Advances in Isotope Geochemistry

Mark Baskaran
Editor

Handbook of Environmental Isotope Geochemistry

Editor
Dr. Mark Baskaran
Wayne State University
Dept. Geology
Detroit Michigan
0224 Old Main Bldg
USA
Baskaran@wayne.edu

ISBN 978-3-642-10636-1 (Printed in 2 volumes)   e-ISBN 978-3-642-10637-8
DOI 10.1007/ 978-3-642-10637-8
Springer Heidelberg Dordrecht London New York

Library of Congress Control Number: 2011935432

© Springer-Verlag Berlin Heidelberg 2011
This work is subject to copyright. All rights are reserved, whether the whole or part of the material is concerned, specifically the rights of translation, reprinting, reuse of illustrations, recitation, broadcasting, reproduction on microfilm or in any other way, and storage in data banks. Duplication of this publication or parts thereof is permitted only under the provisions of the German Copyright Law of September 9, 1965, in its current version, and permission for use must always be obtained from Springer. Violations are liable to prosecution under the German Copyright Law.
The use of general descriptive names, registered names, trademarks, etc. in this publication does not imply, even in the absence of a specific statement, that such names are exempt from the relevant pro-tective laws and regulations and therefore free for general use.

Printed on acid-free paper

Springer is part of Springer Science+Business Media (www.springer.com)

To
  The Founders, Architects and Builders
             Of
  Yesterday, Today and Tomorrow
             Of
  The Field "Isotope Geochemistry"

Endorsement from Prof. Gerald J. Wasserburg

"Environmental Science is concerned with chemical compounds that effect the well being of society-where they come from, how they are transported & where they are deposited. Isotopic geochemistry plays a key role in deciphering this code. The approaches given in this "Hand Book" give a clear view of answers, potential answers & approaches in this field."

G.J. Wasserburg, Caltech
Crafoord Laureate and John D. MacArthur Professor of Geology and Geophysics, Emeritus

Endorsement from Prof. Alex Halliday

"The environment has never before been the focus of such fascination, challenge and global engagement. Yet trying to comprehend it or predict how it might evolve is difficult because of the complexities of the systems. This is a volume of immense scope that provides up to date information on the range of new isotopic tools that are being developed and utilised to assail this issue. It is an invaluable reference work that explains the range of techniques, the archives and the discoveries, and should be of interest to any environmental scientist who wants to explore and understand these same critical parts of Earth's surface."

Alex N. Halliday FRS
Head, Mathematical, Physical and Life Sciences Division
University of Oxford

Preface

Applications of concepts and methods of physics and chemistry to geology resulted in the field of *Isotope Geochemistry* which has revolutionized our understanding of our Earth system and our environment. When isotope techniques applied to understanding environmental changes that have taken place in the anthropocene, the resultant field can be defined as *Environmental Isotope Geochemistry*. The applications of isotopes as tracers and chronometers have permeated not only every sub-branch of geosciences, but also archaeology, anthropology and environmental forensics.

Over the past three to four decades major developments in instrumentation have resulted in high precision and sensitivity in the measurement of a large number of radioactive and stable isotopes that are widely utilized as powerful tools in earth and environmental science. These developments have opened-up new areas of research, resulting in even widening and deepening knowledge of geochemical processes and new discoveries.

The purpose of this two-set volume is to bring together the more recent applications of a much larger number of radioactive and stable isotopes in earth and environmental science, compared to the previous efforts (detailed in Chap. 1), which is necessitated by the rapid developments in the field from the broad expansion of elements studied now and novel applications that have emerged. In this first most-comprehensive edited volume, the 40 chapters that follow cover applications of 115 isotopes from $Z = 1$ to $Z = 95$ (hydrogen to americium; radioactive isotopes of 30 elements, stable isotopes of 27 elements, and both radioactive and stable nuclides of 9 elements) as environmental tracers and chronometers. The topics covered in this Handbook include: the cycling, transport and scavenging of atmospheric constituents; the biogeochemical cycling of inorganic and organic substances in aqueous systems; redox processes; the sources, fate, and transport of organic and inorganic pollutants in the environment; material transport in various Earth's sub-systems (viz., lithosphere, hydrosphere, atmosphere and biosphere); weathering and erosion studies; effective surface exposure ages; sediment dynamics; the chronology of inorganic and organic substances; reconstruction of paleoclimate and paleoenvironment; water mass mixing; tracing both the production and origin of food; tracing the sources of pollutant metals in the human body; archaeology; and anthropology. We anticipate that this handbook will serve as an excellent resource for veteran researchers, graduate students, applied scientists in environmental companies, and regulators in public agencies, reviewing many tried and tested techniques as well as presenting state of the art advances in *Environmental Isotope Geochemistry*.

The field of *Environmental Isotope Geochemistry* is so vast that it is not possible to cover every aspect of this field. The audience in this field includes atmospheric scientists, geologists, hydrologists, oceanographers, limnologists, glaciologists, geochemists, biogeochemists, soil scientists, radiation and health physicists and it is our hope that we have included in-depth chapters in this Handbook that are relevant to everyone of this group.

This idea of editing this handbook was conceived in 2008 and the proposal was submitted to the publisher in 2009 during my sabbatical leave as a *Plummer Visiting Research Fellow* at St. Anne's College, University of Oxford. I owe a special thanks to Don Porcelli for inviting me to Oxford and for the countless discussions we have had during my pleasant stay there. An advisory committee comprised of the following scientists was formed to advise the editor in selecting topics and authors: Per Andersson (Sweden), Joel Blum (USA), Thure Cerling (USA), Brian Gulson (Australia), Kastsumi Hirose (Japan), Gi-Hoon Hong (South Korea), Carol Kendall (USA), Devendra Lal (USA), Don Porcelli (UK), R. Ramesh (India), Henry Schwarcz (Canada), Peter Swarzenski (USA), and Jing Zhang (China). I would like to thank the Editorial Advisory Board members for thoughtful suggestions at various stages of this Handbook. I would like to specially thank Carol Kendall, Gi-Hoon Hong, Henry Schwarcz and Peter Swarzenski for suggesting some key chapters for inclusion in this Handbook that enhanced the overall breadth of coverage. S. Krishnaswami has been very helpful in looking through the original list of topics and came forth with many good suggestions to improve the content of this handbook. My association with him over the past 28 years (first as one of my early advisors in graduate school at the Physical Research Laboratory (PRL), in Ahmedabad, India) has been very enjoyable. I thank student assistant Vineeth Mohan of my department for his editorial assistance with the manuscripts. We thank all of the 82 external reviewers who gave up their time for reviewing all the manuscripts. Finally, I am deeply indebted to all the authors for their relentless efforts in collectively producing a thorough comprehensive two-set volume of articles with breadth and depth for a variety of audiences and their efforts will be highly appreciated by students of yesterday, today and tomorrow.

Wayne State University Mark Baskaran
Detroit, Michigan, USA

Contents

Volume I

Part I Introductory Chapters

1 "Environmental Isotope Geochemistry":
   Past, Present and Future 3
   Mark Baskaran

2 An Overview of Isotope Geochemistry in Environmental Studies 11
   D. Porcelli and M. Baskaran

3 Humans and Isotopes: Impacts and Tracers of Human
   Interactions with the Environment 33
   Karl K. Turekian

Part II Isotopes as Tracers of Continental and Aquatic Processes

4 Lithium Isotopes as Tracers in Marine and Terrestrial
   Environments 41
   K.W. Burton and N. Vigier

5 Meteoric ^{7}Be and ^{10}Be as Process Tracers in the Environment 61
   James M. Kaste and Mark Baskaran

6 Silicon Isotopes as Tracers of Terrestrial Processes 87
   B. Reynolds

7 Calcium Isotopes as Tracers of Biogeochemical Processes 105
   Laura C. Nielsen, Jennifer L. Druhan, Wenbo Yang,
   Shaun T. Brown, and Donald J. DePaolo

8 Natural and Anthropogenic Cd Isotope Variations 125
   M. Rehkämper, F. Wombacher, T.J. Horner, and Z. Xue

9 Stable Isotopes of Cr and Se as Tracers of Redox Processes
   in Earth Surface Environments 155
   Thomas M. Johnson

10 Stable Isotopes of Transition and Post-Transition Metals
   as Tracers in Environmental Studies 177
   Thomas D. Bullen

| 11 | Applications of Osmium and Iridium as Biogeochemical Tracers in the Environment | 205 |

Mukul Sharma

| 12 | Applications of Stable Mercury Isotopes to Biogeochemistry | 229 |

Joel D. Blum

| 13 | Thallium Isotopes and Their Application to Problems in Earth and Environmental Science | 247 |

Sune G. Nielsen and Mark Rehkämper

| 14 | Po-210 in the Environment: Biogeochemical Cycling and Bioavailability | 271 |

Guebuem Kim, Tae-Hoon Kim, and Thomas M. Church

| 15 | Applications of Groundwater Helium | 285 |

J.T. Kulongoski and D.R. Hilton

| 16 | Applications of Short-Lived Radionuclides (^{7}Be, ^{210}Pb, ^{210}Po, ^{137}Cs and ^{234}Th) to Trace the Sources, Transport Pathways and Deposition of Particles/Sediments in Rivers, Estuaries and Coasts | 305 |

J.Z. Du, J. Zhang, and M. Baskaran

| 17 | Radium Isotope Tracers to Evaluate Coastal Ocean Mixing and Residence Times | 331 |

L. Zhang, J. Zhang, P.W. Swarzenski, and Z. Liu

| 18 | Natural Radium and Radon Tracers to Quantify Water Exchange and Movement in Reservoirs | 345 |

C.G. Smith, P.W. Swarzenski, N.T. Dimova, and J. Zhang

| 19 | Applications of Anthropogenic Radionuclides as Tracers to Investigate Marine Environmental Processes | 367 |

G.-H. Hong, T.F. Hamilton, M. Baskaran, and T.C. Kenna

| 20 | Applications of Transuranics as Tracers and Chronometers in the Environment | 395 |

Michael E. Ketterer, Jian Zheng, and Masatoshi Yamada

| 21 | Tracing the Sources and Biogeochemical Cycling of Phosphorus in Aquatic Systems Using Isotopes of Oxygen in Phosphate | 419 |

Adina Paytan and Karen McLaughlin

| 22 | Isotopic Tracing of Perchlorate in the Environment | 437 |

Neil C. Sturchio, John Karl Böhlke, Baohua Gu, Paul B. Hatzinger, and W. Andrew Jackson

| 23 | The Isotopomers of Nitrous Oxide: Analytical Considerations and Application to Resolution of Microbial Production Pathways | 453 |

Nathaniel E. Ostrom and Peggy H. Ostrom

24 Using Cosmogenic Radionuclides for the Determination of Effective Surface Exposure Age and Time-Averaged Erosion Rates 477
D. Lal

25 Measuring Soil Erosion Rates Using Natural (^{7}Be, ^{210}Pb) and Anthropogenic (^{137}Cs, ^{239,240}Pu) Radionuclides 487
Gerald Matisoff and Peter J. Whiting

26 Sr and Nd Isotopes as Tracers of Chemical and Physical Erosion 521
Gyana Ranjan Tripathy, Sunil Kumar Singh, and S. Krishnaswami

27 Constraining Rates of Chemical and Physical Erosion Using U-Series Radionuclides 553
Nathalie Vigier and Bernard Bourdon

Volume II

Part III Isotopes as Tracers of Atmospheric Processes

28 Applications of Cosmogenic Isotopes as Atmospheric Tracers 575
D. Lal and M. Baskaran

29 Uranium, Thorium and Anthropogenic Radionuclides as Atmospheric Tracers 591
K. Hirose

30 Oxygen Isotope Dynamics of Atmospheric Nitrate and Its Precursor Molecules 613
Greg Michalski, S.K. Bhattacharya, and David F. Mase

Part IV Isotopes as Tracers of Environmental Forensics

31 Applications of Stable Isotopes in Hydrocarbon Exploration and Environmental Forensics 639
R. Paul Philp and Guillermo Lo Monaco

32 Utility of Stable Isotopes of Hydrogen and Carbon as Tracers of POPs and Related Polyhalogenated Compounds in the Environment 679
W. Vetter

Part V Isotopes as Tracers in Archaeology and Anthropology

33 Light-Element Isotopes (H, C, N, and O) as Tracers of Human Diet: A Case Study on Fast Food Meals 707
Lesley A. Chesson, James R. Ehleringer, and Thure E. Cerling

34 Stable Isotopes of Carbon and Nitrogen as Tracers for Paleo-Diet Reconstruction 725
H.P. Schwarcz and M.J. Schoeninger

| 35 | Applications of Sr Isotopes in Archaeology | 743 |

N.M. Slovak and A. Paytan

| 36 | Sources of Lead and Its Mobility in the Human Body Inferred from Lead Isotopes | 769 |

Brian L. Gulson

Part VI Isotopes as Tracers of Paleoclimate and Paleoenvironments

| 37 | Dating of Biogenic and Inorganic Carbonates Using ^{210}Pb-^{226}Ra Disequilibrium Method: A Review | 789 |

Mark Baskaran

| 38 | Isotope Dendroclimatology: A Review with a Special Emphasis on Tropics | 811 |

S.R. Managave and R. Ramesh

| 39 | The N, O, S Isotopes of Oxy-Anions in Ice Cores and Polar Environments | 835 |

Joël Savarino and Samuel Morin

| 40 | Stable Isotopes of N and Ar as Tracers to Retrieve Past Air Temperature from Air Trapped in Ice Cores | 865 |

A. Landais

Author Index 887

Subject Index 939

Contributors

W. Andrew Jackson Texas Tech University, Lubbock, TX 79409, USA

M. Baskaran Department of Geology, Wayne State University, Detroit, MI 48202, USA, baskaran@wayne.edu

S.K. Bhattacharya Physical Research Laboratory, Navrangpura, Ahmedabad, India, bhatta@prl.res.in

Joel D. Blum Department of Geological Sciences, University of Michigan, 1100 North University Avenue, Ann Arbor, MI 48109, USA, jdblum@umich.edu

Bernard Bourdon Institute of Geochemistry and Petrology, Clausiusstrasse 25 ETH Zurich, Zurich 8092, Switzerland, Bernard.bourdon@erdw.ethz.ch

Shaun T. Brown Earth Sciences Division, Lawrence Berkeley National Laboratory, Berkeley, CA 94720, USA

Thomas D. Bullen U.S. Geological Survey, MS 420, 345 Middlefield Road, Menlo Park, CA 94025, USA, tbullen@usgs.gov

K.W. Burton Department of Earth Sciences, Durham University, Science Labs, Durham, DH1 3LE, UK, kevin.burton@durham.ac.uk

Thure E. Cerling IsoForensics Inc, Salt Lake City, UT 84108, USA; Department of Biology, University of Utah, Salt Lake City, UT 84112, USA; Department of Geology and Geophysics University of Utah, Salt Lake City, UT 84112, USA, thure.cerling@utah.edu

Lesley A. Chesson IsoForensics Inc, Salt Lake City, UT 84108, USA; Department of Biology University of Utah, Salt Lake City, UT 84112, USA, lesley@isoforeniscs.com

Thomas M. Church College of Earth, Ocean, and Environment, University of Delaware, Newark, DE, USA, tchurch@udel.edu

Donald J. DePaolo Department of Earth and Planetary Science, University of California, Berkeley, CA 94720, USA; Earth Sciences Division Lawrence Berkeley National Laboratory, Berkeley, CA 94720, USA, depaolo@eps.berkeley.edu

N. Dimova UC-Santa Cruz/US Geological Survey, 400 Natural Bridges Drive, Santa Cruz, CA, USA, ndimova@usgs.gov

Jennifer L. Druhan Department of Earth and Planetary Science, University of California, Berkeley, CA 94720, USA; Earth Sciences Division Lawrence Berkeley National Laboratory, Berkeley, CA 94720, USA

J.Z. Du State Key Laboratory of Estuarine and Coastal Research, East China Normal University, Shanghai 200062, China, jzdu@sklec.ecnu.edu.cn

James R. Ehleringer IsoForensics Inc, Salt Lake City, UT 84108, USA; Department of Biology University of Utah, Salt Lake City, UT 84112, USA, ehleringer@biology.utah.edu

Baohua Gu Oak Ridge National Laboratory, Oak Ridge, TN 37831, USA

Brian L. Gulson Graduate School of the Environment, Macquarie University, Sydney, NSW 2109, Australia, bgulson@gse.mq.edu.au

T.F. Hamilton Center for Accelerator Mass Spectrometry, Lawrence Livermore National Laboratory, Livermore, CA 94551-0808, USA, hamilton18@llnl.gov

Paul B. Hatzinger Shaw Environmental, Inc., Lawrenceville, NJ 08648, USA

D.R. Hilton Scripps Institution of Oceanography, University of California, San Diego, La Jolla, CA 92093-0244, USA, drhilton@ucsd.edu

K. Hirose Department of Materials and Life Sciences, Sophia University, 7-1 Kioicho, Chiyodaku, Tokyo 102-8554, Japan, hirose45037@mail2.accsnet.ne.jp

G.-H Hong Korea Ocean Research and Development Institute, Ansan P.O.Box 29 Kyonggi 425-600, South Korea, ghhong@kordi.re.kr

T.J. Horner Department of Earth Sciences, University of Oxford, Oxford OX1 3AN, UK, Tristan.Horner@earth.ox.ac.uk

Thomas M. Johnson Department of Geology, University of Illinois at Urbana-Champaign, Urbana, IL 61801, USA, tmjohnsn@illinois.edu

John Karl Böhlke U.S. Geological Survey, Reston, VA 20192, USA, jkbohlke@usgs.gov

James M. Kaste Department of Geology, The College of William & Mary, Williamsburg, VA 23187, USA, jmkaste@wm.edu

T.C. Kenna Lamont-Doherty Earth Observatory, Columbia University, Palisades, NY 10964, USA, tkenna@ldeo.columbia.edu

Michael E. Ketterer Department of Chemistry and Biochemistry, Box 5698, Northern Arizona University, Flagstaff, AZ 86011-5698, USA, Michael.Ketterer@nau.edu

Guebuem Kim School of Earth and Environmental Sciences, Seoul National University, Seoul, South Korea, gkim@snu.ac.kr

Tae-Hoon Kim School of Earth and Environmental Sciences, Seoul National University, Seoul, South Korea, esutaiki@snu.ac.kr

S Krishnaswami Geosciences Division, Physical Research Laboratory, Ahmedabad 380009, India, swami@prl.res.in

J.T. Kulongoski California Water Science Center, U.S. Geological Survey, San Diego, CA 92101, USA, kulongos@usgs.gov

D. Lal Geosciences Research Division, Scripps Institution of Oceanography, 9500 Gilman Drive, La Jolla, CA 92093-0244, USA; Scripps Institution of Oceanography University of California, San Diego, CA 92093, USA, dlal@ucsd.edu

A. Landais Institut Pierre Simon Laplace/Laboratoire des Sciences du Climat et de l'Environnement, CEA/CNRS/UVSQ, Orme des Merisiers, 91191 Gif sur Yvette, France, amaelle.landais@lsce.ipsl.fr

Z. Liu Key Laboratory of Marine Environment and Ecology (Ocean University of China), Ministry of Education, 238 Songling Road, Qingdao 266003, P.R. China, liuzhe_ecnu@yahoo.com.cn

Guillermo Lo Monaco School of Geology and Geophysics, University of Oklahoma, Norman, OK 73019, USA

S.R. Managave Presently at Department of Earth Sciences, Pondicherry University, R.V. Nagar, Kalapet, Puducherry 605014, shreyasman@gmail.com

David F. Mase Purdue University, 550 Stadium Mall Drive, West Lafayette, IN 47907-1210, USA, dmase@purdue.edu

Gerald Matisoff Department of Geological Sciences, Case Western Reserve University, Cleveland, OH 44106-7216, USA, gerald.matisoff@case.edu

Karen McLaughlin Southern California Coastal Water Research Project, Costa Mesa, CA 92626, USA, karenm@sccwrp.org

Greg Michalski Purdue University, 550 Stadium Mall Drive, West Lafayette, IN 47907-1210, USA, gmichals@purdue.edu

Samuel Morin Institut National des Sciences de l'Univers, CNRS, Grenoble, France; Météo-France CNRM-GAME, Centre d'Études de la Neige, St Martin d'Hères, France, morin.samuel@gmail.com

Sune G. Nielsen Woods Hole Oceanographic Institution, Dept. Geology and Geophysics, 02543 Woods Hole, MA, USA, snielsen@whoi.edu

Laura C. Nielsen Department of Earth and Planetary Science, University of California, Berkeley, CA 94720, USA; Earth Sciences Division, Lawrence Berkeley National Laboratory, Berkeley, CA 94720, USA, lnielsen@berkeley.edu

Nathaniel E. Ostrom Department of Zoology, Michigan State University, 204 Natural Sciences Building, East Lansing, MI 48824, USA, ostromn@msu.edu

Peggy H. Ostrom Department of Zoology, Michigan State University, 204 Natural Sciences Building, East Lansing, MI 48824, USA, ostrom@msu.edu

Adina Paytan University of California, Santa Cruz, CA 95064, USA, apaytan@ucsc.edu

R. Paul Philp School of Geology and Geophysics, University of Oklahoma, Norman, OK 73019, USA, pphilp@ou.edu

D. Porcelli   Department of Earth Sciences, Oxford University, South Parks Road, Oxford OX1 3AN, UK, Don.Porcelli@earth.ox.ac.uk

R. Ramesh   Geosciences Division, Physical Research Laboratory, Navrangpura, Ahmedabad 380009, India, rramesh@prl.res.in

Mark Rehkämper   Department of Earth Science and Engineering, Imperial College, London SW7 2AZ, UK, markrehk@imperial.ac.uk

B. Reynolds   Institute for Geochemistry and Petrology, ETH Zurich, Switzerland, reynolds@erdw.ethz.ch

Joël Savarino   Laboratoire de Glaciologie et Géophysique de l'Environnement, Université Joseph Fourier, St Martin d'Hères, France; Institut National des Sciences de l'Univers CNRS, Grenoble, France, jsavarino@lgge.obs.ujf-grenoble.fr

M.J. Schoeninger   University of California at San Diego, 9500 Gillman Drive, La Jolla, CA 92093, USA, mjschoen@ucsd.edu

H.P. Schwarcz   School of Geography and Earth Sciences, McMaster University, Hamilton ON L8S 4K1, Canada, schwarcz@mcmaster.ca

Mukul Sharma   Radiogenic Isotope Geochemistry Laboratory, Department of Earth Sciences, Dartmouth College, 6105 Fairchild Hall, Hanover, NH 03755, USA, Mukul.Sharma@Dartmouth.edu

Sunil Kumar Singh   Geosciences Division, Physical Research Laboratory, Ahmedabad 380009, India, sunil@prl.res.in

N.M. Slovak   Department of Behavioral Sciences, Santa Rosa Junior College, 1501 Mendocino Avenue, Santa Rosa, CA, USA, nmslovak@yahoo.com

C.G. Smith   US Geological Survey, 600 4th Street South, St. Petersburg, FL, USA, cgsmith@usgs.gov

Neil C. Sturchio   University of Illinois at Chicago, Chicago, IL 60607, USA, sturchio@uic.edu

P.W. Swarzenski   US Geological Survey, 400 Natural Bridges Drive, Santa Cruz, CA 95060, USA, pswarzen@usgs.gov

Gyana Ranjan Tripathy   Geosciences Division, Physical Research Laboratory, Ahmedabad 380009, India, gyana@prl.res.in

Karl K. Turekian   Department of Geology and Geophysics, Yale University, New Haven, CT 06511, USA, karl.turekian@yale.edu

W. Vetter   University of Hohenheim, Institute of Food Chemistry, Garbenstrasse 28, 70599 Stuttgart, Germany, walter.vetter@uni-hohenheim.de

Nathalie Vigier   CRPG-CNRS, Nancy Université, 15 rue Notre Dame des Pauvres, 54500 Vandoeuvre les Nancy, France, nvigier@crpg.cnrs-nancy.fr

Peter J. Whiting   Department of Geological Sciences, Case Western Reserve University, Cleveland, OH 44106-7216, USA, peter.whiting@case.edu

F. Wombacher Institut für Geologie und Mineralogie, Universität zu Köln, Zülpicher Straße 49b 50674, Köln, Germany, fwombach@uni-koeln.de

Z. Xue Department of Earth Science and Engineering, Imperial College London, London SW7 2AZ, UK, z.xue07@imperial.ac.uk

Masatoshi Yamada Environmental Radiation Effects Research Group, National Institute of Radiological Sciences, 4-9-1 Anagawa, Inage, Chiba 263-8555, Japan

Wenbo Yang Department of Earth and Planetary Science, University of California, Berkeley, CA 94720, USA; Earth Sciences Division, Lawrence Berkeley National Laboratory, Berkeley, CA 94720, USA

J. Zhang State Key Laboratory of Estuarine and Coastal Research, East China Normal University, Shanghai 200062, China, jzhang@sklec.ecnu.edu.cn

Jian Zheng Environmental Radiation Effects Research Group, National Institute of Radiological Sciences, 4-9-1 Anagawa, Inage, Chiba 263-8555, Japan, jzheng@nirs.go.jp

Part I
Introductory Chapters

Chapter 1
"Environmental Isotope Geochemistry": Past, Present and Future

Mark Baskaran

1.1 Introduction and Early History

A large number of radioactive and stable isotopes of the first 95 elements in the periodic table that occur in the environment have provided a tremendous wealth of information towards unraveling many secrets of our Earth and its environmental health. These isotopes, because of their suitable geochemical and nuclear properties, serve as tracers and chronometers to investigate a variety of topics that include chronology of rocks and minerals, reconstruction of sea-level changes, paleoclimates, and paleoenvironments, erosion and weathering rates of rocks and minerals, rock-water interactions, material transport within and between various reservoirs of earth, and magmatic processes. Isotopic data have also provided information on time scales of mixing processes in the oceans and atmosphere, as well as residence times of oceanic constituents and gases in the atmosphere. Arguably the most important milestone on the application of isotopes to earth science is the determination of the age of the Earth and our solar system. Isotope-based dating methods serve as the *gold-standard* and are routinely used to validate other non-isotope-based dating methods. Dating of hominid fossils provides a handle to understand the evolution and migration pattern of humans and stable isotope analyses of organic matter, as well as phosphate in bones and teeth in recovered fossils provide evidence for food sources consumed by humans and other animals (e.g., Abelson 1988; Schwarcz and Schoeninger 2011). To put it succinctly, our current understanding of the chronological evolution of the earth, its exterior and interior processes occurring on time scales of minutes to billions of years and the reconstruction of the evolution of human civilization has been developed in great part by the measurement of isotopic ratios.

The field of isotope geochemistry started taking its roots shortly after the discovery of "*radioactivity*" (term coined by Marie Curie) in 1896 by Henri Becquerel (Becquerel 1896; Curie 1898). Within a few years of this remarkable discovery, Rutherford reported an exponential decrease of activity of a radioactive substance with time and introduced the concept of half-lives opening the door for age determination of natural substances containing radioactive elements (Rutherford 1900). The term *isotope* was introduced by Soddy in 1913 (Soddy 1913). The rules governing the transmutation of elements during radioactive decay was simultaneously established by Soddy (1913) and Fajans (1913). Secular equilibrium between radioactive parent and daughter was first described by Rutherford and Soddy (1902). The first radiometric age determination of a geologic sample was made on a sample of pitchblende in 1905 by Rutherford and ages of a variety of other minerals were subsequently made by Strutt (1905) and Boltwood (1907) using the U-He and U-Pb systems. The complete ^{238}U and ^{232}Th chain was established by 1913 and is similar to the one that is in use today (e.g., Ivanovich and Harmon 1992; Henderson 2003). The disequilibria between the members of the U-Th series resulting from differences in geochemical properties of different elements within the chain opened a new field of research to investigate aqueous geochemical processes, rock-water interaction, dating of inorganic precipitates, detrital and biogenic sediments and archaeological objects (e.g., Ivanovich and Harmon

M. Baskaran (✉)
Department of Geology, Wayne State University, Detroit, MI 48202, USA
e-mail: Baskaran@wayne.edu

1992; Bourdon et al. 2003; Krishnaswami and Cochran 2008). One of the key discoveries in this field was made when large fractionations of ^{238}U and ^{234}U were observed in rocks, leachates of rocks and natural waters (Cherdyntsev 1955; Thurber 1962). This disequilibrium is caused by a *nuclear* effect resulting from the displacement of ^{234}U (in some cases release into the surrounding aqueous phase from mineral grain surfaces) by recoil during alpha decay (^{238}U decays to ^{234}Th which is displaced from the original position of ^{238}U due to recoil), in contrast to the fractionation caused by differences in the *geochemical* properties between different members of the decay chain.

Soon after the discovery of radioactivity, Victor Hess (1912) measured the radiation levels in the atmosphere at various altitudes using a Geiger counter (developed in 1908) and reported that the radiation levels increased with altitude. He attributed this to radiation coming from outer space called *cosmic radiation* and now commonly called *cosmic rays*. Cosmic rays comprise of charged particles (including high-energy charged particles) such as protons, alpha particles, electrons, helium, nuclei of other elements and subatomic particles. The high-energy charged particles entering the atmosphere interact with atmospheric constituents (N, O, Ar, etc) and produce a suite of cosmogenic radioactive isotopes, whose half-lives range from less than an hour to millions of years (Lal and Peters 1967; Krishnaswami and Lal 2008; Lal and Baskaran 2011). Some of the cosmogenic isotopes (^{14}C, ^{10}Be, ^{7}Be) have found extensive applications in quantifying processes in earth surface reservoirs such as air-sea exchange, atmospheric mixing, ocean circulation and mixing, scavenging, sediment accumulation and mixing rates in aqueous systems, erosion rates, exposure ages, changes in cosmic ray production rates, and history of human civilization. Among these, ^{14}C has been used universally as the most-robust dating tool and has contributed tremendously to our understanding of human civilization.

The field of stable isotope geochemistry started taking roots with the first set of stable isotope measurements of terrestrial samples made by Murphy and Urey (1932), Nier and Gulbransen (1939), Dole and Slobod (1940) and Urey (1948), much before the discovery of cosmogenic isotopes such as ^{7}Be, ^{10}Be, and ^{14}C. The theoretical foundation of isotopic fractionation was later provided by Urey (1947) and Bigeleisen and Mayer (1947) using the methods of statistical quantum mechanics and statistical thermodynamics. Of the 54 elements (first 82 elements are stable of which promethium and technetium are radioactive; 26 are monoisotopic elements) that have two or more stable isotopes, only six of them (H, C, N, O, S and Si) have been extensively studied, with >10,000 published papers, abstracts and theses published on C and O isotope variations since late 1930s for investigating various near earth and earth-surface processes. Fractionations caused by mass-dependent processes such as isotope-exchange reactions, physical and biological reactions result in variations in the isotopic ratios of these elements. All of these elements form chemical bonds that have a high degree of covalent character and some are found in multiple oxidation states in the environment. In contrast, variations in radiogenic stable isotope ratios, such as Pb (^{206}Pb/^{204}Pb, ^{207}Pb/^{204}Pb and ^{208}Pb/^{204}Pb), Sr (^{87}Sr/^{86}Sr), Ce (^{138}Ce/^{142}Ce), Nd (^{143}Nd/^{144}Nd), Os (^{187}Os/^{186}Os) and Hf (^{176}Hf/^{177}Hf) in natural materials depend on the differences of their initial ratios, the parent concentrations, decay constants, and time elapsed since the solid material was formed.

A natural progression of the light-element stable isotope research is to look for stable isotope fractionation of transition and post-transition elements. Sporadic attempts were made to look for them during 1970's and 1980's, and the initial results appeared to be encouraging. Nonetheless, because of the existing technology at that time and the preconceived notion that mass-dependent fractionations among heavy elements are expected to be negligible, there was no real progress in this area of research (O'Neil 1986). Indeed, mass-dependent isotopic fractionation in transition and post-transition elements are small compared to those in light-elements. However, as will be discussed later, these inferences and conclusions have been challenged and the occurrence of isotope fractionation in transition elements is more of a rule than exception.

The radiogenic isotopes (Pb (^{206}Pb/^{204}Pb, ^{207}Pb/^{204}Pb and ^{208}Pb/^{204}Pb), Sr (^{87}Sr/^{86}Sr), Ce (^{138}Ce/^{142}Ce), Nd (^{143}Nd/^{144}Nd), Hf (^{176}Hf/^{177}Hf) and Os (^{187}Os/^{186}Os) have been widely used as tracers and stratigraphic chronometers in the environment. Pb isotopes have been utilized to trace the sources of transboundary atmospheric pollution (e.g., Bollhöfer and Rosman 2001; Komárek et al. 2008), the sources of local and global Pb pollution in a variety of natural reservoirs that include lake and coastal sediments, snow and ice samples, peat deposits, tree rings, lichens and grasses (e.g., Komárek et al. 2008) and to trace the pathways of lead from the environment in to human bodies (e.g., Gulson 2011).

The dawn of the Atomic Age started with the detonation of the first nuclear weapon in 1945. Subsequent nuclear weapon tests during 1950's (started in 1952) and early 1960's (implementation of Nuclear Test Ban Treaty in 1963) released a large amount of radioactive isotopes, mainly ^{137}Cs, ^{90}Sr, Pu and ^{14}C to the environment. These isotopes have been extensively utilized to investigate environmental processes that have occurred since the 1950's, a period of time that witnessed considerable environmental changes due to anthropogenic activities. Although over 70% of the fission products ^{137}Cs and ^{90}Sr derived from global nuclear weapons tests have already decayed away, several of the long-lived isotopes (^{14}C and transuranics) continue to serve as effective tracers and chronometers in environmental studies.

In this present Anthropocene Era, elements of economic value are mobilized from their respective sources into various Earth's subsystems of the lithosphere, hydrosphere, atmosphere and biosphere. With the increases in population and the spectacular sustained growth of emerging economies over the past 2–3 decades, the demand for Earth's resources have increased exponentially. While several hundreds of millions of people are taken out of poverty as a result of global economic growth, rapid industrialization has resulted in sustained environmental degradation in many developed and emerging economies. For example, in Detroit, Michigan, USA, the average Pb concentration in soil is more than an order of magnitude higher than the average upper crustal value. The Environmental Protection Agency in the USA and many regulatory agencies in the United States and elsewhere have listed the following ten elements as priority pollutants: Cd, Cr, Cu, Pb, Hg, Ni, Se, Ag, Tl and Zn. Except for Pb, high precision measurements of the isotopes of these elements given above for environmental studies are relatively new (<15 years old), and thus, the isotopic ratios of these pollutant elements offer exciting opportunities for future research.

The twentieth century witnessed an explosion of the applications of radioactive and stable isotopes in studies of earth system science. Many of these studies, particularly those made prior to 1980 were discussed in some detail in the first two published volumes entitled *Environmental Isotope Geochemistry* edited by P. Fritz and J.Ch. Fontes (Volume-I, 1980 and Volume-II, 1986). These volumes focused primarily on stable isotopes of light elements (H,C,O,N, and S), Pb, Sr, Cl, U-series, and a suite of noble gases and provided state-of-the-art reviews of their applications in selected areas of earth sciences. In the last 25 years since the second volume was published, there have been major advances in instrumentation for high precision isotope measurements of several elements that include U-Th-series radionuclides (^{238}U, ^{234}U, ^{232}Th, ^{230}Th, ^{231}Pa, ^{226}Ra) using thermal ionization mass spectrometer (TIMS), platinum group elements using negative thermal ionization mass spectrometer (NTIMS) and multiple-collector inductively-coupled plasma mass spectrometer (MC-ICPMS). These developments have opened new areas of research in different areas of earth and environmental science.

The purpose of this volume is to bring together recent applications of a much larger number of radioactive and stable isotopes in earth and environmental sciences. There are a few earlier published volumes wherein selected aspects of these applications pertaining to particular processes or environment were discussed. The earlier edited volumes include: *Reviews in Mineralogy and Geochemistry* Volumes 16, 33, 38, 43, 50, and 52 (16: *Stable Isotopes in High Temperature Geological Processes*; 33: *Boron: Mineralogy Petrology and Geochemistry*; 38: *Uranium: Mineralogy, Geochemistry and the Environment*; 43: *Stable Isotope Geochemistry*; 50: *Beryllium: Mineralogy, Petrology, and Geochemistry*; and 52: *Uranium Series Geochemistry*); similarly, there are volumes on application of U-Th series and fallout isotopes in aqueous systems (e.g., Ivanovich and Harmon 1992; Livingston 2004; Krishnaswami and Cochran 2008). The scope of the present volume is much wider. New applications have become possible because of new understandings of isotope fractionation processes, improvements in chemical separation and purification techniques (in particular developments of actinide-specific resins) and development of high sensitivity instruments for measurements of isotopic ratios. Some of these are briefly discussed below.

1.2 Mass Independent Fractionation

The discovery of chemically-produced mass-independent isotope fractionation opened a new variety of applications, including investigations in paleoclimatology, biologic primary productivity, origin and evolution of life in Earth's earliest environment, and atmospheric chemistry (Young et al. 2002; Thiemens 2006). On the surface of Earth, the kinetic and equilibrium fractionations as well as isotopic exchange reactions result in fractionations that are mass-dependent. For example, when $CaCO_3$ precipitates from a solution, $^{18}O/^{16}O$ fractionation is twice as that of $^{17}O/^{16}O$ fractionation. However, in meteorites, mass-independent isotopic fractionation was observed and was attributed to nucleosynthetic processes (Clayton et al. 1973). Subsequently, it was shown that a chemically produced, mass-independent fractionation of oxygen is possible (Thiemens and Heidenreich 1983). Mass-independent fractionation appears to be caused by a molecular symmetry effect and the asymmetric molecule (e.g., O_3 in the form of $^{16}O^{17}O^{18}O$) undergoes mass-independent fractionation and is enriched in heavy isotopes while O_3 in the form of $^{18}O^{18}O^{18}O$ or $^{16}O^{16}O^{16}O$ is not (Heidenreich and Thiemens 1986). Recent results on Hg indicate that Hg displays mass-independent isotope fractionation during photochemical radical pair reactions, wherein the reactivity of odd and even mass number isotopes differs (summarized in Blum 2011).

1.3 Developments in Instrumentation

It took 30 years since the first publication of ^{14}C dating with beta counting method (Anderson et al. 1947) to achieve the first break-through in instrumentation with the use of cyclotron and tandem accelerators as a high-energy mass spectrometer (commonly denoted as accelerated mass spectrometers, AMS), for the measurement of ^{14}C (cyclotron, Muller 1977; tandem accelerators, Bennett et al. 1977; Nelson et al. 1977), resulting in an order of magnitude higher precision than the beta counting method which in turn resulted in three to four orders of magnitude reduction in the sample size (e.g., Trumbore 2002). Subsequently, AMS has been extensively used for high precision measurements of other cosmogenic radionuclides including ^{10}Be, ^{26}Al, ^{36}Cl and ^{129}I. The second major break-through came in the measurements of U-Th series radionuclides ($^{238}U,^{235}U,^{234}U$, $^{232}Th, ^{230}Th, ^{231}Pa$, ^{226}Ra), using thermal ionization mass spectrometer (TIMS) starting from mid 1980's, and was mainly due to high ionization efficiency achieved on U,Th,Pa, Ra and other elements when the samples were prepared by the graphite-sandwich technique on a single Re filament and the ionization efficiency improved from 10^{-4} to few percent (U and Th isotopes: Chen et al. 1986; Edwards et al. 1987; ^{231}Pa: Pickett et al. 1994; ^{226}Ra: Volpe et al. 1991; Cohen and O'Nions 1991). Concentrations as low as 0.1 µg of ^{238}U and ^{232}Th can now be measured with better than 1–3‰ precision (summarized in Goldstein and Stirling 2003). Such high precision and sensitivity resulted in the possibility of dating very young corals (< 500 years) with $^{230}Th/^{234}U$ dating method, as well as extending the dating limit of $^{230}Th/^{234}U$ method to ~500 kyr (Edwards et al. 1987, 1988). The third major break-through came in the measurement of some of the platinum group elements (PGE, specifically Os, Re, and Ir), using the negatively charged oxides of these elements. Conventional TIMS had a low precision for PGE due to poor ionization efficiency, as these elements have high ionization potential (Os: 8.7 eV, Re: 7.9 eV and Ir: 9.1 eV). The principal ion species of Os, Re, and Ir are negative oxides and negative thermal ionization mass-spectrometry resulted in high ionization efficiencies (2–6% for Os, >20% for Re; Creaser et al. 1991). A higher precision of $\geq \pm 2$‰ (2σ) on the isotopic composition of Os with 4 ng Os have been reported. The fourth major break-through in the mass spectrometry, Inductively Coupled Plasma Mass Spectrometer (ICPMS) came in 1980's, most of them in the initial generation were conventional ICPMS, comprised of quadrupole ICPMS, high-resolution sector field ICPMS (HR-ICPMS), and time of flight ICPMS (TOF-ICPMS). Very high ionization (>90%) is obtained with ICPMS for all elements at high temperature (~6,000°K), including those that have high first ionization potentials (such as the PGE elements listed above). The sample throughput is faster and sample preparation time is significantly less in ICPMS compared to TIMS. It was not until the use of multiple-collector ICPMS (MC-ICPMS) in mid 1990s that combined sector-field ICPMS with a multiple collector detector system, major strides have been made in obtaining high precision measurements of U-Th series radionuclides as well as other transitional and post-transitional elements. This technique emerged as an alternative or in some cases superior to TIMS method (a comparison of the sample size, precision, and sensitivity of TIMS, SIMS and ICPMS (multiple-collector, laser ablation and laser ablation-multiple collector) are given in Goldstein and Stirling (2003)). Although the first U measurements with ICPMS was made nearly 20 years ago (Walder and Freedman 1992; Taylor et al. 1995), only recently high precision measurements have been achieved. High precision measurements of Cu and Zn isotopes were made for the first time in late 1990s using ICPMS equipped with multiple collectors and magnetic sector, although earlier attempt to measure differences in Zn isotopes in environmental samples were not successful due to lack of sensitivity and precision (Rosman 1972; Maréchal et al. 1999). High-precision Tl isotopic measurement was made in 1999 for the first time with MC- ICPMS, with a precision of 0.1–0.2‰ (Rehkämper and Halliday 1999), although earlier attempts with TIMS resulted in relatively large errors (>2‰). It has been assumed all along that the $^{238}U/^{235}U$ atomic ratio to be constant, 137.88, except for uranium in the Oklo natural nuclear fission reactor discovered in 1972 in Gabon, Africa. Measurements of unprecedented high precision were made recently of $^{238}U/^{235}U$ ratios on up to 40 ppm (0.040‰) precision in seawater and other aqueous systems (Stirling et al. 2007). With the advent of MC-ICPMS and improvements in the preparation of gases for introduction into gas-source mass spectrometers, the precision for a large number of isotopes is getting better than 0.050‰.

In radioactive counting, some of the short-lived radionuclides can be measured at very low levels. With a delayed-coincidence counting system, ~3,000 atoms of ^{224}Ra corresponding to 1.1×10^{-18} g of ^{224}Ra (half-life = 3.66 days) or 5.0×10^{-21} mole of ^{224}Ra, can be measured with a precision of ~10% (1σ) (Moore 2008). This technique can be employed to measure some other short-lived radionuclides such as ^{223}Ra (half-life = 11.435 days) and ^{228}Th (half-life = 1.913 year). The precision and sensitivity for some of the short-lived U-Th series (^{210}Po, ^{234}Th, ^{228}Th and ^{227}Th) with radioactive counting methods (in particular with alpha and beta counting instruments) are excellent (Baskaran 2011; Baskaran et al. 2009) and most likely counting instruments will be the method of choice in the foreseeable future.

1.4 Future Forecast for 25 Years from Now

We have come a long way over the past ~100 years in improving the precision of isotopic analyses. In the first stable isotope paper (Murphy and Urey 1932), the relative abundances of the nitrogen and oxygen isotopes on natural samples had a precision of about 10% and now we are at the threshold of reaching a precision of ~10 ppm (10,000 times better precision). Technological advances will continue to drive the new and innovative application of the tracer techniques. For example, 1% uncertainties in ^{14}C measurements in the late 1970's with AMS was considered to be major advance (compared to the beta counting), but over the last 10 years, 3‰ has been the state-of-the-art. Attempts have been made to achieve a precision below 2‰, while reaching 1‰ still remains a challenge (Synal and Wacker 2010). This is probably applicable to other key cosmogenic radionuclides including ^{10}Be, ^{26}Al, ^{36}Cl, and ^{129}I. The conventional paradigm that no significant mass-dependent isotopic fractionations is expected in alkali and alkaline earth elements that commonly bond ionically or elements that are heavy where the mass difference between the heavier and lighter isotope is small is undergoing a major shift. Now the accepted view is that chemical, physical and biological processes that take place at normal environmental conditions result in measurable variations in the isotopic ratios of heavy elements. It behooves us to ask the question: Will the instrumental developments be in the *driver-seat* for the scientific breakthroughs in the applications of isotopes (both radioactive and stable) for earth and environmental studies?

Although the foundations for mass-dependent isotope fractionation resulting from kinetic and equilibrium processes were established in 1940's, the refinements in those theoretical foundations over the past 6 decades are minor. It is likely that when the precision improves by a factor of 5–10 (to 10 ppm level), better understanding of the fractionation mechanisms could result in multiple fractionation laws and could result in the reevaluation of the reference mass fractionation line for lighter stable isotope ratios (e.g., C, O, N). In the case of environmental forensics for organic pollutants, the future research in the source identification and fate and transport of pesticides, herbicides and other POPs, VOCs, and other organic pollutants appears to depend on the analysis of molecular compounds at a higher precision. Compound-specific stable carbon isotope measurements of dissolved chlorinated ethane (PCE and TCE) in groundwater provide evidence for reductive dechlorination of chlorinated hydrocarbons. For example, the $\delta^{13}C$ values were found to be more positive (−18.0‰) in down-gradient wells compared to those in the source region (−25.0 to −26.0‰), paving the way for the quantification of extent of biodegradation between the zones of the contaminant plume (Sherwood Lollar et al. 2001). Multiple compound-specific isotopic analyses ($\delta^{37}Cl$, $\delta^{81}Br$, $\delta^{13}C$, δD) on chlorinated compounds with gas chromatography interfaced to ICPMS with much improved precision and sensitivity could provide a powerful tool in source(s) identification, fate and transport of organic pollutants including emerging contaminants, in aqueous systems.

When the biological fractionation is well understood and the precision is significantly improved from the present limit, then, isotopes of Zn and Cd and certain redox-sensitive elements such as Cr, Cu, Fe, Se, Hg, Tl, and U could provide insight on the biogeochemical processes in marine and lacustrine environments. One of the major concerns in the surface waters is the ever-increasing temporal and spatial extent of harmful algal blooms (HAB). The isotopes of macro- and micro-nutrient elements [macro- (N and P) and micro-(Fe, Cr, Mn Se, Zn, Mo, I); for P, oxygen isotope ratios in phosphate can be used] could serve as effective tracers to investigate the factors and processes that lead to the formation and sustenance of HAB. Fractionation of these transition and post-transition elements caused during smelting operation could result in isotopically light elements in the vapor phase and when the condensation of the vapor phase takes place in the environment, gradient in the isotopic ratios of elements from the source of release to farther distances is expected and the isotopic ratios could provide a tool for tracing industrial sources of these elements (Weiss et al. 2008).

In conclusion, if we were to plot the sheer number of publications or the total funding made available for conducting isotope-related research in earth and environmental science versus time since the 1930's, the slope would suggest that we have every reason to be optimistic. We eagerly await the ground-breaking research that will be made in this field in the next 25 years!

Acknowledgments I thank S. Krishnaswami, Jim O'Neil and Peter Swarzenski for their in-depth reviews which resulted in considerable improvement of this chapter. Some of their suggestions on the past and present status of the work are also included in this revised version.

References

Abelson PH (1988) Isotopes in earth science. Science 242:1357

Anderson EC, Libby WF, Weinhouse S, Reid AF et al (1947) Radiocarbon from cosmic radiation. Nature 105:576–577

Baskaran M (2011) Dating of biogenic and inorganic carbonates using ^{210}Pb-^{226}Ra disequilibrium method – a review. In: Handbook of environmental isotope geochemistry. Springer, Berlin

Baskaran M, Hong GH, Santschi PH (2009) Radionuclide analyses in seawater. In: Wurl O (ed) Practical guidelines for the analysis of seawater. CRC Press, Boca Raton, pp 259–304

Becquerel AH (1896) On the rays emitted by phosphorescent bodies. Comptes Rendus de Seances de l'academie de Sciences 122:501–503

Bennett RP, Beukens RP, Clover HE, Gove RB et al (1977) Radiocarbon dating using electrostatic accelerators: negative ions provide the key. Science 198:508–510

Bigeleisen J, Mayer M (1947) Calculation of equilibrium constant for isotope exchange reactions. J Chem Phys 15:261–167

Blum JD (2011) Applications of stable mercury isotopes to biogeochemistry. In: Advances in isotope geochemistry. Springer, Heidelberg

Bollhofer A, Rosman K (2001) Isotopic source signatures for atmospheric lead; the Northern hemisphere. Geochim Cosmochim Acta 65:1727–1740

Boltwood BB (1907) Note on a new radioactive element. Am J Sci 24:370–372

Bourdon B, Henderson GM, Lundstrom CC, Turner SP (2003) Uranium-series geochemistry (eds) vol 52. Geochemical Society – Mineralogical Society of America, Washington, DC, pp. 656

Chen JH, Edwards RL, Wasserburg GJ (1986) ^{238}U, ^{234}U, and ^{232}Th in seawater. Earth Planet Sci Lett 80:241–251

Cherdyntsev VV (1955) On isotopic composition of radioelements in natural objects, and problems of geochronology. Izv Akad Nauk SSR 175

Clayton RN, Grossman L, Mayeda TK (1973) A component of primitive nuclear composition in carbonaceous meteorites. Science 182:485–488

Cohen AS, O'Nions RK (1991) Precise determination of femtogram quantities of radium by thermal ionization mass spectrometry. Anal Chem 63:2705–2708

Creaser RA, Papanastassiou DA, Wasserburg GJ (1991) Negative thermal ion mass-spectrometry of osmium, rhenium and iridium. Geochim Cosmochim Acta 55:397–401

Curie M (1898) Rays emitted by compounds of uranium and thorium. Comptes Rendus de Seances de l'academie de Sciences 126:1101–1103

Dole M, Slobod RL (1940) Isotopic composition of oxygen in carbonate rocks and iron oxide ores. J Am Chem Soc 62:471–479

Edwards RL, Chen JH, Wasserburg GJ (1987) ^{238}U-^{234}U-^{230}Th-^{232}Th systematic and the precise measurement of time over the past 500,000 years. Earth Planet Sci Lett 81: 175–192

Edwards RL, Chen JH, Wasserburg GJ (1988) Dating earthquakes with high-precision thorium-230 ages of very young corals. Earth Planet Sci Lett 90:371–381

Fajans K (1913) Radioactive transformations and the periodic system of the elements. Berichte der Dautschen Chemischen Gesellschaft 46:422–439

Goldstein SJ, Stirling CH (2003) Techniques for measuring uranium-series nuclides: 1992–2002. Rev Mineral Geochem 52:23–57

Gulson B (2011) Sources of lead and its mobility in the human body Inferred from Lead Isotopes. In: Advances in isotope geochemistry Springer, Heidelberg

Heidenreich BC III, Thiemens MH (1986) A non-mass-dependent oxygen isotope effect in the production of ozone from molecular oxygen: the role of symmetry in isotope chemistry. J Chem Phys 84:2129–2136

Henderson GM (2003) One hundred years ago: the birth of uranium-series science. Rev Mineral Geochem 52:v–x

Ivanovich M, Harmon RS (eds) (1992) Uranium-series disequilibrium. Applications to earth, marine, and environmental sciences, 2nd edn. Clarendon Press, Oxford, p 909

Komárek M, Ettler V, Chrastný V, Mihaljevič M (2008) Lead isotopes in environmental sciences: a review. Environ Int 977 34:562–577

Krishnaswami S, Cochran JK (2008) U-Th series nuclides in aquatic systems, Vol 13 (Radioactivity in the Environment). Elsevier, Amsterdam, pp. 458

Krishnaswami S, Lal D (2008) Cosmogenic nuclides in the environment: a brief review of their applications. In: Gupta H, Fareeduddin (eds) Recent Advances in Earth System Sciences. Geol Soc India Golden Jubilee Volume pp 559–600

Lal D, Baskaran M (2011) Applications of cosmogenic-isotopes as atmospheric tracers. In: Advances in isotope geochemistry. Springer, Heidelberg

Lal D, Peters B (1967) Cosmic-ray produced radioactivities on the earth. Handbuch Phys. 46. Springer, Berlin, p 551

Livingston HD (2004) Marine Radioactivity (In: Radioactivity in the Environment Series), 6. Elsevier, Amsterdam, pp. 310.

Maréchal C, Télouk P, Albarède F (1999) Precise analysis of copper and zinc isotope compositions by plasma-source mass spectrometry. Chem Geol 156:251–273

Moore WS (2008) Fifteen years experience in measuring ^{224}Ra and ^{223}Ra by delayed-coincidence counting. Mar Chem 109:188–197

Muller RA (1977) Radioisotope dating with a Cyclotron. Science 196:489–494

Murphy GM, Urey HC (1932) On the relative abundances of the nitrogen and oxygen isotopes. Phys Rev 41:921–924

Nelson DE, Koertling RG, Stott WR (1977) Carbon-14: direct detection at natural concentrations. Science 198:507–508

Nier AO, Gulbransen EA (1939) Variations in the relative abundance of the carbon isotopes. J Am Chem Soc 61:697–698

O'Neil JR (1986) Theoretical and experimental aspects of isotopic fractionation. Rev Mineral 16:1–40

Pickett DA, Murrell MT, Williams RW (1994) Determination of femtogram quantities of protactinium in geologic samples by thermal ionization mass spectrometry. Anal Chem 66:1044–1049

Rehkämper M, Halliday AN (1999) The precise measurement of Tl isotopic compositions by MC-ICPMS: application to the analysis of geological materials and meteorites. Geochim Cosmochim Acta 63:935–944

Rosman KJR (1972) A survey of the isotopic and elemental abundances of zinc. Geochim Cosmochim Acta 36:801–819

Rutherford E (1900) A radioactive substance emitted from thorium compounds. Philos Mag 49:1–14

Rutherford E, Soddy F (1902) The cause and nature of radioactivity Part 1. Philos Trans R Soc 4:370–396

Schwarcz HP, Schoeninger MJ (2011) Stable isotopes of carbon and nitrogen as tracers for paleo-diet reconstruction. In: Advances in isotope geochemistry. Springer, Heidelberg

Sherwood Lollar B, Slater GF, Sleep B, Witt M et al (2001) Stable carbon isotope evidence for intrinsic bioremediation of tetrachloroethene and trichloroethene at Area 6, Dover Air Force Base. Environ Sci Technol 35:261–269

Soddy F (1913) Radioactivity. In: Annual Reports on the Progress of Chemistry. The Chemical Society, London, pp 262–288

Stirling CH, Andersen MB, Potter E-K, Halliday AN (2007) Low-temperature isotopic fractionation of uranium. Earth Planet Sci Lett 264:208–225

Strutt RJ (1905) On the radio-active minerals. Proc R Soc London 76:88–101

Synal H-A, Wacker L (2010) AMS Measurement technique after 30 years: possibilities and limitations of low energy systems. Nucl Inst Meth Phys Res B 268:701–707

Taylor PDP, De Bievre P, Walder AJ, Entwistle A (1995) Validation of the analytical linearity and mass discrimination correction model exhibited by a Multiple Collector Inductively Coupled Plasma Mass Spectrometer by means of a set of synthetic uranium isotope mixtures. J Anal At Spectrom 10:395–398

Thiemens MH (2006) History and applications of mass-independent isotope effects. Annu Rev Earth Planet Sci 34:217–262

Thiemens MH, Heidenreich JE (1983) The mass-independent fractionation of oxygen: a novel isotope effect and its possible cosmochemical implications. Science 219:1073–1075

Thurber DL (1962) Anomalous ^{234}U/^{238}U in nature. J Geophys Res 67:4518–1523

Trumbore SE (2002) Radiocarbon chronology. In: Noller JS, Sowers JM, Lettis WR (eds) Quaternary geochronology: methods and applications, AGU Reference Shelf 4. pp 41–60

Urey HC (1947) The thermodynamic properties of isotopic substances. J Chem Soc Lond 562–581

Urey HC (1948) Oxygen isotopes in nature and in the laboratory. Science 108:489–496

Volpe AM, Olivares JA, Murrell MT (1991) Determination of radium isotope ratios and abundances in geologic samples by thermal ionization mass spectrometry. Anal Chem 63:913–916

Walder AJ, Freedman PA (1992) Isotopic ratio measurement using a double focusing magnetic-sector mass analyzer with an inductively coupled plasma as an ion-source. J Anal Atom Spec 7:571–575

Weiss DJ, Rehkämper M, Schoenberg R, McLaughlin M et al (2008) Application of nontraditional stable-isotope systems to the study of sources and fate of metals in the environment. Eiviron Sci Technol 42:655–664

Young ED, Gay A, Nagahara H (2002) Kinetic and equilibrium mass-dependent isotope fractionation laws in nature and their geochemical and cosmochemical significance. Geochim Cosmochim Acta 66:1095–1104

Chapter 2
An Overview of Isotope Geochemistry in Environmental Studies

D. Porcelli and M. Baskaran

Abstract Isotopes of many elements have been used in terrestrial, atmospheric, and aqueous environmental studies, providing powerful tracers and rate monitors. Short-lived nuclides that can be used to measure time are continuously produced from nuclear reactions involving cosmic rays, both within the atmosphere and exposed surfaces, and from decay of long-lived isotopes. Nuclear activities have produced various isotopes that can be used as atmospheric and ocean circulation tracers. Production of radiogenic nuclides from decay of long-lived nuclides generates widespread distinctive isotopic compositions in rocks and soils that can be used to identify the sources of ores and trace water circulation patterns. Variations in isotope ratios are also generated as isotopes are fractionated between chemical species, and the extent of fractionation can be used to identify the specific chemical processes involved. A number of different techniques are used to separate and measure isotopes of interest depending upon the half-life of the isotopes, the ratios of the stable isotopes of the element, and the overall abundance of the isotopes available for analysis. Future progress in the field will follow developments in analytical instrumentation and in the creative exploitation of isotopic tools to new applications.

2.1 Introduction

Isotope geochemistry is a discipline central to environmental studies, providing dating methods, tracers, rate information, and fingerprints for chemical processes in almost every setting. There are 75 elements that have useful isotopes in this respect, and so there is a large array of isotopic methods potentially available. The field has grown dramatically as the technological means have been developed for measuring small variations in the abundance of specific isotopes, and the ratios of isotopes with increasing precision.

Most elements have several naturally occurring isotopes, as the number of neutrons that can form a stable or long-lived nucleus can vary, and the relative abundances of these isotopes can be very different. Every element also has isotopes that contain neutrons in a quantity that render them unstable. While most have exceedingly short half-lives and are only seen under artificial conditions (>80% of the 2,500 nuclides), there are many that are produced by naturally-occurring processes and have sufficiently long half-lives to be present in the environment in measurable quantities. Such production involves nuclear reactions, either the decay of parent isotopes, the interactions of stable nuclides with natural fluxes of subatomic particles in the environment, or the reactions occurring in nuclear reactors or nuclear detonations. The isotopes thus produced provide the basis for most methods for obtaining absolute ages and information on the rates of environmental processes. Their decay follows the well-known radioactive decay law (first-order kinetics), where the fraction of atoms, λ (the decay constant), that decay over a period of time is fixed and an intrinsic characteristic of the isotope:

D. Porcelli (✉)
Department of Earth Sciences, Oxford University, South Parks Road, Oxford OX1 3AN, UK
e-mail: Don.Porcelli@earth.ox.ac.uk

M. Baskaran
Department of Geology, Wayne State University, Detroit, MI 48202, USA
e-mail: Baskaran@wayne.edu

$$\frac{dN}{dt} = -\lambda N. \tag{2.1}$$

When an isotope is incorporated and subsequently isolated in an environmental material with no exchange with surroundings and no additional production, the abundance changes only due to radioactive decay, and (2.1) can be integrated to describe the resulting isotope abundance with time;

$$N = N_0 e^{-\lambda t}. \tag{2.2}$$

The decay constant is related to the well-known half-life (t1=2) by the relationship:

$$\lambda = \frac{\ln 2}{t_{1/2}} = \frac{0.693}{t_{1/2}}. \tag{2.3}$$

This equation provides the basis for all absolute dating methodologies. However, individual methods may involve considering further factors, such as continuing production within the material, open system behaviour, or the accumulation of daughter isotopes. The radionuclides undergo radioactive decay by alpha, beta (negatron and positron) or electron capture. The elements that have radioactive isotopes, or isotopic variations due to radioactive decay, are shown in Fig. 2.1.

Variations in stable isotopes also occur, as the slight differences in mass between the different isotopes lead to slightly different bond strengths that affect the partitioning between different chemical species and the adsorption of ions. Isotope variations therefore provide a fingerprint of the processes that have affected an element. While the isotope variations of H, O, and C have been widely used to understand the cycles of water and carbon, relatively recent advances in instrumentation has made it possible to precisely measure the variations in other elements, and the potential information that can be obtained has yet to be fully exploited. The full range of elements with multiple isotopes is shown in Fig. 2.2.

One consideration for assessing the feasibility of obtaining isotopic measurements is the amount of an element available. Note that it is not necessarily the concentrations that are a limitation, but the absolute amount, since elements can be concentrated from whatever mass is necessary- although of course there are considerations of difficulty of separation, sample availability and blanks. For example, it is not difficult to filter very large volumes of air, or to concentrate constituents from relatively large amounts of water, but the dissolution of large silicate rock samples is more involved. In response to difficulties in present methods or the challenges of new applications, new methods for the separation of the elements of interest from different materials are being constantly developed. Overall, analyses can be performed not only on major elements, but even elements that are trace constituents; e.g. very pure materials (99.999% pure) still contain constituents in concentrations of micrograms

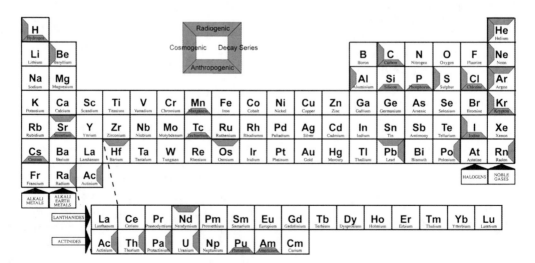

Fig. 2.1 Elements with isotopes that can be used in environmental studies and are anthropogenic, cosmogenic (produced either in the atmosphere or within exposed materials), radiogenic (from production of long-lived isotopes), or that are part of the U- or Th-decay series

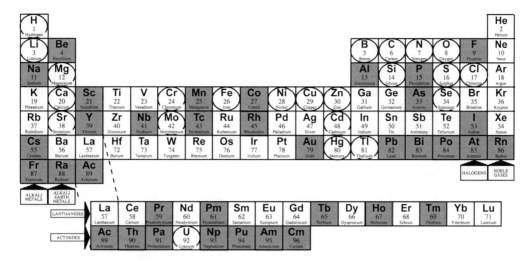

Fig. 2.2 The elements that have stable isotope variations useful for environmental studies. The shaded elements do not have more than one isotope that is stable and not radiogenic. While all others potentially can display stable isotope variations, significant isotopic variations in the environment have been documented for the circled elements

per gram (ppm) and so are amenable to analysis. Further considerations of the abundances that can be measured are discussed below.

The following sections provide a general guide to the range of isotopes available, and the most widespread uses in the terrestrial environment. It is not meant to be exhaustive, as there are many innovative uses of isotopes, but rather indicative of the sorts of problems can be approached, and what isotopic tools are available for particular question. More details about the most commonly used methods, as well as the most innovative new applications, are reported elsewhere in this volume. Section 2.3 provides a brief survey of the analytical methods available.

2.2 Applications of Isotopes in the Environment

In the following sections, applications of isotopes to environmental problems are presented according to the different sources of radioactive isotopes and causes of variations in stable isotopes.

2.2.1 Atmospheric Short-Lived Nuclides

A range of isotopes is produced from reactions involving cosmic rays, largely protons, which bombard the Earth from space. The interactions between these cosmic rays and atmospheric gases produce a suite of radionuclides with half-lives ranging from less than a second to more than a million years (see list in Lal and Baskaran 2011). There are a number of nuclides that have sufficiently long half-lives to then enter into environmental cycles in various ways (see Table 2.1). The best-known nuclide is ^{14}C, which forms CO_2 after its production and then is incorporated into organic matter or dissolves into the oceans. With a half-life of 5730a, it can be used to date material that incorporates this $^{14}CO_2$, from plant material, calcium carbonate (including corals), to circulating ocean waters. Information can also be obtained regarding rates of exchange between reservoirs with different $^{14}C/^{12}C$ ratios, and biogeochemical cycling of C and associated elements. Other nuclides are removed from the atmosphere by scavenging onto aerosols and removed by precipitation, and can provide information on the rates of atmospheric removal. By entering into surface waters and sediments, these nuclides also serve as environmental tracers. For example, there are two isotopes produced of the particle-reactive element Be, ^{7}Be ($t_{1/2} = 53.3$ days) and ^{10}Be (1.4Ma). The distribution of $^{7}Be/^{10}Be$ ratios, combined with data on the spatial variations in the flux of Be isotopes to the Earth's surface, can be used to quantify processes such as stratospheric-tropospheric exchange of air masses, atmospheric circulation, and the removal rate

Table 2.1 Atmospheric radionuclides

Isotope	Half-life	Common applications
^{3}H	12.32a	Dating of groundwater, mixing of water masses, diffusion rates
^{7}Be	53.3 day	Atmospheric scavenging, atmospheric circulation, vertical mixing of water, soil erosion studies
^{10}Be	1.4×10^6a	Dating of sediments, growth rates of Mn nodules, soil erosion study, stratosphere-troposphere exchange, residence time of aerosols
^{14}C	5730a	Atmospheric circulation, dating of sediments, tracing of C cycling in reservoirs, dating groundwater
^{32}Si	140a	Atmospheric circulation, Si cycling in the ocean
^{32}P	25.3 day	Atmospheric circulation, tracing oceanic P pool
^{33}P	14.3 day	
^{35}S	87 day	Cycling of S in the atmosphere
^{39}Ar	268a	Atmospheric circulation and air-sea exchange
^{81}Kr	2.3×10^5a	Dating of groundwater
^{129}I	1.6×10^7a	Dating of groundwater
^{210}Pb	22.3a	Dating, deposition velocity of aerosols, sources of air masses, soil erosion, sediment focusing

Selected isotopes generated within the atmosphere that have been used for environmental studies, along with some common applications. All isotopes are produced by nuclear reactions in the atmosphere induced by cosmic radiation, with the exception of ^{210}Pb, which is produced by decay of ^{222}Rn released from the surface

of aerosols (see Lal and Baskaran 2011). The record of Be in ice cores and sediments on the continents can be used to determine past Be fluxes as well as to quantify sources of sediments, rates of sediment accumulation and mixing (see Du et al. 2011; Kaste and Baskaran 2011). Other isotopes, with different half-lives or different scavenging characteristics, provide complementary constraints on atmospheric and sedimentary processes (Table 2.1).

A number of isotopes are produced in the atmosphere from the decay of ^{222}Rn, which is produced within rocks and soils from decay of ^{226}Ra, and then released into the atmosphere. The daughter products of ^{222}Rn (mainly ^{210}Pb (22.3a) and ^{210}Po (138 days)) that are produced in the atmosphere have been used as tracers to identify the sources of aerosols and their residence times in the atmosphere (Kim et al. 2011; Baskaran 2011). Furthermore, ^{210}Pb adheres to particles that are delivered to the Earth's surface at a relatively constant rate and are deposited in sediments, and its subsequent decay provides a widely used method for determining the age of sediments and so the rates of sedimentation.

A number of isotopes are incorporated into the hydrologic cycle and so provide means for dating groundwaters. This includes the noble gases ^{39}Ar ($t_{1/2}$ = 268a), ^{81}Kr (230ka) and ^{3}H (Kulongoski and Hilton 2011) which dissolve into waters and then provide ideal tracers that do not interact with aquifer rocks and so travel conservatively with groundwater, but are present in such low concentrations that their measurement has proven to be difficult. The isotope ^{129}I (16Ma) readily dissolves and also behaves conservatively: with such a long half-life, however, it is only useful for very old groundwater systems. The readily analyzed ^{14}C also has been used for dating groundwater, but the ^{14}C/^{12}C ratio changes not only because of decay of ^{14}C, but also through a number of other processes such as interaction with C-bearing minerals such as calcium carbonate; therefore, more detailed modelling is required to obtain a reliable age.

General reviews on the use of isotopes produced in the atmosphere for determining soil erosion and sedimentation rates, and for providing constraints in hydrological studies, are provided by Lal (1991, 1999) and Phillips and Castro (2003).

2.2.2 Cosmogenic Nuclides in Solids

Cosmogenic nuclides are formed not just within the atmosphere, but also in solids at the Earth's surface, and so can be used to date materials based only upon exposure history, rather than reflecting the time of formation or of specific chemical interactions. The cosmic particles that have escaped interaction within the atmosphere penetrate into rocks for up to a few meters, and interact with a range of target elements to generate nuclear reactions through neutron capture, muon capture, and spallation (emission of various fragments). From the present concentration and the production rate, an age for the exposure of that surface to cosmic rays can be readily calculated (Lal 1991). A wide range of nuclides is produced, although only a few are produced in detectable amounts, are sufficiently long-lived, and are not naturally present in concentrations that overwhelm additions from cosmic ray interactions. An additional complexity in obtaining ages from this method is that production rates must be well known, and considerable research has been devoted to their determination. These are dependent upon target characteristics, including the concentration of target isotopes, the depth of burial, and the angle of exposure, as well as factors affecting the intensity of the incident cosmic radiation, including altitude and geomagnetic latitude. Also, development of these methods has been coupled to advances in analytical capabilities that have made it possible to measure the small number of atoms involved. It is the high resolution available from accelerator mass spectrometry (see Sect. 2.3.2) that has made it possible to do this in the presence of other isotopes of the same element that are present in quantities that are many orders of magnitude greater.

The most commonly used cosmogenic nuclides are listed in Table 2.2 (see also Fig. 2.3). These include several stable isotopes, ^{3}He and ^{21}Ne, which accumulate continuously within materials. In contrast, the radioactive isotopes ^{10}Be, ^{14}C, ^{26}Al, and ^{36}Cl will continue to increase until a steady state concentration is reached in which the constant production rate is matched by the decay rate (which is proportional to the concentration). While this state is approached asymptotically, in practice within ~5 half-lives concentration changes are no longer resolvable. At this

Table 2.2 Widely used cosmogenic nuclides in solids

Isotope	Primary targets	Half-life	Commonly dated materials
^{10}Be	O, Mg, Fe	1.4 Ma	Quartz, olivine, magnetite
^{26}Al	Si, Al, Fe	705 ka	Quartz, olivine
^{36}Cl	Ca, K, Cl	301 ka	Quartz
^{3}He	O, Mg, Si, Ca, Fe, Al	Stable	Olivine, pyroxene
^{21}Ne	Mg, Na, Al, Fe, Si	Stable	Quartz, olivine, pyroxene

The most commonly used isotopes that are produced by interaction of cosmic rays the primary target elements listed

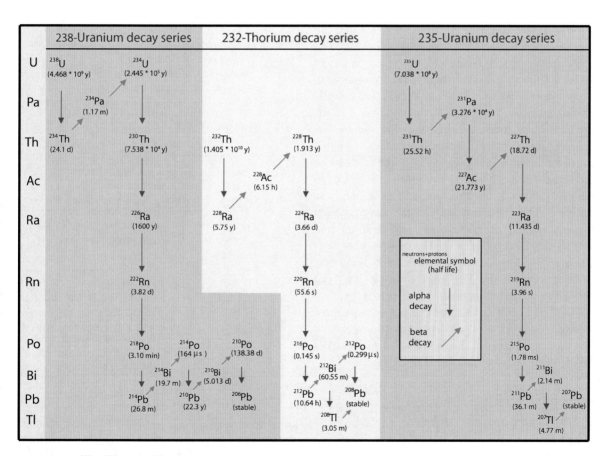

Fig. 2.3 The ^{238}U, ^{235}U, and ^{232}Th decay series. The presence of the series of short-lived nuclides throughout the environment is due to their continuous production by long-lived parents

point, no further time information is gained; such samples can then be assigned only a minimum age.

There have been a considerable number of applications of these methods, which have proven invaluable to the understanding of recent surface events. There are a number of reviews available, including those by Niedermann (2002), which focuses on ^{3}He and ^{21}Ne, and Gosse and Phillips (2001). A full description of the methods and applications is given by Dunai (2010). Some of the obvious targets for obtaining simple exposure ages are lava flows, material exposed by landslides, and archaeological surfaces. Meteorite impacts have been dated by obtaining exposure ages of excavated material, and the timing of glacial retreats has been constrained by dating boulders in glacial moraines and glacial erratics. The ages of the oldest surfaces in dry environments where little erosion occurs have also been obtained. Movement on faults has been studied by measuring samples along fault scarps to obtain the rate at which the fault face was exposed.

Cosmogenic nuclides have also been used for understanding landscape evolution (see review by Cockburn and Summerfield 2004) and erosion rates (Lal 1991). In this case, the production rate with depth must be known and coupled with an erosion history, usually assumed to occur at a constant rate. The concentrations of samples at the surface (or any depth) are then the result of the production rate over the time that the sample has approached the surface due to erosion and was subjected to progressively increasing production. The calculations are somewhat involved, since production rates due to neutron capture, muon reactions, and spallation have different depth dependencies, and the use of several different cosmogenic nuclides can provide better constraints (Lal 2011). Overall, plausible rates have been obtained for erosion, which had hitherto been very difficult to constrain. The same principle has been applied to studies of regional rates of erosion by measuring ^{10}Be in surface material that has been gathered in rivers (Schaller et al. 2001), and this has provided key data for regional landscape evolution studies (Willenbring and von Blanckenburg 2010).

2.2.3 Decay Series Nuclides

The very long-lived nuclides ^{238}U, ^{235}U, and ^{232}Th decay to sequences of short-lived nuclides that eventually lead to stable Pb isotopes (see Fig. 2.3 and Table 2.3). The decay of these parents thus supports a continuous definable supply of short-lived nuclides in the environment that can be exploited for environmental studies, especially for determining ages and rates over a range of timescales from days to hundreds of thousands of years. The abundance of each isotope is controlled by that of its parent, and since this dependence continues up the chain, the connections between the isotopes can lead to considerable complexity in calculating the evolution of some daughters. However, since the isotopes represent a wide range of elements with very different geochemical behaviours, such connections also present a wealth of opportunities for short-term geochronology and

Table 2.3 Decay series nuclides used in environmental studies

Isotope	Half-life	Common applications
^{238}U	4.468×10^{9}a	Dating, tracing sources of U
^{234}U	2.445×10^{4}a	Dating of carbonates, tracing sources of water
^{232}Th	1.405×10^{10}a	Quantifying lithogenic component in aqueous system, atmosphere
^{230}Th	7.538×10^{4}a	Dating, scavenging, ventilation of water mass
^{234}Th	24.1 day	Particle cycling, POC export, rates of sediment mixing
^{228}Th	1.913a	Particle scavenging and tracer for other particulate pollutants
^{227}Th	18.72 day	Particle tracer
^{231}Pa	3.276×10^{4}a	Dating, sedimentation rates, scavenging
^{228}Ra	5.75a	Tracing water masses, vertical and horizontal mixing rates
^{226}Ra	1600a	Dating, water mass tracing, rates of mixing
^{224}Ra	3.66 day	Residence time of coastal waters, mixing of shallow waters
^{223}Ra	11.435 day	Residence time of coastal waters, mixing of shallow waters
^{227}Ac	21.773a	Dating, scavenging
^{222}Rn	3.82 day	Gas exchange, vertical and horizontal diffusion
^{210}Pb	22.3a	Dating (e.g. carbonates, sediments, ice cores, aerosols, artwork); sediment mixing, focusing and erosion; scavenging; resuspension
^{210}Po	138 day	Carbon export, remineralization, particle cycling in marine environment

Radioactive isotopes within the ^{238}U, ^{235}U, and ^{232}Th decay series that have been directly applied to environmental studies. The remaining isotopes within the decay series are generally too short-lived to provide useful environmental information and are simply present in activities equal to those of their parents

environmental rate studies (see several papers in Ivanovich and Harmon 1992; Baskaran 2011).

The abundance of a short-lived nuclide is conveniently reported as an activity ($= N\lambda$, i.e. the abundance times the decay constant), which is equal to the decay rate (dN/dt). In any sample that has been undisturbed for a long period (>~1.5 Ma), the activities of all of the isotopes in each decay chain are equal to that of the long-lived parent element, in what is referred to as secular equilibrium. In this case, the ratios of the abundances of all the daughter isotopes are clearly defined, and the distribution of all the isotopes in a decay chain is controlled by the distribution of the long-lived parent. Unweathered bedrock provides an example where secular equilibrium could be expected to occur. However, the different isotopes can be separated by a number of processes. The different chemical properties of the elements can lead to different mobilities under different environmental conditions. Uranium is relatively soluble under oxidizing conditions, and so is readily transported in groundwaters and surface waters. Thorium, Pa and Pb are insoluble and highly reactive with surfaces of soil grains and aquifer rocks, and adsorb onto particles in the water column. Radium is also readily adsorbed in freshwaters, but not in highly saline waters where it is displaced by competing ions. Radon is a noble gas, and so is not surface-reactive and is the most mobile.

Isotopes in the decay series can also be separated from one another by the physical process of recoil. During alpha decay, a sufficient amount of energy is released to propel alpha particles a considerable distance, while the daughter isotope is recoiled in the opposite direction several hundred Angstroms (depending upon the decay energy and the matrix). When this recoil sends an atom across a material's surface, it leads to the release of the atom. This is the dominant process releasing short-lived nuclides into groundwater, as well as releasing Rn from source rocks. This mechanism therefore can separate short-lived daughter nuclides from the long-lived parent of the decay series. It can also separate the products of alpha decay from those of beta decay, which is not sufficiently energetic to result in substantial recoil. For example, waters typically have (^{234}U/^{238}U) activity ratios that are greater than the secular equilibrium ratio of that found in crustal rocks, due to the preferential release of ^{234}U by recoil.

A more detailed discussion of the equations describing the production and decay of the intermediate daughters of the decay series is included in Appendix 2. In general, where an intermediate isotope is isolated from its parent, it decays according to (2.2). Where the activity ratio of daughter to parent is shifted from the secular equilibrium value of 1, the ratio will evolve back to the same activity as its parent through either decay of the excess daughter, or grow-in of the daughter back to secular equilibrium (Baskaran 2011). These features form the basis for dating recently produced materials. In addition, U- and Th- series systematics can be used to understand dynamic processes, where the isotopes are continuously supplied and removed by physical or chemical processes as well as by decay (e.g. Vigier and Bourdon 2011).

The U-Th series radionuclides have a wide range of applications throughout the environmental sciences. Recent reviews cover those related to nuclides in the atmosphere (Church and Sarin 2008; Baskaran 2010; Hirose 2011), in weathering profiles and surface waters (Chabaux et al. 2003; Cochran and Masque 2003; Swarzenski et al. 2003), and in groundwater (Porcelli and Swarzenski 2003; Porcelli 2008). Recent materials that incorporate nuclides in ratios that do not reflect secular equilibrium (due either to discrimination during uptake or availability of the nuclides) can be dated, including biogenic and inorganic carbonates from marine and terrestrial environments that readily take up U and Ra but not Th or Pb, and sediments from marine and lacustrine systems that accumulate sinking sediments enriched in particle-reactive elements like Th and Pb (Baskaran 2011). The migration rates of U-Th-series radionuclides can be constrained where continuing fractionation between parent and daughter isotopes occurs. These rates can then be related to broader processes, such as physical and chemical erosion rates as well as water-rock interaction in groundwater systems, where soluble from insoluble nuclides are separated (Vigier and Bourdon 2011). Also, the effects of particles in the atmosphere and water column can be assessed from the removal rates of particle-reactive nuclides (Kim et al. 2011).

2.2.4 Anthropogenic Isotopes

Anthropogenic isotopes are generated through nuclear reactions created under the unusual circumstances of high energies and high atomic particle fluxes, either within nuclear reactors or during nuclear weapons explosions. These are certainly of concern as contaminants in the environment, but also provide tools for environmental studies, often representing clear signals from defined sources. These include radionuclides not otherwise present in the environment that can therefore be clearly traced at low concentrations (e.g. ^{137}Cs, ^{239,240}Pu; Hong et al. 2011; Ketterer et al. 2011), as well as distinct pulses of otherwise naturally-occurring species (e.g. ^{14}C). There is a very wide range of isotopes that have been produced by such sources, but many of these are too short-lived or do not provide a sufficiently large signal over natural background concentrations to be of widespread use in environmental studies. Those that have found broader application are listed in Table 2.4.

The release of radioactive nuclides into the environment from reactors, as well as from nuclear waste reprocessing and storage facilities, can occur in a number of ways. Discharges through airborne effluents can widely disperse the isotopes over a large area and enter soils and the hydrological cycle through fallout, as observed for the Chernobyl release in 1986. Discharges of water effluents, including cooling and process waters, can be followed from these point sources in circulating waters, such as from the Sellafield nuclear processing and power plant, which released radionuclides (e.g. ^{137}Cs, ^{134}Cs, ^{129}I) from the western coast of England into the Irish Sea and which can be traced high into the Arctic Ocean (discussed in Hong et al. 2011). Leaks into the ground from storage and processing facilities can also release radionuclides that are transported in groundwater, at locations such as the Hanford nuclear production facility in Washington State. At the Nevada Test Site, Pu was found to have migrated >1 km, facilitated by transport on colloids. Releases by all of these mechanisms at lower levels have been documented around many reactor facilities. Even where levels are too low to pose a health concern, the isotopes can be readily measured and their sources identified. Transuranics are also utilized as tracers for investigating soil erosion, transport and deposition in the environment (Ketterer et al. 2011; Matisoff and Whiting 2011).

As discussed above, cosmic rays entering the atmosphere cause nuclear reactions that produce a number of radioisotopes. Atmospheric bomb testing, which peaked in the late 1950s and diminished dramatically after 1963, produced a spike in the production of some of these radionuclides (e.g. ^{14}C), and high atmospheric concentrations have persisted due to long atmospheric residence times or continuing fluxes from nuclear activities. These higher concentrations can then be used as markers to identify younger materials. The most widely used isotopes in this category are ^{3}H and ^{14}C. There are a number of others, including ^{85}Kr (10.72a), ^{129}I (15.7Ma), and ^{36}Cl (3.0 × 10^{5}a)

Table 2.4 Anthropogenic isotopes most commonly used in environmental studies

Isotope	Half-life	Main sources	Examples of major uses	Notes
^{3}H	12.32a	Bomb testing	Tracing rainwater from time of bomb peak in ground waters and seawater	Background cosmogenic ^{3}H
		Reactors	Ages determined using ^{3}H-^{3}He	
^{14}C	5730a	Bomb testing	Identifying organic and inorganic carbonate materials produced in last 60 years	Background cosmogenic ^{14}C
^{54}Mn	312 day	Bomb testing	Trophic transfer in organisms	No background
^{137}Cs, ^{134}Cs	30.14a	bomb testing, reactors, reprocessing plants	Tracing seawater, soil erosion, sediment dating and mixing, sources of aerosols	No background
^{90}Sr	28.6a	Reactors, bomb testing	Tracing seawater, sources of dust	No background
^{239}Pu	24100a	Reactors, bomb testing	Tracing seawater, identifying sources of Pu, sediment dating, soil erosion, tracing atmospheric dust	No naturally-occurring Pu
^{240}Pu	6560a			
^{241}Pu	14.4a			
^{131}I	8.02 day			
^{129}I	1.57 × 10^{7}a	Bomb testing, reactors	Dating groundwater, water mass movements	Background cosmogenic ^{129}I
^{85}Kr	10.72a	Bomb testing	Dating groundwater	
^{36}Cl	3.0 × 10^{5}a	Bomb testing	Dating groundwater	Background cosmogenic ^{36}Cl
^{241}Am	432.2a	Reactors	Tracing sources of Am	No naturally-occurring Am

The anthropogenic isotopes that can be used as point source tracers, or provide global markers of the time of formation of environmental materials

that have not been as widely applied, partly because they require difficult analyses. For a discussion of a number of applications, see Phillips and Castro (2003).

For the anthropogenic radionuclides that have half-lives that are long compared to the times since their release, time information can be derived from knowing the time of nuclide production, e.g. the present distribution of an isotope provides information on the rate of transport since production or discharge. An exception to this is ^{3}H, which has a half-life of only 12.3 years and decays to the rare stable isotope ^{3}He. By measuring both of these isotopes, time information can be obtained, e.g. ^{3}H-^{3}He ages for groundwaters, which measure the time since rainwater incorporating atmospheric ^{3}H has entered the aquifer (Kulongoski and Hilton 2011).

2.2.5 Radiogenic Isotopes

There are a number of radioactive isotopes with half-lives that are useful for understanding geological timescales, but do not change significantly over periods of interest in environmental studies. However, these isotopes have produced substantial variations in the isotopes of the elements that have daughter isotopes. The resulting isotopic signatures in rocks, sediments, and ores have been exploited in environmental studies. An example is the Rb-Sr system, which is the most commonly used. The long-term decay of ^{87}Rb produces ^{87}Sr, and so the abundance of this isotope, relative to a stable isotope of Sr such as ^{86}Sr, varies according to the equation:

$$\frac{^{87}Sr}{^{86}Sr} = \left(\frac{^{87}Sr}{^{86}Sr}\right)_0 + \frac{^{87}Rb}{^{86}Sr}\left(e^{\lambda_{87}t} - 1\right), \quad (2.4)$$

where the (^{87}Sr/^{86}Sr)$_0$ ratio is the initial isotope ratio and λ_{87} is the decay constant of ^{87}Rb. Clearly, the present ^{87}Sr/^{86}Sr ratio is greater for samples that have greater ages, t, and samples with larger Rb/Sr ratios. The ^{87}Sr/^{86}Sr ratio therefore varies between different rock types and formations. Since Rb is an alkali metal and Sr is an alkaline earth, these elements behave differently in geological processes, creating large variations in Rb/Sr, and so large variations in ^{87}Sr/^{86}Sr. The ^{87}Sr/^{86}Sr ratio has been shown to vary widely in surface rocks, and so any Sr released into soils, rivers, and groundwaters has an isotopic signature that reflects its source. Changes in local sources can also be identified; for example, Keller et al. (2010) identified thawing of permafrost from the changes in stream water ^{87}Sr/^{86}Sr ratio as tills with greater amounts of carbonate deeper in soil profiles were weathered. Strontium isotopes have been used as tracers for identifying populations of migratory birds and fish in streams and rivers (Kennedy et al. 1997; Chamberlain et al. 1997) as well as in archaeology (Slovak and Paytan 2011). Sr isotopes have also been used to trace agricultural products, which have incorporated Sr, along with Ca, from soils incorporating Sr isotope ratios of the underlying rocks. The sources and transport of dust across the globe have been traced using the isotopic compositions of Sr, Nd, and Pb (Grousset and Biscaye 2005). Variations in the Pb isotopic ratios between environmental samples (soil, aerosols, household paint, household dust, and water samples in house) and children's blood have been utilized to identify and quantify the sources of blood Pb in children (Gulson 2011).

Other isotope systems that can be used for similar purposes are listed in Table 2.5. In each case, chemical differences between the chemical behaviour of the parent and daughter isotopes have led to variations in their ratio over geological timescales, and so identifiable variations in the environment can be exploited for tracing. The most obvious differences are often regional, as reflected in the isotopic compositions in rivers, especially for Sr and Nd. More local studies can identify different rocks, or even separate minerals, involved in weathering or contributing to waters (Tripathy et al. 2011). Using this information, groundwater flow paths and the sources of inorganic constituents can be deduced.

2.2.6 Stable Isotopes

In addition to variations in isotope ratios produced by the production or decay of nuclides due to nuclear processes, there are variations in the distribution of other isotopes between different phases or chemical species. These are subtle, as isotopes of an element still behave in fundamentally the same way, but the slight variations in atomic mass result in differences in bonding and kinetics. This is due to differences in vibrational energies reflecting differences in the masses of atoms, since the vibrational frequency of a

Table 2.5 Radiogenic isotopes commonly used in environmental studies

Decay scheme	Half-life	Isotope ratio	Comments
^{87}Rb → ^{87}Sr	4.88×10^{10} a	^{87}Sr/^{86}Sr	Bioarchaeology, erosion rates
^{147}Sm → ^{143}Nd	1.06×10^{11} a	^{143}Nd/^{144}Nd	Regional variations due to age, rates of weathering, source tracking of sediments
^{238}U → ^{206}Pb	4.468×10^{9} a	^{206}Pb/^{204}Pb	Distinguish sources of Pb ores and sources of Pb; tracing sources of Pb to human body
^{235}U → ^{207}Pb	7.038×10^{8} a	^{207}Pb/^{204}Pb	
^{232}Th → ^{208}Pb	1.405×10^{10} a	^{208}Pb/^{204}Pb	
^{176}Lu → ^{176}Hf	3.71×10^{10} a	^{176}Hf/^{177}Hf	Complements Nd/Sm
^{187}Re → ^{187}Os	4.6×10^{10} a	^{187}Os/^{188}Os	Organic-rich sources have distinctly high Re/Os
^{238}U, ^{235}U, ^{232}Th → ^{4}He	As above	^{3}He/^{4}He	Dating by ^{4}He accumulation
			Tracing deep sources of volatiles
^{40}K → ^{40}Ar	1.397×10^{9} a	^{40}Ar/^{36}Ar	Dating

The isotopes that are produced by long-lived parents and so exhibit variations in geological materials that can be used as tracers in environmental studies

molecule is directly proportional to the forces that hold atoms together, such as electron arrangements, nuclear charges and the positions of the atoms in the molecule, and inversely proportional to the masses of the atoms. The result is stable isotope fractionation, which is generally considered only for isotopes that are neither radioactive nor radiogenic, since variations in these isotopes often cannot be readily separated from decay effects. While studies of such fractionations were originally confined to light elements (e.g. H, O, N, S), isotopic variations have now been found even in the heavy stable elements Tl (Rehkamper and Halliday 1999) and U (Stirling et al. 2007), and so across the entire periodic table.

The difference in the ratio of two isotopes of the same element in species or phases A (R_A) and B (R_B) can be described by a fractionation factor α

$$\alpha = \frac{R_A}{R_B}. \qquad (2.5)$$

Fractionation between isotopes of the same element generally appear to be linear; e.g. the ratio of isotopes two mass units apart will be fractionated twice as much as the ratio of elements one mass unit apart (mass dependent fractionation). Since the changes in isotopic values are generally very small, values are typically given as parts per thousand (permil) deviations from a standard; e.g. for example, for O,

$$\delta^{18}O_{sample} = \frac{\left(^{18}O/^{16}O\right)_{sample} - \left(^{18}O/^{16}O\right)_{standard}}{\left(^{18}O/^{16}O\right)_{standard}} \times 10^3. \qquad (2.6)$$

Therefore, results are reported relative to a standard. In some cases, however, several standards are in use, especially when analytical procedures are being developed, making comparisons between datasets difficult. In those cases where smaller variations can be measured precisely, values are given in parts per 10^4 (as ε values). When the isotopic composition of two phases in equilibrium are measured, the difference between them can be related to the fractionation factor by

$$\delta^{18}O_A - \delta^{18}O_B \approx (\alpha - 1) \times 10^3. \qquad (2.7)$$

Therefore, the fractionation factor between two phases can be obtained by measuring the difference in isotopic composition between phases in equilibrium.

Schauble (2004) provides a summary of the general patterns seen in equilibrium isotope fractionation. Fractionation effects are largest for low mass elements, and scale according to the difference in atomic masses of two isotopes relative to the average mass of the element squared, i.e. $\Delta m/m^2$. Heavy isotopes will favour species in which the element has the stiffest bonds. In environmental systems, this includes bonds of metals with the highest oxidation state, those with a bond partner with a high oxidation state, and bonds with anions with low atomic number (including organic ligands).

Fractionation can occur not only by equilibrium partitioning, but also due to kinetic processes. This includes transport phenomena such as diffusion, where lighter isotopes have higher velocities (inversely proportional to the square root of atomic/

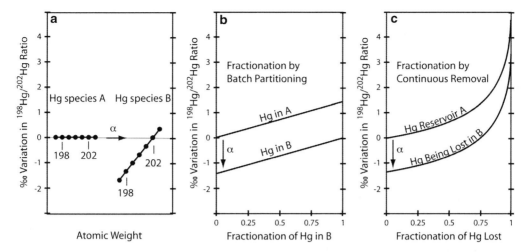

Fig. 2.4 An example of stable isotope fractionation. (**a**) When Hg is partitioned between two phases, isotopic fractionation can occur; the heavier isotopes are preferentially incorporated into phase B, so that $\alpha(^{198}\text{Hg}/^{202}\text{Hg})_A = (^{198}\text{Hg}/^{202}\text{Hg})_B$. Fractionation between all the isotopes is generally linear, i.e. the extent of fractionation between any two isotopes is proportional to the mass difference between them. (**b**) If there is equilibrium batch partitioning between A and B, then there is a difference of α between both phases, and the specific composition of each depends upon mass balance; e.g. if there is very little of phases A, then the ratio of phase B is about equal to the bulk value, while phase A has a value that is α times higher. (**c**) If there is continuous removal of phase B, so that only each increment being removed is in equilibrium with phase A, then this process of Rayleigh distillation will drive the isotopic composition of A to more extreme compositions than a batch process

molecular weight) that generate fractionation over diffusive gradients. This can also be due to reaction kinetics related to the strength of bonding, where lighter isotopes have bonds that can be broken more readily and so can be enriched in reactants. This can be seen in many biological processes, where uptake of elements can involve pathways involving multiple reactions and so considerable fractionation. It is important to note that it is not always possible to determine if equilibrium or kinetic fractionation is occurring; while laboratory experiments often endeavour to ensure equilibrium is reached, it is not always clear in the natural environment. Also, where kinetic fractionation is produced in a regular fashion, for example due to microbial reduction, it may not be easily distinguished from inorganic equilibrium fractionation processes

When the isotopes of an element are distributed between two separate phases or species in equilibrium, then the difference in isotope ratio is determined by (2.7). However, the absolute ratio of each is determined by mass balance according to the simple mixing equation

$$R_{Total} = R_A f_A + R_B(1 - f_A). \quad (2.8)$$

For example, as shown in Fig. 2.4, if an element is partitioned between two phases, and the mixture is almost entirely composed of phase A, then that phase will have the ratio close to that of the total mixture, while the ratio of phase B will be offset by the fractionation factor. If the mixture is almost entirely composed of phase B, it is the composition of phase A that will be offset from the bulk. Intermediate values of R for each phase will reflect the balance between the two phases according to (2.8). Such batch partitioning is more likely to apply where both forms are entirely open to chemical exchange, e.g. two different species in solution.

Often equilibrium is not maintained between a chemical species or phase and the source reservoir. In this case, the isotopic composition of the source evolves according to Rayleigh distillation, in which a small increment of a phase or species is formed in equilibrium with the source, but is immediately removed from further interaction, and so does not remain in equilibrium with the changing source. For example, crystals precipitating from solution are unlikely to continue reacting with the solution. If the source originally had an isotope ratio R_0, and the isotopes of an element of interest are fractionated with material being removed according to a fractionation factor α, then when in total a fraction f of the element has been removed, the integrated effect will be to generate a ratio R_A in the source according to the equation:

$$\frac{R_A}{R_0} = f^{(\alpha-1)}. \quad (2.9)$$

This is illustrated in Fig. 2.4; once a significant amount of an element has been removed, the ratio of the source has changed to a greater degree than by batch partitioning, and more extreme ratios can be obtained when f approaches 1 (most of the element has been removed). As long as the same fractionation factor affects each increment, (2.9) will describe the evolution of the solution, whether the fractionation is due to equilibrium or kinetic fractionation. Such evolution is likely to apply where solid phases are formed from solution, where water-rock interaction occurs, and where biological uptake occurs.

While fractionation between isotopes of an element is generally mass dependent, i.e. proportional to the mass deference between any two isotopes, there are a few circumstances where mass-independent fractionation is observed and a number of theories have been proposed to explain this phenomenon. Examples include O isotopes in stratospheric and tropospheric ozone (due to rotational energies of asymmetric molecules; Gao and Marcus 2001), S in Archean sediments (possibly from S photo-dissociated in the early atmosphere; Farquhar et al. 2000), and Hg isotopes involved in photochemical reactions in the environment (due to nuclear volume and magnetic isotope effects; see Blum 2011). These provide an added dimension for diagnosing the processes affecting these elements.

A wide range of applications has been explored for stable isotope fractionations. Oxygen isotopes vary in precipitation, and have been used as a hydrological tracer, while C, S, and N have been used to trace biogeochemical cycles (see e.g. reviews in Kendall and McDonnell 1998) and tracers of human diet (Chesson et al. 2011; Schwarcz and Schoeninger 2011). Isotopic variations in Li, Mg, and Sr have been related to adsorption and incorporation in clays, and so can be used to understand weathering and trace element migration (Burton and Vigier 2011; Reynolds 2011). Various metals with multiple redox states, including Fe, Cr, U, Mo, Ni, Cu, Zn, Cd, Hg, Se, and Tl, have isotopic variations that have been used to understand redox processes in present and past environments (see the four papers by Bullen 2011; Johnson 2011; Nielsen and Rehkämper 2011; Rehkämper et al. 2011).

Stable isotopes can also be used in understanding environmental contamination. Chemical processing can generate isotopic differences in industrial products that provide a means for identifying sources. For example, C and Cl isotope differences have been found between different manufacturers of chlorinated solvents (van Warmerdam et al. 1995) and this can be used to trace contaminant migration. It has also been shown that degradation of these compounds can fractionate Cl isotopes, so that the extent of degradation can be determined (Sturchio et al. 1998, 2011). As another example, Cr isotopes are fractionated in groundwaters when Cr is reduced to a less soluble species, and so decreases in concentrations due to dilution can be distinguished from decreases due to precipitation (Ellis et al. 2002).

Other examples of documented isotope variations that have been used to understand environmental processes are listed in Table 2.6. Measurements are all made using mass spectrometry. Samples are corrected for instrumental fractionation using either comparison with standards in gas-source mass spectrometry and ICPMS, or using a double spike and TIMS and ICPMS (see below).

2.3 Measurement Techniques

There are two fundamental approaches to measuring the abundance of isotopes. Radioactive isotopes can be measured by quantifying the radiation resulting from decay, and then relating the decay rate to the abundance using (2.1). A number of these counting methods are widely used, especially for short-lived nuclides. Alternatively, atoms of any isotope can be directly measured by mass spectrometry, which separates ions according to mass followed by direct counting. These methods typically involve greater investment in instrumentation, but can often detect more subtle variations. The techniques are discussed further in the following sections.

2.3.1 Counting Methods

There are a number of techniques available for the detection of alpha, beta, and gamma radiation that

Table 2.6 Commonly used stable isotopes

Element	Typical reported ratio	Examples of applications
H	$^2H/^1H$ (D/H)	Tracing both production and origin of food, groundwater sources
Li	$^6Li/^7Li$	Weathering processes, adsorption of Li in aquifers
B	$^{10}B/^9B$	
C	$^{13}C/^{12}C$	Tracing both production and origin of food, paleodiet reconstruction, hydrocarbon exploration, environmental forensics
N	$^{15}N/^{14}N$	N cycling, tracing both production and origin of food
O	$^{18}O/^{16}O$	Tracing both production and origin of food, sulphate and nitrate cycle in the atmosphere and aqueous system, oxidation pathways of compounds, paleoclimate (air temperature, humidity and amounts of precipitation), sources of groundwater
Mg	$^{26}Mg/^{24}Mg$	Lithological studies, adsorption of Mg in aquifers
Si	$^{30}Si/^{28}Si$	Global Si cycle, weathering rates, biogeochemical cycling of Si
S	$^{34}S/^{32}S$	Sulfur cycling in the atmosphere, sources of S in the environment
Cl	$^{37}Cl/^{35}Cl$	Forensics, source tracking of organochlorine and inorganic chlorate, fractionation in deep brines
Ca	$^{44}Ca/^{40}Ca$	Vertebrate paleodiet tracer, biogeochemical pathways of Ca
Cr	$^{53}Cr/^{52}Cr$	Source tracking of Cr contamination, Cr reduction
Fe	$^{56}Fe/^{54}Fe$	Fe redox cycling, Fe incorporation in aquifers
Ni	$^{60}Ni/^{58}Ni$	Sources and transport of Ni; methanogenesis
Cu	$^{65}Cu/^{63}Cu$	Sources and transport of Cu; partitioning of Cu between ligand-bound dissolved and particulate phases
Zn	$^{66}Zn/^{64}Zn$	Transport of Zn along cell walls of xylem; micronutrient transport in plants; pathways of anthropogenic Zn
Se	$^{80}Se/^{76}Se$	Redox reactions
Sr	$^{88}Sr/^{86}Sr$	Sr adsorption in aquifers
Mo	$^{97}Mo/^{94}Mo$	Sources, transport of Mo, paleo-redox changes in aqueous system
Cd	$^{114}Cd/^{110}Cd$, $^{112}Cd/^{110}Cd$	Paleo-ocenography, reconstruction of past nutrient levels in seawater
Os	$^{187}Os/^{188}Os$	Tracking anthropogenic and weathering sources of Os
Mo	$^{97}Mo/^{95}Mo$	Redox reactions
Hg	$^{202}Hg/^{198}Hg$, $^{199}Hg/^{198}Hg$, $^{201}Hg/^{198}Hg$	Forensics, sources tracking, reaction pathways of Hg using both mass-dependent and mass-independent fractionation
Tl	$^{205}Tl/^{203}Tl$	Paleoinput of Fe and Mn, hydrothermal fluid flux
Pb	$^{206}Pb/^{204}Pb$, $^{207}Pb/^{204}Pb$, $^{208}Pb/^{204}Pb$	Source identification of Pb in metals, tracing pathways of Pb to human system, tracing dust
Po	^{210}Po	POC export, remineralization of particles, biogenic particle cycling
U	$^{238}U/^{235}U$	Depleted/enriched uranium identification, redox

can be applied depending upon the mode and energy of decay, as well as the emission spectrum that is produced. In some circumstances, an isotope will decay to another short-lived isotope that can be more readily measured and then related to the abundance of the parent (e.g. ^{210}Pb using ^{210}Po measured by alpha spectrometry or ^{226}Ra using ^{214}Pb or ^{214}Bi measured by gamma spectrometry).

Following is a brief description of the different analytical methods available. For many isotopes, there are several measurement options, and the best method is determined by the type of sample, the abundance of the isotope of interest, and the precision required. A more extensive review is given by Hou and Roos (2008).

2.3.1.1 Alpha Spectrometry

Isotopes that are subject to alpha decay can be measured by detecting and counting the emitted alpha particles. Contributions from the decay of other isotopes can be distinguished by alpha particle energy, although there is overlap between alpha energies emitted by some of the nuclides. However, practical difficulties arise due to the interaction of alpha particles with atoms of other species that are loaded in the source along with the sample. This can cause a spread in particle energies and so reduce the energy resolution of the alpha spectrum, making the data often unusable. Therefore, good chemical separation and purification from other elements is critical, and the isotope of interest must be then deposited as a thin layer to minimize adsorption by the element of interest itself (self-shielding). This is done by electroplating onto metal discs or by sorbing onto thin MnO_2 layers. Accurate determination of the concentration of these isotopes requires careful control of chemical yields during processing as well as of counter efficiency. A detailed review of the chemical procedures adopted for measuring alpha-emitting radionuclides in the U-Th-series is given in Lally (1992).

A number of detectors have been utilized for alpha measurements that include ionization chambers, proportional counters, scintillation detectors (plastic and liquid), and semiconductor detectors. Of these, the semiconductors (in particular, surface barrier and ion-implanted silicon detectors) have been the most widely used detectors due to their superior energy resolution and relatively high counting efficiency (typically 30% for a 450-mm^2 surface-barrier detector). More detailed comparison of the various types of detector material is given in Hou and Roos (2008).

The very low background of alpha detectors (0.8–2 counts for a particular isotope per day), relatively high efficiency (~30% for 450 mm^2 surface area detectors) and very good energy resolution (about 17–20 keV for 5.1 MeV) lead to detection limits of 0.010–0.025 mBq (where Bq, a Becquerel, is a decay per second); this can be related to an abundance using (2.1), so that, e.g. for a 1×10^5 year half-life radionuclide this corresponds to as little as 0.08 femtomoles. While the long preparation times associated with chemical separation, purification, electroplating and counting impose a practical constraint on sample throughput, the relatively low-cost of alpha spectrometers and their maintenance is an advantage.

2.3.1.2 Beta Spectrometry

For isotopes that are subject to beta decay, abundances can be measured by detecting and counting the beta particles (electrons) that are emitted. The most common detector is a Geiger-Müller (GM) counter, which employs a gas-filled tube where atoms are ionized by each intercepted beta particle, and this is converted into an electrical pulse. Background signals from cosmic rays are reduced through shielding and anti-coincidence counters, which subtract counts that correspond to detections of high-energy particles entering the counting chamber.

Although adsorption by surrounding atoms is not as efficient as for alpha particles, it is still significant, and so samples are often deposited as thin sources onto a metal disc to increase counting efficiencies. Also, filters that contain particles collected from air and water samples can be measured directly. Dissolved species of particle-reactive elements (e.g. Th) can be concentrated from large water volumes by scavenging onto precipitated Fe or Mn oxides that are then collected onto filters. These filters can be analyzed directly (Rutgers van der Loeff and Moore 1999).

It should be noted that GM counters do not distinguish between beta particles by energy, and so cannot separate contributions from different isotopes through the energy spectrum. While isotopes can often be isolated by chemical separation, some elements have several isotopes and some isotopes have short-lived beta-emitting daughter isotopes that are continuously produced. In these cases, multiple counting can be done to identify and quantify different isotopes based upon their different half-lives. Also, selective absorbers of various thicknesses between the source and counter can be used to suppress interference from isotopes that produce lower energy beta particles.

The counting efficiency of a GM counter varies depending on the thickness of the source and on counter properties. One of the commonly used GM counters, the Risø Counter (Risø National Laboratory, Roskilde, Denmark), has a typical efficiency of ~40% for most commonly occurring beta–emitting radionuclides. A typical background of this anti-coincidence counter, with 10 cm of lead shielding, is 0.15–0.20 counts per minute, corresponding to ~0.8 mBq. This can be compared to the detection limit for an alpha-emitting radionuclide; the background of a new alpha detector is 0.03–0.09 counts per hour, which corresponds to 0.03–0.08 mBq (at ~30% counting efficiency). Thus, the detection limit in alpha spectrometry is more than an order of magnitude higher than that of beta spectrometry.

The GM counter is not suitable for analyzing isotopes such as ^{3}H and ^{14}C that emit low-energy (~200 keV) electrons, which are adsorbed in the detector window. For these nuclides, a liquid scintillation counter (LSC) is used, in which a sample is dissolved in a solvent solution containing compounds that convert energy received from each beta particle into light, and the resulting pulses are counted. This method has minimal self-adsorption and relatively easy sample preparation, and can be used for all beta-emitters. However, background count rates are substantially higher than in GM counters.

2.3.1.3 Gamma Spectrometry

Alpha and beta-emitting radionuclides often decay to atoms in excited energy states, which then reach the ground state by emitting gamma radiation at discrete characteristic energies. Gamma rays can be detected when they ionize target atoms, so that the resulting charge can be measured. Gamma rays can penetrate a considerable distance in solid material, so that isotopes of interest do not need to be separated from sample material before counting, unless preconcentration from large volumes is required. Therefore, the method has the advantage of being non-destructive and easy to use.

A number of detectors have been used for the detection and quantification of gamma rays emitted from radionuclides of interest. Lithium-doped Ge (Ge(Li)) and Si (Si(Li)) semiconductor detectors have been widely used due to their high-energy resolution. For example, using large-volume Ge crystals with relatively high efficiencies, absolute efficiencies of ~70% have been achieved in Ge-well detectors for small samples (1–2 mL volume) at energies of 46.5 keV (^{210}Pb) and 63 keV (^{234}Th). This efficiency decreases to about 24–30% for 10 mL geometries due to the wider solid angle between the counting vial and Ge-crystal (Jweda 2007). However, the counting efficiency of gamma spectrometry with planar or co-axial detectors is generally quite low (<10%), and is dependent upon detector geometry and gamma energy. An advantage of the gamma counting method is that energy spectra are obtained, so that a number of gamma-emitting radionuclides can be measured simultaneously. The typical energy resolution for the Ge-well detector is high, ~1.3 keV at 46 keV and ~2.2 keV at 1.33 MeV.

The backgrounds in gamma-ray detectors are high compared to alpha and beta detectors, mainly due to cosmic-ray muons and Earth's natural radioactivity. Graded shields (of Cu, Cd, Perspex, etc) have considerably reduced the background mainly arising from environmental radioactivity surrounding the detector, and anti-coincidence techniques have helped to cut down background from muons. Ultra-low background cryostats have also helped to reduce the overall background of the gamma-ray detector. Nonetheless, the detection limit is typically several orders of magnitude higher than for beta and alpha counting.

2.3.2 Mass Spectrometry

For measuring stable isotopes, or as an alternative to counting decays of atoms, it is possible to count the atoms directly. This requires a separation of atoms and this can be achieved by utilizing the fact that the path of charged particles travelling through a magnetic field is diverted along a radius of curvature r that is dependent upon mass and energy:

$$r = \left(\frac{2V}{e H^2}\right)^{1/2} (m/z)^{1/2}, \qquad (2.10)$$

where V is accelerating potential, z is charge of the ion, e is the magnitude of electronic charge, and H is the intensity of the magnetic field. The accelerated ions will traverse a semi-circular path of radius r while within the magnetic field. When all ions of an element have the same charge and subject to the same magnetic field, they are accelerated to the same velocity, so that the path radii are only a function of mass. Ions will therefore exit the magnetic field sorted in beams according to mass. This property therefore dictates the basic features of a mass spectrometer: a source where the element of interest is ionized to the same degree and the atoms are accelerated with as little spread in energy as possible; a magnet with a field that is homogeneous throughout the ion paths; detectors located where each beam can be collected separately; and a flight tube that maintains a vacuum along the ion paths, reducing energy-scattering collisions. These methods generally require separation of the element of interest from other constituents that may produce isobaric interferences (either as separate ions or as molecular species) or interfere with the generation of ions. More sophisticated instruments are equipped with further filters, including energy filters that discriminate between ions based on their energy and so separate ions that otherwise interfere due to scattering.

Mass spectrometers can provide high sensitivity measurements of element concentrations as well. While comparisons with standards to calibrate the instrument are needed, these measurements can be limited by variations between sample and standard analysis. In this case, higher precisions can be obtained using isotope dilution, in which a spike that has the relative abundance of at least one isotope artificially enhanced is mixed with the sample once the sample is dissolved. The resulting isotope composition reflects the ratio of spike to sample, and if the amount of spike that is added is known, then the amount of sample can be calculated with high precision directly from measurement of the isotope composition. A further advantage of this approach is that this isotope composition is not affected by losses during chemical separation or analysis. Spikes can also be used to correct for isotope fractionation that occurs within the instrument. In this case, a double spike, which is enriched in two isotopes, is added. The measured ratio of these two isotopes can then be used to calculate the instrument fractionation, and this correction can be applied to the ratios of other measured stable isotopes to obtain corrected sample values.

There are many designs of magnet geometries, as well as electric fields, in order to improve focusing and resolution. However, for isotope geochemistry there are a number of basic, widely used approaches, as discussed below and in the review articles by Goldstein and Stirling (2003) and Albarede and Beard (2004).

2.3.2.1 Thermal Ionization Mass Spectrometry (TIMS)

In this design, samples are manually loaded onto metal filaments installed within the source. Under vacuum, the metal is resistance-heated, and atoms are evaporated. Some atoms are ionized, and are then accelerated down the flight tube and into the magnetic field. The proportion of atoms that are ionized can be very low depending upon the element, but can be enhanced using various techniques, including the use of different filament metals, loading the sample in a mixture that raises the evaporation temperature, and using a second filament held at much higher temperature to ionize atoms vaporized from the evaporation filament. While TIMS can produce ion beams with a narrow range of energies and generate high precision data, it is time-consuming and the proportion of vaporizing atoms that ionize can be low for elements with high ionization potentials. However, it is the analytical tool of choice for high precision data for some elements. Due to signal variability between samples, concentration data must be obtained using isotope dilution. Also, variable isotope fractionation during analysis is usually corrected for by assuming that the ratio of stable, nonradiogenic isotopes is equal to a standard value and so that there is no natural isotope fractionation. Therefore, detection of natural stable isotope fractionation in the samples requires the use of a double spike.

2.3.2.2 Inductively-Coupled Plasma Source Mass Spectrometry (ICP-MS)

In ICP-MS, samples in solution are introduced into an Ar plasma where ions are efficiently ionized. An ion extraction system then transfers ions from the high-pressure plasma source to the vacuum system of the mass analyzer. However, ions exit the source with a much greater energy spread than in TIMS sources, and precise measurements require additional correction of ion beams. The most recent generation of instruments, with electrostatic fields and additional magnets to focus beams, and techniques such as collision cells to decrease interferences, have achieved high sensitivity and high precision data. These instruments have been found to be versatile and have been used for a wide range of elements. Isotopic fractionation within the instrument can be quite consistent and so corrected for by alternating analyses of samples with those of standards. Corrections can also be made using other elements within the sample with similar masses, which exhibit comparable fractionations. Therefore, it is this method that has been used to document stable isotope variations in many elements. Also, sample throughput is considerably greater than for the TIMS method. Methods for high spatial resolution analyses are advancing using laser ablation for in situ sampling, which is coupled to an ICPMS for on-line analysis. For further information, see Albarede and Beard (2004).

2.3.2.3 Gas Source Mass Spectrometry

Volatile elements, including H, O, S, and the noble gases, are measured using a mass spectrometer that has a gas source, where samples are introduced as a gas phase and are then ionized by bombardment of electrons emitted from a heated metal filament. Fractionation of the sample during introduction, ionization, and analysis is corrected for by monitoring a standard, which is introduced and analyzed under identical conditions and run between samples.

2.3.2.4 Accelerator Mass Spectrometry (AMS)

There are a number of isotopes that have such low concentrations that they are difficult to measure precisely by mass spectrometry. C has a $^{14}C/^{12}C$ ratio of $<1 \times 10^{12}$, and measurement precision is limited by isobaric interferences. The problem is similar for other cosmogenic nuclides such as ^{10}Be, ^{26}Al, ^{36}Cl, and ^{129}I. The development of accelerator mass spectrometry has made it possible to precisely measure as few as 10^4 atoms of ^{14}C, and so has extended the time scale of high precision ^{14}C age dating on substantially smaller samples than by counting methods, as well as making it possible to obtain accurate measurements on the longer-lived cosmogenic nuclides. In accelerator mass spectrometry, negative ions are accelerated to extremely high energies using a tandem accelerator and then passed through a stripper of low pressure gas or a foil, where several electrons are stripped off to produce highly charged positive ions and where molecules are dissociated, thereby eliminating molecular interferences. Subsequent mass discrimination involves the use of an analyzing magnet, with further separation of interfering ions made possible using an electrostatic analyzer (which separates ions by kinetic energy per charge) and a velocity filter. Detection is made by counters that allow determination of the rate of energy loss of the ion, which is dependent upon the element, and so allows further separation of interfering ions. There are a number of AMS facilities worldwide, and each is generally optimally configured for a particular set of isotopes. A more extended review is provided by Muzikar et al. (2003).

Comparison Between Alpha Spectrometry and Mass Spectrometry

Techniques for decay counting as well as for mass spectrometry have been continuously evolving over more than 60 years, with progressively improving precision and sensitivity. For measuring radioactive isotopes, both approaches can be used, although often to different levels of precision. For measuring the abundance of any radioactive isotope, the precision possible by mass spectrometry can be compared to that achievable by decay counting. Considering only the uncertainty, or relative standard deviation, due to counting (Poisson) statistics, then for a total number of atoms N of a nuclide with a decay constant of λ that are counted by a mass spectrometer with an ionization efficiency of η_i and where there is a fraction F of ions counted, the uncertainty is

$$P_{ms} = (N\,F\,\eta_i)^{-1/2}. \qquad (2.11)$$

A reasonable value for F is 20%; the ionization can differ by orders of magnitude between elements. Alternatively, if the decay rate of these atoms is measured in a counting system of counting efficiency η_c and over a time t, then

$$P_c = (N\,\lambda\,t\,\eta_c)^{-1/2}. \qquad (2.12)$$

A reasonable maximum value for t is 5 days, and a typical counting efficiency for alpha counting is 30%. The precision for counting is better as the decay constant increases, and so as the half-life decreases. For an ionization efficiency during mass spectrometry of $\eta_c = 1 \times 10^{-3}$, counting is equally precise for nuclides with $\lambda = 0.02$ year^{-1} ($t_{1/2} = 30$ years). Nuclides with longer half-lives would be more precisely measured by mass spectrometry. While the values of the other parameters may vary, and precisions are often somewhat worse due to factors other than counting statistics, in general isotopes with half-lives of less than $\sim 10^2$ years are more effectively measured by decay counting methods (Chen et al. 1992; Baskaran et al. 2009).

2.4 The Future of Isotope Geochemistry in the Environmental Sciences

Isotope geochemistry has become an essential tool for the environmental sciences, providing clearly defined tracers of sources, quantitative information on mixing, identification of physical and chemical processes, and information on the rates of environmental processes. Clearly, this tool will continue to be important in all aspects of the field, including studies of contamination, resource management, climate change, biogeochemistry, exploration geochemistry, archaeology, and ecology. In addition to further utilization of established methods, new applications will continue to be developed. These will likely include discovering new isotope and trace element characteristics of materials, and defining isotope variations that are diagnostic of different processes. One particular area of great potential is in understanding the role of microbial activity in geochemical cycles and transport.

There will certainly be further development of new isotope systems as isotopic variations in elements other than those now widely studied are discovered. Advances in instrumentation and analytical techniques will continue to improve precision and sensitivity, so that it will be possible to identify and interpret more subtle variations. Methods for greater spatial resolution will also improve with better sample preparation and probe techniques, so that time-dependent variations could be resolved in sequentially-deposited precipitate layers and biological growth bands in corals and plants. New applications will follow the development of new methods for extraction of target elements, greater understanding of controls on particular isotopes, and certainly new ideas for isotope tracers. Isotope variations of new elements, in particular stable isotope variations, will be explored, further widening the scope of isotope geochemistry.

Overall, it is inevitable that there will be increasing integration of isotopic tools in environmental studies.

Appendix 1: Decay Energies of Decay Series Isotopes

Table 2.7 Commonly measured isotopes of the ^{238}U decay series data taken from Firestone and Shirley (1999)

Isotope	Half-life	Decay	Emission energy (MeV)[a]	Yield (%)
^{238}U	4.468 × 10^{9}a	α	4.198 (α_0)	77
			4.151 (α_1)	23
			4.038 (α_2)	0.23
^{234}Th	24.1 day	β$^-$	0.199	70.3
		β$^-$	0.104	19
		γ	0.093	5.58
		γ	0.063	4.85
^{234}Pa	1.17 m	β$^-$	2.290	98.2
		β$^-$	1.500	1.8
^{234}U	2.445 × 10^{5}a	α	4.775 (α_0)	72.5
			4.722 (α_1)	27.5
		γ	0.014	10.4
^{230}Th	7.538 × 10^{4}a	α	4.687 (α_0)	76.3
			4.621 (α_1)	23.4
^{226}Ra	1.6 × 10^{3}a	α	4.784 (α_0)	94.55
			4.601 (α_1)	5.45
		γ	0.186	3.51
^{222}Rn	3.82 day	α	5.489 (α_0)	~100
^{218}Po	3.10 min	α	6.002 (α_0)	~100
^{214}Pb	26.8 min	β$^-$	1.023 (β_0)	~100
		γ	0.3519	35.9
		γ	0.2952	18.5
		γ	0.2420	7.5
^{214}Bi	19.7 min	α	5.621 (α_0)	0.021
		β$^-$	3.272	~100
		γ	~100	0.609
		γ	1.120	15.0
			1.764	15.9
^{214}Po	164 μs	α	7.687 (α_0)	~100
^{210}Pb	22.3 year	β$^-$	0.0635 (β_0)	16
		γ	0.0465	4.25
^{210}Bi	5.013 day	β$^-$	1.163 (β_0)	100
^{210}Po	138.38 day	α	5.304 (α_0)	100
	Stable	–	–	–
^{206}Pb	138.38 day	α	5.304 (α_0)	100

[a]Energies of beta decay are maxima

Table 2.8 Commonly measured isotopes of the ^{235}U decay series data taken from Firestone and Shirley (1999)

Isotope	Half-life	Decay	Emission energy (MeV)[a]	Yield (%)
^{235}U	7.038 × 10^{8}a	α	4.215–4.4596	93.4
		γ	0.1857	57.3
		γ	0.1438	10.96

(*continued*)

Table 2.8 (continued)

Isotope	Half-life	Decay	Emission energy (MeV)[a]	Yield (%)
^{231}Th	25.52 h	β^-	0.3895	100
		γ	0.0256	14.5
		γ	0.0842	6.6
^{231}Pa	3.276×10^4a	α	4.934–5.059	86.5
		α	4.681–4.853	12.3
		γ	0.0274	10.3
		γ	0.300–0.303	4.7
^{227}Ac	21.773a	α (1.38%)	4.941–4.953	87.3
		β (98.62%)	0.0449	~100
^{227}Th	18.72 day	α	5.960–6.038	50.6
			5.701–5.757	37.2
		γ	0.236	12.3
			0.256	7.0
^{223}Ra	11.435 day	α	5.539–5.747	96.8
		γ	0.2695	13.7
^{219}Rn	3.96 s	α	6.819 (α_0)	79.4
			6.553 (α_1)	12.9
			6.425 (α_2)	7.6
^{215}Po	1.78 ms	α	7.386 (α_0)	100
^{211}Pb	36.1 min	β^-	1.373	100
^{211}Bi	2.14 min	α (99.72%)	6.623	83.8
			6.278	16.2
		β^- (0.28%)	0.579	100
		γ	0.351	13.0
^{207}Tl	4.77 min	β^-	1.423	100
^{207}Pb	Stable	–	–	–

[a] Energies of beta decay are maxima

Table 2.9 Commonly measured isotopes of the ^{232}Th decay series data taken from Firestone and Shirley (1999)

Isotope	Half-life	Decay	Emission energy (MeV)[a]	Yield (%)
^{232}Th	1.405×10^{10}a	α	4.013	77.9
			3.954	22.1
^{228}Ra	5.75a	β^-	0.046	100
^{228}Ac	6.15 h	β^-	2.127	100
		γ	0.9112	26.6
			0.9690	16.2
			0.3383	11.3
^{228}Th	1.913a	α	5.423	71.1
			5.340	28.2
^{224}Ra	3.66 day	α	5.685 (α_0)	94.9
			5.449 (α_1)	5.1
		γ	0.2410	3.97
^{220}Rn	55.6 s	α	6.288 (α_0)	99.89
			5.747 (α_1)	0.11
^{216}Po	0.145 s	α	6.778 (α_0)	~100
^{212}Pb	10.64 h	β^-	0.574	100
		γ	0.2386	43.3
			0.3001	3.28

(continued)

Table 2.9 (continued)

Isotope	Half-life	Decay	Emission energy (MeV)[a]	Yield (%)
^{212}Bi	60.55	α (35.94%)	6.051	69.9
			6.090	27.1
		β^-	2.254	100
		γ		
^{212}Po	0.299 μs	α	8.784 (α_0)	100
^{208}Tl	3.05 min	β^-	1.372	100
		γ	0.405	3
			0.832	2.8
^{207}Pb	Stable	–	–	–

[a] Energies of beta decay are maxima

Appendix 2: Decay Series Systematics

Within a decay series, the evolution of the abundance of a daughter radioactive isotope is dependent upon its decay rate as well as production from its radioactive parent. Since the abundance of the parent is in turn dependent upon that of its parent, and so on up the decay chain, the systematics can become complicated, although for most applications simplifying circumstances can be found. The evolution of the abundances of isotopes within a decay series is described by the Bateman equations (see Bourdon et al. 2003 for derivations). For a decay series starting from the long-lived parent N_1 and ending with a stable isotope S,

$$N_1 \rightarrow N_2 \rightarrow N_3 \rightarrow N_4 \rightarrow ....S. \quad (2.13)$$

The long-lived parent evolves according to the basic decay equation:

$$(N_1) = (N_1)_0 e^{-\lambda_1 t}. \quad (2.14)$$

(N_1) is the activity, so that

$$(N_1) = \frac{dN_1}{dt} = \lambda_1 N_1. \quad (2.15)$$

The next two nuclides in the decay series evolves according to the equations

$$(N_2) = \frac{\lambda_2}{\lambda_2 - \lambda_1}(N_1)_0\left(e^{-\lambda_1 t} - e^{-\lambda_2 t}\right) + (N_2)_0 e^{-\lambda_2 t} \quad (2.16)$$

$$(N_3) = \frac{(N_1)_0 \lambda_2 \lambda_3}{\lambda_2 - \lambda_1} \left(\frac{e^{-\lambda_1 t}}{(\lambda_2 - \lambda_1)(\lambda_3 - \lambda_1)} \right.$$
$$\left. + \frac{e^{-\lambda_2 t}}{(\lambda_1 - \lambda_2)(\lambda_3 - \lambda_2)} + \frac{e^{-\lambda_3 t}}{(\lambda_1 - \lambda_3)(\lambda_2 - \lambda_3)} \right)$$
$$+ \frac{(N_2)_0 \lambda_3}{(\lambda_3 - \lambda_2)} \left(e^{-\lambda_2 t} - e^{-\lambda_3 t} \right) + (N_3)_0 e^{-\lambda_3 t}.$$

(2.17)

The equations become increasingly complex, but are rarely necessary for environmental applications. Further, these equations can be greatly simplified under most circumstances.

- For all environmental timescales, the abundance of the long-lived parent (N_1) of the chain is constant, and $e^{-\lambda_1 t} \approx 1$. Further, $\lambda_2 - \lambda_1 \approx \lambda_2$ and $\lambda_3 - \lambda_1 \approx \lambda_3$.
- When a system has been closed for a long time, i.e. when $e^{-\lambda_2 t} \approx e^{-\lambda_3 t} \approx 0$, the above equations reduce to $(N_1 \lambda_1) = (N_2 \lambda_2) = (N_3 \lambda_3)$, so that all the nuclides have the same activity as their parents, i.e. are in secular equilibrium, and the entire chain has the same activity as that of the long-lived parent. This is the state within very long-lived materials, such as unweathered rocks, and is the state all systems evolve towards when nuclides are redistributed.
- The activity of the daughter evolves towards secular equilibrium with its parent according to its half-life (and *not* that of the half life of its parent). This could be grow-in, where the daughter starts with a lower activity, or decay when the daughter starts with a higher activity; in either case it is the difference in the activities that declines at a rate determined by the half-life of the daughter. Therefore, the half-life of the daughter dictates the time-scale for which it is useful.

The linking of all the isotopes in the chain can lead to considerable complexity, since the parent concentration of each nuclide is changing, and so potentially the concentrations of the entire chain must be considered. However, for most applications, this can be greatly simplified by the following considerations:

- When a nuclide is isolated from its parent, then it becomes the head of the decay chain and simply decays away according to the decay equation (2.1); that is, it becomes the top of the chain. An example is ^{210}Pb, which is generated in the atmosphere from ^{222}Rn and then transferred to sediments in fallout, where its activity simply diminishes according to (2.14) with a 22.3a half-life.
- Where the time scale of interest is short compared to the half-life of an isotope, the abundance of that isotope can be considered to be constant. In this case, this nuclide can be considered the head of the decay chain, and the influence of all nuclides higher in the decay series, can be ignored. For example, over periods of several 1,000 years, ^{230}Th ($t_{1/2}$ = 75ka) remains essentially constant, while ^{226}Ra ($t_{1/2}$ = 1.6ka) will grow into secular equilibrium according to (2.16).
- Where the time scale of interest is long compared to the half-life of an isotope, the activity of that isotope can be considered to be equal to that of the parent. For example, in a closed mineral, the ^{222}Rn activity will be equal to that of parent ^{226}Ra after several weeks, and so while the activity of ^{210}Pb will grow-in towards that of its parent ^{222}Rn, this can be represented by measurements of ^{226}Ra. Therefore, shorter-lived intermediate isotopes can be ignored.
- The most common circumstance when (2.17) is required is dating materials that have incorporated U from waters. Since the (^{234}U/^{238}U) ratio of waters is often above that of secular equilibrium, obtaining an age from the grow-in of ^{230}Th requires considering the chain ^{238}U → ^{234}U → ^{230}Th → .

References

Albarede F, Beard B (2004) Analytical methods for non-traditional isotopes. Rev Mineral Geochem 55:113–152

Baskaran M (2011) Po-210 and Pb-210 as atmospheric tracers and global atmospheric Pb-210 fallout: a Review. J Environ Radioact 102:500–513

Baskaran M (2011) Dating of biogenic and inorganic carbonates using ^{210}Pb-^{226}Ra disequilibrium method – a review. In: Baskaran M (ed) Handbook of environmental isotope geochemistry. Springer, Berlin

Baskaran M, Hong G.-H, Santschi PH (2009) Radionuclide analysis in seawater. In: Oliver Wurl (ed) Practical guidelines for the analysis of seawater. CRC Press, Boca Raton, FL, pp 259–304

Blum J (2011) Applications of stable mercury isotopes to biogeochemistry. In: Baskaran M (ed) Handbook of environmental isotope geochemistry. Springer, Berlin

Bourdon B, Turner S, Henderson GM, Lundstrom CC (2003) Introduction to U-series geochemistry. Rev Mineral Geochem 52:1–21

Bullen TD (2011) Stable isotopes of transition and post-transition metals as tracers in environmental studies. In: Baskaran M (ed) Handbook of environmental isotope geochemistry. Springer, Berlin

Burton KW, Vigier N (2011) Lithium isotopes as tracers in marine and terrestrial environment. In: Baskaran M (ed) Handbook of environmental isotope geochemistry. Springer, Berlin

Chabaux F, Riotte J, Dequincey O (2003) U-Th-Ra fractionation during weathering and river transport. Rev Mineral Geochem 52:533–576

Chamberlain CP, Blum JD, Holmes RT, Feng X et al (1997) The use of isotope tracers for identifying populations of migratory birds. Oecologia 109:132–141

Chen JM, Edwards RL, Wasserburg GJ (1992) Mass spectrometry and application to uranium-series disequilibrium. In: Ivanovich M, Harmon RS (eds) Uranium-series disequilibrium. Applications to Earth, Marine, and Environmental Sciences. 2nd Ed. Clarendon Press, Oxford, pp. 174–206

Chesson LA, Ehleringer JR, Cerling TE (2011) Light-element isotopes (H, C, N, and O) as tracers of human diet: a case study on fast food meals. In: Baskaran M (ed) Handbook of environmental isotope geochemistry. Springer, Berlin

Church T, Sarin MM (2008) U and Th series nuclides in the atmosphere: supply, exchange, scavenging, and applications to aquatic processes. In: Krishnaswami S, Cochran JK (eds) U/Th series radionuclides in aquatic systems. Elsevier, Amsterdam, pp 105–153

Cochran JK, Masque P (2003) Short-lived U/Th series radionuclides in the ocean: tracers for scavenging rates, export fluxes and particle dynamics. Rev Mineral Geochem 52:461–492

Cockburn HAP, Summerfield MA (2004) Geomorphological applications of cosmogenic isotope analysis. Prog Phys Geogr 28:1–42

Du JZ, Zhang J, Baskaran M (2011) Applications of short-lived radionuclides (^{7}Be, ^{210}Pb, ^{210}Po, ^{137}Cs and ^{234}Th) to trace the sources, transport pathways and deposition of particles/sediments in rivers, estuaries and coasts. In: Baskaran M (ed) Handbook of environmental isotope geochemistry. Springer, Berlin

Dunai TJ (2010) Cosmogenic nuclides: principles, concepts, and applications in the earth surface sciences. Cambridge University Press, Cambridge

Ellis AS, Johnson TM, Bullen TD (2002) Chromium isotopes and the fate of hexavalent chromium in the environment. Science 295:2060–2062

Farquhar JB, Bao H, Thiemens M (2000) Atmospheric influence of Earth's earliest sulfur cycle. Science 289:756–758

Firestone RB, Shirley VS (1999) Table of isotopes, 8th edn. Update with CD-ROM, Wiley-Interscience

Gao YQ, Marcus RA (2001) Srange and unconventional isotope effects in ozone formation. Science 293:259–263

Goldstein SJ, Stirling CH (2003) Techniques for measuring uranium-series nuclides: 1992–2002. Rev Mineral Geochem 52:23–57

Gosse JC, Phillips FM (2001) Terrestrial in situ cosmogenic nuclides: theory and application. Quatern Sci Rev 20:1475–1560

Grousset FE, Biscaye PE (2005) Tracing dust sources and transport patterns using Sr, Nd, and Pb isotopes. Chem Geol 222:149–167

Gulson B (2011) Sources of lead and its mobility in the human body using lead isotopes. In: Baskaran M (ed) Handbook of environmental isotope geochemistry. Springer, Berlin

Hirose K (2011) Uranium, thorium and anthropogenic radionuclides as atmospheric tracer. In: Baskaran M (ed) Handbook of environmental isotope geochemistry. Springer, Berlin

Hong G-H, Hamilton TF, Baskaran M, Kenna TC (2011) Applications of anthropogenic radionuclides as tracers to investigate marine environmental processes. In: Baskaran M (ed) Handbook of environmental isotope geochemistry. Springer, Berlin

Hou X, Roos P (2008) Critical comparison of radiometric and mass spectrometric methods for the determination of radionuclides in environmental, biological and nuclear waste samples. Anal Chim Acta 608:105–139

Ivanovich M, Harmon RS (1992) Uranium-series disequilibrium. In: Ivanovich M, Harmon RS (eds) Applications to Earth, Marine, and Environmental Sciences. 2nd Ed. Clarendon Press, Oxford, pp. 909

Johnson TM (2011) Stable isotopes of Cr and Se as tracers of redox processes in Earth surface environments. In: Baskaran M (ed) Handbook of environmental isotope geochemistry. Springer, Berlin

Jweda J (2007) Short-lived radionuclides (^{210}Pb and ^{7}Be) as tracers of particle dynamics and chronometers for sediment accumulation and mixing rates in a river system in Southeast Michigan. M.S. Thesis, Department of Geology, Wayne State University, Detroit, Michigan, pp. 167

Kaste JM, Baskaran M (2011) Meteoric ^{7}Be and ^{10}Be as process tracers in the environment. In: Baskaran M (ed) Handbook of environmental isotope geochemistry. Springer, Berlin

Keller K, Blum JD, Kling GW (2010) Stream geochemistry as an indicator of increasing permafrost thaw depth in an arctic watershed. Chem Geol 273:76–81

Kendall C, McDonnell JJ (1998) Isotope tracers in catchment hydrology (edited volume), Elsevier Science B.V., Amsterdam, pp. 839

Kennedy BP, Folt CL, Blum JD, Chamberlain CP (1997) Natural isotope markers in salmon. Nature 387:766–767

Ketterer ME, Zheng J, Yamada M (2011) Source tracing of transuranics using their isotopes. In: Baskaran M (ed) Handbook of environmental isotope geochemistry. Springer, Berlin

Kim G, Kim T-H, Church TM (2011) Po-210 in the environment: biogeochemical cycling and bioavailability. In: Baskaran M (ed) Handbook of environmental isotope geochemistry. Springer, Berlin

Kulongoski JT, Hilton DR (2011) Helium isotope studies of ground waters. In: Baskaran M (ed) Handbook of environmental isotope geochemistry. Springer, Berlin

Lal D (1991) Cosmic ray tagging of erosion surfaces: in situ production rates and erosion models. Earth Planet Sci Lett 104:424–439

Lal D (1999) An overview of five decades of studies of cosmic ray produced nuclides in oceans. Sci Total Environ 237/238:3–13

Lal D (2011) Using cosmogenic radionuclides for the determination of effective surface exposure age and time-averaged erosion rates. In: Baskaran M (ed) Handbook of environmental isotope geochemistry. Springer, Berlin

Lal D, Baskaran M (2011) Applications of cosmogenic isotopes as atmospheric tracers. In: Baskaran M (ed) Handbook of environmental isotope geochemistry. Springer, Berlin

Lally AE (1992) Chemical procedures. In: Ivanovich M, Harmon RS (eds) Uranium-series disequilibrium. Applications to Earth, Marine, and Environmental Sciences. 2nd Ed. Clarendon Press, Oxford, pp. 95–126

Matisoff G, Whiting PJ (2011) Measuring soil erosion rates using natural (^{7}Be, ^{210}Pb) and anthropogenic (^{137}Cs, ^{239,240}Pu) radionuclides. In: Baskaran M (ed) Handbook of environmental isotope geochemistry. Springer, Berlin

Muzikar P, Elmore D, Granger DE (2003) Accelerator mass spectrometry in geologic research. Geol Soc Am Bull 115:643–654

Niedermann S (2002) Cosmic-ray-produced noble gases in terrestrial rocks: dating tools for surface processes. Rev Mineral Geochem 47:731–784

Nielsen SG, Rehkämper M (2011) Thallium isotopes and their application to problems in earth and environmental science. In: Baskaran M (ed) Handbook of environmental isotope geochemistry. Springer, Berlin

Phillips FM, Castro MC (2003) Groundwater dating and residence time measurements: in Treatise on Geochemistry (Holland, H.D., and Turekian, K.K., eds.), Vol. 5, Surface and Ground Water, Weathering, and Soils (Drever, J.I., ed.), Oxford University Press, Oxford, p. 451–497

Porcelli D (2008) Investigating groundwater processes using U- and Th-series nuclides. In: Krishnaswami S, Cochran JK (eds) U-Th series nuclides in aquatic systems, Elsevier, London, pp 105–154

Porcelli D, Swarzenski PW (2003) The behavior of U- and Th-series nuclides in groundwater. Rev Mineral Geochem 52:317–362

Rehkamper M, Halliday A (1999) The precise measurement of Tl isotopic compositions by MC-ICPMS: application to the analysis of geological materials and meteorites- reading the isotopic code. Geochim Cosmochim Acta 63(63):935–944

Rehkämper M, Wombacher F, Horner TJ, Xue Z (2011) Natural and anthropogenic Cd isotope variations. In: Baskaran M (ed) Handbook of environmental isotope geochemistry. Springer, Berlin

Reynolds B (2011) Silicon isotopes as tracers of terrestrial process. In: Baskaran M (ed) Handbook of environmental isotope geochemistry. Springer, Berlin

Rutgers van der Loeff M, Moore WS (1999) Determination of natural radioactive tracers. In: Grasshoff K, Ehrardt M, Kremling K (eds), Weinheim, Germany: Wiley-VCH, pp. 365–397

Schaller M, von Blanckenburg F, Hovius N, Kubik PW (2001) Large-scale erosion rates from in situ-produced cosmogenic nuclides in European river sediments. Earth Planet Sci Lett 188:441–458

Schauble EA (2004) Applying stable isotope fractionation theory to new systems. Rev Mineral Geochem 55:65–111

Schwarcz HP, Schoeninger MJ (2011) Stable isotopes of carbon and nitrogen as tracers for paleo-diet reconstruction. In: Baskaran M (ed) Handbook of environmental isotope geochemistry. Springer, Berlin

Slovak NM, Paytan A (2011) Applications of Sr isotopes in archaeology. In: Baskaran M (ed) Handbook of environmental isotope geochemistry. Springer, Berlin

Stirling CH, Andersen MB, Potter E-K, Halliday AN (2007) Low-temperature isotopic fractionation of uranium. Earth Planet Sci Lett 264:208–225

Sturchio NC, Clausen JL, Heraty LJ, Huang L, Holt BD, Abrajano A Jr (1998) Chlorine isotope investigation of natural attenuation of trichloroethene in an aerobic aquifer. Environ Sci Tech 32:3037–3042

Sturchio NC, Bohlke JK, Gu B, Hatzinger PB, Jackson WA (2011) Isotopic tracing of perchlorate in the environment. In: Baskaran M (ed) Handbook of environmental isotope geochemistry. Springer, Berlin

Swarzenski PW, Porcelli D, Andersson PS, Smoak JM (2003) The behavior of U- and Th-series nuclides in the estuarine environment. Rev Mineral Geochem 52:577–606

Tripathy GR, Singh SK, Krishnaswami S (2011) Sr and Nd isotopes as tracers of chemical and physical erosion. In: Baskaran M (ed) Handbook of environmental isotope geochemistry. Springer, Berlin

van Warmerdam EM, Frape SK, Aravena R, Drimmie RJ, Flatt H, Cherry JA (1995) Stable chlorine and carbon isotope measurements of selected chlorinated organic solvents. Appl Geochem 10:547–552

Vigier N, Bourdon B (2011) Constraining rates of chemical and physical erosion using U-series radionuclides. In: Baskaran M (ed) Handbook of environmental isotope geochemistry. Springer, Berlin

Willenbring JK, von Blanckenburg F (2010) Meteoric cosmogenic beryllium-10 adsorbed to river sediment and soil: applications for Earth-surface dynamics. Earth Sci Rev 98:105–122

Chapter 3
Humans and Isotopes: Impacts and Tracers of Human Interactions with the Environment

Karl K. Turekian

Abstract Radioactive, radiogenic and light stable isotopes have been useful adjuncts in addressing matters of concern to humans including human health, environmental tracers, radioactive geochronometry of the history of humans and forensic uses including reconstructing diets of the past.

3.1 Introduction

Humans are influenced by radioactive and stable isotopes both by affecting them directly and as tracers of processes that affect their weal. Some radioactive isotopes have been injected into the environment from human activity such as nuclear warfare, testing and accidents. Most are naturally produced both from the initial composition of Earth and the subsequent continual production by cosmic rays. The former include U isotopes and Th that produce a range of radioactive daughters as well as ^{40}K accompanying potassium in biological systems and the environment. The latter include radiocarbon as well as a host of other radioactive species discussed in Chaps. 7 and 33. The aim of this chapter is to discuss in narrative fashion the salient ways in which these isotopes from various sources affect human health and the environmental processes that influence human activities.

K.K. Turekian (✉)
Department of Geology and Geophysics, Yale University, New Haven, CT 06511, USA
e-mail: karl.turekian@yale.edu

3.2 Human Health

After Ra was discovered by Marie Curie its various properties as a radiation emitter were utilized mainly for the production of watch faces with luminescent figures and arms. The radiation dose for the wearers was certainly present but the greatest impact was on the workers (mainly women) who painted on the radium solution on the watch faces. Many of these workers eventually came down with radiation-produced diseases (as did Marie Curie) and the mortality rate among watch factory workers became high. The practice continued from 1917 to 1926.

As the result of the increased need for uranium during World War II and thereafter in connection with the nuclear weapons program followed by more peaceful uses, mines developed to obtain uranium ore. One of the daughters of uranium, radon, produced a range of energetic alpha emitting daughters that were adsorbed onto mine dust or cigarette smoke. The inhalation of these aerosols caused an increase in lung cancer rates among uranium miners.

These two processes have been replicated in other situations that arose from related activities. The detonation of nuclear devices into the atmosphere usually penetrated into the stratosphere. As a result long-lived fission isotopes of Sr (^{90}Sr) and Cs (^{137}Cs) made their way slowly into the troposphere and ultimately into the human food chain. After World War II with the increased atmospheric testing lasting until 1962 there was concern about ^{90}Sr entering the human body and lodging in the bone just as Ra had been shown to do. This resulted in an Atomic Energy Commission program called "Project Sunshine" (1953) that was classified for a while. Among the institutions involved in the project was Columbia University and part of my thesis on the geochemistry and biogeochemistry of Sr was supported by this project to obtain the initial estimate of the expected body burden of ^{90}Sr, with a paucity of bone ^{90}Sr data, based on the distribution of Sr in human bones (Turekian and Kulp 1956).

The concern about the perils of cigarette smoking was pursued by Edward Martell of NCAR. He at first reasoned that tobacco leaves stored in curing barns scavenged Rn daughters that were then inhaled by smoking the tobacco. He soon realized rather that smoking produced aerosols that adsorbed short lived and energetically intense Rn daughters in closed environs such as basements (Martell 1983). This was analogous to the lung cancer associated with U miners. Subsequently basements of houses were checked for Rn seeping in from soils through their foundations. The capture by aerosols and especially smoke from cigarettes would burden the lung with radiation and be one additional contributor to the health hazards of smoking.

As a footnote one of the last things that Martell was working on just before he died was on directly estimating the effects of alpha radiation on genetic variation. Feeding fruit flies free of the lethal gene on food laced with ^{228}Th he determined that the lethal gene was generated over time. Rather than concluding that he had discovered the cause of alpha decay cancer formation instead he concluded that the bombardment of skin surfaces by alpha radiation eliminated many cells to allow reformation to eradicate cells that might be altered by chemical and ultraviolet exposure. Although these results have not been confirmed yet by others it provides an interesting possibility of the possible benefits of some radiation.

Lethality and medical harm caused by extreme radiation as during a nuclear blast or accidental exposure around a nuclear facility is the concern for all involved in nuclear activities. Care in the operation of nuclear power plants is exercised with great diligence because of these concerns and the proper disposal of nuclear waste remains a primary consideration in the exploitation of nuclear energy.

3.3 Natural and Artificially Produced Radionuclides as Environmental Tracers

The properties of the atmosphere and oceans and freshwater system of importance to human activities can be studied using the radionuclides discussed above. Specifically the following applications have been made:

(a) The mean residence time of air in the stratosphere was initially determined by heat balance in the atmosphere but confirmed by measurements of the fluxes of the bomb produced nuclides ^{90}Sr and ^{137}Cs each with half lives of about three decades. The concentration in stratospheric air and the flux to Earth's surface provided such information (Peirson et al. 1966). This measurement can also be made using the cosmogenic isotopes such as ^{10}Be. The mean residence time of stratospheric air also provides a method of determining the flux of tropospheric pollutants to the stratosphere. These fluxes provide a measure of the rate of destruction of stratospheric ozone. They also can be used to determine the rate of supply of water responsible for the formation of stratospheric and upper tropospheric cirrus clouds, important as a component of tropospheric heating.

(b) The mean residence time of aerosols in the troposphere has been determined using ^{210}Pb and its daughters, ^{210}Bi and ^{210}Po. The ^{210}Pb is produced in the decay chain of ^{222}Rn which is supplied to the atmosphere from soils (Turekian and Graustein 2003 – This reference will be used in several places in the following discussion because it is a summary paper with appropriate references contained therein). The value obtained for the troposphere is about 5 days based on the standing crop of ^{210}Pb in the troposphere and the precipitation flux of ^{210}Pb. The application of this residence time to aerosols injected into the troposphere from pollution sources and natural elevation of submicron aerosols from fires and dust storms provides a way of determining the extent of transport of those aerosols from point sources.

(c) Sources of potential pollutants can be tracked by using both atmospheric ^{210}Pb and cosmogenic ^{7}Be. The former has its highest concentration near Earth's surface, decreasing with elevation and the latter increases with elevation. Thus the ratio of ^{7}Be to ^{210}Pb provides a unique indication of the source of other components of tropospheric air. Using this property it was discovered that in the free troposphere in the Canary Islands the source of both ozone and nitrate was the upper troposphere and was not related to either biomass burning in Africa or pollution from high population centers in Europe (Turekian and Graustein 2003).

(d) The short-lived isotopes, ^{210}Pb, ^{90}Sr and ^{137}Cs have been used to determine time in salt marsh deposits and lake sediments to track the pollution history of various environments. ^{210}Pb has also been used in salt marshes to determine the rate of rise of sea level over the past 100 years due to global warming and the associated increase in the volume of the oceans because of valley glacial melting, ground water exploitation and thermal expansion. These are clearly ways of monitoring the impact of humans on the environment (Von Damm et al. 1979; McCaffrey and Thomson 1980; Cochran et al. 1998).

(e) Bomb produced tritium in the atmosphere reach the marine environment as water molecules exchange with the surface ocean. In the North Atlantic, where the deep water has its origin at high latitudes, the use of ^{3}H-^{3}He dating to determine the rate of formation of North Atlantic Deep Water and the rate of its transport meridionally has been done using mass spectrometric measurements of ^{3}He (Jenkins 2008). This type of information feeds directly into the understanding of the oceanic "conveyor belt" some consider a critical component in understanding the effects of global warming and fresh water fluxes to the high latitude oceans.

(f) The transport of chemicals in ground waters can be traced using the emission and adsorption properties of members of uranium and thorium decay series. Using the Rn flux to the ground water as a measure of other recoil fluxes of alpha emitters in the uranium and thorium decay chains the retardation factors for a number of elements can be determined. These include Pb, Ra, Th and U. By analogy to these elements responsive to oxidation-reduction conditions as well as acidity and salinity the result can be extended to a range of ionic species including those associated with nuclear materials that might be stored underground and subject to accidental release to ground water (Krishnaswami et al. 1982).

3.4 Radioactive Geochronometry of Human Development

(a) The history of human evolution has been documented using principally the radioactive isotope ^{40}K that decays to ^{40}Ar. Potassium in volcanic materials can be dated using this technique. In some cases the dates can be backed up with U-Pb dating of zircons but the main tool for determining the time scale of hominid evolution especially in east Africa is via the ^{40}K-^{40}Ar dating method. A current chronology of hominid evolution in east Africa and its relation to climate change is reported in a National Research Council report entitled "Understanding climate's influence on human evolution" (2010).

In addition to this method, which has been useful over a five million year time period, travertine deposits (and in some cases bones) have been dated using the ^{234}U-^{238}U-^{230}Th decay chain. This method has been found useful for about 300,000 years (Shen et al. 2004; Shen et al. 2007).

(b) Dating using radiocarbon (^{14}C) with a half-life of 5,730 years has provided important chronological data for human migration and development. Once it had been discovered that the ^{14}C/^{12}C ratio in the atmosphere had not remained constant over time efforts were made to calibrate the time dependent variation using tree rings. These results were only useful to about 10,000 years before present. The problem of extending the information on variability over a longer period of time was resolved by the use of coral, specifically those in a section undergoing submergence off Barbados, for precise dating using the ^{234}U-^{238}U-^{230}Th dating technique alluded to above. Precise dates of corals matched with the radiocarbon content of the synchronously dated coral provided accurate information on the history of variability of atmospheric ^{14}C/^{12}C over the past 50,000 years (Fairbanks et al. 2005).

One of the greatest triumphs of this approach was to determine the migration time scale of Homo sapiens along the European side of the Mediterranean Sea as humans migrated from the near east beginning 49,000 years BP to Spain, reaching there at about 41,000 years BP (Mellars 2006). This migration was accomplished in the interstadial of the last major glaciation when Europe, especially the southern part, was relatively more ice free than during peak glaciation. The rate of migration was about a kilometer a year. Curiously this is the same rate as the spread of agriculture from the near east to northwestern Europe about 7,000 years ago!

3.5 The Forensic Uses of Radiactive and Stable Isotopes

(a) Tracing human diets using carbon isotopes has proven to be a valuable pursuit in anthropology and human history. The ratio of ^{13}C to ^{12}C usually expressed as variations relative to a standard as delta values ($\delta^{13}C$) is different for different food sources used by humans over time and place. For example, the value for C3 plants and animals feeding on them has a $\delta^{13}C$ value of -25. C4 plants such as maize, sorghum and cane sugar have $\delta^{13}C$ values of about -12.5. Sea food generally is also on the less negative side. By using these differences, carbon in fossil human bones can be used to determine the source of food and how it changed over time. For example it was shown that the ancestors of the Viking in Denmark shifted from eating marine animals to agriculture-based products composed of C3 plants about 5,000 BP. In North America the southwestern native population survived on a maize food base seen as a $\delta^{13}C$ value of -12.5 in carbon in their bones. The native populations of the Midwest and to the east were dependent on C3 plants and animals using these plants for nourishment until about 1,000 BP when maize was introduced. This radically change the culture of the Midwest and the northeast and this transition can be seen in the carbon isotopes of fossil human bones where there is a shift in $\delta^{13}C$ from -25 to -12.5.

(b) Oxygen and hydrogen isotopes in textiles and related materials will record the isotopic composition of the water from which the fibers were formed whether vegetable or animal. For example, the paper used in American currency is uniquely made of cotton from Texas fields grown with irrigated water from the Ogallala formation and has a unique oxygen and hydrogen isotopic composition (commonly expressed as $\delta^{18}O$ and δD) (Ehleringer, personal communication). Similarly textiles from the Aswan region of the Nile River has been shown to be partitioned into early locally grown cotton and later imported flax from the Nile delta region because of the diagnostic water isotopic compositions of the upper Nile water source and the modified water at the delta region (Judith Schwartz, personal communication).

(c) A similar use of oxygen and hydrogen isotopes has been used to identify the sources of bacteria in the study of epidemiology (Kreuzer-Martin et al. 2003). Also these isotopes provide definite markers for the sources of drugs such as marijuana which may be important in tracking down the sources in drug trafficking.

(d) There are some times when the issue of authenticity of human artifacts is in question. In those cases it is possible that some of the information contained in radioactive isotopes can be of assistance. One example is the establishment of the true antiquity of an iron object. Before about 1789 all iron was made using charcoal and the carbon alloyed or occluded with the iron would have a radiocarbon signature that actually could be used in chronometric studies (van der Merwe 1969). Any iron artifact manufactured using coke would not have radiocarbon in it thereby distinguishing it from the older material. If there were any question about the authenticity of an iron object purportedly made before the eighteenth century it could be verified with a radiocarbon measurement (Turekian 1977).

Also most contemporarily available lead has been processed from recycled lead exposed to rainfall and therefore impacted by atmospherically derived ^{210}Pb. This was discovered when lead from bombed cathedrals in England during World War II were assayed for radioactivity with the prospect of using "old" lead for radioactive shielding in the Manhattan Project. Since ^{210}Pb has a half-life of 22 years any object made of lead, whether it contained ^{210}Pb or not from any source, would have decayed away whereas any object made using contemporary commercially available lead would be high in ^{210}Pb. This difference has been used to establish the authenticity of lead bearing artifacts of questionable chronological authenticity found in museums. In principle the method can be applied to white lead oxide used in paintings as well as in assaying lead containing alloys in testing for authenticity (Keisch 1968).

(e) Another use of forensics involving isotopes is the tracking of nuclear proliferation, nuclear bomb testing and nuclear fuel processing. These provide important constraints on treaties and environmental considerations. For example each nuclear blast has characteristic plutonium isotope signatures as well as xenon isotope signatures. The measurement of contemporary atmospheric gas samples and aerosols can determine if there has been clandestine testing of a nuclear device. A record of past nuclear activity from a variety of sources can be determined from the Pu isotope signature observed if such historical records as ice cores and sediments. Where spent nuclear fuel has been processed ^{129}I (half-life of 17 million years) is commonly released and if this release is adjacent to the oceans it can act as a tracer for surface ocean circulation (Edmonds et al. 2001).

3.6 Conclusions and Future Directions

Radioactive and stable isotopes have served as hazards, tracers and forensic tools for a number of human encounters. The capacity to make refined measurement of a variety of radioactive and stable isotopes on ever smaller sample sizes will allow the tracing of organic compounds as well as small quantities of pollutants for effective evaluation of human involvement with the environment. Clearly the study of the presence and distribution of these various isotopes has and will continue to have significance for every aspect of human life. The continued exploration for these applications is certain to be made in the future.

References

Cochran JK, Frignani M, Salamanca M et al (1998) Lead-210 as a tracer of atmospheric input of heavy metals in the northern Venice Lagoon. Mar Chem 62:15–29

Edmonds HN, Zhou ZQ, Raisbeck G et al (2001) The distribution and behavior of anthropogenic ^{129}I in water masses ventilating the North Atlantic. J Geophys Res 106:6881–6894

Fairbanks RG, Mortlock RA, Chiu T-C et al (2005) Radiocarbon calibration curve spanning 0 to 50,000 years BP based on paired ^{230}Th/ ^{234}U/ ^{238}U and ^{14}C dates on pristine corals. Quat Sci Res 24:1781–1796

Jenkins WJ (2008) Tritium-helium dating. Encyclopedia of Ocean Sciences (Second Edition) doi:10.1016/B978-012374473-9.00164-8 Elsevier, Editor-in-Chief: John H. Steele, Karl K. Turekian, and Steve A. Thorpe, London

Keisch B (1968) Dating works of art through their natural radioactivity: Improvements and applications. Science 160: 413–415

Kreuzer-Martin HW, Lott MJ, Dorigan J, Ehleringer JR (2003) Microbe forensics: Oxygen and hydrogen stable isotope ratios in *Bacillus subtilis* cells and spores. Proc Nat Acad Sci 100:815–819

Krishnaswami S, Graustein WC, Turekian KK, Dowd JW (1982) Radium, thorium and radioactive lead isotopes in groundwaters: application to the in situ determination of adsorption-desorption constants and retardation factors. Water Resources Res 18:1663–1675

Martell EA (1983) a-radiation dose at bronchial bifurcations of smokers from indoor exposure to radon progeny. Proc Nat Acad Sci 30:1285–1289

McCaffrey RJ, Thomson J (1980) A record of the accumulationof sediment and trace metals in a Connecticut salt marsh. In: Saltzman, B., Ed., Estuarine physics and chemistry: studies in long island sound. Adv Geophys 22:129–164

Mellars P (2006) A new radiocarbon revolution and the dispersal of modern humans in Eurasia. Nature 434:931–934

National Research Council (2010) Understanding climate's influence on human evolution. The National Academies Press, Washington, p 115

Peirson DH, Cambray RS, Spicer GS (1966) Lead-210 and polonium-210 in the atmosphere. Tellus XVIII 2:427–433

Project Sunshine (1953) Worldwide effects of atomic weapons. R-251-AEC (amended) 112 pp. Rand Corporation, copyright 1956

Shen GJ, Cheng H, Edwards RL (2004) Mass spectrometric U-series dating of New Cave at Zhoukoudian, China. J Arch Sci 31:337–342

Shen GJ, Wang W, Cheng H, Edwards RL (2007) Mass spectrometric U-series dating of Laibin hominid site in Guangxi, southern China. J Arch Sci 34:2109–2114

Turekian KK (1977) Physical and chemical approaches to the study of archaeological materials. Discovery (Yale Peabody Museum) 12(2):52–64

Turekian KK and Graustein WC (2003) Natural radionuclides in the atmosphere. In: The Atmosphere (ed. RF Keeling) Vol.4 Treatise on Geochemistry (eds HD Holland and KK Turekian). Elsevier-Pergamon, Oxford, pp 261–279

Turekian KK, Kulp JL (1956) Strontium content of human bones. Science 124:405–407

van der Merwe NJ (1969) The carbon-14 dating of iron. University of Chicago Press, Chicago

Von Damm KL, Benninger LK, Turekian KK (1979) The ^{210}Pb chronology of a core from Mirror Lake, New Hampshire. Limnol Oceanogr 24:434–439

Part II
Isotopes as Tracers of Continental and Aquatic Processes

Chapter 4
Lithium Isotopes as Tracers in Marine and Terrestrial Environments

K.W. Burton and N. Vigier

Abstract The investigation and the use of lithium isotopes as tracers of water-rock interactions at low and high temperature have significantly developed over the last 10 years. This chapter relates our current understanding of lithium isotope and elemental behaviour in the Earth's surface environment. In the introduction, we provide information on the chemical properties and behaviour of lithium, its occurrence and applications. The first section reviews the techniques used for the measurement of Li isotopes. The second section outlines the primary sources of Li in the environment and their potential impact on the hydrological cycle. The third and fourth sections investigate the impact of chemical weathering of continental rocks and oceanic crust, respectively. Finally, the last section assesses marine records of lithium.

4.1 Introduction

Naturally occurring lithium (Li) has two stable isotopes, ^{6}Li and ^{7}Li, (7.52% ^{6}Li and 92.48% ^{7}Li), which have the largest relative mass difference of any isotope pair (~16%) aside from hydrogen and deuterium. Lithium is fluid-mobile, moderately incompatible, and like other alkali metals is only present on Earth in the +1 valence state, so the isotope composition is not directly influenced by redox reactions. Moreover, unlike more established light isotope systems (such as C, O and S), lithium is a trace element and does not form an integral part of hydrological, atmospheric or biological cycles.

Nevertheless, there is significant potential for mass-dependent fractionation, demonstrated by the early experiments of Taylor and Urey (1938) who separated ^{6}Li from ^{7}Li using ion exchange chromatography, suggesting natural chemical exchange processes, such as water-rock interaction, have the potential to generate distinct isotope reservoirs on Earth.

4.1.1 Occurrence of Lithium

In the solar system: Lithium was one of the few elements produced during Big Bang nucleosynthesis, within the first 3 min of the beginning of the Universe (e.g. Spite and Spite 1982; Burles et al. 2001). Lithium is also produced during stellar nucleosynthesis, however, stellar processes tend to destroy lithium, along with beryllium and boron, so their abundances in the solar system are lower than might be expected by comparison with elements with either lower or higher atomic numbers. Stars that achieve temperatures necessary for fusing hydrogen (2.5×10^6 K) rapidly lose their Li through collision of ^{7}Li and a proton, producing two ^{4}He nuclei. Convection in low mass stars mixes the Li into the hotter interior where it will be destroyed (e.g. Korn et al. 2006). Therefore the presence of Li in a star's spectrum is a good indication that it is sub-stellar (that is, a Brown dwarf too low in mass to sustain hydrogen fusion). As a consequence of these stellar processes most of the lithium, and beryllium and boron, in the solar system was formed through cosmic ray (mostly high energy proton) mediated break-up

K.W. Burton (✉)
Department of Earth Sciences, Durham University, Science Labs, Durham, DH1 3LE, UK
e-mail: kevin.burton@durham.ac.uk

N. Vigier
CRPG-CNRS, Nancy Université, 15 rue Notre Dame des Pauvres, 54500 Vandoeuvre-les-Nancy, France

(spallation) of heavier elements residing in interstellar gas and dust (e.g. Olive and Schramm 1992).

On Earth: Lithium is widely distributed on Earth, but does not occur naturally in elemental form due to its very high reactivity. Lithium abundances in mantle rocks range from 0.1 to 7.7 ppm by weight, whereas crustal rocks range from 1 to 80 ppm, with an average abundance of 22 ppm, making Li the 25th most abundant element in the Earth's crust. Lithium is incompatible in most minerals but is preferentially incorporated into biotite, cordierite and alkali-feldspar, and forms a major element in spodumene ($LiAl[Si_2O_6]$) and petalite ($Li[AlSil_4O_{10}]$) characteristic minerals in Li-rich granitic pegmatites, both of which are, at present, the most commercially viable sources of Li. The world's largest reserve base of lithium is located in brines under salt flats, in the Salar de Uyumi area of Bolivia, which are estimated to hold at least 5.4 million tons of Li (US Geological Survey 2009).

4.1.2 Applications of Lithium

High temperature lubricant: Because of its high specific heat capacity, the highest of all solids, lithium is often used as a coolant for heat transfer applications. Thus, the first major application of lithium was as a high temperature lubricant for aircraft engines, and lithium stearate is still used at the present day.

Thermonuclear weapons: Lithium deuteride ($^6Li^2H$) is the fusion fuel used in thermonuclear weapons of the Teller-Ulam design. The lithium deuteride is bombarded with neutrons, heated and compressed, producing tritium from the lithium, which then triggers a reaction between deuterium and helium nucleons.

Nuclear energy: Lithium fluoride (enriched in ^7Li) also forms a constituent of the fluoride salt mixture ($LiF-BeF_2$) into which uranium and thorium are introduced in liquid-fluoride nuclear reactors.

Medical use: Lithium is used medically for the treatment of bipolar disorders (e.g. Schou 1988). How Li^+ works remains a matter of debate – but it is known to elevate levels of serotonin – however, therapeutic levels of Li are only slightly lower than toxic amounts, which affect the muscular and renal system. Lithium isotopes potentially offer a means of quantifying Li dosage, and ^6Li and ^7Li have a differential effect on kidney health (Stoll et al. 2001).

The enrichment of lithium for the nuclear industry has used: (1) ion exchange (Colex separation) of a lithium-mercury amalgam and lithium hydroxide solution, where ^6Li has a greater affinity for mercury than ^7Li (Palko et al. 1976); (2) vacuum distillation, where lithium is heated in a vacuum and gaseous isotopes diffuse at different speeds because of their different atomic weights, and are collected preferentially on a cold surface; (3) electrochemical separation (Umeda et al. 2001). Enrichment factors for ^6Li of 1–8% are typical of these techniques (Symons 1985). One interesting side effect of these industrial enrichment processes is that some commercial standards possess highly anomalous isotope compositions with over 300% enrichment in ^7Li relative to natural terrestrial materials (Qi et al. 1997).

Lithium and its compounds also have a number of industrial applications including high-resistance glass and ceramics, and high strength to weight alloys in aircraft (see also Bach 1985). Of increasing importance is the use of lithium as an anode material in rechargeable batteries, because of the low atomic mass of Li such batteries have a high charge and power-to-weight ratio.

4.2 Analytical Methods

The sampling procedures used for Li isotope analyses of environmental materials, such as river or soil pore waters and particles, are essentially the same as those used for trace element analyses. In brief, waters are filtered at 0.2 or 0.45 μm using frontal filtration techniques and stored in acid pre-cleaned polyethylene or polypropylene bottles. After acidification to a pH of 2, bottles are stored at 4°C (see e.g. Witherow et al. 2010; Kisakurek et al. 2005; Millot et al. 2010a, b, c). River particles, soils and sediments are dissolved in a mixture of concentrated acids (usually HF, HNO_3, HCl, and sometimes $HClO_4$) (e.g. Vigier et al. 2009). Lithium isotopes in such samples are analysed either using TIMS (James and Palmer 2000a), QUAD-ICP-MS (Kosler et al. 2001; Vigier et al. 2008) or MC-ICP-

MS (Millot et al. 2004; Jeffcoate et al. 2004; Hathorne and James 2006). Analysis using MC-ICP-MS enables the measurement of lower level samples, such as carbonates (<1 ppm) and waters (ppb level), with better precisions (<1‰). All of these techniques require the pre-separation of Li from the rest of the sample matrix, in order to avoid interference and matrix effects that are difficult to correct for at the level of precision required. All of these separation techniques use chromatography and cation exchange resins (in particular, the AGW 50X8 resin). Two types of eluant used in recent studies can be found in recent literature: organic based eluant (using a mixture of ethanol or methanol with nitric or hydrochloric acid) (Moriguti and Nakamura 1998; Tomascak et al. 1999; Marriott et al 2004a; Rosner et al. 2007; Witherow et al. 2010; Nishio et al. 2010), and dilute HCl (Chan 1987; You and Chan 1996; James and Palmer 2000a; Hall et al. 2005; Millot et al. 2004). Since Li isotopes significantly fractionate during the elution, a yield of 100% is necessary. This yield can be confirmed by measuring Li content in the sample before and after column elution. However, it is important to note that even a small percentage loss of Li may bias the δ^7Li value by a few per mil. The analysis of a well-characterised secondary standard, having a similar matrix to the sample, provides another means of establishing the accuracy of the entire chemical and analytical procedure. Seawater, rock reference materials and Li solutions doped with various elements are generally used. A review of published δ^7Li values for these reference materials can be found in Carignan et al. (2007) and Jeffcoate et al. (2004).

Lithium isotope analysis using Quad and MC-ICPMS techniques are undertaken following the standard-bracketing method, using L-SVEC as the bracketing standard. This allows the instrumental mass bias and its variation as a function of time to be corrected for. For analyses, the Li fraction is dissolved in highly diluted nitric acid (typically from 0.05 to 0.5 M HNO_3), and passed through either a dual cyclonic/Scott spray chamber or a desolvating system before being injected in the Ar-plasma of the mass spectrometer (see e.g. Millot et al. 2004; Rosner et al. 2007; Hathorne and James 2006). Using MC-ICP-MS, the long-term reproducibility of pure Li solutions is better than 0.3‰ at the 2σ level (Millot et al. 2010a). For reference materials, that additionally need to go through the chromatographic separation procedure, this reproducibility is estimated to be 0.4–0.5‰ for seawater (Millot et al. 2010a, 2004), and 0.6‰ for JB-2 (a basalt standard from Geological Survey of Japan) (see Carignan et al. 2007 for a compilation).

Recently, Li isotopes have also been measured in situ in low level environmental samples (such as carbonates), using the 1270 Ion Microprobe (Chaussidon and Robert 1998; Kasemann et al. 2005; Vigier et al. 2007; Rollion-Bard et al. 2009). The estimated reproducibility of this technique for the analyses of corals and foraminifera tests is ~2‰ (2σ). The instrumental mass bias is corrected for using a carbonate standard having the same matrix and being isotopically homogeneous, and for which the δ^7Li value has been measured with a MC-ICP-MS technique (Vigier et al. 2007; Rollion-Bard et al. 2009).

4.3 Primary Sources of Li in the Environment

4.3.1 δ^7Li of the Oceanic and the Continental Crust

Most of the Li of the silicate Earth is located in the mantle (91.8%). Since Li is incompatible during magmatic processes, the continental crust is enriched in Li compared to the oceanic crust, which contain 7 and 1% of the global Li budget respectively (Teng et al. 2004). Inorganic and biogenic carbonates are particularly depleted in Li (usually 1 ppm in pure carbonate) and this reservoir is minor in the overall Li budget (Hoefs and Sywall 1997).

The average δ^7Li value for MORB estimated by Tomascak et al. (2008) is 3.7 ± 1.9‰ close to that estimated for the undepleted upper mantle (3.5‰; Jeffcoate et al. 2007) (Fig. 4.1). The range of δ^7Li values in MORB is relatively narrow, between 1.5 and 5.6‰, but still significantly larger than that for mantle peridotite (displaying a 1‰ range), considered to represent the residual source of MORB melts. The cause of the measured heterogeneity displayed by MORB remains a matter of debate (cf. Elliott et al. 2006; Tomascak et al. 2008). Crystal fractionation induces only minor Li isotope fractionation

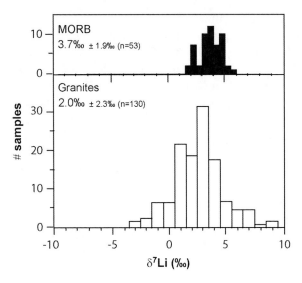

Fig. 4.1 Li isotope compositions of pristine MORB (Tomascak et al. 2008), and granites from the continental crust (Teng et al. 2004, 2008, 2009)

(Tomascak et al. 1999), and diffusion effects may be difficult to preserve in ridge magma chambers with high melt to rock ratio. Incorporation of recycled subduction-metasomatized upper mantle in the MORB source may explain the correlations highlighted with radiogenic isotopes from two localities on the East Pacific Rise (EPR) (Elliott et al. 2006). Alternatively the incorporation of seawater altered components in shallow level magma chambers may explain some of the highest MORB δ^7Li values (Tomascak et al. 2008).

Continental granites of various types (I, A, S) display an average δ^7Li value slightly lighter than that of the mantle (Teng et al. 2008). The Li isotope composition for the continental crust has recently been estimated to be 2.0 ± 2.3‰, and the corresponding range is large (−3.1 to +8.1‰) compared to MORB. This lighter composition is best explained by weathering and water-rock interactions, that have, over geological timescales, successively removed "heavy" lithium from the continents, which has been carried away by river and ground waters to the ocean (see following sections). However, even if shales are significantly enriched in clay minerals and in Li, they display a similar range of δ^7Li to loess and granites (Teng et al. 2004), and the overall role of weathering on the isotope composition of the present-day continental crust still remains to be investigated.

4.3.2 Li Isotope Composition of Natural Waters

The best evidence of fractionation of Li isotopes during low temperature weathering is the systematic enrichment of natural waters in ^{7}Li (river, groundwaters and the ocean), when compared to the drained bedrocks. In rivers, the δ^7Li value of the dissolved phase is systematically higher than the δ^7Li value of suspended or bed sediments (see Fig. 4.2). This contrasted feature is observed everywhere, in mixed lithology basins, but also in small monolithologic basins, such as the basaltic basins of Iceland (Vigier et al. 2009) and the granitic sub-basins of the Orinoco and the Mackenzie (Huh et al. 2001; Millot et al. 2010b). In addition, for all mixed lithology basins studied thus far, most of the dissolved Li (>80%) is estimated to derive from the weathering of silicate lithologies present within the basin (Kisakurek et al. 2005; Millot et al. 2010b).

Atmospheric precipitation (rain and snow) possesses low Li contents, but highly variable δ^7Li values (Millot et al. 2010b, c). The range displayed by rainwaters collected in France is 3–96‰. The highest values are explained by a significant contribution of anthropogenic lithium potentially derived from fertilizers in agricultural areas, suggesting that such pollution may significantly impact the lithium isotope signatures in some rivers, an aspect yet to be investigated.

A recent compilation of δ^7Li values measured in filtered seawater highlighted significant dispersion of the data (28.9–33.4‰) (Carignan et al. 2004). This was surprising since the ocean is expected to be isotopically homogenous as the residence time of lithium in the ocean is long (~1 Ma) (Stoffyn-Egli and Mackenzie 1984; Huh et al. 1998). More recently, improved chemical purification and new generation of MC-ICP-MS instrumentation have enabled a much improved precision, and the most recent estimate of δ^7Li for North Atlantic seawater is 31.2 ± 0.4‰ (e.g. Millot et al. 2010a; Jeffcoate et al. 2004), which corresponds well to the mean value obtained when considering all published data for the ocean. This value for seawater is significantly higher than the δ^7Li values determined for its principal Li sources: the average δ^7Li for oceanic hydrothermal fluids is 9‰, and for river waters is 23.4‰ (Huh et al. 1998). This offset might be partly accounted by a better estimate of Li sources to the

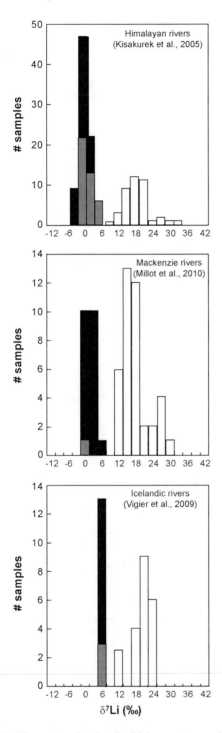

Fig. 4.2 Histograms showing the Li isotope composition of selected river waters (in *white*), suspended sediments (in *grey*) and bedload sands (in *black*). River water δ^7Li values are systematically higher than those measured for river particles and sediments, for both mixed lithology basins (Himalaya, Canada) and monolithologic basins (in Iceland). The range of values measured in river solid phases is narrow and close to the average of the source rocks (see Fig. 4.1)

ocean and their isotope compositions. Also, low temperature seafloor weathering and the formation of ^{6}Li-rich oceanic clays are thought to play a key role in the Li ocean cycle (see the following sections).

4.4 Using Li Isotopes as Tracers of Continental Weathering

Lithium is relatively mobile in the weathering environment, as a result of its low ionic charge, relatively small ionic radius (approximately 0.068 nm; Shannon 1976) and high degree of covalency. Stable isotopes, such as those of lithium, will not be affected by variations in rock and mineral composition that result from the decay of the parent isotope over time, such as for example, for the rubidium-strontium (^{87}Rb-^{86}Sr) or lutetium-hafnium (^{176}Lu-^{176}Hf) isotope systems (e.g. Blum and Erel 1995; Bayon et al. 2006). Nevertheless, there are a number of processes that may lead to the fractionation of ^{6}Li and ^{7}Li during weathering, including:

Primary mineral dissolution: The composition and mineralogy of the parent bedrock determines the initial isotope composition of the waters during mineral dissolution. The primary mineralogy also partly determines which elements are available in solution to form secondary minerals. Previous studies suggest that there is little fractionation of Li isotopes at magmatic temperatures (e.g. Tomascak et al. 1999) and little variation in the composition of coexisting minerals (Chan and Frey 2003). Hence under normal circumstances the preferential weathering of a primary mineral phase should not generate significant differences in the Li isotope composition of waters. Nevertheless, in principle, during diffusion or the incongruent dissolution of primary phases small differences in the bond energies of ions of a different mass may result in isotope fractionation (O'Neil 1986). For lithium breaking the higher energy bonds of ^{6}Li is energetically favourable, potentially enriching any solution with the lighter isotope.

Secondary mineral formation: Lithium is highly hydrated in aqueous solutions relative to most other alkali metals, and ordinarily assumes tetrahedral hydrated coordination (Olsher et al. 1991). In most minerals lithium occupies either tetrahedrally- or octahedrally coordinated sites (Olsher et al. 1991; Wenger and

Armbruster 1991). During the process of crystallization isotope fractionation can occur when an energetic advantage is gained by the incorporation of one isotope over another into a crystallographic site (O'Neil 1986). One basis of ab initio calculations, ^{6}Li is preferentially incorporated into octahedral crystallographic sites of minerals from aqueous solution (Yamaji et al. 2001).

Adsorption onto minerals: Adsorption of aqueous ions onto mineral surface sites is another mechanism that can potentially lead to isotope fractionation. Sorption of Li onto minerals from aqueous solutions at ambient temperatures has been seen in a range of minerals (Taylor and Urey 1938; Anderson et al. 1989) despite early predictions that Li would tend to be the least sorbed of the alkalis, due to its high degree of hydration (Heier and Billings 1970). Sorption and retention of ions from solution depends upon both mineral surface chemistry and the composition of the solution. Early experiments were taken to indicate that the magnitude of mass fractionation on desorption is equivalent to that of adsorption (Taylor and Urey 1938) however, empirical data suggest that this fractionation is not reversible (Comans et al. 1991; James and Palmer 2000a, b). The irreversibility of this process has been attributed to physical adsorption into crystallographic surface sites rather than surface ion exchange sites, in which case the magnitude of this effect will vary among minerals.

The actual secondary minerals formed depends upon elemental supply (saturating solutions in their constituent elements), the stability of those phases (principally controlled by pH and temperature), and their formation kinetics. Lithium is not a nutrient element, and there appears to be little Li isotope fractionation accompanying incorporation into biomass (Lemarchand et al. 2010). Nevertheless, the depletion of other elements, such as K and Mg, due to their incorporation into biomass may inhibit the formation of secondary minerals, indirectly affecting the isotope behaviour of Li (Pogge von Strandmann et al. 2010).

In this section we review how experimental work and studies of soils and rivers yield key information on the influence of weathering processes on Li isotopes, and highlight the utility of Li isotopes as a weathering tracer. Also, studies of Li in the estuarine environment illustrate how the continental weathering signal is transferred to the oceans.

4.4.1 Experimental Work

Primary mineral dissolution: Partial dissolution of basalt in dilute HNO_3 does not result in fractionation of Li isotopes (Pistiner and Henderson 2003) although similar dissolution of granite did result in significant apparent fractionation of Li isotopes. This was attributed to the dissolution of secondary phases in the granite or differences in the isotope composition of primary igneous minerals in this more evolved rock (Pistiner and Henderson 2003). More recently controlled experimental dissolution of basalt glass and olivine at pH and temperatures appropriate to natural conditions (pH 2–4; T = 25°C) indicates that there is no resolvable Li isotope fractionation accompanying dissolution of single phases (Wimpenny et al. 2010b). These experimental observations suggest that primary mineral dissolution does not result in significant fractionation of Li isotopes.

Secondary mineral formation: On the basis of natural observations and experiments the formation of clays, both at high and low temperatures, is thought to be the major process responsible for the fractionation of Li isotopes at the Earth's surface and the high δ^7Li found in waters (see Sect. 4.3). However, there are few experimental studies where Li isotope behaviour has been quantified for the crystallisation of a single phase. Clays are not systematically enriched in Li and the type of clay may influence water isotope signatures. Past work has shown that Li behaviour in soils and waters is strongly linked to that of magnesium (e.g. Huh et al. 1998), suggesting that Mg-bearing primary and secondary phases play a key role in the Li cycle.

Williams and Hervig (2005) studied the Li isotope fractionation of experimentally synthesized illite and smectite under controlled P-T and fluid conditions. It was found that during illitization, Li substitutes into the octahedral sites. The elemental concentration and isotope composition were found to reach a steady state with the onset of nearest neighbour (R1) ordering. However, isotope fractionation for Li of between −5 and −11‰ varied systematically with crystal size, suggesting that kinetic effects may influence Li uptake by clays. More recently tri-octahedral Mg-Li smectites (hectorites) were synthesized at temperatures ranging between 25 and 250°C (Vigier et al. 2008). The Li isotope fractionation factors linked to the incorporation of Li into the octahedral sites, in substitution for

Mg^{2+}, were determined. Experimental $\Delta^7Li_{\text{clay-solution}}$ correlates inversely with temperature, as theory predicts, and ranges between $-1.6 \pm 1.3‰$ at 250°C and $-10.0 \pm 1.3‰$ at 90°C, and then stays relatively constant down to 25°C. The relatively constant isotope fractionation below 90°C was attributed to high concentrations of edge octahedra in low crystallinity smectites. It was also found that the fractionation factor does not depend on the solution composition, nor on the amount of Li incorporated into the clay, at a given temperature. Experimental precipitation of chrysotile from olivine dissolution also results in a significant Li isotope fractionation of the residual fluids (Wimpenny et al. 2010b) although there was insufficient chrysotile for measurement and quantification of the fractionation into the solid phase.

Adsorption: Zhang et al. (1998) found a similar degree of Li isotope fractionation during Li sorption from seawater onto kaolinite and vermiculite, $\alpha = 0.979$ and 0.971, respectively. Pistiner and Henderson (2003) undertook Li sorption experiments on a variety of minerals and found that isotope fractionation does not occur when Li is not structurally incorporated into the solid (physical sorption) for example during outer sphere sorption of Li onto smectite. By contrast, when Li is incorporated by stronger bonds (chemical sorption), an isotope fractionation is observed that is dependent on the chemical structure of the minerals (Anghel et al. 2002). For example, some fractionation was seen to accompany sorption onto ferrihydrate, and significant fractionation occurred during the formation of inner sphere complexes on the surface of gibbsite ($\alpha = 0.986$) (Pistiner and Henderson 2003).

Taken together, these experimental results suggest that secondary mineral formation and adsorption onto secondary minerals are the major processes controlling isotope fractionation during continental weathering. Moreover, while different minerals have variable fractionation factors, the sense of fractionation is always the same, where the lighter isotope ^6Li is preferentially incorporated into the solid phase.

4.4.2 Lithium Behavior Accompanying Soil Formation

Soils represent complex biogeochemical systems, however, given that Li is not significantly fractionated by incorporation into biomass, it can be anticipated that this element will respond primarily to the formation of secondary phases in response to the breakdown of the parent rock and/or external inputs (aeolian or groundwater).

However, early work on soils developed on Hawaiian basalts gave unexpected results. Huh et al. (2004) reported increases in Li concentration, and variable but heavy Li isotope compositions for soils, attributed to the uptake of aerosol lithium from seawater. Whereas, Pistiner and Henderson (2003) found little elemental or isotope variation in a soil developed on a similar basalt, and it was suggested that atmospheric deposits with high δ^7Li may have counterbalanced the preferential release of ^7Li. A laterite soil developed on the Deccan basalt was found to possess Li isotope compositions that were, in general, light relative to the composition of the parent basalt (Kisakurek et al. 2004). However, Li isotopes displayed a strong negative covariation with Li concentration which was best explained by mixing of precursor basalt and aeolian material, rather than simply being due to the presence of Li-enriched secondary phases.

Work on saprolite sequences developed on igneous rocks in South Carolina (USA), a granite and a diabase, demonstrated that the soils can consistently possess a much lighter Li isotope composition than the igneous rocks from which they formed (Rudnick et al. 2004). More recently Lemarchand et al. (2010) studied soils and soil solutions (pore waters) developed on a granite bedrock in the Strengbach catchment (Vosges, France). While there was little variation of the Li isotope composition of the soils with depth, there is a significant variation in the isotope composition of the pore waters, with deep waters possessing heavy isotope compositions and shallow waters light isotope compositions relative to the parent rock (Fig. 4.3). These variations were attributed to dissolution and precipitation processes. In the deepest horizons the precipitation of secondary minerals results in the preferential incorporation of light isotopes, driving the pore waters to heavy values. Whereas in the shallower soil horizons the data suggest that those secondary minerals as well as residual primary phases are actively dissolving shifting the composition of the pore waters to light values.

Overall, soil studies are broadly consistent with the prediction that secondary mineral formation during weathering releases heavy lithium to the hydrosphere,

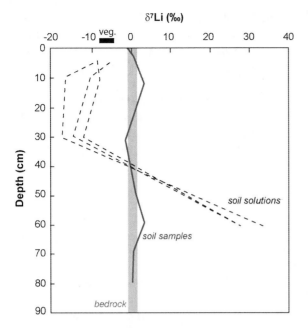

Fig. 4.3 Lithium isotope systematics for an experimental ecosystem located in the Strengbach catchment (Vosges mountains, NE France). The range of δ^7Li values obtained for the local vegetation (spruce needle, root, branches) is shown by the *black bar*. Soil solutions have been sampled at various times of the year (*dashed lines*) and show a similar pattern with depth. Solid phases have δ^7Li values close to those of the bedrock. Modified from Lemarchand et al. (2010)

leaving isotopically light Li behind in the solid phases. However, many of these studies also point to the influence of externally derived Li either from atmospheric input or groundwater, and the breakdown of early formed secondary minerals may also impart a light isotope signal to the pore waters.

4.4.3 Lithium Isotopes in Rivers

Rivers represent the principal means by which the products of continental weathering are transported to the oceans, and exert an important control on the Li isotope composition of seawater. In a study of major rivers (comprising some 30% of the global river discharge) Huh et al. (1998) found that the dissolved phase (filtered to <0.45 μm) possessed a wide range of Li isotope compositions in accord with earlier studies (e.g. Chan et al. 1992; Falkner et al. 1997). No clear relationship between isotope composition and catchment rock type was observed, consistent with a decoupling of δ^7Li and $^{87}Sr/^{86}Sr$ in the same samples, although Li concentrations did show a relationship with the weathering susceptibility of the parent rock type. Lithium concentrations were also found to covary with Mg, but rarely with Si and K, even though both are sourced by the weathering of silicates, in accord with the tendency of Li to be retained in secondary clays, substituting for Mg^{2+}. These results were taken to indicate that the dominant effects on river chemistry are isotope fractionation between waters and secondary minerals and the degree of chemical erosion.

River data for the Orinoco basin shows a conspicuous difference in the δ^7Li of the dissolved load between rivers draining the young Andean mountains (31.0–37.5‰) compared to the shield terrains (13.5–22.8‰) even though the δ^7Li composition of the suspended load is similar between the two (Huh et al. 2001). These differences were attributed not to variations in lithology, but to distinctions in the weathering regimes. In the high-relief Andean mountains weathering is reaction-limited and the dissolved load is high in Li and isotopically heavy. Whereas, in the low-relief stable shield areas weathering is transport limited, Li concentrations in waters are low and isotopically light in proportion to the increasing degree of weathering.

In a study of carbonate and silicate catchments in the Himalayas it was found that most of the dissolved Li is sourced by the dissolution of silicate minerals even in carbonate dominated terrains (Kisakurek et al. 2005). A distinct difference was also found in the δ^7Li isotope composition of both bedload and suspended load reflecting the different composition of the bedrock. However, the isotope composition of the suspended load is nearly always lighter than the corresponding bedload probably reflecting the presence of secondary phases. There is considerable overlap in the δ^7Li composition of the dissolved load from the carbonate-silicate catchments, suggesting that the main control on the Li isotope chemistry of the dissolved load is not the bedrock lithology, but rather weathering reactions involving the formation of secondary minerals.

For rivers in the Mackenzie basin, Canada, it was also found that the Li in the dissolved load is dominated by the weathering of silicates, and that the Li isotope composition is inversely correlated with the relative mobility of Li (when compared with Na) (Millot et al. 2010b). Surprisingly, the most significant ^7Li enrichment in the dissolved phase of the rivers accompanied incipient weathering in the Rocky Mountain and shield areas. The fractionation was not considered to be accounted for by the formation of clays, and was rather attributed to the formation of Fe-oxyhydroxides. By comparison, rivers in the lowland areas are the most depleted in Li, with lower δ^7Li values (although still enriched in ^7Li compared to the bedrock) and this was attributed to equilibrium with clay minerals in aquifers which contribute to the dissolved load of rivers.

Small monolithologic catchments provide a means of understanding and quantifying the effects of weathering processes alone, because there is little variation due to differences in rock type. Over recent years there have been several studies of Li isotope behaviour in rivers draining basalt terrains (Pogge von Strandmann et al. 2006, 2010; Vigier et al. 2009) where the controls on water chemistry are better understood through both natural observations and experimental work. Rivers in Iceland display a wide range of δ^7Li isotope compositions (10.1–43.7‰) that are always heavier than the basalt bedrock and bedload sand (3.1–8.9‰). Whereas, the suspended load in these rivers are invariably lighter than their corresponding bedload (δ^7Li = −1.3–6.0‰) likely due to the presence of secondary phases enriched in ^6Li (Fig. 4.2). Simple mixing between basalt weathering (δ^7Li = 3.9‰), precipitation (rain/snow) which is dominated by seawater aerosols (δ^7Li = 33.3‰) and hydrothermal springs (δ^7Li = 5.5–10.9‰) cannot account for these variations, rather they are better explained by uptake in secondary minerals.

The Li isotope composition of the dissolved phase in the rivers in Iceland shows a dependence upon the weathering regime. In the older basaltic terrains (largely non-glacial) weathering tends to be more reaction-limited, and result in waters with low K/(K + Na) ratios (relative to the parent basalt) and heavy Li isotope compositions. This has been explained by preferential Li and K incorporation into secondary minerals and biomass (K) relative to mobile elements such as Na. Whereas, in the younger, largely glacial catchments, weathering is relatively transport-limited and K/(K + Na) ratios and Li isotope compositions are closer to those of the parent basalt. Moreover, for many rivers in Iceland discharge rates are monitored and chemical erosion rates can be calculated. For these terrains at least Li isotope behaviour can be explained by a simple empirical relationship

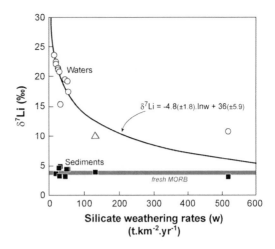

Fig. 4.4 δ^7Li of the dissolved phase (in *white*) and sediments (in *black*) of Icelandic rivers, as a function of weathering or chemical erosion rate, estimated independently from major and trace elements concentrations measured in the dissolved loads (Vigier et al. 2009)

with erosion rate (Fig. 4.4) (cf. Vigier et al. 2009). Put simply, the Li concentrations and isotope compositions reflect the balance of Li derived from the parent rock against that affected by incorporation into secondary phases. When weathering rates are high, Li concentrations of the waters are also high and Li isotope compositions are closer to those of the parent rock, whereas when weathering rates are low, the isotope composition of the waters more strongly reflects incorporation into secondary phases. Finally, Wimpenny et al. (2010a) found that significant fractionation of Li occurs during glacial weathering despite little evidence for the formation of clays in the chemistry of the glacial waters. This was attributed to sub-glacial weathering of sulphide and other Fe-bearing minerals, resulting in the formation of Fe-oxyhydroxides (e.g. Tranter 2003; Raiswell et al. 2006). Leaching experiments suggested that some 65% of the Li in the suspended load is hosted by such Fe-oxyhydroxides, and that this has a light Li isotope composition (relative to the dissolved phases) indicating that this mineral preferentially incorporates ^6Li relative to ^7Li (Wimpenny et al. 2010a).

4.4.4 Estuarine Behaviour of Lithium

Dissolved lithium exhibits conservative behaviour in estuaries, including the well-mixed estuary of the St

Lawrence river, the partially mixed Scheldt (Stoffyn-Egli 1982) and the salt-wedge Mississippi rivers (Colten and Hanor 1984). Lithium also behaves conservatively in the Gulf of Papua, where particulate Li is trapped in aluminosilicate muds on the shelf (Brunskill et al. 2003). Similarly, dissolved Li in estuaries draining dominantly basaltic terrains in Iceland and the Azores show simple elemental and isotopic mixing trends indicating that the Li elemental and isotope signature is transferred to seawater without modification in the estuarine mixing zone (Pogge von Strandmann et al. 2008) (Fig. 4.5).

The Li concentration of suspended particles in the estuary at Borgarfjordur in Iceland is within the range of source basalt, however, remarkably the Li content of the particles and δ^7Li both increase with increasing salinity. This relationship was taken to indicate continued weathering (and precipitation of secondary clays) in the estuarine environment. The particulate data suggests a fractionation factor $\alpha_{mineral-fluid}$ of 0.981 ± 0.003 for the clay-seawater in the estuary, similar to the value of 0.981 reported by Chan et al. (1992) for weathered sea-floor basalts. Moreover, it was concluded that the weathering of such particles in the estuarine environment may significantly impact the Li flux to the oceans, reducing the global riverine flux by between 15 and 25%.

4.5 Li Isotopes as Tracers of Hydrothermal Processes and Ocean Floor Weathering

The isotope and elemental budget of Li in the oceans is controlled by: *input* from continental weathering (via rivers, groundwater and aeolian dust) and from high-temperature hydrothermal fluids; and *removal* through sea-floor alteration, high temperature (hydrothermal) recrystallisation and adsorption. Precipitation of marine carbonates and biogenic silica are minor sinks of Li (<5% total) given their low Li concentrations, and the diffusive flux from sediments is also small (see Table 4.1).

4.5.1 Li Mobility During Ocean-Floor Weathering

During weathering (or alteration) of oceanic basalt Li from seawater is incorporated into secondary phases (in much the same way as during continental weathering) (Fig. 4.6). Data for altered sea-floor basalts indicates that the variations cannot be simply attributed to mixing between pristine basalt and seawater. Rather seawater Li enters the secondary weathering phases with an apparent fractionation

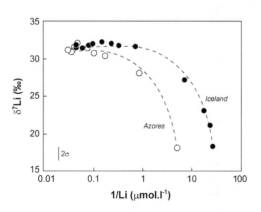

Fig. 4.5 δ^7Li measured in the dissolved phase of two small estuaries on basaltic islands (in the Azores in *white*, and in Iceland in *black*) (Pogge von Strandmann et al. 2008). In both cases, mixing lines between fresh and sea waters strongly suggest a conservative behaviour of lithium within the studied estuaries

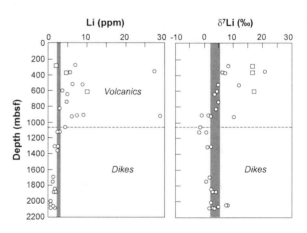

Fig. 4.6 Variation of Li concentration, and δ^7Li values with depth for whole rocks (*white circles*) and smectites (*white squares*) from DSDP/ODP Site 504B (modified from Chan et al. 2002). The range for pristine MORB is shown for comparison (in *light grey*). The uppermost volcanic section is enriched in Li with high δ^7Li values. In contrast, Li is lost from the dyke complex through extraction by hydrothermal fluids with less isotope fractionation

factor α ~ 0.981 (Chan et al. 1992, 2002) indistinguishable from the fractionation factor observed during the weathering of basaltic suspended material in estuaries (Pogge von Strandmann et al. 2008). The relative uniformity of this fractionation at a range of localities suggests either that Li uptake is dominated by a single phase, or else that the phases involved possess similar fractionation factors for Li incorporation. The principal secondary phases in highly weathered basalts are clays, Fe-oxyhydroxides and smectites. Experiments suggest that the fractionation factor accompanying smectite precipitation is <0.990, however, the low-T experiments need to be refined (Vigier et al. 2008). The fractionation factor associated with the precipitation of Fe-oxyhydroxides ranges from 0.978 to 0.999 (Pistiner and Henderson 2003; Chan and Hein 2007; Wimpenny et al. 2010a), although the fractionation factor for zeolite has not yet been determined. Lithium uptake may also be partly controlled by adsorption processes, in a study of Li uptake from seawater onto kaolinite and vermiculite fractionation factors of 0.979 and 0.971, respectively, were determined (Zhang et al. 1998).

4.5.2 Lithium in Hydrothermal Fluids

Seawater circulates through ocean crust of all ages, driven by temperature induced pressure gradients. At mid-ocean ridges the convective flow of seawater is driven by crystallisation and cooling of the magma that upwells from the mantle to form oceanic crust. High-T hydrothermal exchange between basalt and seawater is a significant source of Li to the oceans, although the magnitude of the hydrothermal flux remains poorly quantified, in particular at the ridge flanks (Elderfield and Schultz 1996).

Lithium is enriched in high temperature (>350°C) vent fluids by a factor of 20–50 times that of seawater, and the δ^7Li compositions of fluids from vents in a range of settings yield values ranging from those of MORB to several per mil heavier (Chan and Edmond 1988; Chan et al. 1993, 1994). There is no correlation between δ^7Li and seafloor spreading rate suggesting that mid-ocean ridge hydrothermal fluids derive most of their Li from young, unaltered, basaltic crust, rather than older basalts (Chan et al. 1993). Sediment cover appears to have little effect on the isotope and elemental composition of hydrothermal fluids, suggesting that the isotope signal is sourced by deep processes not linked to surface sediments enriched in Li that is highly susceptible to hydrothermal alteration (Chan et al. 1994).

It has been suggested that oceanic crust preserving a record of the interaction with high-temperature hydrothermal fluids may be uncommon, because of the prevalence of low-temperature weathering. Nevertheless, observations suggest that such hydrothermal alteration preferentially removes heavy Li from the rock, leaving the residual, altered oceanic crustal rock with a lower Li concentration and lighter isotope composition than the parent material (Chan et al. 1992, 1994; You et al. 1995; James et al. 1999).

4.5.3 Lithium in Abyssal Peridotite and Mantle Rocks

In oceanic fracture zones and actively-upwelling mud volcanoes seawater reacts directly with mantle rocks (Decitre et al. 2002; Savov et al. 2005; Benton et al. 2004). Serpentinites (altered mantle rocks) from these studies are enriched in Li relative to unaltered mantle rocks, but possess highly variable δ^7Li isotope compositions, both heavier and lighter than unaltered mantle rocks, although the mean δ^7Li (=+7 ± 5‰) is similar to that of average seafloor vent fluids (+8 ± 4‰). The Li isotope composition of Indian ocean serpentinites has been attributed to the recycling of Li derived from basaltic oceanic crust, rather than from seawater (Decitre et al. 2002). Where the variable isotope compositions of serpentinites result from the interaction of mantle rocks with a hydrothermal fluid, as it cools and reacts along its flow path towards heavier Li isotope compositions (Vils et al. 2009).

4.5.4 Lithium in Marine Pore-Fluids

During diagenesis (compaction and mineral dehydration) pore waters, carrying a chemical signature of fluid-sediment interaction, may be released from marine sediments to seawater. A number of studies have measured the Li isotope and elemental composition of both pore waters and host sediments (e.g. Zhang et al. 1998; Chan and Kastner 2000; James et al. 1999; James and Palmer 2000b) and these data indicate that the pore fluids possess a wide range of Li isotope compositions (δ^7Li values range from less than 20 to >40‰), almost always having a heavier Li isotope composition than their host sediment. Zhang et al. (1998) recognised the importance of NH_4^+ (produced by the breakdown of organic matter) in pore fluids as a counterion for exchange with Li adsorbed on sediments. It was observed that the peak in NH_4^+ concentrations coincided with Li isotope compositions lighter than seawater, suggesting that Li desorbed from sediments is isotopically lighter than the seawater starting composition. Chan and Kastner (2000) documented an inverse correlation between Li concentration and δ^7Li values that was considered to reflect the superimposed effects of the alteration of volcanic ash, taking up Li from the fluids through the formation of smectite, and ion exchange (between Li^+ and NH_4^+).

In a study of active hydrothermal venting through a thick sediment sequence close to the Gorda ridge (James et al. 1999) Li isotope variations were interpreted in terms of initial hydrothermal removal of Li from sediments, followed by uptake of dissolved Li accompanying the crystallisation of secondary minerals, and dilution with seawater. At another site on the Gorda ridge, the Li in pore fluids was estimated to be substantially derived (>65%) from Li initially adsorbed on particles in the water column (James and Palmer 2000a, b). On the basis of comparison between experimental results and natural data it was inferred that the upwelling rate was also a critical control on the Li isotope composition of pore waters, where slow rates of upwelling favour increased fluid-sediment interaction (James et al. 2003).

4.5.5 Lithium Isotope Fractionation During Laboratory Experiments

Laboratory experiments have also been undertaken to quantify Li isotope behaviour during interactions of high temperature fluids with basalt, altered basalts and sediments (Chan et al. 1994; Seyfried et al. 1998; James et al. 2003). All of these studies show that Li is extracted from these rock types even at temperatures <100°C. At temperatures up to 150°C Li is incorporated into clays (such as smectite and illite) during basalt alteration (Seyfried et al. 1998; James et al. 2003) with an affinity that is proportional to the fluid/rock ratio, such that Li is simultaneously added to and removed from fluids. Whereas at higher temperature hydrothermal conditions (~350°C) Li is largely removed from solids, and kept in solution rather than in secondary alteration minerals (Seyfried et al. 1998).

Natural studies of the interaction between seawater and basalts suggest that there is a fractionation of Li isotopes that is strongly temperature dependent (Chan and Edmond 1988; Chan et al. 1992, 1993, 1994; James et al. 1999). High temperature experiments between clinopyroxene and aqueous fluids (500–900°C) also show a temperature dependent isotope fractionation (Wunder et al. 2006). More recent low temperature (25–250°C) experiments have also quantified the temperature dependent fractionation of Li isotopes accompanying seawater-basalt interaction (Millot et al. 2010c), deriving the following relationship between isotope fractionation and temperature:

$$\Delta_{solution-solid} = 7,847/T - 8.093$$

(temperature in Kelvin)

These experiments were taken to indicate that the Li isotope and elemental composition of fluids derived from seawater-basalt interaction are controlled by (1) mixing between Li in seawater, and Li released from the basalt and (2) uptake of Li into secondary minerals (Millot et al. 2010c).

4.6 Marine Li Records

Due mainly to analytical difficulties linked to the particularly low levels of Li in carbonates (<1 ppm), and, in contrast, the high levels of Li (≫10 ppm) in potential contaminants (marine clays and oxides), there are very few published data for δ^7Li in marine carbonates, and data for only two marine records (recorded by foraminifera) have been published thus far. In parallel, experimental studies allow the isotope fractionation during formation of various types of carbonates to be constrained, and some controlling factors to be identified. In parallel, other proxy substrates are investigated, but are not yet fully understood (see Sect. 4.6.3).

Table 4.1 Main marine lithium inputs and outputs

	Li flux (10^{10} g year^{-1})	δ^7Li (‰)	References
Li inputs to the ocean			
Rivers	5.6	+23.5	Huh et al. (1998)
Hydrothermal fluids	4.2 (2–20)	+8.5	Chan et al. (1993), Elderfield and Schultz (1996), and Vigier et al. (2008)
Fluid expulsion at convergent margins	0.06–0.4	+20 to >+40	Zhang et al. (1998), You et al. (1995), Chan and Kastner (2000), and Scholz et al. (2010)
Mud volcanoes	0.2 (0.003–0.5)	+12 to +31	Vanneste et al. (2010), You et al. (2004), and Scholz et al. (2009)
Li outputs			
Uptake by clays	0.7–25	+1 to +14	Stoffyn-Egli and Mackenzie (1984), Chan et al. (1992, 2002, 2006), and Wheat and Mottle (2000)
Carbonate formation	0.2–0.6	+27 to +31	Hoefs and Sywall (1997), Milliman (1993), and Hathorne and James (2006)
Biogenic silica	0.06–1	−1 to +8	Chan et al. (2006)

4.6.1 Foraminiferal Records

The first studies appeared to indicate highly variable Li isotope compositions for foraminifera through time, with δ^7Li values ranging from less than 10‰ up to 40‰ (Kosler et al. 2001; Hoefs and Sywall 1997). However, more recent work indicates that foraminifera preserve a much narrower range of δ^7Li values for the Cenozoic, for a number of different species (Hall et al. 2005; Hathorne and James 2006) (Fig. 4.7). These contrasting results are partly explained by the use of more sensitive MC-ICP-MS, but also by improved cleaning techniques for foraminifera. Marriott et al. (2004a, b) and Vigier et al. (2007) have shown that the cleaning procedure strongly affects both the Li content and isotope composition of the foraminifera. This is because the foraminiferal calcite possesses very low Li contents relative to contaminant phases such as Fe-Mn oxyhydroxides and silicate grains, and there is also a significant difference between the δ^7Li value of foraminiferal calcite and the contaminant phases. It has recently been shown that ion microprobe and laser techniques could be used to avoid contaminant phases, but other difficulties then arise, such as requirement for large shells with thick chamber walls, as well as appropriate calcite standards (Hathorne et al. 2009; Vigier et al. 2007).

The 0.4–35.8 kyr record obtained by TIMS by Hall et al. (2005) suggests that *O. universa* (planktic foraminifera) have δ^7Li values that are close to seawater. Glacial and interglacial samples display similar δ^7Li values to surface sediment samples. This is consistent with the long residence time of Li in the ocean (>1 Myr, Huh et al. 1998), implying that changes in Li sources and sinks in the ocean cannot be resolved in the δ^7Li record over short timescales.

The 0–18 Ma record of Hathorne and James (2006) is based on analysis of several species of planktic foraminifera, separated from sediments from both the Atlantic and the Pacific oceans. The data are globally consistent, and they suggest that the seawater δ^7Li value has been constant for the last 10 Ma. In contrast, the interval from 10 to 18 Ma is marked by a significant decrease in planktic foraminifera δ^7Li. This

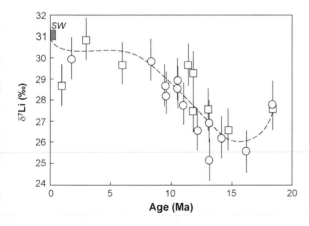

Fig. 4.7 Record of planktic foraminiferal δ^7Li for the past 18 million years (after Hathorne and James 2006). This record is based on different species of foraminifera. Note that there appears to be little difference in the δ^7Li value of samples recovered from the Pacific (*squares*) and Atlantic (*circles*) oceans

decrease was attributed to a change in seawater δ^7Li, due to a change in the rate of continental weathering. Overall, this study illustrates that the Li isotope signature of the ocean may be affected by a change of the isotope signature of river waters. If the river water signature is mainly due to chemical weathering of silicates, then records of past oceanic δ^7Li may yield precious information on the link between climate and silicate weathering.

The same study also highlights the need of constraining, in parallel, changes in the continental Li flux. In principle, the Li flux can be estimated from the Li/Ca ratio measured in foraminifera, although a strong species effect has been demonstrated and must be corrected for (Hathorne and James 2006). This species effect may, at least in part, explain discrepancies between different studies. Significant glacial/interglacial variation in the Li/Ca ratio preserved by planktonic and benthic foraminifera (Hall and Chan 2004; Lear and Rosenthal 2006) has been taken to suggest a role for temperature and/or calcite saturation state on the amount of Li incorporated by these species. In contrast, Delaney and Boyle (1986) found constant Li/Ca ratios for mixed planktonic foraminifera over the last 40 Myr.

4.6.2 Influence of Environmental Parameters on Li Isotope Fractionation

The interpretation of Li isotope signatures measured in foraminifera or other types of carbonates requires knowledge of the isotope fractionation accompanying Li incorporation into the carbonate matrix. Indeed, in order to calculate the ocean δ^7Li from a value measured for an inorganic or a biogenic carbonate, it is necessary first to constrain the isotope fractionation that has occurred during its formation, and second to identify the parameters controlling this fractionation (e.g. T, pH, calcification rate).

The inverse-correlation between Li/Ca and δ^7Li obtained by Hathorne and James (2006) for recent planktonic foraminifera strongly suggests that "species effects" and biomineralization mechanisms are likely to influence both Li isotope fractionation and Li concentration. The influence of environmental parameters on the Li isotope fractionation and Li/Ca ratios has been evaluated for inorganic calcite and aragonite experimentally grown in the laboratory (Marriott et al. 2004a, b). Depending on the experimental conditions, measured Li isotope fractionation between the precipitated calcite and solution is not the same. Despite this discrepancy, no dependency on temperature is found for this fractionation at least between 5 and 30°C (Fig. 4.8). In contrast, Li/Ca inversely correlates with temperature. The Li/Ca ratio recorded by a tropical coral (aragonitic) from the Pacific Ocean varies over the course of the year and displays the same pattern as δ^{18}O, but δ^7Li values remain constant. Marriott et al. (2004a, b) also showed that salinity has no influence on the Li isotope fractionation during experimental formation of calcite and aragonite. The main difference between these forms of calcium carbonate lies in the magnitude of the Li isotope fractionation (relative to the parent solution). Inorganic and biogenic calcite display δ^7Li values that are often close to that of the solution, while aragonite

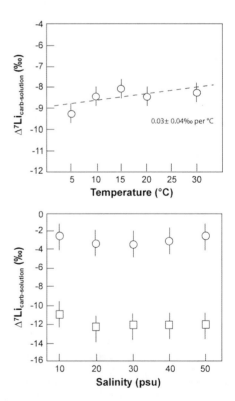

Fig. 4.8 Temperature and salinity dependency on δ^7Li measured in calcite (*circles*) and aragonite (*squares*) grown experimentally (Marriott et al. 2004a, b). In both cases the solution δ^7Li remains constant, at its initial value (0‰)

samples are significantly enriched in the light isotope (generally by more than 10‰).

In brief, thus far, the Li isotope signatures of carbonates seem to be little affected by environmental factors and therefore may be appropriate for determining the evolution of the isotope signature of seawater on geological timescales. However, some parameters still need to be tested such as the calcification rate and/or the role of dissolved CO_3^{2-}.

4.6.3 Other Marine Archives in Development

Two other marine substrates have recently been assessed as potential recorders of the Li and Li isotope composition of seawater: corals and oceanic ferromanganese deposits.

In situ measurements of Li isotopes in calcite and aragonite matrices have recently been developed (Vigier et al. 2007; Rollion-Bard et al. 2009). These studies show first that in situ isotope analyses of aragonitic corals are easier than that of foraminifera tests. They also find that shallow water and deep sea corals, either natural or grown in the laboratory, are isotopically homogeneous at the micron scale (within the 1.1‰ uncertainty of the technique). This implies that the biomineralization process, which is thought to be different for the different skeletal components of the coral (e.g. centres of calcification (COC or EMZ, Early Mineralization Zone) vs. the fibrous parts) does not influence the Li isotope fractionation during skeletal formation. It is also shown that corals grown under highly different pCO_2 and pH conditions display similar isotope signatures. This absence of dependency on environmental conditions strongly suggests that corals can record the evolution of oceanic δ^7Li, with limited disturbance. However, several present-day species display different isotope fractionation relative to seawater, and this aspect remains to be investigated since it is difficult to estimate long term oceanic paleovariations based on a single species.

Recently formed Fe-Mn crusts display highly variable Li contents. In fact, the Li content of Fe-Mn deposits can be used to discriminate among the various types of deposits, because hydrothermal and diagenetic crusts are highly enriched in Li ($\gg 100$ ppm) whereas hydrogenetic Fe-Mn crust that form from seawater on the seafloor display low Li levels (generally <2 ppm) (Chan and Hein 2007; Jiang et al. 2007). Only a few Li isotope data have been published thus far for Fe-Mn crusts. It appears that the crusts incorporate at least two types of Li: loosely bound Li having an isotope signature close to that of seawater (31‰), and more tightly bound Li which can be significantly lighter than the seawater value (Chan and Hein 2007). However, diffusion effects may be significant for Li (Henderson and Burton 1999), and ancient Fe-Mn crusts are unlikely to provide a direct record of the isotope signature of seawater at the time of their formation. More data are certainly needed for confirming these recent observations.

4.7 Conclusion and Future Directions

Studies involving Li isotopes have increased almost exponentially over the past few years. Both experimental and field studies show that significant isotope fractionation of Li occurs during water-rock interactions. Both low and high temperature fluids display δ^7Li values that are significantly different from those of primary or secondary minerals. The major role of silicate lithologies and clay formation in the Li budget has been assessed, both at a regional scale, and at a continental scale. In order to bring more quantitative constraints on present-day and past chemical erosion rates of silicates, using Li isotopes, a better calibration of this fractionation in natural systems is required, in particular during low-temperature rock weathering. Also, materials that record the Li isotope composition of seawater need to be investigated and understood in more detail. The recent developments in mass spectrometry, and in particular the advent of MC-ICP-MS technology, should help the expansion of the use of this promising tool for the study of chemical weathering and erosional processes.

Acknowledgements We thank the two reviewers, an anonymous reviewer and Rachael James, for their constructive comments on this chapter. We are also grateful for the detailed comments of the Editor, Mark Baskaran.

References

Anderson MA, Bertsch PM, Miller WP (1989) Exchange and apparent fixation of lithium in selected soils and clay minerals. Soil Sci 148:46–52

Anghel I, Turin HJ, Reimus PW (2002) Lithium sorption to Yucca Mountain tuffs. Appl Geochem 17:819–824

Bach RO (ed) (1985) Lithium – current applications in science, medicine and technology. Wiley, New York

Bayon G, Vigier N, Burton KW, Brenot A, Carignan J, Chu N-C, Etoubleau J (2006) The control of weathering process on riverine and seawater hafnium isotope ratios. Geology 34:433–436

Benton LD, Ryan JG, Savov IP (2004) Lithium abundance and isotope systematics of forearc serpentinites, conical seamount, Mariana forearc: insights into the mechanics of slab-mantle exchange during subduction. Geochem Geophys Geosyst 8. doi:10.1029/2004GC000708

Blum JD, Erel Y (1995) A silicate weathering mechanism linking increase in marine $^{87}Sr/^{86}Sr$ with global glaciation. Nature 373:415–418

Brunskill GJ, Zagorskis I, Pfitzner J (2003) Geochemical mass balance for lithium, boron, and strontium in the Gulf of Papua, Papua New Guinea (Project TROPICS). Geochim Cosmochim Acta 67:3365–3383

Burles S, Nollett KM, Turner MS (2001) Big bang nucleosynthesis predictions for precise cosmology. Astrophys J 552:L1–L5

Carignan J, Cardinal D, Eisenhauer A, Galy A, Rehkämper M, Wombacher F, Vigier N (2004) A reflection on Mg, Ca, Cd, Li and Si isotopic measurements and related reference materials. Geostand Geoanal Res 28:139–148

Carignan J, Vigier N, Millot R (2007) Three secondary reference materials for Li isotopic measurements: ^{7}Li-N, ^{6}Li-N and LiCl-N. Geostand Geoanal Res 31:7–12

Chan LH (1987) Lithium isotope analysis by thermal ionization mass spectrometry of lithium tetraborate. Anal Chem 59:2662–2665

Chan L-H, Edmond JM (1988) Variation of lithium isotope composition in the marine environment: a preliminary report. Geochim Cosmochim Acta 52:1711–1717

Chan L-H, Frey FA (2003) Lithium isotope geochemistry of the Hawaiian plume: results from the Hawaiian Scientific Drilling Project and Koolau Volcano. Geochem Geophys Geosyst 4:8707

Chan L-H, Hein JR (2007) Lithium contents and isotopic compositions of ferromanganese deposits from the global ocean. Deep Sea Res I Top Stud Oceanogr 54:1147–1162

Chan LH, Kastner M (2000) Lithium isotopic composition of pore fluids and sediments in the Costa Rica subduction zone: implications for fluid processes and sediment contribution to the arc volcanoes. Earth Planet Sci Lett 183:275–290

Chan L-H, Edmond JM, Thompson G, Gillis K (1992) Lithium isotopic composition of submarine basalts: implications for the lithium cycle in the oceans. Earth Planet Sci Lett 108:151–160

Chan L-H, Edmond JM, Thompson G (1993) A lithium isotope study of hot springs and metabasalts from Mid-Ocean ridge hydrothermal systems. J Geophys Res 98:9653–9659

Chan L-H, Gieskes JM, You C-F, Edmond JM (1994) Lithium isotope geochemistry of sediments and hydrothermal fluids of the Guyamas Basin, Gulf of California. Geochim Cosmochim Acta 58:4443–4454

Chan L-H, Alt JC, Teagle DAH (2002) Lithium and lithium isotope profiles through the upper oceanic crust: a study of seawater-basalt exchange at ODP Sites 504B and 896A. Earth Planet Sci Lett 201:187–201

Chan LH, Leeman WP, Plank T (2006) Lithium isotopic composition of marine sediments. Geochem Geophys Geosyst 7:Q06005. doi:10.1029/2005GC001202

Chaussidon M, Robert F (1998) $^7Li/^6Li$ and $^{11}B/^{10}B$ variations in chondrules from the Semarkona unequilibrated chondrite. Earth Planet Sci Lett 164:577–589

Colten VA, Hanor JS (1984) Variations in dissolved lithium in the Mississippi River and Mississippi River Estuary, Louisiana, USA, during Low River stage. Chem Geol 47:85–96

Comans RNJ, Haller M, De Preter P (1991) Sorption of cesium on illite: non-equilibrium behaviour and reversibility. Geochim Cosmochim Acta 55:433–440

Decitre S, Deloule E, Resiberg L, James R, Agrinier P, Mével C (2002) Behaviour of Li and its isotopes during serpentinization of oceanic peridotites. Geochem Geophys Geosyst 3. doi:10.1029/2001GC000178

Delaney ML, Boyle EA (1986) Lithium in foraminiferal shells: implications for high-temperature hydrothermal circulation fluxes and oceanic crustal generation rates. Earth Planet Sci Lett 80:91–105

Elderfield H, Schultz A (1996) Mid-ocean ridge hydrothermal fluxes and the chemical composition of the ocean. Annu Rev Earth Planet Sci 24:191–224

Elliott T, Thomas A, Jeffcoate A, Niu Y (2006) Lithium isotope evidence for subduction enriched mantle in the source of mid-ocean-ridge basalts. Nature 443:565–568

Falkner KK, Chruch M, Measures CI, LeBaron G, Thouron D, Jeandel C, Strodal MC, Gill GA, Mortlock R, Froelich P, Chan LH (1997) Minor and trace element chemistry of Lake Baikal, its tributaries, and surrounding hot springs. Limnol Oceanogr 42:329–345

Hall JM, Chan L-H (2004) Li/Ca in multiple species of benthic and planktonic foraminifera: thermocline, latitudinal, and glacial-interglacial variation. Geochim Cosmochim Acta 68:529–545

Hall JM, Chan L-H, McDonough WF, Turekian KK (2005) Determination of the lithium isotopic composition of planktic foraminifera and its application as a paleo-seawater proxy. Mar Geol 217:255–265

Hathorne EC, James RH (2006) Temporal record of lithium in seawater: a tracer for silicate weathering? Earth Planet Sci Lett 246:393–406

Hathorne EC, James RH, Lampitt RS (2009) Environmental versus biomineralization controls on the intratest variation in the trace element composition of the planktonic foraminifera G. inflata and G. scitula. Paleoceanography 24:PA4204. doi:10.1029/2009PA001742

Heier KS, Billings GK (1970) Lithium. In: Wedepohl KH (ed) Handbook of geochemistry, vol II-1. Springer, Berlin, pp 3-A-1–3-O-1

Henderson GM, Burton KW (1999) Using (234U/238U) to assess diffusion rates of isotope tracers in ferromanganese crusts. Earth Planet Sci Lett 170:169–179

Hoefs J, Sywall M (1997) Lithium isotope composition of quaternary and tertiary biogene carbonates and a global lithium isotope balance. Geochim Cosmochim Acta 61:2679–2690

Huh Y, Chan LH, Zhang L, Edmond JM (1998) Lithium and its isotopes in major world rivers: implications for weathering and the oceanic budget. Geochim Cosmochim Acta 62:2039–2051

Huh Y, Chan L-H, Edmond JM (2001) Lithium isotopes as a probe of weathering processes: Orinoco River. Earth Planet Sci Lett 194:189–199

Huh Y, Chan L-H, Chadwick O (2004) Behaviour of lithium and its isotopes during weathering of Hawaiian basalts. Geochem Geophys Geosyst 5. doi:10.1029/2004GC000729

James RH, Palmer MR (2000a) The lithium isotope composition of international rock standards. Chem Geol 166:319–326

James RH, Palmer MR (2000b) Marine geochemical cycles of the alkali elements and boron: the role of sediments. Geochim Cosmochim Acta 63:3111–3122

James RH, Rudnicki MD, Palmer MR (1999) The alkali element and boron geochemistry of the Escanaba Trough sediment-hosted hydrothermal system: the role of sediments. Earth Planet Sci Lett 171:157–169

James RH, Allen DE, Seyfried WE Jr (2003) An experimental study of alteration of oceanic crust and terrigenous sediments at moderate temperatures (51 to 350°C): insights as to chemical processes in near-shore ridge-flank hysrothermal systems. Geochim Cosmochim Acta 67:681–691

Jeffcoate AB, Elliott T, Thomas A, Bouman C (2004) Precise, small sample size determinations of lithium isotopic compositions of geological reference materials and modern seawater by MC-ICP-MS. Geostand Geoanal Res 28:161–172

Jeffcoate AB, Elliott T, Kasemann SA, Ionov D, Cooper K, Brooker R (2007) Li isotope fractionation in peridotites and mafic melts. Geochim Cosmochim Acta 71:202–218

Jiang X, Lin X, Yao D, Zhai S, Guo W (2007) Geochemistry of lithium in marine ferromanganese oxide deposits. Deep Sea Res I 54:85–98

Kasemann SA, Jeffcoate AB, Elliott T (2005) Lithium isotope composition of basalt glass reference material. Ann Chem 77:5251–5257

Kisakurek B, Widdowson M, James RH (2004) Behaviour of Li isotopes during continental weathering: the Bidar laterite profile. India Chem Geol 212:27–44

Kisakurek B, James RH, Harris NBW (2005) Li and delta Li-7 in Himalayan rivers: proxies for silicate weathering? Earth Planet Sci Lett 237:387–401

Korn AJ, Grundahl F, Richard O, Barklem PS, Mashonkina L, Collet R, Piskunov N, Gustafsson (2006) A probable stellar solution to the cosmological lithium discrepancy. Nature 442:657–659

Kosler J, Kucera M, Sylvester P (2001) Precise measurement of Li isotopes in planktonic foraminiferal tests by quadrupole ICPMS. Chem Geol 181:169–179

Lear CH, Rosenthal Y (2006) Benthic foraminiferal Li/Ca: insights into Cenozoic seawater carbonate saturation state. Geology 34:985–988

Lemarchand E, Chabaux F, Vigier N, Millot R, Pierret M-C (2010) Lithium isotope systematics in a forested granitic catchment (Strengbach, Vosges Mountains, France). Geochim Cosmochim Acta 74:4612–4628

Marriott CS, Henderson GM, Belshaw NS, Tudhope AW (2004a) Temperature dependence of δ^7Li, δ^{44}Ca and Li/Ca incorporation into calcium carbonate. Earth Planet Sci Lett 222:615–624

Marriott CS, Henderson GM, Crompton R, Staubwasser M, Shaw S (2004b) Effect of mineralogy, salinity, and temperature on Li/Ca and Li isotope composition of calcium carbonate. Chem Geol 212:5–15

Milliman JD (1993) Production and accumulation of calcium carbonate in the ocean: budget of a nonsteady state. Glob Biogeochem Cycles 7:927–957

Millot R, Guerrot C, Vigier N (2004) Accurate and high-precision measurement of lithium isotopes in two reference materials by MC-ICP-MS. Geostand Geoanal Res 28:153–159

Millot R, Petelet-Giraud E, Guerrot C, Négrel P (2010a) Multi-isotopic composition (δ^7Li–δ^{11}B–δD–δ^{18}O) of rainwaters in France: origin and spatio-temporal characterization. Appl Geochem 25:1510–1524

Millot R, Vigier N, Gaillardet J (2010b) Behaviour of lithium and its isotopes during weathering in the Mackenzie Basin, Canada. Geochim Cosmochim Acta 74:3897–3912

Millot R, Scaillet B, Sanjuan B (2010c) Lithium isotopes in island arc geothermal systems: Guadeloupe, Martinique (French West Indies) and experimental approach. Geochim Cosmochim Acta 74:1852–1871

Moriguti T, Nakamura E (1998) High-yield lithium separation and precise isotopic analysis for natural rock and aqueous samples. Chem Geol 145:91–104

Nishio Y, Okamura K, Tanimizu M, Ishikawa T, Sano Y (2010) Lithium and strontium isotopic systematics of waters around Ontake volcano, Japan: implications for deep-seated fluids and earthquake swarms. Earth Planet Sci Lett 297:567–576

O'Neil JR (1986) Theoretical and experimental aspects of isotopic fractionation. Rev Mineral 16:1–40

Olive KA, Schramm DN (1992) Astrophysical ^{7}Li as a product of Big Bang nucleosynthesis and galactic cosmic-ray spallation. Nature 360:439–442

Olsher U, Izatt RM, Bradshaw JS, Dalley NK (1991) Coordination chemistry of lithium ion: a crystal and molecular structure review. Chem Rev 91:137–164

Palko AA, Drury JS, Begun GM (1976) Lithium isotope separation factors of some two-phase equilibrium systems. J Chem Phys 64:1828–1837

Pistiner JS, Henderson GM (2003) Lithium-isotope fractionation during continental weathering processes. Earth Planet Sci Lett 214:327–339

Pogge von Strandmann PAE, Burton KW, James RH, van Calsteren P, Gíslason SR, Mokadem F (2006) Riverine behaviour of uranium and lithium isotopes in an actively glaciated basaltic terrain. Earth Planet Sci Lett 251:134–147

Pogge von Strandmann PAE, James RH, van Calsteren P, Gíslason SR, Burton KW (2008) Lithium, magnesium and uranium isotope behaviour in the estuarine environment of basaltic islands. Earth Planet Sci Lett 274:462–471

Pogge von Strandmann PAE, Burton KW, James RH, van Calsteren P, Gíslason SR (2010) Assessing the role of climate on uranium and lithium isotope behaviour in rivers draining a basaltic terrain. Chem Geol 270:227–239

Qi HP, Coplen TB, Wang QZ, Wang YH (1997) Unnatural isotopic composition of lithium reagents. Anal Chem 69:4076–4078

Raiswell R, Tranter M, Benning LG, Siegert M, De'ath R, Huybrechts P, Payne T (2006) Contributions from glacially derived sediment to the global iron (oxyhydr)oxide cycle: implications for iron delivery to the oceans. Geochim Cosmochim Acta 70:2765–2780

Rollion-Bard C, Vigier N, Meibom A, Blamart D, Reynaud S, Rodolfo-Metalpa R, Martin S, Gattuso J-P (2009) Effect of environmental conditions and skeletal ultrastructure on the Li isotopic composition of scleractinian corals, Earth Planet. Sci Lett 286:63–70. doi:10.1016/j.epsl.2009.06.015

Rosner M, Ball L, Peucker-Ehrenbrink B, Blusztajn J, Bach W, Erzinger J (2007) A simplified, accurate and fast method for lithium isotope analysis of rocks and fluids, and δ^7Li values of seawater and rock reference materials. Geostand Geoanal Res 31:77–88

Rudnick RL, Tomascak PB, Njo HB, Gardner LR (2004) Extreme lithium isotopic fractionation during continental weathering revealed in saprolites from South Carolina. Chem Geol 212:45–57

Savov IP, Ryan JG, D'Antonio M, Kelley K, Mattie P (2005) Geochemistry of serpentinized peridotites from the Mariana Forearc Conical Seamount, ODP Leg 125: implications for the elemental recycling at subduction zones. Geochem Geophys Geosyst 6:Q04J15. doi:10.1029/2004GC000777

Scholz F, Hensen C, Reitz A, Romer RL, Liebetrau V, Meixner A, Weise SM, Haeckel M (2009) Isotopic evidence ($^{87}Sr/^{86}Sr$, δ^7Li) for alteration of the oceanic crust at deep-rooted mud volcanoes in the Gulf of Cadiz, NE Atlantic Ocean original research. Geochim Cosmochim Acta 73:5444–5459

Scholz F, Hensen C, DeLange GJ, Haeckel M, Liebetrau V, Meixner A, Reitz A, Romer RL (2010) Lithium isotope geochemistry of marine pore waters – insights from cold seep fluids. Geochim Cosmochim Acta 74:3459–3475

Schou M (1988) Lithium treatment of manic-depressive illness – past, present and perspectives. J Am Med Assoc 259:1834–1836

Seyfried WE Jr, Chen X, Chan L-H (1998) Trace element mobility and lithium isotopic exchange during hydrothermal alteration of seafloor weathered basalt: an experimental study at 350°C, 500 bars. Geochim Cosmochim Acta 62:949–960

Shannon RD (1976) Revised effective ionic-radii and systematic studies of interatomic distances in halides and chalcogenides. Acta Cryst A 32:751–767

Spite M, Spite F (1982) Lithium abundance at the formation of the galaxy. Nature 297:483–485

Stoffyn-Egli P (1982) Conservative behaviour of dissolved lithium in estuarine waters. Estuar Coast Shelf Sci 14:577–587

Stoffyn-Egli P, Mackenzie FT (1984) Mass balance of dissolved lithium in the oceans. Geochim Cosmochim Acta 48:859–872

Stoll PM, Stokes PE, Okamoto M (2001) Lithium isotopes: differential effects on renal function and histology. Bipolar Disord 3:174–180

Symons EA (1985) Lithium isotope separation: a review of possible techniques. Sep Sci Tech 20:633–651

Taylor TI, Urey HC (1938) Fractionation of the lithium and potassium isotopes by chemical exchange with zeolites. J Chem Phys 6:429–438

Teng F-Z, McDonough WF, Rudnick RL, Dalpé C, Tomascak PB, Chappell BW, Gao S (2004) Lithium isotopic composition and concentration of the upper continental crust. Geochim Cosmochim Acta 68:4167–4178

Teng F-Z, Rudnick RL, McDonough WF, Gao S, Tomascak PB, Liu Y (2008) Lithium isotopic composition and concentration of the deep continental crust. Chem Geol 255:47–59

Teng F-Z, Rudnick RL, McDonough WF, Wu F-Y (2009) Lithium isotopic systematics of A-type granites and their mafic enclaves: further constraints on the Li isotopic composition of the continental crust. Chem Geol 262:370–379

Tomascak PB, Tera F, Helz RT, Walker RJ (1999) The absence of lithium isotope fractionation during basalt differentiation: new measurements by multi-collector sector ICP-MS. Geochim Cosmochim Acta 63:907–910

Tomascak PB, Langmuir CH, le Roux P, Shirey SB (2008) Lithium isotopes in global mid-ocean ridge basalts. Geochim Cosmochim Acta 72:1626–1637

Tranter M (2003) Geochemical weathering in glacial and proglacial environments. In: Holland HD, Turekian KK (eds) Treatise on geochemistry. Pergamon, Oxford, pp 189–205

Umeda M, Tuchiya K, Kawamura H, Hasegawa Y, Nanjo Y (2001) Preliminary characterization on Li isotope separation with Li ionic conductors. Fusion Technol 39:654–658

US Geological survey (2009) Mineral commodity summaries 2009. U.S. Geological Survey, 195pp

Vanneste H, Kelly-Gerreyn BA, Connelly DP, James RRH, Haeckel M, Fisher RE, Heeschen K, Mills RA (2010) Spatial variation in fluid flow and geochemical fluxes across the sediment-seawater interface at the Carlos Ribeiro mud volcano (Gulf of Cadiz). Geochim Cosmochim Acta 75(4):1124–1144

Vigier N, Rollion-Bard C, Spezzaferri S, Brunet F (2007) In-situ measurements of Li isotopes in foraminifera. Geochem Geophys Geosyst Q01003. doi:10.1029/2006GC001432

Vigier N, Decarreau A, Millot R, Carignan J, Petit S, France-Lanord C (2008) Quantifying Li isotope fractionation during smectite formation and implications for the Li cycle. Geochim Cosmochim Acta 72:780–792

Vigier N, Gislason SR, Burton KW, Millot R, Mokadem F (2009) The relationship between riverine lithium isotope composition and silicate weathering rates in Iceland. Earth Planet Sci Lett 287:434–441

Vils F, Tonarini S, Kalt A, Seitz H-M (2009) Boron, lithium and strontium isotopes as tracers of seawater–serpentinite interaction at Mid-Atlantic ridge, ODP Leg 209, Earth Planet. Sci Lett 286:414–425

Wenger M, Armbruster T (1991) Crystal chemistry of lithium: oxygen coordination and bonding. Eur J Mineral 3:387–399

Wheat CG, Mottl MJ (2000) Composition of pore and spring waters from Baby Bare: Global implications of geochemical fluxes from a ridge flank hydrothermal system. Geochim Cosmochim Acta 64:629–642

Williams LB, Hervig RL (2005) Lithium and boron isotopes in illite-smectite: the importance of crystal size. Geochim Cosmochim Acta 24:5705–5716

Wimpenny J, James RH, Burton KW, Gannoun A, Mokadem F, Gislason SR (2010a) Glacial effects on weathering processes: new insights from the elemental and lithium isotopic composition of West Greenland rivers. Earth Planet Sci Lett 290:427–437

Wimpenny J, Gisalason SR, James RH, Gannoun A, Pogge von Strandmann PAE, Burton KW (2010b) The behaviour of Li and Mg isotopes during primary phase dissolution and secondary mineral formation in basalt. Geochim Cosmochim Acta 74:5259–5279

Witherow RA, Lyons WB, Henderson GM (2010) Lithium isotopic composition of the McMurdo Dry Valleys aquatic systems. Chem Geol 275:139–147

Wunder B, Meixner A, Romer RL, Heinrich W (2006) Temperature-dependent isotopic fractionation of lithium between clinopyroxene and high-pressure hydrous fluids. Contrib Mineral Petrol 151:112–120

Yamaji K, Makita Y, Watanabe H, Sonoda A, Kanoh H, Hirotsu T, Ooi K (2001) Theoretical estimation of lithium reduced partition function ratio for lithium ions in aqueous solution. J Phys Chem A 105:602–613

You C-F, Chan L-H (1996) Precise determination of lithium isotopic composition in low concentration natural samples. Geochim Cosmochim Acta 60:909–915

You C-F, Chan L-H, Spivack AJ, Gieskes JM (1995) Lithium, boron and their isotopes in sediments and pore waters of Ocean Drilling Program Site 808, Nakai Trough: implications for fluid expulsion in accretionary prisms. Geology 23:37–40

You CF, Gieskes JM, Lee T, Yui TF, Chen HW (2004) Geochemistry of mud volcano fluids in the Taiwan accretionary prism. Appl Geochem 19:695–707

Zhang L, Chan LH, Gieskes JM (1998) Lithium isotope geochemistry of pore waters from Ocean Drilling Program Sites 918 and 919, Irminger Basin. Geochim Cosmochim Acta 62:2437–2450

Chapter 5
Meteoric ^{7}Be and ^{10}Be as Process Tracers in the Environment

James M. Kaste and Mark Baskaran

Abstract ^{7}Be (T$_{1/2}$ = 53 days) and ^{10}Be (T$_{1/2}$ = 1.4 Ma) form via natural cosmogenic reactions in the atmosphere and are delivered to Earth's surface by wet and dry deposition. The distinct source term and near-constant fallout of these radionuclides onto soils, vegetation, waters, ice, and sediments makes them valuable tracers of a wide range of environmental processes operating over timescales from weeks to millions of years. Beryllium tends to form strong bonds with oxygen atoms, so ^{7}Be and ^{10}Be adsorb rapidly to organic and inorganic solid phases in the terrestrial and marine environment. Thus, cosmogenic isotopes of beryllium can be used to quantify surface age, sediment source, mixing rates, and particle residence and transit times in soils, streams, lakes, and the oceans. A number of caveats exist, however, for the general application of these radionuclides as tracers in the environment, as steady deposition and geochemical immobility are not guaranteed in all systems. Here we synthesize and review scientific literature documenting the deposition and behavior of these nuclides at the Earth's surface, focusing on current and potential applications for Earth scientists working to quantify terrestrial and marine processes.

5.1 Introduction

Cosmogenic isotopes of beryllium (Be) form when neutrons and protons spallate oxygen and nitrogen atoms (Fig. 5.1; Lal et al. 1958). The two naturally occurring Be isotopes of use to Earth scientists are the short-lived ^{7}Be (T$_{1/2}$ = 53.1 days) and the longer-lived ^{10}Be (T$_{1/2}$ = 1.4 Ma; Nishiizumi et al. 2007). Because cosmic rays that cause the initial cascade of neutrons and protons in the upper atmosphere responsible for the spallation reactions are attenuated by the mass of the atmosphere itself, production rates of comsogenic Be are three orders of magnitude higher in the stratosphere than they are at sea-level (Masarik and Beer 1999, 2009). Most of the production of cosmogenic Be therefore occurs in the upper atmosphere (5–30 km), although there is trace, but measurable production as oxygen atoms in minerals at the Earth's surface are spallated (in situ produced; see Lal 2011, Chap. 24). After cosmogenic Be is formed in the atmosphere, it is removed by rain, snow, and dry deposition. The near-constant production rate of cosmogenic Be isotopes in the atmosphere (Leya et al. 2000; Vonmoos et al. 2006) and the particle-reactive nature of the Be atom (You et al. 1989) make *meteoric* cosmogenic Be nuclides (sometimes referred to as "garden variety" to differentiate from in situ) a valuable tracer of a wide range of chemical, physical, and biogeochemical processes. Here we focus on the current and potential applications of meteoric ^{7}Be and ^{10}Be for quantifying and tracing natural processes occurring in soil, fluvial, lacustrian, and marine environments.

5.1.1 Production and Deposition of Meteoric ^{7}Be and ^{10}Be

The delivery of meteoric ^{7}Be and ^{10}Be to a particular point on the Earth's surface is controlled by the nuclide production rate in the atmosphere, the region's air mass source, amounts of precipitation, and the

J.M. Kaste (✉)
Department of Geology, The College of William & Mary, Williamsburg, VA 23187, USA
e-mail: jmkaste@wm.edu

M. Baskaran
Department of Geology, Wayne State University, Detroit, MI 48202, USA
e-mail: Baskaran@wayne.edu

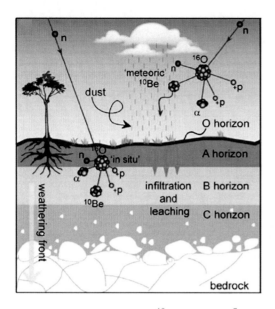

Fig. 5.1 Schematic of cosmogenic ^{10}Be production (^{7}Be forms in a similar manner). Meteoric ^{7}Be and ^{10}Be form in the atmosphere, and are delivered to the Earth's surface by wet and dry deposition. Most cosmogenic ^{7}Be is found on vegetation surfaces and in the O horizon or uppermost 2-cm of soil. Because of its longer half-life, ^{10}Be is commonly found on the surface, throughout the soil profile, and occasionally in saprolite. Reprinted from Willenbring and von Blanckenburg (2010a), Copyright (2010), with permission from Elsevier

efficiency of wet and dry depositional processes. Production rates of cosmogenic Be vary inversely with solar activity, because increased solar output strengthens the Earth's magnetic field which deflects cosmic rays. On timescales of decades, production rates vary by approximately 25% with the 11 year solar cycle, but over the course of hundreds of years, production rates may vary by a factor of two or more due to longer timescale modulations in the activity of the sun (Koch and Mann 1996; Vonmoos et al. 2006; Berggren et al. 2009). Because cosmic rays are deflected towards the poles, production rates of cosmogenic Be are a factor of 3–5 higher in polar air than in equatorial air (Harvey and Matthews 1989; Masarik and Beer 1999), depending on altitude. Polar air masses therefore have higher amounts of cosmogenic Be available for scavenging, but precipitation rates are generally very low here, which ultimately limits the meteoric radionuclide fluxes to the Earth's surface at high latitudes. The strong production gradient of cosmogenic Be with elevation in the atmosphere causes seasonal variability (at least in some latitude belts) in ^{7}Be and ^{10}Be deposition. This has been evident at mid-latitudes, as injections of stratospheric air into the troposphere during the spring season of each year result in higher concentrations of cosmogenic Be in meteoric waters (Husain et al. 1977). Stronger convection in the troposphere during summer months also increases ^{7}Be and ^{10}Be in rainfall as higher air is tapped (Baskaran 1995). Typically, fluxes of meteoric ^{7}Be and ^{10}Be to the Earth's surface scale with precipitation rates more than with latitude, and dry deposition generally is of lesser importance (Olsen et al. 1985; Brown et al. 1989; Wallbrink and Murray 1994; Whiting et al. 2005; Zhu and Olsen 2009), particularly for the shorter-lived ^{7}Be.

While the concentrations of cosmogenic Be in precipitation can vary by a factor of 20 between different storm events at one location, and are strongly dependent on latitude, amounts in rain and snow for each nuclide are commonly on the order of 10^4 atoms g^{-1}, with ^{7}Be/^{10}Be atomic ratios falling between 0.41 and 0.61 in the Southern hemisphere and 0.67 and 0.85 in the Northern hemisphere (Brown et al. 1989; Knies et al. 1994). The production ratio of ^{7}Be/^{10}Be in the atmosphere is projected to be 1.9 (Nishiizumi et al. 1996; Masarik and Beer 1999); ^{7}Be/^{10}Be ratios measured in precipitation will be controlled by residence time in the atmosphere as ^{7}Be decays during air mass or aerosol transport. Snow commonly has higher concentrations of cosmogenic Be isotopes than rain, possibly because of the higher surface area of snowflakes compared with rain droplets (McNeary and Baskaran 2003). Concentrations of cosmogenic Be in precipitation compared from storm to storm are usually inversely proportional to rainfall amount, as aerosols are scavenged during the initial stages of a storm event and larger storms are simply more diluted (Olsen et al. 1985; Todd et al. 1989; Baskaran et al. 1993). For example, Ishikawa et al. (1995) documented a sharp reduction in the concentration of ^{7}Be in precipitation measured over the course of 2-day storm in Japan. During the first ~6 h of the storm, the ^{7}Be content of snowfall was approximately $10^{4.4}$ atoms g^{-1} (4 Bq L^{-1}), but this dropped to <$10^{3.8}$ atoms g^{-1} (<1 Bq L^{-1}) during the latter stages of the event. Resuspended dust can be a significant source of ^{10}Be in rain (Monaghan et al. 1986; Graham et al. 2003; Lal and Baskaran 2011), but is probably a minor source of ^{7}Be to meteoric waters because of its shorter half-life. The ^{10}Be content of continental dust is near that of the average concentrations in regional surface soil, which indicates that topsoil erosion is the likely source of dust over continents (Gu et al. 1996). Lal (2007) estimated the global average value of atmospheric dust to be $2.12(\pm 0.04) \times 10^9$ ^{10}Be atoms g^{-1} dust.

It is presumed that once ^{7}Be or ^{10}Be is produced, it combines with atmospheric oxygen or hydroxyl to form Be oxide (BeO) or hydroxide (BeOH$_x^{2-x}$), molecules that are scavenged by atmospheric moisture. Equilibrium thermodynamics can be used to predict that in slightly acidic cloudwater or rainfall (pH <6), Be speciation will be dominated by Be^{2+} (Vesely et al. 1989; Takahashi et al. 1999). After rainfall hits the surface of the Earth, precipitation having a pH of <6 is usually neutralized rapidly, and cosmogenic Be quickly adsorbs to vegetation and the uppermost layer of soils and sediments (Pavich et al. 1984; Wallbrink and Murray 1996; Kaste et al. 2007). Because of the wide variability in the concentrations of cosmogenic Be isotopes in rainfall within storms, between storms, between seasons, and from year to year, and the potential of dust deposition being significant particularly for ^{10}Be, it takes several years worth of input measurements to accurately generalize deposition rates for a single location (Brown et al. 1989; Baskaran 1995). Furthermore, care must be taken to acidify samples because Be^{2+} can adsorb to plastic sampling funnels or collection bottles (Baskaran et al. 1993). A summary of the volume-weighted ^{7}Be concentrations measured in rainfall and calculated wet deposition fluxes is given in Table 5.1.

5.1.2 General Abundances of ^{7}Be and ^{10}Be at the Earth's Surface

Inventories of meteoric cosmogenic Be on the Earth's surface (in units of atoms (n) m^{-2} or activity (Bq = nλ) m^{-2}) at a particular location are a function of atmospheric deposition rates, exposure age, and soluble (from leaching or "desorption") and particulate (from erosion or sediment deposition) losses or gains. Soluble losses of cosmogenic Be from soils are thought to be negligible in watersheds of near-neutral pH, but can be appreciable in acidic systems (Pavich et al. 1985; Brown et al. 1992b). In most soils on gentle slopes, erosional losses of ^{7}Be are minimal because of its short half-life compared to the residence time of particle-reactive radionuclides in watersheds (or the timescale needed for significant erosion), but ^{10}Be inventory deficiencies are used to calculate steady-state erosion rates (Brown et al. 1995). ^{7}Be in soils may thus be in an approximate radioactive equilibrium with atmospheric deposition, such that decay rates roughly equal deposition rates on a Bq m^{-2} basis (Olsen et al. 1985). However, in areas where there is very strong seasonal variations of precipitation (such as San Francisco, CA and other areas controlled by seasonal monsoons such as East Asia, Southeast Asia, etc.), very high precipitation during certain months may result in transient equilibrium with much higher inventories in soils during wetter periods (Walling et al. 2009).

Generally, concentrations of ^{7}Be and ^{10}Be in soils and sediments near the Earth's surface are in the range of 10^3–10^5 atoms g^{-1} (0.15–15 Bq kg^{-1}), and 10^7–10^9 atoms g^{-1}, respectively (Fig. 5.2). Because of its short half-life, ^{7}Be is confined to vegetation and just the upper few cm of regolith and typically has an exponential decrease of concentration with depth in soils (Wallbrink and Murray 1996; Blake et al. 1999; Whiting et al. 2001; Wilson et al. 2003; Kaste et al. 2007). However, it may be found deeper than 10 cm in sedimentary environments with high deposition rates or where sediment mixing by bioturbation and/or physical mixing is significant (Canuel et al. 1990) or when there are flood events (Sommerfield et al. 1999). Vegetation plays a significant role in scavenging radionuclides from the atmosphere (Russell et al. 1981). A large fraction of the ^{7}Be surface inventory can reside in grasses or the forest canopy (Monaghan et al. 1983; Wallbrink and Murray 1996; Kaste et al. 2002) and decays before it can reach the top of the soil profile or enter into the hydrologic cycle. Surface ^{7}Be inventories typically range between $10^{8.8}$ and $10^{9.8}$ atoms m^{-2} (100–1,000 Bq m^{-2}; Table 5.1) and show some scaling with precipitation patterns (Whiting et al. 2005). Salisbury and Cartwright (2005) took a creative approach in quantifying ^{7}Be deposition along a precipitation gradient by showing that sheep feces sampled on a transect from sea-level to approximately 1,000 m in North Whales had approximately a fivefold gradient in ^{7}Be concentrations. They projected that rainfall rates increased by a factor of approximately three to four along this transect, and because much of the ^{7}Be was deposited directly on vegetation that the sheep ate, fresh feces recorded the deposition signal. An advantage of this technique is that it averages some of the spatial

Table 5.1 Compilation of ^{7}Be fallout fluxes and terrestrial surface inventories

Location	Latitude	Period of collection (month year^{-1})	Annual precipitation (cm)[a]	Average monthly ^{7}Be flux (Bq m^{-2} month^{-1})[a]	Volume-weighted ^{7}Be activity (Bq L^{-1})[a]	References
Precipitation-based collectors						
Bombay (India)	29°0′N	1955–1961, 1963–1965, 1968, 1970	230	106	0.55	Lal et al. (1979)
Galveston, TX	29°18′N	12/1988–2/1992	117 (97–150)	204 (124–322)	1.45–2.58 (2.03)	Baskaran et al. (1993)
College Station, TX	30°35′N	6/1989–2/1992	122 (98–146)	192 (174–208)	1.73–2.07 (1.90)	Baskaran et al. (1993)
Bermuda	32°20′N	9/1977–8/1978	170	238	1.57	Turekian et al. (1983)
Oak Ridge, TN	35°58′N	9/1982–8/1984	127 (110–143)	169 (142–196)	1.60 (1.55–1.65)	Olsen et al. (1986)
Norfolk, VA	36°53′N	1/1983–12/1984	136 (132–141)	173 (167–179)	1.58 (1.50–1.67)	Todd et al. (1989)
Ansan, S. Korea	37°17′N	1/1992–6/1993	112 (106–117)	142	1.08	Kim et al. (1998)
Solomon, MD	38°19′N	3/1986–11/1987	96	189	2.36	Dibb and Rice (1989b)
Onagawa, Japan	38°26′N	4/1987–3/1989	115 (88–142)	150 (118–181)	1.57 (1.54–1.60)	Ishikawa et al. (1995)
Thessaloniki, Greece	40°38′N	1/1987–4/1990	48 (33–65)	51 (40–70)	1.29 (1.10–1.48)	Papastefanou and Ioannidou (1991)
Westwood, NJ	40°59′N	12/1960–8/1961	78	60	0.92	Walton and Fried (1962)
New Haven, CT	41°31′N	3/1977–6/1978	148	315	2.63	Turekian et al. (1983)
Detroit, MI	42°25′N	9/1999–2/2001	76	181	2.87	McNeary and Baskaran (2003)
Geneva, Switzerland	46°16′N	11/1997–11/1998	97	174	2.16	Caillet et al. (2001)
Lake Geneva, Switzerland	46°30′N		120	230	2.30	Dominik et al. (1987)
Lake Zurich, Switzerland	47°22′N		110	223	2.43	Schuler et al. (1991)
Quillayute, Wash.	47°57′N	2/1976–1/1977	270	113	0.50	Crecelius (1981)
Chilton, England	51°26′N	10/1959–9/1960	68	76	1.35	Peirson (1963)
Milford, England	51°42′N	10/1959–9/1960	63	72	1.38	Peirson (1963)
Rijswijk, Netherlands	52°01′N	11/1960–10/1961	93	132	1.70	Bleichrodt and Van Abkoude (1963)
Terrestrial surface inventories		Collection date (month/year)	Median surface inventory (Bq m^{-2})	^{7}Be flux (Bq m^{-2} month^{-1})[b]	Surface type	
SE Queensland, AU	28°01′S	5/2003	440	175	Soil	Doering et al. (2006)
Black Mtn, Australia	35°15′S	9/1988	200	79	Soil + overlying grass	Wallbrink and Murray (1996)
Black Mtn, Australia	35°15′S	9/1988	130	52	Alluvial bare soil	Wallbrink and Murray (1996)
Black Mtn, Australia	35°15′S	5/1989	400	159	Soil + overlying grass	Wallbrink and Murray (1996)
Black Mtn, Australia	35°15′S	5/1989	155	61	Alluvial bare soil	Wallbrink and Murray (1996)
Oak Ridge, TN	35°58′N	7/1984	673	267	Soil + overlying grass	Olsen et al. (1985)
Owens Valley, CA	37°22′N	10/2006	85	34	Soil + grass	Elmore et al. (2008)
Wallops Island, VA	37°55′N	1/1985	673	267	Marsh + overlying grass	Olsen et al. (1985)
Wallops Island, VA	37°55′N	1/1985	107	42	Unvegetated marsh	Olsen et al. (1985)
Mendocino, CA	39°18′N	6/1980	700	278	Above-ground vegetation and litter	Monaghan et al. (1983)
Delaware Marsh	39°27′N	7/1982	207	82	Marsh + overlying grass	Olsen et al. (1985)
Valdivia, Chile	39°44′S	9/2003	573	227	Soil	Schuller et al. (2006)
Valdivia Chile	39°49′S	4/2006	522	207	Recently harvested forest soil	Walling et al. (2009)
Valdivia Chile	39°49′S	10/2006	1,139	452	Recently harvested forest soil	Walling et al. (2009)
Central Idaho	44°50′N	5/1996	139	55	Soil + overlying vegetation	Bonniwell et al. (1999)
Central Maine	44°50′N	12/98	554	220	Bog core + overlying vegetation	Kaste (1999)
Cooke City, MT	45°01′N	6/2000	449	178	Soil	Whiting et al. (2005)
Devon, UK	50°47′N	2/1998	512	203	Soil	Blake et al. (1999)

[a] Numbers in parenthesis denote the range when data are reported for ≥2 years
[b] Assumes steady-state between surface and atmospheric flux: calculated by multiplying the median surface inventory (Bq m^{-2}) by 0.3966 month^{-1}

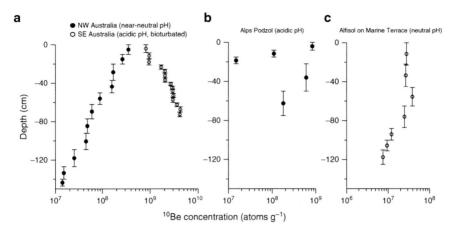

Fig. 5.2 Concentrations of meteoric ^{10}Be with depth in (**a**) a neutral pH soil profile in NW Australia and an acidic soil that was strongly bioturbated in SE Australia (Fifield et al. 2010), (**b**) an acidic Podzol from the Alps (Egli et al. 2010), and (**c**) a near-neutral pH soil developed on a marine terrace in California (Monaghan et al. 1992)

heterogeneity associated with atmospheric deposition on vegetated terrain. Because the half-life of ^{10}Be is ~8 orders of magnitude larger than the residence time of aerosols in the atmosphere, in temperate to tropical regions where the air is stripped of cosmogenic Be isotopes in the initial part of storms, ^{10}Be inventories may not be controlled by rainfall rate as higher rainfall amounts simply dilute concentrations but do not add to the flux (Willenbring and von Blanckenburg 2010a). The fact that small changes in amount of precipitation for a region may not significantly alter depositional fluxes makes it slightly simpler to project steady-state ^{10}Be inventories for a landform.

5.2 Background

5.2.1 General Geochemical Behavior of Be in the Environment

Beryllium has the smallest ionic radius (0.31 Å) of all the metal cations, and exists only in the +2 valence state in natural aqueous solutions (Baes and Mesmer 1976). Its first hydrolysis constant $\{[H^+][BeOH^+]/[Be^{2+}]\}$ is $10^{-5.7}$; hence in meteoric waters where pH is controlled by atmospheric CO_2, there should be a near equal distribution of the divalent cation $[Be^{2+}]$ and its first hydrolysis product $[BeOH^+]$. In acidic precipitation where anthropogenic acids have depressed the pH of rainfall to <5.5, much of the meteoric cosmogenic Be may exist as Be^{2+}, whereas in regions where rainfall pH is buffered by calcite-bearing dust, cosmogenic Be may be in the $BeOH^+$ form. Given the typical pH range of natural waters of 5–9, equilibrium-based thermodynamic models would predict that the relative Be species abundances and the partitioning of cosmogenic Be may vary by orders of magnitude between slightly acidic systems and more alkaline systems. Indeed, experimental studies demonstrate that the partitioning of Be from the aqueous phase to an adsorbed phase on a range of materials is strongly pH dependent (Bloom and Crecelius 1983; Hawley et al. 1986; You et al. 1989). Using 3-week equilibration periods, You et al. (1989) showed a >100-fold variation in the solid-phase partitioning coefficient (K_d in kg L^{-1}) over the pH range of 4–8 for different substrate types (Fig. 5.3).

In freshwater rivers, amounts of dissolved ^{7}Be in the water column are commonly below detection limits (Dominik et al. 1987; Bonniwell et al. 1999), but in marine environments, Be may be characterized as having a limited affinity for suspended matter, as large fraction of the ^{7}Be and ^{10}Be appears to be dissolved (Merrill et al. 1960; Bloom and Crecelius 1983; Kusakabe et al. 1987; Dibb and Rice 1989a; Measures et al. 1996). The mechanism of Be adsorption onto inorganic minerals, including primary silicate minerals, secondary silicates, and iron and aluminum oxyhydroxides is typically considered to be via the formation of a complex between the Be atom and oxygen on

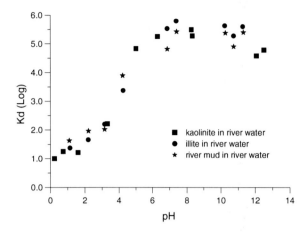

Fig. 5.3 The dependence of the K_D of Be for different materials over varying pH, using equilibration times of 3 weeks and suspended solids concentrations of 0.2 g L^{-1}. Reprinted from You et al. (1989), Copyright (1989), with permission from Elsevier

the surfaces or corners of minerals ("inner sphere adsorption"). The K_d may increase on timescales of weeks to months as adsorption-desorption equilibrium is followed by cationic lattice substitution (Nyffeler et al. 1984; Dibb and Rice 1989a). Like other particle-reactive metals, the K_d varies inversely with suspended solids concentrations below 30 mg L^{-1} (Hawley et al. 1986). While Be can form complexes with organic matter functional groups (daSilva et al. 1996), its partitioning coefficients seem to be significantly higher for inorganic materials compared with algae or seston (Bloom and Crecelius 1983; Dibb and Rice 1989a). Selective chemical extractions of marine sediments (Bourles et al. 1989) and soils (Barg et al. 1997) show that most of the meteoric Be is associated with authigenic phases (secondary Al, Fe, and Mn minerals) and organic matter coatings. Beryllium forms strong complexes with fluoride and humate, which could theoretically affect the strength and rate of aqueous to solid-phase partitioning (Nyffeler et al. 1984; Vesely et al. 1989; Takahashi et al. 1999).

The depth to which meteoric Be penetrates in soils and sediments will be controlled by advection (leaching or percolation) and diffusion-like (mixing) processes. The short half-life and tendency to adsorb to the solid phase causes ^{7}Be to generally be fixed in the vegetation and upper 2-cm of soil (Wallbrink and Murray 1996; Walling et al. 2009), but meteoric ^{10}Be is usually detectable from the soil surface to the C horizon, and in some cases through saprolite (Pavich et al. 1985). In high pH soils that are not intensively mixed, fallout Be exhibit an exponential decline in concentrations with depth (Fig. 5.2a, NW Australia Site) as the affinity of the Be atom for the surface of almost any material (layered clays, secondary oxide minerals, organic matter, etc.) will be very strong. In acidic soil profiles, however, the Be atom may be highly enriched in specific layers that have the most favorable surface sites for adsorption (Fig. 5.2a, b). In most near-neutral pH soils, the concentration of meteoric ^{10}Be in soils has a subsurface concentration maximum in the B horizon layer where iron and aluminum accumulate (Barg et al. 1997; Jungers et al. 2009; Egli et al. 2010); beneath this there is often a general exponential decrease with depth (Fig. 5.2c). Physical soil mixing by organisms, wetting-drying cycles, or freeze-thaw cycles appears to create a layer of homogenized ^{10}Be concentrations (Fifield et al. 2010) as seen in the SE Australia ^{10}Be profile given in Fig. 5.2a.

5.2.1.1 Cosmogenic ^{7}Be and ^{10}Be in Freshwater Environments

The few watershed-scale studies that measure ^{7}Be and/or ^{10}Be in soils and waters have demonstrated that the actual distribution of Be between the solid and aqueous phase is broadly consistent with the thermodynamic predictions described above. In a survey of tropical watersheds in the Orinoco and Amazon basin, Brown et al. (1992b) found that acidic drainage waters had nearly the same concentration of dissolved ^{10}Be as incoming precipitation, indicating very little adsorption to soils and sediments. In more neutral to alkaline waters, concentrations of dissolved ^{10}Be in streamwater are often orders of magnitude smaller than meteoric waters because of adsorption to aluminum and iron oxide coatings on soils and sediments (Brown et al. 1992b; Barg et al. 1997) (Fig. 5.3).

Because of its shorter half-life and tendency to adsorb, the vast majority of ^{7}Be is retained by the vegetation and upper soils of a watershed (Olsen et al. 1986; Cooper et al. 1991; Wallbrink and Murray 1996; Bonniwell et al. 1999). Even during snowmelt events when the storm hydrograph is dominated by new water, most ^{7}Be never reaches the stream (Cooper et al. 1991). There are no reports of ^{7}Be in groundwater above typical detection limits (<10 atoms g^{-1} or 1.5 mBq L^{-1}). Due to the short half-life and high K_ds of ^{7}Be, it is not expected to penetrate very far beyond soil-air interface and hence it is not expected to be present in groundwater. In freshwater rivers, concentrations of dissolved (<0.45 μm) ^{7}Be are usually on the order of undetectable to 10^2 atoms g^{-1} (15 mBq L^{-1}), which is orders of magnitude lower than levels usually found in precipitation. Suspended sediments in streams however can have ^{7}Be concentrations on the order of 10^6 atoms g^{-1} (100–200 Bq kg^{-1}) (Dominik et al. 1987; Bonniwell et al. 1999); export of ^{7}Be from watersheds is evidently dominated by erosion and particulate losses. The concentration of dissolved ^{7}Be in lakes can vary by an order of magnitude over the course of a year as biological productivity varies with seasons. A sub-alpine lake in Bodensee had dissolved concentrations of ^{7}Be fluctuating from approximately $10^{0.8}$–$10^{1.8}$ atoms g^{-1} (1–10 mBq L^{-1}) over the course of the year, which appeared to be controlled by the availability of particulates that are ultimately regulated by biological processes (Vogler et al. 1996).

Cosmogenic ^{10}Be in groundwater has been reported to range from 10^2 to 10^3 atoms g^{-1} (Pavich et al. 1985; McHargue and Damon 1991). Pavich et al. (1985) calculated that the groundwater losses of ^{10}Be from a watershed in Virginia were three orders of magnitude lower than meteoric inputs. In a general survey of North American rivers and the Pearl River Basin in China, Kusakabe et al. (1991) reported dissolved concentrations of ^{10}Be to range from approximately 10^3–$10^{3.7}$ atoms g^{-1}, which was slightly less than the ^{10}Be content of rainfall, but an order of magnitude higher than the amounts found in estuaries. However, dissolved ^{10}Be in the slightly acidic Orinoco River was found to be nearly equivalent to concentrations measured in local precipitation (Brown et al. 1992b). It seems that in acidic watersheds, the exchange sites on mineral surfaces may be dominated by other ions (H$^+$, Al, etc.), which may inhibit Be adsorption. McHargue and Damon (1991) used an average ^{10}Be concentration of 10^3 atoms g^{-1} for freshwaters, the volume of water in the world's rivers and lakes, and the global annual runoff flux to calculate that the residence time of ^{10}Be in the world's freshwaters is approximately 3 years.

5.2.1.2 Cosmogenic ^{7}Be and ^{10}Be in the Marine Environments

After Be radionuclides are delivered to the air-water interface via wet and dry fallout, the ions are removed from the water column primarily by adsorption onto particulate matter. While the removal rate is strongly dependent on the concentration and composition of suspended particulate matter (Dibb and Rice 1989a), the residence time of dissolved Be above the thermocline is on the order of 0.5 years (Fig. 5.4), which

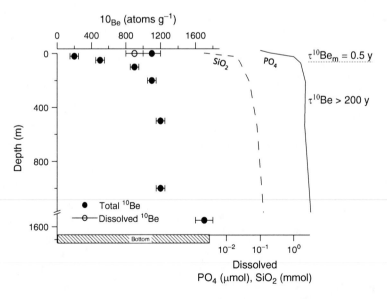

Fig. 5.4 ^{10}Be, phosphate, and silica concentrations in the water column in the San Nicolas Basin (Kusakabe et al. 1982). Above the thermocline in the "mixed layer" (upper ~100 m), high concentrations of suspended particles cause rapid removal of dissolved Be from the water column. As the particles dissolve beneath the thermocline, ^{10}Be is recycled back into the water column where it has a residence time of hundreds of years

is similar to the residence times reported for other particle-reactive atoms like Th and Pb (Kusakabe et al. 1982). The first measurement of ^{7}Be in seawater from the Indian Ocean was reported by Lal et al. (1960). Silker et al. (1968) later demonstrated that ^{7}Be had relatively uniform dissolved concentrations in surface waters of the Atlantic Ocean (5°48′–25°38′N, $n = 7$ samples), ranging from $10^{1.4}$ to $10^{1.7}$ atoms ^{7}Be g^{-1} (40–70 mBq L^{-1}), with a particulate fraction ranging between 7.4 and 14.9% of the total. Kusakabe et al. (1982) found that concentrations of ^{10}Be measured in unfiltered water samples collected from the surface of San Nicolas Basin were indistinguishable from the ^{10}Be measured in filtered water samples (Nucleopore 0.4 μm filter), and concluded that most ^{10}Be was in dissolved form (Fig. 5.4).

Although the monthly depositional fluxes of ^{7}Be have been reported to vary by a factor of ~10, in the water column of the ocean, activities (Bq L^{-1}) vary much less, as the water column integrates the activities over its mean life. The amount of precipitation on the ocean varies with latitude and season, which will affect the depositional fluxes of ^{7}Be and thus their water column inventories. Concentrations of particulate and dissolved ^{7}Be within an estuarine system vary spatially and temporally. For example, the particulate and total (total = dissolved + particulate) ^{7}Be activities in June 2003 in Tampa Bay, Florida varied between $10^{-0.5}$ and $10^{0.4}$ atoms g^{-1} (0.05 and 0.39 mBq L^{-1}) and 10 and $10^{2.5}$ atoms g^{-1} (1.7 and 49 mBq L^{-1}), respectively while the corresponding values varied between $10^{0.3}$ and $10^{1.4}$ atoms g^{-1} (0.3 and 3.7 mBq L^{-1}) and $10^{1.1}$ and $10^{2.3}$ atoms g^{-1} (2 and 28 mBq L^{-1}), respectively in August 2003 (Baskaran and Swarzenski 2007). Particulate ^{7}Be concentrations are generally higher in the spring compared with summer, mainly due to higher amounts of water discharge and precipitation in late spring and early summer months, which results in higher particulate fluxes, resuspension, and scavenging during that time. The total residence time of ^{7}Be (τ_{Be}) is calculated using a simple approach, assuming that the water column is uniformly-mixed:

$$\tau_{Be} = \ln2 \times A_{Be} \times h/I_{Be} \quad (5.1)$$

where A_{Be} is the total activity of ^{7}Be (Bq m^{-3}), I_{Be} is the atmospheric input rate of ^{7}Be (Bq m^{-2} day^{-1}), and h is the mean depth (m) of the well-mixed sampling area. The residence time of dissolved Be in coastal waters varies over an order of magnitude, from <1 to 60 days, depending on the depth of the coastal/estuarine waters and concentrations of suspended particulate matter (Table 5.2). The K_d values in coastal waters may vary over two orders of magnitude, between 7×10^3 and 1.2×10^6 (Olsen et al. 1986; Baskaran and Santschi 1993; Baskaran et al. 1997; Kaste et al. 2002; Jweda et al. 2008). The K_d values are reported to be higher in June than in August, which is related to the variations in the amount of freshwater discharge and the amounts of precipitation that control the extent of resuspension of bottom sediments.

Table 5.2 Residence time of dissolved ^{7}Be in various coastal waters

Location	Residence time of ^{7}Be (d)	References
New York Harbor	8–17	Olsen et al. (1986)
James River Estuary	2–4	Olsen et al. (1986)
Raritan Bay	7–17	Olsen et al. (1986)
Chesapeake Bay	5–52	Dibb and Rice (1989a)
Galveston Bay	0.9–1.8	Baskaran and Santschi (1993)
Sabine-Neches estuary	0.8–10.5	Baskaran et al. (1997)
Tampa Bay	1.6–58.7	Baskaran and Swarzenski (2007)
Clinton River	1.0–60.3	Jweda et al. (2008)
Hudson River Estuary	0.7–9.5	Feng et al. (1999)

The concentration of dissolved ^{10}Be in marine surface waters typically varies between $10^{2.6}$ and $10^{3.2}$ atoms g^{-1}, and the residence time of Be has been generally reported to be shorter than the mixing time of the oceans (Merrill et al. 1960; Frank et al. 2009). Meltwaters from glaciers (both polar and high-altitude) could have higher concentrations of ^{10}Be compared with surrounding waters (e.g., Antarctic ice contains about $10^{4.7}$ atom g^{-1} of ^{10}Be, Raisbeck et al. 1978b). With the exception of one dataset, all data indicate that the water column inventory of ^{10}Be, in the Pacific is higher than that in the Atlantic (Table 5.3), perhaps a result of lower suspended matter concentrations in the Pacific. The transfer of ^{10}Be from the dissolved phase to the solid (adsorbed) phase may be regulated by the recycling of biogenic particles in the oceans (Fig. 5.4).

A wide range of dissolved residence times of ^{10}Be in the open ocean have been reported in the literature. Several approaches have been used to obtain residence times that include the inventories of ^{10}Be and annual removal or depositional input of ^{10}Be, and from the Fe-Mn crust and Mn-nodule measurements (Raisbeck et al. 1978a, 1980; Kusakabe et al. 1982, 1987; Segl et al. 1987; Anderson et al. 1990; Ku et al. 1990; Brown et al. 1992a; Measures et al. 1996; von Blanckenburg et al. 1996; Frank et al. 2002, 2009). A large range of residence time values, if real, has major implications on the changing geochemical processes that lead to varying removal rates of Be^{2+} from the water column. In order to assess the variations, the data from all the open ocean sites were recalculated to determine the inventories of ^{10}Be in the water column (a standard depth of 3,600 m was used for all the basins) and the inventories are given in Table 5.3. Except the Arctic Ocean, where we compared the residence times of ^{10}Be in all four major basins (Nansen, Amundsen, Makarov and Canada Basins), we used the same depositional flux for each of the sites. The vertical profiles of dissolved ^{10}Be from these four basins are given in Fig. 5.5. The errors associated with the concentrations of ^{10}Be in individual depths reported in the publications are propagated to obtain the error associated with the inventory, which we use in the calculation of the errors associated with the residence times. The residence time of ^{10}Be in these major deep basins of the Arctic Ocean ranged from 680 to 830 years (Table 5.3), and given the propagated errors, there is no discernable difference in the residence time of ^{10}Be between the four basins of the Arctic. This observation can be compared to another particle-reactive radionuclide, Th in these major basins. A compilation of ^{230}Th data from these four basins clearly showed that the residence time of Th in the Makarov Basin is the longest (45 ± 1 years) compared to Canada Basin (22 ± 2 years), Amundsen Basin (19 ± 1 years) and Nansen Basin (17 ± 2 years; Trimble et al. 2004). The lack of difference we observe with ^{10}Be could be real or due to using the same atmospheric depositional input of 0.25×10^6 atoms cm^{-2} year^{-1} (Frank et al. 2009) for all four basins. Since the annual removal rate and annual depositional flux data are sparse, we do not know whether there is a real difference between the residence times, but there are real differences between the particle concentrations and particle fluxes between these basins (Trimble et al. 2004).

Our calculated residence time given in Table 5.3 varies by a factor of 2.4 (between 236 and 505 years) between the Atlantic (western North Atlantic, South Atlantic) and Pacific Ocean. Thus, it appears that the

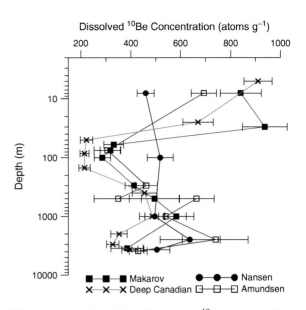

Fig. 5.5 Vertical profiles of dissolved ^{10}Be concentrations from all four major deep basins of the Arctic Ocean. Data are plotted from Frank et al. (2009)

wider range of residence times reported by various authors in the past (factor of ~10) may be due to poorly constrained input term used in their calculations. Lao et al. (1992a) estimated the deposition rate of ^{10}Be for the entire Pacific Ocean to be 1.5×10^6 atoms cm^{-2} year^{-1}, during the Holocene based on dated sediment cores. While the inventories can be measured precisely (with better than 10% precision), the annual depositional (input) flux is more difficult to constrain, as it can significantly vary depending on the amount of precipitation. Multi-year depositional flux measurement could reduce the uncertainty in the depositional flux. Perhaps by quantifying the relationship between ^{10}Be and another fallout isotope that is easier to measure in rainfall or has longer term datasets available (e.g., ^{210}Pb, Preiss et al. 1996; Sheets and Lawrence 1999), the depositional fluxes of ^{10}Be could be better constrained for residence time calculations.

5.2.2 Measuring ^{7}Be and ^{10}Be in Environmental Samples

5.2.2.1 Preconcentration Methods

The concentrations of ^{7}Be in surface water samples are so dilute a preconcentration method is usually needed.

Table 5.3 Concentrations, inventories and residences times of dissolved ^{10}Be in the major world oceans

Sample code	Coordinates	Depth (m)	Surface water ^{10}Be (atoms g^{-1})	^{10}Be inventory (10^8 atoms cm^{-2})	Residence time (years)	References
Arctic ocean						
Nansen Basin	84°16.87'N; 33°39.81'E	4,039	460	1.98 ± 0.30	790 ± 120	Frank et al. (2009)[1]
Amundsen Basin	88°24.48'N; 95°22.78'E	4,400	693	2.08 ± 0.38	830 ± 150	Frank et al. (2009)[1]
Makarov Basin	87°54.97'N; 154°22.50'E	3,985	841	1.70 ± 0.15	680 ± 60	Frank et al. (2009)[1]
Canada Basin	75°12.5'N; 149°57.0'W	3,850	911	1.72 ± 0.24	690 ± 96	Frank et al. (2009)[1]
Atlantic ocean						
Transect at 25°N	25°N; 60°W	>4,500	1,407	4.50 ± 0.22	372 ± 18	Segl et al. (1987)[2]
Transect at 25°N	26°N; 50°W	>4,500	1,671	6.80 ± 0.78	562 ± 64	Segl et al. (1987)[2]
Transect at 25°N	27°N; 44–45°W	>3,319	1,310	5.98 ± 0.70	494 ± 58	Segl et al. (1987)[2]
Transect at 25°N	27°N; 35°W	>4,500	1,416	4.57 ± 0.59	378 ± 49	Segl et al. (1987)[2]
Western North A.	41°32'N; 63°37'W	>3,411	430	2.85 ± 0.18	236 ± 15	Ku et al. (1990)[3]
Western North A.	34°1'N; 63°0'W	>5,178	640	2.94 ± 0.22	243 ± 18	Ku et al. (1990)[3]
South Atlantic	24°40'S; 38°21'W	>3,830	930	3.35 ± 0.14	279 ± 12	Measures et al. (1990)[4]
South Atlantic	24°55'S; 01°00'W	>4,152	585	3.66 ± 0.16	305 ± 13	Measures et al. (1990)[4]
South Atlantic	01°59'S; 04°02'W	>4,850	671	3.12 ± 0.13	260 ± 13	Measures et al. (1990)[4]
Drake Passage	57–63°S; 66–69°'W	>3,700	1,400	4.97 ± 0.41	49 ± 12[a] (279 ± 12)[a]	Kusakabe et al. (1982)
Pacific Ocean	25°00'N; 169°59'E	6,013	1,060	5.74 ± 0.36	879 ± 54[b] (475 ± 29)[b]	Kusakabe et al. (1987)
Pacific Ocean	17°28'N; 117°58'W	3,950	610	5.74 ± 0.33	870 ± 50[b] (475 ± 27)[b]	Kusakabe et al. (1987)
Pacific Ocean	2°46'S; 117°02'W	4,200	810	6.11 ± 0.27	926 ± 41[b] (505 ± 22)[b]	Kusakabe et al. (1987)

The following atmospheric inputs were assumed: 1: 2.5 × 10^5 atoms cm^{-2} year^{-1} (Frank et al. 2009); 2: 1.21 × 10^6 atoms cm^{-2} year^{-1} (Segl et al. 1987); 3: 1.21 × 10^6 atoms cm^{-2} year^{-1} (Monaghan et al. 1986); 4: 1.21 × 10^6 atoms cm^{-2} year^{-1} was assumed

[a]Downward flux of 6.9 × 10^6 atoms cm^{-2} year^{-1} calculated using concentration gradient in ^{10}Be and assuming a vertical diffusivity (Kusakabe et al. 1982); the number in parenthesis is calculated assuming 1.21 × 10^6 atoms cm^{-2} year^{-1} (Monaghan et al. 1986)
[b]Influx value of 6.6 × 10^5 atoms cm^{-2} year^{-1} was assumed (Kusakabe et al. 1987); the number in parenthesis is calculated assuming 1.21 × 10^6 atoms cm^{-2} year^{-1} (Monaghan et al. 1986)

This is particularly important for partitioning studies in lakes, rivers and marine system which require precise measurements in the particulate and dissolved phases. For the extraction of particulate matter, filter cartridges have been widely employed (Silker 1972; Baskaran et al. 1997, 2009b; Feng et al. 1999; Baskaran and Swarzenski 2007). When the dissolved phase (<pore-size of the filter, either absolute cut-off or nominal) is passed through a prefilter, it is possible that some of the dissolved phase may sorb onto the polypropylene/acrylic/glass filters. So far, no studies have been made to quantify this, although recent work indicates that a finite fraction of the dissolved Th is removed by the prefilter (Baskaran et al. 2009b). For preconcentrating the dissolved phase, three common methods are employed that include: (1) evaporation; (2) co-precipitation; and (3) ion-extraction or extraction onto sorbents. Of these three methods, very low ionic strength solutions, including rainwater and freshwaters can be evaporated to reduce the volume (as low as 5–10 mL) that can be directly gamma-counted in Ge-well detectors. Evaporation of seawater samples will result in large amount of salts (~35 g L^{-1} of seawater) and hence it is not the suitable method for seawaters and other waters with high total dissolved solids. Co-precipitation with FeCl$_3$ is one of the most

common methods employed to preconcentrate; detailed procedures are given in Baskaran et al. (2009b). Stable Be carrier is added as yield monitor. Sorbents that have been utilized to extract dissolved radionuclides include $Fe(OH)_3$-impregnated fibers and a bed of aluminum oxide (Lee et al. 1991; Kadko and Olson 1996). Silker (1972) demonstrated that a flow rate of 35 L min^{-1} through a 6.4 mm thick alumina bed removed 70% of the ^7Be from seawater. Kadko and Olson (1996) assumed a constant efficiency of 69 ± 3%, although most cartridge methods for particle-reactive radionuclides (such as Th, Pb, etc.) have been found to have variable extraction efficiencies (Baskaran et al. 2009a). The constant efficiency assumption for Be removal needs to be rigorously tested. Concentrations of ^{10}Be in surface waters of the global ocean vary between 10^2 and 10^3 atoms g^{-1} (Raisbeck et al. 1978a; Kusakabe et al. 1982, 1990; Ku et al. 1990; Frank et al. 2009) and hence preconcentration is required. A minimum of a 2-L water sample ($\sim$0.4–2 × 10^7 atoms) is needed and details on chemical separation and purification are given in Baskaran et al. (2009b). For ^{10}Be analysis, sediments and aerosol samples are digested with concentrated HF, HNO_3, HCl and brought to solution with the addition of ^9Be carrier. Beryllium from this digested solution can be separated and purified following the procedure summarized in Baskaran et al. (2009b). The purified Be in the form of $Be(OH)_2$ is mixed with a small amount of $AgNO_3$ powder and then ashed at 850°C for 6 h and the ashed BeO powder is to prepare the target for the AMS analysis. Although the separation and purification of ^{10}Be is straight forward, isobaric interference from boron isotopes (^{10}B) could affect the measurements of ^{10}Be by AMS and hence care must be exercised in the purification of Be to eliminate B. Earlier studies added ^7Be spike (obtained from a commercial company) for the measurements of ^{10}Be in seawater as a yield monitor which had a high blank levels of ^{10}Be and hence caution needs to be exercised in utilizing ^7Be spike as a yield monitor (Kusakabe et al. 1982).

5.2.2.2 Measurement of ^7Be and ^{10}Be Concentrations

^7Be is most commonly measured in environmental samples using low-background decay counting techniques (Arnold and Al-Salih 1955; Larsen and Cutshall 1981). ^7Be decays to ^7Li via electron capture, with 89.6% of the decays emitting a very low energy x-ray as they go directly to the ground state, and 10.4% of the decays first going to an excited state, which is the fraction that emits the 477.6 keV gamma. Gamma measurements are usually done using shielded scintillation detectors (e.g., NaI crystals) or semiconductor detectors made of Ge(Li) or pure (intrinsic) germanium (Arnold and Al-Salih 1955; Silker 1972; Larsen and Cutshall 1981; Krishnaswami et al. 1982; Murray et al. 1987; Baskaran et al. 1997; Baskaran and Shaw 2001). The absolute detection efficiency of NaI detectors for gamma rays is typically higher than the efficiency of Ge detectors, but NaI detectors have poor energy resolution. The width of the peak at half of the maximum value (full width at half maximum, FWHM of the Gaussian curve) for NaI detectors at 478 keV is commonly ≥30 keV, which is inadequate for resolving the ^7Be gamma from possible emissions from U and Th series isotopes. Most notably, ^{228}Ac (^{232}Th series) has a strong emission (4.74% yield) at 463 keV (Dalmasso et al. 1987), and radon progeny have low yield emissions at 480 and 487 keV (Morel et al. 2004) which can be significant when counting samples high in ^{226}Ra. Separation and purification of the Be atom is therefore necessary when using scintillation detectors, but Ge(Li) and intrinsic Ge detectors have high enough resolution ($\sim$1–2 keV FWHM) in the 478 keV region which allows for the direct analysis of soils, sediments, waters, or filter papers for the ^7Be gamma.

One of the most important considerations that must be made when analyzing for ^7Be is the background of the gamma spectrum, which can control the detection limits. Compton scattering occurs when high energy gammas eject an electron with only a fraction of their full energy, which can create a "count" on the detector's spectrum and a newer lower energy gamma ray that can also create counts on the detector. Thus, high energy photons (>500 keV) from the U and Th decay series, ^{40}K, and cosmic rays thus generate random noise in the region of the gamma spectrum where ^7Be decays. By completely surrounding the scintillation or Ge detectors with 4″ of lead and using detector hardware (preamplifier, wires, etc.) made of ultra-low-background materials, the Compton scattering effect from gammas originating from external sources (the ground, the walls, and the atmosphere) can be minimized. As a general rule, if the detector is completely surrounded by 4″ of lead, external gammas from the ^{238}U, ^{235}U, and ^{232}Th decay series and ^{40}K are effectively attenuated. To minimize the Compton scattering effect from cosmic rays, detectors should be housed in basements of buildings or below ground. A thin copper liner is often used to separate the detector (and sample) from X-rays generated by the lead shielding itself.

In general, the larger the scintillation or Ge Detector (in thickness and surface area), the higher the efficiency for the ^{7}Be detection process, but larger detectors require the thickest shielding. The efficiency of the detector type for absorbing the full energy of the 478 keV ^{7}Be gamma is a function of the detector and the counting geometry. The relationship between efficiency and energy for each detector and counting geometry must be defined to calculate accurate ^{7}Be concentrations. This is commonly done by counting a synthesized mixture of radionuclides that decay over a wide range of energies (40–1,400 keV) in the exact geometry that the unknown samples will be analyzed in, which allows the analyst to construct a relationship between energy (keV) and detection efficiency. Alternately, if certified ^{7}Be Standard Reference Material can be obtained, then, the detector can be calibrated for well-defined geometries and dpm/cpm ratios can be obtained for different geometries and these ratios can be directly used to obtain the activities of samples. Detectors can be planar in form for petri-dish style counting geometries, but large well detectors have the highest efficiencies for small geometry samples (a Ge well detector has an absolute detection efficiency at 478 keV of 21.9 ± 0.3% for 1-mL geometry; Jweda et al. 2008). There have been successful measurements of ^{7}Be in environmental samples by accelerator mass spectrometry, which has a detection limit of ~10^4 atoms (corresponding to 0.09 dpm or 1.5 mBq) which is at least an order of magnitude more sensitive than the counting methods (Nagai et al. 2004). Details on calculation of activities and calibration methods are given in Baskaran et al. (2009b).

A number of accelerators (e.g., Tandem Van De Graaff and other high-voltage Tandem accelerators, cyclotron, etc.) have been utilized to measure ^{10}Be (Raisbeck et al. 1978b; Turekian et al. 1979; Galindo-Uribarri et al. 2007). Details on the differences between these accelerated mass spectrometers are beyond the scope of this article. A detailed methodology on how the Tandem Van De Graaff accelerator is set-up and used for the ^{10}Be measurements is given in Turekian et al. (1979).

5.3 Applications

The near-steady input of ^{7}Be and ^{10}Be and the particle-reactive nature of the Be atom makes meteoric Be nuclides a very useful tracer for quantifying a range of environmental processes operating on timescales from weeks to millennia. Both nuclides can be used to study atmospheric transport and depositional processes (see Lal and Baskaran 2011, Chap. 28). ^{7}Be can be used to quantify a number of short-timescale processes, including the infiltration of particle-reactive elements in soils, overland flow processes, and topsoil erosion (see Matisoff and Whiting 2011, Chap. 25). In lakes, streams, and marine environments, ^{7}Be is a valuable tracer of colloid and particulate dynamics, recent sediment deposition, mixing and focusing, and particle resuspension and transport. Meteoric ^{10}Be has been used to quantify landform age, creep rates, erosion rates, loess accumulation, and it has potential for dating authigenic mineral formation.

5.3.1 Using ^{7}Be and ^{10}Be to Trace Hillslope and Soil Processes

The short half-life of ^{7}Be makes it a valuable tracer of event-scale transport in soil profiles and on hillslopes. Given that ^{7}Be is usually deposited during rainfall events, especially intense rains and thunderstorms, the vertical distribution of this nuclide in soils can be used to infer the initial depth-penetration of other particle-reactive radionuclides and contaminants from a single deposition event. The initial conditions for an advection-diffusion transport model of other fallout isotopes (e.g., ^{210}Pb) can be constrained using the vertical distribution of ^{7}Be (Kaste et al. 2007). Measurements of ^{7}Be in stormwater, for example, can be used to study the fate of atmospherically-deposited particle reactive contaminants (Hg, Pb, etc.) released from a melting snowpack (Cooper et al. 1991). The spatial distribution of ^{7}Be on hillslopes measured after a significant storm can also be used to trace event-scale soil sediment redistribution. If the ^{7}Be distribution is compared with the distribution of other sediment tracers that track longer timescales (e. g., ^{137}Cs), the contribution of single, intense storms to soil movement can be put into perspective with processes operating over the course of years to decades (Walling et al. 1999).

Because of its longer half-life, meteoric ^{10}Be has great potential for quantifying soil and sediment transport processes and authigenic mineral formation rates on soil-mantled hillslopes (Willenbring and von Blanckenburg 2010a). The inventory of meteoric ^{10}Be (I, in units of atoms area^{-1}) at a point on Earth can be expressed as:

$$I = \int_{-z}^{0} C\rho \, dz \qquad (5.2)$$

where $-z$ is the depth in the soil to which meteoric ^{10}Be has penetrated, 0 is the soil surface, C is the concentration of meteoric ^{10}Be (in atoms mass^{-1}), and ρ is the soil density (mass volume^{-1}). In practice, this is measured by collecting samples with sampling depth resolution (dz) on the order of 20 cm to up to a meter (Pavich et al. 1985; McKean et al. 1993; Jungers et al. 2009). The inventory I at a location is governed by ^{10}Be deposition, which is a function of production in the atmosphere and wet + dry depositional processes, age, radioactive decay, particulate losses, and solute losses. Inventories are used to find the age of stable, non-eroding surfaces if the nuclear production rates in the atmosphere and deposition varies around some mean value that can be constrained for time period of interest (Tsai et al. 2008; Willenbring and von Blanckenburg 2010a). However, at many sites, solution and/or erosional losses limit the ^{10}Be inventories (Monaghan et al. 1983). Given an eroding surface where weathering, soil formation, and soil loss are in equilibrium, inventories can be used to calculate steady-state erosion rates (Pavich et al. 1986; Brown et al. 1988). This technique can be extended to study the fate of other particle-reactive elements on landscapes. For example, the loss and accumulation of ^{10}Be over different points of a landform can be used to study how sediment transport processes control the fate and storage of carbon on a hillslope (Harden et al. 2002).

Inventories of meteoric ^{10}Be along points on soil-mantled hillslope profiles often increase with distance from the divide (McKean et al. 1993). This gradient results because points farthest from the divide have traveled the longest distance and resided on the hillslope for a longer duration and thus received more of a ^{10}Be dose. By constructing a linear regression between I values measured for points on a hillslope profile and projected soil particle paths, Jungers et al. (2009) used the rate of inventory change (dI/dx) and an average ^{10}Be input assumption to calculate virtual soil velocities for a hillslope in the Great Smoky Mountains, NC. Downslope soil transport velocities calculated using ^{10}Be inventories can be used to test assumptions about landscape equilibrium. McKean et al. (1993) used this technique to show that the soil creep flux was related to slope, which supported G.K. Gilbert's hypothesis (1877) that hillslopes exist in a dynamic equilibrium with a uniform soil production rate.

Meteoric ^{10}Be may be useful for dating authigenic minerals in soils, which was first suggested by Lal et al. (1991). Dating of secondary minerals in soil profiles could be extremely valuable for quantifying the rate of soil formation, and, using additional information, can be used for putting formation rates into context with erosion rates. Secondary minerals have, for the most part, defied traditional isotopic dating methods, such as U-Th series chronology, because of large uncertainties in defining the initially inherited isotopic composition (Cornu et al. 2009).

Barg et al. (1997) developed a "closed system" model which relied on the ^{10}Be/^{9}Be ratio as a chronometer for modeling the age of clays and iron and aluminum hydroxide minerals. In this model, cosmogenic ^{10}Be and bedrock-derived ^{9}Be equilibrate in the upper soil horizons, and slowly move down the profile by adsorption-desorption reactions. It is assumed that secondary minerals form primarily in the C-horizon, and, at the time of formation ^{10}Be/^{9}Be is locked in, and will change as a function of time. Indeed, the ^{10}Be/^{9}Be values were found to be in a narrower range than the ^{10}Be concentration, and, by using selective chemical extractions that targeted secondary minerals, they found that the highest ^{10}Be/^{9}Be values were found at the soil-bedrock interface, where authigenic mineral formation is projected to take place. This technique needs considerably more development, to account for secondary mineral dissolution and illuviation processes.

5.3.2 Using ^{7}Be as a Tracer in Lake and River Systems

5.3.2.1 ^{7}Be as a Tracer of Metal Scavenging, Colloidal, and Particulate Dynamics in Lakes

When dissolved metals are introduced into lake waters via atmospheric deposition, they can be scavenged by colloids, a process that greatly reduces their potential toxicity to the lake's ecosystem. Honeyman and Santschi (1989) describe a "colloidal pumping" process by which metals are removed from waters in three stages: (1) in the first stage, dissolved trace metals and radionuclides are released into the water column, either from atmospheric deposition or through production from the parents and these species are removed quickly onto colloidal particles through sorption reactions; (2) colloidal-bound trace metals and radionuclides undergo coagulation relatively slowly with the small particle pool and (3) the particles move in the particle size spectrum and are then eventually removed from the water column. In this, the coagulation of colloids is the rate controlling step in trace metals and radionuclide scavenging. Steinmann et al. (1999) modified a steady-state particle scavenging model (Honeyman and Santschi 1989) so that ($\lambda = 0.013$ d^{-1}, the decay constant of ^{7}Be) ^{7}Be could be applied to study mechanisms and rates of metal scavenging from the waters of Lake Lugano (Switzerland, Italy):

$$Be_d \lambda_{ads} = Be_c \lambda + Be_c \lambda_{coag}$$
$$Be_c \lambda_{coag} = Be_p \lambda + Be_p \lambda_{sed}$$
$$F_{sed}^{Be} = Be_p \lambda_{sed} \qquad (5.3\text{-}5.5)$$

By directly measuring the concentrations (in Bq m^{-3}) of dissolved ^{7}Be (Be$_d$; <10 kD), colloidal ^{7}Be (Be$_c$; 10 kD–1 µm), particulate ^{7}Be (Be$_p$; >1 µm), and the sedimentary flux of ^{7}Be (F_{sed}^{Be}; in Bq m^{-3} day^{-1}), the adsorption rate (λ_{ads}) of ^{7}Be onto colloids was calculated to be approximately 0.02–0.005 day^{-1}, corresponding to a dissolved ^{7}Be residence time of 50–200 days. The residence time of colloids (1/λ_{coag}) ranged from a few days to a few weeks, while the residence time of particulates (1/λ_{sed}) was typically less than a week. The authors suggested that colloids <10 kD containing ^{7}Be could be affecting the calculated residence times, and thus adsorption rates of ^{7}Be, and that perhaps the rate limiting step for the removal of Be from the water column was in fact the coagulation of small colloids <10 kD. This application of ^{7}Be was also useful for quantifying the effect of biological processes on trace metal scavenging, as the highest coagulation rates (λ_{coag}) followed algal blooms. While true steady-steady conditions may not be possible on timescales of weeks because of the episodic nature of ^{7}Be deposition, the quantitative modeling approach described above is still useful for constraining the rates of metal adsorption, colloidal coagulation, and particle sedimentation in natural systems (Dominik et al. 1989).

^{7}Be can also be used to identify sources of particulates to the water column in lakes and to examine mixing processes. In Lake Michigan, under isothermal conditions that persist from December to May, the concentration of ^{7}Be on suspended matter is nearly constant with depth (Robbins and Eadie 1991), indicating that the waters are mixed vertically on short timescales.

However, in early summer, the ^{7}Be content of suspended sediments reaches a maxima near the surface of the lake (<20 m) as lake stratification limits the settling of sediment (Robbins and Eadie 1991). A similar stratification effect on the vertical distribution of ^{7}Be has been observed in waters of Lake Zurich (Schuler et al. 1991). In the summer, calcite formation and/or algal blooms can effectively scavenge ^{7}Be from the upper fraction of the water column, reducing ^{7}Be in the epilimnion (Robbins and Eadie 1991; Vogler et al. 1996). In the late fall, lake turnover recharges the suspended sediment pool to lake water again, as ^{7}Be increases again in the epilimnion (Robbins and Eadie 1991; Vogler et al. 1996). These tracer studies have demonstrated how the dissolved to particulate ^{7}Be varies seasonally, and that the scavenging of metals from surface waters can be controlled by biological productivity and/or authigenic mineral formation.

5.3.2.2 ^{7}Be as a Tracer of Sediment Source, Focusing, Resuspension, and Transport in Rivers

Sediment can directly and indirectly degrade river ecosystems, and humans have altered landscapes and sediment fluxes and dynamics in nearly every river system on Earth. The focusing of fine sediment deposition from land use change, in particular, can degrade spawning gravels, and, in some cases, sediments contain particle-reactive contaminants such as Pb, Hg, and PCBs that directly impact organisms. Cosmogenic ^{7}Be can be a useful tracer in fluvial systems, and has been successfully used to trace sediment sources in rivers, residence times, resuspension rates, transit times, and the fate of recently contaminated sediments. In many cases the use of ^{7}Be in fluvial systems relies on the fact that sediments that are buried for a period of more than 6 months will have undetectable ^{7}Be. This "old" sediment can be separated from recently exposed sediment by doing end-member mixing analysis (Bonniwell et al. 1999; Matisoff et al. 2005). In rivers, sediment deposition and resuspension rates can be calculated by projecting the inventories present in bed sediments present from atmospheric deposition, and quantifying depletion from resuspension, or excess, from deposition (Jweda et al. 2008). This technique was used in the Fox River, in Wisconsin, to determine the fate and dynamics of a PCB-contaminated sediment layer (Fitzgerald et al. 2001), much of which was resuspended during high flow despite the impounded nature of the river.

Humans have profoundly altered the flow regimes and sediment supplies of rivers through urbanization, logging, the construction of dams, and the engineering of stream banks (Croke et al. 1999; Magilligan et al. 2003). In recent years, the use of ^{7}Be as a tracer of fine sediment deposition, resuspension, and transport has been expanded to study processes in a wide range of fluvial environments, and to quantify how sediment dynamics might respond to different forcings. The enrichment of the upper few cm of topsoil with ^{7}Be makes it a valuable tracer of sediments derived from surface erosion (Blake et al. 1999, 2002; Wallbrink et al. 1999) and enables a quantitative partitioning of runoff processes between sheetwash and rilling (Whiting et al. 2001). Sediment transport rates within a fluvial system can also be calculated using ^{7}Be. Salant et al. (2007) repeatedly measured point bar and streambed sands in a regulated river immediately downstream of a dam in Vermont, U.S.A. During the winter months, sediment stored behind the dam became depleted in ^{7}Be, and when the gates were opened in the early spring the pulse of depleted "new" sediment was discernable from the surrounding sediments that were supplied by tributaries and previously exposed bars. Sediment transport rates of 30–80 m day^{-1} were calculated using the known starting point of the sands and monthly sampling at fixed points to monitor the pulse of sediment as it moved down river. The ^{7}Be content of bedload sediment in a single river varies by more than a factor of two over the course of a single year, as seasonal changes in ^{7}Be delivery (Olsen et al. 1985) and flow regimes regulate sediment source, grain size, and transport processes. Cosmogenic ^{7}Be may trace sediment substrates and nutrients that are favorable for certain aquatic insects. Svendsen et al. (2009) reported a significant correlation between ^{7}Be concentrations in transitional bed load sediment and benthic community structure at tributary junctions along a mainstem river in Vermont.

There is a strong dependence of grain size on the adsorption of ^{7}Be, and, in rivers, grain size will vary with flow regime which can make the relationship between ^{7}Be and age difficult to reconstruct. To normalize for grain-size effects on ^{7}Be values, models have been developed that utilize the ^{7}Be/^{210}Pb$_{ex}$ ratio for calculating sediment age or fraction of new sediment in suspension. This relies on the fact that ^{210}Pb$_{ex}$, which is the atmospherically-delivered portion of ^{210}Pb in sediments (^{210}Pb in "excess" of that projected to be supported by in situ ^{222}Rn decay) also adsorbs strongly to sediment and organic matter surfaces, and is reported to have partitioning coefficients (K_d) at least as high as Be (Sauve et al. 2000a, b). The change in the ^{7}Be/^{210}Pb$_{ex}$ can thus be a more sensitive indicator of age, since ^{7}Be decays much faster than ^{210}Pb. Others have found a very strong correlation with ^{7}Be/^{210}Pb values and ^{7}Be divided by grain surface area (Fisher et al. 2010). In a study of a coastal plain river in Maine, U.S.A., Fisher et al. (2010) used a constant initial ^{7}Be/^{210}Pb model to measure the timescale of sediment storage behind large woody debris and boulders. They found that in independently-assessed "transport-limited" reaches of the Ducktrap River, sediment sequestration behind in-stream obstructions was >100 days, but in supply limited reaches sediment was replenished more quickly.

5.3.3 Applications of Meteoric ^{7}Be and ^{10}Be as Process Tracers in the Marine Environment

From the time the ^{7}Be and ^{10}Be are produced in the atmosphere until its ultimate disposal, either through radioactive decay or their permanent incorporation into the bottom sediments in oceans, these nuclides often pass through a number of transient reservoirs, such as various layers of atmosphere, precipitation and surface waters. The half-life of ^{7}Be is suitable for studying the vertical eddy diffusion coefficients and associated vertical transport rates in the upper layers of the thermocline (Silker 1972; Kadko and Olson 1996). Silker (1972) utilized the vertical concentration gradients of ^{7}Be in the Atlantic water column to determine the vertical eddy diffusion coefficients within the thermocline, assuming that atmospheric input of ^{7}Be is constant and the horizontal advective transport is negligible. Due to its short mean-life (76.9 days) and the lack of biological removal of Be, the effects of long-range advective transport on the vertical distribution is likely negligible and hence it is reasonable to assume that ^{7}Be will be effective in tracing diffusion at the top of the thermocline. Since ^{7}Be is delivered at the air-water interface, that layer is "tagged" with ^{7}Be constantly and when this layer undergoes sinking, the vertical profiles of ^{7}Be can be utilized to retrieve the record of the seasonal changes in the mixed layer depth (Kadko 2000).

5.3.3.1 Measuring Sediment Accumulation, Mixing Rates, and Sediment Focusing with ^{7}Be

A large number of studies have been conducted utilizing ^{7}Be to identify and quantify flood deposits (Corbett et al. 2007; more discussion are given in Du et al. 2011, Chap. 16). ^{7}Be is also utilized to quantify recent sediment accumulation and mixing rates (Krishnaswami et al. 1980; Clifton et al. 1995; Corbett et al. 2007). The sediment mixing involves both advective (physical particle transport) and diffusive mixing components. A combined effect of advective and diffusive mixing is analogous to eddy diffusion and has been modeled to determine eddy mixing coefficients. In coastal areas, the ^{7}Be profiles in sediments are not often in steady-state due to seasonal variations in the riverine discharge which delivers relatively large amount of ^{7}Be. In marine and lacustrine environments, the sediment inventories at selected sites are often far excessive compared to what is expected and this elevated levels is often caused by physical transport processes of sediments (such as sediment focusing by bottom currents and gravity flows, e.g., slumps and gravity flows). A comparison of the measured sediment inventories (I_{sed}, Bq m^{-2}) to the measured depositional fluxes (or expected) of ^{7}Be (I_{dep}) will yield information on sediment focusing ($I_{sed}/I_{dep} > 1$) or erosion ($I_{sed}/I_{dep} < 1$).

5.3.3.2 Boundary Scavenging and Long-Range Transport of ^{10}Be in Marine Systems

Advective transport of dissolved nuclides to regions of high productivity waters in the margins results in higher suspended particle concentrations and fluxes which result in enhanced scavenging of ^{10}Be and other particle-reactive nuclides (e.g., ^{230}Th, ^{231}Pa, ^{210}Pb, etc.). In ocean margin sediments, the ^{10}Be concentrations were found to be much higher than that expected from the direct atmospheric deposition and this was attributed to boundary scavenging. The deposition rates of ^{10}Be in pelagic red clays were reported to be much less than its global average production rate, whereas in ocean-margin sediments, the ^{10}Be deposition rate exceeds its global average production rate (Brown et al. 1985; Anderson et al. 1990; Lao et al. 1992b). Particle composition (such as opal Si, Fe-Mn oxides, carbonate, terrigenous material such as clays; Luo and Ku 2004a, b; Chase et al. 2003) also plays a significant role on the scavenging of dissolved ^{10}Be. Sharma et al. (1987) reported strong correlation between ^{10}Be and Al in the sediment trap and attributed the aluminosilicate to be the major carrier phase for the scavenging of dissolved ^{10}Be.

Deposition rates of ^{10}Be at margin sites calculated from sedimentary records (= sediment accumulation rates (g cm^{-2} year^{-1}) × concentration of ^{10}Be in surficial sediments (atoms g^{-1})) have been found to be significantly higher than the rate of ^{10}Be supply at the Earth's surface at sites far away from the margins (such as off Northwest Africa and off California, equatorial Pacific off Ecuador, margins in North Atlantic, Brown et al. 1985; Anderson et al. 1990) and is attributed to boundary scavenging. Boundary scavenging exerts influence when the dissolved residence time of Be is greater than the time required for the advective transport to move the water from the center of the basin to the margins. Physical transport processes such as bottom currents and gravity flows (turbidities and slumps) also could lead to increased rates of ^{10}Be accumulation rates, resulting in higher ocean-wide average ^{10}Be deposition rates.

There are limited data on the temporal and spatial variations of ^{10}Be and ^{7}Be isotopes in oceanic particulate matter. Depositional fluxes measured in sediment traps deployed at two sites in the eastern equatorial North Pacific Ocean show that the ^{10}Be flux at traps deployed 50 m above the bottom of the sediment-water interface were up to an order of magnitude higher than that at mid-depth in the water depth column (at 1,565 and 1,465 m with total water depth of 3,100 and 3,600 m). The flux in the bottom trap (4.6×10^6 and 2.4×10^6 atoms cm^{-2} year^{-1}) was significantly higher than the average global depositional flux of 1.2×10^6 atoms cm^{-2} year^{-1} while at intermediate depth the ^{10}Be flux was lower (0.9×10^6 and 0.4×10^6 atoms cm^{-2} year^{-1}) than the global depositional flux (Sharma et al. 1987). This difference was attributed to contribution from lateral input as well as from the resuspended material in the bottom nepheloid layers. A strong correlation between ^{10}Be concentration and Al suggest that aluminosilicate is the major carrier phase of ^{10}Be. Variations in atmospheric dust input may result in variations in the scavenging of ^{10}Be from the water column and thus, the variations in particle fluxes can be linked to the depositional fluxes in the traps.

5.3.3.3 ^{7}Be as a Tracer of Ice Rafted Sediments (IRS) in the Oceans

Concentrations of ^{7}Be in ice-rafted sediments (IRS) in the seasonal ice-cover of the Arctic Ocean have been reported to be relatively high (Masque et al. 2007). When sea ice is formed mostly in shallow waters in the marginal seas of the Arctic Ocean (such as Siberian margins of the Laptev, Kara and Barents Seas), large amounts of fine-grained sediments get incorporated into the sea ice during formation of new ice mainly through suspension freezing of bottom sediments and river-borne sediments (Barnes et al. 1982; Reimnitz et al. 1992; Nurnberg et al. 1994) and are subsequently transported to different parts of the Arctic. During its transit, atmospherically-delivered ^{7}Be is intercepted by the ice cover and the ^{7}Be is added to the IRS during freezing-thawing cycles (Baskaran 2005). The activities of ^{7}Be in IRS have been reported to be highly variable, with values ranging between $10^{4.9}$ and $10^{6.2}$ atoms g^{-1} (13 and 212 Bq kg^{-1}), which is 1–2 orders of magnitude higher than those reported in the coastal sediments. The dissolved ^{7}Be activities in melt ponds were also reported to be much higher (30–60 mBq L^{-1}) than the surface waters in the Arctic (0.6–2.3 mBq L^{-1}) as well as surface water samples collected between 1.0 and 2.0 m from the air-sea interface (2.6 mBq L^{-1}) (Eicken et al. 2002; Kadko and Swart 2004). The concentrations of ^{7}Be in sea ice also were reported to be ~2 orders of magnitude higher than that in the ocean mixed layer below the sea ice and this was attributed to dilution of ^{7}Be with longer depths of the mixed layer compared with the <1 m thick seasonal ice (Cooper et al. 1991). During summer, the snow and upper 0.3–0.7 m of the ice melt off the ice surface. Tracer studies indicate that meltwaters disperse over distances of tens of meters on time scales of days, with much of the meltwater pooling in depressions at the surface (Eicken et al. 2002). IRS accumulates at the surface and in melt pools and are subjected to meltwater flushing (Nurnberg et al. 1994). During this process, the IRS released from melting comes in contact with seawater in meltponds and additional scavenging may take place. It has been estimated that ~60% of the ^{7}Be inventory is found in the water column from the sea ice melt and the remaining ~40% in found in sea ice (Eicken et al. 2002). Based on controlled experiments, it has been estimated that ~15% of the total sediment load is released to the water column during melt (Eicken et al. 2002). During multiple freezing-thawing cycles, the IRS is often pelletized and thus, can undergo sinking very quickly. Cooper et al. (2005) reported ^{7}Be in benthic surface sediments collected as deep as 945 m in Barrow Canyon in the Arctic, but it has not been reported from any other ocean basins. The activities of ^{7}Be in benthic sediments at different water column depths will be useful to quantify the amount of sediments released from melting of sea ice.

5.3.3.4 Using Meteoric ^{10}Be as a Chronological Tool in Marine Sediments

^{10}Be has been extensively utilized to date marine sediments, Mn-nodules, and other authigenic minerals. Prior to the development of AMS techniques, dating of Mn-nodules and deep-sea sediment cores were dated by conventional beta counting technique (Somayajulu 1967; Sharma and Somayajulu 1982). The concentration of ^{10}Be in Mn-nodules is very high (~10^{10} atoms g^{-1}) and hence a very small amount of sample is required (<10 mg). The first measurement of

^{10}Be by AMS basically confirmed the slow growth rates of Mn-nodules observed by conventional counting method (Somayajulu 1967; Turekian et al. 1979). Based on the exponential decay of ^{10}Be with depth on manganese crusts over a period of past 7 Ma, Ku et al. (1982) suggested that the ^{10}Be/^{9}Be ratio has remained constant over this time period. The ^{10}Be concentrations in Mn-nodules extrapolated to zero depth indicate that there is wide range of ^{10}Be values in nodules collected at different basins and varied between 1.5×10^{10} and 6.5×10^{10} atoms g^{-1} (Kusakabe and Ku 1984; Segl et al. 1984a, b; Mangini et al. 1986). The ^{10}Be/^{9}Be ratios in nodules from Atlantic and Pacific only ranged between 1.1×10^{-7} and 1.6×10^{-7}. A comparison of the extrapolated to the surface value (corresponding to present time) of ^{10}Be/^{9}Be ratio to that in the deep waters suggesting that the dissolved Be are well mixed in the bottom waters and that the ratio is well-preserved in authigenic minerals formed out of seawater (Kusakabe et al. 1987). Sediment cores have been dated using certain assumptions that include constant supply of ^{10}Be to the seafloor over the time interval of interest. It has been shown that the flux of ^{10}Be is largely dependent on the sedimentation rate (Tanaka et al. 1982; Mangini et al. 1984). Sediment focusing and boundary scavenging, however, make the assumption of constant supply of ^{10}Be questionable. In the Arctic, the variations in the particle flux (due to variations in the release of ice-rafted sediments) have resulted in highly varying concentrations of ^{10}Be in a vertical profile (Eisenhauer et al. 1994; Aldahan et al. 1997). The authigenic minerals seems to be better candidates for ^{10}Be dating compared to deep sea sediments, as the chemical link between deep seawater and authigenic minerals make this method work well for authigenic minerals. By measuring the ^{10}Be in deep-ocean Fe-Mn crusts, the decay corrected ^{10}Be/^{9}Be ratio was found to be constant over the past 12 Myr and this was attributed to constant global chemical weathering flux over this time period (Willenbring and von Blanckenburg 2010b).

5.3.3.5 Case Study: Dating of Marine Particulate Matter Using ^{7}Be/^{10}Be Ratios as Tracer

As was discussed earlier, once ^{7}Be and ^{10}Be are delivered at the air-sea interface primarily through wet precipitation, during their life-span in the water column, they are scavenged by suspended particulate matter and eventually reach the ocean floor. Due to the short mean-life of ^{7}Be, most of the ^{7}Be undergoes radioactive decay while virtually no ^{10}Be undergoes radioactive decay. Although there are more than an order of magnitude variations on the monthly depositional fluxes of these nuclides, the variations on the ^{7}Be/^{10}Be ratios is very narrow, within a factor of ~2 (Brown et al. 1989; Graham et al. 2003; Heikkila et al. 2008; Lal and Baskaran 2011). When marine particulate matter leaves the surface layer with a finite ^{7}Be/^{10}Be ratio, as it sinks the ratio values will decrease and the rate of decrease can, in principle, be used to date the marine particulate matter, provided additional scavenging of ^{7}Be and ^{10}Be in subsurface layers does not alter the ^{7}Be/^{10}Be ratio. With the detection limit of ^{7}Be of 10^4 atoms using AMS (Raisbeck and Yiou 1988), one can measure ^{7}Be/^{10}Be ratios in particulate matter collected from ~20 L water sample assuming that 0.2 Bq/100 L of ^{7}Be (2.65×10^5 atoms), the average value for a six coastal water systems along the Gulf coast (Baskaran and Santschi 1993; Baskaran et al. 1997; Baskaran and Swarzenski 2007) and ^{10}Be abundance of ~500 atoms g^{-1} (corresponding to ~10^7 atoms of in 20-L sample), particulate matter can be reliably dated.

When particles leave the mixed layer with ^{7}Be/^{10}Be ratio (R_0), at any depth below mixed-layer, the measured ratio is R_d, then the time of transit (or residence time) of particulate matter is given by:

$$t = \tau \ln(R_0/R_d) \qquad (5.6)$$

where τ is the mean-life of ^{7}Be. If we assume that the particles are fresh in the mixed-layer, then, t can be considered as the age of the particulate matter. At depth "d," if there is equilibrium between the particulate matter and dissolved phase, then, the dissolved phase will likely have the same ^{7}Be/^{10}Be ratio as the particulate matter and hence by measuring ^{7}Be/^{10}Be ratio in the dissolved phase, we will be able to determine the "age" of the particulate matter.

5.3.4 Other Applications of Meteoric ^{7}Be and ^{10}Be at the Earth's Surface

The unique source term of the ^{7}Be and ^{10}Be nuclides make them particularly useful for tracing deposition and examining whether minerals are formed from elements derived from the atmosphere or the lithosphere.

Recently, these nuclides were used to study how rock varnish accumulates elements (Moore et al. 2001). Rock varnish specimens shielded from rainfall accumulate detectable ^{7}Be, presumably from dew or dry deposition, but specimens exposed to precipitation accumulate considerably higher doses of ^{7}Be. It was calculated that rock varnish accumulates a few percent of the total deposition of cosmogenic Be nuclides; ultraviolet radiation did not inhibit ^{7}Be accumulation in rock varnish, which suggests that varnish may be formed abiotically (Moore et al. 2001). The use of cosmogenic Be isotopes in conjunction with other fallout radionuclides (^{137}Cs, ^{210}Pb) may be useful for understanding the rates and mechanisms by which atmospheric elements are incorporated into mineral structures (Fleisher et al. 1999; Moore et al. 2001).

5.4 Other Future Directions

5.4.1 Deposition

There is considerable uncertainty on how cosmogenic Be deposition varies spatially over the Earth's surface and how climate might control fluxes to points on land and sea. The meteoric input uncertainty appears to be limiting our ability to quantify and compare the residence time of dissolved Be radionuclides in the oceans (Table 5.3). Climate change will likely result in the global redistribution of rainfall, and changes in the frequency and intensity of rainfall may affect the seasonal depositional fluxes of meteoric Be. However, input functions need to be well constrained in order to effectively use ^{7}Be and ^{10}Be as tracers in the environment, and, in particular, to apply ^{10}Be to quantify ages or steady-state erosion rates on timescales of thousands to millions of years. While there is often an inverse relationship between ^{7}Be and ^{10}Be concentrations in rainfall and storm duration at many sites (Baskaran et al. 1993; Baskaran 1995; Ishikawa et al. 1995), the dependence of annual cosmogenic Be fluxes on precipitation rates has not been well described. If cosmogenic Be is removed from an air mass during the initial parts of a storm, as most datasets indicate, a question remains as to whether or not additional rainfall over some threshold has an additive effect on fluxes. A number of studies have demonstrated that rainfall rates ultimately control the total amounts of tropospheric contaminants and weapons-derived atmospheric ^{137}Cs delivered to the surface (Kiss et al. 1988; Simon et al. 2004). However, Willenbring and Von Blanckenburg (2010a) suggested that at coastal and island settings where the atmospheric transport time is fast and ^{10}Be and ^{7}Be concentrations in rain have a strongly inverse relationship with precipitation rate, annual fluxes should remain constant even if climate were to vary. The controls that climate and annual precipitation have on cosmogenic Be nuclide deposition should be studied further.

There are essentially two methods by which annual atmospheric fluxes can be determined for a region: direct monitoring of wet + dry atmospheric deposition on a weekly to monthly basis, or indirect determination using soil, sediment, and/or ice inventories. Long-term monitoring of cosmogenic Be wet depositional fluxes on landscapes can be costly, and it can be difficult assessing dry fluxes. The use of stable surfaces that record cosmogenic Be inventories has some advantages for calculating fluxes if surface age is known or if steady-state can be assumed. Relating ^{10}Be inventories to atmospheric fluxes would be complex given that surface age, and solute and particulate losses would be difficult to constrain in everything but perhaps ice records, but the use of ^{7}Be for this purpose does not suffer from these complications. The advantage of using surface inventory measurements of ^{7}Be for calculating fluxes is that a single measurement can integrate months of deposition, which works particularly well in temperate climates where rainfall is generally evenly distributed throughout the year. At sites where there is a distinct wet and dry season, multiple samples may be necessary to calculate annual atmospheric fluxes. There is an order of magnitude range of ^{7}Be soil inventories reported for stable soils around the world (Table 5.1). While many of the sites measured for ^{7}Be inventories are of temperate climate, the one arid region dataset falls in the low end of the range, reporting approximately 85 Bq m^{-2} for Owens Valley, CA (Elmore et al. 2008). Salisbury and Cartwright (2005) used ^{7}Be measurements of sheep feces to infer that vegetation along an elevation gradient in North Wales had higher concentrations of ^{7}Be. While they could not calculate atmospheric fluxes from their measurements, their data indicate some relationship between wet precipitation amounts and cosmogenic Be deposition. It would be valuable to have surface ^{7}Be inventory data available

5.4.2 Species and Geochemical Behavior

The geochemical behavior of Be is complex, yet simplifications are often made in order to extend the use of ^{7}Be and ^{10}Be as process tracers. It is evident from experimental work that partitioning strength is generally high ($K_d > 5{,}000$), but dissolved meteoric Be is commonly measurable or even dominates over particle-bound Be in the water columns of lakes and oceans (Silker et al. 1968; Kusakabe et al. 1982, 1987, 1990; Bloom and Crecelius 1983; Sharma et al. 1987; Dibb and Rice 1989a; Dominik et al. 1989; Steinmann et al. 1999). It appears that the availability of particles in a water column may limit the scavenging of meteoric Be radionuclides, but the effects of salinity, time, and particle composition (e.g., algae, organic detritus, Al/Fe/Mn phases) on partitioning strength and kinetics need to be better quantified.

In neutral to alkaline soil environments, Be is expected to form complexes with oxygen on surfaces of secondary clays and oxides and organic matter (Nyffeler et al. 1984; You et al. 1989). However, it is possible that soluble Be complexes could form, inhibiting adsorption and increasing the geochemical mobility of ^{7}Be and ^{10}Be in the environment. Equilibrium thermodynamic approach has been used to predict that Be forms soluble complexes with humic and fulvic acids and fluoride (Vesely et al. 1989; Takahashi et al. 1999) but there are no field-based studies documenting enhanced mobility in the environment from these species. It is also possible that cosmogenic Be could adsorb to very small colloids formed at soils through the breakdown of organic matter and Be could migrate vertically downward. Steinmann et al. (1999) used continuous flow centrifugation and tangential flow ultrafiltration to separate dissolved, colloidal, and particulate ^{7}Be phases in lake water. Despite their substantial efforts to separate truly dissolved ^{7}Be from other phases, they concluded that ^{7}Be might be adsorbed to very fine colloids (<10 kD) that are not easily separated from other phases or size fractions. More theoretical, experimental, and field-based data need to be collected on the distribution of Be species in the environment, and to identify the various mechanisms by which cosmogenic Be can be transported in different systems.

5.4.3 "Dating" of Suspended Particulate Matter with ^{7}Be/^{10}Be

Particulate matter at the air-sea interface is tagged with a finite ^{7}Be/^{10}Be ratio which is likely to remain constant at any season and as the particles undergo sinking, ^{7}Be undergoes radioactive decay. Typical settling rates of a terrigenous particle (density = 2.6 g cm^{-3}) of 4 mm diameter is ~1 m day^{-1} (for 16 mm diameter ~19 m day^{-1}, corresponding to reaching a depth of ~1,460 m over the mean-life of ^{7}Be). Particulate matter from a sediment trap deployed at ~1,500 m can be utilized to determine the settling velocity of particulate matter. ^{7}Be/^{10}Be measurements on size-fractionated particulate matter could provide differences in the settling velocity of particulate matter. The deposition velocities could provide insight on the particle aggregation-disaggregation processes that are taking place in marine environment. It is known that beryllium is biologically inactive and hence could serve as an ideal tracer for the settling of terrigenous particulate matter. In the Arctic, ^{7}Be/^{10}Be ratios in sediment traps from seasonally ice-covered areas could provide insight on the amount of sediments released from sea ice during seasonal melting-freezing cycles. While ^{7}Be in sea ice sediments records the history of the material for <1 year, ^{10}Be data would record a much longer history, and hence the ^{7}Be/^{10}Be ratio may provide information on the time scale and extent of recycling of IRS. There are currently no ^{10}Be data available for sea ice sediments, so more work is needed to develop this potentially valuable tracing tool.

5.5 Concluding Remarks

Meteoric ^{7}Be and ^{10}Be are valuable tracers that can be used to characterize an incredibly wide range of environmental processes. While ^{7}Be can be used to quantify short-timescale processes, including the scavenging of metals and particles from waters, and the origin and fate of fine grained sediments in terrestrial and marine systems, meteoric ^{10}Be can be used to date geomorphic surfaces and authigenic minerals, test quantitative theories about landscape evolution, and study boundary scavenging processes. However, a considerable amount of fundamental work still needs to be done to effectively use these isotopes of beryllium as tools to study Earth systems processes. The atmospheric fluxes of these nuclides and the processes that control deposition need to be characterized better, and the biogeochemical behavior of these isotopes in a wide range of environments needs to be described. We are optimistic that new studies will shed further light on the dynamics of ^{7}Be and ^{10}Be in the environment, and that these tracers will continue to help scientists solve complex environmental problems in the future.

Acknowledgements We thank Jane Willenbring for a thoughtful and thorough review of the earlier version of this manuscript.

References

Aldahan AA, Ning S, Possnert G et al (1997) Be-10 records from sediments of the Arctic Ocean covering the past 350 ka. Mar Geol 144:147–162

Anderson RF, Lao Y, Broecker WS et al (1990) Boundary scavenging in the Pacific Ocean – a comparison of Be-10 and Pa-231. Earth Planet Sci Lett 96:287–304

Arnold JR, Al-Salih H (1955) Beryllium-7 produced by cosmic rays. Science 121:451–453

Baes CF, Mesmer RE (1976) The hydrolysis of cations. Wiley, New York

Barg E, Lal D, Pavich MJ et al (1997) Beryllium geochemistry in soils: evaluation of Be-10/Be-9 ratios in authigenic minerals as a basis for age models. Chem Geol 140:237–258

Barnes PW, Reimnitz E, Fox D (1982) Ice rafting of fine-grained sediment: a sorting and transport mechanism, Beaufort Sea, Alaska. J Sediment Petrol 52:493–502

Baskaran M (1995) A search for the seasonal variability on the depositional fluxes of Be-7 and Pb-210. J Geophys Res Atmos 100:2833–2840

Baskaran M (2005) Interaction of sea ice sediments and surface sea water in the Arctic Ocean: evidence from excess Pb-210. Geophys Res Lett 32:4

Baskaran M, Santschi PH (1993) The role of particles and colloids in the transport of radionuclides in coastal environments of Texas. Mar Chem 43:95–114

Baskaran M, Shaw GE (2001) Residence time of arctic haze aerosols using the concentrations and activity ratios of Po-210, Pb-210 and Be-7. J Aerosol Sci 32:443–452

Baskaran M, Swarzenski PW (2007) Seasonal variations on the residence times and partitioning of short-lived radionuclides (Th-234, Be-7 and Pb-210) and depositional fluxes of Be-7 and Pb-210 in Tampa Bay, Florida. Mar Chem 104:27–42

Baskaran M, Coleman CH, Santschi PH (1993) Atmospheric depositional fluxes of Be-7 and Pb-210 at Galveston and College Station, Texas. J Geophys Res Atmos 98:20555–20571

Baskaran M, Ravichandran M, Bianchi TS (1997) Cycling of Be-7 and Pb-210 in a high DOC, shallow, turbid estuary of south-east Texas. Estuar Coast Shelf Sci 45:165–176

Baskaran M, Swarzenski PW, Biddanda BA (2009a) Constraints on the utility of MnO_2 cartridge method for the extraction of radionuclides: a case study using Th-234. Geochem Geophys Geosyst 10:9

Baskaran M, Hong GH, Santschi PH (2009b) Radionuclide analysis in seawater. In: Wurl O (ed) Practical guidelines for the analysis of seawater. CRC Press, Boca Raton

Berggren AM, Beer J, Possnert G et al (2009) A 600-year annual Be-10 record from the NGRIP ice core, Greenland. Geophys Res Lett 36:5

Blake WH, Walling DE, He Q (1999) Fallout beryllium-7 as a tracer in soil erosion investigations. Appl Radiat Isot 51:599–605

Blake WH, Walling DE, He Q (2002) Using cosmogenic beryllium-7 as a tracer in sediment budget investigations. Geogr Ann Ser A Phys Geogr 84A:89–102

Bleichrodt JF, Van Abkoude ER (1963) On the deposition of cosmic-ray-produced ^{7}Be. J Geophys Res 68:5283–5288

Bloom N, Crecelius EA (1983) Solubility behavior of atmospheric Be-7 in the marine environment. Mar Chem 12:323–331

Bonniwell EC, Matisoff G, Whiting PJ (1999) Determining the times and distances of particle transit in a mountain stream using fallout radionuclides. Geomorphology 27:75–92

Bourles D, Raisbeck GM, Yiou F (1989) Be-10 and Be-9 in marine-sediments and their potential for dating. Geochim Cosmochim Acta 53:443–452

Brown L, Klein J, Middleton R (1985) Anomalous isotopic concentrations in the sea off Southern-California. Geochim Cosmochim Acta 49:153–157

Brown L, Pavich MJ, Hickman RE et al (1988) Erosion of the Eastern United States observed with ^{10}Be. Earth Surf Process Landforms 13:441–457

Brown L, Stensland GJ, Klein J et al (1989) Atmospheric deposition of Be-7 and Be-10. Geochim Cosmochim Acta 53:135–142

Brown ET, Measures CI, Edmond JM et al (1992a) Continental inputs of beryllium to the oceans. Earth Planet Sci Lett 114:101–111

Brown ET, Edmond JM, Raisbeck GM et al (1992b) Beryllium isotope geochemistry in tropical river basins. Geochim Cosmochim Acta 56:1607–1624

Brown ET, Stallard RF, Larsen MC et al (1995) Denudation rates determined from the accumulation of in-situ produced Be-10 in the Luquillo experimental forest, Puerto-Rico. Earth Planet Sci Lett 129:193–202

Caillet S, Arpagaus P, Monna F et al (2001) Factors controlling Be-7 and Pb-210 atmospheric deposition as revealed by sampling individual rain events in the region of Geneva, Switzerland. J Environ Radioact 53:241–256

Canuel EA, Martens CS, Benninger LK (1990) Seasonal-variations in Be-7 activity in the sediments of Cape Lookout-Bight, North Carolina. Geochim Cosmochim Acta 54:237–245

Chase Z, Anderson RF, Fleisher MQ, Kubik PW (2003) Scavenging of ^{230}Th, ^{231}Pa, and ^{10}Be in the Southern Ocean (SW Pacific sector): the importance of particle flux, particle composition, and advection. Deep Sea Res Part II 50:739–768

Clifton RJ, Watson PG, Davey JT et al (1995) A study of processes affecting the uptake of contaminants by intertidal sediments, using the radioactive tracers Be-7, Cs-137, and unsupported Pb-210. Estuar Coast Shelf Sci 41:459–474

Cooper LW, Olsen CR, Solomon DK et al (1991) Stable isotopes of oxygen and natural and fallout radionuclides used for tracing runoff during snowmelt in an arctic watershed. Water Resour Res 27:2171–2179

Cooper LW, Larsen IL, Grebmeier JM et al (2005) Detection of rapid deposition of sea ice-rafted material to the Arctic Ocean benthos using the cosmogenic tracer Be-7. Deep Sea Res 52:3452–3461

Corbett DR, Dail M, McKee B (2007) High-frequency time-series of the dynamic sedimentation processes on the western shelf of the Mississippi River Delta. Cont Shelf Res 27:1600–1615

Cornu S, Montagne D, Vasconcelos PM (2009) Dating constituent formation in soils to determine rates of soil processes: a review. Geoderma 153:293–303

Crecelius EA (1981) Prediction of marine atmospheric deposition rates using total Be-7 deposition velocities. Atmos Environ 15:579–582

Croke J, Hairsine P, Fogarty P (1999) Sediment transport, redistribution and storage on logged forest hillslopes in south-eastern Australia. Hydrol Process 13:2705–2720

Dalmasso J, Maria H, Ardisson G (1987) Th-228 nuclear-states fed in Ac-228 decay. Phys Rev C 36:2510–2527

daSilva J, Machado A, Ramos MA et al (1996) Quantitative study of Be(II) complexation by soil fulvic acids by molecular fluorescence spectroscopy. Environ Sci Technol 30:3155–3160

Dibb JE, Rice DL (1989a) The geochemistry of beryllium-7 in Chesapeake Bay. Estuar Coast Shelf Sci 28:379–394

Dibb JE, Rice DL (1989b) Temporal and spatial-distribution of beryllium-7 in the sediments of Chesapeak Bay. Estuar Coast Shelf Sci 28:395–406

Doering C, Akber R, Heijnis H (2006) Vertical distributions of ^{210}Pb excess, ^{7}Be and ^{137}Cs in selected grass covered soils in Southeast Queensland, Australia. J Environ Radioact 87:135–147

Dominik J, Burrus D, Vernet J-P (1987) Transport of the environmental radionuclides in an alpine watershed. Earth Planet Sci Lett 84:165–180

Dominik J, Schuler C, Santschi PH (1989) Residence times of 234Th and ^{7}Be in Lake Geneva. Earth Planet Sci Lett 93:345–358

Du JZ, Zhang J, Baskaran M (2011) Applications of short-lived radionuclides (^{7}Be, ^{210}Pb, ^{210}Po, ^{137}Cs and ^{234}Th) to trace the sources, transport pathways and deposition of particles/sediments in rivers estuaries and coasts. In: Baskaran M (ed) Handbook of environmental isotope geochemistry. Springer, Heidelberg

Egli M, Brandova D, Bohlert R et al (2010) Be-10 inventories in Alpine soils and their potential for dating land surfaces. Geomorphology 121:378

Eicken H, Krouse HR, Kadko D et al (2002) Tracer studies of pathways and rates of meltwater transport through Arctic summer sea ice. J Geophys Res Oceans 107:20

Eisenhauer A, Spielhagen RF, Frank M et al (1994) Be-10 records of sediment cores from high northern latitudes – implications for environmental and climatic changes. Earth Planet Sci Lett 124:171–184

Elmore AJ, Kaste JM, Okin GS et al (2008) Groundwater influences on atmospheric dust generation in deserts. J Arid Environ 72:1753–1765

Feng H, Cochran JK, Hirschberg DJ (1999) Th-234 and Be-7 as tracers for the sources of particles to the turbidity maximum of the Hudson River estuary. Estuar Coast Shelf Sci 49:629–645

Fifield LK, Wasson RJ, Pillans B et al (2010) The longevity of hillslope soil in SE and NW Australia. Catena 81:32–42

Fisher GB, Magilligan FJ, Kaste JM et al (2010) Constraining the timescale of sediment sequestration associated with large woody debris using cosmogenic ^{7}Be. J Geophys Res Earth Surf 115:F01013

Fitzgerald SA, Klump JV, Swarzenski PW et al (2001) Beryllium-7 as a tracer of short-term sediment deposition and resuspension in the Fox River, Wisconsin. Environ Sci Technol 35:300–305

Fleisher M, Liu TZ, Broecker WS et al (1999) A clue regarding the origin of rock varnish. Geophys Res Lett 26:103–106

Frank M, van der Loeff MMR, Kubik PW et al (2002) Quasi-conservative behaviour of Be-10 in deep waters of the Weddell Sea and the Atlantic sector of the Antarctic Circumpolar Current. Earth Planet Sci Lett 201:171–186

Frank M, Porcelli D, Andersson P et al (2009) The dissolved beryllium isotope composition of the Arctic Ocean. Geochim Cosmochim Acta 73:6114–6133

Galindo-Uribarri A, Beene JR, Danchev M et al (2007) Pushing the limits of accelerator mass spectrometry. Nucl Instrum Meth Phys Res Sect B Beam Interact Mater Atoms 259:123–130

Graham I, Ditchburn R, Barry B (2003) Atmospheric deposition of Be-7 and Be-10 in New Zealand rain (1996–98). Geochim Cosmochim Acta 67:361–373

Gu ZY, Lal D, Liu TS et al (1996) Five million year Be-10 record in Chinese loess and red-clay: climate and weathering relationships. Earth Planet Sci Lett 144:273–287

Harden JW, Fries TL, Pavich MJ (2002) Cycling of beryllium and carbon through hillslope soils in Iowa. Biogeochemistry 60:317–335

Harvey MJ, Matthews KM (1989) ^{7}Be deposition in a high-rainfall area of New Zealand. J Atmos Chem 8:299–306

Hawley N, Robbins JA, Eadie BJ (1986) The partitioning of beryllium-7 in fresh-water. Geochim Cosmochim Acta 50:1127–1131

Heikkila U, Beer J, Alfimov V (2008) Beryllium-10 and beryllium-7 in precipitation in Dubendorf (440 m) and at Jungfraujoch (3580 m), Switzerland (1998-2005). J Geophys Res Atmos 113:10

Honeyman BD, Santschi PH (1989) A brownian-pumping model for oceanic trace-metal scavenging – evidence from Th-isotopes. J Mar Res 47:951–992

Husain L, Coffey PE, Meyers RE et al (1977) Ozone transport from stratosphere to troposphere. Geophys Res Lett 4:363–365

Ishikawa Y, Murakami H, Sekine T et al (1995) Precipitation scavenging studies of radionuclides in air using cosmogenic Be-7. J Environ Radioact 26:19–36

Jungers MC, Bierman PR, Matmon A et al (2009) Tracing hillslope sediment production and transport with in situ and meteoric Be-10. J Geophys Res Earth Surf 114:16

Jweda J, Baskaran M, van Hees E et al (2008) Short-lived radionuclides (Be-7 and Pb-210) as tracers of particle dynamics in a river system in southeast Michigan. Limnol Oceanogr 53:1934–1944

Kadko D (2000) Modeling the evolution of the Arctic mixed layer during the fall 1997 Surface Heat Budget of the Arctic Ocean (SHEBA) project using measurements of Be-7. J Geophys Res Oceans 105:3369–3378

Kadko D, Olson D (1996) Beryllium-7 as a tracer of surface water subduction and mixed-layer history. Deep Sea Res Part I Oceanogr Res Pap 43:89–116

Kadko D, Swart P (2004) The source of the high heat and freshwater content of the upper ocean at the SHEBA site in the Beaufort Sea in 1997. J Geophys Res Oceans 109:10

Kaste JM (1999) Dynamics of cosmogenic and bedrock-derived beryllium nuclides in forested ecosystems in Maine, USA. MSc Thesis, University of Maine

Kaste JM, Norton SA, Hess CT (2002) Environmental chemistry of beryllium-7. In: Grew E (ed) Beryllium: mineralogy, petrology, and geochemistry. Mineralogical Society of America, Washington

Kaste JM, Heimsath AM, Bostick BC (2007) Short-term soil mixing quantified with fallout radionuclides. Geology 35: 243–246

Kim SH, Hong GH, Baskaran M, Park KM, Chung CS, Kim KH (1998) Wet removal of atmospheric ^{7}Be and ^{210}Pb at the Korean Yellow Sea Coast. The Yellow Sea 4:58–68

Kiss JJ, De Jong E, Martz LW (1988) The distribution of fallout Cs-137 in Southern Saskatchewan, Canada. J Environ Qual 17:445–452

Knies DL, Elmore D, Sharma P et al (1994) Be-7, Be-10, and Cl-36 in precipitation. Nucl Instrum Meth Phys Res 92:340–344

Koch DM, Mann ME (1996) Spatial and temporal variability of Be-7 surface concentrations. Tellus Ser B Chem Phys Meteorol 48:387–396

Krishnaswami S, Benninger LK, Aller RC et al (1980) Atmospherically-derived radionuclides as tracers of sediment mixing and accumulation in near-shore marine and lake-sediments – evidence from Be-7, Pb-210, and Pu-239, Pu-240. Earth Planet Sci Lett 47:307–318

Krishnaswami S, Mangini A, Thomas JH et al (1982) Be-10 and Th isotopes in manganese nodules and adjacent sediments – nodule growth histories and nuclide behavior. Earth Planet Sci Lett 59:217–234

Ku TL, Kusakabe M, Nelson DE et al (1982) Constancy of oceanic deposition of Be-10 as recorded in manganese crusts. Nature 299:240–242

Ku TL, Kusakabe M, Measures CI et al (1990) Beryllium isotope distribution in the Western North-Atlantic – a comparison to the Pacific. Deep Sea Res Part I 37:795–808

Kusakabe M, Ku TL (1984) Incorporation of Be isotopes and other trace-metals into marine ferromanganese deposits. Geochim Cosmochim Acta 48:2187–2193

Kusakabe M, Ku TL, Vogel J et al (1982) Be-10 profiles in seawater. Nature 299:712–714

Kusakabe M, Ku TL, Southon JR et al (1987) Distribution of Be-10 and Be-9 in the Pacific Ocean. Earth Planet Sci Lett 82:231–240

Kusakabe M, Ku TL, Southon JR et al (1990) Beryllium isotopes in the ocean. Geochem J 24:263–272

Kusakabe M, Ku TL, Southon JR et al (1991) Be isotopes in river estuaries and their oceanic budgets. Earth Planet Sci Lett 102:265–276

Lal D (2007) Recycling of cosmogenic nuclides after their removal from the atmosphere; special case of appreciable transport of Be-10 to polar regions by aeolian dust. Earth Planet Sci Lett 264:177–187

Lal D (2011) Using cosmogenic radionuclides for the determination of effective surface exposure age and time-averaged erosion rates. In: Baskaran M (ed) Handbook of environmental isotope geochemistry. Springer, Heidelberg

Lal D, Baskaran M (2011) Applications of cosmogenic-isotopes as atmospheric tracers. In: Baskaran M (ed) Handbook of environmental isotope geochemistry. Springer, Heidelberg

Lal D, Malhotra PK, Peters B (1958) On the production of radioisotopes in the atmosphere by comic radiation and their application to meteorology. J Atmos Terr Phys 12:306–328

Lal D, Nijampurkar VN, Rajagopalan G, Somayajulu BLK (1979) Annual fallout of Si-32, Pb-210, Na-22, S-35 and Be-7 in rains in India. Proc Indian Natl Sci Acad 88:29–40

Lal D, Rama T, Zutshi PK (1960) Radioisotopes P-32, Be-7, and S-35 in the atmosphere. J Geophys Res 65:669–674

Lal D, Barg E, Pavich M (1991) Development of cosmogenic nuclear methods for the study of soil-erosion and formation rates. Curr Sci 61:636–640

Lao Y, Anderson RF, Broecker WS et al (1992a) Transport and burial rates of Be-10 and Pa-231 in the Pacific-Ocean during the Holocene period. Earth Planet Sci Lett 113:173–189

Lao Y, Anderson RF, Broecker WS et al (1992b) Increased production of cosmogenic Be-10 during the last glacial maximum. Nature 357:576–578

Larsen IL, Cutshall NH (1981) Direct determination of ^{7}Be in sediments. Earth Planet Sci Lett 54:379–384

Lee T, Barg E, Lal D (1991) Studies of vertical mixing in the southern California Bight with cosmogenic radionuclides P-32 and Be-7. Limnol Oceanogr 36:1044–1053

Leya I, Lange HJ, Neumann S et al (2000) The production of cosmogenic nuclides in stony meteoroids by galactic cosmic-ray particles. Meteor Planet Sci 35:259–286

Luo SD, Ku TL (2004a) On the importance of opal, carbonate, and lithogenic clays in scavenging and fractionating ^{230}Th, ^{231}Pa and ^{10}Be in the ocean. Earth Planet Sci Lett 220:201–211

Luo SD, Ku TL (2004b) Reply to comment on "On the importance of opal, carbonate, and lithogenic clays in scavenging and fractionating ^{230}Th, ^{231}Pa and ^{10}Be in the ocean". Earth Planet Sci Lett 220:223–229

Magilligan FJ, Nislow KH, Graber BE (2003) Scale-independent assessment of discharge reduction and riparian disconnectivity following flow regulation by dams. Geology 31:569–572

Mangini A, Segl M, Bonani G et al (1984) Mass spectrometric Be-10 dating of deep-sea sediments applying the Zurich tandem accelerator. Nucl Instrum Meth Phys Res Sect B Beam Interact Mater Atoms 5:353–358

Mangini A, Segl M, Kudrass H et al (1986) Diffusion and supply rates of Be-10 and Th-230 radioisotopes in 2 manganese encrustations from the South China Sea. Geochim Cosmochim Acta 50:149–156

Masarik J, Beer J (1999) Simulation of particle fluxes and cosmogenic nuclide production in the Earth's atmosphere. J Geophys Res Atmos 104:12099–12111

Masarik J, Beer J (2009) An updated simulation of particle fluxes and cosmogenic nuclide production in the Earth's atmosphere. J Geophys Res Atmos 114:9

Masque P, Cochran JK, Hirschberg DJ et al (2007) Radionuclides in Arctic sea ice: tracers of sources, fates and ice transit time scales. Deep Sea Res Part I 54:1289–1310

Matisoff G, Whiting PJ (2011) Measuring soil erosion rates using natural (^{7}Be, ^{210}Pb) and anthropogenic (^{137}Cs, ^{239,240}Pu) radionuclides. In: Baskaran M (ed) Handbook of environmental isotope geochemistry. Springer, Heidelberg

Matisoff G, Wilson CG, Whiting PJ (2005) The Be-7/Pb-210 (xs) ratio as an indicator of suspended sediment age or fraction new sediment in suspension. Earth Surf Process Landforms 30:1191–1201

McHargue LR, Damon PE (1991) The global beryllium 10 cycle. Rev Geophys 29:141–158

McKean JA, Dietrich WE, Finkel RC et al (1993) Quantification of soil production and downslope creep rates from cosmogenic Be-10 accumulations on a hillslope profile. Geology 21:343–346

McNeary D, Baskaran M (2003) Depositional characteristics of Be-7 and Pb-210 in southeastern Michigan. J Geophys Res Atmos 108:15

Measures CI, Ku TL, Luo S et al (1996) The distribution of Be-10 and Be-9 in the South Atlantic. Deep Sea Res Part I Oceanogr Res Pap 43:987–1009

Merrill JR, Lyden EFX, Honda M et al (1960) The sedimentary geochemistry of the beryllium isotopes. Geochim Cosmochim Acta 18:108–129

Monaghan MC, Krishnaswami S, Thomas JH (1983) Be-10 concentrations and the long-term fate of particle-reactive nuclides in 5 soil profiles from California. Earth Planet Sci Lett 65:51–60

Monaghan MC, Krishnaswami S, Turekian K (1986) The global-average production rate of ^{10}Be. Earth Planet Sci Lett 76:279–287

Monaghan MC, McKean J, Dietrich W et al (1992) Be-10 chronometry of bedrock-to-soil conversion rates. Earth Planet Sci Lett 111:483–492

Moore WS, Liu TZ, Broecker WS et al (2001) Factors influencing Be-7 accumulation on rock varnish. Geophys Res Lett 28:4475–4478

Morel J, Sepman S, Rasko M et al (2004) Precise determination of photon emission probabilities for main X- and gamma-rays of Ra-226 in equilibrium with daughters. Appl Radiat Isot 60:341–346

Murray AS, Marten R, Johnston A et al (1987) Analysis for naturally-occurring radionuclides at environmental concentrations by gamma spectrometry. J Radioanal Nucl Chem Artic 115:263–288

Nagai H, Tada W, Matsumura H et al (2004) Measurement of Be-7 at MALT. Nucl Instrum Meth Phys Res Sect B Beam Interact Mater Atoms 223:237–241

Nishiizumi K, Finkel RC, Klein J et al (1996) Cosmogenic production of Be-7 and Be-10 in water targets. J Geophys Res Solid Earth 101:22225–22232

Nishiizumi K, Imamura M, Caffee MW et al (2007) Absolute calibration of Be-10 AMS standards. Nucl Instrum Meth Phys Res Sect B Beam Interact Mater Atoms 258:403–413

Nurnberg D, Wollenburg I, Dethleff D et al (1994) Sediments in arctic sea ice-implications for entrainment, transport, and release. Mar Geol 119:185–214

Nyffeler UP, Li YH, Santschi PH (1984) A kinetic approach to describe trace-element distribution between particles and solution in natural aquatic systems. Geochim Cosmochim Acta 48:1513–1522

Olsen CR, Larsen IL, Lowry PD et al (1985) Atmospheric fluxes and marsh soil inventories of ^{7}Be and ^{210}Pb. J Geophys Res 90:10487–10495

Olsen CR, Larsen IL, Lowry PD et al (1986) Geochemistry and deposition of ^{7}Be in river-estuarine and coastal waters. J Geophys Res 91:896–908

Papastefanou C, Ioannidou A (1991) Depositional fluxes and other physical characteristics of atmospheric beryllium-7 in the temperate zones (40-degrees N) with a dry (precipitation-free) climate. Atmos Environ A Gen Topics 25:2335–2343

Pavich MJ, Brown L, Klein J et al (1984) Be-10 accumulation in a soil chronosequence. Earth Planet Sci Lett 68:198–204

Peirson DA (1963) ^{7}Be in air and rain. J Geophys Res 68:3831–3832

Pavich MJ, Brown L, Valettesilver N et al (1985) ^{10}Be analysis of a quaternary weathering profile in the Virginia Piedmont. Geology 13:39–41

Pavich M, Brown L, Harden JW et al (1986) ^{10}Be distribution in soils from Merced river terraces, California. Geology 50:1727–1735

Preiss N, Melieres MA, Pourchet M (1996) A compilation of data on lead 210 concentration in surface air and fluxes at the air-surface and water-sediment interfaces. J Geophys Res Atmos 101:28847–28862

Raisbeck GM, Yiou F (1988) Measurement of Be-7 by accelerator mass-spectrometry. Earth Planet Sci Lett 89:103–108

Raisbeck GM, Yiou F, Fruneau M et al (1978a) Be-10 mass spectrometry with a cyclotron. Science 202:215–217

Raisbeck GM, Yiou F, Fruneau M et al (1978b) Measurement of Be-10 in 1,000 year-old and 5,000 year old antarctic ice. Nature 275:731–733

Raisbeck GM, Yiou F, Fruneau M et al (1980) ^{10}Be concentration and residence time in the deep ocean. Earth Planet Sci Lett 51:275–278

Reimnitz E, Marincovich L, McCormick M et al (1992) Suspension freezing of bottom sediment and biota in the northwest passage and implications for arctic-ocean sedimentation. Can J Earth Sci 29:693–703

Robbins JA, Eadie BJ (1991) Seasonal cycling of trace-elements Cs-137, Be-7, and Pu-239+240 in Lake Michigan. J Geophys Res Oceans 96:17081–17104

Russell IJ, Choquette CE, Fang SL et al (1981) Forest vegetation as a sink for atmospheric particulates – quantitative studies in rain and dry deposition. J Geophys Res 86:5247–5363

Salant NL, Renshaw CE, Magilligan FJ et al (2007) The use of short-lived radionuclides to quantify transitional bed material transport in a regulated river. Earth Surf Process Landforms 32:509–524

Salisbury RT, Cartwright J (2005) Cosmogenic Be-7 deposition in North Wales: Be-7 concentrations in sheep faeces in relation to altitude and precipitation. J Environ Radioact 78:353–361

Sauve S, Hendershot W, Allen HE (2000a) Solid-solution partitioning of metals in contaminated soils: dependence on pH, total metal burden, and organic matter. Environ Sci Technol 34:1125–1131

Sauve S, Martinez CE, McBride M et al (2000b) Adsorption of free lead (Pb2+) by pedogenic oxides, ferrihydrite, and leaf compost. Soil Sci Soc Am J 64:595–599

Schuler C, Wieland E, Santschi PH et al (1991) A multitracer study of radionuclides in Lake Zurich, Switzerland. 1. Comparison of atmospheric and sedimentary fluxes of Be-7, Be-10, Pb-210, Po-210, and Cs-137. J Geophys Res Oceans 96:17051–17065

Schuller P, Iroume A, Walling DE et al (2006) Use of beryllium-7 to document soil redistribution following forest harvest operations. J Environ Qual 35:1756–1763

Segl M, Mangini A, Bonani G et al (1984a) Be-10 dating of the inner structure of Mn-encrustations applying the Zurich tandem accelerator. Nucl Instrum Meth Phys Res Sect B Beam Interact Mater Atoms 5:359–364

Segl M, Mangini A, Bonani G et al (1984b) Be-10 dating of a manganese crust from central North Pacific and implications for ocean paleocirculation. Nature 309:540–543

Segl M, Magini A, Beer J et al (1987) ^{10}Be in the Atlantic Ocean, a transect at 25 deg N. Nucl Instr Meth Phys Res B29:332–334

Sharma P, Somayajulu BLK (1982) Be-10 dating of large manganese nodules from world oceans. Earth Planet Sci Lett 59:235–244

Sharma P, Mahannah R, Moore WS et al (1987) Transport of Be-10 and Be-9 in the ocean. Earth Planet Sci Lett 86:69–76

Sheets RW, Lawrence AE (1999) Temporal dynamics of airborne lead-210 in Missouri (USA): implications for geochronological methods. Environ Geol 38:343–348

Silker WB (1972) Horizontal and vertical distributions of radionuclides in North Pacific ocean. J Geophys Res 77:1061

Silker WB, Robertson DE, Rieck HG et al (1968) Beryllium-7 in ocean water. Science 161:879–880

Simon SL, Bouville A, Beck HL (2004) The geographic distribution of radionuclide deposition across the continental US from atmospheric nuclear testing. J Environ Radioact 74:91–105

Somayajulu BLK (1967) Beryllium-10 in a manganese nodule. Science 156:1219

Sommerfield CK, Nittrouer CA, Alexander CR (1999) Be-7 as a tracer of flood sedimentation on the northern California continental margin. Cont Shelf Res 19:335–361

Steinmann P, Billen T, Loizeau JI et al (1999) Beryllium-7 as a tracer to study mechanisms and rates of metal scavenging from lake surface waters. Geochim Cosmochim Acta 63:1621–1633

Svendsen KM, Renshaw CE, Magilligan FJ et al (2009) Flow and sediment regimes at tributary junctions on a - regulated river: impact on sediment residence time and benthic macroinvertebrate communities. Hydrol Process 23:284–296

Takahashi Y, Minai Y, Ambe S et al (1999) Comparison of adsorption behavior of multiple inorganic ions on kaolinite and silica in the presence of humic acid using the multitracer technique. Geochim Cosmochim Acta 63:815–836

Tanaka S, Inoue T, Huang ZY (1982) Be-10 and Be-10/Be-9 in near Antarctica sediment cores. Geochem J 16:321–325

Todd JF, Wong GTF, Olsen CR et al (1989) Atmospheric depositional characteristics of beryllium-7 and lead-210 along the southeastern Virginia coast. J Geophys Res Atmos 94:11106–11116

Trimble SM, Baskaran M, Porcelli D (2004) Scavenging of thorium isotopes in the Canada basin of the Arctic Ocean. Earth Planet Sci Lett 222:915–932

Tsai H, Maejima Y, Hseu ZY (2008) Meteoric Be-10 dating of highly weathered soils from fluvial terraces in Taiwan. Quat Int 188:185–196

Turekian KK, Cochran JK, Krishnaswami S et al (1979) Measurement of Be-10 in manganese nodules using a tandem vandergraaff accelerator. Geophys Res Lett 6:417–420

Turekian KK, Benninger LK, Dion EP (1983) Be-7 and Pb-210 total deposition fluxes at New Haven, Connecticut and at Bermuda. J Geophys Res 88:5411–5415

Vesely J, Benes P, Sevcik K (1989) Occurrence and speciation of beryllium in acidified freshwaters. Water Res 23:711–717

Vogler S, Jung M, Mangini A (1996) Scavenging of Th-234 and Be-7 in Lake Constance. Limnol Oceanogr 41:1384–1393

von Blanckenburg F, O'Nions RK, Belshaw NS et al (1996) Global distribution of beryllium isotopes in deep ocean water as derived from Fe-Mn crusts. Earth Planet Sci Lett 141:213–226

Vonmoos M, Beer J, Muscheler R (2006) Large variations in Holocene solar activity: constraints from Be-10 in the Greenland Ice Core Project ice core. J Geophys Res Space Phys 111:A10105

Wallbrink PJ, Murray AS (1994) Fallout of Be-7 in South Eastern Australia. J Environ Radioact 25:213–228

Wallbrink PJ, Murray AS (1996) Distribution and variability of Be-7 in soils under different surface cover conditions and its potential for describing soil redistribution processes. Water Resour Res 32:467–476

Wallbrink PJ, Murray AS, Olley JM (1999) Relating suspended sediment to its original soil depth using fallout radionuclides. Soil Sci Soc Am J 63:369–378

Walling DE, He Q, Blake W (1999) Use of Be-7 and Cs-137 measurements to document short- and medium-term rates of water-induced soil erosion on agricultural land. Water Resour Res 35:3865–3874

Walling DE, Schuller P, Zhang Y et al (2009) Extending the timescale for using beryllium 7 measurements to document soil redistribution by erosion. Water Resour Res 45:13

Walton A, Fried RE (1962) The deposition of ^{7}Be and ^{32}P in precipitation at north temperate latitudes. J Geophys Res 67:5335–5340

Whiting PJ, Bonniwell EC, Matisoff G (2001) Depth and areal extent of sheet and rill erosion based on radionuclides in soils and suspended sediment. Geology 29:1131–1134

Whiting PJ, Matisoff G, Fornes W (2005) Suspended sediment sources and transport distances in the Yellowstone River basin. Geol Soc Am Bull 117:515–529

Willenbring JK, von Blanckenburg F (2010a) Meteoric cosmogenic Beryllium-10 adsorbed to river sediment and soil: applications for earth-surface dynamics. Earth Sci Rev 98:105–122

Willenbring JK, von Blanckenburg F (2010b) Long-term stability of global erosion rates and weathering during late-Cenozoic cooling. Nature 465:211–214

Wilson CG, Matisoff G, Whiting PJ (2003) Short-term erosion rates from a Be-7 inventory balance. Earth Surf Process Landforms 28:967–977

You CF, Lee T, Li YH (1989) The partitioning of Be between soil and water. Chem Geol 77:105–118

Zhu J, Olsen CR (2009) Beryllium-7 atmospheric deposition and sediment inventories in the Neponset River estuary, Massachusetts, USA. J Environ Radioact 100:192–197

Chapter 6
Silicon Isotopes as Tracers of Terrestrial Processes

B. Reynolds

Abstract Chemical weathering of silicate rocks consumes carbon dioxide during the breakdown of minerals and liberates ions and silicic acid. Weathering reactions can lead to the fractionation of stable silicon (Si) isotopes, which allows variations in Si isotopic ratios to be used to quantify the global Si cycle. Here, I detail the methods and applications of assessing variations in continental Si stable isotope compositions, and show how such studies can elucidate the processes that affect the terrestrial Si cycle. The global Si cycle and its stable isotope variations are strongly affected by biology, with Si uptake by phytoplankton in oceans and lakes, and biological uptake by terrestrial plants, especially modern crops, on land. This biological uptake and subsequent release often leads to a strong seasonal cycle overprinted onto the underlying weathering reactions. Nevertheless, it can be shown that weathering reactions lead to the fractionation of Si into a light Si reservoir stored in soils and sedimentary basins, and a heavy Si reservoir in rivers, lakes and the oceans.

6.1 Introduction

Silicon (Si) is the most abundant element at the Earth's surface after oxygen, with the continental crust containing 25.7 wt% Si. Chemical weathering of silicate rocks leads to the consumption of CO_2. Hence the Si biogeochemical cycle is linked to the global carbon cycle, and chemical weathering can influence the global climate over geological timescales via the greenhouse effect and the carbon cycle.

Global weathering rates or intensities may potentially be quantified using isotope systems, both classical radiogenic isotopes, like Sr, and novel stable isotope variations, like Li (Vigier et al. 2009; Burton and Vigier 2011). Indeed, the enrichment of radiogenic Sr over the Cenozoic, as recorded in marine sediments, has been considered to reflect an increase in continental weathering rates, and a draw-down of CO_2, leading to global cooling via a weakened greenhouse effect (Raymo and Ruddiman 1992). Stable isotope compositions of divalent cations, such as Ca, Mg and Sr, represent relatively new tools for investigating global biogeochemical cycles and variations in weathering intensities over geological timescales (Farkas et al. 2007; Tipper et al. 2008; de Souza et al. 2010). Although the draw-down of CO_2 is primarily due to the weathering of divalent cations (Mg^{2+} and Ca^{2+}) and their subsequent precipitation as carbonates, the weathering reactions also lead to the release of Si. Hence, similarly to the studies of Mg and Ca isotopes, stable Si isotopes may also potentially be used to quantify chemical weathering reactions or processes, as shown here.

The term "weathering" can imply a number of processes. Weathering involves the physical and chemical breakdown of rocks and minerals through numerous processes, but here we focus on the chemical weathering of primary minerals. Two basic approaches for investigating chemical weathering are discussed here: (i) studying the material removed by weathering via quantification of dissolved ions drained from a catchment, and (ii) studying the remnants of weathering retained during soil formation. The chemical reactions of weathering can be simplified and written as various reactions:

B. Reynolds (✉)
Institute for Geochemistry and Petrology, ETH Zurich, Switzerland
e-mail: reynolds@erdw.ethz.ch

Dissolution reaction:

$$CaAl_2Si_2O_8 + 2CO_2 + 8H_2O \rightarrow$$
$$Ca^{2+} + 2Al(OH)_3 + 2Si(OH)_4 + 2HCO_3^-$$

Hydration reaction:

$$CaAl_2Si_2O_8 + CO_2 + 7H_2O \rightarrow$$
$$Ca(CO_3) + 2Al(OH)_3 + 2Si(OH)_4$$

Or silicate to carbonate exchange reactions:

$$CaSiO_3 + CO_2 \rightarrow Ca(CO_3) + SiO_2$$

These reactions neglect the fact that weathering of primary minerals is often accompanied by the formation of secondary minerals such as oxides and clay minerals, or the leaching of cations from interlayer crystallographic sites. Furthermore, the presence of organic ligands or microbial action can strongly affect the weathering reactions in terms of both the rate and extent of reaction. Thus, in natural systems "chemical weathering" may involve numerous reactions, which overall release dissolved ions and Si.

Dissolved Si released from weathering reactions forms silicic acid, which can be written as $Si(OH)_4$ or H_4SiO_4. As silicic acid is fully hydrated, unlike carbonic acid, H_2CO_3, it does not form a meta-stable ion, and remains predominantly an uncharged species at most naturally occurring pHs, shown in Fig. 6.1. However, silicic acid can adsorb to surfaces, due to the presence of charged species (e.g. Hiemstra et al. 2007). Silicic acid also polymerizes, especially at higher concentrations and higher pH, converting mono-silicic acid to polymeric silica, or even larger colloidal silica particles (see Iler 1979). Polymeric silica can remain truly dissolved, passing through colloidal-size filtration. Although polymeric silica does not react with molybdate ions to form coloured complexes (Boltz et al. 1978; Strickland 1952), it can be detected by the difference of sample concentrations between samples left untreated and samples reacted to enhanced dissolution using NaOH or hydrofluoric acid (Iler 1979). The solubility of mono-silicic acid is strongly dependent on the concentration of other ions, with additional salt increasing the solubility of Si (Icopini et al. 2005). The solubility of mono-silicic acid in "pure" water is close to that of quartz (~3 ppm Si (Morey et al. 1962)), but Si concentrations in natural waters are often in much higher, with a solubility close to that of amorphous silica (~100 ppm Si, see Iler (1979)).

It is important to appreciate how observations of stable Si isotope variations can be used to evaluate the terrestrial Si geochemical cycle, and hence the role chemical weathering plays on the release of silicic acid. The terrestrial Si cycle is also strongly affected by biology, with Si uptake and precipitation of biogenic opal (phytoliths) by plants, which can equally lead to stable isotope fractionation. Hence in order to understand the role of chemical weathering, we must also address how biological cycling can influence the terrestrial cycle of Si and its isotopes.

6.1.1 Nomenclature

Variations in the stable isotope compositions are usually given as relative variations in isotope ratios, scaled and expressed with a δ-notation. In terms of quoting stable isotope variations, this standard definition leads to several potential problems in comparing different data. Firstly, the variation given in permil depends upon the isotope ratio used. Secondly, the standard δ definition is actually non-linear (Miller 2002) and a slightly different definition, the linearized form δ' (Hulston and Thode 1965), is mathematically more exact to describe mass-dependent fractionation. This linearization becomes critical when considering correlations between three isotopes with a large range in fractionation (Miller 2002). Thirdly, the scale is only relative and an arbitrary zero point must be defined and the scale calibrated.

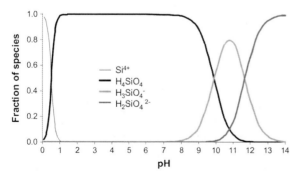

Fig. 6.1 The relative composition of each silicic acid species with variable pH

Silicon has three stable isotopes: ^{28}Si (92.22%), ^{29}Si (4.69%) and ^{30}Si (3.09%). Variations in stable Si isotopes are given relative to the zero reference standard, NBS28 (SRM 8546) (Coplen et al. 2002), and expressed using the standard δ- (and linearized δ'-) notation, in units of permil (‰):

$$\delta^x\text{Si} = 10^3 \left[\frac{(^x\text{Si}/^{28}\text{Si})_{\text{sample}}}{(^x\text{Si}/^{28}\text{Si})_{\text{NBS28}}} - 1 \right] \cong \delta'^x\text{Si}$$

$$\delta'^x\text{Si} = 10^3 \ln \left[\frac{(^x\text{Si}/^{28}\text{Si})_{\text{sample}}}{(^x\text{Si}/^{28}\text{Si})_{\text{NBS28}}} \right]$$

where x is either of the heavier isotopes, 29 or 30. In order to ensure a scale calibration, secondary reference materials of Diatomite and Big Batch were used to inter-compare laboratories and measurement techniques (Reynolds et al. 2007). All terrestrial Si isotope fractionations are mass-dependent, with δ^{30}Si values approximately twice those of δ^{29}Si, hence datasets of δ^{29}Si and δ^{30}Si can be compared assuming δ^{29}Si = 0.51 δ^{30}Si (Reynolds et al. 2007), as shown in Fig. 6.2. Higher δ^{30}Si values reflect an increase in the ^{30}Si/^{28}Si ratio, or enrichment in heavier isotopes making Si heavier; a δ^{30}Si value 1‰ higher is equivalent to an increase in the atomic weight of Si by 2.8 ppm (79 μg mol^{-1}). Current precision does not allow for the determination of differences in the mass-dependent fractionation laws between kinetic and equilibrium fractionation (see Young et al. 2002) to be adequately resolved for the small fractionations (<2‰ δ^{30}Si) and the <0.02‰ difference between the standard and linearized forms of the δ-notation is hence immaterial.

In order to define the degree of isotope fractionation of a process, the isotope fractionation factor (α, or ε in permil) between two substances, A and B, is defined as:

$$\alpha_{A-B} = \frac{(^{30}\text{Si}/^{28}\text{Si})_A}{(^{30}\text{Si}/^{28}\text{Si})_B} = \frac{1 + 1000(\delta^{30}\text{Si})_A}{1 + 1000(\delta^{30}\text{Si})_B}$$

$$= 1 + \frac{^{30}\varepsilon_{A-B}}{10^3}$$

It should be noted that the fractionation factor written as α depends upon the ratio pair used, just like the δ notation, although the distinction is not commonly stated for α. If using a linearized form of the δ-notation (δ'), the fractionation factor in permil can also be written in a number of other ways:

$$10^3 \ln(\alpha_{A-B}) = 10^3 \ln\left[(^{30}\text{Si}/^{28}\text{Si})_A / (^{30}\text{Si}/^{28}\text{Si})_B \right]$$
$$\simeq {}^{30}\varepsilon_{A-B}$$
$$= 10^3 \ln(^{30}\text{Si}/^{28}\text{Si})_A - 10^3 \ln(^{30}\text{Si}/^{28}\text{Si})_B$$
$$= \delta'^{30}\text{Si}_A - \delta'^{30}\text{Si}_B$$
$$\simeq \delta^{30}\text{Si}_A - \delta^{30}\text{Si}_B = \Delta^{30}\text{Si}_{A-B}$$

So the fractionation factor, expressed in permil, is almost equivalent to the difference between the two δ-values, typically written as Δ_{A-B}. Unfortunately, expressing fractionation as either ε or Δ can lead to confusion as both are also used to in isotope geochemistry to denote isotopic anomalies: either radiogenic anomalies such as ε^{143}Nd, or variations from an equilibrium mass-dependent fractionation line such as Δ^{17}O. It is high time that a new standardization is employed such that these confusions can be avoided, and that values given for an element are not dependent upon the isotope ratio used. For example, fractionation could be given as variations in the atomic weight or as variations in ppm per amu, such that data can easily be compared across the mass-spectrum:

$$\text{ppm per amu'} = \varphi = \frac{10^6}{m_1 - m_2} \ln\left[\frac{R_A}{R_B}\right]$$

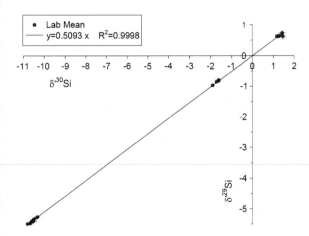

Fig. 6.2 Three isotope plot of δ'^{29}Si versus δ'^{30}Si (error bars are 1 σ_{SD}) for the standard reference materials measured by several international laboratories, adapted from Reynolds et al. (2007). All data fall along a mass-dependent fractionation array with a slope of 0.5093 ± 0.0016, in agreement with kinetic isotope fractionation of Si of 0.5092

where m_1-m_2 denotes the mass difference in amu between the two masses used in the isotope ratio, R (Rudge et al. 2009).

6.2 Materials and Methods

The natural stable isotope variations of Si have been measured for over 50 years, but terrestrial Si isotope variations have been inadequately investigated mainly due to analytical difficulties and poor reproducibility. Conventional analyses involved measurement of gaseous fluorinated silicon (SiF_3^+) in a Nier-type mass spectrometer, using either direct fluorination of purified SiO_2 (Allenby 1954) or by the thermal decomposition of Ba_2SiF_2 (Reynolds and Verhoogen 1953). Careful early studies of silicon in meteorites and lunar material could not resolve any Si isotopic anomalies due to non-mass-dependent processes, with all of the measured Si isotope compositions falling close to a single mass-dependent fractionation line (e.g. Epstein and Taylor 1970). The measured variations in the stable isotope compositions of Si in igneous rocks were small, 1.1‰ (Douthitt 1982). Whilst the $^{30}Si/^{28}Si$ ratios of biogenic opal have a larger range of 6‰ compared to the range for silicate rocks of 3.5‰, assessing the fractionation related to aqueous or biological processes was limited by the inability to measure dissolved Si isotope compositions. Hence, in the following decades little work was done to further investigate Si isotope variations (see Ding 2004).

Recently, there has been a renewed interest in the measurement of silicon isotope variations in natural samples because of the importance of Si in terrestrial and marine biogeochemical cycles (Basile-Doelsch 2006; Brzezinski et al. 2002), and its importance as the most abundant non-volatile element in the solar system (Fitoussi et al. 2009; Georg et al. 2007a). The application of Si isotope variations to the investigation of geochemical cycles has been facilitated by two separate developments. Firstly, the development of new chemical preparation techniques to measure the Si isotope composition of a wide range of samples: dissolved and biogenic Si from marine and freshwater environments (Alleman et al. 2005; André et al. 2006; Cardinal et al. 2005; De La Rocha et al. 2000; Engström et al. 2006; Georg et al. 2006a, b; Reynolds et al. 2006b; Varela et al. 2004), plants and organic-rich soils (Ding et al. 2005; Opfergelt et al. 2006a, b); as well as altered rocks and mineral soils (André et al. 2006; Basile-Doelsch et al. 2005; Georg et al. 2006b; van den Boorn et al. 2006; Ziegler et al. 2005a, b). Secondly, the advancement of Multi-Collector Inductively-Coupled-Plasma Mass-Spectrometry (MC-ICPMS) (Cardinal et al. 2003; De La Rocha 2002; Reynolds et al. 2006c), and recently improved gas source isotope ratio mass spectrometer (IRMS) techniques (Brzezinski et al. 2006), now provide the precision required to better quantify biogeochemical Si isotope fractionations. It has been shown by inter-laboratory comparison that various analytical techniques do not bias the measured stable isotope variations (Reynolds et al. 2007), and there are now several reference materials with well-known $\delta^{30}Si$ values with which to compare datasets, as shown in Table 6.1.

The MC-ICPMS offers two distinct advantages compared to analysis using the IRMS. Firstly, analytical protocols are faster (both chemical processing and analysis time), including repeated analyses; and secondly, much smaller sample sizes are required. Some types of sample preparation for IRMS are complicated, time consuming, and require the use of extremely hazardous chemicals such as BrF_5 or F_2 (De La Rocha et al. 1996; Ding 2004). The MC-ICPMS also has some significant disadvantages that must be addressed. The plasma source induces a much larger mass-dependent fractionation than dual-inlet IRMS, and this fractionation must be corrected for by standard-sample bracketing which can be highly sensitive to "matrix effects" that can induce systematic offsets between samples and standards. Furthermore, the plasma source can produce significant polyatomic isobaric interferences. However, these drawbacks can be addressed with the correct analytical protocols, as discussed below.

Table 6.1 The Si isotope composition of silica reference materials determined by several laboratories with errors indicating 95% confidence limits (Hendry et al. 2010; Reynolds et al. 2007; Savage et al. 2010)

Reference material	$\delta^{30}Si$ value (in ‰)
NBS 28	0.0
Diatomite	+1.25 ± 0.02
Big batch	−10.48 ± 0.04
Terrestrial basalts (e.g. BHVO-1 or -2)	−0.29 ± 0.03
Sponge (LMG08)	−3.35 ± 0.06

6.2.1 Multi-Collector ICP-MS Methods

On MC-ICPMS instruments, all three Si ion beams are measured simultaneously in static mode with Faraday collectors, typically equipped with 10^{11} Ω resistors. In order to ionize Si in an ICP source, the Si must be introduced into the plasma as a dilute aerosol, which is typically done by volatilization of a silicic acid solution. The analyte (sample solution) is turned into small droplets in a nebulizer. This wet aerosol is often then dried by a desolvating unit before being introduced into the plasma in an Ar carrier flow. The transfer of ions from the plasma operating at atmospheric pressure into the high vacuum analyzer of the MS results in considerable loss of ions (>99%) and significant mass-fractionation, which results in a measurement bias towards the heavier isotopes (termed mass-bias). This mass-bias can be corrected for by external doping of another element with a known isotope composition (known as external doping), and/or *relative* Si isotope variations can be measured using a standard-sample-standard bracketing technique that requires identical conditions between samples and standards, and minimal drift in the instrumental mass-bias with time. External doping with Mg isotopes has been demonstrated to be valid for external mass-bias correction for Si isotopes (Cardinal et al. 2003). Failure to ensure identical conditions of samples and standards for the bracketing protocol can easily lead to variations in the mass-bias and thus erroneous determination of relative isotopic compositions. Such mass-bias effects are often referred to as "matrix-effects", although "plasma loading effects" may be a more accurate description since these are usually caused by the presence of additional ions in the sample solutions. Although such effects may be quantified, they may also be dependent upon specific tuning conditions and cannot be easily corrected for. External doping can help determine significant "matrix effects", but it also may also induce effects itself, and may not adequately correct for variable mass-bias at the *sub*-permil level. Thus, in order to directly compare standard and samples without inducing "matrix effects" during ICP-MS analysis, challenging requirements exist for sample purity.

Initially, measurements of Si isotopes using MC-ICPMS used a "wet-plasma" sample introduction technique, with 0.25 M HF (De La Rocha 2002). The analytical precision was poor, because of very low sensitivity, poor temporal stability of instrumental mass-bias and severe memory effects. A refined analytical technique was developed by Cardinal et al. (2003), with a "dry-plasma" sample introduction using a desolvating nebulization system and Mg external doping for mass bias correction. This technique introduced very dilute acids, with HF below 0.01 M, in order to increase sensitivity and stabilize instrumental mass-bias over time. However, the presence of HF can lead to the loss of Si as volatile SiF_4 within the desolvating unit, and it degrades analytical reproducibility (Georg et al. 2006b). Thus, if a desolvating nebulization system is used to increase sensitivity and minimize polyatomic interferences, sample and standard preparation techniques must avoid significantly large amounts of HF within the final analyte solutions.

Under normal operating conditions, in addition to the efficient ionization of Si, interfering polyatomic ions may be produced either within the plasma or in the interface region of the MC-ICPMS where the pressure drops by twelve orders of magnitude. Assuming a typical $^{28}Si^+$ intensity of 60 pA on an instrument equipped with a desolvating nebulizer (Cardinal et al. 2003; Georg et al. 2006b), the polyatomic species $^{14}N_2^+$, $^{14}N_2H^+$ and $^{14}N^{16}O^+$ would typically result in signals corresponding to at least 1, 0.7 and 5% of the $^{28}Si^+$, $^{29}Si^+$ and $^{30}Si^+$ ion beams, respectively (see Fig. 6.3). So, even though the $^{14}N_2^+$ is the largest ion beam, it is the $^{14}N^{16}O^+$ ion beam that results in the largest interference for determination of the Si isotope compositions. For instruments using conventional "wet-plasma" sample introduction systems (Engström et al. 2006), the magnitude of these spectral interferences on the silicon isotopes would be even more pronounced. Although some of these interferences can be corrected for (Reynolds et al. 2006c), the polyatomic interference on mass 30 prevented the accurate measurement of $^{30}Si/^{28}Si$ on the standard MC-ICPMS instruments initially used. However, as polyatomic ions are slightly heavier than atomic ions of the same nominal mass (below 60 amu), the slight difference in mass can be exploited using a higher mass-resolution to separate the Si^+ ion beams from potential polyatomic interferences, as shown in Fig. 6.3. This requires the ability to resolve ion beams that differ by less than 1‰ in mass (usually expressed as requiring a mass-resolution ($m/\Delta m$) in excess of 1,000). Since all polyatomic interferences are heavier than the isobaric Si^+ ion, it is not necessary to "fully resolve" all the ion beams.

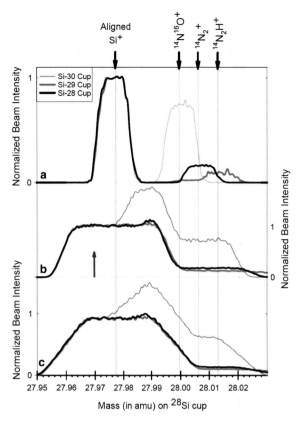

Fig. 6.3 Peak scans for silicon isotopes ^{28}Si (*black line*), ^{29}Si (*thick grey line*) and ^{30}Si (*thin grey line*) using a multi-collector ICP-MS in (**a**) high mass-resolution (**b**) pseudo high mass-resolution, and (**c**) low mass-resolution. Position of Si$^+$ beams and polyatomic interferences are shown relative to m/z on the Si-28 cup. Mass-resolutions shown (full peak width, $m/\Delta m$) are approximately 2,000, 600 and 500, with resolving powers (slope width, $m/\Delta m$) of 6,000, 4,000 and 1,800 for (**a**, **b** and **c**) respectively. In pseudo-high resolution mode, the measurements were carried out on the left side of the peak plateau (indicated by the *arrow*) to avoid the interferences. Interference from SiH$^+$ ions cannot be observed on peak-scans

Instead instruments can be run using a "pseudo" high-resolution mode (Engström et al. 2006), where all the Si ion beams are positioned towards the high mass-side of the collector, such that heavier polyatomic interferences do not enter the Faraday cup collector, as shown schematically in Fig. 6.3.

6.2.2 IRMS Methods

Traditional gas-source mass-spectrometry measures Si ions as SiF$_3^+$ ions, generated from the electron bombardment of gaseous SiF$_4$. The details of this IRMS technique, which has been used for over 50 years, can be found in numerous books and articles (e.g. Taylor 2004), and will not be fully detailed here. Whilst many preparation methods have used hazardous techniques for the fluorination of silica samples (Allenby 1954; De La Rocha et al. 1996; Ding 2004), relatively pure SiF$_4$ can be produced from the thermal decomposition of BaSiF$_6$ (Debievre et al. 1995; Reynolds and Verhoogen 1953) or acidic decomposition of Cs$_2$SiF$_6$ (Brzezinski et al. 2006). Purity is important as volatile fluorides and oxyfluorides are hard to separate from SiF$_4$ (Ding 2004), such as COF$_3^+$ leading to isobaric interferences on masses 85, 86 and 87. As F has only one isotope, no further corrections are required. The use of a dual-inlet system optimizes the standard-sample-standard bracketing technique, with fast switching between samples and standards through a switch-over valve. Typical analytical precisions obtained range from 0.06 to 0.2‰ ($2\sigma_{SD}$) (Ding 2004), and are thus directly comparable to MC-ICPMS.

6.3 Applications

The geochemical cycle of Si is explicitly linked to chemical weathering, but weathering studies often focus on dissolved cations. For the terrestrial and biological cycling of elements, there is extensive literature on Si, but this chapter focuses on Si isotopes, rather than Si per se, and will not review the wealth of information about Si within various sub-disciplines that has been reviewed elsewhere (e.g. Sommer et al. 2006; Struyf et al. 2009). Here is shown how the use of Si isotope can be useful in quantifying the terrestrial Si biogeochemical cycle.

It has been known for a long time that the processes controlling the Si cycle have a measurable impact on the Si isotope composition at the Earth's surface (as reviewed by Douthitt (1982) and shown in Fig. 6.4). Although samples precipitated from solution were shown to be highly fractionated (Douthitt 1982), it was not until the truly dissolved δ^{30}Si was measured from natural waters (De La Rocha et al. 2000), that the importance of aqueous chemistry on δ^{30}Si values was finally realized. This pioneering work showed that "marine and riverine δ^{30}Si are more positive than

δ^{30}Si of igneous rocks, suggesting isotopic fractionation during weathering and clay formation and/or biomineralization" (De La Rocha et al. 2000). Although some work has been done to quantify the global Si isotopic mass-balance (Basile-Doelsch et al. 2005), there has been almost no discussion of the clear difference in δ^{30}Si between mafic (−0.3‰ (Savage et al. 2010)) and granitic rocks (0.0‰), shown in Fig. 6.4. This difference would imply that either there is a fractionation during partial melting and crystallization, or that the progressive recycling of silicate material over Earth's history has lead to the enrichment of the upper continental crust in heavier Si isotopes. That the continental crust has a higher δ^{30}Si value is counterintuitive given that Si in recycled material, such as secondary clay and biogenic opal, is isotopically lighter, as detailed below. Such a scenario would suggest a preferential loss of light Si back into the mantle during crustal recycling, although the nature of this process remains to be explored.

It is important to consider that Si is a nutrient for plants and that the interactions between chemical and biological processes largely occur within soils. In soils Si mainly exists within minerals, including primary minerals, secondary minerals developed through soil formation, and as biogenically precipitated opal (either phytogenic, microbial or protozoic Si precipitates, as shown in Fig. 6.5). In general, knowledge about size, properties, and transformation of these biogenic pools is very scarce for almost all soils (Sommer et al. 2006). However, it has been long recognized that the recycling of Si by plants within soils can be much larger than fluxes associated with chemical weathering (Alexandre et al. 1997). Thus the formation of phytoliths within plants and subsequent dissolution during litter degradation can be a major source of dissolved Si (Farmer et al. 2005). We thus need to assess isotope fractionation in plants to

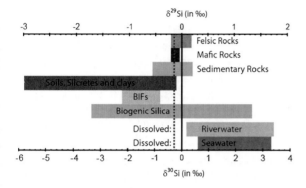

Fig. 6.4 Reported Si isotopic composition of terrestrial samples, expressed as δ^{29}Si and δ^{30}Si. Dissolved Si reservoirs are enriched in the heavier Si isotopes, and secondary precipitated reservoir such as clays and biogenic Si contain light Si (sources: André et al. 2006; Ding et al. 1996; Douthitt 1982; Reynolds et al. 2007; Savage et al. 2010). The *vertical dashed line* represents the estimate of the Bulk Silicate Earth (Savage et al. 2010)

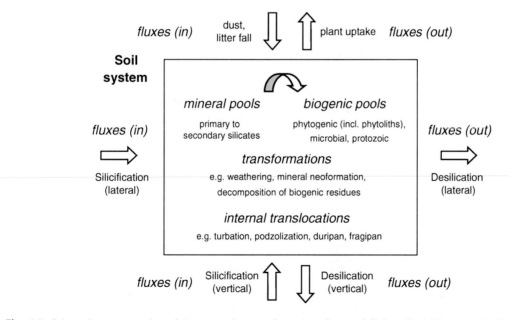

Fig. 6.5 Schematic representation of the reservoirs, transformation, fluxes of Si in soils (without erosion/deposition), from Sommer et al. (2006). Copyright Wiley-VCH Verlag GmbH & Co. KGaA. Reproduced with permission

quantify their impact on the Si isotope compositions of soils, before looking at larger scale catchment-scale studies aimed at quantifying the global Si biogeochemical cycle.

6.3.1 Si Uptake by Plants

Silicon is often one of the most abundant mineral elements in vegetation (besides Ca, Mg, K and P), but is not generally considered as an essential element for higher plants (Marschner 1995) as terrestrial plants rarely experience Si deficiency. The uptake of Si by plants is often passive, but some plants are able to actively enhance or via Si rejection lower Si uptake compared to that of water (Cornelis et al. 2010b; Raven 1983). It has long been recognized that plant uptake and formation of biogenic silica leads to Si isotope fractionation (Douthitt 1982), and these initial studies open interesting applications of isotope geochemistry tools to agricultural or plant sciences. It has been shown that Si is an important nutrient for vascular plants, with beneficial effects for crop growth (Epstein 2003; Mitani and Ma 2005), including crops, such as rice, cereals, and bananas. Silicon helps create the rigidity of cell walls and can increase resistance to pathogens, water stress, and grazing (Ma 2004; Sangster and Hodson 2007).

Within the plant the transpirational loss of water results in the concentration of silicic acid to the saturation value and the subsequent irreversible precipitation of silica, mainly in cell walls (Raven 1983, 2001).

Rice requires high levels of Si for healthy growth, and Si is actively transported against a concentration gradient into the cortical cells and then into the xylem (Ma et al. 2004). Experiments with rice plants grown hydroponically determined the Si uptake and fractionation during growth (Ding et al. 2008a; Sun et al. 2008). Dissolved Si not taken up by the plants evolved to high δ^{30}Si values (>1.4‰) as the plants preferentially incorporated lighter Si isotopes during active Si uptake (Ding et al. 2008a; Sun et al. 2008). However, this evolution to higher δ^{30}Si values with decreasing Si concentrations in the nutrient solutions does not follow a simple Rayleigh-type fractionation curve expected from irreversible uptake with a constant fractionation factor. Instead the data fits a model with a continuous variation in fractionation between the rice plant and nutrient supply between −1.7 and −0.5‰ δ^{30}Si (Ding et al. 2008a). Alternatively, the data could fit an equilibrium fractionation reaction, as shown in Fig. 6.6, with a fractionation of −1.6‰ δ^{30}Si (or −2‰ if fractionation only occurs after about one third of the available Si is taken up, when there is sufficient evapotranspiration through the leaves). However, the equilibrium modeling of the fractionation

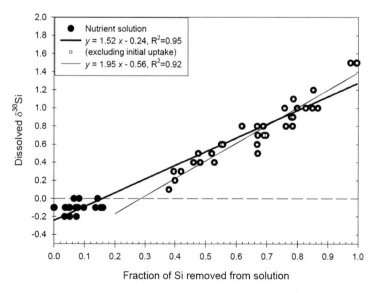

Fig. 6.6 The δ^{30}Si variations for silicic acid in the nutrient solution as a function of the fraction of silicon removed from the solution by the uptake of rice plants grown under hydroponic conditions (Ding et al. 2008a). The apparent linear fit to the data shown here could imply a reversible exchange of Si between plant and solution within uniform fractionation factor. However, such reversible exchange may not be considered appropriate given the active Si uptake by rice plants

presented here would require Si transport from the xylem to the soil solution as well as the active uptake of Si, which is counter to views held by many plant physiologists.

An investigation of the Si isotope composition in Chinese rice plants showed that the Si concentrations and $\delta^{30}Si$ values increased within the plants from the soil and roots through the stems and leaves to the husks (Ding et al. 2005), reaching over 10 wt% silica and an increase in $\delta^{30}Si$ of 0.9‰. Within the rice grains, silica content is low, typically between 0.02 and 0.05%, but the measured $\delta^{30}Si$ values are exceptionally high, ranging from +1.3 to +6.1‰ (Ding et al. 2005). The increasing $\delta^{30}Si$ values are shown to fit a simple model of Rayleigh-type fractionation within the plants as Si is transported from the roots to the grains (Ding et al. 2005). The distribution of $\delta^{30}Si$ values within the plant reflects both a temporal increase in the $\delta^{30}Si$ values of the bulk plant, but also the larger intra-plant fractionation, with $\delta^{30}Si$ variations of 3.5‰ between low values in roots and high values of rice grains (Ding et al. 2008a).

In order to determine the fractionation factor between bulk banana plantlets and source solutions, plants were grown in hydroponics (Opfergelt et al. 2006b), as the exact nature of fractionation during Si uptake by banana plants cannot be understood from field studies. The fractionation factor $^{30}\varepsilon_{plantlet-solution}$ ranged from -0.6 to -1.1‰, with larger differences observed for samples with higher Si uptake (Opfergelt et al. 2006b). It was further shown that there are large variations within each plant in both the Si concentration and $\delta^{30}Si$ value, as observed for rice, with values increasing from the roots, through the midrib to the lamina of the leaves, for both in situ and in vitro plants (Opfergelt et al. 2006a, b). The larger Si isotope fractionation between the external dissolved Si pool and that of the xylem indicates that an active transport mechanism exists that preferentially transport the lighter isotopes, as above.

Despite the large $\delta^{30}Si$ variations within plants, similar plant organs collected from different plant individuals at similar times in the growth cycle should have experienced equivalent Si mass fractionation, thus any differences between these plants may potentially reflect differences in the source $\delta^{30}Si$ composition in the soils (Opfergelt et al. 2008). Variation of $\delta^{30}Si$ values observed in mature banana plants should reflect variations in the $\delta^{30}Si$ values of the nutrient Si, which may be related to the degree of soil weathering (Opfergelt et al. 2008, 2009). Variations of $\delta^{30}Si$ values between bamboo plants in China have also been inferred to reflect variations in the soil, with higher plant $\delta^{30}Si$ values associated with more organic rich soils (Ding et al. 2008b).

6.3.2 Si in Soils

As reviewed by Sommer et al. (2006), the Si cycle within soil systems is complex, with transfer of Si via the dissolved phase between primary minerals, secondary precipitates and plants, as shown in Fig. 6.5. Dissolved Si in the soil is lost via four main pathways: (1) as discussed above, dissolved Si that is taken up by plants is precipitated and return to the soil predominantly as phytoliths, (2) dissolved Si is precipitated as secondary mineral or amorphous complexes, (3) dissolved Si is absorbed to surfaces, like Fe oxyhydroxides, and (4) dissolved Si is washed out of the soils into the river systems. Only the flux of dissolved Si out of the soil is not associated with a Si isotope fractionation, with the other pathways leading to Si isotope fractionation during the development of soils. In addition, Si isotope fractionation within the soils may be induced during dissolution (Demarest et al. 2009; Ziegler et al. 2005b) and by adsorption/desorption reactions (Opfergelt et al. 2009). Initial studies of Si isotope fractionation during soil development (Ziegler et al. 2005a, b), investigated granitic and basaltic soils in tropical climates, where weathering rates are high (White et al. 1998). For the Hawaiian Islands variable lava ages allowed for the construction of a chronosequence of basaltic soil development, and assessment of the processes affecting the Si isotope composition during basaltic weathering (Ziegler et al. 2005b).

On Hawaii, primary minerals are replaced by allophane, a metastable clay which slowly converts to kaolinite. Primary volcanic minerals are almost completely weathered within the first 20,000 years, whilst conversion of allophane to kaolinite requires millions of years, and is thus only seen in the oldest soil profiles (Bern et al. 2010; Ziegler et al. 2005b). During progressive mineral weathering the Si isotope composition of the soil (<2 mm size fraction) changes from the initial -0.5‰ $\delta^{30}Si$ of the unweathered basalt to lower $\delta^{30}Si$ values, down to -2.5‰ $\delta^{30}Si$ in the

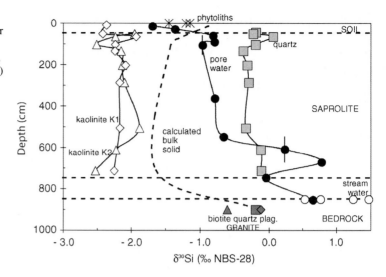

Fig. 6.7 Depth plot of $\delta^{30}Si$ values of different Si phases (*vertical bars* on pore-water data indicates depth interval sampled; *horizontal dotted lines* separate soil, saprolite, and bedrock zones), from Ziegler et al. (2005a)

oldest samples with >20 wt% kaolinite. It has been shown that this progressive enrichment in lighter isotopes is correlated to the net mass loss of Si in the lower soil horizons. It was also noted that the topmost soil may be shifted to lower $\delta^{30}Si$ values due to the recycling of biogenic Si from plants (phytoliths) (Ziegler et al. 2005b). The slightly lower $\delta^{30}Si$ value for the unweathered Hawaiian basalt (−0.5‰ (Ziegler et al. 2005b)) compared to other basalts analyzed (−0.3‰ (Savage et al. 2010)) may indicate some alteration of the unweathered basalt, or simply reflect errors in external reproducibility between groups (Reynolds et al. 2007). Regardless, the development of a clay-rich soil leads to the progressive reduction in bulk soil $\delta^{30}Si$ values (Ziegler et al. 2005b).

The Si isotope compositions of soils have also been investigated for basaltic soils in Cameroon, western Africa (Opfergelt et al. 2009). Just as for the Hawaiian soils, soil $\delta^{30}Si$ values varies from the initial basaltic pumice value of −0.3 to −2.6‰ in the clay rich samples. However, in Cameroon amorphous Si phases were observed, which have higher $\delta^{30}Si$ values, ranging from −0.5 to +0.6‰, and were derived from volcanic ashes and biogenic silica (Opfergelt et al. 2010).

In Puerto Rico, rapid weathering of a quartz diorite intrusion produces a saprolite that has minimal loss of quartz and biotite, but near complete dissolution of plagioclase and hornblende, and formation of secondary clay minerals. It was shown that secondary kaolinites, which precipitated from either transformation via epitaxial overgrowth of biotite or by re-precipitation after dissolution of plagioclase feldspars, have constant $\delta^{30}Si$ values around −2.2‰ (Ziegler et al. 2005b). This is distinctly lower than any primary silicates including biotite (−0.6‰), plagioclase (−0.1‰), and quartz (−0.2‰) in the host diorite, as shown in Fig. 6.7. The mass-fractionation during formation of the kaolinite was calculated to be −2.1‰ from the re-precipitation after dissolution of plagioclase feldspars, and −1.6‰ from the transformation via epitaxial overgrowth of biotite (Ziegler et al. 2005a). However, the $\delta^{30}Si$ difference between co-existing biotite and plagioclase is hard to reconcile with the small potential mass-fractionation between different minerals at higher temperatures of crystallization. Potentially the lower $\delta^{30}Si$ value measured for biotite may be due to some partial alteration of the biotite since crystallization. Unfortunately no systematic studies of the Si isotope fractionation factor between silicic acid and clay minerals have been completed.

Although the mineral soil pool may evolve, and clays incorporate lower $\delta^{30}Si$ values, it has been argued that there are feedback reactions between the elemental cycling by plants and soil mineralogical properties (Lucas 2001; Sommer et al. 2006), such that Si cycling through plants has consequences for soil development itself. It has also been shown that in terrestrial systems the internal recycling of the phytolith pool is larger than the overall Si weathering rate (Conley 2002). As well as the temporal development of soils, it is important to consider their vertical development. Silicon recycling by plants results in the active transport of dissolved Si at depth to biogenic Si at the surface. Furthermore, humic acids from plants can lead to distinctly acidic conditions that can result in the dissolution of clay minerals (Viers et al. 1997). It has been proposed for organic-rich swampy soils in the Congo Basin (Cardinal et al. 2010), that although clays with light Si may be formed at the weathering front at the bottom of the soil profile, the dissolution of clays near the top can provide a source of light Si, resulting in little overall mass-fractionation between the primary minerals and dissolved riverine compositions.

As the Si isotope fractionation associated with plant uptake and mineral precipitation is similar, Si isotopes alone cannot be used to uniquely distinguish between biotic and abiotic cycling. Difficulties in assessing the recycling by plants arise because it is hard to determine the average Si isotope composition of the Si source within the soil for the plant biomass, and the efficiency of recycling of biogenic Si between the soil and plants. However, the behaviour of Si and germanium (Ge), a chemical analog of Si, are different between the two processes, and can be used to constrain the Si pathways within the soil-plant system (Cornelis et al. 2010a). During uptake by plants, Ge/Si ratios are fractionationed; phytoliths contain low Ge/Si ratios, whilst clay minerals contain high Ge/Si ratios, compared to the bedrocks.

The importance of plant-soil recycling compared to the total weathering input and hydrological output may thus be quantified by examining the Ge/Si ratios. Data from an initial Hawaiian study (Derry et al. 2005) was used to infer that most of the hydrological Si flux out of the watershed had passed though the biogenic Si pool, rather than being directly weathered. It was further suggested that "other systems exhibiting strong Si depletion of the mineral soils and/or high Si uptake rates by biomass will also have strong biological control on silica cycling and export" (Derry et al. 2005). Thus it is crucial to evaluate the role of vegatation on the Si biogeochemical cycle.

Within the African soils, higher $\delta^{30}Si$ values were associated with lower Ge/Si ratios (Opfergelt et al. 2010). The low Ge/Si found in the soils must reflect the recycling of phytoliths from banana plants with low Ge/Si at the studied sites, which have also resulted in higher $\delta^{30}Si$ values. Hence, plant material can have a significant impact on the Si isotope composition of soils. It should be noted that there may be mass-fractionation during further dissolution, with preferential dissolution of lighter isotopes (Demarest et al. 2009; Ziegler et al. 2005b).

Mass balance considerations imply that without the build-up of a fractionated Si reservoir, the recycling of Si by plants could not have a large impact on the Si isotope composition supplied to the oceans from the terrestrial cycle, even though they may dominate the fluxes in the global Si cycle. However, the production of particulate Si by plants could strongly affect the difference between the Si isotope composition of dissolved and particulate matter at the global scale. Within the Anthropocene, human cultivation could have an extreme effect on the global Si cycle and the dissolved Si isotope composition supplied to seawater (Conley 2002). For example, the annual discharge of Si from the Yangtze River in China is estimated to be almost 100 Gmol Si (Ding et al. 2004), about 2% of the global riverine discharge of silicic acid, but the amount of Si in rice plants within this catchment is ~500 Gmol (Ding et al. 2004), about 5 times the annual riverine discharge. Thus, most Si must be cycled through the rice plants several times, strongly affecting the seasonal cycle of Si. The annually removal of harvested rice grains contain ~5 Gmol Si, so rice export should have a small effect on the Si budget and the overall Si isotope composition. In summary, it is clear from these early studies that the seasonal growth and uptake by plants and the seasonal degradation of litter should add a strong seasonality to the cycling of Si and the Si isotope composition of soil waters.

6.3.3 Dissolved Si and Catchment-Scale Studies

Assuming that the recycling of Si by plants is at a steady-state, where the inputs to the soil-plant cycle are equal to the outputs, then the recycling of Si by plants will not impact the riverine Si flux, or its Si isotope composition at larger temporal and spatial scales. The total riverine flux is then a function of the silicate weathering within the catchment, and the riverine Si isotope composition should largely reflect abiotic processes. For example, Cardinal et al. (2010) showed that a typical stable phytolith pool does not impact the $\delta^{30}Si$ value of water exported, despite a significant contribution to the recycling of Si in the soil. However, we must consider whether the assumption of steady-state recycling of Si between plants and soils is applicable to catchment scale studies, or the global Si budget. Studies of dissolved Si isotopic composition do indeed show dramatic δ^{30}Si variations, both spatially and temporally (Alleman et al. 2005; Engström et al. 2010; Georg et al. 2006a), but often the cause of the fractionation could not be resolved.

As discussed above, the large uptake of Si by rice plants and bamboo in China could influence the dissolved Si composition of Chinese rivers. An early study of the Yangtze (Changjiang) River (Ding et al. 2004) documented dissolved δ^{30}Si values ranging from +0.7‰ in the uppermost reaches to >3.0‰ in the middle and lower reaches of the river. Such high δ^{30}Si values are hard to reconcile with other published dissolved δ^{30}Si values (+0.8 ± 0.5‰ (De La Rocha et al. 2000)), without invoking large influences of biologically induced fractionation, or a counterbalancing build-up of a light Si reservoir in the Chinese soils (Ding et al. 2004). A weak inverse correlation between Si concentrations and δ^{30}Si values of the Yangtze samples was interpreted to indicate that isotope fractionation must have occurred during removal of dissolved Si by grass and rice plants (Ding et al. 2004). Unfortunately, seasonal data are not yet available to quantify the effect of seasonal uptake by plants on the δ^{30}Si values within this river system.

An investigation of Swiss rivers (Georg et al. 2006a) draining a variety of lithologies showed seasonal variations in Si concentrations and δ^{30}Si values. The δ^{30}Si values ranged from +0.5 to +1.3‰, lower values being associated with high discharge rates, but no obvious differences in the mean δ^{30}Si value between the four rivers studied, despite the difference in sizes and lithologies of their catchments (Georg et al. 2006a). Within one mountainous catchment in the Alps, there is a direct relationship between Si concentration and δ^{30}Si values, with an inverse relationship between Si flux and δ^{30}Si values. This direct relationship between Si concentration and δ^{30}Si values contrasts to the Yangtze River study (Ding et al. 2004), and cannot be driven by the removal of isotopically light dissolved Si. For the small Alpine catchment, it was argued that two distinct components, or reservoirs, of dissolved Si supply the silicic acid to the river: high δ^{30}Si values with high Si concentrations are supplied as a base hydraulic flow from the seepage of soil and/or groundwaters, and low δ^{30}Si values in more Si dilute waters are associated with higher dissolved cation loads, which indicate higher mineral dissolution and less Si fractionation from either clay formation or biological utilization (Georg et al. 2006a).

From the development of the Si isotope composition of Hawaiian soils with age (Ziegler et al. 2005b), there should be a relationship between the age of the catchment and the Si isotope composition of the river waters. However, a catchment scale study of the rivers in Iceland (Georg et al. 2007b) failed to find such a correlation despite the same basaltic bedrock lithology. The biggest difference between Hawaii and Iceland is the climate and vegetation. The colder climate of Iceland does not allow for lush vegetation, and much thinner soils develop. The dissolved δ^{30}Si values of the rivers were found to correlate with the dissolved Ca/Si ratio, which is elevated compared to the basaltic source rocks. It was proposed that, since Ca is not supplied by carbonate weathering on Iceland and is not strongly incorporated into secondary precipitates, increases in the Ca/Si ratio result from the preferential loss of Si. Thus, the Ca/Si ratio can be used to estimate the fraction of Si lost from the dissolved phase. Estimates of Si re-precipitated ranged from 33 to 66%, although in three glacially-fed rivers it appeared that some Ca must have been derived from carbonate weathering. Furthermore, the correlation between δ^{30}Si values and the estimated amount of Si lost could be used to derive an average fractionation factor of −1.6‰ for the processes driving this loss, as shown in Fig. 6.8. This loss was considered to be secondary mineral precipitation and is in agreement with the fractionation factor between water and clays estimated from therotical and field studies (Méheut et al. 2007, 2009; Ziegler et al. 2005b). However, as all sampling occurred in the summer, the influence of seasonal variations in δ^{30}Si values could not be evaluated, and thus the influence of vegetation and soils could not be assessed.

Large seasonal variations in δ^{30}Si values have now been observed in another catchment scale study from the same latitude, in Northern Sweden (Engström et al. 2010). In this granitic catchment of pristine boreal forests, the discharge from the Kalix River is highly seasonal, with minimal flow during the winter due to freezing, and a short period of very high discharge due to the snow melt in the spring. The chemical denudation

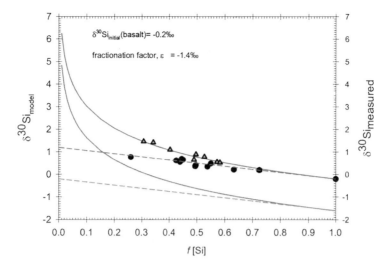

Fig. 6.8 The relation between the riverine Si isotope composition and the fraction dissolved remaining in solution, f_{Si}, computed using the Ca/Si ratios, can be described by open and closed system fractionation behaviour. It appears that rivers collected on the western part of Iceland (*triangles*) follow rather a Rayleigh-type model, whereas rivers collected from eastern and central Iceland (*black dots*) follow an steady-state system fractionation kinetics, from Georg et al. (2007b)

is highest during the high discharge period, with dissolved K, Al and Si concentrations elevated compared to Mg, which can be used to normalize the effect of dilution by snow melt. The $\delta^{30}Si$ values are lowest during this period (~0.7‰), but are still elevated compared to the underlying granitic rocks. However, in late summer, especially in sub-catchments within the Kalix watershed, the $\delta^{30}Si$ values are relatively high (>1.2‰) and associated with a depletion of Si relative to Mg, which was interpreted as a loss of Si and isotope fractionation during the summer due to plant growth, either terrestrial plants or aquatic diatoms. Thus, active Si uptake by plants and algae may influence the seasonal Si isotope composition of rivers. It was argued that lower $\delta^{30}Si$ values could not be due to less secondary clay formation as hypothesized for the Icelandic rivers, given the changes in Al content of the river and the Si/Al ratio of clays (Engström et al. 2010). Potentially, the breakdown of plant material and phytoliths in spring influences the seasonal Si isotope composition of the rivers, as the lowest $\delta^{30}Si$ values are associated with high dissolved organic carbon (DOC) (Engström et al. 2010) as the snow melt leaches the upper, organic-rich soil horizons, and stream waters are dominantly fed from soil water (Land et al. 2000).

The seasonal pattern of chemical weathering observed in the granitic catchment of the Kalix River is also seen in basaltic Siberian rivers (Pokrovsky et al. 2005). Large changes in $\delta^{30}Si$ values are observed, with lowest values during the spring snow melt (~0.7‰) and highest values in the fall (>2‰) (Reynolds et al. 2006a). The lowest $\delta^{30}Si$ values are associated with high DOC, and thus light Si may be supplied from the dissolution of phytoliths, or the leaching of light Si from the soil. However, light Si may also be supplied inorganically from absorption-desorption reactions (Delstanche et al. 2009; Opfergelt et al. 2009), or from precipitation due to freezing (Dietzel 2005). During freezing of water in winter, mono-silicic acid can become super-saturated and fractionate during precipitation, leading to the formation of amorphous opal (Dietzel 2005), similar to phytoliths. The degree to which this phenomenon occurs in nature has yet to be evaluated, so it is impossible to fully quantify the relative roles of organic and inorganic cycling of Si on the isotope composition of Si in rivers.

6.3.4 Si in Groundwater Systems

Since most deep groundwaters systems are fed from the infiltration of surface water, the isotope composition of dissolved Si in groundwaters should have positive $\delta^{30}Si$

values similar to rivers or soil waters. However, measured groundwaters exhibit strongly negative $\delta^{30}Si$ values. Indeed, it has been shown that the $\delta^{30}Si$ values of dissolved Si in groundwater flowing through an artesian basin appear to evolve to lower values during flow (Georg et al. 2009a). The shift towards negative $\delta^{30}Si$ values as Si concentrations increase can readily be explained by the dissolution of previously precipitated clay minerals or silcretes with very negative $\delta^{30}Si$ values ($<-3‰$) (Basile-Doelsch et al. 2005; Georg et al. 2009a). Further evolution to lower $\delta^{30}Si$ values without increasing Si concentrations were explained by the existence of different silcretes within the aquifer with more extreme negative $\delta^{30}Si$ values. Alternatively, continuous precipitation-dissolution or exchange reactions between water and clay minerals within the aquifer system and that potential conversions between kaolinite and smectite result in isotope exchange and re-equilibration between minerals and dissolved species for both Si and Li isotopes.

Since groundwaters have lower $\delta^{30}Si$ values than rivers, an under-estimation of the global subterranean groundwater discharge to the oceans would lead to a bias in the estimated average $\delta^{30}Si$ value supplied to the oceans (Georg et al. 2009b). Since solute concentrations, including silicic acid, are higher in groundwater, it has been estimated that groundwater discharge may constitute up to 40% of the Si flux to the oceans from the Bay of Bengal (Georg et al. 2009b), and thus chemical denudation rates based upon river fluxes alone may dramatically underestimate the total global denudations rates.

6.3.5 Modeling the Si Biogeochemical Cycle

There have now been some attempts at developing models of the Si cycle and its isotope composition for the continental and marine environments (Basile-Doelsch 2006; Basile-Doelsch et al. 2005), as shown in Fig. 6.9, and the oceans themselves (De La Rocha and Bickle 2005; Reynolds 2009; Wischmeyer et al. 2003). Following the observation that river waters and the oceans have higher $\delta^{30}Si$ values than igneous rocks, it was argued that there must be a counterbalancing "negative" reservoir of light Si which

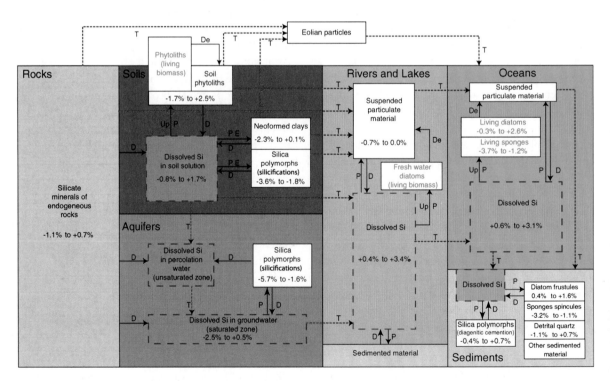

Fig. 6.9 Biogeochemical cycle of Si in continental and marine environments. *D* dissolution; *P* precipitation; *T* transport; *E* epigenesis; *Up* uptake; *De* death. $\delta^{30}Si$ ranges are from the literature and taken from Basile-Doelsch (2006)

develops during weathering, which was initially identified as continental silcretes (Basile-Doelsch et al. 2005). However, this counter-balancing light Si may also be in other continental Si pools, such as phytoliths, clays or biological material (Basile-Doelsch 2006). Above-ground biomass cannot be the light Si counter-balance despite its importance as a Si pool because it does not generally contain light Si. It would appear that the counter-balance must be secondary precipitates stored up in soils and sedimentary basins in the form of clays (Ziegler et al. 2005b) or silcretes (Basile-Doelsch et al. 2005), or adsorption onto Fe oxides (Delstanche et al. 2009).

From modeling the oceanographic Si cycle (Reynolds 2009), it was estimated that the mean $\delta^{30}Si$ value supplied to the ocean was about +0.8‰. Assuming a simple model of isotope fractionation during Si removal with a fractionation factor of -1.6‰ (Georg et al. 2007b), this mean $\delta^{30}Si$ value would imply that about half of the primary weathered Si (with a mean $\delta^{30}Si$ value around zero) is removed from solution and stored on the continents, as estimated for weathering on Iceland (Georg et al. 2007b). This calculation would imply that the continental crust is being enriched in light Si by about 6 Tmol Si per year with a $\delta^{30}Si$ value -0.8‰, which would be equivalent to ~800 million tons of kaolinite per year with a $\delta^{30}Si$ value -2.2‰. This amount of material is the same order of magnitude as the amount of sediment carried by the Mississippi into the Gulf of Mexico per year.

Interestingly, the estimated increase of light Si in the continental crust is in marked contrast to the heavier Si isotopic composition of the granitic continental crust compared to the basaltic mantle. Thus the difference between granitic and basaltic $\delta^{30}Si$ values cannot be due to the continuous recycling of Si via the rock cycle over Earth's history, and the actual cause of the difference is currently enigmatic.

6.4 Future Directions

It seems that the biogeochemical Si cycle is more complex than previously fathomed, with published studies pointing to an intense terrestrial biogeochemical cycling of Si (Struyf et al. 2009). Thus, considerable work is required to assess the terrestrial cycling of Si and its isotopes, in particular the role of plants and siliceous algae. Future field studies will allow the natural cycling of Si to be assessed in more detail, and most importantly, careful laboratory experiments will determine the mechanisms of biogenic silica and mineral silicate formation and associated Si isotope fractionation. Regrettably, the application of Si isotope studies in terrestrial environments is still somewhat hampered by the limited development of sample preparation techniques and the analytical requirements that are still far from routine. Here, I would like to suggest a number of potential research avenues which would help develop this exciting field in the near future.

- For the measurement of Si isotopes by MC-ICPMS, there is currently no technique for separation of silicic acid from anions, without conversion to SiF_6 in hydrofluoric acid. As anions, like sulphur, can cause large "matrix effects" (van den Boorn et al. 2009), a fast and efficient separation of silicic acid from other anions is greatly needed, with existing methods either requiring extensive preparation steps or significant amounts of HF (Engström et al. 2006). If HF cannot be eliminated, a change in the introduction system, with removal of a desolvating unit may be advantageous.
- A change in biomass could lead to significant alteration of the terrestrial Si cycle, such as the anthropogenic transformation from forests to pasture (Conley 2002). Sudden disruption of the plant biomass, like harvesting and forest fires, could also have significant impact on the Si fluxes and isotope composition. As Si is beneficial to plants, we need to understand the molecular mechanism of biogenic opal mineralization and associated isotope fractionation, and the way in which different plants adapt and respond to changes in Si availability.
- Lastly, more studies of river and groundwater systems are required in order to better constrain the global dissolved Si budget, and the role that lithology, climate and vegetation can have on global biogeochemical cycles. However, we should perhaps no longer consider that the global biogeochemical cycles can be accurately assessed from modern observations because, within the modern Anthropocene, human activity now moves more material across the globe than rivers do.

Acknowledgments I would like to thank Mark Baskaran and two anonymous reviewers for their great help in improving this manuscript.

References

Alexandre A, Meunier J-D, Colin F, Koud J-M (1997) Plant impact on the biogeochemical cycle of silicon and related weathering processes. Geochim Cosmochim Acta 61(3): 677–682

Alleman LY, Cardinal D, Cocquyt C, Plisnier P-D, Descy J-P, Kimirei I, Sinyinza D, André L (2005) Silicon isotopic fractionation in Lake Tanganyika and its main tributaries. J Great Lakes Res 31(4):509–519

Allenby RJ (1954) Determination of the isotopic ratios of silicon in rocks. Geochim Cosmochim Acta 5(1):40–48

André L, Cardinal D, Alleman LY, Moorbath S (2006) Silicon isotopes in 3.8 Ga West Greenland rocks as clues to the Eoarchaean supracrustal Si cycle. Earth Planet Sci Lett 245(1–2):162–173

Basile-Doelsch I (2006) Si stable isotopes in the Earth's surface: a review. J Geochem Explor 88(1–3):252–256

Basile-Doelsch I, Meunier JD, Parron C (2005) Another continental pool in the terrestrial silicon cycle. Nature 433(7024): 399–402

Bern CR, Brzezinski MA, Beucher C, Ziegler K, Chadwick OA (2010) Weathering, dust, and biocycling effects on soil silicon isotope ratios. Geochim Cosmochim Acta 74(3): 876–889

Boltz DF, Trudell LA, Potter GV (1978) Silicon. In: Boltz DF, Trudell LA (eds) Colorimetric determination of nonmetals. Wiley, New York, pp 421–462

Brzezinski MA, Pride CJ, Franck VM, Sigman DM, Sarmiento JL, Matsumoto K, Gruber N, Rau GH, Coale KH (2002) A switch from $Si(OH)_4$ to NO_3^- depletion in the glacial Southern Ocean. Geophys Res Lett 29(12):1564. doi:1510.1029/2001GL014349

Brzezinski MA, Jones JL, Beucher CP, Demarest MS, Berg HL (2006) Automated determination of silicon isotope natural abundance by the acid decomposition of cesium hexafluosilicate. Anal Chem 78(17):6109–6114

Burton KW, Vigier N (2011) Lithium Isotopes as Tracers in Marine and Terrestrial Environments. In: Baskaran M (ed) Handbook of Environmental Isotope Geochemistry. Springer, Heidelberg

Cardinal D, Alleman LY, de Jong J, Ziegler K, André L (2003) Isotopic composition of silicon measured by multicollector plasma source mass spectrometry in dry plasma mode. J Anal At Spectrom 18(3):213–218

Cardinal D, Alleman LY, Dehairs F, Savoye N, Trull TW, André L (2005) Relevance of silicon isotopes to Si-nutrient utilization and Si-source assessment in Antarctic waters. Global Biogeochem Cycles 19:2007. doi:10.1029/2004GB002364

Cardinal D, Gaillardet J, Hughes HJ, Opfergelt S, André L (2010) Contrasting silicon isotope signatures in rivers from the Congo Basin and the specific behaviour of organic-rich waters. Geophys Res Lett 37(12):L12403

Conley DJ (2002) Terrestrial ecosystems and the global biogeochemical silica cycle. Global Biogeochem Cycles 16(4): 1121

Coplen TB, Bohlke JK, De Bievre P, Ding T, Holden NE, Hopple JA, Krouse HR, Lamberty A, Peiser HS, Revesz K, Rieder SE, Rosman KJR, Roth E, Taylor PDP, Vocke RD, Xiao YK (2002) Isotope-abundance variations of selected elements – (IUPAC Technical Report). Pure Appl Chem 74(10):1987–2017

Cornelis JT, Delvaux B, Georg RB, Lucas Y, Ranger J, Opfergelt S (2010a) Tracing the origin of dissolved silicon transferred from various soil-plant systems towards rivers: a review. Biogeosci Discuss 7(4):5873–5930

Cornelis JT, Delvaux B, Cardinal D, André L, Ranger J, Opfergelt S (2010b) Tracing mechanisms controlling the release of dissolved silicon in forest soil solutions using Si isotopes and Ge/Si ratios. Geochim Cosmochim Acta 74(14): 3913–3924

De La Rocha CL (2002) Measurement of silicon stable isotope natural abundances via multicollector inductively coupled plasma mass spectrometry (MC-ICP-MS). Geochem Geophys Geosyst 3(8):art. no.-1045

De La Rocha CL, Bickle MJ (2005) Sensitivity of silicon isotopes to whole-ocean changes in the silica cycle. Mar Geol 217(3–4):267–282

De La Rocha CL, Brzezinski MA, DeNiro MJ (1996) Purification, recovery, and laser-driven fluorination of silicon from dissolved and particulate silica for the measurement of natural stable isotope abundances. Anal Chem 68(21): 3746–3750

De La Rocha CL, Brzezinski MA, DeNiro MJ (2000) A first look at the distribution of the stable isotopes of silicon in natural waters. Geochim Cosmochim Acta 64(14): 2467–2477

de Souza GF, Reynolds BC, Kiczka M, Bourdon B (2010) Evidence for mass-dependent isotopic fractionation of strontium in a glaciated granitic watershed. Geochim Cosmochim Acta 74(9):2596–2614

Debievre P, Lenaers G, Murphy TJ, Peiser HS, Valkiers S (1995) The chemical preparation and characterization of specimens for absolute measurements of the molar-mass of an element, exemplified by silicon, for redeterminations of the Avogadro constant. Metrologia 32(2):103–110

Delstanche S, Opfergelt S, Cardinal D, Elsass F, André L, Delvaux B (2009) Silicon isotopic fractionation during adsorption of aqueous monosilicic acid onto iron oxide. Geochim Cosmochim Acta 73(4):923–934

Demarest MS, Brzezinski MA, Beucher CP (2009) Fractionation of silicon isotopes during biogenic silica dissolution. Geochim Et Cosmochim Acta 73(19):5572–5583

Derry LA, Kurtz AC, Ziegler K, Chadwick OA (2005) Biological control of terrestrial silica cycling and export fluxes to watersheds. Nature 433(7027):728–731

Dietzel M (2005) Impact of cyclic freezing on precipitation of silica in $Me-SiO_2-H_2O$ systems and geochemical implications for cryosoils and -sediments. Chem Geol 216(1–2): 79–88

Ding T (2004) Analytical methods for silicon isotope determination. In: de Groot PA (ed) Handbook of stable isotope analytical techniques, vol 1. Elsevier, Amsterdam, pp 523–537

Ding T, Jiang S, Wan D, Li Y, Li J, Song H, Liu Z, Yao X (1996) Silicon Isotope Geochemistry. Geological Publishing House, Beijing, China, p 125

Ding T, Wan D, Wang C, Zhang F (2004) Silicon isotope compositions of dissolved silicon and suspended matter in the Yangtze River, China. Geochim Cosmochim Acta 68(2):205–216

Ding TP, Ma GR, Shui MX, Wan DF, Li RH (2005) Silicon isotope study on rice plants from the Zhejiang province, China. Chem Geol 218(1–2):41–50

Ding TP, Tian SH, Sun L, Wu LH, Zhou JX, Chen ZY (2008a) Silicon isotope fractionation between rice plants and nutrient solution and its significance to the study of the silicon cycle. Geochim Cosmochim Acta 72(23):5600–5615

Ding TP, Zhou JX, Wan DF, Chen ZY, Wang CY, Zhang F (2008b) Silicon isotope fractionation in bamboo and its significance to the biogeochemical cycle of silicon. Geochim Cosmochim Acta 72(5):1381–1395

Douthitt CB (1982) The geochemistry of the stable isotopes of silicon. Geochim Cosmochim Acta 46(8):1449–1458

Engström E, Rodushkin I, Baxter DC, Ohlander B (2006) Chromatographic purification for the determination of dissolved silicon isotopic compositions in natural waters by high-resolution multicollector inductively coupled plasma mass spectrometry. Anal Chem 78(1):250–257

Engström E, Rodushkin I, Ingri J, Baxter DC, Ecke F, Österlund H, Öhlander B (2010) Temporal isotopic variations of dissolved silicon in a pristine boreal river. Chem Geol 271(3–4):142–152

Epstein E (2003) SILICON. Annu Rev Plant Physiol Plant Mol Biol 50(1):641–664

Epstein S, Taylor HP Jr (1970) $^{18}O/^{16}O$, $^{30}Si/^{28}Si$, D/H, and $^{13}C/^{12}C$ studies of lunar rocks and minerals. Science 167(3918, The Moon Issue):533–535

Farkas J, Bohm F, Wallmann K, Blenkinsop J, Eisenhauer A, van Geldern R, Munnecke A, Voigt S, Veizer J (2007) Calcium isotope record of Phanerozoic oceans: implications for chemical evolution of seawater and its causative mechanisms. Geochim Cosmochim Acta 71(21):5117–5134

Farmer VC, Delbos E, Miller JD (2005) The role of phytolith formation and dissolution in controlling concentrations of silica in soil solutions and streams. Geoderma 127(1–2):71–79

Fitoussi C, Bourdon B, Kleine T, Oberli F, Reynolds BC (2009) Si isotope systematics of meteorites and terrestrial peridotites: implications for Mg/Si fractionation in the solar nebula and for Si in the Earth's core. Earth Planet Sci Lett 287(1–2):77–85

Georg RB, Reynolds BC, Frank M, Halliday AN (2006a) Mechanisms controlling the silicon isotopic compositions of river waters. Earth Planet Sci Lett 249(3–4):290–306

Georg RB, Reynolds BC, Frank M, Halliday AN (2006b) New sample preparation techniques for determination of Si isotope composition using MC-ICPMS. Chem Geol 235(1–2):95–104

Georg RB, Halliday AN, Schauble EA, Reynolds BC (2007a) Silicon in the Earth's core. Nature 447(7148):1102–1106

Georg RB, Reynolds BC, West AJ, Burton KW, Halliday AN (2007b) Silicon isotope variations accompanying basalt weathering in Iceland. Earth Planet Sci Lett 261(3–4):476–490

Georg RB, Zhu C, Reynolds BC, Halliday AN (2009a) Stable silicon isotopes of groundwater, feldspars, and clay coatings in the Navajo Sandstone aquifer, Black Mesa, Arizona, USA. Geochim Cosmochim Acta 73(8):2229–2241

Georg RB, West AJ, Basu AR, Halliday AN (2009b) Silicon fluxes and isotope composition of direct groundwater discharge into the Bay of Bengal and the effect on the global ocean silicon isotope budget. Earth Planet Sci Lett 283(1–4):67–74

Hendry KR, Leng MJ, Robinson LF, Sloane HJ, Blusztjan J, Rickaby REM, Georg RB, Halliday AN (2010) Silicon isotopes in Antarctic sponges: an interlaboratory comparison. Antarctic Sci 23(1):34–42

Hiemstra T, Barnett MO, van Riemsdijk WH (2007) Interaction of silicic acid with goethite. J Colloid Interface Sci 310(1):8–17

Hulston JR, Thode HG (1965) Variations in the S^{33}, S^{34}, and S^{36} contents of meteorites and their relation to chemical and nuclear effects. J Geophys Res 70:3475–3584

Icopini GA, Brantley SL, Heaney PJ (2005) Kinetics of silica oligomerization and nanocolloid formation as a function of pH and ionic strength at 25 degrees C. Geochim Cosmochim Acta 69(2):293–303

Iler RK (1979) The Chemistry of silica – solubility, polymerization, colloid and surface properties and biochemistry. Wiley, New York, p 1026

Land M, Ingri J, Andersson PS, Öhlander B (2000) Ba/Sr, Ca/Sr and $^{87}Sr/^{86}Sr$ ratios in soil water and groundwater: implications for relative contributions to stream water discharge. Appl Geochem 15(3):311–325

Lucas Y (2001) The role of plants in controlling rates and products of weathering: importance of biological pumping. Annu Rev Earth Planet Sci 29:135–163

Ma JF (2004) Role of silicon in enhancing the resistance of plants to biotic and abiotic stresses. Soil Sci Plant Nutr 50(1):11–18

Ma JF, Mitani N, Nagao S, Konishi S, Tamai K, Iwashita T, Yano M (2004) Characterization of the silicon uptake system and molecular mapping of the silicon transporter gene in rice. Plant Physiol 136(2):3284–3289

Marschner H (1995) Beneficial mineral elements. In: Mineral nutrition of higher plants, 2nd edn. Academic, London, pp 405–435

Méheut M, Lazzeri M, Balan E, Mauri F (2007) Equilibrium isotopic fractionation in the kaolinite, quartz, water system: prediction from first-principles density-functional theory. Geochim Cosmochim Acta 71(13):3170–3181

Méheut M, Lazzeri M, Balan E, Mauri F (2009) Structural control over equilibrium silicon and oxygen isotopic fractionation: a first-principles density-functional theory study. Chem Geol 258(1–2):28–37

Miller MF (2002) Isotopic fractionation and the quantification of O-17 anomalies in the oxygen three-isotope system: an appraisal and geochemical significance. Geochim Cosmochim Acta 66(11):1881–1889

Mitani N, Ma JF (2005) Uptake system of silicon in different plant species. J Exp Bot 56(414):1255–1261

Morey GW, Fournier RO, Rowe JJ (1962) The solubility of quartz in water in the temperature interval from 25° to 300° C. Geochim Cosmochim Acta 26(10):1029–1040, IN1011, 1041–1043

Opfergelt S, Cardinal D, Henriet C, André L, Delvaux B (2006a) Silicon isotope fractionation between plant parts in banana: in situ vs. in vitro. J Geochem Explor 88(1–3):224–227

Opfergelt S, Cardinal D, Henriet C, Draye X, André L, Delvaux B (2006b) Silicon isotopic fractionation by banana (*Musa* spp.) grown in a continuous nutrient flow device. Plant Soil V285(1):333–345

Opfergelt S, Delvaux B, André L, Cardinal D (2008) Plant silicon isotopic signature might reflect soil weathering degree. Biogeochemistry 91(2):163–175

Opfergelt S, de Bournonville G, Cardinal D, André L, Delstanche S, Delvaux B (2009) Impact of soil weathering degree on silicon isotopic fractionation during adsorption onto iron oxides in basaltic ash soils, Cameroon. Geochim Cosmochim Acta 73(24):7226–7240

Opfergelt S, Cardinal D, André L, Delvigne C, Bremond L, Delvaux B (2010) Variations of $\delta^{30}Si$ and Ge/Si with weathering and biogenic input in tropical basaltic ash soils under monoculture. Geochim Cosmochim Acta 74(1):225–240

Pokrovsky OS, Schott J, Kudryavtzev DI, Dupre B (2005) Basalt weathering in Central Siberia under permafrost conditions. Geochim Cosmochim Acta 69(24):5659–5680

Raven JA (1983) The transport and function of silicon in plants. Biol Rev 58(2):179–207

Raven JA (2001) Chapter 3 Silicon transport at the cell and tissue level. In: Datnoff LE, Snyder GH, Korndörfer GH (eds) Studies in plant science, vol 8. Elsevier, Amsterdam, pp 41–55

Raymo ME, Ruddiman WF (1992) Tectonic forcing of late Cenozoic climate. Nature 359(6391):117–122

Reynolds BC (2009) Modeling the modern marine $\delta^{30}Si$ distribution. Global Biogeochem Cycles 23:GB2015. doi:10.1029/2008GB003266

Reynolds JH, Verhoogen J (1953) Natural variations in the isotopic constitution of silicon. Geochim Cosmochim Acta 3(5):224–234

Reynolds BC, Pokrovsky OS, Schott J (2006a) Si isotopes for tracing basalt weathering in Central Siberia. Geochim Cosmochim Acta 70(18 suppl 1):A528

Reynolds BC, Frank M, Halliday AN (2006b) Silicon isotope fractionation during nutrient utilization in the North Pacific. Earth Planet Sci Lett 244(1–2):431–443

Reynolds BC, Georg RB, Oberli F, Wiechert UH, Halliday AN (2006c) Re-assessment of silicon isotope reference materials using high-resolution multi-collector ICP-MS. J Anal At Spectrom 21(3):266–269

Reynolds BC, Aggarwal J, Andre L, Baxter D, Beucher C, Brzezinski MA, Engstrom E, Georg RB, Land M, Leng MJ, Opfergelt S, Rodushkin I, Sloane HJ, van den Boorn S, Vroon PZ, Cardinal D (2007) An inter-laboratory comparison of Si isotope reference materials. J Anal At Spectrom 22(5):561–568

Rudge J, Reynolds BC, Bourdon B (2009) The double spike toolbox. Chem Geol 265:420–431

Sangster AG, Hodson MJ (2007) Silica in higher plants. In: Evered D, O'Connor M (eds) Ciba Foundation Symposium 121 – silicon biochemistry. Wiley, Chichester, pp 90–111

Savage PS, Georg RB, Armytage RMG, Williams HM, Halliday AN (2010) Silicon isotope homogeneity in the mantle. Earth Planet Sci Lett 295(1–2):139–146

Sommer M, Kaczorek D, Kuzyakov Y, Breuer J (2006) Silicon pools and fluxes in soils and landscapes – a review. J Plant Nutr Soil Sci 169(3):310–329

Strickland JDH (1952) The preparation and properties of silicomolybdic acid III. The combination of silicate and molybdate. J Am Chem Soc 74(4):872–876

Struyf E, Smis A, Van Damme S, Meire P, Conley D (2009) Global Biogeochem Silicon Cycle. SILICON 1(4):207–213. doi:10.1007/s12633-010-9035-x

Sun L, Wu L, Ding T, Tian S (2008) Silicon isotope fractionation in rice plants, an experimental study on rice growth under hydroponic conditions. Plant Soil 304(1):291–300

Taylor BE (2004) Fluorination methods in stable isotope analysis. In: de Groot PA (ed) Handbook of stable isotope analytical techniques, vol 1. Elsevier, Amsterdam, pp 400–472

Tipper ET, Galy A, Bickle MJ (2008) Calcium and magnesium isotope systematics in rivers draining the Himalaya-Tibetan-Plateau region: lithological or fractionation control? Geochim Cosmochim Acta 72(4):1057–1075

van den Boorn SHJM, Vroon PZ, van Belle CC, van der Wagt B, Schwieters J, van Bergen MJ (2006) Determination of silicon isotope ratios in silicate materials by high-resolution MC-ICP-MS using a sodium hydroxide sample digestion method. J Anal At Spectrom 21:734–742

van den Boorn SHJM, Vroon PZ, van Bergen MJ (2009) Sulfur-induced offsets in MC-ICP-MS silicon-isotope measurements. J Anal At Spectrom 24:1111–1114

Varela DE, Pride CJ, Brzezinski MA (2004) Biological fractionation of silicon isotopes in Southern Ocean surface waters. Global Biogeochem Cycles 18:GB1047. doi:10.1029/2003GB002140

Viers J, Dupré B, Polvé M, Schott J, Dandurand J-L, Braun J-J (1997) Chemical weathering in the drainage basin of a tropical watershed (Nsimi-Zoetele site, Cameroon): comparison between organic-poor and organic-rich waters. Chem Geol 140(3–4):181–206

Vigier N, Gislason SR, Burton KW, Millot R, Mokadem F (2009) The relationship between riverine lithium isotope composition and silicate weathering rates in Iceland. Earth Planet Sci Lett 287(3–4):434–441

White AF, Blum AE, Schulz MS, Vivit DV, Stonestrom DA, Larsen M, Murphy SF, Eberl D (1998) Chemical weathering in a tropical watershed, Luquillo mountains, Puerto Rico: I. Long-term versus short-term weathering fluxes. Geochim Cosmochim Acta 62(2):209–226

Wischmeyer A, DelaRocha CL, Maier-Reimer E, Wolf-Gladrow DA (2003) Control mechanisms for the oceanic distribution of silicon isotopes. Global Biogeochem Cycles 17(3):1083. doi:10.1029/202GB002022

Young ED, Galy A, Nagahara H (2002) Kinetic and equilibrium mass-dependent isotope fractionation laws in nature and their geochemical and cosmochemical significance. Geochim Cosmochim Acta 66(6):1095–1104

Ziegler K, Chadwick OA, White AF, Brzezinski MA (2005a) $\delta^{30}Si$ systematics in a granitic saprolite, Puerto Rico. Geology 33(10):817–820

Ziegler K, Chadwick OA, Brzezinski MA, Kelly EF (2005b) Natural variations of $\delta^{30}Si$ ratios during progressive basalt weathering, Hawaiian Islands. Geochim Cosmochim Acta 69(19):4597–4610

Chapter 7
Calcium Isotopes as Tracers of Biogeochemical Processes

Laura C. Nielsen, Jennifer L. Druhan, Wenbo Yang, Shaun T. Brown, and Donald J. DePaolo

Abstract The prevalence of calcium as a major cation in surface and oceanic environments, the necessity of calcium in the functioning of living cells and bone growth, and the large spread in mass between calcium isotopes all suggest that calcium isotope biogeochemistry can be an important avenue of insight into past and present biogeochemical cycling processes. In the following chapter, we review the main areas of research where Ca isotope studies have been pursued and detail recent research results in biogeochemical applications. In marine environments, biogenic fractionation of Ca isotopes during biomineralization produces predictable offsets in some organisms, which facilitate the reconstruction of seawater $\delta^{44/40}$Ca over geologic timescales. In terrestrial studies, observed Ca isotope fractionation between soil and various components of vegetation enables the construction of a local Ca budget and provides a partial explanation for the scale of Ca isotopic variability within a single watershed. The research reviewed in this chapter provides a foundation for future investigations into the macro- and microscopic processes and biochemical pathways dictating the distribution of this essential nutrient using stable Ca isotope ratios.

7.1 Introduction

Calcium is an alkaline earth metal with six naturally occurring isotopes: ^{40}Ca (96.941%), ^{41}Ca, ^{42}Ca (0.647%), ^{43}Ca (0.135%), ^{44}Ca (2.086%), ^{46}Ca (0.004%) and ^{48}Ca (0.187%). By mass, calcium is the fifth most abundant element in both the earth's crust and seawater. The prevalence of calcium as a major cation in surface and oceanic environments, the necessity of calcium in the functioning of living cells and bone growth and the large spread in mass between calcium isotopes all suggest calcium isotope biogeochemistry can be an important avenue of insight into past and present biogeochemical cycling processes. In the following subsections, we review the main research areas where biogeochemical Ca isotope studies have been pursued, and in later sections more details are provided regarding recent research results. Throughout this paper we have renormalized all $\delta^{44/40}$Ca values to a bulk silicate Earth (BSE) standard, which we describe in Section II.

7.1.1 Ca Isotopes and the Marine Calcium Cycle

Calcium isotopes provide a promising means to constrain the global Ca budget over geologic timescales and to identify the causes of variability in the marine calcium cycle (e.g. Zhu and MacDougall 1998; DePaolo 2004; Farkaš et al. 2007). Seawater has a globally homogeneous $\delta^{44/40}$Ca (0.91 ± 0.04‰; Hippler et al. 2003; Farkaš et al. 2007) that is uniform with water depth, due to the long oceanic Ca residence time (~1 million years) relative to oceanic turnover

L.C. Nielsen (✉), J.L. Druhan, W. Yang, and D.J. DePaolo
Department of Earth and Planetary Science, University of California, Berkeley, CA 94720, USA
Earth Sciences Division, Lawrence Berkeley National Laboratory, Berkeley, CA 94720, USA
e-mail: lnielsen@berkeley.edu

S.T. Brown
Earth Sciences Division, Lawrence Berkeley National Laboratory, Berkeley, CA 94720, USA

time (1,500 years; Broecker and Peng 1982). The heavy Ca isotopic composition of seawater relative to global rocks (0.0‰; Simon and DePaolo 2010) and continental water (−0.2‰; Schmitt et al. 2003b) is primarily maintained by biogenic isotope fractionation during calcium carbonate precipitation, where ^{40}Ca is preferentially removed from solution relative to ^{44}Ca.

The concentration and isotopic composition of marine Ca depends upon the balance between the Ca input and output fluxes (e.g. De La Rocha and DePaolo 2000). Calcium enters the ocean as a solute transported by rivers due to continental weathering, by hydrothermal fluids due to metasomatism (Schmitt et al. 2003b; Amini et al. 2008), and by Ca replacement by Mg during dolomitization (Heuser et al. 2005; Böhm et al. 2005; Farkaš et al. 2007; Holmden 2009). These sources are isotopically similar to within ~1‰ (Schmitt et al. 2003b; Fig. 7.1). Riverine input is variable in isotopic composition and unrelated to the source watershed lithology or climate (Zhu and MacDougall 1998; Schmitt et al. 2003b). The magnitude of watershed-scale variability in $\delta^{44/40}$Ca resembles global variability due to mass fractionation during Ca transport through the environment (Tipper et al. 2006). Calcium carbonate precipitation causes the majority of continental Ca isotope fractionation, while preferential uptake of light Ca by vegetation in terrestrial ecosystems also plays a role. Precipitation of carbonate (and minor phosphate) minerals removes Ca from seawater at a rate similar to the input weathering flux (Berner et al. 1983; DePaolo 2004). Average calcium isotopic compositions of pelagic calcite (Skulan et al. 1997; Gussone et al. 2003, 2005) and aragonite differ by over ~0.5‰ due primarily to precipitate mineralogy: biogenic and inorganic aragonite are typically lighter than calcite (Gussone et al. 2003, 2005). Coccolithophores and modern scleractinian corals constitute an important exception to this rule, with average Ca isotopic compositions straddling the calcite and aragonite end members (Böhm et al. 2006). Sources of minor variability in modern marine biogenic carbonate $\delta^{44/40}$Ca include species-dependent as well as systematic intra-species temperature-dependent Ca isotope fractionation (Nägler et al. 2000; Gussone et al. 2003).

Variability in the ratio of Ca weathering to sedimentation fluxes and isotope compositions can affect seawater $\delta^{44/40}$Ca (δ_{sw}) over geologic timescales. Though δ_{sw} can vary with time, the ocean is thought to be currently at steady state with respect to Ca, indicating that the concentration and isotopic composition of Ca are stable over timescales smaller than the Ca oceanic residence time ($\tau = 10^6$ years) (Skulan et al. 1997; De La Rocha and DePaolo 2000; Schmitt et al. 2003a, b). De La Rocha and DePaolo (2000) constructed a simplified model for Ca mass balance in the ocean based on the relative fluxes of Ca into the ocean via continental weathering and out of the ocean via biogenic carbonate sedimentation, where F_w and

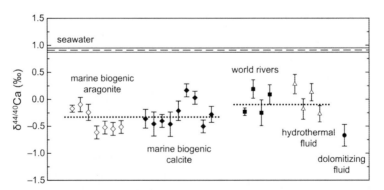

Fig. 7.1 $\delta^{44/40}$Ca of Ca fluxes into and out of seawater. *Dashed line* denotes the average seawater Ca isotopic composition (0.91 ± 0.4‰), with 2σ errors marked as lines above and below. Species average marine biogenic aragonite (Holmden 2005; Gussone et al. 2005; Böhm et al. 2006) and calcite (Gussone et al. 2005; Farkaš et al. 2007; Griffith et al. 2008b; Hippler et al. 2009) $\delta^{44/40}$Ca data are depicted with a *dotted line* through both data sets approximating the average δ_{sed}. Calcium isotopic data for aqueous Ca from major world-rivers, hydrothermal fluids and dolomitizing fluids (Schmitt et al. 2003b; Farkaš et al. 2007) are shown with a *dotted line* approximating the current average δ_w. Ca fluxes into and out of seawater are similar in $\delta^{44/40}$Ca indicating that the ocean is at steady state with respect to Ca

F_{sed} represent weathering and sedimentation fluxes respectively, and δ_w and δ_{sed} represent weathered Ca and sediment Ca isotopic compositions:

$$N_{Ca} \frac{d\delta^{44/40}Ca_{sw}}{dt} = F_w \left(\delta^{44/40}Ca_w - \delta^{44/40}Ca_{sw} \right) - F_{sed}\Delta_{sed},$$

where N_{Ca} is the amount of Ca in the ocean and $\Delta_{sed} = \delta_{sed} - \delta_{sw}$ is average Ca isotope separation (‰) during $CaCO_3$ precipitation. At steady state, $F_w = F_{sed}$ and δ_{sw} is constant in time, so:

$$\delta_{sw} = \delta_w - \Delta_{sed},$$

and

$$\delta_w = \delta_{sed}.$$

Current estimates of δ_{sed} and δ_w are $-0.3 \pm 0.2‰$ and $-0.2 \pm 0.2‰$ relative to BSE, supporting the theory that the ocean is close to steady state with respect to Ca (De La Rocha and DePaolo 2000; Farkaš et al. 2007).

Reconstructions of past δ_{sw} based on marine carbonate oozes (De La Rocha and DePaolo 2000; Fantle and DePaolo 2005, 2007), brachiopod shells, belemnites (Farkaš et al. 2007), rudists (Steuber and Buhl 2006), and foraminifera (Sime et al. 2007; Heuser et al. 2005) form a coherent record of oceanic Ca isotopes throughout the Phanerozoic (Fig. 7.2). Additional reconstructions based on phosphate minerals (Schmitt et al. 2003a; Soudry et al. 2004) are inconsistent with carbonate reconstructions, while those based on marine pelagic barite roughly correspond with carbonate data (Griffith et al. 2008a). A better understanding of marine biogenic and inorganic Ca isotope fractionation during $CaCO_3$ precipitation is necessary to reduce scatter in the reconstructed δ_{sw} record as it currently stands.

Despite a ~0.5‰ spread in the reconstruction, δ_{sw} variability within a given sample type can shed light upon the causes of Ca isotopic shifts with time. Observed long-timescale variation in δ_{sw} could be caused by changes in the ratio of sedimentation to weathering fluxes (De La Rocha and DePaolo 2000; Fantle and DePaolo 2005) or by changes in the Ca isotopic composition of these fluxes (Farkaš et al. 2007). In the first case, inequality of F_w and F_s over a period longer than the residence time of Ca in the ocean (~1 Ma) will result in a quasi steady state δ_{sw}:

$$\delta_{sw}(t) \approx \delta_w(t) - \frac{F_{sed}}{F_w}(t)\Delta_{sed}.$$

The ratio of sedimentation to weathering fluxes (F_{sed}/F_w) approximately ranges between 0.7 and 1.3 (Lasaga et al. 1985), which would cause a ~0.8‰ shift in δ_{sw} (De La Rocha and DePaolo 2000). Observed δ_{sw} variability of ~0.9‰ in fossil oozes of Cretaceous age and younger is consistent with this hypothesis; however, additional analyses of marine biogenic carbonates encompass a range closer to 1.5‰ (Fig. 7.2).

Farkaš et al. (2007) present an alternative hypothesis to explain long-term shifts in the oceanic Ca isotopic composition. They argue that the isotopic composition of the sedimentation flux, δ_{sed}, changes due to transitions in oceanic carbonate mineralogy. Throughout the Phanerozoic, variations in global tectonic activity caused long term alterations in oceanic major element chemistry, specifically Mg/Ca. Periods of enhanced tectonism raised oceanic Mg/Ca, facilitating the precipitation of aragonite relative to

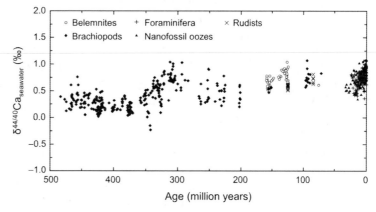

Fig. 7.2 $\delta^{44/40}$Ca of seawater throughout the Phanerozoic from various carbonate data sets (Heuser et al. 2005; Steuber and Buhl 2006; Fantle and DePaolo 2005, 2007; Sime et al. 2007; Farkaš et al. 2007). Seawater Ca isotopic compositions were calculated based on published fractionation factors between the given sediment type and seawater (Fantle and DePaolo 2005; Farkaš et al. 2007). Fantle and DePaolo (2005, 2007) values were first calculated relative to the NIST standard based on their measurement of seawater $\delta^{44/40}$Ca ($0.95 \pm 0.10‰$)

calcite. Based on this hypothesis, "aragonite seas" should correspond with heavy δ_{sw} and "calcite seas" should correspond with light δ_{sw}. Paleozoic, early Mesozoic and Neogene shifts in δ_{sw} are in phase with oscillating "calcite-aragonite seas," as predicted. The relationship between δ_{sw} and dominant mineralogy breaks down in the later Mesozoic and Cenozoic, probably due to the expansion of coccolithophores and scleractinian corals, whose Ca isotopic compositions are decoupled from mineralogy (Farkaš et al. 2007). A third possible explanation for the long-term variability in δ_{sw}, changing δ_w, is still being debated. While some consider major changes in the isotopic composition of the weathering flux to be unlikely based on the current narrow range in riverine $\delta^{44/40}$Ca values (De La Rocha and DePaolo 2000; Tipper et al. 2006), others have not ruled out this possibility (Griffith et al. 2008a; Fantle 2010).

Short-term changes in the calcium isotopic composition of the ocean, those occurring on sub-Ma timescales, are smaller in magnitude and likely caused by different processes than previously discussed. Enhanced silicate weathering due to a period of active volcanism, for example, could cause transient drops in δ_{sw} (Farkaš et al. 2007), as could enhanced carbonate dissolution due to ocean acidification (Payne et al. 2010). Analysis of complementary isotope systems such as sulfur and carbon could confirm or contest proposed mechanisms of these short-term excursions.

Marine Ca isotopic budgets may be applied to paleoclimate reconstructions to investigate aspects of the global carbon cycle. Kasemann et al. (2005) use boron and calcium isotopic analyses of carbonate sediments harboring evidence of Neoproterozoic glaciation in Namibia to refute the snowball earth hypothesis, a hotly debated topic in the Earth sciences. When applied in concert with indicators of paleo-seawater pH such as boron isotopes (δ^{11}B), calcium isotopic variability in marine carbonate sediment can highlight excursions in atmospheric pCO_2 (Kasemann et al. 2005). $\delta^{44/40}$Ca variations are used to estimate seawater [Ca], which in turn may be used to estimate $[CO_3^{2-}]$. With the additional pH indicator provided by B isotopes, the carbonic acid system is fully constrained, and atmospheric pCO_2 may be estimated. The major weakness in this methodology is that absolute Ca fluxes to and from the ocean (F_w and F_{sed}) cannot be fully constrained by calcium isotopic mass balance alone. Additional determinations such as carbonate sedimentation rate are required to generate a more robust estimate of seawater calcium concentrations with time.

Though our understanding of the mechanisms and environmental sensitivity of Ca isotope fractionation during inorganic and biogenic $CaCO_3$ precipitation is still in its infancy, much has already been learned about the global marine Ca budget. As discussed in a later section, investigations into the pathways of Ca transport during biomineralization are beginning to reveal the causes of inter-species variability in $\Delta^{44/40}Ca_{seawater-CaCO_3}$. Constraints on Ca isotopic sensitivity to environmental controls such as T, pH and $[CO_3^{2-}]$ suggest proxy possibilities beyond reconstructing past oceanic δ_{sw}. Non-traditional metal stable isotope systems such as Ca promise to expand typical isotope geochemistry applications and explain the underlying chemical mechanisms of myriad processes observed in nature.

7.1.2 Ca Isotopes and the Terrestrial Calcium Cycle

Although it has already been shown that biological Ca isotope fractionation is an important process within marine biogeochemical Ca cycles, the current understanding of biological influences on terrestrial biogenic Ca isotope fractionation is still under development. From a broad perspective, the flux weighted mean discharge $\delta^{44/40}$Ca value of the largest rivers in the world is −0.2‰ relative to BSE (Fig. 7.1; Schmitt et al. 2003b). In comparison, the total range of values reported for igneous rock is −0.12 to 0.58‰ (Richter et al. 2003). This discrepancy indicates fractionation during the translocation of calcium from primary rock to soil and sedimentary rock to aqueous systems on the continents (e.g. DePaolo 2004). An important component of this isotope separation may be biogenic cycling in higher-order plants, which require calcium as a major nutrient.

Calcium is required to stabilize cell walls, promote enzyme activity, and act as an intracellular messenger and as a counter-cation in the vacuoles of plant cells (Marschner 1995). Unlike other macronutrients incorporated into organic compounds like nitrogen and phosphorous, calcium remains in its elementary cationic form throughout its residence in plants (Amtmann and Blatt 2009). Calcium from the soil solution enters the plant via fine root ends where it is transported through the xylem to shoots, stems and leaves. Uptake from the soil into the xylem can occur at high rates through both apoplastic and symplastic transport protein mediated pathways (White and Broadley 2003). Once taken into the root, the majority of calcium is directly adsorbed to cell walls. The remainder is taken into the cytoplasm where it may be transported to stem and leaf cells, though once incorporated in the phloem it is trapped and cannot undergo recirculation in the plant (von Blanckenburg et al. 2009). Retention of calcium in older growth has direct implications for the calcium signature of surrounding leaf litter and topsoil. Calcium isotope fractionation associated with movement of calcium through a plant will eventually affect the isotopic ratio of the detritus and soil of the ecosystem, suggesting a link between soil $\delta^{44/40}Ca$ and ecosystem evolution.

In the past 5 years, substantial progress has been made in quantifying the isotopic fractionation of calcium associated with uptake, translocation and redeposition by higher-order plants (von Blanckenburg et al. 2009). To date only a few studies directly address the influence of plants and ecosystem dynamics on calcium isotope fractionation (Wiegand et al. 2005; Perakis et al. 2006; Schmitt and Stille 2005; Page et al. 2008; Cenki-Tok et al. 2009; Holmden and Bélanger 2010). Results suggest that fractionation by higher-order plants exhibits a 4‰ range in $\delta^{44/40}Ca$ (Holmden and Bélanger 2010), representing one of the largest terrestrial sources of calcium isotopic variation. In the sections to follow, these studies will be discussed in further detail, and an example of the use of $\delta^{44/40}Ca$ to discern the vital effects in tree rings of phreatophytes in Death Valley, California (Yang et al. 1996), will be presented to illustrate the diversity of applications for calcium stable isotopes in the study of higher-order plants.

7.1.3 Calcium Isotopes in Food Chains

Skulan et al. (1997) were the first to report a systematic trend towards lighter $\delta^{44/40}Ca$ values with increasing trophic level in a marine ecosystem (Fig. 7.3a), and similar trends were subsequently observed in terrestrial systems (Fig. 7.3b; Skulan and DePaolo 1999; Skulan 1999). The marine system shows a $\Delta^{44/40}Ca$ of 3.2‰ between seawater and an orca, separated by four trophic levels. The terrestrial $\Delta^{44/40}Ca$ is slightly larger at 3.5‰ between total soil calcium and a cougar, with three trophic levels of separation. The clear loss of heavy ^{44}Ca relative to light ^{40}Ca across both food chains indicates preferential incorporation of the lighter isotope in the bone structure relative to that available in the organism's diet (Clementz et al. 2003; DePaolo 2004).

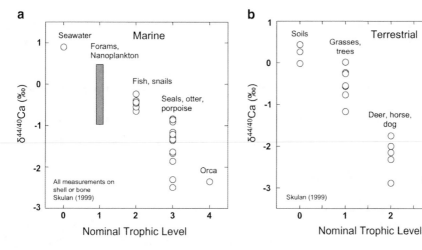

Fig. 7.3 Calcium isotopic composition of materials as a function of marine (**a**) and terrestrial (**b**) trophic levels from DePaolo (2004). $\delta^{44/40}Ca$ of calcium reported relative to the Skulan et al. (1997) reference material. Data represent values reported in Skulan et al. (1997), Skulan and DePaolo (1999) and Skulan (1999). More recent data may be found in Reynard et al. (2010). Reprinted with permission from the author

The fractionation between dietary and bone $\delta^{44/40}Ca$ has several important applications. First, Skulan and DePaolo (1999) present a model for $\delta^{44/40}Ca$ fractionation in organisms based on observations suggesting that the $\delta^{44/40}Ca$ of living tissue is much more transient and responsive to recent diet than bone, which is generally ~1.3‰ lighter. The model incorporates the availability of calcium in the diet and the rate of bone growth and bone loss and indicates healthy adults should have a near constant bone $\delta^{44/40}Ca$ value, while healthy juveniles should show $\delta^{44/40}Ca$ ~1.5‰ lower than the diet source. These findings have lead to subsequent research regarding calcium isotopes as a proxy for calcium availability in vertebrates (Chu et al. 2006; Hirata et al. 2008; Bullen and Walczyk 2009; Heuser and Eisenhauer 2010; Reynard et al. 2010). Second, Clementz et al. (2003) reported similar systematic variation between trophic levels, and showed preservation of this fractionation in bones as old as 15 million years, presenting a novel means of tracking vertebrate paleodiets. Recent developments built upon these studies will be addressed in subsequent sections.

7.2 Analytical Methods and Standards

Ca isotope fractionation can be measured by a variety of techniques including thermal ionization mass spectrometry (TIMS), multi-collector inductively coupled plasma-mass spectrometry (MC-ICP-MS), and in situ by ion microprobe secondary ion mass spectrometry (SIMS; Rollion-Bard et al. 2007). The large spread in Ca isotopic abundances and the large relative mass difference (~20%) make measurement of Ca isotopes complex. Precise determinations of mass dependent Ca isotope composition by TIMS requires the use of a double spike (a tracer solution enriched in two minor isotopes), typically $^{42}Ca/^{48}Ca$ (Russell et al. 1978; Zhu and Macdougall 1998; Skulan et al. 1997), $^{43}Ca/^{48}Ca$ (Nägler et al. 2000; Heuser et al. 2002; Gussone et al. 2003; Hippler et al. 2003) or $^{43}Ca/^{46}Ca$ (Fletcher et al. 1997) to correct for isotopic fractionation in the mass spectrometer. Early TIMS methods employed "peak jumping" or magnetic field stepping, measuring the four masses required for spike subtraction (typically 40, 42, 44 and 48) in sequence on a single collector (Russell et al. 1978; Skulan et al. 1997). This method provided superior external precision compared to the previous generation of multicollector mass spectrometers, but required a long-lived, stable ion beam (Skulan et al. 1997). Discrepancies between the analytical precision based on counting statistics and the external reproducibility of samples and standards were commonly attributed to inter-block focusing. Most researchers now measure Ca isotopes using the new generation TIMS instruments such as the Thermo Triton. The current generation (~1998–2009) of Triton TIMS instruments are only capable of measuring ~17% relative mass difference and thus do not allow for simultaneous measurement of all Ca isotopes. To circumvent this problem most labs measure multiple sequences with a "low mass" scan that measures ^{40}Ca–^{44}Ca and a second "high mass" sequence that measures ^{42}Ca–^{48}Ca. An alternate approach is to avoid measuring either ^{40}Ca or ^{48}Ca (Fantle and Bullen 2009; Holmden 2005). Fantle and Bullen (2009) provide a detailed summary of commonly used cup configurations and double spike compositions. Approximately 50–200 cycles are collected during the course of a sample run, and post-run processing is used to correct data for mass fractionation during analysis. Typical 2σ external reproducibility using TIMS is 0.1–0.2‰ (DePaolo 2004), which is worse than the internal errors generated from counting statistics (0.02–0.05‰). The source of additional error in Ca isotope measurements is still attributed to poor focusing (DePaolo 2004) and reservoir mixing during sample ionization (Fantle and Bullen 2009) though the effect has been minimized in recent years. Despite some shortcomings TIMS is the preferred and most common technique for Ca isotope analysis.

MC-ICP-MS is another widely used method in measuring calcium isotope fractionation. The technique may be employed using either the "cool plasma technique" (CPT; Fietzke et al. 2004) or the "hot plasma technique" (HPT; Halicz et al. 1999). Isobaric interferences from $^{40}Ar^+$ and numerous other sources undermine the utility of MC-ICP-MS for Ca isotope measurements, and the original hot plasma technique required a thorough accounting of and correction for interferences (Halicz et al. 1999). These effects can be greatly diminished by lowering the plasma RF power to ~400 W (Fietzke et al. 2004). MC-ICP-MS analyses enable the simultaneous measurement of various Ca isotope masses as well as sample-standard bracketing, rendering the use of a double spike unnecessary. This method yields poorer precision than TIMS, with 2σ uncertainty in $\delta^{44/40}Ca$ of 0.2–0.3‰ (Fietzke et al. 2004; Hippler et al. 2007) and in $\delta^{44/42}Ca$ of 0.09‰ (Tipper et al. 2006).

A novel ion microprobe technique has recently been developed to measure Ca isotope ratios via SIMS in situ with high (15–20 μm) spatial resolution (Rollion-Bard et al. 2007; Kasemann et al. 2008). Secondary $^{40}Ca^+$ and $^{44}Ca^+$ ions are simultaneously measured on two Faraday cups, providing good internal reproducibility. Interferences from $^{40}K^+$ and $^{88}Sr^{2+}$ on $^{40}Ca^+$ and $^{44}Ca^+$ respectively cannot currently be resolved but must be monitored on mass 41 and 43.5 to ensure no corrections are needed. Uncertainties up to 0.3‰ (2σ) are reported for in situ sample analyses (Rollion-Bard et al. 2007; Kasemann et al. 2008), meaning this method must be further refined for application in systems where minor isotopic variability is expected.

Isotopic analyses of most Ca carbonate minerals require minimal chemical preparation. Other materials with increased concentrations of matrix elements require more extensive separation procedures, which are detailed in the literature (e.g. Skulan and DePaolo 1999). For the TIMS technique, carbonate samples are first dissolved in 2N to 4N ultrapure nitric acid or HCl and centrifuged if detrital material is present. One M acetic acid is more commonly used in lieu of the stronger HCl or HNO_3 to avoid extraction of Ca from any associated detrital clay minerals (Ewing et al. 2008). The Ca double spike is added to an aliquot of sample to correct for mass discrimination during analysis. Optimal spike-sample ratios are discussed elsewhere (Rudge et al. 2009 and references therein). Samples are then dried and re-dissolved in acid. Purification of Ca from the sample matrix may be achieved using a variety of ion chromatography techniques such as cation exchange (Russell et al. 1978; Fantle and DePaolo 2007); and MCI gel (Heuser et al. 2002). Recently the Berkeley lab has started using DGA resin (Eichrom Technologies, normal variety), which efficiently separates Ca from matrix elements in a similar fashion to the better-known Sr-spec resin (Horwitz et al. 2005). Sufficiently pure carbonates may not require chemical purification (Eisenhauer et al. 2004). Purified Ca is dissolved in a small volume of 3N HNO_3 and a few micrograms of dissolved Ca are loaded onto degassed, zone-refined Re filaments and subsequently capped with phosphoric acid.

An alternate technique uses approximately 300 ng of Ca loaded with a Ta_2O_5 activator using the "sandwich technique" for analysis on single Re filaments (activator-sample-activator; Gussone et al. 2003). Sample throughput of ~6–12 per day is possible with this method. Cleaning and preparation of Ca bearing samples is similar for the MC-ICP-MS technique, and ion chromatography is often used to prevent matrix effects and isobaric interferences (Halicz et al. 1999). Once dissolved in acid, samples are diluted with nitric acid solution and introduced into the plasma source through a nebulizer (Halicz et al. 1999; Fietzke et al. 2004). Sample preparation for analysis on the ion microprobe requires that samples be mounted in epoxy, polished to 1 μm using a diamond paste, and coated in gold (Rollion-Bard et al. 2007).

7.2.1 Calcium Isotope Standards: A Proposal

Calcium isotopes are reported in delta notation (‰) as the ratio of the heavy ^{44}Ca over the abundant ^{40}Ca isotope with reference to a known standard value, consistent with the convention of other stable isotope systems (Skulan et al. 1997):

$$\delta^{44/40}Ca = \left(\frac{\left[\frac{^{44}Ca}{^{40}Ca}\right]_{sample}}{\left[\frac{^{44}Ca}{^{40}Ca}\right]_{std}} - 1 \right) \times 1,000.$$

Isotope compositions may also be presented in the $^{44/42}Ca$ notation, which may be converted to $\delta^{44/40}Ca$ to avoid discrepancies due to radiogenic ingrowth of ^{40}Ca (Hippler et al. 2003):

$$\delta^{44/40}Ca = \delta^{44/42}Ca \times \frac{43.956 - 39.963}{43.956 - 41.959}.$$

At the time of writing the most widely used standard reference material for calcium measurements is the NIST synthetic carbonate, SRM 915a (Hippler et al. 2003). The original material, SRM 915a is no longer available and has been superseded by a new reference material SRM 915b, which is 0.72 ± 0.04‰ heavier in $\delta^{44/40}Ca$ than SRM 915a (Heuser and Eisenhauer 2008). Other common standard reference materials include purified calcium salts such as CaF_2

(Russell et al. 1978; Nägler et al. 2000; Gussone et al. 2003) and CaCO$_3$ (Skulan et al. 1997) and seawater (Zhu and Macdougall 1998; Schmitt et al. 2001; Hippler et al. 2003). NIST SRM 915a has a $\delta^{44/40}$Ca of -1.88 ± 0.04‰ relative to seawater (Hippler et al. 2003) and $\delta^{44/40}$Ca of -1.21‰ relative to the Skulan et al. (1997) standard. Stable Ca isotope fractionation between two phases (e.g. i, j) is expressed as a fractionation factor,

$$\alpha_{i-j} = \frac{^{44}Ca/^{40}Ca_i}{^{40}Ca/^{40}Ca_j},$$

or as isotope separation,

$$\Delta^{44}Ca_{i-j} = \delta^{44}Ca_i - \delta^{44}Ca_j \approx 1000\ln(\alpha_{i-j}).$$

Though SRM 915a provides the most reliable interlab working reference standard, it can be argued that the most desirable reference standard for reporting $\delta^{44/40}$Ca values is bulk silicate Earth (BSE). Unlike seawater, this value does not shift through Earth history, except for very small effects due to radioactive decay of ^{40}K, and because it represents the overall bulk composition of the Earth, all deviations from the BSE value have geochemical significance. While SRM 915a is the most widely used laboratory standard, laboratory standards are not the same as standard values used for referencing the delta values for most stable isotope systems. In the case of O isotopes, for example, there are multiple laboratory standards and multiple reference materials. Simon and DePaolo 2010 have reported analyses of silicate materials from the Earth, Mars, and the Moon, and have determined that the bulk silicate value for these three planetary bodies is indistinguishable and identical to that of ordinary chondritic meteorites. Relative to this Bulk Planetary or Chondritic value (BSE), the SRM 915a standard is reported to have $\delta^{44/40}$Ca $= -0.97 \pm 0.04$. Values for mantle clinopyroxene reported by Huang et al. (2010) are consistent with the Simon and DePaolo (2010) results. Thus we will henceforth report all $\delta^{44/40}$Ca values relative to the BSE value, and we propose that the convention be that the standard BSE value of ^{44}Ca/^{40}Ca is exactly 1.00100 times the value of ^{44}Ca/^{40}Ca in SRM 915a. In general practice, it will be sufficiently accurate to convert $\delta^{44/40}$Ca relative to SRM 915a to $\delta^{44/40}$Ca relative to BSE by simply subtracting 1.00 from the former. In this chapter we report all $\delta^{44/40}$Ca values relative to BSE. These $\delta^{44/40}$Ca values differ from those reported by the U.C. Berkeley group through 2010 by 0.2 units of $\delta^{44/40}$Ca.

7.3 Current Research

7.3.1 Calcium Isotopes and Marine Biomineralization

The calcium isotopic composition of carbonate-secreting marine organisms constitutes a promising tool for paoleoclimatologic research, because the shells of certain species preserve paleoenvironmental information such as seawater temperature at the time of formation (e.g. Nägler et al. 2000; Hippler et al. 2009). Some species of planktonic foraminifera such as *Globigerinoides sacculifer* (*G. sacculifer*) and *Neogloboquadrina pachyderma* (sinistral) appear to exhibit strong $\delta^{44/40}$Ca temperature sensitivity on order 0.2‰ per 1°C (Nägler et al. 2000; Gussone et al. 2004; Hippler et al. 2006, 2009); however, this sensitivity in *G. sacculifer* is widely contested (Chang et al. 2004; Sime et al. 2005; Griffith et al. 2008b). Most species of marine organisms, including corals (Böhm et al. 2006), dinoflagellates (Gussone et al. 2010), and foraminifera (Skulan et al. 1997; Gussone et al. 2003; Griffith et al. 2008b) are effectively independent of temperature in calcium isotopic composition, providing a robust means of determining oceanic $\delta^{44/40}$Ca variability over geologic timescales. Understanding the temporal variability in oceanic Ca stable isotope composition has helped to constrain the global Ca budget. Recent research on Ca isotopes in marine calcifying organisms focuses on small but significant vital effects and variability in $\delta^{44/40}$Ca not explained by temperature dependence or inorganic control, because they may contain additional information about the biological pathways of calcification (e.g. Eisenhauer et al. 2009).

Biological mechanisms of calcification vary widely among marine organisms. Corals secreting aragonitic skeletons follow a biocalcification pathway much different from that of foraminifera, resulting in a distinct relationship between $\delta^{44/40}$Ca and growth environment. Though the Ca fractionation behavior of both

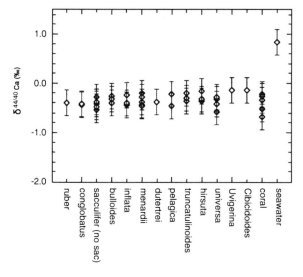

Fig. 7.4 Ca isotopic composition of seawater, corals and foraminifera from Holocene sediments distributed worldwide (Chang et al. 2004). Minimal intra-species variability in the magnitude of $\delta^{44/40}$Ca suggests a similar Ca stable isotope fractionation mechanism between organisms. Reprinted from Chang et al. (2004), Copyright (2004), with permission from Elsevier

foraminifera and corals has been shown to resemble abiotic Ca isotope fractionation during CaCO$_3$ precipitation (Gussone et al. 2003; Lemarchand et al. 2004; Marriott et al. 2004; Böhm et al. 2006; Tang et al. 2008), coral Ca is likely biologically fractionated by active transport through tissues (Böhm et al. 2006), while foraminifera appear to precipitate their shells directly from pH-modified seawater entrained in vacuoles (Erez 2003). Care should be taken when deducing biological Ca isotope fractionation mechanisms based on the experiments of Lemarchand et al. (2004) and Tang et al. (2008), because these groups report directly opposite relationships between $\Delta^{44/40}$Ca and precipitation rate. Despite inter-species differences and the challenge of interpretation, the overall homogeneity in average $\delta^{44/40}$Ca across Holocene marine calcifying organisms (Fig. 7.4) suggests that a universal mechanism may control marine biogenic Ca isotope fractionation.

7.3.1.1 Calcium Isotopes in Coral Skeletons

Though the details of coral calcification are still a major subject of debate, corals are currently thought to actively transport calcium from seawater to the site of calcification by a multi-step process requiring

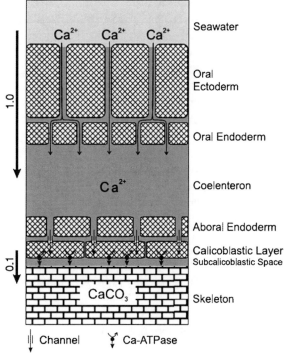

Fig. 7.5 Schematic of calcium transport and skeletal growth in corals. *Numbered arrows* on the *left* of the diagram indicate relative rates of Ca transport; calcium flows into the coelenteron an order of magnitude more quickly than into the calcifying fluid. Reprinted from Böhm et al. (2006), Copyright (2006), with permission from Elsevier

active Ca pumping through numerous tissues (Böhm et al. 2006; Fig. 7.5). Ca^{2+} ions first diffuse from seawater into coral tissue, through the oral ectoderm and endoderm into the coelenteron. There, they are sealed during the day while calcification takes place. Ca then passes from the coelenteron through the aboral endotherm to the calicoblastic layer, where it is actively transported via Ca-ATPase either to the calcifying subcalicoblastic space or possibly from the cell to the skeleton via an organic matrix (Allemand et al. 2004; Cuif and Dauphin 2005). Neither the existence of a subcalicoblastic space (Tambutté et al. 2007; Holbrèque et al. 2009) nor the role of the organic matrix in directing the transport and deposition of Ca ions (Allemand et al. 2004) has definitively been established. From its final reservoir, Ca^{2+} is incorporated into skeletal material (Clode and Marshall 2003; Böhm et al. 2006).

Each step in the coral biogeochemical Ca transport pathway may fractionate stable calcium isotopes, resulting in the preferential incorporation of ^{40}Ca over ^{44}Ca into skeletal material. Unlike inorganic CaCO$_3$ growth, Ca isotope fractionation likely does not occur at the carbonate mineral surface. Instead, it has been suggested that the kinetics of solvated Ca^{2+} dehydration at the cell membrane and of Ca^{2+} ion attachment to proteins such as Ca-ATPase causes the preferential incorporation of light calcium in coral skeletons (Gussone et al. 2006; Böhm et al. 2006). Dehydration kinetics of solvated Ca^{2+} ions is also a possible cause of kinetic, surface-controlled Ca isotope fractionation during inorganic calcite and aragonite precipitation (DePaolo 2011), so it is not surprising that the fractionation factor generated by these very different processes is similar.

The $\delta^{44/40}$Ca of both cultured and open-ocean coral skeletal material is relatively insensitive to seawater temperature, with a T-dependence similar to marine biogenic calcite and aragonite (Fig. 7.6; Böhm et al. 2006):

$$\delta^{44/40}\text{Ca (BSE)} = 0.020(\pm 0.015) \times T(°C) - 0.7(\pm 0.4).$$

Despite the morphological similarity between coral skeletal aragonite and synthetic aragonite precipitated under controlled growth conditions (Holcomb et al. 2009), coral $\delta^{44/40}$Ca values are offset by 0.5‰ from inorganic precipitates and other marine biogenic aragonite (Böhm et al. 2006). The magnitude of Ca isotope fractionation in coral skeletal material, then, is primarily controlled by biological mechanisms, while the T-dependence of fractionation is likely due to geochemical processes that are mechanistically similar in both inorganic and biological systems.

7.3.1.2 Calcium Isotopes in Foraminiferal Biomineralization

Most species of marine calcifying foraminifera exhibit a $\delta^{44/40}$Ca T-dependence similar to that of corals and inorganic calcite and aragonite. Because the calcification mechanism of foraminifera more closely resembles inorganic CaCO$_3$ precipitation (e.g. Erez 2003), it may be more appropriate to consider the rate dependence of Ca isotope fractionation during foraminiferal calcification than during coral skeletal growth. The growth of calcite skeletal material in foraminifera is a two-stage process producing two distinct phases, primary and secondary calcite. These phases differ in both trace element (Erez 2003) and Ca isotopic (Rollion-Bard et al. 2007; Kasemann et al. 2008) composition due to mechanistic differences in the calcification process. Erez (2003) summarizes a working hypothesis of the foraminiferal calcification process as follows. Primary calcite formation is initiated by Ca concentration in a highly soluble mineral phase (also rich in Mg, P and likely S) embedded in the cellular membrane of the endoplasm. These soluble granules provide Ca for the primary calcite crystals, which precipitate as a wall of spherulites directly over an organic matrix. Primary calcite spherulites are Mg-rich, and may reflect the formation of an amorphous calcium carbonate (ACC) precursor (Bentov and Erez 2006). Secondary calcite precipitates adjacent to the primary layer as a low-Mg, crystallographically oriented mineral phase (c-axis perpendicular to skeletal wall). Vacuoles of seawater provide calcium for the secondary phase, so unlike in corals, Ca is not actively transported through cellular membranes and fractionation cannot be attributed to biological pumping. This phase constitutes the bulk of foraminiferal skeletal material, and secondary calcite precipitation resembles inorganic, surface-controlled calcite

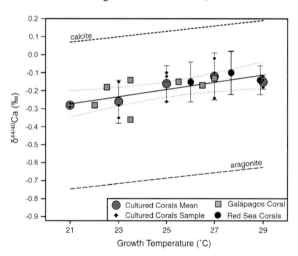

Fig. 7.6 Cultured and open-ocean coral skeletal $\delta^{44/40}$Ca plotted as a function of temperature, compared with similar linear regressions of marine biogenic calcite (braciopods, sclerosponges, and red alga) and aragonite (sclerosponges and pteropods; Böhm et al. 2006). The slope of the T dependence of these organisms is very similar to other marine biogenic calcite and aragonite, but the magnitude of fractionation is offset by ~0.5‰ from each. Reprinted from Böhm et al. (2006), Copyright (2006), with permission from Elsevier

precipitation. Alternating layers of primary and secondary calcite are deposited onto the skeletal wall, causing heterogeneity of both trace elements (Erez 2003) and possibly stable Ca isotopes (Rollion-Bard et al. 2007) within a single shell.

The $\delta^{44/40}$Ca temperature dependence of most studied foraminiferal shell material resembles that of inorganic calcite (Marriott et al. 2004) and aragonite (Gussone et al. 2003). The skeletal Ca isotopic composition typically increases approximately 0.02‰ per 1°C increase in seawater temperature (e.g. Skulan 1997; Gussone et al. 2003; Griffith et al. 2008b). Two exceptions are *N. pachyderma* (sin.) and *G. sacculifer*, which have temperature sensitive Ca isotopic compositions. Linear regression of the temperature sensitivity of Ca isotopic composition for *N. pachyderma* (sin.) for sea surface temperatures greater than 2.0 ± 0.5°C can be expressed as (Hippler et al. 2009):

$$\delta^{44/40}\text{Ca(BSE)} = 0.15(\pm 0.01) \times \text{SST}(°\text{C}) - 1.12(\pm 0.06).$$

The $\delta^{44/40}$Ca-temperature relationship for surface water *G. sacculifer* has similarly been expressed as follows (Hippler et al. 2006):

$$\delta^{44/40}\text{Ca(BSE)} = 0.22(\pm 0.05) \times \text{SST}(°\text{C}) - 5.85,$$

which is very similar to the regression generated by Nägler et al. (2000) for cultured samples. Despite the apparent strength of the temperature signal, the temperature sensitivity of $\delta^{44/40}$Ca in *G. sacculifer* is not agreed upon in the literature. Attempts at T calibration of this species from core-top samples have resulted in a range of interpretations from no T dependence (Sime et al. 2005, 2007), to a weak dependence similar to other foraminifera (Griffith et al. 2008b). Ca isotope fractionation in low-salinity cultures of *G. sacculifer* also appears to be only weakly temperature dependent (Gussone et al. 2009). In general, plankton tow samples exhibit a stronger T dependence than the same species collected from core-top samples (e.g. Griffith et al. 2008b), which may reflect the averaged sampling of organisms accumulated from throughout the water column. This ambiguity calls into question the utility of down-core samples of *G. sacculifer* as a paleotemperature proxy.

7.3.1.3 Marine Calcification, Ca Isotope Fractionation and Precipitation Rate

The temperature dependence of Ca isotope fractionation during inorganic and marine biogenic $CaCO_3$ precipitation is likely caused by the T sensitivity of mineral growth rates due to changing oversaturation (Lemarchand et al. 2004; Gussone et al. 2005; Steuber and Buhl 2006). Increasing temperature alters carbonic acid speciation, shifting phase equilibrium towards the CO_3^{2-} endmember and increasing precipitation rate (e.g. Millero 1995). The rate dependence of Ca isotope fractionation during calcite precipitation is not well understood, and is a subject of recent debate. At issue is whether Ca isotope fractionation during $CaCO_3$ precipitation is equilibrium- (e.g. Marriott et al. 2004; Lemarchand et al. 2004) or kinetically-controlled. Lemarchand et al. (2004) found that the Ca isotopic fractionation factor increases (approaches $\alpha = 1.000$) with increasing rate, while more recent calcite precipitation experiments performed by Tang et al. (2008) generated the opposite relationship at similar growth rates (Fig. 7.7). Both groups argue that the low-rate limit fractionation factor (-1.5 ± 0.25‰ for Lemarchand et al., ~0‰ for Tang et al.) approaches the equilibrium fractionation between calcite and aqueous Ca^{2+}. An equilibrium fractionation factor close to 0‰ is supported by calcite equilibrated with pore fluid Ca^{2+} over tens of millions of years (Fantle and DePaolo 2007), so it follows that marine biogenic Ca isotope fractionation must be kinetically controlled.

Calculated precipitation rates and ranges for marine biogenic calcite and aragonite fall within experimental inorganic Ca carbonate growth rate ranges. A comparison of associated Ca fractionation factors reveals that most biogenic carbonates are consistent with inorganic experimental data, with the exception of *Acopora* coral skeletal material (Fig. 7.7). The range of both surface area normalized growth rates and fractionation factors of an additional foraminiferan, *G. sacculifer*, fall within the experimental region (1.5–3.3 μmol/m²/h, −0.81 to −1.64‰; Carpenter and Lohmann 1992; Erez 1982), but the fractionation factor as a function of growth rate cannot yet be determined for this species. Differences in the Ca coordination environment between $CaCO_3$ polymorphs likely explains the additional offset in magnitude of ~0.6‰ between calcite and aragonite $\delta^{44/40}$Ca. Better constraint of surface-area normalized marine biocalcification rates

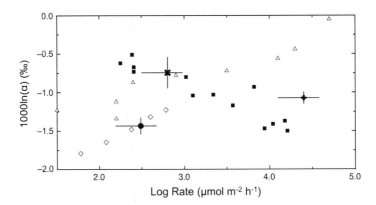

Fig. 7.7 CaCO$_3$ fractionation (1000ln(α)) as a function of precipitation rate for inorganic and marine biogenic calcite and aragonite. Tang et al. (2008) data (*filled squares*) suggest that experimentally precipitated inorganic calcite becomes lighter at faster growth rates, while Lemarchand et al. (2004) data (*open triangles*) follow the opposite trend. Inorganic aragonite (*open diamonds*) appears to follow a similar rate vs. fractionation trend as the latter. Marine biogenic carbonate skeletons from *Ceratoporella nicholsoni* (aragonite; *filled circle*) and *Acanthochaetetes wellsi* (calcite; *x symbol*) (Gussone et al. 2005) are consistent with inorganic precipitates, while the shells of cultured *Acropora* coral (aragonite; + symbol; Böhm et al. 2006) fall off of the expected trends. Error bars signify 2σ uncertainty in 1000ln(α) and ±50% in rate

is necessary before the relationship between shell growth rate and Ca isotope fractionation can be established. Because the growth rates of foraminiferal shells have been shown to depend upon environmental conditions (Kisakürek et al. 2008), vital effects in calcium isotopic composition may eventually provide proxy information for rate-controlling environmental conditions such as seawater pH.

7.3.2 Calcium Isotopes in Bone

Since Skulan et al. (1997) first published evidence for a systematic decrease in $\delta^{44/40}$Ca with increasing trophic level, subsequent studies have broadly fallen into two groups, (1) fractionation between tissue and bone in living and fossilized samples for use as a paleoproxy, and (2) development of $\delta^{44/40}$Ca tests to assess calcium consumption and bone composition in living subjects. Within the first group, Clementz et al. (2003) built upon the studies of Skulan et al. (1997), Skulan and DePaolo (1999) and Skulan (1999), extending analysis to a 15 Ma fossil record of marine species with similar evidence of isotopic enrichment across trophic levels. Fossilized embryos of *Megaclonophycus onustus* and *Parapandorina raphospissa* from as early as the Neoproterozoic have since been analyzed by Komiya et al. (2008) and show a 0.1–0.5‰ depletion in $\delta^{44/40}$Ca relative to host phosphorite and dolostone values, a finding the authors attribute to preservation of the dietary $\delta^{44/40}$Ca signal.

To date the most comprehensive study of fossilized calcium isotopes, and the first to report the $\delta^{44/40}$Ca of human bones, is Reynard et al. (2010). Building upon the techniques outlined in Chu et al. (2006) and work done by Skulan et al. (2007), the Reynard et al. (2010) study reports a survey of $\delta^{44/40}$Ca in both a herd of modern sheep and a variety of fossilized herbivores and humans at five locations ranging in age from ~11 to 3 ka. Results of modern sheep bone show female $\delta^{44/40}$Ca values are higher than males by an average of 0.14 ± 0.08‰. The study presents a modified version of the Skulan and DePaolo (1999) model showing this difference may be attributed to a combination of calcium loss due to lactation and additional bone growth during pregnancy. Fossil data illustrate significant variation between sites as a result of variation in the source values of $\delta^{44/40}$Ca between regions. At all sites, human bone values are lower than those of herbivores (Fig. 7.8) by an average of 0.2‰. The absence of systematic variation between sites predating and postdating herbivore domestication suggests dairy consumption cannot account for the wider variation in $\delta^{44/40}$Ca of human bones relative to those of local fauna. The authors conclude that substantial variation in the bone $\delta^{44/40}$Ca of individuals due to unique physiology may impede the use of calcium isotopes as an archeological tool, but support their use in dietary studies of living organisms. In the study of living

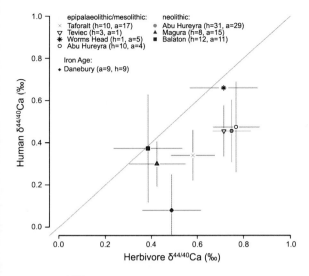

Fig. 7.8 $\delta^{44/40}$Ca for herbivore and human bones from common archeological sites. Error bars show 1σ standard deviation and line represents 1:1 ratio of herbivore to human values. Site locations are listed in the key with corresponding numbers of human (h) and animal (a) samples. Reprinted from Reynard et al. (2010), Copyright (2010), with permission from Elsevier

organisms, calcium isotopes are currently under development as a proxy for internal calcium balance. At the risk of deviating too far from the topic of calcium isotopes in biogeochemistry, we will refer the reader to Skulan et al. (2007), Hirata et al. (2008) and Heuser and Eisenhauer (2010) for further information.

7.3.3 Calcium Isotopes and Higher-Order Plants

The first study to report systematic lowering of $\delta^{44/40}$Ca values between calcium bound in plant material and associated exchangeable soil calcium was conducted in tropical ecosystems of Hawaii (Wiegand et al. 2005). The authors reported fractionation between basaltic soils and ohia trees (*Metrosideros polymorpha*) in forests ranging in soil age from 0.2 to 4.0 Ma (Fig. 7.9). Significant results of this study include (1) evidence for a systematic decrease in $\delta^{44/40}$Ca from roots to leaves, and (2) evidence that younger soils exhibit exchangeable soil pools with high ^{40}Ca, whereas older, more developed soils have higher ^{44}Ca pools, indicating previous recycling of plant material in the upper soil layers. Complementary to this research, studies by Schmitt et al. (2003b) and Schmitt and Stille (2005) reported values of riverine and precipitation $\delta^{44/40}$Ca suggesting that at local scales forests might influence the calcium isotopic value of precipitation by incorporation of canopy calcium sources such as leaf excretions.

Temperate hardwood forests of the Oregon coast range (Perakis et al. 2006), Adirondack Mountains of New York (Page et al. 2008) and Northeastern France (Cenki-Tok et al. 2009) exhibit similar systematic lowering of the $\delta^{44/40}$Ca ratio in trees relative to adjacent exchangeable soil calcium (Fig. 7.9). Soils measured by Perakis et al. (2006) show lower $\delta^{44/40}$Ca values in surface soils relative to soils at 60 cm depth. The authors suggest Douglas fir (*Pseudotsuga menziesii*) growing on this soil add leaf litter with light $\delta^{44/40}$Ca to the upper soil layers, thus recycling calcium to meet nutrient needs and creating a gradient in the soil isotope profile. Page et al. (2008) report similar results in species such as sugar maples, American beech, white ash, American basswood and yellow birch, where again extractable soil calcium $\delta^{44/40}$Ca is lighter in forest floor samples (−1.55 ± 0.61‰ BSE) than soils at depth (−0.14 ± 0.7‰ BSE at 75 cm depth). The study also shows the lowest $\delta^{44/40}$Ca value in both sugar maples and beech are found in the roots, followed by progressive enrichment of ^{44}Ca from stems to leaves. Cenki-Tok et al. (2009) used calcium isotope measurement of stream water, springs, soil solutions, precipitation and vegetation to show that both water-rock interaction and biologic cycling influence aqueous calcium isotopic signatures.

These studies result in the following observations:

1. ^{40}Ca is preferentially removed from the soil solution and incorporated into the plant roots, leaving the adjacent residual soil calcium pool enriched in ^{44}Ca relative to the plant.
2. Once taken into the plant, calcium is fractionated during translocation from root to stems to leaves such that the isotopic value of the leaf has the highest $\delta^{44/40}$Ca value.
3. Soil profiles in well developed forests typically show a gradient in $\delta^{44/40}$Ca with depth from lower values at the surface to greater values at depth, indicating recycling of isotopically light leaf and wood matter in the upper soil layers.

The points listed above are illustrated (Fig. 7.10) by data collected on a simplified system of aspen seedlings grown in a container of mixed basalt and

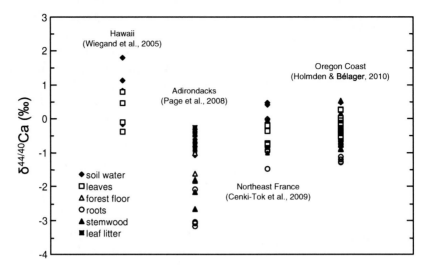

Fig. 7.9 $\delta^{44/40}$Ca of terrestrial material from four recent studies on ecosystem calcium isotope fractionation. Wiegand et al. (2005) report total soil values, exchangeable soil Ca values using a 0.1 N ammonium acetate extraction (shown above as soil water), and digested leaf values (shown). Page et al. (2008) report values from two adjacent catchments on a calcium rich feldspar bedrock including an exchangeable soil Ca value using 0.1 N ammonium acetate (shown above as soil water), separate values of forest floor soil and leaf litter, roots, stems and leaves from sugar maple, beech, white ash and basswood (shown above as root, stemwood and leaves). Cenki-Tok et al. (2009) report values along a hill slope transect including soil solutions obtained from lycimeters, roots, stems and leaves from beech and spruce varieties. Holmden and Bélanger (2010) report values from two toposequences including soil water collected in lycimeters, root, stemwood and foliage from a variety of species. Values for jack pine, trembling aspen, and black spruce are shown above

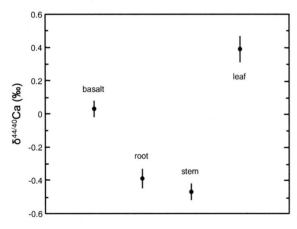

Fig. 7.10 $\delta^{44/40}$Ca for an aspen seedling grown in a potted mixture of basalt and quartz (data reproduced from Holmden and Bélanger 2010)

quartz (Holmden and Bélanger 2010). After 2 months of growth the seedlings were harvested for analysis. Between the basalt mixture and the root the $\Delta^{44/40}$Ca$_{basalt-root}$ is 0.42‰, illustrating preferential incorporation of the lighter ^{40}Ca isotope (observation 1). The values of the root and stem are within error of one another, but between the root and leaf $\Delta^{44/40}$Ca$_{root-leaf}$ is -0.78‰, indicating fractionation during translocation within the plant (observation 2). As described by Wiegand et al. (2005), on fresh parent rock a high ^{40}Ca flux is noted from the soil to the root, but presumably if a soil profile were allowed to develop as this aspen "forest" matured, the $\delta^{44/40}$Ca would show decreasing values from the basalt end-member to surface soil incorporating leaf and wood debris (observation 3).

Holmden and Bélanger (2010) have developed a model based on their study of a forested catchment in Saskatchewan, Canada (Fig. 7.11) towards a more quantitative analysis of the information calcium isotopes contain regarding interaction between vegetation, soil development and rock weathering. The model incorporates two available calcium pools, (1) the exchangeable calcium in the forest floor down to 10 cm, and (2) the exchangeable calcium in the B soil horizon from 10 to 25 cm depth. Fluxes in and out of these pools are soil mineral weathering, atmospheric deposition, leached plant litter, plant uptake, exchange between reservoirs and removal to the groundwater. The atmospheric calcium flux is adjusted to match independent calcium flux data based on ^{87}Sr/^{86}Sr measurements and 80% of annual tree calcium is derived from recycled leaf litter. The steady state solution for

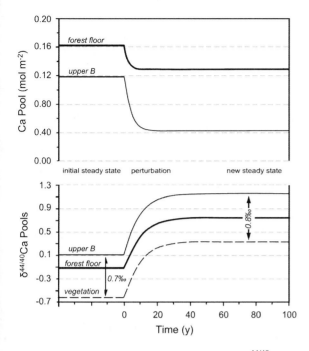

Fig. 7.11 Model calcium concentrations and $\delta^{44/40}$Ca in exchangeable pools. Initial steady state values are shown prior to time = 0, at which point calcium uptake flux by trees is increased 50%. This increase alters the balance of calcium input fluxes from mineral weathering, atmospheric deposition and leaf litter with uptake flux by the trees, leading to a change in the isotopic composition of all exchangeable pools. Reprinted from Holmden and Bélanger (2010), Copyright (2010), with permission from Elsevier

the Boreal forest yields values in agreement with those obtained from independent measurement and soil models. Using the time-dependent form of the model, the authors show an increase in calcium uptake by plants depletes soil calcium concentrations and causes a shift to higher $\delta^{44/40}$Ca values in the vegetation and soil pools (Fig. 7.11), illustrating a relationship between soil isotopic signature and the balance between calcium input (mineral weathering and leaf litter) and removal (root) fluxes. These results illustrate the potential for calcium isotopes as a tracer of water-rock-vegetation interactions and overall ecosystem health.

7.3.4 Case Study: Desert Environments and Vital Effects in Tree Rings

We determined $\delta^{44/40}$Ca variation in the tree rings of exotic phreatophytes in the desert climate of Death Valley, California. These results provide an indication of the behavior of biologically mediated Ca in a desert environment, the variability of Ca isotope fractionation during Ca uptake over several decades of growth, and the relationship between Ca, S, and C isotopes and precipitation variations.

The distribution of the two major types of desert plant species in Death Valley correlate with geology and soil hydrology. Xerophytes dominate on the alluvial fan gravels and have ephemeral water supplies in the vadose zone. The less-abundant phreatophytes grow at the toes of the alluvial fan gravels where their roots can reach the water table to provide a perennial water supply. On the valley floor, the salt-pan is devoid of plants. These desert plants constitute a potential source of information regarding hydrological, ecological and biogeochemical processes in Death Valley (Yang et al. 1996), complementary to abiotic effects observed in the Ca isotopes of arid environments (Ewing et al. 2008).

Among the larger phreatophytes in Death Valley is an exotic perennial, *Tamarix aphylla*, which serves as a dune stabilizer and windbreak (Robinson 1965; Baum 1978). Its taproots can reach down to 30 m depths and sub-superficial side roots may reach 50 m horizontally; the species can store large amounts of water in its roots and undergoes high evapotranspiration. Salt secretion is another typical feature of this tree, which leads to desalinization of deeper soil and increase of salinity in the upper layers. Since *Tamarix aphylla* is a perennial tree growing in desert environments and its roots extend to the water table, it can be a good proxy for modeling the biological calcium isotope fractionation in desert ecological and hydrologic systems.

Chemical analyses of *Tamarix aphylla* show high sulfur concentrations in the tree-rings (4–6 wt%, expressed as sulfate) and X-ray diffraction analyses show that the dominant sulfur compound is CaSO$_4$ (0.15–0.62 H$_2$O) (Yang et al. 1996). The δ^{34}S values of soluble sulfate increase from +13.5 to +18‰ VCDT from the core to the bark of the trunk, which are interpreted as reflecting deeper sulfate sources as the tree grew (Fig. 7.12). The δ^{13}C variations of the tree-ring cellulose (−27.6 to −24.0‰ VPDB) reflect changes in local precipitation, and show that *Tamarix aphylla* undergoes C$_3$ photosynthesis (Yang et al. 1996).

Calcium isotopes were measured in the soluble fraction of 12 tree ring samples from a 50-year-old specimen growing on an alluvial fan in Death Valley (Yang et al. 1996). The $\delta^{44/40}$Ca for the soluble calcium sulfates through the tree-ring section, which

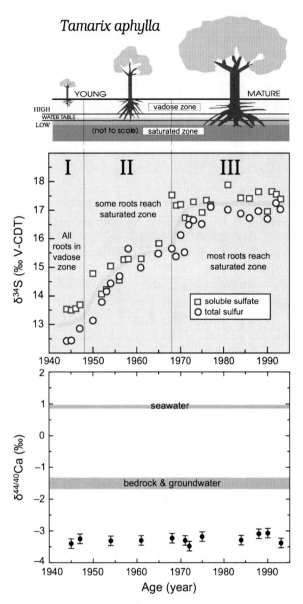

Fig. 7.12 $\delta^{34}S$ and $\delta^{44/40}Ca$ variation with time as recorded in the tree rings of *Tamarix aphylla* from Death Valley, California

cover a time period from 1945 to 1993, have an average value of $\delta^{44/40}Ca = -3.3‰$ (Fig. 7.12). Small variations are observed, from -3.46 to $-3.05‰$; the highest value (for 1990) occurs near the end of an extended drought. The low $\delta^{44/40}Ca$ values are similar to previous measurements of stems from a temperate forest (Page et al. 2008; Fig. 7.9). Our current limited data for the Death Valley hydro-ecological systems indicate that the tree has only one calcium source ($\delta^{44/40}Ca \approx -1.6‰$) from Nevares Spring water, which is isotopically uniform and similar to both local rainfall and limestones ($\delta^{44/40}Ca \approx -1.4‰$). The calcium isotopic fractionation between trunk and source is $\Delta^{44/40}Ca_{trunk-water} = -1.7‰$. This fractionation must be due to transport processes during root uptake of calcium. The slight increase in $\delta^{44/40}Ca$ under drought conditions suggests that when the tree is stressed, there may be less calcium isotope fractionation, either (1) because the calcium is held more tightly in small pores in the soil, or (2) the available calcium pool shrinks such that soil calcium starts to shift to more positive $\delta^{44/40}Ca$ values due to depletion of light calcium by the plant. The accumulating database on calcium isotopes in plants continues to suggest that systematic calcium isotope studies may be fruitful for understanding cation transport in plants, and soil ecological conditions in general.

7.4 Conclusions and Future Directions

In this chapter, we have shown that studies of Ca isotope fractionation shed light on the biogeochemical cycling of Ca isotopes in diverse natural ecosystems spanning oceanic and terrestrial environments. Biogenic fractionation of Ca isotopes during biomineralization produces predictable offsets in some marine organisms, which facilitate the reconstruction of seawater $\delta^{44/40}Ca$ over geologic timescales. Minor variability in Ca isotope fractionation both among and between marine calcifying organisms has begun to shed light on microscopic biomineralization pathways and their sensitivity to environmental change. Ca cycling in terrestrial ecosystems is similarly informed by studies of calcium isotopic fractionation. Observed Ca isotope fractionation between soil and various components of vegetation enables the construction of a local Ca soil budget which complements other data and provides a partial explanation for the scale of Ca isotopic variability within a single watershed. Watershed-scale variability is in turn similar in magnitude to variability on the global scale. The dependence of Ca isotopic composition on trophic level suggests the utility of this isotope system as a tracer of both food source and calcium balance within individual organisms. The research reviewed in this paper provides a foundation for future investigations into the macro- and microscopic processes and biogeochemical pathways dictating the distribution of this essential nutrient.

Potential future applications of Ca isotopes as biogeochemical tracers are diverse. To better constrain variability in the global marine Ca cycle, seawater $\delta^{44/40}$Ca reconstructions may be coupled with other isotopic and trace element indicators to explain short- and long-term excursions (e.g. Farkaš 2007; Payne et al. 2010). The relationship between $CaCO_3$ precipitation rate and microscopic mechanisms of Ca isotope partitioning remains unresolved, so future researchers must constrain the rates of biogenic calcification and determine the sensitivity of both Ca isotopic and trace element compositions to growth rates and environmental variables. An associated challenge will be to improve the small-scale resolution of Ca isotopic analyses and to improve upon the ion probe methods developed by Rollion-Bard et al. (2007) and Kasemann et al. (2008). The development of such a tool may elucidate the microscopic effects of mineralization pathway and perturbations in growth conditions on biomineral Ca isotope composition, which will in turn control global Ca isotopic signatures. Associated investigations of inorganic $CaCO_3$ precipitation to determine surface-controlled kinetic Ca isotope fractionation mechanisms may provide a basis for comparison with biogenic growth to constrain biological Ca transport pathways and biomineralization mechanisms. Studies of metal stable isotopes including calcium embody the cutting edge of research in isotope geochemistry, providing a framework for understanding the mechanisms and pathways controlling biogeochemical processes.

Acknowledgements Research on the biogeochemistry of Ca isotopes by the authors has been supported by the National Science Foundation (NSF EAR-9526997; NSF EAR-9909639; NSF EAR-0838168), a NASA Astrobiology Institute grant (BioMARS; NAI02-0024-0006), and by the Director, Office of Science, Office of Basic Energy Sciences, of the U.S. Department of Energy under Contract No. DEAC02-05CH11231 to the Lawrence Berkeley National Laboratory.

References

Allemand D, Ferrier-Pagès C, Furla P et al (2004) Biomineralisation in reef-building corals: from molecular mechanisms to environmental control. C R Palevol 3:453–467

Amini M, Eisenhauer A, Böhm F et al (2008) Calcium isotope ($\delta^{44/40}$Ca) fractionation along hydrothermal pathways, Logatchev field (Mid-Atlantic Ridge, 14°45′N). Geochim Cosmochim Acta 72:4107–4122

Amtmann A, Blatt M (2009) Regulation of macronutrient transport. New Phytol 181:35–52

Baum BR (1978) The genus *Tamarix*. The Israel Academy of Sciences and Humanities, Jerusalem

Bentov S, Erez J (2006) Impact of biomineralization processes on the Mg content of foraminiferal shells: a biological perspective. Geochem Geophys Geosyst 7:1–11

Berner R, Lasaga A, Garrels R (1983) The carbonate-silicate geochemical cycle and its effect on atmospheric carbon dioxide over the past 100 million years. Am J Sci 283:641–683

Böhm F, Eisenhauer A, Heuser A, Kiessling W, Wallmann K (2005) Calcium isotope fractionation during dolomitization. GeoErlangen, Schriften Deutsch Ges Geowissensch 39.

Böhm F, Gussone N, Eisenhauer A et al (2006) Calcium isotope fractionation in modern scleractinian corals. Geochim Cosmochim Acta 70:4452–4462

Broecker W, Peng T (1982) Tracers in the sea. Eldigo Press, New York

Bullen T, Walczyk T (2009) Environmental and biomedical applications of natural metal stable isotope variations. Elements 5:381–385

Carpenter S, Lohmann K (1992) Sr/Mg ratios of modern marine calcite: empirical indicators of ocean chemistry and precipitation rate. Geochim Cosmochim Acta 56:1837–1849

Cenki-Tok B, Chabaux F, Lemarchand D et al (2009) The impact of water-rock interaction and vegetation on calcium isotope fractionation in soil- and stream waters of a small, forested catchment (the Strengbach case). Geochim Cosmochim Acta 73:2215–2228

Chang V, Williams R, Makishima A et al (2004) Mg and Ca isotope fractionation during $CaCO_3$ biomineralisation. Biochem Biophys Res Commun 323:79–85

Chu N, Henderson G, Belshaw N et al (2006) Establishing the potential of Ca isotopes as proxy for consumption of dairy products. Appl Geochem 21:1656–1667

Clementz M, Holden P, Koch P (2003) Are calcium isotopes a reliable monitor of trophic level in marine settings? Int J Osteoarchaeol 13:29–36

Clode P, Marshall A (2003) Calcium associated with fibrillar organic matrix in the scleractinian coral *Galaxea fascicularis*. Protoplasma 220:153–161

Cuif J, Dauphin Y (2005) The environment recording unit in coral skeletons – a synthesis of structural and chemical evidences for a biochemically driven, stepping-growth process in fibers. Biogeoscience 2:61–73

De La Rocha C, DePaolo D (2000) Isotopic evidence for variations in the marine calcium cycle over the Cenozoic. Science 289:1176–1178

DePaolo D (2004) Calcium isotope variations produced by biological, kinetic, radiogenic and nucleosynthetic processes. In: Johnson C, Beard B, Albarede F (eds) Reviews in mineralogy and geochemistry: geochemistry of the non-traditional stable isotopes, 52. Mineralogical Society of America, Washington

DePaolo D (2011) Theory of isotopic and trace element fractionation during precipitation of carbonate minerals from aqueous solutions: surface reaction control limit. Geochim Cosmochim Acta 75:1039–1056

Eisenhauer A, Nägler T, Stille P et al (2004) Proposal for international agreement on Ca notation resulting from discussions at workshops on stable isotope measurements held in Davos (Goldschmidt 2002) and Nice (EGS-AGU-EUG 2003). Geostand Geoanal Res 28:149–151

Eisenhauer A, Kisakürek B, Böhm F (2009) Marine calcification: an alkali earth metal isotope perspective. Elements 5:365–368

Erez J (1982) Calcification rates, photosynthesis and light in planktonic foraminifera. In: Westbroek P, De Jong E (eds) Biomineralization and biological metal accumulation. Reidel, Dordrecht

Erez J (2003) The source of ions for biomineralization in foraminifera and their implications for paleoceanographic proxies. In: Dove P, De Yoreo J, Weiner S (eds) Reviews in mineralogy and geochemistry: biomineralization, vol 54. Mineralogical Society of America, Washington

Ewing S, Yang W, DePaolo D et al (2008) Non-biological fractionation of stable Ca isotopes in soils of the Atacama desert, Chile. Geochim Cosmochim Acta 72:1096–1110

Fantle M (2010) Evaluating the Ca isotope proxy. Am J Sci 310:194–230

Fantle M, DePaolo D (2005) Variations in the marine Ca cycle over the past 20 millino years. Earth Planet Sci Lett 237:102–117

Fantle M, DePaolo D (2007) Ca isotopes in carbonate sediment and pore fluid from ODP site 807A: the Ca^{2+}(aq)-calcite equilibrium fractionation factor and calcite recrystallization rates in Pleistocene sediments. Geochim Cosmochim Acta 71:2524–2546

Fantle M, Bullen T (2009) Essentials of iron, chromium, and calcium isotope analysis of natural materials by thermal ionization mass spectrometry. Chem Geol 258:50–64

Farkaš J, Böhm F, Wallmann K et al (2007) Calcium isotope record of Phanerozoic oceans: implications for chemical evolution of seawater and its causative mechanisms. Geochim Cosmochim Acta 71:5117–5134

Fietzke J, Eisenhauer A, Gussone N et al (2004) Direct measurement of $^{44}Ca/^{40}Ca$ ratios by MC-ICP-MS using the cool plasma technique. Chem Geol 206:11–20

Fletcher I, Maggi A, Rosman K et al (1997) Isotopic abundance measurements of K and Ca using a wide-dispersion multicollector mass spectrometer and low-fractionation isonisation techniques. Int J Mass Spectrom Ion Process 163:1–17

Griffith E, Paytan A, Caldeira K et al (2008a) A dynamic marine calcium cycle during the past 28 million years. Science 322:1671–1674

Griffith E, Paytan A, Kozdon R et al (2008b) Influences on the fractionation of calcium isotopes in planktonic foraminifera. Earth Planet Sci Lett 268:124–136

Gussone N, Eisenhauer A, Heuser A et al (2003) Model for kinetic effects on calcium isotope fractionation ($\delta^{44}Ca$) in inorganic aragonite and cultured planktonic foraminifera. Geochim Cosmochim Acta 67:1375–1382

Gussone N, Eisenhauer A, Tiedemann R et al (2004) $\delta^{44}Ca$, $\delta^{18}O$ and Mg/Ca reveal Caribbean Sea surface temperature and salinity fluctuations during the pliocene closure of the Central-American gateway. Earth Planet Sci Lett 227:201–214

Gussone N, Böhm F, Eisenhauer A et al (2005) Calcium isotope fractionation in calcite and aragonite. Geochim Cosmochim Acta 69:4485–4494

Gussone N, Langer G, Thoms S et al (2006) Cellular calcium pathways and isotope fractionation in *Emiliania huxleyi*. Geology 34:625–628

Gussone N, Hönisch B, Heuser A et al (2009) A critical evaluation of calcium isotope ratios in tests of planktonic foraminifers. Geochim Cosmochim Acta 73:7241–7255

Gussone M, Zonneveld K, Kuhnert H (2010) Minor element and Ca isotope composition of calcareous dinoflagellate cysts of cultured *Thoracosphaera heimii*. Earth Planet Sci Lett 289:180–188

Halicz L, Galy A, Belshaw N et al (1999) High-precision measurement of calcium isotopes in carbonates and related materials by multiple collector inductively coupled plasma mass spectrometry (MC-ICP-MS). J Anal At Spectrom 14:1835–1838

Heuser A, Eisenhauer A, Gussone N et al (2002) Measurement of calcium isotopes ($\delta^{44}Ca$) using a multicollector TIMS technique. Int J Mass Spec 220:385–397

Heuser A, Eisenhauer A (2008) The calcium isotope composition ($\delta^{44/40}Ca$) of NIST SRM 915b and NIST SRM 1486. Geostand Newsl J Geostand Geoanal 32:311–315

Heuser A, Eisenhauer A (2010) A pilot study on the use of natural calcium isotope ($^{44}Ca/^{40}Ca$) fractionation in urine as a proxy for the human body calcium balance. Bone 45:889–896

Heuser A, Eisenhauer A, Böhm F et al (2005) Calcium isotope ($\delta^{44/40}Ca$) variations of Neogene planktonic foraminifera. Paleoceanography 20:1–13

Hippler D, Schmidtt A, Gussone N et al (2003) Calcium isotopic composition of various reference materials and seawater. Geostand Newsl J Geostand Geoanal 27:13–19

Hippler D, Eisenhauer A, Nägler T (2006) Tropical Atlantic SST history inferred from Ca isotope thermometry over the last 140ka. Geochim Cosmochim Acta 70(90):100

Hippler D, Kozdon R, Darling K et al (2007) Calcium isotopic composition of high-latitude proxy carrier *Neogloboquadrina pachyderma* (sin.). Biogeosci Discuss 4:3301–3330

Hippler D, Kozdon R, Darling K et al (2009) Calcium isotopic composition of high-latitude proxy carrier *Neogloboquadrina pachyderma* (sin.). Biogeosci Discuss 4:3301–3330

Hirata T, Tanoshima M, Suga A et al (2008) Isotopic analysis of calcium in blood plasma and bone from mouse samples by multiple collector-ICP-mass spectrometry. Anal Sci 24:1501–1507

Holbrèque F, Meibom A, Cuif J-P et al (2009) Strontium-86 labeling experiments show spatially heterogeneous skeletal formation in the scleractinian coral *Porites porites*. Geophys Res Lett 36:L04604

Holcomb M, Cohen A, Gabitov R et al (2009) Compositional and morphological features of aragonite precipitated experimentally from seawater and biogenically by corals. Geochim Cosmochim Acta 73:4166–4179

Holmden C (2005) Measurement of $\delta^{44}Ca$ using a ^{42}Ca-^{43}Ca double-spike TIMS technique: summary of investigations. In: Saskachewan Geological Survey, Sask Industry Resources, Misc Rep 1:A-4

Holmden C (2009) Ca isotope study of Ordovician dolomite, limestone, and anhydrite in the Williston Basin: implications for subsurface dolomitization and local Ca cycling. Chem Geol 268:180–188

Holmden C, Bélanger N (2010) Ca isotope cycling in a forested ecosystem. Geochim Cosmochim Acta 74:995–1015

Horwitz E, McAlister D, Bond A et al (2005) Novel extraction of chromatographic resins based on Tetraalkyldiglycolamides: characterization and potential applications. Solvent Extr Ion Exch 23:319–344

Huang S, Farkaš J, Jacobsen S (2010) Calcium isotopic fractionation between clinopyroxene and orthopyroxene from mantle peridotites. Earth Planet Sci Lett 292:337–344

Kasemann S, Hawkesworth C, Pravec A et al (2005) Boron and calcium isotope composition in Neoproterozoic carbonate rocks from Namibia: evidence for extreme environmental change. Earth Planet Sci Lett 231:73–86

Kasemann S, Schmidt D, Pearson P et al (2008) Biological and ecological insights into Ca isotopes in planktic foraminifers as a paleotemperature proxy. Earth Planet Sci Lett 271:292–302

Kisakürek B, Eisenhauer A, Böhm F et al (2008) Controls on shell Mg/Ca and Sr/Ca in cultured planktonic foraminiferan, *Globigerinoides ruber* (white). Earth Planet Sci Lett 273:260–269

Komiya T, Suga A, Ohno T et al (2008) Ca isotopic compositions of dolomite, phosphorite and the oldest animal embryo fossils from the Neoproterozoic in Weng'an. S China Gondwana Res 14:209–218

Lasaga A, Berner R, Garrels R (1985) An improved geochemical model of atmospheric CO_2 fluctuations over the past 100 million years. In: Sundquist E, Broecker W (eds) The carbon cycle and atmospheric CO_2: natural variations archean to present. American Geophysical Union, Washington

Lemarchand D, Wasserburg G, Papanastassiou D (2004) Rate-controlled calcium isotope fractionation in synthetic calcite. Geochim Cosmochim Acta 68:4665–4678

Marschner H (1995) Mineral nutrition of higher plants, 2nd edn. Academic, London

Marriott C, Henderson G, Belshaw N et al (2004) Temperature dependence of δ^7Li, $\delta^{44}Ca$ and Li/Ca during growth of calcium carbonate. Earth Planet Sci Lett 222:615–624

Millero F (1995) Thermodynamics of the carbon dioxide system in the oceans. Geochim Cosmochim Acta 59:661–677

Nägler T, Eisenhauer A, Müller A et al (2000) $\delta^{44}Ca$-temperature calibration on fossil and cultured *Globigerinoides sacculifer*: new tool for reconstruction of past sea surface temperatures. Geochem Geophys Geosyst 1:2000GC000091

Page B, Bullen T, Mitchell M (2008) Influences of calcium availability and tree species on Ca isotope fractionation in soil and vegetation. Biogeochemistry 88:1–13

Payne J, Turchyn A, Paytan A et al (2010) Calcium isotope constraints on the end-Permian mass extinction. Proc Natl Acad Sci USA 107:8543–8548

Perakis S, Maguire D, Bullen T et al (2006) Coupled nitrogen and calcium cycles in forests of the Oregon coast range. Ecosystems 9:63–74

Reynard L, Henderson G, Hedges R (2010) Calcium isotope ratios in animal and human bone. Geochim Cosmochim Acta 74:3735–3750

Richter F, Davis A, DePaolo D (2003) Isotope fractionation by chemical diffusion between molten basalt and rhyolite. Geochim Cosmochim Acta 67:3905–3923

Robinson T (1965) Geological Survey Professional Paper 491-A: introduction, spread and areal extent of saltcedar (*Tamarix*) in the Western States. United States Government Printing Office, Washington

Rollion-Bard C, Vigier N, Spezzaferri S (2007) In situ measurements of calcium isotopes by ion microprobe in carbonates and application to foraminifera. Chem Geol 244:679–690

Rudge J, Reynolds B, Bourdon B (2009) The double spike toolbox. Chem Geol 265:420–431

Russell W, Papanastassiou D, Tombrello T (1978) Ca isotope fractionaion on the Earth and other solar system materials. Geochim Cosmochim Acta 42:1075–1090

Schmitt A, Stille P (2005) The source of calcium in wet atmospheric deposits: Ca-Sr isotope evidence. Geochim Cosmochim Acta 69:3463–3468

Schmitt A, Bracke G, Stille P et al (2001) The calcium isotope composition of modern seawater determined by thermal ionisation mass spectrometry. Geostand Newsl J Geostand Geoanal 25:267–275

Schmitt A, Stille P, Vennemann T (2003a) Variations of the $^{44}Ca/^{40}Ca$ ratio in seawater during the past 24 million years: evidence from $\delta^{44}Ca$ and $\delta^{18}O$ values of Miocene phosphates. Geochim Cosmochim Acta 67:2607–2614

Schmitt A, Chabaux F, Stille P (2003b) The calcium riverine and hydrothermal isotopic fluxes and the oceanic calcium mass balance. Earth Planet Sci Lett 213:503–518

Sime N, De La Rocha C, Galy A (2005) Negligible temperature dependence of calcium isotope fractionation in 12 species of planktonic foraminifera. Earth Planet Sci Lett 232:51–66

Sime N, De La Rocha C, Tipper E et al (2007) Interpreting the Ca isotope record of marine biogenic carbonates. Geochim Cosmochim Acta 71:3979–3989

Simon J, DePaolo D (2010) Stable calcium isotopic composition of meteorites and rocky planets. Earth Planet Sci Lett 289:457–466

Skulan J (1999) Calcium isotopes and the evolution of terrestrial reproduction in vertebrates. PhD Dissertation, University of California, Berkeley

Skulan J, DePaolo D (1999) Calcium isotope fractionation between soft and mineralized tissues as a monitor of calcium use in vertebrates. Proc Natl Acad Sci USA 96:13709–13713

Skulan J, DePaolo D, Owens T (1997) Biological control of calcium isotopic abundances in the global calcium cycle. Geochim Cosmochim Acta 61:2505–2510

Skulan J, Bullen T, Anbar A et al (2007) Natural calcium isotopic composition of urine as a marker of bone mineral balance. Clin Chem 53:1155–1158

Soudry D, Segal I, Nathan Y et al (2004) $^{44}Ca/^{42}Ca$ and $^{143}Nd/^{144}Nd$ isotope variations in Cretaceous-Eocene Tethyan francolites and their bearings on phosphogenesis in the southern Tethys. Geology 32:389–392

Steuber T, Buhl D (2006) Calcium-isotope fractionation in selected modern and ancient marine carbonates. Geochim Cosmochim Acta 70:5507–5521

Tambutté E, Allemand D, Zoccola D et al (2007) Observations of the tissue-skeleton interface in the scleractinian coral *Stylophora pistillata*. Coral Reefs 26:517–529

Tang J, Dietzel M, Böhm F et al (2008) Sr^{2+}/Ca^{2+} and $^{44}Ca/^{40}Ca$ fractionation during inorganic calcite formation: II. Ca isotopes. Geochim Cosmochim Acta 72:3733–3745

Tipper E, Galy A, Bickle M (2006) Riverine evidence for a fractionated reservoir of Ca and Mg on the continents: implications for the oceanic Ca cycle. Earth Planet Sci Lett 247:267–279

von Blanckenburg F, Wiren N, Guelke M et al (2009) Fractionation of metal stable isotopes by higher plants. Elements 5:375–380

White P, Broadley M (2003) Calcium in plants. Ann Bot 92:487–511

Wiegand B, Chadwick O, Vitousek P et al (2005) Ca cycling and isotopic fluxes in forested ecosystems in Hawaii. Geophys Res Lett 32:L11404. doi:10.1029/2005GL022746

Yang W, Spencer R, Krouse H (1996) Stable sulfur isotope hydrogeochemical studies using desert shrubs and tree rings, Death Valley, California, USA. Geochim Cosmochim Acta 60:3015–3022

Zhu P, Macdougall J (1998) Calcium isotopes in the marine environment and the oceanic calcium cycle. Geochim Cosmochim Acta 62:1691–1698

Chapter 8
Natural and Anthropogenic Cd Isotope Variations

M. Rehkämper, F. Wombacher, T.J. Horner, and Z. Xue

Abstract Cadmium is a transition metal with eight naturally occurring isotopes that have atomic mass numbers of between 106 and 116. The large Cd isotope anomalies of meteorites have been subject to investigation since the 1970s, but improvements in instrumentation and techniques have more recently enabled routine studies of the smaller stable Cd isotope fractionations that characterize various natural and anthropogenic terrestrial materials. Whilst the current database is still comparatively small, pilot studies have identified two predominant mechanisms that routinely generate Cd isotope effects – partial evaporation/condensation and biological utilization. Processes that involve evaporation and condensation appear to be largely responsible for the Cd isotope fractionations of up to 1‰ (for $^{114}Cd/^{110}Cd$) that have been determined for industrial Cd emissions, for example from ore refineries. Cadmium isotope measurements hence hold significant promise for tracing anthropogenic sources of this highly toxic metal in the environment. The even larger Cd isotope fractionations that have been identified in the oceans (up to 4‰ for $^{114}Cd/^{110}Cd$) are due to biological uptake and utilization of dissolved seawater Cd. This finding confirms previous work, which identified Cd as an essential marine micronutrient that exhibits a phosphate-like distribution in the oceans. The marine Cd isotope fractionations are of particular interest, as they can be used to study micronutrient cycling and its impact on ocean productivity. In addition, they may also inform on past changes in marine nutrient utilization and how these are linked to global climate, if suitable archives of seawater Cd isotope compositions can be identified.

8.1 Introduction

8.1.1 The Element Cd

The element cadmium (Cd) was discovered in 1817 by Friedrich Strohmeyer of Göttingen (Germany), whilst analyzing a brownish, Cd-rich Zn-oxide or carbonate. The association of Cd with Zn is common in nature and the scarce mineral deposits of greenockite (CdS) and otavite ($CdCO_3$) are generally found in association with more abundant sphalerite (ZnS) and smithsonite ($ZnCO_3$). Hence, industrial recovery of Cd occurs as a by-product of processing Zn and also Pb ores, as galena (PbS) is commonly associated with sphalerite. Zinc metal is first produced from these ores and Cd metal is then isolated from Zn by either distillation or electrolysis. Ultra-pure Cd (99.9999%) is prepared by zone refining (Chizhikov 1966; Aylett 1973; Holleman et al. 1985; Cotton and Wilkinson 1988).

Some key chemical and physical properties of Cd, which is a Group 12 element along with Zn and Hg, are summarized in Table 8.1. Cadmium is a soft, silvery-blue lustrous metal that tarnishes in air but is otherwise stable at room temperature. Cadmium metal is comparatively volatile and produces a monoatomic

---

M. Rehkämper (✉) and Z. Xue
Department of Earth Science and Engineering, Imperial College London, London SW7 2AZ, UK
e-mail: markrehk@imperial.ac.uk

F. Wombacher
Institut für Geologie und Mineralogie, Universität zu Köln, Zülpicher Straße 49b 50674, Köln, Germany

T.J. Horner
Department of Earth Sciences, University of Oxford, Oxford OX1 3AN, UK

Table 8.1 Chemical and physical properties of Cd

Property	
Atomic number	48
Standard atomic weight (g/mol) (Wieser 2006)	112.411
Electron configuration of Cd^0	$[Kr]\,4d^{10}5s^2$
Melting point (°C)	320.9
Boiling point (°C)	767.3
Density of solid at 25°C (g/cm^3)	8.642
Density of liquid at m.p. (g/cm^3)	8.02
Oxidation states	+2, +1 (rare)
Electronegativity[a]	1.69
E^0 for $Cd^{2+} + 2e^- = Cd$ (V)	−0.402
First ionization energy (kJ/mol)	867
Second ionization energy (kJ/mol)	1,625
Atomic radius (pm) (Slater 1964)	155
Ionic radius of Cd^{2+} (pm) (Shannon 1976)	95
Mohrs hardness	2.0

All other data from Chizhikov (1966), Aylett (1973), Holleman et al. (1985) and Cotton and Wilkinson (1988)
[a] Pauling's scale

vapor. The metal also forms many alloys, burns to CdO on heating, and reacts readily with both oxidizing and non-oxidizing acids. Cadmium displays chemical similarities to Zn, with both elements showing a strong preference for the +2 oxidation state but Cd has a more diverse complex chemistry in solution. The electronegativity of Cd (1.69 on Pauling's scale) is similar to V, Cr and Tl, and most Cd compounds (e.g., chlorides, nitrates, sulfates, etc.) therefore have a predominant ionic character. Cadmium and many of its compounds are highly toxic and carcinogenic. This is particularly problematic because Cd accumulates in organisms and is difficult to flush out (Nriagu 1981).

The production of rechargeable Ni-Cd batteries constitutes the most important industrial use of Cd. Other important applications include its use as a corrosion resistant coating for metals, and in the production of pigments and plastic stabilizers. Overall, industrial use of Cd has been decreasing, in account of its toxicity and legislation that governs handling and application of the metal (Chizhikov 1966; Aylett 1973; Holleman et al. 1985; Cotton and Wilkinson 1988).

8.1.2 Cadmium Isotopes and Natural Isotope Variations

The identity of the eight stable or quasi-stable isotopes of Cd with mass numbers of between 106 and 116 was first confirmed in the 1930s, based on the pioneering mass spectrometric analyses of Aston (1935), Nier (1936), and others (Table 8.2). The isotopic abundances that were published by Leland and Nier (1948) are remarkably precise and already identical, within uncertainty, to the most recent reference values published by IUPAC (Böhlke et al. 2005) (Table 8.1). The latter compilation lists the results of Rosman et al. (1980) as the best measurement from a single source.

On Earth, the only significant natural source of Cd isotope variations are stable isotope fractionations. The isotopes ^{113}Cd and ^{116}Cd are long-lived radioactive isotopes that transform to ^{113}In and ^{116}Sn by β^--decay. Both nuclides have extremely long half-lives (of ~9 × 10^{15} a and ~2.6 × 10^{19} a, respectively), however, such that they are considered to be quasi-stable isotopes with essentially constant abundances (Pfennig et al. 1998).

In extraterrestrial materials, Cd isotope compositions can be altered as a result of neutron capture reactions, which are induced by secondary neutrons generated by cosmic rays. The isotope ^{113}Cd has a large neutron capture cross section of 20,600 barns for (epithermal) neutrons with energies of less than about 0.5 eV and produces ^{114}Cd by a (n, γ) reaction (Pfennig et al. 1998). Consequently, the ^{113}Cd/^{114}Cd ratios of lunar soils and surface rocks were observed to differ from the terrestrial value by up to 5‰ (Sands et al. 2001; Schediwy et al. 2006; Wombacher et al. 2008). Such cosmogenic Cd isotope effects are not significant for terrestrial samples (Wombacher et al. 2008), as the Earth's surface completely shields the interior from cosmic rays, whilst the atmosphere severely attenuates the cosmic ray flux to the surface.

As is common in geochemistry, variations in Cd stable isotope compositions are reported using a relative notation and both ε- and δ-values are currently in use. For the ^{114}Cd/^{110}Cd isotope ratio, these are calculated as:

$$\varepsilon^{114/110}Cd = \left[\frac{\left(^{114}Cd/^{110}Cd\right)_{sample}}{\left(^{114}Cd/^{110}Cd\right)_{std}} - 1\right] \times 10,000 \quad (8.1)$$

$$\delta^{114/110}Cd = \left[\frac{\left(^{114}Cd/^{110}Cd\right)_{sample}}{\left(^{114}Cd/^{110}Cd\right)_{std}} - 1\right] \times 1,000. \quad (8.2)$$

Both notations provide the relative deviation of the isotope composition of a sample from a standard

Table 8.2 Isotope ratios and abundances of Cd

Mass number	106	107	108	109	110	111	112	113	114	115	116	117	118
Cadmium isotope ratios $^xCd/^{114}Cd$													
Rosman et al. (1980)[a]	0.044247		0.031374		0.438564	0.44845	0.84324	0.42605			0.259629		
Wombacher et al. (2004)[a]	0.044206		0.031364		0.438564	0.448848	0.843259	0.42645			0.259661		
Prizkow et al. (2007)[b]	0.043314		0.030907		0.433845	0.445155	0.838821	0.425395			0.261562		
Isotope abundances (in %) of Cd and isobaric elements (Böhlke et al. 2005)													
Cd (Böhlke et al. 2005)	1.25		0.89		12.49	12.80	24.13	12.22	28.73		7.49		
Cd (Prizkow et al. 2007)	1.25		0.89		12.47	12.80	24.11	12.23	28.74		7.52		
Pd	27.3		26.5		11.7								
Ag		51.8		48.2									
In										95.7			
Sn							0.97	4.3	0.65	0.36	14.5	7.68	24.2
Isotope abundances (in %) of major molecular interferences (Böhlke et al. 2005)													
$M^{40}Ar^+$	^{66}Zn 27.9	^{67}Zn 4.1	^{68}Zn 18.8	^{69}Ga 60.1	^{70}Ge 20.5 ^{70}Zn 0.6	^{71}Ga 39.9	^{72}Ge 27.4	^{73}Ge 7.8	^{74}Ge 36.5	^{75}As 100	^{76}Ge 7.8 ^{76}Se 9.0	^{77}Se 7.6	^{78}Se 23.6 ^{78}Kr 0.35
$M^{16}O^+$	^{90}Zr 51.5	^{91}Zr 11.2	^{92}Zr 17.2 ^{92}Mo 14.8	^{93}Nb 100	^{94}Zr 17.4 ^{94}Mo 9.25	^{95}Mo 15.9	^{96}Zr 2.80 ^{96}Mo 16.7 ^{96}Ru 5.52	^{97}Mo 9.55	^{98}Mo 24.1 ^{98}Ru 1.88	^{99}Ru 12.7	^{100}Mo 9.63 ^{100}Ru 12.6	^{102}Ru	^{101}Ru 31.6 ^{102}Pd 1.02

Also shown are the isotopic abundances of other elements with isotopes that can produce major isobaric and molecular interferences on Cd

[a] The isotopic analyses utilized internal normalization relative to $^{110}Cd/^{114}Cd = 0.438564$ for mass bias correction

[b] The results are based on analyses of gravimetric mixtures of seven enriched Cd isotope materials. The mass bias correction is based on an absolute measurement of atomic weight;

reference material (std) but they differ in sensitivity by a factor of 10.

A number of recent studies have investigated Cd stable isotope variations in various terrestrial materials and environments. Large Cd isotope effects were detected (1) in seawater, which features $\varepsilon^{114/110}$Cd values as high as +40, presumably as a result of preferential biological uptake of isotopically light Cd by plankton (see Sect. 8.4.4); (2) dust and other products of Cd refining as well as polluted soils from the vicinity of ore processing plants, which feature $\varepsilon^{114/110}$Cd values as low as about −8, probably due to Cd isotope fractionation during partial evaporation/condensation (Sect. 8.4.1). Very large Cd isotope effects from thermal processing were also detected in primitive meteorites (chondrites) with $\varepsilon^{114/110}$Cd values of between about −80 and +160 (Sect. 8.4.2), and even larger fractionations were observed in industrial Cd metal samples, which remained after Cd loss by evaporation (Sect. 8.4.1).

Overall, these results demonstrate that significant Cd isotope variations, of more than 0.5% for ^{114}Cd/^{110}Cd, exist in various terrestrial environments. These appear to be generated primarily by either biological uptake of Cd or partial evaporation and/or condensation of volatile Cd. Hence, this indicates that Cd isotopes can be applied as a tracer of materials that have been affected by such processes. Such applications are further presented and discussed in Sect. 8.4. However, in many environments, such as the silicate Earth (Sect. 8.4.3), Cd isotope variations are commonly very small at less than 0.5 ε for ^{114}Cd/^{110}Cd. The reliable resolution of Cd isotope variations at this level is particularly challenging, as Cd is a trace metal with concentrations of less than 1 μg/g in most terrestrial environments. Hence, highly sensitive and precise techniques of Cd isotope analysis are necessary and these invariably require that Cd is efficiently separated from the sample matrix prior to analysis. Such procedures are discussed in Sect. 8.2.

8.2 Cadmium Isotope Analyses

8.2.1 Chemical Separation of Cd

The acquisition of accurate and precise Cd stable isotope data for geological samples necessitates that Cd is separated from the sample matrix prior to mass spectrometric analysis. Ideally, the separation procedure should provide both highly purified Cd separates and essentially perfect (100%) yields, to ensure the best possible conditions for the isotopic measurements. Near quantitative yields are particularly important, to prevent analytical artifacts that may result from the isotope fractionation that is known to occur during incomplete elution of Cd from columns that utilize anion exchange or Eichrom resins (Wombacher et al. 2003; Cloquet et al. 2005b; Schmitt et al. 2009b). Application of the double spike methodology largely bypasses the latter problem but high yields are nevertheless desirable for optimum sensitivity, because Cd is a trace element in most relevant materials.

All recently published Cd isotope studies apply anion exchange chromatography to affect the separation of Cd from a wide range of natural materials, including meteorites, silicate rocks, sediments, minerals, seawater, and industrial samples. Such methods are particularly advantageous, as Cd is strongly retained on various types of strongly basic anion-exchange resin from (generally dilute) solutions of HCl (Kraus and Nelson 1955), HBr (Andersen and Knutsen 1962; Strelow 1978), and HBr-HNO$_3$ mixtures (Strelow 1978). For examples, Cd displays $D_{S/L}$ values (solid/liquid distribution coefficients) of >100 for anion exchange resin at HCl concentrations from 0.05 to 8 M (Korkisch 1989). Such media, applied at appropriate concentrations, are also suitable for the effective elution of relevant matrix elements, including the major elements of common geological samples (e.g., Na, Mg, Ca, Fe, Al) and problematic trace elements, which can produce isobaric or molecular interferences on Cd during the isotopic analyses (Table 8.2). Following elution of the matrix, Cd can be readily rinsed from the resin columns using mineral acid media that generate low $D_{S/L}$ values, such as (dilute) HNO$_3$ (Faris and Buchanan 1964) or suitable HBr-HNO$_3$ mixtures (Strelow 1978).

Most studies have found that a single pass of the sample through a column with about 1–2 mL of anion-exchange resin is generally adequate to obtain Cd separates, which are sufficiently pure for mass spectrometry (Wombacher et al. 2003; Cloquet et al. 2005b; Shiel et al. 2009). A second pass, using smaller columns with less resin (~100 μL), has been used for further purification of Cd separated from seawater samples, which exhibit particularly low Cd contents and unfavorable Cd/matrix ratios (Ripperger and Rehkämper 2007). Prior to isotopic analyses by MC-ICPMS (multiple collector inductively coupled plasma mass spectrometry) a number of workers have further purified the Cd obtained from the anion exchange separation using columns filled with ~100 μL Eichrom TRU resin (Wombacher et al. 2003; Ripperger and Rehkämper 2007). This procedure serves to effectively isolate Cd from Sn as well as any possible traces of Zr and Mo, which can produce isobaric/molecular interferences on isotopes of Cd (Table 8.2). The Cd purification protocol of Schmitt et al. (2009b) also employs an additional separation step. This is carried out prior to the anion-exchange procedure and removes Fe from the 6 M HCl sample solutions by liquid-liquid extraction into an organic phase.

8.2.2 Cd Isotope Ratio Measurements

8.2.2.1 Thermal-Ionization Mass Spectrometry (TIMS)

The pioneering Cd isotope studies of Rosman and de Laeter (1975) first investigated terrestrial minerals but subsequent work, which revealed the first natural variations in Cd stable isotope compositions focused exclusively on extraterrestrial samples (Rosman and de Laeter 1976, 1978). These analyses were accomplished by TIMS, using a silica gel activator to enhance ionization and an electron multiplier for detection. To correct for the effects of instrumental mass fractionation, the sample data were initially compared to results obtained for a pure Cd standard, whilst all measurements applied the same, rigorous data acquisition protocol (Rosman and de Laeter 1975, 1978). The precision of these data was limited (Table 8.3), however, and subsequent analyses therefore applied a ^{106}Cd-^{111}Cd double spike to correct for instrumental mass fractionation. Application of the double spike improved reproducibility to about ±4 ε/amu (2sd; Table 8.3), for analyses in which Cd ion beams with a total intensity of about 0.1 pA were monitored for several hours using a single electron multiplier (Rosman and de Laeter 1978). More recent Cd isotope measurements from the same laboratory also utilized a ^{106}Cd-^{111}Cd double spike but the application of multiple Faraday cups for data acquisition provided improved reproducibilities of about ±2 ε/amu (Schediwy et al. 2006) (Table 8.3).

The so far most precise method for Cd stable isotope measurements was published by Schmitt et al. (2009b). These workers applied a ^{106}Cd-^{108}Cd double spike and a Thermo Scientific Triton TIMS instrument equipped with Faraday cups, to achieve external reproducibilities of about ±0.07 ε/amu (2sd) for multiple standard measurements (Table 8.3). Analyses of ~100 ng Cd loads typically yielded total Cd ion

Table 8.3 Reproducibility of Cd isotope measurements by various laboratories

Analysts	Method	Instrument	±2sd εCd/amu
Rosman and de Laeter (1975, 1978)	Constant run conditions	TIMS	8–16
Rosman and de Laeter (1978)	^{106}Cd-^{111}Cd DS	TIMS	≤4
Wombacher et al. (2004)	SSB	MC-ICPMS	1.0–1.5
Wombacher et al. (2003)	Ag-, Sb-normalization	MC-ICPMS	0.2–0.8
Cloquet et al. (2005b)	SSB	MC-ICPMS	0.1–0.5
Schediwy et al. (2006)	^{106}Cd-^{111}Cd DS	TIMS	2[a]
Lacan et al. (2006)	Ag normalization	MC-ICPMS	0.1–0.5
Ripperger and Rehkämper (2007)	Ag normalization	MC-ICPMS	0.4
Ripperger and Rehkämper (2007)	^{110}Cd-^{111}Cd DS	MC-ICPMS	0.2–0.3
Gao et al. (2008)	SSB	MC-ICPMS	0.2–0.3
Schmitt et al. (2009b)	^{106}Cd-^{108}Cd DS	TIMS	0.07
Shiel et al. (2009)	Ag normalization	MC-ICPMS	0.2–0.8
Horner et al. (2010)	^{111}Cd-^{113}Cd DS	MC-ICPMS	0.2–0.3
Xue and Rehkämper, unpublished results obtained using techniques adapted from Horner et al. (2010)	^{111}Cd-^{113}Cd DS	MC-ICPMS	0.1–0.2

The reproducibilities are based on the 2sd values obtained for multiple analyses of pure Cd standard solutions and are quoted in units of εCd/amu (modified from Schmitt et al. 2009b). Methods: DS double spike; SSB standard-sample bracketing
[a]Uncertainty quoted for multiple analyses of the USGS rock standard BCR-1

beam intensities of about 240 pA for ~30 min, equivalent to an ionization efficiency of approximately 0.3% (Schmitt et al. 2009b).

8.2.2.2 Multiple Collector Inductively Coupled Plasma Mass Spectrometry (MC-ICPMS)

The study of Wombacher et al. (2003) was the first to apply MC-ICPMS to Cd isotope measurements. The success of this and subsequent MC-ICPMS investigations spurred further interest in Cd isotope research, and established MC-ICPMS as the dominant instrumental technique for such analyses (Tables 8.3 and 8.4).

In principle, Cd isotope measurements are more straightforward by MC-ICPMS than TIMS, because the high first ionization potential of Cd (Table 8.1) is readily overcome by plasma ionization (Rehkämper et al. 2001). There are other factors, however, which render the acquisition of precise and, in particular, accurate Cd stable isotope data by MC-ICPMS challenging. First, the high ionization efficiency of the plasma source implies that the numerous possible spectral interferences on the isotopes of Cd from both isobars and molecular species (see Table 8.2) must be either avoided or reduced to tolerable levels by chemical separation. Second, the instrumental mass bias correction for natural samples, which invariably contain some residual matrix (including remains of ion exchange resin), is readily affected by non-spectral interferences (matrix effects), if carried out by the commonly applied methods of either (1) standard sample bracketing (SSB), whereby the $\varepsilon^{114/110}$Cd values are calculated directly from the "raw" (unprocessed) isotope data of samples, by referencing these to results obtained for "bracketing" analyses of a Cd standard; or (2) "external" normalization, whereby a suitable fractionation law or empirical correlation is applied for mass bias correction, based on the fractionation observed for a constant isotope ratio of an admixed (dopant) element (Rehkämper et al. 2001). In particular, it has been shown for Cd that even very low matrix levels in sample solutions can generate unsystematic mass bias behavior that is not corrected by either of the above techniques, and which can thus lead to poor precision and/or inaccurate results (Wombacher et al. 2003; Shiel et al. 2009).

Problems from both spectral and non-spectral interferences can be avoided through appropriate analytical

Table 8.4 Summary of Cd isotopic data obtained for various intercalibration standards by different laboratories and techniques

Analysts	Method and instrument	BAM-I012 Cd	Münster Cd	Alfa Cd Zürich	MPI JMC Cd	Nancy Spex Cd	NIST 3108 Cd
Wombacher and Rehkämper (2004)	Ag-n, MC-ICPMS	−10.8 ± 1.5	46.5 ± 0.5				
Cloquet et al. (2005b)	SSB, MC-ICPMS		44.3 ± 0.4			−0.5 ± 1.2	
Lacan et al. (2006)	Ag-n, MC-ICPMS		44.3 ± 2.0				
Ripperger and Rehkämper (2007)	DS, MC-ICPMS	−12.4 ± 1.1	46.4 ± 1.2	0.0 ± 0.5			
Ripperger and Rehkämper (2007)	Ag-n, MC-ICPMS	−11.4 ± 1.5	46.0 ± 1.5				
Schmitt et al. (2009a, b)	DS, TIMS	−12.3 ± 0.3	44.8 ± 0.2		2.2 ± 0.2	0.0 ± 0.4	
Gao et al. (2008)	SSB, MC-ICPMS	−12.0 ± 1.2	45.9 ± 1.2			0.1 ± 1.2	
Shiel et al. (2009)	Ag-n, MC-ICPMS	−13.7 ± 2.5	45.0 ± 0.3				
Horner et al. (2010)	DS, MC-ICPMS			0.5 ± 0.4	2.6 ± 0.4		
Wombacher, unpublished results	Ag-n, MC-ICPMS						0.8 ± 0.5
Xue and Rehkämper, unpublished results	DS, MC-ICPMS	−12.2 ± 0.4					1.0 ± 0.4

All results are reported as $\varepsilon^{114/110}$Cd values (±2sd) relative to the JMC Cd Münster zero-epsilon standard
Methods: *SSB* standard-sample bracketing; *Ag-n* Ag normalization; *DS* double spike

protocols including rigorous purification of the samples prior to mass spectrometry. This is not straightforward, however, as the separation of Cd must provide essentially quantitative yields. Any significant loss of Cd during chromatography will result in Cd isotope fractionations (Wombacher et al. 2003; Cloquet et al. 2005b; Schmitt et al. 2009b) and thus generate analytical artifacts. A number of Cd isotope studies have shown that such problems can be overcome successfully either by careful Cd purification and/or the use of double spiking.

The majority of the published MC-ICPMS Cd isotope data were obtained using external normalization to admixed Ag (and in one case Sb) for the correction of mass fractionation (Table 8.3). These investigations quote typical external reproducibilities (±2sd) for multiple standard analyses of about 0.2–0.8 ε/amu (Table 8.3). Using a Nu Plasma MC-ICPMS instrument, Wombacher et al. (2003) reported instrumental sensitivities of about 125×10^{-11} A/ppm (125 V/ppm), equivalent to a transmission efficiency of about 0.09%.

It was also shown that the application of standard sample bracketing for correction of instrumental mass bias yields reproducibilities of about 0.2–1.0 ε/amu for pure Cd standards, similar to the results obtained with external normalization (Table 8.3). Comparative studies have demonstrated, however, that the standard sample bracketing approach is clearly more susceptible to matrix effects that produce inaccurate isotopic data for natural samples than various techniques of external normalization (Wombacher et al. 2003; Lacan et al. 2006; Shiel et al. 2009).

In order to avoid problems associated with matrix effects and to ensure accurate data acquisition even for challenging seawater samples, Ripperger and Rehkämper (2007) employed the double spike approach for mass bias correction in conjunction with analyses by MC-ICPMS. Using a ^{110}Cd-^{111}Cd double spike, they were able to achieve reproducibilities (±2sd) of about 0.2–0.3 ε/amu for multiple standard and sample analyses, which consumed about 40–60 ng of natural Cd (Table 8.3). More recent measurements (Xue and Rehkämper, unpublished results) applied a more favorable ^{111}Cd-^{113}Cd spike and routinely provided an external precision (±2sd) of about 0.1–0.2 ε/amu, which is only slightly worse than the best reported TIMS results (Table 8.3). The latter analyses, conducted with a Nu Plasma MC-ICPMS instrument, consumed about 30 ng of natural Cd at typical sensitivities of about 250–350 V/ppm, equivalent to a transmission efficiency of up to 0.25%.

8.2.2.3 Double Spike Methodology

The double spike methodology involves spiking of samples with a well-characterized solution that is enriched in two isotopes of the target element, relative to the natural isotope abundances. The isotopic fractionation of the natural sample is then determined from mass spectrometric data, which are collected for three isotope ratios (sharing a common denominator) of the sample-spike mixture (Dodson 1963; Wetherill 1964; Hofmann 1971; Galer 1999; Rudge et al. 2009). Double spiking of Cd (and other elements) offers three principle advantages. (1) The procedure of mass bias correction is, in essence, similar to the method of "internal" normalization commonly employed in radiogenic isotope geochemistry. Importantly, internal normalization is known to be more robust towards the generation of analytical artifacts from matrix effects than either standard sample bracketing or external normalization (Rehkämper et al. 2001); (2) If the double spike is added to and equilibrated with a sample prior to the chemical separation of Cd, the approach can accurately correct for the effects of both instrumental and laboratory-induced mass fractionations; (3) As a byproduct of double spiking, precise Cd concentrations can be determined. The main disadvantage of double spiking is that it requires (at least) four non-radiogenic isotopes, which must all be free from spectral interferences (Dodson 1963; Wetherill 1964; Hofmann 1971; Galer 1999; Rudge et al. 2009).

Cadmium has eight stable isotopes (Table 8.2) and a multitude of double spike compositions are thus available, in principle, for the acquisition of stable isotope data. A quantitative evaluation of the most suitable spike composition(s) for precise Cd isotope analyses of natural samples is not straightforward but has been carried out using numerical methods. To this end, both Galer (1999) and Rudge et al. (2009) applied somewhat distinct error propagation models to identify double spike compositions and associated spike/sample ratios, which yield the most favorable uncertainties for the determination of natural isotope fractionations, following the correction of instrumental mass bias. Both analyses focus on the propagation of uncertainties that are scaled to the relative ion beam intensities. A systematic investigation of how the choice of double spike may effect the accuracy of the stable isotope data that are acquired has yet to be carried out, however. This will be difficult in any case, given the numerous sources of systematic error that need to be considered.

However, there are number of well know issues that should be factored into the decision to adopt a particular Cd double spike composition: (1) problematic isobaric and molecular interferences (Table 8.2); (2) possible isotopic variations from sources other than mass dependent isotope fractionation, such as cosmogenic isotope anomalies for samples (Sands et al. 2001; Wombacher et al. 2008) or magnetic odd-even isotope effects during TIMS analyses that apply a silica gel activator (Manhès and Göpel 2003, 2007; Schmitt et al. 2009b); note that such non-mass dependent instrumental fractionations have hitherto not been reported for MC-ICPMS measurements; (3) limitations imposed by the collector configuration and the maximum mass dispersion that can be accommodated by the collectors.

Based on a comprehensive consideration of these factors, Schmitt et al. (2009b) chose a ^{106}Cd-^{108}Cd double spike (measured in conjunction with ^{110}Cd, ^{112}Cd) for TIMS analyses, instead of the ^{106}Cd-^{111}Cd spike that was applied in previous TIMS studies (Rosman and de Laeter 1978; Schediwy et al. 2006) (Table 8.3). A particular advantage of the former composition is that it avoids uncertainties, which are related to the apparent mass independent fractionation behavior of odd-mass isotopes of Cd during TIMS analyses (Manhès and Göpel 2003, 2007; Schmitt et al. 2009b). The error propagation model of Rudge et al. (2009) does not consider such effects and predicts that an optimized ^{106}Cd-^{111}Cd double spike should provide the most precise Cd stable isotope results, with uncertainties that are more than a factor of 2 better than to those obtainable with a ^{106}Cd-^{108}Cd spike.

The ^{110}Cd-^{111}Cd double spike that was used by Ripperger and Rehkämper (2007) with MC-ICPMS, fares even worse in the evaluation of Rudge et al. (2009), but its application to seawater analyses was most severely compromised by isobaric interferences from ^{110}Pd (Table 8.2), which required correction in a separate (and hence undesirable) measurement sequence. More recently, a ^{111}Cd-^{113}Cd double spike (in conjunction with ^{112}Cd, ^{114}Cd) was therefore used for MC-ICPMS analyses (Baker et al. 2010; Horner et al. 2010), as this (1) avoids problematic isobaric interferences from Sn and Pd on ^{116}Cd and ^{110}Cd, respectively, and (2) should provide approximately the same measurement precision (Rudge et al. 2009) as the ^{106}Cd-^{108}Cd double spike of Schmitt et al. (2009b).

8.3 Cadmium Isotope Data

8.3.1 Cadmium Isotope Reference Materials

A straightforward comparison of the ε or δ values obtained for samples by different laboratories is only possible if all data are presented relative to the same "zero-epsilon" or "zero-delta" reference material. Unfortunately, this has not been the case for most published Cd analyses. In fact, many laboratories have utilized different commercially available 1,000 μg/g Cd solutions (generally sold for the calibration of ICP-MS/OES concentration analyses) as in-house zero-epsilon reference standards. Typically, these are designated based on the name of the manufacturer and/or laboratory location, e.g., JMC Cd Münster, Alfa Cd Zürich, MPI JMC Cd (Wombacher et al. 2003; Ripperger and Rehkämper 2007; Schmitt et al. 2009b).

It is obvious that the renormalization of sample data requires a precise cross-calibration between different zero-epsilon standards. Until recently such data were not available, in part because it was assumed that various shelf standards would not exhibit significant Cd isotope variability (Wombacher and Rehkämper 2004). However, a number of analyses have now revealed differences of more than 2 $\varepsilon^{114/110}$Cd for different batches of Cd shelf standards obtained from Alfa Aesar/JMC (e.g., Table 8.4). These differences are probably due to stable isotope fractionation of Cd during the industrial separation and purification of the element. This interpretation is supported by the observation that highly purified samples of Cd metal, which have presumably been processed by zone refining in addition to electrolysis and/or distillation, appear to commonly feature distinctly light Cd isotope compositions. Examples of this are the BAM-I012 reference material (Table 8.4) and Alfa Aesar/JMC Puratronic grade Cd metal (99.9999% purity). Analyses of two different batches of the latter yielded $\varepsilon^{114/110}$Cd values of -14.7 (MPI JMC metal; Schmitt et al. 2009b) and -17.0 (JMC Cd metal Zürich; Ripperger and Rehkämper 2007) relative to the JMC Cd Münster standard.

Due to the significant isotopic variability of commercially available Cd metal and Cd solutions, it would be advantageous for all laboratories to adopt a single zero-epsilon Cd isotope reference material. A suitable material for this purpose should fulfill a number of criteria including: (1) storage and distribution in solution form to ensure isotopic homogeneity; (2) long-term availability; (3) high purity, such that elemental impurities do not generate significant interferences during isotopic analyses; (4) an isotope composition, which closely resembles the bulk silicate Earth (BSE).

The JMC Cd Münster solution, introduced by Wombacher et al. (2003), was used in a number of studies and laboratories (Table 8.4) as zero-epsilon standard with an isotope composition that is essentially identical to the BSE. However, the availability of this material is limited, such that it is not suitable as a long-term reference material. The BAM-I012 Cd isotope standard was produced by the German Federal Institute for Materials Research and Testing (BAM) to address the lack of a suitable zero-epsilon reference material. Subsequent analyses, however, revealed an isotopic composition that is far removed from the BSE value (Tables 8.4 and 8.5) and BAM-I012 Cd was therefore not adopted as zero-epsilon standard by the geochemical community. The continued need for a common zero-epsilon reference led to a concerted search effort by various Cd isotope laboratories, and the NIST SRM 3108 Cd solution was identified as a potentially suitable material (Abouchami et al. 2010b). An inter-laboratory calibration exercise to characterize the isotopic offset between NIST 3108 Cd and other previously applied in-house zero-epsilon reference materials is currently ongoing (Abouchami et al. 2010b).

Recent Cd isotope studies relied on data for additional secondary "fractionated" reference materials, to enable the comparison of results obtained by different laboratories. Such analyses will continue to be of importance for quality control, even if a common zero-epsilon reference is adopted (Abouchami et al. 2010b). The use of the same secondary reference materials by all laboratories is furthermore desirable, to facilitate the inter-laboratory comparison of results obtained with different instruments and techniques. Two particularly suitable materials with fractionated Cd isotope compositions are already available for this purpose – the isotopically heavy Münster Cd solution (aliquots available from F. Wombacher and J. Carignan at CRPG-CNRS in Nancy, France) and BAM-I012 Cd, which is characterized by a light isotope composition (Table 8.4). Both standards have already been analyzed by a number of laboratories and the measurements have yielded remarkably consistent results (Table 8.4), which can serve as benchmarks for future analyses.

8.3.2 Notation and Conversion of Cd Isotope Data

The majority of recent studies reported Cd stable isotope compositions using either an ε- or δ-notation for a specific isotope ratio (8.1) and (8.2). The δ-notation is most commonly applied for many classical and non-traditional stable isotope systems elements but the use of ε values provides a higher sensitivity, which is advantageous for terrestrial Cd isotope studies, as the isotopic variations are often smaller than 0.1‰. Fortunately, the conversion between ε- and δ-values is very straightforward.

In addition, a number of Cd isotope studies have reported fractionations using a ε/amu, δ/amu or even ‰/amu notation (Rosman and de Laeter 1976; Wombacher et al. 2003; Lacan et al. 2006; Schediwy et al. 2006). With this technique, the fractionation (δ or ε value) calculated for any isotope ratio is normalized to a mass difference of one amu. As an example, the calculation of ε/amu values from ^{114}Cd/^{110}Cd utilizes:

$$\varepsilon\text{Cd/amu} = \left[\frac{(^{114}\text{Cd}/^{110}\text{Cd})_{\text{sample}}}{(^{114}\text{Cd}/^{110}\text{Cd})_{\text{std}}} - 1\right] \times 10{,}000/(m_{114} - m_{110})$$

where m_{114} and m_{110} denote the (exact) atomic masses of the Cd isotopes. The advantage of this approach is that εCd/amu (or δCd/amu) can be determined from multiple isotope ratios. Data from multiple isotope ratios or different laboratories (which measure

Table 8.5 Selected Cd concentrations and isotope compositions (reported as $\varepsilon^{114/110}$Cd relative to NIST 3108 Cd) for important meteorite types, Earth reservoirs and terrestrial samples

Sample type or reservoir		Typical concentrations or calculated mean values[a]	References	$\varepsilon^{114/110}$Cd range or mean[a]	References
Chondritic meteorites					
Carbonaceous chondrites	ng/g	650–710 (CI); ~370 (CM, CV)	Wasson and Kallemeyn (1988), Palme and O'Neill (2005), McDonough and Sun (1995), Lodders (2003)	−4.4 to 5.1[b] 3.0 ± 1.4[b]	Baker et al. (2010)
Ordinary chondrites (H,L, LL groups)	ng/g	~10–40; total range of [Cd]: 0.1–1,000	Wasson and Kallemeyn (1988), Wombacher et al. (2008)	−78 to 159	Wombacher et al. (2008)
Bulk silicate Earth (BSE)	ng/g	64; 40; 35 ± 7; 18	Palme and O'Neill (2005), McDonough and Sun (1995), Witt-Eickschen et al. (2009), Yi et al. (2000)	−0.3 ± 0.5	Schmitt et al. (2009a)
Mantle and mantle-derived rocks					
Peridotites	ng/g	40 ± 28; 41 ± 14; 10–40	Heinrichs et al. (1980), McDonough (1990), Witt-Eickschen et al. (2009)		
Mid-ocean ridge basalts (MORB)	ng/g	119 ± 22	Yi et al. (2000)	0.0 ± 0.3	Schmitt et al. (2009a)
Ocean island basalts (OIB)	ng/g	100[c]; 100–160	Heinrichs et al. (1980), Yi et al. (2000)	−0.7 ± 0.3	Schmitt et al. (2009a)
Continental crust and crustal rocks					
Bulk continental crust	ng/g	102; 98; 90	Wedepohl (1995), Taylor and McLennan (1985), Rudnick and Gao (2005)	(−0.1 ± 0.4[d])	Schmitt et al. (2009a)
Upper continental crust	ng/g	100; 100; 80	Wedepohl (1995), Taylor and McLennan (1985), Rudnick and Gao (2005)	−0.1 ± 0.4[d] 0.5 ± 1.0[e]	Schmitt et al. (2009a)
Andesites	ng/g	58	Heinrichs et al. (1980)		
Granites/ granodiorites	ng/g	60–130	Heinrichs et al. (1980)		
Gneisses and granulites	ng/g	95–115	Heinrichs et al. (1980)		
Clastic river and shelf sediments	ng/g	~100–200	Heinrichs et al. (1980)	−4 to +4	Wombacher et al. (2003), Gao et al. (2008)
Loess	ng/g	110–130	Schmitt et al. (2009a)	−0.1 ± 0.4	Schmitt et al. (2009a)
Cd and Cd-rich minerals				−2 to +3	Schmitt et al. (2009a), Wombacher et al. (2003), Shiel et al. (2010)
Marine environment					
Surface ocean (dissolved Cd)	nmol/L	~0.001–0.5	Bruland (1983), de Baar et al. (1994)	~0 to +38	Ripperger et al. (2007), Lacan et al. (2006)
Oceans ≥1 km (dissolved Cd)	nmol/L	~0.2–1.1	Bruland (1983), de Baar et al. (1994)	2.9 ± 0.5	Ripperger et al. (2007)
Cd/Ca ratios of foraminiferal tests	µmol/ mol	<0.01 to >0.25	Boyle (1988), Ripperger et al. (2008)		
Pelagic clays	ng/g	100–350; 420	Heinrichs et al. (1980), Li and Schoonmaker (2005)		
Hydrogenetic Fe-Mn crusts	µg/g	1–20	Horner et al. (2010), Hein et al. (1997)	2.5 ± 1.0	Horner et al. (2010)
Hydrogenetic Fe-Mn nodules	µg/g	1–20	Schmitt et al. (2009a), Hein et al. (1997)	2.2 ± 1.1	Schmitt et al. (2009a)

(continued)

Table 8.5 (continued)

Sample type or reservoir		Typical concentrations or calculated mean values[a]	References	$\varepsilon^{114/110}Cd$ range or mean[a]	References
Hydrothermal Fe-Mn crusts	µg/g	3–20	Schmitt et al. (2009a), Hein et al. (1997)	0.6–1.1	Schmitt et al. (2009a)
Hydrothermal sulfides	µg/g	~100–400	Schmitt et al. (2009a)	−4.5 to 3.9	Schmitt et al. (2009a)
Anthropogenic samples					
Purified Cd and Cd metal				−17 to 4	Shiel et al. (2010), This study, Table 8.3, Ripperger and Rehkämper (2007)
Samples from of Pb-Zn refinery				−7 to +5	Shiel et al. (2010), Cloquet et al. (2006)
Soils polluted by Pb-Zn refinery				−7 to +3	Gao et al. (2008), Cloquet et al. (2006)

[a]Ranges in concentration and $\varepsilon^{114/110}Cd$, based on published data, are given in normal font; calculated or published mean concentrations or $\varepsilon^{114/110}Cd$ values are given in *italics* and are quoted with 1sd, if available
[b]Excludes the anomalous samples Colony, Leoville and Kainsaz, which are characterized by $\varepsilon^{114/110}Cd \approx 10\text{–}40$
[c]Includes continental basalts
[d]Based on analyses of upper crustal loess
[e]Based on ten high-precision analyses of continental sulfides

different Cd isotopes) can therefore be readily compared and applied to calculate, presumably more reliable, averages. A detailed evaluation demonstrates, however, that various Cd isotope ratios (e.g., $^{110}Cd/^{111}Cd$ and $^{113}Cd/^{114}Cd$) will yield slightly different εCd/amu results. These differences reflect: (1) variations in the relative mass difference of the isotopes (e.g., 0.91% for ^{110}Cd and ^{111}Cd and 0.89% for ^{113}Cd and ^{114}Cd), (2) the mechanism and hence mass dependence of mass fractionation (equilibrium vs. kinetic effects), and (3) the absolute extent of isotope fractionation, whereby larger fractionations produce greater differences (Wombacher and Rehkämper 2004). For example, assuming that kinetic mass fractionation is relevant, a value of εCd/amu = 10.0 determined from $^{114}Cd/^{110}Cd$ corresponds to εCd/amu = 9.9 for $^{114}Cd/^{112}Cd$. This difference is small, but it is an inherent drawback of the methodology that may become more serious with future improvements in analytical precision. We suggest that the "per amu" notation is therefore best avoided in the future.

Given that Cd has eight stable isotopes, of which five have relative abundances of more than 10% (Table 8.2), it is not surprising that various studies report stable isotope fractionations using different Cd isotope ratios. Most commonly, ε or δ values are reported for $^{114}Cd/^{110}Cd$ (Cloquet et al. 2005b; Ripperger et al. 2007; Wombacher et al. 2008; Shiel et al. 2009) or $^{112}Cd/^{110}Cd$ (Schmitt et al. 2009a). The former approach yields larger numerical values, which can be advantageous to differentiate between results that differ only slightly, whilst the latter features a 2-amu mass difference that is also commonly applied for the notation of other elements (e.g., Mg, Fe, Cu, Zn).

An accurate comparison of ε-values calculated for different Cd isotope ratios is possible but this requires that the results are recalculated using a suitable mass fractionation law (Wombacher and Rehkämper 2004). In many cases, particularly where the Cd isotope fractionations are large (e.g., from evaporation and biological utilization), the kinetic law (Young et al. 2002) may be most suitable. The $\varepsilon^{k/m}Cd$ value for one isotope ratio $R_{k/m} = {}^kCd/{}^mCd$ is related to the $\varepsilon^{l/m}Cd$ value of a second isotope ratio $R^{l/m} = {}^lCd/{}^mCd$ by:

$$\varepsilon^{k/m}Cd = \left[\left(\varepsilon^{l/m}Cd/10{,}000 + 1\right)^\beta - 1\right] \times 10{,}000. \quad (8.4)$$

For the kinetic law, β is defined as:

$$\beta = \ln(m_k/m_m)/\ln(m_l/m_m), \quad (8.5)$$

where m_k, m_l and m_m refer to the exact atomic masses of the Cd isotopes k, l, and m (Wombacher and Rehkämper 2003). Note that a simplified equation can be used in many instances, whereby

$$\varepsilon^{k/m}Cd \approx \varepsilon^{l/m}Cd \times \beta. \quad (8.6)$$

This simplified technique is accurate to better than ± 0.01 ε (=1 ppm) for the conversion between $\varepsilon^{114/110}Cd$ and $\varepsilon^{112/110}Cd$, for $\varepsilon^{114/110}Cd$ values of 0 ± 50.

8.3.3 Renormalization of Cd Isotope Data

The $\varepsilon^{114/110}Cd$ values that are determined for a sample relative to an in-house zero-epsilon standard can be renormalized to a second reference material using the standard conversion identity (Criss 1999). The relevant equation is formulated here (as an example) for the conversion of an ε value originally determined for a sample (S) relative to JMC Münster Cd, to an ε value relative to NIST 3108 Cd:

$$\begin{aligned}\varepsilon^{114/110}Cd_{S-NIST} &= \varepsilon^{114/110}Cd_{S-JMC} + \varepsilon^{114/110}Cd_{JMC-NIST} \\ &+ \frac{\left(\varepsilon^{114/110}Cd_{S-JMC}\right)\left(\varepsilon^{114/110}Cd_{JMC-NIST}\right)}{10,000}.\end{aligned} \quad (8.7)$$

In practice, the conversion is often carried out using the approximation:

$$\begin{aligned}\varepsilon^{114/110}Cd_{S-NIST} &\approx \varepsilon^{114/110}Cd_{S-JMC} \\ &+ \varepsilon^{114/110}Cd_{JMC-NIST}.\end{aligned} \quad (8.8)$$

The application of this simplified approach is justified in many cases because it introduces systematic errors than are much smaller than the analytical uncertainties.

8.3.4 Conventions Adopted in this Study

In order to directly compare the results obtained by different laboratories in this manuscript, it was useful to (1) adopt a common method of Cd isotope notation and (2) renormalize published isotopic data to a common zero-epsilon reference standard. To this end, results quoted in the present paper are reported using an $\varepsilon^{114/110}Cd$ notation, relative to the NIST SRM 3108 Cd solution (except when otherwise stated). This approach was taken in anticipation of the future adoption of NIST SRM 3108 Cd as the commonly accepted zero-epsilon reference standard (Abouchami et al. 2010b).

The renormalization was carried out based on preliminary results that were obtained for analyses of NIST SRM 3108 Cd, JMC Münster Cd, Alfa Cd Zurich, MPI JMC Cd, BAM-I012 Cd and Münster Cd as part of a current intercalibration effort (Abouchami et al. 2010b). Furthermore utilized were published differences between the in-house zero-epsilon materials of various laboratories and secondary Cd isotope standards. Such data are fortunately available in all recently published Cd isotope studies (Table 8.4). Based on the preliminary results compiled by Abouchami et al. (2010b), the following isotopic differences were adopted in this manuscript for the $\varepsilon^{114/110}Cd$ values of zero-epsilon standards *relative* to NIST SRM 3108 Cd: JMC Münster Cd: -0.8; Alfa Cd Zürich: -0.4; MPI JMC Cd: $+1.5$; Nancy Spex Cd: -0.8; PCIGR-1 Cd: -0.5. Note that these isotopic differences are preliminary results that are subject to (slight) future revisions, which may be adopted once the current intercalibration effort is finalized (Abouchami et al. 2010b). The renormalization of the data was carried out using the standard conversion identity (8.7).

8.4 Cadmium Isotopes: Applications and Results

8.4.1 Industrial and Anthropogenic Samples

8.4.1.1 Cadmium Stable Isotope Fractionation During Evaporation

The variability of Cd isotope compositions in different pure Cd metal samples and standard solutions was mentioned previously (Table 8.4) and this is likely to reflect isotopic fractionations from the purification of

Cd by either (non-quantitative) distillation or zone refining. For the evaporation of volatile Cd (Table 8.1), kinetic theory predicts an isotope fractionation factor α that can be calculated from the square root of the masses (m) of the relevant isotopes or isotopologues. Given that Cd generally appears to evaporate as a monoatomic gas (Chizhikov 1966), the relevant equation is $\alpha_{kin} = (^{114}Cd/^{110}Cd)_{residue} / (^{114}Cd/^{110}Cd)_{vapor} = (m_2/m_1)^{0.5} \approx (114/110)^{0.5} = 1.018$. This is equivalent to a fractionation of $\Delta \approx 18‰$ for $^{114}Cd/^{110}Cd$ during kinetic evaporation, where the "big delta" is given by $\Delta \approx 1{,}000 \ln\alpha$ (Criss 1999).

Cadmium isotope fractionation during evaporation and condensation processes were investigated by Wombacher et al. (2004) using three Cd metal samples that were residues from evaporation of molten Cd in a vacuum chamber at about 10^{-4} mbar. At these conditions, Cd has a boiling point of ~180°C. The samples were obtained from an industrial process, in which Cd vapor is used to coat metal parts with a thin layer of Cd for a high level of corrosion protection. In detail, the analyzed metals (Samples C, D, E of Fig. 8.1) were residues, which remained after about 90–99.5% of the initial Cd had been lost to evaporation, such that they featured highly fractionated (heavy) Cd isotope compositions that differed from the starting material (Sample A of Fig. 8.1) by up to 4.8% for $^{110}Cd/^{114}Cd$ and almost 10% for $^{106}Cd/^{114}Cd$ (Wombacher et al. 2004). At such large absolute fractionations, the isotopic compositions that are expected for pure equilibrium and kinetic evaporation processes (Young et al. 2002) are sufficiently different in three isotope plots (e.g., $^{106}Cd/^{114}Cd$ vs. $^{110}Cd/^{114}Cd$; Fig. 8.1) to be resolved by mass spectrometry. In principle, this approach is, therefore, able to provide fundamental constraints on fractionation mechanisms.

Wombacher et al. (2004) utilized the generalized power law of Maréchal et al. (1999) to evaluate the Cd isotope data obtained for the metal samples and they determined that the evaporation residues displayed a mass dependence intermediate between that expected for equilibrium and kinetic isotope fractionation (Fig. 8.1). They also found that the evaporation process yielded smaller fractionations than expected from $\alpha_{kin} = (m_2/m_1)^{0.5}$. Taken together, these observations were interpreted to reflect hybrid evaporation with kinetic isotope fractionation characterized by $\alpha_{kin} = (m_2/m_1)^{0.5}$ and limited, but still significant, equilibrium isotope partitioning between the vapor phase

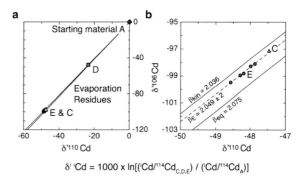

Fig. 8.1 Three isotope plot of Cd data for industrial Cd metal samples (Wombacher et al. 2004). Results are shown in the δ' notation, as defined below the figure (Hulston and Thode 1965), because this produces linear fractionation lines rather than curved trends, which would result from the use of conventional δ or ε values. Note that the δ' values of the fractionated evaporation residues C, D, E are calculated relative to the starting material A. (**a**) Samples C, D, E and the starting material A define a fractionation line in $\delta'^{106}Cd$ vs. $\delta'^{110}Cd$ isotope space. (**b**) Close-up of (**a**) that shows the results obtained for samples C and E. Shown in (**b**) are reference lines for mass-dependent isotope fractionation according to the kinetic and equilibrium fractionation laws, with slopes β_{kin} and β_{equil} calculated as described by Young et al. (2002). Also denoted in (**b**) is the best-fit fractionation line inferred for sample E with slope $\beta_E = \delta'^{106}Cd/\delta'^{110}Cd$. The inferred β values for samples E and C (not shown) are intermediate between those expected for equilibrium and kinetic fractionation. The drift of the instrumental mass discrimination, which is apparent for the repeated measurements of sample E, is negligible compared to the overall magnitude of the isotope fractionation from evaporation

and melt. More recent data have shown, however, that evaporation of Mg also produces an intermediate mass dependence, albeit at very high temperatures (Davis et al. 2005). Given that equilibrium isotope partitioning is temperature sensitive, Wombacher et al. (2008) concluded that equilibrium partitioning of Cd isotopes between vapor and melt (α_{equil}) were overestimated by Wombacher et al. (2004), because evaporation into a vacuum under kinetic conditions is not necessarily characterized by kinetic mass scaling (Young et al. 2002) and $\alpha_{kin} = (m_2/m_1)^{0.5}$. Unidirectional evaporation processes which display (1) fractionations that are about 50% lower than predicted by $\alpha_{kin} = (m_2/m_1)^{0.5}$ and (2) mass-dependencies intermediate between those expected for equilibrium and kinetic isotope effects, are therefore probably the result of chemical interactions in the melt phase (Simon and Young 2007) and/or the melt-vacuum interface (Richter et al. 2002).

8.4.1.2 Metallurgy and Anthropogenic Emissions

The large Cd isotope fractionations that can be produced by evaporation, suggest that isotopic analyses may be useful for tracing Cd pollution, which is released by anthropogenic emissions, including smelting/refining processes that are associated with the recovery of Cd from ores, waste incineration and the burning of coal. Such studies are of particular interest, because Cd is highly toxic and commonly enriched (to levels of up to a few percent) in sulfide ores, which are processed for the recovery of Pb and Zn. So far, the combustion of coal and other fossil fuels have not been investigated using Cd isotope methods, even though they represent the most important anthropogenic source of Cd to the environment (Nriagu and Pacyna 1988). Furthermore, only preliminary results are available for a Cd isotope study of waste incineration. Importantly, this investigation identified variations of up to 10 ε for ^{114}Cd/^{110}Cd, in bulk, leachate and leachate residue samples of fly ash from an urban waste combustor (Cloquet et al. 2005a). These large isotope effects were thought to reflect modification of the Cd isotope composition in the residual gas phase by progressive kinetic condensation.

Three published studies have looked for and identified significant Cd isotope fractionations in various samples from Pb-Zn smelters and refineries, as well as the immediate environments of such plants (Cloquet et al. 2006; Gao et al. 2008; Shiel et al. 2010). Cloquet et al. (2006) investigated the Cd and Pb isotope compositions of soils collected in the vicinity of a Pb-Zn refinery, at distances of up to 4.5 km. Also analyzed were dust from exhaust filters and a slag produced in the plant furnace. The isotopic analyses showed that two dust samples from the refinery exhibited Cd isotope compositions that were lighter in ^{114}Cd/^{110}Cd by ~10 ε compared to the slag (Fig. 8.2). This difference in Cd isotope composition was interpreted to reflect progressive evaporation of Cd, under equilibrium conditions, to form a gas phase enriched in light isotopes, from which the dust condensed on cooling. As the $\varepsilon^{114/110}$Cd value of the slag was thought to represent the last fraction of evaporated Cd in the refinery, an equilibrium solid-gas fractionation factor α of only 1.0001 was derived for the system (Cloquet et al. 2006). This result is significantly lower than the fractionation predicted for

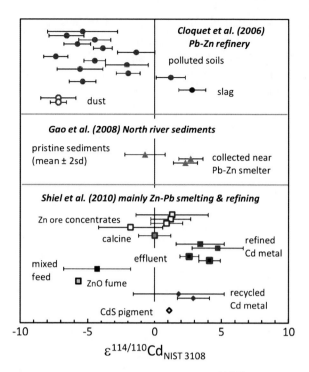

Fig. 8.2 Cadmium isotope data (reported as $\varepsilon^{114/110}$Cd relative to NIST 3108 Cd) for samples from Zn-Pb smelting/refining plants and environmental samples that were affected by emissions from such plants (Cloquet et al. 2006; Gao et al. 2008; Shiel et al. 2010). The uncertainties are shown as reported in the original publications, except for the average calculated for five unpolluted river sediments (Gao et al. 2008)

kinetic evaporation (α_{kin} = 1.0018) and the (presumably overestimated) fractionation factor of α_{equil} = 1.0007 that was inferred for equilibrium evaporation at 180°C (Wombacher et al. 2004). The origin of the low observed α value is not well constrained but may be due to either negligible equilibrium isotope fractionation at high temperature (Cloquet et al. 2006) and/or a lack of isotopic homogenization in the residual materials during the evaporation process.

Cloquet et al. (2006) also found that soil samples from the vicinity of the refinery plant were characterized by high Cd contents and clear differences in Cd isotope compositions with $\varepsilon^{114/110}$Cd ranging from -7.4 to $+1.2$ (Fig. 8.2, Table 8.5). Soils from locations close to the refinery exhibited the lowest $\varepsilon^{114/110}$Cd values and the isotopic variability was thought to reflect contamination of the soils by variable mixtures of dust- and slag-derived Cd. The viability of tracing anthropogenic emissions of Cd to the environment was further corroborated by the work of Gao et al. (2008), who carried out a combined Cd-Pb isotope study of sediments from the North River in China. This investigation found that two samples from the vicinity of a Pb-Zn smelter displayed elevated Cd concentrations and higher $\varepsilon^{114/110}$Cd values (of +2.3 and +2.7) compared to five pristine river sediments, which exhibited $\varepsilon^{114/110}$Cd $\approx$ 0 (Fig. 8.2). The clear difference in Cd isotope compositions was interpreted to reflect isotope fractionation from processing of Cd in the Pb-Zn smelter (Gao et al. 2008).

Shiel et al. (2010) investigated the Cd, Zn and Pb isotope compositions of samples from a Pb-Zn smelter and refining plant in Canada. Cadmium isotope data were obtained for (1) four Zn ore concentrates (primarily sphalerite); (2) calcine (mainly ZnO, obtained from roasting of the sulfide ores); (3) the Pb smelter feed (comprised of Pb ore concentrates and residues from the extraction of Zn from the calcine); (4) CdO-rich ZnO fumes produced during Pb purification; (5) effluents (Zn and Cd-rich waste products from the Zn plant that are recycled to the smelter); and (6) the final refined Cd metal (Fig. 8.2). The Zn ore concentrate samples exhibited Cd isotope compositions ($\varepsilon^{114/110}$Cd $\approx$ -2 to $+1$) similar to those observed for other continental and oceanic sulfides (Fig. 8.3, Table 8.5). The calcine sample also falls within this range, probably due to the near-quantitative Cd yield of the roasting process (Fig. 8.2). The Zn and Cd-rich effluent and the refined Cd metal displayed similar and distinctly positive $\varepsilon^{114/110}$Cd values (of between about +3 and +5), whilst the ZnO fume and the Pb smelter feed are enriched in isotopically light Cd ($\varepsilon^{114/110}$Cd $\approx$ -4 to -6). The origin of these distinct differences, which should essentially balance out to yield the Cd isotope composition of the original ore concentrate, is difficult to trace due to the unknown yields of various production stages. Based on a careful assessment of mass balances, Shiel et al. (2010) concluded that the isotopic fractionation is mainly (but not solely) related to differences in the Cd isotope compositions of various mineralogical phases that are formed during roasting, and which are separated from one another, as they display distinct reactivities during further processing. Shiel et al. (2010) also analyzed Cd metal that was thermally recovered from Ni-Cd batteries and a CdS pigment sample. These materials had inconspicuous Cd isotope compositions that were identical to the range of values determined for Zn ore concentrates and refined Cd metal (Fig. 8.2).

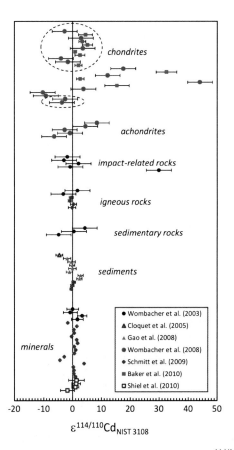

Fig. 8.3 Cadmium stable isotope data (reported as $\varepsilon^{114/110}$Cd relative to NIST 3108 Cd) for various terrestrial sediments and rocks (including a tektite and other samples related to impact events), Zn and Cd minerals, and selected meteorites. The meteorite data are limited to carbonaceous chondrites, enstatite chondrites (of type EH) and selected achondrites (eucrites and SNC meteorites from Mars). The chondrite data encircled by the *hashed line*, denote results for unfractionated carbonaceous chondrites which have not been overprinted by strong parent body metamorphism (Wombacher et al. 2003, 2008)

In summary, it should be considered significant that these pioneering studies of Cd isotope fractionations in industrial and polluted environmental samples were all able to resolve clear differences in $\varepsilon^{114/110}$Cd that are primarily of anthropogenic origin. Importantly, these differences were identified even though individual industrial processes are unlikely to be associated with fractionation factors that exceed values of $\Delta \approx 10$ ε. It was also shown that the isotopic signatures can be applied to interrogate the behavior of Cd during industrial processes. Furthermore, they show significant promise for tracing anthropogenic emissions to and cycling of Cd in polluted environments.

The power of such applications is essentially limited by the precision of the analytical data relative to the observed isotopic variability.

8.4.2 Extraterrestrial Materials

The first systematic studies of Cd isotope compositions, which led to the identification of natural Cd isotope fractionations were carried out by Rosman and de Laeter on meteorite samples in the 1970s (Rosman and de Laeter 1976, 1978) and such investigations continue to draw significant interest. A thorough review of this literature is outside the scope of this contribution but a short synopsis of key results is nonetheless provided, as a reference for the study of terrestrial Cd isotope variations.

Chondrites are primitive meteorites that formed in the early solar system and amongst these, some carbonaceous chondrites feature elemental abundances that most closely resemble the bulk solar system. The majority of carbonaceous chondrites have relatively constant Cd isotope compositions (Baker et al. 2010) with an average of $\varepsilon^{114/110}Cd = +3.0 \pm 2.8$ (2sd). This indicates that the depletion of Cd and other volatile elements in inner solar system bodies did not generate large isotopic fractionations and hence probably did not involve partial kinetic evaporation. In contrast, large Cd stable isotope effects, with $\varepsilon^{114/110}Cd$ values ranging from −78 to +159 are recorded in ordinary and many enstatite chondrites (Fig. 8.3, Table 8.5). The Cd isotope fractionations in these meteorites are thought to be caused by redistribution of Cd during thermal metamorphism in the interiors of the asteroidal parent bodies (Wombacher et al. 2003, 2008). However, there are also many extraterrestrial samples that display Cd isotope compositions, which are very similar or identical to the (unfractionated) carbonaceous chondrites (Fig. 8.3). In particular, this includes achondrites, which are derived from differentiated (and hence formerly molten) parent asteroids. Also essentially indistinguishable from unfractionated carbonaceous chondrites are most terrestrial rocks (Fig. 8.3, Table 8.5) and two samples from the Moon (Wombacher et al. 2003, 2008; Schediwy et al. 2006). Taken together, these results demonstrate that Cd isotope heterogeneities are homogenized within planetary bodies during differentiation and that large stable isotope fractionations do not occur as a consequence of volatile element depletion and parent body differentiation.

8.4.3 Reservoirs and Samples of the Silicate Earth

In solid Earth geochemistry, Cd is generally classified as a trace element that exhibits both lithophile and chalcophile characteristics. High concentrations of Cd are typically found in continental igneous sulfide ore deposits, particularly as greenockite (CdS) associated with sphalerite (ZnS). In silicate rocks, Cd preferentially partitions into Mg-Fe silicate minerals such as biotite, chlorite, pyroxene and amphibole (Heinrichs et al. 1980; Witt-Eickschen et al. 2009). In most igneous, sedimentary and metamorphic rocks of the continental crust and the mantle, Cd furthermore typically exhibits relatively uniform concentrations of about 20–200 ng/g (Table 8.5). This uniformity reflects, in part, that Cd is only moderately incompatible during mantle melting with a partition coefficient that resembles the rare earth elements dysprosium (Dy) or ytterbium (Yb) (Yi et al. 2000; Witt-Eickschen et al. 2009).

Most compositional estimates for the continental crust and its sub-reservoirs (of upper, middle, and lower crust) provide Cd concentrations of ~100 ng/g, with only limited intra-crustal differences (Table 8.5). A notable exception is the recent study of Hu and Gao (2008), which suggests a relatively low Cd content of 60 ng/g for the upper continental crust. For the BSE, four abundance estimates derive mean Cd concentrations of between 18 and 64 ng/g (Table 8.5). Based on the geochemical similarity of Cd to Dy and Yb during mantle melting, Cd should be moderately enriched, by about a factor of 5, in the continental crust (CC) relative to the BSE (McDonough and Sun 1995; Palme and O'Neill 2005). Taken at face value, this conclusion is in accord with only some of the published Cd abundance estimates for silicate Earth reservoirs, whilst the majority of the data are more suggestive of $[Cd]_{CC} \approx (2.5 \pm 1) \times [Cd]_{BSE}$ (Table 8.5). Similarly, the geochemical similarity of Cd to Dy and Yb can also be used to infer the Cd concentration of the depleted MORB source mantle (DMM) as

$[Cd]_{DMM} \approx (0.80 \pm 0.05) \times [Cd]_{BSE}$ (Workman and Hart 2005), which is equivalent to $[Cd]_{DMM} \approx$ 15–50 ng/g.

Despite of the large uncertainty of the BSE estimates, the available data suggest that the silicate Earth is depleted in Cd relative to carbonaceous chondrites (Table 8.5), as a consequence of volatile depletion and potentially also core partitioning. Estimates for the Cd content of the Earth's core are highly speculative, however, due to our limited understanding of Cd partitioning between silicate, metal, and sulfide phases during terrestrial accretion (Lagos et al. 2008). Based on mass balance, McDonough (2005) estimated that about 60–65% of the bulk Earth's Cd resides in the core, to support a concentration of ~150 ng/g in the metal.

At present, only very limited Cd isotope data are available, which characterizes the silicate Earth and explores Cd isotope variations amongst and within its reservoirs. Overall, the variability of $\varepsilon^{114/110}$Cd in rocks and minerals from the crust and mantle is limited with values between −4 and +4 (Fig. 8.3, Table 8.5). This excludes all polluted and industrially processed samples, as well as marine sediments and minerals, which show larger Cd isotope variations. Also excluded is a tektite analyzed by Wombacher et al. (2003), which features a heavy isotope composition ($\varepsilon^{114/110}$Cd = 30) and low Cd content, presumably due to loss Cd by evaporation (Fig. 8.3). Additional analyses of other impact-related rocks did not reveal any significant Cd isotope variations, however (Wombacher et al. 2003).

Apart from the tektite and the marine sediments (which are discussed in Sect. 8.4.4.3), the only samples from the solid Earth to exhibit clearly resolvable Cd isotope variations are some continental mineral deposits (Fig. 8.3, Table 8.5). In particular, no significant variability in Cd isotope compositions is apparent for loess and clastic sediments from rivers and continental margins as well as igneous rocks derived from the crust and mantle (Fig. 8.3, Table 8.5). As such, the present dataset does not provide evidence for systematic Cd isotope fractionations between or within the major reservoirs of the silicate Earth. At present, the isotope compositions of the Earth's mantle and continental crust are therefore best characterized by the limited but precise double spike TIMS data of Schmitt et al. (2009a). These workers analyzed three samples of loess (representative of the continental crust) and two samples each of MORB (mid ocean ridge basalts) and OIB (ocean island basalts) to characterize the mantle (Table 8.5). The results obtained suggest that both the continental crust and the mantle display essentially identical $\varepsilon^{114/110}$Cd values of −0.1 and −0.4, respectively (Table 8.5). These data can be combined (Schmitt et al. 2009a) to derive a precise estimate for the mean isotope composition of the BSE of $\varepsilon^{114/110}$Cd = −0.3 ± 1.0 (2sd; Table 8.5).

The conclusion that Cd isotopes were not fractionated during the formation of the continental crust by the extraction of partial melts from the (originally primitive) mantle (Schmitt et al. 2009a) is reasonable given that high temperature processes are generally not expected to generate large Cd isotope effects. However, further high-precision analyses for additional samples, particularly mantle peridotites, are required to fully address this issue. This is of particular relevance because preliminary data obtained by Schönbächler et al. (2009) hint at a minor difference between the Cd isotope compositions of mantle-derived basalts and ultramafic komatiites/peridotites. Such analyses are furthermore of interest, to resolve whether there is indeed a small but resolvable fractionation in $\varepsilon^{114/110}$Cd between the BSE and chondritic meteorites (Schönbächler et al. 2009). The current Cd isotope database for terrestrial rocks is too imprecise and limited in size to adequately address this question.

8.4.4 Marine Environments

8.4.4.1 Cadmium in the Oceans and Marine Sediments

The geochemistry of Cd in the oceans and marine sediments has attracted significant scientific interest for more than 30 years. This interest is primarily based on the observation that the dissolved Cd concentrations of seawater show a nutrient-like distribution (Fig. 8.4), which very closely resembles the distribution of the marine macronutrient phosphate (Boyle et al. 1976; Bruland 1980). Both Cd and P display low concentrations in surface waters, due to the uptake by marine organisms (Fig. 8.4). After the organisms cease, the organic material sinks through the water column and is subsequently decomposed at depth, so that Cd and P are remineralized into the dissolved form (Bender and Gagner 1976; Boyle et al. 1976; Bruland 1980). Whilst it has been suggested in the past that the nutrient-like behavior of Cd may also reflect passive adsorption of Cd onto (deceased) organic matter (Collier and Edmond 1984; Yee and Fein 2001; Dixon et al. 2006), it has been demonstrated by a number of recent studies that Cd is a marine micronutrient with an important biological function. In particular, it has been shown that Cd can substitute for Zn in the enzyme carbonic anhydrase, which plays an important role in the process of inorganic carbon acquisition (Price and Morel 1990; Cullen et al. 1999; Sunda and Huntsman 2000; Lane et al. 2005; Xu et al. 2008).

Oceanic circulation modifies this simple one-dimensional pattern of Cd and phosphate distribution (Fig. 8.4) to create water masses with distinct chemical signatures. A key feature is that Pacific deepwater, which is characterized by older ventilation ages, has significantly higher Cd and P concentrations than comparatively young Atlantic deepwater (Fig. 8.4). This difference reflects the progressive accumulation of remineralized nutrients, as deepwater masses pass from the Atlantic through the Southern to the Pacific Ocean (Donat and Bruland 1995). Despite of this complexity, the Cd and phosphate data for seawater correlate remarkably well for a large global dataset (Fig. 8.5). A particular feature of this largely linear correlation is the "kink" at intermediate Cd and phosphate abundances, which has been described by some as a consequence of a slightly curved [Cd]-[PO$_4$] trend (Elderfield and Rickaby 2000). The origin of this feature has been the subject of numerous investigations and its interpretation continues to be contentious (Frew and Hunter 1992; de Baar et al. 1994; Elderfield and Rickaby 2000; Cullen 2006).

The particular interest of paleoceanographers in the element Cd is based on the [Cd]-[PO$_4$] correlation of seawater (Fig. 8.5) coupled with the finding that past seawater phosphate contents are generally not recorded in marine sediments (Henderson 2002). In contrast, it was observed that Cd is incorporated into the tests of foraminifera, in proportion to the Cd (and hence phosphate) concentration of the surrounding seawater (Boyle 1981; Ripperger et al. 2008). Subsequent studies have further refined this understanding and developed (1) empirical depth-dependent distribution coefficients for Cd between benthic foraminifera and seawater (Boyle 1992) and (2) temperature dependent corrections for the tests of planktonic foraminifera from surface waters (Rickaby and Elderfield 1999). On this basis, the Cd/Ca data of foraminifera have become an important tool in paleoceanography as proxy of past marine nutrient (phosphate) distribution and utilization (Henderson 2002; Lynch-Stieglitz 2006). Paleoceanographic studies that aim to reconstruct past changes in the marine distribution of Cd

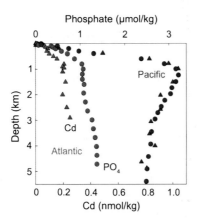

Fig. 8.4 Depth profiles of Cd and phosphate concentrations (shown as *triangles* and *circles*, respectively) for stations in the Atlantic Ocean (in *red*) and Pacific Ocean (in *blue*) (Bruland 1983). The Cd and P contents are well correlated with extremely low surface water abundances and a rapid increase of concentrations at depths of 0–1 km. The higher Cd and phosphate contents of the Pacific deep water are a typical feature of macro- and micronutrient elements

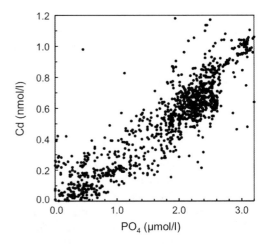

Fig. 8.5 Cadmium vs. phosphate concentrations for a global dataset of open ocean samples. The dataset is essentially identical to that used by (Hendry et al. 2008). The "kink" or slight curvature in the well-defined positive correlation is apparent

have two principal aims. (1) Cadmium is applied as a non-conservative tracer of water masses to trace temporal changes in ocean circulation patterns. (2) The seawater concentration of Cd is a nutrient- and thus carbon-linked property, which can be used to infer variations in both deep and surface water nutrient levels (Boyle 1988, 1992; Rickaby and Elderfield 1999; Elderfield and Rickaby 2000). The Cd data for sedimentary archives thus provides constraints on past changes in ocean productivity and the processes, which drive atmospheric CO_2 variability (Elderfield and Rickaby 2000; Sigman and Boyle 2000).

8.4.4.2 Cadmium Isotopes in the Oceans

At present, the published Cd isotope database for seawater samples is limited to results from only two studies (Lacan et al. 2006; Ripperger et al. 2007). However, there are a number of recent conference abstracts that announce further results from ongoing work, and this highlights the significant current interest in this field of Cd isotope research (Abouchami et al. 2009; Gault-Ringold et al. 2009; Abouchami et al. 2010a, b; Xue et al. 2010a, b).

The study of Ripperger et al. (2007) currently provides the most comprehensive and precise Cd isotope dataset for seawater, with results for 22 samples, representing both surface waters and depth profiles, from eight locations in all major ocean basins. This investigation found large variations in Cd isotope compositions, with $\varepsilon^{114/110}Cd$ raging from ~0 to values as high as $+38 \pm 6$ (Fig. 8.6). The isotopic variability of Cd was found to be restricted almost exclusively to the upper level of the water column (at depths ≤ 500 m) and the largest fractionations were observed for highly Cd depleted surface water samples from the Pacific, which featured Cd contents as low as ~5 pmol/kg (Fig. 8.6). In contrast, nearly uniform isotope compositions were determined for samples from ≥ 900 m depth (Fig. 8.6). The majority of the samples from the study of Ripperger et al. (2007) were furthermore seen to display an inverse relationship between dissolved Cd contents and isotope compositions (Fig. 8.7). A few samples, however, do not follow this broad trend because they exhibit both low Cd contents and nearly unfractionated Cd isotope compositions (Fig. 8.7).

The Cd isotope data of Lacan et al. (2006) are less precise and 28 samples from two depth profiles (in the northwest Pacific Ocean and Mediterranean Sea) display a much smaller range in Cd concentrations and isotope compositions, with $\varepsilon^{114/110}Cd$ values between about -4 and $+2$ (with an uncertainty of ± 3 ε, 2sd). Despite the limited variation, the samples are in accord with the $\varepsilon^{114/110}Cd$ vs. [Cd] trend observed by Ripperger et al. (2007), as the data for the Pacific Ocean depth profile also exhibit a statistically significant correlation between Cd isotope composition and concentration (Lacan et al. 2006). Importantly,

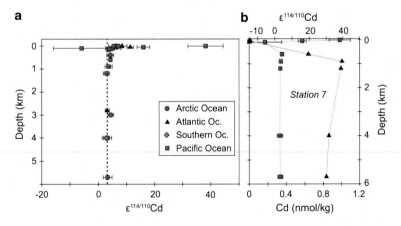

Fig. 8.6 Cd concentrations and isotope data (reported as $\varepsilon^{114/110}Cd$ relative to NIST 3108 Cd) for seawater samples, plotted a function of depth. (**a**) Shows the results for all 22 samples analyzed by Ripperger et al. (2007). Seawater from depths of ≤ 150 m in the photic zone exhibits the largest Cd isotope fractionations, presumably due to isotope fractionation from biological utilization. (**b**) Vertical distribution of [Cd] (*triangles*) and $\varepsilon^{114/110}Cd$ (*squares*) in the central North Pacific Ocean (Ripperger et al. 2007). Note that the samples with the lowest Cd contents exhibit the largest isotopic variability. The uncertainties (2sd) are shown as reported in the original publication

Lacan et al. (2006) also presented Cd isotope data for cultured freshwater phytoplankton and found that the organisms were systematically enriched in isotopically light Cd, presumably due to the kinetic isotope fractionation that results from biological uptake of dissolved seawater Cd.

Based on these findings, Ripperger et al. (2007) concluded that the large Cd isotope variations that were observed in the upper water column are most reasonably accounted for by biological utilization of Cd and associated isotope fractionation. This interpretation is supported by the observation that the seawater Cd data are in accord with closed system Rayleigh fractionation trends for fractionation factors of $\alpha \approx 1.0002–1.0006$ (Fig. 8.7). This result is in general agreement with the isotope fractionation factor of 1.0014 ± 0.0006 (2sd) that was determined for Cd uptake by cultured phytoplankton (Lacan et al. 2006). The small but significant difference in the fractionation factors inferred from seawater and the culturing experiments may reflect vital effects or additional uncharacterized marine processes, which have an impact on Cd isotope compositions (Ripperger

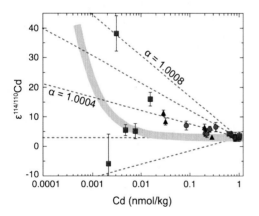

Fig. 8.7 Plot of Cd isotope compositions (reported as $\varepsilon^{114/110}Cd$ relative to NIST 3108 Cd) vs. Cd concentrations for seawater samples (Ripperger et al. 2007). Samples from different ocean basins are denoted by distinct symbols (see legend of Fig. 8.6). The *dashed lines* denote calculated isotope fractionation trends for dissolved seawater Cd, assuming that biological uptake of Cd occurs by closed-system Rayleigh fractionation with fractionation factors α of between 0.9998 and 1.0008. The *broad green line* is the mixing trend expected for addition of Cd-rich deep water ($\varepsilon^{114/110}Cd = 3$, [Cd] = 1 nmol/kg) to Cd-depleted surface water ($\varepsilon^{114/110}Cd = 40$, [Cd] = 0.5 pmol/kg). Note that such mixing produces downward deviations from the main fractionation trend that is exhibited by the majority of the seawater samples. The uncertainties (2sd) are shown as reported in the original publication

et al. 2007). Water mass mixing in particular, may be able to generate large deviations from the main fractionation trend if this involves both Cd-rich and Cd-depleted endmembers, a scenario that may be appropriate for the upwelling of deep waters to the surface ocean. In this case, mixing would generate distinct downward deviations from the main $\varepsilon^{114/110}Cd$ vs. [Cd] trend (as highlighted by the green curve of Fig. 8.7). Mixing processes may hence be responsible for the unusual compositions of samples, which feature both low Cd contents and $\varepsilon^{114/110}Cd$ values of about 0 to +5 (Ripperger et al. 2007).

For the Cd-rich seawater samples obtained from depths ≥900 m in the Arctic, Atlantic, Southern and Pacific Oceans, Ripperger et al. (2007) observed nearly constant and unfractionated Cd isotope compositions, which define a mean deep water value of $\varepsilon^{114/110}Cd = +2.9 \pm 1.0$ (2sd; Fig. 8.8). The sample data of Lacan et al. (2006) are less precise (at about $\pm 3\ \varepsilon$) but this study also reports very similar Cd isotope data for seven Pacific water samples, obtained from depths ≥1 km at a single station, with a slightly lower (but statistically identical) mean value of $\varepsilon^{114/110}Cd = +0.7 \pm 1.4$. The constant Cd isotope compositions of the deep water samples stand in contrast to the Cd contents, which increase systematically along the global deep water pathway, from concentrations of ~0.2–0.3 nM in the Arctic and Atlantic Oceans to values of ~0.8–1.0 nM in the Pacific (Fig. 8.8). These observations are most readily reconciled if the organic matter, which increases the Cd contents of deep water masses by remineralization, also features a nearly constant Cd isotope composition of $\varepsilon^{114/110}Cd \approx +3$. It was, therefore, concluded by Ripperger et al. (2007) that such biomass is most readily produced if the distribution of Cd in surface water is dominated by near-quantitative closed system uptake of Cd by phytoplankton. This interpretation is in accord with the results of previous investigations of marine nutrient cycling (De La Rocha et al. 1998; Elderfield and Rickaby 2000) and the extremely low Cd contents of open ocean surface waters.

Taken together, the currently published Cd isotope database for seawater, which encompasses only 50 samples, shows interesting systematics, including a broad correlation of increasing Cd isotope fractionation with decreasing [Cd]. These results are of particular interest as they provide a tentative foundation for two important applications of Cd isotopes in marine research. (1) The Cd isotope data reveal more detail and complexity in the marine distribution of Cd than can be discerned from concentration data alone. Hence, Cd isotope compositions have the potential to provide new insights into the biogeochemical processes, which govern the cycling of Cd in the oceans. (2) The $\varepsilon^{114/110}$Cd-[Cd] correlation observed for seawater suggests that Cd isotopes could be used in paleoceanography, as a proxy of nutrient utilization. The scarcity of seawater data currently renders such an application premature, however. In addition, it also requires the identification of suitable archives that reliably record the Cd isotope composition of seawater over periods of ka to Ma.

8.4.4.3 Cadmium Isotopes in Marine Sediments

At present, only a limited number of marine deposits have been analyzed for Cd isotope compositions, with a database that encompasses ten samples related to sub-seafloor hydrothermal activity at spreading centers and 50 hydrogenetic ferromanganese nodules and crusts (Fig. 8.9).

The set of hydrothermal samples (Schmitt et al. 2009a) includes seven sulfides with high Cd concentrations (>100 ppm Cd; Table 8.5) from the vent chimneys of either hot black smokers or slightly cooler grey smokers. Furthermore analyzed were three hydrothermal ferromanganese (Fe-Mn) crusts with Cd contents of about 3–20 µg/g (Table 8.5). Such crusts are typically deposited from cool (<100°C) diffuse flows of vent fluids in the distal parts of hydrothermal systems (Hein et al. 1997). With the exception of a single black smoker sulfide from the Lau Basin (a back-arc spreading center), all hydrothermal deposits are from a narrow region at ~10°N of the East Pacific Rise (EPR).

The Cd isotope data for the four black smoker sulfides cluster around a value of $\varepsilon^{114/110}$Cd ≈ 0.5 (Fig. 8.9) and were thought to record the isotope composition of the pristine hydrothermal fluids, from

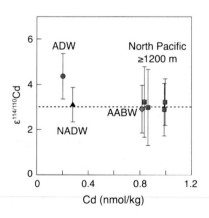

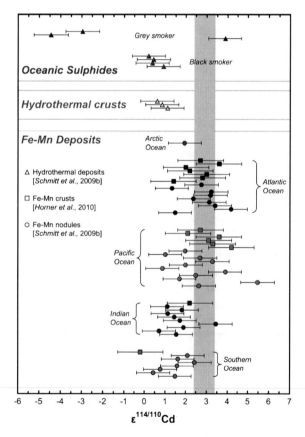

Fig. 8.8 Cd isotope data (reported as $\varepsilon^{114/110}$Cd relative to NIST 3108 Cd) plotted vs. Cd concentrations for water samples from depths of ≥900 m: Arctic deep water (ADW); North Atlantic deep water (NADW); Antarctic bottom water (AABW); and North Pacific water masses from depths ≥1,200 m (Ripperger et al. 2007). These samples display little isotopic variation and define a mean isotope composition, denoted by the *hashed line*, of $\varepsilon^{114/110}$Cd = +2.9 ± 1.0 (2sd) (Ripperger et al. 2007). The uncertainties (2sd) are shown as reported in the original publication

Fig. 8.9 Compilation of Cd isotope data (reported as $\varepsilon^{114/110}$Cd relative to NIST 3108 Cd) for marine sediments and deposits (Schmitt et al. 2009a; Horner et al. 2010). The samples are grouped by deposit type and location (for the hydrogenetic ferromanganese Fe-Mn deposits). The mean Cd isotopic composition of oceanic deep water (Ripperger et al. 2007) of $\varepsilon^{114/110}$Cd = 2.9 ± 1.0, is denoted by the *grey band*. The uncertainties (2sd) are shown as reported in the original publications

which the deposits were precipitated (Schmitt et al. 2009a). The $\varepsilon^{114/110}$Cd values of the chimney sulfides are lower than the average oceanic deep water value derived by Ripperger et al. (2007) ($\varepsilon^{114/110}$Cd ≈ 2.9, Fig. 8.8, Table 8.5) but similar to the bulk silicate Earth estimate of $\varepsilon^{114/110}$Cd ≈ −0.3 (Fig. 8.9, Table 8.5). This may indicate that the isotopic signature of pristine hydrothermal fluids is primarily inherited from Cd that is leached from the basaltic ocean crust, without significant isotope fractionation. The three hydrothermal Fe-Mn crusts display $\varepsilon^{114/110}$Cd values akin to those of the black smoker sulfides and hence these samples were also thought to record the Cd isotope composition of the (pristine) vent fluids, without large fractionation effects (Schmitt et al. 2009a).

The sulfides from a grey smoker (T ≈ 200°C) at the EPR displayed the largest Cd isotope variations of the hydrothermal deposits (Fig. 8.9). These samples displayed $\varepsilon^{114/110}$Cd values that varied between −4.5 and +3.9 and furthermore showed a good (negative) linear correlation with 1/[Cd]. This correlation was speculated to be a consequence of sulfide precipitation in the vent system and associated isotope fractionation (Schmitt et al. 2009a). Similar processes were previously inferred to be responsible for the Zn isotope variability of these samples (John et al. 2008).

Investigations of hydrogenetic Fe-Mn deposits are of particular interest and such samples hence dominate the current Cd isotope database for marine sediments (Fig. 8.9). This interest is based on the observation that the surface layers of hydrogenetic Fe-Mn crusts and nodules faithfully record the isotopic composition of seawater for numerous trace metals. Consequently, time series data obtained for older deposits have been used in paleoceanographic studies, to interrogate temporal changes in the radiogenic (e.g., Nd, Pb) and stable (e.g., Tl, Mo) isotope composition of seawater (Frank 2002; Rehkämper et al. 2002, 2004; Siebert et al. 2003).

The compendium of presently available Cd isotope data encompasses 15 Fe-Mn crusts (Horner et al. 2010) and 37 Fe-Mn nodules (Schmitt et al. 2009a). These two types of deposits differ in that Fe-Mn crusts form on basaltic substrates whereas nodules are precipitated on pelagic sediments. Despite this difference, both crusts and nodules were observed to display similar Cd contents (of about 1–20 μg/g) and isotope compositions, with $\varepsilon^{114/110}$Cd values of between about 0 and +5.5 (Fig. 8.9, Table 8.5). In particular, essentially all hydrogenetic Fe-Mn deposits are characterized by Cd isotope composition that are identical, within error, to the oceanic deep water average of $\varepsilon^{114/110}$Cd = 2.9 ± 1.0 (Ripperger et al. 2007). This result is significant because it implies that Fe-Mn crusts and nodules record the Cd isotope composition of seawater without significant isotopic fractionation (Schmitt et al. 2009a; Horner et al. 2010). This observation is surprising because it contrasts with the behavior of Zn, an element that generally exhibits similar marine biogeochemical characteristics to Cd (de Baar and La Roche 2002). The Zn isotope composition of deep ocean seawater (Bermin et al. 2006) appears to be offset from isotopically heavier Zn in Fe-Mn nodules (Maréchal et al. 2000) by about 5 ε for $^{66}Zn/^{64}Zn$. This fractionation has been corroborated by laboratory studies, which have shown that "heavy" Zn isotopes are preferentially scavenged from solutions by (oxy)hydroxide precipitates (Pokrovsky et al. 2005; Balistrieri et al. 2008; Juillot et al. 2008). Offsets in stable isotope compositions between seawater and hydrogenetic Fe-Mn deposits are also common for other elements, including Mo (Barling et al. 2001) and Tl (Rehkämper et al. 2002; Rehkämper et al. 2004), although the mechanisms of isotopic fractionation are still the subject of ongoing research and may be different for each element.

The absence of isotopic effects during the incorporation of Cd into hydrogenetic Fe-Mn crusts and nodules is a promising result because this simplifies the interpretation of Cd isotope time series data that are obtained for such samples. Based on this finding, it may possible to interpret temporal variations in $\varepsilon^{114/110}$Cd that are identified for Fe-Mn deposits, as reflecting changes in the Cd isotope composition of seawater, rather than differences in the isotopic fractionation between seawater and the Fe-Mn minerals. Preliminary time-resolved Cd isotope data were obtained for a Fe-Mn nodule from the Atlantic Ocean by Schmitt et al. (2009a). Six samples with ages of between 0 and 8 Ma were analyzed and no significant change in $\varepsilon^{114/110}$Cd was detected. This suggests that long-term temporal variations in the Cd isotope composition of oceanic deep water are likely to be small (Schmitt et al. 2009a). However, it is possible that the stable Cd isotope systematics of this nodule were altered by diagenetic processes, such as isotopic equilibration with Cd from pore fluids (Rehkämper et al. 2002).

The time series data for the nodules also do not preclude short-term fluctuations in the Cd isotope composition of the oceans and these may be best recorded in marine carbonates. This conclusion follows from the observation that the carbonate records of Cd/Ca (Boyle and Keigwin 1982; Boyle 1992) and other trace metal isotope systems (Lea 2006; Lynch-Stieglitz 2006) show significant variability on glacial/interglacial timescales. Cadmium isotope results are, however, currently not available for carbonate samples that are commonly employed in paleoclimate studies. This lack of data reflects the challenging nature of the analyses, which are rendered particularly difficult by the low Cd contents of such materials (e.g., foraminiferal tests; Table 8.5). This necessitate that large samples (typically >1 g) are processed for analysis, as at least 1–10 ng of Cd are required for a reasonably precise isotopic measurement. The only carbonates for which Cd isotope data are currently available, are aragonites from laboratory precipitation experiments. These samples were observed to exhibit consistently light Cd isotope signatures, with a fractionation of $-6.4 \pm 0.9\ \varepsilon^{114/110}Cd$ ($\pm 2sd$) relative to the growth solution. It is possible but currently unconstrained whether natural biogenic carbonates exhibit similar Cd isotope effects.

8.4.5 Mass Independent Cd Isotope Fractionations

Most natural stable isotope fractionations scale with the mass differences of the isotopes involved. The ability to acquire more precise isotopic data as a consequence of advances in mass spectrometric techniques and instrumentation has, however, also enabled the identification of stable isotope effects that clearly do not scale with differences in nuclide mass. Such mass independent stable isotope effects were first documented for the light elements O and S (Thiemens 1999, 2006). More recently, the importance of mass independent fractionation for heavier elements was recognized based on results acquired for both natural samples (particularly Hg, Tl and U; Rehkämper et al. 2004; Smith et al. 2005; Xie et al. 2005; Schauble 2007; Stirling et al. 2007; Weyer et al. 2008; Blum 2011; Nielsen and Rehkämper 2011) and laboratory experiments (Hg, U and many other elements; Biegeleisen 1996; Estrade et al. 2009; Fujii et al. 2009a).

For heavier elements, non-mass dependent isotope fractionations are generally thought to be a consequence of either nuclear volume (Biegeleisen 1996; Schauble 2007; Fujii et al. 2009a) or magnetic isotope effects (Buchachenko 2001). Nuclear volume fractionations are also known as nuclear field shift effects and are a consequence of non-mass dependent differences in nuclear volume and shape. Nuclear volume (or the nuclear charge radius) increases with the number of neutrons and this increase is generally slightly smaller for isotopes with odd mass numbers as compared to even mass numbers. Hence, nuclear volume effects are typically expected to generate odd-even isotope fractionation patterns (Biegeleisen 1996; Schauble 2007; Fujii et al. 2009a). In the case of Cd, ^{116}Cd is an exception, however (Fig. 8.10), as the nuclear volume of this even mass isotope displays a relative deficit similar to that observed for the odd mass isotopes ^{111}Cd and ^{113}Cd (Angeli 2004). Nuclear volume effects are most significant for heavier elements but smaller, yet resolvable, effects were also deemed to be reasonable, based on theoretical considerations, for elements that have atomic masses similar to those of Cd, such as Ru and Sn (Knyazev and Myasoedov 2001; Schauble 2007; Fujii et al. 2009a). Magnetic isotope effects stringently distinguish between isotopes with odd and even mass numbers, as they result from hyperfine coupling between the non-zero nuclear spins of odd nuclei and the magnetic moments of electrons (Buchachenko 2001).

Both nuclear volume and magnetic isotope fractionations have been suggested to affect Cd under certain conditions. Magnetic isotope effects, however, only appear to play a role during Cd isotope analyses by TIMS when silica gel is used as an activator. In particular, it was shown for such TIMS measurements that Cd isotope ratios with the odd mass nuclides ^{111}Cd and ^{113}Cd do not display a mass dependent instrumental fractionation relative to ratios that feature only the even mass isotopes ^{110}Cd, ^{112}Cd, ^{114}Cd and ^{116}Cd (Manhès and Göpel 2003, 2007; Schmitt et al. 2009b). Such odd-even isotope anomalies were also observed for Pb and Zn isotope analyses by TIMS (Thirlwall 2000; Doucelance and Manhès 2001; Manhès and Göpel 2003) and ascribed to a magnetic isotope effect because (1) nuclear volume fractionations would also generate anomalies for ^{116}Cd and (2)

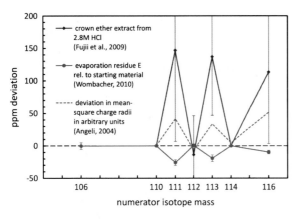

Fig. 8.10 Experimental evidence for nuclear volume Cd isotope effects that was obtained for a crown ether extraction experiment (Fujii et al. 2009b) and an evaporation residue (Wombacher 2010). The relative deviation of a Cd isotope ratio in parts per million (ppm) from the value expected for purely mass dependent stable isotope fractionation is shown on the y-axis. Results are depicted for all Cd isotope ratios from $^{106}Cd/^{114}Cd$ to $^{116}Cd/^{114}Cd$. The data for the odd mass isotopes ^{111}Cd and ^{113}Cd and the even isotope ^{116}Cd deviate from mass dependent behavior. This is in accord with the pattern expected for non-mass dependent nuclear volume effects, which reflect differences in nuclear charge radii (Angeli 2004). The results for the crown ether extract (Fujii et al. 2009b) were recalculated, such that ^{114}Cd is the common denominator; the uncertainties were assessed based on the reported δ-values and are likely to be overestimated. The data for the Cd evaporation residue E (see Fig. 8.1; Wombacher 2010) were obtained with a Neptune MC-ICPMS instrument (Universität Bonn) in 30 min analyses at high ion beam intensities. Shown are the average results (±2sd uncertainties) of three analyses conducted using two different collector settings and sample introduction systems. The generalized power law (with exponent $n = -0.353$) was applied to correct for the large mass dependent fractionation of the Cd, as described in (Wombacher et al. 2004)

comparatively large (permil-level) non-mass dependent effects were observed for ^{111}Cd and ^{113}Cd.

In order to investigate potential nuclear volume isotope effects for Cd, Fujii et al. (2009b) conducted an experimental study of isotope fractionation during extraction of Cd from aqueous HCl solutions using a crown ether dissolved in an immiscible organic solvent. Large Cd isotope fractionations of up to about 50 $\varepsilon^{114/110}Cd$ were generated and resolvable non-mass dependent isotope effects were observed for ^{111}Cd, ^{113}Cd, and ^{116}Cd as a consequence of nuclear field shifts (Fig. 8.10). Fujii et al. (2009b) concluded that between 5 and 30% of the measured Cd isotope fractionations (depending on the isotope ratio evaluated) were contributions from nuclear field shift effects.

Further evidence for non-mass dependent Cd isotope effects was provided by analyses of highly fractionated Cd metal samples that were produced as residues of evaporation (Fig. 8.1, Sect. 8.4.1.1). Inspection of the original measurement results suggests that the isotope fractionation, which is recorded by ratios that include ^{111}Cd and ^{113}Cd, may indeed exhibit ppm level deviations from the trends defined by ^{106}Cd, ^{108}Cd, ^{110}Cd, ^{112}Cd and ^{114}Cd (Wombacher et al. 2004). These differences are, however, barely resolvable given the quoted analytical uncertainties. Wombacher (2010) reanalyzed these Cd metal samples with improved precision and determined small but clearly resolvable non-mass dependent isotope anomalies of -26 ± 5, -19 ± 5 and -10 ± 3 ppm (parts per million) for $^{111}Cd/^{114}Cd$, $^{113}Cd/^{114}Cd$ and $^{116}Cd/^{114}Cd$, respectively (Fig. 8.10). Importantly, the relative magnitude of the measured effects is in accord with, or similar to, the fractionation pattern that is predicted based on published nuclear charge radius data (Fig. 8.10). The interpretation that the observed anomalies indeed reflect nuclear field shift effects from evaporation is furthermore supported by the observation of similar nuclear volume isotope effects for evaporation residues of Hg (Estrade et al. 2009).

The nuclear field shift effects determined for the evaporation of Cd are very small, however, as they contribute only about 0.5% to the total observed isotope fractionation from evaporation of Cd (Wombacher 2010). Hence, these non-mass dependent effects (Fig. 8.10) were only detectable because the absolute extent of mass dependent isotope fractionation from evaporation was extremely large, at up to almost 50‰ for $^{110}Cd/^{114}Cd$ (Fig. 8.1). The mass independent Cd isotope effects determined for TIMS analyses and crown ether extractions are much larger but these results may not be relevant for the generation of isotope anomalies in natural systems. Biological uptake of Cd has been shown to create large isotope fractionations (Figs. 8.6 and 8.7, Sect. 8.4.4.2) and thus provides an alternative pathway for the generation of mass-independent Cd isotope effects. Biological processes, however, are likely to involve bonding environments for Cd that exhibit significant covalent contributions (due to electron sharing) and such conditions are thought to greatly reduce the importance of nuclear field shift effects (Knyazev and Myasoedov 2001; Schauble 2007).

In summary, these observations suggest that mass independent Cd isotope effects may not be significant in nature. It will be important, however, to verify this somewhat speculative conclusion by further high-precision Cd isotope analyses of suitable, highly fractionated samples without use of a double spike. This is particularly significant as (1) non-mass dependent Cd isotope fractionations may enable novel tracer applications and (2) the current techniques of Cd double spike data reduction explicitly assume that any differences Cd isotope compositions arise only from mass dependent effects.

8.5 Future Directions and Outlook

The first targeted search for natural differences in the Cd isotope composition of terrestrial materials was conducted in the 1970s (Rosman and de Laeter 1975). However, no variability could be resolved until more precise mass spectrometric techniques were developed less than a decade ago (Wombacher et al. 2003; Ripperger and Rehkämper 2007; Schmitt et al. 2009b). The current Cd isotope database for terrestrial materials is still relatively small, compared to that available for other non-traditional stable isotope systems, such as Fe, Zn, or Mo (Johnson et al. 2004). However, small natural Cd isotope variations have been identified even for samples of the silicate Earth, such as rocks and minerals from the continental crust (Figs. 8.2 and 8.3). Furthermore, two specific mechanisms have been identified, which routinely generate larger Cd isotope fractionations in nature – these are partial evaporation/condensation (Wombacher et al. 2004; Cloquet et al. 2006) and biological utilization (Lacan et al. 2006; Ripperger et al. 2007). These findings have recently stirred significant interest in Cd isotope studies and an increasing number of laboratories are currently acquiring results.

Due to this interest, it is of particular importance to quickly establish a new zero-ε (or zero-δ) reference material that is universally adopted by all laboratories. This development is needed, to ensure that the results of different laboratories do not feature systematic offsets (which can result from poor intercalibration) but are accurate at the current level of analytical precision (about ±0.1 to 0.3 εCd/amu; Table 8.3). The currently ongoing intercalibration of NIST SRM 3108 Cd (Abouchami et al. 2010b) is promising in this respect and will hopefully lead to the adoption of this standard as a universally accepted zero-ε reference material for Cd isotopes in the near future.

Given the current scarcity of Cd isotope data, it will be important for future studies to expand this database in a systematic manner. Of particular interest will be data acquired at high precision and accuracy, and these are probably best obtained using a double spike approach with either TIMS or MC-ICPMS (Tables 8.3 and 8.4). Additional analyses that are conducted without a double spike also will be important, however, to search for non-mass dependent Cd isotope anomalies. In any case, the samples for such analyses will need to be carefully chosen, as the methods are time-consuming and it is unclear whether future improvements in instrumentation or techniques will be able to significantly increase sample throughput. Regardless of such limitations, there are a number of interesting problems and potential applications, which should (and are very likely to be) investigated in the near future.

Fractionation mechanisms. A striking advantage of the Cd stable isotope system is that essentially all larger isotopic effects are known to be due to either evaporation/condensation or biological utilization. This observation is backed up by a few experimental studies, which clearly demonstrate the effect of these processes on Cd isotope compositions (Wombacher et al. 2004; Lacan et al. 2006). However, in detail, there are many questions that should be addressed. For example, Cd isotope fractionation by biological utilization has only been investigated in experiments with freshwater phytoplankton (Lacan et al. 2006), but such effects are most common and scientifically interesting in marine (seawater) environments. Hence, laboratory studies of Cd uptake by various marine organisms, which characterize potential vital effects and fractionation pathways are particularly desirable. Similarly, it has been observed that isotope fractionations for Cd (and other elements) from partial evaporation are typically much smaller (by about a factor of two) than expected from kinetic theory, which predicts a fractionation factor of $\alpha_{kin} = (m_2/m_1)^{0.5}$ (Wombacher et al. 2004, 2008). The origin of this behavior is currently not understood and should hence be investigated.

Anthropogenic Cd isotope fractionations. An improved understanding of Cd isotope effects from evaporation and condensation is of particular interest, as a number of important anthropogenic fluxes of Cd to the environment involve volatilization. This includes, for example, the burning of coal and urban waste as well as the smelting and refining processes that are applied in the industrial production of Cd. Based on the results obtained to date (Fig. 8.2), combined Cd isotope and concentration measurements hold significant promise for the tracing and quantification of anthropogenic inputs and such studies are of particular relevance due to the high toxicity of Cd (Nriagu 1981). The isotopic tracing approach furthermore appears promising, given (1) the high volatility of Cd, which should aid in the generation of fractionation effects that are larger compared to those observed for the geochemically similar but less volatile element Zn; and (2) the lack of other significant (abiotic) fractionation pathways for Cd isotopes.

Cd isotope fractionation in the silicate Earth. Only few high precision isotopic data are currently available for samples of the Earth's continental crust and mantle. These results indicate very limited natural variability in Cd isotope compositions (Figs. 8.2 and 8.3, Table 8.5) but this may be an artifact of the small number of samples analyzed. Hence, additional high precision measurements are important that target diverse environments, lithologies and samples types, to confirm or revise the present observations. For example, preliminary results obtained for ultramafic samples may indicate that these differ in Cd isotope composition from mafic and felsic rocks, potentially as a result of magmatic processes (Schönbächler et al. 2009).

The primary aim of such analyses will be to better establish the extent of Cd isotope variability in the silicate Earth. Such data will be of importance to environmental scientists, as isotopic tracing of anthropogenic Cd is not feasible without a detailed understanding natural Cd isotope variations in the silicate Earth. Furthermore, the larger database will also provide more robust constraints on the Cd isotope composition of the silicate Earth. Notably, the best current estimate of $\varepsilon^{114/110}$Cd for the bulk silicate Earth (Table 8.5) is based on results for only seven silicate rock samples (Schmitt et al. 2009a). A comparison of the mean terrestrial $\varepsilon^{114/110}$Cd value with meteorite data is important, as this may provide new insights into the processes responsible for the volatile depleted nature of the Earth and other planetary bodies (Palme et al. 1988; Wombacher et al. 2008).

Cd isotopes in marine environments. The small database for Cd isotopes in seawater is currently growing rapidly, as samples from the GEOTRACES research program (Measures 2007) are analyzed. The availability of such data for a wider range of marine locations and conditions is desirable, as this will enhance our understanding of the cycling and biological role of Cd in the oceans.

Further interest will focus on the question whether this larger dataset still exhibits a robust co-variation between Cd concentrations and isotope compositions (Fig. 8.7). Confirmation of this trend is desirable because this would provide a solid basis for paleoceanographic applications. Specifically, a robust $\varepsilon^{114/110}$Cd-[Cd] relationship for seawater would imply that marine records of Cd isotope compositions could be used, either alone or in combination with foraminiferal Cd/Ca ratios, to study past changes in ocean circulation and marine nutrient utilization (Boyle and Keigwin 1982; Boyle 1988; Rickaby and Elderfield 1999). Potential archives that warrant further investigation are Fe-Mn crusts and nodules (Schmitt et al. 2009a; Horner et al. 2010), the hydrogenetic components of pelagic sediments, and marine carbonates. The latter archives are of particular interest, as they can be sampled at sufficient resolution to resolve glacial-interglacial cycles. Hence, they can be used to further interrogate the role of the marine carbon cycle in driving changes in atmospheric CO_2 levels and global climate conditions (Elderfield and Rickaby 2000; Sigman and Boyle 2000). Cadmium isotope analyses of marine carbonates will be very challenging, however, as the low Cd contents of such phases necessitate that relatively large samples are processed for isotopic analyses. Typically, a precise Cd isotope analysis will require about 100–1,000 time more Cd than are necessary for Cd/Ca measurements of carbonates. Nonetheless, such analyses will be of great interest, if they can provide robust records for a marine $\varepsilon^{114/110}$Cd proxy that can be applied to study changes in marine nutrient levels.

The development of such paleoceanographic applications also requires, however, that any isotopic fractionation from incorporation of Cd into the archive is well characterized, ideally by studies of both natural and experimental systems. The presently published data indicate that the incorporation of Cd into marine Fe-Mn oxyhydroxide phases is probably not accompanied by significant Cd isotope fractionations (Schmitt et al. 2009a; Horner et al. 2010), whilst such effects have been detected for carbonate phases (Wombacher et al. 2003). These first results, however, need to be thoroughly verified in further studies. An investigation of the dependence of carbonate mineral fractionation factors on temperature, precipitation rate and other parameters will be particularly important because the Cd/Ca ratios of planktonic foraminiferal tests have been shown to vary strongly with seawater temperature (Rickaby and Elderfield 1999).

Of general interest is, furthermore, an isotopic investigation of the sources and sinks of Cd in the oceans. For the sources, the analyses should ideally encompass rivers and estuarine systems, hydrothermal fluids and atmospheric inputs, including volcanic emissions, which might be expected to exhibit large Cd isotope effects from volatilization. Marine sedimentary deposits represent the main sink of Cd in the oceans, and relatively high Cd concentrations (of >100 ng/g) are commonly found in pelagic clays (Table 8.5) and C-rich sediments from continental margins (Heinrichs et al. 1980). Of particular importance is the application of these data to carry out an independent review of previously published mass balance estimates for the input and output fluxes of marine Cd. In addition, the isotopic results will also provide new constraints on the marine residence time of Cd, which has been estimated at about 18,000–180,000 a by previous investigations (Balistrieri et al. 1981; Martin and Whitfield 1983; Boyle 1988). These results will furthermore be of relevance for paleoceanography, as they inform on whether processes other than nutrient utilization are able to alter the Cd isotope composition of the oceans.

Acknowledgments Wafa Abouchami and Steve Galer from the Max Planck Institute for Chemistry in Mainz (Germany) are thanked for their efforts to identify a suitable common Cd isotope reference material and their encouragement to present data relative to the NIST standard in this manuscript. Constructive formal reviews by Tom Bullen and an anonymous referee as well as the editorial efforts of Mark Baskaran are gratefully acknowledged. F.W. thanks Carsten Münker, Stefan Weyer and Ambre Luguet for support of additional Cd isotope analyses of reference materials and Cd metal samples in the Neptune lab in Bonn and Steve Galer for discussions. M.R. is grateful for helpful advice and contributions provided by Maria Schönbächler.

References

Abouchami W et al (2009) The Cd isotope signature of the Southern Ocean. Eos Trans. AGU 90, Fall Meet. Suppl: Abstract PP13A-1376

Abouchami W et al (2010a) Cd isotopes, a proxy for water mass and nutrient uptake in the Southern Ocean. Eos Trans. AGU 91, Ocean Sci. Meet. Suppl:Abstract CO13A-03

Abouchami W et al (2010b) In search of a common reference material for cadmium isotope studies. 2010 Goldschmidt Conference A2

Andersen T, Knutsen AB (1962) Anion-exchange study. I. Adsorption of some elements in HBr-solutions. Acta Chem Scand 16:849–854

Angeli I (2004) A consistent set of nuclear rms charge radii: properties of the radius surface R (N, Z). At Data Nucl Data 87:185–206

Aston FW (1935) The isotopic constitution and atomic weights of hafnium, thorium, rhodium, titanium, zirconium, calcium, gallium, silver, carbon, nickel, cadmium, iron and indium. Proc R Soc Lond A 149:396–405

Aylett BJ (1973) Group IIB – cadmium. In: Trotman-Dickenson AF (ed) Comprehensive inorganic chemistry. Pergamon, Oxford

Baker RGA et al (2010) The thallium isotope composition of carbonaceous chondrites – new evidence for live ^{205}Pb in the early solar system. Earth Planet Sci Lett 291:39–47

Balistrieri L, Brewer PG, Murray JW (1981) Scavenging residence times of trace metals and surface chemistry of sinking particles in the deep sea. Deep Sea Res 28:101–121

Balistrieri LS et al (2008) Fractionation of Cu and Zn isotopes during adsorption onto amorphous Fe(III) oxyhydroxide: experimental mixing of acid rock drainage and ambient river water. Geochim Cosmochim Acta 72:311–328

Barling J, Arnold GL, Anbar AD (2001) Natural mass-dependent variations in the isotopic composition of molybdenum. Earth Planet Sci Lett 193:447–457

Bender ML, Gagner CL (1976) Dissolved copper, nickel, and cadmium in the Sargasso Sea. J Mar Res 34:327–339

Bermin J et al (2006) The determination of the isotopic composition of Cu and Zn in seawater. Chem Geol 226:280–297

Biegeleisen J (1996) Nuclear size and shape effects in chemical reactions. Isotope chemistry of the heavy elements. J Am Chem Soc 118:3676–3680

Blum JD (2011) Chapter 12 Applications of stable mercury isotopes to biogeochemistry. In: Baskaran MM (ed) Handbook of environmental isotope geochemistry. Springer, Berlin

Böhlke JK et al (2005) Isotopic compositions of the elements, 2001. J Phys Chem Ref Data 34:57–67

Boyle EA, Sclater F, Edmond JM (1976) On the marine geochemistry of cadmium. Nature 263:42–44

Boyle EA (1981) Cadmium, zinc, copper, and barium in foraminifera tests. Earth Planet Sci Lett 53:11–35

Boyle EA, Keigwin LD (1982) Deep circulation of the North Atlantic over the last 200,000 years: geochemical evidence. Science 218:784–787

Boyle EA (1988) Cadmium: chemical tracer of deepwater paleoceanography. Paleoceanography 3:471–489

Boyle EA (1992) Cadmium and δ^{13}C paleochemical ocean distributions during the Stage 2 glacial maximum. Annu Rev Earth Planet Sci 20:245–287

Bruland KW (1980) Oceanographic distributions of cadmium, zinc, nickel, and copper in the North Pacific. Earth Planet Sci Lett 47:176–198

Bruland KW (1983) Trace elements in sea-water. In: Ripley JP, Chester R (eds) Chemical oceanography. Academic Press, London

Buchachenko AL (2001) Magnetic isotope effect: nuclear spin control of chemical reactions. J Phys Chem A 105:9995–10011

Chizhikov DM (1966) Cadmium. Pergamon, Oxford

Cloquet C, Carignan J, Libourel G (2005a) Kinetic isotope fractionation of Cd and Zn during condensation. Eos Trans. AGU 86 Fall Meet. Suppl:Abstract V41F-1523

Cloquet C et al (2005b) Natural cadmium isotopic variations in eight geological reference materials (NIST SRM 2711, BCR 176, GSS-1, GXR-1, GXR-2, GSD-12, Nod-P-1, Nod-A-1) and anthropogenic samples, measured by MC-ICP-MS. Geostand Geoanalyt Res 29:95–106

Cloquet C et al (2006) Tracing source pollution in soils using cadmium and lead isotopes. Environ Sci Technol 40: 2525–2530

Collier R, Edmond JM (1984) The trace element geochemistry of marine biogenic particulate matter. Prog Oceanogr 13:113–199

Cotton FA, Wilkinson G (1988) Advanced inorganic chemistry. Wiley, New York

Criss RE (1999) Principles of stable isotope distribution. Oxford University Press, Oxford

Cullen JT et al (1999) Modulation of cadmium uptake in phytoplankton by seawater CO_2 concentration. Nature 402:165–167

Cullen JT (2006) On the nonlinear relationship between dissolved cadmium and phosphate in the modern global ocean: could chronic iron limitation of phytoplankton growth cause the kink? Limnol Oceanogr 51:1369–1380

Davis AM et al (2005) Isotopic mass-fractionation laws and the initial solar system $^{26}Al/^{27}Al$ ratio. Lunar Planet Sci Conf XXXVI Abstract No:2334

de Baar HJW et al (1994) Cadmium versus phosphate in the world ocean. Mar Chem 46:261–281

de Baar HJW, La Roche J (2002) Trace metals in the oceans: evolution, biology and global change. In: Wefer G, Lamy F, Mantoura F (eds) Marine science frontiers for Europe. Springer, Berlin

De La Rocha CL et al (1998) Silicon-isotope composition of diatoms as an indicator of past oceanic change. Nature 395: 680–683

Dixon JL et al (2006) Cadmium uptake by marine micro-organisms in the English Channel and Celtic Sea. Aquat Microb Ecol 44:31–43

Dodson MH (1963) A theoretical study of the use of internal standards for precise isotopic analysis by the surface ionization technique: part I – General first-order algebraic solutions. J Sci Instrum 40:289–295

Donat JR, Bruland KW (1995) Trace elements in the oceans. In: Salbu B, Steinnes E (eds) Trace elements in natural waters. CRC Press, Boca Raton

Doucelance R, Manhès G (2001) Reevaluation of precise lead isotope measurements by thermal ionization mass spectrometry: comparison with determinations by plasma source mass spectrometry. Chem Geol 176:361–377

Elderfield H, Rickaby REM (2000) Oceanic Cd/P ratio and nutrient utilization in the glacial Southern Ocean. Nature 405:305–310

Estrade N et al (2009) Mercury isotope fractionation during liquid-vapor evaporation experiments. Geochim Cosmochim Acta 73:2693–2711

Faris JP, Buchanan RF (1964) Anion exchange characteristics of elements in nitric acid medium. Anal Chem 36:1157–1158

Frank M (2002) Radiogenic isotopes: tracers of past ocean circulation and erosional input. Rev Geophys 40:1–38

Frew RD, Hunter KA (1992) Influence of Southern Ocean waters on the cadmium–phosphate properties of the global ocean. Nature 360:144–146

Fujii T, Moynier F, Albarède F (2009a) The nuclear field shift effect in chemical exchange reactions. Chem Geol 267: 139–156

Fujii T et al (2009b) Nuclear field shift effect in the isotope exchange reaction of cadmium using a crown ether. Chem Geol 267:157–163

Galer SJG (1999) Optimal double and triple spiking for high precision lead isotopic measurements. Chem Geol 157: 255–274

Gao B et al (2008) Precise determination of cadmium and lead isotopic compositions in river sediments. Anal Chim Acta 612:114–120

Gault-Ringold M et al (2009) Cadmium isotopic compositions and nutrient cycling in the Southern Ocean. Goldschmidt Conference A418

Hein JR et al (1997) Iron and manganese oxide mineralization in the Pacific. In: Nicholson K, Hein JR, Bühn B, Dasgupta S (eds) Manganese mineralization: geochemistry and mineralogy of terrestrial and marine deposits. Special Publications, London

Heinrichs H, Schulz-Dobrick B, Wedepohl KH (1980) Terrestrial geochemistry of Cd, Bi, Tl, Pb, Zn, and Rb. Geochim Cosmochim Acta 44:1519–1533

Henderson GM (2002) New oceanic proxies for paleoclimate. Earth Planet Sci Lett 203:1–13

Hendry KR et al (2008) Cadmium and phosphate in coastal Antarctic seawater: implications for Southern Ocean nutrient cycling. Mar Chem 112:149–157

Hofmann A (1971) Fractionation correction for mixed-isotope spikes of Sr, K, and Pb. Earth Planet Sci Lett 10:397–402

Holleman AF, Wiberg E, Wiberg N (1985) Lehrbuch der Anorganischen Chemie (91–100 ed). de Gruyter, Berlin

Horner TJ et al (2010) Ferromanganese crusts as archives of deep water Cd isotope compositions. Geochem Geophys Geosys 11:Q04001. doi:10.1029/2009GC002987

Hu Z, Gao S (2008) Upper crustal abundances of trace elements: a revision and update. Chem Geol 253:205–221

Hulston JR, Thode HG (1965) Variations in the S^{33}, S^{34}, and S^{36} contents of meteorites and their relation to chemical and nuclear effects. J Geophys Res 70:3475–3484

John SG et al (2008) Zinc stable isotopes in seafloor hydrothermal vent fluids and chimneys. Earth Planet Sci Lett 269:17–28

Johnson CM, Beard BL, Albarède F (Eds.) (2004) Geochemistry of Non-Traditional Stable Isotopes, 454 pp., Mineralogical Society, Washington, DC

Juillot F et al (2008) Zn isotopic fractionation caused by sorption on goethite and 2-lines ferrihydrite. Geochim Cosmochim Acta 72:4886–4900

Knyazev DA, Myasoedov NF (2001) Specific effects of heavy nuclei in chemical equilibrium. Separ Sci Technol 36: 1677–1696

Korkisch J (1989) Handbook of ion exchange resins: their application to inorganic analytical chemistry. CRC Press, Boca Raton

Kraus KA, Nelson F (1955) Anion Exchange Studies of the Fission Products, paper presented at International Conference on the Peaceful Uses of Atomic Energy, United Nations, New York, Geneva, 1956

Lacan F et al (2006) Cadmium isotopic composition in the ocean. Geochim Cosmochim Acta 70:5104–5118

Lagos M et al (2008) The Earth's missing lead may not be in the core. Nature 456:89–92

Lane TW et al (2005) A cadmium enzyme from a marine diatom. Nature 435:42

Lea DW (2006) Elemental and isotopic proxies of past ocean temperatures. In: Elderfield H (ed) The oceans and marine geochemistry. Elsevier, Amsterdam

Leland WT, Nier AO (1948) The relative abundances of the zinc and cadmium isotopes. Phys Rev 73:1206

Li Y-H, Schoonmaker JE (2005) Chemical composition and mineralogy of marine sediments. In: Mackenzie FT (ed) Sediments, diagenesis, and sedimentary rocks. Elsevier, Amsterdam

Lodders K (2003) Solar system abundances and condensation temperatures of the elements. Astrophys J 591:1220–1247

Lynch-Stieglitz J (2006) Tracers of past ocean circulation. In: Elderfield H (ed) The oceans and marine geochemistry. Elsevier, Amsterdam

Manhès G, Göpel C (2003) Heavy stable isotope measurements with thermal ionization mass spectrometry: non mass-dependent fractionation effects between even and uneven isotopes. Geophys Res Abstr 5:10936

Manhès G, Göpel C (2007) Mass-independant fractionationation during TIMS measurements: evidence of nuclear shift effect? Geochim Cosmochim Acta 71:A618

Maréchal CN, Télouk P, Albarède F (1999) Precise analysis of copper and zinc isotopic compositions by plasma-source mass spectrometry. Chem Geol 156:251–273

Maréchal CN et al (2000) Abundance of zinc isotopes as a marine biogeochemical tracer. Geochem Geophys Geosys 1:1999GC000029

Martin JM, Whitfield M (1983) River input of chemical elements to the ocean. In: Wong CS, Boyle EA, Bruland KW, Burton JD, Goldberg ED (eds) Trace metals in seawater. Plenum, New York

McDonough WF (1990) Constraints on the composition of the continental lithospheric mantle. Earth Planet Sci Lett 101:1–18

McDonough WF, Sun S-s (1995) The composition of the Earth. Chem Geol 120:223–253

McDonough WF (2005) Compositional model for the Earth's core. In: Davis AM (ed) Meteorites, comets, and planets. Elsevier, Amsterdam

Measures C (2007) GEOTRACES – an international study of the global marine biogeochemical cycles of trace elements and their isotopes. Chem d Erde Geochem 67:85–131

Nielsen SG, Rehkämper M (2011) Chapter 13 Thallium isotopes and their application to problems in earth and environmental science. In: Baskaran MM (ed) Handbook of environmental isotope geochemistry. Springer, Berlin

Nier AO (1936) A mass-spectrographic study of the isotopes of argon, potassium, rubidium, zinc and cadmium. Phys Rev 50:1041–1045

Nriagu JO (Ed.) (1981) Cadmium in the Environment. Part II: Health Effects, 908 pp., Wiley & Sons, New York

Nriagu JO, Pacyna JM (1988) Quantitative assessment of worldwide contamination of air, water and soils by trace metals. Nature 333:134–139

Palme H, Larimer JW, Lipschutz ME (1988) Moderately volatile elements. In: Kerridge JF, Matthews MS (eds) Meteorites and the early solar system. University of Arizona Press, Tucson

Palme H, O'Neill HStC (2005) Cosmochemical estimates of mantle composition. In: Carlson RW (ed) The mantle and core. Elsevier, Amsterdam

Pfennig G, Klewe-Nebenius H, Seelmann-Eggebert W (1998) Chart of the nuclides. Forschungszentrum Karlsruhe, Karlsruhe

Pokrovsky OS, Viers J, Freydier R (2005) Zinc stable isotope fractionation during its adsorption on oxides and hydroxides. J Coll Interf Sci 291:192–200

Price NM, Morel FMM (1990) Cadmium and cobalt substitution for zinc in marine diatom. Nature 344:658–660

Prizkow W et al (2007) The isotope abundances and the atomic weight of cadmium by a metrological approach. Int J Mass Spectrom 261:74–85

Rehkämper M, Schönbächler M, Stirling CH (2001) Multiple collector ICP-MS: introduction to instrumentation, measurement techniques and analytical capabilities. Geostand Newslett 25:23–40

Rehkämper M et al (2002) Thallium isotope variations in seawater and hydrogenetic, diagenetic, and hydrothermal ferromanganese deposits. Earth Planet Sci Lett 197:65–81

Rehkämper M et al (2004) Cenozoic marine geochemistry of thallium deduced from isotopic studies of ferromanganese crusts. Earth Planet Sci Lett 219:77–91

Richter FM et al (2002) Elemental and isotopic fractionation of type B calcium-, aluminum-rich inclusions: experiments, theoretical considerations, and constraints on their thermal evolution. Geochim Cosmochim Acta 66:521–540

Rickaby REM, Elderfield H (1999) Planktonic foraminiferal Cd/Ca: paleonutrients or paleotemperature? Paleoceanography 14:293–303

Ripperger S, Rehkämper M (2007) Precise determination of cadmium isotope fractionation in seawater by double-spike MC-ICPMS. Geochim Cosmochim Acta 71:631–642

Ripperger S et al (2007) Cadmium isotope fractionation in seawater – a signature of biological activity. Earth Planet Sci Lett 261:670–684

Ripperger S et al (2008) Cd/Ca ratios of in situ collected planktonic foraminiferal tests. Paleoceanography 23:PA3209

Rosman KJR, de Laeter JR (1975) The isotopic composition of cadmium in terrestrial minerals. Int J Mass Spectrom Ion Phys 16:385–394

Rosman KJR, de Laeter JR (1976) Isotopic fractionation in meteoritic cadmium. Nature 261:216–218

Rosman KJR, de Laeter JR (1978) A survey of cadmium isotopic abundances. J Geophys Res 83:1279–1287

Rosman KJR et al (1980) Isotope composition of Cd, Ca, Mg in the Brownfield chondrite. Geochem J 14:269–277

Rudge JF, Reynolds BC, Bourdon B (2009) The double spike toolbox. Chem Geol 265:420–431

Rudnick RL, Gao S (2005) Composition of the continental crust. In: Rudnick RL (ed) The crust. Elsevier, Amsterdam

Sands DG, de Laeter JR, Rosman KJR (2001) Measurements of neutron capture effects on Cd, Sm and Gd in lunar samples with implications for the neutron energy spectrum. Earth Planet Sci Lett 186:335–346

Schauble EA (2007) Role of nuclear volume in driving equilibrium stable isotope fractionation of mercury, thallium, and other very heavy elements. Geochim Cosmochim Acta 71:2170–2189

Schediwy S, Rosman KJR, de Laeter JR (2006) Isotope fractionation of cadmium in lunar material. Earth Planet Sci Lett 243:326–335

Schmitt A-D, Galer SJG, Abouchami W (2009a) Mass-dependent cadmium isotopic variations in nature with emphasis on the marine environment. Earth Planet Sci Lett 277:262–272

Schmitt A-D, Galer SJG, Abouchami W (2009b) High-precision cadmium stable isotope measurements by double spike thermal ionisation mass spectrometry. J Anal At Spectrom 24:1079–1088

Schönbächler M et al (2009) The cadmium isotope composition of the Earth. Geochim Cosmochim Acta 73:A1183

Shannon RD (1976) Revised effective ionic radii and systematic studies of interatomic distances in halides and chalcogenides. Acta Cryst A32:751–767

Shiel AE et al (2009) Matrix effects on the multi-collector inductively coupled plasma mass spectrometric analysis of high-precision cadmium and zinc isotope ratios. Anal Chim Acta 633:29–37

Shiel AE, Weiss D, Orians KJ (2010) Evaluation of zinc, cadmium and lead isotope fractionation during smelting and refining. Sci Total Environ 408:2357–2368

Siebert C et al (2003) Molybdenum isotope records as a potential new proxy for paleoceanography. Earth Planet Sci Lett 211:159–171

Sigman DM, Boyle EA (2000) Glacial/interglacial variations in atmospheric carbon dioxide. Nature 407:859–869

Simon JI, Young ED (2007) Evaporation and Mg isotope fractionation: model constraints for CAIs. Lunar Planet Sci XXXVIII Abstract No:2426

Slater JC (1964) Atmic radii in crystals. J Chem Phys 41:3199–3204

Smith CN et al (2005) Mercury isotope fractionation in fossil hydrothermal systems. Geology 33:825–828

Stirling CH et al (2007) Low-temperature isotopic fractionation of uranium. Earth Planet Sci Lett 264:208–225

Strelow FWE (1978) Distribution coefficients and anion exchange behavior of some elements in hydrobromic-nitric acid mixtures. Anal Chem 50:1359–1361

Sunda WG, Huntsman SA (2000) Effect of Zn, Mn, and Fe on Cd accumulation in phytoplankton: implications for oceanic Cd cycling. Limnol Oceanogr 45:1501–1516

Taylor SR, McLennan SM (1985) The continental crust: its composition and evolution. Blackwell Scientific, Oxford

Thiemens MH (1999) Mass independent isotope effects in planetary atmospheres and the early solar system. Science 283:341–345

Thiemens MH (2006) History and applications of mass-independent isotope effects. Annu Rev Earth Planet Sci 34:217–262

Thirlwall MF (2000) Inter-laboratory and other errors in Pb isotope analyses investigated using a ^{207}Pb-^{204}Pb double spike. Chem Geol 163:299–322

Wasson JT, Kallemeyn GW (1988) Compositions of chondrites. Phil Trans R Soc Lond Ser A 325:535–544

Wedepohl KH (1995) The composition of the continental crust. Geochim Cosmochim Acta 59:1217–1232

Wetherill GW (1964) Isotopic composition and concentration of molybdenum in iron meteorites. J Geophys Res 69:4403–4408

Weyer S et al (2008) Natural fractionation of ^{238}U/^{235}U. Geochim Cosmochim Acta 72:345–359

Wieser ME (2006) Atomic weights of the elements (IUPAC technical report). Pure Appl Chem 78:2051–2066

Witt-Eickschen G et al (2009) The geochemistry of the volatile trace elements As, Cd, Ga, In and Sn in the Earth's mantle: new evidence from in situ analyses of mantle xenoliths. Geochim Cosmochim Acta 73:1755–1778

Wombacher F, Rehkämper M (2003) Investigation of the mass discrimination of multiple collector ICP-MS using neodymium isotopes and the generalised power law. J Anal At Spectrom 18:1371–1375

Wombacher F et al (2003) Stable isotope compositions of cadmium in geological materials and meteorites determined by multiple collector-ICPMS. Geochim Cosmochim Acta 67:4639–4654

Wombacher F, Rehkämper M (2004) Problems and suggestions concerning the notation of Cd stable isotope compositions and the use of reference materials. Geostand Geoanal Res 28:173–178

Wombacher F et al (2004) Determination of the mass-dependence of cadmium isotope fractionation during evaporation. Geochim Cosmochim Acta 68:2349–2357

Wombacher F et al (2008) Cadmium stable isotope cosmochemistry. Geochim Cosmochim Acta 72:646–667

Wombacher F (2010) Mass-independent cadmium isotope fractionation during evaporation. 88th Ann. Meet. German Mineral. Soc. p 65

Workman RK, Hart SR (2005) Major and trace element composition of the depleted MORB mantle (DMM). Earth Planet Sci Lett 231:53–72

Xie Q et al (2005) High precision Hg isotope analysis of environmental samples using gold trap-MC-ICP-MS. J Anal At Spectrom 20:515–522

Xu Y et al (2008) Structure and metal exchange in the cadmium carbonic anhydrase of marine diatoms. Nature 452:56–62

Xue Z et al (2010a) Cadmium isotope constraints on nutrient cycling in the Peruvian oxygen minimum zone. Goldschmidt 2010 Conference A1166

Xue Z et al (2010b) Cadmium isotope fractionation in the Southern Ocean. Eos Trans. AGU 91, Ocean Sci Meet Suppl:Abstract CO15C-12

Yee N, Fein J (2001) Cd adsorption onto bacterial surfaces: a universal adsorption edge. Geochim Cosmochim Acta 65:2037–2042

Yi W et al (2000) Cadmium, indium, tin, tellurium, and sulfur in oceanic basalrs: implications for chalcophile element fractionation in the Earth. J Geophys Res 105:18927–18948

Young ED, Galy A, Nagahara H (2002) Kinetic and equilibrium mass-dependent isotope fractionation laws in nature and their geochemical and cosmochemical significance. Geochim Cosmochim Acta 66:1095–1104

Chapter 9
Stable Isotopes of Cr and Se as Tracers of Redox Processes in Earth Surface Environments

Thomas M. Johnson

Abstract Redox reactions play a central role in the environmental geochemistry of chromium (Cr) and selenium (Se). A small but growing body of research shows that the stable isotope abundances of both elements are altered by these redox reactions. As is observed with nitrate and sulfate, reduction of the higher valence oxoanions to lower valence forms induces isotopic fractionation, with the reaction product enriched in lighter isotopes. The magnitudes of isotope ratio shifts range up to 4.5‰ for $^{53}Cr/^{52}Cr$ during Cr(VI) reduction, and up to 18‰ for $^{82}Se/^{76}Se$ during Se(VI) or Se(IV) reduction, but they vary with reaction mechanism and can be much smaller for environmentally relevant conditions. Other chemical processes generally produce lesser fractionation. Accordingly, Cr and Se isotope ratios have potential as indicators of reduction. Application of Cr isotope data as a quantitative indicator of natural reduction of hexavalent Cr in groundwater appears to be successful. Application of Se isotope data to constrain selenate reduction in wetlands has not been successful because effective fractionations induced by redox reactions are small, but recent studies suggest that reduction in groundwater produces significant fractionation and may be detected using Se isotope measurements.

9.1 Introduction

Chromium and selenium each have multiple valence states in environmental systems, and reduction and oxidation (redox) reactions are of central importance in their geochemistry. These reactions alter the two elements' chemical behavior markedly and can provide energy for biological activity relevant to environmental processes. Redox reactions have the potential to induce large isotopic fractionations because they greatly alter the atoms' bond energies. Decades of C, N, and S isotope studies have exploited redox-related isotope ratio shifts (e.g., Valley and Cole 2001). More recent research has revealed that many heavier elements exhibit similar redox-driven isotopic fractionation. In this chapter, we review research on Cr and Se isotopes together, because of similarities in their isotope systematics and applications.

Decades of theoretical studies of isotopic fractionation, mostly focused on lighter elements, provide some general systematic concepts that aid in understanding redox-driven isotopic fractionation in the Cr and Se isotope systems. We provide only a brief review here; more complete reviews exist, and those treat these concepts in detail (e.g., Valley and Cole 2001). Many studies have predicted equilibrium isotope fractionation, using chemical bond energies derived from spectroscopic data and molecular models (Chacko et al. 2001; Schauble et al. 2004). The most general prediction is that, at isotopic equilibrium, heavier isotopes of an element tend to partition into sites with greater bonding energies (Criss 1999). Greater valence number generally leads to shorter bonds with greater energies, so heavier isotopes tend to be enriched in phases containing higher valence states, relative to

T.M. Johnson
Department of Geology, University of Illinois at Urbana-Champaign, Urbana, IL 61801, USA
e-mail: tmjohnsn@illinois.edu

phases containing lower valence states with which they are in isotopic equilibrium.

At earth surface conditions, however, redox reactions are often far from equilibrium and proceed dominantly in one direction with little back reaction. This absence of chemical equilibrium is often accompanied by an absence of isotopic equilibrium. The most important C and S isotope effects at earth surface temperatures arise from reduction reactions (photosynthesis and sulfate reduction) operating far from equilibrium (Canfield 2001; Hayes 2001). In these cases, *kinetic isotope effects* may occur. In the simplest form of kinetic isotope fractionation, bonds of lighter isotopes are effectively weaker, so their reaction rates tend to be greater than those of heavier isotopes (Bigeleisen 1949; Criss 1999). For example, during sulfate reduction, the weaker bonds in ^{32}S-bearing sulfate cause it to have a greater reaction rate compared to ^{34}S-bearing sulfate (Canfield 2001).

Kinetic isotope fractionation can be complex, and the fractionation factor for a given reaction varies, depending on the reaction mechanism and/or the relative rates of steps within a multi-step reaction. For example, the magnitude of S isotope fractionation induced by microbial sulfate reduction depends on the type of microbial metabolism and, for a single species, on the electron donor availability and other-environmental variables (Kaplan and Rittenberg 1964; Detmers et al. 2001). Furthermore, individual steps within a multi-step reaction can be close to isotopic equilibrium while the overall reaction is not (Hayes 2001). This leads to complex and variable kinetic isotopic effects observed for the overall reaction. A basic understanding of this variability is essential in applying Se and Cr isotopes in geochemical studies.

9.1.1 Chromium Geochemistry and Stable Isotopes

The aqueous geochemistry of Cr was reviewed recently by Ball and Nordstrom (1998); an Eh-pH diagram is presented in Fig. 9.1. Chromium is a group 6 element; the +6 valence is stable under oxidizing conditions. This is also true of S, and thus some insight into Cr isotope geochemistry can be gained from understanding that of S. The chromate (CrO_4^{2-}) and hydrochromate ($HCrO_4^-$) ions, together referred to as Cr(VI), are generally soluble, weakly adsorbing, and mobile (Davis and Olsen 1995). Under weakly to strongly reducing conditions, Cr(VI) tends to reduce to the +3 valence, or Cr(III), which is geochemically similar to ferric iron. It readily precipitates and/or adsorbs to solid surfaces. Importantly, Cr is mobile and bioavailable in its oxidized forms, but generally immobile and non-bioavailable in its reduced forms. Cr(III) can be oxidized to Cr(VI), but oxidation by dissolved oxygen is slow. MnO_2 minerals are thought to be required for rapid oxidation of Cr(III) (Fendorf et al. 1993).

Cr(VI) has been used in several common industrial processes (U.S. Department of Health 2000); groundwater Cr(VI) plumes are present in most industrial

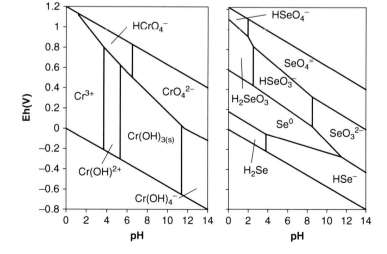

Fig. 9.1 Thermodynamic stability diagrams for Cr and Se after Ball and Nordstrom (1998) and Séby et al. (2001), respectively. Total Cr and Se concentrations are 10^{-6} and 10^{-8} M, respectively

centers of developed countries and occasionally at rural industrial installations. Cr(VI) also occurs naturally, where weathering of ultramafic rocks occurs (Oze et al. 2007; Izbicki et al. 2008). Cr(VI) is known to have various negative health effects whereas Cr(III) is relatively non-toxic (Katz and Salem 1994). The U.S. Environmental Protection Agency's drinking water maximum contaminant level for chromium is 0.10 mg/L (U.S. Environmental Protection Agency 2009). Cr(VI) reduction is often an attractive remediation strategy, as it immobilizes and detoxifies the Cr. In some cases, natural reduction occurs, whereas in other cases, active remediation is attempted by emplacement of reducing agents or stimulation of microbial action (Blowes 2002).

Cr has four stable isotopes, ^{50}Cr, ^{52}Cr, ^{53}Cr, ^{54}Cr, with natural abundances 4.3, 83.8, 9.5, and 2.4%, respectively. All studies to date have measured the ^{53}Cr/^{52}Cr ratio, which provides a mass difference of about 2%.

9.1.2 Selenium Geochemistry and Stable Isotopes

The aqueous geochemistry of Se is reviewed in a recent article (Seby et al. 2001); an Eh-pH diagram is presented in Fig. 9.1. Se is a group 16 element, with valence electron configurations and geochemical properties similar to those of S. Like S, Se is stable in the +6 and −2 valence states. Selenate (SeO_4^{2-}) and selenide (Se^{2-}) species, which are geochemically similar to sulfate and sulfide, are formed. However, Se geochemistry departs from that of S in that the + 4 and 0 valences can be thermodynamically stable (Seby et al. 2001). The + 4 valence forms SeO_3^{2-} and $HSeO_3^{-}$ ions, together referred to as Se(IV), which are highly soluble like Se(VI), but more prone to sorption. The 0 valence state readily forms a solid Se^0 precipitate, even at sub-micromolar concentrations (McNeal and Balistrieri 1989; Seby et al. 2001). Like Cr, Se is highly mobile and bioavailable in its oxidized forms, but is often reduced to insoluble Se(0) and precipitated in moderately reducing environments. Alternatively, under moderately to strongly reducing conditions, metal selenides (e.g., HgSe, CuSe, Ag$_2$Se, FeSe) may form. In Se-contaminated modern environments, reduction may sequester soluble Se in insoluble reduced forms.

Like S, Se is utilized in biological activity. But whereas most systems contain inorganic S well in excess of biological uptake, Se is much less abundant and much of the bioavailable Se in a given system may cycle like a nutrient (Cutter and Cutter 2001), with organic forms more abundant than inorganic forms in some waters and many sediments (Clark and Johnson 2010). The strong organic cycling of Se, combined with the stability of the +4 and 0 valences, causes the environmental geochemistry of Se to differ greatly from that of sulfur.

Se is used by organisms as a nutrient, but at concentrations only one to two orders of magnitude above optimal levels, Se may be toxic and has caused disease, deformity, and/or death among certain animals and, in very rare cases, among humans (Lemly 1998; Skorupa 1998; Zhu et al. 2004). In most cases, Se contamination arises from weathering of Se-rich sedimentary rocks, with human land and water management playing an important role (Presser 1994). Mining activities can release problematic masses of Se in some cases (Stilling and Amacher 2010). The drinking water maximum contaminant level for selenium is 0.050 mg/L (U.S. Environmental Protection Agency 2009).

Se has six stable isotopes, ^{74}Se, ^{76}Se, ^{77}Se, ^{78}Se, ^{80}Se, and ^{82}Se, with natural abundances 0.89, 9.4, 7.6, 23.8, 49.6, and 8.7%, respectively. Although Se is a somewhat heavy element, the mass of ^{82}Se is nearly 8% greater than that of ^{76}Se in the commonly measured ^{82}Se/^{76}Se ratio. This large difference amplifies isotopic fractionation.

9.2 Methods

9.2.1 Notation and the Rayleigh Fractionation Equation

This chapter focuses on mass-dependent fractionation of isotopes, in which chemical processes shift the relative abundances of heavy isotopes of an element relative to its lighter isotopes. These shifts are expressed as ratios of abundances, and by convention that of the heavier isotope is the numerator. For example:

$$R = \frac{^{53}Cr}{^{52}Cr}. \quad (9.1)$$

Because meaningful variations in R are much less than one percent, it is inconvenient to write ratios explicitly (e.g., 0.11338 and 0.11339 may be significantly different ratios). Measured ratios are thus expressed as relative deviations of a measured ratio from that of a standard reference material:

$$\delta^{53}Cr = \frac{R_{Sample}}{R_{Standard}} - 1. \quad (9.2)$$

Typically, delta is expressed in parts per thousand, or per mil. For example, if the measured ratio is greater than that of the standard by a factor of 1.002, then $\delta^{53}Cr = 0.002$, which typically is written as "+2‰." Alternatively, this result can be expressed in parts per million, i.e., "+2,000 ppm." Some workers have used the symbol ε in place of δ to express "parts per 10,000" deviation. This can lead to confusion, as ε has long been used to conveniently express the magnitude of an isotopic fractionation (see below).

It is essential to have a parameter that expresses quantitatively the degree to which isotopes are fractionated by a chemical process. This is done by defining a fractionation factor, which compares the isotope ratio of a reactant to that of a product, or compares two phases in equilibrium:

$$\alpha = \frac{R_{Reactant}}{R_{Product}} \quad (9.3)$$

or

$$\alpha = \frac{R_{Product}}{R_{Reactant}}. \quad (9.4)$$

One can choose either of these two definitions of alpha. The latter may seem a natural choice, but the former has been used for decades by the carbon and sulfur isotope communities because it is more convenient (see below). Both have been used in the Cr and Se isotope literature.

Just as it is more convenient to express measurement results using δ rather than R, the parameter ε is defined to conveniently quantify isotopic fractionation:

$$\varepsilon = \alpha - 1, \quad (9.5)$$

ε is expressed in per mil form, and is a convenient quantity because it is very close to the difference in delta values between the reactant and the product flux:

$$\varepsilon \approx \delta_{reactant} - \delta_{product}, \quad (9.6)$$

when (9.3) is used to define α. For example, if $R_{reactant}$ is greater than $R_{product}$ by a factor of 1.002, we write $\varepsilon = +2‰$.

For normal kinetic isotope effects, $R_{reactant} > R_{product}$ and thus ε is a positive quantity when (9.3) is used to define α. When (9.4) is used to define α, simple kinetic effects lead to ε values that are equal in magnitude but negative in sign. This leads to confusing language, because greater (stronger) isotopic fractionation corresponds to lesser (more negative) ε values. For this reason, studies focusing on "normal" kinetic effects, in which reaction products are depleted in heavy isotopes relative to reactants (Canfield 2001; Hayes 2001) often choose (9.3) as the definition of α. This approach is followed in this chapter.

The magnitude of equilibrium isotope fractionation is often expressed using Δ:

$$\Delta \equiv \delta_A - \delta_B, \quad (9.7)$$

where A and B are the two phases in equilibrium. Like ε, Δ is more convenient than α for routine use.

When kinetic isotope effects occur, the Rayleigh distillation equation is often invoked to provide quantitative relationships between the extent of the reaction and the isotope ratios of the reactant and product. This concept applies rigorously to a closed, well-mixed system in which the reaction products do not interact with the reactant pool (e.g, no back-reaction occurs). When α is defined as $R_{reactant}/R_{product}$, the Rayleigh equation giving the isotopic composition of the remaining reactant is:

$$\delta = (\delta_0 + 1000)f^{\frac{1}{\alpha}-1} - 1000, \quad (9.8)$$

where f is the fraction of reactant remaining. An example of a Rayleigh model is plotted in Fig. 9.2, along with an equilibrium fractionation model, in which the product and reactant remain in full isotopic equilibrium as a reaction proceeds.

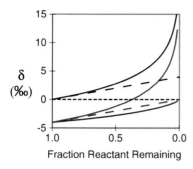

Fig. 9.2 Models for isotopic fractionation as a function of the extent of a reaction. *Solid lines* give values for a Rayleigh model with $\varepsilon = 4‰$; *dashed lines* for an equilibrium model with $\Delta = 4‰$. For both models, black lines give the isotopic composition of the reactant; blue lines give that of the accumulated reaction product. The *red line* gives the isotopic composition of the product flux (the incremental product produced at one instant in time) for the Rayleigh model

9.2.2 Sampling

Water samples must be treated carefully to maintain species in solution and avoid isotopic fractionation during transport and storage. In all cases, samples must be filtered to avoid microbial redox reactions and resulting isotopic shifts. Even after filtration, Cr (VI) may be reduced by abiotic reducing agents such as dissolved organic molecules. Reduction to Cr(III), which causes loss of Cr from solution and alters isotope ratios, is promoted by acidic conditions, so acidification of samples should be avoided (Stollenwerk and Grove 1985). Se(VI) is relatively insensitive to abiotic reduction and acidification of waters to pH < 2 has been recommended for preservation of dissolved speciation (Cheam and Agemian 1980). However, Se(IV) is more prone to reduction via abiotic reduction, and Se(IV) oxidation has been reported for some HCl-bearing solutions (Heninger et al. 1997). It is wise to monitor Se(IV) concentrations, if possible, during sample storage. Refrigeration of samples is highly advantageous for both Se and Cr, as it slows these reactions (Heninger et al. 1997). Storage in the dark is also advisable, to avoid photochemical reactions.

Samples of solid materials can be similarly unstable, as microbial action and ongoing abiotic reactions can greatly change redox speciation and alter isotope ratios during transport and storage. Immediate processing of samples and/or refrigeration, and/or careful exclusion of air may be required.

9.2.3 Mass Spectrometry

Methods for Cr and Se isotope measurements are rather specialized, and a number of analytical issues affect how studies are planned and carried out (e.g., choices of analytical hardware and method dictate minimum usable sample size). For this reason, the major analytical issues are reviewed here with the goal of providing a guide for non-specialists to navigate the analytical choices involved. Sample preparation methods are described after this section, as those methods are determined by the details of mass spectrometry reviewed here.

9.2.3.1 Hardware Choices and Double Spike Methods

Most recent Cr and Se isotope studies employ multicollector inductively coupled plasma mass spectrometry (MC-ICP-MS), in which the analyte is ionized in a plasma, the ion beam passes through a magnetic sector to separate the isotopes, and the individual ions beams are measured simultaneously in an array of Faraday collectors (Albarède and Beard 2004). Single collector ICP-MS instruments do not provide the required precision for the vast majority of Cr and Se applications. Thermal ionization mass spectrometry (TIMS) is a viable option for both elements. Well-developed methods exist, though they generally have lesser sample throughput, compared to MC-ICP-MS.

Both TIMS and MC-ICP-MS instruments exhibit "mass bias." MC-ICP-MS instruments have a strong bias in favor of heavier isotopes (See reviews by Albarède and Beard 2004; Weiss et al. 2008). For example, measured $^{53}Cr/^{52}Cr$ ratios are a few percent larger than true values. With TIMS, the bias is roughly an order of magnitude smaller, but changes over time during analysis. With MC-ICP-MS, the bias is nearly constant over time, and can be determined by analyzing standards before and after each sample (Albarède and Beard 2004). However, this "sample-standard bracketing" approach can lose precision if mass bias variation between standards does not vary smoothly because of instrumental noise or sample matrix effects.

Some Cr and Se MC-ICP-MS studies, and all of the TIMS studies, have employed double isotope tracer, or "double spike" methods to determine instrumental mass bias (Johnson and Beard 1999; Johnson et al. 1999; Clark and Johnson 2008; Schoenberg et al. 2008; Berna et al. 2010). A "spike" solution containing two stable isotopes (e.g., for Cr, ^{54}Cr and ^{50}Cr are used) in a known ratio is mixed thoroughly with each sample (i.e., care must be taken to assure complete isotopic equilibration between spike and sample). The measured ratio of the spike isotopes is then used to monitor mass bias. Although this ratio is partially determined by the sample's contributions to the spike isotope signals, the relative proportions of spike and sample are precisely determined, and the spike and sample contributions to measured signals are separated mathematically. Calculation routines to carry this out are complex, but they are described in the literature and, once implemented, require little effort per sample (Johnson and Beard 1999; Johnson et al. 1999; Schoenberg et al. 2008).

The double spike approach has important advantages: It monitors mass bias during, not before and after, sample measurements and thus can accommodate sample matrix effects and non-systematic shifts in mass bias. Also, it can correct for isotopic fractionation that may occur during sample preparation, provided the spike is mixed with the sample prior to sample preparation steps. Finally, it can provide highly precise concentration measurements if the spike solution's concentration is carefully calibrated. The double spike approach also has some disadvantages: At least four isotopes must be measured rather than two, and this may increase the number of interferences that must be addressed. Furthermore, many elements (e.g., Li, Mg, Si, Cu, Sb, Tl) have fewer than 4 stable isotopes.

A third approach used to correct for MC-ICP-MS mass bias drift is the "elemental spike" approach. In this case, a second element is doped into the analyzed solution; drift in the mass bias is monitored by measuring drift in an isotope ratio of the spike element. For example, Cu can be used as an elemental spike for Fe analyses, with the ^{65}Cu/^{63}Cu ratio monitored to detect mass bias drift (Albarède and Beard 2004). There has been minor use of Sr as an elemental spike for Se (Rouxel et al. 2002), but overall, this approach has not been used for Cr and Se.

9.2.3.2 Cr Mass Spectrometry Methods

Methods used for Cr isotope analysis are closely similar to those used for Fe, with most of the methods development occurring first for Fe measurements, with later adaptation to Cr. A double spike TIMS method for Cr (Ellis et al. 2002) was used in a few studies and, though its usage seems to be waning, it remains a viable option. MC-ICP-MS analysis offers faster sample throughput but is more prone to interfering species with masses close to those of the Cr isotopes. ArC$^+$, ArN$^+$, ClO$^+$, and SO$^+$ interferences can be avoided using high resolution methods (Halicz et al. 2008; Schoenberg et al. 2008) that closely follow methods used for Fe (Weyer and Schwieters 2003). Recent methods and mass spectrometers are capable of resolving small differences in mass, thereby measuring the Cr isotopes while excluding interfering ArC$^+$ and ArN$^+$ ions, which have slightly greater masses. This approach, first developed for Fe isotope ratios (Weyer and Schwieters 2003) requires special instrumentation; not all MC-ICP-MS instruments have the required narrow ion source slit and precise ion optics. Furthermore, the narrow slit decreases signal intensity, so the mass of Cr required for analysis is increased roughly tenfold.

An alternative to this high-resolution approach measures the combined intensity of the Cr and interfering beams, then subtracts the interferences. The interference intensities are determined by measurements on blank solutions. This approach has been used for Fe isotope measurements (Anbar et al. 2000; Belshaw et al. 2000), but so far, no Cr isotope publications have used it. Desolvating nebulizers are often used to reduce Ar-based interferences and increase signal strength (Anbar et al. 2000). Some mass spectrometers are equipped with collision cells that greatly decrease ArN$^+$ and ArO$^+$ intensities (Albarède and Beard 2004).

MC-ICP-MS measurements of Cr isotope ratios can be done with either double spike methods (Schoenberg et al. 2008) or sample-standard bracketing (Halicz et al. 2008). When the ^{54}Cr + ^{50}Cr double spike approach is used, interferences from ^{54}Fe and ^{50}Ti must be dealt with. Clean sample preparation methods can reduce these to small levels, but traces of Fe and Ti usually remain. The masses of the interfering isotopes are so close to those of the Cr isotopes that high mass resolution methods are of no help. Thus, small corrections, calculated using, for example, ^{56}Fe and ^{49}Ti beam intensities, are usually applied. Sample standard bracketing methods also require almost complete purification of the sample's Cr, as any matrix components found in the sample could cause its mass bias to differ relative to that of the more pure bracketing standard solution.

Precision reported in the literature ranges from 0.04 to 0.2‰ (Ellis et al. 2002; Halicz et al. 2008; Schoenberg et al. 2008; Berna et al. 2010). The more precise measurements require about one microgram Cr. With less Cr, analyses can be carried out with diminishing precision as signal strength decreases.

Results are generally reported relative to IRMM-012 or NIST SRM-979. These materials should be isotopically identical, as IRMM-012 is a solution made from SRM-979 material. NIST SRM-3112a may be a useful secondary standard; its δ^{53}Cr value is -0.07‰ relative to IRMM-012 (Schoenberg et al. 2008). Analyses of basaltic rocks, ultramafic rocks, and Cr ores (Schoenberg et al. 2008) fall within a very narrow range close to SRM-979 (-0.124 ± 0.101‰), so all three of these standards are reasonably close to the bulk-earth value.

9.2.3.3 Se Mass Spectrometry Methods

The most effective method for Se isotope analysis at present is MC-ICP-MS, though other methods have been used in the past. Early methods used gas-source mass spectrometry of SeF_6 (Krouse and Thode 1962), but that method requires relatively large amount of Se per analysis and fluorination of the purified Se. It is not currently in use. More recent work successfully employed TIMS methods (Johnson et al. 1999), but that method also requires a relatively large mass of Se (500 ng per analysis) and has slower sample throughput relative to MC-ICP-MS methods.

MC-ICP-MS methods exploit hydride generation, whereby Se(IV) in solution is converted to H_2Se via reaction with a reducing agent and injected into the plasma as a gas. This approach, developed for Se isotopic measurements by Rouxel et al. (2002) has the advantage of delivering nearly 100% of the Se to the plasma, along with only a few other elements and minor amounts of H_2O, H_2, and HCl vapor. This efficient delivery and lack of sample matrix loading in the plasma enables high sensitivity and precise analyses that consume as little as 20 ng Se (Rouxel et al. 2002; Clark and Johnson 2008). This small sample size requirement is essential to the success of many studies, as Se concentrations in many waters and solids are small.

MC-ICP-MS methods for Se must address a large array of interferences; every Se isotope is affected (Rouxel et al. 2002; Clark and Johnson 2008). These include $ArAr^+$, Kr^+, Ge^+, $ArCl^+$, SeH^+, BrH^+, and AsH^+. Although most elements do not follow Se through the hydride generation process into the plasma, Ge and As do. Specific strategies used to address interferences are given in recent publications (Rouxel et al. 2002; Clark and Johnson 2008). High-resolution methods have not been successful to date, in part because they decrease signal strength and thus require more Se per analysis.

Mass bias correction has been accomplished through either sample-standard bracketing (Rouxel et al. 2002; Carignan and Wen 2007) or double spike methods, which include $^{82}Se+^{74}Se$ and $^{77}Se+^{74}Se$ spikes (Johnson et al. 1999; Clark and Johnson 2008). Reported precision for sample-standard bracketing and $^{77}Se+^{74}Se$ double spike methods is ± 0.25 per mil on the $^{82}Se/^{76}Se$ ratio. The $^{82}Se+^{74}Se$ double spike approach produces precision of 0.15‰ on the $^{78}Se/^{76}Se$ ratio (Clark and Johnson 2008). Various in-house standards have been used in Se isotope studies, but a recent publication proposes that the readily available NIST SRM-3149 solution is a suitable interlaboratory standard with an isotopic composition close to that of the bulk earth (Carignan and Wen 2007). Analyses reported relative to the "Merck" and "MH-495" in-house standards used in some early publications can be converted to the SRM-3149 scale using the results of Carignan and Wen (2007).

9.2.4 Separation/Purification Prior to Mass Spectrometry

Sample preparation methods differ greatly between Cr and Se, but in both cases the goal is to separate the analyte from the sample matrix and purify the analyte prior to mass spectrometry. Choice of analytical method dictates the type and amount of impurities that can be tolerated. TIMS methods are usually very sensitive to contaminants; small amounts of organic compounds, for example, can greatly inhibit signal strength. With MC-ICP-MS methods, mass bias can be greatly shifted by contaminants. Double spike methods, and, to a lesser extent, elemental spike methods can often correct for such mass bias shifts. Simple sample-standard bracketing methods cannot.

9.2.4.1 Cr Sample Preparation

Two different ion exchange methods have been used to purify Cr for mass spectrometry. The first approach exploits redox manipulation, which shifts Cr from an anionic form to a cationic form (Ball and Bassett 2000; Ellis et al. 2002; Schoenberg et al. 2008). The sample's Cr, in the form of the Cr(VI) anion, is adsorbed onto anion exchange resin, the sample matrix is flushed from the resin, and the Cr(VI) is reduced to Cr(III), which is then easily eluted. Incomplete recovery of sample Cr, and thus the potential for isotopic fractionation, has been reported for this approach (Schoenberg et al. 2008). Hence, the procedures published to date should not be used if mass spectrometry utilizes the sample-standard bracketing method. However, if a double isotope spike is used, and is added prior to sample preparation, this fractionation is readily removed along with the mass bias (see above). The second approach (Halicz et al. 2008) similarly adsorbs Cr(VI) onto anion exchange resin, elutes it with 2 M HNO_3, converts it to Cr(III), adsorbs the Cr(III) onto cation exchange resin, and finally elutes it in 5 M HNO_3. This method recovers essentially all of the Cr and does not fractionate isotopes (Halicz et al. 2008). However, it is more time consuming, with the second ion exchange step involving passage of a large volume (180 mL) of solution through the exchange column.

9.2.4.2 Se Sample Preparation

Because the hydride generation method that delivers Se to the plasma is selective, many matrix components (e.g., Na, K, Mg, Ca, Cl^-, SO_4^{2-}) can be tolerated in large concentrations. Unfortunately, Ge and As are delivered to the plasma, so they must be removed. ^{74}Ge and ^{76}Ge are isobaric interferences, so only trace amounts of Ge can be tolerated (Rouxel et al. 2002). ^{75}As is not itself an interference, but a small fraction of As forms $^{75}AsH^+$ ions, which interfere with ^{76}Se when large amounts of As are present (Rouxel et al. 2002). Also, oxidants like HNO_3 and H_2O_2, and metals like Hg or Cu that form selenide precipitates, can interfere with formation and delivery of H_2Se.

Selenium can be extracted and purified from a variety of aqueous matrices using cotton with thiol functional groups attached (Rouxel et al. 2002). The thiol groups extract Se(IV) from aqueous solutions; most matrix components pass through. This method affords excellent separation of Se from Ge, but somewhat incomplete separation from As. Separation of Se(VI) and Se(IV) species, or pre-concentration of both species from dilute solutions, can be done by anion exchange (Clark and Johnson 2008). This method can be effective with large-volume sample of fresh water, and Se is efficiently separated from Ge and As. However, high concentrations of sulfate and other anions can overwhelm the resin's capacity; this method is therefore ineffective for seawater and other saline waters.

9.3 Chromium Isotope Systematics and Applications

9.3.1 Cr Isotope Systematics

9.3.1.1 Equilibrium vs. Kinetic Isotopic Fractionation

A few experiments and some theoretical considerations suggest that isotopic equilibration between Cr(VI) and Cr(III) species is slow in environmental systems. Recent experiments (Zink and Schoenberg 2009) indicate negligible exchange between aqueous Cr(VI) and Cr(III) over a time scale of weeks. It thus appears chromium is somewhat similar to sulfur, in that chemical and isotopic disequilibrium between the dominant natural species is common, and the most important isotopic fractionations are kinetically controlled. Possibly, this arises from the fact that multiple electrons must be transferred, and unstable, rare intermediates must react, as the redox transformations occur. This contrasts strongly with the situation with Fe isotopes, where a single electron transfer amounts to an isotopic exchange reaction between the dominant Fe(III) and Fe(II). The issue of Cr isotopic equilibration is not fully settled, though, as Cr(VI) and Cr(III) often coexist for decades or longer, in the presence of solid phases and dissolved constituents that might catalyze reactions. Knowledge of the degree to which isotopic equilibrium occurs in various types of system is needed to choose appropriate models with which data can be quantitatively interpreted. Figure 9.2 shows a pure Rayleigh model (no reactant-product interaction), along with a pure equilibrium model (full reactant-product equilibrium); in a real system partial equilibrium could lead to intermediate relationships between these extremes.

Although isotopic equilibrium may not occur in most Cr isotope studies, equilibrium fractionation factors provide important guidance toward understanding Cr isotope systematics. One study (Schauble et al. 2004) provides estimates of equilibrium ^{53}Cr-^{52}Cr isotope fractionation factors using measured vibrational spectra of various Cr species and models derived from both empirical and ab initio force field methods. Reported values of Δ for equilibrium between CrO_4^{2-} and $\{Cr(H_2O)_6\}^{3+}$ complexes are 7‰ at 0°C and 6‰ at 30°C, with the ^{53}Cr/^{52}Cr ratio being greater for CrO_4^{2-}. Substitution of Cl^- for O^{2-} ligands in $\{Cr(H_2O)_6\}^{3+}$ complexes leads to additional fractionation; Δ is about 4‰ for $Cr(H_2O)^{3+}$-$CrCl_6^{3-}$ equilibrium, with the ^{53}Cr/^{52}Cr ratio being lesser in $CrCl_6^{3+}$. This fractionation is fairly large but will not be observed in most practical settings, where Cr(III) does not normally include more than one or two Cl^- ligands in its first coordination sphere. In general we expect potential Cr isotopic fractionation up to several per mil when Cr redox reactions occur, with much smaller fractionations occurring if ligands are exchanged.

9.3.1.2 Cr(VI) Reduction

Because Cr(VI) reduction is a critically important process in contaminated groundwater settings, this reaction has been the focus of most research on Cr isotope systematics. Unlike nitrate and sulfate reduction, Cr(VI) reduction may proceed via a variety of abiotic processes in addition to microbial ones. All reduction mechanisms studied to date have been found to induce Cr isotope fractionation, with the lighter isotopes enriched in the product. The magnitude of fractionation varies considerably; results of the studies done to date are given in Table 9.1. In the microbial reduction study, weaker isotopic fractionation was observed when the cells were supplied with a greater concentration of electron donor and reduced Cr(VI) more rapidly. Sulfate reducing bacteria have similar variability in ϵ (Canfield 2001). Abiotic and microbial reduction have overlapping ranges of the isotopic fractionation factor.

9.3.1.3 Cr(III) Oxidation

Cr(III) oxidation is a key step in the natural generation of mobile Cr(VI) from Cr-rich minerals, in which Cr occurs predominantly as Cr(III) (Oze et al. 2007; Izbicki et al. 2008). Similarly, Cr(III) generated by remediation of Cr(VI) contamination could later be remobilized if subsurface conditions become more oxidizing. To date, a few determinations of isotopic fractionation induced by oxidation of aqueous Cr(III) by Mn oxides have been reported. Bain and Bullen (2005) reported, in a conference abstract, variable ϵ values during this process, with the Cr(VI) product being enriched in the lighter isotope at times, but evolving toward enrichment in heavier isotopes over time. A second study, also presented only in a conference abstract (Villalobos-Aragon et al. 2008) reported a consistent fractionation, with the product enriched in the heavier isotope by +1.1‰. This result cannot be explained as a simple kinetic isotope effect, which would yield isotopically light reaction products. The multi-step oxidation of Cr(III) must include considerable back-reaction or partial approach toward isotopic equilibrium at one or more of the steps. For example, if isotopic equilibrium between Cr(III) and an ephemeral Cr(IV) intermediate species were to occur within the overall reaction chain, the more oxidized Cr(IV) should be enriched in the heavier isotope. This could cause the final Cr(VI) product to be enriched in the heavier isotope as observed. This issue is not settled, but at present preliminary evidence suggests Cr(III) oxidation involves isotopic fractionation, probably with enrichment of the heavier isotope in the Cr(VI) product.

9.3.1.4 Precipitation and Adsorption

Adsorption of Cr(VI) onto Al_2O_3 or Geothite does not induce significant fractionation (Ellis et al. 2004), presumably because Cr(VI)-O bonds within the tetrahedral CrO_4^{2-} complex are not affected much. Similarly, little isotopic fractionation occurs during

Table 9.1 Isotopic fractionation caused by Cr transformations

Study	Reaction	Mechanism or reactant	ε (‰)[a] $^{53}Cr/^{52}Cr$
Ellis et al. (2002)	Cr(VI) to Cr(III)	Magnetite	3.5
Kitchen et al. (2004)	"	Fe(II)	4.3[b]
Kitchen et al. (2004)	"	Fulvic, humic, mandelic acids	3.0
Sikora et al. (2008)	"	*Shewanella oneidensis* MR-1	1.8–4.5
Ellis et al. (2002)	"	Pond and estuary sediment	3.3, 3.5
Berna et al. (2010)	"	Aquifer sediment	2.4, 3.1
Bain and Bullen (2005)	Cr(III) oxidation	Mn oxides	2.5 to −0.7
Villalobos-Aragon et al. (2008)	"	Mn oxides	−1.1
Ellis et al. (2004)	Cr(VI) adsorption	Goethite, alumina	<0.04

[a]The fractionation factor is defined here so that a positive value indicates enrichment of the reaction product in the lighter isotope
[b]Two of these experiments were flawed, and the reported pH-dependence reported was incorrect (unpublished results demonstrate this)

precipitation of PbCrO$_4$, which has $\delta^{53}Cr$ only 0.1‰ greater than the parent solution (Schoenberg et al. 2008). Based on studies presently available, it appears redox reactions will control $\delta^{53}Cr$ values, whereas non-redox processes will not affect them much. This follows the general pattern observed with redox-sensitive elements; redox reactions cause large changes in the local bonding environments, whereas non-redox reactions tend to have much smaller affects (Kendall 1998; Canfield 2001; Hayes 2001; Beard and Johnson 2004; Asael et al. 2009).

9.3.2 Cr Isotope Ratios in Rocks, Natural Dissolved Cr(VI), and Anthropogenic Contaminant Sources

Recent analyses of an array of mantle rocks, mafic igneous rocks, and one granite revealed a very narrow range of $\delta^{53}Cr$ values, with a mean of −0.12‰ and a standard deviation of 0.05‰ (Schoenberg et al. 2008). This provides a good estimate for the composition of the bulk earth, as Cr-rich mantle rocks are the major Cr reservoir on earth. An array of 9 marine sediments analyzed in the same study yielded slightly greater values (−0.06 ± 0.05 1σ). A study focusing on 0.55–3.75 billion-year-old banded iron formation rocks found that most of their $\delta^{53}Cr$ values clustered around bulk-earth values, but during some time periods, they shifted upward, scattering between bulk-earth values and 4.96‰ (Frei et al. 2009). The authors interpreted these excursions as a response to increases in atmospheric oxygen about 2.4 and 0.75 billion years ago. Although a complete survey of rock values does not yet exist, it appears that most rocks cluster close to bulk-earth values, with notable exceptions related to Cr redox processes.

Cr isotope data on naturally produced Cr(VI) in groundwater are consistent with enrichment of the heavier isotope in the Cr(VI) released by weathering. Cr(VI) released from ultramafic rocks into groundwater in certain areas of the Mojave desert, USA has $\delta^{53}Cr$ values of 0.7‰ to about 3.1‰, whereas the source rocks are close to 0.0‰ (Izbicki et al. 2008). Values >3.1‰ are also found in the area, but are most likely related to Cr(VI) reduction (see below). Naturally sourced Cr(VI) in the eastern Snake River Plain aquifer of Idaho, USA has a $\delta^{53}Cr$ range of 0.79–2.45‰ (Raddatz et al. 2011). The emerging pattern is of mildly elevated $\delta^{53}Cr$ values in natural Cr(VI) in groundwater, but the data are still sparse.

Industrial Cr supplies and reagent Cr appear to be restricted to near-zero values (Ellis et al. 2002; Schoenberg et al. 2008; Berna et al. 2010). This is expected, as Cr ores are usually igneous in origin and little Cr is lost during processing of the ore. Cr plating bath solutions, a common source of Cr contamination, appear to be slightly positive (Ellis et al. 2002; Berna et al. 2010), with solutions sampled at three facilities ranging from 0.29 to 0.83‰. There may be some potential to distinguish natural dissolved Cr(VI), with its tendency toward positive $\delta^{53}Cr$ values, from industrial contamination with near-zero values (Izbicki et al. 2008), but one would need to consider potential overlap in the values and possible elevation of $\delta^{53}Cr$ by Cr(VI) reduction.

9.3.3 Cr Isotope Applications

9.3.3.1 Detection of Chromate Reduction: Berkeley Example

Detection and quantification of groundwater Cr(VI) reduction has been the major environmental application of Cr isotopes to date. Because reduction renders the contaminant immobile and nearly non-toxic, reduction, either naturally driven or artificial, is commonly used to remediate Cr(VI) plumes (Blowes 2002). Given the complexity of groundwater systems, with sorption, dilution, and advection processes operating, it is often difficult, slow, and expensive to detect slow natural reduction via Cr(VI) concentration measurements alone.

Cr isotope data can provide a more direct indication of Cr(VI) reduction that is unaffected by sorption, dilution, and advection (Ellis et al. 2002; Berna et al. 2010). Qualitatively, if $\delta^{53}Cr$ of Cr(VI) in a groundwater plume is greater than that of the contaminant source, partial reduction of the Cr(VI) must have occurred. Quantitatively, one can invoke the Rayleigh model (9.8) to approximate the evolution of $\delta^{53}Cr$ as a function of the extent of reduction. However, the classic Rayleigh model is built upon an assumption of a closed system, whereas a "packet" of groundwater seeping from a contaminant source area to a sampled well is not closed and exchanges mass with neighboring packets via hydrodynamic dispersion and diffusion. The effects of longitudinal dispersion have been shown to affect interpretation of isotopic data to a minor or moderate extent (Abe and Hunkeler 2006). If Cr(VI) reduction is restricted to fine-grained lenses within a more permeable matrix, a large decrease in the effective value of ε is expected (Berna et al. 2010). Rayleigh models thus have known flaws, but can provide good approximations if properly applied and/or modified.

A recent study of an urban Cr(VI) plume in Berkeley, CA, USA (Berna et al. 2010) provides a good example of the use of $\delta^{53}Cr$ to quantify Cr(VI) reduction. $\delta^{53}Cr$ was measured over several years' time on dissolved Cr(VI) from an array of 15 monitoring wells and a few temporary "direct push" sampling points. The original contaminant Cr(VI), released decades ago, was assumed to have a $\delta^{53}Cr$ value close to 0.83‰, the value of the present-day Cr plating bath on site. All groundwater $\delta^{53}Cr$ measurements were greater than that, ranging from 1.26 to 6.27‰ (Fig. 9.3). The values are lowest close to the plume source and increase consistently with distance traveled from it. This indicates reduction occurs as Cr(VI) migrates through the aquifer, and the extent of reduction increases with distance traveled.

A quantitative interpretation of these data was developed using a modified Rayleigh fractionation model (Berna et al. 2010). Two laboratory experiments with Cr(VI) reduction by aquifer materials from the site yielded ε values of −2.4 and −3.1‰ respectively. Other experiments revealed that reduction is restricted to greenish lenses of fine-grained reduced material, whereas the dominant coarser, brownish materials did not reduce Cr(VI). In light of that finding, a modified Rayleigh model was constructed with $\varepsilon = -1.37‰$, one half the mean value of the two experiments. This smaller value was used to account for a presumed diffusive barrier between reduction sites in the greenish layers and the main mass of Cr(VI) in the coarser layers. The resulting model is reasonably close to the $\delta^{53}Cr$ vs. Cr(VI) concentration trend defined by samples along the plume axis (Fig. 9.4). Samples from the lateral fringes of the plume have lower concentrations relative to this trend. This is expected, as transverse dispersion dilutes contaminant Cr(VI) in those areas.

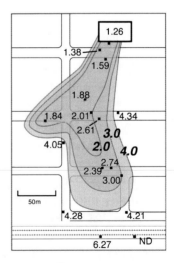

Fig. 9.3 Map view of $\delta^{53}Cr$ (per mil, relative to SRM-979) in a groundwater Cr(VI) plume in Berkeley, California, USA (Berna et al. 2010). *Large bold* numbers give per mil values of the contour lines

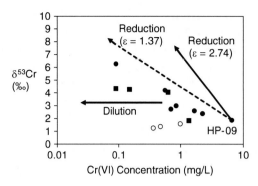

Fig. 9.4 Relationship between $\delta^{53}Cr$ and Cr concentration for the Berkeley Cr(VI) plume. *Solid circles* represent locations close to the plume axis. *Squares* represent locations toward the lateral fringes, where dilution is strong. *Open circles* represent locations in an upgradient area where concentrations are decreasing because the contaminant source has been removed. Rayleigh fractionation models follow straight lines as shown; dilution (e.g., transverse dispersion) shifts waters to the left. The data do not conform to a Rayleigh model constructed using the mean ε value derived from experiments with site sediments. Data from samples taken near the plume axis are close to a modified Rayleigh model with an ε value equal to 1.37‰, one half the experimental value

The results of this study lead to important implications for the fate of Cr(VI) in the system. The estimated extent of Cr(VI) reduction at the most distal sampling point, using the Rayleigh model with $\varepsilon = 1.37$‰, was 98%. This indicates that the downgradient fringe of the plume was not an advancing contaminant front, but rather the location where Cr(VI) reduction reached completion. This suggests remediation via natural attenuation is an effective strategy at the site. Furthermore, $\delta^{53}Cr$ of most of the monitoring wells were shown to be invariant over several years' time, suggesting that the extents of reduction at these points are stable and the Cr(VI)-reducing power of the aquifer is not waning (Berna et al. 2010).

9.3.3.2 Detection of Chromate Reduction and/or Source Identification when Multiple Sources are Present

Two other studies have used Cr isotope data to detect reduction in groundwater systems. One investigated the fate of dissolved Cr(VI) in parts of the Mojave desert, USA (Izbicki et al. 2008), where both natural and industrial sources exist. In uncontaminated areas, the $\delta^{53}Cr$ values of native dissolved Cr(VI) are invariably greater than those inferred for the parent rocks: Values in waters where Cr(VI) reduction is thought to be negligible ranged from 0.7 to 3.1‰, with a mean of 1.7‰ and a standard deviation of 0.4‰. Despite the potential for elevated $\delta^{53}Cr$ values in the natural background Cr, Cr(VI) reduction could be inferred in some areas. Trends of increasing $\delta^{53}Cr$ and decreasing Cr(VI) concentration along groundwater flow paths were strongly suggestive of reduction, with up to half the originally present Cr(VI) lost from solution.

The Izbicki et al. (2008) study was also successful in distinguishing sources of Cr contamination. Monitoring wells near a facility where chemical milling of metal parts was conducted contained elevated Cr(VI) concentrations and low $\delta^{53}Cr$ values suggestive of contamination by industrial Cr with near-zero values. One well yielded $\delta^{53}Cr$ values of -0.1‰, which is clearly outside the range of native Cr and within the expected range for industrial Cr. However, for several other wells, $\delta^{53}Cr$ values were not so low and data interpretation was not so simple because of the presence of native Cr(VI) in the aquifer. A mixing model was constructed to calculate Cr(VI) concentrations and $\delta^{53}Cr$ values for mixtures between the native Cr(VI) and the inferred range of contaminant Cr(VI). Waters from some of the wells conform to this mixing model, with varying fractions of the sampled Cr(VI) sourced from the facility. As an alternative interpretation, $\delta^{53}Cr$ and Cr(VI) concentration data from most wells in the vicinity of the facility conformed to a model where most Cr(VI) was sourced from the facility and was subject to variable amounts of Cr(VI) reduction. Overall, this study demonstrated the potential to distinguish anthropogenic Cr from naturally occurring Cr, at least in the study area. But it also revealed the complexity that arises when mixing of two sources and Cr(VI) reduction are both potentially involved.

The second study investigated Cr(VI) in the eastern Snake River Plain aquifer of Idaho, USA (Raddatz et al. 2011). Results for naturally sourced Cr(VI) were similar to those of the Mojave desert study: Mean $\delta^{53}Cr$ was 1.15‰; standard deviation was 0.46. Once again, it appears that Cr(VI) released by weathering is enriched in the heavier isotope relative to the parent material, which in this case is the basaltic host rock of the aquifer. Contaminant Cr(VI) had near-zero $\delta^{53}Cr$ values. The main goal of the study was to detect any reduction of the contaminant. Some contaminated wells showed near-zero $\delta^{53}Cr$ values indicative of negligible Cr(VI) reduction along some flow paths. Other contaminated wells showed elevated $\delta^{53}Cr$ values ranging from to 0.78 to 1.79‰. Interpretation of these data was complicated by the presumed presence of naturally sourced Cr(VI) with similar $\delta^{53}Cr$ values in the waters. A mixing model was used to define possible combinations of Cr(VI) and $\delta^{53}Cr$ in mixtures of contaminant and natural Cr(VI), without reduction. Comparison with the data revealed that several contaminated wells had $\delta^{53}Cr$ values greater than plausible mixtures of contaminant and natural Cr(VI). This indicated that small amounts of reduction occurred, but only along the shallowest flow paths in the aquifer.

9.3.4 Outlook for Environmental Applications of Cr Isotope Data

The studies published to date strongly suggest that $\delta^{53}Cr$ data can provide an effective means of detecting Cr(VI) reduction. In systems where contaminant Cr(VI) concentrations are much greater than the natural background, elevated $\delta^{53}Cr$ values relative to contaminant source values are indicative of Cr(VI) reduction. At present, this approach seems quite robust, barring the unlikely discovery of some unforeseen phenomenon other than Cr(VI) reduction that causes large increases in $\delta^{53}Cr$. However, one must be aware that in waters where the contaminant does not completely overwhelm the natural background, interpretation is more complicated. One should determine the $\delta^{53}Cr$ range of the background, then use mixing calculations to determine if elevated $\delta^{53}Cr$ values are indeed indicative of reduction rather than mixing alone.

Quantitative determination of reduction using $\delta^{53}Cr$ data is more challenging, and more research must be done to address certain issues. The fractionation factor for Cr(VI) reduction depends on the reduction mechanism and environmental conditions, and a full understanding of its variability does not yet exist. Rayleigh fractionation calculations provide an imperfect model of reaction within groundwater systems, and it appears this can greatly affect the calibration of $\delta^{53}Cr$ increase as a function of the extent of reduction.

The use of $\delta^{53}Cr$ data as source indicators is less well developed, but the emerging picture suggests that anthropogenic Cr(VI) tends to have lower $\delta^{53}Cr$ values than that of natural background Cr(VI). This tendency is based on few studies at present and must be explored in future work. If indeed such a contrast exists at a particular site, one must be careful to consider that reduction of the contaminant Cr(VI) can increase its $\delta^{53}Cr$ and render the source contrast useless. If the goal is to "fingerprint" Cr from multiple anthropogenic plumes, one should not expect them to have distinct "signatures" a priori because industrial Cr contamination seems to have near-zero $\delta^{53}Cr$ values as a general rule. However, there are likely processes, such as Cr(VI) reduction in soils prior to release into groundwater, that could generate distinctive $\delta^{53}Cr$ values that can be traced. Such usage of Cr isotope analyses would necessarily be developed on a site-by-site basis and has been explored little at this time.

9.4 Selenium Isotope Systematics and Applications

9.4.1 Se Isotope Systematics

Selenium has 4 stable valence states in natural systems, so its isotopic systematics are more complex than those of Cr. However, the general pattern of isotopic fractionation, with the greatest fractionation induced by reduction and smaller effects due to oxidation and non-redox processes such as adsorption, is similar. Here we discuss the magnitude of fractionation for the $^{82}Se/^{76}Se$ ratio (Table 9.2), as it is the most commonly reported ratio with the newer analytical methods and will likely be the one used in most future studies. Some of the fractionation factors have been converted from other originally reported ratios, such as $^{80}Se/^{76}Se$.

9.4.1.1 Equilibrium Isotope Fractionation

Attainment of isotopic equilibrium in Se redox reactions is probably extremely slow under most

Table 9.2 Kinetic isotopic fractionation caused by Se transformations

Study	Reaction	Mechanism	ε (‰)^{82}Se/^{76}Se
Rees and Thode (1966)	Se(VI) to Se(IV)	8 M HCl, 25°C	18
Johnson and Bullen (2003)	"	"Green rust"	11.1[a]
Ellis et al. (2003)	"	Sediment slurry (microbial)	3.9–4.7[a]
Clark and Johnson (2008)	"	"	1.8
Kirk et al. (2009)	"	*Dechloromonas sp.*	2.4
Herbel et al. (2000)	"	Bacterial cultures	1.7–7.2[a]
Krouse and Thode (1962), Rees and Thode (1966), Webster (1972), Rashid and Krouse (1985)	Se(IV) to Se(0)	NH$_2$OH or ascorbic acid	15.0–19.2
Ellis et al. (2003)	"	Sediment slurry (microbial)	8.4[a]
Herbel et al. (2000)	"	Bacterial cultures	9.0–13.7[a]
Johnson (2004)	Se(IV) to Se(VI)	NaOH + H$_2$O$_2$	< 0.5[a]
Johnson et al. (1999)	Se(0) oxidation	Incubated soil	< 0.5[a]
Herbel et al. (2002)	Plant uptake	Wetland plants	< 1.5[a]
Hagiwara (2000)	Algal uptake	*C. reinhardtii*	1.5–3.9[a]
Johnson et al. (1999)	Se volatilization	Cyanobacteria	< 1.7[a]
"	"	Soil (microbes)	< 0.9[a]
Schilling et al. (2010)	"	Fungi	1.0–4.0
"	Se(IV) adsorption	Fe(OH)$_3$·nH$_2$O	0.8[a]

[a]Converted from original measured ratio using $\varepsilon_{82/76} = 1.48\, \varepsilon_{80/76}$

environmental conditions, because chemical disequilibrium is observed in nature (White et al. 1991) and, as with Cr, redox reactions involve multiple electron transfers. Furthermore, the similarities between Se and S suggest that the predominantly kinetic interpretation used for S isotope studies at environmental temperatures should apply to Se also. However, experiments have not yet been done to measure rates of Se isotopic equilibration.

Although isotopic equilibrium may not be relevant in most Se isotope studies, equilibrium fractionation factors provide guidance toward understanding Se isotope systematics. Krouse and Thode (1962) used vibrational spectra and theoretical calculations to estimate fractionation factors for several Se species. Their estimate for Δ for SeO$_4$–H$_2$Se equilibrium at 25°C is 33‰. In comparison, Tudge and Thode (1950) estimated $\Delta = 75$‰ for SO$_4^-$–H$_2$S equilibrium. These results suggest the potential for significant Se isotope fractionation, but it also indicates fractionation for a given Se redox reaction will tend to be much smaller than that of an analogous S redox reaction. Furthermore, because the +4 and 0 valences of Se are stable, reactions involving the entire eight-electron difference between +6 and −2 valences are not relevant. The actual relevant reactions involve two or four electrons transferred, have less bonding contrast, and exhibit lesser isotopic fractionation.

9.4.1.2 Se(VI) Reduction to Se(IV)

As with sulfate, abiotic reduction of Se(VI) is kinetically sluggish at environmental temperatures. One environmentally relevant abiotic Se(VI) reduction process is known; Se(VI) can be reduced to Se(IV) and ultimately to Se(0) by "green rust," a layered ferric+ferrous oxide with anions occupying interlayer spaces (Myneni et al. 1997). Johnson and Bullen (2003) performed room-temperature laboratory experiments to obtain an ε value of 11.1‰ ± 0.3‰ for the Se(VI) reduction reaction. Less relevant experiments have been carried out which use 8 N HCl as the reductant of Se(VI) (Table 9.2).

Selenate can be reduced readily by certain microbes, and generally proceeds stepwise, with Se(IV) as a potentially stable product. This differs from sulfate reduction, where a single microbe may convert sulfate to sulfide. Se isotope fractionation factors were reported by Herbel et al. (2000) for Se(VI) reduction by growing, anaerobic batch cultures of *Bacillus arsenicoselenatis* strain E1H, a haloalkaliphile, and *Sulfurospirillum barnesii* strain SES-3, a freshwater bacterium. Fractionation factors varied in these experiments. In the *S. barnesii* experiment, isotopic fractionation was nearly nonexistent in the earliest stages, then increased to $\varepsilon \approx 6$‰, whereas with *B. arsenicoselenatis* $\varepsilon \approx 7$‰. Washed cell suspensions of *S. barnesii* yielded ε ≈ 1.6‰. In all the experiments, high initial concentrations of electron donor (e.g., 20 mmol/L lactate) and Se(VI) (e.g., 10 mmol/L) were used. So although the results clearly indicate microbial Se(VI) reduction can induce isotopic fractionation, they cannot be directly applied to natural settings. Furthermore, the variability observed in the fractionation factor suggests that environmental conditions will to some extent control the magnitude of fractionation, as is observed with sulfate reduction (Canfield 2001).

In an attempt to obtain fractionation factors relevant to natural settings, Ellis et al. (2003) performed batch incubation experiments with unamended, anaerobic sediments from three different Se-impacted surface water systems in California, USA. Starting Se(VI) concentrations were environmentally relevant, ranging from 230 nmol/L to 100 μmol/L. The results yielded a narrow range in the fractionation factor, with ε ranging from 3.9 to 4.7‰. Clark and Johnson (2008) performed a similar incubation of sediment from a Se-impacted wetland at Sweitzer Lake in Colorado, USA, with 30 μmol/L initial Se(VI), and obtained a much smaller value (1.8‰). Kirk et al. (2009) also reported a relatively small value of 2.4‰ for incubations of phosphate mine waste rock from Idaho, USA, inoculated with an enrichment culture of native microbes and supplied with 150 μmol/L initial Se(VI). In all of these studies, Se(VI) reduction was microbially mediated. In general, the results suggest that microbial Se(VI) reduction in natural settings induces little isotopic fractionation, relative to that observed for the abiotic reaction. This raises the possibility of distinguishing microbial Se(VI) reduction from abiotic reduction using observed fractionation factors in a field setting, but such an approach would require detailed knowledge of the factors relevant for each site. For example, a large fractionation generated by a reaction may be manifested as a relatively small apparent fractionation if the measured Se is separated from the reduction sites by a diffusive barrier (see below).

9.4.1.3 Se(IV) Reduction

Abiotic Se(IV) reduction is not kinetically inhibited as are selenate and sulfate reduction. For example, the strong reducing agent BH_4^- instantaneously reduces Se(IV) to H_2Se at room temperature but does not react with Se(VI) quickly. Se(IV) can be reduced by several reducing agents such as dissolved ascorbic acid or SO_2, and presumably a variety of naturally occurring organic reductants can reduce Se(IV) to Se(0) in nature. Microbes may also reduce Se(IV) (Blum et al. 1998).

Several studies have explored Se isotope fractionation during abiotic Se(IV) reduction. Krouse and Thode (1962), Rees and Thode (1966), and Rashid and Krouse (1985) reported ε values ranging from 15.0 to 19.2‰, for reduction by ascorbic acid and hydroxylamine. The latter authors noted that their data were not consistent with a simple Rayleigh-type fractionation process; they modeled the reaction as a two step process, with the rate constant of the second step two orders of magnitude smaller than the first, and kinetic isotope effects of 4.8 and 13.2‰ for the first and second steps. Data from all three studies were fit. This highlights the need to consider the potential variability of fractionation factors for kinetic processes, even ones that are abiotic and apparently simple.

Se isotope fractionation induced by microbial reduction of Se(IV) to Se(0) appears to be somewhat smaller than that induced by abiotic reduction. Herbel et al. (2000) reported fractionation factors for pure cultures of selenate respiring bacteria *B. arsenicoselenatis* and *S. Barnesii* and also *B. selenitireducens*, which grows with Se(IV) as an electron acceptor but not with Se(VI). The cultures were actively growing during the experiments, and initial concentrations were 10 mmol/L Se(IV) and lactate. Results yielded a range of ε from 9.0 to 13.7‰. These experiments demonstrate the potential for relatively large isotopic fractionation, but were conducted rather far from natural conditions. The Ellis et al. (2003) study described above obtained a slightly smaller value, 8.4‰, for two unamended sediment slurry experiments, with Se(IV) concentrations ranging from 0.24 to 100 μmol/L. Overall, it appears that Se(IV) reduction imparts a much larger isotopic fractionation than does reduction of Se(VI) to Se(IV).

9.4.1.4 Se Oxidation and Adsorption

Isotopic effects induced by Se oxidation reactions have not yet been studied fully. The only experiments reported so far are oxidation of Se(IV) to Se(VI) in a 0.8-M NaOH + 3% H_2O_2 solution. No isotopic shift was detected within the 0.2‰ uncertainty of the $^{80}Se/^{76}Se$ measurements (Johnson 2004). As with Cr (III) oxidation, a simple kinetic isotope effect is not observed, presumably because of the competing tendencies of simple kinetic effects, which lead to enrichment of lighter isotopes in the oxidized product, and back reaction or equilibrium effects, which promote enrichment of the product in heavier isotopes. Oxidation of Se(0) and Se(-II) have not yet been studied isotopically. However, S isotopic fractionation induced by sulfide oxidation is relatively weak (Canfield 2001) and it is likely that Se follows suit.

In one adsorption experiment, Johnson et al. (1999) found Se(IV) adsorbed onto hydrous ferric oxides to be slightly (0.8‰) enriched in the lighter isotope relative to the coexisting solution. Apparently, Se(IV) adsorption, at least with this surface, does not change the local bonding environment of Se(IV) much. This fits the general pattern set by adsorption of sulfate (Van Stempvoort et al. 1990) and chromate (Ellis et al. 2004), which both show small or unmeasurable effects.

9.4.1.5 Se in Biota

Assimilation of Se(VI) and Se(IV) by algae induces minor Se isotope fractionation. Hagiwara (2000) found that the fresh water alga, *Chlamydomonas reinhardtii*, grown in Se(VI)- and Se(IV)-bearing media, was enriched in the lighter isotope by 1.7–3.9‰. Clark and Johnson (2010) reported a much smaller offset (mean of 0.59‰ in the same direction) for phytoplankton recovered from Sweitzer Lake, Colorado, relative to the mean composition of waters in the lake. This result is consistent with the minor S isotopic fractionation (0.9–2.8‰) induced by algal assimilation of sulfate (Kaplan et al. 1963).

Uptake of Se into higher plants also induces generally small but significant Se isotope fractionations. Herbel et al. (2002) determined the isotopic compositions of Se in five macrophytes from an artificial wetland receiving water with ca. 320 nmol/L Se(VI). The plants' Se isotope ratios varied little and the mean value was slightly (1.1‰) less than that of the surface waters. In a more extensive survey of roots, stems, and leaves from several plants in a Se(VI)-impacted wetland, plant tissues' $\delta^{82}Se$ values generally ranged from identical to the median wetland water Se(VI) value to 1.8‰ less than that, with a few outliers up to 4.2‰ less (Clark and Johnson 2010). Once again, Se isotope systematics are similar to those of sulfur; Trust and Fry (1992) found higher plants to be an average of 1.5‰ enriched in the lighter isotope relative to the sulfate assimilated.

Se can be methylated and volatilized by algae, plants, and fungi. Johnson et al. (1999) reported that volatile Se generated by cyanobacterial mats and incubated soils was enriched in the lighter isotope by 0.0–1.7‰ relative to the supplied selenate. Volatile Se species recovered from four water samples from a Se-impacted lake in Colorado, USA were enriched in the lighter isotope by 0.0–1.5‰ (Clark and Johnson 2010) relative to the lake's selenate. In contrast to those results, pure cultures of the fungus *Alternaria alternata* have been reported to produce volatile Se species enriched in the lighter isotope by as much as 4.0‰ (Schilling et al. 2011).

Se isotope ratios in zooplankton and fish have been studied at the Sweitzer Lake field site (Clark and Johnson 2010). Two zooplankton samples had $\delta^{82}Se$ values 0.2‰ and 0.5‰ less than coexisting phytoplankton. Livers and muscle tissue from four fish were, respectively, 0.5 and 1.3‰ less than coexisting phytoplankton. This limited data set suggests that animals largely reflect their food sources, with only minor Se isotopic offsets.

9.4.2 Se Isotope Ratios in Rocks and Waters

Rouxel et al. (2002) reported $\delta^{82}Se$ for several igneous rocks and four iron meteorites. These all lie within 0.6‰ of SRM-3149. An array of recent marine and estuarine sediments, and unweathered sedimentary rocks from a variety of geologic time periods fall within 1.6‰ of zero (Hagiwara 2000; Rouxel et al. 2002; Zhu et al. 2008; Clark and Johnson 2010; Mitchell et al. 2010). Weathered organic-rich shale exhibits greater variation, with values spanning the range −13‰ to +13‰ in a single outcrop (Hagiwara 2000; Wen et al. 2011; Zhu et al. 2008). This variation is thought to reflect near-surface redox cycling during the weathering process.

Se in water appears to be somewhat more variable and weakly positive in δ^{82}Se, but few studies have been published to date. Several analyses of surface waters generated by agricultural activities in the San Joaquin Valley, California, USA range from 0.9 to 2.9‰ (Herbel et al. 2002), 4 water samples from the northern arm of the San Francisco estuary, California ranged from 0.5 to 3.3‰ (Johnson et al. 2000), and surface waters from the Sweitzer Lake site in western Colorado, USA range from 0.6 to 4.4‰ (Clark and Johnson 2010). Wastewaters from four oil refineries near the northern reach of San Francisco bay range from 1.2 to 4.5‰ (Johnson et al. 2000).

This lack of significant isotopic fractionation was interpreted to be the result of an absence of direct Se (VI) or Se(IV) reduction and dominance of Se cycling via plant uptake, but more recent research suggests that the isotopic contrasts generated by reduction reactions may be quite small in surface water systems. Clark and Johnson (2008) investigated isotopic fractionation in sediments from the Sweitzer Lake site (Fig. 9.5). Unamended site sediment incubated with Se(VI)-spiked site water showed rather small isotopic fractionation for microbial Se(VI) reduction ($\varepsilon = 1.8$‰).

9.4.3 Se Isotope Applications

Most Se-related environmental research has focused on wetland environments, where the greatest environmental impacts have occurred (Presser 1994). Se isotope research followed this emphasis, with the goal of providing a new means to detect and perhaps quantify selenate reduction in wetlands and other surface water environments. This reaction is of primary importance, as it leads to sequestration of the contaminant. Se isotope studies in groundwater or vadose zone settings have developed more slowly.

9.4.3.1 Detection of Selenate Reduction in Wetlands and Lakes

Early studies of isotopic fractionation by Se(VI) and Se(IV) reduction led to a working hypothesis that these reactions would produce strong Se isotope ratio shifts and would thus be readily detected via isotopic analyses. Johnson et al. (2000) reported that δ^{82}Se values of reduced Se found in estuarine sediments of the northern arm of San Francisco Bay, California, USA, were only about 3‰ less than the overlying waters. Herbel et al. (2002) found a similar offset of less than 2.0‰ in an artificial wetland receiving selenate-rich agricultural drain waters in the San Joaquin Valley of California, USA. More importantly, although about 75% of the selenate input to the wetland was removed by biogeochemical processes, the outlet water was not enriched in the heavier isotopes relative to the inflow.

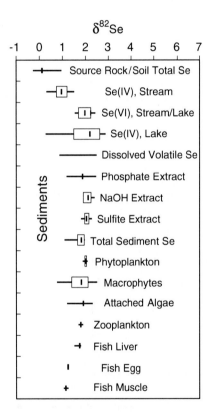

Fig. 9.5 Se isotope data from Sweitzer Lake, Colorado, USA (Clark and Johnson 2010). Mean values are given by the vertical tick marks. *Boxes* indicate the 25th and 75th quartiles of measurements, when sufficient data are available. The horizontal lines indicate the total range of measurements. The observed range in the isotopic composition of dissolved Se(VI) and the other major Se reservoirs in the system, is narrow. Se(VI) released into the environment is slightly enriched in heavier isotopes relative to the parent rock, but little isotope fractionation is manifested within the lake/wetland system, despite the presence of reducing sediments. Phosphate, NaOH, and sulfite extracts are designed to recover adsorbed Se, labile organic compounds, and elemental Se, respectively

Additional experimental results and theory presented by Clark and Johnson (2008) reveal a phenomenon that diminishes the isotopic shifts generated by reactions in surface water-sediment systems. This phenomenon applies whenever any isotope ratios in surface waters are used to detect reactions hosted by sediments, not just those of selenium. Here we review the concept briefly. Figure 9.6 provides concentration and isotopic profiles for a simple case. When reduction occurs inside sediments, isotopic fractionation occurring at the sites of reduction is not fully manifested in the surface waters above. The cause of this can be described as a "reservoir effect"; in the pore spaces of the sediment, reduction causes the dissolved Se(VI) to become enriched in the heavier isotopes (Fig. 9.6). This counteracts the tendency for the reaction to enrich the product in lighter isotopes. Although the pore space is connected by diffusion to surface water above by, diffusion is generally not fast enough to erase the reservoir effect. Bender (1990) and Brandes and Devol (1997) developed mathematical models for this phenomenon, and found that the isotopic effect observed in the surface water can be described using an effective fractionation factor. For a given system, the effective ε value is smaller than the intrinsic ε value (the micro-scale fractionation induce by the reaction itself) by a factor of two or more, depending on diffusive communication between water and sediment. Clark and Johnson (2008) performed a series of laboratory experiments that confirmed the predictions of these models and yielded an effective ε value of only 0.6‰ for an intact Sweitzer lake core.

It thus appears that Se isotope data may not provide a simple indicator of Se(VI) reduction in wetlands and other surface systems. The effective fractionation is small in the Sweitzer Lake case and is no greater than that occurring with plant and phytoplankton uptake. This may be the case elsewhere. For example, the lack of large isotopic contrast between reduced Se in sediments and that in the overlying water in the San Francisco Bay and artificial wetland settings (Johnson et al. 2000; Herbel et al. 2002) may reflect small effective fractionation for Se(VI) reduction and not a lack of dissimilatory microbial reduction as the original studies concluded.

The data from Sweitzer Lake (Clark and Johnson 2010) provide additional support for the view that effective fractionation is small (Fig. 9.5). The study provided a comprehensive survey of Se isotope data in the wetland/lake system and found a remarkably small range of Se isotope ratios. $\delta^{82}Se$ values for the vast majority of samples, including dissolved Se(VI) and Se(IV), extracted sediment components, plants, phytoplankton and higher plants, fall between 1.5 and 4.1‰ (relative to NIST SRM-3149); the Se(VI) input had a consistent 3.4‰ value. Incubations of site sediment showed that Se(VI) reduction was rapid (Clark and Johnson 2008), suggesting that it occurs in the system, but the associated effective fractionation is small.

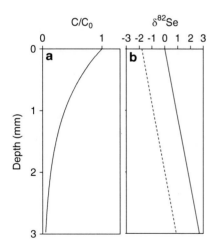

Fig. 9.6 Illustration of reservoir effects in a sediment-water system (Bender 1990; Clark and Johnson 2008). Concentration (**a**) and isotopic composition (**b**) of dissolved Se(VI) are given as a function of depth below the sediment/water interface. C_0 is the concentration in the overlying water. The *solid* and *dashed lines* in (**b**) give the isotopic compositions of Se(VI) in pore water and of reduced Se accumulating in the sediment, respectively. $\delta^{82}Se$ increases with increasing depth because isotopic fractionation enriches pore water Se(VI) in heavier isotopes. The magnitude of isotopic fractionation (ε) at any given point is 1.8‰. However, the average $\delta^{82}Se$ value of the accumulated reduced Se is offset from the overlying water by much less than 1.8‰, and thus the effective isotopic fractionation (ε = 0.9‰ for this particular model) is much smaller than the intrinsic fractionation

9.4.3.2 Reduction of Se in Subsurface Waters

There are few published Se isotope data from groundwater or vadose systems at this time. Whole-rock samples from an outcrop in Hubei Province, China that contains extremely high Se concentrations causing human Se poisoning decades ago (Zhu et al. 2004) have been analyzed for Se isotope ratios (Wen et al. 2011; Zhu et al. 2008). In contrast to the lack of significant isotopic variation in surface water settings, δ^{82}Se values vary strongly, ranging from -13.2 to $+11.4‰$ over a short distance in the outcrop. Wen et al. (2011) concluded that redox reactions, and potential repeated redox cycling of Se must have occurred at the site and caused the Se enrichment. Zhu et al. (2008) reported a relative lack of Se isotope variation in drill core samples from greater than 50 m depth, confirming that the isotopic variability was indeed caused by near-surface processes. This demonstrates that Se redox reactions occurring in shallow groundwater or vadose zone water can impart large isotopic fractionation.

Data from soil and weathered Mancos shale of western Colorado, USA, along with surface waters derived from the same rock unit, also suggest Se isotope ratios are shifted by subsurface redox reactions (Clark and Johnson 2010). Se(VI) in surface water was enriched in heavier isotopes by about 3‰ relative to the rock (Fig. 9.5). This is suggestive of redox reactions causing some fractionation during weathering.

9.4.4 Outlook for Environmental Applications of Se Isotope Data

At present, it appears that Se isotope ratios are not likely to provide a robust indicator of Se redox reactions in wetlands and other surface water systems. Despite the potential for large Se isotope variations to be induced by Se(VI) and Se(IV) reduction reactions in general, the intrinsic fractionations of the actual reduction mechanisms operating in wetland and lake sediments appear to be small. Additionally, when reduction is restricted to underlying sediments, the effective fractionation observed in surface waters is even smaller. We suggest the outlook for tracing Se redox reactions in these systems is poor, because the small redox-driven isotopic variations are no larger than those caused by a variety of other processes, such as assimilation of Se by plants. However, the lack of large isotopic fractionations observed in surface water systems suggests the possibility that Se from different sources may be traced in surface water systems, as their isotopic "signatures" may be affected little by redox reactions.

The outlook for groundwater applications appears much better. Data are sparse at this time, but it appears Se isotopic fractionation can be large in subsurface systems. Se isotope data could possibly be used to detect reduction of contaminant Se(VI) or Se(IV) in groundwater just as Cr isotope data are used to detect Cr(VI) reduction. For example, Se released from mine wastes or fly ash may be immobilized in the subsurface by microbial reduction (Kirk et al. 2009). It is likely that δ^{82}Se measurements of dissolved Se(VI) will provide a means to detect and possibly quantify reductive immobilization of the Se.

References

Abe Y, Hunkeler D (2006) Does the Rayleigh equation apply to evaluate field isotope data in contaminant hydrogeology? Environ Sci Technol 40:1588–1596

Albarède F, Beard BL (2004) Analytical methods for non-traditional isotopes. In: Johnson CM, Beard BL, Albarede F (eds) Geochemistry of non-traditional stable isotopes. Mineralogical Society of America, Washington

Anbar AD, Roe JE, Holman ES, Barling J, Nealson KH (2000) Non-biological fractionation of iron isotopes. Science 288:126–128

Asael D, Matthews A, Oszczepalski S, Bar-Matthews M, Halicz L (2009) Fluid speciation controls of low temperature copper isotope fractionation applied to the Kupferschiefer and Timna ore deposits. Chem Geol 262:147–158

Bain DJ, Bullen TD (2005) Chromium isotope fractionation during oxidation of Cr(III) by manganese oxides. Geochim Cosmochim Acta 69(10) Suppl. 1:A212

Ball JW, Bassett RL (2000) Ion exchange separation of chromium from natural water matrix for stable isotope mass spectrometric analysis. Chem Geol 168:123–134

Ball JW, Nordstrom DK (1998) Critical evaluation and selection of standard state thermodynamic properties for chromium metal and its aqueous ions, hydrolysis species, oxides, and hydroxides. J Chem Eng Data 43:895–918

Beard BL, Johnson CM (2004) Fe isotope variations in the modern and ancient earth and other planetary bodies. In: Johnson CM, Beard BL, Albarede F (eds) Geochemistry of non-traditional stable isotopes. Mineralogical Society of America, Washington

Belshaw NS, Zhu XK, Guo Y, O'Nions RK (2000) High precision measurement of iron isotopes by plasma source mass spectrometry. Int J Mass Spectrom 197:191–195

Bender ML (1990) The δ^{18}O of dissolved o_2 in seawater: a unique tracer of circulation and respiration in the deep sea. J Geophys Res 95:22243–22252

Berna EC, Johnson TM, Makdisi RS, Basu A (2010) Cr stable isotopes as indicators of Cr(VI) reduction in groundwater: a detailed time-series study of a point-source plume. Environ Sci Technol 44:1043–1048

Bigeleisen J (1949) The relative reaction velocities of isotopic molecules. J Chem Phys 17:675–678

Blowes DW (2002) Tracking hexavalent Cr in groundwater. Science 295:2024–2025

Blum JS, Bindi AB, Buzzelli J, Stolz J, Oremland RS (1998) Bacillus arsenicoselenatis, sp. nov., and Bacillus selenitireducens, sp. nov.: two haloalkaliphiles from Mono Lake, California that respire oxyanions of selenium and arsenic. Arch Microbiol 171:19–30

Brandes JA, Devol AH (1997) Isotopic fractionation of oxygen and nitrogen in coastal marine sediments. Geochim Cosmochim Acta 61:1793–1801

Canfield DE (2001) Biogeochemistry of sulfur isotopes. In: Valley JW, Cole DR (eds) Stable isotope geochemistry. Mineralogical Society of America, Washington

Carignan J, Wen HJ (2007) Scaling NIST SRM 3149 for Se isotope analysis and isotopic variations of natural samples. Chem Geol 242:347–350

Chacko T, Cole DR, Horita J (2001) Equilibrium oxygen, hydrogen, and carbon isotope fractionation factors applicable to geologic systems. In: Valley JW, Cole DR (eds) Stable isotope geochemistry. Mineralogical Society of America, Washington

Cheam V, Agemian H (1980) Preservation and stability of inorganic selenium-compounds at ppb levels in water samples. Analyt Chim Acta 113:237–245

Clark SK, Johnson TM (2008) Effective isotopic fractionation factors for solute removal by reactive sediments: a laboratory microcosm and slurry study. Environ Sci Technol 42:7850–7855

Clark SK, Johnson TM (2010) Selenium stable isotope investigation into selenium biogeochemical cycling in a lacustrine environment: Sweitzer lake, Colorado. J Environ Qual 39: 2200–2210

Criss RE (1999) Principles of stable isotope distribution. Oxford University press, New York, 254 pp

Cutter GA, Cutter LS (2001) Sources and cycling of selenium in the western and equatorial Atlantic Ocean. Deep-Sea Res Part II 48:2917–2931

Davis A, Olsen RL (1995) The geochemistry of chromium migration and remediation in the subsurface. Ground Water 33:759–768

Detmers J, Brüchert V, Habicht KS, Kuever J (2001) Diversity of sulfur isotope fractionations by sulfate-reducing prokaryotes. Appl Eviron Microbiol 67:888–894

Ellis AS, Johnson TM, Bullen TD (2002) Cr isotopes and the fate of hexavalent chromium in the environment. Science 295:2060–2062

Ellis AS, Johnson TM, Bullen TD, Herbel MJ (2003) Stable isotope fractionation of selenium by natural microbial consortia. Chem Geol 195:119–129

Ellis AS, Johnson TM, Bullen TD (2004) Using chromium stable isotope ratios to quantify Cr(VI) reduction: lack of sorption effects. Environ Sci Technol 38:3604–3607

U.S. Environmental Protection Agency (2009) National Primary Drinking Water Regulations. http://water.epa.gov/drink/contaminants/index.cfm Accessed 23 Nov 2010

Fendorf SE, Zasoski RJ, Burau RG (1993) Competing metal ion influences on chromium(III) oxidation by birnessite. Soil Sci Soc Am J 57:1508–1515

Frei R, Gaucher C, Poulton SW, Canfield DE (2009) Fluctuations in Precambrian atmospheric oxygenation recorded by chromium isotopes. Nature 461:250–253

Hagiwara Y (2000) Selenium isotope ratios in marine sediments and algae – a reconaissance study. M.S. Thesis, University of Illinois at Urbana-Champaign, Urbana, IL

Halicz L, Yang L, Teplyakov N, Burg A, Sturgeon R, Kolodny Y (2008) High precision determination of chromium isotope ratios in geological samples by MC-ICP-MS. J Analyt Atom Spectrom 23:1622–1627

Hayes JM (2001) Fractionation of carbon and hydrogen isotopes in biosynthetic processes. In: Valley JW, Cole DR (eds) Stable isotope geochemistry. Mineralogical Society of America, Washington

Heninger I, PotinGautier M, deGregori I, Pinochet H (1997) Storage of aqueous solutions of selenium for speciation at trace level. Fresenius J Analyt Chem 357:600–610

Herbel MJ, Johnson TM, Oremland RS, Bullen TD (2000) Selenium stable isotope fractionation during bacterial dissimilatory reduction of selenium oxyanions. Geochim Cosmochim Acta 64:3701–3709

Herbel MJ, Johnson TM, Tanji KK, Gao S, Bullen TD (2002) Selenium stable isotope ratios in agricultural drainage water systems of the western San Joaquin Valley, CA. J Environ Qual 31:1146–1156

Izbicki JA, Ball JW, Bullen TD, Sutley SJ (2008) Chromium, chromium isotopes and selected trace elements, western Mojave Desert, USA. Appl Geochem 23:1325–1352

Johnson TM (2004) A review of mass-dependent fractionation of selenium isotopes and implications for other heavy stable isotopes. Chem Geol 204:201–214

Johnson CM, Beard BL (1999) Correction of instrumentally produced mass fractionation during isotopic analysis of Fe by thermal ionization mass spectrometry. Int J Mass Spectrom 193:87–99

Johnson TM, Bullen TD (2003) Selenium isotope fractionation during reduction by Fe(II)-Fe(III) hydroxide-sulfate (green rust). Geochim Cosmochim Acta 67:413–419

Johnson TM, Herbel MJ, Bullen TD, Zawislanski PT (1999) Selenium isotope ratios as indicators of selenium sources and oxyanion reduction. Geochim Cosmochim Acta 63: 2775–2783

Johnson TM, Bullen TD, Zawislanski PT (2000) Selenium stable isotope ratios as indicators of sources and cycling of selenium: Results from the northern reach of San Francisco Bay. Environ Sci Tech 34:2075–2079

Kaplan IR, Rittenberg SC (1964) Microbial fractionation of sulfur isotopes. Gen Microbiol 43:195–212

Kaplan IR, Emery KO, Rittenberg SC (1963) The distribution and isotopic abundance of sulphur in recent marine sediments off southern California. Geochim Cosmochim Acta 27:297–331

Katz SA, Salem H (1994) The biological and environmental chemistry of chromium. VCH Publishers, New York

Kendall C (1998) Tracing nitrogen sources and cycling in catchments. In: Kendall C, McDonnell JJ (eds) Isotope tracers in catchment hydrology. Elsevier, Amsterdam

Kirk LB, Childers SE, Peyton B, McDermott T, Gerlach R, Johnson TM (2009) Geomicrobiological control of selenium solubility in subsurface phosphate overburden deposits. Geochim Cosmochim Acta 73(13) Suppl. 1:A661

Kitchen JW, Johnson TM, Bullen TD (2004) Chromium Stable Isotope Fractionation During Abiotic Reduction of Hexavalent Chromium. Eos Trans AGU 85(47) Fall Meet. Suppl.: Abstract V51A-0519

Krouse HR, Thode HC (1962) Thermodynamic properties and geochemistry of isotopic compounds of selenium. Can J Chem 40:367–375

Lemly AD (1998) Pathology of selenium poisoning in fish. In: Frankenberger WT Jr, Engberg RA (eds) Environmental chemistry of selenium. Marcel Dekker, New York

McNeal JM, Balistrieri LS (1989) Geochemistry and occurrence of selenium; an overview. In: Jacobs LW (ed) Selenium in agriculture and the environment. SSSA Special Publication. Soil Science Society of America, Madison, WI

Mitchell K, Mason PRD, Johnson TM, Lyons TW, Van Cappellen P (2010) Selenium isotope variation during oceanic anoxic events. Geochim Cosmichim Acta 74(12) Suppl. 1:A714

Myneni SCB, Tokunaga TK, Brown GE Jr (1997) Abiotic selenium redox transformations in the presence of Fe (II, III) oxides. Science 278:1106–1109

Oze C, Bird DK, Fendorf S (2007) Genesis of hexavalent chromium from natural sources in soil and groundwater. Proc Natl Acad Sci USA 104:6544–6549

Presser TS (1994) The Kesterson effect. Environ Manage 18:437–454

Raddatz AL, Johnson TM, McLing TL (2011) Cr stable isotopes in Snake River Plain aquifer groundwater: evidence for natural reduction of dissolved Cr(VI). Environ Sci Technol 45:502–507

Rashid K, Krouse HR (1985) Selenium isotopic fractionation during SeO_3^{2-} reduction to Se^0 and H_2Se. Can J Chem 63:3195–3199

Rees CB, Thode HG (1966) Selenium isotope effects in the reduction of sodium selenite and of sodium selenate. Can J Chem 44:419–427

Rouxel O, Ludden J, Carginan J, Marin L, Fouquet Y (2002) Natural variations of Se isotopic composition determined by hydride generation multiple collector inductively coupled plasma mass spectrometry. Geochim Cosmochim Acta 66: 3191–3199

Schauble E, Rossman GR, Taylor HP (2004) Theoretical estimates of equilibrium chromium-isotope fractionations. Chem Geol 205:99–114

Schilling K, Johnson TM, Wilcke W (2011) Isotope fractionation of selenium during fungal biomethylation by *Alternaria alternata*. Environ Sci Technol 45:2670–2676

Schoenberg R, Zink S, Staubwasser M, von Blanckenburg F (2008) The stable Cr isotope inventory of solid earth reservoirs determined by double spike MC-ICP-MS. Chem Geol 249:294–306

Seby F, Potin-Gautier M, Giffaut E, Borge G, Donard OFX (2001) A critical review of thermodynamic data for selenium species at 25 degrees C. Chem Geol 171:173–194

Sikora ER, Johnson TM, Bullen TD (2008) Microbial mass-dependent fractionation of chromium isotopes. Geochim Cosmochim Acta 72:3631–3641

Skorupa JP (1998) Selenium poisoning of fish and wildlife in nature: lessons from twelve real-world-examples. In: Frankenberger WT Jr, Engberg RA (eds) Environmental chemistry of selenium. Marcel Dekker, New York

Stilling LL, Amacher MC (2010) Kinetics of selenium release in mine waste from the Meade Peak Phosphatic Shale, Phosphoria Formation, Wooley Valley, Idaho, USA. Chem Geol 269:113–123

Stollenwerk KG, Grove DB (1985) Reduction of hexavalent chromium in water samples acidified for preservation. J Environ Qual 14:396–399

Trust BA, Fry B (1992) Stable sulphur isotopes in plants: a review. Plant Cell Environ 15:1105–1110

Tudge AP, Thode HG (1950) Thermodynamic properties of isotopic compounds of sulfur. Can J Res B28:567–578

U.S. Department of Health (2000) Toxicological Profile for Chromium. U.S. Department of Health and Human Services, Atlanta

Valley JW, Cole DR (eds) (2001) Stable isotope geochemistry. Mineralogical Society of America, Washington

Van Stempvoort DR, Reardon EJ, Fritz P (1990) Fractionation of sulfur and oxygen isotopes in sulfate by soil sorption. Geochim Cosmochim Acta 54:2817–2826

Villalobos-Aragon A, Ellis AS, Johnson TM, Bullen TD, Glessner JJ (2008) Chromium stable isotope fractionation during transport: sorption and oxidation experiments. 2008 Geological Society of America Annual Meeting abstracts, p. 239

Webster CL (1972) Selenium isotope analysis and geochemical applications. Ph.D Dissertation, Colorado State University, Fort Collins, CO

Weiss DJ et al (2008) Application of nontraditional stable-isotope systems to the study of sources and fate of metals in the environment. Environ Sci Technol 42:655–664

Wen HJ, Carignan J (2011) Selenium isotopes trace the source and redox processes in the black shale-hosted Se-rich deposits in China. Geochim Cosmochim Acta 75:1411–1427

Weyer S, Schwieters JB (2003) High precision Fe isotope measurements with high mass resolution MC-ICPMS. Int J Mass Spectrom Ion Processes 226:355–368

White AF, Benson SM, Yee AW, Wollenberg HA, Flexser S (1991) Groundwater contamination at the Kesterson reservoir, California.2. Geochemical parameters influencing selenium mobility. Water Resour Res 27:1085–1098

Zhu J, Zuo W, Liang X, Li S, Zheng B (2004) Occurrence of native selenium in Yutangba and its environmental implications. Appl Geochem 19:461–467

Zhu JM, Johnson TM, Clark SK (2008) Selenium isotope variations in weathering zones of Se-rich carbonaceous rocks at Yutangba, China. Geochim Cosmochim Acta 72(12) Suppl. 1:A1102

Zink S, Schoenberg R, Staubwasser M (2010) Isotopic fractionation and reaction kinetics between Cr(III) and Cr(VI) in aqueous media. Geochim Cosmochim Acta 74:5729–5745

Chapter 10
Stable Isotopes of Transition and Post-Transition Metals as Tracers in Environmental Studies

Thomas D. Bullen

Abstract The transition and post-transition metals, which include the elements in Groups 3–12 of the Periodic Table, have a broad range of geological and biological roles as well as industrial applications and thus are widespread in the environment. Interdisciplinary research over the past decade has resulted in a broad understanding of the isotope systematics of this important group of elements and revealed largely unexpected variability in isotope composition for natural materials. Significant kinetic and equilibrium isotope fractionation has been observed for redox sensitive metals such as iron, chromium, copper, molybdenum and mercury, and for metals that are not redox sensitive in nature such as cadmium and zinc. In the environmental sciences, the isotopes are increasingly being used to understand important issues such as tracing of metal contaminant sources and fates, unraveling metal redox cycles, deciphering metal nutrient pathways and cycles, and developing isotope biosignatures that can indicate the role of biological activity in ancient and modern planetary systems.

10.1 Introduction

The transition metals form a geochemically and isotopically complex and interesting group of elements clustered in the center of the periodic table. The term *transition metal*, according to the International Union of Pure and Applied Chemistry (IUPAC), refers to an element whose atom has an incomplete d sub-shell or which can give rise to cations with an incomplete d sub-shell. As shown in Fig. 10.1, according to this definition transition metals occur in Groups 1B and 3B-8B of the periodic table, and the transition metals across the periodic table have progressively increasing d-shell electron occupancy up to a maximum of 9. A characteristic of transition metals is that they generally exhibit two or more oxidation states, which as a consequence often leads to differences in the stable isotope composition of the chemical compounds and aqueous species they form. Transition metals for which significant variation of stable isotope composition in natural materials has been demonstrated include chromium (Cr), iron (Fe), nickel (Ni), copper (Cu), molybdenum (Mo) and silver (Ag).

The above definition of *transition metal* specifically excludes the Group 2B elements zinc (Zn), cadmium (Cd) and mercury (Hg), which are often referred to as *post-transition metals*. These metals have the d shell filled and thus generally occur as cations only in the +2 oxidation state. However, it is useful to include the post-transition metals in any discussion of transition metals, as they provide interesting similarities and contrasts in chemical behavior, are subject to change in redox state either naturally (e.g. Hg) or under anthropogenic influence (e.g. Zn, Cd) and display significant variation of stable isotope composition in natural materials.

Given this broad perspective, a description of the isotope systematics of the transition and post-transition metals is given here, including general isotopic distribution, examples of processes leading to stable isotope fractionation, and chemical properties pertinent to the discussion of stable isotope variability observed in natural materials. Rather than providing a comprehensive treatment of these issues, the intent here is to provide a general understanding of how the transition metal stable isotopes behave in Earth's hydrosphere

T.D. Bullen (✉)
U.S. Geological Survey, MS 420, 345 Middlefield Road, Menlo Park, CA 94025, USA

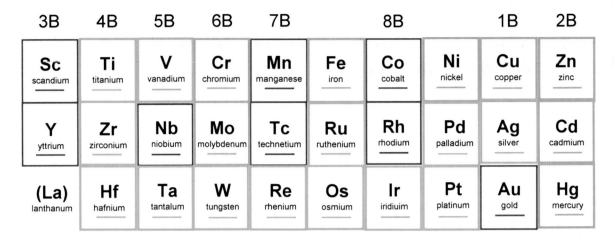

Fig. 10.1 A portion of the Periodic Table of the Elements, showing the transition and post-transition metals. Elements with *green borders* are those for which significant variability of stable isotope composition has been observed in natural materials. Elements with *blue borders* are those that have more than one stable isotope and thus the potential to have variability of stable isotope composition in natural materials, but such variability has yet to be rigorously demonstrated. Elements with *red borders* have only one stable isotope (no stable isotope in the case of Tc)

and atmosphere that are the venue of environmental research. Familiarity with the material in this section will help readers who are not well versed in stable isotope geochemistry to better appreciate the specific applications discussed in the later part of the chapter.

10.1.1 Chromium (Cr)

Chromium is an important trace element, particularly in ultramafic and mafic rock minerals, and is a major element in several minerals such as chromite ($FeCr_2O_4$) of the spinel group, uvarovite ($Ca_3Cr_2(SiO_4)_3$) of the garnet group, and crocoite ($PbCrO_4$). Chromium has numerous industrial applications, including use in electroplating, in leather tanning, as pigment in paint, and as an anticorrosion agent. Chromium is stable at the Earth's surface in two oxidation states: Cr^{6+} (Cr(VI)), commonly referred to as hexavalent Cr, and Cr^{3+} (Cr(III)), commonly referred to as trivalent Cr. Cr(VI) is a strong oxidant, and occurs mainly as the toxic, water-soluble oxyanions chromate (CrO_4^{2-}) and bichromate ($HCr_2O_7^{1-}$). Cr(III) is nontoxic and is insoluble in water.

Chromium has four stable isotopes: ^{50}Cr (4.35%), ^{52}Cr (83.79%), ^{53}Cr (9.50%) and ^{54}Cr (2.37%); note that for Cr and other elements discussed below that relative percentages of the isotopes are the values recommended by IUPAC to represent average terrestrial abundances. Igneous rocks have a very restricted range of $^{53}Cr/^{52}Cr$ which on average is about 0.12‰ (parts per thousand) less than that of NIST (National Institute of Standards and Technology) 979 Cr metal (Schoenberg et al. 2008). Presently the measured range of $^{53}Cr/^{52}Cr$ in natural materials is ~6‰, essentially reflecting the range measured for naturally occurring Cr(VI) in groundwater (Izbicki et al. 2008). Both inorganic and microbially mediated reduction of Cr(VI) to Cr(III) have been shown to result in isotope fractionation, with breakage of the strong Cr-O bond in chromate and bichromate appearing to be the rate limiting step (Ellis et al. 2002; Sikora et al. 2008). During reduction of Cr(VI), the product Cr(III) formed at any step of the process is 3–4‰ lighter than coexisting remnant Cr(VI). The fact that the oxidized species, Cr(VI), is heavier than the reduced species, Cr(III), is consistent with observations for other redox-sensitive elements such as selenium (Johnson et al. 2000; Johnson 2011) and mercury (Bergquist and Blum 2009; Blum 2011). While oxidation of Cr(III) to Cr(VI) is also likely to cause stable isotope fractionation (Schauble 2004), this phenomenon has yet to be rigorously demonstrated. Sorption of chromate and bichromate on Fe- and Al-oxyhydroxides has only a negligible instantaneous effect on the distribution of Cr isotopes, with preferential partitioning of light Cr isotopes onto sorption sites. Sorption effects may be magnified at reaction fronts such as the leading edge of a Cr contaminant plume in groundwater (Ellis et al. 2004).

10.1.2 Iron (Fe)

Iron is the most abundant element on Earth and the fourth most abundant element in the Earth's crust. Iron is the most widely used metal in industry, and thus is pervasive in the environment. Iron is stable at the Earth's surface in two oxidation states: Fe^{2+} (Fe(II)), commonly referred to as ferrous Fe, and Fe^{3+} (Fe(III)), commonly referred to as ferric Fe. Fe(II) is a strong reductant, is water-soluble, and is an important constituent of Fe minerals formed under reducing conditions (e.g. magnetite, siderite, pyrite). Fe(III) is only sparingly water-soluble and is an important constituent of Fe minerals formed under oxidizing conditions (e.g. ferrihydrite, goethite, hematite). Iron metal has a wide variety of applications in metallurgy such as construction of automobiles and ships, structural components of buildings, and machinery in general. Iron compounds such as Fe(III)-chloride and Fe(II)-sulfate are likewise widely used in industry for a variety of purposes. Iron plays an important role in biology and is an essential nutrient for nearly all living organisms. For example, Fe combines with oxygen to form hemoglobin and myoglobin, the two main oxygen transport proteins in vertebrates. Iron can serve as both electron donor and electron acceptor for a wide range of microorganisms, and thus Fe redox reactions have the potential to support substantial microbial populations in soil and sedimentary environments (Weber et al. 2006). Iron has been viewed as a limiting nutrient for phytoplankton growth in the oceans (Martin and Fitzwater 1988). Iron transformations and cycling processes are critical determinants of plant, forest and agricultural ecosystem function. Iron itself is generally non-toxic, but dissolved Fe(II) does impart an unpleasant taste to water and Fe(III) staining of plumbing fixtures can be a nuisance.

Like Cr, Fe has four stable isotopes: ^{54}Fe (5.58%), ^{56}Fe (91.95%), ^{57}Fe (2.18%) and ^{58}Fe (0.30%). Igneous rocks have a very restricted range of $^{56}Fe/^{54}Fe$. Presently the measured range of $^{56}Fe/^{54}Fe$ in natural materials is ~5‰, with pyrites from Precambrian black shales spanning this range (−3.6 to +1.2‰ relative to igneous rocks; Severmann and Anbar 2009). Both inorganic and microbially-mediated oxidation of Fe(II) to Fe(III) and microbial reduction of Fe(III) to Fe(II) have been shown to result in stable isotope fractionation (Bullen et al. 2001; Balci et al. 2006; Beard et al. 1999). Perhaps most important in this regard is the experimentally and theoretically well-established isotope exchange equilibrium between Fe (II)- and Fe(III)-hexaquo complexes of ~3‰ (Johnson et al. 2002; Anbar et al. 2005), in which the oxidized species, Fe(III), is heavier than the reduced species, Fe(II), as is the case for Cr. In addition, an important role for sorption of Fe onto mineral surfaces has been recognized, with heavy Fe species observed to populate the sorption sites after equilibrium is attained (Beard et al. 2010).

10.1.3 Nickel (Ni)

Nickel is used in many industrial and consumer products, including stainless steel and other metal alloys, coins, rechargeable batteries, and electroplating. Nickel is an important trace element in ultramafic and mafic rocks. Nickel is stable at the Earth's surface in three oxidation states: primarily as Ni^{2+} (Ni(II)), but also as Ni^{1+} (Ni(I)) and Ni^{3+} (Ni(III)). Nickel is an important enzyme cofactor, for example in urease, which assists in the hydrolysis of urea.

Nickel has five stable isotopes: ^{58}Ni (68.1%), ^{60}Ni (26.2%), ^{61}Ni (1.1%), ^{62}Ni (3.6%) and ^{64}Ni (0.9%). In a recent study of Ni stable isotope variations related to methanogenesis, Cameron et al. (2009) found that methanogens cultured in the laboratory preferentially assimilated light Ni from the media, leaving the residue enriched in heavy Ni isotopes. Fractionation of as much as 1.6‰ for $^{60}Ni/^{58}Ni$ was observed. In contrast, little variability was observed among meteorites, basalts and continental sediments, suggesting a lithologic baseline for comparison of this potential biomarker of methanogenesis.

10.1.4 Copper (Cu)

Copper has a wide range of industrial applications, including use for electrical wires, in metal alloys, in boat paint to prevent biologic fouling, in water and gas supply pipes, and in coins. Copper is stable at the Earth's surface in two oxidation states: Cu^{1+} (Cu(I)), which is the common form in sulfide minerals such as chalcopyrite ($CuFeS_2$), chalcocite (Cu_2S), and enargite (Cu_3AsS_4); and Cu^{2+} (Cu(II)), which is the common form in aqueous solution. Copper can be toxic to animals, due in large part to its ability to accept and donate single electrons readily as it changes between oxidation states. This is thought to catalyze the production of reactive radical ions leading to the condition known as oxidative stress.

Cu has two stable isotopes: ^{63}Cu (69.17%) and ^{65}Cu (30.83%). In natural materials, the measured range of $^{65}Cu/^{63}Cu$ is approximately 9‰ for solid samples and 3‰ for water samples (Larson et al. 2003; Borrok et al. 2008). In one study of possible Cu stable isotope fractionation processes, abiotic oxidative dissolution of Cu-bearing sulfide minerals resulted in the product aqueous Cu(II) being ~1.4‰ heavier in terms of the $^{65}Cu/^{63}Cu$ ratio than Cu in chalcopyrite and ~3.0‰ heavier than Cu in chalcocite. Involving the microbe *Acidithiobacillus ferrooxidans* in the dissolution process reduced the fractionation observed in both cases (Mathur et al. 2005).

Copper isotopes are beginning to provide important new insights into how Cu behaves in Earth's rivers and oceans, as well as how the isotopes may be fractionated. Vance et al. (2008) reported $^{65}Cu/^{63}Cu$ ratios for riverine water, estuarine water and particulates, and open ocean water samples. They observed that dissolved Cu in rivers had $^{65}Cu/^{63}Cu$ of from 0.02 to 1.45‰ greater than that of the NIST 976 Cu standard, and for the most part greater than that of Cu in crustal rocks (+0.16‰ relative to NIST 976). In the estuarine samples, $^{65}Cu/^{63}Cu$ of dissolved Cu was 0.8–1.5‰ greater than that of Cu associated with particulate material. Vance et al. (2008) interpreted this isotopic contrast to reflect isotopic partitioning of the weathered pool of Cu between a light fraction adsorbed to particulates and a heavy dissolved fraction dominated by Cu bound to strong organic complexes, providing a mechanism to deliver heavy dissolved Cu to the oceans. Dissolved Cu in open ocean water samples was even heavier than that in the river waters, suggesting a continuation of the fractionation between dissolved and particulate Cu fractions particularly in the surface ocean.

10.1.5 Zinc (Zn)

The post-transition metal Zn is used in electroplating, metal alloys, pigment in paint, agricultural fertilizers and pesticides, galvanized roofing material, and television-screen phosphors. Zinc is stable at the Earth's surface in one oxidation state, Zn^{2+} (Zn(II)). The large pool of Zn^0 metal produced industrially is manufactured specifically for purposes that take advantage of its high oxidation potential relative to other metals (e.g. as an anti-fouling and anti-corrosive agent on the submerged portion of ship hulls and propeller shafts). Zinc is an important enzyme cofactor, for example in alcohol dehydrogenase in humans, and is generally thought to be non-toxic.

Zinc has five stable isotopes: ^{64}Zn (48.63%), ^{66}Zn (27.90%), ^{67}Zn (4.10%), ^{68}Zn (18.75%) and ^{70}Zn (0.62%). In natural materials, the measured range of $^{66}Zn/^{64}Zn$ is approximately 2‰ (Cloquet et al. 2008). Maréchal et al. (1999) published the first high-quality measurements of zinc and copper isotope compositions in a variety of minerals and biological materials. Since this pioneering work the field of Zn isotope biogeochemistry has rapidly expanded into a variety of environmental applications. For example, Zn has a relatively low boiling point (~910°C) compared with that of Cr or Cu, within the operating temperature range of ore smelters. Thus Zn can evaporate from ore during smelting, which would favor escape of light Zn isotopes in the exhaust as demonstrated by Mattielli et al. (2009), and retention of heavy Zn isotopes in the slag residue as shown by Sivry et al. (2009). Kavner et al. (2008) demonstrated that electroplated Zn, reduced to metallic state from a large pool of aqueous Zn(II), is isotopically light compared to the parent solution, potentially resulting in a large pool of isotopically light electroplated Zn in the environment. In addition, Weiss et al. (2007) demonstrated that Zn adsorption due to complexation by organic compounds is selective for the heavy Zn isotopes, resulting for example in peat samples that have heavier Zn isotope compositions than any potential natural source. On the other hand, Borrok et al. (2008) pointed out that Zn uptake by microorganisms is selective for the light Zn isotopes. Thus there are several fractionation processes for Zn that might modify its isotope composition in the environment.

10.1.6 Molybdenum (Mo)

Molybdenum is used in the production of high-strength steel alloys, as pigments and catalysts, and in a variety of high temperature applications such as lubricants, electrical contacts and industrial motors. Molybdenum is stable at the Earth's surface in several oxidation states, the most stable being Mo^{4+} (Mo(IV)) and Mo^{6+} (Mo(VI)). Molybdenum is an essential enzyme cofactor in nearly all organisms, with particular importance for nitrogen fixation, nitrate reduction and sulfite oxidation. Molybdenum is soluble in strongly alkaline water such as seawater, forming molybdates (MoO_4^{2-}). Molybdates bear structural similarities to chromates, yet are weaker oxidants. In fact, although Mo lies directly below Cr in the periodic table, Mo compounds show more similarity to those of tungsten (W), which lies directly below Mo in the periodic table.

Molybdenum has seven stable isotopes: ^{92}Mo (14.84%), ^{94}Mo (9.25%), ^{95}Mo (15.92%), ^{96}Mo (16.68%), ^{97}Mo (9.55%), ^{98}Mo (24.13%) and ^{100}Mo (9.63%). In natural materials, the measured range of $^{97}Mo/^{95}Mo$ is ~3‰, with pelagic clays extending to −1‰ and suboxic sediment pore fluids extending to +2‰ relative to a modeled "bulk earth" composition as defined by Mo ores (Barling et al. 2001; Siebert et al. 2003; Anbar 2004; Arnold et al. 2004). Molybdenum isotopes are fractionated during adsorption to ferromanganese oxides in oxidizing environments (Barling et al. 2001; Barling and Anbar 2004) and precipitation of Fe-Mo-S solids in reducing environments (Poulson et al. 2006; Neubert et al. 2008), and the extent of fractionation differs in these two extreme conditions. Thus Mo isotopes can provide information about local redox conditions.

10.1.7 Cadmium (Cd)

The post-transition metal Cd is used primarily in Ni-Cd batteries, metal alloys, electroplating, pigment in paint, and electronic components. Cadmium is stable at the Earth's surface primarily in the Cd^{2+} (Cd(II)) oxidation state. Cadmium has a strong geological affinity to Zn, but unlike Zn can be toxic to humans due to its effective substitution for calcium in bone and its tendency to concentrate in the kidneys and liver where it may become carcinogenic.

Cadmium has eight stable isotopes: ^{106}Cd (1.25%), ^{108}Cd (0.89%), ^{110}Cd (12.49%), ^{111}Cd (12.80%), ^{112}Cd (24.13%), ^{113}Cd (12.22%), ^{114}Cd (28.73%) and ^{116}Cd (7.49%). In terrestrial materials, the measured range of $^{114}Cd/^{110}Cd$ is approximately 0.5‰; considering anthropogenically influenced materials extends that range to ~1.5‰ (Ripperger and Rehkämper 2007; Schmitt et al. 2009a, b; Shiel et al. 2010). Cadmium has an even lower boiling point (~760°C) than Zn, and like Zn, Cd in ores likely evaporates during smelting favoring escape of light Cd isotopes in the exhaust, as recently demonstrated by Cloquet et al. (2006a). Cadmium behaves like Zn during electroplating (Kavner et al. 2008), potentially resulting in a large pool of isotopically light Cd in the environment. Although Cd is discussed in this chapter mainly in a multi-tracer context, Rehkämper et al. (2011) provide a thorough discussion of the application of Cd isotopes in environmental studies.

10.1.8 Mercury (Hg)

The post-transition metal Hg is used in thermometers and other scientific apparatus, amalgam for dental restoration, Hg vapor lamps, cosmetics, and liquid mirror telescopes. Historically, Hg has been widely used in gold mining operations. Mercury is stable at the Earth's surface as soluble Hg^{2+} (Hg(II)) and as Hg^+ (Hg(I)) in the form of monomethylmercury (CH_3Hg^+). In addition, and unlike the other transition and post-transition metals, Hg has a stable gaseous form (Hg^0) at environmental temperatures with a residence time in the atmosphere of approximately 1 year (Schroeder and Munthe 1998). Mercury and its compounds (e.g. monomethylmercury (MeHg), mercuric chloride) are extremely toxic and can be readily ingested through inhalation or absorption through the skin and mucous membranes.

Mercury has seven stable isotopes: ^{196}Hg (0.15%), ^{198}Hg (9.97%), ^{199}Hg (16.87%), ^{200}Hg (23.10%), ^{201}Hg (13.18%) ^{202}Hg (29.86%) and ^{204}Hg (6.87%). Based on the compilation of existing high-precision Hg isotope data reported by Bergquist and Blum (2009), the measured range of $^{202}Hg/^{198}Hg$ in natural materials resulting from *mass dependent fractionation* (MDF) between coexisting Hg pools is approximately 7‰. This is remarkable, considering that measurable fractionation of Hg isotopes was generally thought to be non-existent as late as the 1990s. In addition, Hg stable isotopes exhibit considerable *mass independent fractionation* (MIF), in which the odd-numbered isotopes ^{199}Hg and ^{201}Hg behave differently than the even-numbered isotopes in certain chemical reactions such as photochemical reduction (Bergquist and Blum 2007), leading to significant enrichments and depletions of the odd-numbered isotopes relative to the even-numbered isotopes in environmental Hg pools.

Although Hg is included in the discussion here, Blum (2011) provides a thorough discussion of the theory and application of Hg isotopes in environmental studies. In his chapter, Blum (2011) points out that most environmental applications of Hg isotopes are concerned with either (1) interpreting the isotopic composition of MeHg occurring in sediments and organisms, or (2) interpreting the isotopic composition of Hg(II) associated with sediments or deposited onto the land surface or into a water body. In addition, the author stresses the importance of redox reactions as the main cause of isotope fractionation in the Hg system and the role that the isotopes can play in helping to increase our understanding of the chemical bonding environments that lead to various Hg fractions in sediments and other natural materials.

10.1.9 Other Transition Metals Potentially Having Useful Isotopic Variability

There are several other transition metals, outlined in blue in Fig. 10.1, that have multiple stable isotopes and thus the potential for stable isotope variability in nature, but which have yet to be rigorously explored in terms of isotope systematics. Examples of these elements that would be especially useful in environmental studies are titanium (Ti), vanadium (V), silver (Ag), tungsten (W) and rhenium (Re). For example, as pointed out by Severman and Anbar (2009), Re (^{185}Re (37.4%), ^{187}Re (62.6%); common oxidation states Re^{2+}, Re^{4+}, Re^{6+}, Re^{7+}) is removed from solution into sediments only under anoxic or suboxic conditions, making it a suitable tracer for moderately reducing conditions. In contrast, V (^{50}V (0.25%), ^{51}V (99.75%); common oxidation states V^{2+}, V^{3+}, V^{4+}, V^{5+}) is scavenged from solution by manganese oxides and Fe-oxyhydroxides, making it a suitable tracer for oxic conditions. Both potential redox-tracing applications assume that sedimentation promotes isotope fractionation between sediment and residual aqueous pools of Re or V, but such fractionation has yet to be demonstrated.

Silver, in addition to its historic value as a precious metal, is widely used in dentistry, photographic materials, mirrors and optics, control rods in nuclear reactors, clothing (as an anti-fungal agent) and in certain medicines. Silver has the highest electrical conductivity of any element and the highest thermal conductivity of any metal and thus is widely used in electrical contacts and conductors. Silver is stable at the Earth's surface primarily in three oxidation states: Ag^0 or native silver, and Ag^{1+} and Ag^{2+}, the common forms in a wide variety of Ag-containing compounds and minerals. Silver has two stable isotopes: ^{107}Ag (51.84%) and ^{109}Ag (48.16%). In an initial study of Ag stable isotope variation in sediment and domestic and industrial sludge samples, Luo et al. (2010) reported a small but significant range of variation in ^{109}Ag/^{107}Ag of 0.1‰ that bracketed the value for their standard. They found a different value for Ag from a fish liver (+0.28‰ relative to their standard) that may indicate a biological influence. However, at this stage Ag is only beginning to be explored in terms of stable isotope variability.

Tungsten provides an especially intriguing but as yet untested candidate for stable isotope variation in natural materials at the Earth's surface. Tungsten is used in many high-temperature applications, such as light bulbs and welding processes, and in lubricants, abrasives, ceramic glazes, and heavy metal alloys. The most common oxidation state of W at the Earth's surface is W^{6+}, but W exhibits all oxidation states from W^{2-} to W^{6+}. Tungsten has five stable isotopes: ^{180}W (0.12%), ^{182}W (26.50%), ^{183}W (14.31%), ^{184}W (30.64%) and ^{186}W (28.43%); ^{180}W is radioactive, but with a half-life of 1.8×10^{18} years it is essentially stable. While there is substantial literature describing ^{182}W variations in meteorites, there have been essentially no published systematic studies of low-temperature processes involving W. However, Irisawa and Hirata (2006) reported that W isotope ratios for sediment reference materials differed significantly from values for igneous rocks, hinting at the possibility that low-temperature redox and/or sorption processes may fractionate the W isotopes.

The potential for Ti to have significant variability of stable isotope composition in natural materials at the Earth's surface is probably less than that for W, V and Re. This is due to the fact that the geochemistry of Ti is dominated by the Ti^{4+} oxidation state, and Ti is not known to play a large role in biological processes such as enzyme stabilization. Regardless, titanium has five stable isotopes spanning a large mass range: ^{46}Ti (8.0%), ^{47}Ti (7.3%), ^{48}Ti (73.8%), ^{49}Ti (5.5%) and ^{50}Ti (5.4%). If our experience with exploration of stable isotope variability of other elements dominated by a single oxidation state may serve as a guide, it is not unrealistic that a thorough reconnaissance study of Ti or other unexplored transition metal stable isotope systematics in environmental materials (e.g. nanoparticles, pigments, electroplated steel) would reveal unexpected surprises.

10.2 Methodology

Although the transition and post-transition metals represent a geochemically diverse group of elements, the details of the analytical procedures required for measurement of their isotope compositions are quite uniform. Once a laboratory establishes a protocol for analysis of one of these metals, it can be relatively straightforward to add more of the remaining metals to the repertoire. On the other hand, transition and post-transition metals are everywhere, in field and laboratory equipment, and care must be taken to maintain strict cleanliness during sample collection, processing and measurement.

10.2.1 Sampling Protocols

As a general rule, the sampling protocol for transition or post-transition metal isotope analysis of water is identical to that for standard dissolved cation analysis. The sample is collected using plastic tubing for pumps or plastic containers for dipping and placed into plastic containers that have been rinsed several times with the water being collected. Pre-cleaning of the plastic containers with nitric acid may be required if container "blanks", obtained by allowing distilled water to sit in the container for an extended period of time, reveal significant metal content. The sample needs to be filtered, the pore-size of the filter being determined by the question being asked (e.g. is there isotopic fractionation between colloidal and non-colloidal size fractions?). Generally an in-line filter with 0.1 μm pore size provides sufficient filtration, but this may cause retention of some of the colloidal fraction as well (size: 1 nm–1 μm). Alternatively, the sample can be filtered after initial collection using a plastic syringe fitted with a disposable filter cartridge. Care must be taken to prevent possible redox changes of the redox-sensitive species in the oxic environment. Use of an in-line filter minimizes potential formation of Fe- and Mn-oxyhydroxides that can scavenge the metals from solution. Additionally, collection and filtration can be carried out in a portable nitrogen-atmosphere glove box. The sample should be acidified with exceptionally clean nitric or hydrochloric acid, to a pH of approximately 2.0. Additional precautions such as wearing ultra-clean over-garments, gloves and boots may be required, particularly when metal concentrations are expected to be extremely small.

Knowledge of metal concentrations in the sample prior to collection for isotope analysis is useful in order to ensure that sufficient metal is collected to meet the requirements of the analytical technique. As a general rule of thumb, sufficient material to provide 1–10 μg of the target metal should be collected when possible. The amount of metal required for isotope analysis ranges from as little as 5 ng to as much as 1 μg, depending on factors such as ionization efficiency in the mass spectrometer and resolution mode employed (see below). Moreover, knowledge of procedural blank levels is useful for deciding what kinds of plastic tubing and containers to use and what protective measures to take. A worthwhile exercise is to take a supply of the cleanest distilled laboratory water available to the field site and to treat the distilled water as a sample, using all intended plastic tubing and containers, filters and other collection equipment, and techniques. One or two field blanks should be collected during each sampling session, and the blank should be measured for concentration of the metal of interest prior to investing time in sample preparation and mass spectrometry.

Collection of solid samples such as soils, plant tissues and landfill materials requires common sense more than anything. Solid samples should be collected into plastic bags, and care should be taken to ensure that sampling equipment (e.g. shovels, augers, stem-wood borers) do not contain the metal of interest. Particular care should be taken to avoid sampling equipment with painted surfaces as the paint is likely to have transition or post-transition metals as pigments.

10.2.2 Sample Preparation

Prior to mass spectrometry, the metal of interest must be isolated from the rest of the sample matrix and particularly from elements that would produce isobaric interferences on the target masses for analysis. For example, in preparing a sample for Cr isotope analysis, all Ti, V and Fe must be removed, as each of these metals has an isotope at the same mass as a Cr isotope. Similarly, some elements form double-charged ions in the mass spectrometer that are detected at ½ of their mass due to the direct dependence of ion flight path through the magnetic sector of the mass spectrometer on mass:charge ratio of the ion. For example, Ba^{2+} ions can be formed in both multi-collector inductively coupled plasma mass spectrometry (MC-ICP-MS) and thermal ionization mass spectrometry (TIMS), the two common methods for analysis of metal stable isotope composition. $^{132}Ba^{2+}$ would be detected at mass 66, $^{134}Ba^{2+}$ would be detected at mass 67, and $^{136}Ba^{2+}$ would be detected at mass 68, all directly overlapping the Zn isotopes of the same masses. Thus, a goal of the chemistry is to produce as pure a target metal fraction as possible, preferably if not assuredly with quantitative recovery.

Most chemical procedures for purification of the transition and post-transition metals involve some form of ion exchange chromatography, making use of different affinities of the metals for the ion exchange resins in different strengths of acid media. For example, Wombacher et al. (2003) developed a method to purify Cd from geological materials and meteorites. The digested sample is dissolved in 3 M HCl and loaded directly onto an exchange column containing AG1-X8 strong acid anion exchange resin. The column is rinsed with set volumes of 0.5, 1, 2 and finally 8 M HCl in order to remove the majority of matrix elements. Following this series of rinses, all that remains on the resin are Ag, Zn, tin (Sn) and Cd. Additional rinsing with 8 M HCl removes the Ag, which can be retained for isotope analysis if desired (although quantitative recovery must be demonstrated). The rinse solution is then changed to a 0.5 M HNO_3–0.1 M HBr mixture, which removes Zn and some of the Sn from the column; this fraction can be retained for further processing for Zn isotope analysis, if desired. The rinse solution is then changed to 2 M HNO_3, which removes the Cd and the remainder of the Sn. This product solution is then taken to dryness, re-dissolved in 6 M HCl and loaded onto a small column containing a resin (TRU Spec, Eichrom Industries) that retains the Sn but not the Cd, which is collected in a beaker. The product solution contains only Cd, and is ready for isotopic determination. This example demonstrates the versatility of the ion exchange resin method, and new element-specific resins are continually being developed as the need for new applications is encountered.

Procedures for chemical purification of the other transition and post-transition metals follow a similar logic and most are able to achieve quantitative recovery, thereby avoiding isotope fractionation on the resin column which has been demonstrated to occur for several elements such as Fe (Anbar et al. 2000). References providing detailed descriptions of purification techniques for the transition and post-transition metals discussed here include: Cr, Ellis et al. (2002; supplementary material); Fe, Beard and Johnson (1999); Ni, Cameron et al. (2009); Cu and Zn, Borrok et al. (2007); Mo, Barling et al. (2001); Cd, Wombacher et al. (2003); Ag, Luo et al. (2010); Ti, Das and Pobi (1990); V, Pelly et al. (1970); W, Sahoo et al. (2006); Re, Liu et al. (1998). Note that sample preparation of Hg for isotope analysis requires an entirely different approach, as discussed in Blum (2011).

10.2.3 Instrumentation

As mentioned above, the stable isotope composition of the transition and post-transition metals is measured using either MC–ICP–MS or TIMS. A generalized schematic of these mass spectrometers is given in Fig. 10.2, which illustrates essential aspects of their

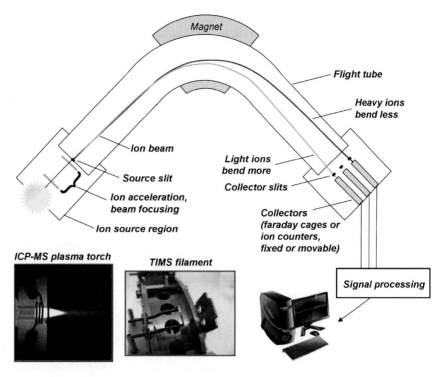

Fig. 10.2 Generalized schematic of the important aspects of multi-collector mass spectrometers used for determination of metal stable isotope compositions. Key features are discussed in the text. Modified after Bullen and Eisenhauer (2009)

function (Bullen and Eisenhauer 2009). All instruments consist of three main components: ion source, ion optics (ion acceleration, ion beam focusing, isotope mass separation) and ion collection. Ions are produced by thermal processes in the source region: in MC-ICP-MS, the sample is injected into the source region either as a nebulized liquid aerosol or as ablated solid particulate in a gaseous stream which passes through a high-temperature plasma (up to 10,000°K; see photo insert in Fig. 10.2) at near-atmospheric pressure. At this high temperature, atoms of the sample material are very efficiently ionized. In TIMS, the sample is deposited on a wire filament (see photo insert in Fig. 10.2), placed in a high vacuum (~10^{-8} torr), and the filament is heated to the temperature required for ion production of the particular element (typically 800–1,700°C). The ions produced in either instrument then enter the optics region, where they are accelerated through a large potential (up to 10 kV) and focused electronically to create a discrete ion beam. In MC-ICP-MS, the vacuum must be substantially improved over this short optics region via efficient mechanical and electrical "ion getter" vacuum pumps. Beyond the ion optics region, an exceptional vacuum (~10^{-8} torr) must be maintained in both instruments.

The accelerated and focused ion beam then passes through a slit into the flight tube where the different isotopes of the element are dispersed by mass using an electromagnet. The width of the slit determines the ability of the mass spectrometer to partially resolve sample ions and extraneous molecules of nearly similar mass ("isobaric interferences"). For example, since the plasma in ICP-source instruments is predominantly composed of Ar, a narrow slit must be used to resolve ^{56}Fe from ^{40}Ar^{16}O. In TIMS instruments, ^{56}Fe must be resolved from organic molecules with mass ~56 atomic mass units. In general, a narrower slit provides greater resolving power but reduced signal intensity.

The fundamental power of MC-ICP-MS and TIMS instruments to disperse the isotopes of an element rests in the ability of the magnet to differentially divert the flight paths of the various isotopes according to their individual energies. The ion flight paths are bent along the axis of the flight tube depending on the ratio of ion mass to charge: light ions are bent more (green path in diagram), heavy ions are bent less (red path in diagram). For any particular isotope, a restricted range of ion energies leaving the magnetic sector is required to create a concentrated, flat-topped signal that is ideal for detection and measurement. In TIMS, ions of a particular isotope produced at the source filament are essentially mono-energetic, and once the ion beam is focused in the source region mass (i.e. energy) dispersion can be effectively achieved in the magnetic sector. In contrast, in MC-ICP-MS ions of a particular isotope leave the source region having a fairly wide energy distribution due to numerous processes occurring in the source region. Thus in most MC-ICP-MS instruments, the poly-energetic ion beam must first pass through an electrostatic discriminator (not shown) which disperses the ion beam in exactly the opposite sense to that which follows in the magnetic sector. As a result of exacting electromechanical design, the ion beam that ultimately passes out of the magnetic sector has a balanced and much smaller energy spread that allows the flat-topped signal to develop. In a few MC-ICP-MS instruments, the ion beam first passes through a "collision cell" containing molecules of a gas such as Ar where collisions of the metal ions with the gas molecules substantially reduce the kinetic energies of those ions, which can then be reaccelerated in a nearly mono-energetic condition for passage through the magnetic field as in TIMS.

Ion detection collectors are configured to perfectly intersect the resulting flight paths of the component isotope ion beams of the element. In some instruments, collectors can be moved perpendicularly to the ion paths until desired coincidence is achieved for a particular element. In other instruments, the collector positions are fixed and the trajectories of the ion beams themselves are adjusted along the final section of the flight tube by further electronic focusing in order to achieve coincidence. Collectors are either Faraday cages, credit-card sized, thin rectangular boxes open on one narrow end to receive and trap ions on the internal surfaces resulting in a measurable total ion current on the internal surfaces, or ion counters that record individual ion collisions with the counter surfaces. Faraday cages may have slits positioned along the ion flight path prior to the point of ion entry into the cage in order to better shape the ion beam for optimum signal detection. Channel-specific total ion currents detected over a certain period of time are then processed through a series of electronic conversions into signals that can be interpreted by computer software as relative isotope abundances for the element.

Each instrument has its advantages in certain applications, but as a general rule the stable isotope composition of all the transition and post-transition metals discussed in this chapter can be most efficiently determined using MC-ICP-MS. The high ionization efficiency of the plasma source used to generate ions in MC-ICP-MS compared to that of the metal filament source used in TIMS results in improved isotope measurement of very small quantities (i.e. tens of nanograms) of elements with high ionization potentials, such as Zn (e.g. Chen et al. 2008), Mo (e.g. Barling and Anbar 2004), and Hg (e.g. Blum and Bergquist 2007). In addition, MC-ICP-MS instruments typically have the capability to sufficiently resolve measureable portions of metal ion beams from "isobaric" polyatomic interfering species having nearly the same mass (e.g. ^{56}Fe vs. ^{40}Ar^{16}O generated in the Ar plasma). Some interfering species, primarily oxides, can also be removed from the ion stream to a large extent by desolvating the sample prior to introduction to the plasma. Alternatively, in some instruments polyatomic molecules can be disaggregated in the collision cell apparatus described above due to collisions of the molecules with atoms of the gas supplied to the cell. In MC-IPC-MS, uncertainty due to signal noise resulting from the inherent instability of the plasma source is eliminated by measuring all ion beams simultaneously with the multicollector array. An additional strength of MC-ICP-MS is that samples can be introduced using laser ablation equipment, in which layers of atoms are sputtered off the surface of a solid material and transported by a gas stream to the plasma for ionization. Excellent references on the general topic of isotope analysis by MC-ICP-MS are Albarède and Beard (2004), Rehkämper et al. (2004) and Wieser and Schwieters (2005).

On the other hand, TIMS has been used for more than 50 years and has long been the workhorse for metal isotope research. TIMS provided much of the data on transition and post-transition metal stable isotope systematics prior to the proliferation of MC-ICP-MS in research labs and continues to provide excellent isotope measurements for more readily ionized elements such as Cr (e.g. Ellis et al. 2002) and Fe (Bullen and Amundson 2010) and even for elements that are difficult to ionize such as Cd (Schmitt et al. 2009a, b). TIMS is less affected by isobaric interferences than MC–ICP–MS because there is no need for the Ar gas and because the sample is ionized under high vacuum rather than in an aqueous solution or gas stream at atmospheric pressure. TIMS likewise benefits from simultaneous multicollector ion beam detection (Fantle and Bullen 2009). Typical external precision (i.e. comparison of total procedural replicates for natural samples) is at the 0.1‰ level for both TIMS and MC–ICP–MS. A disadvantage of both instruments is a sizable footprint in the laboratory and high costs for purchase and maintenance.

10.2.4 Data: Acquisition, Reduction, Accuracy, and Standard Reference Materials

Isotope fractionation of the ion beam produced in both TIMS and MC-ICP-MS, which is an unavoidable characteristic of both analytical platforms, means that the relative proportion of isotopes of an element reaching the detectors in the mass spectrometer is not identical to that of the sample. This instrument-induced fractionation, or "mass bias", is often larger than the isotopic variation between samples. In MC-IPC-MS, light ions of any element are preferentially discriminated against and removed from the ion stream at the sampling cones situated just beyond the plasma source, resulting in relatively heavy raw measured ratios that tend to stay very constant over the course of a measurement. In TIMS, light ions of any element are preferentially ionized off the filament surface, causing the raw measured ratio to be light at the outset and progressively heavier over the course of a measurement. Thus it is essential to provide some means to correct for isotope discrimination so that measured isotope compositions of samples can be compared to those of standards and other samples.

The classic approach to correct for isotope discrimination in TIMS is referred to as the *double spike technique*. An excellent treatment of the theoretical aspects of the double spike technique is given by Johnson and Beard (1999), and a practical overview of the technique is provided by Rudge et al. (2009). This technique requires that the element have at least four naturally-occurring isotopes, two that can be used as isotopes for the double spike and two that can be used to calculate a ratio that provides information on the relative natural isotope composition of a sample. The double spike is a mixture of two typically minor isotopes of the element that have each been industrially enriched in their proportion relative to the other isotopes. There are now several commercial suppliers of highly enriched (i.e. typically >90% pure) stable isotopes of most multi-isotopic elements, and thus it is relatively straightforward to create a double spike using any combination of two isotopes of an element. The double spike mixture, which has been carefully characterized for isotope composition and concentration, is added to a sample prior to analysis thus imparting an unnatural isotope composition to the sample-double spike mixture. Following analysis, the measured ratio of the two isotopes used to create the double spike can be compared to the known, carefully characterized value to obtain a "mass bias factor" that quantifies isotope fractionation that has occurred during analysis. The measured isotope composition of the sample-double spike mixture can then be corrected based on this mass bias factor, and the double spike component of the mixture can be mathematically subtracted from the fractionation-corrected composition to reveal the natural ratios of the two or more isotopes not used for the double spike as the residual.

For example, the $^{53}Cr/^{52}Cr$ natural isotope ratio of a sample can be determined using a double spike consisting of a mixture of nearly pure ^{50}Cr and ^{54}Cr, the two minor isotopes of Cr. Of the elements discussed here, Cr, Fe, Ni, Zn, Cd, Mo, Hg, Ti, and W are suitable for analysis by the double spike technique while Ag, Re and V are not. An important advantage of the double spike technique is that the double spike solution can be added to the sample prior to purification chemistry, and thus, without any additional effort, can be used to correct for isotope fractionation that might occur during chemical purification prior to analysis (e.g. due to non-quantitative recovery of the metal from the resin).

The nature of MC-ICP-MS allows application of two additional techniques for correction of isotope discrimination in the instrument during analysis, *standard-sample-standard bracketing* and *internal standardization*. As the name implies, in standard-sample-standard bracketing a standard solution is analyzed directly before and after a sample, and the interpolated isotope ratio of the standard at the time of the sample analysis is taken as the reference value. This is an incredibly straightforward approach, but assumes that the sample and standard behave similarly in the plasma (e.g. have identical matrices). With internal standardization, an element with a known isotope composition and similar fractionation behavior to the element being analyzed is added to the sample solution prior to aspiration into the plasma (e.g. Zr as an internal standard for Mo isotope analysis, Cu as an internal standard for Fe or Zn isotope analysis). The isotope ratio measured for the internal standard can be compared to its known value to quantify machine-induced fractionation and a correction can then be applied to the isotope ratio of the target element on a scan by scan basis. Using internal standardization, a plot of the measured isotope ratios of the target element vs. those of the internal standard element typically defines a mass dependent fractionation relationship that provides the basis for the correction (Longerich et al. (1987); Maréchal et al. (1999)).

The power of the standard-sample-standard bracketing and internal standardization techniques is that they can be used for isotopic analysis of any element having two or more isotopes. On the other hand, the double spike technique required for TIMS stable isotope analysis can likewise be used for MC-ICP-MS isotope analysis of elements having four or more isotopes, and is in essence a special case of internal standardization. The advantage of using a double spike for MC-ICP-MS analysis is that quantitative yield during sample purification is not required as it is for standard-sample-standard bracketing and internal standardization approaches (Siebert et al. 2001).

Differences in transition and post-transition metal stable isotope composition between materials tend to be small, a few parts per thousand at most, so that differences in absolute isotope ratio are difficult to conceptualize. For example, say that the $^{53}Cr/^{52}Cr$ ratio determined by TIMS for a native groundwater sample is 0.114000, while the value for the Cr isotope standard is 0.113392. To better visualize this small difference, most researchers in the field have adopted the delta notation used for gas isotope ratio comparison so that ratios can be compared on a per mil (‰) basis. Using the Cr example:

$$\delta^{53/52}Cr = 1000 * (^{53}Cr/^{52}Cr_{groundwater}$$
$$- ^{53}Cr/^{52}Cr_{standard})/^{53}Cr/(^{52}Cr_{standard})$$
$$= 1000 * (0.114000 - 0.113392)/0.113392$$
$$= +5.36.$$

δ values are preceded by a plus or minus sign to denote ratios that are, respectively, heavier or lighter than the reference material, which in this case is NIST 979 Cr metal. There are two important variants of the δ notation: (1) Δ_{A-B} may be used to represent the difference in δ values between two samples, A and B; and (2) workers may use the ε notation, equal to 10 times δ, or parts per 10,000, in cases where differences in isotope composition between samples are extremely small. Caution is advised as there are additional uses of both Δ and ε in isotope science terminology, and some researchers report isotope ratio variations on a per atomic mass unit basis.

Standard reference materials (SRM) provide the basis for inter- and intra-laboratory comparison of isotope ratio measurements. While the mass spectrometers used in different laboratories might produce different absolute values for an SRM, the ratios obtained on samples are comparable through use of the delta scale described above, calculated relative to the SRM. Most standards used in metal stable isotope research are supplied by two sources: the Institute for Reference Materials and Measurements (IRMM, Belgium) and the National Institute for Standards and Technology (NIST, USA). Both organizations have gone to great lengths to produce large quantities of metal isotope SRM, which have been rigorously assessed through round-robin analyses at a number of isotope laboratories. A good SRM must (1) be in abundant supply and available for distribution, (2) be homogeneous in isotopic composition across the supply, (3) have an isotopic composition similar to or bordering the range to be measured in samples, and (4) have an agreed-upon value. Ideally, for a given metal it is preferable to have at least two well-characterized SRM that bracket the range of isotope ratios anticipated in samples. This is rarely the case, and thus most researchers rely on a single SRM that is used widely by other laboratories, together with routinely analyzed internal standards used to monitor consistency of laboratory processing and mass spectrometer function. An important point is that all isotope research should be conducted using widely available SRM in order to allow other laboratories to make a check of accuracy.

Commonly used SRM include: Cr, NIST 979; Fe, IRMM-014; Ni, NIST 986; Cu, NIST 976; Zn, JMC (Johnson Matthey Corporation) 3-0749-L and IRMM-3702; Mo, NIST 3134 (proposed; presently there is no internationally accepted SRM (Wen et al. 2010), but acceptance by the isotope community requires a period of inter-laboratory comparison to validate the

choice); Cd, JMC lot 502552A and NIST 3108; Ag, NIST 978a. Note that there are currently no internationally accepted or proposed SRM for Ti, V, W or Re isotope analysis.

IRMM and NIST typically supply an SRM for metal stable isotope analysis in the form of a purified metal or salt, both of which are easy to dissolve and use routinely. However, some researchers choose to use natural samples as standards, provided those samples meet the criteria listed above. The advantage of using a natural material as an SRM is that it must follow the same chemical purification steps as samples, thus assuring integrity of the chemical procedures. For some metals, there are both natural and synthetic SRM. For Fe stable isotope research, for example, there are two widely available SRM: specific basalts and other igneous rocks, which have been shown to have uniform isotope composition (Beard et al. 2003), and IRMM-014, a synthetic iron standard. The isotope composition of IRMM-014 is now generally considered to be the zero point for the delta scale; on this basis, igneous rocks have on average a $\delta^{56/54}$Fe value of +0.09‰. Unfortunately, Fe lacks SRM having isotope compositions at the light and heavy extremes of the natural spectrum.

10.3 Metal Stable Isotope Fractionation Processes

Isotope fractionation is a set of processes which can divide an isotopically homogeneous pool of an element into multiple fractions that have different complementary isotope compositions. For a given element, observed fractionation may be either mass dependent, scaling approximately as the mass difference between isotope pairs, or mass independent, where one isotope behaves non-systematically with respect to the others. By far, most of the stable isotope variability of the transition and post-transition metals observed to date has resulted from mass dependent processes.

There are two main categories of isotope fractionation processes that affect the transition and post-transition metals, kinetic isotope effects and equilibrium isotope exchange. Kinetic isotope effects result from uni-directional physical and chemical reactions, during which the light isotope is always favored in the product phase. Examples of processes resulting in kinetic isotope effects are evaporation, where light isotopes of metals such as Zn and Cd escape more efficiently at the liquid-air interface than heavy isotopes (e.g. Cloquet et al. 2006a, b), electroplating, where light isotopes of metals from the plating bath are preferentially deposited onto a metal substrate (Kavner et al. 2008), and diffusion (Richter et al. 2009). A special class of kinetic isotope effects arises for elements having differences in specific isotope reaction rates, for example where the odd-numbered isotopes of Hg behave non-systematically with respect to the even-numbered isotopes (Bergquist and Blum 2009; Blum 2011). Isotope equilibrium results from protracted exchange between reaction products and reactants, and at equilibrium the products may be either lighter or heavier than the reactants depending on factors such as metal ion coordination, packing density and bond strength. Examples of processes resulting in isotope equilibrium are the partitioning of relatively heavy Fe into $[Fe(III)(H_2O)_6]_{(aq)}^{3+}$ compared to Fe in coexisting $[Fe(II)(H_2O)_6]_{(aq)}^{2+}$ (Johnson et al. 2002; Welch et al. 2003), and the exchange of Fe between hematite and $[Fe(III)(H_2O)_6]_{(aq)}^{3+}$ leading to identical Fe isotope compositions of these two phases (i.e. no isotope fractionation). This latter example is interesting in that hematite formed initially in synthesis experiments has light Fe relative to coexisting $[Fe(III)(H_2O)_6]_{(aq)}^{3+}$ due to kinetic isotope effects that are subsequently reversed (Skulan et al. 2002). Note that isotope equilibrium between liquid and vapor is likewise possible, as in the case of Hg (Estrade et al. 2009). Excellent discussions of the topic of kinetic and equilibrium isotope fractionation, with examples pertinent to environmental studies, are provided by Clark and Fritz (1997) and Kendall and McDonnell (1998).

Redox reactions, whether strictly inorganic or microbially mediated, consistently result in significant metal stable isotope fractionation, and the reduced species of the metal redox pair is generally isotopically lighter than the oxidized species. Thus, determination of the isotope composition of redox-sensitive metals has proven useful for understanding redox dynamics in paleo- and present-day environmental systems (Severmann and Anbar 2009; Bergquist and Blum 2009; Bullen and Walczyk 2009). On the other hand, the stable isotope composition of metals which occur naturally in only one redox state (e.g. Cd, Zn) but which may fractionate during complexation with redox-sensitive species such as sulfur (Schauble 2003) or with organic matter (Weiss et al. 2007) can likewise be indicative of redox conditions.

Sorption of metals onto the surfaces of minerals in sediments and soils is an environmentally important process that can help to moderate aqueous concentrations of those metals and aid in metal contaminant remediation. Isotope fractionation associated with sorption has been recognized for several of the transition and post-transition metals, but is highly element specific. For example, sorption of isotopically light Mo onto ferromanganese oxides in the oxic oceans is viewed as having been an important isotope fractionation process for Mo over much of Earth history (Wasylenki et al. 2008). In contrast, sorbed Fe(II) on goethite was found to be isotopically heavy compared to $[Fe(II)(H_2O)_6]_{(aq)}^{3+}$ in a series of carefully conducted equilibrium isotope exchange experiments reported by Beard et al. (2010). In the case of Cr, only a negligible isotope fractionation was observed between sorbed and aqueous CrO_4^{2-} and $HCr_2O_7^-$ when either γ-Al_2O_3 or goethite was used as the sorption substrate (Ellis et al. 2004).

Coordination environment of the metals in liquids and solids is an additional important determinant of stable isotope fractionation. In general, for a given system chemistry, higher coordination favors the lighter isotopes. Using the system Fe-O as an example, Fe in hematite (Fe_2O_3) is in octahedral coordination while Fe in magnetite (Fe_3O_4) is in both octahedral and tetrahedral coordination within their respective oxygen framework. The result is a higher average coordination number for Fe in hematite than for Fe in magnetite. As expected, when the two phases coexist at or near equilibrium, Fe in hematite is isotopically lighter than Fe in magnetite by approximately 3‰, as has been reported for coexisting magnetite-hematite pairs in banded iron formations (Johnson et al. 2003). The difference in metal ion coordination environment in coexisting solid-liquid pairs is likewise an important determinant of stable isotope fractionation. For example, theoretical calculations of equilibrium isotope fractionation predict that aqueous Fe(II), which is in octahedral coordination within a "hexaquo" inner hydration sphere ($[Fe(II)(H_2O)_6]_{(aq)}^{2+}$), should be heavier than Fe(II) in coexisting siderite ($FeCO_3$), in which the Fe(II) is in more open octahedral coordination with oxygen of the carbonate groups (Polyakov and Mineev 2000; Schauble 2004). Siderite synthesis experiments confirm the sense of this theoretical fractionation, but observed magnitudes are less than those predicted by theory (Wiesli et al. 2004). Gaining a better understanding of aqueous metal coordination environments and the implications for stable isotope fractionation is one of the major challenges facing the field.

10.4 Applications of Metal Isotopes to Environmental Studies

The transition and post-transition metals are subject to numerous potential isotope fractionation processes over a wide range of environmental conditions, and thus their isotope compositions may provide a unique record of those processes that can be used to understand present-day and paleo-environmental systems. In this section, several examples are given that illustrate the spectrum of environmental issues that can be addressed using these novel stable isotope tracers as process recorders. These include using the isotopes to trace metal contaminants, to identify metal redox processes, to assess plant nutrition cycles and understand processes within plants, and to serve as biomarkers. Although familiarity with the general topic of stable isotope biogeochemistry will help the reader to appreciate the details of the following examples, the introductory material and references in this and other chapters of this book should help those not familiar with the field to better understand why the isotopes are useful and how they work in these specific cases.

10.4.1 Tracing Metal Contaminant Sources and Sinks

The ability of isotope signatures to trace the sources of and processes affecting metal contaminants in groundwater and the atmosphere appears to be receiving the most attention from research groups focused on environmental applications of transition and post-transition metal stable isotopes. Of metal contaminants in groundwater, Cr(VI) is arguably the best understood in terms of isotope systematics based on laboratory experiments, and Cr stable isotopes have been used to study the fate of toxic Cr contamination at numerous field sites. Of the potential metal contaminants in the atmosphere, the most progress has been made in understanding Hg sources and sinks as discussed by Blum (2011), but Zn, Cd and Cr are likewise receiving considerable attention and are discussed here.

10.4.1.1 Tracing Sources of Chromium in Groundwater

As described above in the overview of Cr, reduction of Cr(VI) to Cr(III) is accompanied by a large isotope fractionation. Thus, Cr stable isotopes should be useful for monitoring Cr(VI) reduction in a contaminant plume, for distinguishing between Cr(VI) reduction and advective mixing with Cr-free groundwater (two mechanisms that can decrease Cr(VI) concentrations in the contaminant plume), and potentially for distinguishing between industrial and natural sources of Cr(VI). Sources of Cr used for industrial purposes are likely to be Cr ore deposits and minerals, which consistently have $\delta^{53/52}$Cr values close to 0‰ relative to NIST 979 (Ellis et al. 2002; Schoenberg et al. 2008). In contrast, naturally occurring Cr in groundwater displays a range of values ($\delta^{53/52}$Cr: +1.0 to +5.8‰; Ellis et al. 2002; Izbicki et al. 2008). These values likely reflect fractionation during oxidation of Cr(III) on manganese oxides and fractionation during subsequent reduction of Cr(VI) along groundwater flowpaths following mobilization of the highly soluble Cr(VI) from mineral surface sorption sites. Thus under favorable conditions the natural and industrial Cr pools at a contaminated site may have different $\delta^{53/52}$Cr values.

In order to test the usefulness of Cr stable isotopes for Cr contamination source assessment and process identification at a well-studied contaminated site, a pilot study (CH2MHill 2007) was carried out at the Pacific Gas & Electric (PG&E) Compressor Facility in Hinkley, California, made famous in the movie "Erin Brockovich." From 1952 to 1966, waste water containing Cr(VI) used as an anti-corrosive in the cooling towers leaked into the underlying sandy aquifer, creating a plume of Cr(VI)-contaminated groundwater that by 2000 had migrated nearly 2 miles downgradient from the facility. Groundwater samples from immediately adjacent to the facility, from within the plume, and of regional groundwater collected well away from the plume were analyzed for Cr stable isotope composition (Bullen 2007). As shown in Fig. 10.3, the data plot along a well-organized trend, with near-facility (contaminant) and regional groundwaters defining the extremes, while samples from the contaminant plume plot in between. There is no question that in this particular case, the Cr stable isotopes clearly distinguish the natural and anthropogenic Cr(VI) end-members.

It is less clear how the samples collected from the contaminant plume obtained their compositions in $\delta^{53/52}$Cr-Cr concentration space. For example, as shown in Fig. 10.3 the plume sample data all lie well to the low-Cr side of model trends describing both abiotic and microbially-mediated Cr(VI) reduction (Ellis et al. 2002; Sikora et al. 2008) and well to the high-$\delta^{53/52}$Cr side of the model trend for advective mixing between near-facility contaminant and regional groundwaters, suggesting that neither Cr(VI) reduction nor advective mixing alone is able to account for those compositions. However, some combination of Cr(VI) reduction and advective mixing could account for the plume sample compositions, and a scenario that is at least consistent with field relations (Bullen 2007) is shown in Fig. 10.3. In this scenario, Cr(VI) at the front and margins of the migrating contaminant plume would come into contact with the limited reductant (e.g. ferrous iron, organic matter) in this aquifer, allowing small amounts of Cr(VI) reduction to occur and driving remnant Cr(VI) compositions along the model Cr(VI) reduction vectors. At the same time, advective mixing with regional groundwaters at the plume front and margins would shift remnant Cr(VI) compositions to the low-Cr side of the model Cr(VI) reduction vectors (e.g. along trajectories similar to the lines labeled "plume front Cr(VI) reduction-mixing" in Fig. 10.3). After 40 years of plume migration, one could imagine an array of developed plume front and margin groundwater compositions similar to region "FM" in Fig. 10.3. This range of plume front and margin groundwater compositions would result from specific evolutionary trends that depend on the relative amount of Cr(VI) reduction and advective mixing with regional groundwater that occurred at any particular position along the plume front or margin. In this scenario, the actual plume sample array would develop by advective mixing between these hypothesized plume front and margin groundwaters and the contaminated groundwater

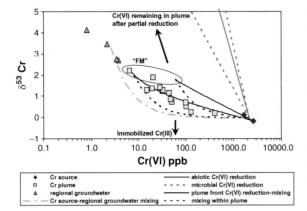

Fig. 10.3 $\delta^{53/52}Cr$ and Cr(VI) concentration data for groundwater at the Hinkley, CA site "Cr source" samples were collected at the PG&E compressor facility, "Cr plume" samples were collected downgradient of the compressor facility within the Cr(VI) plume. "Regional groundwater" samples were collected just upgradient, just down gradient, and well away from the plume. The "Cr source-regional groundwater mixing" curve shows mixing between Cr source and average regional groundwater. The "abiotic Cr(VI) reduction vector" is from Ellis et al. (2002); the "microbial Cr(VI) reduction" vectors are from Sikora et al. (2008), and show the range of fractionation observed in their experiments. The field labeled "FM" represents plume Front and Margin samples produced through combined Cr(VI) reduction and advective mixing with regional groundwater as described in the text. The "mixing within plume" curves show mixing between "FM"-type and "Cr source" groundwaters. The arrow labeled "Cr(VI) remaining in plume following partial reduction" shows the direction that homogeneous Cr(VI) reduction within the plume will move residual Cr(VI) compositions on this diagram. The arrow labeled "immobilized Cr(III)" shows that Cr(III) produced during Cr(VI) reduction will have negative $\delta^{53/52}Cr$. The uncertainty of the data for all samples is less than the size of the symbol

within the plume body itself. Note that extremely careful, discrete depth sampling at the plume front and margins would be necessary to confirm the existence and determine the exact shape of the "FM" region hypothesized in Fig. 10.3, but such an effort would be useful for supporting the validity of this proposed scenario.

In their detailed time-series study of a point-source Cr contamination plume emanating from a Cr plating facility in Berkeley, California, Berna et al. (2010) observed a similar trend for plume groundwater samples in $\delta^{53/52}Cr$-Cr concentration space, such that the plume samples had intermediate compositions between vectors describing Cr(VI) reduction and advective mixing with Cr-free groundwater. Although they concluded that Cr(VI) reduction and advective mixing were important processes determining the composition of the plume groundwater samples, they likewise pointed to the potential importance of heterogeneous Cr(VI) reduction in the aquifer. Their laboratory incubation experiments had demonstrated that Cr(VI) reduction was limited to greenish, fine-grained lenses that are embedded in non-reducing aquifer materials. They proposed that total reduction of small pools of Cr(VI) in the greenish lenses would decrease the amount of Cr(VI) measured in a downgradient sampling well, but would not change the $\delta^{53/52}Cr$ value of the remaining Cr(VI) transported through the non-reducing aquifer materials. As a result, the $\delta^{53/52}Cr$ value of groundwater sampled along the plume is not as high and the $\delta^{53/52}Cr$ value of Cr(III) immobilized on the aquifer sediment is not as low as would be predicted for a system where reduction occurs homogeneously throughout the aquifer. As at Hinkley, careful discrete depth sampling would help to determine the influence of aquifer heterogeneity on Cr(VI) reduction effectiveness and measured $\delta^{53/52}Cr$ values.

10.4.1.2 Sources of Zinc in River Water

As described above in the overview on Zn, there are several processes that may result in Zn stable isotope fractionation including biological incorporation, abiotic adsorption, chemical diffusion and industrial processing. Although Zn is not a highly toxic metal it does provide a useful index of environmental contamination because it is commonly associated with toxic metals such as Cd and Pb. Thus Zn stable isotopes may be useful for distinguishing natural and anthropogenic sources of metals in both hydrologic and atmospheric systems.

An interesting example of the usefulness of Zn isotopes to trace contaminant sources is the work of Chen et al. (2008, 2009) who studied Zn isotope systematics of the Seine River of France. The Seine River is one of the most anthropogenically impacted rivers in Europe, with high relative concentrations of metals (Zn, Cu, Pb, Ni, etc.) in dissolved and suspended loads compared to large rivers of the world (Chen et al. 2009). For their study, these workers sampled both the dissolved and suspended particulate matter (SPM) fractions of Seine River waters along the length of the river and, during both high and low flow periods, in the center of the city of Paris. This sampling strategy allowed these workers to test for a Zn isotope signal of industrial and urban activity, since most industry is concentrated in and downstream of Paris.

Chen et al. (2009) observed an overall negative correlation between $\delta^{66/64}$Zn of SPM and the Zn enrichment factor (EF), which describes the enrichment of Zn in a sample relative to the natural background and is defined as:

$$EF = (Zn/Al)_{sample}/(Zn/Al)_{background}$$

where $(Zn/Al)_{background}$ was estimated using average concentrations of Zn and Al in uncontaminated forest sediments and pre-historical deposits as assessed by Thevenot et al. (2007). Specifically, as shown in Fig. 10.4a, SPM samples with lesser $\delta^{66/64}$Zn and greater Zn EF were those collected within Paris during the low flow periods and from downstream of Paris. Chen et al. (2009) identified two potential anthropogenic contaminant sources, Zn in roof runoff from buildings in Paris and waste water treatment plant effluent, each having low $\delta^{66/64}$Zn and high Zn EF that could explain the low flow Zn isotope signal. In contrast, background Zn from upstream sources was determined to have elevated $\delta^{66/64}$Zn and low Zn EF based on analyses of bedrock from the region (the composition for bedrock granite from the region is shown in Fig. 10.4a) and the compositions of SPM from upstream of Paris.

Chen et al. (2009) observed additional complexity when they considered $\delta^{66/64}$Zn of SPM and $\delta^{66/64}$Zn of dissolved Zn together (Fig. 10.4b). Seine Basin transect and Paris low flow samples defined a linear array that could be explained by mixing between natural and anthropogenic end-members. The natural end-member (Zn_{nat}) was inferred to have $\delta^{66/64}$Zn of SPM similar to that of bedrock granite (+0.33‰) and $\delta^{66/64}$Zn of dissolved Zn similar to that of Cretaceous chalk from the region (+0.90‰), highlighting the fact that Zn in SPM and the dissolved phase can come from different sources. Some Paris high flow samples required an additional component, again having $\delta^{66/64}$Zn of SPM similar to that of bedrock granite, but having $\delta^{66/64}$Zn of dissolved Zn reflecting a mixture of Zn from Cretaceous chalk (the dissolved component of Zn_{nat}) and the dissolved component of Zn from roof runoff (Zn_{rr}). Chen et al. (2009) suggested that urban Zn storm runoff (Zn_{urb}) adds considerable dissolved Zn but only minor SPM associated with zinc roofing materials to the river.

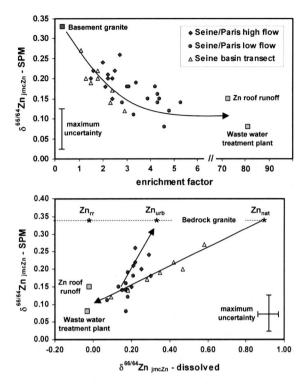

Fig. 10.4 Zn isotope systematics of the Seine River, modified after Chen et al. (2008, 2009). (**a**) $\delta^{66/64}$Zn of suspended particulate matter (SPM) vs. the Zn enrichment factor (defined in text). Curved vector describes mixing between a natural end-member, basement granite, and likely anthropogenic end-members, Zn roof runoff and waste water treatment plant effluent. Note break in enrichment factor scale. (**b**) $\delta^{66/64}$Zn of SPM vs. $\delta^{66/64}$Zn of dissolved Zn. Symbols are the same as in Fig. 10.4a. Zn isotope compositions of basin transect and Paris low flow samples can be explained as mixtures of a natural end-member (Zn_{nat}, having $\delta^{66/64}$Zn of SPM dominated by bedrock granite and $\delta^{66/64}$Zn of dissolved Zn dominated by Cretaceous chalk), and an anthropogenic end-member having Zn isotope composition similar to that of Zn roof runoff and waste water treatment plant effluent. Paris high flow samples require an additional "urban" end-member (Zn_{urb}, having $\delta^{66/64}$Zn of SPM dominated by bedrock granite and $\delta^{66/64}$Zn of dissolved Zn reflecting a mixture of Cretaceous chalk Zn and Zn from roof runoff (Zn_{rr}). Errors are equal to or less than maximum uncertainty bars shown

10.4.1.3 Tracing Sources of Metals in Dust Using a Multi-Tracer Approach

Inhalation of dust particles is potentially a primary delivery mechanism of toxic metals to humans. Moreover, incorporation of deposited dust in soil can allow the metals to become available to plants, providing an additional source of metal loading to crops. There are many possible sources of transition and post-transition metals that could supply the metal content of dust. The major contributors are likely to include metal refining and smelting, coal combustion fly ash and residues, petroleum burning and particularly vehicle emissions, municipal waste incineration, scrap metal deterioration, and wood burning. Early approaches for identifying dust sources relied on tracers such as concentration ratios of metals and surrogates such as strontium isotopes that might be able to tie the dust back to specific geologic or regional sources. While there have been numerous examples of success with this approach, environmental isotope scientists have been exploring whether the stable isotope composition of the transition and post-transition metals might be more directly useful for fingerprinting the sources of the metals in dust.

Of the potential sources for these metals listed above, metal refining and smelting have received the most attention, mainly for Zn and Cd, pointing to the unique physical characteristics of these two post-transition metals. As noted above in the overview on these elements, their boiling points are sufficiently low that they can both be evaporated in the smelter furnace, resulting in a kinetic isotope fractionation that partitions light metal into the vapor that escapes the factory stack. For Zn, light compositions in refinery stack emissions have been documented by Mattielli et al. (2009), while heavy compositions in residual slag compositions have been documented by Sivry et al. (2009). For Cd, the same sense of isotope fractionation was observed for smelter exhaust and slag by Cloquet et al. (2006a). Mercury similarly has a low boiling point within the range of smelter and waste incinerator furnace temperatures, and partitioning of light Hg into stack emissions has been proposed (Estrade et al. 2010). Thus the association of light Zn, Cd, and Hg, along with unfractionated compositions relative to the ore sources for other metals having far greater boiling points (such as Cr and Cu) could implicate a metal refining smelter exhaust stack source for those metals in dust samples.

The Zn and Cd stable isotope compositions of the other potential sources of metals in dust listed above are less constrained, although some reconnaissance values for Zn are available in the literature. Cloquet et al. (2006b) reported $\delta^{66/64}$Zn values for particulates trapped on bus air filters and emitted in urban waste incinerator flue gas from Metz, France, which spanned a small range from +0.04 to +0.19‰ relative to the JMC-Lyon Zn standard. On the other hand, lichens collected from around Metz that had high Pb concentrations and Pb isotope signatures consistent with a leaded gasoline source likewise had high Zn concentrations and negative $\delta^{66/64}$Zn (to −0.2‰), suggesting that light Zn might be associated with road traffic circulation. Sivry et al. (2008) reported $\delta^{66/64}$Zn of +0.72‰ for a coal ash from southwest France, which is consistent with the association of heavy Zn with organic materials recognized by Weiss et al. (2007).

In their study of sources of metals observed in dust samples collected in Paris, France, Widory et al. (2010) have expanded this emerging isotope database for the sources and specifically have measured Zn, Cd, and Cr stable isotope compositions for each of their samples. Important aspects of their data, shown in Fig. 10.5a and b, include: (1) a broad range of $\delta^{114/110}$Cd for particulates in vehicle exhaust; (2) the tendency of organic-associated materials such as fuel oil and coal to have positive $\delta^{66/64}$Zn; and (3) the positive $\delta^{53/52}$Cr values of the coal combustion samples, perhaps reflecting the particular redox history of the coal deposit. These data highlight the potential variability of the anthropogenic signal and show the utility of the multi-isotope approach.

Although the influence of dust from deteriorating metal in scrapyards has yet to be rigorously quantified in terms of either metal flux or stable isotope composition, there is potentially an important isotope signal resulting from electroplating as noted above. Electroplating is a ubiquitous process that can produce an isotopically light pool for all the transition and post-transition metals (e.g. Kavner et al. 2008), and recognition of light isotope compositions for a suite of metals such as Zn, Cd and Cr in a set of dust samples may point to such sources. Cadmium and Zn isotope compositions of an electroplated metal nut and Cr and Zn isotope compositions of an electroplated metal bolt, measured by the author, are included in Fig. 10.5a and b to illustrate the potential signal.

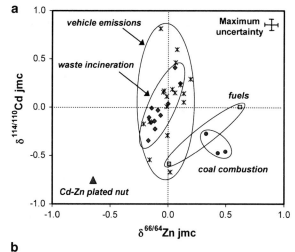

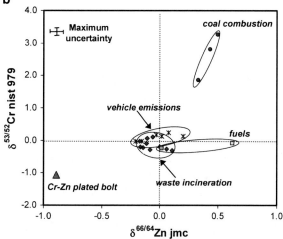

Fig. 10.5 (a) Cd-Zn and (b) Cr-Zn isotope systematics of suspected sources of metals in dust from northern France. Measurements were made on acid leachates of particles collected at municipal waste incinerators and coal combustion plants, from vehicle exhaust pipe filters, on acid leachates of plated nuts and bolts, and on total digests of fuels. Modified from Widory et al. (2010) with additional unpublished data from author

The obvious need to understand the potential influence of deteriorating electroplated metal stockpiles on the isotope composition of atmospheric dust, perhaps mostly to monitor the global transport of potentially toxic metals, makes this an important avenue for future research.

10.4.2 Metal Redox Cycles

The common association of changes in redox status of the transition and post-transition metals with isotope fractionation between oxidized and reduced states points to the potentially powerful application of the isotopes as tracers of redox cycles in natural systems. Isotopes can be used, for example, to distinguish between biological and geochemical processes, such as microbial reduction and inorganic oxidation, and physical processes, such as advective mixing and diffusion, to explain chemical and isotopic composition gradients in hydrologic systems.

Teutsch et al. (2009) used Fe isotopes to study the Fe cycle across the oxic-anoxic boundary of meromictic Lake Nyos, Cameroon, the site of a sudden catastrophic release of CO_2 in 1986 that killed about 1,700 people. To avoid the recurrence of catastrophic outgassing of CO_2 that is supplied by groundwater to the lake bottom, the lake is now artificially degassed by a tube that drains deep water to the surface (Kling et al. 2005). The degassing system spreads deep water with concentrations of dissolved Fe(II) in a high jet over the lake, causing immediate oxidation of the Fe(II) to Fe(III)-oxyhydroxide particulate which then settles into the lake. Concentrations of dissolved Fe(II) in the oxic surface waters are negligible, but sharply increase as particles settle across the oxic-anoxic boundary at depth. Partial reduction of the Fe(III)-rich particles at this boundary releases Fe(II) to solution, which then diffuses upwards into the oxic zone where it is re-oxidized to form Fe(III)-oxyhydroxides which settle back to the oxic-anoxic boundary, and the cycle continues.

Depth profiles sampled in the lake revealed a sharp increase in both dissolved Fe(II) concentrations and $\delta^{57/54}$Fe values in a thin zone across the oxic-anoxic boundary. Using a calibrated one dimensional reaction-transport model that considered the isotopic signatures of dissolved Fe(II) and the settling of Fe(III)-rich particles, the settling fluxes of the particles and the dissolved Fe(II) concentration profiles, Teutsch et al. (2009) showed that the sharp increase in $\delta^{57/54}$Fe values is caused by isotopic fractionation associated with dissimilatory Fe(III) reduction across the oxic-anoxic boundary. They further showed that the continued shift toward even greater $\delta^{57/54}$Fe values below the oxic-anoxic boundary is due to vertical mixing of Fe-rich lake bottom water that has high-$\delta^{57/54}$Fe values, perhaps due to precipitation of siderite. In this study, isotopic analyses coupled to modeling proved invaluable for identifying the processes responsible for the broad range of $\delta^{57/54}$Fe values observed in this lake.

Isotopes can likewise be used to assess paleoredox conditions in water bodies, for example by looking at variations of the redox-sensitive Mo isotope composition of sediments. In oxic settings, Mo isotopes are fractionated during adsorption of Mo to ferromanganese oxides, with light isotopes preferentially associated with the solids. Experimental work under oxidizing conditions has shown that $\delta^{97/95}$Mo of aqueous Mo in seawater is 2.7‰ greater than that of adsorbed Mo on sediment (Barling et al. 2001; Barling and Anbar 2004; Wasylenki et al. 2008). As waters progress from oxic to suboxic to anoxic to euxinic (i.e. sulfidic) conditions, MoO_4^{2-} is converted to MoS_4^{2-} which readily adsorbs to particles allowing aqueous Mo to be more effectively removed to the sediment (Poulson et al. 2006; Neubert et al. 2008). The net result is that the isotopic contrast between aqueous and sorbed Mo is greatest at slightly reducing conditions and low sedimentation rates, but decreases with increasingly reducing and sulfidic conditions due to increasingly more quantitative removal of Mo from solution.

This concept was used by Siebert et al. (2006) to account for differences in the average Mo isotope composition of sediment profiles from continental margin settings. These workers recognized a co-variation between the Mo isotope composition of the sediments and the rate of both authigenic Mo accumulation and organic carbon oxidation and burial under reducing conditions. Malinovsky et al. (2007) tried the same approach to understand paleoredox cycles in lakes from northern Sweden and northwestern Russia, where variations of Mo isotope composition in the sediments could be attributed to both redox status of the water column and isotopically variable input of Mo into the lakes due to Mo isotopic heterogeneity of bedrock in the drainage basins. This latter study points to the need for rigorous source term characterization, particularly as study site size decreases from global to local scale.

An area of applied research where transition and post-transition metal stable isotopes are increasingly being used is to understand sedimentary and low temperature hydrothermal metal redox cycling, where both inorganic and biological redox processes can cause isotope fractionation. For example, Asael et al. (2007) used Cu stable isotopes to understand Cu redox cycling and mass transfer functions during sequential cycles of low temperature alteration of igneous copper porphyries, marine sedimentary diagenesis and epigenetic mobilization of Cu in sandstones at a site of historic Cu mining in southern Israel. As predicted by theory and experiment, they found that $\delta^{65/63}$Cu values of Cu(I)-sulfides in the stratiform sediment-hosted Cu deposits are significantly lower than those of coexisting Cu(II)-carbonates and hydroxides. Coupling the Cu isotope composition of the minerals to a mass balance model revealed that the main Cu reservoir of the ore body is a sandstone-shale sequence and that the importance of Cu-sulfide reservoirs is small, consistent with field observations. Thus most of the Cu transport occurred in relatively oxidized conditions, the knowledge of which provides a novel tool for exploration in the future.

10.4.3 Plant Nutrition and Processes Within Plants

Numerous transition and post-transition metals play important chemical roles in plant nutrition (Marschner 1995). For example, of the metals discussed in this chapter, Fe is essential for chlorophyll development and function, assists in energy transfer functions and redox reactions, is an important constituent of enzymes and participates in nitrogen fixation. Zinc is important for carbohydrate and protein metabolism, pollen formation, and general disease resistance, and is a structural constituent or regulatory co-factor of numerous plant enzymes. Molybdenum, Cu and Ni each play additional important physiological roles in plant nutrition, while excesses can lead to a variety of toxic effects on plant vitality. Using the stable isotopes of these metals as indicators of metal sources and cycling processes, and perhaps to reveal the onset of metal toxicity, is an emerging field in isotope biogeoscience.

As pointed out by von Blanckenburg et al. (2009), the metabolic processes that control the behavior of metals in plants can be envisaged as a gigantic geochemical pump that continuously moves metals between reservoirs. As plants move metals from soils into roots and along the transpiration stream, the metals are cycled through a variety of chemical species via processes that can lead to isotope fractionation. During the past several years, for the metals discussed here stable isotope fractionation in higher plants has been demonstrated for Fe and Zn (Weiss et al. 2005; Guelke and von Blanckenburg 2007; Viers et al. 2007; Moynier et al. 2008; Kiczka et al. 2010). This is an intriguing pair of metals to consider, as the isotope fractionations observed in plants are largely controlled by redox status for Fe, and by complexation chemistry for Zn.

Plants have developed two different strategies for transporting Fe from soils into roots. Roots of dicotyledons and nongraminaceous monocotyledons ("strategy I plants", e.g. pea, bean) release protons into the rhizosphere, promoting dissolution of Fe(III) precipitates, and activating membrane-bound enzymes that reduce chelate-bound Fe(III). The reduced Fe is then transported across the root plasma membrane by unspecific metal transporters (Briat et al. 2007). Partial reduction of an Fe(III)-solid produces aqueous Fe(II) with an $^{56}Fe/^{54}Fe$ ratio that is approximately 1–1.5‰ less than that in the remaining Fe(III)-solid (Johnson et al. 2004). As a consequence, the Fe(II) that crosses the root plasma membrane, and thus the pool of Fe in a strategy I plant, is lighter than the soil Fe pool. In contrast, roots of graminaceous plant species ("strategy II plants", e.g. maize, wheat) can release phytosiderophores into the rhizosphere, which form complexes with Fe(III) (Romheld and Marschner 1986; Kraemer et al. 2006). These Fe(III)-complexes are then transported across the root plasma membrane by a specialized class of transport proteins, but due to lack of reduction, the pool of Fe(III) in a strategy II plant has the same Fe isotope composition as the soil Fe pool.

The results of two studies of Fe isotope variability in plants are shown in Fig. 10.6. In a variety of strategy I plants studied by Guelke and von Blanckenburg (2007), $\delta^{56/54}Fe$ values of tissues decreased from soils to stems, from stems to leaves and from leaves to seeds. In contrast, all tissues in their strategy II plants had similar $\delta^{56/54}Fe$ values, which they attributed to differences in the way that Fe is translocated in the two plant types. These results are consistent with the different Fe uptake and transport processes proposed above for strategy I and strategy II plants. Kiczka et al. (2010), working with strategy I and II plants collected from a recently deglaciated site, identified several reaction points along the translocation stream of both plant types where Fe(III) reduction and thus isotope fractionation might occur, including between the root symplast and apoplast, between the leaf cytoplasm and the xylem, and between the phloem and the leaf cytoplasm, at each step moving lighter Fe along the transpiration stream. However, in contrast to the results of Guelke and von Blanckenburg (2007), Kiczka et al. (2010) observed similar Fe isotope patterns in their strategy I and II plants, with $\delta^{56/54}Fe$ values generally increasing from root stele to stem to leaf. To explain this discrepancy, they pointed to previous studies suggesting that release of siderophores by strategy II plants into the rhizosphere is suppressed under Fe-rich soil conditions similar to

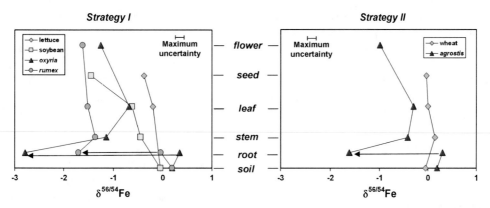

Fig. 10.6 $\delta^{56/54}Fe$ values for a variety of plant tissues. Values for lettuce and soybean (Strategy I) and wheat (Strategy II) are from Guelke and von Blanckenburg (2007), values for *oxyria digya* and *rumex scutatus* (Strategy I) and *agrostis givantea* (Strategy II) are from Kiczka et al. (2010). In each diagram, the arrows between root tissues point from cortex to stele, and demonstrate the large Fe isotope fractionation that occurs across the plasma membrane. Complex Fe isotope variations within individual plants result from a variety of possible isotope fractionation processes which can occur along the transpiration stream

those at their field site (Marschner 1995), and suggesting that strategy I and II plants mobilize and transport Fe similarly under such conditions (Charlson and Shoemaker 2006).

In the case of Zn, dissolution of Fe(III) solids in the soil by protons released from plant roots results in release of Zn that is adsorbed on the Fe(III) solids. Subsequent uptake of Zn by roots depends only on the induction of specific Zn transporters, which belong to the same protein family as the Fe transporters (Kramer et al. 2007). Likewise, the release of phytosiderophores benefits the acquisition of Zn (Suzuki et al. 2006). Apparently the transport of Fe and Zn to plant roots is closely linked. Zn isotope variability within plants appears to be fairly straightforward and similar for strategy I and strategy II plants, based on several laboratory and field based studies (Weiss et al. 2005; Viers et al. 2007; Moynier et al. 2008). Roots tend have the heaviest Zn, probably reflecting the preferential adsorption of heavy Zn onto the root and plaque surfaces. However it should be noted that the possibility of Zn isotope contrast between the root cortex and stele of plants has yet to be explored, as has been done for Fe isotopes (Kiczka et al. 2010). Shoots have similar to lesser $\delta^{66/64}$Zn values than bulk roots, and leaves invariably contain the lightest Zn in the plants. These observations are consistent with light Zn isotopes being transported preferentially during nutrient exchange along the cell walls of the xylem although it is unclear at this stage whether this is a kinetic or an equilibrium process (von Blanckenburg et al. 2009).

There clearly remains much to be done to understand the causes of transition and post-transition metal stable isotope fractionation in plants, and certainly the remaining metals deserve at least a thorough reconnaissance for a variety of plant species and field situations. To date, much of the work has aimed at explaining observed variations of isotope composition in terms of known plant processes. With greater understanding, future work will gradually shift toward using the isotopes to identify as yet unknown or unconstrained processes in plants.

10.4.4 Biosignatures

One of the most enduring hopes and goals of metal stable isotope biogeochemistry is to be able to use the isotopes as biosignatures, by developing either single or multi-tracer tests that could be applied to a natural system to decipher a biological influence on its origins. A recurring theme in metal stable isotope studies of biological systems is that biological activities that result in significant isotope fractionation generally involve preferential processing of lighter isotopes from the metal source pool into and within the biological materials (e.g. Zhu et al. 2002; Bermin et al. 2006; Johnson et al. 2008; Eisenhauer et al. 2009). Indeed, in the early days of the transition and post-transition metal stable isotope data explosion (i.e. the late 1990s), the general thinking was that most metal stable isotope fractionation observed in nature would be due to biological processing (e.g. Beard and Johnson 1999). This thinking led to pioneering attempts to constrain experimentally an Fe isotope biosignature (Beard et al. 1999), but likewise inspired early attempts to demonstrate the importance of inorganic processes that could fractionate Fe isotopes both in the laboratory (e.g. Anbar et al. 2000) and in natural settings (e.g. Bullen et al. 2001). We now realize that ranges of metal stable isotope fractionation caused by inorganic processes can exceed and overlap those produced by biological processing, making development of robust metal stable isotope biosignature tools a challenge but also an exceptional research opportunity.

Given the early and persistent focus on the isotope systematics of Fe, it is not surprising that the greatest progress toward development of a metal stable isotope biosignature has been for Fe, at least for early Earth environments. As stressed by Johnson et al. (2008), the fluxes of mobilized Fe must have been far greater in the Archean and early Proterozoic because of the enormous amount of Fe-rich sediments (e.g. banded iron formations) that were deposited in that time span. The Fe isotope compostion of those sediments is highly variable compared to igneous rocks, implying that the processes responsible for fractionating Fe must have been especially active and variable in extent during that time span. Because microbial dissimilatory Fe reduction (DIR) appears to be the most important process capable of producing large quantities of isotopically light Fe, by an order of magnitude or more compared to inorganic processes, DIR is likely to have been the dominant process affecting Fe during that time span.

In this case, the effectiveness of the Fe isotope biosignature depended on knowledge of the relative sizes of the reduced and oxidized Fe pools during early

Earth history. If we are going to be able to use Fe isotopes to detect a record of DIR in other environments (e.g. in the search for life on Mars and other planetary bodies), we must be able to assess the size of those pools in those environments as well. Confidence in the approach will increase as more terrestrial field sites are studied. A promising example from modern Earth comes from the work of Severmann et al. (2008), who used Fe isotopes to support their contention that DIR coupled to Fe(II) oxidation in oxic shelf sediments can account for the large amount of isotopically light Fe produced on the shelf and shuttled to the euxinic basin of the Black Sea.

Biosignatures based on the other transition and post-transition metal isotope systems have yet to be explored as deeply as that based on Fe, but there are promising signs for success. For example, Borrok et al. (2008) documented a diel cycle for dissolved Zn concentrations and $\delta^{66/64}Zn$ values in Prickly Pear Creek, which drains a former mining district in the Boulder Batholith, Montana, USA. They found that dissolved Zn concentrations and $\delta^{66/64}Zn$ values were inversely correlated, with Zn at low concentrations in the daytime having elevated $\delta^{66/64}Zn$ due to sequestration of light Zn in bed sediments. Biological uptake of light Zn was invoked to explain this relationship, since adsorption of Zn onto amorphous Fe(III) oxides prevalent in bed sediments preferentially concentrates heavy Zn onto sorption sites (Balistrieri et al. 2008). As discussed above, Cameron et al. (2009) demonstrated that methanogenic bacteria cultured in the laboratory preferentially assimilate and retain isotopically light Ni, while a wide range of geologic materials have uniform Ni isotope composition. Nickel isotopes are thus likely to become an important biosignature of methanogenesis. And, as pointed out by von Blanckenburg et al. (2009), plant tissues recycled into soil generally have different metal stable isotope composition than the growth medium, thus the upper soil may retain a time-integrated metal isotope fingerprint of plant activity. New biosignatures are likely to emerge as well understood biological systems are studied.

materials. Arguably, much of the work that has been done to date with the transition and post-transition metal stable isotopes has resulted in a better understanding of the isotope systems themselves rather than the materials that have been measured. Many observed variations in isotope composition have been explained in terms of well understood physical, chemical and biological processes, but isotope analyses have so far rarely revealed some unanticipated or poorly understood set of processes. On the other hand, useful applications pertinent to environmental studies are beginning to emerge: isotopic characteristics of plant homeostasis are being established, multi-tracer approaches for identifying metal sources are being developed, biosignatures are being rigorously constrained through analysis of well understood geological and biological materials, and the isotopic signatures of redox reactions are being characterized. As understanding of metal stable isotope systematics increases, it will become easier to think of novel ways that isotopes can enhance understanding of complex processes in environmental systems.

What should be clear from this chapter is that the field of transition and post-transition metal stable isotope biogeochemistry is really just beginning to mature, and that opportunities are abundant. Major advances over the next several years are likely to include: improving the routine precision and reproducibility of total procedural replicate measurements of samples to the few hundredths of a per mil level; developing robust multi-tracer approaches to identify metal sources and pathways for forensic purposes; looking for the unexpected isotope fractionation of metals such as Ti and Ni in environmental systems; expanding the list of robust biosignatures, including the fingerprint of anthropogenic activities; exploring the metal stable isotope signals of global climate change. This is only a partial list, but one where success is likely to be realized through well-crafted interdisciplinary efforts, careful selection of field sites, and clever laboratory- and field-based experiments.

10.5 Future Directions

As pointed out by Severmann and Anbar (2009), the field of metal stable isotope biogeochemistry emerged primarily as an analytical pursuit driven by advances in mass spectrometry applied to interesting geological

References

Albarède F, Beard B (2004) Analytical methods for non-traditional isotopes. In: Johnson CM, Beard BL, Albarède F (eds) Geochemistry of non-traditional stable isotopes.

Reviews of Mineralogy and Geochemistry, vol 55. Mineralogical Society of America, Washington, pp 113–152

Anbar AD (2004) Molybdenum stable isotopes: observations, interpretations and directions. Rev Mineral Geochem 55: 425–454

Anbar AD, Roe JE, Barling J, Nealson KH (2000) Nonbiological fractionation of iron isotopes. Science 288:126–128

Anbar AD, Jarzecki AA, Spiro TG (2005) Theoretical investigation of iron isotope fractionation between $Fe(H_2O)_6^{3+}$ and $Fe(H_2O)_6^{2+}$: implications for iron stable isotope geochemistry. Geochim Cosmochim Acta 69:825–837

Arnold GL, Anbar AD, Barling J, Lyons TW (2004) Molybdenum isotope evidence for widespread anoxia in mid-Proterozoic oceans. Science 304:87–90

Asael D, Matthews Bar-Matthews M, Halicz L (2007) Copper isotope fractionation in sedimentary copper mineralization (Timna Valley, Israel). Chem Geol 243:238–254

Balci N, Bullen TD, Witte-Lien K, Shanks WC, Motelica M, Mandernack KW (2006) Iron isotope fractionation during microbially stimulated Fe(II) oxidation and Fe(II) precipitation. Geochim Cosmochim Acta 70:622–639

Balistrieri LS, Borrok DM, Wanty RB, Ridley WI (2008) Fractionation of Cu and Zn isotopes during adsorption onto amorphous Fe(III) oxides: experimental mixing of acid rock drainage and pristine river water. Geochim Cosmochim Acta 72:311–328

Barling J, Anbar AD (2004) Molybdenum isotope fractionation during adsorption by manganese oxides. Earth Planet Sci Lett 217:315–329

Barling J, Arnold GL, Anbar AD (2001) Natural mass-dependent variations in the isotopic composition of molybdenum. Earth Planet Sci Lett 193:447–457

Beard BL, Johnson CM (1999) High precision iron isotope measurements of terrestrial and lunar materials. Geochim Cosmochim Acta 63:1653–1660

Beard BL, Johnson CM, Cox L, Sun H, Nealson KH, Aguilar C (1999) Iron isotope biosignatures. Science 285:1889–1892

Beard BL, Johnson CM, Skulan JL, Nealson KH, Cox L, Sun H (2003) Application of Fe isotopes to tracing the geochemical and biological cycling of Fe. Chem Geol 195:87–117

Beard BL, Handler RM, Scherer MM, Wu L, Czaja AD, Heimann A, Johnson CM (2010) Iron isotope fractionation between aqueous ferrous iron and goethite. Earth Planet Sci Lett 295:241–250

Bergquist BA, Blum JD (2007) Mass-dependent and –independent fractionation of Hg isotopes by photoreduction in aquatic systems. Science 318:417–420

Bergquist BA, Blum JD (2009) The odds and evens of mercury isotopes: applications of mass-dependent and mass-independent isotope fractionation. Elements 5:353–357

Bermin J, Vance D, Archer C, Statham PJ (2006) The determination of the isotopic composition of Cu and Zn in seawater. Chem Geol 226:280–297

Berna EC, Johnson TM, Makdisi RS, Basu A (2010) Cr stable isotopes as indicators of Cr(VI) reduction in groundwater: a detailed time-series study of a point-source plume. Environ Sci Technol 44:1043–1048

Blum JD (2011) Applications of stable mercury isotopes to biogeochemistry. In: Baskaran M (ed) Handbook of Environmental Isotope Geochemistry, Advances in Isotope Geochemistry, DOI:10.1007/978-3-642-10637-8_10, © Springer-Verlag Berlin Heidelberg 2011

Blum JD, Bergquist BA (2007) Reporting of variations in the natural isotopic composition of mercury. Anal Bioanal Chem 388:353–359

Borrok DM, Wanty RB, Ridley WI, Wolf R, Lamothe PJ, Adams M (2007) Separation of copper, iron and zinc from complex aqueous solutions for isotopic measurement. Chem Geol 242:400–414

Borrok DM, Nimick DA, Wanty RB, Ridley WI (2008) Isotopic variations of dissolved copper and zinc in stream waters affected by historical mining. Geochim Cosmochim Acta 72:329–344

Briat JF, Curie C, Gaymard F (2007) Iron utilization and metabolism in plants. Curr Opin Plant Biol 10:276–282

Bullen TD (2007) Chromium stable isotopes as a new tool for forensic hydrology at sites contaminated with anthropogenic chromium. In: Bullen TD, Wang Y (eds) Water-rock interaction: proceedings of the 12th international symposium on water-rock interaction, vol 1. Taylor & Francis, London, pp 699–702

Bullen TD, Amundson R (2010) Interpreting Ca and Fe stable isotope signals in carbonates: a new perspective. Proceedings of the 13th International Symposium on Water-Rock Interaction (WRI-13), Guanajuato, Mexico. Taylor & Francis, London

Bullen TD, Eisenhauer AE (2009) Metal stable isotopes in low-temperature systems: a primer. Elements 5:349–352

Bullen TD, Walczyk T (2009) Environmental and biomedical applications of natural metal stable isotope variations. Elements 5:381–385

Bullen TD, White AF, Childs CW, Vivit DV, Schulz MS (2001) Demonstration of significant abiotic iron isotope fractionation in nature. Geology 29:699–702

Cameron V, Vance D, Archer C, House C (2009) A biomarker based on the stable isotopes of nickel. Proc Natl Acad Sci 106:10944–10948

CH2MHill (2007) Groundwater background study report, Hinkley Compressor Station, Hinkley, California. Internal Report to Pacific Gas & Electric Corporation, February 2007

Charlson DV, Shoemaker RC (2006) Evolution of iron acquisition in higher plants. J Plant Nutr 29:1109–1125

Chen JB, Gaillardet J, Louvat P (2008) Zinc isotopes in the Seine River waters, France: a probe of anthropogenic contamination. Environ Sci Technol 42:6494–6501

Chen JB, Gaillardet J, Louvat P, Huon S (2009) Zinc isotopes in the suspended load of the Seine River, France; isotopic variations and source determination. Geochim Cosmochim Acta 73:4060–4076

Clark I, Fritz P (1997) Environmental isotopes in hydrogeology. Lewis Publishers, Boca Raton, 328p

Cloquet C, Carignan J, Libourel G, Sterckeman T, Perdrix E (2006a) Tracing source pollution in soils using cadmium and lead isotopes. Environ Sci Technol 40:2525–2530

Cloquet C, Carignan J, Libourel G (2006b) Isotopic composition of Zn and Pb atmospheric depositions in an urban/periurban area of northeastern France. Environ Sci Technol 40: 6594–6600

Cloquet C, Carignan J, Lehmann MF, Vanhaecke F (2008) Variation in the isotopic composition of zinc in the natural environment and the use of zinc isotopes in biogeosciences: a review. Anal Bioanal Chem 390:451–463

Das J, Pobi M (1990) Separation of titanium, iron and aluminium on a chelating resin with benzoylphenylhydroxylamine group and application to bauxite and clay. Fresenius' J Anal Chem 336:578–581

Eisenhauer A, Kisakurek B, Bohm F (2009) Marine calcification: an alkali earth metal isotope perspective. Elements 5:365–368

Ellis A, Johnson TM, Bullen TD (2002) Chromium isotopes and the fate of hexavalent chromium in the environment. Science 295:2060–2062

Ellis A, Johnson TM, Bullen TD (2004) Using chromium stable isotope ratios to quantify Cr(VI) reduction: lack of sorption effects. Environ Sci Technol 38:3604–3607

Estrade N, Carignan J, Sonke JE, Donard OFX (2009) Mercury isotope fractionation during liquid-vapor evaporation experiments. Geochim Cosmochim Acta 73:2693–2711

Estrade N, Carignan J, Donard OFX (2010) Isotope tracing of atmospheric mercury sources in an urban area of northeastern France. Environ Sci Technol 44:6062–6067

Fantle MS, Bullen TD (2009) Essentials of iron, chromium and calcium isotope analysis of natural materials by thermal ionization mass spectrometry. Chem Geol 258:50–64

Geulke M, von Blanckenburg F (2007) Fractionation of stable iron isotopes in higher plants. Environ Sci Technol 41:1896–1901

Irisawa K, Hirata T (2006) Tungsten isotopic analysis of six geochemical reference materials using multiple collector–ICP-mass spectrometry coupled with rhenium-external correction technique. Geochim Cosmochim Acta 70:A279

Izbicki JA, Ball JW, Bullen TD, Sutley SJ (2008) Chromium, chromium isotopes and selected trace elements, western Mojave Desert, USA. Appl Geochim 23:1325–1352

Johnson TM (2011) Stable isotopes of Cr and Se as tracers of redox processes in earth surface environments. In: Baskaran M (ed) Handbook of environmental isotope geochemistry. Springer, Heidelberg

Johnson CM, Beard BL (1999) Correction of instrumentally produced mass fractionation during isotopic analysis of Fe by thermal ionization mass spectrometry. Int J Mass Spectrom 193:87–99

Johnson TM, Bullen TD, Zawislanski PT (2000) Selenium stable isotope ratios as indicators of sources and cycling of selenium: results from the northern reach of San Francisco Bay. Environ Sci Technol 34:2075–2079

Johnson CM, Skulan JL, Beard BL, Sun H, Nealson KH, Braterman PS (2002) Isotopic fractionation between Fe(III) and Fe(II) in aqueous solutions. Earth Planet Sci Lett 195:141–153

Johnson CM, Beard BL, Beukes NJ, Klein C, O'Leary JM (2003) Ancient geochemical cycling in the Earth as inferred from Fe isotope studies of banded iron formations from the Transvaal Craton. Contrib Mineralog Petrol 144:523–547

Johnson CM, Beard BL, Roden EE, Newman DK, Nealson KH (2004) Isotopic constraints on biogeochemical cycling of Fe. In: Johnson CM, Beard BL, Albarede F (eds) Geochemistry of non-traditional stable isotopes. Mineralogical Society of America Reviews in Mineralogy & Geochemistry 55:359–408

Johnson CM, Beard BL, Roden EE (2008) The iron isotope fingerprints of redox and biogeochemical cycling in modern and ancient Earth. Annu Rev Earth Planet Sci 36:457–493

Kavner A, John SG, Sass S, Boyle EA (2008) Redox-driven stable isotope fractionation in transition metals: Application to Zn electroplating. Geochim Cosmochim Acta 72:1731–1741

Kendall C, McDonnell JJ (1998) Isotope tracers in catchment hydrology. Elsevier, Amsterdam, 839p

Kiczka M, Wiederhold JG, Kraemer SM, Bourdon B, Kretzschmar R (2010) Iron isotope fractionation during Fe uptake and translocation in alpine plants. Environ Sci Technol 44:6144–6150

Kling GW, Evans WC, Tanyileke G, Kusakabe M, Ohba T, Yoshida Y, Hell JV (2005) Degassing Lakes Nyos and Nonoun: defusing certain disaster. Proc Natl Acad Sci USA 102:14185–14190

Kraemer SM, Crowley DE, Kretzschmar R (2006) Geochemical aspects of phytosiderophore-promoted iron acquisition by plants. Adv Agron 91:1–46

Kramer U, Talke IN, Hanikenne M (2007) Transition metal transport. FEBS Lett 581:2263–2272

Larson PB, Maher K, Ramos FC, Chang Z, Gaspar M, Meinert LD (2003) Copper isotope ratios in magmatic and hydrothermal ore-forming environments. Chem Geol 201:337–350

Liu Y, Huang M, Masuda A, Inoue M (1998) High-precision determination of osmium and rhenium isotope ratios by in situ oxygen isotope ratio correction using negative thermal ionization mass spectrometry. Int J Mass Spectrom Ion Processes 173:163–175

Longerich HP, Fryer BJ, Strong DF (1987) Determination of lead isotope ratios by inductively coupled plasma-mass spectrometry (ICP-MS). Spectrochimjjica Acta 42B:39–48

Luo Y, Dabek-Zlotorzynska E, Celo V, Muir DCG, Yang L (2010) Accurate and precise determination of silver isotope fractionation in environmental samples by multi-collector-ICPMS. Anal Chem 82:3922–3928

Malinovsky D, Hammarlund D, Ilyashuk B, Martinsson O, Gelting J (2007) Variations in the isotopic composition of molybdenum in freshwater lake systems. Chem Geol 236:181–198

Maréchal CN, Télouk P, Albarède F (1999) Precise analysis of copper and zinc isotopic compositions by plasma-source mass spectrometry. Chem Geol 156:251–273

Marschner H (1995) Mineral nutrition of higher plants, 2nd edn. Academic, London, p 889

Martin JH and Fitzwater SE (1988) Iron-deficiency limits phytoplankton growth in the Northeast Pacific Subarctic. Nature 331:341–343

Mathur R, Ruiz J, Titley S, Liermann L, Buss H, Brantley SL (2005) Cu isotopic fractionation in the supergene environment with and without bacteria. Geochim Cosmochim Acta 69:5233–5246

Mattielli N, Petit JCJ, Deboudt K, Flament P, Perdrix E, Taillez A, Rimetz-Planchon J, Weis D (2009) Zn isotope study of atmospheric emissions and dry depositions within a 5 km radius of a Pb–Zn refinery. Atmos Environ 43:1265–1272

Moynier F, Pichat S, Pons M-L, Fike D, Balter V, Albarède F (2008) Isotopic fractionation and transport mechanisms of Zn in plants. Chem Geol 267:125–130

Neubert N, Nägler TF, Böttcher ME (2008) Sulfidity controls molybdenum isotope fractionation into euxinic sediments: evidence from the modern Black Sea. Geology 36:775–778

Pelly IZ, Lipschutz ME, Balsiger H (1970) Vanadium isotopic composition and contents in chondrites. Geochim Cosmochim Acta 34:1033–1036

Polyakov VB, Mineev SD (2000) The use of Mossbauer spectroscopy in stable isotope geochemistry. Geochim Cosmochim Acta 64:849–865

Poulson RL, Siebert C, McManus J, Berelson WM (2006) Authigenic molybdenum isotope signatures in marine sediments. Geology 34:617–620

Rehkämper M, Wombacher F, Aggarwal JK (2004) Stable isotope analysis by multiple collector ICP-MS. In: de Groot PA (ed) Handbook of stable isotope analytical techniques. Elsevier, Amsterdam, pp 692–725

Rehkämper M, Wombacher F, Horner TJ, Xue Z (2011) Natural and anthropogenic Cd isotope variations. In: Baskaran M (ed) Handbook of Environmental Isotope Geochemistry, Advances in Isotope Geochemistry, DOI 10.1007/978-3-642-10637-8_10, © Springer-Verlag Berlin Heidelberg 2011

Richter FM, Dauphas N, Teng F-Z (2009) Non-traditional fractionation of non-traditional isotopes: Evaporation, chemical diffusion and Soret diffusion. Chem Geol 258:92–103

Ripperger S, Rehkämper M (2007) Precise determination of cadmium isotope fractionation in seawater by double-spike MC-ICPMS. Geochim Cosmochim Acta 71:631–642

Romheld V, Marschner H (1986) Evidence for a specific uptake system for iron phytosiderophores in roots of grasses. Plant Physiol 80:175–180

Rudge JF, Reynolds BC, Bourdon B (2009) The double spike toolbox. Chem Geol 265:420–431

Sahoo YV, Nakai S, Ali A (2006) Modified ion exchange separation for tungsten isotopic measurements from kimberlite samples using multi-collector inductively coupled plasma mass spectrometry. Analyst 131:434–439

Schauble EA (2003) Modeling zinc isotope fractionations. EOS Trans AGU 84:F232

Schauble EA (2004) Applying stable isotope fractionation theory to new systems. In: Johnson CM, Beard BL, Albarède F (eds) Geochemistry of non-traditional stable isotopes, reviews in mineralogy and geochemistry, vol 55. Mineralogical Society of America and Geochemical Society, Washington, D.C., pp 65–111

Schmitt A-D, Galer SJG, Abouchami W (2009a) High-precision cadmium stable isotope measurements by double spike thermal ionization mass spectrometry. J Anal At Spectrom 24:1079–1088

Schmitt A-D, Galer SJG, Abouchami W (2009b) Mass-dependent cadmium isotopic variations in nature with emphasis on the marine environment. Earth Planet Sci Lett 277:262–272

Schoenberg R, Zink S, Staubwasser M, von Blanckenburg F (2008) The stable Cr isotope inventory of solid Earth reservoirs determined by double spike MC-ICP-MS. Chem Geol 249:294–306

Schroeder WH, Munthe J (1998) Atmospheric mercury – an overview. Atmos Environ 32:809–822

Severmann S, Anbar AD (2009) Reconstructing paleoredox conditions through a multitracer approach: the key to the past is the present. Elements 5:359–364

Severmann S, Lyons TW, Anbar A, McManus J, Gordon G (2008) Modern iron isotope perspective on the benthic iron shuttle and the redox evolution of ancient oceans. Geology 36:487–490

Shiel AE, Weiss D, Orians KJ (2010) Evaluation of zinc, cadmium and lead isotope fractionation during smelting and refining. Sci Total Environ 408:2357–2368

Siebert C, Nägler TF, Kramers JD (2001) Determination of molybdenum isotope fractionation by double-spike multicollector inductively coupled plasma mass spectrometry. Geochem Geophys, Geosyst 2:1032

Siebert C, Nagler TF, von Blanckenburg F, Kramers JD (2003) Molybdenum isotope records as a potential new proxy for paleoceanography. Earth Planet Sci Lett 211:159–171

Siebert C, McManus J, Bice A, Poulson R, Berelson WM (2006) Molybdenum isotope signatures in continental margin marine sediments. Earth Planet Sci Lett 241:723–733

Sikora ER, Johnson TM, Bullen TD (2008) Microbial mass-dependent fractionation of chromium isotopes. Geochim Cosmochim Acta 72:3631–3641

Sivry Y, Riotte J, Sonke JE, Audry S, Schäfer J, Viers J, Blanc G, Freydier R, Dupré B (2008) Zn isotopes as tracers of anthropogenic pollution from Zn-ore smelters: the Riou Mort–Lot River system. Chem Geol 255:295–304

Skulan JL, Beard BL, Johnson CM (2002) Kinetic and equilibrium Fe isotope fractionation between aqueous Fe(III) and hematite. Geochim Cosmochim Acta 66:2995–3015

Suzuki M et al (2006) Biosynthesis and secretion of mugineic acid family phytosiderophores in zinc-deficient barley. Plant J 48:85–97

Teutsch N, Schmid M, Muller B, Halliday AN, Burgmann H, Wehrli B (2009) Large iron isotope fractionation at the oxic-anoxic boundary in Lake Nyos. Earth Planet Sci Lett 285:52–60

Thevenot DR, Moilleron R, Lestel L, Gromaire MC, Rocher V, Cambier P, Bonte P, Colin JL, de Ponteves C, Maybeck M (2007) Critical budget of metal sources and pathways in the Seine River basin (1994–2003) for Cd, Cr, Cu, Hg, Ni, Pb and Zn. Sci Total Environ 375:180–203

Vance D, Archer C, Bermin J, Perkins J, Statham PJ, Lohan MC, Elwood MJ, Mills RA (2008) The copper isotope geochemistry of rivers and the oceans. Earth Planet Sci Lett 274:204–213

Viers J, Oliva P, Nonell A, Gélabert A, Sonke JE, Freydler R, Gainville R, Dupré B (2007) Evidence of Zn isotopic fractionation in a soil plant system of a pristine tropical watershed (Nsimi, Cameroon). Chem Geol 239:124–137

Von Blanckenburg F, von Wirén N, Guelke M, Weiss DJ, Bullen TD (2009) Fractionation of metal stable isotopes by higher plants. Elements 5:375–380

Wasylenki LE, Rolfe BA, Weeks CL, Spiro TG, Anbar AD (2008) Experimental investigation of the effects of temperature and ionic strength on Mo isotope fractionation during adsorption to manganese oxides. Geochim Cosmochim Acta 72:5997–6005

Weber KA, Achenbach LA, Coates JD (2006) Microorganisms pumping iron: anaerobic microbial oxidation and reduction. Nat Rev Microbiol 4:752–764

Weiss DJ, Mason TFD, Zhao FJ, Kirk GJD, Coles BJ, Horstwood MSA (2005) Isotopic discrimination of zinc in higher plants. New Phytol 165:703–710

Weiss DJ, Rausch N, Mason TFD, Coles BJ, Wilkinson JJ, Ukonmaanaho L, Arnold T, Nieminen TM (2007) Atmospheric deposition and isotope biogeochemistry of zinc in ombrotrophic peat. Geochim Cosmochim Acta 71: 3498–3517

Welch SA, Beard BL, Johnson CM, Braterman PS (2003) Kinetic and equilibrium Fe isotope fractionation between aqueous Fe(II) and Fe(III). Geochim Cosmochim Acta 67:4231–4250

Wen H, Carignan J, Cloquet C, Zhu X, Zhang Y (2010) Isotopic delta values of molybdenum standard reference and prepared solutions measured by MC-ICP-MS: proposition for delta zero and secondary references. J Anal At Spectrom 25: 716–721

Widory D, Petelet-Giraud E, LeBihan O, LeMoullec Y, Quetel C, Snell J, Van Bocxstaele M, Hure A, Canard E, Joos E, Forti L, Bullen T, Johnson T, Fiani E (2010) Metals in atmospheric particles: can isotopes help discriminate potential sources? Atmospheric Pollution, Special Edition:75–82

Wieser ME, Schwieters JB (2005) The development of multiple collector mass spectrometry for isotope ratio measurements. Int J Mass Spectrom 242:97–115

Wiesli RA, Beard BL, Johnson CM (2004) Experimental determination of Fe isotope fractionation between aqueous Fe(II), siderite and "green rust" in abiotic systems. Chem Geol 211:343–362

Wombacher F, Rehkamper M, Mezger K, Munker C (2003) Stable isotope compositions of cadmium in geological materials and meteorites determined by multiple-collector ICPMS. Geochim Cosmochim Acta 67:4639–4654

Zhu XK, Guo Y, Williams RJP, O'Nions RK, Matthews A, Belshaw NS, Canters GW, deWaal EC, Weser U, Burgess BK, Salvato B (2002) Mass fractionation processes of transition metal isotopes. Earth Planet Sci Lett 200:47–62

Chapter 11
Applications of Osmium and Iridium as Biogeochemical Tracers in the Environment

Mukul Sharma

Abstract Osmium (Os) and Iridium (Ir) and are among the rarest elements on the surface of the earth and ones whose applications in modern industry are quite limited. However, their environmental burden has been increasing as they occur in nature with other platinum group elements, which have a wide variety of industrial, chemical, electrical and pharmaceutical applications. This review traces the development of the analytical techniques used to precisely measure Os and Ir concentrations and Os isotope composition in the environmental samples from their roots in geochemical and cosmochemical investigations. We then examine the distribution of Os and Ir in natural samples and review recent literature applying these elements as biogeochemical tracers. The primary environmental applications of Os and Ir arise from the fact that these elements are extremely rare on the surface of the earth and their introduction into the environment leads to an increase in their concentration in surface materials. In addition, a unique isotope fingerprint is present for Os introduced into the environment in that it comes from ores mined primarily in South Africa and Russia. In particular, we examine studies where (a) Os isotopes have been utilized to track the dispersal of platinum group elements from automobile catalysts and also to assess dispersal of Os itself and (b) Ir has been introduced as an intentional tracer to evaluate soot contribution from burning of fossil fuels.

11.1 Introduction

The intent of this paper is to outline advances over the last 20 years in the detection and accurate measurements of Os and Ir concentrations and Os isotopes in natural samples and their applications to environmental issues. Development of environmental applications for these elements has followed cosmo- and geo-chemical applications. This paper gives a brief outline of the studies done in these collateral areas of research and then presents a review of the analytical techniques used to measure Os and Ir. Following this we discuss distribution of Os and Ir in natural samples and their environmental applications. Os and Ir are highly siderophile or chalcophile, i.e., they strongly prefer metal or sulfide phases over silicate minerals. During the formation of the earth's core these elements were extracted nearly quantitatively from the mantle into the core. Moreover, during mantle melting and magma formation, both Os and Ir are highly compatible. Consequently, while Os and Ir are highly enriched in meteorites/cosmic dust (average ~600 ng/g; Os/Ir ~ 1, Barker and Anders 1968; Chen et al. 2005; Walker et al. 2002a) they are depleted in the upper continental crust (average ~30 pg/g; Os/Ir ~ 1, Esser and Turekian 1993b; Peucker-Ehrenbrink and Jahn 2001). They are among the rarest elements in seawater with concentrations of 6–14 and 0.08–3.6 pg/kg for Os and Ir, respectively.

Osmium has seven naturally occurring isotopes: ^{184}Os (0.02%), ^{186}Os (1.56%), ^{187}Os (1.6%), ^{188}Os (13.3%), ^{190}Os (26.4%) and ^{192}Os (41.0%). The molar abundances given are nominal as ^{187}Os and ^{186}Os are radiogenic and produced from the decay of ^{187}Re (β-decay, $t_{1/2}$ = 41.2 billion years) and ^{190}Pt (α-decay, $t_{1/2}$ = 494.4 billion years), respectively.

M. Sharma (✉)
Radiogenic Isotope Geochemistry Laboratory, Department of Earth Sciences, Dartmouth College, 6105 Fairchild Hall, Hanover, NH 03755, USA
e-mail: Mukul.Sharma@Dartmouth.edu

Early interest in the ^{187}Re-^{187}Os isotope system stemmed from using it as a geochronometer for dating of Re-bearing ore-minerals (Hintenberger et al. 1954) and iron meteorites (Herr et al. 1961) and as a cosmochronometer in evaluating the age of the Universe (Clayton 1964; Truran 1998). The ^{187}Re-^{187}Os isotope system is unique among long-lived geochronometers as both parent and daughter are siderophile-chalcophile. This has permitted dating of geological samples that could not be dated using traditional isotope systems (^{87}Rb-^{87}Sr, ^{147}Sm-^{143}Nd, U-Th-Pb) that involve lithophile elements. The ^{187}Re-^{187}Os isotope system has been successfully utilized in numerous studies dating meteorites and terrestrial rocks and mineral deposits (e.g., Shen et al. 1996; Stein et al. 2004; Yang et al. 2009; Walker et al. 2002a). It has also been used in assessing the evolution of the planetary mantles (see e.g., Walker 2009). The environmental applications of the ^{187}Re-^{187}Os isotope system are based on using the ^{187}Os/^{188}Os ratio as a tracer of mantle derived Os and other platinum group elements (PGE). The fundamental systematics for such applications has been worked out in studies that have attempted to assess meteorite impacts and changes in continental weathering. It is based on the large Os isotope ratio difference between meteorites (^{187}Os/^{188}Os = 0.13) and continental crustal rocks (^{187}Os/^{188}Os = 1.3), which makes Os isotopes a highly sensitive tracer of extraterrestrial and/or continental material.

A number of investigations have utilized Os isotopes to trace bolide impacts throughout geological time (Gordon et al. 2009; Koeberl et al. 2004; Lee et al. 2003; Luck and Turekian 1983; Meisel et al. 1995; Paquay et al. 2009; Peucker-Ehrenbrink et al. 1995; Schmitz et al. 2001) and to evaluate the size of the impactor (Paquay et al. 2008) and also to assess the temporal variations in the cosmic dust flux (Dalai and Ravizza 2006; Dalai et al. 2006; Esser and Turekian 1988; Marcantonio et al. 1999; Peucker-Ehrenbrink 1996; Peucker-Ehrenbrink and Ravizza 2000a). Another set of studies has utilized Os isotopes to determine the extent to which the meteorite component is present in target rocks (Frei and Frei 2002; Gelinas et al. 2004; Koeberl et al. 1994a, b, c; Lee et al. 2006; McDonald et al. 2007) or in distal ejecta (e.g., tektites) from the impact (Koeberl and Shirey 1993, 1997).

Variations in the ^{187}Os/^{188}Os ratios in the oceans have been used to infer changes in continental weathering resulting from the uplift of mountain ranges (Pegram et al. 1992; Pegram and Turekian 1999; Peucker-Ehrenbrink 1996; Sharma et al. 1999) and waxing and waning of ice ages (Burton et al. 2010; Oxburgh 1998; Oxburgh et al. 2007). They have also been used to trace flood basalt volcanism (Ravizza and Peucker-Ehrenbrink 2003; Robinson et al. 2009).

The environmental studies of ^{187}Re-^{187}Os isotope system began with the discovery of anthropogenic Os in estuarine sediments from Long Island Sound that was inferred to be derived from effluents from hospitals where OsO$_4$ is used as a fixative for electron microscopy (Esser and Turekian 1993a; Williams and Turekian 2002). High concentrations of anthropogenic Os were also found in sewage discharge from Boston suggesting that Os isotopes could be used to assess municipal sewage dispersal from major cities adjoining estuaries (Ravizza and Bothner 1996). Recent environmental studies have focused on the role of automobile catalytic converters and industrial refining of ores as sources of anthropogenic osmium (Chen et al. 2009; Poirier and Gariepy 2005; Rauch et al. 2004, 2005, 2006, 2010; Rodushkin et al. 2007a, b). Pre-historic refining of sulfide ores in the Iberian Peninsula, Spain has also been traced using Os isotopes (Rauch et al. 2010).

The production of ^{186}Os over geologic time is minor as ^{190}Pt is a minor isotope of Pt (0.013%) with an extremely long half-life (Walker et al. 1997). The ^{190}Pt-^{186}Os system has been successfully utilized to date meteorites and old terrestrial rocks (e.g., Cook et al. 2004; Puchtel et al. 2009). It has also been used to study core-mantle interaction (see Walker 2009 and references therein). Variations of the order of a few tens of parts per million in ^{186}Os/^{188}Os ratios have been observed between different lithologies (ultramafic rocks and black shales) that upon erosion contribute Os to the oceans (McDaniel et al. 2004). The ^{190}Pt-^{186}Os system has not been developed for environmental applications so far.

Iridium has two naturally occurring stable isotopes: ^{191}Ir (37.3%) and ^{193}Ir (62.7%). So far, no variations in ^{193}Ir/^{191}Ir ratios have been documented in nature; the expected variations in this ratio are about 1‰ or less (E. Schauble, personal communication). Due to its low abundance on the earth's surface and high concentration in meteorites Ir concentrations in marine sediments and ice cores have been utilized to assess bolide impacts (Alvarez et al. 1980) and long term accretion of cosmic dust (Gabrielli et al. 2004a; Karner et al. 2003; Kyte and Wasson 1986). It has also been used to examine fall out from volcanic eruptions (Gabrielli et al. 2008). In environmental studies Ir has been introduced as an intentional tracer to track soot emissions from coal fired plants, public buses and diesel sanitation truck fleets (Heller-Zeisler et al. 2000; Suarez et al. 1998; Wu et al. 1998). Introduction of Ir in latest automobile catalytic converters has led to a study of its abundance in urban areas (Iavicoli et al. 2008).

and Meisel (2002). Here we will concentrate only on the determination of Os and Ir. Table 11.1 provides a comparison of the detection limits of various procedures used in estimating the concentrations of Os and Ir in sediments, rocks, and waters; with the exception of the study by Meisel and Moser (2004) all studies that have measured Os concentration have also measured Os isotope composition.

11.2 Analytical Techniques

Several excellent reviews have appeared over the years dealing with different aspects of chemical separation and measurements of the PGEs. These include: Barefoot and Van Loon (1999), Meisel and Moser (2004), Pearson and Woodland (2000) and Reisberg

11.2.1 Osmium

11.2.1.1 Determination of Os in Solid Samples

Early osmium concentration measurements were made using Neutron Activation Analysis (NAA; Barker and Anders 1968; Bate and Huizenga 1963;

Table 11.1 Comparison of various procedures used to measure Os and Ir

Sample	Element	Preconc. procedure	Measurement	Detection limit (pg)	References
Sediments	Os	NiS/distillation	ID-N-TIMS	13.4	Ravizza and Pyle (1997)
Rocks	Os	Carius tube/solvent extraction	ID-N-TIMS	≤5	Cohen and Waters (1996) and Shen et al. (1996) and Shirey and Walker (1996)
Rocks	Os	Carius tube/sparging	ID-ICP-MS	<2	Meisel and Moser (2004)
Rocks	Os	HF + HCl + EtOH/distill.	SIMS	<5	Walker (1988)
Rocks	Os	HF + HBr/CrO_3 + Br_2	N-TIMS	0.06	Birck et al. (1997)
Seawater	Os	CrO_3 + Br_2/microdistill	N-TIMS	0.022	Levasseur et al. (1998)
Seawater	Os	H_2SO_4 + H_2O_2 distill	N-TIMS	0.120	Woodhouse et al. (1999)
Seawater	Os	Carius tube/distillation	N-TIMS	0.0027	Chen and Sharma (2009)
Groundwater	Os	Carius tube/Br_2 extract.	N-TIMS	0.069	Paul et al. (2009)
Rain	Os	CrO_3 + Br_2/microdistill	N-TIMS	0.0203	Levasseur et al. (1999)
Rain and snow	Os	Carius tube/distillation	N-TIMS	0.0027	Chen and Sharma (2009)
Sediments and rocks	Ir	Na-peroxide/anion exch.	GFAAS	30	Hodge et al. (1986)
sediments	Ir	Acid digest/anion exch.	ICP-MS	6	Colodner et al. (1993)
Rocks	Ir	Na-peroxide/Te-ppt	RNAA	4	Morcelli and Figueiredo (2000)
Rocks	Ir	Na peroxide/Te-ppt	ICP-MS	300	Enzweiler et al. (1995)
Rocks	Ir	NiS digestion	INAA	790	Tanaka et al. (2008)
Rocks	Ir	NiS digestion	ID-ICP-MS	15.6	Ravizza and Pyle (1997)
Rocks	Ir	NiS-anion exch.	ID-ICP-MS	<1,565	Rehkamper and Halliday (1997)
Rocks	Ir	Carius/anion exch.	ID-ICP-MS	<3	Rehkamper et al. (1998)
Rocks	Ir	Carius/sparging	ID-ICP-MS	<2	Meisel and Moser (2004)
Rocks	Ir	NiS digest/Te-ppt	ICP-MS	40	Pattou et al. (1996)
Aerosols	Ir	None	INAA	0.06	Heller-Zeisler et al. (2000)
Aerosols	Ir	Acid digest/anion exch.	RNAA	0.09	Heller-Zeisler et al. (2000)
Aerosols	Ir	Acid digest/anion exch.	ID-N-TIMS	0.3	Heller-Zeisler et al. (2000)
Ice-core dust	Ir	Filtration from water	INAA	<0.001	Karner et al. (2003)
Snow	Ir	Evaporation	ICP-MS	<0.001	Gabrielli et al. (2004a, b)
Seawater	Ir	Anion exch.	INAA	0.2	Li et al. (2007)
Seawater	Ir	Anion exch.	ID-N-TIMS	0.005	Anbar et al. (1997)

Morgan 1965), which utilized γ-ray detection of ^{185}Os (646 keV) and ^{191}Os (130 keV) and permitted Os concentration determination in nanogram-sized samples. Interest in the geochemical and cosmochemical applications of the ^{187}Re-^{187}Os isotope system required precise measurements of Os isotopes in samples containing 10^{-9}–10^{-12} g/g of Os. However, a high ionization potential of Os (8.7 eV) meant that it could not be analyzed using conventional positive Thermal Ionization Mass Spectrometry. This difficulty led to the development of isotope ratio measurement techniques using secondary ion mass spectrometry (SIMS, Luck and Allegre 1983), resonance ionization mass spectrometry (RIMS, Walker and Fassett 1986), and accelerator mass spectrometry (AMS, Fehn et al. 1986; Teng et al. 1987). The SIMS and AMS techniques involved generation of Os ions by sputtering of a target containing Os with high energy molecular ion beams and required specialized equipment (ion probe and accelerator mass spectrometer, respectively). The RIMS technique involved resonance ionization of evaporated Os atoms with laser and had the advantage of utilizing a thermal ionization mass spectrometer platform for ion detection and measurement. These techniques have been supplanted by the development of Inductively Coupled Plasma-Mass Spectrometry (ICP-MS) and negative Thermal Ionization Mass Spectrometry (N-TIMS).

Routine ICP-MS measurements for Os isotopes involve introducing the sample in solution into plasma through a nebulizer and spray chamber assembly and suffer from low sensitivity. The most sensitive of ICP-MS techniques for Os isotope measurement involve generating volatile osmium tetraoxide (OsO_4) from sample and introducing ("sparging") it directly into the plasma (Hassler et al. 2000; Russ and Bazan 1987a, b). This procedure has the significant advantage of not requiring prior chemical separation of Os. The preferred method for Os isotope analysis is N-TIMS using OsO_3^- (Creaser et al. 1991; Volkening et al. 1991). This technique can analyze very small amounts of Os to a high precision but requires Os pre-concentration and purification (Table 11.1). Purified Os in solution is loaded and dried on high purity Pt filaments following which Ba-hydroxide is loaded and dried; the latter provides electrons needed for negative thermal ionization.

Advances in the chemical separation of Os from solids and its purification have gone hand in hand with the development of mass spectrometry. For SIMS measurements high level Os samples (Laurite Ru(Os, Ir)S_2, Fe-Mn nodules) were dissolved in Ni vessels, the resulting OsO_4 was distilled and purified using ion exchange chromatography and loaded on an Al disc (Luck and Allegre 1983; Palmer and Turekian 1986; Turekian and Luck 1984). RIMS measurements were conducted with Os separated by dissolving powdered rock samples in Teflon bombs with HF + HCl + ethanol. Ethanol was used to keep Os from getting oxidized to volatile OsO_4 during this stage. After dissolution, Os was oxidized and purified by distillation (Walker 1988). This technique was also used for N-TIMS work. Ravizza and Bothner (1996) used the NiS fire essay method in combination with isotope dilution N-TIMS to measure Os concentration and isotope composition in marine sediments. The NiS method was initially developed for neutron activation analysis to permit PGE concentration determination in large samples (Hoffman et al. 1978). This method takes advantage of the chalcophile character of the PGEs and the immiscibility of a sulfide melt to separate the PGEs from the silicate rocks. Powdered solid samples are mixed with a flux containing sodium tetraborate, sodium carbonate, Ni powder, sulfur powder and silica and fused at 1,000°C. The PGEs are concentrated into NiS beads, which are then dissolved in hydrochloric acid, leaving Ni into the solution and isolating insoluble sulfide compounds of the PGEs (Hoffman et al. 1978; Tanaka et al. 2008). Ravizza and Bothner (1996) filtered the PGE sulfides and then dissolved them in a combination of H_2SO_4 + $Cr^{VI}O_3$. The resulting volatile OsO_4 was distilled and further purified using ion exchange. Birck et al. (1997) pioneered a technique in which a powdered sample with tracer is attacked first by HF and HBr in a Teflon vessel. The resulting solution is dried and further dissolved using HNO_3 in the presence of $Cr^{VI}O_3$ and Br_2(liq). The resulting OsO_4 is extracted in Br_2(liq), reduced in HBr and further purified using microdistillation.

Memory effects caused by OsO_4 diffused into Teflon at high temperature can be severe. This observation and the requirement that dissolution procedure has to be carried out in a closed-system to achieve sample-tracer equilibration provided the basic incentive to develop the Carius tube technique (Cohen and Waters 1996; Shen et al. 1996; Shirey and Walker 1995) in which samples are dissolved in sealed thick-walled glass vessels (Carius tubes). Typical digestion temperatures used in the Carius tube method range from 200 to 240°C and are dictated by the bursting pressure of the glass tube, which depends on the glass thickness and internal diameter of the tube. This technique has been subsequently modified to use a high pressure asher (HPA-S), where a confining pressure is applied with dry nitrogen and the operating temperature can be raised to 300°C for a relatively thin-walled and higher-capacity quartz glass tube. During high temperature attack with Lefort aqua regia, Os is leached out from the sample and Os-tracer equilibrate as they get converted to OsO_4, which can then be distilled or extracted using CCl_4 or $CHCl_3$ or Br_2 (liq). No further purification is needed if Os isotopes are measured using an ICP-MS (see e.g., Meisel and Moser 2004). For N-TIMS, Os is further purified using microdistillation, which essentially removes all organics from the sample (Birck et al. 1997). Removal of organics is required to achieve a high ionization efficiency of 10–20%, which has permitted precise isotope ratio measurements samples containing as little as 100×10^{-15} g of Os (see below).

11.2.1.2 Determination of Os in Natural Waters

Suzuki et al. (1998) used an aqueous solution of an Os-thiourea complex with GFAAS to measure Os concentrations in waste water from electron microscopy laboratories, which use OsO_4(aq) to fix cells for microscopy. A number of attempts have been made to determine the osmium concentration and isotope composition of seawater. Its low abundance in seawater and possible multiple oxidation states frustrated initial efforts to precisely measure its concentration and isotope composition. Koide et al. (1996) used isotope dilution RIMS to obtain the osmium concentration and isotope composition of seawater. A 25 L sample of seawater was spiked with ^{190}Os tracer and Os was separated using anion exchange chromatography and purified using distillation. Sharma et al. (1997) reduced seawater and tracer Os by bubbling SO_2(g) and then coprecipitated Os with iron oxyhydroxide in 4–10 L samples. Os was extracted from iron oxyhydroxide using the carius tube technique followed by CCl_4 extraction, microdistillation, and measurement using N-TIMS. These techniques were found to not provide a quantitative yield for Os leading to the development of three new techniques. Levasseur et al. (1998) oxidized seawater and tracer Os by heating 50 g samples at 80°C with a mixture of $Cr^{VI}O_3$ and Br_2(liq); the resulting OsO_4 was extracted with Br_2(liq) and purified. Woodhouse et al. (1999) oxidized Os in 1.5-L seawater samples by adding a mixture of H_2SO_4 and H_2O_2 and distilling OsO_4, which was trapped in HBr and further purified. Os isotopes were measured using N-TIMS. Sharma et al. (1999) modified the Carius tube technique to remove Os from 50 to 70 g of seawater. The samples with tracer and $Cr^{VI}O_3$ were heated at 180°C for more than 40 h; the resulting OsO_4 was distilled and trapped in HBr. It was further purified and measured using N-TIMS. A recent study by Chen and Sharma (2009) has, however, demonstrated that oxidation of all species of Os present in seawater requires much higher temperatures (300°C). In fact as much as 30% of Os in seawater is not extracted by any of the previous techniques. That Os occurring in other natural waters (such as groundwaters) is present in multiple oxidation states has also been inferred in a recent study by Paul et al. (2009) who oxidize waters in a HPA-S using a mixture of H_2O_2 and H_2SO_4 and extract the resulting OsO_4 using Br_2(liq). High temperature oxidation of Os seems to be the only way to provide quantitative estimates of the Os concentrations in waters.

11.2.2 Iridium

11.2.2.1 Determination of Ir in Solid Samples

The classical technique to determine Ir is NAA. Despite the development of other more sensitive methods (see below) the NAA continues to be extensively used on its own (Instrumental Neutron Activation Analysis, INAA) or in combination with chemical purification of Ir from a pre-irradiated sample (Radiochemical Neutron Activation Analysis, RNAA). Neutron irradiation of ^{191}Ir results in the production of ^{192}Ir, which has a half life of 73.83 days and decays via emitting γ rays with energies of 295, 308, 316.51 and 468.07 keV; the last two rays are more intense than the first two and are usually used for concentration determination. The γ rays are detected by high purity Ge-detectors and Ir amounts estimated using sensitivities obtained from concurrently irradiated Ir-standards. Background interference from samples depends on the type of sample being analyzed (rock, water, etc.). It may come mainly from ^{46}Sc, ^{51}Cr, ^{60}Co, ^{65}Zn, ^{76}Se, and ^{124}Sb. Different approaches have been employed to remove the background and to improve detection limit:

1. Electronic suppression of background: This technique involves essentially reducing the background by simultaneously detecting 316.51 and 468.07 keV rays, whose emission is near-coincident, by a pair of detectors; background interference is substantially reduced except from that produced from Compton interactions of high energy coincident pairs of gamma rays from ^{46}Sc and ^{60}Co. This background can also be removed by independent detection of Compton-gamma-rays (Karner et al. 2003; Michel et al. 1990).
2. Preconcentration of Ir using the NiS fire essay method (see previous section for description) followed by neutron irradiation and measurement. The PGE sulfides are then filtered, irradiated and analyzed.
3. Chemical purification following irradiation and after adding Ir carrier:
   (a) Irradiated sample with Ir carrier is sintered with sodium peroxide and then dissolved in HCl. The resulting solution is filtered and Ir co-precipitated with Te metal and analyzed (Morcelli and Figueiredo 2000; Nogueira and Figueiredo 1995).
   (b) Iridium separation using anion exchange chromatography. During the ion exchange chromatography advantage is taken of extremely high affinity of Ir^{IV} to anion exchange resin AG-1 in dilute HCl. Dissolved sample mixed with Ir carrier is loaded on to a column of pre-cleaned AG-1 resin. The interfering nuclides are removed by eluting the column in dilute HCl and water. Since Ir^{IV} can be reduced on the column and mobilized an oxidizing reagent (Ce^{4+}) is mixed with the sample to ensure quantitative retention of Ir on the column. Following this Ir can be retained on the column and resin separated and analyzed or eluted and further purified before analysis (Heller-Zeisler et al. 2000; Keays et al. 1974).

Iridium separated from marine sediments and organisms has been measured using Graphite Furnace Atomic absorption Spectrometry (GFAAS) (Hodge et al. 1986). Samples were sintered with sodium peroxide and then dissolved in HCl and spiked with ^{192}Ir tracer. Chlorine gas was bubbled through the solution, which was then passed through a column of AG-1X8 resin. Iridium retained on the column was recovered using hot nitric acid and further purified prior to analysis. Iridium was determined with the spectrometer's monochromator set on 266.5 nm. Measured Ir concentrations were corrected for loss during separation and purification by determining the loss of ^{192}Ir activity in the final solution.

With the development of Inductively Couple Plasma-Mass Spectrometry (ICP-MS) external calibration or isotope dilution techniques have been successfully employed to accurately and precisely measure Ir concentrations in geochemical samples (Colodner et al. 1993; Date et al. 1987; Gregoire 1988; Gros et al. 2002; Jackson et al. 1990; Meisel and Moser 2004; Sengupta and Gregoire 1989). The ICP-MS technique appears to have successfully replaced atomic absorption spectrometry for the determination of Ir. The technique involves generation of ions by introducing solution carrying elements of interest in to plasma using by means of a nebulizer and spray chamber, followed by acceleration of ions through a Quadrupole filter/magnetic sector and their detection. The technique usually requires pre-concentration of Ir. A procedure involving sintering the sample with sodium peroxide followed by co-precipitation of tracer (enriched ^{191}Ir) and sample Ir with Te and subsequent measurement using ICP-MS was developed in the mid 1990s (Enzweiler et al. 1995). Ir precipitation is not quantitative but not required for this method as it uses isotope dilution to determine the concentration. For any isotope dilution method it is imperative that the sample and tracer fully equilibrate so that losses during subsequent processing do not impact the result. It was found that reproducible results are obtained when spike solution is mixed to strongly alkaline solution resulting from the dissolution of fusion cake.

As NiS fire essay method provides nearly quantitative recovery of Ir (e.g., Hoffman et al. 1978), it has been used as a pre-concentration technique prior to ICP-MS determination (Gros et al. 2002; Plessen and Erzinger 1998). A variation of this technique involves isotope dilution analysis in which Ir-tracer is mixed to the sample prior to the fire essay (Ravizza and Pyle 1997). This permits achievement of isotopic equilibrium between the sample and tracer Ir within the immiscible NiS melt needed to obtain accurate concentration determination. While NiS fire essay is simple and can be used with a large volume of samples, it suffers from high procedural blanks owing to the large amounts of reagents needed (Table 11.1; see also Pearson and Woodland 2000) and its Re yield is not quantitative. These issues along with the desire to measure other PGE and especially Re and Os isotopes in the same sample aliquots (see above) have led to the development of a digestion procedure in which samples mixed with tracers are decomposed with aquaregia at high temperature in Carius tubes (Rehkamper et al. 1998). Under these conditions Ir leached from the sample equilibrates with the tracer. It can then be separated from other elements using anion exchange column chromatography and solvent extraction and measured on an ICP-MS. The procedural blanks for the Carius tube/HPA-S methods are low permitting precise determination in samples with pg/g of Ir (Meisel and Moser 2004; Pearson and Woodland 2000; Rehkamper and Halliday 1997; Rehkamper et al. 1998; Shinotsuka and Suzuki 2007; Table 11.1).

11.2.2.2 Determination of Ir in Natural Waters

Gabrielli et al. (2004a, b) measured Ir concentrations in an ice core from Greenland using external calibration with a high resolution magnetic sector ICP-MS. This procedure involved pre-concentration of Ir by sub-boiling evaporation of water followed by mass spectrometry. Seawater and river water have complex matrices and require pre-concentration procedures. Fresco et al. (1985) removed Ir from 1.5 L sea-water by reducing it with magnesium and then used neutron activation to measure Ir concentration. Hodge et al. (1986) used anion exchange chromatography and GFAAS to measure Ir in seawater. This method required 100 L of seawater for a single measurement. Chlorine gas was bubbled in samples spiked with ^{192}Ir tracer to convert any Ir^{III} to Ir^{IV} (cf. Yi and Masuda 1996a, b). The water was then passed through a column of AG1-X8 resin. Iridium was removed with hot nitric acid and further purified and measured using GFAAS as described above. Anbar et al. (1996, 1997) used highly sensitive isotope dilution Negative Thermal Ionization Mass Spectrometry (ID-NTIMS) to determine Ir in seawater and river water. This procedure requires 2–4 L of sample and involves equilibrating tracer and water Ir by bubbling Cl_2 gas into water. Ir is then separated from water by (1) using anion exchange chromatography or (2) co-precipitation with ferric hydroxide followed by ion exchange chromatography. It is further purified and measured as IrO_2^- ions in a thermal ionization mass spectrometer (cf. Creaser et al. 1991; Walczyk and Heumann 1993). Recently, Li et al. (2007) have determined Ir concentrations in 1–2 L of seawater using anion exchange chromatography and NAA. Seawater samples were loaded on to a column of pre-cleaned AG-1 resin. The interfering nuclides were removed by eluting the column in dilute HCl and water. To ensure quantitative retention Ir^{IV} on the column Cl_2 was bubbled in the sample and eluting solutions (Li et al. 2007). Following this, the resin containing Ir was ashed, irradiated, and analyzed.

11.3 Distributions Among Natural Samples

Here I briefly give the distribution of Os and Ir in solid earth, meteorites, ores, waters and marine sediments and discuss the resulting systematics that form the basis of the environmental applications in Sect. 11.4.

11.3.1 Distribution in Solid Earth and Meteorites

The distribution of Os and Ir in rocks reflects their siderophile and chalcophile character (Table 11.2). They are highly enriched in iron meteorites over stony meteorites. Metal-silicate equilibrium partitioning data suggest that differentiation of earth into a metal core and silicate mantle depleted these elements in the mantle as they were quantitatively sequestered in the core (see Shirey and Walker 1998 and references therein). The higher than expected PGE concentrations in the mantle suggest that following the core formation some PGE (including Os and Ir) were accreted as a "late veneer". The $^{187}Os/^{188}Os$ ratio of the primitive upper mantle indicates that the late veneer was likely either ordinary or enstatite

Table 11.2 Os and Ir concentrations, Os/Ir ratios, and $^{187}Os/^{188}Os$ ratios in natural samples

	[Os][a]	[Ir][a]	Os/Ir	$^{187}Os/^{188}Os$	References
Iron meteorites (IVB)	27.3 μg/g	24.1 μg/g	1.1	0.1240	Walker et al. (2008)
Carbonaceous chondrite	636 ng/g	599 ng/g	1.1	0.1257	Horan et al. (2003)
Ordinary chondrite	685 ng/g	617 ng/g	1.1	0.1281	Horan et al. (2003)
Cosmic dust	1,060 ng/g	–	~1	0.125	Measurement of a single unmelted 21 μg sample of a micrometeorite (Chen and Sharma, unpublished data)
Primitive mantle	3.4 ng/g	3.2 ng/g	1.1	0.1290	McDonough and Sun (1995)
Continental mineral aerosol (upper continental crust)	0.031 ng/g	0.022 ng/g	1.4	1.26	Esser and Turekian (1993b) and Peucker-Ehrenbrink and Jahn (2001)
PGE-sulfide ores	4.6 μg/g	6.3 μg/g	0.7	0.15–0.20	Chen et al. (2009) and Naldrett (2004)
Base-metal sulfide ores	0.17 μg/g	0.15 μg/g	0.8	0.15 to >2	Chen et al. (2009) and Naldrett (2004)
Chromite ores	0.049 μg/g	0.041 μg/g	1.2	0.12–0.14	Ahmed and Arai (2002, 2009), Büchl et al. (2004), Chen et al. (2009), Economou-Eliopoulos (1996), Pedersen et al. (1993), Prichard and Lord (1993), Prichard et al. (2008), and Zaccarini et al. (2004)
Pelagic sediment	0.16 ng/g	0.40 ng/g	0.4		Koide et al. (1991)
Fe-Mn nodules/crusts	0.80 ng/g	1.8 ng/g	0.3		Koide et al. (1991)
Pelagic carbonate	0.065 ng/g	0.022 ng/g	3		Ravizza and Dalai (2006)
Black shale	4 ng/g	1 ng/g	5		Jiang et al. (2007), Paquay et al. (2009), and Pasava (1993)
Seawater	6–14 fg/g	0.08–3.6 fg/g	68	1.05 (deep) < 1.05 (surface)	Chen and Sharma (2009)
River water	3–84 fg/g	0.56–3	7–14	1.45	Anbar et al. (1996), Chen et al. (2006), Gannoun et al. (2006), Levasseur et al. (1999), Martin et al. (2001), Sharma et al. (1999, 2007a), Sharma and Wasserburg (1997), and Turekian et al. (2007)
Groundwater	20–191 fg/g	–	?	0.96–2.8	Paul et al. (2010)
Rainwater/snow	0.3–23 fg/g	0.1–5 fg/g	?	0.16–0.40	Chen et al. (2009) and Gabrielli et al. (2004a)

[a]Concentrations in the table vary about 11 orders of magnitude.
Note: 1 μg/g = 10^{-6} g/g, 1 ng/g = 10^{-9} g/g, and 1 fg/g = 10^{-15} g/g = pg/kg

chondrites (Meisel et al. 1996). Os and Ir are highly compatible elements during melting of the earth's mantle, i.e., they do not partition into the melts that form the crust. As a result, these elements are extremely depleted in the continental crust. Also, the Os/Ir ratio of the continental crust is about the same as the mantle and meteorites (~1). However, during mantle melting Re is incompatible resulting in an average $^{187}Re/^{188}Os$ ratio of the continental material of ~48, which is much higher than that of the mantle and meteorites (~0.36). As a consequence, the time-integrated average $^{187}Os/^{188}Os$ ratio of continental crust is about ten times higher than that of the mantle and meteorites (Table 11.2).

Osmium and Ir are highly enriched in cosmic dust, which is produced as a result of collision of asteroids in the asteroidal belt and travels toward the sun due to the Poynting-Robertson light effect. Approximately, 40×10^6 kg of cosmic dust accretes annually on earth (Love and Brownlee 1993) providing a nearly constant background flux of Os and Ir. Lal and Jull (2003) have pointed out that a relatively large flux of the PGE also comes from the fragmentation and vaporization of meteorites as they enter the earth's atmosphere. This "meteoric smoke" likely gets concentrated in the polar-regions leading to an enhanced apparent flux of the PGE in polar ice (Gabrielli et al. 2006). Osmium and Ir concentrations and Os isotopes in deep marine sediments have been used to assess cosmic dust flux and its variations over long-term (Peucker-Ehrenbrink and Ravizza 2000a and references therein).

11.3.1.1 Distribution in Ore Deposits

Relatively high concentrations of Os and Ir in PGE-sulfide and chromitite ores (Table 11.2) reflect enrichment during crystallization in magma chambers. An excellent recent review of the literature on the PGE bearing magmatic sulfide deposits is by Naldrett (2004). The largest PGE deposits are in the 2 Ga Bushveld Complex of South Africa, where mixing of two magmas engendered nearly simultaneous chromite crystallization and segregation of an immiscible sulfide-rich melt. Scavenging of the PGE by sulfide melts and also by chromites led to unusually high concentrations of the PGEs, which are found in the Merensky Reef and UG-2 ore bodies (Naldrett 2004). Noril'sk Ni-Cu sulfide ore deposits in Russia are the second largest source of PGE (Jones 1999). They are associated with 251 Ma Siberian Flood Basalt Province. Here magma contamination and consequent segregation of immiscible sulfide melts along dykes that fed lavas to the surface has led to crystallization of PGE-enriched Ni-Cu sulfide ores (Naldrett 2004). Modest enrichments of the PGE are always associated with magmatic sulfides (Table 11.2) and quite possibly with other types of sulfide deposits (sedimentary-exhalative deposits, Cu and Mo porphyry deposits; see e.g., Sotnikov et al. 2001). The Re/Os ratios of sulfide ores are quite variable and as a result depending on their age, the $^{187}Os/^{188}Os$ ratios of the ores also vary from ~0.15 to >2 (Chen et al. 2009).

Relatively modest PGE-enrichments are also found all over the world in podiform chromitite bodies associated with ultramafic rocks within ophiolite complexes (e.g., Ahmed and Arai 2002; Ahmed et al. 2009; Büchl et al. 2004; Economou-Eliopoulos 1996; Pedersen et al. 1993; Prichard and Lord 1993; Prichard et al. 2008; Zaccarini et al. 2004). Here chromite encloses primarily laurite and also metal clusters of Ru-Os-Ir alloys. The Re/Os ratios of chromites are extremely low and therefore their $^{187}Os/^{188}Os$ ratios reflect the time when these ores were derived from the mantle (Walker et al. 2002b).

11.3.2 Distribution in Natural Waters

The speciation and oxidation states of Os and Ir in waters are not well constrained. Thermodynamic data predicts that Os should exist as either OsO_4^0 (Byrne 2002) or $H_3OsO_6^-$ (Palmer et al. 1988) in seawater, both of which are in the fully oxidized (Os^{VIII}) state. However, experimental observations indicate that Os in seawater does not solely reside in its fully oxidized state because great effort is required to fully oxidize seawater Os. The rich organometallic chemistry of Os suggests that it may also be present in a reduced inert organic form in seawater (Chen and Sharma 2009; Levasseur et al. 1998). Os is abundant in both oxidizing and reducing marine sediments (see Table 11.2) and its redox behavior seems to impact its distribution in seawater-sediment system. A recent X-ray absorption near-edge structure (XANES) study shows that Os incorporated in reducing sediment is mainly trivalent and that removed in oxic sediment is tetravalent (Yamashita et al. 2007). In comparison, the speciation of Ir in seawater is dominated by Ir^{III} oxidation state forming anionic chloro-complex $IrCl_6^{3-}$ (Anbar 1996 and references therein). Recent work suggests that Ir^{III} in seawater may form mixed ligand complexes (Byrne 2002). The principal species of Ir^{III} in seawater is tentatively assigned to be $IrCl_a(OH)_b^{3-(a+b)}$. Iridium may also be stabilized in seawater by the formation of stable organic complexes (Anbar et al. 1996). Iridium also has a large first hydroxide binding constant and is thus likely to be very reactive toward Fe-Mn oxyhydroxides and clays (Anbar 1996). Oxidation of Ir from trivalent to tetravalent state may occur during mineral formation (Hodge et al. 1986).

The differences in the redox behavior and particle reactivity appear to control the removal and fractionation of Os and Ir in the oceans. These elements are concentrated in anoxic black shales as well as oxic chemical precipitates (Fe-Mn nodules/crusts) (Table 11.2). Remarkably, however, there is a substantial fractionation between Os and Ir in these sediments, i.e., the average Os/Ir ratio of black shales is ~15 times higher than that of the Fe-Mn nodules/crusts. This suggests that geochemical behaviors of these elements in seawater column are vastly different and much unlike their behaviors during planetary formation and differentiation! This observation can be further examined by calculating the ratio of the dissolved to suspended particulate fluxes entering the oceans via rivers. Assuming that the concentrations in pelagic sediments are the same as those transported by rivers and the suspended particulate flow rate from rivers is 15.5×10^{15} g year^{-1} (Martin and Meybeck 1979), the ratio of dissolved to suspended particulates fluxes to the oceans are 0.1 and 0.01 for Os and Ir, respectively. Note that there is a large uncertainty in the riverine Ir flux as only three rivers have been analyzed so far (Anbar et al. 1996). Nonetheless we can infer that only a small fraction of continental Os and Ir are transported to the oceans in dissolved form and that subtle changes in the Ir concentrations of oceanic sediments (e.g., by reduction of oxic marine sediments, which would release Ir) could potentially impact its seawater concentrations, potentially providing a flux that is significant compared to the riverine Ir flux.

The Os concentration (=[Os]) in seawater varies from 6 to 14 of fg/g (Table 11.2; Fig. 11.1). As discussed above, only the latest analytical technique appears to provide a quantitative yield of the Os concentration in seawater. Using these limited data we can infer that the behavior of Os in the oceans is nonconservative (Fig. 11.1a). The surface water of the Atlantic, the Pacific, and the Southern Ocean is less concentrated in Os than the deep oceans. At a depth of 2,000 m the water in the eastern Pacific is more concentrated in Os than the Atlantic. Depth profiles suggest a nutrient like behavior: Os is depleted near the surface and becomes enriched at ~500 m; at depths of 700–1,000 m it seems to decrease.

The Atlantic and Pacific waters from a depth of 2,000 m have a $^{187}Os/^{188}Os$ ratio of 1.05, which is identical to a previously established ratio of 1.05 ± 0.01 for the deep oceans (Levasseur et al. 1998; Sharma et al. 1997; Woodhouse et al. 1999). However, deeper water, in the eastern Pacific is even more radiogenic, $^{187}Os/^{188}Os = 1.08 \pm 0.01$ (2,440 m, Chen and Sharma 2009) and 1.09 ± 0.02 (2,778 m, Woodhouse et al. 1999). While these studies suggest that the deep oceans do not have a homogeneous $^{187}Os/^{188}Os$ ratio, a lot more data from the deep oceans are needed to confirm this observation. The $^{187}Os/^{188}Os$ ratio of the waters from surface to a depth of about 1,000 m appears to be quite variable and always lower than the deep waters (Fig. 11.1b). Os isotopic composition of seawater is therefore not homogeneous. Chen et al. (2009) have surmised that the Os isotope composition of surface waters is impacted by recent anthropogenic Os contributions (see below). More detailed depth profiles are needed to confirm this behavior.

The Ir concentration (=[Ir]) in seawater varies from 0.1 to ~3.5 of fg/g (Table 11.2; Fig. 11.2), with the exception of a sample from the Pacific coast (Point Loma, San Diego, CA) that yielded [Ir] of ~10 fg/g ($N = 6$, Fresco et al. 1985). Hodge et al. (1986) analyzed a sample from the Pier at Scripps Institution of Oceanography (La Jolla, CA) and found [Ir] to be 1.3 fg/g ($N = 4$). Whether the observed large variations in the concentration are real or analytical issues is not clear as no cross-calibrations have been performed. Since the samples analyzed by the above two studies came from within a few km of each other a concentration difference of a factor of ~8 by natural variations is hard to imagine. To examine this issue further depth profiles from two more recent studies are plotted in Fig. 11.2. It is intriguing that seawater analyzed using RNAA (Arctic and Bering Sea,

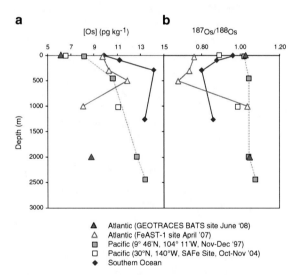

Fig. 11.1 Seawater profiles of Os isotopic compositions and concentrations from the Atlantic, Pacific, and Southern Oceans. Os concentrations are variable across ocean basins and are lowest at the surface and becoming enriched at ~500 m depth (**a**). At a depth of 2,000 m, the eastern North Pacific has a much higher concentration than the Atlantic. While the GEOTRACES sample collected at 2,000 m has an $^{187}Os/^{188}Os$ ratio of 1.052 ± 0.004 ($N = 8$) and confirms earlier results from the Sargasso Sea, the Os isotope compositions of samples at shallower levels vary between basins and with depth (**b**). Moreover, surface samples seem to have lower $^{187}Os/^{188}Os$ ratios than those for deeper waters. Figure modified from Chen and Sharma (2009)

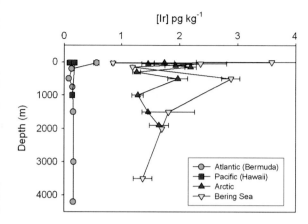

Fig. 11.2 Seawater profiles of Ir concentrations from the Atlantic, Pacific, Arctic and Bering Sea. The Ir concentrations from (**a**) the Atlantic and the Pacific Oceans were measured using negative thermal ionization mass spectrometry (Anbar et al. 1996, 1997) and (**b**) the Arctic Ocean and Bering Sea were measured using neutron activation analysis (Li et al. 2007)

$N = 17$) yields Ir concentrations that are, on average, ~1.8 fg/g and about ten times more than those obtained using isotope dilution (Pacific and Atlantic, $N = 11$). Since all of the analyzed samples come from open oceans the data indicate a large variability in the Ir concentration in different ocean basins. Alternatively, there are unresolved analytical issues with either or both techniques. These issues will need to be investigated in future. For the discussion below we will assume that the data are accurate.

Only three seawater samples from depths of 25, 500 and 3,000 m in the Sargasso Sea have been analyzed for both Os and Ir concentrations in the oceans (Os data: M. Sharma unpublished data; Ir data from Anbar et al. 1997). These yield Os/Ir ratios of 10, 81 and 41, respectively. The Os concentrations were analyzed by heating seawater with tracer and $Cr^{IV}O_3$ to 180°C for more than 40 h (see above). Since this method systematically underestimates Os concentration (Chen and Sharma 2009) it is possible to estimate the "true" Os/Ir ratios, which yields an average Os/Ir ratio of ~68 for the Sargasso Sea. An average Os/Ir ratio for the Arctic can also be estimated. Average Os concentration of four samples from the Arctic (86°N, 85°E, Edmonds et al. 2003) is ~7 fg/g (M. Sharma, unpublished data). Combined with Ir concentration data by Li et al. (2007) gives an Os/Ir ratio for the Arctic to be ~4. This ratio is likely higher as the Os concentration data were obtained by heating the samples at 180°C instead of 300°C. We estimate that the Os/Ir ratio of the Arctic Ocean is ~6. The Os/Ir ratios in the seawater thus appear to be variable and remarkably fractionated, compared to the three main sources of these elements (continents, mantle and cosmic dust/meteoric smoke; Table 11.2). As discussed above, there is a large fractionation of Os and Ir between oxic and anoxic sediments, with Os/Ir ratio of anoxic sediments being much higher than the oxic sediments. Evidence of a recent (last 50 years) large-scale increase in reducing conditions has been found in all major Arctic Ocean basin sediments (Gobeil et al. 2001) suggesting that during this time the rainout rate of Ir in the Arctic has been significantly reduced. Moreover if there is reduction of oxidized sediments taking place, it would release Ir as shown by Colodner et al. (1992). That Os/Ir ratios are rather high in the oxic waters of the Sargasso Sea and low in the Arctic thus appears consistent with the fractionation observed in the sediments.

11.3.2.1 Mass Balance

Details of Os and Ir mass balances have been discussed in a number of papers (Os: Burton et al. 2010; Levasseur et al. 1999; Peucker-Ehrenbrink 2002; Peucker-Ehrenbrink and Ravizza 2000b; Sharma et al. 1997, 2007b, Ir: Anbar et al. 1996). The $^{187}Os/^{188}Os$ ratio of the deep oceans is thought to reflect a balance between inputs from the continents ($^{187}Os/^{188}Os$ ratio ~ 1.26) and from a combination of the mantle and cosmic sources ($^{187}Os/^{188}Os$ ratio ~ 0.13). Assuming that the $^{187}Os/^{188}Os$ ratio of the deep oceans is rather homogeneous (=1.05), the continents currently supply 81% of Os dissolved in the oceans. Using the available data that represent approximately 39% of the annual river flux, the estimated average riverine [Os] = 8.5 fg/g (=45 fmol/kg) and $^{187}Os/^{188}Os = 1.45$. Assuming that the estuaries trap about 15% of riverine Os (Turekian et al. 2007), we find that the rivers supply about 270 kg (=1,423 mol) of Os annually to the oceans. The groundwater and the continental mineral aerosol are also assumed to provide continental Os to the oceans. However, a recent study finds that the Os groundwater flux from Bangladesh delta is negligible (Paul et al. 2010). Approximately, 450×10^{12} g of continental mineral dust, which is generated primarily in the Sahara and Gobi deserts, is deposited on the ocean surface (Jickells et al. 2005). Its dissolution in the waters is critical in providing essentially micro-nutrients (such as Fe) to the oceans. Assuming that the dust contains about 31 pg g^{-1} of Os, its *complete* dissolution would provide ~5% of riverine Os flux to the oceans. The leachable Os associated with continent-derived dust to the oceans would be smaller still (cf. Williams and Turekian 2004).

The isotope mass balance of the deep oceans dictates that 19% of Os must come from mantle and/or cosmic sources. The low and high temperature hydrothermal alteration, and cosmic dust/meteoric smoke have been postulated to provide Os to the oceans. Detailed studies of high and low temperature alteration along and away from the Juan de Fuca ridge have indicated a negligible mantle contribution via hydrothermal alteration of sea floor basalt (Sharma et al. 2000, 2007b). The extent to which dissolution of cosmic dust/meteoric smoke contributes Os to the seawater could be large but remains un-quantified (Sharma et al. 2007b). Additional sources of mantle derived Os include volcanic dust and submarine alteration of abyssal peridotites. Analyses of rain from around the world, however, indicate a large anthropogenic Os source and no (or negligible) input of volcanic Os to the oceans (Chen et al. 2009). Alteration of peridotites in the oceans could potentially deliver a large amount of mantle derived Os to the oceans, provided that Os is mobilized during the alteration (see Burton et al. 2010; Sharma et al. 2007b; Cave et al. 2003).

The mean residence time of Os in the oceans ($\bar{\tau}_{Os}^{SW}$) is given as: $\bar{\tau}_{Os}^{SW} = \frac{N_{Os}^{SW}}{(1-f_{Os}^{E})(QC_{Os}^{C}/f_{Os}^{C})}$ where N_{Os}^{SW} is the total amount of Os in the oceans ($=66 \times 10^6$ mol), f_{Os}^{E} is the fraction of Os trapped in estuaries ($=0.15$), Q is the annual run-off ($=3.74 \times 10^{16}$ kg year^{-1}), C_{Os}^{C} is the average concentration of Os in rivers ($=45 \times 10^{-15}$ mol kg^{-1}) and $f_{Os}^{C} = 0.81$, is the fraction of continent-derived Os in the oceans. These yield a $\bar{\tau}_{Os}^{SW} = 37 \pm 14(2\sigma)$ kyr (Sharma et al. 2007a). This is much longer than the ocean mixing time scale of ~1,500 years.

The $\bar{\tau}_{Os}^{SW}$ can also be estimated from the relaxation time, which can be determined from the recovery rate of $^{187}Os/^{188}Os$ ratio following an excursion. Excursions to lower $^{187}Os/^{188}Os$ ratios associated with glacial maxima have been observed in the rapidly accumulating Fe-Mn sediments on the flanks of the East Pacific Rise (EPR) and in the anoxic sediments from the Cariaco Basin (Oxburgh 1998, 2001). They are also found in sub-oxic to anoxic sediments from the Santa Barbara Basin (Williams and Turekian 2002) and Japan Sea (Dalai et al. 2005). Finally, these excursions have been found in the planktic foraminifera recovered from ODP site 758 in the northern Indian Ocean (Burton et al. 2010). Excursions to lower $^{187}Os/^{188}Os$ ratios associated with glacial maxima followed by upswing to higher $^{187}Os/^{188}Os$ ratios during the interglacials were used to estimate $\bar{\tau}_{Os}^{SW}$. The estimated relaxation time for the EPR sediments $= 8–12$ kyr (Oxburgh 1998) and for the Cariaco Basin $= 3–4$ kyr (Oxburgh 2001). The $\bar{\tau}_{Os}^{SW}$ thus estimated is much less than that determined from mass balance considerations, However, not *all* high resolution Os records display the glacial-interglacial variations: no excursion in $^{187}Os/^{188}Os$ ratios has been observed in a deep sea core from the Cape Basin for the past 100,000 years (core: TNO57-21, Zylberberg et al. 2006), in sediments containing dominantly hydrogeneous Os from ODP Site 806 on the Ontong Java plateau (Dalai and Ravizza 2006), and in early Pleistocene sediments from ODP Site 849 in the eastern equatorial Pacific (Dalai and Ravizza 2010). These observations suggests that the observed glacial-interglacial variations in the marine Os isotope ratios are not global and that the residence times calculated using relaxation times likely reflect different rain-out rates of Os in different basins (see Sharma et al. 2007a).

All of the above-mentioned sources of Os could also be the sources for Ir. In addition, reduction of oxic sediments could also be a source of Ir. However, a detailed understanding of Ir budget in the oceans is not yet possible due to lack of data. Direct measurements of a relatively small river in Sweden (Kalixälven) that is unperturbed by human activities, transport of water through the Baltic Sea, the Ir concentration in the Pacific and Atlantic oceans and its burial in sediments have been used to obtain a rough picture of the Ir budget of the oceans (Anbar et al. 1996). Average Kalixälven concentration is 0.63 fg/g. Assuming this is the concentration in all the rivers gives an annual riverine input of 23.5 kg year^{-1} ($=122$ mol year^{-1}). Following analogy from the Baltic Sea about 75% of riverine Ir input gets trapped in the estuaries (Anbar et al. 1996). This indicates that the riverine input to the deep oceans is only ~5.9 kg year^{-1}. The Os/Ir ratio of this input is likely between 24 and 48. In comparison, the Ir input from cosmic dust/meteoric smoke to the oceans is between 7 and 61 kg year^{-1} (Sharma et al. 2007b). The extent to which Ir from cosmic dust is dissolved in the oceans is unknown. Similarly, dissolution of continental mineral dust could supply up to 9.9 kg year^{-1}. The hydrothermal contribution of Ir is unknown and is likely small primarily due to removal by sulfide precipitation (see Sharma et al. 2001). The amount of Ir released due to reduction of oxic sediments is also unknown. Anbar et al. (1996) estimated lower and upper limits for the mean residence time of (Ir = $\bar{\tau}_{Ir}^{SW}$) in the oceans to be between 2,500 and 21,000 years.

Assuming that amount of Ir released due to reduction of oxic sediments is negligible we can estimate $\bar{\tau}_{Ir}^{SW}$ using $\bar{\tau}_{Os}^{SW}$ and Os/Ir ratios of the inputs and ocean. Defining $\rho^k = \left(\text{Ir}/^{188}\text{Os}\right)^k$ for the ratio of dissolved Ir and ^{188}Os provided by source "k", the ratio of the residence times of Os and Ir in seawater (SW) can be given as (see Sharma et al. 1997):

$$\frac{\bar{\tau}_{Os}^{SW}}{\bar{\tau}_{Ir}^{SW}} = \frac{\rho^C}{\rho^{SW}} f_{188Os}^C + \frac{\rho^H}{\rho^{SW}} f_{188Os}^H + \frac{\rho^D}{\rho^{SW}} f_{188Os}^D \quad (11.1)$$

where f_{188Os}^k denotes fractional contributions of ^{188}Os from continents (C), hydrothermal fluids (H), and cosmic dust/meteoric smoke (D). As $\rho^H \approx \rho^D$, using the data given in Table 11.2 the above equation gives: $\bar{\tau}_{Ir}^{SW} = \frac{\rho^{SW}\bar{\tau}_{Os}^{SW}}{f_{188Os}^C \rho^C + 0.19} = \frac{0.124 \times (37{,}000 \pm 14{,}000)}{0.81\rho^C + 0.19}$. Using the Os/Ir input of the riverine source to be either 24 or 48, we find that 9 kyr $< \bar{\tau}_{Ir}^{SW} <$ 14 kyr.

11.4 Applications

The primary environmental applications of Os and Ir arise from the fact that these elements are extremely rare on the surface of the earth and their introduction into the environment leads to an increase in their concentration in surface materials. In addition, a unique isotope fingerprint is present for Os introduced into the environment in that it comes from ores mined primarily in South Africa and Russia. The following are examples of environmental applications of Os and Ir.

11.4.1 Platinum Group Element Dispersal Due to Ore Processing and Use in Automobiles

As mentioned above, the PGEs are highly siderophile and therefore concentrated in sulfide ores. They are also quite abundant in chromite ores where they may exist as minerals (such as laurite, osmiridium, etc.). Industrial refining of ores to obtain PGEs and their usage in a wide variety of industrial, chemical, electrical and pharmaceutical applications has increased their burden on the environment. In particular, the production of Pt, Pd, and Rh, has increased exponentially over the last 40 years following the requirement in the US and other countries that all new automobiles have to be equipped with catalytic converters that reduce SO_2, CO and NO_x emissions. There are over 500 million automobiles operating around the world and a majority of them has catalytic converters containing (depending on the make and size) 2–15 g of Pt and lesser amounts of Ru and Rh. The catalytic converters have a ceramic substrate with a honeycomb structure to permit the passage and reaction of exhaust gases with the PGEs that are impregnated on the honeycomb with the help of the washcoat, which is a highly porous alumina coating providing high surface area for the catalytic reactions. During the life-time of an automobile physical and chemical stresses on the catalytic converters cause small quantities of the PGEs, including Os, Ir, and Ru that are present as impurities, to emit together with particles from the washcoat. So while the urban air quality has substantially improved following the introduction of automobile catalytic converters, the PGE concentrations in the environment have also increased. Indeed several studies have documented elevated levels of PGEs in air, street dust, soil, vegetation, and organisms (see reviews by Ravindra et al. 2004 and Wiseman and Zereini 2009). In addition, snow from relatively pristine central Greenland has registered an increase in the PGE concentrations, which have been attributed to the large scale atmospheric contamination of these elements by catalytic converters (Barbante et al. 2001).

Osmium isotopes have been utilized to (a) track the dispersal of the PGEs associated with particulates and (b) assess dispersal of Os *itself* from either direct utilization of OsO_4 in a unique research application or indirect spreading of this compound due to industrial processing and refining of PGE and other ores. Approximately 92% of the PGEs produced in the world come from the Merensky reef in South Africa and from the intrusions in the Noril'sk region, Russia (Jones 1999). The $^{187}Os/^{188}Os$ ratios of ores from these deposits lie between 0.15 and 0.20 (see references in Chen et al. 2009; Table 11.2). In comparison, the Os isotope ratios of dominant natural sources of Os on the earth surface are much more radiogenic: the average $^{187}Os/^{188}Os$ ratio of the continental mineral dust examined so far is 1.05 ± 0.23 (Esser and Turekian 1993b; Peucker-Ehrenbrink and Jahn 2001; Table 11.2). If the Os isotope composition (Γ) of a given environmental sample is assumed to be a binary mixture of a lithogenic and an anthropogenic end-member, fractional anthropogenic contribution of Os ($f_{Os}^{Anthropogenic}$) in the sample can be computed as follows:

$$f_{Os}^{Anthropogenic} = \frac{\Gamma_{sample} - \Gamma_{lithogenic}}{\Gamma_{Anthropogenic} - \Gamma_{lithogenic}}. \quad (11.2)$$

Moreover, assuming that the ratio of Os to other PGEs is similar to that in the continental mineral dust, fractional anthropogenic contribution of the PGEs ($f_{Anthropogenic}^{PGE}$) in the sample can also be estimated from:

$$f_{Anthropeogenic}^{PGE} = 1 - \frac{[PGE]_{UCC} / [PGE]_{Sample}}{[Os]_{UCC} / [Os]_{Sample}(1 - f_{Anthropogenic}^{Os})}. \quad (11.3)$$

Here the subscript UCC = Upper Continental Crust. Utilizing this approach Rauch et al. (2004) analyzed a set of samples from an ombrotrophic peat bog (Thoreau's bog) in Concord, Massachusetts for Os, Ir, Pt, Pd, Rh, and Al concentrations and Os isotopes. The samples were dated using ^{210}Pb and were found to have accumulated from 1980 through 2002. Rauch et al. (2004) determined that in all samples the PGEs were enriched over those estimated from PGE/Al ratios and inferred that almost all (i.e., >99%) Pt, Pd, Rh, and Ir and ~94% of Os were of non-lithogenic origin. Assuming that non-lithogenic Os fraction in the peat samples was derived from catalytic converter exhaust and rainwater and that the rain Os isotope composition was similar to that of the continental mineral dust or ocean (~1.05), Rauch et al. (2004) estimated that the anthropogenic and rainwater fractions represent, on average, 43 and 57% of non-lithogenic Os, respectively. That is, a total of 96.5% of Os (=94% + 43% of the remaining 6%) in Thoreau's bog was derived from catalytic exhaust and the rest from the rain.

A recent study, however, shows that the rainwater on the US eastern seaboard has an $^{187}Os/^{188}Os$ ratio of ~0.2 (Chen et al. 2009). Since the average $^{187}Os/^{188}Os$ ratio of the Thoreau's bog samples is ~0.6 an alternative explanation using (11.1) would be that only ~53% Os is rain and catalytic converter derived and the rest comes from lithogenic sources with an average $^{187}Os/^{188}Os$ ratio of 1.05. What fraction of rain and catalytic converter derived Os can be attributed to coming as particulates with Pt, Pd and Rh? It is likely negligible! This is because (1) Os is present only as an impurity in Pt, Pd and Rh and (2) under the oxidizing conditions present in a section of the catalytic converter Os would be lost as OsO_4. It is therefore likely that that bulk of Os in Thoreau's bog comes from atmospheric precipitation and does not provide any insight into the concentration of Pt, Pd and Rh that are automobile exhaust derived. If true, (11.2) cannot be used as the source of anthropogenic Os in the sample is decoupled from that of the other PGEs.

The above interpretation is consistent with another study done by Rauch et al. (2005) where they analyzed Pt, Pd, Rh, and Os concentrations and Os isotope compositions in samples of 10 μm diameter airborne particles (PM_{10}) collected from various sites in the city of Boston. They found significant correlations between Pt, Pd and Rh concentrations that also showed no correlation with Os. In fact the Pt/Os ratio varies widely from 9 to 12,000 with a median of 1,000. The Os isotope compositions of all 28 samples analyzed are highly radiogenic ($^{187}Os/^{188}Os$ = 0.3–2.8; median = 1.12). Intriguingly, there are four samples with Pt/Os ratios ranging from 9 to 40 that yield $^{187}Os/^{188}Os$ ratios with an average of 0.5 ± 0.2 (2σ, $N = 4$). Since the average Pt/Os ratio of PGE ores is about 50, such low Pt/Os ratios are not expected for the catalytic converters, which use purified Pt. The isotope ratio in these samples nonetheless indicates a source that is quite different from the lithogenic source. These observations suggest that catalytic converters provide little Os to the PM_{10} particles in Boston and possibly elsewhere. Nonetheless further studies are needed to evaluate the relationship between Pt, Pd, and Rh and Os isotopes for different size fractions in urban air.

Attempts to use osmium isotopes in identifying and tracing anthropogenic Os pollution have been more successful. While examining the estuarine sediments of New Haven Harbor and the Long Island Sound (LIS) for Os Esser and Turekian (1993a) discovered that the top 15 cm of the sediments had Os isotope compositions that were identical to those of the ores from South Africa. They then traced the source of the anthropogenic Os to the New Haven sewage sludge and suggested that the primary source of anthropogenic Os was fixative solutions used to stain tissue for electron microscopy in medical laboratories. Williams et al. (1997) and Turekian et al. (2007) further extended this study to examine the anthropogenic Os from New York City medical facilities transported by the East River, which is the western source of water to the LIS. They found that there is a net transport of about 0.4–1 mol of anthropogenic Os per year from the East River into the LIS and that bulk of anthropogenic Os coming into the sound is retained there.

Studies in other estuaries have also been promising suggesting that osmium isotopes may track the influence of particle-borne contaminants, delivered by sewage, on sediments in estuarine environments. Ravizza and Bothner (1996) compared Os isotopes and Ag concentrations in sediments collected from Boston Harbor, Massachusetts Bay, and Cape Cod Bay and found that discharge of sewage into Boston Harbor has affected sediments in Cape Cod Bay, about 70 km away! Helz et al. (2000) examined sediments from a number of sites in the Chesapeake Bay region, including the sewage discharge sites from the city of Baltimore. They observed anthropogenic Os in sludge from the city's principal wastewater treatment plant and inferred it to be derived from medical facilities in the city. Baltimore was a major world supplier of chromium compounds and Isaac Tyson's Baltimore Chrome Works operated from 1827 through 1986. However, Helz et al. (2000) did not find any anthropogenic Os derived from processing of chromite ores to extract Cr. This is actually not an unexpected result as the first step in extraction of chromium from chromite involves high temperature oxidation of Cr^{III} to Cr^{VI}. This process would convert all Os present in the ore to OsO_4, which would then be released in the air and transported to long distances (see below).

Helz et al. (2000) also found that a site in the mid-Chesapeake Bay region at a distance of >50 km from any known anthropogenic sources displays the strongest anthropogenic Os signal in their data set. They noted that the transport of particles to this site overrode the northward flowing bottom currents in the estuary and that finding anthropogenic Os at this site would suggest that other particle-borne substances, including hazardous ones, could be dispersed broadly in this estuary. Interestingly this site received anthropogenic Os as early as 1949. Could this come from chromium industry? A more detailed analysis of several sediment cores especially in the northern part of the Bay combined with an understanding of chromium production rates in Baltimore, past wind trajectories and also water circulation would be needed to assess the distribution of Os derived from roasting of chromite ores.

Osmium bearing PGE minerals and nuggets reside within chromites, which can have tens of ng of Os per g. Consideration of the primary reaction involved in the extraction of Cr from chromite ores (e.g., Burke et al. 1991) shows that it would also release OsO_4.

$$4\ FeCr_2O_4 + 8\ Na_2 + 7\ O_2 \rightarrow 8\ Na_2CrO_4 + 2\ Fe_2O_3 + 2\ CO_2.$$

Here insoluble chromite ($FeCr_2O_4$) containing trivalent Cr is converted to water soluble sodium chromate (Na_2CrO_4) containing hexavalent Cr. To fully oxidize Cr(III) finely crushed chromite ore is mixed with soda ash and lime and roasted in rotary kilns at 1,100–1,150°C. The mix does not fuse but the molten soda ash reacts with the chromite to form sodium chromate. That chromite hosted Os minerals released as a result of the breakdown of the chromite lattice would also be attacked under highly oxidizing conditions and volatile OsO_4 produced has been indicated by recent experimental data from Yokoyama et al. (2010). Since the kiln does not trap evolving CO_2 it is likely that OsO_4 would be carried away with CO_2 escaping from the kiln.

In a pioneering study examining the Os isotopes in humus, plants, mushrooms, mosses and lichens as well as soils Rodushkin et al. (2007a, b) demonstrated the impact of OsO_4 transport to the region surrounding a chromium producing plant. They found that mosses and lichens collected in rural northern Sweden had Os concentrations that were tenfold higher than those reported for particulate Os in dust samples from urban centers in the US. In addition, they also noted that from southwest to northeast transect of the sampling area along Sweden's eastern coastline the Os concentrations were increasing with decreasing $^{187}Os/^{188}Os$ ratios indicating a local anthropogenic source in the northeast. They were then able to trace the source to Kemi chromite ore, which is being processed to obtain chromium and ferrochromium in Torneå, Finland. From the data it appears that OsO_4 from Torneå smelter was transported to a distance of 100+ km southwest in Sweden. An important outcome of this study is the demonstration of the utility of using mosses and lichens in absorbing OsO_4 from the ambient atmosphere. Lichens have been used in the past to distinguish between local natural, local anthropogenic and background sources of lead (Simonetti et al. 2003). Their utilization to assess the extent of OsO_4 deposition in major urban centers should shed light on the issue of atmospheric Os loading from catalytic converters.

During the industrial processing and refining of sulfide ores of base metals (Cu, Ni, Pb, Zn) and PGE, OsO_4 is generated during the converting phase, where sulfur is removed as SO_2 by blasting hot air/oxygen through the molten (900–1,250°C) matte that contains mainly metal plus sulfur. The processing of sulfide ores would therefore constitute the single largest source of worldwide OsO_4 emission. However, environmental laws restricting the release of SO_2 from base metal smelters appear to have inadvertently eliminated a large fraction of OsO_4 from the environment (Chen et al. 2009). This is because a majority of nations requires that the base metal sulfide industry scrubs the SO_2 produced from exhaust stream. Scrubbing appears to nearly quantitatively remove OsO_4 (Abisheva et al. 2001). In the first stage of scrubbing the SO_2 is cleaned off any dust. Most of OsO_4 is trapped either by the wash sulfuric acid or the slime produced. Any remaining OsO_4 in the washed SO_2 is likely sequestered during subsequent treatments that attempt to recover sulfur as (1) sulfuric acid, (2) gypsum, (3) liquefied SO_2, or (4) elemental sulfur.

High concentrations of SO_2 in the flue gas stream are needed to make sulfur recovery economically feasible, which is typically possible with modern arc furnace technology used by base-metal industry (Goonan 2005). In contrast, the PGE ore processing facilities do not scrub SO_2 as its low concentration in furnace gases and intermittent production from the converter make scrubbing technically challenging and uneconomical. Instead the SO_2 is discharged into the atmosphere using tall stacks (Jones 1999). So it seems likely only PGE ore processing plants release OsO_4 into the environment. Indeed the Os isotope composition of snow and rain collected worldwide implicates OsO_4 released from PGE ore processing as providing the single largest Os burden to the atmosphere (Chen et al. 2009). Moreover, the deposition of anthropogenic Os over the last several decades has lowered the Os isotopic composition of oceans surface by about 10% compared to that of the deep oceans (Chen et al. 2009; Chen and Sharma 2009).

11.4.2 Distribution of Soot from Diesel Engines

This novel application uses Ir as an intentional tracer to track soot particles generated during burning of fossil fuel. Soot particles contain poly-aromatic hydrocarbons (PAH), which are known mutagenic and carcinogenic agents. Soot emissions from motor vehicles especially those employing diesel engines constitute the single largest source of PAHs in the US. The key issue then is the extent to which urban and surrounding rural areas are affected by such emissions. The answer depends on the aerodynamic size of the soot particles generated and its evolution with time. Since soot in a given location may be derived from multiple sources (power plants, cars and trucks, barbecues, etc.) the basic problem is how to trace soot from a given source. Iridium as an intentional particulate tracer was shown to be useful in studies of soot transport, deposition, and source attribution in two studies (Suarez et al. 1998; Wu et al. 1998). Advantage was taken of the fact an inexpensive commercially available Ir compound (2,4-pentanedionate containing 38.07% of Ir) is readily soluble in organic solvents (i.e., toluene) that in turn are miscible in diesel fuel.

In the first study, diesel fuel used by the sanitation trucks in the city of Baltimore was tagged with Ir^{III} octahedral pentanedionate dissolved in toluene

(Suarez et al. 1998). Size sorted aerosol samples were collected from several sites in the city and surrounding areas. Samples were also collected from a sanitation vehicle's tail-pipe. The samples were analyzed for Ir and Fe concentrations using neutron activation analysis. Iridium concentrations over the background and Iridium to iron ratios were used to assess the extent to which Ir introduced in the diesel fuel was associated with the soot particles. It was found that the tracer Ir concentration in the tail-pipe emission of single sanitation truck showed a skewed distribution with particle size, peaking between 0.1 and 0.2 µm – 50% of Ir was found to lie between 0.1 and 0.2 µm and the remainder on larger particles. The peak in Ir also corresponded to the peak in organic carbon confirming that Ir is a good surrogate for soot emitting from diesel engines. The Ir concentration – particle size distribution spectra for aerosol samples collected on days when the sanitation trucks were running was found to be distinctly different from that of the single truck and also from the background: a large proportion of Ir was found associated with 2.4 µm. The authors concluded that the ambient spectra probably reflected aerosol particle populations from a range of new and old diesel sanitation trucks operating over a wide range of on-road conditions.

In the second study another important question examined was that of the soot loading by public bus exhausts on school-going children who live in urban areas as opposed to those living in suburbs (Wu et al. 1998). Diesel supplies for the public buses in the city of Baltimore were spiked with Ir[III] octahedral pentanedionate dissolved in toluene and student exposure to soot emitted from the buses evaluated by analyzing Ir concentrations on the filters borne by individuals. It was found that the highest exposure of diesel soot occurred when an individual was getting on and off the vehicle. However, the Ir data demonstrate that the public bus derived soot contributed only 2–10% of the total carbon exposure to an individual.

11.5 Future Directions

These are exciting times to understand environmental transport of Os and Ir. Improvements in clean laboratory chemistry and measurement techniques are opening new avenues to examine increasingly lower level samples for Os and Ir concentrations and Os isotope composition. Although much has been learned at the observational level about the dispersal of Os and Ir, many unsolved problems exist and much remains to be discovered. First of all the natural distributions of Os and Ir especially their mass balances in modern oceans need to be worked out in detail. In spite of the mostly phenomenological studies to date, there remains a dearth of information about fundamental mechanisms underlying the transport of Os and Ir nature and in environment. The basic issue of whether Ir isotopes fractionate should be resolved with the help of modern mass spectrometry. A lot more work is also needed to assess the atmospheric burden of Os from base-metal sulfide smelting beginning from the industrial revolution to about 1980s when modern SO_2 scrubbing techniques began to be employed. Utilization of the PGEs is expected to continue to increase through this century and consequently the environmental burden of Os and Ir is also slated to rise. The processes that make these elements bioavailable and the extent to which they are toxic to life are topics of immediate importance (e.g., Ravindra et al. 2004; Smith et al. 1974; Wiseman and Zereini 2009). For example, recent work with rats shows that ingestion of soluble Ir could impact immune system (Iavicoli et al. 2010). Perhaps the biggest unknown in the assessment of transport and bioavailability of anthropogenic Os and Ir is the extent to which these elements occur as nanoparticles in automobile exhausts. It will be important in future studies to seek answers to the following questions related to environmental dispersal of the PGE:

1. How are Pt, Pd and Rh dispersals from catalytic converters related to the distributions of anthropogenic Ir and Os? In particular, analysis of size-segregated aerosol samples collected from major urban areas would be useful and so would be the data obtained directly from automobile exhausts.
2. What is the residence time of Os in the atmosphere? What are the scavenging mechanisms that remove Os from the atmosphere? Of interest here would be to investigate the observed/inferred differences between the path lengths of OsO_4 emitted by chromite smelters and PGE-ore processing plants.
3. What is the role of local sources in impacting the Os isotope composition of snow and rain?
4. Beginning from the industrial revolution what was the impact of the base metal sulfide ore refining on the atmospheric burden of the atmosphere? Both ice-core records and ombrotrophic peat records will be useful to address this question.

Acknowledgements I would like to thank Mark Baskaran for inviting me to write this review. I gratefully acknowledge reviews from Karl Turekian, Mark Baskaran, and an anonymous reviewer that led to considerable improvement of this paper. Over the years I have benefited from discussions with several individuals on various aspects of the geochemistry of Ir and Os. These span from professors, when I was a graduate student and a post doc, to my own graduate students and fellow researchers in the field. I would like to thank them all for generously giving me insights into this fascinating field, and into the art of making and evaluating the challenging measurements involved. I am especially grateful to Ariel Anbar, Gerhard Bruegmann, Cynthia Chen, Udo Fehn, Dimitri Papanastassiou, Bernhard Peucker-Ehrenbrink, Matthieu Roy-Barman, Mark Rehkamper, Karl Turekian, Rich Walker, and G.J. Wasserburg.

References

Abisheva ZS, Zagorognyaya AN, Bukurov TN (2001) Recovery of radiogenic osmium-187 from sulfide copper ores in Kazakhstan. Platinum Met Rev 45:132–135

Ahmed AH, Arai S (2002) Unexpectedly high-PGE chromitite from the deeper mantle section of the northern Oman ophiolite and its tectonic implications. Contrib Mineralog Petrol 143:263–278

Ahmed AH, Arai S, Abdel-Aziz YM et al (2009) Platinum-group elements distribution and spinel composition in podiform chromitites and associated rocks from the upper mantle section of the neoproterozoic Bou Azzer ophiolite, Anti-Atlas, Morocco. J Afr Earth Sci 55:92–104

Alvarez LW, Alvarez W, Asaro F et al (1980) Extraterrestrial cause for the Cretaceous-tertiary extinction – experimental results and theoretical interpretation. Science 208:1095–1108

Anbar AD (1996) I. Rhenium and Iridium in natural waters. II. Methyl bromide: ocean sources, ocean sinks, and climate sensitivity. III. CO_2 stability and heterogeneous chemistry in the atmosphere of Mars. Division of Geological and Planetary Sciences, California Institute of Technology, Pasadena

Anbar AD, Wasserburg GJ, Papanastassiou DA et al (1996) Iridium in natural waters. Science 273:1524–1528

Anbar AD, Papanastassiou DA, Wasserburg GJ (1997) Determination of iridium in natural waters by clean chemical extraction and negative thermal ionization mass spectrometry. Anal Chem 69:2444–2450

Barbante C, Veysseyre A, Ferrari C et al (2001) Greenland snow evidence of large scale atmospheric contamination for platinum, palladium, and rhodium. Environ Sci Technol 35:835–839

Barefoot RR, Van Loon JC (1999) Recent advances in the determination of the platinum group elements and gold. Talanta 49:1–14

Barker JL, Anders E (1968) Accretion rate of cosmic matter from iridium and osmium contents of deep-sea sediments. Geochim Cosmochim Acta 32:627

Bate GL, Huizenga JR (1963) Abundances of ruthenium, osmium and uranium in some cosmic and terrestrial sources. Geochim Cosmochim Acta 27:345–360

Birck JL, Roy Barman M, Capmas F (1997) Re-Os isotopic measurements at the femtomole level in natural samples. Geostandards Newsl 20:19–27

Büchl A, Brugmann G, Batanova VG (2004) Formation of podiform chromitite deposits: Implications from PGE abundances and Os isotopic compositions of chromites from the Troodos complex, Cyprus. Chem Geol 208:217–232

Burke T, Fagliano J, Goldoft M et al (1991) Chromite ore processing residue in Hudson County, New-Jersey. Environ Health Perspect 92:131–137

Burton KW, Gannoun A, Parkinson IJ (2010) Climate driven glacial-interglacial variations in the osmium isotope composition of seawater recorded by planktic foraminifera. Earth Planet Sci Lett 295:58–68

Byrne RH (2002) Inorganic speciation of dissolved elements in seawater: the influence of pH on concentration ratios. Geochem Trans 3:11–16

Cave RR, Ravizza GE, German CR et al (2003) Deposition of osmium and other platinum-group elements beneath the ultramafic-hosted rainbow hydrothermal plume. Earth Planet Sci Lett 210:65–79

Chen C, Sharma M (2009) High precision and high sensitivity measurements of osmium in seawater. Anal Chem 81:5400–5406

Chen C, Taylor S, Sharma M (2005) Iron and osmium isotopes from stony micrometeorites and implications for the Os budget of the ocean. In: Lunar planetary science conference XXXVI. LPI, Houston

Chen C, Sharma M, Bostick BC (2006) Lithologic controls on osmium isotopes in the rio orinoco. Earth Planet Sci Lett 252:138–151

Chen C, Sedwick PN, Sharma M (2009) Anthropogenic osmium in rain and snow reveals global-scale atmospheric contamination. Proc Natl Acad Sci U S A 106:7724–7728

Clayton DD (1964) Cosmoradiogenic chronologies of nucleosynthesis. Astrophys J 139:637–663

Cohen AS, Waters FG (1996) Separation of osmium from geological materials by solvent extraction for analysis by thermal ionisation mass spectrometry. Anal Chim Acta 332:269–275

Colodner DC, Boyle EA, Edmond JM et al (1992) Postdepositional mobility of platinum, iridium and rhenium in marine-sediments. Nature 358:402–404

Colodner DC, Boyle EA, Edmond JM (1993) Determination of rhenium and platinum in natural-waters and sediments, and iridium in sediments by flow-injection isotope-dilution inductively coupled plasma-mass spectrometry. Anal Chem 65:1419–1425

Cook DL, Walker RJ, Horan MF et al (2004) Pt-Re-Os systematics of group IIAB and IIIAB iron meteorites. Geochim Cosmochim Acta 68:1413–1431

Creaser RA, Papanastassiou DA, Wasserburg GJ (1991) Negative thermal ion mass-spectrometry of osmium, rhenium, and iridium. Geochim Cosmochim Acta 55:397–401

Dalai TK, Ravizza G (2006) Evaluation of osmium isotopes and iridium as paleoflux tracers in pelagic carbonates. Geochim Cosmochim Acta 70:3928–3942

Dalai TK, Ravizza G (2010) Investigation of an early Pleistocene marine osmium isotope record from the eastern equatorial Pacific. Geochim Cosmochim Acta 74:4332–4345

Dalai TK, Suzuki K, Minagawa M et al (2005) Variations in seawater osmium isotope composition since the last glacial maximum: a case study from the Japan Sea. Chem Geol 220:303–314

Dalai TK, Ravizza GE, Peucker-Ehrenbrink B (2006) The late Eocene Os-187/Os-188 excursion: chemostratigraphy, cosmic dust flux and the early Oligocene glaciation. Earth Planet Sci Lett 241:477–492

Date AR, Davis AE, Cheung YY (1987) The potential of fire assay and inductively coupled plasma source-mass spectrometry for the determination of platinum group elements in geological-materials. Analyst 112:1217–1222

Economou-Eliopoulos M (1996) Platinum-group element distribution in chromite ores from ophiolite complexes: implications for their exploration. Ore Geol Rev 11:363–381

Edmonds HN, Michael PJ, Baker ET et al (2003) Discovery of abundant hydrothermal venting on the ultraslow-spreading Gakkel Ridge in the Arctic. Nature 421:252–256

Enzweiler J, Potts PJ, Jarvis KE (1995) Determination of platinum, palladium, ruthenium and iridium in geological samples by isotope-dilution inductively-coupled plasma-mass spectrometry using a sodium peroxide fusion and tellurium coprecipitation. Analyst 120:1391–1396

Esser BK, Turekian KK (1988) Accretion rate of extraterrestrial particles determined from osmium isotope systematics of pacific pelagic clay and manganese nodules. Geochim Cosmochim Acta 52:1383–1388

Esser BK, Turekian KK (1993a) Anthropogenic osmium in coastal deposits. Environ Sci Technol 27:2719–2724

Esser BK, Turekian KK (1993b) The osmium isotopic composition of the continental-crust. Geochim Cosmochim Acta 57:3093–3104

Fehn U, Teng R, Elmore D et al (1986) Isotopic composition of osmium in terrestrial samples determined by accelerator mass-spectrometry. Nature 323:707–710

Frei R, Frei KM (2002) A multi-isotopic and trace element investigation of the cretaceous-tertiary boundary layer at Stevns Klint, Denmark – inferences for the origin and nature of siderophile and lithophile element geochemical anomalies. Earth Planet Sci Lett 203:691–708

Fresco J, Weiss HV, Phillips RB et al (1985) Iridium in seawater. Talanta 32:830–831

Gabrielli P, Barbante C, Plane JMC et al (2004a) Meteoric smoke fallout over the Holocene epoch revealed by iridium and platinum in Greenland ice. Nature 432:1011–1014

Gabrielli P, Varga A, Barbante C et al (2004b) Determination of Ir and Pt down to the sub-femtogram per gram level in polar ice by ICP-SFMS using preconcentration and a desolvation system (vol 19, pg 831, 2004). J Anal At Spectrom 19:831–837

Gabrielli P, Plane JMC, Boutron CF et al (2006) A climatic control on the accretion of meteoric and super-chondritic iridium-platinum to the Antarctic ice cap. Earth Planet Sci Lett 250:459–469

Gabrielli P, Barbante C, Plane JMC et al (2008) Siderophile metal fallout to Greenland from the 1991 winter eruption of Hekla (Iceland) and during the global atmospheric perturbation of Pinatubo. Chem Geol 255:78–86

Gannoun A, Burton KW, Vigier N et al (2006) The influence of weathering process on riverine osmium isotopes in a basaltic terrain. Earth Planet Sci Lett 243:732–748

Gelinas A, Kring DA, Zurcher L et al (2004) Osmium isotope constraints on the proportion of bolide component in Chicxulub impact melt rocks. Meteorit Planet Sci 39:1003–1008

Gobeil C, Sundby B, Macdonald RW et al (2001) Recent change in organic carbon flux to arctic ocean deep basins: evidence from acid volatile sulfide, manganese and rhenium discord in sediments. Geophys Res Lett 28:1743–1746

Goonan TG (2005) Flows of selected materials associated with world copper smelting. In: USGS Open-File Report 2004-1395, Reston

Gordon GW, Rockman M, Turekian KK et al (2009) Osmium isotopic evidence against an impact at the Frasnian-Famennian boundary. Am J Sci 309:420–430

Gregoire DC (1988) Determination of platinum, palladium, ruthenium and iridium geological-materials by inductively coupled plasma mass-spectrometry with sample introduction by electrothermal vaporization. J Anal At Spectrom 3:309–314

Gros M, Lorand JP, Luguet A (2002) Analysis of platinum group elements and gold in geological materials using NiS fire assay and Te coprecipitation; the NiS dissolution step revisited. Chem Geol 185:179–190

Hassler DR, Peucker-Ehrenbrink B, Ravizza GE (2000) Rapid determination of Os isotopic composition by sparging OsO_4 into a magnetic-sector icp-ms. Chem Geol 166:1–14

Heller-Zeisler SF, Borgoul PV, Moore RR et al (2000) Comparison of INAA and RNAA methods with thermal-ionization mass spectrometry for iridium determinations in atmospheric tracer studies. J Radioanal Nucl Chem 244:93–96

Helz GR, Adelson JM, Miller CV et al (2000) Osmium isotopes demonstrate distal transport of contaminated sediments in Chesapeake Bay. Environ Sci Technol 34:2528–2534

Herr W, Hoffmeister W, Hirt B et al (1961) Versuch zur datierung von eisenmeteoriten nach der rhenium-osmium-methode. Z Naturforsch 16:1053–1058

Hintenberger H, Herr W, Voshage H (1954) Radiogenic osmium from rhenium-containing molybdenite. Phys Rev 95:1690–1691

Hodge V, Stallard M, Koide M et al (1986) Determination of platinum and iridium in marine waters, sediments, and organisms. Anal Chem 58:616–620

Hoffman EL, Naldrett AJ, Vanloon JC et al (1978) Determination of all platinum group elements and gold in rocks and ore by neutron-activation analysis after preconcentration by a nickel sulfide fire-assay technique on large samples. Anal Chim Acta 102:157–166

Horan MF, Walker RJ, Morgan JW et al (2003) Highly siderophile elements in chondrites. Chem Geol 196:5–20

Iavicoli I, Carelli G, Bocca B et al (2008) Environmental and biological monitoring of iridium in the city of Rome. Chemosphere 71:568–573

Iavicoli I, Fontana L, Marinaccio A et al (2010) Iridium alters immune balance between T helper 1 and T helper 2 responses. Hum Exp Toxicol 29:213–219

Jackson SE, Fryer BJ, Gosse W et al (1990) Determination of the precious metals in geological-materials by inductively coupled plasma mass-spectrometry (ICP-MS) with nickel sulfide fire-assay collection and tellurium coprecipitation. Chem Geol 83:119–132

Jiang SY, Yang JH, Ling HF et al (2007) Extreme enrichment of polymetallic Ni-Mo-PGE-Au in lower Cambrian black shales of south China: an Os isotope and PGE geochemical investigation. Palaeogeogr Palaeoclimatol Palaeoecol 254:217–228

Jickells TD, An ZS, Andersen KK et al (2005) Global iron connections between desert dust, ocean biogeochemistry, and climate. Science 308:67–71

Jones RT (1999) Platinum smelting in South Africa. S Afr J Sci 95:525–534

Karner DB, Levine J, Muller RA et al (2003) Extraterrestrial accretion from the GISP2 ice core. Geochim Cosmochim Acta 67:751–763

Keays RR, Ganapath R, Laul JC et al (1974) Simultaneous determination of 20 trace-elements in terrestrial, lunar and meteoritic material by radio-chemical neutron-activation analysis. Anal Chim Acta 72:1–29

Koeberl C, Shirey SB (1993) Detection of a meteoritic component in Ivory-Coast tektites with Rhenium-Osmium isotopes. Science 261:595–598

Koeberl C, Shirey SB (1997) Re-Os isotope systematics as a diagnostic tool for the study of impact craters and distal ejecta. Palaeogeogr Palaeoclimatol Palaeoecol 132:25–46

Koeberl C, Reimold WU, Shirey SB (1994a) Saltpan impact crater, South-Africa – geochemistry of target rocks, breccias, and impact glasses, and osmium isotope systematics. Geochim Cosmochim Acta 58:2893–2910

Koeberl C, Reimold WU, Shirey SB et al (1994b) Kalkkop crater, Cape Province, South-Africa – confirmation of impact origin using osmium isotope systematics. Geochim Cosmochim Acta 58:1229–1234

Koeberl C, Sharpton VL, Schuraytz BC et al (1994c) Evidence for a meteoritic component in impact melt rock from the Chicxulub structure. Geochim Cosmochim Acta 58:1679–1684

Koeberl C, Farley KA, Peucker-Ehrenbrink B et al (2004) Geochemistry of the end-permian extinction event in Austria and Italy: no evidence for an extraterrestrial component. Geology 32:1053–1056

Koide M, Goldberg ED, Niemeyer S et al (1991) Osmium in marine-sediments. Geochim Cosmochim Acta 55:1641–1648

Koide M, Goldberg ED, Walker R (1996) The analysis of seawater osmium. Deep Sea Res II 43:53–55

Kyte FT, Wasson JT (1986) Accretion rate of extraterrestrial matter – iridium deposited 33 to 67 million years ago. Science 232:1225–1229

Lal D, Jull AJT (2003) Extra-terrestrial influx rates of cosmogenic isotopes and platinum group elements: realizable geochemical effects. Geochim Cosmochim Acta 67:4925–4933

Lee CTA, Wasserburg GJ, Kyte FT (2003) Platinum-group elements (PGE) and rhenium in marine sediments across the cretaceous-tertiary boundary: constraints on Re-PGE transport in the marine environment. Geochim Cosmochim Acta 67:655–670

Lee SR, Horton JW, Walker RJ (2006) Confirmation of a meteoritic component in impact-melt rocks of the Chesapeake Bay impact structure, Virginia, USA – evidence from osmium isotopic and PGE systematics. Meteorit Planet Sci 41:819–833

Levasseur S, Birck JL, Allegre CJ (1998) Direct measurement of femtomoles of osmium and the Os-187/Os-186 ratio in seawater. Science 282:272–274

Levasseur S, Birck JL, Allegre CJ (1999) The osmium riverine flux and the oceanic mass balance of osmium. Earth Planet Sci Lett 174:7–23

Li SH, Chai ZF, Mao XY (2007) Iridium in the Bering Sea and Arctic Ocean studied by neutron activation analysis. J Radioanal Nucl Chem 271:125–128

Love SG, Brownlee DE (1993) A direct measurement of the terrestrial mass accretion rate of cosmic dust. Science 262:550–553

Luck JM, Allegre CJ (1983) Re-187-Os-187 systematics in meteorites and cosmochemical consequences. Nature 302:130–132

Luck JM, Turekian KK (1983) Os-187/Os-186 in manganese nodules and the cretaceous-tertiary boundary. Science 222:613–615

Marcantonio F, Turekian KK, Higgins S et al (1999) The accretion rate of extraterrestrial He-3 based on oceanic Th-230 flux and the relation to Os isotope variation over the past 200,000 years in an Indian ocean core. Earth Planet Sci Lett 170:157–168

Martin JM, Meybeck M (1979) Elemental mass-balance of material carried by major world rivers. Mar Chem 7:173–206

Martin CE, Peucker-Ehrenbrink B, Brunskill G et al (2001) Osmium isotope geochemistry of a tropical estuary. Geochim Cosmochim Acta 65:3193–3200

McDaniel DK, Walker RJ, Hemming SR et al (2004) Sources of osmium to the modern oceans: new evidence from the Pt-190-Os-186 system. Geochim Cosmochim Acta 68:1243–1252

McDonald I, Peucker-Ehrenbrink B, Coney L et al (2007) Search for a meteoritic component in drill cores from the Bosumtwi impact structure, Ghana: platinum group element contents and osmium isotopic characteristics. Meteorit Planet Sci 42:743–753

McDonough WF, Sun S-S (1995) Composition of the earth. Chem Geol 120:223–253

Meisel T, Moser J (2004) Reference materials for geochemical PGE analysis: new analytical data for Ru, Rh, Pd, Os, Ir, Pt and Re by isotope dilution ICP-MS in 11 geological reference materials. Chem Geol 208:319–338

Meisel T, Krahenbuhl U, Nazarov MA (1995) Combined osmium and strontium isotopic study of the cretaceous-tertiary boundary at Sumbar, Turkmenistan – a test for an impact vs a volcanic hypothesis. Geology 23:313–316

Meisel T, Walker RJ, Morgan JW (1996) The osmium isotopic composition of the earth's primitive upper mantle. Nature 383:517–520

Michel HV, Alvarez WA, Alvarez LW (1990) Geochemical studies of the cretaceous-tertiary boundary in ODP holes 689b and 690c. In: Scientific results, Proceedings of the ocean drilling program. Scientific Results, Proc Ocean Drill Prog 113:159–168

Morcelli CPR, Figueiredo AMG (2000) Determination of iridium at sub ng levels in geological materials by RNAA. J Radioanal Nucl Chem 244:619–621

Morgan JW (1965) Simultaneous determination of rhenium and osmium in rocks by neutron activation analysis. Anal Chim Acta 32:8

Naldrett AJ (2004) Magmatic sulfide deposits. Springer, Berlin, p 727

Nogueira CA, Figueiredo AMG (1995) Determination of platinum, palladium, iridium and gold in selected geological reference materials by radiochemical neutron-activation analysis – comparison of procedures based on aqua regia leaching and sodium peroxide sintering. Analyst 120:1441–1443

Oxburgh R (1998) Variations in the osmium isotope composition of sea-water over the past 200,000 years. Earth Planet Sci Lett 159:183–191

Oxburgh R (2001) Residence time of osmium in the oceans. Geochem Geophys Geosyst 2:2000GC000104

Oxburgh R, Pierson-Wickmann AC, Reisberg L et al (2007) Climate-correlated variations in seawater Os-187/Os-188 over the past 200,000 yr: evidence from the cariaco basin, venezuela. Earth Planet Sci Lett 263:246–258

Palmer MR, Turekian KK (1986) Os-187/Os-186 in marine manganese nodules and the constraints on the crustal geochemistries of rhenium and osmium. Nature 319:216–220

Palmer MR, Falkner KK, Turekian KK et al (1988) Sources of osmium isotopes in manganese nodules. Geochim Cosmochim Acta 52:1197–1202

Paquay FS, Ravizza GE, Dalai TK et al (2008) Determining chondritic impactor size from the marine osmium isotope record. Science 320:214–218

Paquay FS, Goderis S, Ravizza G et al (2009) Absence of geochemical evidence for an impact event at the Bolling-Allerød/Younger Dryas transition. Proc Natl Acad Sci U S A 106:21505–21510

Pasava J (1993) Anoxic sediments an important environment for PGE; an overview. Ore Geol Rev 8:425–445

Pattou L, Lorand JP, Gros M (1996) Non-chondritic platinum-group element ratios in the earth's mantle. Nature 379:712–715

Paul M, Reisberg L, Vigier N (2009) A new method for analysis of osmium isotopes and concentrations in surface and subsurface water samples. Chem Geol 258:136–144

Paul M, Reisberg L, Vigier N et al (2010) Dissolved osmium in Bengal plain groundwater: implications for the marine Os budget. Geochim Cosmochim Acta 74:3432–3448

Pearson DG, Woodland SJ (2000) Solvent extraction/anion exchange separation and determination of PGEs (Os, Ir, Pt, Pd, Ru) and Re-Os isotopes in geological samples by isotope dilution ICP-MS. Chem Geol 165:87–107

Pedersen RB, Johannesen GM, Boyd R (1993) Stratiform platinum-group element mineralizations in the ultramafic cumulates of the Leka ophiolite complex, central Norway. Econ Geol Bull Soc Econ Geol 88:782–803

Pegram WJ, Turekian KK (1999) The osmium isotopic composition change of Cenozoic sea water as inferred from a deep-sea core corrected for meteoritic contributions. Geochim Cosmochim Acta 63:4053–4058

Pegram WJ, Krishnaswami S, Ravizza GE et al (1992) The record of sea-water Os-187/Os-186 variation through the Cenozoic. Earth Planet Sci Lett 113:569–576

Peucker-Ehrenbrink B (1996) Accretion of extraterrestrial matter during the last 80-million-years and. Geochim Cosmochim Acta 60:3187–3196

Peucker-Ehrenbrink B (2002) Comment on "residence time of osmium in the oceans" by Rachel Oxburgh. Geochem Geophys Geosyst 3:1

Peucker-Ehrenbrink B, Jahn BM (2001) Rhenium-osmium isotope systematics and platinum group element concentrations: loess and the upper continental crust. Geochem Geophys Geosyst 2:26

Peucker-Ehrenbrink B, Ravizza G (2000a) The effects of sampling artifacts on cosmic dust flux estimates: a reevaluation of nonvolatile tracers (Os, Ir). Geochim Cosmochim Acta 64:1965–1970

Peucker-Ehrenbrink B, Ravizza G (2000b) The marine osmium isotope record. Terra Nova 12:205–219

Peucker-Ehrenbrink B, Ravizza G, Hofmann AW (1995) The marine Os-187/Os-186 record of the past 80-million years. Earth Planet Sci Lett 130:155–167

Plessen HG, Erzinger J (1998) Determination of the platinum-group elements and gold in twenty rock reference materials by inductively coupled plasma-mass spectrometry (ICP-MS) after pre-concentration by nickel sulfide fire assay. Geostandards Newsl 22:187–194

Poirier A, Gariepy C (2005) Isotopic signature and impact of car catalysts on the anthropogenic osmium budget. Environ Sci Technol 39:4431–4434

Prichard HM, Lord RA (1993) An overview of the PGE concentrations in the Shetland ophiolite complex. In: Prichard HM, Alabaster T, Harris NBW, Neary CR (eds) Magmatic processes and plate tectonics, vol 76. Geological Society, London

Prichard HM, Neary CR, Fisher PC et al (2008) PGE -rich podiform chromitites in the Al'ays ophiolite complex, Saudi Arabia: an example of critical mantle melting to extract and concentrate PGE. Econ Geol 103:1507–1529

Puchtel IS, Walker RJ, Brandon AD et al (2009) Pt-Re-Os and Sm-Nd isotope and HSE and REE systematics of the 2.7 Ga Belingwe and Abitibi komatiites. Geochim Cosmochim Acta 73:6367–6389

Rauch S, Hemond HF, Peucker-Ehrenbrink B (2004) Source characterisation of atmospheric platinum group element deposition into an ombrotrophic peat bog. J Environ Monit 6:335–343

Rauch S, Hemond HF, Peucker-Ehrenbrink B et al (2005) Platinum group element concentrations and osmium isotopic composition in urban airborne particles from Boston. Massachusetts Environ Sci Technol 39:9464–9470

Rauch S, Peucker-Ehrenbrink B, Molina LT et al (2006) Platinum group elements in airborne particles in Mexico city. Environ Sci Technol 40:7554–7560

Rauch S, Peucker-Ehrenbrink B, Kylander ME et al (2010) Anthropogenic forcings on the surficial osmium cycle. Environ Sci Technol 44:881–887

Ravindra K, Bencs L, Van Grieken R (2004) Platinum group elements in the environment and their health risk. Sci Total Environ 318:1–43

Ravizza GE, Bothner MH (1996) Osmium isotopes and silver as tracers of anthropogenic metals in sediments from Massachusetts and Cape Cod Bays. Geochim Cosmochim Acta 60:2753–2763

Ravizza G, Dalai TK (2006) The potential of Ir and Os isotopes as point paleoflux tracers. Geochim Cosmochim Acta 70:A520

Ravizza G, Peucker-Ehrenbrink B (2003) Chemostratigraphic evidence of Deccan volcanism from the marine osmium isotope record. Science 302:1392–1395

Ravizza G, Pyle D (1997) PGE and Os isotopic analyses of single sample aliquots with NiS fire assay preconcentration. Chem Geol 141:251–268

Rehkamper M, Halliday AN (1997) Development and application of new ion-exchange techniques for the separation of the platinum group and other siderophile elements from geological samples. Talanta 44:663–672

Rehkamper M, Halliday AN, Wentz RF (1998) Low-blank digestion of geological samples for platinum-group element analysis using a modified carius tube design. Fresenius J Anal Chem 361:217–219

Reisberg L, Meisel T (2002) The Re-Os isotopic system: a review of analytical techniques. Geostandards Newsl 26:249–267

Robinson N, Ravizza G, Coccioni R et al (2009) A high-resolution marine Os-187/Os-188 record for the late Maastrichtian: distinguishing the chemical fingerprints of Deccan volcanism and the K-Pg impact event. Earth Planet Sci Lett 281:159–168

Rodushkin I, Engstrom E, Sorlin D et al (2007a) Osmium in environmental samples from northeast Sweden. Part II. Identification of anthropogenic sources. Sci Total Environ 386:159–168

Rodushkin I, Engstrom E, Sorlin D et al (2007b) Osmium. In environmental samples from northeast Sweden – part I. Evaluation of background status. Sci Total Environ 386:145–158

Russ GP, Bazan JM (1987a) Isotopic ratio measurements with an inductively coupled plasma source-mass spectrometer. Spectrochim Acta B 42:49–62

Russ GP, Bazan JM (1987b) Osmium isotopic ratio measurements by inductively coupled plasma source-mass spectrometry. Anal Chem 59:984–989

Schmitz B, Tassinari M, Peucker-Ehrenbrink B (2001) A rain of ordinary chondritic meteorites in the early Ordovician. Earth Planet Sci Lett 194:1–15

Sengupta JG, Gregoire DC (1989) Determination of ruthenium, palladium and iridium in 27 international reference silicate and iron-formation rocks, ores and related materials by isotope-dilution inductively-coupled plasma mass-spectrometry. Geostandards Newsl 13:197–204

Sharma M, Wasserburg GJ (1997) Osmium in the rivers. Geochim Cosmochim Acta 61:5411–5416

Sharma M, Papanastassiou DA, Wasserburg GJ (1997) The concentration and isotopic composition of osmium in the oceans. Geochim Cosmochim Acta 61:3287–3299

Sharma M, Wasserburg GJ, Hofmann AW et al (1999) Himalayan uplift and osmium isotopes in oceans and rivers. Geochim Cosmochim Acta 63:4005–4012

Sharma M, Wasserburg GJ, Hofmann AW et al (2000) Osmium isotopes in hydrothermal fluids from the Juan de Fuca ridge. Earth Planet Sci Lett 179:139–152

Sharma M, Polizzotto M, Anbar AD (2001) Iron isotopes in hot springs along the Juan de Fuca ridge. Earth Planet Sci Lett 194:39–51

Sharma M, Balakrishna K, Hofmann AW et al (2007a) The transport of osmium and strontium isotopes through a tropical estuary. Geochim Cosmochim Acta 71:4856–4867

Sharma M, Rosenberg EJ, Butterfield DA (2007b) Search for the proverbial mantle osmium sources to the oceans: hydrothermal alteration of mid-ocean ridge basalt. Geochim Cosmochim Acta 71:4655–4667

Shen JJ, Papanastassiou DA, Wasserburg GJ (1996) Precise Re-Os determinations and systematics of iron meteorites. Geochim Cosmochim Acta 60:2887–2900

Shinotsuka K, Suzuki K (2007) Simultaneous determination of platinum group elements and rhenium in rock samples using isotope dilution inductively coupled plasma mass spectrometry after cation exchange separation followed by solvent extraction. Anal Chim Acta 603:129–139

Shirey SB, Walker RJ (1995) Carius tube digestion for low-blank rhenium-osmium analysis. Anal Chem 67:2136–2141

Shirey SB, Walker RJ (1998) The Re-Os isotope system in cosmochemistry and high-temperature geochemistry. Annu Rev Earth Planet Sci 26:423–500

Simonetti A, Gariepy C, Carignan J (2003) Tracing sources of atmospheric pollution in western canada using the pb isotopic composition and heavy metal abundances of epiphytic lichens. Atmospheric Environment 37:2853–2865

Smith IC, Carson BL, Ferguson TL (1974) Osmium: an appraisal of environmental exposure. Environ Health Perspect 8:201–213

Sotnikov VI, Berzina AN, Economou-Eliopoulos M et al (2001) Palladium, platinum and gold distribution in porphyry Cu +/− Mo deposits of Russia and Mongolia. Ore Geol Rev 18:95–111

Stein HJ, Hannah JL, Zimmerman A et al. (2004) A 2.5 Ga porphyry Cu-Mo-Au deposit at Malanjkhand, central India: Implications for late Archean continental assembly. Precambrian Research 134:189–226

Suarez AE, Caffrey PF, Borgoul PV et al (1998) Use of an iridium tracer to determine the size distribution of aerosol emitted from a fleet of diesel sanitation trucks. Environ Sci Technol 32:1522–1529

Suzuki T, Miyada M, Ohta K et al (1998) Determination of osmium in waste water by graphite furnace atomic absorption spectrometry. Mikrochim Acta 129:259–263

Tanaka N, Oura Y, Ebihara M (2008) Determination of iridium and gold in rock samples by using pre-concentration neutron activation analysis. J Radioanal Nucl Chem 278:603–606

Teng RTD, Fehn U, Elmore D et al (1987) Determination of Os isotopes and Re/Os ratios using AMS. Nucl Instrum Meth Phys Res B 29:281–285

Truran JW (1998) The age of the universe from nuclear chronometers. Proc Natl Acad Sci U S A 95:18–21

Turekian KK, Luck JM (1984) Estimation of continental Os-187/Os-186 values by using Os-187/Os-186 and Nd-143/Nd-144 ratios in marine manganese nodules. Proc Natl Acad Sci U S A Phys Sci 81:8032–8034

Turekian KK, Sharma M, Gordon GW (2007) The behavior of natural and anthropogenic osmium in the Hudson river-Long Island Sound estuarine system. Geochim Cosmochim Acta 71:4135–4140

Volkening J, Walczyk T, Heumann KG (1991) Osmium isotope ratio determinations by negative thermal ionization mass-spectrometry. Int J Mass Spectrom Ion Process 105:147–159

Walczyk T, Heumann KG (1993) Iridium isotope ratio measurements by negative thermal ionization mass-spectrometry and atomic-weight of iridium. Int J Mass Spectrom Ion Process 123:139–147

Walker RJ (1988) Low-blank chemical-separation of rhenium and osmium from gram quantities of silicate rock for measurement by resonance ionization mass-spectrometry. Anal Chem 60:1231–1234

Walker RJ (2009) Highly siderophile elements in the Earth, Moon and Mars: update and implications for planetary accretion and differentiation. Chem Erde Geochem 69:101–125

Walker RJ, Fassett JD (1986) Isotopic measurement of subnanogram quantities of rhenium and osmium by resonance ionization mass-spectrometry. Anal Chem 58:2923–2927

Walker RJ, Morgan JW, Beary ES et al (1997) Applications of the Pt-190-Os-186 isotope system to geochemistry and cosmochemistry. Geochim Cosmochim Acta 61:4799–4807

Walker RJ, Horan MF, Morgan JW et al (2002a) Comparative Re-187-Os-187 systematics of chondrites: implications regarding early solar system processes. Geochim Cosmochim Acta 66:4187–4201

Walker RJ, Prichard HM, Ishiwatari A et al (2002b) The osmium isotopic composition of convecting upper mantle deduced from ophiolite chromites. Geochim Cosmochim Acta 66:329–345

Walker RJ, McDonough WF, Honesto J et al (2008) Modeling fractional crystallization of group IVB iron meteorites. Geochim Cosmochim Acta 72:2198–2216

Williams GA, Turekian KK (2002) Atmospheric supply of osmium to the oceans. Geochim Cosmochim Acta 66:3789–3791

Williams GA, Turekian KK (2004) The glacial-interglacial variation of seawater osmium isotopes as recorded in Santa Barbara basin. Earth Planet Sci Lett 228:379–389

Williams G, Marcantonio F, Turekian KK (1997) The behavior of natural and anthropogenic osmium in long island sound, an urban estuary in the eastern US. Earth Planet Sci Lett 148:341–347

Wiseman CLS, Zereini F (2009) Airborne particulate matter, platinum group elements and human health: a review of recent evidence. Sci Total Environ 407:2493–2500

Woodhouse OB, Ravizza G, Kenison-Falkner K et al (1999) Osmium in seawater: vertical profiles of concentration and isotopic composition in the eastern Pacific Ocean. Earth Planet Sci Lett 173:223–233

Wu CC, Suarez AE, Lin ZB et al (1998) Application of an Ir tracer to determine soot exposure to students commuting to school on Baltimore public buses. Atmos Environ 32:1911–1919

Yamashita Y, Takahashi Y, Haba H et al (2007) Comparison of reductive accumulation of Re and Os in seawater – sediment systems. Geochim Cosmochim Acta 71:3458–3475

Yang G, Hannah JL, Zimmerman A et al. (2009) Re-Os depositional age for Archean carbonaceous slates from the southwestern superior province: Challenges and insights. Earth Planet Sci Lett 280:83–92

Yi YV, Masuda A (1996a) Isotopic homogenization of iridium for high sensitivity determination by isotope dilution inductively coupled plasma mass spectrometry. Anal Sci 12:7–12

Yi YV, Masuda A (1996b) Simultaneous determination of ruthenium, palladium, iridium, and platinum at ultratrace levels by isotope dilution inductively coupled plasma mass spectrometry in geological samples. Anal Chem 68:1444–1450

Yokoyama T, Alexander CMO, Walker RJ (2010) Osmium isotope anomalies in chondrites: results for acid residues and related leachates. Earth Planet Sci Lett 291:48–59

Zaccarini F, Pushkarev EV, Fershtater GB et al (2004) Composition and mineralogy of PGE-rich chromitites in the Nurali lherzolite-gabbro complex, southern Urals, Russia. Can Mineralog 42:545–562

Zylberberg DR, Goldstein SL, Sharma M (2006) A 100ky record of the osmium isotopic composition of seawater. In: Eos transactions, vol 87(52), Fall meeting supplement, American Geophysical Union

Chapter 12
Applications of Stable Mercury Isotopes to Biogeochemistry

Joel D. Blum

Abstract The application of mercury (Hg) stable isotopes to problems in environmental chemistry is relatively new, but is a rapidly expanding field of investigation. Interest in this isotope system has been stimulated by concerns regarding the global distribution of mercury, its tendency to be methylated and bioaccumulated in the environment, and the severe health affects associated with this toxic metal. Mercury displays mass-dependent isotope fractionation during most biotic and abiotic chemical transformations. Additionally, mercury displays mass-independent fractionation, mainly during photochemical radical pair reactions, wherein the reactivity of odd and even mass number isotopes differs. The combination of mass-dependent and mass-independent fractionation provides a new and much needed probe into the reaction pathways and history of mercury in biogeochemical systems, and also provides an isotopic fingerprint of various mercury sources to the environment.

12.1 Introduction

Mercury is a geochemically unique element that has attracted a large amount of research interest because it is globally distributed and forms methylmercury (MeHg), which is a potent neurotoxin, in the environment. Once formed, MeHg is highly bioaccumulative and poses a significant health risk to humans and to a wide variety of wildlife. The speciation of Hg in the environment is controlled by microbial activity, dark abiotic reactions, and photochemical reactions. Hg speciation dramatically affects its mobility, which ranges from stable solid phases (e.g., HgS) to a globally distributed atmospheric gas [Hg(0)]. Speciation also strongly affects Hg toxicity and bioaccumulation, which are greatly enhanced when Hg is in the methylated form (e.g., Morel et al. 1998).

Mercury biogeochemistry is of great importance in both atmospheric and aqueous systems. It is estimated that present-day anthropogenic atmospheric emissions of Hg are about 3 times higher than natural emissions (Mason et al. 1994). Anthropogenic emissions are dominated by coal combustion, cement production and waste incineration, whereas primary natural emissions are dominated by volcanic and hydrothermal systems (US EPA 1997). There is also a growing amount of previously deposited natural and anthropogenically-derived Hg that can be re-emitted due to biomass burning and reduction to Hg(0) in both aquatic and terrestrial systems (Gustin et al. 2008).

Details of the speciation of Hg in the atmosphere are not well known, but Hg speciation is generally reported as three operationally defined forms: gaseous elemental mercury [$Hg(0)_{gas}$], reactive gaseous mercury [$Hg(II)_{gas}$] and particulate mercury [$Hg(II)_{part}$]. $Hg(0)_{gas}$ has a relatively long atmospheric residence time of ~1 year, but once it is photochemically oxidized to $Hg(II)_{gas}$ it becomes extremely reactive and is rapidly deposited in precipitation or attached to particles and deposited as $Hg(II)_{part}$ (Schroeder and Munthe 1998). Hg is distributed and deposited globally even to remote and otherwise pristine areas including the Arctic (e.g., Lindberg et al. 2002; Douglas et al. 2008).

In aqueous solutions Hg can exist in the reduced form as dissolved $Hg(0)_{aq}$, or in the oxidized Hg(II) and MeHg(II) forms associated with a wide range of dissolved inorganic and organic ligands and with

J.D. Blum (✉)
Department of Geological Sciences, University of Michigan, 1100 North University Avenue, Ann Arbor, MI 48109, USA
e-mail: jdblum@umich.edu

particle surfaces. Microbial, photochemical, and dark abiotic reactions are known to reduce Hg(II) in solution to Hg(0)$_{aq}$. In reducing environments some sulfate and iron-reducing bacteria can methylate Hg(II) to MeHg. Microbial and photochemical processes can also degrade MeHg back to Hg (II) or Hg(0)$_{aq}$ (e.g., Fitzgerald and Lamborg 2003). Once it is produced, MeHg can enter both aquatic and terrestrial foodwebs. It is strongly bio-accumulated and can reach toxic levels in piscivorous fish, mammals and birds as well as insectivorous birds (Evers et al. 2007). The main pathway for human mercury exposure is through the consumption of piscivorous marine and fresh water fish (Sunderland 2007). To present a graphical picture of environmental Hg isotope fractionations associated with transitions in Hg speciation, several of these transitions are shown as a flow-chart in Fig. 12.1. The small but rapidly growing literature on Hg isotope fractionation along these pathways is keyed to the diagram and will be discussed below.

Mercury has seven stable isotopes with the following approximate abundances for the National Institute of Standards and Technology (NIST) Standard Reference Material (SRM)-3133: ^{196}Hg = 0.155%, ^{198}Hg = 10.04%, ^{199}Hg = 16.94%, ^{200}Hg = 23.14%, ^{201}Hg = 13.17%, ^{202}Hg = 29.73%, ^{204}Hg = 6.83% (Blum and Bergquist 2007; Fig. 12.2a). These abundances are referenced to the certified ^{205}Tl/^{203}Tl ratio of 2.38714 for the NIST SRM-997 because absolute abundances for the Hg isotopes in NIST SRM-3133 have not been accurately determined. Institute for National Measurement Standards of the National Research Council Canada reference material NIMS-1 was recently certified for isotopic abundances at nearly indistinguishable values to those reported above (Meija et al. 2010). Because ^{196}Hg has a very low abundance and ^{204}Hg suffers from an isobaric interference with ^{204}Pb, the ^{202}Hg/^{198}Hg ratio is used to describe mass-dependent stable isotope fractionation (MDF) relative to the NIST SRM-3133. Hg isotopic compositions relative to NIST SRM-3133 are reported as:

$$\delta^{xxx}Hg(‰) = \{[(^{xxx}Hg/^{198}Hg)_{unknown}/(^{xxx}Hg/^{198}Hg)_{SRM3133}] - 1\} \times 1,000$$

Mass-independent isotope fractionation (MIF) has also been observed for the odd isotopes of Hg and is reported as Δ^{199}Hg and Δ^{201}Hg. These values are the differences between the measured δ^{199}Hg and δ^{201}Hg

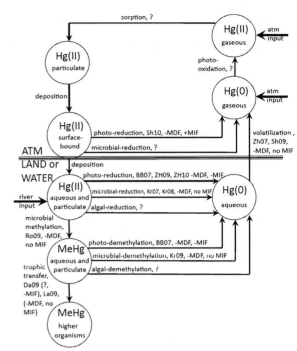

Fig. 12.1 Flow-chart showing some of the major biogeochemical transformations of Hg speciation in the environment and their associated Hg isotope fractionations. Atmospheric and riverine inputs of Hg to the global biogeochemical system are shown as *input arrows*. For those transformations where experimental fractionation studies have been carried out, the positive (+) or negative (−) mass dependent (MDF) and mass independent (MIF) fractionation of the product relative to the reactant is indicated. References to the publications relevant to each transformation are also given with the following abbreviations: BB07 = Bergquist and Blum (2007); Da09 = Das et al. (2009); Kr07,08,09 = Kritee et al. (2007, 2008, 2009); La09 = Laffont et al. (2009); Ro09 = Rodriguez-Gonzalez et al. (2009); Sh09,10 = Sherman et al. (2009, 2010); Zh07 = Zheng et al. (2007); ZH09,10 = Zheng and Hintelmann (2009, 2010b)

values and those predicted based on the measured δ^{202}Hg and the kinetic mass-dependent fractionation law derived from transition state theory. For variations of less than ~10‰, these values were approximated by Blum and Bergquist (2007) to be:

$$\Delta^{199}Hg = \delta^{199}Hg - (\delta^{202}Hg \times 0.2520)$$

$$\Delta^{200}Hg = \delta^{200}Hg - (\delta^{202}Hg \times 0.5024)$$

$$\Delta^{201}Hg = \delta^{201}Hg - (\delta^{202}Hg \times 0.7520)$$

$$\Delta^{204}Hg = \delta^{204}Hg - (\delta^{202}Hg \times 1.493)$$

The magnitude of mass-dependent and mass-independent fractionation will be described in this

12 Applications of Stable Mercury Isotopes to Biogeochemistry

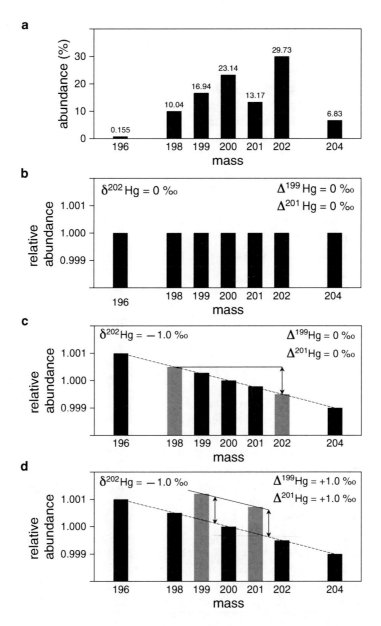

Fig. 12.2 Schematic diagrams illustrating (**a**) the average abundance of the Hg isotopes, (**b**) the abundance of the Hg isotopes relative to a standard reference material with no fractionation, (**c**) with negative mass-dependent fractionation, and (**d**) with negative mass-dependent and positive mass-independent fractionation

chapter primarily with δ^{202}Hg and Δ^{201}Hg (Fig. 12.2b–d). In addition to variations in the magnitude of these isotope ratios, trends in the ratios of the isotope ratios have been shown to be diagnostic of certain chemical transformation mechanisms (Bergquist and Blum 2007). The most diagnostic "ratio of ratios" that have been described in the literature are the Δ^{199}Hg/Δ^{201}Hg and Δ^{201}Hg/δ^{202}Hg ratios and these will be discussed extensively below.

12.2 Materials and Methods

12.2.1 Sampling and Sample Size

The unusual chemistry of mercury and the low concentration of this element in many environmental samples of interest necessitates special attention to sampling protocols. Sampling must be done under trace metal clean conditions and sample bottles must be acid leached in a multi-step cleaning procedure (e.g., Johnson et al. 2008). Unlike other trace metal clean preparations, to obtain low Hg blank levels, it is necessary to leach materials in the strong oxidant BrCl rather than just HCl and HNO_3. Sample sizes needed for Hg isotopic analysis scale with concentration, which can vary by over eight orders of magnitude in environmental samples. The total amount of Hg that is needed for a high-precision isotope measurement varies according to specific sample introduction systems and mass spectrometer sensitivities. Nevertheless the range typically quoted is 10 and 20 ng. With smaller amounts of Hg than 10–20 ng the precision of isotope ratio measurements can be considerably compromised.

12.2.2 Pre-Treatment

A variety of sample preservation protocols are needed and these differ depending on the material that is sampled. In general, solid samples such as rock and sediment can be stored at room temperature in clean plastic or glass containers. Vegetation, fish tissues and other organic samples should be stored in clean plastic or glass containers and either kept frozen or freeze-dried and refrigerated. Water samples require the most specialized attention and must be stored in either glass, Teflon or fluorinated polyethylene bottles. Traditional polyethylene and polypropylene bottles used for other trace elements are much more porous to $Hg(0)_{vapor}$ than Teflon or fluorinated polyethylene bottles and storage of water samples in these containers can result in Hg loss from the sample or gain from the ambient atmosphere. Depending on what fraction of Hg in water is of interest to a particular study, water samples are often filtered prior to further treatment. Samples are then acidified and oxidized typically with 1% BrCl to keep Hg stable in solution as Hg(II). Trace-metal clean reagents must be purged with Hg-free N or Ar to achieve the lowest Hg blank levels.

For the highest precision isotope analyses Hg should be quantitatively separated from samples and concentrated to produce ~4 g of a solution with ~4 ng/g Hg(II) for analysis by mass spectrometry. Liquid samples can be reduced with $SnCl_2$ and the produced Hg(0) can be sparged into an oxidizing solution of $KMnO_4$ or amalgamated onto gold by passing it through a gold-coated bead trap. Solid samples can be digested in acids and then diluted as long as Hg is kept fully oxidized in solution to avoid loss of Hg as Hg(0). Alternatively, solid samples can be combusted in a two-stage furnace to fully volatilize Hg as Hg(0) (Smith et al 2005; Biswas et al 2008) and the Hg(0) can be collected by bubbling into an oxidizing solution of $KMnO_4$ or by amalgamation onto gold. Hg(II) in $KMnO_4$ or diluted acidic solutions can then be reduced with $SnCl_2$, sparged from solution and bubbled into a smaller volume of $KMnO_4$ solution to purify and concentrate the Hg. Similarly, gold traps can be slowly heated in a flow of Hg-free N or Ar and bubbled into $KMnO_4$ solution to yield a solution of ~4 ng/g Hg. It is essential to measure the yield of the Hg extraction and concentration procedure for each sample, because incomplete yield (<90%) can produce significant levels of mass-dependent Hg isotope fractionation.

12.2.3 Instrument Used for Analyses

High precision stable Hg isotope analysis of environmental samples was made possible by two developments. The first was the advent of multiple collector inductively coupled plasma mass spectrometry (MC-ICP-MS) and the second was the development of continuous flow cold vapor introduction systems for MC-ICP-MS. The first study to combine these features and obtain high precision isotope ratios was Lauretta et al. (2001), who used a continuous flow cold vapor generator (CFCVG; Klaue and Blum 1999) fed directly into a VG P-54 MC-ICP-MS. Evans et al (2001), Hintelmann and Ogrinc (2003), Hintelmann and Lu (2003), Xie et al. (2005), Krupp and Donard (2005), Epov et al. (2008) and Sonke et al. (2008) developed methods for measurement of transient signals on MC-ICP-MS but the precision of this approach has not attained the level of performance of CFCVG. Jackson et al. (2004) coupled CFCVG to a Finnigan Neptune MC-ICP-MS but lacked optimal correction for instrumental mass bias. The continuous flow method was adapted to the Nu-Plasma MC-ICP-MS (e.g., Smith et al. 2005; Bergquist and Blum 2007) and the Thermo Neptune MC-ICP-MS (e.g., Foucher and Hintelmann 2006; Zheng et al. 2007; Zambardi et al. 2009; Estrade et al. 2010) with optimal mass bias

12.2.4 Detection Limit, Precision and Accuracy of the Method

The accuracy of Hg isotope ratio measurement is optimized by measuring isotope ratios relative to a standard reference material, which is analyzed before and after each unknown sample. Sample-standard bracketing alone can achieve good mass bias correction, but for the most precise Hg isotope measurements Tl is introduced to the Hg vapor stream as a dry aerosol and the ^{205}Tl/^{203}Tl ratio of NIST SRM-997 is first used for mass bias correction of the Hg isotope ratios, followed by sample-standard bracketing. The optimal conditions for analysis include complete separation of Hg from sample matrix, analysis of bracketing standards at the same concentration and in the same matrix solution as samples, careful monitoring of background signals, and performance of on-peak blank corrections. Under these conditions analytical precision of ±0.08‰ (2SD) for δ^{202}Hg and ±0.05‰ (2SD) for Δ^{201}Hg and Δ^{199}Hg are achievable (e.g., Blum and Bergquist 2007).

12.3 Discussion and Applications

The environmental isotope geochemistry of Hg is a topic of widespread interest because Hg contamination is globally pervasive and found in virtually all ecosystems and their various components. The integration of Hg isotopes into Hg research is in its earliest stages, but is growing at a very rapid rate. Published studies generally fall into one or more of the following three categories: (1) laboratory experimental studies to measure Hg isotope fractionation during biotic and abiotic chemical transformations of Hg, (2) measurements of Hg isotopic compositions in the environment to gain insight into biogeochemical transformations that are occurring or (3) measurements of Hg isotopic compositions in the environment and their use to fingerprint Hg sources. For the purpose of organizing this chapter I will discuss the published literature on experimental investigations first and then delve into applications of Hg isotopes to provide insight into environmental processes and sources of environmental Hg contamination. Like any rapidly developing field there are numerous meeting abstracts with important developments appearing on a monthly basis. In this review I have restricted discussion to only peer-reviewed articles published in research journals. I have also focused this review on applications of Hg isotope measurements to environmental problems rather than on the physical chemistry of Hg isotope fractionation or the characterization of geological reservoirs for Hg isotopes. These topics were recently reviewed by Bergquist and Blum (2009). In an attempt to organize the growing and somewhat complex literature on experimental Hg fractionation studies I have summarized study results that are relevant to environmental systems on a diagram showing important biogeochemical transformation of Hg in the atmosphere, on land and in aquatic systems (Fig. 12.1).

12.3.1 Experimental Studies

12.3.1.1 Mechanisms of Hg Isotope Fractionation

The environmental isotope geochemistry of Hg is a very rich topic owing to the widespread occurrence of both MDF and MIF. MDF is fractionation of isotopes in proportion to the differences in the masses of the isotopes and this has been observed to occur for Hg in most equilibrium and kinetic processes that have been studied. MIF is fractionation of isotopes that is not in proportion to the differences in the masses of the isotopes and MIF of Hg appears to be associated with two different mechanisms, the nuclear volume effect (NVE) and the magnetic isotope effect (MIE). The nuclear volume effect is an isotope fractionation caused because nuclear volume and nuclear charge radius do not scale linearly with mass. This isotope effect has been predicted by quantum mechanical calculations (Schauble 2007) and was recently observed in laboratory experiments (Estrade et al. 2009; Zheng and Hintelmann 2010a). The magnitude of the observed MIF due to the NVE is generally quite small (<0.2‰ for Δ^{199}Hg and Δ^{201}Hg) and results in Δ^{199}Hg/Δ^{201}Hg ratios of ~1.5–2 (Estrade et al. 2009; Zheng and Hintelmann 2010a). The MIE is a kinetic fractionation that appears to occur primarily during photochemical radical pair reactions (e.g., Buchachenko 2009). The magnetic spin of the odd isotopes enhances triplet to singlet and singlet to triplet intersystem crossing. As a result, reaction products can be either enriched or depleted in the odd isotopes of Hg depending on whether singlet or triplet radical pairs are produced photochemically. In laboratory experiments with natural sunlight, photochemical reduction/degradation of Hg(II) and MeHg in aqueous solution with natural DOC have been shown to produce reduced Hg reaction products with depletions of odd isotopes (−MIF; Bergquist and Blum 2007) whereas photochemical reduction of Hg-halogen compounds in snow have been shown to produce reduced Hg with enrichments of odd isotopes (+MIF; Sherman et al. 2010).

12.3.1.2 Biotic Experiments

Experimental studies of fractionation by higher organisms have not yet been performed, but experimental investigations have been carried out to quantify fractionation factors for some of the most important microbially mediated Hg transformations. Kritee et al. (2007, 2008, 2009) determined fractionation factors for Hg transformations mediated by organisms that express the mer-A and mer-B enzymes that reduce Hg(II) and demethylate MeHg, respectively. In both cases the Hg(0) that was produced had lower δ^{202}Hg than the starting Hg(II) or MeHg. Fractionation factors varied depending on the organisms, temperature, growth rate, and other experimental conditions. In no case did any of the microbial experiments produce detectable levels of MIF. A recent study of fermentative methylation by a sulphate reducing bacteria (Rodriguez-Gonzalez et al. 2009) showed that the MeHg product had lower δ^{202}Hg than the Hg(II) starting material (−MDF) and also found no evidence for biotic MIF during methylation.

12.3.1.3 Abiotic Experiments

The first abiotic Hg isotope fractionation experiments used natural sunlight to reduce Hg(II) and MeHg from aqueous solutions in the presence of DOC in the form of Suwannee River fulvic acid (Bergquist and Blum 2007). These experiments were, however, carried out at Hg/DOC ratios much higher than those observed in natural waters. The product, Hg(0), was found to have lower δ^{202}Hg than the reactant (−MDF), as in the case of microbial reduction, but unlike microbial reduction it was also depleted in the odd isotopes of Hg (−MIF). The isotopic trends with increasing amounts of reduction were different for Hg(II) reduction vs. MeHg reduction. The ratio of MDF/MIF (Δ^{201}Hg/δ^{202}Hg) was ~1.5 for Hg(II) reduction and ~3.0 for MeHg reduction. Additionally, the Δ^{199}Hg/Δ^{201}Hg ratio was 1.00 for Hg(II) reduction and 1.36 for MeHg reduction (Bergquist and Blum 2007). Yang and Sturgeon (2009) studied Hg(II) photoreduction using formic acid and a UV light-source and observed −MDF but did not observe significant MIF. The experimental conditions were quite different, but the specific factor(s) responsible for the lack of MIF during this study is not clear.

Photoreduction of Hg(II) was explored in detail in a series of laboratory experiments that tested the importance of the Hg/DOC ratio and the effect of different ligands on Hg isotope fractionation (Zheng and Hintelmann 2009, 2010b). The experiments of Zheng and Hintelmann (2009) differed from those of Bergquist and Blum (2007) in several ways, including the use of a filtered Xe lamp instead of natural sunlight, the use of natural lake water as the source of DOC, and the use of a range of Hg/DOC ratios that approached values found in natural systems. These experiments showed −MDF and −MIF, but also demonstrated a range of Δ^{199}Hg/Δ^{201}Hg ratios from 1.31 at high Hg/DOC, to a lower value of 1.19 at lower and more realistic values of Hg/DOC. The authors suggested that this range represented differing behavior of Hg bonded predominantly to reduced sulfur groups at low Hg/DOC and to oxygen groups at higher Hg/DOC.

Zheng and Hintelmann (2010b) expanded upon earlier work by studying fractionation during photochemical reduction of Hg(II) by individual low molecular weight organic acids that either contained, or did not contain, reduced sulfur groups. When reduced sulfur groups (e.g., cysteine) were present the authors observed fractionation similar to that observed using natural DOC (−MDF and −MIF). However, in the absence of reduced sulfur groups (e.g., serine) the reduction products were enriched in odd isotopes (+MIF), suggesting that opposite magnetic isotope effects were operating. This +MIF behavior is similar to that observed during the photochemical reduction of Hg(II) associated with halogens in Arctic snow (Sherman et al. 2010, see below).

Based on studies of Hg isotopes in natural hydrothermal systems, Smith et al. (2005) and Sherman et al. (2009) suggested that aqueous Hg(0) released from hydrothermal solutions would have lower δ^{202}Hg than Hg in solution (−MDF). This was demonstrated experimentally and quantified by Zheng et al. (2007) who showed that Hg(0) volatilization from an aqueous phase to a gas phase resulted in −MDF and no MIF. Yang and Sturgeon (2009) also showed that reduction of Hg(II) with $SnCl_2$ and sodium tetraethylborate ($NaBH_4$) resulted in −MDF and no MIF. In addition, Yang and Sturgeon (2009) inorganically ethylated Hg (II) using $NaBH_4$ and found −MDF but no MIF.

Estrade et al. (2009) showed that evaporation of $Hg(0)_{vapor}$ from pure metallic Hg(0) caused −MDF and a small degree of +MIF (<0.2‰). The Δ^{199}Hg/Δ^{201}Hg ratio for this MIF was ~2, suggesting that it was the result of the nuclear volume effect rather than the magnetic isotope effect (Estrade et al. 2009). Zheng and Hintelmann (2010a) studied the abiotic dark reduction of Hg(II) by DOC and $SnCl_2$ and found evidence for −MDF and a small (<−0.4‰) amount of −MIF in the reactant with a Δ^{199}Hg/Δ^{201}Hg ratio of 1.5–1.6. They also attributed this small amount of MIF to the nuclear volume effect.

12.3.1.4 Summary of Experimental Studies

As a way of organizing and summarizing the many results of experimental Hg fractionation experiments I have compiled some of the most important results on Fig. 12.3. Figure 12.3a is a plot of MDF vs. MIF (δ^{202}Hg vs. Δ^{201}Hg) with the value for SRM-3133 at the origin. In most environmental applications of Hg isotopes investigators are interested in either (1) interpreting the isotopic composition of MeHg occurring in sediments or organisms, or (2) interpreting the isotopic composition of Hg(II) associated with sediments or deposited onto the land surface or into a water body. Thus I have designed Fig. 12.3 to show the direction of the evolution of the isotopic composition of Hg as it might be expected to move through the biogeochemical cycle. If we begin with Hg(II) with an isotopic composition of SRM-3133, the arrows display the trajectory of isotopic evolution of the residual Hg(II) in the system during microbial reduction, microbial methylation, and photochemical reduction. If we begin with MeHg with an isotopic composition of SRM-3133, the arrows display the trajectory of isotopic evolution of the residual MeHg in the system during microbial and photochemical reduction (Fig. 12.3a). Thus if a set of samples have been affected by the same process, but to varying degrees, we might expect the sample analyses to plot along a line approximated by the lines shown on Fig. 12.3a.

In those processes where MIF occurs, the Δ^{199}Hg/Δ^{201}Hg ratio may be indicative of the reaction mechanism(s) that have affected Hg(II) or MeHg. As in Fig. 12.3a, b shows the direction of evolution of the isotopic composition of Hg as it might be expected to move through the biogeochemical cycle. If we begin with Hg(II) or MeHg with an isotopic composition of SRM-3133, the arrows display the trajectory of isotopic evolution of the residual Hg(II) or MeHg in the system during photochemical reduction (Fig. 12.3b). Thus, as in Fig. 12.3a, if a set of samples have been affected by the same photochemical process, but to varying degrees, we might expect them to plot along a line approximated by those shown on Fig. 12.3b.

12.3.2 Studies in the Environment

12.3.2.1 Fresh Water Sediments and Food Webs

Mercury isotope research on fresh water fisheries began inauspiciously with the publication of stable isotope ratios for sediments and food web animals measured using a quadrupole ICP-MS (Jackson 2001a). Unfortunately the data were not of high enough precision to draw conclusions regarding natural Hg isotope variation and this fact was soon pointed out in a comment by Hintelmann et al. (2001) for which there was also a reply (Jackson 2001b). A subsequent paper by Jackson et al. (2004) used MC-ICP-MS to analyze Hg isotopes in a sediment core from an Arctic lake and documented mass-dependent changes in the Hg isotopic composition with depth. These variations were attributed to microbial activities linked to oxidation–reduction reactions in the lake.

Bergquist and Blum (2007) measured Hg isotopes in fish from small New England lakes and from Lake Michigan with previously unprecedented analytical

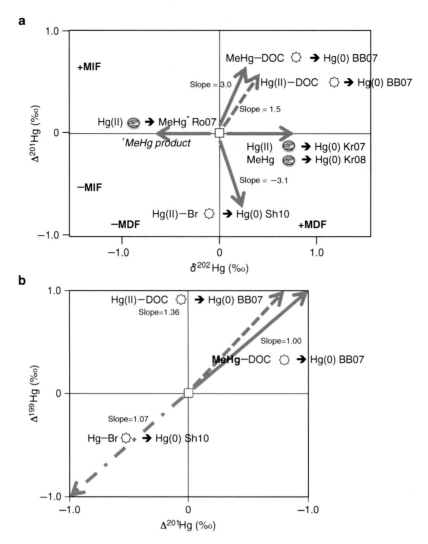

Fig. 12.3 Schematic diagram showing the direction of isotopic evolution of Hg(II) or MeHg beginning with an isotopic composition of NIST SRM-3133 and undergoing photochemical reduction (*sun symbols*), microbial reduction (*microbe symbols*), and microbial methylation (*microbe symbols*). Abbreviations for references to each study are given in the Fig. 12.1 caption. (**a**) Slopes of reaction pathways on a plot of δ^{202}Hg vs. Δ^{199}Hg, and (**b**) slopes of reaction pathways on a plot of Δ^{201}Hg vs. Δ^{199}Hg. "DOC" indicates experiments were carried out in the presence of dissolved organic carbon

precision. In reporting these data, Bergquist and Blum (2007) used an isotopic nomenclature suggested by Blum and Bergquist (2007) that standardized the reporting of Hg isotopic compositions relative to a commonly available standard (NIST SRM-3133). Using fractionation factors determined in the laboratory for photochemical reduction of Hg(II) and MeHg they argued that large variations in Δ^{199}Hg and Δ^{201}Hg in fish tissues resulted from photochemical degradation of MeHg in the water column prior to its entry into aquatic foodwebs. They also provided a means of estimating the proportion of MeHg that had been degraded and suggested that the Δ^{199}Hg/Δ^{201}Hg ratio might be diagnostic of reduction of Hg (II) vs. MeHg. Bergquist and Blum (2007) concluded that MeHg produced in each of three different locations in northern Lake Michigan had experienced differing amounts of photochemical demethylation (up to 76%) and that this was reflected in the Δ^{199}Hg and Δ^{201}Hg of fish tissues.

Subsequent studies of Hg isotopes in fish tissues have all measured positive MIF, but have come to conflicting conclusions as to the origin of this fractionation. Jackson et al. (2008) measured Hg isotopes in sediment, crustaceans, forage fish and predator fish from two boreal lakes and Lake Ontario. Based on correlations between MeHg concentration and Hg isotopic composition the authors argued that MIF of Hg occurred during microbial Hg methylation. More recent studies of isotope fractionation during microbial Hg methylation (Rodriguez-Gonzalez et al. 2009) have found no evidence for the occurrence of MIF during methylation. Das et al. (2009) measured Hg isotopes in fish and a zooplankton sample from a Florida lake and based on a correlation of MIF with $\delta^{15}N$ concluded that in vivo Hg MIF occurring in fish provided the simplest explanation for their data. However, a mechanism for this fractionation has not yet been identified and theoretical arguments have been made that suggest organismal in vivo fractionation is unlikely (Kritee et al. 2009). Laffont et al. (2009) measured Hg isotopes in fish and in human hair from a population eating those fish on a daily basis. They found no evidence for in vivo MIF of Hg in humans, but did find evidence for in vivo MDF of Hg in human hair by ~2‰ (as measured using $\delta^{202}Hg$). Finally, Gantner et al. (2009) measured Hg isotopes in sediments, zooplankton, invertebrates and fish from ten Arctic lakes. Each of these organisms displayed MIF and there was no clear pattern of Hg MIF during trophic transfer; in some lakes zooplankton MIF values even exceeded those of fish. The authors concluded that differences in MIF in fish were most likely caused by MeHg photoreduction in the water column and that MDF largely reflected regional differences due to atmospheric and/or terrestrial inputs of Hg to lakes.

12.3.2.2 Marine Sediments and Food Webs

Mercury isotopes were first measured in marine sediments by Gehrke et al. (2009) in a study of mid-Pleistocene (955 kyr) Mediterranean sediment cores. Large fluctuations in Hg concentration were observed in alternating layers of organic rich and organic poor sediments (sapropels) but the Hg isotopic composition of these sediments remained relatively constant at $\delta^{202}Hg = -0.91‰$ (± 0.15, 1SD, $n = 5$) and $\Delta^{201}Hg = 0.04‰$ (± 0.03, 1SD, $n = 5$). These measurements provided the first estimate of a pre-anthropogenic marine Hg isotope composition. Senn et al. (2010) measured modern marine sediments in coastal waters in the Gulf of Mexico (GOM) and found Hg isotope ratios of $\delta^{202}Hg = -1.00‰$ (± 0.11, 1SD, $n = 6$) and $\Delta^{201}Hg = -0.04‰$ (± 0.03, 1SD, $n = 6$). The GOM $\delta^{202}Hg$ values are indistinguishable from the Mediterranean sapropels and the $\Delta^{201}Hg$ values are only slightly higher.

To address the question of whether coastal vs. oceanic species of fish in the Gulf of Mexico obtain Hg from the same or different sources, Senn et al. (2010) investigated Hg isotopes and $\delta^{15}N$ of fish tissues collected in the GOM. Coastal species had distinct and non-overlapping values of $\delta^{202}Hg$ and $\Delta^{201}Hg$ compared to oceanic species and this isotopic contrast along with $\delta^{15}N$ values was used to argue for isotopically distinct MeHg sources. It was proposed that oceanic MeHg was produced in the open ocean and then underwent substantial photodegradation before entering the base of the oceanic food web. This study is described in more detail below as a case study.

12.3.2.3 Point-Source Contamination of Sediments

Several studies have explored the use of Hg isotopes to "fingerprint" point sources of Hg from historic Hg mines and trace them into the surrounding environment. Early studies determined that there was substantial variability in the $\delta^{202}Hg$ of cinnabar (HgS) ores from various mining districts, ore deposits and volcanic/hydrothermal systems (Hintelmann and Lu 2003; Smith et al. 2005, 2008; Sherman et al. 2009; Zambardi et al. 2009), and explored MDF of Hg isotopes during ore formation. Stetson et al. (2009) showed that the roasting of HgS ore to produce metallic mercury imparted a MDF signature that allows isotopic distinction between roasted ore mine waste (calcine) and un-roasted ore. This provides a means of isotopically differentiating between calcine and background Hg in stream runoff.

Foucher et al. (2009) isotopically characterized cinnabar ores and highly contaminated river sediments from a Hg mining region in Slovenia and then showed that the $\delta^{202}Hg$ value of Hg in downstream river sediments and in sediments in the Adriatic Sea could be used to determine the proportion of Hg from the mine. Feng et al. (2010) performed a similar study but used Hg isotopes to fingerprint metallic Hg that was used as a catalyst in industrial acid production. The authors linked Hg in industrial runoff to Hg in contaminated sediments from a downstream reservoir and confirmed this result by showing a contrasting isotopic composition in sediment from an unaffected reservoir.

12.3.2.4 Atmospheric Processes

Mercury is unique as a trace metal pollutant because it has a common volatile form [Hg(0)] with a long (~1 year) atmospheric lifetime, and is thus distributed and deposited globally. There is great interest in understanding sources of mercury to the atmosphere, sources of Hg in wet and dry deposition to the Earth's surface, and the atmospheric redox reactions that lead to the emission and deposition of Hg. These are complex scientific problems that are analytically challenging due to the difficulties of measuring Hg isotope ratios at low atmospheric concentrations. Biswas et al. (2008) measured Hg isotopes in coals, soils and Arctic peat deposits to evaluate sources and sinks for atmospheric Hg. Most coal deposits had negative values of δ^{202}Hg and Δ^{201}Hg but they varied widely isotopically, offering the possibility that Hg emissions from coal might be distinguishable from global background values in certain coal producing regions. Soils and peat deposits tend to be enriched in Hg concentration near the surface due to deposition from anthropogenic sources and Biswas et al. (2008) found that surface organic soils and peats also had negative values of δ^{202}Hg and Δ^{199}Hg, possibly due to deposition from the atmosphere. Ghosh et al. (2008) reported Δ^{201}Hg (but not δ^{202}Hg) in peat cores and Spanish moss and also found negative (or near 0) values of Δ^{201}Hg. Carignan et al. (2009) investigated Hg isotopes in tree lichens, which have been widely used as biomonitors of atmospheric metal deposition. This study also found consistently negative values of δ^{202}Hg and Δ^{201}Hg, suggesting that these values might be indicative of atmospheric Hg isotopic composition. This is consistent with the hypothesis that biotic, abiotic, and aqueous photochemical reduction of Hg(II) together produce Hg(0) with negative values of δ^{202}Hg and Δ^{201}Hg and that this Hg is ultimately deposited to the Earth's surface and sequestered by various forms of organic matter.

Recent measurements of the isotopic composition of Hg in atmospheric samples (Sherman et al. 2010; Gratz et al. 2010) have, however, complicating the view that lichens, peat and organic soils are representative of the isotopic composition of atmospheric Hg. Measurements of Hg(0) in the atmosphere and Hg(II) deposited with snowfall at Barrow, Alaska yielded δ^{202}Hg and Δ^{201}Hg values close to 0 (Sherman et al. 2010). Measurements of atmospheric Hg(0) and Hg(II) at Dexter, MI showed average δ^{202}Hg and Δ^{201}Hg values close to 0 for Hg(0) and values that were slightly negative and positive for δ^{202}Hg and Δ^{201}Hg, respectively, in Hg(II) in precipitation (Gratz et al. 2010). These data raise the question of why lichens, mosses and soils have negative δ^{202}Hg and Δ^{201}Hg, whereas atmospheric values are 0 or positive. It appears that either the atmospheric Hg isotopic composition is very heterogeneous and samples of Hg(0) have not captured the same values as organic materials exposed to the atmosphere, or there is isotope fractionation that shifts values for vegetation relative to atmospheric deposition. Recent experiments by Sherman et al. (2010) showed that photochemical reduction of Hg(II) from snow produced Hg(0) with a higher Δ^{201}Hg (and Δ^{199}Hg) than that of the Hg remaining in snow. This fractionation is in the opposite direction of that caused by the photochemical reduction of Hg(II) associated with organic matter in aqueous solution (Bergquist and Blum 2007). Zheng and Hintelmann (2010b) also recently demonstrated experimentally that photochemical reduction of Hg(II) can produce Hg(0) with either positive or negative Δ^{201}Hg depending on the ligand to which the Hg(II) was originally attached. Thus, although it remains to be tested rigorously, these data open up the possibility that vegetation acquires negative Δ^{201}Hg due to the loss of Hg(0) by photochemical reduction (Gratz et al. 2010).

12.3.2.5 Case Study from Gulf of Mexico Fisheries

The most important pathway for human exposure to Hg is via the consumption of marine fish (Sunderland 2007), yet there is still considerable uncertainty concerning the sources of Hg to different fisheries and the factors that control the concentrations of Hg in these fisheries (e.g., Sunderland et al. 2009). The Gulf of Mexico (GOM) is a highly productive marine region and a major fishery (Chesney et al. 2000). Fish are harvested in the GOM from both the shallow-water (coastal) region, within tens of kilometers of shore, and from open-ocean (oceanic) regions in deep water hundreds of kilometers from shore. Senn et al. (2010) used N and Hg isotopes in these fisheries to gain insight into the sources and pathways of Hg to these coastal and oceanic fish. This study provides a particularly illustrative example and is used here with some supplementary data to demonstrate how the results of experimental Hg isotope fractionation studies can be linked to environmental data sets to obtain insight into Hg sources and biogeochemical cycling.

Muscle tissue from coastal and oceanic species of fish was collected and additional samples were collected from fish species that are transitional in their habitat (Senn et al. 2010). Samples were analyzed for Hg concentration and the isotopic composition of Hg and N. A plot of $\delta^{15}N$ vs. Hg concentration revealed two separate, but parallel, trends for coastal vs. oceanic fish indicating disconnected foodwebs based in nitrogen from continental runoff in the coastal areas and open-ocean nitrogen sources for the oceanic fish, with transitional species occupying positions intermediate between these sources (Senn et al. 2010). Figure 12.4a is a plot of $\delta^{202}Hg$ vs. $\Delta^{201}Hg$ for the coastal, transitional, and open ocean fish as well as Hg isotope data for the top 2 cm of coastal sediment cores. Sediments have highly negative $\delta^{202}Hg$ but $\Delta^{201}Hg$ values close to 0. These values are consistent with geological sources of Hg (Gehrke et al. 2009) as well as with Hg used in industrial applications (Smith et al. 2008), or a combination of these sources. Published Hg isotope values for atmospheric deposition in the GOM are not yet available, but values for the isotopic composition of precipitation from a rural site in Michigan (Dexter, MI; Gratz et al. 2010) that is relatively unaffected by point sources is plotted on Fig. 12.4c, and provides an estimate of the expected isotopic composition of atmospheric Hg in the GOM.

Coastal fish from a variety of species have $\delta^{202}Hg$ values ranging from the sediment values (−0.82 to −1.11‰) up to a much higher value of −0.03‰. However, unlike the sediments, the coastal fish display MIF with $\Delta^{201}Hg$ values of 0.25–0.50‰. Oceanic fish (yellowfin and bluefin tuna) have much higher, and non-overlapping, values of both $\delta^{202}Hg$ and $\Delta^{201}Hg$ compared to coastal fish. Small yellowfin tuna (<16 kg) have the highest $\Delta^{201}Hg$ values observed in the GOM. Large yellowfin tuna from the Pacific Ocean near Hawaii, which were collected in 1998 and studied by Kraepiel et al. (2003), also display high $\Delta^{201}Hg$ values (Blum unpublished; Fig. 12.4). Transitional species of fish in the GOM have both $\delta^{202}Hg$ and $\Delta^{201}Hg$ values intermediate between coastal and open ocean species – but with Spanish mackerel displaying more coastal values consistent with their largely coastal range and bluerunner displaying closer to deep water values consistent with their tendency to spend part of their life-cycle in oceanic waters. Because the GOM fish tissues display MIF, it is instructive to also plot the data as $\Delta^{199}Hg$ vs. $\Delta^{201}Hg$ (Fig. 12.4b). All of the fish samples from the GOM plot on a single line with slope of 1.20 ± 0.02. Samples of GOM coastal sediment and yellowfin tuna from the Pacific Ocean also plot very close to this line, although with somewhat greater scatter.

The main purpose of presenting this case study is to demonstrate how one can interpret the trends of biogeochemical data (Fig. 12.4a, b) within the context of the available experiment results. From Fig. 12.4a we know that the Hg in all fish tissues has undergone significant amounts of both MDF and MIF. Based on their experimental results, Bergquist and Blum (2007) argued that a slope of ~1.36 in fish tissues would result from photochemical demethylation of MeHg in the water column prior to entering the food web, whereas a slope of ~1.00 would result from photochemical reduction of Hg(II) prior to methylation. The authors measured a $\Delta^{199}Hg/\Delta^{201}Hg$ slope of 1.28 in burbot fish from Lake Michigan and concluded that photochemical degradation of MeHg was the dominant reaction producing MIF. However, with new experimental data from Zheng and Hintelmann (2009, 2010b) showing that the slope of $\Delta^{199}Hg/\Delta^{201}Hg$ varies for photochemical reduction of Hg(II) depending on the Hg/DOC ratio and ligand–Hg binding, it appears that these data may be more complex to interpret than previously realized. Additional data on the $\Delta^{199}Hg/\Delta^{201}Hg$ slope during Hg(II) reduction and MeHg degradation under conditions specifically relevant to marine photochemical reactions will be needed to form more definitive conclusions. Nevertheless, empirical observations of $\Delta^{199}Hg/\Delta^{201}Hg$ slopes in natural materials show values of ~1.0 for processes that involve Hg(II) photoreduction (Biswas et al. 2008; Ghosh et al. 2008; Carignan et al. 2009; Sherman et al. 2010) whereas values in aquatic foodwebs, which

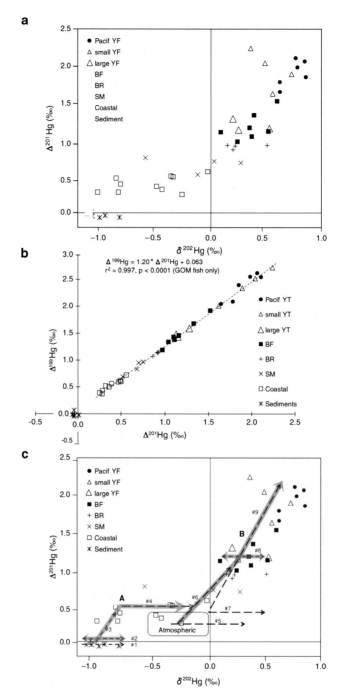

Fig. 12.4 (a) Plot of data for δ^{202}Hg vs. Δ^{201}Hg for Gulf of Mexico coastal sediments and both coastal and oceanic fish muscle tissues (Senn et al. 2010), as well as Pacific Ocean oceanic fish muscle tissues (Blum, unpublished). YT is yellowfin tuna; BF is blackfin tuna; BR is bluerunner; SM is Spanish mackerel; "coastal" includes red snapper, grey snapper, bay anchovy, speckled trout, red drum, flounder and menhaden (b) Plot of Δ^{201}Hg vs. Δ^{199}Hg for the same samples as in (a). Linear regression is for all GOM fish samples but excludes Pacific yellowfin tuna and sediments. (c) Same plot as in (a) but with the addition of a field for regional background precipitation from a remote site (Dexter, MI) from Gratz et al. (2010). *Shaded arrows* indicate inferred isotopic evolution of the MeHg ultimately bio-accumulated in coastal (A) and oceanic (B) fish in the GOM. *Dashed arrows* with numbers represent isotope fractionation pathways expected for nine different processes that are discussed in the text

uniquely involve MeHg, display slopes of ~1.2–1.3 (Bergquist and Blum 2007; Gantner et al. 2009; Laffont et al. 2009; Senn et al. 2010).

We return now to the relationship between δ^{202}Hg and Δ^{201}Hg for the GOM samples using Fig. 12.4c, which is a schematic illustration of how various Hg transformations known to occur in the aquatic environment would be expected to affect the isotopic values of the GOM samples. Isotope values for GOM sediments, which contain Hg predominantly in the form of Hg(II), display no MIF and have δ^{202}Hg values consistent with eroded geological materials or Hg used in industrial processes. A reasonable assumption based on previous studies of marine systems is that Hg in these reduced sediments is methylated by sulfate reducing bacteria, incorporated into the base of the coastal foodweb and then bioaccumulated in coastal fish (e.g., Fitzgerald and Lamborg 2003). Microbial methylation would drive the produced MeHg to lower δ^{202}Hg values and microbial demethylation would increase δ^{202}Hg values in remaining MeHg. However, production of significant MIF of Hg isotopes appears to require photochemistry. Because MIF is observed in fish but not in sediments it is most likely that photochemical demethylation produces the observed MIF. This MIF would be accompanied by MDF and might follow the trajectories on Fig. 12.3a. Based on the experimental results of Bergquist and Blum (2007), the MIF observed in coastal fish would represent demethylation of ~15% of the originally produced MeHg prior to incorporation into the food web. Microbial degradation of MeHg before entry into the food web would also drive the residual MeHg to higher values of δ^{202}Hg and could account for some of the large variability in the δ^{202}Hg values for coastal fish. Although some MDF probably also occurs during trophic transfer of MeHg through the food web, trophic transfer to coastal fish does not appear to be a major driver of the range in δ^{202}Hg values in coastal fish for three reasons. First, some fish have δ^{202}Hg values that are the same as the sediment values; second, fish of the same species display nearly the full range in coastal fish δ^{202}Hg values (Senn et al. 2010); and third, there is no correlation between coastal fish δ^{202}Hg values and δ^{15}N, which is an indicator of trophic position (Senn et al. 2010).

The next question Senn et al. (2010) address is why the Hg isotopic composition of oceanic fish is different than coastal fish. One possible means by which oceanic fish might acquire Hg is by feeding on coastal fish (Fitzgerald et al. 2007). The higher δ^{202}Hg values for oceanic fish could possibly be explained by trophic transfer, but the higher Δ^{201}Hg are inconsistent with this hypothesis. Instead MeHg was most likely photochemically demethylated to a greater extent than in the coastal ecosystem, prior to being incorporated into the oceanic food web. Based on the experimental results of Bergquist and Blum (2007) the MIF observed in oceanic fish would represent demethylation of ~40–60% of the produced MeHg. Microbial demethylation in the open ocean could push δ^{202}Hg values in residual MeHg to higher values, and trophic transfer could also push δ^{202}Hg values to higher values, but the observed higher Δ^{201}Hg values require higher degrees of photochemical MeHg degradation. Thus it is possible that MeHg was produced in the coastal zone and transported to the open ocean with additional photo-demethylation occurring along the way, or that Hg was deposited to the open ocean as Hg(II) through the atmosphere and was methylated and subsequently demethylated in the open ocean.

Several recent studies have argued that open ocean methylation can and does occur (e.g., Sunderland et al. 2009; Monperrus et al. 2007). Based on δ^{15}N values, which indicated contrasting N sources to the coastal and oceanic food webs, Senn et al. (2010) argued that the most likely scenario for the pathway of Hg into GOM open ocean fish was deposition of atmospheric Hg and methylation within the open ocean. Although we do not yet know the isotopic composition of Hg deposited from the atmosphere in the GOM, based on atmospheric Hg isotopic composition data from Michigan (Gratz et al. 2010), photochemical demethylation of Hg that was deposited from the atmosphere as Hg(II) and methylated in the open ocean, corresponds well to the range of values observed in oceanic fish (Fig. 12.4c). Senn et al. (2010) also observed that small yellowfin tuna (<16 kg) from the GOM display higher Δ^{201}Hg than large yellowfin tuna, which suggests that they obtain MeHg from a food web in which a higher degree of photochemical demethylation has occurred prior to entry into that food web. Graham et al. (2007) showed that juvenile yellowfin tuna near the Hawaiian Islands feed in the shallow mixed layer and that as they mature they shift to prey in deeper, colder water, which could provide an explanation of the Δ^{201}Hg data for small yellowfin tuna from the GOM. Large yellowfin tuna from near the Hawaiian Islands have higher Δ^{201}Hg than large yellowfin tuna from the GOM (Blum, unpublished), suggesting a larger degree of photochemical demethylation of the Hg in the foodweb of the large Hawaiian yellowfin tuna.

As a summary of the GOM fisheries case study, trajectories are plotted for likely Hg speciation transitions on a δ^{202}Hg vs. Δ^{201}Hg diagram with the GOM data shown in the background (Fig. 12.4c). This conveniently illustrates the possible pathways for the evolution of the Hg isotopic composition within this biogeochemical system. The range of sediment Hg isotope values could be due to contrasting sources of Hg to sediment and could be shifted to higher δ^{202}Hg, but not higher Δ^{201}Hg, by microbial reduction (path #1). As Hg in sediments is methylated, MeHg can shift to higher and lower δ^{202}Hg by microbial methylation and demethylation (path #2), respectively, and to higher δ^{202}Hg and Δ^{201}Hg by photochemical demethylation (path #3). Coastal fish show about the same amount of photo-demethylation but a wide range in δ^{202}Hg; this could be due to varying amounts of microbial demethylation (path #4) or a mixture of MeHg originating from multiple sources such as sediment and atmospheric deposition. The overall pathway that best explains the Hg isotope values of the coastal fish is shown with a gray arrow and labeled "A".

Atmospheric Hg(II) deposited to the ocean surface could be shifted to higher δ^{202}Hg by microbial reduction (path #5) or to higher δ^{202}Hg and Δ^{201}Hg by photochemical reduction (path #6) following the trajectories shown. Transfer of MeHg through the food web could cause an increase in δ^{202}Hg, but not Δ^{201}Hg (path #7). If oceanic fish fed on coastal fish we would expect some increase in δ^{202}Hg due to trophic transfer following path #7, but no change in Δ^{201}Hg. If, however, Hg(II) was deposited to the ocean, photochemically reduced (path #6), microbially methylated and demethylated in the water column (path #8), and then photochemically demethylated (path #9), this could produce the isotopic patterns observed in oceanic fish. This pathway is consistent with the pathway proposed by Monperrus et al. (2007) for production of oceanic MeHg and is shown as the thick gray arrow and labeled "B" on Fig. 12.4c.

12.4 Future Directions

The first set of papers attempting to apply Hg isotope measurements to biogeochemical problems appeared in 2001, and after approximately 6 years of working through the analytical details of how to produce high precision isotope ratio measurements, the field of Hg isotope biogeochemistry began to flourish in 2007. Since that time several research groups have applied Hg isotope measurements to a wide variety of research problems and at last count there were ~20 papers published predominantly on applications of Hg isotopes and another ~30 on experiments and method development related to Hg isotope biogeochemistry. The rate of growth of this field of research has been rapid, and will continue to grow quickly as the Hg biogeochemistry research community realizes the wide utility of this new approach to the study of Hg in the environment. Like all new stable isotope systems the main challenges will be in understanding all of the various fractionation mechanisms and their fractionation factors, and learning how to separate isotopic variation due to different sources of Hg to the environment from isotopic variation imparted by chemical processes occurring in the environment. However, Hg is unique among stable isotopic systems in the multitude of isotope fractionation signals that are built into the isotopic system. In addition to the basic mass-dependent and mass-independent fractionations recorded as δ^{202}Hg and Δ^{201}Hg, there are several other diagnostic ratios that hold important additional information including the Δ^{199}Hg/Δ^{201}Hg and Δ^{199}Hg/δ^{202}Hg ratios.

With continued laboratory experimentation to better refine the understanding of Hg fractionation mechanisms and higher precision measurements of Hg isotope ratios in environmental samples, there will be an important place for Hg isotopes in future studies of virtually all aspects of Hg biogeochemistry. The areas where we may see the greatest breakthroughs are perhaps those areas where we have made the least progress using concentration analyses of Hg species alone. These include the detailed understanding of many complex multi-step chemical pathways such as: (1) the atmospheric chemistry that leads to the deposition of Hg(0) from the atmosphere as Hg(II)$_{gas}$ and Hg(II)$_{part}$, (2) the chemical bonding environments in sediments that leads to various Hg fractions with chemical behaviors ranging from very recalcitrant to readily bioavailable and (3) the biological pathways that begin with methylation and demethylation and transfer Hg through the food web to top predators. Hg isotopes will undoubtedly also serve as fingerprints of sources of Hg contamination, but it is anticipated that the most significant breakthroughs will come from the insight that Hg isotopes add to our understanding of the chemical processes and pathways of Hg in the environment.

Acknowledgements I wish to thank all of the past and present members of my Hg isotope research group for their persistence, dedication and intellectual contributions to helping develop an important new isotopic system. I also thank J. Sonke and an anonymous reviewer for providing helpful comments that improved the manuscript. This chapter was written while the author was a CIRES visiting fellow at the University of Colorado, and this program is gratefully acknowledged.

References

Bergquist BA, Blum JD (2007) Mass-dependent and -independent fractionation of Hg isotopes by photo-reduction in aquatic systems. Science 318:417–420

Bergquist BA, Blum JD (2009) Mass dependent and independent fractionation of mercury isotopes. Elements 5:353–357

Biswas A, Blum JD, Bergquist BA et al (2008) Natural mercury isotope variation in coal deposits and organic soils. Environ Sci Technol 42:8303–8309

Blum JD, Bergquist BA (2007) Reporting of variations in the natural isotopic composition of mercury. Anal Bioanal Chem 233:353–359

Buchachenko AL (2009) Mercury isotope effects in the environmental chemistry and biochemistry of mercury-containing compounds. Russ Chem Rev 78:319–328

Carignan J, Estrade N, Sonke JE et al (2009) Odd isotope deficits in atmospheric Hg measured in lichens. Environ Sci Technol 43:5660–5664

Chesney EJ, Baltz DM, Thomas RG (2000) Louisiana estuarine and coastal fisheries and habitats: perspectives from a fish's eye view. Ecol Appl 10:350–366

Das R, Salters VJM, Odom AL (2009) A case for in vivo mass-independent fractionation of mercury isotopes in fish. Geochem Geophys Geosyst 10:Q11012

Douglas TA, Sturm M, Simpson WR, Blum JD, Alvarez-Aviles L, Keeler GJ, Perovich DK, Biswas A, Johnson K (2008) The influence of snow and ice crystal formation and accumulation of mercury deposition to the arctic. Environ Sci Technol 42:1542–1551

Epov VN, Rodriguez-Gonzalez P, Sonke JE et al (2008) Simultaneous determination of species-specific isotopic composition of Hg by gas chromatography coupled to multicollector ICPMS. Anal Chem 80:3530–3538

Estrade N, Carignan J, Sonke JE et al (2009) Mercury isotope fractionation during liquid-vapor evaporation experiments. Geochim Cosmochim Acta 73:2693–2711

Estrade N, Carignan J, Sonke JE et al (2010) Measuring Hg isotopes in bio-geo-environmental reference materials. Geostandard Geoanal Res 34:79–93

Evans RD, Hintelmann H, Dillon PJ (2001) Measurement of high precision isotope ratios for mercury from coals using transient signals. J Anal At Spectrom 16:1064–1069

Evers DC, Han Y-J, Driscoll CT et al (2007) Biological mercury hotspots in the northeastern United States and southeastern Canada. BioScience 57:29–43

Feng XB, Foucher D, Hintelmann H et al (2010) Tracing mercury contamination sources in sediments using mercury isotope compositions. Environ Sci Technol 44:3363–3368

Fitzgerald WF, Lamborg CH (2003) Geochemistry of mercury in the environment. In: Lollar BS (ed) Treatise on geochemistry, vol 9. Elsevier, New York, pp 107–148

Fitzgerald WF, Lamborg CH, Hammerschmidt CR (2007) Marine biogeochemical cycling of mercury. Chem Rev 107:641–662

Foucher D, Hintelmann H (2006) High-precision measurement of mercury isotope ratios in sediments using cold-vapor generation multi-collector inductively coupled plasma mass spectrometry. Anal Bioanal Chem 384:1470–1478

Foucher D, Ogrinc N, Hintelmann H (2009) Tracing mercury contamination from the Idrija mining region (Slovenia) to the Gulf of Trieste using Hg isotope ratio measurements. Environ Sci Technol 43:33–39

Gantner N, Hintelmann H, Zheng W et al (2009) Variations in stable isotope fractionation of Hg in food webs of Arctic lakes. Environ Sci Technol 43:9148–9154

Gehrke GE, Blum JD, Meyers PA (2009) The geochemical behavior and isotopic composition of Hg in a mid-Pleistocene Mediterranean sapropel. Geochim Cosmochim Acta 73:1651–1665

Ghosh S, Xu YF, Humayun M et al (2008) Mass-independent fractionation of mercury isotopes in the environment. Geochem Geophys Geosyst 9:Q03004

Graham BS, Grubbs D, Holland K et al (2007) A rapid ontogenetic shift in the diet of juvenile yellowfin tuna from Hawaii. Mar Biol 150:647–658

Gratz LE, Keeler GJ, Blum JD, Sherman LS (2010) Isotopic composition and fractionation of mercury in Great Lakes precipitation and ambient air. Environ Sci Technol 44:7764–7770

Gustin MS, Lindberg SE, Weisberg PJ (2008) An update on the natural sources and sinks of atmospheric mercury. Appl Geochem 23:482–493

Hintelmann H, Lu SY (2003) High precision isotope ratio measurements of mercury isotopes in cinnabar ores using multi-collector inductively coupled plasma mass spectrometry. Analyst 128:635–639

Hintelmann H, Ogrinc N (2003) Determination of stable mercury isotopes by ICP/MS and their application in environmental studies. ACS Symp Ser 835:321–338

Hintelmann H, Dillon P, Evans RD et al (2001) Comment: variations in the isotope composition of mercury in a freshwater sediment sequence and food web. Can J Fish Aquat Sci 58:2309–2311

Jackson TA (2001a) Variations in the isotope composition of mercury in a freshwater sediment sequence and food web. Can J Fish Aquat Sci 58:185–196

Jackson TA (2001b) Reply: variations in the isotope composition of mercury in a freshwater sediment sequence and food web. Can J Fish Aquat Sci 58:2312–2316

Jackson TA, Muir DCG, Vincent WF (2004) Historical variations in the stable isotope composition of mercury in Arctic lake sediments. Environ Sci Technol 38:2813–2821

Jackson TA, Whittle DM, Evans MS et al (2008) Evidence for mass-independent and mass-dependent fractionation of the stable isotopes of mercury by natural processes in aquatic ecosystems. Appl Geochem 23:547–571

Johnson KP, Blum JD, Keeler GJ et al (2008) A detailed investigation of the deposition and re-emission of mercury in arctic snow during an atmospheric mercury depletion event. J Geophys Res. doi:10.1029/2008JD009893

Klaue B, Blum JD (1999) Trace analysis of arsenic in drinking water by inductively coupled plasma mass spectrometry: high resolution versus hydride generation. Anal Chem 71:1408–1414

Kraepiel AML, Keller K, Chin HB et al (2003) Sources and variations of mercury in tuna. Environ Sci Technol 37:5551–5558

Kritee K, Blum JD, Johnson MW et al (2007) Mercury stable isotope fractionation during reduction of Hg(II) to Hg(0) by mercury resistant bacteria. Environ Sci Technol 41:1889–1895

Kritee K, Blum JD, Barkay T (2008) Constraints on the extent of mercury stable isotope fractionation during reduction of Hg(II) by different microbial species. Environ Sci Technol 42:9171–9177

Kritee K, Barkay T, Blum JD (2009) Mass dependent stable isotope fractionation of mercury during microbial degradation of methylmercury. Geochim Cosmochim Acta 73:1285–1296

Krupp EA, Donard OFX (2005) Isotope ratios on transient signals with GC-MC-ICP-MS. Int J Mass Spectrom 242:233–242

Laffont L, Sonke JE, Maurice L et al (2009) Anomalous mercury isotopic compositions of fish and human hair in the Bolivian Amazon. Environ Sci Technol 43:8985–8990

Lauretta DS, Klaue B, Blum JD et al (2001) Mercury abundances and isotopic composition in the Murchison (CM) and Allende (CV) carbonaceous chondrites. Geochim Cosmochim Acta 65:2807–2818

Lindberg SE, Brooks S, Lin C-J et al (2002) Dynamic oxidation of gaseous mercury in the troposphere at polar sunrise. Environ Sci Technol 36:1245–1256

Mason RP, Fitzgerald WF, Morel FMM (1994) The biogeochemical cycling of elemental mercury – anthropogenic influences. Geochim Cosmochim Acta 58:3191–3198

Meija J, Yang L, Sturgeon RE et al (2010) Certification of natural isotopic abundance inorganic mercury reference material NIMS-1 for absolute isotopic composition and atomic weight. J Anal At Spectrom 25:384–389

Monperrus M, Tessier E, Amouroux D et al (2007) Mercury methylation, demethylation and reduction rates in coastal and marine surface waters of the Mediterranean Sea. Mar Chem 107:49–63

Morel FMM, Kraepiel AML, Amyot M (1998) The chemical cycle and bioaccumulation of mercury. Annu Rev Ecol Syst 29:543–566

Rodriguez-Gonzalez P, Epov VN, Bridou R et al (2009) Species-specific stable isotope fractionation of mercury during Hg(II) methylation by an anaerobic bacteria (*Desulfobulbus propionicus*) under dark conditions. Environ Sci Technol 43:9183–9188

Schauble EA (2007) Role of nuclear volume in driving equilibrium stable isotope fractionation of mercury, thallium, and other very heavy elements. Geochim Cosmochim Acta 71:2170–2189

Schroeder WH, Munthe J (1998) Atmospheric mercury – an overview. Atmos Environ 32:809–822

Senn DB, Chesney EJ, Blum JD et al (2010) Stable isotope (N, C, Hg) study of methylmercury sources and trophic transfer in the northern Gulf of Mexico. Environ Sci Technol 44:1630–1637

Sherman LS, Blum JD, Nordstrom DK et al (2009) Mercury isotopic composition of hydrothermal systems in the Yellowstone Plateau volcanic field and Guaymas Basin sea-floor rift. Earth Planet Sci Lett 279:86–96

Sherman LS, Blum JD, Johnson KP et al (2010) Use of mercury isotopes to understand mercury cycling between Arctic snow and atmosphere. Nat Geosci 3:173–177

Smith CN, Klaue B, Kesler SE et al (2005) Mercury isotope fractionation in fossil hydrothermal systems. Geology 33:825–828

Smith CN, Kesler SE, Blum JD et al (2008) Isotope geochemistry of mercury in source rocks, mineral deposits and spring deposits of the California Coast Range, USA. Earth Planet Sci Lett 269:398–406

Sonke JE, Zambardi T, Toutain JP (2008) Indirect gold trap-MC-ICP-MS coupling for Hg stable isotope analysis using a syringe injection interface. J Anal At Spectrom 23:569–573

Stetson SJ, Gray JE, Wanty RB et al (2009) Isotopic variability of mercury in ore, mine-waste calcine, and leachates of mine-waste calcine from areas mined for mercury. Environ Sci Technol 43:7331–7336

Sunderland EM (2007) Mercury exposure from domestic and imported estuarine and marine fish in the US seafood market. Environ Health Perspect 115:235–242

Sunderland EM, Krabbenhoft DP, Moreau JW et al (2009) Mercury sources, distribution, and bioavailability in the North Pacific Ocean: insights from data and models. Global Biogeochem Cycles 23:GB2010. doi:10.1029/2008GB003425

US Environmental Protection Agency (U.S. EPA) (1997) Mercury study report to congress 2. EPA-452/R-97-003. Office of Air Quality Planning and Standards, Office of Research and Development, Washington, DC

Xie QL, Lu SY, Evans D et al (2005) High precision Hg isotope analysis of environmental samples using gold trap-MC-ICP-MS. J Anal At Spectrom 20:515–522

Yang L, Sturgeon R (2009) Isotopic fractionation of mercury induced by reduction and ethylation. Anal Bioanal Chem 393:377–385

Zambardi T, Sonke JE, Toutain JP et al (2009) Mercury emissions and stable isotopic compositions at Vulcano Island (Italy). Earth Planet Sci Lett 277:236–243

Zheng W, Hintelmann H (2009) Mercury isotope fractionation during photoreduction in natural water is controlled by its Hg/DOC ratio. Geochim Cosmochim Acta 73:6704–6715

Zheng W, Hintelmann H (2010a) Nuclear field shift effect in isotope fractionation of mercury during abiotic reduction in the absence of light. J Phys Chem A 114:4238–4245

Zheng W, Hintelmann H (2010b) Isotope fractionation of mercury during its photochemical reduction by low-molecular-weight organic compounds. J Phys Chem A 114:4246–4253

Zheng W, Foucher D, Hintelmann H (2007) Mercury isotope fractionation during volatilization of Hg(0) from solution into the gas phase. J Anal At Spectrom 22:1097–1104

Chapter 13
Thallium Isotopes and Their Application to Problems in Earth and Environmental Science

Sune G. Nielsen and Mark Rehkämper

Abstract This paper presents an account of the advances that have been made to date on the terrestrial stable isotope geochemistry of thallium (Tl). High precision measurements of Tl isotope ratios were only developed in the late 1990s with the advent of MC-ICP-MS and therefore we currently only have limited knowledge of the isotopic behavior of this element. Studies have revealed that Tl isotopes, despite their heavy masses of 203 and 205 atomic mass units, can fractionate substantially, especially in the marine environment. The most fractionated reservoirs identified are ferromanganese sediments and low temperature altered of oceanic crust. These display a total isotope variation of about 35 ε^{205}Tl-units, which is over 50 times the analytical reproducibility of the measurement technique. The isotopic variation can be explained by invoking a combination of conventional mass dependent equilibrium isotope effects and the nuclear field shift isotope fractionation, but the specific mechanisms are still largely unaccounted for.

Thallium isotopes have been applied to investigate paleoceanographic processes in the Cenozoic and there is some evidence to suggest that Tl isotopes may be utilized as a monitor of Fe and Mn supply to the water column over million year time scales. In addition, Tl isotopes can be used to calculate the magnitude of hydrothermal fluid circulation through ocean crust. Such calculations can be performed both for high and low temperature fluids. Lastly, it has been shown that marine ferromanganese sediments can be detected in mantle-derived basalts with Tl isotopes (Nature 439:314–317), which confirms that marine sediments subducted at convergent plate margins can be recycled to the surface possibly via mantle plumes.

13.1 Introduction

The element thallium (Tl) was first discovered by William Crookes in 1861 and named after its green spectral line (thallus is latin for budding twig). Historically, the primary use has been as a rodent poison, but it has more recently also been employed as a component in photocells and infrared optical materials. The distribution of Tl in natural environments on Earth is controlled in part by its large ionic radius, which is akin to the alkali metals potassium (K), rubidium (Rb) and cesium (Cs) (Wedepohl 1974; Shannon 1976; Heinrichs et al. 1980). The large ionic radius renders Tl highly incompatible during igneous processing and hence it in the continental crust (Shaw 1952). Another important aspect of Tl chemistry is the ability to take on two different valence states: Tl^+ and Tl^{3+}. The oxidized form makes Tl generally more particle-reactive in solution than the alkali metals, and results in strong adsorption onto authigenic ferromanganese phases and to a lesser extent clay minerals in aqueous environments (Matthews and Riley 1970; McGoldrick et al. 1979; Hein et al. 2000). The higher particle-reactivity means that rivers and particularly the oceans have lower ratios between Tl and the alkali metals compared to the continental crust (Bruland 1983; Flegal and Patterson 1985; Nielsen et al. 2005). Thallium is also concentrated in some sulfide minerals

S.G. Nielsen (✉)
Woods Hole Oceanographic Institution, Dept. Geology and Geophysics, 02543 Woods Hole, MA, USA
e-mail: snielsen@whoi.edu

M. Rehkämper
Department of Earth Science and Engineering, Imperial College, London SW7 2AZ, UK

(Shaw 1952; Heinrichs et al. 1980) but it is currently unclear exactly what influence sulfide minerals have on the elemental and isotopic distribution of Tl on Earth.

Thallium has two isotopes with atomic masses 203 and 205 (Table 13.1). This equates to a relative mass difference of <1%. Considering that stable isotope fractionation theory states that the magnitude of isotope fractionation should scale with the relative mass difference of the isotopes (Urey 1947) one would not expect large stable isotope effects for Tl. Hence, it may be considered surprising that stable isotope investigations for Tl were commenced in the first place. However, the first attempts to measure Tl isotope ratios in natural materials were not aimed at terrestrial materials. The main reason behind the first Tl isotope studies was the search for potential radiogenic isotope variations due to decay of the now-extinct radioactive isotope ^{205}Pb (Anders and Stevens 1960; Ostic et al. 1969). Lead-205 decays to ^{205}Tl with a half-life of 15.1 Myr (Pengra et al. 1978), whereas ^{203}Tl is stable and has no radioactive precursor. Therefore, it was believed that variable ^{205}Tl/^{203}Tl ratios would reveal the former presence of ^{205}Pb when the solar system formed. A number of studies between 1960 and 1994 all failed to register any resolvable Tl isotope variation for some selected terrestrial and a large number of extraterrestrial materials (Anders and Stevens 1960; Ostic et al. 1969; Huey and Kohman 1972; Chen and Wasserburg 1987, 1994). The failure to detect any Tl isotope variation was due to the relatively large errors (>2‰) associated with the thermal ionization mass spectrometry (TIMS) measurements used at that time. The breakthrough came in 1999 when the first high precision Tl isotope measurements by multiple collector inductively coupled plasma mass spectrometry (MC-ICPMS) were published (Rehkämper and Halliday 1999). This technique provided a reduction in the uncertainty of more than an order of magnitude with errors reported at about 0.1–0.2‰ (Rehkämper and Halliday 1999). As with the previous Tl isotope investigations, Rehkämper and Halliday (1999) were ultimately motivated by the search for ^{205}Pb. However, analyses of a number of terrestrial samples that were meant to define the Tl isotope composition of Earth, which could then be compared to extraterrestrial findings, revealed large Tl isotope variation in excess of 15 times the analytical reproducibility. In principle, it is not possible to distinguish between ^{205}Tl/^{203}Tl variation caused by decay of ^{205}Pb and stable isotope fractionation because Tl has only two isotopes. However, the observed variation was clearly unrelated to radioactive ^{205}Pb and thus was inferred to be the result of stable isotope fractionation.

In this contribution we summarize the knowledge that has been accumulated since 1999 on the stable isotope geochemistry of Tl in various terrestrial environments. We will divide this into sections covering the continental and marine domains, respectively. We will also review the most notable applications of Tl stable isotopes in geochemical research and outline what future studies are necessary in order to make Tl isotopes a more quantitative tracer in Earth sciences.

13.2 Materials and Methods

13.2.1 Mass Spectrometry

As outlined in the introduction, it was mainly the advent of MC-ICPMS that facilitated the development of high-precision Tl isotope ratio measurements. The principal difference to the previous TIMS measurements was the ability to correct for instrumental isotope fractionation that occurs during the measurement (mass bias or mass discrimination). Stable isotope compositions in general and two-isotope systems in particular are difficult to measure by TIMS because isotope fractionation during volatilization from the filament is mass dependent just like the stable isotope variation that is attempted to be measured. As such, it is very difficult to determine precise stable isotope ratios by TIMS without the use of a double spike (Rudge et al. 2009). However, double spiking can only be performed for elements with four or more

Table 13.1 Physical properties of thallium (Nriagu 1998)

Melting point	577 K
Molar weight	204.38 g
Density	11.85 g/cm^3
Valence states	Tl$^+$ Tl^{3+}
Redox potentials (V)	
Tl$_{(s)}$ → Tl$^+$ + e$^-$	+0.336
Tl$^+$ → Tl^{3+} + 2e$^-$	−1.28
Ionic radius (Tl$^+$)	1.50 Å
Ionic radius (Tl^{3+})	0.89 Å
Isotopes	^{203}Tl ^{205}Tl

isotopes (Rudge et al. 2009) and this is the reason why the early Tl isotope studies by TIMS yielded relatively large uncertainties. The great advantage of MC-ICPMS when measuring Tl isotopes (or any two-isotope system) is that, even though the overall magnitude of instrumental mass discrimination is much larger than for TIMS, it is possible to monitor it independently during a measurement and thus correct for it much more precisely than is possible with TIMS. This mass bias correction can be performed because the sample is introduced into the mass spectrometer as a solution (or a desolvated aerosol). Into this solution can be admixed a separate element with a known isotope composition (for Tl this element is Pb) and by assuming that the mass bias incurred for the two elements are proportional, isotope ratios can be determined very accurately and precisely (Rehkämper and Halliday 1999; Nielsen et al. 2004). In practice (as is the case for all other stable isotope systems) isotope compositions are best determined by referencing to a standard. For Tl this standard is the NIST 997 Tl metal, and isotope compositions are reported as:

$$\varepsilon^{205}Tl = 10^4 \times \left(^{205}Tl/^{203}Tl_{sample} - ^{205}Tl/^{203}Tl_{NIST997}\right) / \left(^{205}Tl/^{203}Tl_{NIST997}\right)$$

Note that the terminology used for Tl isotope ratios is slightly different to standard stable isotope ratio conventions, which generally use the δ-notation (variations in parts per 1,000). The reason for this is that the Tl isotope system was originally developed as a cosmochemical radiogenic isotope system, which by convention mostly is reported using the ε-notation (variations in parts per 10,000). Hence, it was deemed sensible to keep the original notation in order to make comparisons between cosmochemical and terrestrial data easier. In addition, the analytical uncertainty on and overall variability of stable Tl isotope ratios make the ε-notation very convenient as most data are thereby shown in whole digits and only a single figure behind the decimal point is needed.

13.2.2 Chemical Separation of Thallium

Prerequisites to obtaining precise and accurate stable isotope ratios by MC-ICPMS are the complete separation of the element of interest from the sample matrix as well as 100% recovery. This is important because residual sample matrix in the purified sample can result in isotope effects either present as instabilities that lead to large uncertainties or as reproducible isotopic offsets leading to precise but inaccurate data (Pietruszka and Reznik 2008; Poirier and Doucelance 2009; Shiel et al. 2009).

The technique for separating Tl from sample matrix when performing isotopic analyses of geologic materials was initially developed by Rehkämper and Halliday (1999). The technique has been modified slightly (Nielsen et al. 2004, 2007; Baker et al. 2009) from the original recipe, but the fundamentals have remained unchanged. Only the elution procedure has been optimized in order to remove matrix elements most efficiently (Fig. 13.1). The separation technique relies on the fact that the trivalent Tl ion produces anionic complexes with the halogens (the technique uses either Cl^- or Br^-) in acidic solutions that partition very strongly to anion exchange resins. Conversely univalent Tl does not form strong anionic complexes and thus does not partition at all to anion exchange resins. Therefore, samples are prepared in oxidizing media by adding small amounts of water saturated in Br_2 to the samples already digested and dissolved in hydrochloric acid. This process ensures that all Tl is in the trivalent state, which will adsorb onto the anion exchange resin prepared in a quartz or teflon column. The sample matrix can then be eluted in various acidic media as long as Br_2 is present. Lastly, Tl is stripped from the resin by elution with a reducing solution that converts Tl to the univalent state. The reducing solution used is 0.1 M hydrochloric acid in which 5% by weight of SO_2 gas has been dissolved. As SO_2 is not stable in solution for long periods of time it is important to make this solution fresh before performing the chemical separation of Tl.

13.2.3 Measurement Uncertainties and Standards

As with most stable isotope measurements by MC-ICPMS, the smallest uncertainties are obtained for pure standard solutions. The most commonly used secondary standard for Tl is a pure 1,000 ppm standard solution for ICP-MS concentration analyses that was originally purchased from Aldrich. Over a period of

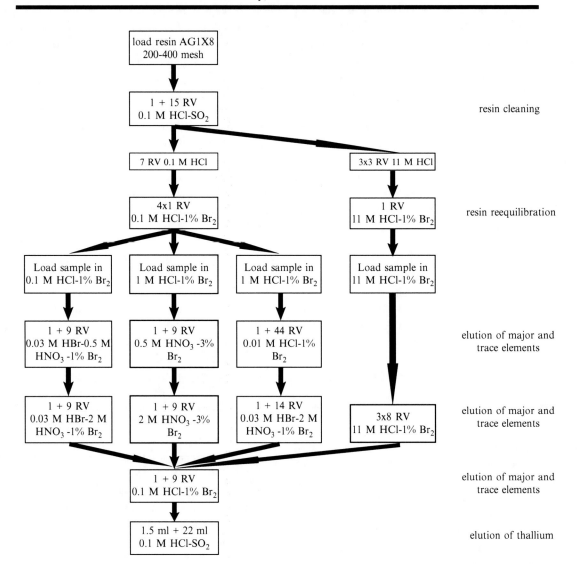

Fig. 13.1 Anion exchange separation procedures for Tl. The recipes can be scaled to any amount of resin (*RV* resin volume) depending on sample size, though large samples require a second 100 μl resin column to ensure that Tl is sufficiently pure. Four different elution procedures are outlined here (Rehkämper and Halliday 1999; Nielsen et al. 2004; Baker et al. 2009), where the procedure in dilute HCl and HNO$_3$ without HBr is the method currently in use in Oxford

more than 10 years this standard has been measured against NIST 997 Tl on six different mass spectrometers with an overall average of ε^{205}Tl = −0.81 ± 0.33 (2sd, n = 133). This uncertainty, however, is not representative of how well samples can be reproduced mainly because small amounts of sample matrix invariably degrade the measurement precision compared to a pure metal standard even when measuring at the same ion beam intensity. This matrix effect generally leads to a relationship between sample concentration and measurement uncertainty with the smallest uncertainties obtained for samples with the highest concentrations (Table 13.2, (Nielsen et al. 2004, 2006a, 2007; Baker et al. 2009; Prytulak and Nielsen, unpublished data)).

In cases where the amount of sample is limited it can become an important issue how many ions can be measured for a given amount of Tl. Over the last 10

Table 13.2 Data for geologic reference materials

Standard	Description	ε^{205}Tl	n	Error[a]	Tl concentration (ng/g)	References
Nod P1	Ferromanganese nodule	0.5	1	0.5	146,000	Rehkämper et al. (2002)
Nod A1[b]	Ferromanganese nodule	10.7	6	0.5	108,000	Rehkämper et al. (2002); Nielsen et al. (2004)
AGV-2	Andesite	−3.0	8	0.6	269	Prytulak and Nielsen, unpublished data; Baker et al. (2009)
BCR-2	Columbia river basalt	−2.5	4	0.4	257	Prytulak and Nielsen, unpublished data
BHVO-2	Hawaii basalt	−1.8	6	0.6	18	Prytulak and Nielsen, unpublished data; Baker et al. (2009)
BIR-1	Iceland basalt	1.1	6	1.2	1.3	Nielsen et al. (2007)
NASS-5	seawater	−5.0	1	1.0	0.0094	Nielsen et al. 2004
TSW[c]	Coastal seawater, Tenerife	−8.7	3	1.0	0.0075	Nielsen et al. (2004)
RRR[c]	Rhine river water	−6.4	4	0.8	0.0036	Nielsen et al. (2004)
Allende	Carbonaceous chondrite	−3.1	8	0.5	55	Baker et al. (2010b)

[a]Errors are either 2sd of the population of separate sample splits processed individually ($n \geq 3$) or estimated based on repeat measurements of similar samples ($n = 1$)
[b]Isotope composition reported for multiple analyses of one large 300 mg aliquot dissolved in 6 M HCl
[c]RRR and TSW are internal laboratory standards that were analyzed multiple times in order to assess reproducibility for samples with such low concentrations of Tl

years MC-ICPMS instruments have been developed to achieve increased transmission (i.e. the fraction of the ions introduced into the machine that reach the collector) and the most recent have values of >1% for Tl and Pb. Although this efficiency may appear low, it enables Tl isotope analyses on samples as small as 1 ng, without compromising counting statistics notably, and an external precision of better than ±1 ε-unit is common (Nielsen et al. 2004, 2006a, 2007; Baker et al. 2009). Smaller sample sizes down to 200 pg can still be analysed on regular Faraday collectors, although precision is significantly degraded to ±3 ε- units (Nielsen et al. 2006a, 2007, 2009b).

13.3 Thallium Isotope Variability in Crustal Environments

13.3.1 Igneous Processes

As discussed previously, Tl generally follows the alkali metals K, Rb and Cs during melting and fractional crystallisation (Shaw 1952; Heinrichs et al. 1980) resulting in higher Tl concentrations in the continental crust (~0.5 ppm) than the mantle (~0.0035 ppm) (Shaw 1952; Wedepohl 1974; Heinrichs et al. 1980; McDonough and Sun 1995; Rudnick and Gao 2003). Though stable isotope fractionation has been documented in igneous systems for relatively heavy elements such as iron (Williams et al. 2004, 2009), the small relative mass difference between the isotopes of Tl in combination with little or no redox chemistry for Tl in igneous systems, favour the prediction that only very little Tl isotope fractionation may occur during magmatic processing. Several studies found that the isotope composition of the average continental crust, represented by loess, is indistinguishable from the mantle, with both exhibiting a value of ε^{205}Tl $= -2.0 \pm 0.5$ (Nielsen et al. 2005, 2006b, c, 2007). The uniform isotope composition of the continental crust is further supported by data obtained for an ultrapotassic dike from the Tibetan Plateau (ε^{205}Tl $= -2.3$; Nielsen and Rehkämper, unpublished results), which represents a melt originating from the sub-continental lithospheric mantle (Williams et al. 2001). Hence, there is no evidence to suggest that melting or fractional crystallisation impart any significant Tl isotope fractionation. Thallium also partitions into some sulphide minerals (Shaw 1952; Heinrichs et al. 1980) and commercial Tl extraction is performed almost exclusively from sulphide ore minerals. So far, there is only one study that investigates Tl isotope fractionation during sulphide ore genesis (Baker et al. 2010a). Here it is shown that the Cu-mineralisation of large porphyry Cu systems in the Collahuasi Formation of Chile is not characterized by Tl enrichments or distinct Tl isotope fingerprints. However, some of the associated hydrothermal processes that mobilised alkali metals also affected Tl concentrations and, to some extent, isotope compositions ($\varepsilon^{205}Tl = -5$ to 0). This study therefore highlights that Tl may retain its lithophile behaviour even in systems that experience sulphide mineralisation (Baker et al. 2010a). Hence, it is unknown at present exactly under what conditions Tl becomes chalcophile and if Tl isotopes would fractionate substantially under such circumstances. However, it might be expected that at least for magmatic ore formation processes the high temperatures in combination with the small relative mass difference of the Tl isotopes would probably not result in significant isotope fractionation.

13.3.2 Volcanic Degassing

Due to its low boiling point (Table 13.1), Tl is significantly enriched in volcanic gasses and particles compared with the geochemically analogous alkali metals (Patterson and Settle 1987; Hinkley et al. 1994; Gauthier and Le Cloarec 1998; Baker et al. 2009). Consequently, volcanic plumes provide a large Tl flux from igneous to surface environments on Earth, with an estimated flux to the oceans of 370 Mg/year (Rehkämper and Nielsen 2004; Baker et al. 2009). The behaviour of Tl isotopes in volcanic systems was recently investigated by Baker et al. (2009), who identified significant isotope variations. The most likely form of isotope fractionation to occur during degassing is kinetic isotope fractionation, where the light isotope is enriched in the gas phase. The extent of kinetic isotope fractionation between two phases is determined by the relative magnitude of the atomic or molecular velocities, where $\alpha = (m_1/m_2)^\beta$ and β varies from 0.5 (in the case of a vacuum) down to values approaching 0 (Tsuchiyama et al. 1994). Hence, even though the relative mass difference between the two Tl isotopes is small, the maximum kinetic $\alpha_{liq-vap}$ is 1.0049, or 49 ε-units. An isotope composition difference of this magnitude is larger than the entire stable Tl isotope variation recorded on Earth and would hence be readily detectable. However, contrary to the expectation of a gas phase enriched in light isotopes, Baker et al. (2009) found no systematic enrichment of either isotope in volcanic emanations compared with average igneous rocks. Despite large isotopic differences between individual samples, the isotope compositions of 34 samples of gas condensates and particles from six separate volcanoes showed an overall average of $\varepsilon^{205}Tl = -1.7 \pm 2.0$. This value is indistinguishable from average igneous rocks (Nielsen et al. 2005, 2006b, c, 2007). The Tl isotope variation of volcanic emanations was interpreted to reflect the complex evaporation as well as condensation processes that occur in volcanic edifices (Baker et al. 2009), and which eventually produce no net isotope fractionation between Tl in the magma and the Tl transported into the atmosphere and surface environments on Earth.

13.3.3 Riverine Transport of Tl

Stable isotope fractionation during weathering is often monitored via measurements of isotope ratios in the dissolved and particulate phases in rivers. For example, elements like lithium and molybdenum have been shown to display significant isotope composition variation between different rivers and rocks of the continental crust (Huh et al. 1998; Archer and Vance 2008; Pogge von Strandmann et al. 2010). Numerous kinetic and equilibrium processes can potentially affect the stable isotope budgets of rivers, and as such it is difficult to predict isotope fractionation during weathering. Nielsen et al. (2005) measured the Tl isotope compositions of dissolved and particulate components for a number of major and minor rivers and found that these generally display values similar to those observed for continental crust. The average value for dissolved riverine Tl is $\varepsilon^{205}Tl = -2.5 \pm 1.0$ (Nielsen et al. 2005), with particulate matter ($\varepsilon^{205} Tl = -2.0 \pm 0.5$) being indistinguishable from the dissolved phase (Fig. 13.2, Table 13.3). These data strongly imply that there is little or no Tl isotope fractionation associated with continental weathering processes.

Natural unpolluted Tl abundances in rivers are generally very low and vary between 1 and 10 pg/g (Cheam 2001; Nielsen et al. 2005). An estimated global average riverine concentration of 6 ± 4 pg/g (or 30 ± 20 pmol/l) results in a flux to the oceans of 230 Mg/year. In the study of Nielsen et al. (2005) three rivers exhibited Tl isotope compositions significantly lighter ($\varepsilon^{205}Tl = -6$ to -4) than average continental crust (Fig. 13.2). These lighter compositions were interpreted to reflect the somewhat different Tl isotope composition of marine carbonates that were a main constituent of the drainage areas for these rivers.

Table 13.3 Thallium isotope and concentration data for rivers

Sample	ε^{205}Tl	Tl (ng/kg)
Amazon	−2.3	16.4
Danube	−6.7	16.4
Doubs	−5.5	3.36
Eder	−4.2	1.93
Kalix	−1.6*	1.31
Nahe	−2.5	7.03
Nidda	−2.7	1.67
Nidder	−2.6	2.85
Nile	±0.0*	3.13
Rhine Rueun	−6.4	3.61
Rhine Laufenbg.	−3.0	4.04
Rhine Speyer	−2.8	5.35
Rhine Bingen	−2.9	6.71
Rhone	−2.7	6.54
Volga	−1.1*	1.60

Sample locations are given in Nielsen et al. (2005)
*Uncertainties for these samples were estimated at 1.5 epsilon units due to smaller amounts of Tl available. The remaining samples have uncertainties of about 1 epsilon unit.

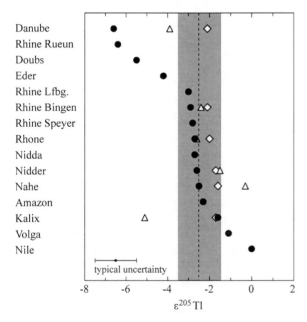

Fig. 13.2 Isotope compositions of riverine dissolved and particulate Tl. The particulate Tl is divided into a labile and detrital fraction. Figure modified from Nielsen et al. (2005)

The more negative ε^{205}Tl values for marine carbonates were expected, based on the current Tl isotope composition measured for seawater (ε^{205}Tl = −5.5 ± 1.0; Rehkämper et al. 2002; Nielsen et al. 2004, 2006c)). However, marine carbonates including carbonate oozes (Rehkämper et al. 2004), corals (Rehkämper, unpublished data) and foraminifera (Nielsen, unpublished data) exhibit very low Tl concentrations and therefore weathering of marine carbonates will only have a strong influence on rivers predominantly draining such lithologies, and will not strongly affect the total global budget of Tl transported by rivers to the oceans.

13.3.4 Anthropogenic Mobilisation of Tl

To date no study has been carried out to investigate the Tl isotope effects incurred from industrial pollution. Industrial purification of Tl is tightly restricted because of the toxicity of Tl. Therefore, the main emissions of Tl to the environment are not from Tl production but stem from coal burning (Nriagu 1998; Cheam et al. 2000) and zinc smelting (Nriagu 1998). Thallium is present in relatively high concentrations in coal and zinc ore (~1 ppm, (Sager 1993)) and is enriched in the waste products from industrial processing of these resources. There are many examples of adverse environmental effects from Tl pollution caused by industry (Lis et al. 2003; Xiao et al. 2003, 2004; Sasmaz et al. 2007) and hence there may be some potential for using Tl isotopes to monitor industrial pollution. Coal burning and zinc smelting both involve significant heating, during which gases and particulates may escape into the natural environment. One might therefore expect kinetic Tl isotope fractionation to occur and isotopically light Tl could potentially be a signature of Tl pollution, as has been found for volatile Cd (Cloquet et al. 2006). However, if volcanic emissions (see Sect. 13.3.2) can be used as an analogue for these processes, industrial waste pollution may not necessarily carry a very distinct Tl isotope signature.

A secondary issue of potentially tracing Tl pollution using isotope data, is the unknown initial composition of the material. In many cases, it may be reasonable to infer that the isotope composition of continental crust (ε^{205}Tl = −2) will be the dominant "signature" of coal and zinc ore. However, since Tl is mainly associated within these deposits as a result of its affinity to sulfides (Sager 1993; Nriagu 1998), it will be important to investigate Tl isotope variations in sulfides to verify this assumption.

13.4 Thallium Isotopes in the Marine Environment

13.4.1 The Isotope Composition of Seawater

In the oceans, Tl is a conservative low-level trace element with an average dissolved concentration of 13 ± 1 pg/g (64 ± 5 pmol/l) (Flegal and Patterson 1985; Schedlbauer and Heumann 2000; Rehkämper and Nielsen 2004; Nielsen et al. 2006c), which is slightly higher than abundances observed in river water (see Sect. 13.3.3). Based on a thorough review of the marine input and output fluxes of Tl, Rehkämper and Nielsen (2004) concluded that the oceans are currently at steady state and that Tl has a residence time of ~21 kyr, which is consistent with a number of previous studies (Flegal and Patterson 1985; Flegal et al. 1989). With an inferred residence time that is more than an order of magnitude longer than the oceanic mixing time and a conservative distribution, Tl should exhibit an invariant isotope composition in the oceans. Analyses of Arctic, Atlantic and Pacific seawater confirm this prediction (Rehkämper et al. 2002; Nielsen et al. 2004, 2006c) but the current dataset is limited to results for seven samples from only three locations.

It may perhaps be somewhat surprising that the oceans are significantly depleted in ^{203}Tl compared to the continental crust and the mantle (See Chap. 3) with an average value of $\varepsilon^{205}Tl = -5.5 \pm 1.0$ (Rehkämper et al. 2002; Nielsen et al. 2004, 2006c). Hence, it is required either that the marine sources of Tl are isotopically light compared to the crust and mantle or that the outputs are fractionated towards heavy values compared with seawater itself. Rehkämper and Nielsen (2004) showed that the most significant marine inputs for Tl are from rivers, high-temperature hydrothermal fluids, mineral aerosols, volcanic emanations and sediment pore water fluxes at continental margins. In contrast, there are only two important marine Tl sinks, namely Tl adsorption by the authigenic phases of pelagic clays and uptake of Tl during low-temperature alteration of oceanic crust. The relative magnitudes and isotope compositions of these fluxes are summarized in Table 13.4. In the following sections, we outline the main observations that have led to the assessment for these fluxes.

13.4.2 Thallium Isotopes in Marine Input Fluxes

The riverine and volcanic input fluxes have already been discussed in previous sections and hence here we will focus on high-temperature hydrothermal fluids, mineral aerosols and sediment pore water fluxes from continental margins.

13.4.2.1 High Temperature Hydrothermal Fluids

In hydrothermal systems where temperatures exceed ~150°C it has been shown that Tl behaves much like the alkali metals Rb and K (Metz and Trefry 2000), and is leached from the oceanic crust by the circulating fluids. This results in end-member high temperature hydrothermal fluids exhibiting Tl concentrations almost 500 times higher than ambient seawater (Metz and Trefry 2000; Nielsen et al. 2006c) with a best estimate of 4.5 ± 2.2 ppb (22 ± 11 nmol/l) (Nielsen et al. 2006c). Thus hydrothermal fluids represent a significant Tl flux to the oceans with an estimated range of 9–200 Mg/year (Nielsen et al. 2006c). Due to the relatively high temperatures involved (300–400°C) isotope fractionation is not expected to be significant. This was confirmed by Nielsen et al. (2006c), who determined the Tl isotope composition of hydrothermal fluids from the East Pacific Rise and Juan de Fuca Ridge. All samples had Tl isotope compositions identical to that of average MORB, thus supporting the interpretation that extraction of Tl from the oceanic crust is not associated with isotope fractionation. The chemical and isotopic behavior inferred for Tl from hydrothermal fluids is further in accord with the results of a study of oceanic crust from ODP Hole 504B altered by high temperature hydrothermal fluids. The latter work revealed Tl concentrations for basalts/dikes that were much lower than expected for depleted ocean crust whilst the Tl isotope compositions were identical to average MORB (Nielsen et al. 2006c).

13.4.2.2 Mineral Aerosols

There are no direct investigations of Tl abundances or isotope compositions for mineral aerosols deposited in the oceans. Based on studies of windborne loess sediments deposited on land, Nielsen et al. (2005) concluded that the average abundance of Tl in dust deposited in the oceans is about 490 ± 130 ng/g. The main uncertainty in determining the Tl flux to the oceans is from estimating the fraction of Tl that is released from the dust into seawater, following deposition and partial dissolution. By comparison with a number of other elements, Rehkämper and Nielsen (2004) concluded that about 5–30% of the Tl transported in aerosol particles would dissolve in seawater, resulting in an annual Tl flux of 10–150 Mg/year.

It may be reasonable to assume that the bulk Tl isotope composition of the material transported to the ocean is identical to loess, and thereby also the continental crust, with $\varepsilon^{205}\text{Tl} = -2$ (Nielsen et al. 2005). However, it is unknown whether dust dissolution is associated with any isotope fractionation. Schauble (2007) has shown that significant equilibrium Tl isotope fractionations will be produced primarily by chemical reactions that involve both valence states, Tl^+ and Tl^{3+}. Based on the strongly correlated behaviour of Tl, Rb, Cs and K in the continental crust, univalent Tl should be dominant in mineral aerosols.

Thermodynamic calculations of the valence state of Tl in seawater also predict that all Tl is univalent in this reservoir (Nielsen et al. 2009a). In addition, continental weathering processes, which are ultimately controlled by aqueous dissolution of silicates and are therefore in many ways analogous to the leaching of dust in seawater, induce no detectable Tl isotope fractionations (Nielsen et al. 2005). Hence, we infer that isotope fractionation is unlikely to be significant during the dissolution of mineral aerosols in seawater.

13.4.2.3 Benthic Fluxes from Continental Margins

It has long been known that pore waters, which seep into the oceans from reduced continental margin sediments, are rich in manganese (Elderfield 1976; Sawlan and Murray 1983). Since Tl is strongly associated with Mn oxides (Koschinsky and Hein 2003), Rehkämper and Nielsen (2004) inferred that such pore waters may be an important source of dissolved marine Tl. However, there are currently no data available to constrain either the average Tl concentration or isotope

Table 13.4 The Tl mass balance of the oceans with estimated source and sink fluxes

	Range of Tl flux estimates (Mg/year)	Best estimate (Mg/year)	%	$\varepsilon^{205}\text{Tl}$	References
Marine input fluxes					
Rivers	76–380	230	27	−2.5	Rehkämper and Nielsen (2004); Nielsen et al. (2005)
Hydrothermal fluids	9–200	30	4	−2	Nielsen et al. (2006c)
Subaerial volcanism	42–700	370	43	−2	Baker et al. (2009)
Mineral aerosols	10–150	50	6	−2	Rehkämper and Nielsen (2004); Nielsen et al. (2005)
Benthic fluxes from continental margins	5–390	170	19	0	Rehkämper and Nielsen (2004), this study
Total input flux	140–1,820	850	100	−1.8	
Marine output fluxes					
Pelagic clays	240–450	310	36	+10	Rehkämper and Nielsen (2004); Rehkämper et al. (2004), this study
Altered ocean crust	225–1,985	540	64	−8.5	Rehkämper and Nielsen (2004); Nielsen et al. (2006c), this study
Total output flux	430–1,530	850	100	−1.8	
	Mass of Tl (Mg)	Steady-state residence time		$\varepsilon^{205}\text{Tl}$	
Global oceans	1.75 (±0.14) × 10[7][a]	21,000 year		−5.5	Rehkämper and Nielsen (2004); Nielsen et al. (2006c); Rehkämper et al. (2002)

[a] For a global ocean system with 1.348 × 10^{21} kg, the Tl mass in the oceans is equivalent to an average seawater concentration of 65 ± 5 pmol/kg or 13 ± 1 ng/kg (Rehkämper and Nielsen 2004)

composition of these benthic fluxes. An estimate was therefore derived indirectly, based on (1) Tl/Mn ratios observed in ferromanganese nodules that were known to have precipitated from pore waters (Rehkämper et al. 2002), combined with (2) estimates for benthic Mn fluxes (Sawlan and Murray 1983; Heggie et al. 1987; Johnson et al. 1992). Taken together, these data yield a Tl flux of 5–390 Mg/year, and a best estimate of 170 Mg/year (Rehkämper and Nielsen 2004) which constitutes a substantial fraction of the total Tl flux to the oceans (Table 13.4).

We can use two distinct approaches to estimate the average Tl isotope composition of pore waters. The first utilizes published Tl isotope data for the various components that may supply pore waters with Tl. The second relies on the Tl isotope compositions of sediments that are precipitated from pore waters. The application of these approaches is summarized below.

There are three principle components, which could be mobilized and incorporated into sedimentary pore waters. (1) Labile Tl associated with riverine particles (Nielsen et al. 2005). (2) Thallium adsorbed onto clay minerals from seawater (Matthews and Riley 1970). (3) Thallium bound to authigenic Mn-oxides that precipitated from seawater as part of deep sea red clays (Rehkämper et al. 2004). All three will be present in continental margin sediments in various proportions depending on sedimentation rate and proximity to estuaries, where high sedimentation rates would tend to dilute the Mn-oxides as it is assumed that authigenic precipitation should remain fairly constant. Hence, assuming that there is no isotope fractionation associated with the release of Tl into pore waters, the isotope composition of each component can define bounds on the composition of pore waters.

The labile components in riverine particles have been shown to be isotopically similar to the continental crust and river waters (Fig. 13.2) and thus is characterized by $\varepsilon^{205}Tl = -2$ (Nielsen et al. 2005).

Matthews and Riley (1970) showed that Tl is readily adsorbed onto some clay minerals (in particular illite) where most Tl is exchanged with K. Because of the required charge balance for adsorption reactions, clay adsorption is unlikely to be associated with any Tl reduction/oxidation processes and should thereby exhibit only minimal or no isotope fractionation (Schauble 2007). Since Tl adsorption to clay minerals would take place primarily within the marine environment, this component should be characterized by $\varepsilon^{205}Tl = -5.5$.

Thallium that is precipitated with Mn oxides onto sedimentary particles has been shown to have a significantly heavier isotope composition than the seawater from which the mineral forms (Rehkämper et al. 2002, 2004). The origin of this isotope fractionation is discussed in Sect. 13.5 below. Theoretically, the pure MnO_2 mineral to which Tl is bound should have approximately the same isotope composition as the surfaces of Fe-Mn crusts, which display $\varepsilon^{205}Tl = +13$ (Fig. 13.4). However, leaching experiments conducted on shelf sediments with HCl and hydroxylamine hydrochloride indicate that the labile Tl on sediment particles features somewhat lower $\varepsilon^{205}Tl$ values of between +3 to +7 (Rehkämper et al. 2004; Nielsen et al. 2005). These lower values could reflect contributions from a component other than Fe-Mn oxides, for example Tl adsorbed to clay minerals that are characterized by negative $\varepsilon^{205}Tl$-values. Nevertheless, it appears reasonable to infer that the Tl associated with marine authigenic Mn in continental margin sediments exhibits $\varepsilon^{205}Tl \approx +4$ to +10.

In summary, the isotopic data available for these three components indicate that pore waters from continental margin sediments are likely to feature $\varepsilon^{205}Tl$ values of about −4 to +6. This range can be compared with the compositions of two different sedimentary archives that form from pore waters. These are diagenic ferromanganese nodules (which are inferred to originate primarily from pore waters) and early diagenetic pyrite. Rehkämper et al. (2002) analyzed two Mn nodules from the Baltic Sea and these exhibited $\varepsilon^{205}Tl = -0.2$ and -5.2. Nielsen et al. (2010) measured Tl isotopes in pyrites younger than 10 Myr from continental shelf sediments in the Northeast Pacific and the Caribbean and found $\varepsilon^{205}Tl$ values of between −1 and +2. Importantly, both ferromanganese nodules and pyrites thus have Tl isotope compositions that are in accord with the range of values inferred for pore waters based on their constituent components. Taken together (and assuming that there is negligible isotope fractionation between pyrite/Fe-Mn nodules and pore waters) these constraints suggests that benthic fluxes from continental shelf sediments are characterized by $\varepsilon^{205}Tl \sim -3$ to +3, with a best estimate of $\varepsilon^{205}Tl \sim 0$.

13.4.3 Thallium Isotope Compositions of Marine Output Fluxes

13.4.3.1 Thallium Adsorption by the Authigenic Phases of Pelagic Clays

There is a significant enrichment of Tl in pelagic clays compared with continental shelf sediments (Matthews and Riley 1970; Heinrichs et al. 1980; Rehkämper et al. 2004). This enrichment is caused by the adsorption of Tl onto hydrogenetic Fe-Mn oxy-hydroxides and clay minerals (Matthews and Riley 1970; McGoldrick et al. 1979; Heinrichs et al. 1980). Two independent methods of calculating the Tl flux associated with these authigenic fluxes agree very well and indicate an annual flux of 200–410 Mg/year, with a best estimate of 270 Mg/year (Rehkämper and Nielsen 2004). Adding the smaller flux of Tl incorporated into pure Fe-Mn deposits of ~40 Mg/year (Rehkämper and Nielsen 2004) yields a total marine authigenic Tl flux of 310 Mg/year (Table 13.4). As discussed in Sect. 13.4.2.3, the isotope composition of the Tl associated with authigenic phases is best characterized by a mixture of Tl bound to Mn oxides that displays $\varepsilon^{205}Tl \sim +13$ and Tl adsorbed onto clay minerals, which likely exhibits $\varepsilon^{205}Tl \sim -5.5$. Due to the strong Mn enrichment of deep sea pelagic clays, it can be assumed that the majority of authigenic Tl in pelagic clays originates from Mn oxides and thus may have an isotope composition more akin to Fe-Mn crusts. We can attempt to quantify this by performing an isotope mass balance calculation for the pelagic clays reported in Rehkämper et al. (2004). The bulk isotope composition of these samples is $\varepsilon^{205}Tl = +3$ to $+5$. About 50% of the Tl in pelagic clays is thought to be of detrital origin (Rehkämper et al. 2004), which is characterized by $\varepsilon^{205}Tl = -2$ (Nielsen et al. 2005). In order to account for the reported bulk isotope compositions, the authigenic component would thus have to display $\varepsilon^{205}Tl \sim +8$ to $+12$, and this implies that the authigenic Tl of pelagic clays features an isotope composition that is slightly lighter than pure Fe-Mn deposits. Based on the above considerations we can calculate that ~75–95% of the authigenic Tl in pelagic clays originates from Mn-oxides, whilst the remainder is bound to clay minerals.

13.4.3.2 Thallium Uptake During Low Temperature Hydrothermal Alteration

It is well-known that alteration minerals produced by interaction between cold (<100°C) seawater and MORB are highly enriched in Tl compared to pristine oceanic crust (McGoldrick et al. 1979; Alt 1995; Jochum and Verma 1996). This process leads to an enrichment of Tl in the volcanic section of oceanic crust (on average the uppermost 500–600 m) that has been shown to be affected by low temperature (low-T) alteration. Estimating an average annual flux of Tl into the oceanic crust would be relatively simple if the average Tl concentration of this material could be determined. This is not a straightforward exercise, however, because of the heterogeneous distribution of alteration minerals in low-T altered crust and the inferred enrichment factors for Tl (compared to its abundance in pristine oceanic crust) therefore vary widely, between values of 1 and 200 (Jochum and Verma 1996; Teagle et al. 1996; Nielsen et al. 2006c). In general, the highest Tl enrichments are detected towards the top of the oceanic crust (Fig. 13.3). Following a conservative approach, based partially on the assumption that Cs and Tl are added to the ocean crust in equal proportions (Rehkämper and Nielsen 2004; Nielsen et al. 2006c), it was proposed that, on average, the upper 600 m of oceanic crust contains 200 ± 150 ng/g of Tl (Nielsen et al. 2006c). This number is equivalent to an element flux of 225–1,985 Mg/year, with a best estimate of ~1,000 Mg/year. The best estimate value, however, is significantly larger than all marine Tl inputs combined (Table 13.4).

Given the uncertainties, it may thus be more reasonable to apply a mass balance approach (which assumes that marine Tl is at steady state) to estimate the Tl flux into low-T altered ocean crust. With a total input flux of 850 Mg/year (Table 13.4) and an authigenic output flux of 310 Mg/year, the marine Tl budget can be balanced with a low-T alteration output of 540 Mg/year, which is well within the range of values estimated based on Tl concentration data available for low-T altered ocean crust (Table 13.4).

The isotope composition of the low-T alteration output is equally difficult to determine. Nielsen et al. (2006c) observed that, in general, the upper oceanic crust is significantly enriched in isotopically light Tl

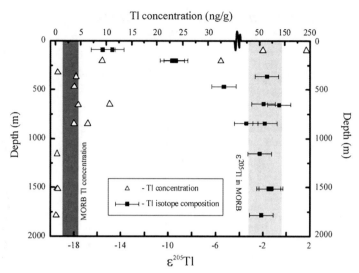

Fig. 13.3 Thallium concentrations (*triangles*) and isotope compositions (*black squares* with error bars) versus depth for samples from ODP Hole 504B. The *dark* and *light shaded areas* denote the average Tl concentration and isotope composition of MORB. Note the change of scale for Tl concentrations higher than 40 ng/g. Figure modified from Nielsen et al. (2006c)

with the shallowest samples displaying the lightest isotope compositions (Fig. 13.3). This isotopic signature was interpreted to reflect closed system isotope fractionation during uptake of Tl from seawater, which gradually becomes more depleted in Tl as it penetrates deeper into the oceanic crust. As shown by a Rayleigh fractionation model in which an isotope fractionation factor of $\alpha = 0.9985$ was applied (Nielsen et al. 2006c), the isotope composition of the total Tl deposited during low-T alteration depends critically on the fraction of Tl that is extracted from the seawater before it is re-injected into the oceans as a low-T hydrothermal fluid. Stripping the fluid of all the Tl originally present would imply an isotope composition for the low-T alteration flux that is identical to seawater ($\varepsilon^{205}Tl = -5.5$), whilst lower degrees of depletion result in lighter isotope compositions.

Based on the isotope fractionation observed for rocks from ODP Hole 504B, it was estimated that extraction of ~50% of the original seawater Tl by ocean crust alteration would yield an average isotope composition of $\varepsilon^{205}Tl \sim -18$ for the Tl deposited during low-T alteration. The application of this approach to obtain a global estimate for the average Tl isotope composition of the low-T alteration flux is fraught with many uncertainties, however, as the database is currently limited to rocks from just one section of altered ocean crust and Tl concentration measurements for three low-T fluids from the Juan de Fuca ridge (Nielsen et al. 2006c). In principle, average altered upper ocean crust may be characterized by an $\varepsilon^{205}Tl$ value of between about -18 to -6. Hence it may again be more appropriate to apply an isotope mass balance approach to obtain a more accurate result. With an isotope composition of $\varepsilon^{205}Tl \sim -1.8$ for the combined marine Tl input fluxes and an authigenic output flux of $\varepsilon^{205}Tl \sim +10$, mass balance dictates that the altered basalt flux is characterized by $\varepsilon^{205}Tl \sim -8.5$ (Table 13.4). This result is certainly within the range of reasonable values but based on Rayleigh fractionation modelling it requires that more than 90% of seawater Tl is removed by alteration processes, which is not entirely supported by the data obtained for ODP Hole 504B (Nielsen et al. 2006c). It is unlikely, however, that the results obtained for Hole 504B are representative for global average ocean crust alteration processes.

13.5 Causes of Thallium Isotope Fractionation

In general, Tl isotope variations on Earth are fairly limited, with only a few select environments displaying significant deviations from the bulk Earth value of $\varepsilon^{205}Tl = -2$ (Nielsen et al. 2005, 2006b, c, 2007). However, the overall magnitude of Tl isotope variation in natural environments on Earth now exceeds 35 $\varepsilon^{205}Tl$-units (Rehkämper et al. 2002, 2004; Nielsen et al. 2006c; Coggon et al. 2009). This variability is comparable to the extent of fractionation observed for

^{56}Fe/^{54}Fe (Anbar 2004), which has a relative mass difference almost four times that of Tl. It would appear that this observation is somewhat contrary to what is predicted from classical stable isotope fractionation theory and it is therefore important to understand the fundamental processes responsible for Tl isotope fractionation.

There are two principle mechanisms by which most stable isotope fractionations are generated – a kinetic route that is associated with unidirectional processes and an equilibrium pathway, which acts during chemical exchange processes. Both mechanisms should scale with the relative mass difference between the two isotopes of interest. As mentioned in Sect. 13.3.2, kinetic isotope fractionation processes are theoretically capable of generating substantial Tl isotope fractionation and there is evidence that such fractionations occur from data acquired for samples from volcanic plumes and some meteorites (Nielsen et al. 2006a; Baker et al. 2009, 2010b). However, the large Tl isotope fractionations observed between seawater and Fe-Mn oxy-hydroxides and for low-T altered oceanic crust are more likely to reflect equilibrium fractionation effects (Rehkämper et al. 2002, Nielsen et al. 2006c).

The larger-than-expected Tl isotope variations for equilibrium reactions were recently shown by Schauble (2007) to be partially caused by the so-called nuclear field shift isotope fractionation mechanism (Bigeleisen 1996). In short, the original equilibrium isotope exchange equation of Bigeleisen and Mayer (1947) had five components, of which four were deemed negligible. However, based on some unusual isotope fractionation effects observed for uranium (Fujii et al. 1989a, b), it was concluded that the nuclear field shift term may be important in some cases (Bigeleisen 1996). Interestingly, nuclear field shift isotope fractionation is not mass dependent but scales positively with the size of the nucleus and hence is also largest for the heavy elements. As Tl has only two isotopes, it is not possible to distinguish nuclear field shift effects from mass dependent isotope fractionation. However, the calculations that were carried out for Tl predict that an equilibrium system with aqueous dissolved Tl$^+$ and Tl^{3+} will feature both regular mass dependent and nuclear field shift isotope effects, with isotope fractionation factors that act in the same direction (Schauble 2007). When combined, these two components can furthermore reproduce the approximate magnitude of Tl isotope variation observed on Earth (Schauble 2007). The calculations are particularly relevant for the isotope compositions determined for Fe-Mn crusts and low-T altered basalts as these represent the heaviest and lightest reservoirs, respectively, found to date. The main requirement for substantial equilibrium Tl isotope fractionation to take place, is a chemical exchange reaction that involves two valence states of Tl (Schauble 2007), and these could be Tl^0, Tl$^+$ or Tl^{3+}. However, the calculations of Schauble (2007) also imply that the largest isotope effects are expected if Tl^{3+} is present.

Based on these theoretical considerations, experimental work is currently being conducted to investigate the specific mechanism for Tl isotope fractionation during adsorption onto hydrogenetic Fe-Mn crusts (Fig. 13.4) and other Fe-Mn sediments (Nielsen et al. 2008; Peacock et al. 2009). The majority of Tl in these deposits is associated with MnO$_2$ minerals and the isotope fractionation is therefore likely to occur at or in such phases. Preliminary interpretations of EXAFS/XANES spectra for Tl sorbed onto Mn-oxides have shown that the MnO$_2$ phase hexagonal birnessite has the capacity to oxidize Tl to Tl^{3+}, following adsorption as a univalent ion (Bidoglio et al. 1993; Peacock et al. 2009). This reaction appears to be associated with isotope fractionation whereas adsorption of Tl onto other MnO$_2$ mineral structures (Na-birnissite and todorokite), which do not have the same oxidation potential, is not associated with significant isotope effects (Nielsen et al. 2008; Peacock et al. 2009). These results are consistent with the Tl isotope data obtained for natural hydrogenetic and hydrothermal Fe-Mn crusts, which are dominated by H-birnissite and todorokite, respectively (Rehkämper et al. 2002; Nielsen et al. 2008; Peacock et al. 2009). Considering the theoretical predictions (Schauble 2007), these preliminary observations are able to account for at least some of the Tl isotope variability observed in the marine environment. However, the current experiments are still unable to reproduce the magnitude of isotope fractionation observed in nature and it is therefore still unresolved exactly how the large isotope fractionation observed between seawater and Mn-oxides is produced (Rehkämper et al. 2002). Future work will hopefully shed more light on this issue.

Even more enigmatic than the Tl isotope fractionation of Mn oxides are the isotope fractionation effects found in low-T altered ocean crust. In this case, Tl is extracted from seawater circulating through the rocks with an isotope fractionation factor of about α = 0.9985 (Nielsen et al. 2006c; Coggon et al. 2009). It is conceivable that these fractionations reflect kinetic isotope effects, for example as a result of the more rapid diffusion of the light isotopes from the hydrothermal fluid to the alteration minerals that concentrate Tl. However, it is unclear whether diffusion rates can be sufficiently fast, to enable the large isotope effects observed (Nielsen et al. 2006c; Coggon et al. 2009). If an equilibrium reaction is responsible, Tl^{3+} is likely to be involved because large equilibrium isotope fractionations are not expected for reactions without this species (Schauble 2007). Simple models of Tl speciation in seawater predict that only Tl^+ is present (Nielsen et al. 2009a), which implies that the hydrothermal processes that deposit Tl in the oceanic crust produce Tl^{3+}. This inference strongly contradicts observations, which demonstrate that low-T hydrothermal alteration generally occurs at conditions that are more reducing than those prevalent in open ocean seawater (Alt et al. 1996). Additionally, Tl^{3+} has a much lower aqueous solubility than Tl^+ (Nriagu 1998) and this implies that trivalent Tl should be deposited during hydrothermal alteration. However, isotope fractionation calculations indicate that oxidized Tl should be enriched in ^{205}Tl (Schauble 2007), which is at odds with the light Tl isotope signatures observed in altered basalts.

Alternatively, if we ignore the inferred univalent nature of dissolved oceanic Tl, it may be hypothesized that seawater consists of a mixture of uni- and trivalent Tl-species that are isotopically fractionated from each other by ~35 $\varepsilon^{205}Tl$-units (Schauble 2007). This configuration would be consistent with oxidized and isotopically heavy Tl adsorbing onto Mn-oxides whilst reduced and isotopically light Tl would partition into basalt alteration minerals. Together, these fractionated species could thus create the isotope difference observed between these two reservoirs (Rehkämper et al. 2002; Nielsen et al. 2006c; Coggon et al. 2009). In order to reproduce the isotope compositions observed for Fe-Mn crusts ($\varepsilon^{205}Tl$ ~ +13, (Rehkämper et al. 2002)) and altered basalts ($\varepsilon^{205}Tl$ ~ −20 (Coggon et al. 2009)), mass balance then dictates that seawater with an average isotope composition of $\varepsilon^{205}Tl = -5.5$ would need be characterized by a ratio of $Tl^{3+}/Tl^+ \approx 0.8$. Some analytical studies have in fact claimed that significant levels of Tl^{3+} are present in natural waters (Lin and Nriagu 1999), but such results are questionable as the sample preparation procedures may have altered the speciation of Tl. Schedlbauer and Heumann (2000) furthermore identified methylated Tl^{3+} in seawater but these analyses yielded variable Tl^{3+}/Tl^+ ratios of 1 to <0.05, which would preclude the globally invariant Tl isotope compositions observed for Fe-Mn crusts (Fig. 13.4).

In summary, the large overall Tl isotope variability observed in the marine environment is most likely produced by a combination of conventional mass dependent and nuclear field shift equilibrium isotope fractionation processes. However, contributions from kinetic isotope effects cannot be ruled out at present. In order to reproduce the magnitude of equilibrium isotope fractionation observed for natural samples, reduction-oxidation processes, in which oxidized Tl^{3+} plays a central role, are predicted to be important. Experimental studies of Tl isotope fractionation are

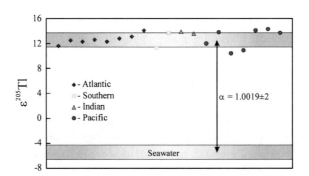

Fig. 13.4 Thallium isotope compositions determined for the growth surfaces of hydrogenetic Fe-Mn crusts and seawater. The Fe-Mn crusts precipitate directly from seawater and hence there is an isotope fractionation between these two reservoirs of ~19 $\varepsilon^{205}Tl$-units. Data from (Rehkämper et al. 2002; Nielsen et al. 2004, 2006c)

only in their infancy, however, and we still have much to learn about the processes that govern Tl isotope fractionation in natural environments.

13.6 Applications of Tl Isotopes to Problems in Earth Science

13.6.1 Studies of Tl Isotopes in Fe-Mn Crusts

Hydrogenetic Fe-Mn crusts grow on hard substrates that experience little or no regular detrital sedimentation, for example on seamounts where ocean currents prevent gravitational settling of particles (Hein et al. 2000). They precipitate directly from the ambient water mass that surrounds them and feature growth rates of a few mm/Myr (Segl et al. 1984, 1989; Eisenhauer et al. 1992). This implies that sample thicknesses in excess of 10 cm could represent a continuous seawater record through the entire Cenozoic (the last ~65 Myr). Extensive investigations of Fe-Mn crusts over the last 15 years have been conducted to infer changes in the radiogenic isotope compositions of various elements in ocean bottom waters (Lee et al. 1999; Frank 2002). Depending on the marine residence time for the element investigated, the isotopic variability has mainly been interpreted as reflecting changes in either ocean circulation patterns or the marine source and/or sink fluxes (Burton et al. 1997; van de Flierdt et al. 2004).

The globally uniform Tl isotope ratios observed for the surfaces of Fe-Mn crusts (Fig. 13.4), imply that there is a consistent equilibrium isotope fractionation between seawater and Tl incorporated into the Fe-Mn crusts. Time-dependent variations of Tl isotope compositions in Fe-Mn crusts can be interpreted to reflect either changes in the isotope fractionation factor between seawater and Fe-Mn crust or the Tl isotope composition of seawater. In principle, both interpretations are possible, but several lines of reasoning currently favor the latter explanation (Rehkämper et al. 2004; Nielsen et al. 2009a).

Two studies have determined Tl isotope depth profiles for several Fe-Mn crusts, and both identified large systematic changes in Tl isotope compositions (Rehkämper et al. 2004; Nielsen et al. 2009a). The first study produced low-resolution depth profiles for a number of samples. The largest Tl isotope variations were observed for the early Cenozoic (Rehkämper et al. 2004), but the low sampling density and uncertainties in the models, which were applied to infer the sample growth rates, made it difficult to determine the exact time and duration of this change. It was furthermore argued that the Fe-Mn crusts record variations in the Tl isotope composition of seawater that were caused by changes in the marine input and/or output fluxes of this element (Rehkämper et al. 2004). The second study generated high resolution Tl isotope time series for two Fe-Mn crusts. Improved age models were applied, which resolved a single large shift in Tl isotope composition, which occurred between ~55 and ~45 Ma (Fig. 13.5; (Nielsen et al. 2009a)). Based on an improved understanding of the marine input and output fluxes of Tl and their resepcteive isotope compositions, it was furthermore proposed that the large shift in the ε^{205}Tl value of seawater reflects a decrease in the amount of authigenic Fe-Mn oxyhydroxides that were deposited with pelagic sediments in the early Eocene (Nielsen et al. 2009a).

It is difficult to assess the underlying mechanism responsible for this global change in Fe-Mn oxyhydroxide precipitation. The strong co-variation of the Tl isotope curve with the sulfur (S) isotope composition of seawater (Fig. 13.5) may imply, however, that the same mechanism is driving the jump in the isotopic evolution of both stable isotope systems, even though S isotopes are known to be unaffected (at least directly) by changes in Fe-Mn oxides precipitation. Baker et al. (2009) proposed that the inferred high Fe-Mn oxyhydroxide precipitation rates for the Paleocene (~65–55 Ma) may be explained by increased deposition of Fe- and Mn-rich volcanic ash particles in the oceans. Such volcanic activity would also supply isotopically light S and this could explain the relatively low δ^{34}S value for seawater at this time. The changes in the Tl and S isotope compositions of the oceans between ~55 and ~45 Ma (Fig. 13.5) would then be controlled by diminishing volcanic activity (Wallmann 2001). An alternative model proposes that Fe-Mn precipitation is controlled by biological utilization and burial of Fe and Mn with organic carbon (Nielsen et al. 2009a). High organic carbon burial rates would lead to diminished Fe-Mn precipitation rates as less Fe and Mn are available in the water column. Simultaneously, the increased organic carbon

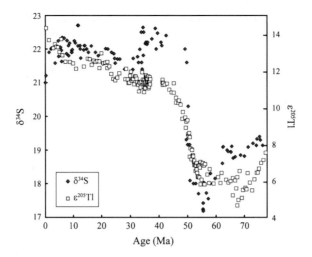

Fig. 13.5 Thallium isotope data for a Fe-Mn crust from the Pacific Ocean and the S isotope seawater curve plotted for the last 75 Myr. The S isotope data are from Paytan et al. (1998; 2004), with ages based on the age model of (Kurtz et al. 2003). The chronology of the Tl isotope curve was determined based on Os isotope data (Burton 2006). Figure modified from Nielsen et al. (2009a)

burial would result in higher rates of sedimentary pyrite burial, which draws isotopically light S out of seawater (Berner 1984).

In summary, the results of both initial paleoceanographic studies thus indicate that it may be possible to utilize Tl isotopes as a proxy for the Fe and Mn supply to the oceans further back in time. However, much work is still needed to confirm whether these preliminary interpretations are indeed correct.

13.6.2 Calculation of Hyrothermal Fluid Fluxes Using Tl Isotopes in the Ocean Crust

Hydrothermal fluids are expelled from the seafloor (i) at high temperature on mid ocean ridge axes, as fueled by the magmatic energy from the crystallization and cooling of fresh ocean crust to ~300–400°C and (ii) at lower temperatures on the ridge flanks, as the ocean crust cools slowly over millions of years. These hydrothermal fluxes play pivotal roles in controlling seawater chemistry, but the magnitude of the high temperature water flux at mid-ocean ridge axes remains widely disputed, whilst the volume of low temperature vent fluids expelled at ridge flanks is virtually unconstrained.

As discussed in Sects. 13.4.2.1 and 13.4.3.2, Tl displays distinct behavior during high and low temperature hydrothermal alteration of the ocean crust. High-T fluids effectively leach Tl from the cooling rocks whereas low-T fluids deposit Tl into the upper part of the oceanic crust.

For the high-T hydrothermal fluid flux (F_{hT}), Nielsen et al. (2006c) constructed the following mass balance equation:

$$F_{hT} \times [Tl]_{hT} = F_{oc\ leach} \times [Tl]_{oc} \times f_{Tl\ leach}$$

where $[Tl]_{hT}$ is the average Tl concentration of the vent fluids, $F_{oc\ leach}$ is the mass flux of ocean crust that is leached by high-T fluids, $[Tl]_{oc}$ is the Tl content of the crust prior to leaching, and $f_{Tl\ leach}$ is the fraction of Tl leached from the rocks during alteration. Based on data obtained for samples from ODP Hole 504B (Nielsen et al. 2006c) and high-T hydrothermal fluids (Metz and Trefry 2000; Nielsen et al. 2006c) each of these parameters could be constrained and the equation solved for F_{hT}. This calculation yielded a high temperature hydrothermal water flux of 0.17–2.93 × 10^{13} kg/year, with a best estimate of 0.72 × 10^{13} kg/year (Nielsen et al. 2006c). This fluid flux, however, only accounts for about 5–80% of the heat available at mid-ocean ridge axes from the crystallization and cooling of the freshly formed ocean crust. Based on this, it was inferred that some energy at mid ocean ridge axes is lost via conduction and/or through the circulation of intermediate temperature hydrothermal fluids that do not alter the chemical budgets of Tl in the ocean crust (Nielsen et al. 2006c).

For the low-T hydrothermal fluid circulation (F_{lT}), the following mass balance equation was shown to apply (Nielsen et al. 2006c):

$$F_{vz} \times \left([Tl]_{avz} - [Tl]_{pvz}\right) = F_{lT} \times [Tl]_{sw} \times f_{upt}$$

where F_{vz} is the mass flux of ocean crust that is affected by low-T alteration, $[Tl]_{avz}$, $[Tl]_{pvz}$, and $[Tl]_{sw}$ are the Tl concentrations of the altered volcanic zone basalts, their pristine equivalents and seawater, respectively. The fraction of Tl that is removed from seawater by basalt weathering is denoted by f_{upt}. Again, samples from ODP Hole 504B (Nielsen et al. 2006c) combined with literature data (Teagle et al. 1996) enabled an assessment of each parameter and yielded an estimated low-T hydrothermal fluid flux at ridge flanks of $0.2-5.4 \times 10^{17}$ kg/year (Nielsen et al. 2006c). Using the ridge flank power output of 7.1 TW (Mottl 2003), it was calculated that such fluids have an average temperature anomaly of only about 0.1–3.6°C relative to ambient seawater, which is lower than most flank fluids sampled to date. It is therefore unclear how representative these fluids are of average low-T ocean crust alteration processes.

In order to improve the utility of Tl mass balance calculations to constrain hydrothermal fluid fluxes it will be essential to obtain more data on altered ocean crust from a number of locations. Such analyses, which are currently in progress, will provide improved constraints on the behavior of Tl during hydrothermal processes and thus ultimately yield more reliable Tl-based estimates of global hydrothermal water fluxes (Coggon et al. 2009).

13.6.3 Thallium Isotopes as a Tracer for Ocean Crust Recycling in the Mantle

One of the central hypotheses of mantle geochemistry over the last 30 years has been the proposed link between ocean crust recycling at convergent plate margins and the upwelling of mantle plumes (Hofmann and White 1982) that generate, for example, ocean island basalt (OIB) magmatism. Much work has gone into tracing sediments and altered ocean crust with radiogenic isotopes (e.g. ^{87}Rb-^{87}Sr, ^{238,235}U-^{207,206}Pb, ^{147}Sm-^{143}Nd and ^{176}Lu-^{176}Hf) which display variations as a function of time and the fractionation of the parent and daughter elements (Hofmann 1997). Whilst it has become clear that these radiogenic isotope ratios are generally different in OIB compared to the average upper mantle, direct evidence for the presence of marine sediments and/or altered ocean crust in OIB has been elusive. This reflects that the radiogenic isotope and trace element data, which have been used to infer and trace sediment recycling in the mantle, are generally open to alternative interpretations. Stable isotope tracers (e.g. oxygen (O) and lithium (Li)) have therefore been employed as they may trace specific processes such as ocean crust alteration (Alt et al. 1986; Chan et al. 2002; Burton and Vigier 2011). However, light isotopes like O and Li may also fractionate during processes occurring at mantle temperatures (Marschall et al. 2007; Williams et al. 2009) and their mantle concentrations are relatively high, such that the isotope composition of a mantle source would not be readily affected by admixing of sediment and/or ocean crust (Elliott et al. 2004; Thirlwall et al. 2004). Seen in this perspective, Tl isotopes could be a near-perfect tracer of ocean crust recycling for the following reasons:

1. Thallium isotopes do not appear to fractionate during high temperature processing and the isotope composition of (normal) upper mantle material is well defined at $\varepsilon^{205}Tl = -2$ (see Sect. 13.3.2).
2. The isotope composition of low-T altered upper ocean crust and Fe-Mn sediments differ significantly ($\varepsilon^{205}Tl = -8$ and $\varepsilon^{205}Tl = +10$, respectively) from the average upper mantle value (see Sects. 13.4.3.1 and 13.4.3.2).
3. Altered basalts and Fe-Mn sediments have much higher Tl concentrations than average upper mantle material (see Sects. 13.4.3.1 and 13.4.3.2).

Taken together, these characteristics should enable the identification of even very minor contributions of altered basalts and/or Fe-Mn sediments to a mantle source. This inference assumes, of course, that the recycled materials do not loose their distinct Tl signatures during subduction related processes. Bulk addition of <1% of low-T altered basalt and just a few parts per million by weight of pure Fe-Mn sediment into a normal mantle material should result in discernible Tl isotope anomalies that are correlated with systematically lower Cs/Tl ratios (Fig. 13.6). The Cs/Tl data can be used to identify admixing of oceanic crust and sediments because this ratio should not be significantly altered by mantle melting or fractional crystallization due to the similar compatibility of the two elements (Nielsen et al. 2006b).

To date, OIB samples from the Azores, Iceland and Hawaii have been investigated (Nielsen et al. 2006b, 2007). As it is unclear whether the Azores basalts were affected by post-eruptional alteration (Nielsen et al. 2007), these data will not be discussed here. Samples from Hawaii exhibit the most convincing Tl isotope evidence for presence of sediments in the mantle source (Fig. 13.6). About 10 ppm of pure Fe-Mn sediment would be sufficient to explain the heavy isotope compositions recorded in these rocks. It is unlikely that the Tl isotope variation originates from anything else than Fe-Mn sediment, but it is uncertain if this component was acquired by the melts during magma ascent via assimilation of modern marine sediments or if it is a feature of the mantle source. The heaviest Tl isotope compositions are, however, also characterized by the least radiogenic Pb isotope compositions (Nielsen et al. 2006b) which would argue for an old age of the sediment contaminant.

The relatively straightforward interpretation of the Tl isotope data for Hawaii could be considered a "smoking gun" for the presence of recycled ocean crust material in the Hawaiian mantle plume. However, results obtained for of a suite of lavas from Iceland strongly indicate that there is some way to go before we can confidently apply Tl isotopes as a unique tracer of crustal recycling processes (Fig. 13.6). The Iceland samples, which include rocks from all major eruption centers, exhibit a complete uniformity of the Tl isotope composition but a large variability in Cs/Tl ratios is apparent. The isotopic invariance is perhaps not so surprising given that the thickness of subducted oceanic lithosphere is >30 km. Thallium isotope anomalies will be situated in only the uppermost ~500–1,000 m, whilst the remainder of the oceanic crust is expected to be isotopically identical to the mantle. The mantle-like Tl isotope signatures of the Iceland basalts hence do not argue against the presence of recycled ocean crust in the plume source. Of more concern is that the Iceland samples also display variable Cs/Tl ratios. The large range demonstrates that processes other than the addition of Fe-Mn sediments and

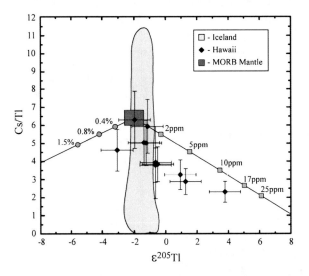

Fig. 13.6 Cs/Tl ratios of primitive basalts from Hawaii (Nielsen et al. 2006b) and Iceland (Nielsen et al. 2007) plotted versus Tl isotope composition. Mixing lines between pristine mantle (*large pink square*), Fe-Mn oxyhydroxides and low-T altered MORB are also shown. The mantle is assumed to be characterized by $\varepsilon^{205}Tl = -2$ (Nielsen et al. 2006b) and Cs and Tl concentrations of 7.7 ppb (Hofmann and White 1983) and 1.2 ppb (Nielsen et al. 2006b), respectively. For the Fe-Mn oxyhydroxides, the Tl concentration and isotope composition are assumed to be 100 µg/g and $\varepsilon^{205}Tl = +10$, akin to values of modern Fe-Mn crusts and nodules (Hein et al. 2000; Rehkämper et al. 2002). The Cs content of Fe-Mn oxyhydroxides is about 500 ng/g (Ben Othmann et al. 1989). Altered MORB is assumed to be characterized by $\varepsilon^{205}Tl = -15$ and Tl and Cs concentrations of about 200 ng/g (Nielsen et al. 2006c). Error bars denote 2sd uncertainties

low-T altered basalts may alter this ratio. It is conceivable that this could include igneous processes, such as partitioning of Tl and or Cs into sulfides and/or phyllosilicates as well as mobilization by magmatic fluids (Nielsen et al. 2007).

It is also prudent to consider two other aspects important for the application of Tl isotopes in mantle geochemistry. First, it is likely that the Tl isotope composition of seawater has not remained constant over time and the isotope signatures of altered basalts and Fe-Mn sediments are therefore also expected to exhibit temporal variability. Any such variability will act to obscure the systematic mixing trends, which are produced by contamination of the ambient mantle with these endmember compositions. Second, it is unclear when the oceans became sufficiently oxic to support the precipitation of Fe-Mn oxyhydroxides, but this probably occurred after ~2.4 Ga (Canfield 1998). Sediments that were recycled more than 2.4 billion years ago are therefore unlikely to be enriched in Tl from Fe-Mn sediments and hence would most probably exhibit ε^{205}Tl ≈ -2. The isotope fractionation mechanism for altered basalts is yet to be determined, so it is unclear whether such basalts would be isotopically fractionated in ancient environments and thus if ocean crust recycling would be able to alter the Tl isotope composition of the mantle.

13.7 Future Directions and Outlook

High precision measurements of Tl isotope ratios have only been possible for little more than a decade and it is therefore not surprising that our knowledge of the isotopic behavior of this element is limited. To date, studies have revealed that Tl isotopes fractionate substantially in some terrestrial environments, despite their heavy masses. Most of the variation occurs in the marine realm with ferromanganese sediments and low temperature hydrothermal alteration displaying a total isotope variation of about 35 ε^{205}Tl-units (Rehkämper et al. 2002, 2004; Nielsen et al. 2006c, 2009a; Coggon et al. 2009). The isotopic variation can most likely be accounted for by invoking a combination of conventional mass dependent equilibrium isotope fractionation and nuclear field shift isotope effects (Bigeleisen 1996; Schauble 2007), but the specific mechanisms are still largely unaccounted for.

Thallium isotopes have thus far mainly been applied to investigate past and present marine environments. Firstly, several papers have focused on paleoceanographic processes in the Cenozoic and it appears that Tl isotopes may be utilized as a monitor of Fe and Mn supply to the water column over million year time scales (Rehkämper et al. 2004; Baker et al. 2009; Nielsen et al. 2009a). Secondly, Tl isotopes can be used to calculate the magnitude of hydrothermal fluid circulation through ocean crust. Such calculations can be performed both for high and low temperature fluids (Nielsen et al. 2006c). Thirdly, it has been shown that marine ferromanganese sediments can be detected in mantle-derived basalts (Nielsen et al. 2006b), which confirms that marine sediments subducted at convergent plate margins can be recycled to the surface via mantle plumes (Hofmann and White 1982).

Apart from the obvious overall necessity of expanding the Tl isotope and concentration database for all terrestrial environments in order to gain a better understanding of the geochemical distribution and behavior of this element, there are a few crucial investigations that are needed to advance the utility of Tl isotopes as quantitative tracers of past and present geological processes.

First of all, we must understand the mechanisms that govern the two major Tl isotope effects observed on Earth and which yield highly fractionated Tl in Fe-Mn sediments and low-T altered basalts. This knowledge will not only help expand our appreciation of the physico-chemical processes that cause very heavy isotopes to fractionate, but also enable us to better utilize the Tl isotope system to quantify low-T hydrothermal fluid flow (Nielsen et al. 2006c) and generate more accurate models for the causes of the Tl isotope variations observed in the marine environment over time (Baker et al. 2009; Nielsen et al. 2009a). Efforts are currently in progress that may eventually shed light on the origin of these Tl isotope fractionations (Nielsen et al. 2008; Coggon et al. 2009; Peacock et al. 2009), but there is probably still a long way to go before constraints from experiments and theory can be combined in comprehensive models that are consistent with observations on natural systems.

Secondly, it will be important to refine the three applications outlined in Sect. 13.6. For the paleoceanographic studies this will include a detailed determination of the magnitude and isotope compositions of the most uncertain marine fluxes. These are benthic pore waters, adsorption processes on pelagic clays and low-T hydrothermal alteration. A complete understanding of the modern marine Tl cycle is a prerequisite for good models of past Tl isotope variations. Another aspect of the oceanic Tl isotope evolution that has yet to be investigated, are short-term fluctuations, for example on glacial-interglacial time-scales. Since the marine residence time of Tl is ~21 kyr (Table 13.4), it should be feasible to observe perturbations of the Tl cycle that happen on geologically rapid time-scales, although it will be vital to identify geological archives that accurately monitor seawater Tl isotope variations with millennial resolution.

The use of Tl isotopes as a tool in mantle geochemistry is currently limited by scant data. Thallium isotope and concentration data are presently available only for a few OIB and we also lack knowledge concerning how Tl is cycled through subduction zones. Two recent studies that attempted to address such questions through Tl isotope analyses of cratonic eclogites (Nielsen et al. 2009b) and island arc lavas (Prytulak and Nielsen, unpublished data) were unable to conclusively constrain the behavior of Tl and further investigations of rocks from subduction related environments are thus necessary. Of particular interest to such studies are the distinct Tl isotope compositions of key arc components, particularly marine sediments and altered basalts. This implies, for example, that Tl isotope data for arc lavas may provide information on the sources of Tl and, more generally, element mobility and transport in subduction zone environments.

Lastly, there are a number of essentially unexplored but potentially useful applications of Tl isotopes. For example, anthropogenic Tl pollution from power plants and zinc smelting might be characterized by distinct Tl isotope fingerprints that reflect kinetic isotope fractionation. The isotope fractionations that may result from the partitioning of chalcophile Tl (Shaw 1952) into sulfide phases has also been largely neglected to date. A Tl isotope study of rocks from the vicinity of a Cu porphyry deposit showed that Tl may retain its lithophile behavior even in sulphide-rich environments (Baker et al. 2010a). Some sulfides, however, are known to be enriched in Tl (Heinrichs et al. 1980), but apart from this, little is known about the association of Tl with sulfide minerals. Hence, it may be of value to investigate Tl isotope fractionation during partitioning into sulfides. Such studies may yield information about the conditions at which Tl behaves as a lithophile or chalcophile element and, potentially, help improve our understanding of fluid mobility in ore systems.

References

Alt JC (1995) Subseafloor processes in mid-ocean ridge hydrothermal systems. In: Humphris SE, Lupton JE, Mullineaux LS, Zierenberg RA (eds) Seafloor hydrothermal systems, physical, chemical, and biological interactions. AGU, Washington

Alt JC, Muehlenbachs K, Honnorez J (1986) An oxygen isotopic profile through the upper kilometer of the oceanic crust, DSDP Hole 504B. Earth Planet Sci Lett 80:217–229

Alt JC, Teagle DAH, Bach W et al (1996) Stable and strontium isotopic profiles through hydrothermally altered uppper oceanic crust, hole 504B. Proc ODP Sci Results 148:57–69

Anbar AD (2004) Iron stable isotopes: beyond biosignatures. Earth Planet Sci Lett 217:223–236

Anders E, Stevens CM (1960) Search for extinct lead 205 in meteorites. J Geophys Res 65:3043–3047

Archer C, Vance D (2008) The isotopic signature of the global riverine molybdenum flux and anoxia in the ancient oceans. Nat Geosci 1(9):597–600

Baker RGA, Rehkämper M, Hinkley TK et al (2009) Investigation of thallium fluxes from subaerial volcanism-implications for the present and past mass balance of thallium in the oceans. Geochim Cosmochim Acta 73(20):6340–6359

Baker RGA, Rehkämper M, Ihlenfeld C et al (2010a) Thallium isotope variations in an ore-bearing continental igneous setting: Collahuasi Formation, Northern Chile. Geochim Cosmochim Acta 74(15):4405–4416

Baker RGA, Schonbachler M, Rehkamper M et al (2010b) The thallium isotope composition of carbonaceous chondrites – new evidence for live Pb-205 in the early solar system. Earth Planet Sci Lett 291(1–4):39–47

Ben Othmann D, White WM, Patchett J (1989) The geochemistry of marine sediments, island arc magma genesis, and crust-mantle recycling. Earth Planet Sci Lett 94:1–21

Berner RA (1984) Sedimentary pyrite formation – an update. Geochim Cosmochim Acta 48(4):605–615

Bidoglio G, Gibson PN, Ogorman M et al (1993) X-ray-absorption spectroscopy investigation of surface redox transformations of thallium and chromium on colloidal mineral oxides. Geochim Cosmochim Acta 57(10):2389–2394

Bigeleisen J (1996) Nuclear size and shape effects in chemical reactions. Isotope chemistry of the heavy elements. J Am Chem Soc 118(15):3676–3680

Bigeleisen J, Mayer MG (1947) Calculation of equilibrium constants for isotopic exchange reactions. J Chem Phys 15 (5):261–267

Bruland KW (1983) Trace elements in seawater. In: Riley JP, Chester R (eds) Chemical oceanography. Academic, London

Burton KW (2006) Global weathering variations inferred from marine radiogenic isotope records. J Geochem Explor 88:262–265

Burton KW, Vigier N (2011) Chapter 4 Lithium Isotopes as Tracers in Marine and Terrestrial Environments. In: Baskaran M (ed) Handbook of environmental isotope geochemistry. Springer, Heidelberg

Burton KW, Ling HF, Onions RK (1997) Closure of the Central American Isthmus and its effect on deep-water formation in the North Atlantic. Nature 386(6623):382–385

Canfield DE (1998) A new model for Proterozoic ocean chemistry. Nature 396(6710):450–453

Chan LH, Alt JC, Teagle DAH (2002) Lithium and lithium isotope profiles through the upper oceanic crust: a study of seawater-basalt exchange at ODP Sites 504B and 896A. Earth Planet Sci Lett 201:187–201

Cheam V (2001) Thallium contamination of water in Canada. Water Qual Res J Can 36(4):851–877

Cheam V, Garbai G, Lechner J et al (2000) Local impacts of coal mines and power plants across Canada. I. Thallium in waters and sediments. Water Qual Res J Can 35:581–607

Chen JH, Wasserburg GJ (1987) A search for evidence of extinct lead 205 in iron meteorites. LPSC XVIII:165–166

Chen JH, Wasserburg GJ (1994) The abundance of thallium and premordial lead in selected meteorites – the search for ^{205}Pb. LPSC XVV:245

Cloquet C, Carignan J, Libourel G et al (2006) Tracing source pollution in soils using cadmium and lead isotopes. Environ Sci Technol 40(8):2525–2530

Coggon RM, Rehkamper M, Atteck C et al (2009) Constraints on hydrothermal fluid fluxes from Tl geochemistry. Geochim Cosmochim Acta 73(13):A234–A234

Eisenhauer A, Gogen K, Pernicka E et al (1992) Climatic influences on the growth-rates of Mn Crusts during the late quaternary. Earth Planet Sci Lett 109(1–2):25–36

Elderfield H (1976) Manganese fluxes to the oceans. Mar Chem 4(2):103–132

Elliott T, Jeffcoate A, Bouman C (2004) The terrestrial Li isotope cycle: light-weight constraints on mantle convection. Earth Planet Sci Lett 220:231–245

Flegal AR, Patterson CC (1985) Thallium concentrations in seawater. Mar Chem 15:327–331

Flegal AR, Sanudo-Wilhelmy S, Fitzwater SE (1989) Particulate thallium fluxes in the northeast Pacific. Mar Chem 28:61–75

Frank M (2002) Radiogenic isotopes: tracers of past ocean circulation and erosional input. Rev Geophys 40:art. no.-1001

Fujii Y, Nomura M, Okamoto M et al (1989a) An anomalous isotope effect of U-235 in U(IV)-U(VI) chemical exchange. Z Naturforsch 44(5):395–398

Fujii Y, Nomura M, Onitsuka H et al (1989b) Anomalous isotope fractionation in uranium enrichment process. J Nucl Sci Technol 26(11):1061–1064

Gauthier PJ, Le Cloarec MF (1998) Variability of alkali and heavy metal fluxes released by Mt. Etna volcano, Sicily, between 1991 and 1995. J Volcanol Geotherm Res 81:311–326

Heggie D, Klinkhammer G, Cullen D (1987) Manganese and copper fluxes from continental margin sediments. Geochim Cosmochim Acta 51(5):1059–1070

Hein JR, Koschinsky A, Bau M et al (2000) Cobalt-rich ferromanganese crusts in the Pacific. In: Cronan DS (ed) Handbook of marine mineral deposits. CRC Press, Boca Raton

Heinrichs H, Schulz-Dobrick B, Wedepohl KH (1980) Terrestrial geochemistry of Cd, Bi, Tl, Pb, Zn and Rb. Geochim Cosmochim Acta 44:1519–1533

Hinkley TK, Lecloarec MF, Lambert G (1994) Fractionation of families of major, minor and trace-metals across the melt vapor interface in volcanic exhalations. Geochim Cosmochim Acta 58(15):3255–3263

Hofmann AW (1997) Mantle geochemistry-the message from oceanic volcanism. Nature 385:219–229

Hofmann AW, White WM (1982) Mantle plumes from ancient oceanic crust. Earth Planet Sci Lett 57:421–436

Hofmann AW, White WM (1983) Ba, Rb and Cs in the Earth's mantle. Z Naturforsch 38:256–266

Huey JM, Kohman TP (1972) Search for extinct natural radioactivity of Pb-205 via thallium-isotope anomalies in chondrites and lunar soil. Earth Planet Sci Lett 16:401–412

Huh Y, Chan LH, Zhang L et al (1998) Lithium and its isotopes in major world rivers: implications for weathering and the oceanic budget. Geochim Cosmochim Acta 62:2039–2051

Jochum KP, Verma SP (1996) Extreme enrichment of Sb, Tl and other trace elements in altered MORB. Chem Geol 130:289–299

Johnson KS, Berelson WM, Coale KH et al (1992) Manganese flux from continental-margin sediments in a transect through the oxygen minimum. Science 257(5074):1242–1245

Koschinsky A, Hein JR (2003) Acquisition of elements from seawater by ferromanganese crusts: solid phase association and seawater speciation. Mar Geol 198:331–351

Kurtz AC, Kump LR, Arthur MA et al (2003) Early Cenozoic decoupling of the global carbon and sulfur cycles. Paleoceanography 18(4):14.1–14.14

Lee D-C, Halliday AN, Hein JR et al (1999) Hafnium isotope stratigraphy of ferromanganese crusts. Science 285:1052–1054

Lin T-S, Nriagu J (1999) Thallium speciation in the Great Lakes. Environ Sci Technol 33:3394–3397

Lis J, Pasieczna A, Karbowska B et al (2003) Thallium in soils and stream sediments of a Zn-Pb mining and smelting area. Environ Sci Technol 37(20):4569–4572

Marschall HR, Pogge von Strandmann PAE, Seitz HM et al (2007) The lithium isotopic composition of orogenic eclogites and deep subducted slabs. Earth Planet Sci Lett 262(3–4):563–580

Matthews AD, Riley JP (1970) The occurrence of thallium in sea water and marine sediments. Chem Geol 149:149–152

McDonough WF, Sun S-S (1995) The composition of the Earth. Chem Geol 120:223–253

McGoldrick PJ, Keays RR, Scott BB (1979) Thallium – sensitive indicator of rock-seawater interaction and of sulfur saturation of silicate melts. Geochim Cosmochim Acta 43:1303–1311

Metz S, Trefry JH (2000) Chemical and mineralogical influences on concentrations of trace metals in hydrothermal fluids. Geochim Cosmochim Acta 64:2267–2279

Mottl MJ (2003) Partitioning of energy and mass fluxes between mid-ocean ridge axes and flanks at high and low temperature. In: Halbach PE, Tunnicliffe V, Hein JR (eds) Energy and mass transfer in marine hydrothermal systems. Dahlem University Press, Berlin

Nielsen SG, Goff M, Hesselbo SP, et al (2011) Thallium isotopes in early diagenetic pyrite - a paleoredox proxy? Geochim Cosmochim Acta, In Press

Nielsen SG, Rehkämper M, Baker J et al (2004) The precise and accurate determination of thallium isotope compositions and concentrations for water samples by MC-ICPMS. Chem Geol 204:109–124

Nielsen SG, Rehkämper M, Porcelli D et al (2005) The thallium isotope composition of the upper continental crust and rivers – an investigation of the continental sources of dissolved marine thallium. Geochim Cosmichim Acta 69:2007–2019

Nielsen SG, Rehkämper M, Halliday AN (2006a) Large thallium isotopic variations in iron meteorites and evidence for lead-205 in the early solar system. Geochim Cosmochim Acta 70:2643–2657

Nielsen SG, Rehkämper M, Norman MD et al (2006b) Thallium isotopic evidence for ferromanganese sediments in the mantle source of Hawaiian basalts. Nature 439:314–317

Nielsen SG, Rehkämper M, Teagle DAH et al (2006c) Hydrothermal fluid fluxes calculated from the isotopic mass balance of thallium in the ocean crust. Earth Planet Sci Lett 251(1–2):120–133

Nielsen SG, Rehkämper M, Brandon AD et al (2007) Thallium isotopes in Iceland and Azores lavas – implications for the role of altered crust and mantle geochemistry. Earth Planet Sci Lett 264:332–345

Nielsen SG, Peacock CL, Halliday AN (2008) Investigation of thallium isotope fractionation during sorption to Mn oxides. Geochim Cosmochim Acta 72(12):A681

Nielsen SG, Mar-Gerrison S, Gannoun A et al (2009a) Thallium isotope evidence for increased marine organic carbon export in the early Eocene. Earth Planet Sci Lett 278:297–307

Nielsen SG, Williams HM, Griffin WL et al (2009b) Thallium isotopes as a potential tracer for the origin of cratonic eclogites? Geochim Cosmochim Acta 73:7387–7398

Nriagu J (ed) (1998) Thallium in the environment. Wiley, New York

Ostic RG, Elbadry HM, Kohman TP (1969) Isotopic composition of meteoritic thallium. Earth Planet Sci Lett 7(1):72–76

Patterson CC, Settle DM (1987) Magnitude of lead flux to the atmosphere from volcanos. Geochim Cosmochim Acta 51(3):675–681

Paytan A, Kastner M, Campbell D et al (1998) Sulfur isotopic composition of Cenozoic seawater sulfate. Science 282(5393):1459–1462

Paytan A, Kastner M, Campbell D et al (2004) Seawater sulfur isotope fluctuations in the cretaceous. Science 304(5677):1663–1665

Peacock CL, Moon EM, Nielsen SG et al (2009) Oxidative scavenging of Tl by Mn oxide birnessite: sorption and stable isotope fractionation. Geochim Cosmochim Acta 73(13):A1003

Pengra JG, Genz H, Fink RW (1978) Orbital electron capture ratios in the decay of ^{205}Pb. Nucl Phys A302:1–11

Pietruszka AJ, Reznik AD (2008) Identification of a matrix effect in the MC-ICP-MS due to sample purification using ion exchange resin: an isotopic case study of molybdenum. Int J Mass Spectrom 270(1–2):23–30

Pogge von Strandmann PAE, Burton KW, James RH et al (2010) Assessing the role of climate on uranium and lithium isotope behaviour in rivers draining a basaltic terrain. Chem Geol 270(1–4):227–239

Poirier A, Doucelance R (2009) Effective correction of mass bias for rhenium measurements by MC-ICP-MS. Geostand Geoanal Res 33(2):195–204

Prytulak J, Nielsen SG (unpublished data) Thallium isotope budget of the Mariana arc and influence of subducted slab components. Chem Geol. In Prep

Rehkämper M, Halliday AN (1999) The precise measurement of Tl isotopic compositions by MC-ICPMS: application to the analysis of geological materials and meteorites. Geochim Cosmochim Acta 63:935–944

Rehkämper M, Nielsen SG (2004) The mass balance of dissolved thallium in the oceans. Mar Chem 85:125–139

Rehkämper M, Frank M, Hein JR et al (2002) Thallium isotope variations in seawater and hydrogenetic, diagenetic, and hydrothermal ferromanganese deposits. Earth Planet Sci Lett 197:65–81

Rehkämper M, Frank M, Hein JR et al (2004) Cenozoic marine geochemistry of thallium deduced from isotopic studies of ferromanganese crusts and pelagic sediments. Earth Planet Sci Lett 219:77–91

Rudge JF, Reynolds BC, Bourdon B (2009) The double spike toolbox. Chem Geol 265(3–4):420–431

Rudnick RL, Gao S (2003) Composition of the continental crust. In: Holland HD, Turekian KK (eds) Treatise on geochemistry. Pergamon, Oxford

Sager M (1993) Determination of arenic, cadmium, mercury, stibium, thallium and zinc in coal and coal fly-ash. Fuel 72(9):1327–1330

Sasmaz A, Sen O, Kaya G et al (2007) Distribution of thallium in soil and plants growing in the keban mining district of Turkey and determined by ICP-MS. At Spectrosc 28(5):157–163

Sawlan JJ, Murray JW (1983) Trace-metal remobilisation in the interstitial waters of red clay and hemipelagic marine sediments. Earth Planet Sci Lett 64(2):213–230

Schauble EA (2007) Role of nuclear volume in driving equilibrium stable isotope fractionation of mercury, thallium, and other very heavy elements. Geochim Cosmochim Acta 71(9):2170–2189

Schedlbauer OF, Heumann KG (2000) Biomethylation of thallium by bacteria and first determination of biogenic dimethylthallium in the ocean. Appl Organometal Chem 14:330–340

Segl M, Mangini A, Bonani G et al (1984) Be-10-Dating of a manganese crust from Central North Pacific and implications for ocean palaeocirculation. Nature 309(5968):540–543

Segl M, Mangini A, Beer J et al (1989) Growth rate variations of manganese nodules and crusts induced by paleoceanographic events. Paleoceanography 4(5):511–530

Shannon RD (1976) Revised effective ionic radii and systematic studies of interatomic distances in halides and chalcogenides. Acta Crystallogr A32:751–767

Shaw DM (1952) The geochemistry of thallium. Geochim Cosmochim Acta 2:118–154

Shiel AE, Barling J, Orians KJ et al (2009) Matrix effects on the multi-collector inductively coupled plasma mass spectrometric analysis of high-precision cadmium and zinc isotope ratios. Anal Chim Acta 633(1):29–37

Teagle DAH, Alt JC, Bach W et al (1996) Alteration of upper ocean crust in a ridge-flank hydrothermal upflow zone: mineral, chemical, and isotopic constraints from hole 896A. Proc ODP Sci Results 148:119–150

Thirlwall MF, Gee MAM, Taylor RN et al (2004) Mantle components in Iceland and adjacent ridges investigated using double-spike Pb isotope ratios. Geochim Cosmochim Acta 68(2):361–386

Tsuchiyama A, Kawamura K, Nakao T et al (1994) Isotopic effects on diffusion in MgO melt simulated by the molecular-dynamics (Md) method and implications for isotopic mass fractionation in magmatic systems. Geochim Cosmochim Acta 58:3013–3021

Urey HC (1947) The thermodynamic properties of isotopic substances. J Chem Soc May:562–581

van de Flierdt T, Frank M, Halliday AN et al (2004) Tracing the history of submarine hydrothermal inputs and the significance of hydrothermal hafnium for the seawater budget-a combined Pb-Hf-Nd isotope approach. Earth Planet Sci Lett 222:259–273

Wallmann K (2001) Controls on the Cretaceous and Cenozoic evolution of seawater composition, atmospheric CO_2 and climate. Geochim Cosmochim Acta 65(18):3005–3025

Wedepohl KH (1974) Handbook of geochemistry. Springer, Berlin

Williams H, Turner S, Kelley S et al (2001) Age and composition of dikes in Southern Tibet: new constraints on the timing of east-west extension and its relationship to postcollisional volcanism. Geology 29(4):339–342

Williams HM, McCammon CA, Peslier AH et al (2004) Iron isotope fractionation and the oxygen fugacity of the mantle. Science 304:1656–1659

Williams HM, Nielsen SG, Renac C et al (2009) Fractionation of oxygen and iron isotopes in the mantle: implications for crustal recycling and the source regions of oceanic basalts. Earth Planet Sci Lett 283:156–166

Xiao TF, Boyle D, Guha J et al (2003) Groundwater-related thallium transfer processes and their impacts on the ecosystem: southwest Guizhou Province. China Appl Geochem 18 (5):675–691

Xiao TF, Guha J, Boyle D et al (2004) Environmental concerns related to high thallium levels in soils and thallium uptake by plants in southwest Guizhou. China Sci Total Environ 318 (1–3):223–244

Chapter 14
Po-210 in the Environment: Biogeochemical Cycling and Bioavailability

Guebuem Kim, Tae-Hoon Kim, and Thomas M. Church

Abstract As the heaviest element of Group 6A, ^{210}Po has a unique biogeochemistry in the environment that challenges our understanding. This chapter provides an overview of the research on ^{210}Po in the atmosphere as well as in marine and other aqueous environments. Excess atmospheric ^{210}Po has been attributed to external sources, such as volcanic emissions, resuspension of soil humus, incursion of stratospheric air, sea spray from the oceanic surface micro-layer, plant exudates including evapotranspiration, anthropogenic emissions (e.g., emission from coal combustion), and bio-volatilization through the formation of dimethyl polonide. Most of these sources have been qualitatively identified, yet they remain difficult to quantify. In the aqueous environment, ^{210}Po is efficiently accumulated in plankton and bacteria and is biomagnified through the food webs, relative to its grandparent ^{210}Pb, causing ^{210}Po to be largely deficient in the euphotic zone. Globally, ^{210}Po deficiency increases as ocean productivity decreases in the upper 1,000 m through biological transfer to the upper trophic levels. Smaller ^{210}Po deficiencies in the productive areas of the ocean appear to be related to relatively active bacterial remineralization. Unusually high activities of ^{210}Po are often found in the suboxic and anoxic waters in association with S, Mn, and Fe redox cycles. As many details of these processes remain elusive and under debate, we propose additional studies that should be conducted.

G. Kim (✉) and T.-H. Kim
School of Earth and Environmental Sciences, Seoul National University, Seoul, South Korea

T.M. Church
College of Earth, Ocean, and Environment, University of Delaware, Newark, DE, USA

14.1 Introduction

Naturally occurring ^{210}Po ($t_{1/2}$ = 138 days) is the decay product of ^{210}Pb ($t_{1/2}$ = 22.3 years) via ^{210}Bi ($t_{1/2}$ = 5 days) in the ^{238}U decay series and widely distributed in the earth's crust, rivers, oceans, and the atmosphere. ^{210}Po is highly radioactive, with alpha particle energy of 5.30 MeV. Although alpha particles are not sufficiently energetic to pass through a person's outer skin, they easily penetrate the unprotected lining and pass through living cells when released within the lungs. Thus, ^{210}Po and other radon daughters in the inhaled air may cause lung cancer, especially when individuals are exposed to high activities of point-sources in regions such as confined spaces associated with radioactive materials. In addition, it could pose a risk to human health when it enters the body through food consumption.

The ratios of ^{210}Po/^{210}Pb vary in the earth's surface, aqueous environment, and atmosphere because of their different chemical reactivity, biological enrichments, and volatility. In the atmosphere, ^{210}Po produced from ^{210}Pb is minor (e.g., ^{210}Po/^{210}Pb < 0.1), as measured in precipitation or fallout particles. However, a large excess (not produced in the troposphere) of ^{210}Po is introduced by a variety of sources, including volcanic emissions, resuspension of top soils, incursion of stratospheric air, sea spray from the oceanic surface micro-layer, bio-volatilization of ^{210}Po from the ocean, and anthropogenic emissions (Vilenskii 1970; Poet et al. 1972; Turekian et al. 1974, 1977; Moore et al. 1976; Bacon and Elzerman 1980; Lambert et al. 1985; Baskaran 2011). In the aquatic environment, ^{210}Po displays a strong biogeochemical cycling, relative to its parent ^{210}Pb, and exhibits a large deficiency in the euphotic zone because of its preferential removal by biota and rapid regeneration in the subsurface layer because of preferential remineralization from sinking biogenic debris.

Concentration factors for ^{210}Po in the pelagic ecosystem increase through the food web from phytoplankton and finally to fish (15×10^4, shrimp), although they are only about $12–23 \times 10^2$ for ^{210}Pb (e.g., Fisher et al. 1983; Stewart and Fisher 2003; Stewart et al. 2005; Waska et al. 2008). Uranium mining operations often result in elevated levels of ^{210}Po in freshwater fish (average 208 mBq g^{-1}), especially in the viscera, while ^{210}Pb concentrates in the bone (Carvalho 1988; Carvalho et al. 2007). Po is effectively taken up by bacteria and dispersed between the cell walls, cytoplasm, and high-molecular-weight proteins in a manner similar to sulfur (Fisher et al. 1983; Cherrier et al. 1995; LaRock et al. 1996; Momoshima et al. 2001; Kim et al. 2005a). Thus, the volatility of Po could be due to the formation of bio-volatile species such as dimethyl polonide (Hussain et al. 1995). Given its high affinity for bacteria, it is not surprising that extremely high and unsupported ^{210}Po activity was found in some sulfide-bearing shallow groundwater in central Florida (Harada et al. 1989).

Thus, studying ^{210}Po in the environment is important in many aspects: (1) it can serve as a tracer for sulfur group elements such as Se and Te in aqueous environments, (2) its bio-accumulation and transfer through terrestrial and marine food chains are of great health concerns for human beings, and (3) the occurrence of excess ^{210}Po in the atmosphere is also important for tracing atmospheric emissions and in enclosed areas for human health. In this chapter, we review recent advances in our understanding of the fate and cycling of ^{210}Po in the atmospheric and aqueous environments.

14.2 Analytical Methods

The analytical methods for measuring ^{210}Po and ^{210}Pb in the same sample are relatively straightforward (e.g., Sarin et al. 1992; Kim et al. 1999; Masque et al. 2002; Stewart et al. 2007; IAEA 2009; Baskaran et al. 2009; GEOTRACES User Manual, Church and Baskaran, http://www.ldeo.columbia.edu/res/pi/geotraces/documents/GEOTRACESIPYProtocols-Final.pdf). The source of ^{210}Po is prepared by spontaneous self-deposition onto a silver disc and following simple pre-concentration procedures to separate ^{210}Pb. Then, ^{210}Pb is measured, via ingrown ^{210}Po, after storing the sample for more than 3 months. Some researchers determine ^{210}Pb activities by beta counting of its daughter, ^{210}Bi, following a specific separation and source preparation in order to obtain ^{210}Pb data without delay (e.g., Nozaki and Tsunogai 1973). Also, some researchers conduct the separation of ^{210}Po from ^{210}Pb via specialized resins before auto-plating in order to eliminate any interference in the alpha counting (e.g., IAEA 2009). However, in this chapter we describe the simplest and most common method for ^{210}Po and ^{210}Pb analyses in water samples. Although we do not include the method for solid samples, the methods after sample dissolution and subsequent Fe(OH)$_3$ precipitation are the same.

In general, approximately 10–20 L of water is required for the analysis of dissolved or total ^{210}Po and ^{210}Pb in seawater, river water, or lake water. Because the activity of ^{210}Pb in particulate matter is approximately one order of magnitude lower than that in solution, the volume of water required for the analysis of particulate ^{210}Po and ^{210}Pb is approximately 3–5 times higher than that required for the analyses of ^{210}Po and ^{210}Pb in the solution form. Although the uncertainties in measuring the sample volume can easily be reduced in the laboratory, special tools, such as an electronic balance, may be required to weigh the samples in order to reduce the uncertainties onboard ships.

The separation between the dissolved and particulate phases can be performed using the membrane or cartridge filters with a pore size of 0.4–0.8 μm immediately after sample collection. From the GEOTRACE intercalibration results (GEOTRACES User Manual), the difference between the particulate ^{210}Po and ^{210}Pb concentrations using 0.4- or 0.8-μm filters was found to be insignificant, but the choice of the filter materials is important – Supor 0.4–0.8 μm filter cartridges (e.g. Acropak 500) have been found to be reliable. The filtered water samples are stored in acid-cleaned plastic containers with acidification to a pH lower than 2. Because the intercalibration showed higher than 20% uncertainties for 10–20 L water volumes for both nuclides, water volumes of at least 50 L are recommended for particulate ^{210}Po and ^{210}Pb measurements.

The water samples are transferred into plastic buckets or cubitainers, and then spiked with NIST-traceable

^{209}Po, 25 mg of ancient Pb carrier (with a negligible ^{210}Pb activity), and 70 mg of Fe^{3+} carrier for 10–20 L volume. Following stirring with rods or bubbling with N$_2$ gas for 10–30 min for equilibration of the spikes and carriers, the samples are allowed to stand for an additional hours to ensure complete equilibration (Kim et al. 1999). The Po and Pb are co-precipitated with Fe(OH)$_3$ at pH 7–8 using ammonia solution, with vigorous stirring during the first 10 min, and the precipitate is then allowed to settle for hours. The supernatant is siphoned off, and the precipitates are collected using a centrifuge or Whatman 54 quantitative-grade paper. The precipitates on the filter paper are dissolved by adding approximately 3 mL of 6 M HCl and are then rinsed thoroughly using about 40 mL of deionized water to bring the samples for plating to approximately 0.5 M HCl. About 200 mg of ascorbic acid is added to reach a colorless solution (Fe is fully reduced). The Po in the solution is spontaneously plated onto a silver disc (the reverse side of which is covered by a neutral cement or plastic film) by swirling the disc in water at temperature at approximately 90°C.

After the self-plating of Po onto the disc, the solution is dried and then dissolved in 5 mL of 9 M HCl. The Pb is separated from Po using a preconditioned 9 M HCl anion-exchange column. The remaining Po is adsorbed onto the column, while Pb passes through the column. To the collected Pb solution, a known amount of ^{209}Po spike is added, and the sample is stored in a clean plastic bottle for more than 3 months. The ^{210}Pb activity is measured by determining the ingrown activity of ^{210}Po. The chemical yield of Pb from all the procedures is determined by measuring the stable Pb recovery for an aliquot of the ^{210}Pb solution. The ^{210}Po activity is determined using alpha spectroscopy.

The time of sampling, anion column separation, and alpha counting should be recorded to conduct the appropriate corrections for the ingrowth and decay of ^{210}Po and ^{210}Pb from sampling to counting (IAEA 2009). The activity of ^{210}Po is calculated by the ^{210}Po/^{209}Po counting ratios multiplied by the activity of the added ^{209}Po spike. The counts of each Po peak should be carefully corrected for background because Po background builds up easily owing to its volatile nature. The blank activities should be further corrected for the ^{210}Po and ^{210}Pb activities.

14.3 ^{210}Po in the Atmosphere

In the atmosphere, ^{222}Rn daughters, ^{210}Pb, ^{210}Bi, and ^{210}Po, grow in during the residence of aerosols because they are adsorbed readily onto particles. Thus, their disequilibria, ^{210}Pb/^{222}Rn, ^{210}Bi/^{210}Pb, and ^{210}Po/^{210}Pb, have been used to determine the residence times of aerosols (e.g., Turekian et al. 1977; Carvalho 1995a; Church and Sarin 2008). Amongst these pairs, the residence times (10–300 days) of aerosols based on ^{210}Po/^{210}Pb disequilibria are generally much longer than those based on the other pairs (2–20 days) and aerosol deposition models (e.g., Kim et al. 2005a). Thus, the various possible sources for the excess ^{210}Po (the observed ^{210}Po activity minus the ^{210}Po activity produced from ^{210}Pb in the troposphere) have been reported in many studies. The suggested primary sources include the input of aged aerosols from the stratosphere (residence time of about 1 year) or resuspended dust from topsoil in which the ^{210}Po/^{210}Pb activity ratios are close to 1. Poet et al. (1972) and Moore et al. (1976) suggested that ^{210}Po mixed down from the stratosphere was 0.2–7% of that in the troposphere. At Lamto, on the west coast of the African continent, soil dusts from Sahara desert contributed to very high ^{210}Po/^{210}Pb ratios, ranging from 0.30 to 0.54, in the winter atmosphere (Nho et al. 1996). In this chapter, we introduce recent progress in finding the sources of preferential ^{210}Po inputs into the atmosphere via geophysical or biogeochemical processes.

First, high ^{210}Po can be introduced into the atmosphere by volcanic activities because of its high volatility relative to ^{210}Pb. The activity ratios of ^{210}Po/^{210}Pb were found to be between 5 and 40 in volcanic plumes from Mt. Etna (Lambert et al. 1985) and up to ~600 at the Stromboli volcano (Gauthier et al. 2000). For this reason, the ^{210}Po/^{210}Pb ratios in freshly erupted pyroclastic or lava flows are close to zero (Bennett et al. 1982; Gill et al. 1985; Rubin et al. 1994). Although quite patchy in space and time, the input of ^{210}Po from volcanic emissions can account for more than one-half of the global budget of ^{210}Po in the atmosphere (Lambert et al. 1979, 1982). In northern Taiwan, Su and Huh (2002) observed higher ^{210}Po/^{210}Pb, derived from pre-eruption gas emission from the Philippines' Mayon volcano. Based on these observational results, they suggested that ^{210}Po might serve as a precursor of volcano eruptions worldwide if a global network can be properly designed.

Second, the source of excess ^{210}Po activities in the atmosphere can be from sea spray. The sea-surface microlayer has high activity of ^{210}Po relative to ^{210}Pb. The secretion from phytoplankton concentrated in the surface microlayer, where the ^{210}Po/^{210}Pb activity ratios are as much as 3.8 (Bacon and Elzerman 1980), could increase the ^{210}Po/^{210}Pb activity ratios in the atmosphere by sea-sprays. Heyraud and Cherry (1983) showed that the ^{210}Po/^{210}Pb ratios were in the ranges of 1–3 and 2–80 for the surface microlayer and neuston, respectively. Kim et al. (2005a) showed that the seawater fraction of ^{210}Po in aerosol samples from Busan, a harbor city in Korea, explained the unusually high excess of ^{210}Po activities. This suggests that a small fraction of the sea spray can result in a large excess of ^{210}Po in the coastal atmosphere.

Third, bio-volatile ^{210}Po from the eutrophic coastal region could result in high excess ^{210}Po in the atmosphere. Hussain et al. (1995) documented that bio-volatile Po (i.e., dimethypolonide) can be formed in natural water together with other species, such as DMS, DMSe, and MMHg, through theoretical calculations and laboratory experiments. Similarly, Momoshima et al. (2001) reported that the formation and emission of volatile Po compounds occurred associated with the biological activity of microorganisms in culture medium as well as in natural seawater. Kim et al. (2000) observed that the increase in the excess ^{210}Po in the Chesapeake Bay air was dependent on the wind speed over a threshold of 3 m s^{-1} (mean), similar to other gases (i.e., CO_2, SF_6, and DMS) (Fig. 14.1). They showed that the simultaneously measured activity ratios of ^{7}Be/^{210}Pb and ^{210}Pb/^{222}Rn argued against either higher-altitude air or continental soils as the source of this excess. Thus, Kim et al. (2000) suggested that the source of excess ^{210}Po could be from surrounding coastal seawater through the gaseous air-sea exchange of volatile biogenic species (e.g., dimethyl polonide). They found that less than 10% of the total ^{210}Po in the Chesapeake or Delaware Bay was required to be volatile to account for the excess ^{210}Po in the observed air. Hussain et al. (1993) showed nearly threefold higher activity ratios of ^{210}Po/^{210}Pb in aerosols in and around the mid-Atlantic region in the summer than that in the fall and winter, even through the winter winds are stronger, as an indication of an additional bio-volatile ^{210}Po source in the marine environment during summer.

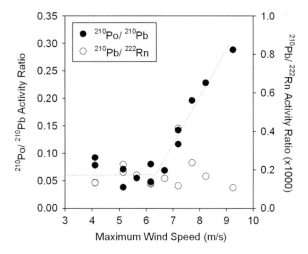

Fig. 14.1 Plots of maximum wind speed vs. ^{210}Po/^{210}Pb activity ratio for aerosol samples collected in Chesapeake Bay, USA, during 10–24 August 1995 (adapted from Kim et al. 2000). Wind speeds were measured on board hourly

Forth, plant exudates, which are submicron particles, can be important sources of excess ^{210}Po in the atmosphere (Moore et al. 1976). Plant exudates are formed by the condensation of volatile organic materials released from the plant surfaces. Stress conditions such as leaf expansion during active growth or by normal weathering due to wind action and abrasion may lead to loss of these exudates (Moore et al. 1976). Moore et al. (1976) predicted that plant exudates contribute to almost 40% of the natural and anthropogenic ^{210}Po fluxes in the atmosphere over the continental USA, on the basis of a weak correlation between submicron aerosols and vegetation density, without actual ^{210}Po data. Since this process seems to be quantitatively important, more direct measurements are necessary in the future.

Fifth, anthropogenic ^{210}Po input can result from its high volatility. Moore et al. (1976) found that up to 7% of the total ^{210}Po flux to the atmosphere in Boulder, Colorado, USA could be from anthropogenic sources. These can be produced from phosphate fertilizer dispersion, a by-product of gypsum or lead refinement, cement and other metal production, and fossil-fuel burning. As such, Carvalho (1995b) reported increased ^{210}Po/^{210}Pb ratios in precipitation from the industrial emission of ^{210}Po in Lisbon, Portugal. Kim et al. (2005a) documented that excess ^{210}Po in the metropolitan areas of Seoul, Korea, originates mostly from anthropogenic sources. They observed a strong correlation between non-sea-salt (nss) SO_4^{2-} and excess ^{210}Po, although there was no correlation between nss-SO_4^{2-} and its parent ^{210}Pb, suggesting that both anthropogenic SO_4^{2-} and excess ^{210}Po are controlled mainly by the same factor (Fig. 14.2). Based on this correlation and the δ^{34}S values, they concluded that the major source for excess ^{210}Po in Seoul precipitation is anthropogenic, likely from the burning of fossil fuels such as coals and petroleum oils, biomass burning, and/or high-temperature incineration. Gaffney et al. (2004) showed that the contribution of soils or coal-fired power plants to the ^{210}Po/^{210}Pb activity ratio in aerosols could be successfully evaluated by measuring the activity ratios for both fine (<1 μm) and coarse (>1 μm) fractions.

Nevertheless, ^{210}Po/^{210}Pb disequilibria have been successfully used for estimating the aerosol residence times in some isolated areas (i.e., Arctic haze). Baskaran and Shaw (2001) observed that the ^{210}Po/^{210}Pb ratios in Arctic haze aerosols varied between 0 and 0.177; the residence times of the aerosols calculated from these ratios were between 0 and 39 days. In addition, the aerosol residence times calculated from the ^{210}Po/^{210}Pb ratios at Centerton, New Jersey, USA, were equal to or slightly shorter than those calculated using ^{210}Bi/^{210}Pb ratios for the same aerosol samples; this indicated the negligible input of excess ^{210}Po in this area. The mean deposition velocity of the aerosols using ^{210}Po was found to be 2.2 cm s^{-1}; this velocity was higher than that reported for ^{210}Pb at the same site (McNeary and Baskaran 2007). The aforementioned difference could be attributed to the difference in the scavenging behaviors of these two nuclides. Thus, to understand the scavenging mechanisms for ^{210}Po- and ^{210}Pb-laden aerosols and to validate the residence time calculation methods using Rn daughters, more extensive measurements must be performed on all Rn daughter pairs under different environmental conditions and for different particle sizes.

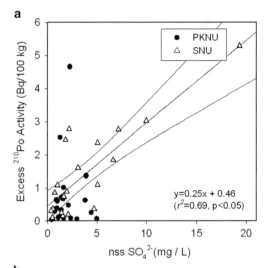

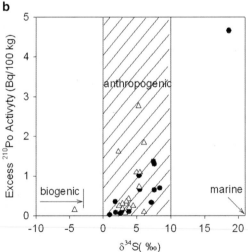

Fig. 14.2 Plots of excess ^{210}Po activities vs. (**a**) non-sea-salt SO_4^{2-} and (**b**) δ^{34}S in precipitation from Seoul (SNU, Korea) and Busan (PKNU, Korea) stations (adapted from Kim et al. 2005a). A 95% confidence interval is shown for the regression (**a**). In general, the δ^{34}S value is approximately 21‰ for sea spray, 5‰ on average for volcanic emissions, and lower than −4‰ for most of the biogenic emissions (Pichlmayer et al. 1998; Rees 1970). Anthropogenic sulfates exhibit a wide range of values from 0 to 10‰

14.4 ^{210}Po in the Ocean

In the surface ocean, ^{210}Po is generally deficient relative to its parent ^{210}Pb due to preferential removal by biota, where it is in near equilibrium or excess below the surface mixed layer due to rapid regeneration from sinking organic matter (Shannon et al. 1970; Bacon et al. 1976; Thomson and Turekian 1976; Cochran et al. 1983; Chung and Finkel 1988; Bacon et al. 1988). Sarin et al. (1999) showed that the dissolved ^{210}Po activities in the surface waters of the equatorial and South Atlantic Ocean are about a third of the equilibrium concentrations. Based on these disequilibria, they calculated that the residence times of dissolved ^{210}Po in the surface water at the equatorial and southern sites were about 73 and 130 days, respectively, with longer residence times for the intermediate depths (100–500 m). The observed ^{210}Po-^{210}Pb disequilibria showed a significant positive correlation ($r^2 = 0.61$) with the POC concentrations, suggesting that the ^{210}Po deficiency in surface water is associated mainly with biological removal from surface waters (Sarin et al. 1999). Therefore, a larger deficiency of ^{210}Po would be expected in the more productive areas of the ocean, where the population of sinking particles is larger. However, Kim (2001) showed that ^{210}Po deficiencies decrease as ocean productivity increases for the globally available ^{210}Po data in the upper 1,000 m (Fig. 14.3). The removal of ^{210}Po through the 500 m depth was about an order of magnitude higher in the oligotrophic ocean than in the productive areas of the ocean.

Nozaki et al. (1990), for the first time, pointed out an unusually large deficiency of Po in the oligotrophic Philippine Sea. They explained this phenomenon with a 2–3-fold larger focusing of atmospheric ^{210}Pb, relative to the model-based estimate, and subsequent ventilation into the deeper ocean. However, Kim (2001) posited that episodic atmospheric focusing could not significantly affect on the large deficiency of ^{210}Po in the upper oligotrophic Sargasso Sea or Philippine Sea due to the following reasons: (1) the effect of a sudden increase in the atmospheric input of ^{210}Pb (with negligible ^{210}Po) cannot be significant because the water residence times in the upper 500 m are longer than 10 years in the major oceans, allowing almost 100% ^{210}Po ingrowth; (2) the directly measured annual atmospheric deposition of ^{210}Pb during the study period was less than 10% of the deficiency of the ^{210}Po inventory in the 0–700 m layer in the Sargasso-Sea; (3) the largest deficiency of ^{210}Po was found in the subsurface layer (Fig. 14.3) rather than in the surface layer; and (4) the large deficiencies are perennial in the Sargasso Sea based on bimonthly measurements (Kim 2001). Kim (2001) showed that the horizontal transport of waters from the ocean margins also could not be the source of this large ^{210}Po deficiency water because ^{210}Po deficiencies are generally smaller in coastal waters due to rapid remineralization.

Alternatively, Kim (2001) suggested that the large deficiencies of ^{210}Po are likely due to biological removal from the total (dissolved plus particulate) pool by cyanobacteria and subsequent transfer to nekton (via grazing), which is unavailable to normal oceanographic water sampling. He showed the following evidence supporting this hypothesis: (1) the sediment trap based fluxes of ^{210}Po (Bacon et al. 1985) through 3,000 m in the Sargasso Sea were an order of magnitude lower than the calculated ^{210}Po removal flux through 3,000 m; (2) the calculated ^{210}Po/^{210}Pb export flux ratio in the upper Sargasso Sea (0–500 m) is about 20, which is similar to that in zooplankton (about 20), but much higher than that in sediment trap samples at 500 m in the Sargasso Sea (about 2–3);

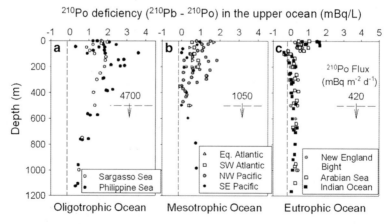

Fig. 14.3 Total (dissolved + particulate) deficiency of ^{210}Po relative to ^{210}Pb in the upper ocean and removal fluxes of ^{210}Po at 500 m (adapted from Kim 2001). The oceanic locations plotted are divided into three categories depending on biological productivity, (**a**) oligotrophic-very low productive ocean, (**b**) mesotrophic-intermediate productive ocean, and (**c**) eutrophic-relatively high productive ocean. The abbreviations Eq., SW, NW, SE represent equatorial, south western, north western, and south eastern, respectively

(3) in the oligotrophic oceans, such as the Sargasso Sea, bacteria constitute the dominant biomass of the microflora and are partially predated upon by protozoa and higher trophic levels; (4) the particulate fraction (15–75%) in the oligotrophic Sargasso Sea is much higher than that in the productive areas of the ocean (e.g. the Bay of Bengal, <15%); and (5) the largest deficiency of ^{210}Po was found in the subsurface layer, where bacterial production peaks. More evidently, the relative proportion of particulate ^{210}Po was considerably higher during the summer of 1997 in the Sargasso Sea when the maximum nitrogen-fixation occurred, showing a good correlation (r = 0.85) between the particulate to dissolved ratios of ^{210}Po and N_2 fixation rates by *Trichodesmium* (size: ~0.5 by 3 mm) (Fig. 14.4). Chung and Wu (2005) also found a large deficiency of Po in the northern South China Sea, adjacent to the Philippine Sea. In this region, the mean value of the total and the dissolved ^{210}Po to ^{210}Pb activity ratios was approximately 0.6 for the entire water column. Because this large ^{210}Po deficiency could not be explained by the scavenging of sinking particulates on the basis of sediment trap results, they concluded that the deficiency is associated with biological removal and transfer to the upper trophic levels.

In contrast, in the eutrophic ocean, the ecosystem is supported mainly by macro-phytoplankton (i.e., diatoms), which are less efficient in taking up ^{210}Po relative to bacteria. The enriched Po in diatoms could be rapidly regenerated by attached bacteria and transferred to abundant free-living bacteria, non-sinking fine particles (0.3–0.6 μm diameter), in a manner similar to dissolved organic carbon (Kim 2001). In general, in the eutrophic ocean, the turnover times of bacterial organic carbon (>95% are free-living bacteria) are about fourfold faster, and the amount of free-living bacteria is an order of magnitude larger, relative to the oligotrophic ocean. This suggests that the Po could reside in the non-sinking organic pool for a much longer time in the eutrophic ocean, potentially with other sulfur group elements.

Recent studies have suggested that ^{210}Po is a potentially good tracer for particulate organic carbon (POC) export (Shimmield et al. 1995; Friedrich and Rutgers van der Loeff 2002; Murray et al. 2005; Stewart et al. 2010), similar to ^{234}Th. Friedrich and Rutgers van der Loeff (2002) reported that the observed fractionation of ^{234}Th and ^{210}Po on particles, dependent on particle

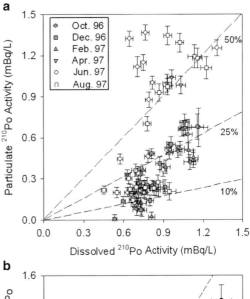

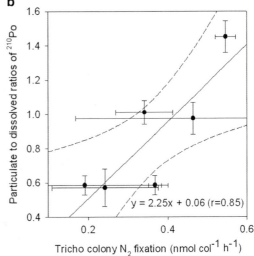

Fig. 14.4 Enrichment of polonium into cyanobacteria (adapted from Kim 2001). (**a**) A plot of dissolved vs. particulate ^{210}Po in the surface (0–100 m) Sargasso Sea between 1996 and 1997, showing higher enrichment of ^{210}Po to particulate matter during summer (the numbered percentages represent the proportion of particulate ^{210}Po), (**b**) a correlation between Trichodesmium N_2 fixation rates in the surface ocean and relative activities of particulate to dissolved ^{210}Po ratios in surface ocean (upper 20 m). Errors for Trichodesmium N_2 fixation rates are from the standard deviation of the average N_2 fixation in Puffs and Tufts for each month, and errors for ^{210}Po are based on 1σ counting error propagation. The *dotted lines* indicate 95% confidence intervals

composition (POC/biogenic silica ratio), is in accordance with the known preference of ^{210}Po for cytoplasm in the Antarctic Circumpolar Current. On this basis, they suggested that the utilization of the two tracers together would enable a more detailed interpretation of POC fluxes than would be possible by using ^{234}Th alone. Murray et al. (2005) found that the POC export fluxes calculated from ^{210}Po were much more variable than those calculated from ^{234}Th because of the more variable correction factors for advection in the central equatorial Pacific. Stewart et al. (2010) found a very good correlation (>80%) between the relative fraction of ^{210}Po and POC in size-fractionated particles, with a better relationship during non-bloom conditions at the Bermuda Atlantic Time-series Study (BATS) site. They suggested that ^{210}Po traces POC export fluxes more accurately in low-export seasons than during high-export seasons, such as during a spring bloom. In the Bellingshausen Sea in Antarctica, the export production measured using ^{210}Po was considerably lower than that measured using ^{234}Th, suggesting that a better understanding of radionuclide uptake and recycling in conjunction with POC is necessary to trace export production (Shimmield et al. 1995). Careful consideration is thus needed with regard to the use of ^{210}Po as a POC tracer because ^{210}Po uptake rates appear to be largely dependent on ecosystem structures and microbial roles.

14.5 ^{210}Po in Suboxic and Anoxic Waters

The activity of ^{210}Po in oxic waters of major oceans ranges from 1 to 5 mBq L^{-1}. Similarly, in the oligotrophic Crystal Lake in Wisconsin, USA, the mean annual total concentration of ^{210}Po is 1.6 ± 0.7 mBq L^{-1}. However, ^{210}Po activities are higher, up to 17 mBq L^{-1}, in seasonally anoxic ponds such as Pond B in South Carolina, USA (Kim et al. 2005b), and Bickford Pond in Massachusetts, USA (Benoit and Hemond 1990), as well as in permanently anoxic seawater such as Framvaren Fjord in Norway (Swarzenski et al. 1999) (Table 14.1). In the permanently anoxic Jellyfish Lake in Palau, the maximum activity of ^{210}Po was 133 mBq L^{-1}, among the highest found in natural waters, except for some sulfide-bearing shallow groundwater (Harada et al. 1989).

Kim et al. (2005b) showed that, in the seasonally oxic environment, Pond B, the activity of ^{210}Po increases sharply from the surface to the bottom layer, with a maximum activity of 14 mBq L^{-1}, in the summer, while it is vertically uniform and low in winter (Fig. 14.5). In Pond B, the bottom layer becomes anoxic from May to October, and the concentrations of Fe and Mn increase sharply from the surface to the bottom layer during this period, similar to the ^{210}Po pattern, due to the dissolution of metal oxides and oxyhydroxides of Fe and Mn in the reducing bottom sediments (Kim et al. 2005b). It has been reported that Po (IV) is insoluble and is reduced to Po (II) near the potential at which Mn (IV) is reduced to Mn (II) (Benoit and Hemond 1990; Balistrieri et al. 1995). Thus, the summer profile of ^{210}Po resulted from its diffusion from bottom sediments under reducing conditions, together with Fe and Mn, in contrast to its much lower winter patterns as a consequence of efficient co-precipitation with Fe and Mn oxides. However, ^{210}Pb did not show such seasonal variations and was vertically uniform in Pond B, similar to Cs, Al, Na, and Cu, those are not sensitive to redox conditions (Kim et al. 2005b). Based on the seasonal changes in the ^{210}Po budget, Kim et al. (2005b) showed that the actual amount of labile ^{210}Po inputs to the water column is a small fraction (<5%) of the ^{210}Po pool in the surface bottom sediment. However, the activities of total ^{210}Po in the seasonally anoxic Sammamish Lake (0.07–0.47 mBq L^{-1}) in Washington, USA, were much lower than those in Pond B (Balistrieri et al.

Table 14.1 The activity of ^{210}Po in various aquatic environments (after Kim et al. 2005b)

Location	Water type	Depth (m)	^{210}Po (mBq L^{-1})	References
Atlantic Ocean	Oxic seawater	0–4,200	1–2	Kim and Church (2001)
Pacific Ocean	Oxic seawater	0–5,420	1–5	Nozaki et al. (1997)
Indian Ocean	Oxic seawater	0–5,587	1–2	Obota et al. (2004)
North Sea	Oxic seawater	5–45	1–5	Zuo and Eisma (1993)
Framvaren Fiord	Permanently anoxic seawater	0–30	0–17	Swarzenski et al. (1999)
Jellyfish Lake	Permanently anoxic seawater	0–30	1–133	Kim et al. (2005b)
Lake Sammamish	Seasonally anoxic freshwater	0–32	0–3	Balistrieri et al. (1995)
Crystal Lake	Seasonally anoxic freshwater	0–20	1–5	Talbot and Andren (1984)
Bickford Pond	Seasonally anoxic freshwater	0–12	1–17	Benoit and Hemond (1990)
Pond B	Seasonally anoxic freshwater	0–12	1–14	Kim et al. (2005b)

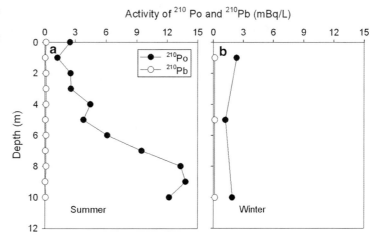

Fig. 14.5 Vertical profiles of ^{210}Po and ^{210}Pb activity in the Pond B (South Carolina, USA) water column in the (**a**) summer and (**b**) winter (adapted from Kim et al. 2005b)

1995). In this pond, the activities of both ^{210}Po and ^{210}Pb increased with depth. The researchers suggested that during oxic periods, ^{210}Pb behavior was linked to Fe cycling, whereas ^{210}Po behavior was more closely linked to the cycling of Mn. In anoxic conditions, both nuclides were influenced by sulfur cycling.

In the permanently anoxic Jellyfish Lake, an order of magnitude higher inventory of ^{210}Po (11 Bq cm^{-2}), compared to Pond B, was observed (Kim et al. 2005b). They showed that this high inventory is associated with an order of magnitude higher ^{210}Pb fluxes to the bottom sediment, compared with the atmospheric input fluxes. In Jellyfish Lake, about 2% of ^{210}Po produced from the ^{210}Pb in the bottom sediments would account for the observed excess in the water column. The benthic fluxes of ^{210}Po to the overlying water column may be controlled either by diffusion or by pore-water advection due to tidal pumping of groundwater and subsequent diffusion upward with the vertical eddy diffusivity of ~0.2 cm^2 s^{-1} between 15 and 30 m.

In the permanently anoxic fjord, Framvaren, in Norway, ^{210}Po and ^{210}Pb were highly enriched at the O$_2$/H$_2$S interface layer, where an active community of microbes, such as anoxygenic phototrophs (e.g., Chromatium, Chlorobium sp.), thrives (Swarzenski et al. 1999). The observed dissolved ^{210}Po enrichment at this depth was likely controlled by a phase transformation (particulate→dissolved), which could be microbially mediated. They suggested that the dissolved ^{210}Po and ^{210}Pb were sequestered efficiently by particulates in the oxic surface layer, and this process occurred rapidly relative to the kinetics of phase redistribution because enrichment in particulate ^{210}Po did not occur. In anoxic bottom waters, on the other hand, the dissolved ^{210}Po activities were lost from the water column to freshly (co)precipitated sulfide minerals (Swarzenski et al. 1999).

A close link between ^{210}Po and the sulfur cycle has also been observed in anoxic marine sediments (Connan et al. 2009). In the pore waters of anoxic marine sediment from the roads of Cherbourg (France) in the central English Channel, ^{210}Po activity exhibited large excess in the uppermost layer (~20 mBq L^{-1}) and lower than 3 mBq L^{-1} in the deeper layer, relative to almost constant ^{210}Pb activities (2.4–3.8 mBq L^{-1}) over the entire depth (Connan et al. 2009). ^{210}Po and ^{210}Pb activities in the pore water were higher than those in seawater, suggesting that sediments act as a source of both nuclides in seawater. The distributions of ^{210}Po activities in the pore waters correlates well with those obtained for Fe, Mn, and SO$_4^{2-}$, exhibiting lower ^{210}Po in the deeper pore water by the formation of a metal sulfide (Bangnall 1957; LaRock et al. 1996). Thus, they concluded that early diagenetic process influences the solid-liquid distribution of ^{210}Po in marine sediments.

14.6 ^{210}Po in Groundwater

In groundwater, the ^{210}Po level rarely exceeds 5 mBq L^{-1} since it is rapidly adsorbed onto particles. However, some previous studies (Harada et al. 1989; Neto and Mazzilli 1998; Ruberu et al. 2007; Seiler 2010) have reported unusually high ^{210}Po activities in groundwater. The natural enrichment of ^{210}Po in groundwater may depend to a large extent on the supply of ^{222}Rn from rocks and sediments, the physicochemical conditions (i.e., pH) of groundwater, and the occurrence of fine particles (i.e., colloids) that can retain Po for a long time in solution. Monitoring of the ^{210}Po levels in groundwater is particularly important for drinking water because Po is highly toxic and it is difficult to predict the exact level of this nuclide on the basis of general geochemical information.

Harada et al. (1989) studied the geochemistry of ^{210}Po in a well containing acidic water (pH < 5) with extraordinarily high ^{210}Po levels (~17 Bq L^{-1}); in this well water, the activity of ^{210}Po was markedly higher than that of its parents, ^{210}Pb and ^{210}Bi, while the ^{222}Rn activity was moderately high. They found that most of the Po species present in this well was in a form that could not be co-precipitated with iron hydroxide scavenging; during sulfide oxidation, the soluble phase took a few days to convert to the particulate phase. Thus, Harada et al. (1989) speculated that Po cycling in such an extreme environment is related to the sulfur cycle, which in turn may be influenced by sulfur bacteria.

Seiler (2010) found that the activity of ^{210}Po ranged from 0.37 to 6,590 mBq L^{-1} in the groundwater in Lahontan Valley, northern Nevada, USA. In this region, wells with high ^{210}Po activities had low dissolved oxygen concentrations (less than 0.1 mg L^{-1}) and commonly had pH greater than 9. In Lahontan Valley, the ^{210}Po activity varied very slightly over a given period that was several times the half-life of this nuclide, indicating the continuous production of ^{210}Po in the aquifer. In this region, the basin-fill sediments were derived mainly from the erosion of uranium-enriched granite rocks in the Sierra Nevada by streams. Thus, radioactive decay of naturally occurring uranium in the sediments was proposed as the ultimate source of ^{210}Po in this region (Seiler 2010). On the basis of the results obtained for the Lahontan valley and those obtained for Florida and Maryland, Seiler (2010) concluded that natural ^{210}Po enrichment may occur in acidic, neutral, or basic groundwater, irrespective of the presence of ^{222}Rn.

In groundwater wells across the state of California, USA, the ^{210}Po activity ranged from 0.3 to 555 mBq L^{-1} (Ruberu et al. 2007), which was always lower than the ^{210}Pb activity (4–1,500 mBq L^{-1}). According to Ruberu et al. (2007), bacteria may play a role in the preferential removal of ^{210}Po, relative to ^{210}Pb, by bio-volatilization. In the Brazilian springs (Vilela and São Bento), which are fed by sandstone aquifers, the ^{210}Po activity ranged from 120 to 400 mBq L^{-1}, with the ^{210}Po/^{210}Pb activity ratios ranging from 0.5 to 0.7 (Neto and Mazzilli 1998). Neto and Mazzilli (1998) stated that the high level of ^{210}Po and ^{210}Pb in these spring waters could be attributed to the low pH of the water (4.6–5.4) and the uraniferous nature of the sandstones (0.1–0.2% U_3O_8). Although there was no correlation between ^{210}Po and ^{210}Pb, a weak correlation existed between the ^{210}Po level and the pH. For these spring waters, the calculated radiation doses corresponding to the ingestion of ^{210}Pb and ^{210}Po by the human kidney and bone surface were up to 5.9 and 1.9 mSv year^{-1}, respectively. This is an issue of concern from the viewpoint of human health.

14.7 Enrichment of ^{210}Po in Marine Organisms

^{210}Po has a similar chemical property to Se and Te (Nieboer and Richardson 1980), but ^{210}Po functions differently in living organisms (Carvalho 1988; Carvalho and Fowler 1994; Barceloux 1999; Chasteen and Bentley 2003; Moroder 2005; Waska et al. 2008). Although ^{210}Po is radiotoxic even in small concentrations, ^{210}Po is assimilated preferentially over ^{210}Pb by marine organisms (Heyraud and Cherry 1979; Cherry and Heyraud 1982; Fisher et al. 1983; Carvalho 1988, 1990; Carvalho and Fowler 1994). Thus, the radiation dose from ^{210}Po alone can be very high in human beings who consume seafood. In the ocean, ^{210}Po activities in phytoplankton range from 40 to 450 Bq kg^{-1} (dry wt) (Shannon et al. 1970; Heyraud and Cherry 1979; Cherry et al. 1987), and those in zooplankton range from 100 to 2,000 Bq kg^{-1} (dry wt) (Shannon et al. 1970; Kharkar et al. 1976; Heyraud 1982). For marine fishes and benthic animals, ^{210}Po activities were found to be from 50 to 900 Bq kg^{-1} (dry wt) for the entire body, with exceptionally elevated concentrations in some organs, e.g. the hepatopancreas of crustaceans and in the liver, gonads, and bones of fish (Beasley et al. 1973; Heyraud and Cherry 1979; Carvalho 1988; Cherry et al. 1989). ^{210}Po activities in small mesopelagic fish can reach ~800 Bq kg^{-1} (wet wt) on a whole-body basis (Carvalho 1988).

Table 14.2 Bioconcentration factors for ^{210}Po, Se, and Te (after Waska et al. 2008)

Sample type	^{210}Po	Se	Te
Phytoplankton	7.6×10^{4a}	1.6×10^{4b}	$600–4.2 \times 10^{4c}$
Zooplankton	8.3×10^{5a}	3.4×10^{4b}	$8–20^{c}$
Loligo sp.			
Muscle	1.9×10^{4d}	2.3×10^{4b}	
Gills	–	7.2×10^{4b}	
Hepatopancreas	2.3×10^{6d}	1.3×10^{5b}	
Todarodes pacificus			
Muscle	6.4×10^{3e}	2.0×10^{4e}	5.8×10^{3e}
Stomach	2.6×10^{4e}	6.2×10^{4e}	5.9×10^{3e}
Gills	2.4×10^{5e}	8.4×10^{4e}	1.6×10^{4e}
Hepatopancreas	1.4×10^{6e}	1.1×10^{5e}	2.2×10^{4e}

[a]Suh et al. (1995)
[b]Liu et al. (1987)
[c]Nolan et al. (1991)
[d]Heyraud and Cherry (1979)
[e]Waska et al. (2008)

Concentration factors for ^{210}Po in the pelagic ecosystem increase through the food web from phytoplankton to zooplankton and finally to fish (Table 14.2). The pattern of ^{210}Po accumulation into marine organisms is similar to that for Se (Heyraud and Cherry 1979; Liu et al. 1987; Skwarzec 1988; Chatterjee et al. 2001), which is much more effective than its parent ^{210}Pb (concentration factors from 1.2 to 2.3×10^3 in fish, Cherry and Heyraud 1981, 1982; Carvalho 1988; Stewart and Fisher 2003). The sulfur-group elements may follow the same metabolic pathways, such as biomethylation (Hussain et al. 1995; Chasteen and Bentley 2003; Kim et al. 2000). A ^{210}Po-sulfur covalent bond may be formed in the hepatopancreas of marine invertebrates (Heyraud et al. 1987).

In order to elucidate the relationship between the chemically similar S, Se, Te, and ^{210}Po in marine invertebrates, Waska et al. (2008) measured the distribution patterns of these elements in the pacific flying squid, *Todarodes pacificus*. They found no relationship between chalcogen concentrations and morphological parameters (mantle length, body weight, and sex), but gills showed slightly elevated levels of Se and ^{210}Po, perhaps due to the absorption and uptake of these elements over the gill surface (Table 14.2). All four chalcogens had their highest concentrations in the hepatopancreas and their lowest concentrations in the muscle tissues (Waska et al. 2008). Waska et al. (2008) demonstrated that bioconcentration factors (BCFs), based on reference seawater values and "internal BCFs" based on the enrichment in the hepatopancreas compared with that in muscle tissues, for ^{210}Po showed a distinct ranking of tissues in the order of muscle < stomach < gills < hepatopancreas, whereas the BCFs for Se and Te in the various tissues were of similar magnitudes. They suggested that the distribution patterns of Se and Te resemble those of essential trace elements such as Zn and Cu, whereas ^{210}Po is partitioned in a manner similar to toxic heavy metals such as Cd and Ag.

14.8 Future Directions

The source of excess ^{210}Po activities in the atmosphere has not been understood thoroughly, especially for plant exudates, anthropogenic sources, and biovolatilization. Thus, more extensive studies are necessary to determine the sources of ^{210}Po, its scavenging mechanisms, and the feasibility of using ^{210}Po as a tracer for aerosol residence time determination. Because the occurrence of radioactive ^{210}Po in air is of great concern, especially when other toxic chemical constituents accompany it in the urban atmosphere, one should determine the detailed sources, mechanisms, and magnitude of anthropogenic ^{210}Po.

In the ocean, Po as a tracer for Se and Te may be taken up efficiently by bacteria (i.e., cyanobacteria) in the oligotrophic ocean and transferred to higher trophic levels (i.e., nekton). In productive areas of the ocean, sulfur group elements seem to reside in the subsurface ocean for much longer periods because they are taken up by abundant free-living bacteria (non-sinking fine particles). Although this hypothesis sheds new light on the global marine biogeochemical cycling of sulfur-group elements in association with microbial activity, more extensive studies are necessary to find direct evidence, together with Se and Te. Although these complicated biogeochemical characteristics hamper the use of ^{210}Po as a tracer for the export of particulate organic matter, it may serve as a tracer of nitrogen fixation in the ocean if a more direct link is demonstrated. In addition, the role of colloids should be investigated in order to fully understand the mechanisms of the biogeochemical cycling of ^{210}Po in the ocean.

In the suboxic and anoxic waters, the direct measurements of ^{210}Po transport from bottom sediments should be conducted using tools such as benthic chambers. In addition, both microbial and chemical roles for the unusually high enrichment of ^{210}Po in the suboxic and anoxic environments should be further investigated because different regions show different enrichment patterns and biogeochemical cycling. Because ^{210}Po can naturally be highly enriched in anoxic environments to a very high level, studies on the bio-concentrations and bio-magnification of ^{210}Po in these extreme environments are highly recommended. The mechanisms of ^{210}Po enrichment in groundwater has not been elucidated as yet, and the observed results differ with the study region and geochemical conditions. Since high levels of ^{210}Po in drinking groundwater remain a cause of concern from the human-health perspective, extensive studies must be carried out to determine the ^{210}Po level in groundwater as well as the groundwater geochemistry (i.e., Eh, pH, and DO) and health effects of ^{210}Po.

In order to better understand the enrichment of ^{210}Po by marine organisms and to evaluate the human uptake rates of ^{210}Po through marine organisms, we have to quantify the ^{210}Po enrichment for all species for different environmental regions in the future. Many more studies are still needed on the detoxification processes that occur at a molecular level in invertebrates, which could differ depending on the species and its ecological niche.

References

Bacon MP, Elzerman AW (1980) Enrichment of ^{210}Pb and ^{210}Po in the sea-surface microlayer. Nature 284:332–334
Bacon MP, Spencer DW, Brewer PG (1976) Pb-210/Ra-226 and Po-210/Pb-210 disequilibria in seawater and suspended particulate matter. Earth Planet Sci Lett 32:277–296
Bacon MP, Huh CA, Fleer AP et al (1985) Seasonality in the flux of natural radionuclides and plutonium in the deep Sargasso Sea. Deep Sea Res 32:273–286
Bacon MP, Belastock RA, Tecotzky M et al (1988) Lead-210 and polonium-210 in ocean water profiles of the continental shelf and slope south of New England. Cont Shelf Res 8:841–853
Balistrieri LS, Murray JW, Paul B (1995) The geochemical cycling of stable Pb, ^{210}Pb, and ^{210}Po in seasonally anoxic Lake Sammamish, Washington, USA. Geochim Cosmochim Acta 59:4845–4861
Bangnall KW (1957) Chemistry of the rare radioelements. Academic, New York
Barceloux DG (1999) Selenium. Clin Toxicol 37:145–172
Baskaran M (2011) Po-210 and Pb-210 as atmospheric tracers and global atmospheric Pb-210 fallout: a review. J Envion Radioact 102:500–513
Baskaran M, Shaw GE (2001) Residence time of arctic haze aerosols using the concentrations and activity ratios of ^{210}Po, ^{210}Pb and ^{7}Be. J Aerosol Sci 32:443–452
Baskaran M, Hong GH, Santschi PH (2009) Radionuclide analysis in seawater. In: Wurl O (ed) Practical guidelines for the analysis of seawater. CRC Press, Boca Raton, pp 259–304
Beasley TM, Eagle RJ, Jokela TA (1973) Polonium-210, lead-210 and stable lead in marine organisms. Q Summ Rep Hlth Saf Lab 273:2–36
Bennett JT, Krishnaswami S, Turekian KK et al (1982) The uranium and thorium decay series nuclides in Mt. St. Helens effusive. Earth Planet Sci Lett 60:61–69
Benoit G, Hemond HF (1990) ^{210}Po and ^{210}Pb remobilization from lake sediments in relation to iron and manganese cycling. Environ Sci Technol 24:1224–1234
Carvalho FP (1988) Polonium-210 in marine organisms: a wide range of natural radiation dose domains. Radiat Prot Dosim 24:113–117
Carvalho FP (1990) Contribution a a l'etude du cycle du polonium-210 et du plomb-210 dans l' environnement. These de Doctorat, Universitè de Nice-Sophia Antipolis
Carvalho FP (1995a) ^{210}Pb and ^{210}Po in sediments and suspended matter in the Tagus estuary, Portugal. Local enhancement of natural levels by wastes from phosphate ore processing industry. Sci Total Environ 159:201–214
Carvalho FP (1995b) Origins and concentrations of ^{222}Rn, ^{210}Pb, ^{210}Bi and ^{210}Po in the surface air at Lisbon, Portugal, at the Atlantic edge of the European continental landmass. Atmos Environ 29(15):1809–1819
Carvalho FP, Fowler SW (1994) A double-tracer technique to determine the relative importance of water and food as sources of polonium-210 to marine prawns and fish. Mar Ecol Prog Ser 103:251–264
Carvalho FP, Oliveira JM, Lopes I et al (2007) Radionuclides from past uranium mining in rivers of Portugal. J Environ Radioact 98:298–314
Chasteen TG, Bentley R (2003) Biomethylation of selenium and tellurium: microorganisms and plants. Chem Rev 103:1–25
Chatterjee A, Bhattacharya B, Das R (2001) Temporal and organ-specific variability of selenium in marine organisms from the eastern coast of India. Adv Environ Res 5:167–174
Cherrier J, Burnett WC, LaRock PA (1995) The uptake of polonium and sulfur by bacteria. Geomicrobiol J 13:103–115
Cherry RD, Heyraud M (1981) Polonium-210 content of marine shrimp: variation with biological and environmental factors. Mar Biol 65:165–175
Cherry RD, Heyraud M (1982) Evidence of high natural radiation doses in certain mid-water oceanic organisms. Science 218:54–56
Cherry MI, Cherry RD, Heyraud M (1987) Polonium-210 and lead-210 in Antarctic marine biota and sea water. Mar Biol 96:441–449
Cherry RD, Heyraud M, James AG (1989) Diet prediction in common clupeoid fish using polonium-210 data. J Environ Radioact 10:47–65
Chung Y, Finkel R (1988) ^{210}Po in the western Indian Ocean: distributions, disequilibria, and partitioning between

dissolved and particulate phases. Earth Planet Sci Lett 88: 232–240

Chung Y, Wu T (2005) Large ^{210}Po deficiency in the northern South China Sea. Cont Shelf Res 25:1209–1224

Church TM, Sarin MM (2008) U- and Th-series nuclides in the atmosphere: supply, exchange, scavenging, and applications to aquatic processes. In: Krishnaswami S, Cochran JK (eds) Chapter 2 radioactivity in the environment, Elsevier, Amsterdam, vol 13, p 11–47

Cochran JK, Bacon MP, Krishnaswami S (1983) Pb-210, Po-210 distributions in the central and eastern Indian Ocean. Earth Planet Sci Lett 65:433–452

Connan O, Boust D, Billon G et al (2009) Solid partitioning and solid–liquid distribution of ^{210}Po and ^{210}Pb in marine anoxic sediments: roads of Cherbourg at the northwestern France. J Environ Radioact 100:905–913

Fisher NS, Burns KA, Cherry RD et al (1983) Accumulation and cellular distribution of ^{241}Am, ^{210}Po, and ^{210}Pb in two marine algae. Mar Ecol Prog Ser 11:233–237

Friedrich J, Rutgers van der Loeff M (2002) A two-tracer (^{210}Po-^{234}Th) approach to distinguish organic carbon and biogenic silica export flux in the Antarctic circumpolar current. Deep Sea Res I 49:101–120

Gaffney JS, Marley NA, Cunningham MM (2004) Natural radionuclides in fine aerosols in the Pittsburgh area. Atmos Environ 38:3191–3200

Gauthier PJ, Le Gloarec MF, Condomines M (2000) Degassing processes at Stromboli volcano inferred from short-lived disequilibria (^{210}Pb-^{210}Bi-^{210}Po) in volcanic gases. Earth Planet Sci Lett 102:1–19

Gill J, Williams R, Bruland K (1985) Eruption of basalt and andesite lava degasses ^{222}Rn and ^{210}Po. Geophys Res Lett 12 (1):17–20

Harada K, Burnett WC, LaRock PA et al (1989) Polonium in Florida groundwater and its possible relationship to the sulfur cycle and bacteria. Geochim Cosmochim Acta 53: 143–150

Heyraud M (1982) Contribution a l'etude du polonium-210 et du plomb-210 dans les organisms marins et leur environnement. These de Dotorat d'Etat, Univ, Paris VI

Heyraud M, Cherry RD (1979) Polonium-210 and Lead-210 in marine food chains. Mar Biol 52:227–236

Heyraud M, Cherry RD (1983) Correlation of ^{210}Po and ^{210}Pb enrichments in the sea-surface microlayer with neuston biomass. Cont Shelf Res 1:283–293

Heyraud M, Cherry RD, Dowdle EB (1987) The subcellular localization of natural ^{210}Po in the hepatopancreas of the rock lobster (Jasus lalandii). J Environ Radioact 5:249–260

Hussain N, Church TM, Burnett WC (1993) Volatile polonium in Floridan groundwater. AGU Abstract H22D-5. Fall meeting, San Francisco

Hussain N, Ferdelman TG, Church TM (1995) Bio-volatilization of polonium: results from laboratory analyses. Aquat Geochem 1:175–188

IAEA (2009) A procedure for the determination of Po-210 in water samples by alpha spectrometry. In: IAEA analytical quality in nuclear applications No. IAEA/AQ/12. International Atomic Energy Agency, Vienna

Kharkar AG, Thomson J, Turekian KK et al (1976) Uranium and thorium decay series nuclides in plankton from the Caribbean. Limmol Oceanogr 21:294–299

Kim G (2001) Large deficiency of polonium in the oligotrophic ocean's interior. Earth Planet Sci Lett 192:15–21

Kim G, Church TM (2001) Seasonal biogeochemical fluxes of ^{234}Th and ^{210}Po in the upper Sargasso Sea: influence from atmospheric iron deposition. Global Biogeochem Cy 15(3): 651–661

Kim G, Hussain N, Church TM et al (1999) A practical and accurate method for the determination of ^{234}Th simultaneously with ^{210}Po and ^{210}Pb in seawater. Talanta 49: 851–858

Kim G, Hussain N, Scudlark JR et al (2000) Factors influencing the atmospheric depositional fluxes of stable Pb, ^{210}Pb, and ^{7}Be into Chesapeake Bay. J Atmos Chem 36:65–79

Kim G, Hong YL, Jang J et al (2005a) Evidence for anthropogenic ^{210}Po in the urban atmosphere of Seoul, Korea. Environ Sci Technol 39:1519–1522

Kim G, Kim SJ, Harada K et al (2005b) Enrichment of excess ^{210}Po in anoxic ponds. Environ Sci Technol 39:4894–4899

Lambert G, Buisson A, Sanak J (1979) Modification of the atmospheric polonium-210 to lead-210 ratios by volcanic emissions. J Geophys Res 84(C11):6980–6986

Lambert G, Ardouin B, Polian G (1982) Volcanic output of long-lived radon daughter. J Geophys Res 87:11103–11108

Lambert G, Le C, Ardouin MF et al (1985) Volcanic emission of radionuclides and magma dynamics. Earth Planet Sci Lett 76:185–192

LaRock PL, Hyun JH, Boutelle S (1996) Bacterial mobilization of polonium. Geochim Cosmochim Acta 60:4321–4328

Liu DL, Yang YP, Hu MH (1987) Selenium content of marine food chain organisms from the coast of China. Mar Environ Res 22:51–165

Masque P, Sanchez-Cabeza JA, Bruach JM et al (2002) Balance and residence times of ^{210}Pb and ^{210}Po in surface waters of the Northwestern Mediterranean Sea. Cont Shelf Res 22: 2127–2146

McNeary D, Baskaran M (2007) Residence times and temporal variations of ^{210}Po in aerosols and precipitation from southeastern Michigan, United States. J Geophys Res 112: D04208. doi:10.1029/2006JD007639

Momoshima N, Song LX, Osaki S et al (2001) Formation and emission of volatile polonium compound by microbial activity and polonium methylation with methylcobalamin. Environ Sci Technol 35:2956–2960

Moore HE, Martell EA, Poet SE (1976) Source of polonium-210 in atmosphere. Environ Sci Technol 10:586–591

Moroder L (2005) Isosteric replacement of sulfur with other chalcogens in peptides and proteins. J Peptide Sci 11: 187–214

Murray JW, Paul B, Dunne J et al (2005) ^{234}Th, ^{210}Pb, ^{210}Po and stable Pb in the central equatorial Pacific: tracers for particle cycling. Deep Sea Res I 52:2109–2139

Neto AN, Mazzilli B (1998) Evaluation of ^{210}Po and ^{210}Pb in some mineral spring waters in Brazil. J Environ Radioact 41: 11–18

Nho EY, Ardouin B, Le Cloarec MF et al (1996) Origins of ^{210}Po in the atmosphere at Lamto, Ivory coast: biomass burning and Saharan dusts. Atmos Environ 30(20):3705–3714

Nieboer E, Richardson DHS (1980) The replacement of the nondescript term "heavy metals" by a biologically and chemically significant classification of metal ions. Environ Pollut 1:3–26

Nolan C, Whitehead N, Teyssié JL (1991) Tellurium speciation in seawater and accumulation by marine phytoplankton and crustaceans. J Environ Radioact 13:217–233

Nozaki Y, Tsunogai S (1973) A simultaneous determination of lead-210 and polonium-210 in sea water. Anal Chim Acta 64:209–216

Nozaki Y, Ikuta N, Yashima M (1990) Unusually large ^{210}Po deficiency relative to ^{210}Pb in the Kurashio current of the East China and Philippine Seas. J Geophys Res 95:5321–5329

Nozaki Y, Zhang J, Takeda A (1997) ^{210}Pb and ^{210}Po in the equatorial Pacific and the Bering Sea: the effects of biological productivity and boundary scavenging. Deep Sea Res II 44:2203–2220

Obota H, Nozaki Y, Alibo DS et al (2004) Dissolved Al, In, and Ce in the eastern Indian Ocean and the Southeast Asian Seas in comparison with the radionuclides ^{210}Pb and ^{210}Po. Geochim Cosmochim Acta 68:1035–1048

Pichlmayer F, Schoner W, Seibert P et al (1998) Stable isotope analysis for characterization of pollutants at high elevation alpine sites. Atmos Environ 32:4075–4086

Poet SE, Moore HE, Martell EA (1972) Lead-210, bismuth-210, and polonium-210 in the atmosphere; accurate ratio measurement and application to aerosol residence time determination. J Geophys Res 77:6515–6527

Rees CE (1970) The sulfur-isotope balance of the ocean, an improved model. Earth Planet Sci Lett 7:366–370

Ruberu SR, Liu YG, Perera SK (2007) Occurrence and distribution of ^{210}Pb and ^{210}Po in selected California groundwater wells. Health Phys 92:432–441

Rubin KH, Macdougall JD, Perfit MR (1994) ^{210}Po-^{210}Pb dating of recent volcanic eruption on the sea floor. Nature 368:841–844

Sarin MM, Bhushan R, Rengarajan R et al (1992) The simultaneous determination of ^{238}U series nuclides in seawater: results from the Arabian Sea and Bay of Bengal. Indian J Mar Sci 21:121–127

Sarin MM, Kim G, Church TM (1999) ^{210}Po and ^{210}Pb in the South-equatorial Atlantic: distribution and disequilibria in the upper 500 m. Deep Sea Res II 46:907–917

Seiler RL (2010) ^{210}Po in Nevada groundwater and its relation to gross alpha radioactivity. Ground Water. doi:10.1111/J.1745-6584.2010.00688

Shannon LV, Cherry RD, Orren MJ (1970) Polonium-210 and lead-210 in the marine environment. Geochim Cosmochim Acta 34:701–711

Shimmield GB, Ritchie GD, Fileman TW (1995) The impact of marginal ice zone processes on the distribution of ^{210}Pb, ^{210}Po, and ^{234}Th and implications for new production in the Bellingshausen Sea, Antarctica. Deep Sea Res II 42:1313–1335

Skwarzec B (1988) Accumulation of ^{210}Po in selected species of Baltic fish. J Environ Radioact 8:111–118

Stewart GM, Fisher NS (2003) Experimental studies on the accumulation of polonium-210 by marine phytoplankton. Limnol Oceanogr 48:1193–1201

Stewart GM, Fowler SW, Teyssie JL et al (2005) Contrasting transfer of polonium-210 and lead-210 across three trophic levels in marine plankton. Mar Ecol Prog Ser 290:27–33

Stewart GM, Cochran JK, Xue J et al (2007) Exploring the connection between ^{210}Po and organic matter in the northwestern Mediterranean. Deep Sea Res I 54:415–427

Stewart GM, Moran SB, Lomas MW (2010) Seasonal POC fluxes at BATS estimated from ^{210}Po deficits. Deep Sea Res I 57:113–124

Su CC, Huh CA (2002) Atmospheric ^{210}Po anomaly as a precursor of volcano eruptions. Geophys Res Lett 29(5). doi:10.1029/2001GL013856

Suh HL, Kim SS, Go YB et al (1995) ^{210}Po Accumulation in the Pelagic Community of Yongil Bay, Korea (Korean). J Korean Fish Soc 28:219–226

Swarzenski PW, McKee BA, Sørensen K et al (1999) ^{210}Pb and ^{210}Po, manganese and iron cycling across the O$_2$/H$_2$S interface of a permanently anoxic fjord: Framvaren, Norway. Mar Chem 67:199–217

Talbot RW, Andren AW (1984) Seasonal variations of ^{210}Pb and ^{210}Po concentrations in an oligotrophic lake. Geochim Cosmochim Acta 48:2053–2063

Thomson J, Turekian KK (1976) ^{210}Po and ^{210}Pb distribution ocean water profiles from the eastern South Pacific. Earth Planet Sci Lett 32:297–303

Turekian KK, Kharkar DP, Thomson J (1974) The fates of ^{210}Pb and ^{210}Po in the ocean surface. J Rech Atmos 8:639–646

Turekian KK, Nozaki Y, Benninger LK (1977) Geochemistry of atmospheric radon and radon products. Ann Rev Earth Planet Sci 5:227–255

Vilenskii VD (1970) The influence of natural radioactive atmospheric dust in determining the mean stay time of lead-210 in the troposphere. Atmos Ocean Phys 6:307–310

Waska H, Kim S, Kim G et al (2008) Distribution patterns of chalogens (S, Se, Te, ^{210}Po) in various tissues of a squid, *Todarodes pacificus*. Sci Total Environ 392:218–224

Zuo Z, Eisma D (1993) ^{210}Pb and ^{210}Po distributions and disequilibrium in the coastal and shelf waters of the southern North Sea. Cont Shelf Res 13:999–1022

Chapter 15
Applications of Groundwater Helium

J.T. Kulongoski and D.R. Hilton

Abstract Helium abundance and isotope variations have widespread application in groundwater-related studies. This stems from the inert nature of this noble gas and the fact that its two isotopes – helium-3 and helium-4 – have distinct origins and vary widely in different terrestrial reservoirs. These attributes allow He concentrations and ^{3}He/^{4}He isotope ratios to be used to recognize and quantify the influence of a number of potential contributors to the total He budget of a groundwater sample. These are atmospheric components, such as air-equilibrated and air-entrained He, as well as terrigenic components, including in situ (aquifer) He, deep crustal and/or mantle He and tritiogenic ^{3}He. Each of these components can be exploited to reveal information on a number of topics, from groundwater chronology, through degassing of the Earth's crust to the role of faults in the transfer of mantle-derived volatiles to the surface. In this review, we present a guide to how groundwater He is collected from aquifer systems and quantitatively measured in the laboratory. We then illustrate the approach of resolving the measured He characteristics into its component structures using assumptions of endmember compositions. This is followed by a discussion of the application of groundwater He to the types of topics mentioned above using case studies from aquifers in California and Australia. Finally, we present possible future research directions involving dissolved He in groundwater.

J.T. Kulongoski (✉)
California Water Science Center, U.S. Geological Survey, San Diego, CA 92101, USA
e-mail: kulongos@usgs.gov

D.R. Hilton
Scripps Institution of Oceanography, University of California San Diego, La Jolla, CA 92093-0244, USA

15.1 Introduction

Studies of the noble gas characteristics of groundwater have long been used to provide information about climatic conditions during periods of aquifer recharge (e.g. Mazor 1972; Andrews and Lee 1979). In this respect, the key attributes of the noble gases are their inert behavior coupled with their well defined solubility characteristics as a function of temperature and salinity. However, to fully exploit noble gases in this manner, groundwater residence times must also be known in order to place an absolute chronology against climatic variations. Ironically, the least useful noble gas for paleo-climate studies – helium, due to the relative insensitivity of its Henry's Law coefficient to temperature change – is one of the most useful gases for providing chronological information, over both short and long timescales (e.g. Andrews and Lee 1979; Marine 1979; Torgersen 1980; Heaton and Vogel 1981). Thus, helium has often been considered as a stand-alone tracer to the other noble gases as it has found widespread application in groundwater-related research, mostly unrelated to climate studies. This is the approach adopted in this review – we emphasize the role of helium in groundwater – and direct the reader to other publications (e.g. Ballentine et al. 2002; Ballentine and Burnard 2002; Kipfer et al. 2002; Phillips and Castro 2003) which provide background on the utility of the other noble gases in groundwater.

We begin with an overview of the collection and measurement approaches commonly used to produce He data on groundwater samples. We provide details of field and laboratory techniques and protocols for the production of both precise and accurate He isotope and concentration data. By adopting reasonable estimates of potential endmember compositions which contribute to observed (i.e., measured) values, we show how He variations can be resolved to yield quantitative information on the different sources comprising the He record. Such information can indeed be used to provide chronological information on the residence time of the groundwater since the time of recharge (i.e., since last equilibration with the atmosphere in the unsaturated zone). However, groundwater He data go far beyond this application, and we include discussion on other issues such as degassing fluxes from the crust and the influence of faults on the transfer of mantle-derived volatile to the atmosphere – topics which are amenable to study through exploiting He measurements of groundwater. We end with a section on future prospects of groundwater He studies, particularly in light of the increasing realization that groundwater He concentrations and its isotope composition can vary on both short and long timescales, and that documenting this variability has exciting potential for understanding fluid characteristics and movement through the crust especially in response to external perturbations, including seismic disturbances.

15.2 Materials and Methods

This section gives an overview of the procedures involved in producing helium data on groundwater samples. It includes a brief description of sample collection methods as well as laboratory procedures used to prepare He for isotopic and abundance measurement. It concludes with a note on data quality, nomenclature, and units.

15.2.1 Sample Collection

Collection of groundwaters for dissolved He analysis involves exploiting actively-pumping commercial (i.e., supply) wells, observation wells, natural springs or seeps. In all cases, the essential prerequisite during sampling is to avoid, or at least minimize, air contamination. In the case of wells, it is important that the well volume is purged prior to sampling so that only aquifer water is collected for analysis. This precaution involves sampling only after a given time period equivalent to purging approximately three volumes of the well. This safeguard is particularly important for observation wells where a pump has to be introduced to sample the groundwater. However, even for apparently continuously-pumping wells, caution must be exercised that sampling does not take place immediately following the start of a new pumping cycle as air can enter the well when the pumping cycle is interrupted or shut-off.

For natural springs and groundwater seeps, air contamination is avoided by inserting sampling tubing as far as possible into the "eye" of the spring, i.e., at the point where water leaves the ground. It must remain submerged and well below the surface level of the water for the entire sampling procedure. The tubing most commonly used is clear and flexible (e.g. PVC or PTFE – polytetrafluoroethylene), and its role is to transfer the groundwater into and out of the sampling vessel. In the case of wells, the same type of tubing can be connected to the well-head directly, usually via a sampling port, or to the outlet of the pump in the case of small, narrow-diameter pumps commonly used for observation wells. In all cases, prolonged flushing with the (water) sample of interest should remove, and can be seen to remove (in the case of transparent plastic tubing), any air bubbles which may have adhered to the sides of the tubing.

There are different types of sampling vessel which are used for the collection and storage of groundwaters for transfer back to the laboratory. The most common type is oxygen-free, annealed copper tubing, which is readily available at plumbers' supply stores. A length of ~30 cm copper tube with a diameter of ~9–10 mm contains ~12 g water which is more than sufficient for He analysis. The copper tube is connected at each end with the plastic tubing and the system is flushed with the groundwater of interest. Water pressure flowing through the copper tubing can be reduced through innovative use of Y-connectors at the up-flow end of the copper – this can help with sealing of the copper tubes. Capturing the sample in the copper tubing can be achieved by use of two hinged knife-edged clamps (a.k.a. refrigeration clamps) which crimp the copper to form a tight seal – where pressure is maintained on the seal by the clamps, or through cold welding of the copper using specific field-portable welding tools. In the case of particularly gas-rich waters, a pressure release valve can be incorporated at the out-flow end of the plastic tubing to help ensure that any dissolved gases remain in solution.

The alternative type of vessel used in some cases for sample collection is a glass flask, which incorporates either one or two stopcock valves. A single stopcock vessel is evacuated to high vacuum and has a Y-inlet which facilitates flushing of the sample past the stopcock prior to opening the valve and capturing the sample. The other type of glass vessel has a stopcock at both ends and the water is flushed through the entire vessel before the stopcock valves are closed in turn. The essential feature of the glass sample vessels that ensures an accurate measurement of any dissolved helium is the type of glass employed. A number of commercially-available glasses (e.g. Corning-1720, Monax, AR-glass) have low He-permeability characteristics and provide secure sample containers over extended time periods (months to years). Pyrex and Duran glass vessels are not suitable for He work, even over short time periods (days to weeks), due to their high He diffusivity (Norton 1957) although they have been used successfully for gases other than He, including the heavier noble gases.

15.2.2 Sample Extraction and Preparation

Prior to measurement, helium must be purified and isolated from both the water phase and the other dissolved gases. This clean-up procedure usually takes place on a dedicated ultra-high vacuum (UHV) system to which the various sampling devices can be attached. Following opening of the seal or valve of the sample container, the water is captured in a bulb and any water vapor is condensed on a trap through the use of a slurry of dry ice and acetone (approximately $-78°C$). The remaining gases can be purified further through a combination of exposure of the gas phase to traps both with and without activated charcoal held at liquid nitrogen temperature (to condense CO_2 and Xe and adsorb Ar and Kr, respectively) and active gas reaction, using an active metal getter pump and/or hot titanium sponge. The remaining gas should be dominated by He and Ne. Depending upon the configuration of the line, the gas can be expanded into a glass breakseal for transfer to a mass spectrometer (e.g. Torgersen 1980; Stute et al. 1992; Kulongoski and Hilton 2002) or, if the preparation line is interfaced directly to the mass spectrometer preparation line, inlet directly for further processing before measurement (e.g. Andrews and Lee 1979; Stute et al. 1995; Beyerle et al. 2000; Furi et al. 2009; Stanley et al. 2009). Further processing usually involves cryogenic separation of He from Ne utilizing an activated charcoal trap attached externally to the cooling stage of a refrigeration unit using He as the coolant. Temperatures ~35 K or lower are usually sufficient to adsorb Ne completely onto the trap whilst He remains in the gas phase and thus ready for inlet into the measurement device.

15.2.3 Instrumentation

Most He isotope and He concentration data are produced using magnetic sector noble gas mass spectrometers, incorporating Nier-type ion sources (which use high-sensitivity electron bombardment for ionization) and a sufficiently large radius flight tube (often with extended geometry) to achieve a resolution of 600 or more without undue restriction on the source slit width. In this way, it is possible to completely separate $^3He^+$ from the interfering HD^+-T^+ doublet at mass 3 and thus record accurate $^3He/^4He$ ratios. With magnetic sector mass spectrometers of lower resolution, a correction is usually applied for the tailing effect of the HD^+ peak into the $^3He^+$ measurement position. Given the large difference between the relative abundance of ^3He and ^4He in most groundwaters (around six orders of magnitude if $^3He/^4He$ ratios are air-like), magnetic sector machines must employ a multiplier device for the small $^3He^+$ ion beams while $^4He^+$ can be measured using a conventional Faraday cup collector.

An alternative means of producing He data is to use quadrupole mass spectrometers (a.k.a. residual gas analyzers) which employ a potential field distribution between source and detector to allow transmittance of selected e/m ratios and deflect others to ground. With this type of device, the mass resolution is severely compromised (compared to sector machines) so that it is not possible to separate $^3He^+$ from interfering species at mass 3: therefore, quadrupole mass spectrometers are useful for measuring ^4He abundances only and $^3He/^4He$ ratio information must be gained elsewhere. It is important to note that these instruments do have sufficient resolution to separate peaks at adjacent masses (e.g. mass 3 and mass 4) so are frequently used to measure the ^{3}He/^{4}He ratio of gases which have been spiked with a known amount of the minor isotope, ^{3}He. In this way, accurate information on sample ^{4}He abundances can be gained by the isotope dilution technique which overwhelms natural variations in ^{3}He (and any HD-T) with spike-derived ^{3}He (e.g. Poole et al. 1997; Kulongoski and Hilton 2002; Hamme and Emerson 2004; Sano and Takahata 2005). Original (sample) ^{3}He/^{4}He information is not recoverable by this technique.

15.2.4 Measurement Protocols and Standards

Once helium has been purified, the usual measurement protocol upon inlet into the mass spectrometer involves either simultaneous collection of the species of interest, i.e., ^{3}He and ^{4}He, or peak jumping whereby masses are measured in sequential fashion by adjustment of a controlling parameter (e.g. magnetic field) allowing greater time to be spent on smaller ion beams, i.e., ^{3}He. Instrument configuration will determine which mode of measurement is possible. In the case of quadrupole-based instruments, peak jumping (a.k.a. mass scanning) is the only possible mode. Data collected on peaks of interest are normally extrapolated to the time of sample inlet in an attempt to minimize discrimination effects caused by differences in ionization efficiencies of various species and/or mass-dependent collector biases. This approach allows the extrapolated ^{3}He/^{4}He ratio or ^{4}He abundance to be determined.

An essential prerequisite for the production of accurate ^{3}He/^{4}He and He abundance data is the preparation of standards which are used to calibrate and optimize instrument performance as well as standardize raw sample results. In the first instance, peak shape and peak resolution issues can be resolved through use of standards as well as ensuring that machine sensitivity is adequate for the measurements to be undertaken. Secondly, sample measurements need to be quoted relative to a standard of known isotopic composition and He abundance. In the case of magnetic sector instruments which employ different types of detectors, determining the relative detector gain is an inherent challenge. This is particularly acute for helium isotope measurements because of the necessity of using a multiplier detector for ^{3}He, and a Faraday cup for ^{4}He. In this case, the measured ^{3}He/^{4}He ratio is a purely arbitrary value and specific to the device used and dependent upon source and/or detector conditions operating at the time of measurement. However, knowledge of the multiplier gain is unnecessary as long as measured ^{3}He/^{4}He values are compared to a standard of known He isotopic composition. Air, collected under known conditions of temperature, pressure and relative humidity, is often the standard of choice given its well known (present-day) ^{3}He/^{4}He ratio (1.399 ± 0.013 ($\times 10^{-6}$); Mamyrin et al. 1970) and He abundance (5.24 ± 0.05 ppm volume fraction dry air; Ozima and Podosek 1983). Thus, by measuring aliquots of air He, similar in size to samples and prepared under identical laboratory conditions, and taking account of any variations in collector linearity, it is possible to normalize sample ^{3}He/^{4}He and He abundances to that of the standard giving a measure of the relative enrichment/depletion of ^{3}He relative to ^{4}He, or just ^{4}He, compared to the standard. For example, sample ^{3}He/^{4}He ratios (R) are routinely quoted relative to the air ^{3}He/^{4}He ratio (R_A) with the absolute value known only through multiplying by the ^{3}He/^{4}He of air (1.4×10^{-6}). Thus, if the sample ^{3}He/^{4}He is given as $0.1R_A$, its absolute value is 1.4×10^{-7}.

15.2.5 Data Quality, Nomenclature, and Units

Helium data – isotope ratios and abundances – should be quoted with realistic estimates of the uncertainty on the measurements. The factors that constitute the quoted uncertainty include the measurement statistics on individual measurements of sample and normalizing standard ^{3}He/^{4}He ratios (primarily a measure of the uncertainty on the ^{3}He measurement), the reproducibility of standard measurements (^{3}He/^{4}He and He abundance) over the course of the measurement run, and blank contributions. The latter is rarely an issue in groundwater studies provided sufficient volumes of water are used for He extraction (i.e., $> \sim 10$ cm^3). Other effects that could contribute to the uncertainty include the linearity of detector response – measurements on standards and samples should be made on approximately the same amounts of

He – and whether He and Ne were separated prior to measurement. Although hardly an issue with the most recent data, there are reports in the older literature (e.g. Rison and Craig 1983) cautioning that the presence of Ne can have an effect on measured ^{3}He/^{4}He ratios.

As stated previously, ^{3}He/^{4}He ratios (R) are usually quoted relative to air (R_A). Occasionally, however, ^{3}He/^{4}He data are reported as absolute values (e.g. 1×10^{-7}). It should be clear from the previous discussion that such values involve selection of the air ^{3}He/^{4}He value – it is worthwhile checking that the air value adopted is explicitly stated to reassure the user when comparing such data to other data quoted in the more conventional manner. We also point out that there are a number of ways that He concentrations are reported. The concentration of He in a groundwater should be reported in stated units (atoms, moles or cm^3 STP) per unit mass (usually gram) of water. For example, the concentration of ^{4}He in groundwater could be reported as 4.2×10^{-6} cm^3 STP g^{-1} H$_2$O. Prefixes such as "n" (nano-) – in the form ncm^3 STP – are potentially confusing but are frequently seen: they imply units of 10^{-9} cm^3 STP.

Finally, we note that in addition to He concentration, it is usual to report the Ne concentration in the same sample – in the following section it will become clear why the Ne measurement is so important. Neon concentrations are usually given as either a stand-alone concentration (e.g. atoms Ne g^{-1} H$_2$O) or as a He/Ne ratio. In the former case, it is worthwhile to check that it is the total Ne concentration that is quoted as opposed to the ^{20}Ne concentration – the major isotope of Ne and one usually selected for measurement. In the latter case of reporting Ne as a He/Ne ratio, it should be explicit in the data table if the value is the absolute ratio or the air-normalized value, i.e., the sample He/Ne ratio divided by the same ratio in air (air He/Ne = 0.288, (Weiss 1971)). Given the means of producing such data by mass spectrometric instrumentation, it is likely to be the latter.

15.3 Applications of Helium Isotopes

Helium is a useful and well-exploited tracer in groundwater-related studies because of its unique isotopic characteristics and inert nature. Helium concentrations in groundwater often exceed the expected solubility equilibrium values as a result of radiogenic production and release of He from aquifer material into associated pore waters. By measuring the abundance of in situ produced He, it is possible to determine the residence time of the groundwater in the aquifer, over both short and long timescales (Andrews and Lee 1979; Marine 1979; Torgersen 1980; Kipfer et al. 2002).

Complications to He-derived chronologies arise with the introduction of He into the groundwater from extraneous sources. Contributions to the extraneous He content can include (a) release of stored He from aquifer material, either by weathering (Torgersen 1980) or diffusion (Solomon et al. 1996), (b) diffusion of He out of confining strata (Andrews and Lee 1979), (c) a basal flux of He from the deeper crust (Heaton 1984; Torgersen and Clarke 1985; Torgersen and Clarke 1987; Torgersen 2010), and/or (d) a mantle He flux usually associated with faults, volcanic activity, crustal extension, or magmatic intrusion (Torgersen 1993; Kennedy et al. 1997; Kulongoski et al. 2005; Sano and Nakajima 2008; Pik and Marty 2009). In all these cases, He can migrate into the aquifer resulting in observed He excesses. In turn, this can lead to overestimation of groundwater residence times (ages).

Means of resolving and quantifying the extraneous sources of helium rely upon careful scrutiny of the helium isotope and concentration record together with associated properties of the aquifer matrix. The weathering and/or diffusive release of He from aquitards and aquifer minerals generally occurs within ~50 Ma (Solomon et al. 1996), such that sediments >50 Ma will no longer be a significant source of extraneous helium. The basal He flux (J_0) may be identified and quantified if the age structure of the groundwater in the aquifer has been established by alternative techniques (e.g. ^{14}C, ^{36}Cl, or hydrodynamically calculated residence times) (e.g. Andrews and Lee 1979; Torgersen and Clarke 1985; Lehmann et al. 2003; Kulongoski et al. 2008). The subsurface addition of He (^{3}He or ^{4}He) will change the ^{3}He/^{4}He ratio from that of air-equilibrated water ~1 R_{eq} (where R_{eq} = ^{3}He/^{4}He of air-equilibrated water = ~1.4×10^{-6}; Clarke et al. 1976), thereby providing a means to identify the source of the extraneous He (e.g. Clarke et al. 1969; Tolstikhin 1975). In the case of mantle helium, its contribution may be identified and quantified based on measured ^{3}He/^{4}He ratios and observed shifts from established crustal production rates (e.g. O'Nions and Oxburgh 1983; Kennedy et al. 1997;

Kulongoski et al. 2003). Consequently, helium isotope ratios have found utility in many areas of contemporary research associated with the occurrence of mantle volatiles in groundwater, including quantifying mantle volatiles fluxes, identifying transport mechanisms, and tracing extant tectonic processes (e.g. Oxburgh et al. 1986; O'Nions and Oxburgh 1988; Torgersen 1993; Kennedy et al. 1997; Kulongoski et al. 2005; Kennedy and Van Soest 2007; Crossey et al. 2009).

15.3.1 Helium Components

In order to effectively use helium as a tracer in groundwater systems, it is first necessary to distinguish the various sources of helium which comprise the total dissolved He content of a water sample. The measured sample He concentration may represent superimposition of several He components, which, upon resolution, can provide insight into the origin of the various He contributions to the system. As described by Kulongoski et al. (2008), the sources of ^{3}He and ^{4}He include terrigenic sources, such as in situ production (He_{is}) from the decay of U, Th, and Li in the aquifer material, a mantle flux (He_m), a deep crustal flux (He_{dc}), and tritiogenic helium-3 (^3He_t) from the decay of tritium in groundwater. Atmospheric sources include air-equilibrated helium (He_{eq}) and dissolved-air bubbles (He_a). The He mass-balance equation describing total He dissolved in groundwater (Torgersen 1980; Stute et al. 1992; Castro et al. 2000) resolves the measured $^3He_s/^4He_s$ ratio of the sample (R_s) into its components:

$$R_s = \frac{^3He_s}{^4He_s}$$
$$= \frac{^3He_{eq} + {}^3He_a + {}^3He_{is} + {}^3He_{dc} + {}^3He_m + {}^3He_t}{^4He_{eq} + {}^4He_a + {}^4He_{is} + {}^4He_{dc} + {}^4He_m}$$
(15.1)

In this equation, ^3He_s and ^4He_s are the helum-3 and helium-4 concentrations measured in a groundwater sample.

The different sources of He have distinct $^3He/^4He$ ratios. End-member $^3He/^4He$ ratio values include air: $R_A = \sim 1.4 \times 10^{-6}$ (Clarke et al. 1976), mantle: $R_m = 1.1 \times 10^{-5} = 8\ R_A$ (Graham 2002), and in situ and deep crustal sources: $R_{is} \sim R_{dc} \sim 3 \times 10^{-8} = 0.02\ R_A$ (Mamyrin and Tolstikhin 1984). These endmembers are used to distinguish between and resolve the various components that comprise the measured $^3He/^4He$ ratio of a sample.

15.3.1.1 Atmospheric Helium Components in Groundwater

The atmospheric He components include He_{eq} and He_a. The contribution of atmospheric He components to the total (measured) He in a groundwater sample may be calculated using the measured concentrations of Ne, He, and the He/Ne ratio of each sample, along with an estimate of the recharge temperature. The recharge temperature can be derived from the concentration of the noble gases other than He (e.g. Mazor 1972; Stute et al. 1992; Weyhenmeyer et al. 2000; Kipfer et al. 2002; Kulongoski et al. 2009) or from the average temperature measured at the recharge area covering the period of groundwater recharge (e.g. Stute and Schlosser 1993; Kulongoski et al. 2005; Kulongoski et al. 2008).

The amount of air-entrained He in each groundwater sample may be calculated assuming that the dissolved Ne of a sample (Ne_s) represents the sum of the Ne from air-equilibration (Ne_{eq}) and from air-bubble entrainment (Ne_a), and that there is no fractionation in the resulting air-entrained He/Ne ratio (Torgersen 1980). The Ne content resulting from air-equilibration can be calculated from knowledge of the temperature at recharge, and is subtracted from the measured Ne concentration to give the amount of Ne due to air-bubble entrainment. The Ne_a is then used to calculate the amount of air-entrained He (He_a), using the known ratio of He to Ne ratio of air ($He_a/Ne_a = 0.288$) (Weiss 1971).

He and Ne oversaturation in groundwater, i.e., relative to air-equilibrated water at a given temperature, may reflect overpressure effects during groundwater infiltration and/or rapid fluctuations in the water table (e.g. Stute and Talma 1998; Beyerle et al. 2003; Kulongoski et al. 2004). This latter possibility is thought to result from large and rapid fluctuations in the water table of shallow aquifers during flood events, resulting in dissolution of trapped air bubbles in the unsaturated zone by increased water pressure (e.g. Kulongoski et al. 2003; Ingram et al. 2007; Kulongoski et al. 2009).

15.3.1.2 Excess (Terrigenic) Helium: Radiogenic, Deep Crustal, Tritiogenic, and Mantle Components in Groundwater

The sum of the in situ produced, deep crustal, tritiogenic, and mantle He components is often referred to as "excess" or "terrigenic" helium (He_{ex}). Note that although tritium has an atmospheric source, it decays to ^{3}He (^{3}H $t_{1/2} = 12.32$ years) within the groundwater system so the daughter product is included as a terrigenic component. The contribution of each of these components to the total He can be determined using measured ^{4}He concentrations and assumed ^{3}He/^{4}He ratios which are characteristic of the different sources (endmembers). First, however, excess helium-3 ($^3He_{ex}$) and helium-4 ($^4He_{ex}$) are calculated by subtracting the concentrations of He due to air equilibration (He_{eq}) and air-bubble entrainment (He_a) from the measured He concentration of the sample (He_s):

$$^4He_{ex} = {}^4He_s - {}^4He_{eq} - {}^4He_a \quad (15.2)$$

Similarly, $^3He_{ex}$ is calculated using $^4He_{eq}$ and ^4He_a concentrations multiplied by the ^{3}He/^{4}He ratios characteristic of the different sources (endmembers):

$$^3He_{ex} = ({}^4He_s \times R_s) - ({}^4He_{eq} \times R_{eq}) \\ - ({}^4He_a \times R_a) \quad (15.3)$$

Equations (15.2) and (15.3) can be combined to yield R_{ex} which represents the excess $^3He_{ex}/^4He_{ex}$ ratio.

15.3.2 Resolution of Helium Components

The total dissolved He of a sample may be separated into its various components by transforming the He mass-balance equation (15.1) (e.g. Torgersen 1980; Stute et al. 1992):

$$\frac{{}^3He_s - {}^3He_a}{{}^4He_s - {}^4He_a} = \\ \frac{{}^3He_{eq} + {}^3He_{is} + {}^3He_{dc} + {}^3He_m + {}^3He_t}{{}^4He_{eq} + {}^4He_{is} + {}^4He_{dc} + {}^4He_m} \quad (15.4)$$

$$= \frac{({}^4He_{eq} \times R_{eq}) + ({}^4He_{is} \times R_{is}) + ({}^4He_{dc} \times R_{dc}) + ({}^4He_m \times R_m) + {}^3He_t}{{}^4He_s - {}^4He_a} \quad (15.5)$$

R_{ex} may then be defined from (15.2) and (15.3):

$$R_{ex} = \frac{({}^4He_{is} \times R_{is}) + ({}^4He_{dc} \times R_{dc}) + ({}^4He_m \times R_m)}{{}^4He_{is} + {}^4He_{dc} + {}^4He_m} \quad (15.6)$$

$$= \frac{({}^4He_{is} \times R_{is}) + ({}^4He_{dc} \times R_{dc}) + ({}^4He_m \times R_m)}{{}^4He_{ex}} \quad (15.7)$$

Substituting (15.7) into (15.5), gives the linear equation ($Y = mX + b$):

$$\underbrace{\frac{{}^3He_s - {}^3He_a}{{}^4He_s - {}^4He_a}}_{Y} = \underbrace{\left(R_{eq} - R_{ex} + \frac{{}^3He_t}{{}^4He_{eq}}\right)}_{m}$$

$$\times \underbrace{\frac{{}^4He_{eq}}{{}^4He_s - {}^4He_a}}_{X} + \underbrace{R_{ex}}_{b} \quad (15.8)$$

In (15.8), Y is the measured ^{3}He/^{4}He ratio corrected for air-bubble entrainment (R_c), X is the fraction of ^{4}He in water resulting from air equilibration with respect to the total ^{4}He in the sample, corrected for air-bubble entrainment, and $b = R_{ex}$ is the isotopic ratio of non-atmospheric excess-He (e.g. Weise and Moser 1987; Castro et al. 2000).

In order to identify the addition of extraneous He in a given suite of groundwater samples, it is useful to plot Y vs. X (as defined above) for each groundwater sample. This helium-isotope-evolution plot presents the evolution of the groundwater systems from recharge conditions, where, after correction for air-bubble entrainment, all of the dissolved He is from air-equilibration [$^4He_{eq}/({}^4He_s - {}^4He_a) \cong 1$], to conditions dominated by crustal and/or mantle contributions, in which air-equilibrated He is a small fraction of the total He [$^4He_{eq}/({}^4He_s - {}^4He_a) < 0.05$]. This Y vs. X plot is used to estimate: (1) R_{ex} (the Y-axis intercept) which is the isotopic ratio of non-atmospheric excess-He (including deep crustal and mantle fluxes), and (2) the possible contribution of tritiogenic ^{3}He (related to the gradient "m") which is diagnostic of groundwater recharged since 1953.

A linear regression of the dataset provides the excess-helium ratio (R_{ex}), which is the Y-intercept. The gradient (m) of the regression line represents the decay of tritium and subsequent production of ^{3}He in the groundwater samples. When using this plot, it is a useful reference to include the evolutionary trajectory representing the addition of radiogenic He (with the ^{3}He/^{4}He ratio of ~0.02 R_A) to air-saturated water (ASW). If samples plot along the radiogenic He-ASW trajectory, this indicates that the flux of He into such groundwater systems is dominated by radiogenic "crustal" ^{4}He production, which is typical of stable continental platforms (e.g. Torgersen and Clarke 1985; Torgersen and Clarke 1987; Ballentine et al. 2002; Kulongoski et al. 2008). If samples plot above the radiogenic He-ASW trajectory, then additional ^{3}He has been added to the groundwater, and may be explained by: (1) a contribution of ^{3}He from anomalous concentrations of lithium in crustal materials and production of ^{3}He via the ^{6}Li(n, α)^{3}H (β^-)^{3}He reaction, (2) mixing of older groundwater with young (nuclear-bomb produced) tritiated water, and/or (3) a contribution from a mantle-derived He flux (Table 15.1).

One or all of these sources of extraneous ^{3}He, however, may be discounted on the basis of aquifer composition or regional geology. The production of ^{3}He from the reaction involving thermal neutron capture by lithium (^{6}Li(n, α)^{3}H (β^-)^{3}He) within the aquifer is negligible if the lithium concentration in the aquifer is low (Andrews 1985). Groundwater that recharged prior to extensive nuclear weapons testing in the 1960s will not contain excess ^{3}He from the decay of nuclear bomb-derived ^{3}H. Indeed, relatively young, nuclear-bomb tritiated water with elevated ^{3}He/^{4}He ratios would plot towards the right-hand side of the helium-isotope-evolution plot (i.e., ^{4}He$_{eq}$/(^{4}He$_s$−^{4}He$_a$) → 1). If the sample ^{3}He/^{4}He ratio (R_s) reflects radiogenic helium, i.e., no extraneous ^{3}He is measured, then a mantle-derived He contribution may be discounted.

15.3.2.1 Case Studies: Uluru Basin, Australia and Mojave Desert, USA

An example of the use of the helium-isotope-evolution plot is shown in Fig. 15.1, where helium results from groundwater samples from the Uluru Basin, Northern Territory, Australia (Kulongoski et al. 2008) and the Eastern Morongo Basin, California, USA (Kulongoski et al. 2005) are plotted together. As a reference, the evolutionary trajectory representing the addition of radiogenic helium (0.02 R_A) to air saturated water (ASW) (line labeled "a") is also included, as is a line representing the addition of 10% mantle helium and ^{3}He$_t$ from the decay of 4.8 TU of tritium (line labeled "d").

A linear regression of the Uluru Basin dataset (red squares) gives the equation Y = 17.5X + 0.07 ($r^2 = 0.99$) and is labeled as line "b". The intercept (of line "b") with the Y-axis is 0.07×10^{-7} (0.005 R_A), which is consistent with in situ radiogenic production of He. The Uluru groundwater samples consistently plot along the radiogenic He-ASW trajectory line "a", from which we conclude that ^{3}He$_m$, ^{3}He$_t$, and ^{3}He$_{is}$ contribute negligible amounts of helium-3 to the Uluru Basin groundwater system, and the Uluru groundwater basin is dominated by production of radiogenic "crustal" He. Uluru Basin R_{ex} values are also consistent with the He production ratio of between 2.2 and 3.3×10^{-8} (0.016–0.024 R_A) calculated by Torgersen and Clarke (1987) for the Hooray Sandstone in eastern Australia. The gradient (m = 17.5) of line "b" reflects the complete decay of approximately 1.8 tritium units (TU) (1 TU = 1 atom of T in 10^{18} atoms of H; upon decay, 1 TU yields 2.5×10^{-15} cm^3 STP ^{3}He g^{-1} H$_2$O); this represents the background tritium value in the Uluru Basin.

In contrast, all of the Eastern Morongo Basin (EMB) samples (blue circles) plot above the radiogenic He–ASW trajectory (line "a"). A linear regression of the EMB data produces the following equation Y = 11.8X + 4.4, ($r^2 = 0.76$), and is labeled line "c" in Fig. 15.1. Significantly, the intercept of line "c" with the Y-axis occurs at 0.32 R_A. This value (R_{ex}) is considerably greater than expected for in situ radiogenic production of helium. As stated above, there are three possibilities to explain the observation of the apparently high value of the ^{3}He/^{4}He excess: (1) a contribution of ^{3}He from the ^{6}Li(n, α)^{3}H (β^-)^{3}He reaction in crustal materials, (2) mixing of older groundwater with younger nuclear-bomb tritiated water, and/or (3) a contribution from a mantle-derived helium flux.

The production of ^{3}He from the reaction involving thermal neutron capture by lithium (^{6}Li(n, α)^{3}H (β^-)^{3}He) may be dismissed due to the unrealistically

Table 15.1 Helium-4 and helium-3 fluxes estimated for different regions

Region		cm^3 STP ^{4}He cm^{-2} a^{-1} (10^{-6})	cm^3 STP ^{3}He cm^{-2} a^{-1}(10^{-13})	References
Global flux estimate (including oceans)		0.40	44.8	Torgersen and Clarke (1985)
Continental crust		3.30	0.67	O'Nions and Oxburgh (1983)
Australia	Great Artesian Basin, Australia	3.60	2.24	Torgersen and Clarke (1985)
	Amadeus Basin, Australia	0.30–0.300	nd	Kulongoski et al. (2008)
Africa	Auob sandstone, Namibia	2.20	nd	Heaton (1984)
	Kalahari Desert, Botswana	1.33	nd	Selaolo et al. (2000)
	Lake Nyos, Cameron	35,212	2,758,285	Sano et al. (1990)
	Central South Africa	1.33	nd	Heaton (1984)
Asia	Northern Taiwan	2.91–3.14	44.8–89.7	Sano et al. (1986)
	South Caspian Sea	2.23	nr	Peeters et al. (2000)
	Central Caspian Sea	6.93	nr	Peeters et al. (2000)
	Black Sea	1.53	nr	Top and Clarke (1983)
	South Lake Bakail	4.26	nr	Hohmann et al. (1998)
	North Lake Bakail	12.8	nr	Hohmann et al. (1998)
Europe	Great Hungarian Plain, Hungary	0.067–0.54	2.0–5.83	Stute et al. (1992)
	Great Hungarian Plain, Hungary	9.41	85.2	Martel et al. (1989)
	Molasse Basin, Austria	0.22	nr	Andrews (1985)
	Lake Laacher See, Switzerland	1,174	87,091	Aeschbach-Hertig et al. (1996)
	Lake Alpnach, Switzerland	1.29	nr	Aeschbach-Hertig et al. (1996)
	Lake Lugano, Swizerland	5.63	Nr	Aeschbach-Hertig et al. (1996)
	Paris Basin, France	0.45	0.45	Marty et al. (1993)
	Paris Basin, France	4.48	4.48	Castro et al. (1998)
North America	Mojave Desert, USA	3.0–10.0	0.20–14.2	Kulongoski et al. (2003)
	Eastern Morongo Basin, USA	0.003–30.0	0.45–135	Kulongoski et al. (2005)
	Crater Lake, USA	64.6	6,397	Collier et al. (1991)
	Pacific Ocean	0.12	46.9	Well et al. (2001)

nd not detected; *nr* not reported

large concentrations of lithium (>850 ppm Li) which would be necessary in aquifer rock to produce the observed ^{3}He/^{4}He ratio (0.32 R$_A$). Assuming a lithium content of ~50 ppm, an upper limit for a typical sedimentary lithology (Andrews 1985; Ballentine and Burnard 2002), the calculated ^{3}He/^{4}He production ratio in the rock would be 2.59 × 10^{-8} or 0.02 R$_A$ (Kulongoski et al. 2003).

For groundwater residence times >50 years, no excess nuclear-bomb tritium is expected. The groundwaters sampled in this study have helium and/or ^{14}C-derived residence times significantly greater than 50 years. The gradient of the linear-regression of EMB data, 11.8, is consistent with complete decay of 4.8 tritium units (TU); this represents the background tritium value in the EMB. If the background tritium value is taken as 4.8 TU, this would produce 1.2 × 10^{-14} cm^3 STP ^{3}He g^{-1} H$_2$O. For EMB ground-water samples older than 50 years, ^{3}He concentrations are significantly higher, ranging from 2 to 1,150 × 10^{-14} cm^3 STP ^{3}He g^{-1} H$_2$O (Table 15.1): therefore, the decay of nuclear "bomb" produced tritium should represent a negligible contribution to the total measured ^{3}He.

By default, we conclude that the high R$_{ex}$ in the EMB groundwaters results from an influx of mantle-derived helium-3 (^{3}He$_m$). Using the regression value of 0.32 R$_A$ we can estimate that the average contribution of mantle He is 4.6% of the total helium, assuming a simple binary mixture of mantle (R$_m$ ~ 8 R$_A$) and crustal (R$_{dc}$ ~ 0.02 R$_A$) components. Although the sub-continental lithospheric mantle have a lower end-member composition (e.g. 6 R$_A$; Dunai and Porcelli 2002), this would make a small difference only to the calculated fraction of mantle-derived He in any given sample.

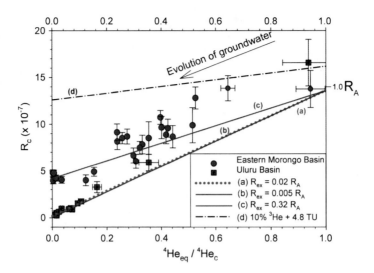

Fig. 15.1 Measured ^{3}He/^{4}He ratios corrected for air-bubble entrainment (R_c), vs. the relative amount of ^{4}He derived from solubility (He$_{eq}$) with respect to total ^{4}He, corrected for air-bubble entrainment (He$_c$). Lines (*a*), (*b*), and (*c*) represent the evolution of the ^{3}He/^{4}He ratio in water samples from atmospheric equilibrium values to excess ratios (see (15.8) in the text). Line (*d*) represents an addition of 10% mantle helium plus 4.8 TU of tritiogenic ^{3}He

15.3.3 Excess Helium and the Utility of Helium Isotopes in Groundwaters

In this section, we report on the utility of He studies of groundwater systems. The list is by no means exhaustive but is meant to illustrate recent applications of He geochemistry to issues of groundwater chronology, crustal degassing, and the role of faulting in loss of mantle-derived volatiles to the atmosphere. The approach is predicated on successfully resolving observed He isotope and abundance variations into component structures, as discussed above.

15.3.3.1 Radiogenic Helium Produced In Situ: Chronological Applications

Radiogenic helium is produced by α-decay of U- and Th-series nuclides in crustal minerals, and accumulates in groundwater. The accumulation rate (^{4}He$_{sol}$) of radiogenic ^{4}He in groundwater can be quantified based on the U and Th content and the porosity of the aquifer (e.g. Andrews and Lee 1979; Torgersen 1980; Stute et al. 1992; Castro et al. 2000):

$$^4\text{He}_{sol} = \rho \times \Lambda \times \{1.19 \times 10^{-13}[\text{U}] + 2.88 \times 10^{-14}[\text{Th}]\} \times \frac{(1-\phi)}{\phi} \quad (15.9)$$

In (15.9) ^{4}He$_{sol}$ is the rate of ^{4}He accumulation (the ^{4}He solution rate in cm^3 STP He g^{-1} H$_2$O a^{-1}); Λ is the fraction of He produced in the rock that is released into the water phase – this is assumed to be unity (Andrews and Lee 1979); and ρ is the bulk density of the aquifer material. The ^{4}He production rates from U and Th decay are = 1.19×10^{-13} cm^3 STP ^{4}He μgU^{-1} a^{-1} and 2.88×10^{-14} cm^3 STP ^{4}He μgTh^{-1} a^{-1}, respectively (Andrews and Lee 1979). The U and Th concentrations ([U] and [Th]) in the aquifer material (ppm) and the fractional effective porosity (ϕ) of the aquifer material are measured parameters.

If the accumulation of excess He (^{4}He$_{ex}$) in groundwater can be attributed solely to in situ production, then the He age of the groundwater may be estimated by dividing the excess ^{4}He (^{4}He$_{ex}$) by the solution rate (^{4}He$_{sol}$) (e.g. Andrews and Lee 1979; Marine 1979). However, we caution that excess He may also be introduced to groundwater reservoirs from a basal flux, in which case it is necessary to account for this He flux in order to calculate realistic groundwater ages (Heaton 1984; Torgersen and Clarke 1985). This topic is considered next. However, we note that in addition to its utility for calibrating He residence times, the flux of He from either continental or mantle degassing also may be used to model constraints on the chemical and thermal evolution of the Earth (e.g. O'Nions and Oxburgh 1988), and/or examine the transport mechanism of volatiles through the crust (e.g. Torgersen 1993; Kulongoski et al. 2003; 2005).

15.3.3.2 Deep Crustal Helium: Crustal Fluxes

If a basal flux contributes He to a groundwater system, it may be quantified if the groundwater age is known by an independent method (e.g. C-14 chronology). Conversely, the He groundwater age may be determined if the basal flux is known. The corrected groundwater residence time (τ_{corr}) can be calculated from the following equation (e.g. Heaton 1984; Torgersen and Clarke 1985):

$$\tau_{corr} = \frac{^4He_{ex}}{\left(\frac{J_0}{\phi z_0 \, \rho_w} + {}^4He_{sol}\right)} \quad (15.10)$$

In (15.10), $^4He_{ex}$ has units of cm^3 STP He g^{-1} H$_2$O, J_0 (cm^3 STP He cm^{-2} a^{-1}) is the basal flux, ϕ is the effective fractional porosity of the aquifer, z_0 is the depth (cm) at which this flux enters the aquifer, and ρ_w is the density of water (~1 g cm^{-3}). In many cases, z_0 may be estimated as the distance from the middle of the perforations in the well casing to the interface between the aquifer and the basement formation (Kulongoski et al. 2003; 2005).

In order to determine the basal flux of He (J_0) and improve the sensitivity of the ^{4}He chronometer, it is useful to plot the ^{4}He residence times (age) vs. ages derived by alternative methods (such as ^{14}C, ^{36}Cl, or hydrologically derived age). Different values of the basal flux are assumed in order to search for concordance between ^{4}He and other age-dating techniques. Values of 3×10^{-8} cm^3 He STP cm^{-2} a^{-1} represent a low basal flux of He, in comparison with fluxes recorded from various aquifers worldwide (see Table 15.1). Groundwater chronometry studies typically investigate aquifers with simple flow systems in which groundwater residence time increases along a confined flow path. Calibration of the flow systems may be made by comparing ^{4}He and ^{14}C or ^{36}Cl-derived ages.

Observations of high He concentrations in the basal layer of aquifers (e.g. the Great Artesian Basin; Torgersen and Clarke 1985), coupled with numerical models of groundwater flow incorporating advective mass transport (Bethke et al. 1999), indicate that the distribution of helium in aquifer systems must also have vertical structure, owing to the basal He flux, in addition to horizontal "plug-piston" flow (Zhao et al. 1998). Introduction of the basal flux may occur by diffusion from crystalline basement followed by entrainment and upward migration, particularly as the discharge region is approached (Bethke et al. 2000).

Case Study: Alice Springs, Australia

The following example shows how He can be utilized for groundwater dating purposes. Helium residence times (ages) were calculated for a suite of samples from Alice Springs, Australia assuming $\phi = 0.15$, $\rho = 2.6$ g cm^{-3}, [U] = 1.7 ppm, [Th] = 6.1 ppm (15.9), and $He_{sol} = 3.93 \times 10^{-12}$ cm^3 STP He g^{-1} H$_2$O a^{-1} (Kulongoski et al. 2008). In Fig. 15.2, we plot calculated mean ages derived from ^{4}He, ^{14}C and ^{36}Cl for a range of basal fluxes of ^{4}He, from zero to 30×10^{-8} (units of cm^3 STP He cm^{-2} a^{-1}). The straight lines (1:1 reference line) would indicate concordance in ages between the various chronometers.

The He ages of the youngest Alice Springs samples show concordance between ^{4}He-^{14}C-^{36}Cl derived ages when zero or low J_0-values (0–3 $\times$ 10^{-8} units of cm^3 STP He cm^{-2} a^{-1}) are adopted (Fig. 15.2). This observation agrees with findings by other workers that relatively young groundwaters (<50 ka) are dominated by in situ He production, rather than a basal flux (e.g. Torgersen and Clarke 1985). However, a larger basal He flux (30 $\times$ 10^{-8} cm^3 STP He cm^{-2} a^{-1}) would be required for the ^{4}He ages of the older groundwaters to approximate the ^{14}C ages. ^{36}Cl-derived ages do not require a larger basal flux for agreement with ^{4}He ages, suggesting that they may also overestimate groundwater ages. The presence of a basal flux in the majority of the groundwater samples suggests that these aquifers accumulate He from a diffusive process rather than from a localized (deep) source, such as faults. It would appear that the stable cratonic continental platform of central Australia, therefore, may limit any trans-crustal flux or mantle-derived volatiles into aquifers of the region.

15.3.3.3 Tritiogenic Helium-3: Chronological Applications

The use of tritium (^{3}H), combined with its daughter isotope, helium-3 (^{3}He$_t$), to study groundwater flow

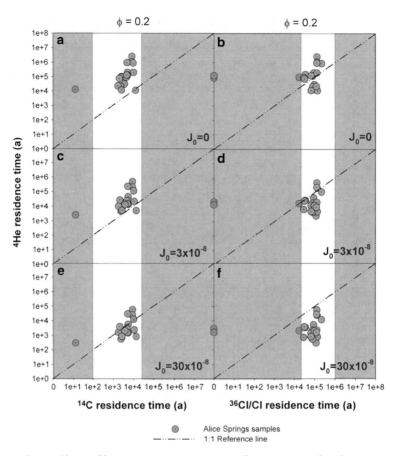

Fig. 15.2 Alice Springs ^{4}He vs. ^{14}C and ^{36}Cl/Cl groundwater residence times. Plots of corrected ^{4}He Alice Springs groundwater residence times (porosity $\phi = 0.2$) vs. ^{14}C derived and ^{36}Cl/Cl derived residence times, for three basal fluxes ($J_o = 0$, 3×10^{-8}, and 30×10^{-8} cm^3 STP He cm^{-2} a^{-1}). Un-shaded areas represent the windows of applicability for ^{14}C (**a, c, e**) and ^{36}Cl/Cl (**b, d, f**). From Kulongoski et al. (2008)

was first proposed by Tolstikhin and Kamenskiy (1969). However, widespread adoption of the ^{3}H–^{3}He method occurred only following the analytical advance of measuring ^{3}H by the ^{3}He in-grow method (Clarke et al. 1976). Early studies of marine and lacustrine systems made use of this technique (e.g. Jenkins and Clarke 1976; Kipfer et al. 2002; Schlosser and Winckler 2002). A decade later, the calculation of groundwater residence times (ages) using ^{3}H–^{3}He dating was made possible when the fraction of ^{3}He derived directly from tritium decay (^{3}He$_t$) could be successfully resolved from other sources (e.g. Takaoka and Mizutani 1987; Poreda et al. 1988; Schlosser et al. 1988; Schlosser et al. 1989). Determination of water residence times in lakes and groundwaters by ^{3}H–^{3}He dating is now used extensively to study transport and mixing processes.

The ^{3}H–^{3}He method provides a quantitative determination of groundwater residence time based on the accumulation of the inert gas ^{3}He$_t$, the β-decay product of tritium decay. As mentioned previously, tritium (^{3}H) is a short-lived radioactive isotope of hydrogen with a half-life of 12.32 years. ^{3}H is produced naturally in the atmosphere from the interaction of cosmogenic radiation with nitrogen (Craig and Lal 1961), by above-ground nuclear explosions, and by the operation of nuclear reactors. Following the release of large amounts of tritium into the atmosphere from nuclear-bomb testing in the 1950s and 1960s, the International Atomic Energy Agency (IAEA) implemented a global monitoring program of tritium in precipitation, with the result that tritium has been used extensively as a tracer in hydrology studies (International Atomic Energy Association 1983; 1986; Hong et al. 2011).

As tritium forms an intrinsic part of a water molecule within a groundwater system, the daughter product (^3He_t) remains dissolved in solution following decay of the parent. Thus, the tritium-helium clock starts when ^3He_t begins accumulating from the decay of ^3H. The time that the water has resided in the aquifer can be calculated by determining ^3He_t. This technique, based on the ratio $^3H/^3He_t$, is more precise than using tritium alone because it does not depend on the tritium input function or the atmospheric tritium content at the time of recharge. Instead, the ^3H–^3He method is based on the assumption that groundwater is a closed system, and the $^3He_t/^3H$ ratio increases predictably as a function of time. Thus, the calculated ^3H–^3He age only depends upon the initial ^3H activity and measurement of its daughter product ^3He_t, which must be resolved from the total ^3He dissolved in groundwater (Takaoka and Mizutani 1987; Poreda et al. 1988; Schlosser et al. 1988; Schlosser et al. 1989).

15.3.3.4 Mantle-Derived Helium: Identification, Quantification, and Geologic Significance

The contribution of mantle-derived He to the total He in a groundwater sample may be identified by resolving the measured He concentration into its various components, and then quantified by assuming a binary mantle-crust mixture with end-member compositions (mantle = 8 R_A, crust = 0.02 R_A). In this way, it is possible to compute the absolute concentrations of He (^3He and ^4He) contributed by each end-member. The crustal and mantle helium contributions can provide information about the evolution of the groundwater since recharge as well as associated geological processes.

High $^3He/^4He$ ratios (i.e., values greater than crustal production ratios), indicative of a mantle-derived input of He, are often found in close proximity to specific geological features such as volcanic systems and major faults, or associated with regions of high heat flow such as regions of crustal extension or where there is magmatic activity. In such cases, mantle-derived ^3He measured in groundwater may be used to explore mantle-crust interaction processes including mechanisms of volatile transfer into and through the crust (Table 15.1). For example, Kennedy et al. (1997) showed that high $^3He/^4He$ ratios (up to ~4 R_A) in groundwater proximal to the San Andreas Fault (SAF) in California could be used to estimate the transport rate of mantle fluids through the crust. By assuming that measured values reflect the dilution of mantle helium (R_m = 8 R_A) with crustal (radiogenic) helium (R_{dc} = 0.02 R_A), they calculated that an upward flow rate of 1–10 mm year^{-1} is necessary to maintain observed ratios assuming steady-state production and release of crustal He. This range was considered a lower limit, since hydrodynamic dispersion, the effects of mixing between fault fluids and radiogenic crustal fluids, or episodic flow events were not taken into account. For a crustal thickness of 30 km, and a flow rate of 10 mm year^{-1}, it would require nearly 3 Ma for mantle fluids to reach the surface. Over the same time period, He would be expected to move only ~1 m by diffusion through bulk granite (D_{He} ~ 5 × 10^{-7} m^2 a^{-1}; (Ballentine and Burnard 2002)) or ~300 m by diffusion through groundwater (D_{He} ~ 0.03 m^2 a^{-1}; (Ballentine and Burnard 2002)), thus emphasizing the important role of advection in the transport of mantle helium through the crust.

Advective flow of (deep) crustal fluids containing mantle-derived He, particularly in the vicinity of faults, implies that the Eastern California Shear Zone (ECSZ) extends to considerable depths, possibility into the lowermost crust. Episodic fracturing and subsequent fluid flow is a likely mechanism to enable the transfer of mantle volatiles through the brittle-ductile boundary – estimated at a depth between 10 and 15 km for typical crustal thermal gradients – and towards the surface (Nur and Walder 1990; Rice 1992; Kennedy et al. 1997). On the other hand, if faults in the ECSZ are limited to relatively shallow depths, by a mid-crustal decollement for example (e.g. Webb and Kanamori 1985; Jones et al. 1994), then their role would be to transport volatiles from the decollement to the surface. In this case, faults would act as conduits, and fault activity would regulate the release of accumulated ^3He_m (and $^4He_{dc}$) into the shallow crust.

Case Study: The Influence of Faults and Seismic Activity in the East Morongo Basin (EMB)

In this section, we review the role of faults – their surface distribution and activity – in controlling the distribution of ^3He_m and $^4He_{dc}$ concentrations in groundwaters of the Eastern Morongo Basin (EMB) (Kulongoski et al. 2005).

The presence of faults, particularly if they lead to greater permeability in the fault zone, should act to channel He fluxes to the surface. If this is the case, then wells in close proximity to faults should have groundwater with associated high ^3He_m (and $^4He_{dc}$) fluxes as a result of relatively rapid transport of helium and other mantle and/or deep-crustal derived volatiles via the fault zone. In the EMB, wells located near the Emerson, Elkins, and Lavic Lake-Bullion Faults have relatively high ^3He_m concentrations; however, wells sampled near the Hidalgo-Surprise Springs Fault Zone (Fig. 3), show no evidence of high concentrations of ^3He_m. This leads to the conclusion that the presence of faults alone does not necessarily lead to enhanced transport of deep crustal and mantle fluids.

Intermittent fault activity or episodic fault rupture (Nur and Walder 1990) may be a contributory factor why ^3He_m fluxes vary between sites either close to one another or close to faults. In this scenario, pore fluid pressure in the fault zone could increase until it reaches the level of the least compressive stress in the crust, whereby it would induce hydraulic fracturing and fluid release: this would then be followed by a drop in pore pressure and resealing of the system (Nur and Walder 1990). Local hydro-fracturing, induced by fluid pressure increases (e.g. from porosity reduction, dehydration, vertical fluid fluxes), could create local inter-connected networks allowing fluid flow (Miller and Nur 2000). Such localized networks may explain variations in observed ^3He_m accumulation and mantle flux rates throughout the EMB. It should be noted that significant time periods (Ma) are necessary for advective transport of ^3He_m from the mantle to the shallow crust; therefore, it is unlikely that an individual rupture event would result in instantaneously high ^3He_m groundwater concentrations, unless these faults directly tapped a shallow reservoir containing ^3He_m (Sano 1986; Gulec et al. 2002). However, the cumulative activity (slip) along faults may facilitate transfer and transport of volatiles from the mantle to the crust over geologic time scales, resulting in a heterogeneous distribution of ^3He_m, as observed in the EMB. In this respect, it is informative to consider the recent seismic record for the EMB region.

Recent seismic activity in the EMB includes the rupture of the Emerson Fault during the 1992 Landers earthquake, and rupture of the Lavic Lake-Bullion Fault (LLBF) during the 1999 Hector Mine earthquake. Significantly, the highest concentrations of ^3He_m are observed at sites near the terminus of the 1999 rupture of the LLBF and in wells near the Emerson Fault (Fig. 15.3). These observations are consistent with the model of episodic hydrofracture-induced flow enhancing the transfer of mantle volatiles to the shallow crust.

Groundwaters in the EMB that contain relatively high concentrations of ^3He_m are also observed to have high concentrations of crustal (radiogenic) helium ($^4He_{dc}$). The relation between high $^4He_{dc}$ and ^3He_m concentrations suggests a similar transport mechanism for the two components through the crust (Ballentine et al. 1991), and may be attributed to increased permeability through the active fault zone. Transmission of mantle and deep crustal fluid/gas through fault zones by episodic fracturing would then represent a plausible mechanism for enhanced helium transport through the crust (Kulongoski et al. 2005).

15.4 Future Prospects

It is becoming increasingly clear that the presence of crustal faulting plays an important role in the transfer of mantle-derived He, and other volatiles through the crust (see discussion in Hilton (2007)). Faults serve as high permeability conduits for fluids as they traverse the crust-mantle transition zone. In turn, this implies that such fluids must exist at extremely high pressures – in order to maintain open pathways during volatile transport. As discussed above, little is known concerning the nature of the flow regime – be it continuous or episodic – so it is difficult to relate spatial variability of groundwater He_m or He_{dc} signals to the location of high permeability zones such as faults. Recent studies have sought to address this issue by adapting or developing instrumentation to collect groundwater and other fluid samples over prolonged time periods to assess temporal variability in the He isotope and abundance record.

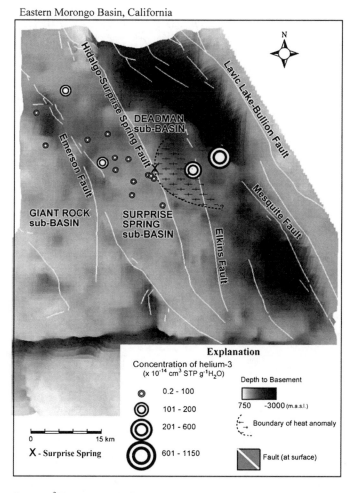

Fig. 15.3 Concentration of excess ^{3}He measured in Eastern Morongo Basin, California groundwater samples. Also shown is the gravimetric depth to basement in *grey-shading*, regional faults, and region of high thermal anomaly, from Kulongoski et al. (2005)

15.4.1 Submarine Studies Using Fluid Flow Meters

The Extrovert Cliffs site in Monterey Bay, California is situated close to several active strike-slip faults, including the Monterey and San Gregorio fault zones, and has many cold seeps located at ~1,000 m depth below sea level. By adapting instrumentation designed to measure diffuse fluid fluxes on the seafloor (Tryon et al. 2001), Furi et al. (2009) were able to capture dissolved gases in emanating cold seep fluids over deployment intervals up to 3 weeks.

As illustrated in Fig. 15.4, the measured helium and neon concentrations (^{4}He$_m$ and Ne$_m$) were found to show remarkable fluctuations, on time-scales of only a few hours. In addition, the isotopic composition of the dissolved He also varied. For example, measured and terrigenic (crust + mantle) ^{3}He/^{4}He ratios increased from 1.2 to 1.5 R$_A$ within just 2.5 h (e.g. at $t = 455$ h). A maximum increase from 0.9 to 1.75 R$_A$ is observed in the data record with the lower time-resolution (at $t = 517$ h). As discussed by Furi et al. (2009), temporal variations in dissolved gas characteristics (isotopes and concentrations) can be influenced by interaction with a hydrocarbon phase within the aquifer. In spite of this complicating process, it was still possible to resolve the He signal into component structures associated with air-equilibration, excess-air entrainment, and terrigenic fluxes (both crustal and mantle-derived). Indeed, the mantle He contribution is mostly 20–30% (up to 2.3 R$_A$), with a short duration spike of 60% mantle He.

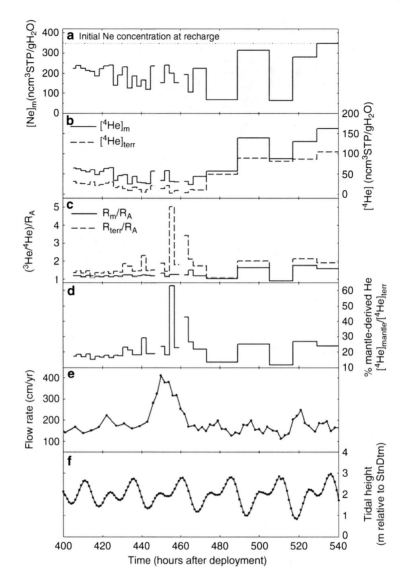

Fig. 15.4 Temporal variation of (**a**) measured Ne concentration, (**b**) measured ^{4}He and calculated terrigenic ^{4}He concentrations, (**c**) measured ^{3}He/^{4}He ratio (R_m/R_A) and calculated R_{terr}/R_A ratio, and (**d**) the fraction of mantle He in seep fluids collected with CAT meter No. 2. Also shown are (**e**) the seep flow rate and (**f**) the tidal height record. Time (in hours) is given relative to the start of sampling on 4 August 2004, at 12 p.m. local time. From Furi et al. (2009)

Significantly, the correlation of high ^{3}He/^{4}He ratios with a transient pulse in the flow rate at Extrovert Cliff (Fig. 15.4e) indicates that the release of mantle-derived fluids is episodic and cannot be explained solely by hydrostatic pressure changes due to the variable tidal load (Fig. 15.4f). In this respect, the controlling factor is likely episodic fracturing in the San Gregorio or Monterey Bay fault zones. This is consistent with the notion of continuous ascent of mantle-derived fluids through the crust, increasing pore fluid pressure until it overcomes the least principle stress, resulting in hydrofracturing and release of fluids (e.g. Nur and Walder 1990).

15.4.2 On-Land Studies: SPARTAH Instrumentation

Barry et al. (2009) reported a new approach to sampling groundwater from land-based groundwater and geothermal wells through development of a new instrument for the continuous collection of water samples over prolonged periods, up to ~6 months. The instrument is dubbed SPARTAH (Syringe Pump Apparatus for the Retrieval and Temporal Analysis of Helium). Motivation for the development of SPARTAH came, in part, through He monitoring studies in seismically-active regions of the crust (e.g. Hilton 1996; De Leeuw et al. 2010) whereby collection of samples took place at regular frequencies of ~3 months. However, in some instances (e.g. Italiano et al. 2001), variations in the monitored He record could occur at comparable or shorter frequencies and so would be missed by adoption of a fixed (and regular) sampling regimen.

The SPARTAH device consists of an off-the-shelf programmable syringe pump equipped with a high-power stepper motor capable of extremely accurate and low withdrawal rates (between 0.0001 mL/h and 220 mL/min). The pump is operated in withdrawal mode whereby fluid is accurately and smoothly drawn into the syringe at a user-defined rate. Well water is drawn into the coils at a constant rate set by the syringe pump, and is independent of the pumping rate of the well. The only stipulation is that the well operates continuously so that air is not entrained into the coils. At any time during a deployment, the tubing can be isolated by closing shut-off valves, and disengaged (for transfer to the laboratory) or replaced to extend the duration of sampling.

SPARTAH was successfully tested at two locations: Selfoss, Iceland and San Bernardino, California (Fig. 15.5). These test deployments demonstrated that well fluids could be drawn smoothly, accurately, and continuously into the Cu-tubing and be time-stamped through user-determined operating parameters. As discussed by Barry et al. (2009), the use of SPARTAH has the potential to revolutionize studies relying on time-series records of dissolved He variations. It has a number of advantages over periodic sampling. Briefly, these include long, maintenance-free deployment times (6 months or more) with continuous sampling, user defined sample resolutions (i.e., sections of coil representing a fixed number of hours or days can be selected for analytical work), and accurate time-stamping of water aliquots due to constant withdrawal rates. Finally, it should be noted that the use of SPARTAH is not restricted to He or any other dissolved gas. The Cu-tubing can be sectioned and processed for various trace elements, stable isotopes or

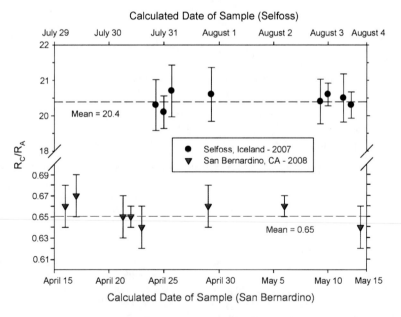

Fig. 15.5 Helium isotope variations captured in the SPARTAH Cu-coils at Selfoss, Iceland (1-week deployment) and San Bernardino, California (1-month deployment). *Note*: helium is mainly mantle-derived in Selfoss but is predominantly crustal in origin at San Bernardino. Individual error bars are given at the 2σ level. The mean ^{3}He/^{4}He values for Selfoss (*black*) and San Bernardino (*red*) are illustrated by *dashed lines* (Barry et al. 2009)

any other intrinsic water property of interest. Thus, innovative analytical processing could lead to contemporary records of different tracers, enabling assessments of relative sensitivities to external perturbations. Deployment of SPARTAH at strategically-placed locations can lead to more detailed and accurate assessments of the responses of He and other geochemical parameters to external forcing factors, such as earthquakes.

Acknowledgements We thank Mark Baskaran for the invitation to contribute this review chapter. We acknowledge funding from the National Science Foundation, U.S. Geological Survey, and Scripps Institution of Oceanography.

References

Aeschbach-Hertig W, Kipfer R, Hofer M et al (1996) Quantification of Gas Fluxes From the Subcontinental Mantle - the Example of Laacher See, a Maar Lake in Germany. Geochim Cosmochim Acta 60:31–41
Andrews JN (1985) The isotopic composition of radiogenic helium and its use to study groundwater movement in confined aquifers. Chem Geol 49:339–351
Andrews JN, Lee DJ (1979) Inert gases in groundwater from the Bunter Sandstone of England as indicators of age and paleoclimatic trends. J Hydrol 41:233–252
Ballentine CJ, Burnard P (2002) Production of noble gases in the continental crust. In: Porcelli D et al (eds) Noble gases in geochemistry and cosmochemistry. Mineralogical society of America, Washington, 47:481–538
Ballentine CJ, O'Nions RK, Oxburgh ER et al (1991) Rare gas constraints on hydrocarbon accumulation, crustal degassing and groundwater flow in the Pannonian basin. Earth Planet Sci Lett 105:229–246
Ballentine CJ, Burgess R, Marty B (2002) Tracing fluid origin, transport and interaction in the curst. In: Porcelli D et al (eds) Noble gases in geochemistry and cosmochemistry. Mineralogical society of America, Washington, 47:539–614
Barry PH, Hilton DR, Tryon MD et al (2009) A new syringe pump apparatus for the retrieval and temporal analysis of helium in groundwaters and geothermal fluids. Geochem Geophys Geosyst 10:Q05004
Bethke CM, Zhao X, Torgersen T (1999) Groundwater flow and the ^{4}He distribution in the Great Artesian Basin of Australia. J Geophys Res 104:12999–13011
Bethke CM, Torgersen T, Park J (2000) The "age" of very old groundwater; insights from reactive transport models. J Geochem Explor 69–70:1–4
Beyerle U, Aeschbach-Hertig W, Imboden DM et al (2000) A mass spectrometric system for the analysis of noble gases and tritium from water samples. Environ Sci Technol 34:2042–2050
Beyerle U, Rueedi J, Leuenberger M et al (2003) Evidence for periods of wetter and cooler climate in the Sahel between 6 and 40 kyr BP derived from groundwater - art. no. 1173. Geophys Res Lett 30:1173
Castro MC, Goblet P, Ledoux E et al (1998) Noble gases as natural tracers of water circulation in the Paris Basin 2. Calibration of a groundwater flow model using noble gas isotope data. Water Resources Research 34:2467–2483
Castro MC, Stute M, Schlosser P (2000) Comparison of ^{4}He ages and ^{14}C ages in simple aquifer systems: implications for groundwater flow and chronologies. Appl Geochem 15:1137–1167
Clarke WB, Beg MA, Craig H (1969) Evidence for primordial He (super 3) in the earth. Eos, Transactions, American Geophysical Union 50:222
Clarke WB, Jenkins WJ, Top Z (1976) Determination of tritium by mass spectrometric measurement of He-3. Int J Appl Radiat Isot 27:515–522
Collier R, Dymond J, McManus J (1991) Studies of hydrothermal processes in Crater Lake, OR. College of Oceanography Report #90 7
Craig H, Lal D (1961) The production rate of natural tritium. Tellus 13:85–105
Crossey LJ, Karlstrom KE, Springer AE et al (2009) Degassing of mantle-derived CO_2 and He from springs in the southern Colorado Plateau region – neotectonic connections and implications for groundwater systems. Bull Geol Soc Am 121:1034–1053
de Leeuw GAM, Hilton DR, Gulec N et al (2010) Regional and temporal variations in $CO_2/^3$He, ^{3}He/^{4}He and d^{13}C along the North Anatolian Fault Zone, Turkey. Appl Geochem 25:524–539
Dunai TJ, Porcelli D (2002) Storage and transport of noble gases in the subcontinental lithosphere. Rev Mineral Geochem 47:371–409
Furi E, Hilton DR, Brown KM et al (2009) Helium systematics of cold seep fluids at Monterey Bay, California, USA: Temporal variations and mantle contributions. Geochem Geophys Geosyst 10:Q08013
Graham DW (2002) Noble gas isotope geochemistry of mid-ocean ridge and ocean island basalts: Characterization of mantle source reservoirs. In: Porcelli D et al. (eds) Noble gases in geochemistry and cosmochemistry. Mineralogical Society of America, Washington, 47:247–318
Gulec N, Hilton DR, Mutlu H (2002) Helium isotope variations in Turkey: relationship to tectonics, volcanism and recent seismic activities. Chem Geol 187:129–142
Hamme RC, Emerson SR (2004) Measurement of dissolved neon by isotope dilution using a quadrupole mass spectrometer. Mar Chem 91:53–64
Heaton THE (1984) Rates and sources of ^{4}He accumulation in groundwater. Hydrol Sci J 29:29–47
Heaton THE, Vogel JC (1981) "Excess air" in groundwater. J Hydrol 50:201–216
Hilton DR (1996) The helium and carbon isotope systematics of a continental geothermal system – results from monitoring studies at Long Valley caldera (California, USA). Chem Geol 127:269–295
Hilton DR (2007) Geochemistry: the leaking mantle. Science 318:1389–1390
Hohmann R, Hofer M, Kipfer R et al (1998) Distribution of helium and tritium in Lake Baikal. Journal of Geophysical Research: Oceans 103:12823–12838

Hong GH, Hamilton TF, Baskaran M, Kenna TC (2011) Applications of anthropogenic radionuclides as tracers to investigate marine environmental processes. In: Baskaran M (ed) Handbook of environmental isotope geochemistry. Springer, Heildelberg, 47:this volume

Ingram RGS, Hiscock KM, Dennis PF (2007) Noble gas excess air applied to distinguish groundwater recharge conditions. Environ Sci Technol 41:1949–1955

International Atomic Energy Association (1983) Environmental isotope data No. 7: World survey of isotope concentration in precipitation (1976-1979). Technical Report Series, #206

International Atomic Energy Association (1986) Environmental isotope data No. 8: World survey of isotope concentration in precipitation (1980-1983). Technical Report Series, #264. 184

Italiano F, Martinelli G, Nuccio PM (2001) Anomalies of mantle-derived helium during the 1997-1998 seismic swarm of Umbria-Marche, Italy. Geophys Res Lett 28:839–842

Jenkins WJ, Clarke WB (1976) The distribution of ^{3}He in the western Atlantic Ocean. Deep-Sea Res 23:481–494

Jones DL, Graymer R, Wang C et al (1994) Neogene transpressive evolution of the California coast ranges. Tectonics 13:561–574

Kennedy BM, Van Soest MC (2007) Flow of mantle fluids through the ductile lower crust: helium isotope trends. Science 318:1433–1436

Kennedy BM, Kharaka YK, Evans WC et al (1997) Mantle fluids in the San Andreas fault system, California. Science 278:1278–1281

Kipfer R, Aeschbach-Hertig W, Peeters F et al (2002) Noble gases in lakes and ground waters. In: Porcelli D et al (eds) Noble gases in geochemistry and cosmochemistry. Mineralogical Society of America, Washington, 47:615–700

Kulongoski JT, Hilton DR (2002) A quadrupole-based mass spectrometric system for the determination of noble gas abundances in fluids. Geochem Geophys Geosyst 3:U1–U10

Kulongoski JT, Hilton DR, Izbicki JA (2003) Helium isotope studies in the Mojave Desert, California: implications for ground-water chronology and regional seismicity. Chem Geol 202:95–113

Kulongoski JT, Hilton DR, Selaolo ET (2004) Climate variability in the Botswana Kalahari from the Late Pleistocene to the present day. Geophys Res Lett 31:L10204

Kulongoski JT, Hilton DR, Izbicki JA (2005) Source and movement of helium in the eastern Morongo groundwater Basin: the influence of regional tectonics on crustal and mantle helium fluxes. Geochim Cosmochim Acta 69:3857–3872

Kulongoski JT, Hilton DR, Cresswell RG et al (2008) Helium-4 characteristics of groundwaters from Central Australia: Comparative chronology with chlorine-36 and carbon-14 dating techniques. J Hydrol 348:176–194

Kulongoski JT, Hilton DR, Izbicki JA et al (2009) Evidence for prolonged El Nino-like conditions in the Pacific during the Late Pleistocene: a 43 ka noble gas record from California groundwaters. Quatern Sci Rev 28:2465–2473

Lehmann BE, Love A, Purtschert R et al (2003) A comparison of groundwater dating with ^{81}Kr, ^{36}Cl and ^{4}He in four wells of the Great Artesian Basin, Australia. Earth Planet Sci Lett 211:237–250

Mamyrin BA, Anufriev GS, Kamenskiy IL et al (1970) Determination of the isotopic composition of atmospheric helium. Geochem Int 7:498–505

Mamyrin BA, Tolstikhin IN (1984) Helium isotopes in nature. Elsevier, Amsterdam, 274

Marine IW (1979) The use of naturally occurring helium to estimate groundwater velocities for studies of gologic storage of radioactive waste. Water Resour Res 15:1130–1136

Martel DJ, Deak J, Dovenyi P et al (1989) Leakage of Helium from the Pannonian Basin. Nature 342:908–912

Marty B, Torgersen T, Meynier V et al (1993) Helium isotope fluxes and groundwater ages in the Dogger Aquifer, Paris Basin. Water Resources Research 29:1025–1035

Mazor E (1972) Paleotemperatures and other hydrological parameters deduced from noble gases dissolved in groundwaters; Jordan Rift Valley, Israel. Geochim Cosmochim Acta 36:1321–1336

Miller SA, Nur A (2000) Permeability as a toggle switch in fluid-controlled crustal processes. Earth Planet Sci Lett 183:133–146

Norton FJ (1957) Permeation of Gases through Solids. J Appl Phys 28:34–39

Nur A, Walder J (1990) Time-dependent hydraulics of the Earth's crust. In: The role of fluids in crustal processes. National Academy Press, Washington, 113–127

O'Nions RK, Oxburgh ER (1983) Heat and helium in the Earth. Nature 306:429–431

O'Nions RK, Oxburgh ER (1988) Helium, volatile fluxes and the development of continental crust. Earth Planet Sci Lett 90:331–347

Oxburgh ER, O'Nions RK, Hill RI (1986) Helium isotopes in sedimentary basins. Nature 324:632–635

Ozima M, Podosek FA (1983) Noble gas geochemistry. Cambridge University Press, Cambrigde

Peeters F, Kipfer R, Achermann D et al (2000) Analysis of deep-water exchange in the Caspian Sea based on environmental tracers. Deep-Sea Res Pt I-Oceanog Res 47:621–654

Phillips FM, Castro MC (2003) Groundwater dating and residence time measurements. In: Holland HD et al. (eds) Treatise on Geochemistry, Surface and Ground Water, Weathering, and Soils. Oxford University Press, Oxford, 5:451–497

Pik R, Marty B (2009) Helium isotopic signature of modern and fossil fluids associated with the Corinth rift fault zone (Greece): implication for fault connectivity in the lower crust. Chem Geol 266:67–75

Poole JC, McNeill GW, Langman SR et al (1997) Analysis of noble gases in water using a quadrupole mass spectrometer in static mode. Appl Geochem 12:707–714

Poreda RJ, Cerling TE, Salomon DK (1988) Tritium and helium isotopes as hydrologic tracers in a shallow unconfined aquifer. J Hydrol 103:1–9

Rice JR (1992) Fault stress, pore pressure distribution, and the weakness of the San Andreas fault. In: Evans B et al. (eds) Fault mechanics and transport properties of rocks, International Geophysics Series. Academic Press, London, 51:475–503

Rison W, Craig H (1983) Helium isotopes and mantle volatiles in Loihi Seamount and Hawaiian Islands basalts and xenoliths. Earth Planet Sci Lett 66:407–426

Sano Y (1986) Helium flux from the solid earth. Geochem J 20:227–232

Sano Y, Nakajima J (2008) Geographical distribution of ^{3}He/^{4}He ratios and seismic tomography in Japan. Geochem J 42:51–60

Sano Y, Takahata N (2005) Measurement of noble gas solubility in seawater using a quadrupole mass spectrometer. J Oceanogr 61:465–473

Sano Y, Wakita H, Huang CW (1986) Helium flux in a continental land area estimated from ^{3}He/^{4}He ratio in northern Taiwan. Nature 323:55–57

Schlosser P, Winckler G (2002) Noble gases in ocean waters and sediments. In: Porcelli D et al. (eds) Noble gases in geochemistry and cosmochemistry. Mineralogical Society of America, Washington, 47:701–730

Schlosser P, Stute M, Dorr H et al (1988) Tritium/^{3}He dating of shallow groundwater. Earth Planet Sci Lett 89:353–362

Schlosser P, Stute M, Sonntag C et al (1989) Tritiogenic He-3 in shallow groundwater. Earth Planet Sci Lett 94:245–256

Selaolo ET, Hilton DR and Beekman HE (2000) Groundwater recharge deduced from He isotopes in the Botswana Kalahari. In: Sililo O et al. (eds.) Groundwater: past achievements and future challenges. Proceedings of the 30th International Association of Hydrogeologists Congress, Cape Town, RSA. A.A. Balkema Publishers, Rotterdam, Netherlands, 313–318

Solomon DK, Hunt A, Poreda RJ (1996) Source of radiogenic helium 4 in shallow aquifers – implications for dating young groundwater. Water Resour Res 32:1805–1813

Stanley RHR, Baschek B, Lott III DE et al (2009) A new automated method for measuring noble gases and their isotopic ratios in water samples. Geochem Geophys Geosyst 10:Q05008

Stute M, Schlosser P (1993) Principles and applications of the noble gas paleothermometer. In Climate Change in Continental Isotopic Records 78:89–100

Stute M, Talma AS (1998) Glacial temperatures and moisture transport regimes reconstructed from noble gases and O-18, Stampriet aquifer, Namibia. In Isotope techniques in the study of environmental change. IAEA, Vienna, 307–318

Stute M, Sonntag C, Deak J et al (1992) Helium in deep circulating groundwater in the Great Hungarian Plain – flow dynamics and crustal and mantle helium fluxes. Geochim Cosmochim Acta 56:2051–2067

Stute M, Clark JF, Schlosser P et al (1995) A 30,000 yr continental paleotemperature record derived from noble gases dissolved in groundwater from the San Juan Basin, New Mexico. Quat Res 43:209–220

Takaoka N, Mizutani Y (1987) Tritiogenic ^{3}He in groundwater in Takaoka. Earth Planet Sci Lett 85:74–78

Tolstikhin IN (1975) Helium isotopes in the Earth's interior and in the atmosphere; a degassing model of the Earth. Earth Planet Sci Lett 26:88–96

Tolstikhin IN, Kamenskiy IL (1969) Determination of groundwater ages by the T-3He method. Geochem Intl 6:810–811

Top Z, Clarke WB (1983) Helium, neon, and tritium in the Black Sea. J Mar Res 41:1–17

Torgersen T (1980) Controls on pore-fluid concentrations of ^{4}He and ^{222}Rn and the calculation of ^{4}He/^{222}Rn ages. J Geochem Explor 13:57–75

Torgersen T (1993) Defining the role of magmatism in extensional tectonics: He-3 fluxes in extensional basins. J Geophys Res 98:16257–16269

Torgersen T (2010) Continental degassing flux of ^{4}He and its variability. Geochem Geophys Geosyst 11:Q06002

Torgersen T, Clarke W (1985) Helium accumulation in groundwater: I. An evaluation of sources and continental flux of crustal ^{4}He in the Great Artesian basin, Australia. Geochim Cosmochim Acta 49:1211–1218

Torgersen T, Clarke WB (1987) Helium accumulation in groundwater; III, limits on helium transfer across the mantle-crust boundary beneath Australia and the magnitude of mantle degassing. Earth Planet Sci Lett 84:345–355

Tryon M, Brown K, Dorman L et al (2001) A new benthic aqueous flux meter for very low to moderate discharge rates. Deep Sea Res I 48:2121–2146

Webb TH, Kanamori H (1985) Earthquake focal mechanisms in the eastern Transverse Ranges and San Emigdio Mountains, southern California and evidence for a regional decollement. Seismological Soc Am Bull 75:737–757

Weise S, Moser H (1987) Groundwater dating with helium isotopes. In: Techniques in water resources development. IAEA, Vienna, 105–126

Weiss RF (1971) Solubility of helium and neon in water and seawater. J Chem Eng Data 16:235–241

Well R, Lupton J, Roether W (2001) Crustal helium in deep Pacific waters. Journal of Geophysical Research C: Oceans 106:14165–14177

Weyhenmeyer CE, Burns SJ, Waber HN et al (2000) Cool glacial temperatures and changes in moisture source recorded in Oman groundwaters. Science 287:842–845

Zhao X, Fritzel TLB, Quinodoz HAM et al (1998) Controls on the distribution and isotopic composition of helium in deep ground-water flows. Geology 26:291–294

Chapter 16
Applications of Short-Lived Radionuclides (^{7}Be, ^{210}Pb, ^{210}Po, ^{137}Cs and ^{234}Th) to Trace the Sources, Transport Pathways and Deposition of Particles/Sediments in Rivers, Estuaries and Coasts

J.Z. Du, J. Zhang, and M. Baskaran

Abstract Natural and anthropogenic radioisotopes can be used to determine not only the mixing and diffusion processes of water masses but also the sources and sedimentary dynamics of particles in aquatic systems such as rivers, estuaries and oceans. Particle-reactive radionuclides that are derived from atmospheric deposition and/or the decay from their parent nuclides in aqueous system, can be used to determine the removal rates of suspended particulate matter, sediment focusing/erosion, sediment resuspension rates and sediment accumulation and mixing rates. They can be also used as analogs for tracing the transport and fates of other particle-reactive contaminants, such as PCBs and PAH. In this chapter, we focus on various applications of short-lived radionuclides (i.e., ^{7}Be, ^{210}Pb, ^{210}Po, ^{137}Cs and ^{234}Th) as tracers for particle and sediment dynamics to quantify several river, estuary and coastal oceanic processes with their concerned timescales ranging from a few days to about 100 years.

16.1 Introduction

Estuaries and coasts are commonly described as complex reactors at land-sea interface where not only significant transformations of river and stream-borne suspended sediments but also the biogeochemical cycling of key nutrients take place (Swarzenski et al. 2001; Swarzenski et al. 2003). Whether estuarine and coastal processes of water masses and soluble materials, or of particle/sediment and particle-reactive materials, we can use a suite of applicable radionuclides to trace these processes in different time scales that are comparable to their half-lives. Based on the geochemical behavior, most radionuclides that occur in the environment can be broadly classified into two major categories: (i) the particle-reactive (i.e., non-conservative tracers); and (ii) water soluble (i.e., conservative tracers). The particle-reactive radioactive radionuclides are sorbed onto biogenic and terrigenous particulate matter, including colloidal material and are removed from the water column. In coastal waters, most of the particle-reactive radionuclides are derived either from direct atmospheric fallout (e.g., ^{7}Be), riverine discharge (e.g., ^{137}Cs), or in situ production from parent nuclides (e.g., ^{234}Th) and are commonly associated with fine particles. The amount of particle-reactive radionuclides discharged through submarine groundwater discharge is likely negligible, compared to other sources (e.g., soluble radionuclides) in the coastal waters. Once-sorbed, the fate of the particle-reactive radionuclides become the fate of the particulate matter, although under some cases, there could be desorption of the radionuclides. In general, the adsorption is faster than the desorption. Owing to irreversible character of solid-water partitioning, particle-reactive radionuclides can serve as a set of powerful tracers to quantify the sedimentary dynamics including removal rates of suspended particulate matter (Baskaran and Swarzenski 2007; Jweda et al. 2008), sediment focusing/erosion, sediment resuspension rates, sediment accumulation and mixing rates, and sediment transport (e.g., Feng et al. 1999a, b; Giffin and Corbett 2003; Lima et al. 2005;

J.Z. Du (✉) and J. Zhang
State Key Laboratory of Estuarine and Coastal Research, East China Normal University, Shanghai 200062, China
e-mail: jzdu@sklec.ecnu.edu.cn; jzhang@sklec.ecnu.edu.cn

M. Baskaran
Department of Geology, Wayne State University, Detroit, MI 48202, USA
e-mail: baskaran@wayne.edu

Du et al. 2010). The particle-reactive radionuclides also provide analog information on other particle-reactive organic (e.g., PCBs, PAHs, and other derivatives from petroleum and coal hydrocarbons, synthetic organics, etc. (Gustafsson et al. 1997a, b; Gustafsson and Gschwend 1998)), and inorganic contaminants (e.g., biogenic silica, heavy metals, industrial substrates derived from industrial pollution, etc., Broecker et al. 1973; Buesseler et al. 2001b; Rutgers van der Loeff et al. 2002; Weinstein and Moran, 2005). Organic-rich coastal waters could be ideal sites to investigate the fate of these nuclides and other metals due to metal-organic complexation. The removal rates of particle-reactive radionuclides are also relevant in the use of these radionuclide-based chronologies. Such chronologies are needed to investigate the historical variations on the inputs of organic and inorganic pollutants.

A large number of papers have been published with ^{234}Th as a tracer in aqueous system, with more than 250 papers since the first work by Bhat et al. (1969) and the application and future use are reviewed in Waples et al. (2006). Kaste et al. (2002) reviewed the environment chemistry of ^{7}Be in the world (also see Kaste and Baskaran 2011). ^{7}Be and ^{234}Th can be used in the determination of sedimentation rates in short time scale, such as sediment deposition by flood (Sommerfield et al. 1999; Palinkas et al. 2005; Huh et al. 2009). Sediment redistribution parameters such as the settling and resuspension rates can be evaluated using ^{7}Be/^{210}Pb$_{xs}$, and ^{234}Th$_{xs}$/^{7}Be activity ratios. Moreover, the depth profiles of ^{7}Be, ^{210}Pb$_{xs}$ and ^{137}Cs have been utilized as effective tracers in the study of soil erosion (Walling et al. 1999; also Matisoff and Whiting 2011). In this chapter, we review the utility of particle-reactive radionuclides as tracers for particle and sediment dynamics to quantify several coastal oceanic processes, with special emphasis on the utility of ^{7}Be, ^{234}Th, ^{210}Po, ^{210}Pb and ^{137}Cs.

16.1.1 Sources of Isotopes: ^{7}Be, ^{137}Cs, ^{210}Pb, ^{210}Po and ^{234}Th

The nuclear properties of these nuclides are listed in Table 16.1. ^{7}Be is continuously and globally produced in an endothermic reaction when cosmic-ray-produced neutrons and protons disintegrate the atomic nuclei of atmospheric nitrogen and oxygen at an average rate of 8.1×10^2 atoms m^{-2} s^{-1} (UNSCEAR 1982, 2000). The production rate of ^{7}Be (and other cosmogenic radionuclides) varies with latitude and altitude. About two thirds of the production takes place in the stratosphere and the remaining one third in the troposphere (Lal and Peters 1967; Lal and Baskaran 2011). The residence time of ^{7}Be in the stratosphere with respect to removal by scavenging is about 1 year, while that in the troposphere is about 10 days. At the mid-latitudes (about 40–50°), there is an intrusion of stratospheric air, primarily during spring, which transfers large amounts of ^{7}Be and other cosmogenic radionuclides into the troposphere (Lal and Baskaran 2011). Moreover, the high production rate of

Table 16.1 Nuclear properties of ^{7}Be, ^{137}Cs, ^{210}Pb, ^{210}Po and ^{234}Th[a]

Nuclide	Atomic number	Decay mode	Energy of main peak (MeV)	Intensity (%)	Half-life	Analysis method	Sources
^{7}Be	4	EC[b]	Gamma-0.4776	10.52	53.3 days	Gamma	Cosmogenic
		EC	–	89.48			(810 atoms m^{-2} s^{-1})
^{137}Cs	55	Beta	Beta-1.176	94.4	30.0 years	Beta	Human nuclear activity
		Beta	Beta-0.514	5.6			
			Gamma-0.6617	94.4		Gamma	
^{210}Pb	82	Beta	Beta-0.0631	16	22.3 years	Beta-^{210}Bi	U-decay series
		Beta	Beta-0.16	84		Alpha-^{210}Po	
			Gamma-0.0465	4.25		Gamma	
^{210}Po	84	Alpha	Alpha-5.304	100	138.4 days	Alpha-^{210}Po	U-decay series
^{234}Th	93	Beta	Beta-0.1985	70.3	24.1 days	Beta	U-decay series
		Beta	Beta-0.1035	26.8			
		Beta	Beta-0.0833	2.9			
			Gamma-0.0629	4.85		Gamma	
			Gamma-0.0926	5.58			

[a]Data reproduced from http://atom.kaeri.re.kr
[b]Electron capture

^{7}Be in the atmosphere is observed during solar minimums during the 11-year sunspot cycle (Masarik and Beer 1999).

Most of the ^{137}Cs present in the oceans was derived from global nuclear fallout, with peak fallout in 1963, riverine input from the erosion of soils and nuclear waste sites, discharge of nuclear effluents from close-in and long-range transport and the Chernobyl accident (more details are given in Hong et al. 2011). Detonation of nuclear weapon tests in the atmosphere starting from 1952 until the cessation of nuclear weapons testing released a large volume of anthropogenic radionuclides to the atmosphere.

There are two components of ^{210}Pb in the Earth's surface. In natural minerals and soils, ^{210}Pb in general is in secular equilibrium with ^{226}Ra, a decay product in the ^{238}U decay series, which is then supplied to rivers along with detrital particles and is termed as the supported ^{210}Pb. An additional source of ^{210}Pb is primarily derived from the atmospheric deposition. ^{222}Rn is constantly released from surface soils, rocks and minerals with a global average rate of 1 atom cm^{-2} s^{-1} from the land surface, while this rate is about two orders of magnitude lower from the ocean (Turekian and Graustein 2003). The atmospheric ^{222}Rn undergoes radioactive decay and eventually produces ^{210}Pb. ^{210}Pb thus formed in the atmosphere is rapidly adsorbed onto aerosol particles, which reach soils and aquatic systems through wet and dry atmospheric depositions. Once delivered at the air-sea interface via wet and dry precipitation, the ^{210}Pb is quickly removed by suspended particulate matter and eventually reaches the sediment-water interface. The concentration of ^{210}Pb in the surface waters thus primarily depends upon the temporally and geographically varying inputs of atmospheric deposition. Subsequently, most of the ^{210}Pb in water column is removed by sorption on to particulate matter.

Polonium, an oxygen-sulfur group element, is used as a tracer of elements fundamental to organic matter. It is a particle-reactive element with a wide range of K_d values, 10^3–10^5 cm^3 g^{-1} (Carvalho 1997; Nozaki et al. 1997; Hong et al. 1999; Kim and Yang 2004). Most of the dissolved ^{210}Po in the upper waters is from the decay of ^{210}Pb and is ultimately derived from atmospheric fallout with the ^{210}Po/^{210}Pb activity ratio in depositional flux < 0.1 (e.g., Hussain et al. 1998). More details can be found in Kim and Church (2011).

^{234}Th is a naturally occurring, highly particle-reactive radionuclide produced from the decay of ^{238}U. Due to its high particle-reactivity, Th-sorbed particulate matter eventually reaches the seabed. The sorbed component of ^{234}Th is called excess ^{234}Th (^{234}Th$_{xs}$), while the lattice-bound ^{234}Th that is in secular equilibrium with ^{238}U is called parent-supported ^{234}Th. The production rate of ^{234}Th is thus dependent on the concentration of its dissolved parent nuclide ^{238}U. In estuaries and coastal areas, U generally behaves conservatively, although non-conservative behavior has been reported at low salinity waters, and thus, ^{238}U activity increases with salinity and thus production rate of ^{234}Th also increases with salinity.

16.1.2 Factors Affecting the Concentrations and Behavior of ^{7}Be, ^{137}Cs, ^{210}Pb, ^{210}Po and ^{234}Th in Estuaries and Coasts

The depositional flux of ^{7}Be to Earth's surface is highly variable, depending upon the amount and frequency of precipitation, latitude, seasons and the nature and height of the cloud cover. The ^{7}Be specific activity in precipitation has been reported to vary widely. For example, ^{7}Be in precipitation samples collected over a period of 3 years (1989–1991) in Galveston, TX varied between 0.09 and 20.7 Bq L^{-1} (Baskaran et al. 1993; activities in global precipitation is summarized in Kaste and Baskaran 2011). Most of the ^{7}Be produced in the stratosphere do not readily reach the troposphere except during spring, when the seasonal thinning of the troposphere occurs at mid-latitudes, resulting in air exchange between the stratosphere and troposphere (e.g., Kim et al. 1998). Therefore, ^{7}Be flux to the earth surface depends on latitude. Several studies indicated that additional meteorological factors, such as seasonal variability in the amount and frequency of rainfall and the vertical mixing of the lower and upper troposphere, could explain the seasonal variability of ^{7}Be flux (Baskaran 1995; Benitez-Nelson and Buesseler 1999; Su et al. 2003; Du et al. 2008). During heavy flood events in rivers, ^{7}Be discharge is very high compared to normal flow periods, and hence, the distribution of particulate ^{7}Be is affected by the volume of water discharge in estuaries and coastal areas.

At present, there is no global fallout of ^{137}Cs derived from nuclear weapons testing. However, the resuspension of very fine ^{137}Cs-laden dust particles could contribute a very small amount to the ocean surface waters. Rivers discharge ^{137}Cs through suspended particulate and bed load as well as in the dissolved phase. Most of this ^{137}Cs is derived from the erosional input from the watershed (Walling and Quine 1993; Matisoff and Whiting 2011). ^{137}Cs in the oceanic water column behaves as a soluble nuclide and has properties that make it useful as a water mass tracer (Bowen and Roether 1973; Murray et al. 1978; Smith et al. 1990). Note that ^{137}Cs is particle-reactive in freshwater systems due to low potassium (K) concentrations, while in marine systems, the presence of large amounts of K compete with Cs, and hence, the geochemical behavior of Cs is different. Very little of the ^{137}Cs delivered to the ocean has reached the sedimentary column. More details on the sources and distribution of ^{137}Cs in marine environment are given in Hong et al. (2011).

^{210}Pb, similar to ^{7}Be, is also primarily derived from atmospheric deposition. The atmospheric deposition is also highly variable, depending on the amounts and frequency of precipitation as well as latitude (Baskaran 2011; Kaste and Baskaran 2011). The amount of ^{210}Pb resulting from the decay of ^{226}Ra in coastal areas is negligible compared to the atmospheric depositional flux. After the deposition of ^{7}Be and ^{210}Pb at the air-sea interface in the coastal waters, these nuclides are quickly removed onto suspended particulate matter. For example, Baskaran and Santschi (1993) reported that about 80 % of ^{7}Be became associated with particulates within an hour after a heavy rainfall event. ^{210}Pb binds irreversibly to soil particles, and is generally immobile in sediments (Koide et al. 1972; Robbins and Edgington 1975; Goldberg et al. 1977). Hawley et al. (1986) reported partition coefficients, K_d for ^{7}Be and ^{210}Pb greater than 10^4, indicating that these radionuclides are sufficiently particle-bound that they can be used to trace erosion, particle transport and deposition (Whiting et al. 2005). Annual depositional fluxes of ^{210}Pb are considerably constant over longer periods of time, although they vary with the amounts of precipitation (Baskaran et al. 1993). The removal of ^{210}Pb from the upper ocean is driven by the flux of biogenic material from surface productivity (Fisher et al. 1988; Moore and Dymond, 1988). The atmospheric depositional flux of ^{210}Po is very small compared to ^{210}Pb, with a ^{210}Po/^{210}Pb activity ratio of <0.1. One could potentially utilize ^{210}Po/^{210}Pb ratios in suspended particulate matter to trace the pathways and extent of resuspension of deposited sediment. Polonium has a strong affinity for biogenic material and gets incorporated into organic matter, and thus, Po is expected to trace biogenic organic matter much more closely than Th or Pb (Fisher et al. 1983; Cherrier et al. 1995; LaRock et al. 1996).

The contrasting geochemical difference between insoluble Th and soluble U in the water column and the relatively short residence times of ^{234}Th in surface waters and in the euphotic zone make ^{234}Th a powerful tracer for quantifying spatial and temporal dynamics of particles and other particle-reactive species. Short-lived thorium isotopes are used to examine a large range of water column and seabed processes in estuarine environments (Waples et al. 2006; Rutgers van der Loeff et al. 2006). The earliest observations regarding the disequilibria between ^{234}Th and its parent ^{238}U revealed that the removal rate of ^{234}Th decreased with distance from shore (Bhat et al. 1969). This finding established the particle-reactive nature of thorium and laid the foundation for the use of short-lived thorium isotopes as tracers for geochemical and sedimentary processes in estuarine and coastal environments. The utilization of ^{234}Th as a tracer in estuarine environments is divided into two categories: (1) quantifying the residences time of ^{234}Th (dissolved and particulate, particulate Th is the same as the particulate matter) in the water column, other particle-reactive species with respect to the scavenging and removal from the water column, and (2) quantifying rates of sedimentary processes such as mixing, deposition and transport.

In this chapter, we review the important applications of these short-lived radionuclides in the coastal environments as tracers to quantify the removal rates of suspended particulate matter, particle cycling, sediment dynamics and utility of these tracers as proxies for other contaminants sorbed on to sedimentary particulate matter. These tracers have provided useful information in quantifying the fluxes of particle-sorbed PCBs, PAHs and trace metals (Gustafsson et al. 1997a, b; Weinstein and Moran 2005).

16.2 Methodology for Sample Collection and Laboratory Analysis

Appropriate sampling and analytical methods are all key factors in obtaining high quality data. The sampling and treatment of samples, the laboratory analysis of radioisotopes and quality management are especially important.

16.2.1 Sampling and Treatment

Compared to other chemical analyses (i.e., heavy metals and organic composition), the unintended contamination of samples for the analysis of short-lived radionuclides by material released from the research vessel or by the sampling apparatus is generally negligible. The particle-reactive radionuclides have a strong tendency to sorb onto the walls of the sampling containers, and hence, acidification of the samples immediately after collection is required.

For large-volume surface water sampling, different methods are available. Water samples can be collected either through the seawater supply hose in the ship or through a submersible pump. The seawater hose in the ship tends to accumulate Th and has been reported to release the daughter products of Th (e.g., ^{224}Ra released from the sorbed ^{228}Th). Caution must be exercised for the possible release of radionuclides from the seawater supply hose. Generally, it is necessary to collect large sample volumes (usually 0.02 M^3 to several M^3) to obtain good counting statistics within an acceptable counting time. Sometimes there are few options except having a large volume of water samples (e.g., ^{7}Be). There are a few methods that can be used for large volume sampling (Rutgers van der Loeff and Moore 1999; Baskaran et al. 2009a).

1. *Rosette sampling* (up to 30 L) is the most convenient method and provides the exact depth profile and for other parameters.
2. *Gerard sampling* (up to 400 L). The bottles must be rinsed using the yo-yo technique to obtain water from the desired depth (Roether 1971). There is a serious risk of exchange of water during recovery (Broecker et al. 1986) due to the loose closure of the lids, which should be checked with indirect methods (e.g., silicate measurements). Radionuclide sampling with 250-L PVC bottles is described in Nozaki (1993).
3. *In situ pumps* combined with (MnO_2-coated) absorbers form the only alternative for subsurface sampling volumes of >400 L for ^{234}Th and ^{210}Pb (Buesseler et al. 1992, 1995; Baskaran et al. 1993; Rutgers van der Loeff and Berger 1993; Rutgers van der Loeff and Moore 1999; Baskaran and Santschi, 2002). This method is ship time-consuming, and there is a risk of loss of particles during the recovery and of particles sticking to the baffles in front of the filters. The simultaneous operation of multiple pumps at multiple depths in the winch will reduce the overall ship time needed for vertical profile sampling. We found a reassuring correspondence between particulate ^{234}Th profiles obtained with Rosette casts and with in situ pumps (Du et al., unpublished data).
4. *Surface seawater supply* is a convenient way to sample surface water. However, there is a risk of artifacts, especially at insufficient throughput through long lines, and clogging inputs, which can affect both the dissolved and particulate phases. Sewage discharge from ship also can pose problems affecting the quality of data.

For volumes less than 30 L, except for the in situ pumps, large-volume filtration normally has to be performed on board. Radionuclides of the dissolved phase (^{7}Be, ^{210}Pb, ^{210}Po and ^{234}Th) are usually preconcentrated onboard by coprecipitation with hydroxides and sulfates of Fe, Ba or Mn such as $Fe(OH)_3$, $Mg(OH)_2$, MnO_2, $BaSO_4$, $PbSO_4$ or Co-APDC (Rutgers van der Loeff and Moore 1999; Buesseler et al. 2001a, b; Benitez-Nelson et al. 2001; Elsinger and Moore 1983; details on various sorbents used in literature are summarized in Baskaran et al. 2009a, b). If quantitative recovery cannot be guaranteed, yield tracers must be added such as ^{209}Po for ^{210}Po, stable Pb carrier for ^{210}Pb, stable Be carrier for ^{7}Be and ^{230}Th or ^{229}Th for ^{234}Th, respectively (Harvey and Lovett 1984).

For the measurements of both dissolved and particulate radionuclides, prefilters followed by MnO_2-coated fibers connected in series are used to obtain particulate matter (prefilter) and dissolved radionuclides (Th, Pb) from the water column. The prefilter is usually ashed at 550°C for polypropylene fiber material and can be utilized for the measurements of ^{7}Be, ^{234}Th, ^{210}Pb and ^{137}Cs but not for particulate ^{210}Po because Po is volatile at such high temperatures. The efficiency of extraction, E, for any nuclide can then be calculated from the ratio of the activities A and B measured in the first and second cartridges: E = 1 −B/A, (so called the "cartridge formula"). The assumption that both cartridges extract radionuclides uniformly (i.e., constant extraction efficiency by both filters A and B) becomes especially critical at lower efficiencies (Baskaran et al. 1993). Furthermore, a recent study indicates that some of the sorbed Th can desorb during filtration, thus adding uncertainty to the uniform extraction efficiency (Baskaran et al. 2009b). The formula cannot be used if there is a risk of desorption of MnO_2 with associated activity from the first to the second cartridge, either by physical entrainment of MnO_2 particles or by MnO_2 reduction in anoxic water masses. The schematic of the procedures is shown in Fig. 16.1. Detailed methods on the preparation of Mn-fiber and MnO_2-cartridges can be found in literature (Moore 1976; Buesseler et al. 1992; Rutgers van der Loeff and Moore 1999; Baskaran et al. 1993; 2009a, b). For preconcentrating ^{137}Cs from the water sample, a cotton fiber filter impregnated with copper ferrocynanide has been utilized (Buesseler et al. 1990; Baskaran et al. 2009a).

Surface sediments and sediment cores are usually collected by push core from a small boat with a PVC coring device outfitted with a one-way check valve and a 4-in.-diameter clear acrylic tube (core tube).

16.2.2 Laboratory Analysis of Radioisotopes

Alpha spectrometry, beta counting and gamma spectrometry are the commonly used methods for measuring the activity of ^{7}Be, ^{210}Pb, ^{210}Po, ^{137}Cs and ^{234}Th (Buesseler et al. 2001a; Rutgers van der Loeff and Moore, 1999; Trimble et al. 2004; Baskaran et al. 2009a). *Gamma spectrometry* allows the simultaneous measurements of ^{7}Be, ^{210}Pb, ^{234}Th and ^{137}Cs without chemical purification and is the ideal method for non-destructive techniques (Buesseler et al. 1992). However, inherent (Compton) background and self-absorption by the sample, energy- and geometry-dependent detector efficiency and low gamma branching ratios for some of the radionuclides can reduce the sensitivity and accuracy of the method. The overall sensitivity is generally low compared to other techniques (Rutgers van der Loeff and Moore 1999).

Alpha spectrometry is characterized by good isotope separation, uniform (i.e., energy- and isotope-independent) detector efficiency and very low background, typically on the order of 0.001–0.003 counts per minute for the 4–8 MeV energy range of a new surface barrier detector, increasing with time as a result of the accumulation of recoil products from measured samples (Rutgers van der Loeff and Moore 1999). Alpha spectrometry allows precise measurements at low activities and easy calibration with yield tracers. Alpha particles have a very short range. They are absorbed or slowed down by interactions with sample material or by impurities on the source

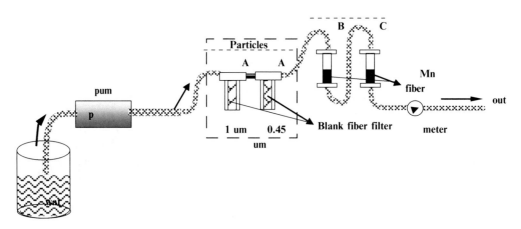

Fig. 16.1 Schematic diagram of the large volume method for ^{234}Th collection in the water column (modified from Baskaran et al. 2009a)

which is normally a flat planchet with the purified element plated on it. The best results in terms of efficiency and energy resolution are obtained with clean sources prepared from well-purified samples.

In the case of Beta Spectroscopy, the energy of the beta particles (electrons) has continuous energy distribution and hence identification of beta-emitting radionuclides using beta spectrometer is not possible. Since the beta particles have continuous energy spectrum extending from zero up to the maximum energy, E_{max}, the thickness of source will affect the beta particles with different energies differently. Thus, in most cases, chemical separation and purification of radionuclides is required.

16.2.2.1 Be-7

The decay of ^{7}Be can proceed directly to the Li ground state or to the first excited state in ^{7}Li with the branching ratio of 89.56 and 10.44%, respectively. The latter then proceeds to the ^{7}Li ground state by emitting gamma rays at 477.6 keV. Because the energy of gamma rays emitted from ^{7}Be is high, most of the samples are measured by γ-spectrometry (Olsen et al. 1986; Kim et al. 1999; 2000).

16.2.2.2 Po-210 and Pb-210

After a set of chemical procedures, ^{210}Po can be plated on a silver disc and assayed in an alpha spectrometer, more details can be found in Mathews et al. (2007). The ^{210}Pb concentration can be measured using a low-energy (46.3 keV) gamma line of ^{210}Pb. However, the low branching ratio (4.25%; Porcelli and Baskaran 2011) requires that a large volume of water sample be processed for ^{210}Pb (Baskaran and Santschi 2002). ^{210}Po and ^{210}Pb measurements in natural samples are routinely conducted using α-spectrometry, first measuring ^{210}Po (called in situ ^{210}Po) and then keeping the sample for a period of 6 months to 2 years for the in-growth of ^{210}Po from ^{210}Pb. The second ^{210}Po (parent-supported) measurement provides the data on the concentration of ^{210}Pb (Sarin et al., 1992; Church et al., 1994; Radakovitch et al., 1998; Hong et al., 1999; Fridrich and Rutgers van der Loeff 2002; Masque et al., 2002; Stewart et al., 2007). ^{210}Pb can also be measured by the beta counting method because the beta-ray emitted by ^{210}Pb has lower energy (β_{max} = 63 keV); these low energy betas particles need to be suppressed, and ^{210}Bi beta (β_{max} = 1.16 MeV) is used to assay ^{210}Pb activity. So when beta counting is used, we can get the concentration of ^{210}Pb through the measurement of ^{210}Bi (half-life: 5.0 days). Multiple counting over a period of 30–35 days (6–7 half-lives of ^{210}Bi) with a beta counter will ensure data quality.

16.2.2.3 Cs-137

^{137}Cs concentrations are most commonly measured using gamma-ray spectrometry when ^{137}Cs decays to the excited state of ^{137}Ba by beta decay and its subsequent transition to ground state. ^{137}Cs can also be directly measured using the beta (β_{max} = 1.176 MeV) counting method after separation and purification (Yoshizawa 1958; Baskaran et al. 2009a).

16.2.2.4 Th-234

^{234}Th can be measured by γ-spectrometry. The two suitable peaks for quantifying ^{234}Th are 63.2 keV (branching ratio 4.85%; Porcelli and Baskaran 2011) and 92.4 + 92.8 keV (branching ratio ~5.6%). These low branching ratios result in low gamma counting rates (e.g., Rutgers van der Loeff et al. 2006). Unlike γ-spectrometry, beta counting has an excellent counting efficiency for ^{234}Th. Apparent efficiencies reach almost 100% when the beta emissions for ^{234}Th (weak beta at E_{max} = 0.27 MeV) and its daughter ^{234m}Pa (strong beta at E_{max} = 2.19 MeV) are combined, and the sample is reduced into a near weightless source via electroplating on a stainless steel disk (Waples et al. 2003). A five-position beta counter manufactured by RISØ National Laboratories (Roskilde, Denmark) with a 10 cm lead shield has been most commonly used at sea and in the laboratory and has background values of 0.15–0.20 cpm for a source of 25-mm diameter. Pates et al. (1996) developed a method for the determination of ^{234}Th in seawater by LSS. A major advantage of using LSS is the avoidance of separate alpha and beta measurements to determine ^{234}Th activities. After the addition of ^{230}Th as a yield tracer, Th is coprecipitated with Fe(OH)$_3$ and purified through ion exchange chromatography to remove Fe and U. Details can be seen in a number of published references (Smith et al. 2004; Kersten et al. 1998; Foster and Shimmield 2002). However, beta counting is not isotope specific. Beta counting is the preferred method for ^{234}Th, and can be used with non-destructive techniques (for particulate samples on filters) and can be used on-board ship. For particulate and sediment samples, chemical separation and purification are required to prevent other beta-emitting radionuclides from interfering with the counting.

Table 16.1 show the nuclear properties of ^{7}Be, ^{137}Cs, ^{210}Pb, ^{210}Po and ^{234}Th, including decay modes, energies of the main peaks, intensities, half-lives, analytical methods and sources of these nuclides.

16.2.3 Quality Assurance and Quality Control

Each of the collection, handling, storage and preservation steps for samples should be validated to ensure that there are interferences with the measurements of isotopes interest. In cases where are there interferences, they need to be minimized and quantified. This quality control involves evaluation of the blanks and standard reference material (SRMs) containing certified levels of the relevant analytes (to determine analytical precision and/or for calibration of instruments). In general, blank samples should accompany each batch of sample containers to the field sampling site and be subjected to the same handling protocols (e.g., opening, closing, preservation) as the actual sample containers. A list of SRMs commonly utilized is given in Baskaran et al. (2009a). The analysis of natural radionuclides from small-volume water samples, usually less than 20 L, can be collected with CTD rosettes. Sorption to the walls of sample bottles could reduce the dissolved concentration of the nuclide considerably, and thus care must be taken.

To ensure that the analytical results are accurate, quality assurance (QA) measurements are often necessary. Generally, the QA for nuclide analyses include the counting of a known radioactive source in a predetermined geometry for a specific count time. After data collection, data analysis is performed to determine that the critical parameters such as peak position, resolution and efficiency are within the acceptable limits. It is noted that the QA measurement is not a substitute for calibration. Moreover, proficiency testing (PT) programs can provide another means for laboratories to assess their ability to perform a particular method and may be an appropriate part of some laboratory approval programs. Currently there are few PT programs for radionuclide analysis, such as GEOTRACES program website (http://www.usgeotraces.org/) and "world-wide open proficiency test" of IAEA (http://nucleus.iaea.org/rpst/index.htm).

16.2.3.1 Be-7

Getting a certified reference standard for ^{7}Be measurements is not an easy task due to its short half-life and the expense of producing it. However, calibration of the gamma ray spectrometer can be commonly carried out by establishing the energy vs. absolute efficiency curve. In particular, the two gamma lines with very little interference, 352 keV (^{214}Pb) and 609 keV (^{214}Bi), can be used to evaluate the absolute efficiency of ^{7}Be at 476 keV (Kaste et al. 2002). The intercalibration of samples collected from natural waters is recommended because it would help to validate the ^{7}Be measurement. Moreover, in samples with high ^{232}Th concentrations, one should be cautious of the 478.3 keV decay energy of ^{228}Ac (Kaste et al. 2002). To our knowledge, there is no intercalibration attempts have been carried out for this nuclide. It is noted that the efficiency of extracting Be in water onto Fe-coated cartridges (i.e., absolute efficiency in two cartridges connected in series) has been assumed to be constant (Kadko and Olson, 1996). If there is desorption of Be from the first cartridge, then the assumption of constant efficiency might be questionable sometimes.

16.2.3.2 Pb-210

A known amount of primary standards of ^{210}Po and ^{210}Pb (c.f. a list of SRMS is given in Baskaran et al. 2009a) should be used to check the precision of the methodology. In particular, ^{210}Po can be volatile if it is in the reduced state, and hence, its loss during the chemical procedure (if any) needs to be carefully monitored.

16.2.3.3 Po-210

Most of the SRMs for ^{210}Po are the same as ^{210}Pb because ^{210}Po and ^{210}Pb are in secular equilibrium (Baskaran et al. 2009a).

16.2.3.4 Cs-137

Adding a known amount of ^{134}Cs with certified value to standard seawater (e.g., 20 L), following the chemical procedures and comparing the activities to the certified values will ensure that the methodology and counting procedures are working satisfactorily. Meanwhile, for the small-volume seawater samples, it is important to know the background level of ^{137}Cs. The ^{137}Cs activity in CsCl was found to be less than 0.03 mBq g^{-1} based on extremely low-background gamma spectrometry. The ^{137}Cs activity in AMP was found to be less than 0.008 mBq g^{-1} (Hirose et al. 2005).

16.2.3.5 Th-234

The best standard for the intercalibration of ^{234}Th techniques is pristine deep-ocean water (1,000 m), in which the activities are expected to be in equilibrium. Earlier studies, however, have shown that some occasional disequilibrium does occur (c.f. Baskaran et al. 2003), and it is not clear whether some physical and/or biogeochemical processes that lead to this disequilibrium are operating or whether this is due to problems associated with the analytical procedures. To be safe, the best method is to collect a large-volume sample, filter it through a 0.45-μm cartridge filter, acidify the filtrate to pH < 1 and store it for about 6 months, which will ensure the ^{234}Th/^{238}U activity ratio of 1.0.

When the small-volume (2–5 L) beta counting method is used, other beta emitters (because beta counting is not isotope specific) of the U-Th decay-series radionuclides (such as betas from the decay of ^{228}Ac, ^{214}Pb, ^{214}Bi, ^{210}Bi, ^{212}Bi, ^{211}Pb, ^{211}Bi, and ^{40}K) could affect the accuracy of the measurements. This problem is more serious when beta counting is conducted without purification of the sample (such as the direct counting of suspended particulate material or MnO$_2$ precipitates; Rutgers van der Loeff et al. 2006). Beta counting of electroplated sources may be ideal for beta counting (uniform thickness) because non-uniform thickness resulting from uneven loading of particles could alter the self-absorption correction factor. Applying appropriate self-absorption corrections could be challenging because low-energy betas from ^{234}Th (E$_{max}$ = 0.27 MeV) and high-energy betas from its progeny, ^{234m}Pa (E$_{max}$ = 2.19 MeV), will be absorbed to varying extents (unless a Mylar is used to cut-off all ^{234}Th betas) in different sample thicknesses. More details can be found in Rutgers van der Loeff et al. (2006). Some of the glass fiber filters have been found to contain detectable natural radioactivity within the filter material, and the use of such filters for low-volume water filtration has resulted in relatively high uncertainty in the ^{234}Th data (Buesseler et al. 2001a, b).

16.3 Application

Figure 16.2 shows the conceptual input and output fluxes of ^{234}Th, ^{210}Po, ^{210}Pb, ^{137}Cs and ^{7}Be in an estuary/coast and its adjacent sea. Note that some

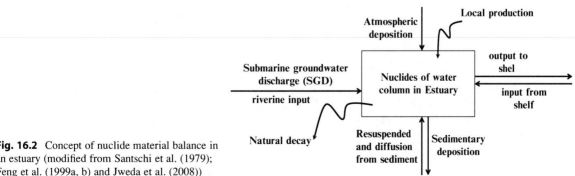

Fig. 16.2 Concept of nuclide material balance in an estuary (modified from Santschi et al. (1979); Feng et al. (1999a, b) and Jweda et al. (2008))

of the input terms for radionuclides are negligible (e.g., submarine groundwater discharge (SGD)-derived ^{137}Cs and ^{7}Be). The unique sources of ^{234}Th, ^{210}Po/^{210}Pb, ^{137}Cs and ^{7}Be make them ideal candidates for quantifying particle dynamics and sediment sources and their fate.

16.3.1 Quantifying the Freshly Delivered and Resuspended Sediments in Coastal and Estuarine Zones

Freshly delivered sediments at a given site in the coastal area predominantly contain the ^{7}Be/^{210}Pb ratio (A^p_{Be}/A^p_{Pb}) of precipitation, while resuspended sediments will have the ratio similar to the fine particulate matter of the upper layers of the bottom sediments. The specific activity ratio can potentially fingerprint unique particle sources; for example, the A^p_{Be}/A^p_{Pb} ratio may prove particularly useful as a storm-seabed process tracer (Allison et al. 2005). In this case, from the measured ^{7}Be/^{210}Pb ratio in the atmospheric deposition and the measured A^p_{Be}/A^p_{Pb} ratio in the particulate matter, bottom sediments and freshly delivered sediments, one can differentiate between resuspension of bed sediments and freshly delivered sediments from the river system. This ratio is also useful for calculating the relative amount of resuspended bed sediments collected in the estuary.

16.3.1.1 ^{7}Be/^{210}Pb$_{xs}$ as Tracer to Identify Sediment Sources and Age

If we assume that the mixtures of suspended sediment in an estuarine system can come from three potential sources: (i) the river system bringing riverine suspended sediment with a finite ^{7}Be/^{210}Pb ratio; (ii) suspended sediments brought from the coastal and marine areas; and (iii) sediments from the estuary. Wilson et al. (2007) measured the A^p_{Be}/A^p_{Pb} ratio in each source materials and observed that the riverine ^{7}Be/^{210}Pb$_{xs}$ ratio was greater than the estuarine sediment ^{7}Be/^{210}Pb$_{xs}$. The A^p_{Be}/A^p_{Pb} ratio in resuspended sediments collected in a sediment trap (subscript S) is given by Wilson et al. 2007:

$$(A^p_{Be}/A^p_{Pb})_S = (A^p_{Be}/A^p_{Pb})_B \times c + (A^p_{Be}/A^p_{Pb})_D \times (1-c), \quad (16.1)$$

where c is the fraction of the resuspension of bed sediments in the efflux, subscript B represents the resuspended bottom sediments and subscript D represents the combined ratio of the recently delivered sediment from both the riverine and adjacent marine system. All of the other terms and symbols used in this equation and the rest of the equations in this article are defined in Appendix. If the suspended sediments in the coastal area and river have the same ^{7}Be/^{210}Pb$_{xs}$ ratio, then both terms can be combined, which results in (16.1). Using (16.1), the efflux of sediments from the bed, c, can be determined.

Furthermore, A^p_{Be}/A^p_{Pb} ratios can also be used as tracers of the "freshness" or "age" of suspended sediments in an estuary. If it is assumed that the suspended sediments in the estuarine water column is from a single riverine source, the decreases in the A^p_{Be}/A^p_{Pb} ratio could arise because of two factors: low A^p_{Be}/A^p_{Pb} ratio due to aging of the sedimentary particulate matter or reduction in this ratio from the dilution of ^{7}Be-rich sediment by ^{7}Be-deficient but not ^{210}Pb$_{xs}$-dead source. Sometimes, both of these cases can exist (Matisoff et al. 2005).

In the first case, it is possible to calculate the age of the sediment from the ^{7}Be/^{210}Pb$_{xs}$ ratio if the value of that ratio in the precipitation is known (Matisoff et al. 2005; Matisoff and Whiting 2011):

$$\frac{A^p_{Be}}{A^p_{Pb}} = \left(\frac{A^p_{Be}}{A^p_{Pb}}\right)_0 \exp[(-(\lambda_{Be} - \lambda_{Pb}) \cdot t] \quad (16.2)$$

where subscript "0" represents the "fresh" and/or original suspended sediment. The age (t) of the sediment is given by

$$t = \frac{1}{\lambda_{Be} - \lambda_{Pb}} \left[\ln\left(\frac{A^p_{Be}}{A^p_{Pb}}\right)_0 - \ln\left(\frac{A^p_{Be}}{A^p_{Pb}}\right) \right]. \quad (16.3)$$

If "new" sediment is defined here as sediment particles that have a A^p_{Be}/A^p_{Pb} ratio equal to that of the precipitation, i.e., "newly tagged sediment", the sediment particle age is the time since the ^{7}Be tag was delivered. Most likely, however, soil particles eroded from the landscape and delivered into suspension will have a A^p_{Be}/A^p_{Pb} ratio that is less than that in the precipitation because of very low ^{7}Be/^{210}Pb ratios in soils (due to differences in the mean-lives of these nuclides). The watershed residence times of ^{7}Be and ^{210}Pb are much longer than their mean lives (e.g., Ravichandran et al. 1995), and hence, it is likely that A_{Be}^p/A_{Pb}^p ratio in a river is primarily controlled by the ^{7}Be/^{210}Pb ratio in the precipitation. Equation (16.3) is a simplified case. In the real world, there could be multiple sources of sediments with distinctly different A_{Be}^p/A_{Pb}^p ratios, such as fine particulate matter derived from the offshore region, resuspension of sediments with different activity ratios at the sampling sites, and different activity ratios in the bed load and suspended load of the river, etc. Alternatively, the A_{Be}^p/A_{Pb}^p ratio in suspended sediment can also decrease by the addition of ^{7}Be-deficient sediment from deep soil erosion or the entrainment of old (i.e., ^{7}Be-dead) bottom sediments. In this case, the percentage of new sediment in suspension is directly proportional to the A_{Be}^p/A_{Pb}^p ratio (Matisoff et al. 2005):

$$\text{\%'new' sediment} = 100 \times \left(\frac{A_{Be}^p}{A_{Pb}^p}\right) \Big/ \left(\frac{A_{Be}^p}{A_{Pb}^p}\right)_0 \quad (16.4)$$

The percentage of new sediment can also be determined from the age of the sediment:

$$\text{\%'new' sediment} = 100 \times \exp(-(\lambda_{Be} - \lambda_{Pb})t) \quad (16.5)$$

16.3.1.2 ^{234}Th$_{xs}$/^{7}Be as a Tracer of Particle Dynamics and Sediment Transport in a Partially Mixed Estuary

Using the mass-balance approach for Th and ^{7}Be proposed by Santschi et al. (1979), Feng et al. (1999a, b) elaborated this approach to determine source and dynamics within the turbidity maximum of the Hudson Estuary using the ^{234}Th$_{xs}$/^{7}Be ratio. In this formulation, the advective transport of particles through the estuary is assumed to be negligible, and transport of particulate ^{234}Th$_{xs}$/^{7}Be owing to temporal variations of the river flow, tidal mixing and resuspension are also not included in the approach.

In Feng et al. (1999a), local mass balances for dissolved and particulate ^{234}Th can be written as below (definitions of each terms are given in Appendix):

$$\frac{\partial A_{Th}^d}{\partial t} = \lambda_{Th} \times A_U - \lambda_{Th} \times A_{Th}^d - k_{Th} \times A_{Th}^d \quad (16.6)$$

and

$$\frac{\partial A_{Th}^p}{\partial t} = k_{Th} \times A_{Th}^d - \lambda_{Th} \times A_{Th}^p - \kappa_{Th}^p \times A_{Th}^p$$
$$+ \frac{\phi_{Th}^{resus}}{h} \times A_{Th}^{sed} \quad (16.7)$$

Under steady-state conditions for the dissolved and particulate ^{234}Th, $\partial A_{Th}^d/\partial t = \partial A_{Th}^p/\partial t = 0$.

$$\lambda_{Th} \times A_U - \lambda_{Th} \times A_{Th}^d - k_{Th} \times A_{Th}^d = 0 \quad (16.8)$$

and

$$k_{Th} \times A_{Th}^d - \lambda_{Th} \times A_{Th}^p - \kappa^p \times A_{Th}^p + \frac{\phi_{Th}^{resus}}{h} \times A_{Th}^{sed} = 0 \quad (16.9)$$

From (16.8) and (16.9), particulate ^{234}Th can be expressed as

$$A_{Th}^p = \frac{k_{Th} \times \left(\frac{\lambda_{Th} \times A_U}{\lambda_{Th} + k_{Th}}\right) + \frac{\phi_{Th}^{resus}}{h} \times A_{Th}^{sed}}{\lambda_{Th} + \kappa_{Th}^p} \quad (16.10)$$

Similar to ^{234}Th, one can set up the mass balance for dissolved and particulate ^{7}Be. Assuming steady-state conditions and re-arranging various terms, the particulate ^{7}Be can be calculated from the following formula:

$$A_{Be}^p = \frac{k_{Be} \times \left(\frac{I_{Be}}{h(\lambda_{Be} + k_{Be})}\right) + \frac{\phi_{Be}^{resus}}{h} \times A_{Be}^{sed}}{\lambda_{Be} + \kappa_{Be}^p} \quad (16.11)$$

Combining the particulate ^{234}Th and ^{7}Be activities, the particulate ^{234}Th/^{7}Be ratio can be obtained from (16.10) and (16.11):

$$\frac{A_{Th}^P}{A_{Be}^P} = \left(\frac{\lambda_{Be} + \kappa^p}{\lambda_{Th} + \kappa^p}\right)$$
$$\times \left[\frac{\left(\frac{k_{Th}}{\lambda_{Th} + k_{Th}}\right) \times \lambda_{Th} \times A_U + \frac{\phi_{Th}^{resus}}{h} \times A_{Th}^{sed}}{\left(\frac{k_{Be}}{\lambda_{Be} + k_{Be}}\right) \times \frac{I_{Be}}{h} + \frac{\phi_{Be}^{resus}}{h} \times A_{Be}^{Sed}}\right]$$
$$(16.12)$$

Generally, the rather short residence time of Th and Be in the estuaries imply that the scavenging of ^{234}Th and ^{7}Be onto particles is more rapid than radioactive decay ($k_{Th} > \lambda_{Th}$ and $k_{Be} > \lambda_{Be}$). Assuming that the resuspension fluxes of Th and Be are negligible relative to ^{234}Th production and ^{7}Be atmospheric input, we can evaluate (16.12) under two special cases, depending on the residence time of particles in the water column.

In case one, the residence time of particles in the water column is short relative to the half-lives of ^{234}Th and ^{7}Be (i.e., $k_P > \lambda_{Th}$ and $k_P > \lambda_{Be}$). Then (16.12) becomes

$$\frac{A_{Th}^P}{A_{Be}^P} = \frac{\lambda_{Th} \times A_U}{\frac{I_{Be}}{h}} \quad (16.13)$$

Equation (16.13) states that the ^{234}Th/^{7}Be ratio in the particulate matter approaches the ratio of their supply rates if there is rapid scavenging of Th and Be from the dissolved phase with a negligible resuspension contribution from the bottom sediments.

In case two, the residence time of particles is relatively long compared to the half-lives (i.e., $k_P \ll \lambda_{Th}$, $k_P \ll \lambda_{Be}$), and the ^{234}Th$_{xs}$/^{7}Be approaches the inventory ratio. Then (16.12) becomes

$$\frac{A_{Th}^P}{A_{Be}^P} = \frac{\lambda_{Be} \times A_U}{\frac{I_{Be}}{h}} \quad (16.14)$$

In most of the cases, the A_U increases linearly with salinity in estuaries because the ^{238}U concentration in seawater is significantly higher than the average value of ^{238}U in the world rivers.

In the plots of ^{34}Th$_{xs}$/^{7}Be vs. salinity (Fig. 16.3), Feng et al. (1999a) determined both case one (16.13) and case two (16.14) conditions can be dominant with the seasonal differences in the Hudson Estuary.

16.3.1.3 Quantifying the Advection and Resuspension of Particulate Matter in the Turbidity Maximum Zone Using ^{234}Th$_{xs}$/^{7}Be as a Tracer

The ^{234}Th$_{xs}$/^{7}Be ratios can also be applied to quantify the sources of suspended particles in the turbidity maximum of an estuary. In estuaries flood tides result in the transport of suspended particulate matter. This advectional component can be quantified using change in ^{234}Th$_{xs}$/^{7}Be ratios (Feng et al. 1999b). The total fluxes of ^{234}Th and ^{7}Be in the turbidity maximum are given by (definitions of each terms are given in Appendix):

$$\varphi_{Th}^{total} = \varphi_{Th}^{prod} + \varphi_{Th}^{resus} + \varphi_{Th}^{adv} \quad (16.15)$$

and

$$\varphi_{Be}^{toal} = \varphi_{Be}^{prod} + \varphi_{Be}^{resus} + \varphi_{Be}^{adv} \quad (16.16)$$

It is assumed that the observed variations in SPM concentrations result both from local bottom resuspension and advective transport and that the observed variation in the SPM flux is the sum of the following:

$$\phi_{part}^{total} = \phi_{part}^{resus} + \phi_{part}^{adv} \quad (16.17)$$

As discussed earlier, ϕ_{part}^{total} can be calculated from the slope of the increasing portion of the SPM variation with time, and an average value can be used.

$$\phi_{part}^{resus} = \phi_{part}^{total} - \phi_{part}^{adv} \quad (16.18)$$

Then, combining (16.15) and (16.16), we get

$$\frac{\varphi_{Th}^{toal}}{\varphi_{Be}^{toal}} = \frac{\varphi_{Th}^{prod} + \varphi_{Th}^{resus} + \varphi_{Th}^{adv}}{\varphi_{Be}^{prod} + \varphi_{Be}^{resus} + \varphi_{Be}^{adv}} \quad (16.19)$$

Combining (16.17), (16.18) and (16.19), one gets

$$\frac{\varphi_{Th}^{toal}}{\varphi_{Be}^{Itotal}} = \frac{\lambda_{Th} A_U + \left(\frac{A_{Th}^{sed}}{A_{Be}^{sed}}\right) \times (\phi_{part}^{total} - \phi_{part}^{adv}) \times A_{Be}^{sed} + \varphi_{Th}^{prod} + \left(\frac{A_{Th}^{adv}}{A_{Be}^{adv}}\right) \times \phi_{part}^{adv} \times A_{Be}^{adv}}{\frac{I_{Be}}{h} + (\phi_{part}^{total} - \phi_{part}^{adv}) \times A_{Be}^{sed} + \phi_{part}^{adv} \times A_{Be}^{adv}} \quad (16.20)$$

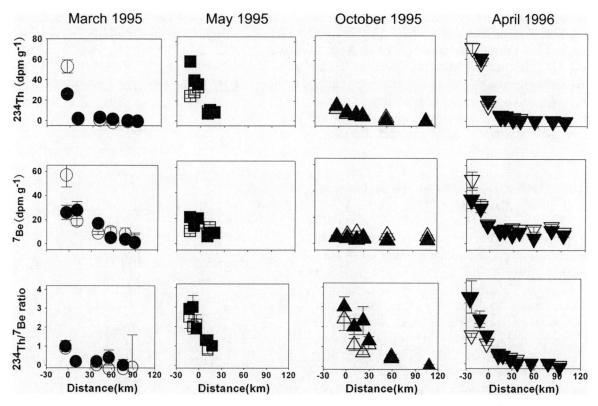

Fig. 16.3 Particulate ^{234}Th and ^{7}Be specific activities and ^{234}Th$_{xs}$/^{7}Be activity ratios vs. distance in the Hudson River estuary. Surface and bottom water samples were collected in March, 1995 (circle), May, 1995 (square), October, 1995 (triangle) and April, 1996 (inverted triangle), respectively. The *open symbols* represent surface water and the *filled symbols* represent bottom water (data plotted from Feng et al. (1999a))

The key parameters for (16.20) are the ^{234}Th/^{7}Be flux ratio in the turbidity maximum zone ($\phi_{Th}^{total}/\phi_{Be}^{total}$) and the ^{234}Th$_{xs}$/^{7}Be activity ratios in the resuspended ($A_{Th}^{sed}/A_{Be}^{sed}$) and advected ($A_{Th}^{adv}/A_{Be}^{adv}$) sediments.

In the turbidity maximum zone, ($A_{Th}^{adv}/A_{Be}^{adv}$) is taken to be the average activity ratio measured from particles collected at the boundary near the estuary. $A_{Th}^{sed}/A_{Be}^{sed}$ can be obtained from the ratio of the average excess ^{234}Th and ^{7}Be activities in the bottom sediments. Bottom sediments are initially considered as representative of locally resuspended sediment, and ^{234}Th and ^{7}Be activities are used in the 0–0.5 cm depth interval of these sediments to estimate the average ^{234}Th$_{xs}$/^{7}Be activity ratio in resuspended particles (Feng et al. 1999b). Then the parameter ϕ_{part}^{adv} can be calculated using (16.20). As an example of the application of this model, Feng et al. (1999b) estimated that the lateral advection from the high salinity region corresponded to 31% of the particles, while the remaining 69% was attributed to the local resuspension from bottom sediments.

16.3.2 Sediment Focusing and Erosion Using ^{210}Pb, ^{7}Be and ^{234}Th as Tracers

Excess ^{234}Th can be used as a tracer of sediment sources from offshore and higher salinity waters in contrast to ^{7}Be, which is supplied to the estuary at a higher rate from the discharge of riverine water. Thus, the combined use of ^{234}Th and ^{7}Be can help to distinguish sediment sources from the supply of new material from the rivers (i.e., high ^{7}Be) and resuspension and/or mixed sediments from the surface seabed (i.e., high ^{234}Th) (Corbett et al. 2004). In shallow estuaries, where the seabed is disturbed by tidal and wind-driven resuspension, the ^{234}Th profiles and inventories within the seabed can yield valuable insights about the frequency and intensity of resuspension (Giffin and Corbett, 2003). Moreover, over longer time scales, the sediment focusing and erosion can be investigated using the inventories of ^{210}Pb in the water column and bottom sediments (Cochran et al. 1990; Thomson et al. 1993; Baskaran and Santschi, 2002).

16.3.2.1 The Residence Times and Scavenging Effectiveness for ^{210}Pb

In the box-model (Fig. 16.2), the major inputs (^{210}Pb) to the coastal environment are included of: (i) atmospheric fallout (I_{Pb}) and (ii) the production from its grandparent, ^{226}Ra ($\lambda_{Pb}A_{Ra}$). The major outputs from the box are (i) the decay of ^{210}Pb ($\lambda_{Pb}A_{Pb}$) and (ii) its removal by scavenging ($\kappa_{Pb}A_{Pb}$). In this simple approach, additional sources/sinks (such as lateral advection from and to shelf, diffusion, etc.) have been assumed to be negligible. Evidence for this assumption comes from the measured sediment inventory of ^{210}Pb in several sediment cores. The measured inventory in a large portion of the sediment cores is about the same as that expected from the production rate in the overlying water column, implying that there is no net lateral advectional input to the site or net lateral input is offset by sediment erosion (Baskaran and Santschi 2002).

The mass balance of total ^{210}Pb in the water column can be written as (definitions of each terms are given in Appendix):

$$\lambda_{Pb}A_{Ra} + I_{Pb}/h = \lambda_{Pb}A_{Pb} + \kappa_{Pb}A_{Pb} \quad (16.21)$$

From (16.21), the residence times ($\tau_{Pb} = 1/\kappa_{Pb}$ for total ^{210}Pb) with regards to scavenging can be calculated as follows:

$$\tau_{Pb} = 1/\kappa_{Pb}$$
$$= A_{Pb}/[\lambda_{Pb}(A_{Ra} + (I_{Pb}/(h\lambda_{Pb})) - A_{Pb})] \quad (16.22)$$

To determine the residence time of ^{210}Pb in the entire water column, (16.22) can be rewritten in terms of the inventory of ^{210}Pb as

$$\tau_{Pb} = A_{Pb}\lambda_{Pb}/[\lambda_{Pb}A_{Ra} + (I_{Pb}/h) - \lambda_{Pb}A_{Pb}] \quad (16.23)$$

The term $[\lambda_{Pb}A_{Ra} + (I_{Pb}/h)]$ is the total supply of ^{210}Pb. From the mass balance considerations, one can have

Total supply $[(A_{Ra}h + (I_{Pb}/\lambda_{Pb})]$
 − Measured inventory $[hA_{Pb}\lambda_{Pb}]$
 = Deficiency $[\lambda_{Pb}hA_{Ra} + (I_{Pb}) - \lambda_{Pb}hA_{Pb}]$.
$$(16.24)$$

Equation (16.23) thus reduces to

$$\tau_c = (\text{Total supply} - \text{Deficiency})/(\lambda_{Pb} \times \text{Deficiency}) \quad (16.25)$$

The scavenging effectiveness (SE) is defined as (Cochran et al. 1990; Thomson et al. 1993; Baskaran and Santschi, 2002) follows:

$$SE = Deficency/Total\,supply \quad (16.26)$$

$$\tau_c = [(1/SE - 1) \times 1/\lambda_{Pb}]. \quad (16.27)$$

Using (16.26) and (16.27), the residence times and SE for ^{210}Pb for the entire water column can be determined.

16.3.2.2 ^{234}Th as a Tracer of the Fate of Particle-Reactive Species

Using a two-dimensional model, McKee et al. (1984) calculated the residence times of ^{234}Th in the coastal waters. The supply and removal of ^{234}Th are assumed to be in steady state, and the ^{234}Th activity measured in surface waters is assumed to be representative of the average water-column ^{234}Th activity at that station during the ~100 days (i.e., 4–5 times of ^{234}Th's half life) prior to collection. ^{234}Th sediment inventories reveal that the supply and removal of ^{234}Th are two-dimensional. This result is consistent with the idea of a well-mixed water column during winter. Vertical mixing tends to homogenize (to some extent) the concentrations of suspended sediments and the activities on suspended sediment within the water column. Therefore, the ^{234}Th values measured in surface waters are a reasonable indication of the entire water column ^{234}Th activity during this time of year. Meanwhile, it is assumed that the adsorption of ^{234}Th is irreversible (negligible desorption) because of the short time scales for scavenging and removal in near-shore environments. In addition, the horizontal ^{234}Th flux is modeled as an advective transport (multiplying the first derivative of ^{234}Th, with respect to x, by an advection coefficient). The horizontal flux can also be modeled as diffusive transport. The governing equations for this model are shown in Table 16.2.

To implement this model, measurements of ^{234}Th were made in surface waters and in the seabed at each station. v and k_{Th} can be evaluated using (16.28–16.30). In the case of the Changjiang Estuary, the residence times of particle-active species from near-shore to offshore areas are around 0.3–4 days with respect to scavenging, and 0.5–11.0 days with respect to the removal to the seabed (McKee et al. 1984). This approach can be extended to other particle-reactive contaminant species that have K_d values similar to that of ^{234}Th.

16.3.3 Tracing the Pathways of ^{7}Be and ^{210}Pb Following a Heavy Thunderstorm

During rainstorm events, large amounts of ^{210}Pb and ^{7}Be, along with increase in the river discharge, can be flushed into coastal areas. It was reported that most of the ^{7}Be (74–86% of the total) as well as ^{210}Pb (80–89%) were associated with particles even less than an hour after the pulse rain input stopped injecting ^{7}Be and ^{210}Pb into the water. Initial ^{7}Be and ^{210}Pb concentrations per gram of suspended matter were high when the particle concentration was at a maximum (Fig. 16.4). Meanwhile, there is extensive particle resuspension because of strong winds and high tides. Sommerfield et al. (1999) introduced the application of ^{7}Be as a tracer of food sedimentation. At steady state, and assuming no lateral transport, the riverine flux and atmospheric deposition flux are balanced by radioactive decay in the water column (λA_{Be}) and by removal to the sediments (φ_{Be}), i.e., the mass balance of ^{7}Be in coastal region is expressed as follows:

$$\varphi_{Be}^{river} + I_{Be}/h = \varphi_{Be} + \lambda_{Be} A_{Be} \quad (16.31)$$

Using (16.31), two cases are examined. First, the presence of ^{7}Be in shelf sediments is expected to undergo rapid removal from the water column through scavenging and deposition prior to radioactive decay (i.e., $\varphi_{Be} \gg \lambda A_{Be}$) and/or loss through advective transport (Fig. 16.2). This process is likely facilitated by a lateral (i.e., riverine influx) source of ^{7}Be and the rapid deposition of ^{7}Be-laden sediment during floods. In the second case, when rainfall and river runoff are negligible, radioactive decay of ^{7}Be in the water column (e.g., supported by dry fallout from atmosphere) will generally exceed the removal to the seabed (i.e., $\varphi_{Be} \ll \lambda A_{Be}$) due to the much longer residence time of ^{7}Be in the water column.

16.3.4 ^{7}Be and ^{210}Pb as Tracers of Particle Dynamics in the River

By modeling the distributions of particulate ^{7}Be and ^{210}Pb$_{xs}$, the sediment resuspension rate can be quantified. The mass balance equations for the particulate ^{7}Be and ^{210}Pb$_{xs}$ can be written as follows (Jweda et al. 2008; definitions of each terms are given in Appendix):

$$k_{Be} A_{Be}^d + \varphi_{Be}^{river} + \phi_{Be}^{resus} A_{Be}^{resus}/h \\ = \varphi_{Be}^{out} + \lambda_{Be} A_{Be}^p + \phi_{Be}^{resus} A_{Be}^{trap}/h \quad (16.32)$$

Table 16.2 The governing equations for this model (x is the horizontal dimension of the box (in cm); the other horizontal dimension is assumed to have unit length; "in"/"out" are the input/output of ^{234}Th in the box)

Advection-input + Production = Advection-output + Decay + Removal
Total
$hvA_{Th}^{in} + hx\lambda A_U^t = hvA_{Th}^{out} + hx\lambda A_{Th} + hx\varphi_{Th}$ (16.28)
Dissolved
$hvA_{Th}^{d-in} + hx\lambda A_U^d = hvA_{Th}^d + hx\lambda A_{Th}^d + hxk_{Th} A_{Th}^d$ (16.29)
Particulate
$hvA_2^{p-in} + hx(\lambda A_U^p + k_{Th} A_{Th}^d) = hvA_{Th}^{p-out} + hx\lambda A_{Th}^p + hx\varphi_{Th}$ (16.30)

Fig. 16.4 Particulate $^{210}Pb_{ex}$, ^{7}Be and particle concentrations plotted against time after a heavy pulsed rain input into Galveston (Texas) coastal water (data taken from Baskaran and Santschi 1993)

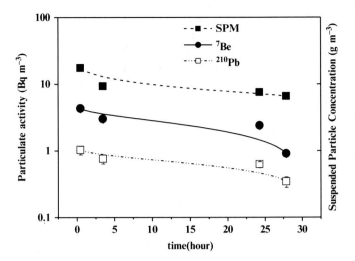

and

$$k_{Pb}A^d_{Pb} + \lambda_{Pb}A^p_{Rn} + \varphi^{river}_{Pb} + \phi^{resus}_{Pb}A^{resus}_{Pb}/h$$
$$= \varphi^{river}_{Pb} + \lambda_{Pb}A^p_{Pb} + \phi^{resus}_{Pb}A^{trap}_{Pb}/h \quad (16.33)$$

where A^p_{Rn} is the activity of ^{222}Rn (Bq cm^{-3}) adsorbed onto particulate matter.

The activities of the resuspension material are considered to be those of the nuclides in the upper 1 cm of the bottom sediments because this upper layer is likely the most frequently resuspended. The importance of the $\lambda_{Pb}A^p_{Rn}$ term in (16.39) is evaluated by Jweda et al. (2008).

Assuming that the ^{7}Be and $^{210}Pb_{xs}$ activities of the upper layer of the bottom sediment are equal to those of the resuspended sediment, $A^{resus}_{Be} = A^{sed}_{Be}$ and $A^{resus}_{Pb} = A^{sed}_{Pb}$, then the mass balance equations for particulate ^{7}Be and $^{210}Pb_{xs}$ yield sediment resuspension rates (g cm^{-2} day^{-1}) using ^{7}Be and $^{210}Pb_{xs}$, respectively:

$$\phi^{resus}_{Be} = h \times \frac{k_{Be}A^d_{Be} - \lambda_{Be}A^p_{Be}}{A^{trap}_{Be} - A^{resus}_{Be}} \quad (16.34)$$

and

$$\phi^{resus}_{Pb} = h \times \frac{k_{Pb}A^d_{Pb} - \lambda_{Pb}A^p_{Pb}}{A^{trap}_{Pb} - A^{resus}_{Pb}}. \quad (16.35)$$

As an example, Jweda et al. (2008) reported that the sediment resuspension rates in a river system in southeast Michigan calculated using ^{7}Be and $^{210}Pb_{xs}$ activities to vary between 0.50 and 1.34 g cm^{-2} year^{-1} with average of 0.83 ± 0.34 g cm^{-2} year^{-1} (n = 10) and 0.16 to 1.49 g cm^{-2} year^{-1} with average of 0.38 ± 0.38 g cm^{-2} year^{-1} (n =14), respectively.

16.3.5 ^{137}Cs and $^{210}Po/^{210}Pb$ as Tracers of Historical Records in Sediment Cores

The activity $A^{sed}_{Pb}(0)$ of ^{210}Pb in freshly deposited surface sediments can be expressed by the following equation:

$$A^{sed}_{Pb}(0) = \varphi^{sed}_{Pb}(0)/\phi^{sed}(0) \quad (16.36)$$

where $\varphi^{sed}_{Pb}(0)$ is the sedimentation flux of ^{210}Pb (Bq cm^{-2} year^{-1}), and $\phi^{sed}(0)$ is the mass accumulation flux of sediment (g cm^{-2} year^{-1}). It is assumed that the there is a constant mass and ^{210}Pb flux to the sediment. The activity of $A^{sed}_{Pb}(t)$ at any depth in the core (H) corresponding to any time can be written as a Constant Accumulation (CA) model (Appleby and Oldfield 1978):

$$A^{sed}_{Pb}(t) = \varphi^{sed}_{Pb}(t)\exp(-\lambda_{Pb}t)/\phi^{sed}(t)$$
$$= A^{sed}_{Pb}(0)(\exp(-\lambda_{Pb}t)) \quad (16.37)$$

In the CA model, the sedimentary accumulation rate (v, cm year^{-1}) can be written as

$$v = -\lambda_{Pb}/k_{sl} \quad (16.38)$$

where k_{sl} is the slope in the plot of $\ln A_{Pb}^{sed}(t)$ vs. the core depth (in cm). Then the sediment core age (t, year) corresponding to any depth "H" depth (cm) is as follows:

$$t = \frac{H}{v} = -\frac{H \cdot k_{sl}}{\lambda_{Pb}} \quad (16.39)$$

If there is no physical and/or biological mixing, then the plot of $\ln A_{Pb}^{sed}(t)$ vs. the core depth should show a linear relationship. However, biological and/or physical mixing is very common in riverine and coastal sediments. In such cases, the accumulation rate can be expressed by

$$v = -\lambda_{Pb}/k_{sl} + k_{sl}D, \quad (16.40)$$

where D is the biological mixing coefficient (cm^2 year^{-1}). Comparing (16.40) to (16.38), it is seen that the deposition rate will be over-estimated if the biological mixing is significant but is ignored. A combination of ^{210}Pb and ^{239,240}Pu or ^{137}Cs can be used to delineate the sediment mixing and accumulation rates (e.g., Zuo et al. 1990).

^{137}Cs has also been used to date sediment cores. The maximum activity of ^{137}Cs in sediment cores is generally observed in the layer of sediment corresponding to 1963 because of the peak nuclear weapon testing were conducted in early 1960s (i.e.,1962–1963; Hong et al. 2011). Hirose et al. (2008) reported the atmospheric deposition flux of ^{137}Cs in Tsukuba (Japan) during the period 1957–2005 (Fig. 16.5). Two peaks were observed corresponding to calendar years 1963 and 1986. If the peak activity of ^{137}Cs is observed in a layer at the core depth H, the average sedimentation rate v from 1963 to the year the sediment core was collected (y) (y > 1963) for sampling can be evaluated as

$$v = \frac{H}{y - 1963} \quad (16.41)$$

It is pertinent to point out two-thirds of the peak activity deposited in 1963 has decayed away by 2010, leaving only one-third of the peak activity. Furthermore, diffusion of ^{137}Cs has led to smearing of this peak, making it more challenging to date sediment cores with this one-layer dating method. In April 1986, the Chernobyl nuclear accident made a secondary peak of ^{137}Cs activity in sediment cores from some regions (e.g., Europe). However, while this activity was very dominant in the Scandinavian Arctic, including countries around Baltic Sea, the Chernobyl-derived fallout was negligible in the Alaskan Arctic and the contiguous United States. In any case, if there is a Chernobyl-derived radiocesium peak (Fig. 16.4), then, the average deposition rate v from 1986 to the calendar year (y > 1986) for sampling can be similarly expressed as

$$v = \frac{H}{y - 1986} \quad (16.42)$$

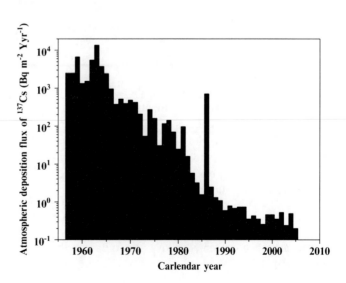

Fig. 16.5 Annual ^{137}Cs deposition (Bq m^{-2}) observed at Tsukuba (Japan) during the period 1957–2005 (data plotted from Hirose et al. (2008))

16.3.6 Case Studies: Utilizing ^{7}Be as a Tracer to Quantify the Redistribution of PCBs and Other Contaminants

The unique properties of ^{7}Be, such as particle-reactivity, a relatively short half-life and easily measurable rate of delivery to the surface of the Earth, make this isotope a powerful tracer to quantify seasonal variations and episodic pulses of sediment delivery such as hurricanes, major flood events, etc. in coastal areas (Canuel et al. 1990; Sommerfield et al. 1999; Allison et al. 2000). Furthermore, the K_d of ^{7}Be ($\sim10^4$–10^5) is comparable to most hydrophobic contaminants, including PCBs, and hence, ^{7}Be can be used as a surrogate tracer to quantify the rates of processes that affect the cycling of these contaminants (Fitzgerald et al. 2001). As discussed earlier, once ^{7}Be is introduced into the water column, it is removed quickly onto surface of particles. More than 70% of the ^{7}Be introduced into the coastal water column by direct atmospheric deposition was removed from the water column in less than 1 day (Baskaran and Santschi 1993), and hence, this tracer is useful for quantifying particle resuspension and deposition. If typically upper 10–15 cm of sediment cores is collected at two different times, T_1 and T_2, and the inventories of ^{7}Be at T_1 and T_2 are I_1 and I_2, respectively, then the residual inventory (I_r) before the second sampling is given by the following equation:

$$I_r = I_1 \exp(-\lambda t) = I_2 - I_d + I_{fe} \quad (16.43)$$

where t is the time elapsed between successive samplings, I_d (Bq cm^{-2}) is the depositional flux between T_1 and T_2 (measured flux) and I_{fe} (Bq cm^{-2}) is the inventory resulting from sediment focusing (+ve) or erosion (−ve). If $I_2 > (I_1 \exp(-\lambda t) + I_d)$, then it represents newly deposited ^{7}Be-laden sediments. When $I_2 < (I_1 \exp(-\lambda t) + I_d)$, then there is sediment resuspension and removal from that site. The term $I_d + I_{fe}$ corresponds to the new inventory, and the sediment accumulation rate (ϕ^{sed}, g cm^{-2} day^{-1}) can be calculated using the terms as follows:

$$\phi^{sed} = (I_d + I_{fe})/A_{Be}^{sed}/(1.44(\ln(2)/\lambda_{Be}) \quad (16.44)$$

The approach presented above is similar to the one proposed by Canuel et al. (1990). The standard technique for calculating the mass accumulation rate from the slope of the best-fit line of the semi-log plot of ^{7}Be activity plotted against mass depth assumes that the flux (both mass and ^{7}Be) is constant, which may be questionable in sediments derived from storm and flood events (Canuel et al. 1990; Sommerfield et al. 1999). It is important to note that the presence of ^{7}Be indicates sediments deposited in less than ~6 months, and hence, this approach cannot be extended for sediments beyond approximately 1 year.

From the I_{fe} calculated as described above and the mean activity of ^{7}Be in the surficial sediments (A_{Be}^{sed}), one can calculate the depositional or erosional flux (ϕ^{er}, g cm^{-2} day^{-1}) as below

$$\phi^{sed}(\phi^{er}) = (I_d + I_{fe})/A_{Be}^{sed}/(1.44(\ln(2)/\lambda_{Be}) \quad (16.45)$$

If C_{PCB} (µg g^{-1} dry wt) is the concentration of PCB in the surficial layer, then the depositional or erosional flux of PCB (ϕ_{PCB}, µg cm^{-2} day^{-1}) due to resuspension of sediments during episodic events (e.g., storm, flood and other events) can be calculated as

$$\phi_{PCB} = \phi^{sed}(\phi^{er}) \times C_{PCB}) \quad (16.46)$$

Knowing the area of the contaminated region, one can calculate the total loading of PCB from the bottom into the water column. Utilizing this approach, Fitzgerald et al. (2001) estimated the total reintroduction of PCBs from sediments to the water column over a period of 40 days to be about 10 kg, which can be compared to the estimated annual flux of PCBs from the Fox River to the Green Bay of 200 kg year^{-1}. This approach can be applied to other contaminants (i.e., both organic and inorganic) that have K_d values similar to that of ^{7}Be (e.g., Hg, Pb, etc). This approach can also be utilized to quantify the amounts of contaminated sediment resuspension during freight movements in relatively shallow water systems.

Moreover, similar to the application of ^{7}Be, ^{234}Th is one of the powerful particulate nuclides to trace the PCB and other contaminant transport in aquatic environment. As mentioned above, combination of (16.6) and (16.7), the activity of ^{234}Th in the surface marine water can be a balance between its formation via radioactive decay of ^{238}U, its radioactive decay and removal by particulate matter.

$$\frac{\partial A_{Th}}{\partial t} = \lambda_{Th} \times A_U - \lambda_{Th} \times A_{Th} - \kappa^p_{Th} \times A^p_{Th} \quad (16.47)$$

Based on the assumption of steady state, (16.47) can be rewritten as:

$$\kappa^p_{Th} = \lambda_{Th}(A_U - A_{Th})/A^p_{Th} \quad (16.48)$$

If the concentration of particulate pollutants (e.g., PCBs/PAHs) (mol m^{-3}) in the surface water is known, then, averaged fluxes for PCBs/PAHs can be calculated as follows (definitions of each terms are given in Appendix):

$$\phi_{PCB/PAH} = \kappa^p_{Th} \cdot C_{PCB/PAH} \quad (16.49)$$

Using above equation, Gustafsson et al. (1997a, b) estimated that the vertical fluxes of PCB/PAH in the northwestern Atlantic Ocean in 1993–1994 (Figs. 16.6 and 16.7). It was found that fluxes both of PCB and PAH decreased with distance from coast. But, considering the areal extent, there are only 5% total PCB in the 20-km-wide coastal region closest to the continental source and 20% on the continental shelf. It means that most of PCB would be transported to open sea via lateral particulate matter advection. Similarly, only around 10% of PAH was deposited in coastal waters while the remaining 90% was exported farther offshore on the continental shelf and to the open sea.

Definitely, pollutants with similar properties to PCBs/PAHs can also be evaluated using this approach.

16.4 Future Directions

Although many authors have reported the measurement of particle-active radionuclides (^{210}Pb, ^{7}Be and ^{234}Th) as one of the useful methods for tracing the sources, transport and deposition of particles/sediments in estuaries and coasts, there are still many interesting questions that are unanswered and many exciting avenues for future research remain unexplored. These applications to quantify dynamic processes in aquatic environment are still complex because there is no generally accepted model for temporal and spatial variations of these nuclides in different areas. For example, even in the same region, the atmospheric deposition fluxes of ^{7}Be and ^{210}Pb vary spatially and temporally. The watershed erosional input of these nuclides depends on several factors, such as relief, mineralogy, amounts of precipitation, etc. This component is critically needed for constraining the input functions to the coastal and estuarine zones. Colloidal material plays a significant role in the transport of these particle-reactive radionuclides, and their removal rate depends on the complexation capacity of these colloidal materials. The relative importance of colloidal material and suspended particulate material need to be quantified. The behavior of U, whether conservative or non-conservative, plays a key role on the production rates of ^{234}Th in the estuarine and coastal regions, and hence the phases of U that behave as non-conservative components must be determined. For ^{234}Th, there is an

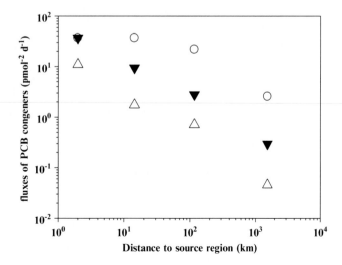

Fig. 16.6 Vertical fluxes into the subsurface ocean of three PCB congeners over coastal to pelagic regimes of the western North Atlantic Ocean (*open circles*, tetrachlorobiphenyl; *filled inverted triangles*, hexachlorobiphenyl; *open triangles*, octachlorobiphenyl) (data plotted from Gustafsson et al. (1997a))

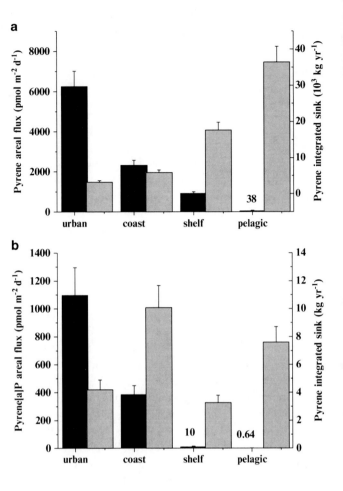

Fig. 16.7 Surface-ocean fluxes and area-integrated mass removal rates of pyrene (**a**) and benzo[a]pyrene (**b**) for four marine regimes in the western North Atlantic Ocean. *Dark bars* represent PAH fluxes and *grey bars* denote areally integrated PAH sinks. The uncertainties represent one standard deviation of the propagated analytical variabilities (data plotted from Gustafsson et al. (1997b))

obvious lack of information about the comparisons of ^{234}Th (and/or ^{238}U) behavior in the world estuaries.

Higher-frequency sampling would greatly improve our understanding of the fate of these nuclides themselves and the material they trace (suspended particle and particle-reactive chemicals) in dynamic estuarine environments. The distribution of these nuclides in size-fractionated particulate matter (both micro- and macro particulate matter) would provide insight into the scavenging behavior of the particulate matter. Characterizing the particulate matter (organic/inorganic/surface area/charge, etc.) will provide much needed information on the processes that control the removal of particle-reactive radionuclides and other particle-reactive contaminants in coastal and estuarine areas.

Acknowledgments This work was funded by the Natural Science Foundation of China (41021064; 40976054) and the Ministry of Science and Technology of PR China (2011CB409801). We thank the two reviewers (Chih-An Huh and Brent A. McKee) for their critical reviews of the earlier version of this manuscript.

Appendix: Descriptions of symbols and abbreviation used in this chapter

Symbols	Meaning	Unit
$A_{Th}/A_{Be}/A_{Pb}$	Activity of excess ^{234}Th/^{7}Be/excess ^{210}Pb in the water column	Bq cm^{-3}
$A_{Th}^p/A_{Be}^p/A_{Pb}^p$	Activity of excess ^{234}Th/^{7}Be/excess ^{210}Pb in the particulate phase	Bq g^{-1} or Bq cm^{-3}
$A_{Th}^d/A_{Be}^d/A_{Pb}^d$	Activity of ^{234}Th/^{7}Be/^{210}Pb in the dissolved phase	Bq cm^{-3}

(*continued*)

Symbols	Meaning	Unit	Symbols	Meaning	Unit
$A_{Th}^{sed}/A_{Be}^{sed}/A_{Pb}^{sed}$	Activity of excess ^{234}Th/^{7}Be/excess ^{210}Pb in surficial bottom sediments subject to resuspension	Bq g^{-1}	$\varphi_{Th}^{adv}/\varphi_{Be}^{adv}/\varphi_{Pb}^{advs}$	Advective flux of particulate excess ^{234}Th and ^{7}Be and excess ^{210}Pb	Bq cm^{-3} day^{-1}
$A_{Th}^{adv}/A_{Be}^{adv}/A_{Pb}^{sdv}$	Activity of excess ^{234}Th/^{7}Be/excess ^{210}Pb on advected particles	Bq g^{-1}	$\varphi_{Th}^{prod}/\varphi_{Be}^{prod}/\varphi_{Pb}^{prod}$	Supply flux of excess ^{234}Th and ^{7}Be and excess ^{210}Pb from ^{238}U decay or the atmosphere (^{7}Be, ^{210}Pb) followed by rapid scavenging fluxes	Bq cm^{-3} day^{-1}
$A_{Th}^{resus}/A_{Be}^{resus}/A_{Pb}^{resus}$	Activity of excess ^{234}Th/^{7}Be/excess ^{210}Pb on resuspension particles from sea bed	Bq g^{-1}	$\varphi_{Th}^{total}/\varphi_{Be}^{total}/\varphi_{Pb}^{total}$	Total supply flux of particulate excess ^{234}Th/^{7}Be/excess ^{210}Pb into turbidity maximum zone	Bq cm^{-3} day^{-1}
$A_{Th}^{trap}/A_{Be}^{trap}/A_{Pb}^{trap}$	Activity of excess ^{234}Th/^{7}Be/excess ^{210}Pb on settling particles collected by sediment trap	Bq g^{-1}	$\phi_{Th}^{resus}/\phi_{Be}^{resus}/\phi_{Pb}^{resus}$	Sediment resuspension flux by excess ^{234}Th and ^{7}Be and excess ^{210}Pb from the sea bed	g cm^{-2} day^{-1}
$\lambda_{Th}/\lambda_{Be}/\lambda_{Pb}$	Decay constant of ^{234}Th/^{7}Be/^{210}Pb	day^{-1},			
$k_{Th}/k_{Be}/k_{Pb}$	Rate constant for the scavenging of dissolved ^{234}Th/^{7}Be/^{210}Pb onto particles	day^{-1},	$\phi_{part}^{resus}/\phi_{part}^{adv}/\phi_{part}^{total}$	Particle flux from resuspension/advective input/total input	g cm^{-3} day^{-1}
$\kappa_{Th}/\kappa_{Be}/\kappa_{Pb}$	Removal rate constant of excess ^{234}Th/^{7}Be/excess ^{210}Pb in the water column	day^{-1}	ϕ^{sed}/ϕ^{er}	Sedimentary deposition/erosion flux	g cm^{-2} day^{-1}
			$\tau_{Pb} = 1/\kappa_{Pb}$	Residence times for total ^{210}Pb in the water column	Day
$\varphi_{Th}/\varphi_{Be}/\varphi_{Pb}$	Removal flux of excess ^{234}Th and ^{7}Be and excess ^{210}Pb from water column	Bq cm^{-3} day^{-1}	v	Horizontal flux constant	cm day^{-1}
			$\varphi_{Be}^{river}/\varphi_{Pb}^{river}$	Riverine input particulate ^{7}Be and excess ^{210}Pb	Bq cm^{-3} day^{-1}
$\varphi_{Th}^{p}/\varphi_{Be}^{p}/\varphi_{Pb}^{p}$	Removal flux of particulate excess ^{234}Th and ^{7}Be and excess ^{210}Pb from water column to the sea bed	Bq cm^{-3} day^{-1}	$\varphi_{Be}^{out}/\varphi_{Pb}^{out}$	Riverine output flux of particulate ^{7}Be and excess ^{210}Pb	Bq cm^{-3} day^{-1}
			$C_{PCB/PAH}$	Concentration of PCB/PAH in mixing layer of water column	g m^{-3}
$\kappa_{Th}^{p}/\kappa_{Be}^{p}/\kappa_{Pb}^{p}$	Removal rate constant of particulate excess ^{234}Th/^{7}Be/excess ^{210}Pb in the water column	day^{-1}	$\phi_{PCB/PAH}$	Vertical flux of PCB/PAH from mixing layer of water column	g cm^{-3} day^{-1}
			AMP	Amaniamolybdenum (III) phosphate	
I_{Be}/I_{Pb}	^{7}Be/^{210}Pb atmospheric deposition flux	Bq cm^{-2} day^{-1}	SPM	Suspended particle matter	
A_U/A_{Ra}	Activity of ^{238}U/^{226}Ra in the water column	Bq cm^{-3}	PCB	Polychlorinated biphenyl	
h	Water depth	cm	PAH	Polycyclic aromatic hydrocarbons	
$\varphi_{Th}^{resus}/\varphi_{Be}^{resus}/\varphi_{Pb}^{resus}$	Resuspension flux of particulate excess ^{234}Th and ^{7}Be and excess ^{210}Pb from the sea bed	Bq cm^{-3} day^{-1}	LSS	Liquid scintillation spectrometry	

(*continued*)

References

Allison MA, Kineke GC, Gordon ES et al (2000) Development and reworking of a seasonal flood deposit on the inner continental shelf off the Atchafalaya River. Cont Shelf Res 20:2267–2294

Allison MA, Sheremet A, Goñic MA (2005) Storm layer deposition on the Mississippi–Atchafalaya subaqueous delta generated by Hurricane Lili in 2002. Cont Shelf Res 25:2213–2232

Appleby P, Oldfield F (1978) The calculation of lead-210 dates assuming a constant rate of supply of unsupported lead-210 to the sediment. Catena 5:1–8

Baskaran M (1995) A search for the seasonal variability on the depositional fluxes of ^{7}Be and ^{210}Pb. J Geophys Res 100:2833–2840

Baskaran M (2011) Po-210 and Pb-210 as atmospheric tracers and global atmospheric Pb 210 fallout: a review. J Environ Radioactiv 102:500–513

Baskaran M, Santschi PH (1993) The role of particles and colloids in the transport of radionuclides in coastal environments of Texas. Mar Chem 43:95–114

Baskaran M, Santschi PH (2002) Particulate and dissolved ^{210}Pb activities in the shelf and slope regions of the Gulf of Mexico waters. Cont Shelf Res 22:1493–1510

Baskaran M, Swarzenski PW (2007) Seasonal variations on the residence times and partitioning of short-lived radionuclides (^{234}Th, ^{7}Be and ^{210}Pb) and depositional fluxes of ^{7}Be and ^{210}Pb in Tampa Bay, Florida. Mar Chem 104:27–42

Baskaran M, Coleman CH, Santschi PH (1993) Atmospheric depositional fluxes of ^{7}Be and ^{210}Pb at Galveston and College Station, Texas. J Geophys Res 98:20255–20571

Baskaran M, Swarzenski PW, Porcelli P (2003) Role of colloidal material in the removal of ^{234}Th in the Canada Basin of the Arctic Ocean. Deep Sea Res I 50:1353–1373

Baskaran M, Hong GH, Santschi PH (2009a) Radionuclide analysis in seawater, Chapter 13. In: Wurl O (ed) Practical guidelines for the analysis of seawater. CRC press, Boca Raton

Baskaran M, Swarzenski PW, Biddanda BA (2009b) Constraints on the utility of MnO$_2$ cartridge method for the extraction of radionuclides: a case study using ^{234}Th. Geochem Geophys Geosyst 10, Q04011, doi:10.1029/2008GC002340

Benitez-Nelson CR, Buesseler KO (1999) Phosphorus-32, phosphorus-33, beryllium-7 and lead-210: atmospheric fluxes and utility in tracing stratosphere/troposphere exchange. J Geophys Res 104:11745–11754

Benitez-Nelson CR, Buesseler KO, Rutgers van der Loeff M et al (2001) Testing a new small-volume technique for determining ^{234}Th in seawater. J Radioanal Nucl Chem 248:795–799

Bhat SG, Krishnaswami S, Lal D et al (1969) ^{234}Th/^{238}U ratios in the ocean. Earth Planet Sci Lett 5:483–491

Bowen VT, Roether W (1973) Vertical distributions of strontium-90, cesium-137 and tritium about 45° north latitude in the Atlantic. J Geophys Res 78:6277–6285

Broecker WS, Kaufman A, Trier RM (1973) The residence time of thorium in surface sea water and its implication regarding the fate of reactive pollutants. Earth Planet Sci Lett 20:35–44

Broecker WS, Patzert WC, Toggweiler JR et al (1986) Hydrography, chemistry and radioisotopes in the Southeast-Asian Basins. J Geophys Res 91:14345–14354

Buesseler KO, Casso SA, Hartman MC, Livingston HD (1990) Determination of fission-products and actinides in the Black Sea following the Chernobyl accident. J Radioanal Nucl Chem 138:33–47

Buesseler KO, Cochran JK, Bacon MP et al (1992) Determination of thorium isotopes in seawater by non-destructive and radiochemical procedures. Deep Sea Res 39:1103–1114

Buesseler KO, Andrews JA, Hartman MC et al (1995) Regional estimates of the export flux of particulate organic carbon derived from thorium-234 during the JGOFS EqPac program. Deep Sea Res II 42:777–804

Buesseler KO, Benitez-Nelson CR, Rutgers van der Loeff M et al (2001a) A comparison of methods with a new small-volume technique for thorium-234 in seawater. Mar Chem 74:15–28

Buesseler KO, Ball L, Andrews J et al (2001b) Upper ocean export of particulate organic carbon in the Arabian Sea derived from thorium-234. Deep Sea Res II 45:2461–2467

Canuel EA, Martens CS, Benninger LK (1990) Seasonal variations in ^{7}Be activity in the sediments of Cape Lookout Bight, North Carolina. Geochim Cosmochim Acta 54:237–245

Carvalho PF (1997) Distribution, cycling and mean residence time of ^{226}Ra, ^{210}Pb and ^{210}Po in the Tagus estuary. Sci Total Environ 196:151–161

Cherrier J, Burnett WC, LaRock PA (1995) Uptake of polonium and sulfur by bacteria. Geomicrobiol J 13:103–115

Church TM, Hussain N, Ferdelman TG et al (1994) An efficient quantitative technique for the simultaneous analyses of radon daughters Pb-210, Bi-210 and Po-210. Talanta 41:243–249

Cochran JK, McKibbin-Vaughan T, Dornblaser MM et al (1990) ^{210}Pb scavenging in the North Atlantic and North Pacific Oceans. Earth Planet Sci Lett 97:332–352

Corbett R, McKee B, Duncan D (2004) An evaluation of mobile mud dynamics in the Mississippi River deltaic region. Mar Geol 209:91–112

Du J, Zhang J, Zhang J et al (2008) Deposition patterns of atmospheric ^{7}Be and ^{210}Pb in Coast of East China Sea, Shanghai, China. Atmos Environ 42:5101–5109

Du J, Wu Y, Huang D et al (2010) Use of ^{7}Be, ^{210}Pb and ^{137}Cs tracers to the transport of surface sediments of the Changjiang Estuary, China. J Mar Syst 82:286–294

Elsinger RJ, Moore WS (1983) ^{224}Ra, ^{228}Ra and ^{226}Ra in Winyah Bay and Delaware Bay. Earth Planet Sci Lett 64:430–436

Feng H, Cochran JK, Hirschberg DJ (1999a) ^{234}Th and ^{7}Be as tracers for the transport and dynamics of suspended particles in a partially mixed estuary. Geochim Cosmochim Acta 23:2487–2505

Feng H, Cochran JK, Hirschberg DJ (1999b) ^{234}Th and ^{7}Be as tracers for the sources of particles to the turbidity maximum of the hudson river estuary. Estuar Coast Shelf Sci 49:629–645

Fisher NS, Burns KA, Cherry RD et al (1983) Accumulation and cellular distribution of ^{241}Am, ^{210}Po, and ^{210}Pb in two marine algae. Mar Eco Prog Ser 11:233–237

Fisher NS, Cochran JK, Krishnaswami S et al (1988) Predicting the oceanic flux of radionuclides on sinking biogenic debris. Nature 355:622–625

Fitzgerald SA, Klump JV, Swarzenski PW, Mackenzie RA, Richards KD (2001) Beryllium-7 as a Tracer of Short-Term Sediment Deposition and Resuspension in the Fox River, Wisconsin. Environ Sci Technol 35:300–305

Foster JM, Shimmield GB (2002) Th-234 as a tracer of particle flux and POC export in the northern North Sea during a coccolithophore bloom. Deep Sea Res II 49:2965–2977

Fridrich J, Rutgers van der Loeff M (2002) A two-tracer (^{210}Po-^{234}Th) approach to distinguish organic carbon and biogenic silica export flux in the Antarctic circumpolar current. Deep Sea Res I 49:101–20

Giffin D, Corbett R (2003) Evaluation of sediment dynamics in coastal systems via short-lived radioisotopes. J Mar Syst 42:83–96

Goldberg ED, Gamble E, Griffin JJ et al (1977) Pollution history of Narragansett Bay as recorded in its sediments. Estuar Coast Mar Sci 5:549–561

Gustafsson Ö, Gschwend PM, Buesseler KO (1997a) Settling removal rates of PCBs into the Northwestern Atlantic derived from ^{238}U-^{234}Th disequilibria. Environ Sci Technol 31(12):3544–3550

Gustafsson Ö, Gschwend PM, Buesseler KO (1997b) Using Th-234 disequilibria to estimate the vertical removal rates of polycyclic aromatic hydrocarbons from the surface ocean. Mar Chem 57(1–2):11–23

Gustafsson Ö, Gschwend PM (1998) The Flux of Black Carbon to Surface Sediments on the New England Continental Shelf. Geochim Cosmochim Acta 62:465–472

Harvey BR, Lovett MB (1984) The use of yield tracers for the determination of alpha emitting actinides in the marine environment. Nucl Instrum Methods Phys Res 223:224–234

Hawley N, Robbins JA, Eadie BJ (1986) The partitioning of beryllium-7 in fresh water. Geochim Cosmochim Acta 50:1127–1131

Hirose K, Aoyama M, Igarashi Y et al (2005) Extremely low background measurements of ^{137}Cs in seawater samples using an underground facility (Ogoya). J Radioanal Nucl Chem 263:349–353

Hirose K, Igarashi Y, Aoyama M (2008) Analysis of the 50-year records of the atmospheric deposition of long-lived radionuclides in Japan. Appl Radiat Isot 66:1675–1678

Hong GH, Park SK, Baskaran M et al (1999) Lead-210 and Polonium-210 in the winter well-mixed turbid waters in the mouth of the yellow sea. Cont Shelf Res 19:1049–1064

Hong GH, Hamilton TF, Baskaran M, Kenna TC (2011) Applications of anthropogenic radionuclides as tracers to investigate marine environmental processes. In: Baskaran M (ed) Handbook of environmental isotope geochemistry. Springer, Heidelberg

Huh CA, Liu JT, Lin HL et al (2009) Tidal and flood signatures of settling particles in the Gaoping submarine canyon (SW Taiwan) revealed from radionuclide and flow measurements. Mar Geol 267:8–17

Hussain N, Church TM, Veron AJ (1998) Radon daughter disequilibria and lead systematics in the western North Atlantic. J Geophys Res 103(D3):16059–16071

Jweda, J (2007) Short-lived radionuclides (^{210}Pb, ^{7}Be and ^{137}Cs) as tracers of particle dynamics and chronometers for sediment accumulation and mixing rates in a river system in Southeast Michigan. M.S. Thesis, Department of Geology, Wayne State University, pp 167.

Jweda J, Baskaran M, van Hees Ed (2008) Short-lived radionuclides (^{7}Be and ^{210}Pb) as tracers of particle dynamics in a river system in southeast Michigan. Limnol Oceanogr 53:1934–1944

Kadko D, Olson D (1996) Be-7 as a tracer of surface water subduction and mixed layer history. Deep Sea Res I 43:89–116

Kaste JM, Baskaran M (2011) Meteoric ^{7}Be and ^{10}Be as process tracers in the environment. In: Baskaran M (ed) Handbook of environmental isotope geochemistry. Springer, Heidelberg

Kaste JM, Norton SA, Hess CT (2002) Environmental chemistry of beryllium-7. Rev Mineral Geochem 50:271–289

Kersten M, Thomsen S, Priebsch W et al (1998) Scavenging and particle residence times determined from ^{234}Th/^{238}U disequilibria in the coastal waters of Mecklenburg Bay. Appl Geochem 13:339–347

Kim G, Church TM (2011) Po-210 in the environment: biogeochemical cycling and bioavailability. In: Baskaran M (ed) Handbook of environmental isotope geochemistry. Springer, Heidelberg

Kim Y, Yang HS (2004) Scavenging of ^{234}Th and ^{210}Po in surface water of Jinhae Bay, Korea during a red tide. Geochem J 38:505–513

Kim SK, Hong GH, Baskaran M et al (1998) Wet removal of atmospheric ^{7}Be and ^{210}Pb at the Korean Yellow Sea Coast. The Yellow Sea 4:58–68

Kim G, Alleman LY, Church TM (1999) Atmospheric depositional fluxes of trace element, ^{210}Pb, and ^{7}Be to the Sargasso sea. Global Biogeochem Cycle 13(4):1183–1192

Kim G, Hussain N, Scudlark JR et al (2000) Factors influencing the atmospheric depositional fluxes of stable Pb, ^{210}Pb, and 7Be into Chesapeake Bay. J Atmos Chem 36:65–79

Koide M, Soutar A, Goldberg ED (1972) Marine geochronology with ^{210}Pb. Earth Planet Sci Lett 14:2–446

Lal D and Peters B (1967) Cosmic ray produced radioactivity on the Earth. In: Encyclopedia of physics. Springer, Berlin

Lal D, Baskaran M (2011) Applications of cosmogenic-isotopes as atmospheric tracers. In: Baskaran M (ed) Handbook of environmental isotope geochemistry. Springer, Heidelberg

Larock P, Hyun JH, Boutelle S et al (1996) Bacterial mobilization of polonium. Geochim Cosmochim Acta 60(22):4321–4328

Lima AL, Hudeny JB, Reddy CM et al (2005) High-resolution historical records from Pettaquamscutt River basin sediments: 1. ^{210}Pb and varve chronologies validate record of ^{137}Cs released by the Chernobyl accident. Geochim Cosmochim Acta 69:1803–1812

Masarik J, Beer J (1999) Simulations of particle fluxes and cosmogenic nuclide production in the earth's atmosphere. J Geophys Res D104:12099–13012

Masque P, Sanchez-Cabeza JA, Bruach JM et al (2002) Balance and residence times of ^{210}Pb and ^{210}Po in surface waters of the Northwestern Mediterranean Sea. Cont Shelf Res 22:2127–2146

Matisoff G, Whiting PJ (2011) Measuring soil erosion rates using natural (^{7}Be, ^{210}Pb) and anthropogenic (^{137}Cs, ^{239,240}Pu) radionuclides. In: Baskaran M (ed) Handbook of environmental isotope geochemistry. Springer, Heidelberg

Matisoff G, Wilson CG, Whiting PJ (2005) The ^{7}Be/^{210}Pb$_{xs}$ ratio as an indicator of suspended sediment age or fraction new sediment in suspension. Earth Surf Process Landforms 30:1191–1201

Matthews KM, Kim CK, Martin P (2007) Determination of ^{210}Po in environmental materials: a review of analytical methodology. Appl Radiat Isot 65:267–279

McKee BA, DeMaster DJ, Nittrouer CA (1984) The use of ^{234}Th/^{238}U disequilibrium to examine the fate of particle-reactive species on the Yangtze continental shelf. Earth Planetary Sci Lett 68:431–442

Moore WS (1976) Sampling ^{228}Ra in the deep ocean. Deep Sea Res 23:647–651

Moore WS, Dymond J (1988) Correlation of ^{210}Pb removal with organic carbon fluxes in the Pacific Ocean. Nature 331:339–341

Murray CN, Kautsky H, Hoppenheit M (1978) Actinide activities in water entering the northern North Sea. Nature 276:225–230

Nozaki Y (1993) Actinium-227: a steady state tracer for the deep-sea basin-wide circulation and mixing studies. In: Teramoto Y (ed) Deep ocean circulation: physical and chemical aspects. Amsterdam, Elsevier

Nozaki Y, Zhang J, Takeda A (1997) ^{210}Pb and ^{210}Po in the equatorial Pacific and the Bering Sea: the effects of biological productivity and boundary scavenging. Deep Sea Res II 44:2203–2220

Olsen CR, Larsen IL, Lowry PD et al (1986) Geochemistry and deposition of ^{7}Be in river-esturine and coastal waters. J Geophys Res 91(Cl):896–908

Palinkas CM, Nittrouer CA, Wheatcroft RA et al (2005) The use of ^{7}Be to identify event and seasonal sedimentation near the Po River delta, Adriatic Sea. Mar Geol 222–223:95–112

Pates JM, Cook GT, MacKenzie AB et al (1996) Determination of ^{234}Th in marine samples by liquid scintillation spectrometry. Anal Chem 68:3783–3788

Porcelli D, Baskaran M (2011) An overview of isotope geochemistry in environmental studies. In: Baskaran M (ed) Handbook of environmental isotope geochemistry. Springer, Heidelberg

Radakovitch O, Cherry RD, Heyraud M et al (1998) Unusual Po-210/Pb-210 ratio in the surface water of the Gulf of Lions. Oceanol Acta 21:459–468

Ravichandran M, Baskaran M, Santschi PH et al (1995) Geochronology of sediments in the Sabine-Neches estuary, Texas, USA. Chem Geol 125:291–306

Robbins JA, Edgington DN (1975) Determination of recent sedimentation rates in Lake Michigan using ^{210}Pb and ^{137}Cs. Geochim Cosmochim Acta 39:285–304

Roether W (1971) Flushing of the Gerard-Ewing large-volume water sampler. J Geophys Res 76:5910–5912

Rutgers van der Loeff M, Berger GW (1993) Scavenging of ^{230}Th and ^{231}Pa near the Antarctic Polar Front in the South Atlantic. Deep Sea Res I 40:339–357

Rutgers van der Loeff M, Moore WS (1999) The analysis of natural radionuclides in seawater. In: Grasshoff K, Ehrhadt MG, Kremling K (eds) Methods of seawater analysis, 3rd edn. Chemie, Weinheim

Rutgers van der Loeff M, Meyer R, Rudels B, Rachor E (2002) Resuspension and particle export in the benthic nepheloid layer in and near Fram Strait in relation to faunal abundances and Th-234 depletion. Deep Sea Res I 49:1941–1958

Rutgers van der Loeff M, Sarin MM, Baskaran M (2006) A review of present techniques and methodological advances in analyzing ^{234}Th in aquatic systems. Mar Chem 100:190–212

Santschi PH, Li YH, Bell J (1979) Natural radionuclides in the water of Narragansett Bay. Earth Planet Sci Lett 45:201–213

Sarin MM, Bhushan R, Remgarajan R et al (1992) The simultaneous determination of ^{238}U series nuclides in seawater: results from the Arabian Sea and Bay of Bengal. Indian J Mar Sci 21:121–127

Smith JN, Ellis KM, Jones EP (1990) Caesium-137 transport into the Arctic Ocean through Fram Strait. J Geophys Res 95(C2):1693–1701

Smith KJ, Vintro LL, Mitchell PI et al (2004) Uranium-thorium disequilibrium in north-east Atlantic waters. J Environ Radioact 74:199–210

Sommerfield CK, Nittrouer CA, Alexander CR (1999) ^{7}Be as a tracer of flood sedimentation on the northern California continental margin. Cont Shelf Res 19:335–361

Stewart G, Cochran JK, Xue J et al (2007) Exploring the connection between ^{210}Po and organic matter in the northwestern Mediterranean. Deep Sea Res I 54:415–427

Su CC, Huh CA, Lin FJ (2003) Factors controlling atmospheric fluxes of ^{7}Be and ^{210}Pb in northern Taiwan. Geophys Res Lett 30(19):2018. doi:10.1029/2003GL018221

Swarzenski PW, Reich CD, Spechler RM et al (2001) Using multiple geochemical tracers to characterize the hydrogeology of the submarine spring off Crescent Beach, Florida. Chem Geol 179:187–202

Swarzenski PW, Porcelli D, Andersson Per S et al (2003) The behavior of U- and Th-series nuclides in the Estuarine environment. Rev Mineral Geochem 52:577–606

Thomson J, Colley S, Anderson R et al (1993) ^{210}Pb in the sediments water column of the Northeast Atlantic from 47 to 59°N along 20°W. Earth Planet Sci Lett 115:75–87

Trimble SM, Baskaran M, Porcelli D (2004) Scavenging of thorium isotopes in the Canada Basin of the Arctic Ocean. Earth Planet Sci Lett 222:915–932

Turekian KK, Graustein WC (2003) Natural radionuclide in the atmosphere. Treatise Geochem 4:261–279

UNSCEAR (1982) Ionizing radiation: sources and biological effects. United Nations, New York

UNSCEAR (2000) Sources and effects of ionizing radiation. United Nations, New York

Walling DE, Quine TA (1993) Use of Cs-137 as a tracer of erosion and sedimentation: handbook for the application of the Cs-137 technique. University of Exeter, Exeter

Walling DE, He Q, Blake W (1999) Use of ^{7}Be and ^{137}Cs measurements to document short- and medium-term rates of water-induced erosion and agricultural land. Water Resour Res 35:3865–3874

Waples JT, Orlandini KA, Weckerly KM et al (2003) Measuring low concentrations of ^{234}Th in water and sediment. Mar Chem 80:265–281

Waples JT, Benitez-Nelson CR, Savoye N et al (2006) An introduction to the application and future use of ^{234}Th in aquatic systems. Mar Chem 100:166–189

Weinstein SE, Moran SB (2005) Vertical flux of particulate Al, Fe, Pb and Ba from the upper ocean determined using ^{234}Th/^{238}U disequilibrium. Deep Sea Res I 52:1477–1488

Whiting PJ, Matisoff G, Fornes W et al (2005) Suspended sediment sources and transport distances in the Yellowstone River basin. GSA Bullet 117:515–529

Wilson CG, Matisoff G, Whiting PJ (2007) The use of ^{7}Be and ^{210}Pb$_{ex}$ to differentiate fine suspended sediment sources in South Slough, Oregon. Estuar Coast 30(2):348–358

Yoshizawa Y (1958) Beta and gamma ray spectroscopy of ^{137}Cs. Nucl Phys 5:122–140

Zuo ZZ, Eisma D, Berger GW (1990) Determination of sediment accumulation and mixing rate in the Gulf of Lion, the Mediterranean Sea. Water Poll Res Rep 20:469–499

Chapter 17
Radium Isotope Tracers to Evaluate Coastal Ocean Mixing and Residence Times

L. Zhang, J. Zhang, P.W. Swarzenski, and Z. Liu

Abstract The identification and provenance of unique coastal water masses is essential in near-shore biogeochemical studies. Water mass mixing and residence times impact water quality and can play a role in the evolution of algal blooms. Such information is thus critical for resource managers who have an interest in understanding the source and fate of contaminants and their eventual fate in the coastal ocean. If mixing is important for quantitatively assessing the amount of exchange, the water residence time or water age is important to assess the rate of this exchange. An understanding of water mass residence times is useful to examine time scales of contaminant discharge and to evaluate transport phenomena.

This review summarizes the scientific significance, measurement approaches, and models to evaluate coastal water mixing and residence times using radium isotopes. Each method or model described here is valid, although each has its own advantages and disadvantages. Examples of mixing among different end-members are given as case studies. All approaches presented here demonstrate the utility of radium isotopes for the evaluation of water mass mixing and residence times.

L. Zhang (✉) and J. Zhang
State Key Laboratory of Estuarine and Coastal Research, East China Normal University, 3663 Zhongshan Road North, Shanghai 200062, P.R. China
e-mail: 1979_zhanglei@163.com; jzhang@sklec.ecnu.edu.cn

P.W. Swarzenski
US Geological Survey, 400 Natural Bridges Drive, Santa Cruz, CA 95060, USA
e-mail: pswarzen@usgs.gov

Z. Liu
Key Laboratory of Marine Environment and Ecology (Ocean University of China), Ministry of Education, 238 Songling Road, Qingdao 266003, P.R. China
e-mail: liuzhe_ecnu@yahoo.com.cn

17.1 Introduction

There are four isotopes of radium in the natural environment with a board range of half-lives that extend from 3.66 days to 1,600 years, i.e. ^{226}Ra ($t_{1/2}$ = 1,600 year), ^{228}Ra ($t_{1/2}$ = 5.75 year), ^{223}Ra ($t_{1/2}$ = 11.4 day) and ^{224}Ra ($t_{1/2}$ = 3.66 day). Radium is an alkaline earth metal and each isotope is produced by the decay of their radiogenic parents, which belongs to members of the three U/Th decay series (see Fig. 2.2 in Chap. 2). They have different half-lives but similar physicochemical properties. Figure 17.1 is a biogeochemical behavior model for radium isotopes in a coastal environment. In river water, radium isotopes are strongly adsorbed onto particles, but they will rapidly be released into seawater with an increase in ion strength on a short timescale (Moore 1981; Key et al. 1985; Moore and Scott 1986). In dissolved form, radium isotopes move with a water body and are transported in the ocean by advective and diffusive mixing processes, while irreversibly removed from water column via radioactive decay. It has been recently documented that submarine groundwater discharge is an important source of radium in coastal waters (Moore 1999; Krest et al. 1999; Yang et al. 2002; Burnett et al. 2006). Based on their properties and different half-lives, Ra is of particular interest in oceanographic studies because it may lead to an understanding of various mobilization processes in the ocean and has great potential as tracers to evaluate geophysical and environmental processes in the geosphere (Moore 1996; Rama and Moore 1996). Numerous publications have demonstrated the suitability of radium to evaluate mixing dynamics of various water masses and residence time in estuaries, coastal environments, and the

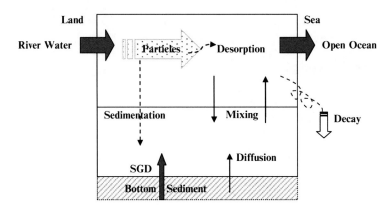

Fig. 17.1 A model budget for radium isotopes in a coastal region. SGD is the abbreviation of submarine groundwater discharge. Assume Ra inputs include river discharge, desorption from riverine particles, diffusion from bottom sediment and open ocean (Moore et al. 1986; Moore and Todd 1993; Kim et al. 2005; Moore 2006). inputs from SGD; Ra outputs include mixing with the open ocean and radioactive decay (Kim et al. 2005; Burnett et al. 2006)

17.2 Methods

17.2.1 Gamma Spectrometry

In oceanographic observations, temperature and salinity are measured *in-situ* by a CTD profiler, while for radium isotopes different methods exist for measurements. Gamma spectrometry needs relatively less chemical processing after sample collection, but relative large sample volumes. Large volume (>200 L) water samples can be pumped directly through polypropylene cartridge pre-filters to remove suspended particles followed by two MnO_2 impregnated cartridges connected in a series to concentrate radium (Baskaran et al. 1993, 2009). After the filtration, the MnO_2-fibers are detached from the cartridge and washed with distilled water to remove salts. In the laboratory, the pre-filtered and MnO_2 impregnated cartridges are ashed at 550°C for more than 6 h in a muffle furnace. After cooling, the ashes are weighed and transferred into a plastic vials sealed with epoxy for determination by Gamma-spectrometry. The extraction efficiency of radium from seawater is calculated from relative efficiency of two MnO_2-fibers. Assuming no difference in extraction efficiency exists between two MnO_2-fiber cartridges, the extraction efficiencies (η) can be calculated by equation $\eta = 1 - A_2/A_1$, where A_1 and A_2 are measured radium isotope activities of the first and second cartridge filters (Hong et al. 2002). However, the extraction of Ra onto MnO_2-fibers at filtration rates of 5–10 L/min is not quantitative. Thus, the large-volume filtration is useful to obtain activity ratios of Ra (such as $^{226}Ra/^{228}Ra$, $^{224}Ra/^{226}Ra$ and $^{223}Ra/^{226}Ra$) and a separate aliquot of 20 L is required for precise ^{226}Ra measurements. From the precise measurement of ^{226}Ra and activity ratios of Ra, specific activity of ^{223}Ra, ^{224}Ra and ^{228}Ra can be determined. Obviously, in most circumstances, the data of short-lived isotopes are difficult to acquire, and the samples volume required for determination are large. So the small volume methods are predominant in near 10 years.

17.2.2 ^{226}Ra by Alpha Spectrometry

For seawater samples, about 20 L water samples are sufficient for this method. After collection, samples are filtered immediately through a filter (e.g. pore size = 0.45 μm). The resulting pre-filtered waters are stored in pre-cleaned polyethylene containers, and ^{229}Th-^{225}Ra yield tracers are added after acidification. Samples are thoroughly mixed for ca. 10 min and then allowed to stand for ca. 6 h for spikes to equilibrate. Briefly, concentrated NH_4OH, $KMnO_4$ and $MnCl_2$ solutions are added to form a suspension of MnO_2. The sample is stirred for 30 min, and then allowed to settle. The precipitate is separated from the supernatant and taken back to the lab for further treatment. In the laboratory, the resultant precipitate is re-dissolved in concentrated HNO_3 and H_2O_2 to make a 0.1-M HNO_3 solution. Detailed procedures for Ra separation and purification are described in Hancock and Martin (1991). The purified Ra is evaporated to a low volume, brought to dryness at low heat, dissolved in 0.5 mL 0.1 M HNO_3, and finally transferred to a deposition cell with 1 mL 0.05 M HCl and 9 mL ethanol. Ra is electro-deposited onto a stainless steel disc for alpha measurement. One can readily obtain alpha spectrometry data for ^{226}Ra, ^{224}Ra, and ^{223}Ra, but for ^{228}Ra one needs to wait more than 1 year to allow for sufficient ingrowth of ^{228}Th.

17.2.3 Combination of Delayed Coincidence System with Gamma Spectrometry (RaDeCC)

One can obtain short-lived Ra isotope data quickly using a delayed coincidence counting system. Moore and Arnold (1996) used a portable delayed coincidence counter to obtain ^{224}Ra and ^{223}Ra, after which samples were measured using gamma ray spectrometry to assess the activities of ^{226}Ra and ^{228}Ra. The advantage of RaDeCC is the very low detection limit. For example, ^{224}Ra concentrations in the range of 8 mBq m^{-3} will have 1σ uncertainty of ±10% and thus, total number of ^{224}Ra as low as 3,000 atoms can be detected (Moore 2007). For ^{223}Ra, activities as low as 3 mBq m^{-3} can be detected (Moore 2007). Seawater samples sizes are generally 50–100 L, while 20 L groundwater samples are usually sufficient. Radium is extracted from the samples by passing the water through a column filled with about 50 g of manganese-coated acrylic fiber (Moore 1976). Each Mn fiber sample is partially dried with a stream of air and placed in closed loop air circulation system (Moore and Arnold 1996). Using helium as a transport gas, the radiogenic daughters of ^{219}Rn and ^{220}Rn generated by ^{223}Ra and ^{224}Ra decay are used to quantify ^{223}Ra and ^{224}Ra. After the ^{223}Ra and ^{224}Ra measurements, the Mn fibers are leached with HCl to quantitatively remove the long-lived Ra isotopes. The Ra is co-precipitated with $BaSO_4$, and the precipitate is aged for 3 weeks to determine ^{226}Ra and ^{228}Ra in a ultra low background gamma detector.

17.3 Water Mass Mixing and Residence Times Using Radium as Tracers

17.3.1 Study of Water Mixing

The identification of various water masses in the coastal ocean is of critical importance in near-shore oceanography, where oceanographic characteristics are largely determined by the mixing of source water masses (Pickard and Emery 1990; Vilibic and Orlic 2002). The study of water mixing processes can be traced back to the early twentieth century, when the method of classical temperature and salinity data analysis was first established by Helland-Hansen (1916), and later globally applied (e.g. Sverdrup et al. 1946). In 1927, the temperature and salinity (T-S) diagram was used to study the mixing of two water masses by Jacobsen (1927), and later for the analysis of mixing among three water masses at the edge of the continental shelf in the North Atlantic Ocean (Miller 1950; Dulaiova and Burnett 2007). With regards to more complex mixing conditions, for example among four parcels over a shelf, Chen et al. (1995) elaborated a method based on temperature-salinity diagram analysis and series of assumptions, which could give quantitative results at specific conditions and that method had been applied to marginal seas of NW Pacific Ocean. However, this method fails in some circumstances and the intrinsic assumptions are invalid in terms of mathematics and reasonable derivation in oceanography (Zhang et al. 2007). Based on an original theory of equations, in the case when the number of tracers (i.e. in the case of $n = 2$) is less than that of source end-members minus one (i.e. 3) will not produce a definite solution. Other tracers are required ultimately to resolve problems with exact solutions, when the end-members of water mixing are >3.

In the 1980s, a new water mass analysis technique, the so-called Optimum Multi-parameter (OMP) analysis, with the aid of multiple tracers, was first introduced to characterize water masses in the sea

(Tomczak 1981). Generally, the governing equations describing mixing processes among different water masses can be described by matrix theory:

$$\begin{pmatrix} 1 & 1 & 1 & 1 \\ T_{11} & T_{12} & \cdots & T_{1n} \\ T_{21} & T_{22} & \cdots & T_{2n} \\ \vdots & \vdots & \ddots & \vdots \\ T_{m1} & T_{m2} & \cdots & T_{mn} \end{pmatrix} \begin{pmatrix} f_1 \\ f_2 \\ \vdots \\ f_n \end{pmatrix} = \begin{pmatrix} 1 \\ T_1^{obs} \\ T_2^{obs} \\ \vdots \\ T_m^{obs} \end{pmatrix}, \quad (17.1)$$

where f_i ($i = 1, 2, \cdots, n$) is the fraction for the $i-th$ water mass, T_{ij} is the $j-th$ value of tracer for the $i-th$ water mass, T_j^{obs} is the $j-th$ tracer value for a given water sample. By solving these matrix equations, the mixing proportions f_i can be expressed as:

$$\begin{pmatrix} f_1 \\ f_2 \\ \vdots \\ f_n \end{pmatrix} = \begin{pmatrix} 1 & 1 & 1 & 1 \\ T_{11} & T_{12} & \cdots & T_{1n} \\ T_{21} & T_{22} & \cdots & T_{2n} \\ \vdots & \vdots & \ddots & \vdots \\ T_{m1} & T_{m2} & \cdots & T_{mn} \end{pmatrix}^{-1} \begin{pmatrix} 1 \\ T_1^{obs} \\ T_2^{obs} \\ \vdots \\ T_m^{obs} \end{pmatrix}. \quad (17.2)$$

This method solves a set of linear mixing equations with the contribution of the each water type to the given oceanic water as variables and the hydrographic properties as the parameters of the system. OMP analysis has been successfully applied to shelf seas where all hydrographic properties can be considered as a first approximation to be conservative (Maamaatuaiahutapu et al. 1992, 1994; Klein and Tomczak 1994), and is still undergoing further development (Poole and Tomczak 1999; Henry-Edwards and Tomczak 2006).

Stable isotopes, including $\delta^{18}O$ and $\delta^{13}C$, nutrient and radioactive isotopes (such as ^{134}Cs, ^{137}Cs and ^{125}Sb) have been used to track mixing of water masses (e.g. Liu et al. 1995; Bailly 1996; Lakey and Krothe 1996; Katz et al. 1997; Lee and Krothe 2001; Zhang 2002; Hong et al. 2011). However, the concentration of $\delta^{18}O$ in typical marine waters is rather stable with limited variation (e.g. $0 \pm 1‰$), if used as tracer will introduce considerable uncertainty (Chen 1985; Liu et al. 1995). The $\delta^{13}C$ and nutrients are not strictly conservative because of air-sea exchange and participation in metabolic regulations, and therefore, are unsuitable in water source discrimination when biological processes are important factors relative to the mixing processes. In order to improve the traceability of the discrimination of source waters, Ra nuclides can be chosen as additional tracers because of its conservative behavior and non-particle reactive nature in the ocean (Moore 2000). In the early 1980s, ^{226}Ra was used as tracer to analyze the mixing between two water parcels in north Pacific (Chung and Craig 1980). Moore (2003) developed a three-end-member mixing model that used radium isotopes to estimate the fractions of ocean water. Furthermore, off western Florida waters from a shallow aquifer and from a deep aquifer were separated using Ra isotopes (Moore et al. 2006). Recently, Zhang et al. (2007) used temperature, salinity and ^{226}Ra as passive tracers to evaluate the mixing proportions among four end-members in the East China Sea.

17.3.2 Study of Residence Time

Trajectory, in terms of a Lagrangian view, is used to describe the path of a moving water parcel through space. Let t denote time, $\vec{r}(t)$ represent the position vector of the fluid parcel at time t, and $\vec{u}(\vec{r}(t), t)$ describe the velocity at the time t and position $\vec{r}(t)$. From the notion of trajectory, the relation between $\vec{r}(t)$ and $\vec{u}(\vec{r}(t), t)$ can be expressed as:

$$\vec{u}(\vec{r}(t), t) = \frac{d\vec{r}(t)}{dt}. \quad (17.3)$$

However, it is rather difficult to understand transport processes by only investigating magnitude and direction of flow (Deleersnijder et al. 2001). Therefore, auxiliary parameters are needed, which are closely related to the flow field and can be used to quantify their functioning of material transport in seawater (Deleersnijder et al. 2001). Water age and residence time are two such variables. Residence time is the complement of age (Takeoka 1984), and provides the timescale available for components to accumulate in the water column (Moore et al. 2006). It should be pointed out that, in some studies, residence time is also taken as the time span of the water parcel (e.g. Bolin and Rodhe 1973), which is different from the above concept. In order to facilitate better understanding, a

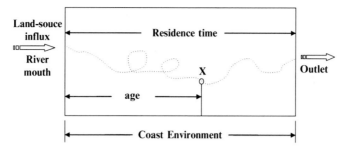

Fig. 17.2 Water age and residence time schematic. The hollow dot represents a water parcel. From river mouth, the water parcel enters the water body, and leaves the water body from outlet. The point X location means water age of the water parcel and the duration for the water parcel from entering to leaving the water body is residence time (Bolin and Rodhe 1973)

schematic diagram is drawn in Fig. 17.2. In such a view, water age (which is zero when a water parcel enters a water body) and residence time that is the duration for a water parcel from entering to leaving the water body of interest is analogous to that of human age that is zero when people were born and lifespan (i.e. the duration for people from birth to death) (Bolin and Rodhe 1973). Water age is regarded as the time a water parcel has spent since entering the control volume through one of its boundaries, where its age is prescribed to be zero (Delhez et al. 1999). Hence, the age $a[\vec{r}(t), t]$ of water parcel in the domain of interest must progress at the same pace (17.4), since the age is measured by the elapsed time.

$$\frac{dt}{dt} = 1 = \frac{d}{dt} a[\vec{r}(t), t]. \quad (17.4)$$

Residence time is defined as "the time it takes for any water parcel to leave a given water body through its outlet to the sea" (Zimmerman 1976; Monsen et al. 2002). Hence, residence time is the remainder of the life time of a water parcel considered (17.5)

$$RT(\vec{r}(t), t) = a[\vec{r}(t^{out}), t^{out}] - a[\vec{r}(t), t], \quad (17.5)$$

where $RT(\vec{r}(t), t)$ is the residence time, $a[\vec{r}(t), t]$ is the water age at the time of t considered, and $a[\vec{r}(t^{out}), t^{out}]$ is the age the water parcel reaches the outlet and leaves (i.e. life time).

17.3.2.1 Conservative Ra Without Additional Input

This model assumes there is no additional input of radium beyond the nearshore. The coastal ocean is usually an area for significant radium inputs because radium is released into a water column from riverine particles as well as bottom sediments (Moore 1981; Key et al. 1985; Moore and Scott 1986). In exclusion of the coastal zone, desorption from sediment and pore-water diffusion are the only other inputs (Moore 1969; Li and Chan 1979), and this contribution can be ignored in some circumstances, i.e. strong thermocline will prevent the approach from bottom layer (Moore 2000). If the radioactive isotope with the half-life of $T \log 2$ is contained in a water parcel, where the T is the half-life of the isotope, the concentration within the fluid parcel of this radioactive constituent decreases as:

$$C_r[\vec{r}(t^{end}), t^{end}] = C_r[\vec{r}(t^{ini}), t^{ini}] \exp\frac{t^{ini} - t^{end}}{T}, \quad (17.6)$$

where $C_r[\vec{r}(t^{ini}), t^{ini}]$ and $C_r[\vec{r}(t^{end}), t^{end}]$ is the activity of a radionuclide at initial time t^{ini} in $\vec{r}(t^{ini})$ and arrival time t^{end} in $\vec{r}(t^{end})$, respectively.

Then, combining (17.5) and (17.6) yields

$$RT[\vec{r}(t^{end}), t^{end}] = t^{end} - t^{ini}$$
$$= T \log \frac{C_r[\vec{r}(t^{ini}), t^{ini}]}{C_r[\vec{r}(t^{end}), t^{end}]}, \quad (17.7)$$

or

$$RT[\vec{r}(t^{end}), t^{end}] = t^{end} - t^{ini}$$
$$= T \log \frac{C_r[\vec{r}(t^{ini}), t^{ini}] / C_p[\vec{r}(t^{ini}), t^{ini}]}{C_r[\vec{r}(t^{end}), t^{end}] / C_p[\vec{r}(t^{end}), t^{end}]}, \quad (17.8)$$

where $RT[\vec{r}(t^{end}), t^{end}]$ is the residence time for the water parcel considered, and $C_p[\vec{r}(t), t]$ is the concentration of a passive conservative constituent. The constituent is neither produced nor destroyed. That is to say, the concentration $C_p[\vec{r}(t), t]$ within the fluid parcel remains constant as time progresses (i.e. $C_p[\vec{r}(t^{end}), t^{end}] = C_p[\vec{r}(t^{ini}), t^{ini}]$).

17.3.2.2 Non-Conservative Ra with Additional Input

In estuaries, salt marshes, near-shore coastal waters, and even on the shelf where radium additions are continuous, the above-mentioned activities of a radioactive constituent at some initial time are not constant. The model described in Sect. 17.3.2.1 thus becomes not applicable. Considering the additions and mixing between fresh and ocean water, a different approach has been proposed (Nozaki et al. 1991). Assuming that the decay of isotopes does not measurably affect the activities of long-lived ^{226}Ra and ^{228}Ra on the timescale of estuarine mixing, and biological uptake/release is negligible, the activity of radium isotopes can be affected by the following three factors: (1) Ra_R; dissolved radium in the river freshwater end-member, (2) Ra_S; dissolved radium in saline water end-member, and (3) Ra_{ex}; "excess" dissolved radium in observed station. The "excess" here means a supply of radium from other sources except the mixture of river water and sea water end-members, including, for example, desorption from particles, diffusion from sediment (e.g. sediment-water exchange) and submarine groundwater (Yang et al. 2002; Moore 2006). Thus, the budget of radium isotopes can be expressed by (Nozaki et al. 1989, 1991)

$$^{226}Ra_{obs} = f\,^{226}Ra_R + (1-f)\,^{226}Ra_S + ^{226}Ra_{ex} \quad (17.9)$$

and

$$^{228}Ra_{obs} = f\,^{228}Ra_R + (1-f)\,^{228}Ra_S + ^{228}Ra_{ex}, \quad (17.10)$$

where f is the fraction of river water, subscript *obs* is the observed activity of radium for a given station.

Assuming that the net effect of evaporation and precipitation is negligible, the surface water at salinity S is the mixture of the river freshwater and the saline water. The f is given by

$$f = (S_S - S)/(S_S - S_R), \quad (17.11)$$

where S_R, S_S, S is the salinity of the river water end-member, sea water end-member, and measured salinity in the sample, respectively. At steady state, the residence time of excess dissolved radium (τ_w) can be expressed using the results of $^{226}Ra_{ex}$ and $^{228}Ra_{ex}$.

$$^{226}Ra_{ex} = F^{226}Ra\,\tau_w \quad (17.12)$$

$$^{228}Ra_{ex} = (F^{228}Ra - \lambda_{228}\,^{228}Ra_{obs})\tau_w, \quad (17.13)$$

where $F^{226}Ra$ and $F^{228}Ra$ are the total fluxes of excess radium supplied from particles and sediments, and λ_{228} is the decay constant of ^{228}Ra. The ratio of (17.12) and (17.13) can be expressed as:

$$\left(\frac{^{228}Ra}{^{226}Ra}\right)_{ex} = \frac{F^{228}Ra}{F^{226}Ra} - \left(\frac{\lambda_{228}}{F^{226}Ra}\right){}^{228}Ra_{obs}. \quad (17.14)$$

The $(^{228}Ra/^{226}Ra)_{ex}$ is a linear function of the $^{228}Ra_{obs}$, and the slope is the ratio of the decay constant of ^{228}Ra and total flux of excess ^{226}Ra (i.e. slope $= \frac{\lambda_{228}}{F^{226}Ra}$). Then the slope and y-intercept allow us to calculate the absolute values of $F^{226}Ra$ and $F^{228}Ra$. In this manner the residence time for each sample can be calculated.

In certain circumstances, the uncertainties are large enough to affect the resulting linear relationship. If that's the case, it is difficult to obtain the available results of the slope mentioned above. Moore et al. (2006) proposed a mass balance approach which assumed a system under steady state conditions. Radium additions, including fluxes from sediment, river and groundwater would be balanced by losses. For short-lived radium isotopes, losses are due to mixing and radioactive decay, but for long-lived radium radioactive decay would be negligible and so affected only by mixing processes.

In the case of ^{224}Ra mass balance, the equation can be expressed as:

$$F^{224}Ra = I^{224}Ra(\lambda_{224} + 1/\tau), \quad (17.15)$$

where $F^{224}Ra$ is the total flux of ^{224}Ra to the system, $I^{224}Ra$ is the inventory of ^{224}Ra in the system, λ_{224} is the decay constant for ^{224}Ra, and τ is the apparent age of water in the system. A similar equation can be written for ^{226}Ra, with radioactive decay ignored:

$$F^{226}Ra = I^{226}Ra(1/\tau). \quad (17.16)$$

Dividing (17.15) by (17.16), we can get:

$$F(^{224}Ra/^{226}Ra) = I^{224}Ra(\lambda_{224} + 1/\tau)/[I^{226}Ra(1/\tau)]. \quad (17.17)$$

This equation can be rearranged and solved for τ:

$$\tau = [F(^{224}Ra/^{226}Ra) - I(^{224}Ra/^{226}Ra)]/ \\ I(^{224}Ra/^{226}Ra)\lambda_{224} \quad (17.18)$$

Except for the above-mentioned radiochemical tracers, there have been a number of physical, chemical and modeling techniques used to estimate the residence time of coastal waters, including the tidal prism model, fresh water budgets, dye tracers and numerical models (e.g. Hougham and Moran 2007). Hydrodynamic models are based upon a large number of parameters including tidal range and prism, wind speed and direction, currents and geometry of the reservoir. The acquisition of data has inherent spatio-temporal limitations. Compared with real-time measurements, numerical models can provide information with much higher spatio-temporal resolution. In addition, advection and diffusion processes can be separated in numerical models. However, it can be difficult to separate these two terms from observations alone. All models need observational data for the validation of the model. If real-time measurements are not available, uncertainties associated with the modeled residence time could become quite large, therefore diminishing the utility of the model (Rapaglia et al. 2009). If, for example, the velocity of the current and the turbulent stream are dynamic, the model will yield the correct residence time only if these parameters are known for the period under investigation (Pougatch et al. 2007; Rapaglia et al. 2009). Due to the same chemical properties and different half-lives, the ratio of different radium isotopes could be considered as a good "inner-clock" to study residence time on different time scales. Water flow or separation will result in the variation of the radium activity, while the ratios could thus in effect offset water fractionation and turbulence.

17.3.3 Case Studies: Application of Radium Isotopes

17.3.3.1 Water Mixing

Three-End-Member Mixing

Moore et al. (2006) used ^{228}Ra and salinity as two tracers to analyze the mixing among three end members at Okatee Estuary; namely, (1) water from Port Royal Sound (PRS, here represented by the Colleton River), (2) water from the Okatee River, and (3) groundwater. Equations for water, salt and ^{228}Ra could be expressed as:

$$\begin{pmatrix} 1 & 1 & 1 \\ ^{228}Ra_S & ^{228}Ra_R & ^{228}Ra_{GW} \\ S_S & S_R & S_{GW} \end{pmatrix} \begin{pmatrix} f_S \\ f_R \\ f_{GW} \end{pmatrix} = \begin{pmatrix} 1 \\ ^{228}Ra_M \\ S_M \end{pmatrix}, \quad (17.19)$$

where f_S, f_R and f_{GW} are the fractions of PRS, Okatee River, and groundwater end-members, respectively; $^{228}Ra_S$, $^{228}Ra_R$ and $^{228}Ra_{GW}$ are the ^{228}Ra activities in PRS, Okatee River and groundwater end-members, respectively; S_S, S_R and S_{GW} are the salinities in the PRS, Okatee and groundwater end-members, respectively. $^{228}Ra_M$ and S_M are the measured ^{228}Ra concentration and salinity, respectively.

These equations may be solved for the fractions of each end-member (Moore et al. 2006):

$$f_S = \frac{\left(\frac{^{228}Ra_M - ^{228}Ra_R}{^{228}Ra_{GW} - ^{228}Ra_R}\right) - \left(\frac{S_M - S_R}{S_{GW} - S_R}\right)}{\left(\frac{^{228}Ra_S - ^{228}Ra_R}{^{228}Ra_{GW} - ^{228}Ra_R}\right) - \left(\frac{S_S - S_R}{S_{GW} - S_R}\right)} \quad (17.20)$$

$$f_{GW} = \frac{S_M - S_R - f_S(S_S - S_R)}{S_{GW} - S_R} \quad (17.21)$$

$$f_R = 1 - f_S - f_{GW}. \quad (17.22)$$

The results of this mixing using data from different dates are shown in Fig. 17.3.

Four-End-Member Mixing

Zhang et al. (2007) used temperature, salinity and ^{226}Ra as tracers to quantify the contribution among four

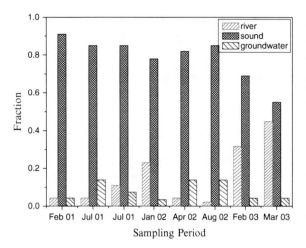

Fig. 17.3 Fraction of river, sound, and groundwater based on a three end-member mixing model (Moore et al. 2006)

end-members mixing in the East China Sea. Figure 17.4 shows the schematic of mixing among the four water masses; namely, Kuroshio Surface Water (KSW), Kuroshio Sub-surface Water (KSSW), Changjiang Diluted Water (CDW), and Taiwan Strait Warm Water (TSWW).

In this example, there were three variables: temperature (T), salinity (S) and radium (Ra) as tracers to analyze the mixing among four end members in that study. Based on the (17.1), the following equations could be obtained:

$$\begin{pmatrix} 1 & 1 & 1 & 1 \\ T_A & T_B & T_C & T_D \\ S_A & S_B & S_C & S_D \\ Ra_A & Ra_B & Ra_C & Ra_D \end{pmatrix} \begin{pmatrix} f_A \\ f_B \\ f_C \\ f_D \end{pmatrix} = \begin{pmatrix} 1 \\ T_{obs} \\ S_{obs} \\ Ra_{obs} \end{pmatrix}, \quad (17.23)$$

where T_i, S_i, Ra_i ($i = $ A, B, C, D) are the variable parameters of temperature, salinity and concentrations of Ra isotopes, respectively, for source water type i. Here, A, B, C and D represented the four actual source water masses, namely, the Kuroshio Surface Water, Kuroshio Sub-surface Water, Changjiang Diluted Water and Taiwan Strait Warm Water. $T_{obs}, S_{obs}, Ra_{obs}$ are the measured values of temperature, salinity and Ra concentration, and f_i the corresponding fractions of A, B, C and D for the observed stations, respectively. By solving these equations, the proportions of individual source water components in the ECS water mixed among four end-members could be estimated through exact solutions:

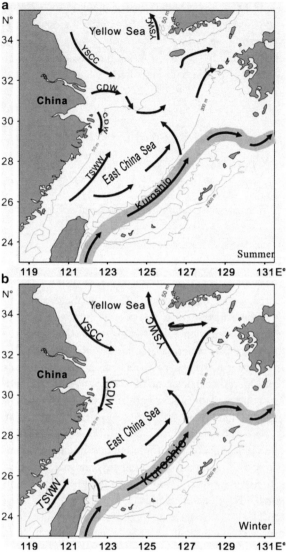

Fig. 17.4 Schematic of mixing among multiple water masses in the East China Sea. The four dominant source waters include: Kuroshio Surface Water (KSW), Kuroshio Sub-surface Water (KSSW), Changjiang Diluted Water (CDW), and Taiwan Strait Warm Water (TSWW) (**a**) In summer; (**b**) In winter

$$\begin{pmatrix} f_A \\ f_B \\ f_C \\ f_D \end{pmatrix} = \begin{pmatrix} 1 & 1 & 1 & 1 \\ T_A & T_B & T_C & T_D \\ S_A & S_B & S_C & S_D \\ Ra_A & Ra_B & Ra_C & Ra_D \end{pmatrix}^{-1} \begin{pmatrix} 1 \\ T_{obs} \\ S_{obs} \\ Ra_{obs} \end{pmatrix}. \quad (17.24)$$

Mixing proportions of the four water masses are shown in Fig. 17.5. In this manner, the contribution of each of the source water masses could be evaluated directly.

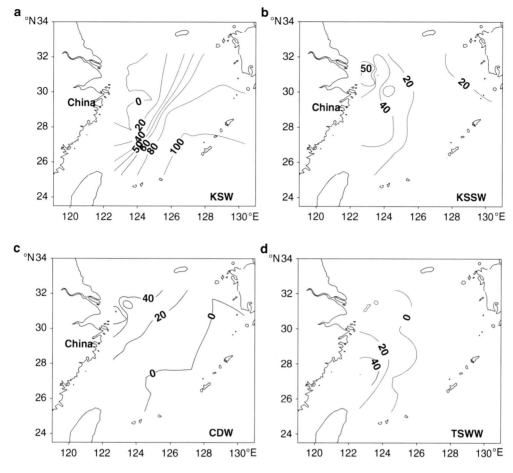

Fig. 17.5 Horizontal distributions of mixing proportion lines of the four water end-members at the surface. (**a**) Kuroshio surface water (KSW); (**b**) Kuroshio sub-surface water (KSSW); (**c**) Changjiang diluted water (CDW); (**d**) Taiwan strait warm water (TSWW)

17.3.3.2 Residence Time

Conservative Ra Model

^{223}Ra and ^{224}Ra have been used to evaluate near-shore water residence times (Dulaiova et al. 2006; Swarzenski and Izbicki 2009). In a study by Moore (2000) it is assumed that a strong pycnocline isolates the surface water from the bottom water (waters are strongly stratified), and that lateral transport can be ignored. The change in concentration of radium activity (A) with time (t) as a function of distance offshore (Z) may be expressed as a balance of advection and diffusion:

$$\frac{dA}{dt} = K_h \frac{\partial^2 A}{\partial Z^2} - \omega \frac{\partial A}{\partial Z} - \lambda A, \quad (17.25)$$

where K_h is the eddy diffusion coefficient, and ω is the advective velocity. For the long-lived radium isotopes, the decay term again can be neglected. If the offshore distribution is dominated by diffusion with constant K_h, a plot of activity vs. distance will be a straight line connecting the end members. The above equation reduces to

$$\frac{dA}{dt} = K_h \frac{\partial^2 A}{\partial Z^2} - \lambda A. \quad (17.26)$$

Solving the (17.26), we can obtain

$$A = A_1 e^{\sqrt{\frac{\lambda}{K_h}}Z} + A_2 e^{-\sqrt{\frac{\lambda}{K_h}}Z}. \quad (17.27)$$

A_1 and A_2 are constant.
In first case, the boundary conditions are:

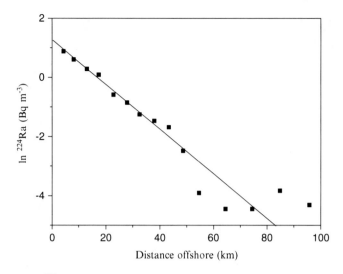

Fig. 17.6 The distance-averaged ln^{224}Ra as a function of distance offshore. The lines are based on the best fit to samples collected within 50 km of shore. The slope is -0.0121 Bq m^{-3} km^{-1} with R$^2 = 0.986$ (Moore 2000)

$A = A_0$ at $Z = 0$,
$A \to 0$ as $Z \to \infty$.

If K_h is constant and the system is in steady state,

$$A_Z = A_0 e^{-\sqrt{\frac{\lambda}{K_h}}Z}, \quad (17.28)$$

where A_Z is the activity at distance Z from the coast, A_0 is the activity at distance 0 from the coast and λ is the decay constant. Based on the above equation, the distance Z can be expressed as:

$$Z = \sqrt{\frac{K_h}{\lambda}} \cdot \ln\frac{A_0}{A_z}. \quad (17.29)$$

A plot of ln ^{224}Ra as a function of distance from the coast may be used to estimate K_h, i.e. Fig. 17.6 (Moore 2000). In second case, the boundary conditions are:
$A = A_0$ at $Z = 0$,
$A \to A_L$ as $Z = Z_L$
Combined with (17.27), we obtain:

$$A_0 = A_1 + A_2$$

and

$$A_L = A_1 e^{\sqrt{\frac{\lambda}{K_h}}Z_L} + A_2 e^{-\sqrt{\frac{\lambda}{K_h}}Z_L}. \quad (17.30)$$

The A_1 and A_2 can be solved as follow

$$A_1 = \frac{A_L - A_0 e^{-\sqrt{\frac{\lambda}{K_h}}Z_L}}{e^{\sqrt{\frac{\lambda}{K_h}}Z_L} - e^{-\sqrt{\frac{\lambda}{K_h}}Z_L}}$$

and

$$A_2 = \frac{A_0 e^{\sqrt{\frac{\lambda}{K_h}}Z_L} - A_L}{e^{\sqrt{\frac{\lambda}{K_h}}Z_L} - e^{-\sqrt{\frac{\lambda}{K_h}}Z_L}}. \quad (17.31)$$

Combined with $Sinh x = \frac{e^x - e^{-x}}{2}$, the concentration of radium activity (A) is expressed as (Chung and Kim 1980):

$$A = \frac{1}{Sinh\left(\sqrt{\frac{\lambda}{K_h}} \cdot Z_L\right)} \left(A_L Sinh\left(\sqrt{\frac{\lambda}{K_h}} \cdot Z\right) \right.$$
$$\left. + A_0 Sinh\left[\sqrt{\frac{\lambda}{K_h}} \cdot (Z_L - Z)\right] \right). \quad (17.32)$$

Non-Conservative Ra Model

Kim et al. (2005) used the method (described in Sect. 17.3.2.2) to estimate the average water residence time in Yellow Sea and adjacent ocean. Figure 17.7 shows the $(^{228}Ra/^{226}Ra)_{ex}$ vs. ^{228}Ra activity; the intercept and slope can be calculated by this relationship. The average water residence times of ~4.9 year is obtained using (17.12).

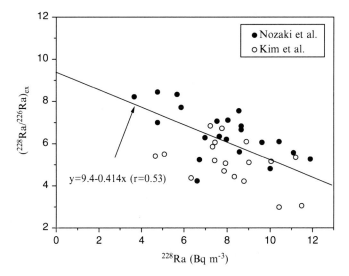

Fig. 17.7 A plot of $(^{228}Ra/^{226}Ra)_{ex}$ vs. ^{228}Ra activity. The data for the *closed circles* are from Nozaki et al. (1991) and those for the *open circles* are from Kim et al. (2005). The slope and intercept of this plot are used for calculating Ra fluxes in the study area using (17.14)

17.4 Future Directions

River-dominated coastal waters and the adjacent continental shelf are important marine regions where the continuous exchange of water and chemical constituents bear the geochemical signals entrained between the continent and the open oceans. These waters occupy less than one fifth of the surface area of the world ocean, yet they play a major role in terms of global biogeochemical cycles (Walsh 1991; Iseki et al. 2003). Thus, water mass mixing and exchange rates will continue to be of great interest.

To protect our natural environment, we should resolve to understand the "three-how's" that define the behavior of a contaminant: "how" many contaminants are flowing into a near-shore, coastal system, "how" long will the contaminants reside there, and "how" far will the impact of the contaminants be? By knowing the source and fate of contaminants in the near-shore, one can more realistically make informed decisions as to their management. For mixing among four unique water masses, a study cannot cease at the relative mixing proportions, but should include an understanding of absolute contributions. That will help evaluate the degree of influence for source contaminants. After establishing the contributions from source water masses, we then can utilize appropriate tracers to assess the ages of different coastal source waters.

Acknowledgments We thank Willard S. Moore of University of South Carolina for his discussions on this topic. PWS thanks the USGS Coastal and Marine Program for continued support. The application example in the East China Sea was supported by Chinese Ministry of Science and Technology (2006CB400601).

References

Bailly BP (1996) Mapping of water masses in the North Sea using radioactive tracers. Endeavour 20:2–7

Baskaran M, Murray DJ, Santschi PH et al (1993) A method for rapid in situ extraction and laboratory determination of Th, Pb, and Ra isotopes from large volumes of seawater. Deep Sea Res 40:849–865

Baskaran M, Hong GH, Santschi PH (2009) Radionuclide analyses in seawater. In: Wurl O (ed) Practical guidelines for the analysis of seawater. CRC Press, Boca Raton, pp 259–304

Bolin B, Rodhe H (1973) A note on the concepts of age distribution and transit time in natural reservoirs. Tellus 25:58–63

Burnett WC, Aggarwal PK, Bokuniewicz H et al (2006) Quantifying submarine groundwater discharge in the coastal zone via multiple methods. Sci Total Environ 367:498–543

Chen CT (1985) Preliminary observations of oxygen and carbon dioxide of the winter time Bering Sea marginal ice zone. Cont Shelf Res 4:465–483

Chen CTA, Ruo R, Pai SC et al (1995) Exchange of water masses between the East China Sea and the Kuroshio off northeastern Taiwan. Cont Shelf Res 15:19–39

Chung YC, Craig H (1980) ^{226}Ra in the Pacific Ocean. Earth Planet Sci Lett 49:267–292

Chung YC, Kim K (1980) Excess ^{222}Rn and the benthic boundary layer in the western and southern Indian Ocean. Earth Planet Sci Lett 49:351–359

Deleersnijder E, Campin JM, Delhez EJM (2001) The concept of age in marine modeling. I. I. Theory and preliminary model results. J Mar Syst 28:229–267

Delhez EJM, Campin JM, Hirst AC et al (1999) Toward a general theory of the age in ocean modeling. Ocean Model 1:17–27

Dulaiova H, Burnett WC (2007) Evaluation of the flushing rates of Apalachicola Bay, Florida via natural geochemical tracers. Mar Chem. doi:10.1016/j.marchem.2007.09.001

Dulaiova H, Burnett WC, Wattayakorn G, Sojisuporn P (2006) Are groundwater inputs into river-dominated areas important? The Chao Phraya River – Gulf of Thailand. Limnol Oceanogr 51:2232–2247

Hancock GJ, Martin P (1991) Determination of Ra in environmental samples by alpha particle spectrometry. Appl Radiat Isot 42:63–69

Helland-Hansen B (1916) Nogen hydrografiske metoder. Forh Skand Naturf Mote 16:357–359

Henry-Edwards A, Tomczak M (2006) Detecting changes in Labrador Sea Water through a water mass analysis of BATS data. Ocean Sci 2:19–25

Hong GH, Zhang J, Chung CS (2002) Impact of interface exchange on the biogeochemical processes of the yellow and East China seas. Bum Shin Press, Korea

Hong G-H et al (2011) Applications of anthropogenic radionuclides as tracers to investigate marine environmental processes. In: Baskaran M (ed) Handbook of environmental isotope geochemistry. Springer, Berlin

Hougham AL, Moran SB (2007) Water mass ages of coastal ponds estimated using ^{223}Ra and ^{224}Ra as tracers. Mar Chem 105:194–207

Iseki K, Okamura K, Kiyomoto K (2003) Seasonality and composition of downward particulate fluxes at the continental shelf and Okinawa Trough in the East China Sea. Deep Sea Res 50:457–473

Jacobsen JP (1927) Eine graphsche Methode zur Bestimmung des Vermischungs-Koeffzienten in Meere. Gerlands Beitr Geophsik 16:404

Katz BG, Coplen TB, Bullen TD et al (1997) Use of chemical and isotopic Tracers to characterize the interactions between ground water and surface water in mantled karst. Ground Water 35:1014–1028

Key RM, Stallard RF, Moore WS et al (1985) Distribution and flux of ^{226}Ra and ^{228}Ra in the Amazon River estuary. J Geophys Res 90:6995–7004

Kim G, Ryu JW, Yang HS et al (2005) Submarine groundwater discharge (SGD) into the Yellow Sea revealed by ^{228}Ra and ^{226}Ra isotopes: implications for global silicate fluxes. Earth Planet Sci Lett 237:156–166

Klein B, Tomczak M (1994) Identification of diapycnal mixing through Optimum Multi-parameter analysis: 2. Evidence for unidirectional diapycnal mixing in the front between North and South Atlantic Central Water. J Geophys Res 99:25275–25280

Krest JM, Moore WS, Rama (1999) ^{226}Ra and ^{228}Ra in the mixing zones of the Mississippi and Atchafalaya Rivers: indicators of groundwater input. Mar Chem 64:129–152

Lakey BL, Krothe NC (1996) Stable isotopic variation of storm discharge from a Perennial karst spring, Indiana. Water Resour Res 32:721–731

Lee ES, Krothe NC (2001) A four-component mixing model for water in a karst terrain in south-central Indiana, USA. Using solute concentration and stable isotopes as tracers. Chem Geol 179:129–143

Li YH, Chan LH (1979) Adsorptions of Ba and ^{226}Ra from river borne sediments in the Hudson estuary. Earth Planet Sci Lett 43:343–350

Liu B, Phillips F, Hoines S et al (1995) Water movement in desert soil traced by hydrogen and oxygen isotopes, chloride, and chlorine-36, southern Arizona. J Hydrol 168:91–110

Maamaatuaiahutapu K, Garcon VC, Provost C et al (1992) Brazil-Malvinas Confluence: water mass composition. J Geophys Res 97:9493–9505

Maamaatuaiahutapu K, Garcon VC, Provost C et al (1994) Spring and winter water mass composition in the Brazil-Malvinas Confluence. J Mar Res 52:397–426

Miller AR (1950) A study of mixing processes over the edge of the continental shelf. J Mar Res 9:145–160

Monsen NE, Cloern JE, Lucas LV (2002) A comment on the use of flushing time, residence time, and age as transport time scales. Limnol Oceanogr 47:1545–1553

Moore WS (1969) The measurements of ^{228}Ra and ^{228}Th in sea water. J Geophys Res 74:694–704

Moore WS (1976) Sampling radium-228 in the deep ocean. Deep Sea Res 23:647–651

Moore WS (1981) Radium isotopes in the Chesapeake Bay. Estuar Coast Shelf Sci 12:713–723

Moore WS (1996) Large ground water inputs to coastal waters revealed by ^{226}Ra enrichments. Nature 380:612–614

Moore WS (1999) The subterranean estuary: a reaction zone of ground water and sea water. Mar Chem 65:111–125

Moore WS (2000) Determining coastal mixing rates using radium isotopes. Cont Shelf Res 20:1993–2007

Moore WS (2003) Sources and fluxes of submarine groundwater discharge delineated by radium isotopes. Biogeochemistry 66:75–93

Moore WS (2006) Radium isotopes as tracers of submarine groundwater discharge in Sicily. Cont Shelf Res 26:852–861

Moore WS, Arnold R (1996) Measurement of ^{223}Ra and ^{224}Ra in coastal waters using a delayed coincidence counter. J Geophys Res 101:1321–1329

Moore WS (2007) Fifteen years experience in measuring ^{224}Ra and ^{223}Ra by delayed-coincidence counting. Mar Chem. doi:10.1016/j.marchem.2007.06.015

Moore DG, Scott MR (1986) Behavior of ^{226}Ra in the Mississippi River mixing zone. J Geophys Res 91:14317–14329

Moore WS, Todd JF (1993) Radium isotopes in the Orinoco estuary and eastern Caribbean Sea. J Geophys Res 98:2233–2244

Moore WS, Sarmiento JL, Key RM (1986) Tracing the Amazon component of surface Atlantic water using ^{228}Ra, salinity and silica. J Geophys Res 91:2574–2580

Moore WS, Blanton JO, Joye SB (2006) Estimates of flushing times, submarine groundwater discharge, and nutrient fluxes to Okatee Estuary, South Carolina. J Geophys Res 111:1–14

Nozaki Y, Kasemuspaya V, Tsubota H (1989) Mean residence time of the shelf water in the East China and Yellow Seas determined by ^{228}Ra/^{226}Ra measurements. Geophys Res Lett 16:1297–1300

Nozaki Y, Tsubota H, Kasemsupaya V et al (1991) Residence times of surface water and particle-reactive ^{210}Pb and ^{210}Po in the East China and Yellow seas. Geochim et Cosmochim Acta 55:1265–1272

Pickard GL, Emery WJ (1990) Descriptive physical oceanography: an introduction, 5th edn. Pergamon Press, New York

Poole R, Tomczak M (1999) Optimum multi-parameter analysis of the water mass structure in the Atlantic Ocean thermocline. Deep Sea Res 46:1895–1921

Pougatch K, Salcudean M, Gartshore I et al (2007) Computational modeling of large aerated lagoon hydraulics. Water Res 41:2109–2116

Rama, Moore WS (1996) Using the radium quartet to estimate water exchange and ground water input in salt marshes. Geochim et Cosmochim Acta 60:4645–4652

Rapaglia J, Ferrarin C, Zaggia L et al (2009) Investigation of residence time and groundwater flux in Venice Lagoon: comparing radium isotope and hydrodynamical models. J Environ Radioact. doi:10.1016/j.jenvrad.2009.08.010

Sverdrup HU, Johnson MW, Fleming RH (1946) The oceans, their physics, chemistry and general biology. Prentice-Hall, Inc, New York

Swarzenski PW, Izbicki JA (2009) Examining coastal exchange processes within a sandy beach using geochemical tracers, seepage meters and electrical resistivity. Estuar Coast Shelf Sci 83:77–89

Takeoka H (1984) Fundamental concepts of exchange and transport time scales in a coastal sea. Cont Shelf Res 3:322–326

Tomczak M (1981) A multi-parameter extension of temperature/salinity diagram techniques for the analysis of non-isopycnal mixing. Prog Oceanogr 10:147–171

Vilibic I, Orlic M (2002) Adriatic water masses, their rates of formation and transport through the Otranto Strait. Deep Sea Res 49:1321–1340

Walsh JJ (1991) Importance of continental margins in the marine biogeochemical cycling of carbon and nitrogen. Nature 50:53–55

Yang HS, Hwang DW, Kim G (2002) Factors controlling excess radium in the Nakdong River estuary, Korea: submarine groundwater discharge versus desorption from riverine particles. Mar Chem 78:1–8

Zhang J (2002) Biogeochemistry of Chinese estuaries and coastal waters: nutrients, trace metals and biomarkers. Reg Environ Change 3:65–76

Zhang L, Liu Z, Zhang J et al (2007) Reevaluation of mixing among multiple water masses in the shelf: an example from the East China Sea. Cont Shelf Res 27:1969–1979

Zimmerman JTF (1976) Mixing and flushing of tidal embayments in the Western Dutch Wadden Sea, Part I: distribution of salinity and calculation of mixing time scales. Nethl J Sea Res 10:149–191

Chapter 18
Natural Radium and Radon Tracers to Quantify Water Exchange and Movement in Reservoirs

C.G. Smith, P.W. Swarzenski, N.T. Dimova, and J. Zhang

Abstract Radon and radium isotopes are routinely used to quantify exchange rates between different hydrologic reservoirs. Since their recognition as oceanic tracers in the 1960s, both radon and radium have been used to examine processes such as air-sea exchange, deep oceanic mixing, benthic inputs, and many others. Recently, the application of radon-222 and the radium-quartet (^{223,224,226,228}Ra) as coastal tracers has seen a revelation with the growing interest in coastal groundwater dynamics. The enrichment of these isotopes in benthic fluids including groundwater makes both radium and radon ideal tracers of coastal benthic processes (e.g. submarine groundwater discharge). In this chapter we review traditional and recent advances in the application of radon and radium isotopes to understand mixing and exchange between various hydrologic reservoirs, specifically: (1) atmosphere and ocean, (2) deep and shallow oceanic water masses, (3) coastal groundwater/benthic pore waters and surface ocean, and (4) aquifer-lakes. While the isotopes themselves and their distribution in the environment provide qualitative information about the exchange processes, it is mixing/exchange and transport models for these isotopes that provide specific quantitative information about these processes. Brief introductions of these models and mixing parameters are provided for both historical and more recent studies.

18.1 Introduction

18.1.1 Hydrologic Cycle

Roughly 97% of the Earth's water resides as salt water in the world's oceans, while the remaining ~3% is sequestered as freshwater in ice caps and glaciers, groundwater, and surface water features such as lakes, rivers, and wetlands (Table 18.1; Fig. 18.1). This water is also the primary transport agent for sediment, solutes, and heat, and the most essential compound for sustaining life on Earth. As a result of human dependencies on these unique characteristics, water and its transfer through the Earth systems has been an active area of human interest for at least 12,000 years (Miller 1982; Hardcastle 1987; Moore et al. 1995). As the Earth is a closed system with respect to water, a global water mass balance (Table 18.1) can be visualized as a group of cyclic processes, such as evaporation and precipitation, that exchange water between three large-scale reservoirs (i.e. ocean, atmosphere, and land). This model represents the water or hydrologic cycle (Fig. 18.2). The purpose of this chapter is to review how the radioisotopes of radium and radon have been utilized as tracers to quantify several hydrologic processes, including water and material exchange across the three main reservoirs and their respective sub-reservoirs. Specific exchange mechanisms addressed in this chapter span

C.G. Smith (✉)
US Geological Survey, 600 4th Street South, St. Petersburg, FL, USA
e-mail: cgsmith@usgs.gov

P.W. Swarzenski
US Geological Survey, 400 Natural Bridges Drive, Santa Cruz, CA, USA
e-mail: pswarzen@usgs.gov

N.T. Dimova
UC-Santa Cruz/US Geological Survey, 400 Natural Bridges Drive, Santa Cruz, CA, USA
e-mail: ndimova@usgs.gov

J. Zhang
State Key Laboratory of Estuarine and Coastal Research, East China Normal University, Shanghai 2000062, China
e-mail: jzhang@sklec.ecnu.edu.cn

Table 18.1 List of major reservoirs and sub-reservoirs within the hydrologic cycle and there combined contribution to total water budget and freshwater budget (values obtained from Maidment 1993)

	Volume (km^3)	Percentage of total	Residence time (year)
Water in land areas	47,971,710	3.5	–
Lakes	176,400	0.013	–
Rivers	2,120	0.00015	4
Marshes	11,470	0.00083	–
Groundwater and soil moisture	23,416,500	1.7	20,000
Biological water	1,120	0.00008	–
Icecaps and glaciers	24,364,100	1.8	–
Atmosphere	12,900	0.0009	0.02
Oceans	1,338,000,000	96.5	3,000
Total	1,385,984,610	100	–

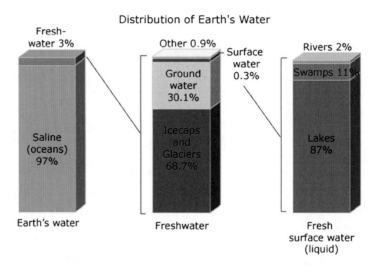

Fig. 18.1 Bar graph showing the two predominant fractions of Earth's water: fresh and saline. The freshwater component is further broken into various fractions or storage (sub)reservoirs. Note, variations in percentages listed in this graph and those given in Table 18.1 reflect general uncertainty in the division of water into its various forms on Earth. (http://ga.water.usgs.gov/edu/earthwherewater.html)

from interstitial fluids to the atmosphere, and include groundwater-coastal water-, riverine -estuarine-, coastal ocean-, and ocean-atmospheric-exchange.

18.1.2 Geochemical and Radiochemical Qualities of Radium-Isotopes

Radium (Ra) is the heaviest (atomic number 88) of the alkaline earth metals, placing it at the base of the second column of the periodic table of elements. Like other alkaline earth metals, radium is a divalent cation (Ra^{2+}) and occurs in aquatic environments as either a dissolved cation or as a complex with some anionic ligand. The ionic radius of Ra is 1.40 Å and its ionic potential (Z/r) is 1.43 (Railsback 2003). In solution, radium behaves very similar to barium (Ba) due to similar ionic potentials (Ra = 1.43; Ba = 1.48) and substitutes readily for Ba in many minerals (e.g. barite, BaSO$_4$). The geochemical behavior of Ra and Ba in coastal-marine environments has been studied for over 50 years (e.g. Koczy 1958; Turekian and Johnson 1966; Hanor and Chan 1977; Li and Chan 1979; Key et al. 1985; Moore and Dymond 1991; Webster et al. 1995; Gonneea et al. 2008). In freshwater environments (i.e. ground/surface water), Ra occurs as a dissolved cation that commonly is adsorbed onto negatively-charged clays, mineral coatings, and/or particulate organic matter (Beneš et al. 1984; Langmuir and Riese 1985; Beneš et al. 1986). In contrast, in marine environments Ra remains dissolved (Ra^{2+}) or forms complexes with inorganic

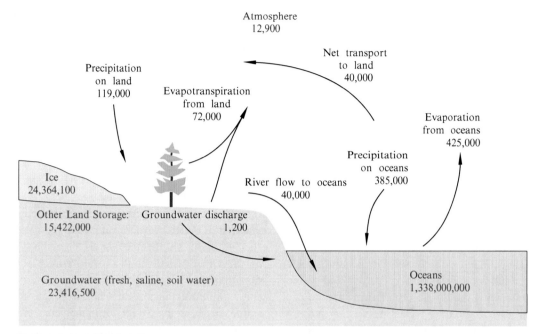

Fig. 18.2 Diagrammatic representation of the hydrologic cycle, which shows the predominant (sub) reservoirs (units of km^3) and the exchange pathways (units of km^3 year^{-1}) among each. Figure modified from Winter et al. (1998); values obtained from Maidment (1993)

(e.g. RaCl$^+$, RaSO$_4^0$, RaOH$^+$) and organic anionic ligands (Lauria et al. 2004). As Ra is transported through estuaries (i.e. mixing between fresh and saline waters) desorption and surface exchange play an important role in its distribution in the water between adsorbed and dissolved phases. This dissimilar behavior in fresh and marine (saline) environments allows Ra to be used to study water mass mixing processes, solute exchange, and fluid exchange between the terrestrial and marine environment (i.e. the land and ocean reservoirs).

There are four naturally occurring isotopes of Ra ($^{223, 224, 226, 228}$Ra), which have varying half-lives that span from just a few days to over 1,600 year ($t_{1/2}$ ^{223}Ra = 11.4 day, ^{224}Ra = 3.66 day, ^{226}Ra = 1,600 year, and ^{228}Ra = 5.75 year) (Fig. 2.3, Chap. 2). The production of $^{223, 224, 226, 228}$Ra is from alpha decay of the strongly particle reactive, sediment-derived, $^{227, 228, 230, 232}$Th parents that precede them in the U/Th decay series. The contrasting chemical behavior between Ra and Th generally leads to excess levels of Ra relative to its Th parent in waters in close proximity to sediment sources (i.e. aquifers, estuaries, pelagic ocean). Also, because the half-life of each Ra isotope is substantially different while each isotope is physiochemically identical, collectively the "Ra-quartet" has been successfully used to examine complex coastal mixing processes and benthic exchange (e.g. Moore 2000a, b).

18.1.3 Geochemical and Radiochemical Qualities of Radon-Isotopes

Radon (Rn) is the heaviest of the noble gases with an atomic number of 86. Because of its balanced outer electron shell, it is chemically inert and behaves conservatively in nearly all chemical reactions in nature (except weak van der Waals bonds). Radon is ubiquitous in the natural environment and found in the atmosphere, trapped in mineral grains, and also dissolved in solution. At standard temperature and atmospheric conditions, the gas solubility constant of Rn in water is 10$^{-2.03}$ M (K = 10$^{19.06}$ Bq m^{-3}), suggesting that the possibility of saturation is low. There are two isotopes – radon-222 (^{222}Rn or radon[1]) and radon-220 (^{220}Rn or thoron) that are relevant to environmental and hydrologic studies. Both ^{222}Rn and ^{220}Rn are short-lived ($t_{1/2}$ = 3.82 days and 55.6 s, respectively), and are produced by the alpha decay of a parent Ra isotope (^{226}Ra and ^{224}Ra, respectively) (Fig. 2.3, Chap. 2). Similar to the Th-Ra parent-daughter relationship, dissimilar geochemical behaviors of Ra and Rn provide for a unique system where the daughter elements can easily decouple, and depending on the environment and processes (turbulent mixing, groundwater flow), either excess or deficiencies of the daughter isotopes relative to their radiogenic parents are likely. To date, ^{222}Rn has been more widely applied to environmental processes than short-lived ^{220}Rn; however, continued improvements in in situ Rn detection (see Methodology section on Planar silicon detectors) has opened the possibility of using thoron as a powerful hydrologic tracer. Due to its utility more so than its ease of measurement, ^{222}Rn has been used at almost every possible (sub)reservoir interface within the hydrologic cycle (e.g. Broecker et al. 1967; Clements and Wilkening 1974; Hammond et al. 1977; Gruebel and Martens 1984; Hoehn et al. 1992; Bertin and Bourg 1994; Cable et al. 1996a).

18.2 Analysis of Rn and Ra

18.2.1 Radon and Thoron

A variety of analytical techniques exist for quantifying radon and thoron activity; most methods measure particle emission from the Rn isotope (α) and/or daughter isotopes (α/β). For example, ^{222}Rn decay to ^{210}Pb results in three α–particles emissions (Fig. 2.3), all of which are equally detectable. Although other methods exits, the three most commonly used radioanalytical techniques for the measuring Rn in natural waters are the Rm extraction line with Lucas cell system, liquid scintillation counting (LSC) methods, and portable radon-in-air detectors.

One of the earliest and still actively used techniques for measuring Rn and Rn emanation (Ra) is the Rn extraction line with Lucas cell system. The method builds off the closed-vessel experiments by the Curies, Friedrich Ernst Dorn, and Ernest Rutherford that lead to the discovery of a gaseous release from Ra (i.e. radium emanation or radon) and thorium (thorium emanation or thoron) minerals (Rutherford 1906). Detailed procedures are provided in a number of studies and references therein (e.g. Broecker et al. 1967; Key et al. 1979; Mathieu et al. 1988; Cable et al. 1996b; Smith et al. 2008; Peterson et al. 2009b). The system has one of the lowest minimum detection limits/activities (MDL or MDA) and the highest detection efficiencies of all the techniques described in this chapter. Other benefits to this measurement technique are supported ^{222}Rn (^{226}Ra) can be measured on the same sample and the method is non-destructive so that sample can be recovered for additional analysis. However, the sample collection and analysis require substantial preparation and user interaction.

Another common detection method for radon (^{222}Rn) in water is liquid scintillation counting; this technique is widely used in groundwater and pore water studies (e.g. Kitto 1994; Freyer et al. 1997; Purkl and Eisenhauer 2004; Smith et al. 2008) and procedures generally follow Prichard and Gesell (1977). LSC has several advantages that stem from collection to analysis, including rapid and easy sample collection, small sample volumes, high sample throughput (automatic sample changers and pre-set count times), and limited (required) user interaction/involvement during measurements. In addition, modern commercial LSC instruments have a high efficiency (280–340%). Drawbacks of LSC determination of radon are that only total radon is measured, high background can limit MDA, and energy resolution is generally poor. However, improvements to pulse-shape discrimination (PSD) circuitry, coincidence circuitry, and additional high-purity shielding have improved on many of these problems with liquid scintillation analyzers.

More recently, portable radon-in-air monitors coupled with air-water exchangers have been used to provide accurate, near real-time in-situ measurement of Rn dissolved in water (e.g. Burnett and Dulaiova 2003). Commercially available systems are available and are of three general types: (1) portable scintillation or "Lucas" cells and PMTs (e.g. AB-4/AB-4A

---

[1] From this point forward, radon refers to the radionuclide ^{222}Rn unless otherwise indicated.

Portable Radiation Monitors from Pylon Electronics, Inc.); (2) ionization chambers (e.g. TBM-IC-RN by Technical Associates Nuclear Instrumentation); or (3) solid-state (semiconductor) detectors (e.g. RAD7 by Durridge Company, Inc). Critical to these instruments is the extraction of Rn from water into air; this has been accomplished by either sparging a discrete water sample with ambient air (e.g. Lee and Kim 2006) in a closed air loop or by aspirating the water sample into a closed air loop (Burnett et al. 2001; Burnett and Dulaiova 2003; Dulaiova et al. 2005; Dimova et al. 2009). Both techniques have been rigorously tested and verified in both laboratory and field settings (e.g. Burnett et al. 2001; Burnett and Dulaiova 2003; Dulaiova et al. 2005; Lee and Kim 2006). The combined portability of the solid-state detectors and the two quantitative extraction techniques have allowed these systems to be used to map Rn in surface waters (e.g. Dulaiova et al. 2005) as well as provided a more autonomous system for measuring discrete samples, potentially replacing the traditional "radon line" and LSC techniques for discrete samples (Stringer and Burnett 2004; Lee and Kim 2006).

18.2.2 Radium

As with radon and thoron, numerous techniques have been developed to quantify ^{223,224,226,228}Ra isotopes in natural waters (Ivanovich and Murray 1992). Those highlighted in this chapter, which are by no means exhaustive, represent the bulk of the techniques currently used to address the questions concerning fluid and material exchange. Traditional alpha-, beta, or gamma-emissions measurements of appropriate parent or daughter nuclides as well as liquid scintillation (Rutgers van der Loeff and Moore 1999), Rn emanation (Butts et al. 1988), coincidence detection (i.e. α-β, γ-γ) (Antovic and Svrkota 2009), and mass spectrometry (Cohen and O'Nions 1991; Ollivier et al. 2008) have been used for determination of Ra-isotopes.

Each of the four Ra isotopes can be very precisely quantified using alpha spectrometry; however, all require antecedent wet-chemistry that is laborious. Thus, indirect ingrowth scintillation methods of Rn-daughters are common. Because of the low activity of ^{223,224,228}Ra in seawater, most of the radioanalytical techniques require pre-concentration of Ra from large volumes of water. Acrylic fiber impregnated with MnO_2 (i.e. Mn-fiber) or some equivalent (e.g. manufactured MnO_2 cartridge) originally described by Moore and Reid (1973) is the most common pre-concentration media in use today for oceanographic and coastal studies (Bourquin et al. 2008). ^{223,224}Ra isotopes adsorbed on to the Mn-fiber are commonly measured using a delayed coincidence counting method (commonly referred to as RaDeCC) first established by Moore and Arnold (1996). Principles of the RaDeCC system were pioneered by Giffin et al. (1963) who worked on a delayed coincidence method for actinon (^{219}Rn) and thoron (^{220}Rn). Rama et al. (1987) developed a similar recirculating, α-particle measurement system for ^{224}Ra that relied on the ingrowth of ^{220}Rn. However, determination of low-level ^{220}Rn generally required long counting periods and interference from ingrowth ^{222}Rn and its daughter isotopes greatly affected the precision of the instrument. The RaDeCC system can also be used to measure ^{226}Ra on Mn-fiber after an extended 30-day ingrowth period (Waska et al. 2008; Peterson et al. 2009b). Kim et al. (2001) and Dimova et al. (2007) utilized a commercial radon-in-air monitor (RAD7-Durridge Company, Inc) that uses electrostatic attraction to a silicon semiconductor detector to quantify Ra (-224 and -226 specifically) via Rn-ingrowth. One potential problem with these techniques is the reliance on the pre-concentration of Ra onto Mn-fiber. While numerous studies have shown near-quantitative extraction (i.e. nearly 100% adsorption of Ra onto the fiber), a number of other studies have found that the Ra sorption to MnO_2 is sensitive to pH (Bourquin et al. 2008; Dimova et al. 2008), salinity (Moon et al. 2003), and flow-rate (Moon et al. 2003).

Radium-isotopes can also be measured by γ-spectrometry. Each gamma ray photon has a discrete energy, and this energy is characteristic of the source isotope (Michel et al. 1981; Moore 1984; Krishnaswami et al. 1991; Dulaiova and Burnett 2004; van Beek et al. 2010). This measurement technique, while offering the benefit of being minimally destructive and requiring substantially less laboratory time, suffers from usually being less sensitive than standard α- or β-spectrometry methods due to interferences by other gamma-emitting isotopes. As a result, this method is usually applicable to samples containing relatively high activities of ^{226,228}Ra.

Measurement of Ra by thermal ionization- and inductively coupled plasma-mass spectrometry (TIMS and ICP-MS, respectively) has increased significantly over the last 20 years (Cohen and O'Nions 1991; Larivière et al. 2003; Ollivier et al. 2008). TIMS has been used to precisely measure the longer-lived ^{226,228}Ra isotopes in a number of field studies where overall activity and/or excess activity were low (Ollivier et al. 2008) and sample volumes were limited (Cohen and O'Nions 1991). However, sample preparation for TIMS is laborious and generally involves the pre-concentration and purification of Ra to remove barium (Ghaleb et al. 2004). Generally speaking, sample preparation times for TIMS are comparable to other radiometric techniques (e.g. α-spectrometry). The advent and progression of high-resolution ICP-MS (HR-ICP-MS) and multiple collector ICP-MS (MC-ICP-MS) have improved the precision and throughput (relative to TIMS) of water samples for Ra analysis (e.g. Varga 2008; Sharabi et al. 2010), often reducing the analytical time for ^{226}Ra to minutes. However, isobaric interferences at the 228 mass often occur with ICP-MS and have limited the detection and quantification of ^{228}Ra via this method (Hou and Roos 2008).

18.3 Application of Radon and Radium to Fluid and Material Exchange Among Hydrologic Reservoirs

18.3.1 Air-Sea Exchange

The exchange of dissolved gases and fluid across the air-sea boundary has been an active area of research for over 60 years (Redfield 1948). However, the importance of the boundary flux has had a substantial revival and progression over the last 20 years due to increased concerns of how the emission of anthropogenic-derived CO_2 and other greenhouse gases will affect global climate and ocean biogeochemistry. While this transfer is not directly facilitated by water (i.e. water is not a "transport agent"), the exchange between two important hydrologic reservoirs (ocean and atmosphere) is the focal point of this globally significant topic, and ^{222}Rn/^{226}Ra disequilibrium has provided much insight to exchange rates and processes governing exchange (Peng et al. 1979; Fanning et al. 1982; Kawabata et al. 2003; Schery and Huang 2004; Hirsch 2007). This section provides a brief overview of air-sea exchange and how Rn and Ra isotopes have been used to advance this study. A more comprehensive, updated review of air-sea exchange and its quantification can be found in Liss et al. (1988), Wanninkhof et al. (2009), and references therein.

18.3.1.1 ^{222}Rn/^{226}Ra Disequilibrium in the Surface Ocean

In 1965, Broecker (1965) suggested that the "discontinuities" (i.e. disequilibrium) between ^{222}Rn and its parent ^{226}Ra in the surface ocean could be used to provide estimates of air-sea exchange, mixing between the surface and deep ocean, and sediment-water (benthic) exchange. Radon's predictable source/sink terms in the surface ocean, negligible activity in the atmosphere, and its ease of measurement on board ships (using the radon-line and Lucas cell method) made it an attractive tracer to quantify air-sea exchange rates; however, some obstacles for sampling had to be overcome (Peng et al. 1979).

In the surface ocean, the ^{222}Rn/^{226}Ra disequilibrium originates from the evasion of radon across the air-sea interface. Thus, a near surface water column mass balance of ^{222}Rn ($A_{Rn_{water}}$) can be used to estimate evasion rates or gas transfer velocities (k, m day^{-1}) in the open ocean using (Fig. 18.3)

$$F_{Rn} = k(A_{Rn_{water}} - \alpha A_{Rn_{air}}), \quad (18.1)$$

where F_{Rn} is the flux of radon across air-sea interface (Bq m^{-2} day^{-1}), α is the Ostwald solubility coefficient (-), and $A_{Rn_{air}}$ is the ^{222}Rn activity in the air (Bq m^{-3}).

Integrating the radon deficit over the thickness of the surface mixed layer provides a deficit inventory of radon (I_{def}) which is proportional to the efflux of radon due to atmospheric evasion (assuming steady-state mixing)

$$F_{^{222}Rn} = \lambda I_{def} = \lambda \sum_{z=z_{base}}^{z=0} (A_{Rn}(z) - A_{Ra}(z))\Delta z, \quad (18.2)$$

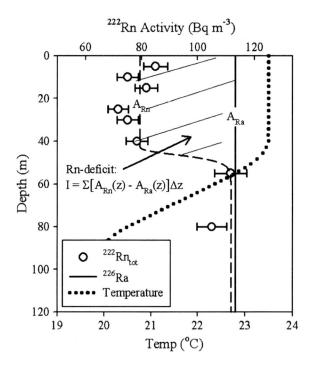

Fig. 18.3 Example of how earlier research where radon was used to evaluate air-sea exchange using a version of the Stagnant Boundary Layer model. The radon deficit is integrated over the upper surface mixed layer (as denoted by both radon and temperature) and used to compute the gas transfer or piston velocity (18.3). Radon data are taken from GEOSECS station number 57 as published by Broecker and Peng (1974) (units were modified from dpm 100 L^{-1} to Bq m^{-3})

where λ is the decay constant of ^{222}Rn (0.181 day^{-1}), $A_{Rn}(z)$ and $A_{Ra}(z)$ are the activities (Bq m^{-3}) of ^{222}Rn and ^{226}Ra in the water column, respectively at depth z (m), and Δz is the sample interval (m). Substituting (18.2) into (18.1) and assuming that radon in air is negligible, gas transfer velocity (k) is simply

$$k = \lambda I_{def} / A_{Rn_{water}}. \qquad (18.3)$$

Broecker et al. (1967) provided a detail proof of concept with data collected from Pacific Ocean (ocean stretch between Hawaii and Japan); however, the data left uncertainties as to the validity of transposing radon based evasion rates to quantifying CO$_2$ air-sea exchange. Broecker and Peng (1971) used a similar approach at the Bomex site in the Atlantic Ocean; the radon-derived surface boundary layer (SBL) thickness and radon gas transfer velocity were estimated to be 62 μm and 1.9 m day^{-1}, respectively (Table 18.2). Peng et al (1974) followed up with a north Pacific survey at the Papa site; estimates of SBL thickness and radon gas transfer velocity were 20 μm and 3.6 m day^{-1} (Table 18.2). Then as part of GEOSECS, Peng et al. (1979) used radon-deficit to quantify gas exchange rates at over 100 sites within the Atlantic and Pacific Oceans, which they used to derive a global gas transfer rate of 2.9 m day^{-1} and a SBL thickness of 36 μm. The radon-based global gas transfer rate (Peng et al. 1979) agreed favorably well (within 20%) with radiocarbon-based fluxes (Broecker and Peng 1974). Smethie et al. (1985) revisited the transect of Broecker and Peng (1974) and found that radon-estimates of gas exchange rates were within error of those provided by Broecker and Peng (1974) a decade earlier using bomb carbon-14 (^{14}C). Many of these early estimates of radon evasion rates or gas transfer velocities did not specifically address the effects of prolonged storminess or variable wind conditions. In the western North Pacific, Kawabata et al. (2003) estimated gas transfer velocities that ranged from 2.1 to 30.2 m day^{-1}; which is 1 to 10× the global average proposed by Peng et al. (1979). The higher velocities were related to periods of extended storminess and high apparent windspeeds, which are not well represented by the steady-state assumption implied in (18.3).

A number of other scientists argued that the gas transfer velocities obtained by Peng et al. (1979) were not consistent with direct estimates of air-sea CO$_2$ fluxes (e.g. Jones and Smith 1977; Smith and Jones 1985; Liss et al. 1988; Wanninkhof et al. 2009). Peng et al. (1979) admitted to some uncertainty in their own data, as the k values they obtained from the radon-deficit method did not demonstrate a discernible trend with wind speed. The relationship between gas exchange rates and wind speed had already been verified in the laboratory experiments (Broecker et al. 1978). The discrepancy was later shown to reflect the dissimilarity in the time scale of atmospheric forcings (i.e. hours to days) relative to the equilibration of ^{222}Rn in the surface mixed layer (days). That is to say, the ^{222}Rn distributions measured at some point in time in the surface mixed layer is not only a reflection of decay and evasion given the current wind stress, but also had a hysteresis effect from transient atmospheric forcing(s) (Liss et al. 1988; Kawabata et al. 2003; Wanninkhof et al. 2009). If this previous

Table 18.2 Summary table of studies where ^{222}Rn and ^{226}Ra disequilibrium in the surface mixed to determine gas transfer velocities, the general location of each site, temperature (Temp., °C), molecular diffusion coefficient (D_m, cm^2 s^{-1}), gas transfer velocity (v or k, m day^{-1}), stagnant film thickness (z*, μm), and gas transfer velocity normalized to 20°C (v_{20} or k_{20}, m day^{-1})

Location	Temp (°C)	D_m (cm^2 s^{-1})	k or v (m day^{-1})	z* (mm)	k_{20} or v_{20} (m day^{-1})	Author
Atlantic Ocean – Bomex	25	1.36E-05	1.91	62	1.6	Broecker and Peng (1971)
North Pacific Ocean – Papa	5	8.10E-06	3.6	20	5.2	Peng et al. (1974)
90-locations Atlantic and Pacific Oceans	20	1.20E-05	2.86	36	2.8	Peng et al. (1979)
18-locations: S. Polar Atlantic, Pacific and Global Oceans	20	1.20E-05	4.1	25	4.0	Peng et al. (1979)
19-locations: S. Temperate Atlantic and Pacific Oceans	20	1.20E-05	2.9	36	2.8	Peng et al. (1979)
24-locations: Equatorial Atlantic and Pacific Oceans	20	1.20E-05	2.1	49	2.0	Peng et al. (1979)
19-locations: N. Temperate Atlantic and Pacific Oceans	20	1.20E-05	2.6	40	2.5	Peng et al. (1979)
10-locations: N. Polar Atlantic and Pacific Oceans	20	1.20E-05	2.6	40	2.5	Peng et al. (1979)
Theoritical for 40 m thick mixed zone	0	7.00E-06	1.5	40	2.5	Broecker and Peng (1982)
Theoritical for 40 m thick mixed zone	24	1.40E-05	3	40	2.5	Broecker and Peng (1982)
Pee Dee River, SC	19.6	1.15E-05	2.7	37	2.8	Elsinger and Moore (1983)
San Francisco Bay – 6 year average	–	–	1.0	–	–	Hartman and Hammond (1984)
26-locations Tropical Atlantic Ocean	26.2	1.42E-05	3.7	34	3.0	Smethie et al. (1985)
Bering Sea*	–	2.74E-07	2.4	42	100.8	Glover and Reeburgh (1987)
Barents Sea – Summer	–	–	1.8	–	3.5	Fanning and Torres (1991)
Barents Sea – Winter	–	–	1.1	–	2.0	Fanning and Torres (1991)
western North Pacific Ocean	20	1.20E-05	9.4	11	9.1	Kawabata et al. (2003)
Upper Gulf of Thailand	–	–	0.38	–	–	Dulaiova and Burnett (2006)

forcing were on different intensity and frequency scales, then the assumption of a steady-state ^{222}Rn distribution would be invalid and likewise the applicability of the model to estimate gaseous transfer.

Kawabata et al. (2003) modified the radon-based gas transfer velocity model to account for sporadic gas exchange. By allowing the radon to exchange with the atmosphere sporadically, (i.e. once or twice per time allotted), the influence of local and regional wind-forcing via storms is more readily represented in a model. Using this approach, the radon activity ($A_{Rn(t)}$; Bq m^{-3}) at any point in time after a sporadic removal and given a prescribed initial gas transfer velocity and mixed layer (i.e. residence time) can be represented by:

$$A_{Rn(t)} = \left[I_{Rn(t-\Delta t)}e^{-\lambda_{Rn}\Delta t} + A_{Ra}h\left(e^{-\lambda_{Ra}\Delta t} - e^{-\lambda_{Rn}\Delta t}\right)\right]/h. \quad (18.4)$$

where; A_{Ra} is the activity of ^{226}Ra in the surface mixed layer (Bq m^{-3}); λ_{Rn} and λ_{Ra} are the decay constants for ^{222}Rn and ^{226}Ra (day^{-1}), respectively; $I_{Rn(t-\Delta t)}$ is the inventory of ^{222}Rn in the surface mixed layer (Bq m^{-2}) for time $t-\Delta t$; h is the thickness of the surface mixed layer (m). The transient gas transfer velocity, $k(t)$ (m day^{-1}) is

$$k(t) = h\lambda_{Rn}\left(\frac{A_{Ra}}{A_{Rn(t)}} - 1\right). \quad (18.5)$$

This model is not capable of modeling specific wind events but simply addresses how large-scale disturbances influence gas transfer velocities obtained following an event. Kawabata et al. (2003) found that using this approach and assuming a 7-day period between storms, that the steady-state overestimated gas transfer velocities by less than 20% relative to the sporadic model. Model improvement such as these, which investigate transient impacts to traditionally steady-state assumptions, could represent the next generation of natural tracer estimates of gas transfer between the ocean and atmosphere as well as a means to further refine existing empirical relationships between gas transfer velocities and wind-speed (e.g. MacIntyre et al 1995; Raymond and Cole 2001; Wanninkhof et al. 2009).

Despite the uncertainties and misnomers associated with the radon-derived fluxes, what the radon-deficit method did was start a revelation and progression in air-sea exchange research. For example, Jones and Smith (1977) using eddy correlation technique (i.e. correlation of $pCO2$ in the atmosphere and measured wind-speed) to provide some of the first direct measurements of CO_2 flux. Roether and Kromer (1978) and Kromer and Roether (1983) refined atmospheric and aqueous Rn sampling and analyzed the uncertainties with radon-deficit method. While radon is not as widely used to quantify air-sea exchange as it once was; mass balance models are still used, namely the dual-deliberate isotopic methods (e.g. ^{3}He and SF_6) highlighted by Wanninkhof et al. (1993). Also, the atmospheric radon flux and its parameterization as defined here is one of the key parameters in estimating benthic fluid discharge to the coastal zone as will be described in Sect. 18.3.3 (Cable et al. 1996a, b; Corbett et al. 1997; Burnett and Dulaiova 2003). Modern coastal groundwater studies often use empirical relationships between gas transfer velocities and wind speed (e.g. MacIntyre et al. 1995) and measured radon gradients, temperature, and salinity to estimate atmospheric radon flux. However, re-examination of measured air-sea radon gradients in different coastal settings and different wind forcing could further refine existing empirical relationships between gas transfer and wind velocities.

18.3.2 Mixing in the Deep and Coastal Ocean

Radium-226 has a mean lifetime, σ, of 2,309 year (σ = decay constant^{-1}) that is conveniently in line with the residence time of water in the oceans (~3,200 year; Broecker and Peng 1982; Maidment 1993; Schlesinger 1997). Further, its distribution in the water column is closely tied to the upward diffusion from bottom sediment (Holland and Kulp 1954; Cochran 1980; Cochran et al. 1983; Cochran 1992; Cochran and Kadko 2008) and removal associated with particle scavenging and biological uptake (Broecker and Peng 1982; Ku and Luo 1994; Stewart et al. 2008). Thorium, as the immediate radiogenic parent of Ra, has a very low solubility in seawater that renders the supply of ^{226}Ra from the in situ decay of ^{230}Th mostly negligible. Rivers continuously supply both particulate and dissolved Ra to the ocean and this component usually consists of up to 10% of the total oceanic ^{226}Ra inventory. Global hydrothermal vents contribute another ~5% Ra. As a consequence, ^{226}Ra is ideally suited as a tracer of large scale oceanic mixing, and was used in the seminal work of Koczy (1958) and colleagues in the Baltic Sea. Since then, all the major ocean basins have been examined using ^{226}Ra, and too lesser extent ^{228}Ra, to gain information about water exchange rates across the principal thermoclines and to obtain vertical eddy diffusion coefficients of water mass transport phenomena (Table 18.3).

Table 18.3 Summary of horizontal (K_h) and vertical (K_z) eddy diffusivity coefficients obtained from ^{226}Ra and ^{228}Ra models for various ocean basins

Location	Isotope	K_h (cm^2 s^{-1})	K_z (cm^2 s^{-1})	Source
Eastern and Central Indian Ocean	^{226}Ra	$10^{4.00}$–$10^{8.00}$	$10^{-2.00}$–$10^{1.00}$	Ku and Luo (1994)
Arabian Sea	^{228}Ra/^{226}Ra	$10^{6.11}$–$10^{6.49}$	$10^{-0.49}$–$10^{0.18}$	Somayajulu et al. (1996)
North Pacific	^{226}Ra		$10^{-0.13}$–$10^{-0.01}$	Chan et al. (1976)
Pacific	^{226}Ra	$10^{5.70}$–$10^{7.00}$	~$10^{-2.00}$	Chung and Craig (1980)
East Atlantic	^{226}Ra		$10^{-1.00}$–$10^{1.10}$	Schlitzer (1987), Rhein and Schlitzer (1988)
North Atlantic	^{226}Ra	$10^{7.70}$–$10^{7.90}$		Sarmiento et al. (1982)
Southern Ocean	^{226,228}Ra		$10^{0.18}$–$10^{0.75}$	van Beek et al. (2008)
Southern Ocean	^{223,224}Ra; ^{226,228}Ra	$10^{0.75}$–$10^{1.59}$	$10^{0.18}$–$10^{2.00}$	Charette et al. (2007)

A typical ^{226}Ra profile in the ocean decreases upward through the water column as the distance from sediment/water interface increases. The concave-shape of ^{226}Ra profiles in the open ocean lead to very simple models for vertical exchange across thermoclines and lateral mixing (e.g. Koczy 1958). These simplified models eventually gave way to more sophisticated 2-D advection and diffusion models capable of resolving lateral variations in mixing coefficients (e.g. Ku and Luo 1994) as well as production and consumption associated with particle scavenging and biological uptake (Stewart et al. 2008). A detailed description of the various model types can be found in Ku and Luo (2008). Incorporating results from a variety of model-types, horizontal and vertical diffusivities (K_h and K_z, respectively) estimated by ^{226}Ra and ^{228}Ra for different ocean basins span four orders of magnitude ($K_h = 10^4 - 10^8$ cm^2 s^{-1} and $K_z = 10^{-2} - 10^2$ cm^2 s^{-1}) (Chan et al. 1976; Chung and Craig 1980; Sarmiento et al. 1982; Schlitzer 1987; Rhein and Schlitzer 1988; Ku and Luo 1994; Somayajulu et al. 1996; Charette et al. 2007; van Beek et al. 2008). Overall, despite simplification to a 1-D case, many of the earlier values reported in the literature are still within the uncertainty of the more robust modern models and estimates determined using other radiotracers (e.g. ^{3}H and ^{14}C) (Ku and Luo 1994). For example, Ra-based K_z values bracket the 0.3 cm^2 s^{-1} vertical mixing coefficient for the global ocean that Toggweiler et al. (1991) derived using a steady-state radiocarbon model.

Lateral transport processes in the water column can result in large discrepancies between diffusive seafloor fluxes and water-column inventories for long-lived ^{226}Ra. In contrast, a reasonable balance between a water-column inventory and a diffusive flux across the sediment/water interface usually is maintained for ^{228}Ra due to its shortened half-life ($t_{1/2} = 5.7$ year). This attribute allows ^{228}Ra to be particularly useful to examine water mixing in regions of the coastal ocean as well as the deep sea, on timescales of up to 30 year. A simple, steady-state 1-D diffusion model has been developed to model the observed seaward exponential decrease of ^{228}Ra in surface waters (Kaufman et al. 1973; Yamada and Nozaki 1986; Moore et al. 1986). Nearshore, ^{228}Ra is often enriched relative to ^{226}Ra and a systematic change in the isotopic ratio of ^{228}Ra/^{226}Ra as a function of distance away from the shoreline can yield information on water-mass mixing (horizontal eddy diffusivity) as well as water-mass sources (Moore 1996). Similarly, the two shorter-lived Ra-isotopes (^{223,224}Ra) have also been used to quantify coastal mixing rates (Moore 2000a, b; Burnett et al. 2008; Swarzenski and Izbicki 2009). Moore (2000a) observed that the natural logarithm of the activity of ^{223,224}Ra decreased linearly offshore in the south Atlantic Bight along the southeastern United States coastline. Assuming the water mass becomes isolated from the source and mixes conservatively offshore with losses only by natural radioactive decay, then the slope of the log-linear relationship between Ra-activity and distance offshore could be used to compute a coastal (diffusive) mixing rate (K_h)

$$\ln Ra_x = \ln Ra_0 - xm_1 \quad (18.6a)$$

$$m_1 = \sqrt{\frac{\lambda}{k_h}}, \quad (18.6b)$$

where Ra_x and Ra_0 are the activity of the Ra isotopes at distance x offshore and at the shoreline, respectively and λ is the decay constant of the Ra isotope. Alternatively, Burnett et al. (2008) used the activity ratio of ^{224}Ra/^{223}Ra to compute an average K_h

$$\ln \left(\frac{^{224}Ra}{^{223}Ra}\right)_x = \ln \left(\frac{^{224}Ra}{^{223}Ra}\right)_0 - xm_2 \quad (18.7a)$$

$$m_2 = \sqrt{\frac{(\lambda_{224} - \lambda_{223})}{K_h}}. \quad (18.7b)$$

The underlying assumptions of this model are: (1) diffusive-transport dominates, (2) all Ra input occurs at the shoreline (i.e. water mass is disconnected from benthic sources or stratified), and (3) Ra losses are through mixing and/or decay only (Moore 2000a). The activity ratio of ^{224}Ra/^{223}Ra can also be used to estimate apparent water mass ages in the coastal zone, using an equation similar to (18.8)

$$\left(\frac{^{224}Ra}{^{223}Ra}\right)_{sample} = \left(\frac{^{224}Ra}{^{223}Ra}\right)_{source} \frac{e^{-\lambda_{224}t}}{e^{-\lambda_{223}t}}, \quad (18.8)$$

where the subscripts *sample* and *source* reflect an offshore water column sample and the expected source of Ra, (e.g. river, estuarine, or groundwater) respectively (Moore 2000b).

As stated earlier, the Ra-quartet has been used in a variety of coastal settings to establish mixing rates and residence times (Table 18.3; also see Zhang et al. 2011). Moore and Todd (1993) used ^{224}Ra and ^{228}Ra/^{226}Ra to compute apparent ages of water in the Orinoco Estuary and the eastern Caribbean Sea. Based on excess ^{224}Ra data, fresher waters discharged from the Gulf of Paria took anywhere from 0.9 to 9.8 days to thoroughly mix in the Caribbean Sea. Using these apparent ages, current speeds on the order of 15–33 cm s^{-1} were needed to account for the mixing observed (Fig. 18.4). Along the coast of Santa Barbara, CA, USA, Swarzenski and Izbicki (2009) noted much slower mixing rates and current velocities. Based on the short-lived ^{223,224}Ra, they estimated mixing rates on the order of $10^{4.12}$ to $10^{4.46}$ cm^2 s^{-1}. Along the west Florida shelf (USA), Smith et al. (2010) reported mixing rates on the order of $10^{5.43}$–$10^{6.26}$ cm^2 s^{-1}; they noted that the mixing rate was strongly dependent on the offshore extent for which mixing was considered. For example, rates obtained from ^{224}Ra increased an order magnitude when the extent of offshore mixing was extended from 5 to 15 km (Okubo 1971). This distance from shore sensitivity is inevitably tied to the assumptions originally implied when the model was derived (Okubo 1971); namely that the water mass is truly separated from the benthic inputs and completely stratified, such that radioactive decay and mixing must control offshore distribution of Ra. Also in the examples presented (Fig. 18.4b, c), mixing rates derived from ^{224}Ra are larger than those derived from ^{223}Ra. This difference in theory is related to differences in half-lives of the two isotopes and the temporal integration assumed by applying the isotopes to simple mixing equations. Given (18.6b), a similar slope in the natural log of ^{223}Ra and ^{224}Ra would produce mixing rates that differ by a factor of 3 (Charette et al. 2007). The impact of these assumptions on mixing rates is also apparent when considering multiple data sets. For example, mixing rates compiled from 13 different locations incorporating a variety of geologic and hydrodynamic settings vary by over seven orders of magnitude (Table 18.4). While it is not unreasonable that significantly different sites could influence mixing over seven orders of magnitude, it is worth considering how refinement of the model assumptions (i.e. benthic inputs, onshore/offshore advection, etc) might influence the consistency of coastal mixing rates.

Hancock et al. (2006) tested the sensitivity of the Ra-mixing model to benthic inputs with data collected along the Great Barrier Reef and an adjacent lagoon. They found that by including benthic inputs mixing rates decreased by as much as 70% relative to the more

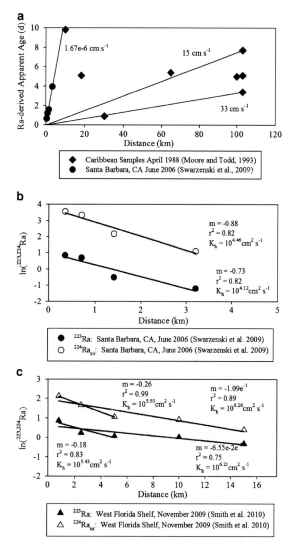

Fig. 18.4 Figure showing various plots of the Ra-derived apparent ages (**a**) or natural log of the activity of short-lived Ra isotopes (^{223,224}Ra) (**b**, **c**) vs. distance offshore compiled from data from Caribbean Sea (Moore and Todd 1993); Santa Barbara, California (Swarzenski and Izbicki 2009); and west Florida shelf (Smith et al. 2010)

Table 18.4 Summary of horizontal eddy diffusivity coefficients (K_h) obtained from different combinations of ^{223,224,226,228}Ra models for various coastal ocean sites

Location	Isotopes/ratios	K_x (cm^2 s^{-1})	Source
Tokyo Bay, Japan and adjacent Shelf	^{226,228}Ra	$10^{5.60}$–$10^{7.60}$	Yamada and Nozaki (1986)
Long Island Sound, NY, USA	^{224}Ra	$10^{4.70}$–$10^{5.70}$	Torgersen et al. (1996)
South Atlantic Bight, USA	^{223,224,226,228}Ra	$10^{6.56}$–$10^{6.62}$	Moore (2000a)
northeastern Gulf of Mexico, USA	^{223}Ra, ^{224}Ra	$10^{4.43}$, $10^{5.08}$	Moore (2003)
Eckernforder Bay, Baltic Sea, Germany	^{224}Ra	$10^{4.0}$–$10^{5.70}$	Purkl and Eisenhauer (2004)
Huntington Beach, CA, USA	^{223}Ra, ^{224}Ra	$10^{3.08}$, $10^{3.41}$	Boehm et al. (2006)
West Neck Bay, NY, USA	^{223}Ra	$10^{4.87}$–$10^{4.99}$	Dulaiova et al. (2006)
Great Barrier Reef, Australia	^{223,224}Ra	$10^{6.35}$–$10^{6.47}$	Hancock et al. (2006)
Ubatuba and Flamengo Bay, Brazil	^{224}Ra/^{223}Ra	$10^{5.47}$	Burnett et al. (2008)
Drake Passage, Southern Ocean/Antarctic	^{224}Ra	$10^{8.80}$	Dulaiova et al. (2009)
Santa Barbara, California	^{223}Ra, ^{224}Ra	$10^{4.12}$, $10^{4.46}$	Swarzenski and Izbicki (2009)
northeastern Gulf of Mexico, USA	^{223}Ra, ^{224}Ra	$10^{4.99}$, $10^{5.01}$	Santos et al. (2009)
Eastern Gulf of Mexico, USA	^{223,224}Ra	$10^{5.43}$–$10^{6.26}$	Smith et al. (2010)

commonly used, simple 1-D mixing model (Hancock et al. 2006). Take for example Fig. 18.4b or c, the slope of radium isotopes offshore remains constant despite the physical process(es) used to model them. For the simple mixing model (Moore 2000a) only mixing (unknown), supply at the shoreline (constant and assumed to be known), and decay (known) are considered, the slope of the offshore gradient implies that turbulent mixing transports the shoreline Ra-signal offshore before it decays away. However, if there were offshore Ra inputs, then turbulent mixing could occur at a slower rate and the same slope could be maintained. Continued refinement of Ra mixing models is occurring, and Zhang et al. (2011) provides derivations for a number of these models. Despite the remaining uncertainties, Ra continues to provide reasonable first-order and sometimes better estimates of coastal mixing that are generally unattainable without sophisticated numerical physical models or large oceanographic surveys.

18.3.3 Benthic Fluxes and Groundwater Discharge

18.3.3.1 Benthic Fluxes and Exchange

Active exchange of fluid across the sediment – water interface is a well-known vector where dissolved constituents in the water column are transported into the sediment and pore water constituents are released into the water column (e.g. classic textbooks by Berner 1980 and Boudreau 1997). The rate of exchange across the sediment – water interface will ultimately determine the residence time of the fluid in the pore space and influence the transient behavior of chemical reactions that occur between the water and sediment. Quantifying these exchange rates has proven difficult as the size of sediment and pore space vary greatly from one basin to another, and the processes that drive the exchange are numerous (e.g. advection, dispersion, bioturbation, bio-irrigation, etc). Broecker (1965) first suggested radon as a tracer for examining such exchange and interaction. In a symposium note, Broecker (1965) combined the pore water radon data with a simple model of diffusion and a first-order decay equation as a lower boundary condition for water column mass balance of ^{222}Rn in the deep ocean. Numerous other works have built off of this radon pore water profile approach to examine mixing and exchange processes in coastal and marine environments (e.g. Hammond et al. 1977; Hammond and Fuller 1979; Martens et al. 1980; Martin and Sayles 1987; Cable et al. 1996b; Martin et al. 2006, and Smith et al. 2008). In order to quantify rates of exchange, some form of the transient advection-dispersion-reaction equation is used to model the ^{222}Rn pore water distribution:

$$\frac{\partial \varphi C}{\partial t} = \frac{\partial}{\partial z}\left(D_s \varphi \frac{\partial C}{\partial z}\right) - \frac{\partial}{\partial z}(v \varphi C) - \alpha(z)\varphi(C - C_{sw}) + \lambda P - \lambda C, \quad (18.9)$$

where φ is porosity (–), C is the pore water concentration of ^{222}Rn (atoms m^{-3}), t is time (day), z is depth (positive downward) (m), D_s is bulk sedimentary diffusion/dispersion coefficient (m^2 s^{-1}), v is seepage velocity (m s^{-1}), $\alpha(z)$ is a depth-dependant (non-local) mass transfer coefficient (s^{-1}), C_{sw} is the concentration of ^{222}Rn (atoms m^{-3}) in the flushing water of the upper zone, λ is the decay constant for ^{222}Rn (2.09 × 10^{-6} s^{-1}), and P is the concentration of ^{222}Rn emanated from the sediments and produced by dissolved ^{226}Ra (atoms m^{-3}). On the RHS of (18.9), the first term describes diffusive transport by molecular or dispersive processes; the second term describes advective exchange by pressure gradients (e.g. SGD, wave-induced circulation); the third term describes non-local mixing of porewater with the overlying water column (bioirrigation or non-local advective exchange); the fourth term describes radon production; and the fifth term radon decay. Unknown parameters (e.g. v or $\alpha(x)$) can be estimated by numerically solving (18.9) and minimizing the error between measured and modeled pore water ^{222}Rn (e.g. Meile et al. 2001; Smith et al. 2008).

The above model can be used to evaluate both excess and deficit radon profiles relative to supported fractions. Numerous studies have observed pore water ^{222}Rn deficiencies relative to production from sediment and dissolved ^{226}Ra (e.g. Hammond et al. 1977; Martin and Sayles 1987; Martin et al. 2006; Cable and Martin 2008). In these scenarios (e.g. Martin and Sayles 1987), diffusion or advective exchange occur at time-scale faster than radon production, causing in a radon deficits in the upper 10–20 cm. The rate of advective flow or biological-processes (i.e. bioturbation or bioirrigation) is quantified using either a numerical or analytical solution to (18.9). In comparison, Smith et al. (2008) found excess radon in shallow pore waters collected along the east coast of Florida. While it was hypothesized that the same shallow mixing processes that cause deficit radon profiles were occurring at this site, additional advective inputs from submarine groundwater discharge delivered radon-enriched groundwater from depth. Smith et al. (2008) reiterated the importance of including non-local processes and more importantly heterogeneous production when considering nearshore pore-water and radon transport.

18.3.3.2 Submarine Groundwater Discharge

The discharge of coastal groundwater is an obvious but hard to quantify pathway for water and associated material transport to the sea (Fig. 18.5). Armed with new geochemical methods for determining submarine groundwater discharge rate measurements, coastal scientists are now assessing the volumetric (Webster et al. 1994; Moore 1996; Burnett et al. 2003, 2006; Cable et al. 1996a, b; Swarzenski and Izbicki 2009) and ecological significance of submarine groundwater discharge (SGD) (Johannes 1980). Observations suggest that the discharge of groundwater into coastal waters may have important environmental consequences because groundwater often carries elevated concentrations of select nutrients, trace elements, radionuclides, and organics. If a coastal groundwater resource has somehow become compromised by salt-water

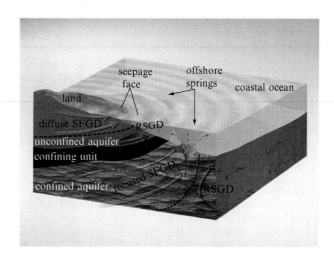

Fig. 18.5 Idealized representation of submarine groundwater discharge, expressed here as discrete springs as well as near-shore diffusive fluxes

intrusion or anthropogenic perturbations, the submarine discharge of this groundwater may directly contribute to the environmental degradation of receiving coastal waters. In this section we briefly explore the unique application of select U/Th series radioisotopes to the study of submarine groundwater discharge.

For almost as long as ^{226}Ra has been utilized as an oceanic water mass tracer, a growing suite of diverse geochemical elements have been explored to quantify rates of fluid exchange across the sediment/water interface. This group of tracers includes: ^{87}Sr, ^{3,4}He, ^{3}H, Fe/Mn, CH_4, and select U/Th-series isotopes (Cable et al. 1996a, b; Swarzenski 2007; Charette et al. 2008). It is the U/Th-series radionuclides of Ra (e.g. Moore 1996; Rama and Moore 1996; Charette et al. 2001) and Rn (e.g. Cable et al. 1996b; Burnett and Dulaiova 2006; Swarzenski et al. 2007) that have been used extensively to quantify rates of SGD and to identify discharging aquifers. The application of Ra and Rn as SGD tracers is built upon a simple box model that quantifies the removal and supply terms (Fig. 18.6) for a respective radionuclide in a coastal water column (Charette et al. 2008). This balance between supply and removal terms can be expressed as follows:

$$\frac{\partial A}{\partial t} = \left[\frac{A_s - A_0}{T_r}\right] - \left[\frac{QA_r + A_{desorb}}{V}\right]$$
$$- \left[\varphi D_s \left(\frac{\partial^2 A_{sed}}{\partial z_{sed}^2}\right) + \omega\left(\frac{\partial A}{\partial z}\right)\right]$$
$$+ \left[\frac{k(A_s + A_{atm})}{z_{wc}}\right] + \lambda A_s, \quad (18.10)$$

where the first term on the RHS addresses oceanic exchange (A_s and A_o = activity of sample and the adjacent ocean, respectively, Bq m^{-3}; T_r = water residence time, day), the second term quantifies riverine inputs (Q = river discharge, m^3 s^{-1}; A_r = fresh, riverine activity, Bq m^{-3}; A_{desorb} = suspended sediment activity available for desorption, Bq m^{-3}), the third addresses benthic flux (φ = sediment porosity, unitless; D_s = molecular diffusion coefficient, m^2 s^{-1}; ω = advective pore water flux, m s^{-1}), the fourth term accounts for atmospheric evasion (k = piston or gas transfer velocity, m s^{-1} and z_{wc} = depth of the water column, m), and the last term, λA, accounts for radioactive decay. By measuring the radionuclide activity of the discharging groundwater, A_{gw}, and assuming steady-state conditions, one can rearrange the above mass balance equation (after Burnett et al. 2006; Charette et al. 2008) to obtain an SGD rate in terms of m^3 m^{-2} s^{-1} (m s^{-1}):

$$SGD = z_w \left(\left[\frac{A_s - A_0}{T_r}\right] - \left[\frac{QA_r + A_{desorb}}{V}\right]\right.$$
$$\left. - \left[\varphi D_s \left(\frac{\partial^2 A_{sed}}{\partial z_{sed}^2}\right)\right] + \left[\frac{k(A_s + A_{atm})}{z_{wc}}\right] + \lambda A\right) \Big/ A_{gw}.$$
$$(18.11)$$

Burnett et al. (2007) carefully examined the sensitivity of this ^{222}Rn water column mass balance model for SGD to variations in atmospheric evasion, mixing losses, and groundwater end-member. Of these, the groundwater endmember (A_{gw}) has the most obvious

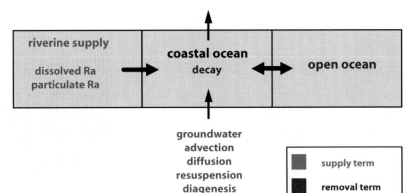

Fig. 18.6 Simplified box model for the application of select U/Th-series isotopes in submarine groundwater discharge studies. Removal and source terms correspond to terms described in (18.11)

influence with a 1:1 inverse relationship with SGD rate. More complicated are the effects of atmospheric evasion (fourth bracketed term on RHS of (18.11)), which generally account for less than 25% of the total radon flux (Burnett et al. 2007). Dulaiova and Burnett (2006) and Burnett et al. (2007) found that for wind speeds less than 10 m s^{-1}, empirical relationships (see section MacIntyre et al. 1995) provide reliable estimates of evasion rates. Import and export of ^{222}Rn into the model via tidal mixing (first bracketed term on RHS of (18.11)) and riverine discharge/offshore mixing (second bracketed term on RHS of (18.11)) are site dependant. Accurate representation of these terms in the model often requires direct or indirect measurements of tidal exchange and/or river discharge. One approach has been to physically instrument the study site with a velocity meter (e.g. ADCP or ADV) and pressure transducer, then compute fluxes using recorded velocity fields, tidal stage, and a representative end-member ^{222}Rn concentration (i.e. ocean, riverine, or estuarine) (Peterson et al. 2009a). When Ra studies are conducted in parallel with Rn studies, ^{223,224}Ra-based mixing coefficients (i.e. (18.4) and (18.5)) can be combined with ^{222}Rn gradients to compute a uni-directional offshore flux (Burnett et al. 2007, 2008). When no other data are available, average negative ^{222}Rn fluxes (i.e. change in inventory over time) are used as representative mixing loss (Burnett et al. 2007). Negative fluxes are generally averaged over a finite time window, usually the local tidal cycle. Refinement in these model parameters will inevitably improve the accuracy of SGD estimates and its role as a coastal vector.

18.3.3.3 Groundwater Discharge into Lakes

Groundwater discharge to a lake can play a major role in its water budget and can also impact its ecological health and general trophic structure (Hayashi and Rosenberry 2002). In the case of seepage lakes, groundwater input is the only substantial source of nutrients besides direct atmospheric deposition. The success of using radon as a groundwater tracer in marine environments in the last few decades has motivated hydrologists to expend its application as a groundwater tracer in fresher water systems, such as, rivers and lakes. While the models developed to interpret radon data for coastal environments can be complex as corrections for tidal mixing with offshore low-radon waters need to be applied (Burnett and Dulaiova 2003; Burnett et al. 2006), lakes often present a much simpler situation. A single-box model for quantifying groundwater discharge into a lake based on a surface water time-series record of radon inventories was recently developed by Dimova and Burnett (2011). Similar to the coastal SGD radon approach, the lake model for groundwater-surface water interaction is based on a ^{222}Rn water column mass balance. For small seepage lakes, the radon budget is essentially a balance between inflows from groundwater and losses by decay and atmospheric evasion. Thus, this lake radon model requires only equations for radon flux and corrections for radioactive decay. Atmospheric losses through the water-gas interface are evaluated using an empirical relationship for the piston or gas transfer velocity (k; see Sect. 18.3.1 for background), correlating wind speed and water temperature, as presented in MacIntyre et al. (1995).

$$k(600) = 0.45 \times u_{10}^{1.6} \times (Sc/600)^{-0.5}, \quad (18.12)$$

where u_{10} is the wind speed at 10 m height above the water surface and Sc is the Schmidt number ($Sc = v/D_m$). In this case k (i.e. $k(600)$) is normalized to CO_2 at 20°C in freshwater. Although the equation is not strictly derived for radon gas transfer, the relationship could be and has shown to be adequate for evaluating $k(600)$. The diffusion of SF_6 from water surface has similar physics as the diffusion of noble gases, its molecule is non-polar and the molecular weights are not too different. Calculated fluxes based on this equation and another empirical relationship developed specifically for Rn (Hartman and Hammond 1984) showed only 20% difference (Dimova and Burnett 2011). Experience shows that D_m values vary between $10^{-5.086}$ to $10^{-4.824}$ cm^2 s^{-1} depending on temperature variations. For the same range of temperature and molecular diffusion, Sc calculated were between $10^{2.713}$ and $10^{3.236}$. Main controls include temperature, salinity, and density (respectively kinematic viscosity). Both v and D_m can be adjusted for temperature changes by well-known relationships. The temperature dependence of the molecular diffusion can be expressed by the equation (Peng et al. 1974; Ullman and Aller 1982):

$$-\log D_m = (980/T_w) + 1.59, \quad (18.13)$$

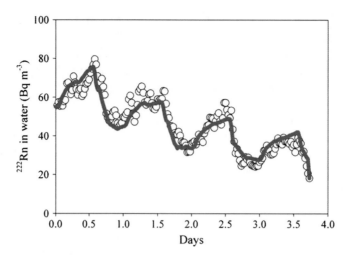

Fig. 18.7 An observed surface-water ^{222}Rn time series in Lake Haines, Florida that can be used to model rates of lacustrine groundwater discharge (modified from Dimova and Burnett 2011)

where T_w is the temperature of the water in K. The kinematic viscosity is a simple ratio of the absolute viscosity, μ, to the density (ρ) of the water at a given temperature:

$$v = \mu/\rho. \quad (18.14)$$

To adjust the absolute viscosity for the differences in the water temperature (T_w in K) the following relationship (Seeton 2006) can be used:

$$\mu = A \times 10^{B(T_w - C)}, \quad (18.15)$$

Where the coefficients in this equation are: $A = 2.414 \times 10^{-5}$ kg s^{-1} m^{-1} (or Pa s); $B = 247.8$ K; and $C = 140$ K.

To calculate the water density required in (18.15) as a function of temperature, an empirical relationship for freshwater as derived by Gill (1982) was applied:

$$\begin{aligned}\rho =\ & 999.842598 + 6.793952 \times 10^{-2} \times T_w \\ & - 9.09529 \times 10^{-3} \times T_w^2 + 1.001685 \\ & \times 10^{-4} \times T_w^3 - 1.120083 \times 10^{-6} \times T_w^4 \\ & + 6.536332 \times 10^{-9} \times T_w^5.\end{aligned} \quad (18.16)$$

Thus, based on information for wind speed and temperature alone, one can model the trends in the radon concentration in the lake by adjusting the radon flux (via groundwater) to balance the observed trends. The minimum root mean square (RMS) error can be used as a statistical measure to establish best fit between the modeled and observed radon concentration time-series (Fig. 18.7). As in the previously discussed SGD models, one can convert this estimate of the steady-state flux to a water flux by simply dividing by the measured radon-in-groundwater activity. Such a simple, lacustrine groundwater discharge model based solely on Rn offers hydrologists a new tool to easily quantify ground water/surface water exchange.

18.4 Summary and Future Directions

Radium and radon isotopes have proven to be valuable tracers of reservoir exchanges, allowing scientists to quantify fluxes from the ocean to the atmosphere, from land to the ocean, and within the coastal zone. To date, Ra and/or Rn isotopes have been used at almost every hydrological reservoir boundary. Given the diversity of current applied Ra and Rn research, it is difficult to foresee how these radioisotopes will be used to examine additional hydrologic processes. However, improved collection procedures (long-term deployments and in situ measurement) and detection of the short-lived isotopes (i.e. ^{223,224}Ra and $^{219, 220}$Rn) will inevitably open the door to quantifying processes that occur at much shorter (e.g. advective exchange across rippled bedforms) and longer time-scales (e.g. atmospheric evasion during event and periodic forcing) and smaller spatial-scales (influence of pore-scale diffusion and grain size on current Rn and Ra source terms) than are currently being examined. These improvements will not only benefit research at new "scales" but also allow scientists to refine current estimates of (hydrologic) reservoir exchange. For example, many of the models that use Rn and Ra to quantify submarine groundwater discharge are relatively young (<15 years) and make various assumptions (e.g. steady-state, uni-directional mixing, isolation of water mass, specific boundary conditions) to simplify the computational aspect of the problem. Thus future research should focus on examining these assumptions in detail, in hopes to removing assumptions outright and having an all-inclusive model. Similarly, the advancement personal computing power over the last 20 years has opened the door to a new application of Ra and Rn data sets, new and old. In some instances this is happening (e.g. Ra in circulation models); however, this approach is not widespread.

Acknowledgments We would like to thank Christopher Conaway and Nancy Prouty of the U.S. Geological Survey for reviewing this manuscript prior to publication; their comments and suggestions added greatly to this document. We would also like to thank William Burnett for his comments and suggestions that improved the quality of this manuscript. Any use of trade names is for descriptive purposes only and does not imply endorsement by the U.S. Geological Survey.

References

Antovic N, Svrkota N (2009) Measuring the radium-226 activity using a multidetector gamma-ray coincidence spectrometer. J Environ Radioactiv 100(10):823–830

Beneš P, Strejc P, Lukavec Z (1984) Interaction of radium with fresh-water sediments and their mineral components. 1 ferric hydroxide and quartz. J Radioanal Nucl Chem 82 (2):275–285

Beneš P, Borovec Z, Strejc P (1986) Interaction of radium with fresh-water sediments and their mineral components .3 muscovite and feldspar. J Radioanal Nucl Chem Art 98 (1):91–103

Berner RA (1980) Early diagenesis: a theoretical approach. Princeton University Press, Princeton

Bertin C, Bourg CM (1994) Radon-222 and chloride as natural tracers of the infiltration of river water into an alluvial aquifer in which there is significant river/groundwater mixing. Environ Sci Technol 28:794–798

Boehm AB, Paytan A, Shellenbarger GG et al (2006) Composition and flux of groundwater from a California beach aquifer: implications for nutrient supply to the surf zone. Cont Shelf Res 26(2):269–282

Boudreau BP (1997) Diagenetic models and their implementation: modelling transport and reactions in aquatic sediments. Springer, New York

Bourquin M, van Beek P, Reyss JL et al (2008) Comparison of techniques for pre-concentrating radium from seawater. Mar Chem 109(3–4):226–237

Broecker HC, Petermann J, Siems W (1978) The influence of wind on CO_2-exchange in a wind/wave tunnel, including the effects of monolayers. J Mar Res 36:595–610

Broecker WS (1965) An application of natural radon to problems in ocean circulation. In: Ichiye T (ed) Symposium on diffusion in oceans and fresh waters. Lamont-Doherty Geological Observatory, Palisades

Broecker WS, Peng TH (1971) Vertical distribution of radon in Bomex area. Earth Planet Sci Lett 11(2):99–108

Broecker WS, Peng TH (1974) Gas exchange rates between air and sea. Tellus XXVI(1–2):21–35

Broecker WS, Peng TH (1982) Tracers in the sea. Lamont-Dohert Geological Observatory, Palisades

Broecker WS, Li YH, Cromwell J (1967) Radium-226 and Radon-222: concentration in Atlantic and Pacific Oceans. Science 158(3806):1307–1310

Burnett WC, Dulaiova H (2003) Estimating the dynamics of groundwater input into the coastal zone via continuous radon-222 measurements. J Environ Radioactiv 69(1–2): 21–35

Burnett WC, Dulaiova H (2006) Radon as a tracer of submarine groundwater discharge into a boat basin in Donnalucata, Sicily. Cont Shelf Res 26:862–873

Burnett WC, Kim G, Lane-Smith D (2001) A continuous monitor for assessment of Rn-222 in the coastal ocean. J Radioanal Nucl Chem 249(1):167–172

Burnett WC, Bokuniewicz H, Huettel M et al (2003) Groundwater and pore water inputs to the coastal zone. Biogeochemistry 66(1–2):3–33

Burnett WC, Aggarwal PK, Aureli A et al (2006) Quantifying submarine groundwater discharge in the coastal zone via multiple methods. Sci Total Environ 367(2–3):498–543

Burnett WC, Santos IR, Weinstein Y et al (2007) Remaining uncertainties in the use of Rn-222 as a quantitative tracer of submarine groundwater discharge. In: Sanford W, Langevin C, Polemio M, Povinec P (eds) A new focus on groundwater-seawater interactions. IAHS Publ, Perugia, 312, pp 125–133

Burnett WC, Peterson R, Moore WS et al (2008) Radon and radium isotopes as tracers of submarine groundwater discharge – results from the Ubatuba Brazil SGD assessment intercomparison. Estuar Coast Shelf Sci 76(3):501–511

Butts J, Todd JF, Lerche I et al (1988) A simplified method for Ra-226 determinations in natural-waters. Mar Chem 25 (4):349–357

Cable JE, Martin JB (2008) In situ evaluation of nearshore marine and fresh pore water transport into Flamengo Bay, Brazil. Estuar Coast Shelf Sci 76(3):473–483

Cable JE, Burnett WC, Chanton JP et al (1996a) Estimating groundwater discharge into the northeastern Gulf of Mexico using Radon-222. Earth Planet Sci Lett 144:591–604

Cable JE, Bugna GC, Burnett WC et al (1996b) Application of ^{222}Rn and CH_4 for assessment of groundwater discharge to the coastal ocean. Limnol Oceanogr 41(6):1347–1353

Chan LH, Edmond JM, Stallard RF et al (1976) Radium and barium at GEOSECS stations in the Atlantic and Pacific. Earth Planet Sci Lett 32(2):258–267

Charette MA, Buesseler KO, Andrews JE (2001) Utility of radium isotopes for evaluating the input and transport of groundwater-derived nitrogen to a Cape Cod estuary. Limnol Oceanogr 46(2):465–470

Charette MA, Gonneea ME, Morris PJ et al (2007) Radium isotopes as tracers of iron sources fueling a Southern Ocean phytoplankton bloom. Deep Sea Res II 54(18–20):1989–1998

Charette MA, Moore WS, Burnett WC (2008) Uranium- and thorium-series nuclides as tracers of submarine groundwater discharge. In: Krishnaswami S, Cochran JK (eds) Radioactivity in the environment, Vol 13. Elsevier, Oxford

Chung Y, Craig H (1980) ^{226}Ra in the Pacific Ocean. Earth Planet Sci Lett 49(2):267–292

Clements WE, Wilkening MH (1974) Atmospheric pressure effects on ^{222}Rn transport across the earth-air interface. J Geophys Res 79(33):5025–5029

Cochran JK (1980) The flux of ^{226}Ra from deep-sea sediments. Earth Planet Sci Lett 49(2):381–392

Cochran JK (1992) The oceanic chemistry of the uranium and thorium series nuclides. In: Ivanovich M, Harmon RS (eds) Uranium series disequilibrium: applications to environmental problems. Claredon Press, Oxford

Cochran JK, Bacon MP, Krishnaswami S et al (1983) ^{210}Po and ^{210}Pb distributions in the central and eastern Indian Ocean. Earth Planet Sci Lett 65(2):433–452

Cochran JK, Kadko DC (2008) Uranium- and thorium-series radionuclides in marine groundwaters. In: Krishnaswami S, Cochran JK (eds) Radioactivity in the environment, Vol 13. Elsevier, Oxford

Cohen AS, O'Nions RK (1991) Precise determination of femtogram quantities of radium by thermal ionization mass spectrometry. Anal Chem 63(23):2705–2708

Corbett DR, Burnett WC, Cable PH et al (1997) Radon tracing of groundwater input into Par Pond, Savannah River Site. J Hydrol 203(1–4):209–227

Dimova N, Burnett WC, Horwitz EP et al (2007) Automated measurement of ^{224}Ra and ^{226}Ra in water. Appl Radiat Isot 65(4):428–434

Dimova N, Dulaiova H, Kim G et al (2008) Uncertainties in the preparation of ^{224}Ra Mn fiber standards. Mar Chem 109(3–4):220–225

Dimova N, Burnett WC, Lane-Smith D (2009) Improved automated analysis of Radon (^{222}Rn) and Thoron (^{220}Rn) in natural waters. Environ Sci Technol 43(22):8599–8603

Dimova NT, Burnett WC (2011) Evaluation of groundwater discharge into small lakes based on the temporal distribution of radon-222. Limnol Oceanogr 56(2):486–494

Dulaiova H, Burnett WC (2004) An efficient method for γ-spectrometric determination of radium-226,228 via manganese fibers. Limnol Oceanogr Meth 2:256–261

Dulaiova H, Burnett WC (2006) Radon loss across the water-air interface (Gulf of Thailand) estimated experimentally from ^{222}Rn-^{224}Ra. Geophys Res Lett 33(5):L05606

Dulaiova H, Peterson R, Burnett WC et al (2005) A multi-detector continuous monitor for assessment of Rn-222 in the coastal ocean. J Radioanal Nucl Chem 263(2):361–365

Dulaiova H, Burnett WC, Chanton JP et al (2006) Assessment of groundwater discharges into West Neck Bay, New York, via natural tracers. Cont Shelf Res 26(16):1971–1983

Dulaiova H, Ardelan MV, Henderson PB et al (2009) Shelf-derived iron inputs drive biological productivity in the southern Drake Passage. Global Biogeochem Cycles 23:GB4014, doi:10.1029/2008GB003406

Elsinger RJ, Moore WS (1983) Ra-224, Ra-228, and Ra-226 in Winyah Bay and Delaware Bay. Earth Planet Sci Lett 64(3):430–436

Fanning KA, Torres LM (1991) ^{222}Rn and ^{226}Ra: indicators of sea-ice effects on air-sea gas exchange. Polar Res 10(1):51–58

Fanning KA, Breland JA, Byrne RH (1982) Radium-226 and Radon-222 in the Coastal Waters of West Florida: high concentrations and atmospheric degassing. Science 215(4533):667–670

Freyer K, Treutler HC, Dehnert J et al (1997) Sampling and measurement of radon-222 in water. J Environ Radioactiv 37(3):327–337

Ghaleb B, Pons-Branchu E, Deschamps P (2004) Improved method for radium extraction from environmental samples and its analysis by thermal ionization mass spectrometry. J Anal At Spectrom 19(7):906–910

Giffin C, Kaufman A, Broecker WS (1963) Delayed coincidence counter for the assay of actinon and thoron. J Geophys Res 68:1749–1757

Gill AE (1982) Atmosphere-ocean dynamics. Academic, San Diego

Glover DM, Reeburgh WS (1987) Radon-222 and radium-226 in southeastern Bering Sea shelf waters and sediment. Cont Shelf Res 7(5):433–456

Gonneea ME, Morris PJ, Dulaiova H et al (2008) New perspectives on radium behavior within a subterranean estuary. Mar Chem 109(3–4):250–267

Gruebel KA, Martens CS (1984) Radon-222 tracing of sediment-water chemical transport in an estuarine sediment. Limnol Oceanogr 29(3):587–597

Hammond DE, Fuller C (1979). The use of Radon-222 to estimate benthic exchange and atmospheric exchange rates in San Francisco Bay. San Fransisco Bay: the urbanized estuary The Pacific Division of the American Association for the Advancement of Science (California Academy of Sciences, San Fracncisco, CA). pp 213–229

Hammond DE, Simpson HJ, Mathieu G (1977) Radon-222 distribution and transport across the sediment-water interface in the Hudson River estuary. J Geophys Res 82:3913–3920

Hancock GJ, Webster IT, Stieglitz TC (2006) Horizontal mixing of Great Barrier Reef waters: offshore diffusivity determined from radium isotope distribution. J Geophys Res 111(C12):C12019

Hanor JS, Chan LH (1977) Non-conservative behavior of barium during mixing of Mississippi River and Gulf of Mexico Waters. Earth Planet Sci Lett 37(2):242–250

Hardcastle BJ (1987) Wells ancient and modern; an historical review. Q J Eng Geol 20(3):231–238

Hartman B, Hammond DE (1984) Gas exchange rates across the sediment-water and air-water interfaces in south San Francisco Bay. J Geophys Res 89(C3):3593–3603

Hayashi M, Rosenberry DO (2002) Effects of ground water exchange on the hydrology and ecology of surface water. Ground Water 40(3):309–316

Hirsch AI (2007) On using radon-222 and CO_2 to calculate regional-scale CO_2 fluxes. Atmos Chem Phys 7(14): 3737–3747

Hoehn E, von Gunten HR, Stauffer F et al (1992) Radon-222 as a groundwater tracer: a laboratory study. Environ Sci Technol 26:734–738

Holland HD, Kulp LJ (1954) The transport and deposition of uranium, ionium and radium in rivers, oceans and ocean sediments. Geochim Cosmochim Acta 5(5):197–213

Hou X, Roos P (2008) Critical comparison of radiometric and mass spectrometric methods for the determination of radionuclides in environmental, biological and nuclear waste samples. Anal Chim Acta 608(2):105–139

Ivanovich M, Murray A (1992) Spectroscopic methods. In: Ivanovich M, Harmon RS (eds) Uranium-series disequilibrium: applications to environmental problems. Clarendon Press, Oxford

Johannes RE (1980) The ecological significance of the submarine discharge of groundwater. Mar Ecol-Prog Ser 3: 363–373

Jones EP, Smith SD (1977) A first measurement of sea-air CO_2 flux by eddy correlation. J Geophys Res 82(37):5990–5992

Kaufman A, Trier RM, Broecker WS et al (1973) Distribution of ^{228}Ra in the World Ocean. J Geophys Res 78(36):8827–8848

Kawabata H, Narita H, Harada K et al (2003) Air-sea gas transfer velocity in stormy winter estimated from radon deficiency. J Oceanogr 59(5):651–661

Key RM, Brewer RL, Stockwell JH et al (1979) Some improved techniques for measuring radon and radium in marine-sediments and in seawater. Mar Chem 7(3):251–264

Key RM, Stallard RF, Moore WS et al (1985) Distribution and flux of Ra-226 and Ra-228 in the Amazon River Estuary. J Geophys Res Oceans 90:6995–7004

Kim G, Burnett WC, Dulaiova H et al (2001) Measurement of ^{224}Ra and ^{226}Ra activities in natural waters using a radon-in-air monitor. Environ Sci Technol 35:4680–4683

Kitto ME (1994) Characteristics of liquid scintillation analysis of radon in water. J Radioanal Nucl Chem 185(1):91–99

Koczy FF (1958) Natural radium as a tracer in the ocean. In: Proceedings of the second UN international conference on the peaceful uses of atomic energy. IAEA 18:351–357

Krishnaswami S, Bhushan R, Baskaran M (1991) Radium isotopes and ^{222}Rn in shallow brines, Kharaghoda (India). Chem Geol 87(2):125–136

Kromer B, Roether W (1983). Field measurement of air-sea exchange by the radon deficit method during JASIN 1978 and FGGE 1979. Meteor Forsch Ergebnisse A/B24:55–75

Ku T-L, Luo S (1994) New appraisal of Radium-226 as a large-scale oceanic mixing tracer. J Geophys Res 99(C5): 10255–10273

Ku T-L, Luo S (2008) Ocean circulation/mixing studies with decay-series isotopes. In: Krishnaswami S, Cochran JK (eds) Radioactivity in the environment, Vol 13. Elsevier, Oxford

Langmuir D, Riese AC (1985) The thermodynamic properties of radium. Geochim Cosmochim Acta 49(7):1593–1601

Larivière D, Epov VN, Evans RD et al (2003) Determination of radium-226 in environmental samples by inductively coupled plasma mass spectrometry after sequential selective extraction. J Anal At Spectrom 18(4):338–343

Lauria DC, Almeida RMR, Sracek O (2004) Behavior of radium, thorium and uranium in groundwater near the Buena Lagoon in the Coastal Zone of the State of Rio de Janeiro, Brazil. Environ Geol 47(1):11–19

Lee J-M, Kim G (2006) A simple and rapid method for analyzing radon in coastal and ground waters using a radon-in-air monitor. J Environ Radioactiv 89(3):219–228

Li Y-H, Chan L-H (1979) Desorption of Ba and ^{226}Ra from river-borne sediments in the Hudson estuary. Earth Planet Sci Lett 43(3):343–350

Liss PS, Heimann M, Roether W (1988) Tracers of air-sea gas exchange [and discussion]. Philos Trans R Soc Lond A 325 (1583):93–103

MacIntyre S, Wannikhof R, Chanton JP (1995) Trace gas exchange across the air-water interface in freshwater and coastal marine environments. In: Matson PA, Hariss RC (eds) Biogenic trace gases: measuring emissions from soil and water. Blackwell Science, Oxford

Maidment DR (1993) Hydrology. In: Maidment DR (ed) Handbook of hydrology. New York, McGraw-Hill

Martens CS, Kipphut GW, Klump JV (1980) Sediment-water chemical exchange in the coastal zone traced by in situ Radon-222 flux measurements. Science 208(4441):285–288

Martin JB, Cable JE, Jaeger J et al (2006) Thermal and chemical evidence for rapid water exchange across the sediment-water interface by bioirrigation in the Indian River Lagoon, Florida. Limnol Oceanogr 51(3):1332–1341

Martin WR, Sayles FL (1987) Seasonal cycles of particle and solute transport processes in nearshore sediments: ^{222}Rn/^{226}Ra and ^{234}Th/^{238}U disequilibrium at a site in Buzzards Bay, MA. Geochim Cosmochim Acta 51:927–943

Mathieu G, Biscaye P, Lupton R et al (1988) System for measurement of ^{222}Rn at low level in natural waters. Health Phys 55:989–992

Meile C, Koretsky CM, Cappellen PV (2001) Quantifying bioirrigation in aquatic sediments: an inverse modeling approach. Limnol Oceanogr 46:164–177

Michel J, Moore WS, King PT (1981) γ-ray spectrometry for determination of radium-228 and radium-226 in natural waters. Anal Chem 53(12):1885–1889

Miller R (1982) Public health lessons from prehistoric times. World Water 5:22–25

Moon DS, Burnett WC, Nour S et al (2003) Preconcentration of radium isotopes from natural waters using MnO_2 Resin. Appl Radiat Isot 59(4):255–262

Moore JE, Zaporozec A, Mercer JW (1995) Groundwater – a primer. Childress J (ed) AGI environmental awareness series. American Geological Institute, Arlington

Moore WS (1984) Radium isotope measurements using germanium detectors. Nucl Instrum Methods Phys Res 223:407–411

Moore WS (1996) Large groundwater inputs to coastal waters revealed by Ra-226 enrichments. Nature 380(6575): 612–614

Moore WS (2000a) Determining coastal mixing rates using radium isotopes. Cont Shelf Res 20(15):1993–2007

Moore WS (2000b) Ages of continental shelf waters determined from ^{223}Ra and ^{224}Ra. J Geophys Res 105(C9):22117–22122

Moore WS (2003) Sources and fluxes of submarine groundwater discharge delineated by radium isotopes. Biogeochemistry 66:75–93

Moore WS, Reid DF (1973) Extraction of radium from natural-waters using manganese-impregnated acrylic fibers. J Geophys Res 78(36):8880–8886

Moore WS, Dymond J (1991) Fluxes of Ra-226 and Barium in the Pacific-Ocean - the importance of boundary processes. Earth Planet Sci Lett 107(1):55–68

Moore WS, Todd JF (1993) Radium isotopes in the Orinoco Estuary and eastern Caribbean Sea. J Geophys Res Oceans 98(C2):2233–2244

Moore WS, Arnold R (1996) Measurement of Ra-223 and Ra-224 in coastal waters using a delayed coincidence counter. J Geophys Res Oceans 101(C1):1321–1329

Moore WS, Sarmiento JL, Key RM (1986) Tracing the Amazon component of surface Atlantic Water using Ra-228, salinity and silica. J Geophys Res Oceans 91(C2):2574–2580

Okubo A (1971) Oceanic diffusion diagrams. Deep Sea Res Oceanogr Abstr 18(8):789–802

Ollivier P, Claude C, Radakovitch O et al (2008) TIMS measurements of ^{226}Ra and ^{228}Ra in the Gulf of Lion, an attempt to quantify submarine groundwater discharge. Mar Chem 109(3–4):337–354

Peng TH, Takahash T, Broecker WS (1974) Surface radon measurements in North Pacific Ocean station Papa. J Geophys Res 79(12):1772–1780

Peng TH, Broecker WS, Mathieu GG et al (1979) Radon evasion rates in the Atlantic and Pacific Oceans as determined during the Geosecs program. J Geophys Res 84(C5):2471–2486

Peterson RN, Santos IR, Burnett WC (2009a) Evaluating groundwater discharge to tidal rivers based on a Rn-222 time-series approach. Estuar Coast Shelf Sci 86(2):165–178

Peterson RN, Burnett WC, Dimova N et al (2009b) Comparison of measurement methods for radium-226 on manganese-fiber. Limnol Oceanogr Meth 7:196–205

Prichard HM, Gesell TF (1977) Rapid measurements of Rn-222 concentration in water with a commerical liquid scintillation counter. Health Phys 33:577–581

Purkl S, Eisenhauer A (2004) Determination of radium isotopes and Rn-222 in a groundwater affected coastal area of the Baltic Sea and the underlying sub-sea floor aquifer. Mar Chem 87(3–4):137–149

Railsback LB (2003) An earth scientist's periodic table of the elements and their ions. Geology 31(9):737–740

Rama, Moore WS (1996) Using the radium quartet for evaluating groundwater input and water exchange in salt marshes. Geochim Cosmochim Acta 60(23):4645–4652

Rama, Todd JF, Butts JL et al (1987) A new method for the rapid measurement of 224Ra in natural waters. Mar Chem 22(1):43–54

Raymond PA, Cole JJ (2001) Gas exchange in rivers and estuaries: choosing a gas transfer velocity. Estuaries 24(2):312–317

Redfield AC (1948) The exchange of oxygen across the sea surface. J Mar Res 7:347–361

Rhein M, Schlitzer R (1988) Radium-266 and barium sources in the deep east atlantic. Deep Sea Res A 35(9):1499–1510

Roether W, Kromer B (1978) Field determination of air-sea gas exchange by continuous measurement of radon-222. Pure Appl Geophys 116(2):476–485

Rutgers van der Loeff MM, Moore WS (1999) Determination of natural radioactive tracers. In: Grasshoff K, Kremling K, Ehrhardt M (eds) Methods of Seawater Analysis, 3rd Ed. Wiley-VCH, Weinheim

Rutherford E (1906) Radioactive transformation. Charles Scribner's Sons, New York

Santos IR, Burnett WC, Dittmar T et al (2009) Tidal pumping drives nutrient and dissolved organic matter dynamics in a Gulf of Mexico subterranean estuary. Geochim Cosmochim Acta 73(5):1325–1339

Sarmiento JL, Rooth C, Broecker W (1982) Radium-228 as a tracer of basin wide processes in the abyssal ocean. J Geophys Res Oceans 87:9694–9698

Seeton CJ (2006) Viscosity-temperature correlation for liquids. Tribol Lett 22:67–78

Schery SD, Huang S (2004) An estimate of the global distribution of radon emissions from the ocean. Geophys Res Lett 31(19):L19104

Schlesinger WH (1997) Biogeochemistry: an analysis of global change. Academic Press, Boston

Schlitzer R (1987) Renewal rates of East-Atlantic Deep-Water estimated by inversion of C-14 data. J Geophys Res Oceans 92(C3):2953–2969

Sharabi G, Lazar B, Kolodny Y et al (2010) High precision determination of Ra-228 and Ra-228/Ra-226 isotope ratio in natural waters by MC-ICPMS. Int J Mass Spectrom 294(2–3):112–115

Smethie WM, Takahashi T, Chipman DW (1985) Gas-exchange and CO_2 flux in the tropical Atlantic Ocean determined from Rn-222 and pCO_2 measurements. J Geophys Res Oceans 90(Nc4):7005–7022

Smith CG, Swarzenski PW, Reich C (2010) Examining the source and magnitude of submarine groundwater discharge along the west Flroida shelf, USA. In: Proceedings of the ASLO/NABS Joint Summer Meeting, Santa Fe, NM

Smith CG, Cable JE, Martin JB et al (2008) Evaluating the source and seasonality of submarine groundwater discharge using a Radon-222 pore water transport model. Earth Planet Sci Lett 273(3–4):312–322

Smith SD, Jones EP (1985). Evidence for wind-pumping of air-sea gas-exchange based on direct measurements of CO_2 fluxes. J Geophys Res Oceans 90(Nc1):869–875

Somayajulu BLK, Sarin MM, Ramesh R (1996) Denitrification in the eastern Arabian Sea: evaluation of the role of continental margins using Ra isotopes. Deep Sea Res II 43(1):111–117

Stewart GM, Fowler SW, Fisher NS (2008) The bioaccumulation of U- and Th-series radionuclides in marine organisms. In: Krishnaswami S, Cochran JK (eds) Radioactivity in the environment, Vol 13. Elsevier, Oxford

Stringer C, Burnett WC (2004) Sample bottle design improvements for radon emanation analysis of natural waters. Health Phys 87:642–646

Swarzenski PW (2007) U/Th series radionuclides as coastal groundwater tracers. Chem Rev 107(2):663–674

Swarzenski PW, Izbicki JA (2009) Coastal groundwater dynamics off Santa Barbara, California: combining geochemical tracers, electromagnetic seepmeters, and electrical resistivity. Estuar Coast Shelf Sci 83(1):77–89

Swarzenski PW, Reich C, Kroeger KD et al (2007) Ra and Rn isotopes as natural tracers of submarine groundwater discharge in Tampa Bay, Florida. Mar Chem 104(1–2):69–84

Toggweiller JR, Dixon K, Broecker WS (1991) The Peru upwelling and the ventilation of the South Pacific thermocline. J Geophys Res 96:20467–20497

Torgersen T, Turekian KK, Turekian VC et al (1996) Ra-224 distribution in surface and deep water of Long Island Sound: sources and horizontal transport rates. Cont Shelf Res 16(12):1545–1559

Turekian KK, Johnson DG (1966) The barium distribution in sea water. Geochim Cosmochim Acta 30(11):1153–1174

Ullman WJ, Aller RC (1982) Diffusion coefficients in nearshore marine sediments. Limnol Oceanogr 27(3):552–556

van Beek P, Souhaut M, Reyss JL (2010) Measuring the radium quartet (^{228}Ra, ^{226}Ra, ^{224}Ra, ^{223}Ra) in seawater samples using gamma spectrometry. J Environ Radioactiv 101(7):521–529

van Beek P, Bourquin M, Reyss JL et al (2008) Radium isotopes to investigate the water mass pathways on the Kerguelen Plateau (Southern Ocean). Deep Sea Res II 55(5–7):622–637

Varga Z (2008) Ultratrace-level radium-226 determination in seawater samples by isotope dilution inductively coupled plasma mass spectrometry. Anal Bioanal Chem 390(2):511–519

Wanninkhof R, Asher W, Weppernig R et al (1993) Gas transfer experiment on Georges Bank using 2 volatile deliberate tracers. J Geophys Res Oceans 98(C11):20237–20248

Wanninkhof R, Asher WE, Ho DT et al (2009) Advances in quantifying air-sea gas exchange and environmental forcing*. Ann Rev Mar Sci 1(1):213–244

Waska H, Kim S, Kim G et al (2008) An efficient and simple method for measuring Ra-226 using the scintillation cell in a delayed coincidence counting system (RaDeCC). J Environ Radioactiv 99(12):1859–1862

Webster IT, Hancock GJ, Murray AS (1994) Use of radium isotopes to examine pore-water exchange in an estuary. Limnol Oceanogr 39:1917–1927

Webster IT, Hancock GJ, Murray AS (1995) Modelling the effect of salinity on radium desorption from sediments. Geochim Cosmochim Acta 59:2469–2476

Winter TC, Harvey JW, Franke OL et al (1998) Ground water and surface water: a single resource, U.S. Geological Survey Circular 1139. U.S. Geological Survey, Denver

Yamada M, Nozaki Y (1986) Radium isotopes in coastal and open ocean surface waters of the Western North Pacific. Mar Chem 19(4):379–389

Zhang L, Zhang J, Swarzenski PW, Liu Z (2011) Chapter 19 Radium isotope tracers to evaluate coastal ocean mixing and residence times. In: Baskaran M (ed) Handbook of environmental isotope geochemistry. Springer, Berlin

Chapter 19
Applications of Anthropogenic Radionuclides as Tracers to Investigate Marine Environmental Processes

G.-H. Hong, T.F. Hamilton, M. Baskaran, and T.C. Kenna

Abstract Since the 1940, anthropogenic radionuclides have been intentionally and accidentally introduced into the environment through a number of activities including nuclear weapons development, production, and testing, and nuclear power generation. In the ensuing decades, a significant body of research has been conducted that not only addresses the fate and transport of the anthropogenic radionuclides in the marine environment but allows their application as tracers to better understand a variety of marine and oceanic processes. In many cases, the radionuclides are derived entirely from anthropogenic sources and the release histories are well constrained. These attributes, in conjunction with a range of different geochemical characteristics (e.g., half-life, particle affinity, etc.), make the anthropogenic radionuclides extremely useful tools. A number of long-lived and largely soluble radionuclides (e.g., ^{3}H, ^{14}C, ^{85}Kr, ^{90}Sr, ^{99}Tc, ^{125}Sb, ^{129}I, ^{134}Cs, ^{137}Cs) have been utilized for tracking movement of water parcels in horizontal and vertical directions in the sea, whereas more particle-reactive radionuclides (e.g., ^{54}Mn, ^{55}Fe, ^{103}Ru, ^{106}Ru, Pu isotopes) have been utilized for tracking the movement of particulate matter in the marine environment. In some cases, pairs of parent-daughter nuclides (e.g., ^{3}H-^{3}He, ^{90}Sr-^{90}Y and ^{241}Pu-^{241}Am) have been used to provide temporal constraints on processes such as the dynamics of particles in the water column and sediment deposition at the seafloor. Often information gained from anthropogenic radionuclides provides unique/complementary information to that gained from naturally occurring radionuclides or stable constituents, and leads to improved insight into natural marine processes.

19.1 Introduction

Sustained atmospheric nuclear testings and bomb explosions from 1945 to 1980 in the Equatorial Pacific (Bikini Atoll, Christmas Island, Enewetak Atoll, Johnson Atoll), northern temperate latitudes (Algeria, Japan, Kapustin Yar, Lop Nor, New Mexico, Nevada Test Site, Semipalantinsk, Totsk), polar-north (Nova Zemlya), Southern Hemisphere (Fangatufa Atoll, Malden Island, Maralinga/EMU Test ranges, Monte Bello Islands, Mururoa Atoll) contaminated the entire surface of the earth including the ocean with a suite of anthropogenic radionuclides (Hamilton 2004), ranging from short-lived to long-lived radionuclides (Table 19.1). The oceanic inventory of some selected fallout anthropogenic radionuclides is listed in Table 19.2. Other source terms including effluents from nuclear waste reprocessing plants, nuclear power plants and nuclear weapons production facilities, accidents and losses involving nuclear materials, the burn-up of nuclear powered satellites in the atmosphere have all contributed to the anthropogenic radionuclide contamination (Hong et al. 2004; Linsley et al. 2004). Notably, releases from European nuclear fuel reprocessing facilities (primarily from Sellafield, UK and La Hague, France) have been documented in

G.-H. Hong (✉)
Korea Ocean Research and Development Institute, Ansan P.O.Box 29 Kyonggi 425–600, South Korea
e-mail: ghhong@kordi.re.kr

T.F. Hamilton
Center for Accelerator Mass Spectrometry, Lawrence Livermore National Laboratory, Livermore, CA 94551–0808, USA

M. Baskaran
Department of Geology, Wayne State University, Detroit, MI 48202, USA

T.C. Kenna
Lamont-Doherty Earth Observatory, Columbia University, Palisades, NY 10964, USA

Table 19.1 List of anthropogenic radionuclides produced and globally dispersed, with their half-lives, decay mode and K_d. (a) Atmospheric nuclear weapon testing and discharge from reprocessing plants. (b) Radionuclides released from nuclear power plants in the United States[1]

(a)

Radio-nuclide (UNSCEAR 2000)	Half-life (UNSCEAR 2000; Browne and Firestone 1986)	$K_d{}^a$	Decay mode (Browne and Firestone 1986)	Daughter (Browne and Firestone 1986)	Dominant form in seawater Byrne (2002) and in the surface of the earth (Bruland 1983; Emsley 1989) (nuclear reactions for the production of radionuclides) (Hou and Roos 2008)
H-3	12.32 a	1×10^0	β	$_2$He-3 (stable)	H_2O (gas, liquid, solid), biogenic matter, also produced by the natural process ($^2H(n, \gamma)^3H$; $^3He(n, p)^3H$; $^6Li(n, \alpha)^3H$)
C-14	5730 a	2×10^3	β	$_7$N-14 (stable)	Gas CO_2, liquid, H_2CO_3, $HCO_3{}^-$, $CO_3{}^{2-}$, organic matter, biogenic or nonbiogenic (Ca,Mg or other metal) CO_3, also produced by the natural process $^{14}N(n, p)^{14}C$; $^{13}C(n, \gamma)^{14}C$; $^{17}O(n, \alpha)^{14}C$
Mn-54	312.5 d	2×10^8	EC, γ	$_{24}$Cr-54 (stable)	Ion Mn^{2+} and $MnCl^+$, solid MnO_2, $MnCO_3$, earth's crustal material, manganese nodule in the bottom of the sea, ferromanganese oxides (Fe, Mn)Ox, biogenic matter. ($Cr^{53}(d, n)Mn^{54}$) (Kafalas and Irvine 1956)
Fe-55	2.74 a	2×10^8	EC	$_{25}$Mn-54 (n, γ), $_{25}$Mn-55 (stable)	$Fe(OH)_3$, organic complex, contained in the phytoplankton cell, earth's crustal material
Se-79	2.95×10^5 a (Bienvenu et al. 2007)	1×10^3	β	$_{35}$Br-79 (stable)	Se, Se^{2+}, $SeO_4{}^{2-}$, $SeO_3{}^{2-}$, $HSeO_3{}^-$, earth's crustal material ($^{78}Se(n, \gamma)^{79}Se$; $^{235}U(n, f)^{79}Se$)
Kr-85	10.72 a	1×10^0	β	$_{37}$Rb-85 (stable)	Noble gas, also produced by the natural process
Sr-89	50.55 d	2×10^2	β	$_{39}$Y-91	Sr^{2+}, largely soluble in seawater, forms $SrSO_4$ (celestite) by Aacantharia (protozoa) in upper 400 m, earth's crustal material
Sr-90	28.6 a	2×10^2	β	$_{39}$Y-90 (2.67 d) to $_{40}$Zr-90 (stable)	$^{88}Sr(n, \gamma)^{89}Sr$, ($^{235}U(n, f)^{89}Sr$; $^{88}Sr(n, \gamma)^{89}Sr$, $^{235}U(n, f)^{90}Sr$)
Y-91	58.51 d	7×10^6	β	$_{40}$Zr-91 (stable)	$YCO_3{}^+$, YOH^{2+}, Y^{3+}, particle reactive, earth's crustal material
Zr-93	1.53×10^6 a	7×10^6	β	$_{41}$Nb-93	$Zr(OH)_4{}^0$, $Zr(OH)_5{}^-$, earth's crustal material
Zr-95	64.03 d	7×10^6	β, γ	$_{41}$Nb-95(34.97 d) to $_{42}$Mo-95 (stable)	
Tc-99	2.13×10^5 a	1×10^2	β	$_{44}$Ru-99 (stable)	$TcO_4{}^-$, earth's crustal material ($^{235}U(n, f)^{99}Tc$; $^{98}Mo(n, \gamma)^{99}Mo(\beta)^{99}Tc$)
Ru-103	39.25 d	1×10^3	β, γ	$_{45}$Rh-106 (29.8 s) to $_{46}$Pd-106	Earth's crustal material
Sb-125	2.73 a	4×10^3	β, γ	$_{52}$Te-125 (stable)	$Sb(OH)_5{}^0/Sb(OH)_6{}^-$, Earth's crustal material
I-129	1.57×10^7 a	2×10^2	β, γ	$_{54}$Xe-129	$IO_3{}^-$, I^- Dissolved in the water, incorporated into the seaweeds, lichens, grass, bovine thyroids ($^{129}Xe(n, p)^{129}I$; $^{235}U(n, f)^{129}I$; $^{127}I(2n, \gamma)^{129}I$) ($^{130}Te(n, \gamma)$ ^{131}Te (β^-) ^{131}I)
I-131	8.02 d	2×10^2	β, γ	$_{54}$Xe-131(sable)	
Cs-137	30.14 a	2×10^3	β, γ	$_{56}$Ba-137 (stable)	Cs^+, earth's crustal material ($^{235}U(n, f)^{137}Cs$)

(continued)

Table 19.1 (continued)

(a)

Radionuclide (UNSCEAR 2000)	Half-life (UNSCEAR 2000; Browne and Firestone 1986)	$K_d{}^a$	Decay mode (Browne and Firestone 1986)	Daughter (Browne and Firestone 1986)	Dominant form in seawater Byrne (2002) and in the surface of the earth (Bruland 1983; Emsley 1989) (nuclear reactions for the production of radionuclides) (Hou and Roos 2008)
Ba-140	12.75 d	9×10^3	β, γ	$_{57}$La-140 (1.68 d) to $_{58}$Ce-140 (stable)	Ba^{2+}, $BaSO_4$, earth's crustal material
Ce-141	32.50 d	7×10^7	β, γ	$_{59}$Pr-141 (stable)	$CeCO_3{}^+$, Ce^{3+}, $CeCl^{2+}$, earth's crustal material
Ce-144	284.9 d	7×10^7	β, γ	$_{59}$Pr-144 (7.2 min) to $_{60}$Nd-144 (2.1×10^5 a) to $_{58}$Ce-140	
Sm-151	90 a	5×10^5	β	$_{63}$Eu-151 (stable)	$SmCO_3{}^+$, Sm^{3+}, $SmSO_4{}^+$, earth's crustal material
Eu-155	4.96 a	2×10^6	β, γ	$_{64}$Gd-155	$EuCO_3{}^+$, Eu^{3+}, $EuOH^{2+}$, $Eu(CO_3)_2{}^-$, earth's crustal material
Np-237	2.144×10^6 a	1×10^3	α	$_{91}$Pa-233 (27.01 d)	Trace in uranium mine, ^{238}U(n, 2n)^{237}U→^{237}Np; ^{235}U(n, γ)^{236}U(n, γ)^{237}U→^{237}Np
Pu-238	87.74 a	1×10^5	α	$_{92}$U-234	Soluble Pu(V), particulate Pu(IV) ^{235}U(n, γ)^{236}U(n, γ)^{237}U(β$^-$)^{237}Np(n, γ)^{238}Np(β$^-$)^{238}Pu, ^{238}U(n, 2n)^{237}U (β$^-$)^{237}Np(n, γ)^{238}Np(β$^-$)^{238}Pu
Pu-239	24,100 a	1×10^5	α, γ	$_{92}$U-235	^{238}U(n, γ)^{239}U(β$^-$)^{239}Np(β$^-$)^{239}Pu
Pu-240	6560 a	1×10^5	α, γ	$_{92}$U-236 (2.32×10^7 a)	^{238}U(n, γ)^{239}U(β$^-$)^{239}Np(β$^-$)^{239}Pu(n, γ)^{240}Pu
Pu-241	14.4 a	1×10^5	β	$_{95}$Am-241 (432.75 a)	^{238}U(n, γ)^{239}U(β$^-$)^{239}Np(β$^-$)^{239}Pu(n, γ)^{240}Pu(n, γ)^{241}Pu
Am-241	432.75 a	2×10^6	α	^{237}Np	^{241}Pu(β$^-$)^{241}Am

(b)

Noble gases	^{41}Ar, ^{85}Kr, ^{85m}Kr, ^{87}Kr, ^{131m}Xe, ^{133}Xe, ^{133m}Xe, ^{138}Xe
Others	^{3}H, ^{24}Na, ^{51}Cr, ^{54}Mn, ^{55}Fe, ^{56}Mn, ^{57}Co, ^{58}Co, ^{59}Fe, ^{60}Co, ^{65}Zn, ^{89}Sr, ^{90}Sr, ^{92}Sr, ^{95}Nb, ^{95}Zr, ^{97}Nb, ^{97}Zr, ^{99}Mo, ^{99m}Tc, ^{103}Ru, ^{106}Ru, ^{110m}Ag, ^{113}Sn, ^{122}Sb, ^{124}Sb, ^{131}I, ^{132}I, ^{132}Te, ^{133}I, ^{134}Cs, ^{134}I, ^{135}I, ^{136}Cs, ^{137}Cs, ^{138}Cs, ^{139}Ba, ^{140}Ba, ^{140}Pa, ^{141}Ce, ^{143}Ce, ^{144}Ce, ^{187}W, ^{239}Np

$^a K_d$ (dimensionless) = [concentration per unit mass of particulate (kg/kg or Bq/g dry weight)]/[Concentration per unit mass of water (kg/kg or Bq/kg)]. Value in parenthesis indicate that data area insufficient to calculate K_ds and was chosen to be equal to the K_ds of periodically adjacent elements (IAEA 2004)

Scandinavian waters and the northern portions of the North Atlantic (Irish, North, Norwegian, Barents and Greenland Seas) (Gray et al. 1995; Lindahl et al. 2005). Chernobyl derived contamination has been documented in the Mediterranean Sea and elsewhere (e.g., Buesseler and Livingston 1996; Livingston and Povinec 2000; Noureddine et al. 2008; Papucci et al. 1996). Former nuclear weapons program facilities in Siberia (e.g., the Techa and the Tom tributaries of the River Ob, the River Yenisey, Mayak plant explosion in 1957, Karachi Lake in 1967 reported in Vakulovsky 2001) continues to release anthropogenic radionuclides to the Arctic Ocean (e.g., Cooper et al. 1999; Kenna and Sayles 2002). While there is no more direct deposition of fallout to the ocean (ceased since early 1980), a secondary pathway for global fallout nuclides reaching the oceans, such as continental run-off, release of anthropogenic radionuclides from estuarine processes, and atmospheric deposition of continental dust of previously deposited debris, becomes important in some ocean regions.

The anthropogenic radionuclides are the ubiquitous global contaminants as some of them descended from the stratosphere and landed at sea (Tables 19.1 and 19.2). Therefore, they have received great attention from the radiological protection purposes in both the

Table 19.2 Oceanic inventory of fission products and transuranium elements originating from globally dispersed debris, and local and regional deposition from atmospheric nuclear tests, including estimates of regional fallout from Pacific Ocean tests sites and deposition from SNAP-9A decay corrected to 1 January 2000 (reproduced from Hamilton 2004 with some modification, used with permission)

Radio-nuclide	Arctic ocean		Atlantic Ocean		Indian Ocean		Pacific Ocean		Total Oceanic Inventory	
	PBq	Kg	PBq	Kg	PBq	Kg	PBq	Kg	PBq	Kg
^{3}H									8000[a]	22
^{14}C									130[a]	3.9
^{90}Sr	2.0	0.40	51	10	22	4	114	23	189	37
^{99}Tc	0.001	1.8	0.03	46	0.01	20	0.07	110	0.11	178
^{129}I	0.000005	0.72	0.0012	18	0.000050	7.7	0.003	43	0.0005	69
^{137}Cs	3.2	1.0	81	25	34	11	182	57	300	93
^{237}Np	0.0003	11.3	0.007	263	0.002	94	0.02	888	0.03	1,256
^{238}Pu	0.002	0.0032	0.13	0.2	0.11	0.17	0.50	0.78	0.73	1.16
^{239}Pu	0.054	23	1.4	591	0.58	250	4.5	2960	6.5	2820
^{240}Pu	0.036	4.2	0.90	106	0.38	45	4.0	477	5.4	632
$^{239+240}$Pu	0.090	28	2.3	697	0.96	295	8.6	2436	12	3,456
^{241}Pu	0.17	0.046	4.3	1.1	1.8	0.48	24	6.2	30	7.9
^{242}Pu	0.00001	0.091	0.0003	2.3	0.0002	1.0	0.003	22	0.004	26
^{241}Am	0.04	0.29	0.92	7.2	0.39	3.1	3.7	29	5.1	40

[a]Taken from Povinec et al. (2010)

terrestrial and marine environment for the past 60 years. As anthropogenic radionuclides are introduced into the sea, they behave almost identical to their stable counterpart (Pu does not have stable nuclide) chemical elements (Table 19.1). Moreover, changes in the level and distribution of specific radionuclides and isotopic ratios in the oceans through radioactive decay and/or transport dynamics provide internal clocks (tracers) of many oceanographic processes including fluxes and input history (e.g., Bowen et al. 1980). The utility of these anthropogenic radionuclides was immediately recognized by the marine science community to apply them to understand the natural marine processes. Often the natural radionuclides present in the earth surface were simultaneously utilized to understand marine biogeochemical processes. The dynamics of water mass movement, biological particle formation, sorption-desorption reactions and decomposition processes, settling rates of particulate matter through the water column, and ultimate deposition of radionuclides onto the seafloor in various ocean basins are of prime interest in ocean geochemistry. In this connection, Broecker (1974) observed that "During the International Geophysical Year (July 1957–December 1958) the atomic technology boom that occurred during the Second World War finally reached the seas. Since then, the field has seen spectacular growth. Great advances have been made in our understanding of the substances dissolved in the sea and buried in the sediments and their utilization as guides to the nature of both past and present processes within the sea."

The application of anthropogenic radionuclides has been, however, often limited by the techniques available for sampling and analysis. Earlier analytical methods of most anthropogenic radionuclides required relatively large volumes of seawater (approximately several 100 L), followed by preconcentration and subsequent radiochemical processing and measurements using analysis resulting from radioactive decay using alpha, beta, and gamma ray spectrometers and mass spectrometric atom counting. However, the developments in instrumentation and technology in sample collection, preconcentration, and analysis have reduced the sample size as well as the time involved in processing and measurement of samples for many radionuclides. Readers are advised to consult the sampling and analytical protocols, such as Baskaran et al. (2009), for the individual anthropogenic radionuclide concerned.

A large number of studies of anthropogenic radionuclides in the ocean have contributed either directly or indirectly to the knowledge on the rates, pathways of advection, and physical mixing of ocean water, quantification of marine particle sinking rate and fate in the ocean interior and burial processes in the sea floor. Numerous studies have applied these man-made tracers to study processes such as velocities of ocean current systems and their mixing rates, particle cycling and transport, sediment accumulation and mixing rates, and pore waters dynamics as well as biological processes (see review articles in Livingston and Povinec 2002; Sholkovitz 1983 and references therein). Here we have collated previous researches that utilize anthropogenic radionuclides to understand processes of circulation and mixing of ocean water, and transport and fate of the particulate matter in the ocean. Application of anthropogenic radionuclides to tracing material transport in the atmosphere, soil, sedimentation dynamics in estuaries, and transuranics are also presented in Chaps. 25 (Matisoff and Whiting), 16 (Du et al.), and 20 (Ketterer et al.), respectively, in this volume.

19.2 Principles of Application

In most cases, the ocean input history of anthropogenic radionuclides is relatively well known; therefore a large number of radionuclides were utilized to trace their carrier phases in the sea as time markers. In particular, a large number of network stations were monitored around the globe by the former Environmental Measurements Laboratory, U.S. Department of Energy and ^{90}Sr fallout were measured over 30 years (from 1952 onwards). Using a constant ratio between ^{90}Sr and other nuclides (such as ^{137}Cs/^{90}Sr, ^{239,240}Pu/^{90}Sr, etc.) in the nuclear-weapons testing-derived fallout, the history of atmospheric fallout of most of anthropogenic radionuclides was documented. Upon reaching the surface of the earth's, each nuclide behaves similar to their stable counterpart chemical elements (except those of plutonium and technetium, as plutonium and ^{99}Tc have no corresponding stable isotopes). The anthropogenic radionuclide carrier phases in the sea are water (dissolved phase), dissolved organic matter, suspended particulate matter, bottom sediments, and biota. The carrier phase of each radionuclide is determined by the chemistry of prevailing redox and acid–base conditions in situ. Chemical forms of radionuclides influence their solubility, cell-membrane transport and bioavailability, adsorptive behavior onto particles, oceanic residence times, and volatility in the sea. In some cases, a pair or multiple anthropogenic radionuclides originated from a particular source could serve to trace the provenance of the carrier phase material. The daughter products of some of the anthropogenic radionuclides have different particle-affinity than that of their parents and the disequilibrium between the daughters and parents may be used to trace the dynamics of particle formation, sinking rates in the ocean interior and deposition rates on the seafloor. Numerical modeling techniques can be used to improve our understanding of many different environmental processes, and may apply to pollution control on a local, regional and/or planetary scale including climate change predictions. Anthropogenic radionuclides are playing an even important role in helping test the validity of these models by providing ground-truth measurement data on rates and fluxes of carrier phases such as air and water. These measurements are on the input history of radionuclides tracers and their evolution through space and time.

19.2.1 Isotopic Composition of Different Radionuclide Contaminant Sources

The transuranic composition and relative abundance of fission products produced in nuclear explosions are proportional to the duration and intensity of neutron irradiation as well as the isotopic composition of the initial material. More intense the neutron flux (e.g., high versus low burn-up fuels or high versus low explosive yield of an atomic weapon) will lead to a higher proportion of heavier isotopes of transuranics and higher yields of fission products in the irradiated material. For example, fallout debris derived from high yield weapons tests will have a higher ^{240}Pu/^{239}Pu ratio and contain more ^{137}Cs relative to fallout from a low yield weapons tests. Levels of contamination originating from the reprocessing facilities will differ based on the nature and burn-up characteristics of the fuel (e.g., low burn-up fuel from the production of weapons-grade plutonium or high burn-up fuel resulting from nuclear power generation).

Numerous studies have used the source specific signatures of nuclear contaminants to reconstruct radionuclide time histories and delineate inputs from multiple sources. Much of the available isotopic information documents the isotopic signatures of various sources as they are recorded in different environmental samples (e.g. soils, sediments, biota, ice and water).

Table 19.3 Activity ratios of selected pair of anthropogenic radionuclides useful for marine environmental applications (taken from Hong et al. 2004, used with permission). (a) Activity ratios of selected pair of global fallout radionuclides. (b) Activity and atom ratios of anthropogenic radionuclides originated from various sources other than global fallout. Activity ratios are given at the year 2000

(a)

Radionuclides	Activity ratio	Reference
$^{3}H/^{90}Sr$	299	Aarkrog (2003)
$^{14}C/^{90}Sr$	0.3424	Aarkrog (2003)
$^{137}Cs/^{90}Sr$	1.52	Aarkrog (2003)
$^{238}Pu/^{239+240}Pu$	0.030	Baskaran et al. (1995, 2000)[a]
$^{239}Pu/^{90}Sr$	0.0105	Aarkrog (2003)
$^{240}Pu/^{90}Sr$	0.007	Aarkrog (2003)
$^{241}Pu/^{90}Sr$	0.2283	Aarkrog (2003)[a]
$^{241}Pu/^{239+240}Pu$	13–14	
$^{240}Pu/^{239}Pu$	0.6672 (activity ratio)	Hirose et al. (2001)
	0.182 (atom ratio)	Kelly et al. (1999); Kim et al. (2004)
$^{239+240}Pu/^{137}Cs$	0.12 (activity ratio)	Baskaran et al. (1995)

(b)

Source	$^{240}Pu/^{239}Pu$ (atom ratio)	$^{238}Pu/^{239+240}Pu$ (activity ratio)	$^{241}Am/^{239+240}Pu$ (activity ratio)	$^{137}Cs/^{134}Cs$ (activity ratio)	$^{137}Cs/^{90}Sr$ (activity ratio)	$^{239+240}Pu/^{137}Cs$ (activity ratio)	$^{238}U/^{235}U$ (atom ratio)
Chernobyl	0.40 (Warneke et al. 2002)	0.5 (Hirose et al. 2001)		2 (Aoyama et al. 1991)	12.1 (Aoyama et al. 1991)		
Effluent from Selafield	0.242 (Lee et al. 2001)	0.27 ± 0.02 (Baskaran et al. 1995)					
dumped reactors in Kara Sea		0.25–0.46 (Baskaran et al. 1995)					
Coral rock and top soil (Mururoa Atoll) (1997)		0.54 ± 0.16 (Irlweck and Hrnecek 1999)	1.69 ± 0.51				
Loose coral rocks		30.7 ± 8.6 (Irlweck and Hrnecek 1999)	5.0 ± 1.2				
Reactor							
MAGNOX reactor (GCR)						0.23[b] (Warneke et al. 2002)	15–32 before burn up (Warneke et al. 2002)
Pressurized heavy water reactor (PHWR)						0.41[b] (Warneke et al. 2002)	
Advanced gas-cooled reactor (AGR)						0.57[b] (Warneke et al. 2002)	
Pressure tube boiling water reactor (RBMK)						0.67[b] (Warneke et al. 2002)	

19 Applications of Anthropogenic Radionuclides as Tracers

Sample	Value 1	Value 2	Value 3 (Reference)
Boiling water reactor (BWR)	0.40[b] (Warneke et al. 2002)		
Pressurized water reactor (PWR)	0.43[b] (Warneke et al. 2002)		
Natural uranium			137.88 (Warneke et al. 2002)
Weapon grade uranium			<0.1 (Warneke et al. 2002)
Depleted uranium			250–500 (Warneke et al. 2002)
Weapon production	0.01–0.07 Warneke et al. 2002		
Ivy/Mike shot Eniwetok Island	0.36 (Koide et al. 1985)		
Rongelap atoll	0.276 (Muramatsu et al. 2001)	0.28	
Bikini atoll	0.306 (Muramatsu et al. 2001)	0.12	
Enewetak atoll (Runit Island)	0.065 (Muramatsu et al. 2001)	34.7	
Enewetak atoll (Aej Island)	0.254 (Muramatsu et al. 2001)	1.06	
Marshall Island	0.303 (Muramatsu et al. 2001)	0.20	
Semipalantinsk, Chenaya Guba, Nevada test sites soils	0.03–0.08 (Muramatsu et al. 2001)		
Nagasaki, Japna	0.037 (Yoshida and Muramatsu 2003)		
Mururoa and Fangataufa, sediments	0.035–0.05 (Chiappini et al. 1999)		

[a] Decay corrected to the year of 1 January 2000 using Aarkrog (1988) for the northern hemisphere
[b] After fuel burn-up

A list of useful activity and atom ratios of selected pairs of anthropogenic radionuclides in marine environmental applications are shown in Table 19.3.

19.2.2 Radioactive Fallout from Nuclear Weapons Tests

Atmospheric and aboveground testing of nuclear weapons is significant because it results in the injection of radioactive material into the stratosphere and troposphere; the subsequent deposition of this material on the planet's surface is termed fallout. Fallout can generally be divided into two types: global fallout and local (or close-in) fallout. Global fallout occurs when an explosion of sufficient yield occurs, and the debris is injected into the stratosphere. The deposition pattern of global fallout exhibits a latitudinal dependence with maxima at mid-latitudes and minima at the poles and equator. This is due to the fact that large volumes of air exit the stratosphere via the tropopause discontinuity in the mid-latitudes. Since interhemispheric-stratospheric exchange of materials occurs on longer time scales than materials exchanged between the stratosphere and troposphere, most global fallout is deposited within its hemisphere of origin (Perkins and Thomas 1980).

It is estimated that about 6,500, 4,300, and 40 TBq (4×10^{13} Bq) of ^{239}Pu, ^{240}Pu, and ^{237}Np, respectively have been released globally by the surface and atmospheric weapons tests conducted between 1945 and 1980 (Lindahl et al. 2005). Due to the long half-lives of these radionuclides, these values have not changed substantially. The estimates for ^{137}Cs reached a maximum during the mid to late 1960 of 460 PBq (4.6×10^{17} Bq). Due to its relatively short half-life, this value will have decreased to about 170 PBq by 2010. Of these total, approximately 76% was deposited in the northern hemisphere, nearly all being deposited between 0° N and 70°. Using a value of 55% of oceanic areal coverage, it is estimated that 2,500, 1,600, and 15 TBq and 71 PBq of ^{239}Pu, ^{240}Pu, ^{237}Np, and ^{137}Cs, respectively have been deposited to the marine areas of the Northern Hemisphere (Lindahl et al. 2005; UNSCEAR 2000).

While the initial pathway of fallout nuclides to the marine environment was direct deposition, a secondary pathway for global fallout nuclides reaching the oceans is continental run-off and related estuarine processes and tropospheric resuspension of previously deposited debris that may serve to modify the global fallout isotopic signatures (e.g., Linsalata et al. 1985; Shlokovitz and Mann 1987; Hamilton et al. 1996; see discussion below about the different geochemical behavior of the radionuclides of interest).

19.2.3 Discharges from the Nuclear Fuel Reprocessing Plant

A number of nuclear fuel reprocessing plants are located at the coast and they discharge radioactive wastes into the sea (Hu et al. 2010). The Sellafiled and La Hague in the northern Europe are of global significance in terms of ocean process tracers. The Sellafield Nuclear reprocessing plant has been discharging liquid radioactive wastes containing plutonium isotopes, ^{237}Np, and ^{137}Cs (with some amounts of ^{134}Cs) to the Irish Sea. It has been estimated that 610 and 9.5 TBq (1 TBq = 10^{12} Bq) of ^{239,240}Pu and ^{237}Np, respectively and ~20 PBq (1 PBq = 10^{15} Bq) of ^{137}Cs (decay corrected to 2010) have been released since the plant began operating in 1952 (Assinder 1999; Beasley et al. 1988; Gray et al. 1995; Kuwabara et al. 1996). The reported atom ratios of ^{240}Pu/^{239}Pu and ^{237}Np/^{239}Pu in discharges from Sellafield weighted over the operating period of the plant are 0.242 and 1.69, respectively. The La Hague Plant located in the west of Cherbourg, France began operating in 1966. Its radioactive waste discharge peaked in the late 1970 to early 1980. The cumulative discharge from La Hague between 1967 and 1995 was 33, 1,600, 654, 1.3 and 2.5 TBq of ^{60}Co, ^{90}Sr, ^{137}Cs, ^{238}Pu, $^{239+240}$Pu, respectively (Cundy et al. 2002). The 1997 sample of effluent showed a relatively high ^{240}Pu/^{239}Pu of 0.34 (Ketterer and Szechenyi 2008). La Hague also discharged ^{129}I as much as 1,640 kg for the period of 1975–1997 (Raisbeck and Yiou 1999).

19.2.4 Chernobyl Derived Contamination

It is estimated that the total activity of ^{239,240}Pu and ^{137}Cs released to the environment as a result of the accident at Chernobyl (April 1986) was 0.055 and 85 PBq (24 PBq in 2010), respectively. Kirchner and Noack (1988) estimate that the ^{240}Pu/^{239}Pu and ^{237}Np/^{239}Pu atom composition in the reactor core at the time of the accident were 0.56 ± 0.16 and 0.023 ± 0.006, respectively. Although there are no published values for Chernobyl derived ^{237}Np, ^{240}Pu/^{239}Pu atom ratios of around 0.4 have been determined in soils near the facility (Krey et al. 1986; Muramatsu et al. 2000). Several studies of the Chernobyl accident have shown that the deposition pattern of radioactivity was highly variable over Europe and that the isotopic composition of this material varied throughout the period of the accident (Buesseler and Livingston 1996; Krey et al. 1986; Livingston et al. 1988). Chernobyl-derived ^{137}Cs was estimated to have added 5 and 3 PBq into the Baltic and Black Seas, respectively (Livingston and Povinec 2000) and 3 PBq in the Mediterranean Sea (Papucci et al. 1996).

19.3 Applications of Selected Anthropogenic Radionuclides

19.3.1 Tritium (^{3}H)

One of the most significant applications of anthropogenic radionuclides to climate studies up to the present is the utilization of the bomb tritium distribution in the North Atlantic (Fig. 19.1). With its presence in intermediate and deep waters, it directly confirmed that the deep water forms in the North Atlantic as postulated by the box model of global ocean conveyor-belt circulation (Broecker 1974). Before the era of nuclear weapons testing, the world's inventory of cosmic ray-produced tritium was estimated to be ~7 kg, however, by the time of the moratorium on widespread testing in 1963 an additional amount of the order of 100 kg had been introduced largely to the northern hemisphere. Both natural and anthropogenic tritium is rapidly transferred to the surface ocean as HTO (^{1}H^{3}HO) via direct precipitation and gas exchange. Tritium decays to ^{3}He with 12.43 years half-life. Bomb-produced tritium has considerable potential as a tracer for oceanic circulation and for the study of processes with time-scales of less than 100 years due to its short half-life. Tritium is normally reported in Tritium unit (1 TU = 1 × 10^{-18} atoms of ^{3}H per atom of hydrogen or 1 tritium atom in 10^{18} hydrogen atoms; 1 TU = 3.19 pCi/L = 118 Bq m^{-3}). In subsurface waters, the ^{3}He thus formed cannot escape, so that the combined measurements of tritium and the in situ grown ^{3}He enables tritium–^{3}He "dating", the age is the period since the parcel of water left the surface mixed layer (Roether et al. 1999)

The tritium/^{3}He age, τ_{He-3}, is calculated using (19.1),

$$\tau_{He-3} = T_{1/2}/\ln 2 \times \ln(1 + [^3He_{-tri}]/[^3H]) \quad (19.1)$$

where $T_{1/2}$ is a half life of tritium (^{3}H) and ^{3}He$_{-tri}$ is tritiogenic ^{3}He.

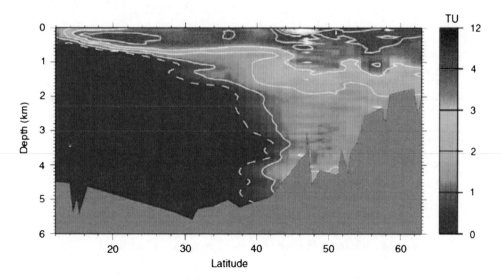

Fig. 19.1 A North Atlantic tritium meridional section in the Denmark Strait to the Central Sargasso Sea taken in the early 1980. The extensive penetration of bomb tritium in the Norwegian Sea (60°N) is clearly visible (Jenkins 2001, used with permission)

Due to radioactive decay, the bomb-produced tritium inventory has reduced from 113,000 PBq at 1963 to 8,000 PBq in 2010 or only about 4 times larger than the natural fallout level. The reprocessing plants-originated ^3H oceanic inventory is estimated to be 45 PBq in 2010 (Povinec et al. 2010). And after cessation of bomb-derived ^3H global fallout in the early 1980, the water circulation and mixing appears to influence ^3H distribution in the ocean more than the atmospheric input in the past.

19.3.2 Radiocarbon (^{14}C)

Radiocarbon is also produced naturally in the atmosphere by nuclear reaction of cosmic ray-produced neutrons with atmospheric nitrogen. The ^{14}C production rate of 2.2 atoms cm^{-2} s^{-1} is balanced by its disintegration by beta decay. During its mean lifetime of 8,200 years radiocarbon can penetrate the active carbon reservoirs through chemical reactions of carbonic acid formation and plant photosynthesis (Fig. 19.2). The production rate of ^{14}C has not been constant with time, and neither have the rates of processes that distribute ^{14}C among various reservoirs. In addition to these natural perturbations, the emission of CO_2 to the atmosphere by fossil fuel combustion has measurably reduced the atmospheric $^{14}C/^{12}C$ ratio.

The large input of ^{14}C into the upper atmosphere resulting from nuclear weapon testing became measurable in 1954. At the time of the implementation of the test ban treaty in 1963, the number of nuclear-weapons-derived ^{14}C atoms in the atmosphere was roughly equal to the number of cosmogenic ^{14}C atoms. This excess ^{14}C has decreased to ~10% of the cosmogenic ^{14}C inventory as of the year 2000 (Broecker 2003). This decrease is mainly due to removal of ^{14}C by the exchange with ocean ΣCO_2 and terrestrial biospheric carbon and dilution by the addition of ^{14}C free fossil-fuel-derived CO_2 molecules to the atmosphere (Broecker 2003). And the current atmospheric and biotic mass activities of ^{14}C are close to levels observed prior to atmospheric nuclear weapons testing (Yim and Caron 2006).

As it is difficult to measure absolute ^{14}C concentrations, it is conventional to express ^{14}C determinations as the per mil difference between the specific activity of the sample and 0.95 times the activity of a standard carbon sample ($A_{NBS\,Std}$), with the 'modern' is defined as the year of 1950. Thus

$$\delta^{14}C = (A_{sample} - 0.95 A_{NBS\,Std})/(0.95 A_{NBS\,Std}) \times 1,000$$
(19.2)

The principal modern radiocarbon standard ($A_{NBS\,Std}$) is NIST oxalic acid I ($C_2H_2O_4$), made from a crop of 1955 sugar beets. Ninety-five percent of the activity

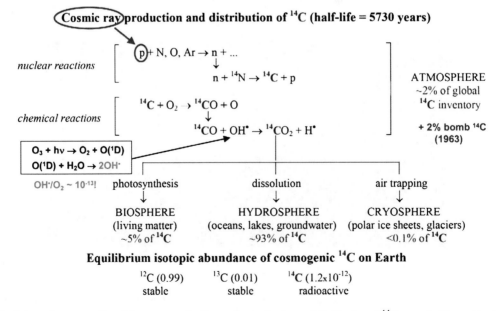

Fig. 19.2 Schematic presentation of the processes leading to the production and distribution of ^{14}C on earth. The sudden increase of ^{14}C in the atmosphere by nuclear weapons testing in the early 1960 is also indicated. (Kutschera 2010, used with permission)

of oxalic acid I from the year 1950 is equal to the measured activity of the absolute radiocarbon standard which is 1890 wood (chosen to represent the pre-industrial atmospheric $^{14}CO_2$), corrected for radioactive decay to 1950.

Furthermore, it is necessary to allow for ^{14}C differences produced by isotopic fractionation. This is achieved by use of the $^{13}C/^{12}C$ ratio, which is itself also expressed as enrichment:

$$\delta^{13}C(‰) = [(^{13}C/^{12}C)_{sample} - (^{13}C/^{12}C)_{standard}] / [(^{13}C/^{12}C)_{standard}] \times 1{,}000 \tag{19.3}$$

Normalized ^{14}C enrichments are then given by the formula:

$$\Delta^{14}C = \delta^{14}C - 2(\delta^{13}C + 25)(1 + \delta^{14}C/1{,}000) \text{ ‰} \tag{19.4}$$

19.3.2.1 Validating Global Ocean Carbon Model

The distribution of bomb-produced ^{14}C in the ocean has been summarized on the basis of radiocarbon measurements made during GEOSECS (Geochemical Ocean Sections Study), TTO (Transient Tracers in the Ocean), and SAVE (South Atlantic Ventilation Experiment) ocean survey programs. The inventory of bomb ^{14}C and the mean penetration depth of this tracer in the water column for the Atlantic (1972–1973), Pacific (1973–1974), and Indian (1977–1978) oceans have been published from the GEOSECS results. To eliminate the time difference, these bomb ^{14}C inventories are normalized to 1 January 1975. This represents the global spatial distribution of bomb ^{14}C tracer, which is required for calibration of ocean models, especially when these models are to be used for estimating the oceanic uptake of CO_2. In addition, results obtained from expeditions during later years from TTO in the northern and tropical Atlantic (1981–1982), and SAVE in the southern Atlantic (1987–1988) are also published by Broecker et al. (1995). This information depicts the temporal variations of the bomb ^{14}C distribution in the Atlantic Ocean. The evolution of bomb ^{14}C inventory in the ocean with time is another valuable piece of information for verifying the models of the global ocean carbon cycle (Peng et al. 1998).

19.3.2.2 Tracing Dissolved Organic Carbon Sinking in the Sea

According to Beaupre and Druffel (2009), dissolved organic carbon (DOC), is largely derived from the autochthonous production in the sun-lit surface ocean, and is the largest reservoir of reduced carbon in the ocean with its magnitude of about 685×10^{15} g C and is comparable to the carbon in the atmosphere in the form of CO_2. The 4,000–6,000 year ^{14}C ages of deep ocean DOC suggest that a significant portion cycles on longer time scales and ages during deep water transit. The processes that produce these old ages remain unknown. As a tracer of time and carbon sources, the ^{14}C content of marine DOC is a powerful tool for potentially constraining many of these uncertainties. They were able to infer the sinking of DOC from the surface ocean to depths of about 450 m on time scale of months based on the time series observations of $\Delta^{14}C$ of DOC between 1991 and 2004, and the magnitude and synchronicity of major $\Delta^{14}C$ anomalies (Beaupre and Druffel 2009).

19.3.2.3 Dating Marine Samples for the Recent Past ~60 Years with High Accuracy

Living organisms take up radiocarbon through the food chain and via metabolic processes. This provides a supply of ^{14}C that compensates for the decay of the existing ^{14}C in the organism, establishing equilibrium between the ^{14}C concentration in living organisms and that of the atmosphere. When an organism dies, this supply is cut off and the ^{14}C concentration of the organism starts to decrease by radioactive decay at a rate determined by the radiocarbon half-life. This rate is independent of other physical and environmental factors. The time t elapsed since the organism was originally formed can be determined from (19.5):

$$t = T_{1/2}/\ln 2 \times \ln(N_t/N_o) \tag{19.5}$$

where $T_{1/2}$ is the radiocarbon half-life, N_o is the original ^{14}C concentration in the organism and N_t is its residual ^{14}C concentration at time t (Hua 2009). This method has been utilized extensively for the climate proxies and dating older objects in archaeology. However, we would like to highlight the importance of bomb-derived ^{14}C as the anthropogenic ^{14}C

overwhelmed the naturally produced ^{14}C by masking its natural variability, and thus allowed dating objects of the recent past ~60 years with much greater accuracy than for ^{14}C age-dating conducted during the pre-bomb period, e.g., determination of age-depth model for a salt marsh (Marshall et al. 2007). As another example, accurate measurements of the age of fish provide valuable information for the sustainable management of fish stock in the sea (for dating of fish otoliths, see Chap. 37). Piner et al. (2006) collected otoliths from bocassio rockfish off the coast of Washington State of USA and determined their birth years and inferred that they can live at least 37 years. The age-structured stock assessment is helpful for the fish mangers to evaluate the sustainability of the fish populations in the region.

19.3.2.4 Tracing Source of Organic Matter in Estuary

Organic matter is one of the controlling factors determining the fertility and environmental quality of estuaries and coastal oceans. Organic matter may have its origin from in situ primary production, resuspension of the bottom sediment, and terrestrial detritus discharged from surface runoff through rivers and streams. The advantage of ^{14}C determinations over the use of ^{13}C as a marker for the source of organic material is the fact that ^{14}C age of biomass of short living organisms is, unlike-δ^{13}C, the same for all organic constituents, because the effect of isotopic fractionation is removed by the normalization procedure used in ^{14}C age determination as described above. For example, Megens et al. (2001) were able to elucidate that particulate organic matter in the southern North Sea during winter is mainly derived from the resuspension from the bottom sediment by utilizing bomb-^{14}C signal.

19.3.3 Manganese (^{54}Mn)

^{54}Mn, ^{58}Co, ^{60}Co, ^{134}Cs and ^{137}Cs are among the common beta/gamma emitting radionuclides discharged under normal operating conditions by many nuclear facilities. For instance in 1995, these five isotopes accounted for about 68% of the non-tritium low-level radioactive liquid wastes from French 1,300 MW pressurized reactors. In addition to this radioecological aspect, the three elements selected present a special interest from a biological standpoint. Cs is biochemically analogous to K while Mn and Co are classified among the ten vital elements for life. Co is vital to many enzymatic systems and to the formation of noble molecules, such as vitamin B-12. Mn is a coactivator of such enzymes as transferases and decarboxylases, and is a constituent of several metalloenzymes, including pyruvate carboxylase and superoxide dismutase. Therefore, these radionuclides could be used to study metal physiology, e.g., trophic transfer factors, in biological organisms in the marine areas adjacent to the nuclear waste discharge facilities (Baudin et al. 2000).

19.3.4 Iron (^{55}Fe)

Introduction, formation, decomposition, dissolution, and sinking of particulate matter is largely responsible for the vertical segregation of biophilic chemical elements in the sea. The main aspects of particulate matter have been the size, settling rate, and physical, chemical and biological compositions. During the 1960, ^{55}Fe constituted one of the major radioactive isotopes present in atmospheric fallout. Although ^{55}Fe, which has a 2.4 year half-life, decays exclusively through electron capture and emits a very weak 5.9 keV X-ray, this isotope is of biological interest because Fe is an essential element for plant growth and is absorbed by red blood cells of animals. ^{55}Fe fallout from atmospheric weapons detonations was largely associated with aerosols as an amorphous oxide or as extremely small particulate species attached to the surfaces of large aerosol particles. The ^{55}Fe contained in these aerosols was more readily solubilized and became available to marine organisms than the stable iron in geological matrix of soil (Weimer and Langford 1978). Massic ^{55}Fe activity was utilized to obtain Fe-laden particle dynamics in the Pacific Ocean. Lal and Somayajulu (1977) found that ^{14}C-laden biogenic calcareous particles (~ 6 μm diameter) sank faster than ^{55}Fe-labeled small particles (~ 1 μm diameter) sinking to the depths of 2,500 m in the Pacific Ocean. And ^{55}Fe-labeled particles sank faster than Pu isotope-labeled particles in the North Pacific Ocean (Livingston et al. 1987). These studies indicated that particulate carrier phase may be specific to each metallic element. Recently the role of iron in the photosynthesis of marine plant, hence, its influence on the climate change has drawn considerable attention. Its role on the sequestration of atmospheric CO_2 received extensive interests from both scientific and commercial community (e.g., Betram 2010). In this context, biogeochemistry of iron in the sea could be elucidated using ^{55}Fe as a tracer at sites where it is released.

19.3.5 Cobalt (^{58}Co and ^{60}Co)

Controlled low level radioactive waste release from routine operation of nuclear power plants could be monitored using ^{60}Co and other radionuclides (^{137}Cs, ^{134}Cs) and their spatial gradients could be used for estimating the extent of discharge plume in the receiving water body and sediment budget (Olsen et al 1981). Cutshall et al. (1986) used ^{60}Co and ^{152}Eu to trace the Columbia River derived sediment in Quinault Canyon, Washington, USA by utilizing unusual, once-through, open-loop cooling system at the Hanford nuclear facility. Donoghue et al. (1989) have used ^{134}Cs to estimate sediment trapping behind the river dams located below the nuclear reactors as the affinity of cesium for sediment particles, illite mineral in particular, in the freshwater are very high and desorption does not occur. The presence of ^{60}Co in the bottom sediment was used to confirm a nuclear submarine reactor accident occurred in 1985 in Sterlok bay, Peter the Great Bay off Vladivostok (Tkalin and Chaykovskaya 2000).

19.3.6 Krypton (^{85}Kr)

^{85}Kr concentrations in the atmosphere and ocean have been increasing steadily since 1945 as a result of atmospheric release from nuclear power plants, plutonium production and nuclear waste processing facilities, and in atmospheric nuclear weapons tests. ^{85}Kr has a great potential as a tracer for ocean ventilation and water mass formation. There are two basic reasons for this: (1) ^{85}Kr enters the ocean by gas exchange with an equilibration time of 1 month or less. Hence nearly the entire ocean surface will be in equilibrium with the atmosphere, and the surface water concentration as a function of time at any location in the ocean can be accurately calculated from the documented atmospheric history (Winger et al. 2005; Kemp 2008) and krypton solubility data; and (2) Krypton is an inert gas and ^{85}Kr is absolutely conservative in seawater except for its radioactive decay. Smethie et al. (1986) were able to show that Norwegian Sea Deep Water forms from a mixture of Greenland Sea Deep Water and Eurasian Basin Deep Water. They estimated the volume transports for exchange between the surface and deep Greenland Sea and for exchange between the deep Greenland and deep Norwegian seas. They also estimated the residence time of water in the deep Greenland Sea with respect to exchange with surface water. As ^{85}Kr is introduced from the air, deep water sample for the gas analysis may be checked with its presence as an indicator of air contamination during sampling (Schlosser et al. 1995).

19.3.7 Strontium (^{89}Sr and ^{90}Sr)

Most of the ^{90}Sr fallout in surface waters is derived from global fallout. The interest in ^{90}Sr stems from the following: (1) ^{90}Sr is a high-yield product of U and Pu fission; (2) ^{90}Sr is a tracer for stable Sr, and Sr has similar properties as Ca; (3) ^{90}Sr is the most widely and carefully monitored fallout radionuclide in precipitation, aerosols, and soils. Indeed, the bomb test fallout of ^{137}Cs is calculated based on the monitored ^{90}Sr fallout, assuming constancy of the ^{90}Sr/^{137}Cs ratio, and (4) its moderate half-life of 28.9 years (Baskaran et al. 2009). In the year of 2000, total oceanic inventory of ^{90}Sr was estimated to be 189 PBq (37 kg) (Hamilton 2004) and it constitutes ca. 0.3×10^{-9} g ^{90}Sr/g Sr in the world ocean based on the average Sr concentration (8.7×10^{-5} mole/kg) and world ocean volume (1.37×10^{21} L) (Broecker and Peng 1982). ^{90}Sr largely resides in the water column with subsurface peak at ca. 300 m (Fig. 19.6) and very small fractions (0.02–0.04%) are buried in the sediment in the deep Pacific Ocean Basin (Lee et al. 2005). ^{90}Sr concentration in the surface 1 km depth decreased approximately 30% over the past 24 years (Fig. 19.3). As the environmental half-life of ^{90}Sr in the central NW Pacific Ocean is ca. 15 years (Povinec et al. 2005), ^{90}Sr would decrease to 0.1 Bq m^{-3} by 2060 (if there is no additional sources in the future), which would be difficult to measure with 50 L of seawater sample using oxalate precipitation method (Baskaran et al. 2009).

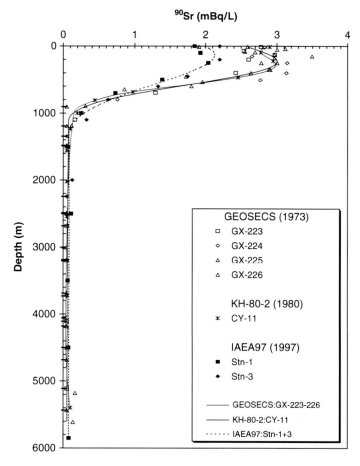

Fig. 19.3 ^{90}Sr concentration profiles in the central NW Pacific Ocean in 1997 (IAEA 1997) along with earlier GEOSECS (1973), KNORR (1978), Hakuho Maru (1980) measurements (Povinec et al. 2003, used with permission)

The residence time of stable Sr is ~ 5×10^6 year in the ocean, and several orders of magnitude longer than the global ocean turnover time (~ 1,500 years). The removal from the sea can be made through the formation of $SrCO_3$ (strontianite), $SrSO_4$ (celestite), and incorporated into Barite ($BaSO_4$) as impurities. $SrCO_3$ is formed when Sr is incorporated into coralline skeletons at sea. The sodium carbonate fortified seawater from which coral skeleton precipitates has a Sr/Ca ratio close to that of seawater (~0.8×10^{-2}), but it is depleted in Mg and Ba (Gaetani and Cohen 2006). Recently, it was shown that $SrCO_3$ mol percentage in coralline algae (rhodoliths) in coastal waters off Scotland closely followed the in situ temperature (Kamenos et al. 2008). Therefore, ^{90}Sr might be utilized as a paleothermometer for corals and other biogenic carbonates proxies.

Celestite is formed as major skeletons and cysts of acantharians, abundant planktonic protists. Settling skeletons of dead acantharians and acantharian cysts are readily dissolved because the ocean is undersaturated with respect to $SrSO_4$. Most oceanic water is undersaturated with respect to barite ($BaSO_4$), yet barite particles are ubiquitous throughout the water column in the ocean. Barite formation in the water column and its accumulation in sediments are closely related to export production of carbon. Sr is incorporated into barite precipitation (average 36.6 mmol Sr/mol Ba) (Paytan et al. 2007). However, the mechanism of barite formation has not been clearly understood yet. Three main hypotheses that have been proposed for barite formation in the oceanic water column include: (1) Barite is formed in microenvironments in which sulfate is enriched due to organic matter oxidation;

(2) A thermodynamically driven barite formation process in which the dissolution of acantharian celestite (SrSO$_4$), which is enriched in Ba (BaSO$_4$/SrSO$_4$), creates barium rich microenvironments conducive to barite precipitation; and (3) Barite is formed by Ba enrichment rather than SO$_4^{2-}$ enrichments. It is reported that these proposed mechanisms for barite formation are not mutually exclusive. Acantharian dissolution within microenvironments appears to lead to BaSO$_4$ supersaturation and subsequent barite formation (Bernstein and Byrne 2004).

In case of a relatively fresh fission product mixture, another radiostrontium, ^{89}Sr with a half-life of 50.5 days, can also be present. As the fission yields of both ^{89}Sr and ^{90}Sr are known, their activity ratio can be applied in the dating of the fission product formation. According to UNSCEAR (2000), the ^{89}Sr/^{90}Sr activity ratio should be 188 just after a nuclear explosion (Paatero et al. 2010)

19.3.7.1 Quantifying Particle Removal

The disequilibrium between the soluble parent isotope ^{90}Sr ($t_{1/2} = 29.1$ years) and its particle-reactive daughter ^{90}Y ($t_{1/2} = 64$ h) has been used to estimate particle settling rates in freshwater systems (Orlandini et al. 2003) and in estuaries and the coastal ocean. Because of the short half-life (64 h) of ^{90}Y, this method could-be used to trace faster processes than what is possible using ^{238}U ($t_{1/2} = 4.468 \times 10^9$ years)–^{234}Th ($t_{1/2} = 24.1$ days) disequilibria. However, in order to utilize ^{90}Y–^{90}Sr pair as a particle-cycling tracer, the particulate ^{90}Y (^{90}Sr likely negligible) and dissolved ^{90}Y needs to be separated immediately after sample collection and hence immediate filtration is required. The mathematical expressions to relate the measurements of ^{90}Sr–^{90}Y disequilibrium to the physical dynamics of particles and particle-reactive species are similar to expressions developed and applied to uranium–thorium disequilibria in the sea (19.6).

$$\lambda_p N_p = \lambda_d N_d + k N_d \quad (19.6)$$

Here λ_p and λ_d are the decay constants of the parent (here ^{90}Sr) and the daughter nuclide (^{90}Y), respectively. The N terms are the atom concentrations of the parent and daughter nuclides in the water column. The k term is the first-order net removal rate coefficient (nonradioactive) for the particle-active daughter nuclide. In (19.6), the advection (horizontal and vertical) and diffusion are neglected. Solving for the removal coefficient (residence time = 1/k) and converting to activities (A, expressed in Bq m^{-3}) by using the appropriate decay constants gives the following:

$$k = \lambda_d (A_p - A_d)/A_d \quad (19.7)$$

19.3.7.2 Tracing Water Mass Movement

Water column distribution of ^{90}Sr, ^{99}Tc, ^{129}I, ^{137}Cs, ^{238}Pu, ^{239}Pu, and ^{241}Am over time was used to estimate horizontal advection rates of various water masses in various parts of the ocean, for example, in the Arctic Ocean (Livingston et al. 1984), Adriatic Sea (Franić 2005), Norwegian Sea (Yiou et al. 2002), and Sulu Sea (Yamada et al. 2006).

Also, ^{90}Sr activity in the river varies depending upon the watershed soil and denudation characteristics in its watershed, and could serve as a source indicator in the estuary. For example, ^{90}Sr activity in the Danube River was 15–30 times lower than that of Dneiper River in the north-west Black Sea. Based on this information, Stokozov and Buesseler (1999) constructed a water mixing model using ^{90}Sr and salinity as water mass tracers for the northwest Black Sea. Present distribution of ^{90}Sr also serves as the benchmark to assess future spreading of radionuclides for specific release scenarios. Gao et al. (2009) modeled the difference between the present-day and the 2 × atmospheric CO_2-warming scenario runs for the accidental releases of ^{90}Sr in the Ob and Yenisey rivers and indicated that more of the released ^{90}Sr would be confined to the Arctic Ocean in the global warming run, particularly in the coastal, non-European part of the Arctic Ocean.

19.3.8 Ruthenium (^{103}Ru and ^{106}Ru)

^{103}Ru and ^{106}Ru fallout radionuclides released from the Chernobyl accident along with other gamma-emitting fallout radionuclides, ^{134}Cs, ^{137}Cs, ^{110m}Ag, were used to quantify water mass movement in the seas. Carlson and Holm (1992) measured the concentrations of these nuclides in marine plant, *Fucus vesiculosus* L. in the

Baltic Sea area following the Chernobyl accident. The activity ratios of $^{106}Ru/^{137}Cs$ and $^{144}Ce/^{137}Cs$ in sediment trap and suspended particles were utilized to distinguish the particles laden with Chernobyl Cs ($^{134}Cs/^{137}Cs = 0.5$) and earlier global fallout ($^{134}Cs/^{137}Cs = 0.0$) in the Black Sea (Buesseler et al. 1990).

19.3.9 Antimony (^{125}Sb)

Antimony-125 is a conservative tracer. A large set of ^{125}Sb data collected from the English Channel and southern North Sea between 1987 and 1994 have been utilized to validate hydrodynamic model in this region (du Bois and Dumas 2005; du Bois et al. 1995). For the field validation with tracers, the coverage of the spatial and temporal tracer concentrations with high accuracy is required. The ideal water mass tracers should show conservative behavior in the water mass, i.e., neither fixed by environmental components (sediment, living species) nor modified during its stay in seawater, and when subsequently diluted, it must be measurable at several hundred or thousand kilometers from its point of discharge. Good tracers must have only one or a set of well-defined and characterized input functions and their flow must be well tracked. In this context, anthropogenic radionuclides released by nuclear fuel reprocessing plants fully meet these specifications if their half-life is long enough compared to the transit-times in the ocean basin. Since 1960, large-scale studies have been chiefly concerned with ^{137}Cs, a point source tracer due to the discharge from the nuclear fuel reprocessing plant at Sellafield on the Irish Sea. Other tracers, namely ^{134}Cs, ^{90}Sr and ^{99}Tc, ^{125}Sb and $^{239+240}Pu$ have also been used to monitor the transport of water masses. For example, dispersion of a water from the Rhone River into the coastal Mediterranean Sea was modeled using ^{125}Sb, ^{137}Cs and $^{239+240}Pu$ (Periáñez 2005).

19.3.10 Iodine (^{129}I)

In the terrestrial and marine environment, natural levels of ^{129}I (cosmogenic origin) have been overwhelmed by a build-up of "new" ^{129}I, a product of the nuclear age. Much of this new ^{129}I has entered the ocean and is now found in its upper layers. Using ^{129}I as a point source tracer due to the discharge from the nuclear fuel reprocessing plants, it has been employed as an oceanographic water mass tracer to determine transit time scale based on the horizontal concentration gradients from the point of discharge. ^{129}I has also been used as a tracer for monitoring nuclear activities, including nuclear safeguard investigations. Furthermore, the differences in the $^{129}I/^{137}Cs$ and $^{129}I/^{99}Tc$ activity ratios of reprocessed and unprocessed nuclear wastes are also utilized as markers to distinguish water masses because of their unique chemical properties (e.g., solubility, volatility) and high sensitivity of detection. The activity ratio $^{129}I/^{137}Cs$ was used to distinguish between accidental or deliberate discharges of these two types of radioactive wastes to the ocean (e.g., Raisbeck and Yiou 1999).

During primary production, iodine is incorporated in marine organic matter and migrates through the food chain. The $^{129}I/^{127}I$ atom ratio in marine organics therefore reflects the value found in the ocean's photic zone when the organic matter formed. Because the $^{129}I/^{127}I$ atom ratio in any well-mixed marine basin has increased rapidly since the advent of the nuclear age, establishing the buildup pattern of ^{129}I in that basin's surface waters would allow us to ''date'' the time of formation of any organic matter in the euphotic zone, provided we can obtain an adequate amount of iodine from samples (Schink et al. 1995). Measurement of this ratio currently requires the use of accelerator mass spectrometer (AMS).

19.3.11 Cesium (^{134}Cs and ^{137}Cs)

Cesium is an alkali metal existing as the Cs^+ ion in the oceans. Similar to other alkali metals (e.g., potassium), cesium is conservative in seawater (Brewer et al. 1972). In terrestrial environment, Cs is strongly associated with soil and sediment particles. K_d values for Cs in freshwater environments are large and on the order of $1-5 \times 10^5$. This value decreases substantially as particulate matter is delivered to the oceans via rivers and estuaries. As salinity increases, so too does competition for sediment sorption sites from other cationic species such as K^+, resulting in desorption of cesium. In pelagic environments, K_d is observed to be significantly lower and has been reported to vary between 4×10^2 and 2×10^4 L kg^{-1} (IAEA 2004), with lower end in the Black Sea (Topcuoğlu et al. 2002).

In the year of 2000, total oceanic inventory of ^{137}Cs was estimated to be 300 PBq (93 kg) (Hamilton 2004) and it constitutes ca. 0.2×10^{-12} g ^{137}Cs/g Cs in the world ocean based on the average Cs concentration of 2.3×10^{-9} mole/kg and world ocean volume of 1.37×10^{21} L (Broecker and Peng 1982). ^{137}Cs largely resides in the water column with subsurface peak at ca. 200 m and very small fractions (0.01–0.12%) are buried in the sediment in the deep Pacific Ocean Basin (Lee et al. 2005). ^{137}Cs concentration in the surface 1 km depth decreased approximately 40% over the past 24 years (Fig. 19.4). As the environmental half-life of ^{137}Cs in the central NW Pacific Ocean is ca. 24 years (Povinec et al. 2005), ^{137}Cs would decrease to ca. 0.1 Bq m^{-3} by 2108, which would be difficult to measure with current AMP precipitation method (Baskaran et al. 2009).

19.3.11.1 Utilizing ^{134}Cs/^{137}Cs Activity Ratios for Estimating Water Transit Time in the Arctic

One of the most significant point source discharges of anthropogenic radionuclides in the ocean is the Sellafield reprocessing plant (formerly Windscale) UK into the Irish Sea since 1952 in the North Atlantic. The reprocessing waste from La Hague, north-west France discharging into the English Channel since 1966 is relatively small compared to Sellafield (Kershaw and Baxter 1995). The contribution of La Hague to the marine inventory of ^{137}Cs, ^{90}Sr, ^{99}Tc and Pu was estimated to be 2.3, 12.2, 12.6 and 0.4%, respectively, of the Sellafield releases (Kershaw and Baxter 1995). The discharge of anthropogenic radionucldies resulted in substantial increase in their inventories in the North Atlantic and its marginal seas. The soluble radionuclides (^{90}Sr, ^{99}Tc, ^{129}I, ^{134}Cs, ^{137}Cs) released from Sellafield are carried northwards out of the Irish Sea

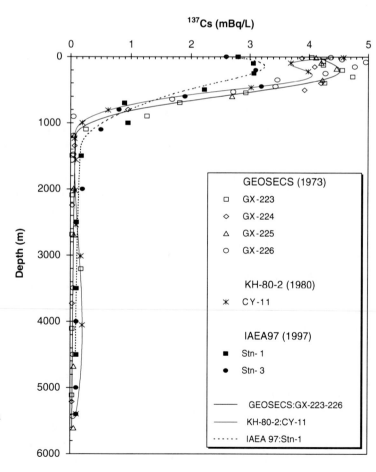

Fig. 19.4 ^{137}Cs profiles in the central NW Pacific Ocean in 1997 (IAEA 1997) along with earlier GEOSECS (1973), KNORR (1978), Hakuho Maru (1980) measurements (Povinec et al. 2003, used with permission)

via the North Channel, and flow around the coast of Scotland into the North Sea and then in the Norwegian Coastal Current (NCC). This NCC branches off northern Norway with one branch passing through eastwards into the Barents Sea (Vakulovsky 1987) and other current becomes the West Spitzbergen Current, passing through the Fram Strait into the Nansen Basin (Holm et al. 1983; Kautsky 1988; Smith et al. 1990; Kershaw and Baxter 1995). Using $^{134}Cs/^{137}Cs$ ratios at the source, one can determine the transit time of water masses from the discharge point to the Arctic Ocean. If R_0 and R_t are the $^{134}Cs/^{137}Cs$ activity ratios at the discharge point in Sellafield and at a point in the middle of Nansen Basin, then,

$$R_t = R_0 \exp(\lambda_{137} - \lambda_{134})t \qquad (19.8)$$

From this, the transit time (t) can be calculated as follows:

$$\tau = 1/(\lambda_{134} - \lambda_{137}) \times \ln(R_0/R_t) \qquad (19.9)$$

λ_{134} and λ_{137} are the decay constants of ^{134}Cs and ^{137}Cs, respectively. There is no other background ^{134}Cs, but there are background ^{137}Cs derived from the global fallout. At the time of sampling (1970), the background ^{137}Cs due to global fallout was in the range of 3–5 Bq m^{-3} at the corresponding latitudes, whereas levels at >120 Bq m^{-3} occurred in the northern Scottish waters. Equation (19.9) assumes the following: (1) the decrease in $^{134}Cs/^{137}Cs$ is only due to radioactive decay as the water mass moves; (2) there is no preferential removal of ^{134}Cs or ^{137}Cs; and (3) the change in $^{134}Cs/^{137}Cs$ activity ratio due to mixing with ^{134}Cs depleted water is negligible and the ^{137}Cs contribution from the global fallout to the sample collected is negligible. Using the $^{134}Cs/^{137}Cs$ ratios, the transit times were calculated to be 5–6, 7, and 7–9 years in North Cape, Svalbard, East Greenland, respectively (Kautsky 1988). This range of transit time is comparable to the advective time for transport around the perimeter of the Arctic Ocean from the Santa Anna Trough to the southern Canada Basin (~6,000 km) of 7.5 years obtained using H-3-He-3 and chlorofluorocarbon data (Mauldin et al. 2010). These values agree well with the values reported based on ^{137}Cs and ^{99}Tc (summarized in Kershaw and Baxter 1995). The transit time from La Hague to the Arctic is expected to be 3–4 years, about 2 years shorter than that reported for Sellafield (Kershaw and Baxter 1995).

19.3.11.2 Tracing Deep Water Formation Originated from the Regional Climate Change

The understanding of the deep water formation or intermediate water formation in the ocean is very important to understand the oceanic response due to climate change and vice versa. Sinking of surface water (formation of intermediate and deep water) is largely induced by the intense evaporation and cooling forced by regional climate, and it was generally studied using hydrographic observation such as temperature, salinity and dissolved oxygen. However, these hydrographic variables provide no direct information on the timing and duration of the sinking events. A time series measurement of ^{90}Sr and ^{137}Cs in deep waters could give direct information on the deep water formation. In the Mediterranean Sea, deep water formation occurring in the eastern part of the sea shifted from the southern Adriatic Sea to the Aegean Sea largely during 1989–1995 with exceptional severe winter in 1991–1993, known as Eastern Mediterranean Transient (EMT). The deep water formation in the Aegean Sea was relaxed partially by 1995. The deep penetration of the Aegean surface water was confirmed using time series depth distributions of ^{137}Cs in the water column collected over the period of 1975–1999 by Delfanti et al. (2003). They also utilized the ^{90}Sr-poor water formed during Chernobyl accident in 1986. In the Eastern Mediterranean Sea, ^{137}Cs concentration at 2,000 m depth was found to be ca. 1 and 3 Bq m^{-3} in 1977 and 1995, respectively. The activity ratio of $^{137}Cs/^{90}Sr$ in the surface water of the Mediterranean was 1.54 and 12 (in 1986) for the previous global fallout contained water and Chernobyl fallout contained water, respectively. This difference in the $^{137}Cs/^{90}Sr$ activity ratio is used to distinguish the water mass laden with ^{90}Sr-poor Chernobyl originated ^{137}Cs (high $^{137}Cs/^{90}Sr$ ratio) in 1986 from ^{90}Sr rich previous global fallout ^{137}Cs (low $^{137}Cs/^{90}Sr$ ratio) in the early 1960.

Miayo et al. (2000) found the timing of the deep water formation in a large marginal sea in the Northwest Pacific Ocean, East Sea (Sea of Japan) using historical ^{137}Cs profiles collected over the period of

1976–1996. They found that ^{137}Cs concentrations in the upper 200 m layer decreased with time, while those in the deeper layers below 1,500 m depth sharply increased during the period of 1985–1995, accompanied with the increase of the ^{137}Cs inventory. They further identified the area of deep water formation in this period being in the deep Japan Basin off the Peter the Great Bay, Vladivostok, Russia.

19.3.11.3 Tracing Ice Rafted Sediment in the Polar Seas

Sea ice in the polar region is one of the most important transport vectors of terrestrial material to the interior of the ocean. Terrestrial material is incorporated into the sea ice in the coastal areas when it forms and released at farther distances in the Arctic when it melts. In some cases, sea ice melts and the ice-rafted sediments (IRS) are retained in the melt ponds and during early fall, it refreezes. About 15% of the IRS are released during one freeze-melt cycle (details given in Kaste and Baskaran, Chap. 5 in this volume). From the measurements of ^{137}Cs in sea ice sediments, it was found that there is no enrichment of ^{137}Cs or Pu while one to two orders of magnitude enrichment was found for other atmospherically-delivered radionuclides such as ^{7}Be and ^{210}Pb (Meese et al. 1997; Cooper et al. 1998; Landa et al. 1998; Baskaran 2005; Masque et al. 2007). The ^{239}Pu/^{240}Pu atom ratios in the IRS were found to be ~0.18, suggesting that most of the Pu were derived from the global fallout (Masque et al. 2007).

19.3.12 Neptunium (^{237}Np)

Np is also an actinide element, but it differs from plutonium in that it is most likely present as the highly soluble Np(V) as NpO$_2^+$ (Keeney-Kennicutt and Morse 1984). While it can be present as Np(IV), it is proposed that Np (IV) in this form is rapidly oxidized to Np(V) under normal seawater conditions (McCubbin and Leonard 1997; Pentreath et al. 1986). K$_d$ value of 1×10^3 kg L^{-1} has been reported for Np in coastal sediments. In the open ocean, ^{237}Np is thought to behave conservatively. Livingston et al. (1988) used fallout and Sellafield derived ^{137}Cs and ^{90}Sr to study ventilation and circulation processes in the Mediterranean and the Arctic Seas and documented their input into the North Atlantic Deep Water. ^{237}Np may have similar applications, with added benefits of a significantly longer half-life (2.14 × 10^6 years) and measurement by ICP-MS which allows lower levels of detection as compared to traditional radio-counting methods for ^{137}Cs and ^{90}Sr (i.e., gamma and beta counting). Furthermore, with the isotopic composition of the main sources of contamination well characterized, water column ^{237}Np inventories and ^{237}Np/^{239}Pu and ^{237}Np/^{137}Cs inventory ratios may be useful for identifying non-fallout sources of contamination and providing additional insight into water column scavenging processes.

19.3.13 Plutonium (^{239}Pu and ^{240}Pu)

Plutonium is an actinide element. Although Pu can exist in four oxidation states in natural environment, it is thought to exist in the oceans predominantly in two oxidation states, the particle reactive Pu(IV) as Pu(OH)$_4$ and the relatively soluble Pu(V) as PuO$_2^+$ (Sholkovitz 1983; McMahon et al. 2000). In many publications, the Pu(III) and Pu(IV) are grouped together as the reduced forms having average K$_d$ values of 2.5 × 10^6 kg L^{-1} and Pu(V) and Pu(VI) are grouped together as the oxidized forms having average K$_d$ values of 1.5 × 10^4 kg L^{-1} (Nelson and Lovett 1978).

Much of the published plutonium data is based on alpha spectrometry, which cannot resolve ^{240}Pu (5.168 MeV (73.5%), 5.123 MeV (26.4%)) from ^{239}Pu (5.157 MeV (73.3%), 5.144 MeV (15.1%), 5.106 MeV (11.5%)) and is usually reported as the sum of the ^{239}Pu and ^{240}Pu activity (i.e., ^{239,240}Pu). The average ^{239,240}Pu/^{137}Cs activity ratio of the global fallout is well documented to be 0.026 ± 0.01 (decay corrected to 1 Jan 2010) (Koide et al. 1977; Koide et al. 1975; Krey et al. 1976). More sensitive mass spectrometric techniques (e.g., ICP-MS and TIMS) have the ability to resolve the different isotopes of plutonium (details given in Chap. 18). Kelley et al. (1999) have shown that the atom ratios ^{240}Pu/^{239}Pu, ^{237}Np/^{239}Pu, and ^{241}Pu/^{239}Pu in soils contaminated by global fallout exhibit relatively little variation globally. Their average ratios in global fallout for the Northern Hemisphere are 0.180 ± 0.014, 0.480 ± 0.070, and (2.44 ± 0.35) × 10^{-3}, respectively (Chap. 20). Several studies have documented lower plutonium isotopic ratios in debris originating from low-yield nuclear tests conducted at the Nevada Test Site to have an average ^{240}Pu/^{239}Pu ratio of about 0.035 (Buesseler and Sholkovitz 1987; Krey et al. 1976; Perkins and Thomas 1980).

In the year of 2000, total oceanic inventory of ^{239}Pu and ^{240}Pu are estimated to be 6.5 PBq (2,820 kg) and 5.4 PBq (632 kg), respectively (Hamilton 2004). $^{239+240}$Pu largely resides in the water column with subsurface peak at ca. 600 m depth (Fig. 19.5) and substantial fractions (30–112%) are buried in the sediment in the deep Pacific Ocean Basin (Lee et al. 2005). The $^{239+240}$Pu profiles in the central NW Pacific Ocean indicate that the subsurface maximum of Pu concentration over the past 24 years decreased by a factor of ca. 4 and the subsurface peaks moved to deeper water from ca. 450 to ca. 800 m and they were getting less sharp and more wide (Fig. 19.6). However, the stations located in the west-ward flowing North Equatorial Current (NEC) with very low biological activity showed no temporal changes in the vertical profiles of $^{239+240}$Pu over the last 24 years (Povinec et al. 2003). The persistence of subsurface Pu peak appears to be correlated with the water density gradient (Wong et al. 1992). As the environmental half–life of $^{239+240}$Pu in the central NW Pacific Ocean is ca. 7 years (Povinec et al. 2005), $^{239+240}$Pu would decrease to less than 1 mBq m^{-3} by 2035, which would be difficult to measure with a conventional 60 L water volume and alpha spectrometry (Baskaran et al. 2009). Similar observation was made for the Santa Monica and San Padro Basins off California by Wong et al. (1992).

19.3.13.1 Tracing Provenances and Fates of Nuclear Fallout in the Ocean

Buesseler and Sholkovitz (1987) used Pu to document the input of both global stratospheric fallout and tropospheric fallout (derived from tests conducted at the Nevada Test Site) to North Atlantic sediments. They found the ^{240}Pu/^{239}Pu atom ratios ranged from ~ 0.18 on the shelf to ~ 0.10 at 5,000 m in the solid phase and

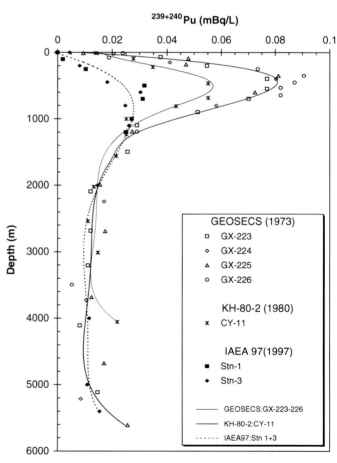

Fig. 19.5 ^{239,240}Pu profiles in the central NW Pacific Ocean in 1997 (IAEA 1997) along with earlier GEOSECS (1973), KNORR (1978), Hakuho Maru (1980) measurements (Povinec et al. 2003, used with permission)

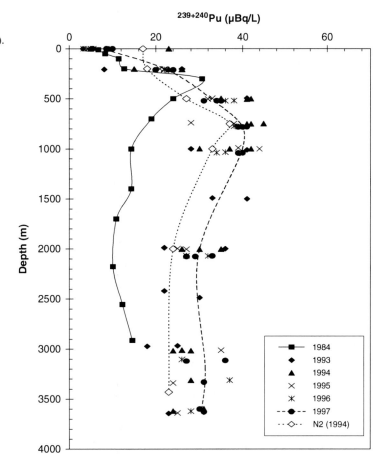

Fig. 19.6 Depth distribution of dissolved $^{239+240}$Pu concentrations over 1984–1997 in the Sea of Japan (East Sea) (from IAEA 2005). The time-series measurements were not done in a single fixed station

~ 0.18 in the pore water phases of bottom sediments collected from a transect between Woods Hole and Bermuda in the 1950 and early 1960. Based on the Pu isotopic composition, they were able to infer that there have been at least two distinct sources of fallout Pu in this region: one is global fallout (^{240}Pu/^{239}Pu ~ 0.18) and another tropospheric fallout from the Nevada Tests (^{240}Pu/^{239}Pu ~ 0.035), in which Pu derived from the Nevada Tests fallouts have been much more efficiently removed from the water column to deep-sea sediments and they were more refractory relative to Pu from global fallout.

19.3.13.2 Tracing Subtropical Water Movement in the Pacific Ocean

Although the entire surface of the Earth has been contaminated with radioactive fallout due to the atmospheric nuclear weapons tests conducted from 1945 to the early 1980, radioactive contamination in the Pacific Ocean is different from the rest of the world ocean. The earlier nuclear tests were dominated by the U.S. at the Pacific Proving Ground (close-in fallout, atom ratio of ^{240}Pu/^{239}Pu >0.3) in the 1950, whereas the later tests were largely by the former Soviet Union (stratospheric fallout, atom ratio of ^{240}Pu/^{239}Pu ~ 0.18) in 1961 and 1962. Stratospheric fallout from the atmospheric nuclear tests virtually ceased in the early 1980. Most of the terrestrial Pu contaminations in the world are largely originating from the former USSR nuclear tests. However, the Pu signal in seawater collected in early 2000 from the Pacific Ocean revealed that it was decoupled from that of the adjacent terrestrial environment and the data from the Pacific water has shown that US signal was found even off the Aleutian Chain, several thousand kilometers away from the Proving Ground.

The seawater in the Pacific Ocean had received close-in fallout in the 1950 and the stratospheric fallout in the 1960. The earlier work (e.g., Noshkin et al. 1975; Buesseler 1997) showed that the ^{239}Pu/^{240}Pu atom ratios in the Pacific surface seawater had changed from the U.S. signal of >0.3 in the 1950 to a former Soviet signal in the 1960 and returned to the U.S. signal in the early 2000. No observations have been made between the late 1970 and the early 2000, which could be filled using samples from coral skeletons or archived seawater.

The Pu isotopic signature in the region is a potent and unique tracer to distinguish among the different sources of Pu, as well as to understand the variations of ^{239}Pu/^{240}Pu ratios through time (Ketterer et al., Chap. 20). The US Proving Grounds in the Pacific continue to export Pu with high ^{240}Pu/^{239}Pu ratio, mainly originated from the nuclear weapons tests in 1950, through the leaching of contaminated carbonate species deposited in the ocean. The massive (dome-shaped) coral *Porites* found throughout the northern tropical Pacific Ocean will allow us to reconstruct the Pu signal in surface waters spanning the last 50 years. By determining atom ratios of ^{240}Pu/^{239}Pu recorded in coral cores and dating the corals using the ^{210}Pb/^{226}Ra disequilibria (details on dating of corals is given in Chap. 37), we may be able to accomplish a better understanding of the mixing and circulation of waters between the Warm Pool and adjacent water masses in the Northwest Pacific Ocean. However, this application has not been fully realized yet (Hong et al. 2004).

19.3.13.3 Tracing Material Exchange Between Ocean and its Margins

The dissolved ^{230}Th activity concentration increases with depth in agreement with the balance between the in situ production from the dissolved ^{234}U and the scavenging of ^{230}Th on sinking particles generally in the ocean. However, a significant fraction of dissolved ^{231}Pa (daughter of ^{235}U) and some dissolved ^{230}Th were found to be exported from area of low particle flux to area of high particle flux where ^{230}Th and ^{231}Pa are efficiently scavenged by settling particulate matter to the sea floor. This process became widely known in the early 1980 as "boundary scavenging" because ocean boundaries/margins are generally areas with high biological productivity, high particle flux and high sediment accumulation. The concept of boundary scavenging was first applied to horizontal transport of ^{210}Pb (Spencer et al. 1981; Baskaran and Santchi 2002; Roy-Barman 2009). And enhanced deposition of fallout-derived Pu isotopes in ocean-margin sediments has also been reported (e.g. Koide et al. 1975).

In some ocean margins where large rivers are absent, e.g., East Sea (Sea of Japan), the oceanic $^{239+240}$Pu is effectively scavenged through this boundary scavenging process and the deep waters become enriched with $^{239+240}$Pu over time (Fig. 19.6) and the marginal sea is destined to continue to serve as a filter for all particle-reactive chemical materials introduced from incoming oceanic waters or atmosphere. This aspect should be investigated more fully in the future.

19.3.13.4 Identifying Subsurface Water Upwelling Areas in the Sea

Although both ^{137}Cs and $^{239+240}$Pu radionuclides were injected into the ocean surface by similar processes, the oceanic behavior of plutonium and ^{137}Cs differ from each other. ^{137}Cs, most of which exists in a dissolved form, moves as water mass in the ocean, whereas oceanic behavior of $^{239+240}$Pu is controlled both by water movement and adsorption onto the particles formed in situ or introduced and subsequently sinking to the deeper waters. Therefore, in the upper 1,000 m depths, ^{137}Cs concentration decreases exponentially with depth with surface maximum or subsurface maxima where surface water subducts. $^{239+240}$Pu has increased with depth toward the subsurface maximum which lies about 500–800 m depth. Hirose et al. (2009) found that activity ratio of $^{239+240}$Pu/^{137}Cs ($R_{Pu/Cs}$) increases with depth in the upper 100–1,000 m depth as follows:

$$R_{Pu/Cs} = R_{Pu/Cs,0} \exp(\lambda Z) \qquad (19.10)$$

Where λ is the constant related to ocean biogeochemical process of water movement (downwelling and upwelling) and particle formation and destruction and Z is the water depth. They further noted that a half-generation depth of $^{239+240}$Pu, $Z_h = \ln2/\lambda$, corresponds to the half-decrease depth of the particulate organic matter in the water column. As the ocean primary productivity depends on the upwelling intensity of the nutrient-rich deep water, they found that Z_h is deep (~ 300 m) in the eutrophic subarctic Pacific and shallow (~ 100 m) in the subtropical North Pacific.

19.3.13.5 Constraining Sediment Accumulation and Mixing Rates

Particle reactive anthropogenic radionuclides (Table 19.1) has been utilized to determine sediment accumulation and mixing rates in the seafloor based on the input history and their half-lives as they often are adsorbed to the sediment irreversibly in the oceanic environment. Bomb-derived fallout radionuclide (^{137}Cs, $^{239+240}$Pu) peaks its concentration in 1963 in the Atlantic Ocean. However $^{239+240}$Pu peaked in 1954 in the Equatorial-North Pacific Ocean (Hamilton 2004) and the Antarctic Region (Koide et al. 1985) due to the close-in fallout in the Equatorial Pacific. Sediment accumulation rates may be calculated by assuming a fallout maximum in 1963 in the rapid sediment accumulating coastal ocean (e.g., Santschi et al. 2001) or sediment mixing rates may be calculated in the deep ocean basin where sediment accumulation rate is slower than mixing rate (e.g., Pacific Ocean, Cochran 1985).

19.3.14 Americium (^{241}Am)

Measurements of ^{241}Pu ($t_{1/2}$ = 14.3 years) and its daughter product, ^{241}Am ($t_{1/2}$ = 433 years) in the sediment sample can be used to establish the timing of the source term. Assuming that an initial activity of ^{241}Am in fallout debris is zero, the time elapsed since the Pu was placed, t, can be given by the radioactive decay equation governing the time-dependent activities A_{Pu-241} and A_{Am-241}, respectively (19.11).

$$t = 1/(\lambda_{Pu-241} - \lambda_{Am-241})$$
$$\times \ln[1 + (\lambda_{Pu-241} - \lambda_{Am-241})A_{Am-241}/ \quad (19.11)$$
$$(\lambda_{Am241} - \lambda_{Pu-241})]$$

where λ is the radioactive decay constant (0.693/$t_{1/2}$) for a given isotope. In this way, Smith et al. (1995) reaffirmed that the underwater nuclear tests were conducted in Chernaya Bay to be in 1955 and 1957 (t obtained using (19.11) 36.6 ± 1.2 years can be compared to the time elapsed between 1955 and 1994 (or 1957 and 1994)).

19.4 Future Directions

A major portion of the remaining water soluble radionuclides (such as ^{90}Sr and ^{137}Cs) still exist in the water column with maximum in the surface or subsurface layer and hence these radionuclides can be utilized to study the geochemical pathways of their stable counterparts (e.g., pathways of Sr using ^{90}Sr as a tracer). About 60–70% of the total inventories of these nuclides derived from weapons testing have undergone radioactive decay, but the remaining portions continue to penetrate deeper water column in the ocean. Formation of barite minerals and the chemical reactions linked with such formation can be studied by using ^{90}Sr as a tracer.

Particle reactive anthropogenic radionuclides, e.g., Pu isotopes, continue to be transferred from the water column to the bottom sediments as enhanced in the ocean margins. Therefore, sediment inventories of these radionuclides would provide valuable temporal constraints on the boundary scavenging process occurring between ocean proper and its margins.

Slow movement of waters in the ocean, e.g., downwelling of surface water (deep water formation, drawdown of atmospheric CO_2), upwelling of subsurface water (supplying nutrient rich water to the surface, evasion of CO_2 to the atmosphere), spillover of deep water between basins are generally difficult to address using hydrographic variables such as temperature, salinity and stable isotopes of hydrogen and oxygen, where they have no time information. The current understanding of the spatial and temporal distribution in the sea and marine biogeochemistry of the well-studied anthropogenic radionuclides could serve to detect the slow movement of waters in the ocean. As the changes in slow movement of waters in the ocean are induced by the regional or global climate change, selected anthropogenic radionuclides, reviewed here, warrant a further investigation in a globally coordinated manner. The recently initiated GEOTRACES (an international study of the marine biogeochemical cycles of trace elements and their isotopes) could serve a useful forum for this purpose (http://www.ldeo.columbia.edu/res/pi/geotraces/).

Acknowledgements The authors are grateful to Dr. Katusmi Hirose and an anonymous reviewer for providing valuable comments on the manuscript. This work was partially supported by Korea Ocean Research and Development PM55861 (GHH), was performed under the auspices of the U.S. Department of Energy by Lawrence Livermore National Laboratory in part under Contract W-7405-Eng-48 and in part under Contract DE-AC52-07NA27344 (TFH), Wayne State's Board of Governor's Distinguished Faculty Fellowship (MB). Copyright permission to reproduce here was generously granted for Table 19.2 and Figs. 19.1–19.5 from Elsevier Limited, UK and Table 19.3 from Terra Scientific Publishing Company, Japan.

References

Aarkrog A (1988) Worldwide data on fluxes of ^{239,240}Pu and ^{238}Pu to the ocean. In Inventories of selected radionuclides in the oceans. IAEA-TECDOC-481, 103–137

Aarkrog A (2003) Input of anthropogenic radionuclides into the world ocean. Deep-Sea Res II 50:2597–2602

Aoyama M, Hirose K, Sugimura Y (1991) The temporal variation of stratospheric fallout derived from the Chernobyl accident. J Environ Radioact 13:103–115

Assinder DJ (1999) A review of the occurrence and behavior of neptunium in the Irish Sea. J Environ Radioact 44:353–347

Baskaran M (2005) Interaction of sea ice sediment and surface sea water in the Arctic Ocean: evidence from excess ^{210}Pb. Geophys Res Lett: 32:L12601. doi:10.1029/2004GL022191

Baskaran M, Asbill S, Santschi PH, Davis T, Brooks JM, Champ MA, Makeyev V, Khlebovich V (1995) Distribution of ^{239,240}Pu and ^{238}Pu concentrations in sediments from the Ob and Yenisey rivers and the Kara Sea. Appl Radiat Isot 46:1109–1119

Baskaran M, Asbill S, Schwantes J, Santschi PH, Champ MA, Brooks JM, Adkins D, Makeyev V (2000) Concentrations of ^{137}Cs, ^{239,240}Pu, and ^{210}Pb in sediment samples from the Pechora Sea and biological samples from the Ob, Yenisey Rivers and Kara Sea. Mar Pollut Bull 40:830–838

Baskaran M, Hong GH, Santschi PH (2009) Radionuclide analyses in seawater. In: Wurl O (ed) Practical guidelines for the analysis of seawater. CRC, Boca Raton, pp 259–304

Baskaran M, Santschi PH (2002) Particulate and dissolved ^{210}Pb activities in the shelf and slope regions of the Gulf of Mexico waters. Cont Shelf Res 22:1493–1510

Baudin JP, Adam C, Garnier-Laplace J (2000) Dietary uptake, retention and tissue distribution of ^{54}Mn, ^{60}Co and ^{137}Cs in the rainbow trout (*Onchoryhynchus Mikiss* Walabaum). Water Resour 34:2869–2878

Beasley TM, Cooper LW, Grebmeier JM, Aagard K, Kelley JM, Kilius LR (1988) ^{237}Np/^{129}I atom ratios in the Arctic Ocean: has ^{237}Np from western European and Russian fuel reprocessing facilities entered the Arctic Ocean? J Environ Radioact 39:255–277

Beaupre SR, Druffel ERM (2009) Constraining the propagation of bomb-radiocarbon through the dissolved organic carbon (DOC) pool in the northeast Pacific Ocean. Deep-Sea Res I 56:1717–1726

Bernstein RE, Byrne RH (2004) Acantharians and marine barite. Mar Chem 86:45–50

Betram C (2010) Ocean iron fertilization in the context of the Kyoto protocol and the post-Kyoto process. Energy Policy 38:1130–1139

Bienvenu P, Cassette P, Andreoletti G, Be M-M, Comte J, Lepy M-C (2007) A new determination of ^{79}Se half-life. Appl Radiat Isot 65:355–364

Bowen VT, Noshkin VE, Livingston HD, Volchock HL (1980) Fallout radionuclides in the Pacific Ocean: vertical and horizontal distributions, largely from GEOSECS stations. Earth and Planet Sc Lett 49:411–434

Brewer PG, Spencer DW, Robertson DE (1972) Trace element profiles from the GEOSECS-II test station in the Sargasso Sea. Earth Planet Sci Lett 16:111–116

Broecker WS (1974) Chemical oceanography. Harcourt Brace Jovanovich, New York, 214p

Broecker WS (2003) Radiocarbon. In: Holland HD, Turekian KK (eds) Treatise in geochemistry, the atmosphere, vol. 4. Elsevier-Pergamon, Oxford, pp 245–260

Broecker WS, Peng TH (1982) Tracers in the sea. Eldigio, New York, p 690

Broecker WS, Southerland S, Smethie W, Peng TH, Ostlund G (1995) Oceanic radiocarbon: separation of the natural and bomb components. Global Biogeochem Cycles 9:263–288

Browne E, Firestone RB (1986) Table of radioactive isotopes. Wiley-Interscience, New York

Bruland KW (1983) Trace elements in seawater. In: Riley JP, Chester R (eds) Chemical oceanography, vol 8. Academic, London, pp 157–221

Buesseler KO (1997) The isotopic signature of allout plutonium in the North Pacific. J Environ Radioact 36:69–83

Buesseler KO, Livingston HD (1996) Natural and man-made radionuclides in the Black Sea. In: Guéguéniat P, Germain P, Métivier H (eds) Radionuclides in the oceans: Inputs and Inventories, Les Editions de physique, Europe Media Duplication S.A. Les Ulis, pp 199–217

Buesseler KO, Livingston HD, Honjo S, Hay BJ, Konuk T, Kempe S (1990) Scavenging and particle deposition in the southwestern Black Sea – evidence from Chernobyl radiotracers. Deep-Sea Res 37:413–430

Buesseler KO, Sholkovitz ER (1987) The geochemistry of fallout plutonium in the North Atlantic: II. ^{240}Pu/^{239}Pu ratios and their significance. Geochim Cosmochim Acta 51:2623–2637

Byrne RH (2002) Inorganic speciation of dissolved elements in seawater: the influence of pH on concentration ratios. Geochem Trans 3:11–16

Carlson L, Holm E (1992) Radioactivity in *Fucus vesiculosus L.* from the Baltic Sea following the Chernobyl accident. J Environ Radioact 15:231–248

Chiappini R, Pointurier R, Milies-Lacroix JC, Lepetit G, Hemet P (1999) ^{240}Pu/^{239}Pu isotopic ratios and $^{239+240}$Pu total measurements in surface and deep waters around Muroroa and Fangataufa atolls compared with Rangiroa atoll (French Polynesia). Sci Total Environ 237(238):269–276

Cochran JK (1985) Particle mixing rates in sediments of the eastern equatorial Pacific: evidence from ^{210}Pb, ^{239,240}Pu and ^{137}Cs distributions at MANOP sites. Geochim Cosmochim Acta 49:1195–1210

Cooper LW, Larsen IL, Beasley TM, Dolvin SS, Grebmeier JM, Kelley JM, Scott M, Johnson-Pyrtle A (1998) The distribution of radiocesium and plutonium in sea ice-entrained Arctic sediments in relation to potential sources and sinks. J Environ Radioact 39:279–303

Cooper LW, Beasley T, Aagaard K, Kelley JM, Larsen IL, Grebmeier JM (1999) Distributions of nuclear fuel-reprocessing tracers in the Arctic Ocean: indications of Russian river influence. J Mar Res 57:715–738

Cundy AB, Croudace IW, Warwick PE, Oh JS, Haslett SK (2002) Accumulation of COGEMA-La Hague-derived reprocessing wastes in French salt marsh sediments. Environ Sci Technol 36:4990–4997

Cutshall NH, Larsen IL, Olsen CR, Nittrouer CA, DeMaster DJ (1986) Columbia River sediment in Quinault Canyon, Washington – Evidence from artificial radionuclides. Mar Geol 71:125–136

Delfanti R, Klein B, Papucci C (2003) Distribution of ^{137}Cs and other radioactive tracers in the eastern Mediterranean: relationship to the deepwater transient. J Geophys Res 108 (C9):8108. doi:10.1029/2002JC001371

Donoghue JF, Bricker OP, Oslen CR (1989) Particle-borne radionuclides as tracers for sediment in the Susquehanna River and Chesapeake Bay. Estuarine Coastal Shelf Sci 29:341–360

du Bois PB, Dumas F (2005) Fast hydrodynamic model for medium-and long-term dispersion in seawater in the English Channel and southern North Sea, qualitative and quantitative validation by radionuclide tracers. Ocean Modell 9:169–210

du Bois PB, Salomon JC, Gandon R, Guegueniat P (1995) A quantitative estimate of English Channel water fluxes into the North Sea from 1987 to 1992 based on radiotracer distribution. J Mar Syst 6:457–481

Emsley J (1989) The elements. Clarendon, Oxford, p 256

Franić Z (2005) Estimation of the Adriatic Sea water turnover time using fallout ^{90}Sr as a radioactive tracer. J Mar Syst 57:1–12

Gaetani GA, Cohen AL (2006) Element partitioning during precipitation of aragonite from seawater: a framework for understanding paleoproxies. Geochim Cosmochim Acta 70:4617–4634

Gao Y, Drange H, Johannessen OM, Pettersson LH (2009) Sources and pathways of ^{90}Sr in the North-Arctic region: present day and global warming. J Environ Radioact 100:375–395

Gray J, Jones SR, Smith AD (1995) Discharges to the environment from the Sellafield site, 1951–1992. J Radiol Prot 15:99–131

Hamilton TF, Milliès-Lacroix J-C, Hong GH (1996) ^{137}Cs(^{90}Sr) and Pu isotopes in the Pacific Ocean: sources and trends. In: Guéguéniat P, Germain P, Métivier H (eds) Radionuclides in the oceans. Inputs and Inventories. Les Editions de Physique, Les Ulis, pp 29–58

Hamilton TF (2004) Linking legacies of the cold war to arrival of anthropogenic radionuclides in the oceans through the 20th century. In: Livingston HD (ed) Marine radioactivity. Elsevier, Amsterdam, pp 23–78

Hirose K, Aoyama M, Povinec PP (2009) ^{239,240}Pu/^{137}Cs ratios in the water column of the North Pacific: a proxy of biogeochemical process. J Environ Radioact 100:258–262

Hirose K, Igarashi Y, Aoyama M, Miyao T (2001) Long-term trends of plutonium fallout observed in Japan. In: Kudo A (ed) Plutonium in the environment. Elsevier Science, Amsterdam, pp 251–266

Holm E, Persson BRR, Hallstadius L, Aarkrog A, Dahlgaard H (1983) Radiocesium and transuranium elements in the Greenland and Barents Seas. Oceanolog Acta 6:457–462

Hong GH, Baskaran M, Povinec PP (2004) Artificial radionuclides in the western North Pacific: a review. In: Shiyomi M, Kawahata H, Koizumi H, Tsuda A, Awaya Y (eds) Global environmental change in the ocean and on land. Terrapub, Tokyo, pp 147–172

Hou X, Roos P (2008) Critical comparison of radiometric and mass spectrometric methods for the determination of radionuclides in environmental, biological and nuclear waste samples. Anal Chim Acta 608:105–139

Hua Q (2009) Radiocarbon: a chronological tool for the recent past. Quat Geochronol 4:378–390

Hu QH, Weng JQ, Wang JS (2010) Sources of anthropogenic radionuclides in the environment: a review. J Environ Radioact 101:426–437

IAEA (2004) Sediment distribution coefficients and concentration factors for biota in the marine environment. Technical reports series No. 422, International Atomic Energy Agency, Vienna, p 93

IAEA (2005) Worldwide marine radioactivity studies (WOMARS). Radionuclide levels in oceans and seas. IAEA-TECDOC-1429. International Atomic Energy Agency, Vienna, p 187

Irlweck K, Hrnecek E (1999) ^{241}Am concentration and ^{241}Pu/$^{239(240)}$Pu ratios in soils contaminated by weapons-grade plutonium. Akamemiai Kiado, Budapest, pp 595–599

Jenkins WJ (2001) Tritium-helium dating. Encyclopedia of Ocean Sciences 6:3048–3056

Kamenos NA, Cusack M, Moore PG (2008) Coralline algae are global paleothermometers with bi-weekly resolution. Geochim Cosmochim Acta 72:771–779

Kafalas P, Irvine J Jr (1956) Nuclear excitation functions and thick target yields: (Cr^{+d}). Phys Rev 104:703–705

Kautsky H (1988) Determination of distribution processes, transport routes and transport times in the North Sea and the northern Atlantic using artificial radionuclides as tracers. In: Guary JC, Guegueniat P, Pentreath RJ (eds) Radionuclides: a tool for oceanography. Elsevier Applied Science, London, pp 271–280

Kelley JM, Bond LA, Beasley TM (1999) Global distribution of Pu isotopes and ^{237}Np. The Sci Total Environ 237(238):483–500

Keeney-Kennicutt WL, Morse JW (1984) The interaction of Np (V)O$_2^+$ with common mineral surfaces in dilute aqueous solutions and seawater. Mar Chem 15:133–150

Kemp RS (2008) A performance estimate for the detection of undeclared nuclear-fuel reprocessing by atmospheric ^{85}Kr. J Environ Radioact 99:1341–1348

Kenna TC, Sayles FL (2002) The distribution and history of nuclear weapons related contamination in sediments from the Ob River, Siberia as determined by isotopic ratios of plutonium and neptunium. J Environ Radioact 60:105–137

Kershaw P, Baxter A (1995) The transfer of reprocessing wastes from north-west Europe to the Arctic. Deep-Sea Res II 42:1413–1448

Ketterer ME, Szechenyi SC (2008) Determination of plutonium and other transuranic elements by inductively coupled plasma mass spectrometry: a historical perspective and new frontiers in the environmental sciences. Spectrochim Acta Part B 63:719–737

Kim CK, Kim CS, Chang BU, Choi SW, Chung CS, Hong GH, Hirose K, Igarashi Y (2004) Plutonium isotopes in seas around the Korean Peninsula. Sci Total Environ 318:197–209

Kirchner G, Noack CC (1988) Core history and nuclide inventory of the Chernobyl core at the time of accident. Nucl Saf 29:1–5

Koide M, Berine KK, Chow TJ, Goldberg ED (1985) The $^{240}Pu/^{239}Pu$ ratio, a potential geochronometer. Earth Planet Sci Lett 72:1–8

Koide M, Goldberg ED, Herron MM, Langway CC Jr (1977) Transuranic depositional history in South Greenland firn layers. Nat 269:137–139

Koide M, Griffin JJ, Goldberg ED (1975) Records of plutonium fallout in marine and terrestrial samples. J Geophys Res 80(30):4153–4162

Krey PW, Hardy EP, Pachucki C, Rourke F, Coluzza J, Benson WK (1976) Mass isotopic composition of global fall-out plutonium in soil: Transuranium nuclides in the environment, San Francisco, CA, USA, pp. 671–678

Kutschera W (2010) AMS and climate change. Nucl Instrum Methods Phys Res B 268:693–700

Kuwabara J, Yamamoto M, Assinder DJ, Komura K, Ueno K (1996) Sediment profiles of ^{237}Np in the Irish Sea: estimation of the total amount of ^{237}Np from Sellafield. Radiochim Acta 73:73–81

Lal D, Somayajulu BLK (1977) Particulate transport of radionuclides ^{14}C and ^{55}Fe to deep waters in the Pacific Ocean. Limnol Oceanogr 22:55–59

Landa E, Reimnitz E, Beals D, Pockowski J, Rigor I (1998) Transport of ^{137}Cs and $^{239,240}Pu$ by ice rafted debris in the Arctic Ocean. Arct 51:27–39

Lee SH, Gastaud J, La Rosa JJ, Liong Wee Kwong L, Povinec PP, Wyse E, Fitfield LK, Hausladen PA, Di Tada LM, Santos GM (2001) Analysis of plutonium isotopes in marine samples by radiometric, ICP-MS and AMS techniques. J Radioanal Nucl Chem 248:757–764

Lee SH, Povinec PP, Wyse E, Pham MK, Hong GH, Chung CS, Kim SH, Lee HJ (2005) Distribution and inventories of ^{90}Sr, ^{137}Cs, ^{241}Am and Pu isotopes in sediments of the Northwest Pacific Ocean. Mar Geol 216:249–263

Lindahl P, Roos P, Holm E, Dahlgaard H (2005) Studies of Np and Pu in the marine environment of Swedish – Danish waters and the North Atlantic Ocean. J Environ Radioact 82:285–301

Linsalata P, Simpson HJ, Olsen CR, Cohen N, Trier RM (1985) Plutonium and radiocesium in the water column of the Hudson River estuary. Environ Geol Water Sci 7:193–204

Linsley G, Sjoblom K-L, Cabianca T (2004) Overview of point sources of anthropogenic radionuclides in the oceans. In: Livingston HD (ed) Marine radioactivity, vol 6, Radioactivity in the environment. Elsevier, Amsterdam, pp 109–138

Livingston HD, Buesseler KO, Izdar E, Konuk T (1988) Characteristics of Chernobyl fallout in the southern Black Sea. In: Guary J, Guegueniat P, Pentreath RJ (eds) Radionuclides: a tool for oceanography. Elsevier, Essex, UK, pp 204–216

Livingston H, Kupferman SL, Bowen VT, Moore RM (1984) Vertical profile of artificial radionuclide concentrations in the Central Arctic Ocean. Geochim Cosmochim Acta 48:2195–2203

Livingston HD, Mann DR, Casso SA, Schneider DL, Surprenant LD, Bowen VT (1987) Particle and solution phase depth distributions of transuranics and ^{55}Fe in the North Pacific. J Environ Radioact 5:1–24

Livingston HD, Povinec PP (2000) Anthropogenic marine radioactivity. Ocean Coastal Manage 43:689–712

Livingston HD, Povinec PP (2002) A millennium perspective on the contribution of global fallout radionuclides to the ocean science. Health Phys 82:656–668

Marshall WA, Gehrels WR, Garnett MH, Freeman SPHT, Maden C, Xu S (2007) The use of 'bomb spike' calibration and high-precision AMS ^{14}C analyses to date salt-marsh sediments deposited during the past three centuries. Quat Res 68:325–337

Masque P, Cochran JK, Hirschberg DJ, Dethleff D, Hebbeln D, Winkler A, Pfirman S (2007) Radionuclides in Arctic sea ice: tracers of sources, fates and ice transit time scales. Deep-Sea Res 154:1289–1310

Mauldin A, Schlosser P, Newton R, Smithie WM Jr, Bayer R, Rhein M, Peter JE (2010) The velociy and mixing time scale of the Arctic Ocean Boundary Current estimated with transient tracers. J Geophys Res 115:C08002. doi:10.1029/2009JC005965

McCubbin D, Leonard KS (1997) Laboratory studies to investigate short-term oxidation and sorption behabiour of neptunium in artificial and natural seawater solutions. Mar Chem 56:107–121

McMahon CA, Leon Vintro L, Mitchell PI, Dahlgaard H (2000) Oxidation-state distribution of plutonium in surface and subsurface waters at Thule, northwest Greenland. Appl Radiat Isot 52:697–703

Meese DA, Reimnitz E, Tucker WB III, Gow AJ, Bischoff J, Darby D (1997) Evidence for radionuclide transport by sea ice. Sci Total Environ 202:267–278

Megens L, van der Plight J, de Leeuw JW (2001) Temporal variations in ^{13}C and ^{14}C concentrations in particulate organic matter from the southern North Sea. Geochim Cosmochim Acta 65:2899–2911

Miayo T, Hirose K, Aoyama M, Igarashi Y (2000) Trace of the recent deep water formation in the Japan Sea deduced from historical ^{137}Cs data. Geophys Res Lett 27:3731–3734

Muramatsu Y, Rühm W, Yoshida S, Tagami K, Uchida S, Wirth E (2000) Concentrations of ^{239}Pu and ^{240}Pu and their isotopic ratios determined by ICP-MS in soils collected from the Chernobyl 30-km zone. Environ Sci Technol 34:2913–2917

Muramatsu Y, Hamilton T, Uchida S, Tagami K, Yoshida S, Rosbinson W (2001) Measurement of $^{240}Pu/^{239}Pu$ isotopic ratios in soils from the Marshall Islands using ICP-MS. Sci Total Environ 278:151–159

Nelson DM, Lovett MB (1978) Oxidation state of plutonium in the Irish Sea. Nat 276:599–601

Noshkin VE, Wong KM, Eagle RJ, Gatrousis C (1975) Transuranics and other radionuclides in Bikini lagoon. Concentration data retrieved from aged coral sections. Limnol Oceanogr 20:729–742

Noureddine A, Benkrid M, Maoui R, Menacer M, Boudjenoun R, Kadi-Hanifi M, Lee SH, Povinec PP (2008) Radionuclide tracing of water masses and processes in the water column and sediment in the Algerian Basin. J Environ Radioact 99(8):1224–1232

Olsen CT, Larsen IL, Cutshall NH, Donoghue JF, Bricker OP, Simpson HJ (1981) Reactor released radionuclides in Susquehanna River sediments. Nat 294:242–245

Orlandini KA, Bowling JW, Pinder JE III, Penrose WR (2003) ^{90}Y-^{90}Sr disequilibrium in surface waters: investigating short-term particle dynamics by using a novel isotope pair. Earth Planet Sci Lett 207:141–150

Paatero J, Saxen R, Buyukay M, Outola L (2010) Overview of strontium-89,90 deposition measurements in Finland 1963–2005. J Environ Radioact 101:309–316

Papucci C, Charmasson S, Delfanti R, Gasco C, Mitchell P, Sanchez-Cabeza JA (1996) Time evolution and levels of man-made radioactivity in the Mediterranean Sea. In: Guéguéniat P, Germain P, Métivier H (eds) Radionuclides in the oceans. Inputs and inventories. Les Edition de Physique. Institut de Protection et de Surete Nuclaire, Les Ulis cedex A, pp 177–197

Paytan A, Averyt K, Faul K, Gray E, Thomas E (2007) Barite accumulation, ocean productivity, and Sr/Ba in barite across the Paleocene-Eocene thermal maximum. Geol 35:1139–1142

Peng T-H, Key RM, Ostlund HG (1998) Temporal variations of bomb radiocarbon inventory in the Pacific Ocean. Mar Chem 60:3–13

Pentreath RJ, Harvey BR, Lovett M (1986) Speciation of fission and activation products in the environment. In: Bulman RA, Cooper JR (eds) Chemical speciation of transuranium nuclides discharged into the marine environment. Elsevier, London, pp 312–325

Periáñez R (2005) Modeling the dispersion of radionuclides by a river plume: application to the Rhone River. Cont Shelf Res 25:1583–1603

Perkins RW, Thomas CW (1980) Worldwide fallout. In: Hanson WC (ed) Transuranic elements in the environment US DOE/TIC-22800. Office of Health and Environmental Research, Washington DC, pp 53–82

Piner KR, Wallace JR, Hamel OS, Mikus R (2006) Evaluation of ageing accuracy of bocaccio (*Sebastes paucispinis*) rockfish using bomb radiocarbon. Fish Res 77:200–206

Povinec PP, Aarkrog A, Buesseler KO, Delfanti R, Hirose K, Hong GH, Ito T, Livingston HD, Nies H, Noshkin VE, Shima S, Togawa O (2005) ^{90}Sr, ^{137}Cs and $^{239,240}Pu$ concentration surface water time series in the Pacific and Indian Oceans – WOMARS results. J Environ Radioact 81:63–87

Povinec PP, Lee SH, Liong Wee Kwong L, Oregioni B, Jull AJT, Kieser WE, Morgenstern U, Top Z (2010) Top Z (2010) Tritium, radiocarbon, ^{90}Sr and ^{129}I in the Pacific and Indian Oceans. Nucl Instrum Methods Phys Res B 268:1214–1218

Povinec PP, Livingston HD, Shima S, Aoyama M, Gastaud J, Goroncy I, Hirose K, Hyunh-Ngoc L, Ikeuchi Y, Ito T, La Rosa J, Kwong LLW, Lee SH, Moriya H, Mulsow S, Oregioni B, Pettersson H, Togawa O (2003) IAEA'97 expedition to the NW Pacific Ocean – results of oceanographic and radionuclide investigations of the water column. Deep-Sea Res II 50:2607–2637

Raisbeck GM, Yiou F (1999) ^{129}I in the oceans: origins and applications. Sci Total Environ 237(238):31–41

Roether W, Beitzel V, Sultenfu J, Putzka A (1999) The eastern Mediterranean tritium distribution in 1987. J Mar Syst 20:49–61

Roy-Barman M (2009) Modelling the effect of boundary scavenging of thorium and protactinium profiles in the ocean. Biogeosci 6:3091–3107

Santschi PH, Presley BJ, Wade TL, Garcia-Romero B, Baskaran M (2001) Historical contamination of PAHs, PCBs, DDTs, and heavy metals in Mississippi River Delta, Galveston Bay and Tampa Bay sediment cores. Mar Environ Res 52:51–79

Schink DR, Santschi PH, Corapcioglu O, Fehn U (1995) Prospects for "iodine-129 dating" of marine organic matter using AMS. Nucl Instrum Methods Phys Res B 99:524–527

Schlosser P, Bonisch G, Kromer B, Loosli HH, Buhler R, Bayer R, Bonani G, Koltermann P (1995) Mid-1980s distribution of tritium, ^3He, ^{14}C and ^{39}Ar in the Greenland/Norwegian Seas and the Nansen Basin of the Arctic Ocean. Prog Oceanogr 35:1–28

Sholkovitz ER (1983) The geochemistry of plutonium in fresh and marine water environments. Earth Sci Rev 19:95–161

Shlokovitz ER, Mann DR (1987) $^{239,240}Pu$ in estuarine and shelf waters of the north-eastern United States. Estuarine Coast Shelf Sci 25:413–434

Smith JN, Ellis KM, Jones EP (1990) Caesium-137 transport into the Arctic Ocean through Fram Strait. J Geophys Res 95 (C2):1693–1701

Smith JN, Elis KM, Naes K, Dahle S, Matishov D (1995) Sedimentation and mixing rates of radionuclides in Barents Sea sediments off Novaya Zemlya. Deep-Sea Res II 42:1471–1493

Smethie WM Jr, Ostlund HG, Loosli HH (1986) Ventilation of the deep Greenland and Norwegian seas: evidence from krypton-85, tritium, carbon-14 and argon-39. Deep-Sea Res 33:675–703

Spencer DW, Bacon MP, Brewer PG (1981) Models of the distribution of ^{210}Pb in a section across the north Equatorial Atalntic Ocean. J Mar Res 39:119–138

Stokozov NA, Buesseler KO (1999) Mixing model for the northwest Black Sea using ^{90}Sr and salinity as tracers. J Environ Radioact 43:173–186

Tkalin AV, Chaykovskaya EL (2000) Anthropogenic radionuclides in Peter the Great Bay. J Environ Radioact 51:229–238

Topcuoğlu S, Güngör N, Kirbaşaoğlu C (2002) Distribution coefficients (K_d) and desorption rates of ^{137}Cs and ^{241}Am in Black Sea sediments. Chemos 49:1367–1373

UNSCEAR (2000) United Nations Scientific Committee on the Effects of Atomic Radiation (UNSCEAR) Report to the general assembly with scientific annexes. Annex C: exposures from man-made sources of radiation. http://www.unscear.org. Accessed April 2010

Vakulovsky S (1987) Determination of distribution processes, transport routes and transport times in the North Sea and the northern North Atlantic using artificial radionuclides as tracers. In: Guary JC, Guegueniat P, Pentreath RJ (eds) Radionuclides: a tool for oceanography. Elsevier Applied Science, London, pp 271–280

Vakulovsky S (2001) Radiation monitoring in Russia at the turn of the millennium. J Eniron Radioact 55:219–220

Warneke T, Croudace IW, Warwick PE, Taylor RN (2002) A new ground-level fallout record of uranium and plutonium isotopes for northern temperate latitudes. Earth Planet Sci Lett 203:1047–1057

Weimer W, Langford JC (1978) Iron-55 and stable iron in oceanic aerosols: forms and availability. Atmos Environ 12:1201–1205

Winger K, Feichter J, Kalinowski MB, Sartorius H, Schlosser C (2005) A new compilation of the atmospheric ^{85}krypton inventories from 1945 to 2000 and its evaluation in a global transport model. J Environ Radioact 80:183–215

Wong KM, Jokela TA, Eagle RJ, Brunk JL, Noshkin VE (1992) Radionuclide concentrations, fluxes, and residence times at Santa Monica and San Pedro Basins. Prog Oceanogr 30: 353–391

Yamada M, Wang Z-L, Zhen J (2006) The extremely high ^{137}Cs inventory in the Sulu Sea: a possible mechanism. J Environ Radioact 90:163–171

Yim MS, Caron F (2006) Life cycle and management of carbon-14 from nuclear power generation. Prog Nucl Energy 48:2–36

Yiou F, Raisbeck GM, Christensen GC, Holm E (2002) ^{129}I/^{127}I, ^{129}I/^{137}Cs and ^{129}I/^{99}Tc in the Norwegian coastal current from 1980 to 1988. J Environ Radioact 60:61–71

Yoshida S, Muramatsu Y (2003) Determination of U and Pu isotopes in environmental samples by inductively coupled plasma mass spectrometry. http://www.nirs.go.jp/report/nene/h13/05/85.htm. Accessed April 2010

Chapter 20
Applications of Transuranics as Tracers and Chronometers in the Environment

Michael E. Ketterer, Jian Zheng, and Masatoshi Yamada

Abstract The transuranic elements (TRU) Np, Pu, Am, and Cm have prominently emerged as powerful tracers of earth and environmental processes, applicable to the recent, post nuclear-era timescale. Various long-lived isotopes of these elements are found in the earth's surface environment, almost exclusively as a result of nuclear weapons production, testing, or nuclear fuel cycle activities. A globally recognizable signal, of consistent composition, from stratospheric fallout derived from 1950–1960 above-ground weapons tests is itself useful in tracing applications; in specific local/regional settings, stratospheric fallout is mixed with or dominated by other TRU sources with contrasting isotopic signatures. Both decay-counting and MS approaches have been utilized to measure the concentrations and isotopic ratios of TRU and are useful as discriminators for source characterization, provenance, and apportionment. Examples include the activity ratios ^{238}Pu/$^{239+240}$Pu, ^{241}Am/$^{239+240}$Pu, and ^{241}Pu/$^{239+240}$Pu; atom ratios such as ^{240}Pu/^{239}Pu, ^{237}Np/^{239}Pu, ^{241}Pu/^{239}Pu, and ^{242}Pu/^{239}Pu are also used in this context. Of the TRU elements, Pu is by far the most widely studied; accordingly, this chapter mainly emphasizes the use of Pu activities and/or atom ratios as tracers and/or chronometers. Nevertheless, Pu is sometimes measured in combination with one or more isotopes of other elements. The TRU elements offer several prominent applications in environmental/geochemical tracing: (1) chronostratigraphy of sediments and related recent Holocene deposits; (2) using fallout TRU as quantitative probes of soil erosion, transport and deposition; (3) investigating water mass circulation, the transport and scavenging of particulate matter, and tracking the marine geochemical behavior of the TRU elements themselves in the marine environment; and (4) studies of the local/regional transport, deposition and inventories of non-fallout TRU in the surficial environment.

20.1 Introduction

20.1.1 The Elements Np, Pu, Am, and Cm

The transuranic elements (also known as transuranium elements, abbreviated herein as TRU) have atomic numbers (Z) greater than 92. None of these elements are stable; all of their nuclides undergo radioactive decay to form other isotopes, eventually reaching $Z = 82$ (Pb) or $Z = 83$ (Bi). Elements with up to $Z = 103$ (Np–Lw) are members of the 5f actinide series, and the named elements with $Z = 104$–112 are known as transactinides. The elements with $Z > 98$ (Cf) have not been produced in significant quantities and have never been used other than in limited research settings. Isotopes of the elements Np through Cm ($Z = 93$–96, respectively) are most important in terms of their environmental distribution and occurrence; hence, this chapter focuses upon the elements Np ($Z = 93$), Pu ($Z = 94$), Am ($Z = 95$) and Cm ($Z = 96$). Of these, Pu is by far the most prominent in terms of its importance as an environmental tracer, and in

M.E. Ketterer (✉)
Department of Chemistry and Biochemistry, Box 5698, Northern Arizona University, Flagstaff, AZ 86011-5698, USA
e-mail: Michael.Ketterer@nau.edu

J. Zheng and M. Yamada
Environmental Radiation Effects Research Group, National Institute of Radiological Sciences, 4-9-1 Anagawa, Inage, Chiba 263-8555, Japan

terms of numbers of published studies pertaining to its environmental distribution.

Neptunium has two long-lived, α-emitting isotopes, ^{236}Np ($t_{1/2} = 1.54 \times 10^5$ years) and ^{237}Np ($t_{1/2} = 2.14 \times 10^6$ years). Plutonium has six prominent isotopes: ^{238}Pu ($t_{1/2} = 87.74$ years), ^{239}Pu ($t_{1/2} = 24,110$ years), ^{240}Pu ($t_{1/2} = 6,561$ years), ^{241}Pu ($t_{1/2} = 14.29$ years), ^{242}Pu ($t_{1/2} = 373,000$ years) and ^{244}Pu ($t_{1/2} = 8.3 \times 10^7$ years). The isotopes ^{238}Pu, ^{239}Pu, ^{240}Pu, ^{242}Pu and ^{244}Pu all decay by emission of α particles, while ^{241}Pu undergoes β decay to produce α-emitting ^{241}Am ($t_{1/2} = 432.6$ years). ^{241}Am also generates 59 keV gamma photons coincident with its α decay. In addition to ^{241}Am, Am has a second long-lived α emitting isotope, ^{243}Am ($t_{1/2} = 7,370$ years), and a meta-stable isotope, ^{242m}Am ($t_{1/2} - 141$ years). The element Cm has seven prominent isotopes: ^{243}Cm, ^{244}Cm, ^{245}Cm, ^{246}Cm, ^{247}Cm, ^{248}Cm, and ^{250}Cm; these all undergo α decay with respective half-lives of 29.1, 18.1, 8,500, 4,730, 1.56×10^7, 340,000, and 9,000 years.

The half-lives of the isotopes of Np, Pu, Am, and Cm are too short to have survived decay over the Earth's 4.5 billion year history, though these elements' abundances were probably similar to Th and U at the time of Earth accretion (Taylor 2001). The one exception is ^{244}Pu, of which a few primordial atoms per kg crust are expected to still be present (Taylor 1995; Wallner et al. 2004). The astrophysical r-process is the main production mechanism for ^{244}Pu and other heavy isotopes. ^{244}Pu is also thought to be present in interstellar medium (ISM) grains found in the inner solar system (Ofan et al. 2006). ISM material continuously enters the solar system and the Earth's atmosphere as a source of "live", recently produced ^{244}Pu. The ISM grains are thought to have elemental and isotopic compositions resembling the early Earth; hence, ISM should contain many radionuclides (^{244}Pu and others) with half-lives much less than the age of the Earth. Ofan et al. (2006) estimate that the steady-state flux of ^{244}Pu is about 0.1 atom cm^{-2} year^{-1}. In deep-sea sediments with a sedimentation rate of ca. 1 μm/year, the ^{244}Pu flux would generate a concentration of about 10^5 atoms/kg in young deposits.

The presence of small amounts of ^{237}Np and ^{239}Pu in Nature is well established; these nuclides are continuously produced from ongoing neutron capture processes, and hence natural ^{239}Pu is most abundant in U ores. Curtis et al. (1999) determined ^{239}Pu/^{238}U atom ratios of 2.4×10^{-12} to 4.4×10^{-11} for ores from the Cigar Lake uranium deposit. In an older study, Myers and Lindner (1971) determined a ^{237}Np/^{238}U atom ratio of 2.16×10^{-12} in a pitchblende ore. Both ^{237}Np and ^{239}Pu, along with ^{236}U, are produced in Nature by neutron capture on ^{235}U or ^{238}U, with neutrons originating from (α, n) reactions, spontaneous and induced fission, and by cosmic rays (Hotchkis et al. 2000). These neutron capture processes are much less probable in the bulk Earth's crust (ca. 3 mg ^{238}U/kg) than in U ores but still produce a crustal concentration of ~2×10^{-14} g ^{239}Pu/kg (Taylor 1995).

Oxidation-reduction processes are important in determining the environmental behavior of the TRU elements. The elements Am and Cm exist only in the +3 oxidation state under normal environmental conditions (Choppin 2007), while Np and Pu have more complex redox behavior. Np can exist in the +4, +5, and +6 states, with the +5 state being most common in oxic natural waters and sediment- or soil-water-multiphase environments. Pu has analogous behavior with mixtures of the +4 and +5 states being prevalent; under strongly reducing (anoxic) conditions, Pu can also exist in the +3 oxidation state. Given the high charge and small ionic radii of these species, all of the TRU elements tend to be present in colloidal rather than true solute forms in natural waters (Choppin 2007), and all display rather significant propensity to become associated with particle phases (namely, particle reactivity). The sorption coefficient, K_d, is a commonly encountered operational parameter that is used to describe the partitioning of the element between solid-adsorbed and solution phases ($K_d = C_{solid}/C_{solution}$) which has units of mL/g. Accordingly, a high value of K_d is indicative of strong tendency towards particle association, and hence, relative immobility as a true solute, while a low K_d indicates significant solubility and the substance's tendency to be transported in solution form. For the TRU elements, a reported K_d is mainly limited to use in the specific environmental circumstances for which it was measured, and may obviously be greatly affected by factors such as the oxidation state/speciation of the element, the environmental temperature, the characteristics of the aqueous phase (pH, ionic strength, and presence of complexants), and the physical and chemical properties of the solid phase (surface area, particle size distribution, mineral composition, etc.).

Despite these limitations, K_d values have been calculated for TRU elements in many different environments. As a case in point, in seawater as the aqueous phase, and with fine-grained clay sediments from the Irish Sea, Nelson and Lovett (1978) determined K_d values in the range of 10^3–10^4 for Pu in the +5 and +6 states, while tetravalent Pu exhibited a higher K_d in the range of 10^6. This result obviously conveys that Pu will be less mobile and more strongly associated with the sediment when present in the reduced +4 state. Hence, calculations and models performed using K_d values should be used as a starting point, or guide for the expected tendencies of the element of interest, and some *a priori* knowledge or actual measurements in the environment of interest is essential. In general, however, most K_d measurements of TRU imply the strong association of these elements with solid phases in many different freshwater and marine environments, which lends well to their use as tracers of particle transport in the environment (*vide infra*).

20.1.2 Sources, Release Histories, and Isotope Compositions

Aside from the above-mentioned natural occurrences of TRU in the Earth surface environment, these elements have become widely distributed in the Earth's surface environment as a result of anthropogenic production in the last six decades. The production mode is again neutron capture in a fission bomb, or a nuclear reactor that is specifically designed to produce electrical power or to transmute ^{238}U into ^{239}Pu for weapons. As a result, essentially all of the TRU in the Earth's surface environment is of anthropogenic origin. The uppermost 30 cm of undisturbed crust, for example, contains ^{237}Np and ^{239}Pu at levels ~10^4–10^5 times greater than the natural concentrations, and no evidence to date has been identified for the natural sources for Am, Cm, or other Pu isotopes besides ^{239}Pu and ^{244}Pu. The main source term for Np, Pu and Am is global fallout from 1950's and 1960's aboveground nuclear weapons tests; chronologies and locations of these tests are described in Carter and Moghissi (1977). Americium (as ^{241}Am) is not itself initially formed in the nuclear tests, but rather it accumulates over time, even during stratospheric residence timeframes, due to its in-growth from the decay of relatively short-lived ^{241}Pu. Curium, however, is neither produced in significant amounts by nuclear tests, nor do any Cm isotopes form by in-growth from fresh testing debris; instead, Cm isotopes are mainly derived from power reactors (Schneider and Livingston 1984; Holm et al. 2002).

In the 30-year span 1945–1975, 801 nuclear detonations were conducted by the United States, former Soviet Union, United Kingdom, Republic of France, People's Republic of China, and India, with a cumulative yield of about 325 Mt TNT (Carter and Moghissi 1977). Several additional Chinese low-yield atmospheric tests were conducted at Lop Nor in the late 1970 (Roy et al. 1981) with the last Chinese atmospheric detonation having occurred in 1980 (Bennett 2002). These tests released ~1.5×10^{16} Bq of $^{239+240}Pu$ into the global environment, corresponding to ~4,000 kg ^{239}Pu, 700 kg ^{240}Pu, and $\geq$kg quantities of several other TRU isotopes. After release of these vaporized nuclides into the stratosphere, the TRU and fission products condensed to form very fine aerosols of 10–20 nm diameter (Harley 1980); these fine particles returned to the troposphere and thereafter, became associated with naturally occurring particulate matter, ultimately being removed as wet and dry deposition. In soils, TRU have become associated with specific phases such as iron/manganese oxides and humic acids.

Besides atmospheric weapons tests, other important sources include nuclear reactor accidents (e.g., the 1986 Chernobyl accident), nuclear fuel re-processing facilities (e.g., Sellafield, UK), atmospheric disintegration of satellites with onboard ^{238}Pu-powered thermoelectric generators (e.g., the 1964 SNAP accident), and releases from facilities devoted to the manufacturing or handling of weapons components (e.g., Rocky Flats).

Atmospheric tests can be categorized into two main groups. The first consists of "low-yield" tests that released the bulk of the radionuclide yield into the lower atmosphere, where it was rapidly deposited on a local/regional scale as "tropospheric fallout"; and the second group is "high-yield" tests that injected large amounts of material into the stratosphere, where it was later deposited on a hemispheric scale as "stratospheric" or "global" fallout. The atmospheric deposition of Pu and associated fission products such as ^{137}Cs ($t_{1/2} = 30.07$ years) and ^{90}Sr ($t_{1/2} = 28.78$ years) occurred mainly in the 1950's and 1960's,

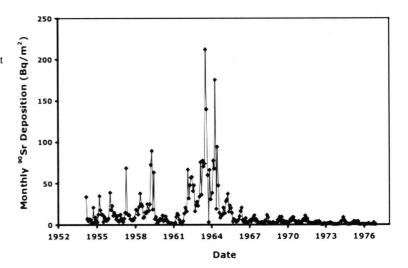

Fig. 20.1 Monthly deposition of ^{90}Sr at New York City, USA in the 1954–1976 timeframe. Source: Environmental Measurements Laboratory, US Department of Homeland Security, www.eml.st.dhs.gov/databases/fallout, accessed 19 May 2010

with a pronounced and well-known maximum in 1963, followed by a sharp decrease in deposition. Figure 20.1 depicts the temporal history of ^{90}Sr fallout deposition from atmospheric tests, with prominent 1963 peak, at New York City. A strong seasonality is evident, with the fallout delivery from the stratosphere to the troposphere and surface peaking in the late spring and early summer (Bennett 2002). The decrease in deposition coincided with the 1963 Limited Test Ban Treaty, signed by the USA, former Soviet Union, and United Kingdom. Comparatively minor amounts of fallout are evident in the Fig. 20.1 profile as a result of late 1960's to early 1970's tests conducted by France and China, non-signatories to the Test Ban Treaty. Residence times in the atmosphere are typically less than 1 month for nuclides introduced into the troposphere, while residence times are 0.7–3.4 years for material injected into the stratosphere from different types of tests conducted at different latitudes (Bennett 2002).

In addition to the temporal and seasonal dependencies of fallout deposition, the accumulated inventories of fallout also vary widely versus location on the Earth's surface. The highest inventories are found at mid-latitudes of the Northern Hemisphere, with lower inventories at the Equator and poles; less total fallout deposition occurred in the Southern Hemisphere. The Southern Hemisphere was the location of fewer tests, and the stratospheric air masses are not well-mixed between hemispheres (Bennett 2002); Northern Hemisphere inventories of all fallout nuclides are ~5-fold greater than those in the Southern Hemisphere. Besides its latitude dependency, fallout depositional inventories also exhibit variation resulting from climate, as atmospheric precipitation is the chief removal mechanism for fallout deposition. For equivalent latitudes, greater fallout inventories are observed in climates with higher mean annual precipitation; consequently, locations with marine climates have larger inventories than desert or humid continental climate patterns (Beck and Bennett 2002). In montane areas with vertical climates, significant local differences in fallout inventory versus altitude can be observed as a result of large differences in annual precipitation (Ulsh et al. 2000).

An authoritative estimate of the isotopic composition of stratospheric fallout Pu is given in Kelley et al. (1999), a study that used samples from 54 worldwide locations. Kelley et al. (1999) reported depositional inventories (10^{12} atoms ^{239}Pu/m^{2}) and atom ratios (^{237}Np/^{239}Pu, ^{240}Pu/^{239}Pu, ^{241}Pu/^{239}Pu, and ^{242}Pu/^{239}Pu); several important trends are apparent. First, regional effects, originating from locations such as the USA's Nevada Test Site (NTS), are important in some areas; secondly, slight differences in isotopic composition exist between Northern and Southern Hemisphere deposition. Since Pu is synthetic in origin, its isotope composition varies widely according to its production mechanism; hence, this intrinsic variance serves as the basis for a wide variety of isotopic tracing possibilities in atmospheric, marine, hydrologic, and crustal environments. The atom ratio ^{240}Pu/^{239}Pu is the most readily measured fingerprint; the ^{237}Np/^{239}Pu atom ratio and minor Pu isotope ratios provide additional insights and discriminatory power. Figure 20.2

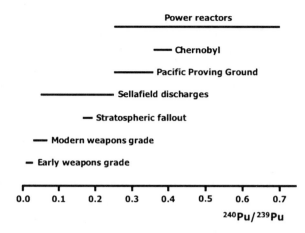

Fig. 20.2 Characteristic ranges for ^{240}Pu/^{239}Pu in stratospheric fallout and other prominent sources of TRU in the surface environment

depicts ^{240}Pu/^{239}Pu atom ratios for a number of prominent sources in the global environment. The Pu isotopic composition can therefore provide a "fingerprint" of different sources and is useful in studies of the movement of a specific point or regional source. In these studies, sources other than global fallout are identified and apportioned via their isotopic contrast versus the globally distributed baseline of stratospheric fallout.

In a manner analogous to source provenance with atom ratios, activity ratios are also useful in discrimination between stratospheric fallout and other local/regional sources of TRU. The most important activity ratio in this context is ^{238}Pu/$^{239+240}$Pu, which is useful in distinguishing many prominent Pu sources (Baskaran et al. 1995). This ratio ranges considerably between relatively low values of <0.05 in both weapons-grade and stratospheric fallout Pu, to much higher values in Pu originating from reactor sources (nuclear fuel reprocessing, Chernobyl, Kara Sea dumped reactor). More details on ^{238}Pu/$^{239+240}$Pu source discrimination with reference to the Arctic are given in Baskaran et al. (1995).

"Weapons-grade" Pu consists mainly of ^{239}Pu, with ^{240}Pu/^{239}Pu ~0.03–0.07 being typical for fission weapons and triggers for thermonuclear devices. Releases from low-yield fission tests, such as at the Nevada Test Site, exhibit ^{240}Pu/^{239}Pu very similar to the Pu present in the original device. The low neutron fluxes in typical kiloton-range fission tests generate only small amounts of neutron capture products such as ^{237}Np, the heavier Pu isotopes, and ^{241}Am. In contrast, the high-yield (thermonuclear) tests, accounting for the great majority of the worldwide TRU inventory, produced Pu that has undergone significant neutron activation in the high neutron flux environment, thereby accounting for the greater proportions of ^{240}Pu, ^{241}Pu, and ^{242}Pu. For stratospheric fallout in the 31–70°N latitude band, resulting from the cumulative average of all testing, average atom ratios ($\pm 2\sigma$) are as follows: ^{237}Np/^{239}Pu $= 0.48 \pm 0.07$; ^{240}Pu/^{239}Pu $= 0.180 \pm 0.014$; ^{241}Pu/^{239}Pu $= 0.00194 \pm 0.00028$ (decay-corrected to 1 January 2000); and ^{242}Pu/^{239}Pu $= 0.00387 \pm 0.00071$. Plutonium that has been subjected to lengthy reactor irradiation (i.e., "burn-up") also contains relatively larger abundances of the heavier isotopes. As is seen in Fig. 20.2, the Pu isotopic composition from power reactors can vary widely, being characteristic of the neutron flux and reactor conditions.

A specific global source of ^{238}Pu is the 1964 atmospheric burn-up of the radioisotope thermoelectric generator onboard the US satellite SNAP-9A. Upon re-entry to the atmosphere, the satellite incinerated and released ~6 × 10^{14} Bq of ^{238}Pu into the southern Hemisphere stratosphere (Harley 1980). The vast majority of the SNAP-9A ^{238}Pu was subsequently deposited in the southern Hemisphere, since limited mixing occurs between the northern and southern Hemisphere stratospheric air masses. The SNAP ^{238}Pu is now blended with stratospheric fallout Pu, and as a result, the northern and southern Hemispheres exhibit ^{238}Pu/$^{239+240}$Pu signatures of ~0.025 and 0.25, respectively (Sam et al. 2000). The 1978 re-entry of the Soviet satellite Cosmos-954 over northern Canada led to its mechanical destruction upon impact; large debris was recovered by clean-up teams; however, high-altitude balloon sampling conducted by the USA demonstrated minimal presence of debris in the atmosphere (Krey et al. 1979).

Another well-known situation where TRU source mixing is observed is in marine systems. Nuclear fuel re-processing facilities such as Sellafield (UK) and La Hague (France) historically released large quantities of TRU and fission products into Northern Hemisphere oceans (Hong et al. 2011); some of these nuclides have remained in the water column, though large amounts have become associated with sediments in coastal areas through particle scavenging processes. In the Pacific Ocean, extensive mixing occurs between directly deposited stratospheric (global) fallout, and close-in regional fallout from the Pacific Proving Ground (PPG). PPG fallout was deposited mainly from US tests in the 1952–1958 pre-moratorium timeframe (Koide et al. 1985; Hong et al. 2011). As shown in Fig. 20.2, ^{240}Pu/^{239}Pu is higher in PPG fallout than stratospheric fallout; this isotopic difference is very useful towards tracing the transport of dissolved Pu in the North Pacific (e.g., Zheng and Yamada 2004).

20.2 Techniques for Sample Collection, Processing, and Analysis

20.2.1 Sampling and Sample Size

TRU are commonly determined in environmental media such as soils, sediments, water, atmospheric aerosols, atmospheric deposition, snow pack, ice cores, and biota. Procedures for sampling these media are very similar to methods used in other types of geochemical analysis. TRU are mainly found in surficial soils and sediments; in undisturbed soils, essentially the entire inventory is concentrated in the top 30 cm (Kelley et al. 1999). Grab and composite soil samples are obtained using trowels, augers, and shallow coring tools. Boulyga et al. (2003) collected soil cores to a depth of 20 cm, subsequently sectioned into 1 cm intervals, in order to study the depth distribution of Pu and ^{241}Am in soils from Belarus.

In sediments, the depositional TRU inventory may be focused in a shallow band of <5 cm or distributed across several meters of sediment, depending upon the sedimentation conditions. Coring techniques are best selected based on the water depth and sediment thickness to be recovered. Ketterer et al. (2004a) utilized a "Mini-Mackereth" corer based upon the design of Mackereth (1969) to recover a 120 cm core from 170 m water depth in Loch Ness; detectable levels of ^{239}Pu, ^{240}Pu, and ^{241}Am were found in the top 20 cm of this core. For situations requiring coring to greater depths, other devices such as piston corers (e.g., Nesje 1992) are appropriate.

Determinations of TRU in soil samples require enough mass to isolate and measure (by decay counting or mass spectrometry techniques) a sufficient number of detector events. The sensitivity of a mass spectrometric determination is governed by numbers of analyte atoms; decay-counting analyses are constrained by the numbers of decays that can be recorded during a reasonable counting period. The requisite sample mass may be in the mg range for high activity soils or sediments, particularly where sensitive techniques such as thermal ionization mass spectrometry (TIMS) are used. Kelley et al. (1999) used TIMS to determine ^{237}Np and Pu isotopes, including low-abundance ^{241}Pu and ^{242}Pu in 1 g of surficial soil. Zheng and Yamada (2006a) used SF-ICP-MS to determine ^{239}Pu and ^{240}Pu in 0.1–1 g samples of settling marine particles. In other cases, sample sizes of 10–100 g are utilized to improve detection of low concentrations of the analyte; Agarande et al. (2001) used sector field inductively coupled plasma mass spectrometry (SF-ICPMS) for ^{241}Am determinations, using up to 100 g for soils and sediments containing <1 Bq/kg (massic activity) of ^{241}Am from stratospheric fallout. For the alpha spectrometric determination of Pu, Am and Cm isotopes, Mietelski and Was (1997) used soil aliquots of 20–40 g.

In other situations, the available amount of analyte is not the basis for determining sample size; rather, one is concerned with analyzing a singe aliquot with sufficient mass to contain a statistical representation of high-activity "hot" particles. Maxwell (2008) described the preparation of 100–200 g aliquots to meet this need for determinations of Pu, Am, and Cm in soils.

Preservation is not required for soils and sediments, since the transuranic elements are particle-reactive and non-volatile; however, collection of water samples typically requires preservation by acidification. In a demonstrative example of this methodology, Yamada et al. (2006) collected 250 L seawater samples, which were acidified to pH <2 with HCl and treated with 3 g of Fe^{3+} (aq) carrier before co-precipitation of Pu with $Fe(OH)_3$. The precipitate was separated aboard ship prior to further processing in a land-based laboratory. The determination of TRU in seawater and other natural waters typically requires very large volume-samples, as the $^{239+240}$Pu activity ranges from 1–10 mBq/m^3 in the open oceans up to >100 mBq/m^3 in semi-enclosed waters near specific sources such as the Irish Sea (Lindahl et al. 2010).

20.2.2 Sample Pre-Treatment Instrumental Methods of Analysis

^{241}Am in solid samples at ≥Bq/kg activity is amenable to direct gamma spectrometric determination

(Warwick et al. 1996); with this exception, most other TRU analytical procedures require destructive radiochemical analyses. The following is a typical sequence of steps: (1) weighing of a homogenous sample aliquot; (2) addition of a spike isotope, also referred to as a yield tracer; (3) sample treatment with reagents suitable to dissolve the entire sample or quantitatively extract the target analyte from the matrix; (4) separation of the analyte from the dissolved sample matrix, and (5) analyte determination by decay-counting or mass spectrometric techniques. A long-lived spike isotope for ^{237}Np is not commercially available, though Kelley et al. (1999) used a reactor-produced ^{236}Np ($t_{1/2} = 1.54 \times 10^5$ years) spike in their TIMS work. Other workers have used freshly prepared ^{239}Np ($t_{1/2} = 2.36$ days) counted separately by gamma spectrometry (Morris et al. 2000), or ^{242}Pu as a pseudo-spike isotope, the latter taking advantage of the rather similar chemical behavior of Np(IV) and Pu(IV) in separations (Chen et al. 2002). Pu determinations entail using 2.86 years half-life ^{236}Pu spike (Mietelski and Was 1995) and in some cases, ^{244}Pu (Kelley et al. 1999), though ^{242}Pu has been most widely used and is available as certified solutions from the US's National Institute of Standards and Technology (NIST) and the European Union's Institute for Reference Materials and Measurements (IRMM). ^{236}Pu and ^{244}Pu would be the ideal spikes to use in alpha spectrometry (AS) and mass spectrometry (MS), respectively, as these isotopes are essentially absent in most samples and their presence does not interfere with analyte isotope measurements. Some tracing studies entail determination of the sample's indigenous ^{242}Pu/^{239}Pu; in this case, the preparation of both ^{242}Pu-spiked and un-spiked sub-samples is required. ^{243}Am is widely available as certified solutions and is the routine spike used for determination of ^{241}Am by AS and MS. For measurements of Cm, $^{243+244}$Cm is usually measured as a single un-resolved activity peak by AS versus a ^{243}Am pseudo-spike, taking advantage of the very similar separation behavior of Am and Cm (Mietelski and Was 1997).

In many types of solid environmental samples, TRU can be quantitatively extracted by leaching with mineral acids (HNO_3 and/or HNO_3-HCl) since these elements are not incorporated into the crystal lattices of naturally occurring minerals. TRU from stratospheric fallout origin are readily recovered in this manner, which simplifies chemical pre-treatment. A representative example is the procedure followed by Liao et al. (2008), in which sediment samples were leached with 8 M HNO_3 on a 200°C hot plate for 4 h. Acid leaching is simplified by first dry-ashing of the samples at 450–600°C to thoroughly remove organic matter. Nevertheless, some samples contain TRU in physical/chemical forms not recoverable by simple acid leaching; in this event, more rigorous treatment with HF-containing acid mixtures or fusion with alkaline salts is preferred. Sill (1975) first advocated the routine use of high-temperature fusions for complete dissolution of "refractory" TRU contained within fused vitreous silicate particles or in oxide particles previously subjected to high-temperature calcination. Cizdziel et al. (2008) used HNO_3-HF-H_3BO_3 dissolutions on ~3 g sub-samples to dissolve Pu contained in vitreous silicates originating from the NTS. Kenna (2002) used HNO_3-HF dissolutions for determination of Np and Pu isotopes in ~10 g environmental samples prior to MS determinations.

In the analysis of water samples, isolation and purification of the TRU elements is generally a multi-stage process. A typical sequence consists of filtration, preservation by acidification, addition of spike isotopes and carriers, precipitation of a carrier such as $Fe(OH)_3$ or MnO_2, collection of the TRU-containing precipitate, dissolution of the precipitate, and a subsequent chemical separation of the TRU from the re-dissolved precipitate. Examples of specific procedures for determination of TRU in large-volume water samples are given elsewhere (Chiappini et al. 1999; Yamada et al. 2006; Baskaran et al. 2009).

Following dissolution or leaching, TRU are isolated by preparative separation techniques. In many cases, a targeted TRU element can be separated from other TRU elements, or other more abundant actinides such as Th and U, by taking advantage of differences in redox behavior. For example, U can be maintained as U(VI) under conditions where Pu is stable as Pu(IV); the latter is easily pre-concentrated on anion exchange resins from nitrate solutions while U(VI) is not retained under the same conditions. The preparative steps are intended to remove the sample matrix and any components interfering directly with measurement of the targeted isotope(s), and to enhance the concentration of the target element. Chemical purity (for MS) and obtaining a thin deposit for minimal peak tailing, in AS, are often more paramount considerations than the chemical yield. Co-precipitation with

Fe(OH)$_3$, MnO$_2$, or calcium oxalate are typical initial separation steps in analysis of water samples (Yamada et al. 2006; Becker et al. 2004), and sometimes solid samples (Maxwell 2008). Separations are generally based upon ion-exchange schemes with anion or cation resin columns, or selective solid-phase extractants (Ketterer and Szechenyi 2008). The development of commercially available extraction resins for Np, Pu and Am/Cm has greatly simplified TRU separation processes, allowing for the use of small columns and reagent volumes with minimal waste generation (Thakkar 2002).

20.2.3 Instrumental Methods of Analysis

Alpha spectrometry is the classical approach used for decades in the determination of Pu, Am and Cm (for a recent review see Vajda and Kim 2010), though the method is less practical for longer-lived isotopes (e.g., ^{237}Np and ^{244}Pu). In AS, the element of interest is prepared as a thin-film deposit by microprecipitation with NdF$_3$, or electrodeposition (Baskaran et al. 2009; Vajda and Kim 2010); the source is placed into an evacuated chamber with a planar Si detector. Equipped with a multi-channel analyzer, the alpha spectrum (Fig. 20.3) is acquired over sufficient counting time to collect adequate counts for the analyte and spike isotopes (typically hours–weeks). The main advantages of AS include its use of relatively simple and inexpensive hardware, the ability to re-measure previously prepared sources, and the ability to determine selected isotopes such as ^{238}Pu and ^{241}Am that have shorter half-lives or serious isobaric interferences in MS. Besides the long acquisition times, the other major drawback in the AS determination of Pu is its inability to routinely resolve ^{239}Pu and ^{240}Pu, precluding using ^{240}Pu/^{239}Pu as a powerful isotopic fingerprint, though this is sometimes possible in high-activity samples by spectral de-convolution methods (Montero and Sanchez 2001). With AS methods, discriminatory activity ratios routinely studied as source tracers include ^{238}Pu/$^{239+240}$Pu (Mietelski and Was 1995) and ^{241}Am/$^{239+240}$Pu. AS remains the preferred approach for determinations of ^{238}Pu, ^{241}Am and $^{243+244}$Cm, and it appears unlikely that MS will become competitive in the future for measurements of these isotopes.

^{241}Pu is a beta-emitting isotope not directly detectable by alpha spectrometry, though it has been measured by alpha-counting the ingrown ^{241}Am after a 1–2 year delay (Mietelski et al. 1999). Alternatively, ^{241}Pu is directly measured by beta scintillation techniques, enabling source discrimination and mixing calculations using ^{241}Pu/$^{239+240}$Pu activity ratios (Mietelski et al. 1999).

Though it has long had a niche presence in specialized studies (Hardy et al. 1980; Koide et al. 1985), MS has recently enjoyed a tremendous renaissance for low-level measurements of TRU in environmental samples (Lariviere et al. 2006; Ketterer and Szechenyi 2008; Lindahl et al. 2010). Mass spectrometry is inherently advantageous in determinations of long-lived radionuclides having $t_{1/2} \geq 1,000$ years (Lariviere et al. 2006). Several MS techniques have been widely used in TRU determinations, including TIMS, ICPMS, resonance ionization MS (RIMS), accelerator MS (AMS), and secondary ionization MS (SIMS); the role of different MS techniques is discussed in more detail elsewhere (Lariviere et al. 2006; Ketterer and Szechenyi 2008). Though sample preparation is still required, the typical MS determination is performed in a few minutes versus hours–days for decay-counting methods. A mass spectrum depicting the detection of ^{237}Np, ^{239}Pu and ^{240}Pu in a stratospheric fallout-containing surface soil is shown in Fig. 20.4.

At this point in time, quadrupole and sector field ICPMS (Q-ICPMS, SF-ICPMS) dominate all the others in numbers of studies, while the remaining MS techniques are nonetheless invaluable in specialized

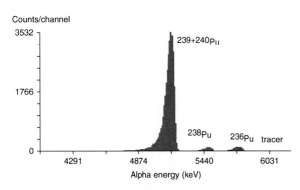

Fig. 20.3 Alpha spectrum of Pu isotopes prepared from 20 g of a surface soil sample from southern Poland; dissolution method per Mietelski and Was (1995). Energy width = 7 keV per channel; measurement time = 1.0 days; 0.0837 Bq ^{236}Pu tracer was added at the onset of the preparation

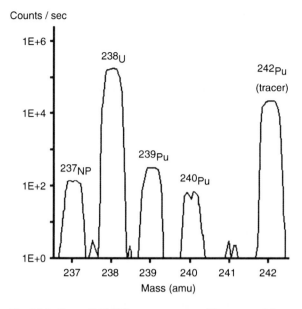

Fig. 20.4 Sector field ICP mass spectrum of Pu extracted from 50 g of a surface soil sample from Chihuahua, Mexico; acid-leaching and Pu separation per Ketterer et al. (2002). Mass spectral data were collected by averaging 50 one-second sweeps in the depicted mass range; the ^{242}Pu peak originates from 50 pg of added yield tracer; ^{238}U is the amount remaining after removal of >99.999% of the originally leached U content

situations. ICPMS has achieved "workhorse" status because of its widespread availability in labs worldwide; instruments have relatively low capital cost (<$150,000 US for a basic Q-ICPMS). ICPMS is highly compatible with automated introduction of liquid samples, and has modest operator skill requirements. Sample throughput, after preparation of separated sample aliquots, can surpass 100 samples/day. The contrast between ICPMS and AS is apparent by considering the determination of 1 pg of recovered ^{239}Pu, corresponding to 2.5×10^9 atoms and an α decay rate of 2.3 mBq. With an AS counting efficiency of 30% (Vajda and Kim 2010), approximately 17 days are required to collect 1,000 detector events, yielding a relative counting error of 3.1%. However, even with a relatively atom counting efficiency of 0.01% typical of Q-ICPMS, approximately 50,000 atom counts can be registered (in a 20% duty cycle peak-jump experiment with five isotopes sequentially counted) in a matter of several minutes time. Significantly better atom counting efficiencies are routine in SF-ICPMS, and specialized TIMS methods have achieved ionization efficiencies of 2% for ^{237}Np and ~5% for Pu isotopes (Kelley et al. 1999). The main drawbacks of MS versus decay-counting techniques are permanent contamination of the instrument, high hardware costs, and difficulties with determination of some specific isotopes that have relatively short half-lives, such as ^{238}Pu. Mass spectrometric techniques, particularly ICPMS and SIMS, can be constrained by interfering isobaric polyatomic ions, which are not resolved under low-resolution conditions where sensitivity is optimum. In ICPMS, isobars such as PbCl$^+$ (Nygren et al. 2003) can generally be removed by suitable preparative separations prior to the determination step; the well-known ^{238}U^{1}H interference on ^{239}Pu is not problematic if U is sufficiently eliminated beforehand (Ketterer et al. 2004a). In AMS, isobars and abundance sensitivity effects are completely eliminated by high-energy charge-exchange reactions, and definitive detection of transuranic isotopes is performed in an interference-free manner (Fifield 2008). AMS, however, is very capital- and operator skill-intensive, with only a few systems currently in use worldwide for TRU determinations in niche applications. The main advantage of SIMS is its imaging capability and the possibility of performing elemental and isotopic analyses of individual particles, which represents an important capability in nuclear safeguards applications (Pöllänen et al. 2006).

20.2.4 Figures of Merit

Some representative detection limits for MS determinations of Pu in environmental samples are given in Table 20.1. In general, these detection limits are far superior to AS, which affords detection limits of ca. 0.1–1 mBq/sample (Vajda and Kim 2010). Total propagated uncertainties for analytical results of <5% are achieved relatively easily, and can be better than 1% in some instances. ICPMS or TIMS with multiple ion counting systems (Taylor et al. 2001) potentially offers excellent precision for ratio measurements, both of the sample isotope ratios, and sample-spike atom ratios for isotope dilution determinations. Though these MC-ICPMS systems are expensive and not widely used, they have potential for producing outstanding accuracy if all sources of systematic error (namely isobars, mass discrimination, and detector cross-calibration effects) can be accounted for. Taylor et al. (2001) illustrated the determination of

Table 20.1 Representative detection and performance limits for determination of Pu in environmental samples

Technique	Detection limits	Comments	Reference
TIMS	10 fg $^{239+240}$Pu	Early TIMS work; ice cores	Koide et al. (1985)
TIMS	~0.01 fg for ^{241}Pu, ^{242}Pu	Precise ratio measurements in ~1 g soils; total dissolution with HNO$_3$/HF; ^{244}Pu spike	Kelley et al. (1999)
RIMS	~0.4–4 fg for each isotope	Soil and sediment analysis; alkali fusion for total dissolution	Nunnemann et al. (1998)
AMS	<0.1 fg ^{239}Pu	Ice cores; analysis constrained by 2–5 fg procedural blanks for ^{240}Pu and ^{239}Pu	Olivier et al. (2004)
Q-ICPMS[a]	20 fg $^{239+240}$Pu	Ratio measurements in Pu solution SRM's, IAEA-135 sediment	Godoy et al. (2007)
SF-ICPMS[a]	5 fg ^{237}Np, ^{239}Pu, ^{240}Pu, ^{241}Pu	10 g samples (LOD 0.5 fg/g); total dissolution with HNO$_3$/HF; ^{236}Np and ^{242}Pu spikes; sediments and soils	Kenna (2002)
	0.07 fg ^{242}Pu	Analysis of settling marine particles 0.03–0.50 g	Zheng and Yamada (2006a)
MC-ICPMS	<5 fg ^{242}Pu	<10% RSD achieved for ^{242}Pu/^{239}Pu ratio measurements with 5 fg ^{242}Pu; extracts from sediments	Taylor et al. (2001)

[a]More discussion and examples of ICPMS detection limits for Pu are given in a review by Kim et al. (2007)

Table 20.2 Standard reference materials with certified activities of TRU in environmental sample matrices

Material[a]	Certified activities	Uncertified (information values)
NIST 4350b, Columbia River Sediment	^{238}Pu, $^{239+240}$Pu, ^{241}Am	^{240}Pu/^{239}Pu atom ratio
NIST 4353a, rocky flats soil #2	^{238}Pu, $^{239+240}$Pu	^{241}Am, ^{240}Pu/^{239}Pu, ^{241}Pu/^{239}Pu
NIST 4354, lake sediment	^{238}Pu, $^{239+240}$Pu, ^{241}Am	
NIST 4357, ocean sediment	^{238}Pu, $^{239+240}$Pu	^{241}Am, ^{237}Np
IAEA 375, soil		^{238}Pu, $^{239+240}$Pu, ^{241}Am
IAEA 381, Irish sea water	^{237}Np, ^{238}Pu, ^{239}Pu, ^{240}Pu, $^{239+240}$Pu, ^{241}Am	^{240}Pu/^{239}Pu atom ratio[b]
IAEA 384, Fangataufa Sediment	^{238}Pu, $^{239+240}$Pu, ^{241}Am	^{239}Pu, ^{240}Pu, ^{241}Pu
IAEA 385, Irish Sea Sediment	^{238}Pu, $^{239+240}$Pu, ^{241}Am	^{239}Pu, ^{240}Pu, ^{241}Pu

[a]NIST National Institute of Standards and Technology; IAEA International Atomic Energy Agency (Povinec et al. 2002)
[b]A recommended ^{240}Pu/^{239}Pu value of 0.22 ± 0.03 is given by IAEA. Zheng and Yamada (2007) reported ^{240}Pu/^{239}Pu = 0.2315 ± 0.0008

^{240}Pu/^{239}Pu in 100 fg Pu with 0.7% RSD, and with a precision of better than 0.15% RSD with 3 pg Pu (the latter quantity corresponds to a mixture of 6.1 × 10^9 atoms ^{239}Pu and 1.4 × 10^9 atoms ^{240}Pu, or 5.6 mBq ^{239}Pu and 4.7 mBq ^{240}Pu). Clearly, the combination of high transmission efficiency along with simultaneous counting of multiple ion beams at 100% duty cycle is responsible for these outstanding attributes.

Many different standard reference materials are readily available from NIST or IAEA for assay of activities of ^{238}Pu, $^{239+240}$Pu, and ^{241}Am (a list of reference materials is given in Baskaran et al. 2009). Materials are available that contain nuclides from stratospheric fallout and/or other sources in a range of activities; some examples of commonly used assay standards are given in Table 20.2. Reference materials certified for activities of one or more Cm isotopes are lacking, owing largely to the paucity of environmental Cm studies. Materials certified for ^{237}Np activity and/or ^{237}Np/^{239}Pu (along with the availability of ^{236}Np from a commercial or governmental agency source) represent an additional need.

Since many studies are focusing on TRU in seawater, particularly Pu, a standard reference material (IAEA 381, Irish Sea Water) has become available. However, this material contains grossly elevated levels of radionuclides originating from the Sellafield nuclear fuel reprocessing facility; for instance, $^{239+240}$Pu is certified at 0.0137 ± 0.0012 Bq/kg, or about 14 Bq/m^3, which is 1,000–10,000-fold higher than typical Pu activities in open ocean water. Though the availability of a seawater standard certified for baseline Pu activities, along with atom ratios, would be greatly beneficial to the marine radiochemistry community, the preparation of this material would be exceedingly difficult.

Certified Pu atom ratio standards are available from the US Department of Energy's New Brunswick Labs as CRM 136 and 137 (12 and 18% ^{240}Pu, respectively), though NBL is, at present, unwilling to distribute these CRM's in quantities smaller than 250 mg of Pu metal. The European Union's Institute for Reference Materials and Measurements (IRMM) now sells 1 mg sets of solid Pu nitrate materials certified for ^{239}Pu/^{242}Pu atom ratios; however, these materials are less useful for evaluating data for the environmentally significant ratios ^{240}Pu/^{239}Pu and ^{241}Pu/^{239}Pu. Also lacking are natural matrix reference materials certified for Pu atom ratios, though numerous referee analyses of Pu assay SRM's have been published (e.g., Muramatsu et al. 2001; Zheng and Yamada 2007; Ketterer and Szechenyi 2008). In conclusion, critical needs are the development of a spike for mass spectrometric determination of ^{237}Np, and the development of standards for Pu isotopic compositions, in small-quantity solutions, and particularly in actual environmental media.

20.3 Uses of TRU as Tracers and Chronometers

Several prominent areas of applicability exist for using TRU in the context of environmental/geochemical tracing. These may be broadly classified as (1) chronostratigraphy of sediments and related recent Holocene deposits; (2) using fallout TRU as quantitative probes of soil erosion, transport and deposition; (3) investigating water mass circulation, the transport and scavenging of particulate matter, and tracking the marine geochemical behavior of the TRU elements themselves in the marine environment; and (4) studies of the local/regional transport, deposition and inventories of non-fallout TRU in the surficial environment. Each of these is considered separately, again with the perspective that Pu is the most prominent analytical target in all four applications. In some instances, e.g. (1) and (2), one focuses on the globally deposited stratospheric fallout signal, which has a well-known source term. In principle, it should be sufficient to determine activity of one or more TRU nuclides; however, in practice, source provenance with ratios such as ^{240}Pu/^{239}Pu (Fig. 20.2) is essential.

20.3.1 Chronostratigraphy of Recent Deposits

Layered, cumulative deposits, including aquatic sediments, peats, and ice cores require reliable chronology in studies of phenomena such as anthropogenic pollution and climate change. Th-U dating and ^{14}C ($t_{1/2} = 5,730$ years) are effective for longer timescales, though the well-known excess ^{210}Pb method is broadly applicable to these deposits for the past 100–200 years (Baskaran and Santschi 2002). ^{210}Pb ($t_{1/2} = 22.2$ years) is a model-dependent dating method that assumes a constant rate of supply or constant initial concentration of the excess (unsupported) ^{210}Pb. In contrast, the delivery of stratospheric fallout has a well-defined history (refer to Fig. 20.1) and a signal recognizable worldwide, even in the Southern Hemisphere. Chronostratigraphic dating pinpoints one or more specific event-associated dates in the layered deposits, and thus is an excellent compliment to ^{210}Pb chronology. Chronostratigraphy based upon stratospheric fallout has long been performed using ^{137}Cs; in freshwater aquatic sediments, Cs is associated with fine-grained clay sediments, intercalating into smectites and illites (Ritchie and McHenry 1990). In an ideal depositional setting, fallout accumulation would precisely coincide with the depositional history (Fig. 20.1); however, in real systems, the fallout nuclides appear in the sediment profile as a result of mixing of direct atmospheric deposition with previously deposited fallout transported in from the catchment basin. These mixing processes, along with time-averaging of material that can remain suspended for long period of times, have the effect of dampening the precise monthly and annual deposition patterns apparent in Fig. 20.1. As a result, one is usually only able to pinpoint three specific dates from the sediment record: (1) fallout onset (1952–1954), (2) maximum activity (1963), and (3) the date of core collection. In most cases, the pre-moratorium and post-moratorium peaks are not separately resolved, and frequently, post-deposition disturbances of the sediment record exact additional loss of resolution or distortion of the record. Post-deposition mixing, whether by physical turbation processes, or by bioturbation, tends to introduce ambiguity in the accurate detection of the onset date marker, although in many of these cases a clear 1963 maximum is still evident. Indeed, an evaluation of the quality/integrity of the un-disturbed sediment record can be garnered from examining or modeling the sediment activity profile versus the atmospheric deposition source term. Cores exhibiting erratic activity profiles or un-interpretable chronostratigraphies tend to be of limited utility in studies where an accurate timeline is required as part of investigations of other recent phenomena such as pollution or climate histories.

Although ^{137}Cs chronostratigraphy has proven quite successful, it is limited by analytical constraints (i.e., the need to gamma-count many individual samples for hours–days) as well as an appreciable loss of activity due to decay (about two-thirds of the originally deposited fallout ^{137}Cs inventory has decayed away by the year 2010). In regions affected by Chernobyl fallout, including many parts of northern Europe and the former Soviet Union, an additional, larger 1986 ^{137}Cs signal is also present. ^{137}Cs is also less successful as a date marker in marine systems, where K_d for Cs is much lower than in freshwater systems due to competition between Cs and high concentrations of other alkali cations. As alternatives to Cs, any of the TRU similarly associated with stratospheric fallout could be used, and there is now substantial interest in using $^{239+240}$Pu in this application. Jaakkola et al. (1983) first used $^{239+240}$Pu activities, measured by AS, in a chronostratigraphic study of sediments in Finnish lakes, demonstrating close agreement between the $^{239+240}$Pu and ^{137}Cs profiles. The labor involved and the low throughput of AS for $^{239+240}$Pu activity measurements render this nuclide less attractive if measured by AS; however, ICPMS-based $^{239+240}$Pu activity measurements can be performed rapidly and at low cost. In a proof-of-principle study, Ketterer et al. (2002) used Q-ICPMS to determine a $^{239+240}$Pu chronology at Old Woman Creek (Ohio, USA); 20 g samples were leached with 16 M HNO$_3$, and the resulting $^{239+240}$Pu profile was in excellent agreement with ^{137}Cs. Other studies have used SF-ICPMS (e.g., Ketterer et al. 2004a; Liao et al. 2008; Reynolds et al. 2010), Q-ICPMS (Schiff et al. 2010) or AMS (Tims et al. 2010) and in many cases, excellent results were realized with small sample aliquots of <1 g.

Global inventories of TRU are dominated by stratospheric fallout; nevertheless, it is a pre-condition of chronostratigraphy to be able to demonstrate, unequivocally, that the source term (Fig. 20.1) is precisely what is being measured in a particular setting. This confirmation of source cannot be performed in conventional ^{137}Cs chronostratigraphy, which is based upon activity measurements of a single radionuclide. Advantageously, mass spectrometric measurements of Pu can confirm or disprove the stratospheric fallout origin of Pu through examining the ^{240}Pu/^{239}Pu atom ratio (Fig. 20.2). An example of this vetting of the Pu source is shown in Fig. 20.5 for a small arctic lake in southern Alaska (Schiff et al. 2010). The Pu activity record indicates the onset of $^{239+240}$Pu detection at 8.50–9.00 cm and a $^{239+240}$Pu activity maximum at 5.20–5.40 cm. All depth intervals for which adequate ^{240}Pu/^{239}Pu ratio counting statistics were realized, in fact, exhibit Pu with atom ratios within measurement error of the stratospheric fallout composition (Kelley et al. 1999).

An important advantage of using $^{239+240}$Pu chronostratigraphy in preference to ^{137}Cs is its applicability to marine sediments. Even though ^{210}Pb chronology is applicable in marine settings as well, it is nevertheless very instructive to validate the results using a second, independent tracer such as Pu. In studies of shelf sediments at Poverty Bay (New Zealand), Miller and Kuehl (2009) used $^{239+240}$Pu activities, measured in acid-leached samples by SF-ICPMS. Despite the lower inventories of stratospheric fallout in the Southern Hemisphere, Pu was readily measured in these sediments and the resulting $^{239+240}$Pu activity profiles validated the ^{210}Pb-inferred chronologies. Sanders et al. (2010) successfully determined sediment accumulation rates and $^{239+240}$Pu penetration depths in intertidal mangrove mudflats at a coastal location in southeastern Brazil.

20.3.2 Soil Erosion, Transport, and Deposition

The erosion, transport, and re-deposition of soil are of paramount concern in agriculture as well as the earth sciences. These processes have been studied for decades, and the importance of land-use and tillage practices in minimizing erosion rates is now recognized (Matisoff and Whiting 2011). For obtaining a quantitative understanding of these processes, ^{137}Cs has been used since the 1960's, along with ^{210}Pb and several cosmogenic isotopes (Matisoff and Whiting 2011). Such studies are based upon known source

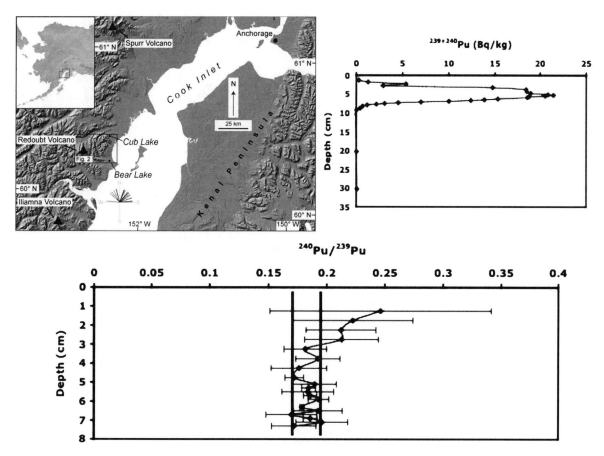

Fig. 20.5 $^{239+240}$Pu chronstratigraphy in Bear Lake, Alaska. The location of the lake is shown in the top left panel. $^{239+240}$Pu is first detected in the 8.5–9.0 cm depth interval (ca. 1952) and the peak activity is located at 5.2–5.4 cm, ascribed to 1963/1964. The ^{240}Pu/^{239}Pu atom ratios, which could be measured in many of the samples with <0.5 g dry sediment, indicate that the Pu is derived from stratospheric fallout

timing with an initially uniform deposition of an isotope within a small-scale experimental area. A "reference inventory" (Bq/m^2) is measured at "undisturbed" locations within or adjoining the study area, and loss or gain of soil by erosion or sedimentation is inferred via comparative measurements of inventories at each specific location versus the reference inventory. ^{137}Cs is well-known in this context, as Cs strongly associates with soil particles (particularly clays) and can be counted using the 661.62 keV gamma photon peak. These studies require an accurate two-dimensional picture of the inventory loss or gain, which is related to the erosion rate via various models (Ritchie and McHenry 1990). As such, a detailed inventory map of even a small field site requires many samples (i.e., hundreds), and the rate of data generation is constrained by numbers of available gamma counters. In Chernobyl-affected locations, the use of stratospheric fallout ^{137}Cs is complicated by a second 1986 source term that was deposited non-uniformly as tropospheric fallout; gamma spectrometric de-convolution of the 1963 and Chernobyl fallout terms using ^{134}Cs/^{137}Cs activity ratios is no longer possible now that Chernobyl-produced ^{134}Cs ($t_{1/2} = 2.06$ years) has decayed to non-detectable levels. However, TIMS was used by Lee et al. (1993) to measure ^{137}Cs/^{135}Cs; the long-lived ^{135}Cs isotope ($t_{1/2} = 2.3 \times 10^6$ years) potentially allows for source discrimination between stratospheric fallout and other ^{137}Cs components of the soil.

TRU isotopes, especially $^{239+240}$Pu, are attractive alternatives to ^{137}Cs, as these isotopes are also associated with soil particles, though they may be present in different soil mineral phases than Cs. Huh and Su (2004) compared the application of ^{137}Cs and $^{239+240}$Pu in Taiwan soils, finding that Cs and Pu conveyed very similar information about soil inventories and transport. When combined with MS (rather than AS) measurements, it is apparent that Pu has significant under-utilized potential as a tracer of soil re-distribution. Further, the distribution of TRU from the Chernobyl accident is much more constrained than is the case with ^{137}Cs; the TRU isotopes are specifically associated with non-volatile fuel particles (Mietelski and Was 1995, 1997), as opposed to Cs, which was volatilized in the reactor accident and became widespread over Europe and Russia. Once again, MS, through measurements of ^{240}Pu/^{239}Pu and/or other ratios, routinely allows evaluation of source provenance so that a specific input function (namely, stratospheric fallout) can be confirmed, or de-convoluted if mixed sources are identified.

Kaste et al. (2006) compared ^{137}Cs and $^{239+240}$Pu depth profiles in soils from the Konza Prairie site (Kansas, USA), an un-disturbed tall-grass prairie. The activity profiles were very similar, leading to the inference that $^{239+240}$Pu could serve the same purpose as a soil erosion tracer. Everett et al. (2008) used AMS to determine $^{239+240}$Pu activities in soils from the Herbert River catchment area in northeastern Queensland (Australia); again, $^{239+240}$Pu and ^{137}Cs conveyed similar information about erosion and re-distribution of soils in this Southern Hemisphere location. In a study focusing on aeolian transport of surface soils in the steppes of west Texas, Van Pelt et al. (in preparation) compared ^{137}Cs and $^{239+240}$Pu activities in soils and transported dusts, again demonstrating the rather similar inferences possible for these two systems (Fig. 20.6).

20.3.3 Transport and Scavenging in Marine Systems

The oceans have received TRU, mainly through global fallout, both as direct atmospheric deposition and as material transported from the continents. Once again, the greatest emphasis in published studies has been on Pu. Sholkovitz (1983) summarized early studies on Pu marine geochemistry. Subsequent to Sholkovitz's review, it is now increasingly apparent that the behavior of Pu in the oceans is very complex. More recent reviews by Skipperud (2004) and Lindahl et al. (2010) describe the behavior of Pu in the arctic marine environment, with prominent mention of the significance of atom ratio data obtained by mass spectrometry.

Classic alpha spectrometric $^{239+240}$Pu activity measurements in marine studies often do not provide adequate insight into marine Pu sources. Nevertheless, marine Pu transport depends dramatically upon its source and hence physicochemical speciation, with the oxidation state, size fractionation, and chemical composition of Pu-bearing particles all inevitably determining the resulting behavior. Therefore, MS can play a vital role in these studies, both in terms of measuring activities of very small quantities of Pu, and additionally, in isotopic analysis for source discrimination.

The Pacific Ocean has received significant local/regional fallout from 1946 to 1958 tests at the US's Pacific Proving Grounds and from 1966 to 1974 French tests at Mururoa and Fangataufu. The first ocean-wide study of Pu in the Pacific Ocean originated with the GEOSEC sampling program in the early

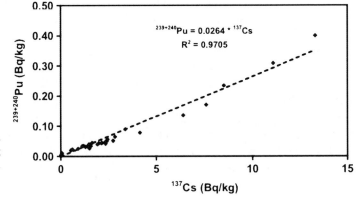

Fig. 20.6 Comparison of $^{239+240}$Pu and ^{137}Cs activities in soils from west Texas, USA. ^{137}Cs was determined by gamma spectrometry (24 h counting times); $^{239+240}$Pu was determined by SF-ICPMS in 50 g sub-samples leached with 100 mL of 16 M HNO$_3$ for 16 h at 80°C

1970's, some 10–15 years after the major input of fallout Pu (Bowen et al. 1980). Since then, Pu activities and the element's behavior in the water column and sediments of the Pacific have been intensively studied. In the South Pacific, Chiappini et al. (1999) determined ^{240}Pu/^{239}Pu atom ratios and $^{239+240}$Pu activities in surface and deep waters in the vicinity of Mururoa and Fangataufa atolls. The measurements were obtained using nominal 500 L sample volumes, with a co-precipitation/purification procedure being first used to prepare an alpha spectrometric source; thereafter, the source deposit was dissolved and analyzed by high-sensitivity Q-ICPMS. Evidence for a localized effect from 1966 to 1974 French tests was manifested as ^{240}Pu/^{239}Pu ratios of 0.07–0.10 in the upper 500 m of the water column. In the North Pacific, a pervasive signal is observed for PPG fallout mixed with stratospheric fallout. The TIMS study of Buesseler (1996) identified ^{240}Pu/^{239}Pu atom ratios in the range 0.19–0.34 that were ascribed to tropospheric fallout from the PPG; deep water ^{240}Pu/^{239}Pu atom ratios were systematically higher than global fallout. Buesseler (1996) demonstrated that PPG-derived Pu was more rapidly removed from surface waters than Pu from stratospheric fallout; corals exhibited ^{240}Pu/^{239}Pu > 0.20 for 1955–1961 growth bands, but ^{240}Pu/^{239}Pu was congruent with stratospheric fallout in 1962–1964 growth. Buesseler (1996) also found evidence for elevated ^{240}Pu/^{239}Pu ratios in surface sediments as far north as 40°N. Subsequent work by many groups has confirmed and elaborated upon these findings. Various studies have now documented the presence of PPG fallout in many North Pacific locations, including the Korean Peninsula (Kim et al. 2004), the Southern Okinawa Trench (Lee et al. 2004), the Japan Sea (Zheng and Yamada 2005), the Sea of Okhotsk (Zheng and Yamada 2006b), and the Sulu and South China Seas (Dong et al. 2010). In these cases, it is apparent that Pu is being transported as a solute via ocean currents; in the pelagic zone, particulate matter is present at very low concentrations, and there is little opportunity for particle scavenging. However, once the currents reach near-shore locations, Pu is scavenged from the water column and appears as a sediment-associated component (Zheng and Yamada 2006c). Yamada and Aono (2002) reported large particle-associated fluxes of $^{239+240}$Pu on the East China Sea continental margins, which indicated that episodic lateral transport of particles was significant for $^{239+240}$Pu delivery on the continental slope in the East China Sea.

The near-shore removal of Pu from the water column is quite apparent in the sediment records from Sagami Bay from the eastern margin of Japan (Zheng and Yamada 2004). This can best be seen by simultaneously evaluating the ^{240}Pu/^{239}Pu atom ratios alongside the $^{239+240}$Pu activity profile (Fig. 20.7). A $^{239+240}$Pu activity peak is observed in Core

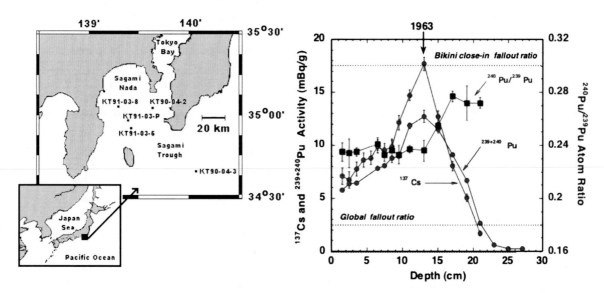

Fig. 20.7 Sediment core location from Sagami Bay near Tokyo, Japan; vertical profiles for $^{239+240}$Pu activity, ^{240}Pu/^{239}Pu, and ^{137}Cs activity in Core KT-91-03-8. The vertical lines shown for comparison versus the ^{240}Pu/^{239}Pu data are stratospheric fallout (0.18) and Bikini close-in fallout (0.30). Source: Zheng and Yamada (2004)

KT-91-03-8 at a depth of 12–14 cm, which coincides with the ^{137}Cs activity peak; both the $^{239+240}$Pu and ^{137}Cs peaks are ascribed to 1963. However, at a depth of 16–18 cm, preceding the 1963 peak and probably reflecting mid-late 1950's sedimentation, a maximum ^{240}Pu/^{239}Pu ratio of 0.277 ± 0.004 is observed. In post-1963 sediments, a rather uniform ^{240}Pu/^{239}Pu ratio of ~0.23 is observed. These patterns are definitely produced via mixing of stratospheric fallout Pu (^{240}Pu/^{239}Pu ~0.18; Kelley et al. 1999) with higher ratio fallout from the PPG (^{240}Pu/^{239}Pu ~0.30), the latter being transported with the North Equatorial Current and Kuroshio Current. The mixing of PPG source Pu and stratospheric fallout Pu has been observed in the water column by Bertine et al. (1986). They found an average value of ^{240}Pu/^{239}Pu of 0.23 in two seawater profiles from the North Pacific Ocean; ^{240}Pu/^{239}Pu was nearly invariant with depth, indicating that Pu has been homogenized in the water column for a time period of the past several decades.

Although the PPG close-in fallout Pu could be removed more rapidly from surface waters than Pu from stratospheric fallout (Buesseler 1996), high contributions of PPG-origin Pu are still observed in the surface waters in the Sulu and Indonesian Seas (39%) and in the South China Sea (42%) after six decades of their input (Yamada et al. 2006). A recent SF-ICP-MS study of Yamada and Zheng (2010) found anomalous increases of $^{239+240}$Pu inventory in water columns of Yamato and Tsushima Basins in the Japan Sea; $^{239+240}$Pu inventory in the water columns increased almost two times in a 10-year time scale, and constant ^{240}Pu/^{239}Pu atom ratios of ca. 0.24 were observed in seawaters from surface down to 3,000 m deep, indicating a continuous input and accumulation of Pu with PPG fallout signal in the Japan Sea.

In addition to the Pacific Ocean, other studies have examined the marine transport of Pu from specific point sources. This discrimination is possible using AS data due to the contrast between ^{238}Pu/$^{239+240}$Pu in stratospheric fallout (~0.025) versus Sellafield emissions (0.20–0.30). Baskaran et al. (1995) summarized the ^{238}Pu/$^{239+240}$Pu activity ratios in nuclear effluents from Sellafield and La Hague in dissolved and suspended particulate phases, surficial sediments, and terrestrial samples in the Arctic. Based upon a plot of the ^{238}Pu versus $^{239+240}$Pu activities for 82 surface sediments obtained from the Ob and Yenisey River deltas and the Kara Sea, Baskaran et al. (1996) obtained a slope of 0.034 ± 0.003, resembling the Northern Hemisphere fallout activity ratio, and concluded that there is virtually no detectable input from either the European effluents nor from the dumped nuclear reactors in the Kara Sea.

Masqué et al. (2003) measured $^{239+240}$Pu activities and ^{240}Pu/^{239}Pu atom ratios in bottom sediments from the Fram Strait of the Arctic Ocean; the low ^{240}Pu/^{239}Pu atom ratios were ascribed to transport of Pu from sources in the Kara Sea and Novaya Zemlya towards the North Atlantic by sea ice. Kershaw et al. (1999) used AS to demonstrate the transport of Sellafield-derived Pu in seawater samples from the northeast Atlantic. Along the northern Scottish coast, the highest activity sample contained 73 mBq/m^3 $^{239+240}$Pu and a Sellafield-dominated ^{238}Pu/$^{239+240}$Pu activity ratio of 0.19. Despite the long-range transport of Pu into seas well north of the Arctic Circle, Kershaw et al. (1999) concluded that the vast majority of Sellafield Pu and ^{241}Am reside in Irish Sea sediments.

In marine and coastal estuarine systems, $^{239+240}$Pu activities have been widely used to delineate sediment mixing processes, sediment inventories and export, and average sedimentation rates. Ravichandran et al. (1995) used $^{239+240}$Pu along with excess ^{210}Pb to investigate sediment deposition processes in the Sabine-Neches estuary of Texas, USA; the sediment mixing coefficients were determined using $^{239+240}$Pu activity profiles, which revealed that mixing rates were relatively small. The average sedimentation rates, determined via the location of the 1963 fallout activity maxima, were quite different among four cores, thus indicating substantial differences in sediment deposition processes at different locations within the estuary. Relatively low core inventories of $^{239+240}$Pu versus terrestrial deposition indicated the export of substantial quantities of Pu from the estuary to the continental shelf. The $^{239+240}$Pu activity profiles yielded better resolution of sedimentation rates than excess ^{210}Pb activity profiles (Ravichandran et al. 1995). Clearly, this and other examples illustrate the utility of Pu as a tracer of marine sedimentation processes in coastal environments.

One overarching conclusion that can be drawn from a synthesis of many different marine studies is that the water column inventories and isotope signatures of TRU in a specific location cannot necessarily be rationalized and anticipated as easily as is the case in the terrestrial environment. While the *initial* deposition at any given point may have stemmed from stratospheric fallout and/or close-in (tropospheric) debris, it is clear that oceanic currents play a strong role in transport and re-distribution of the initial inventory over vast distances within decadal timescales. Though post-depositional transport complicates the interpretation of spatially resolved data, the results can often be utilized advantageously to obtain new insights into complex marine transport and scavenging processes (Livingston and Povinec 2002).

20.3.4 Studies of Local/Regional TRU Sources in the Surficial Environment

Relatively limited inferences regarding sources and mixing processes can be inferred by solely considering activities, as surficial TRU activities can vary widely depending upon latitude, rainfall, topography, and erosion/deposition processes. It is commonplace to study the characteristics of specific point or local/regional sources and their mixing with ubiquitous stratospheric fallout through use of atom and/or activity ratios. Once again, Pu is the most prominently studied element, though various studies have considered Np and Am, and in some cases, Cm, alongside Pu data. Now that the use of mass spectrometry has become widespread in these studies, the most commonly interpreted discriminatory ratio is ^{240}Pu/^{239}Pu (refer to Fig. 20.2). Many sources that mix with stratospheric fallout have contrasting values of ^{240}Pu/^{239}Pu as well as other atom/activity ratios.

TRU with contrasting fingerprints is evident from a large variety of sources, though stratospheric fallout dominates globally. The 1950's US Marshall Islands tests also generated significant higher-ratio close-in fallout, which is prominently evident in the Pacific (refer to Sect. 20.3.3). Table 20.3 lists various prominent additional sources of TRU in the environment; in terms of a global mass balance, stratospheric fallout is by far the largest, accounting for ~4,000 kg of ^{239}Pu along with associated amounts of other TRU isotopes (Harley 1980). Among the other sources, accurate mass balances are not always known, though many of these probably amount to <10 kg ^{239}Pu. While the smaller quantities associated with many different local/regional sources (Table 20.3) may be of minor significance in the global sense, in local settings, these sources may overwhelm the inventories and surface activities of TRU from stratospheric fallout. Many of the releases shown in Table 20.3 consist of weapons-grade Pu with low ^{240}Pu/^{239}Pu ratios of 0.02–0.07, originating from low-yield tests (Trinity, NTS, Semipalatinsk-21, Lop Nor, Reggane, Maralinga, Mururoa/Fangataufa), weapons deployment (Nagasaki), weapons-grade Pu production reactors (Hanford, Savannah River, Mayak, and Sellafield/Windscale), or weapons component fabrication processes (Rocky Flats, Los Alamos, Krasnoyarsk). The low-yield tests produced fallout resembling the starting weapons-grade material because very little production of heavier isotopes occurred in the low neutron fluxes of these kiloton-range fission tests. In contrast, Pu from stratospheric fallout reflects a composite of many high-yield (megaton) tests where intense neutron fluxes generated significant quantities of heavier isotopes by neutron capture processes. Similarly, releases of high burn-up material from re-processing of power reactor fuel (Sellafield, La Hague, West Valley) or from power reactor accidents (Chernobyl) consists of TRU with enhanced abundances of heavier isotopes such as ^{240}Pu, ^{241}Pu, ^{242}Pu, and heavier elements such as Am and Cm not originally present in the fuels.

It is commonplace in isotope geochemistry to investigate mixing processes through a three-isotope, common denominator plot of C/A versus B/A, where A, B and C represent individual isotopes. These atom ratio mixing plots are readily used to pinpoint the mixing of two specific end-members; binary mixtures of the end-members alone plot along a "mixing line" segment, while third sources usually, but do not necessarily have to, appear as deviants from the mixing line. A well-known example is the use of the atom ratios ^{240}Pu/^{239}Pu, ^{241}Pu/^{239}Pu, ^{242}Pu/^{239}Pu, and ^{237}Np/^{239}Pu in investigations of mixing between stratospheric fallout and low-ratio regional fallout from the NTS (Kelley et al. 1999; Cizdziel et al. 2008). Besides having lower ^{240}Pu/^{239}Pu, the NTS regional source is also characterized by much lower ^{241}Pu/^{239}Pu, ^{242}Pu/^{239}Pu, and ^{237}Np/^{239}Pu. Kelley et al. (1999) used TIMS to measure all four of these ratios in a suite of soils from worldwide locations; for several samples from Nevada and Utah in the western US, mixed ratio fingerprints were evident, though ratios in samples from more distant US locations agreed well with the global stratospheric fallout

Table 20.3 Local/regional sources of TRU in the surficial environment

Source	Location	Type of release
Nevada Test Site	Nevada, USA	1
Trinity Site	New Mexico, USA	1
Semipalatinsk-21	Kurchatov, Kazakhstan	1
Lop Nor	Sinkiang, China	1
Maralinga	South Australia	1
Nagasaki	Nagasaki, Japan	1a
Reggane	French Sahara (now Algeria)	1
Mururoa/Fangataufa	French Polynesia	1
Chernaya Bay	Novaya Zemlya, Siberia, Russia	2
Amchitka Island	Alaska, USA	2
Mayak	Chelyabinsk, Russia	3
Hanford	Washington, USA	3
Savannah River Site	South Carolina, USA	3
Sellafield	Cumbria, UK	3, 4
La Hague	Normandy, France	4
West Valley	New York, USA	4
Los Alamos	New Mexico, USA	5
Krasnoyarsk Complex	Krasnoyarsk, Russia	5
Rocky Flats	Colorado, USA	5
Chernobyl	Chernobyl, Ukraine	6
Palomares	Spain	7
Thule	Greenland	7

1 = Atmospheric releases from above-ground, low-yield nuclear tests; local/regional deposition; resembles weapons-grade Pu. 1a = low-yield deployed weapon; localized deposition resembling weapons-grade Pu. 2 = Underwater, low-yield nuclear tests; local/regional deposition in marine environment; resembles weapons-grade Pu. 3 = Plutonium production reactors for weapons programs; atmospheric and/or aquatic releases resembling weapons-grade Pu. 4 = Atmospheric and/or aquatic releases from re-processing of spent nuclear fuel from commercial or research-grade power reactors; higher burn-up Pu containing heavier isotopes. 5 = Industrial facilities for Pu metallurgy and fabrication of weapons components; atmospheric and/or aquatic releases of material resembling weapons-grade Pu. 6 = Local/regional release of "hot" fuel particles from a 1986 reactor accident; high burn-up Pu with elevated abundances of heavier isotopes. 7 = Aircraft accidents resulting in the mechanical destruction of weapons and local dispersal of particulate Pu in the surface environment

signatures. The inclusion of ^{237}Np/^{239}Pu provides additional dimensionality in the data, as ^{237}Np is produced through either successive neutron capture of ^{235}U or by (n, 2n) processes of ^{238}U. Moreover, ^{237}Np/^{239}Pu is more sensitive to subtle differences in nuclear device construction (i.e., how much ^{235}U was used). The ^{237}Np/^{239}Pu fingerprint could also be influenced by different post-release chemical behavior of Np and Pu in environmental systems, and might be expected to vary significantly in discharges from nuclear fuel re-processing.

The ^{241}Pu/^{239}Pu ratio also generates interesting possibilities for chronometry; as ^{241}Pu decays with a half-life of 14.29 years, the ^{241}Pu/^{239}Pu ratio steadily decreases. In a setting where "aged" Pu is mixed with more recently separated Pu, the latter will exhibit much higher ^{241}Pu/^{239}Pu that is an obvious trend outlier from an established mixing process between two aged sources. Alternatively, similar inferences could be drawn through activity comparisons of the ^{241}Pu parent with its daughter, ^{241}Am (Pöllänen et al. 2006).

The powerful inferences available from these mixing plots can be seen in Fig. 20.8. Figure 20.8 depicts data from Kelley et al. (1999); all points shown represent samples from the continental US. The Eureka, Nevada location, labeled "1" in the top panel, reasonably approximates the NTS end-member, while the encircled group "5" consists of soils from more distant locations, comprised exclusively of stratospheric fallout. The points labeled 2, 3, and 4 from locations in Utah, California and Texas, respectively, contain binary mixtures of stratospheric fallout and NTS debris. The same inferences result from using any of these three ratios versus ^{240}Pu/^{239}Pu, though in other binary or ternary mixing situations ^{237}Np/^{239}Pu could behave differently than ^{241}Pu/^{239}Pu and ^{242}Pu/^{239}Pu.

For ^{241}Pu/^{239}Pu, the ratios are calculated for a reference date of 1 January 2000; a decade later, all of these ratios would plot along a similar line with 62% of the 2000 slope. Kelley et al. (1999) discuss mixing calculations for quantitative apportionment of binary sources using individual or pairs of ratios; obviously, the relative contribution of the two sources is a direct linear function of the position along the mixing line between the two end members.

Other published examples of two-component TRU source mixing have used two-component mixing diagrams in a similar manner. Kenna and Sayles (2002) used ^{237}Np/^{239}Pu versus ^{240}Pu/^{239}Pu to investigate mixing of multiple sources of TRU in sediments of the Ob River Delta in Siberia; in this case, ratios were observed that are not solely attributable to simple two-component mixing, as the samples do not all plot along a single line segment. In a study of mixing between stratospheric fallout and Chernobyl fallout Pu in northeastern Poland, Ketterer et al. (2004b) demonstrated a clear case of two-component mixing between these sources on a plot of ^{241}Pu/^{239}Pu versus ^{240}Pu/^{239}Pu; the inferences agreed well with source apportionment previously conducted using ^{238}Pu/$^{239+240}$Pu activity ratios (Mietelski and Was 1995).

The requisite measurements of ^{237}Np/^{239}Pu, ^{241}Pu/^{239}Pu and ^{242}Pu/^{239}Pu by mass spectrometry each present analytical challenges, particularly in small-mass samples and where complete dissolution of all silicates must be performed. For ^{237}Np/^{239}Pu, the atom ratios are reasonably high and present no great difficulty, aside from the limitation that a ^{236}Np spike is not available. As time progresses, more and more ^{241}Pu is decaying away, and ^{241}Pu/^{239}Pu measurements will become impractical for scenarios such as Fig. 20.8 in succeeding decades. Satisfactory measurements of ^{241}Pu/^{239}Pu will be possible long into the future, however, for situations like the

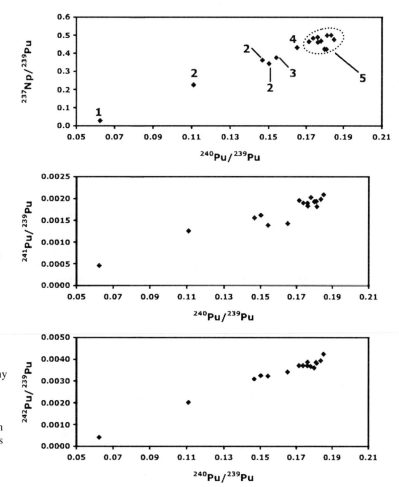

Fig. 20.8 Common-denominator mixing plots of ^{237}Np/^{239}Pu, ^{241}Pu/^{239}Pu, and ^{242}Pu/^{239}Pu versus ^{240}Pu/^{239}Pu, demonstrating two-component mixing between tropospheric fallout from the NTS, and stratospheric fallout. Point 1 (Eureka, Nevada soil) is a reasonable approximation of the NTS end-member. Group 5 represents soils from distant locations in the continental US lacking any NTS influence; isotope ratios from these samples are congruent with stratospheric fallout in other northern Hemisphere locations. Points 2, 3, and 4 are soils from Utah, California and Texas; these samples are described by two-component mixing between NTS and stratospheric fallout. Source: Kelley et al. (1999)

Chernobyl-stratospheric fallout mixing in Poland (Ketterer et al. 2004b), as the Chernobyl source term presently still contains ~3 atom percent ^{241}Pu.

20.4 Future Directions

The number and scope of studies of environmental TRU has rapidly expanded over the past decade. An important change has been that many more individuals and academic institutions are now involved in a field that has long been studied mainly among government organizations such as the US Department of Energy's National Labs. The widespread availability of MS (particularly ICPMS) has helped to stimulate much of this new interest. The use of $^{239+240}$Pu activities and inventories in studies of recent Earth processes, such as erosion and sedimentation, has become routine via MS, while previously the use of Pu this context would have been totally impractical with AS. In this light, $^{239+240}$Pu will probably replace ^{137}Cs as the predominant tracer of these processes.

Most researchers studying TRU sources in the environment will probably favor MS techniques (particularly ICPMS), though AS will remain an important technique for specific measurements such as ^{238}Pu, ^{241}Am, and Cm isotopes. The role of MS in developing a better understanding of Pu distribution and transport in the oceans is quite pronounced (Lindahl et al. 2010). Mass spectrometry already is well-established and an essential tool in nuclear forensics; the role of MS in these endeavors can only be expected to expand. The aforementioned application areas are by no means, fully explored or saturated, and many opportunities for new tracing applications exist that could be tapped with the present well-developed measurement technology. Without question, the upcoming years and decades will generate many exciting and unanticipated new applications of TRU in studies of earth and environmental processes.

Acknowledgments The authors thank J.W. Mietelski for providing the alpha spectrum shown in Fig. 20.3. The authors are indebted to J.W. Mietelski, T.C. Kenna, one anonymous reviewer, and the editorship of M. Baskaran for constructive criticisms of the manuscript. MEK acknowledges ICPMS instrumentation support from Intel Corp., NSF MRI Award No. CHE0118604, and the State of Arizona Technology Research and Innovation Fund. MEK also owes thanks for nearly two decades of rich and productive interactions with many students and collaborators, and is gratefully indebted to JAK for perspective and inspiration towards facing apparent difficulties in life.

References

Agarande M, Benzoubir S, Bouisset P et al (2001) Determination of ^{241}Am in sediments by isotope dilution high resolution inductively coupled plasma mass spectrometry (ID HR ICP-MS). Appl Radiat Isot 55:161–165

Baskaran M, Santschi PH (2002) Particulate and dissolved ^{210}Pb activities in the shelf and slope regions of the Gulf of Mexico waters. Continent Shelf Res 22:1493–1510

Baskaran M, Asbill S, Santschi P et al (1995) Distribution of ^{239,240}Pu and ^{238}Pu concentrations in sediments from the Ob and Yenisey Rivers and the Kara Sea. Appl Radiat Isot 46:1109–1119

Baskaran M, Asbill S, Santschi P et al (1996) Pu, ^{137}Cs and excess ^{210}Pb in Russian Arctic sediments. Earth Planet Sci Lett 140:243–257

Baskaran M, Hong G-H, Santschi PH (2009) Radionuclide analysis in seawater. In: Wurl O (ed) Practical guidelines for the analysis of seawater. CRC, Boca Raton, FL

Beck HL, Bennett BG (2002) Historical overview of atmospheric nuclear weapons testing and estimates of fallout in the continental United States. Health Phys 82:591–608

Becker JS, Zoriy M, Halicz L et al (2004) Environmental monitoring of plutonium at ultratrace level in natural water (Sea of Galilee – Israel) by ICP-SFMS and MC-ICP-MS. J Anal At Spectrom 19:1–6

Bennett BG (2002) Worldwide dispersion and deposition of radionuclides produced in atmospheric tests. Health Phys 82:644–655

Bertine KK, Chow TJ, Koide M et al (1986) Plutonium isotopes in the environment: some existing problems and some new ocean results. J Environ Radioact 3:189–201

Boulyga SF, Zoriy M, Ketterer ME et al (2003) Depth profiling of Pu, ^{241}Am and ^{137}Cs in soils from southern Belarus measured by ICP-MS and a and γ spectrometry. J Environ Monit 5:661–666

Bowen VT, Noshkin VE, Livingston HD et al (1980) Fallout radionuclides in the Pacific Ocean: vertical and horizontal distributions, largely from GEOSEC stations. Earth Planet Sci Lett 49:411–434

Buesseler KO (1996) The isotopic signature of fallout plutonium in the north Pacific. J Environ Radioact 36:69–83

Carter MW, Moghissi AA (1977) Three decades of nuclear testing. Health Phys 33:55–71

Chen Q, Dahlgaard H, Nielsen SP et al (2002) ^{242}Pu as tracer for simultaneous determination of ^{237}Np and ^{239,240}Pu in environmental samples. J Radioanal Nucl Chem 253: 451–458

Chiappini R, Pointurier F, Millies-Lacroix JC et al (1999) ^{240}Pu/^{239}Pu isotopic ratios and $^{239+240}$Pu total measurements in surface and deep waters around Mururoa and Fangataufa atolls compared with Rangiroa atoll (French Polynesia). Sci Total Environ 237(238):269–276

Choppin GR (2007) Actinide speciation in the environment. J Radioanal Nucl Chem 273:695–703

Cizdziel JV, Ketterer ME, Farmer D et al (2008) ^{239,240,241}Pu fingerprinting of plutonium in western US soils using ICPMS: solution and laser ablation measurements. Anal Bioanal Chem 390:521–530

Curtis D, Fabryka-Martin J, Dixon P et al (1999) Nature's uncommon elements: plutonium and technetium. Geochim Cosmochim Acta 63:275–285

Dong W, Zheng J, Guo QJ et al (2010) Characterization of plutonium in deep-sea sediments of the Sulu and South China Seas. J Environ Radioact 101:622–629

Everett SE, Tims SG, Hancock GJ et al (2008) Comparison of Pu and ^{137}Cs as tracers of soil and sediment transport in a terrestrial environment. J Environ Radioact 99:383–393

Fifield LK (2008) Accelerator mass spectrometry of the actinides. Quat Geochron 3:276–290

Godoy MLDP, Godoy JM, Roldão LA (2007) Application of ICP-QMS for the determination of plutonium in environmental samples for safeguards purposes. J Environ Radioact 97:124–136

Hardy EP, Volchok HL, Livingston HD et al (1980) Time pattern of off-site plutonium deposition from Rocky Flats plant by lake sediment analyses. Environ Int 4:21–30

Harley JH (1980) Plutonium in the environment – a review. J Radiat Res 21:83–104

Holm E, Roos P, Aarkrog A et al (2002) Curium isotopes in Chernobyl fallout. J Radioanal Nucl Chem 252:211–214

Hong G-H (2011) Applications of anthropogenic radionuclides as tracers to investigate marine environmental processes. In: Baskaran M (ed) Handbook of environmental isotope geochemistry. Springer, Heidelberg

Hotchkis MAC, Child D, Fink D et al (2000) Measurement of ^{236}U in environmental media. Nucl Instrum Meth Phys Res B 172:659–665

Huh CA, Su CC (2004) Distribution of fallout radionuclides (^{7}Be, ^{137}Cs, ^{210}Pb and ^{239,240}Pu) in soils of Taiwan. J Environ Radioact 77:87–100

Jaakkola T, Tolonen K, Huttunen P et al (1983) The use of fallout ^{137}Cs and ^{239,240}Pu for dating of lake sediments. Hydrobiology 103:15–19

Kaste JM, Heimsath AM, Hohmann M (2006) Quantifying sediment transport across an undisturbed prairie landscape using cesium-137 and high resolution topography. Geomorpholgy 76:430–440

Kelley JM, Bond LA, Beasley TM (1999) Global distribution of Pu isotopes and ^{237}Np. Sci Total Environ 237(238):483–500

Kenna TC (2002) Determination of plutonium isotopes and neptunium-237 in environmental samples by inductively coupled plasma mass spectrometry with total sample dissolution. J Anal At Spectrom 17:1471–1479

Kenna TC, Sayles FL (2002) The distribution and history of nuclear weapons related contamination in sediments from the Ob River, Siberia as determined by isotopic ratios of plutonium and neptunium. J Environ Radioact 60:105–137

Kershaw PJ, McCubbin D, Leonard KS (1999) Continuing contamination of north Atlantic and Arctic waters by Sellafield radionuclides. Sci Total Environ 237(238):119–132

Ketterer ME, Szechenyi SC (2008) Determination of plutonium and other transuranic elements by inductively coupled plasma mass spectrometry: a historical perspective and new frontiers in the environmental sciences. Spectrochim Acta B 63:719–737

Ketterer ME, Watson BR, Matisoff G et al (2002) Rapid dating of recent aquatic sediments using Pu activities and ^{240}Pu/^{239}Pu as determined by quadrupole inductively coupled plasma mass spectrometry. Environ Sci Technol 36:1307–1311

Ketterer ME, Hafer KM, Jones VJ et al (2004a) Rapid dating of recent sediments in Loch Ness: inductively coupled plasma mass spectrometric measurements of global fallout plutonium. Sci Total Environ 322:221–229

Ketterer ME, Hafer KM, Mietelski JW (2004b) Resolving Chernobyl vs. global fallout contributions in soils from Poland using Plutonium atom ratios measured by inductively coupled plasma mass spectrometry. J Environ Radioact 73:183–201

Kim CK, Kim CS, Chang BU et al (2004) Plutonium isotopes in seas around the Korean Peninsula. Sci Total Environ 318:197–209

Kim CS, Kim CK, Martin P et al (2007) Determination of plutonium concentrations and isotope ratio by inductively coupled plasma mass spectrometry: a review of analytical methodology. J Anal At Spectrom 22:827–841

Koide M, Bertine KK, Chow TJ et al (1985) The ^{240}Pu/^{239}Pu ratio, a potential geochronometer. Earth Planet Sci Lett 72:1–8

Krey PW, Leifer R, Benson WK et al (1979) Atmospheric burn-up of the Cosmos-954 reactor. Science 205:583–585

Lariviere D, Taylor VF, Evans RD et al (2006) Radionuclide determination in environmental samples by inductively coupled plasma mass spectrometry. Spectrochim Acta B 61:877–904

Lee T, Teh-Lung K, Hsiao-Ling L et al (1993) First detection of fallout Cs-135 and potential applications of ^{137}Cs/^{135}Cs ratios. Geochim Cosmochim Acta 57:3493–3497

Lee S, Huh C, Su C et al (2004) Sedimentation in the Southern Okinawa Trough: enhanced particle scavenging and teleconnection between the Equatorial Pacific and western Pacific margins. Deep Sea Res I 51:1769–1780

Liao H, Zheng J, Wu F et al (2008) Determination of plutonium isotopes in freshwater lake sediments by sector-field ICP-MS after separation using ion-exchange chromatography. Appl Radiat Isot 66:1138–1145

Lindahl P, Lee S, Worsfold P et al (2010) Plutonium isotopes as tracers for ocean processes: a review. Mar Environ Res 69:73–84

Livingston HD, Povinec PP (2002) A millennium perspective on the contribution of global fallout radionuclides to ocean science. Health Phys 82:656–668

Mackereth FJH (1969) A short core sampler for sub-aqueous deposits. Limnol Oceanogr 14:145–151

Masqué P, Cochran JK, Hebbeln D et al (2003) The role of sea ice in the fate of contaminants in the Arctic Ocean: plutonium atom ratios in the Fram Strait. Environ Sci Technol 37:4848–4854

Matisoff G, Whiting PJ (2011) Measuring soil erosion rates using natural (7Be, 210Pb) and anthropogenic (137Cs, 239,240Pu) radionuclides. In: Baskaran M (ed) Handbook of environmental isotope geochemistry. Springer, Heidelberg

Maxwell SL III (2008) Rapid method for determination of plutonium, americium and curium in large soil samples. J Radioanal Nucl Chem 275:395–402

Mietelski JW, Was B (1995) Plutonium from Chernobyl in Poland. Appl Radiat Isot 46:1203–1211

Mietelski JW, Was B (1997) Americium, curium and rare earths radionuclides in forest litter samples from Poland. Appl Radiat Isot 48:705–713

Mietelski JW, Dorda J, Was B (1999) Pu-241 in samples of forest soil from Poland. Appl Radiat Isot 51:435–447

Miller AJ, Kuehl SA (2009) Shelf sedimentation on a tectonically active margin: a modern sediment budget for poverty continental shelf, New Zealand. Mar Geol 270:175–187

Montero PR, Sanchez AM (2001) Plutonium contamination from accidental release or simply fallout: study of soils at Palomares (Spain). J Environ Radioact 55:157–165

Morris K, Butterworth JC, Livens FR (2000) Evidence for the remobilization of Sellafield waste radionuclides in an intertidal salt Marsh, West Cumbria, U.K. Estuar Coast Shelf Sci 51:613–625

Muramatsu Y, Hamilton T, Uchida S et al (2001) Measurement of $^{240}Pu/^{239}Pu$ isotopic ratios in soils from the Marshall Islands using ICP-MS. Sci Total Environ 278:151–159

Myers WA, Lindner M (1971) Precise determination of the natural abundance of ^{237}Np and ^{239}Pu in Katanga pitchblende. J Inorg Nucl Chem 33:3233–3238

Nelson DM, Lovett MB (1978) Oxidation state of plutonium in the Irish Sea. Nature 276:599–601

Nesje A (1992) A piston corer for lacustrine and marine sediments. Arctic Alpine Res 24:257–259

Nunnemann M, Erdmann N, Hasse H-U et al (1998) Trace analysis of plutonium in environmental samples by resonance ionization mass spectroscopy (RIMS). J Alloy Comp 271–273:45–48

Nygren U, Rodushkin I, Nilsson C et al (2003) Separation of plutonium from soil and sediment prior to determination by inductively coupled plasma mass spectrometry. J Anal At Spectrom 18:1426–1434

Ofan A, Ahmad I, Greene JP et al (2006) Development of a detection method for ^{244}Pu by resonance ionization mass spectrometry. New Astron Rev 50:640–643

Olivier S, Bajo S, Fifield LK et al (2004) Plutonium from global fallout recorded in an ice core from the Belukha Glacier, Siberian Altai. Environ Sci Technol 38:6507–6512

Pöllänen R, Ketterer ME, Lehto S et al (2006) Multi-technique characterization of a nuclear bomb particle from the Palomares accident. J Environ Radioact 90:15–28

Povinec PP, Badie C, Baeza A et al (2002) Certified reference material for radionuclides in seawater IAEA-381 (Irish Sea Water). J Radioanal Nucl Chem 251:369–374

Ravichandran M, Baskaran M, Santschi PH et al (1995) Geochronology of sediments in the Sabine-Neches estuary, Texas, USA. Chem Geol 125:291–306

Reynolds RL, Mordecai JS, Rosenbaum JG et al (2010) Compositional changes in sediments of subalpine lakes, Uinta Mountains (Utah): evidence for the effects of human activity on atmospheric dust inputs. J Paleolimnol 44:161–175

Ritchie JC, McHenry JR (1990) Application of radioactive fallout cesium-137 for measuring soil erosion and sediment accumulation rates and patterns: a review. J Environ Qual 19:215–233

Roy JC, Turcotte J, Cote JE et al (1981) The detection of the 21st Chinese nuclear explosion in eastern Canada. Health Phys 41:449–454

Sam AK, Ahamed MMO, Khangi FE et al (2000) Plutonium isotopes in sediments from the Sudanese coast of the Red Sea. J Radioanal Nucl Chem 245:411–414

Sanders CJ, Smoak JM, Sanders LM et al (2010) Intertidal mangrove mudflat $^{240+239}Pu$ signatures, confirming a ^{210}Pb geochronology on the southeastern coast of Brazil. J Radioanal Nucl Chem 283:593–596

Schiff CJ, Kaufman DS, Wallace KL et al (2010) An improved proximal tephrochronology for Redoubt Volcano, Alaska. J Volcanol Geoth Res 193:203–214

Schneider DL, Livingston HD (1984) Measurement of curium in marine samples. Nucl Instrum Meth Phys Res 223:510–516

Sholkovitz ER (1983) The geochemistry of Pu in fresh and marine water environments. Earth Sci Rev 19:95–161

Sill CW (1975) Some problems in measuring plutonium in the environment. Health Phys 29:619–626

Skipperud L (2004) Plutonium in the arctic marine environment – a short review. Sci World J 4:460–481

Taylor DM (1995) Environmental plutonium in humans. Appl Radiat Isot 46:1245–1252

Taylor DM (2001) Environmental plutonium – creation of the universe to twenty-first century mankind. In: Kudo A (ed) Plutonium in the environment. Elsevier Science, Amsterdam

Taylor RN, Warneke T, Milton JA et al (2001) Plutonium isotope ratio analysis at femtogram to nanogram levels by multicollector ICP-MS. J Anal At Spectrom 16:279–284

Thakkar AH (2002) A rapid sequential separation of actinides using Eichrom's extraction chromatographic material. J Radioanal Nucl Chem 252:215–218

Tims SG, Pan SM, Zhang R (2010) Plutonium AMS measurements in Yangtze River estuary sediment. Nucl Instrum Meth Phys Res B 268:1155–1158

Ulsh B, Rademacher S, Whicker FW (2000) Variations of ^{137}Cs depositions and soil concentrations between alpine and montane soils in northern Colorado. J Environ Radioact 47:57–70

Vajda N, Kim C (2010) Determination of Pu isotopes by alpha spectrometry: a review of analytical methodology. J Radioanal Nucl Chem 283:203–223

Van Pelt RS, Ketterer ME, Zobeck T et al. Anthropogenic radioisotopes to estimate rates of soil redistribution by wind (manuscript in preparation)

Wallner C, Faestermann T, Gerstmann U et al (2004) Supernova produced and anthropogenic 244Pu in deep sea manganese encrustations. New Astron Rev 48:145–150

Warwick PE, Croudace IW, Carpenter R (1996) Review of analytical techniques for the determination of americium-241 in soils and sediments. Appl Radiat Isot 47:627–642

Yamada M, Aono T (2002) Large particle flux of $^{239+240}Pu$ on the continental margin of the East China Sea. Sci Total Environ 287:97–105

Yamada M, Zheng J (2010) Temporal variability of $^{240}Pu/^{239}Pu$ atom ratio and $^{239+240}Pu$ inventory in water columns of the Japan Sea. Sci Total Environ 408:5951–5957

Yamada M, Zheng J, Wang Z (2006) ^{137}Cs, $^{239+240}Pu$ and $^{240}Pu/^{239}Pu$ atom ratios in the surface waters of the western

North Pacific Ocean, eastern Indian Ocean and their adjacent seas. Sci Total Environ 366:242–252

Zheng J, Yamada M (2004) Sediment core record of global fallout and bikini close-in fallout Pu in Sagami Bay, Western Northwest Pacific Margin. Environ Sci Technol 38:3498–3504

Zheng J, Yamada M (2005) Vertical distributions of $^{239+240}$Pu activities and ^{240}Pu/^{239}Pu atom ratios in sediment cores: implications for the sources of Pu in the Japan Sea. Sci Total Environ 340:199–211

Zheng J, Yamada M (2006a) Inductively coupled plasma-sector field mass spectrometry with a high-efficiency sample introduction system for the determination of Pu isotopes in settling particles at femtogram levels. Talanta 69:1246–1253

Zheng J, Yamada M (2006b) Determination of Pu isotopes in sediment cores in the Sea of Okhotsk and the NW Pacific by sector field ICP-MS. J Radioanal Nucl Chem 267:73–83

Zheng J, Yamada M (2006c) Plutonium isotopes in settling particles: transport and scavenging of Pu in the western northwest Pacific. Environ Sci Technol 40:4103–4108

Zheng J, Yamada M (2007) Precise determination of Pu isotopes in a seawater reference material using ID-SF-ICP-MS combined with two-stage anion-exchange chromatography. Anal Sci 23:611–615

Chapter 21
Tracing the Sources and Biogeochemical Cycling of Phosphorus in Aquatic Systems Using Isotopes of Oxygen in Phosphate

Adina Paytan and Karen McLaughlin

Abstract Phosphorous (P) is an essential nutrient for all living organisms and when available in surplus could cause eutrophication in aquatic systems. While P has only one stable isotope, P in most organic and inorganic P forms is strongly bonded to oxygen (O), which has three stable isotopes, providing a system to track phosphorus cycling and transformations using the stable isotopes of O in phosphate (PO_4), $\delta^{18}O_p$. This isotope system has only recently been utilized in aquatic environments. Available data obtained from different settings indicate that $\delta^{18}O_p$ of dissolved phosphate in aquatic systems can be applied successfully for identifying sources and cycling of phosphate in a broad range of environments. Specifically, work to date indicates that $\delta^{18}O_p$ is useful for deciphering sources of phosphate to aquatic systems if these sources have unique isotopic signatures and if phosphate cycling within the system is limited compared to input fluxes. In addition, because various processes are associated with distinct fractionation effects, the $\delta^{18}O_p$ tracer can be utilized to determine the degree of phosphorous cycling within the biomass and shed light on the processes imprinting the isotopic signatures. As a better understanding of the systematics of and various controls on $\delta^{18}O_p$ is gained, it is expected that $\delta^{18}O_p$ would be extensively applied in research geared to understand phosphorous dynamics in many environments.

A. Paytan (✉)
University of California, Santa Cruz, Santa Cruz, CA 95064, USA
e-mail: apaytan@ucsc.edu

K. McLaughlin
Southern California Coastal Water Research Project, Costa Mesa, CA 92626, USA

21.1 Introduction

Phosphorus (P, atomic number 15, relative atomic mass 30,9738) is a multivalent nonmetal element of the nitrogen group. Although 23 isotopes of phosphorus are known (all possibilities from ^{24}P up to ^{46}P), only ^{31}P is stable. Two radioactive isotopes of phosphorus have half-lives which make them useful for scientific experiments. ^{32}P has a half-life of 14.26 days and ^{33}P has a half-life of 25.34 days. Phosphorous compounds (organic and inorganic) are found with phosphorous oxidation states ranging from -3 to $+5$, however the most common oxidation states are $+5$, $+3$ and -3. Phosphorous abundance in earth's crust is 1,050 ppm by weight (730 ppm by moles) and the abundance in the solar system is 7 ppm by weight (300 ppb by moles) (Emsley 2000). Due to its high reactivity, phosphorus does not occur as a free element in nature, but it is found in many different minerals (e.g. apatite) and organic compounds (e.g. DNA, RNA, ATP, phospholipids) essential for all living cells. It is produced commercially from calcium phosphate (phosphate rock). Large deposits of phosphate rock are located in the Middle East, China, Russia, Morocco and the United States of America. Based on 2010 estimates, at the current rate of consumption, the supply of phosphorus is estimated to run out in about 300 years. Peak P consumption will occur in 30 years and reserves will be depleted in the next 50–100 years (Vaccari 2009).

Phosphorus, being an essential plant nutrient, is predominantly used as a constituent of fertilizers for agriculture. Phosphorus is also used as a precursor for various chemicals, in particular the herbicide glyphosate and to make organophosphorus compounds which have many applications, including in plasticizers, flame retardants, pesticides, extraction agents, and water treatment. It is an important component in steel production, utilized in the making of special glasses and fine china, a component in some laundry detergents, baking powder, matchbook strikers, flares, and for military use in incendiary bombs and grenades.

Phosphorus is a key element in all known forms of life. Inorganic phosphorus in the form of phosphate (PO_4^{3-}) plays a major role in biological molecules such as DNA and RNA where it forms part of the structural framework of these molecules. Living cells also use phosphate to transport cellular energy in the form of adenosine triphosphate (ATP). Nearly every cellular process that uses energy obtains it in the form of ATP. ATP is also important for phosphorylation, a key regulatory event in cells. Phospholipids are the main structural components of all cellular membranes and calcium phosphate salts assist in stiffening bones. Due to its biological role phosphorous is an essential macromineral (nutrient) for terrestrial plants and for marine phytoplankton, algae, and sea-grasses. In ecological terms, phosphorus is often a limiting nutrient in many environments; i.e. the availability of phosphorus governs the rate of growth of many organisms. Indeed, it has been suggested that phosphorous availability may limit primary productivity in some aquatic systems (Bothwell 1985; Hecky and Kilham 1988; Howarth 1988; Karl and Tien 1997; Karl et al. 2001; Krom et al. 1991; Wu et al. 2000), and may be co-limiting in others (Nicholson et al. 2006; Sundareshwar et al. 2003). However, at times an excess of phosphorus can be problematic causing eutrophication and algal blooms (Sharp 1991; Smith and Kalff 1983; Smith 1984).

Agricultural expansion over the next 50 years is expected to be accompanied by a 2.4- to 2.7-fold increase in nitrogen (N)- and phosphorus (P)-driven eutrophication of terrestrial, freshwater, and near-shore marine environments (Tilman et al. 2001). Much of the P from fertilizer and animal waste enters surface waters and eventually also groundwater (Carpenter et al. 1998) and these nutrient loads can stimulate large scale macroalgal and/or phytoplankton blooms in receiving waters (Beman et al. 2005; Rabalais et al. 2002). Phosphorus enrichment in aquatic systems can cause diverse problems such as harmful algal blooms, anoxia, fish kills, and loss of habitat and biodiversity (Carpenter et al. 1998; Tilman et al. 2001). Thus, identifying and understanding phosphorous input and cycling and the effects phosphorous limitation or enrichment may have on aquatic ecosystems are of critical importance to management and restoration efforts.

Phosphorous is continuously and rapidly cycled in aquatic environments. Figures 21.1 and 21.2 represent the global biogeochemical cycle of P in the ocean and lakes, respectively and Fig. 21.3 illustrates the various pools and processes involved in the P cycle in aquatic systems.

21.1.1 Stable Isotope Use to Study P Sources and Cycling

Monitoring P sources and transformations in natural environments using stable isotopes has been difficult to do because, in contrast to C, N, O and S, P has only one stable isotope (^{31}P) thus the use of P stable isotope tracing is not an option. Although radioactive P isotopes (^{32}P, ^{33}P) can and have been used for investigation of P transformations in aquatic systems (Benitez-Nelson and Buessler 1998, 1999; Benitez-Nelson and Karl 2002; Lal et al. 1988; Lal and Lee 1988; Lee et al. 1991) there are many complications involved with this procedure. The use of natural stable isotope signatures has advantages as this approach does not perturb the system (e.g. by adding phosphate) and integrates processes over longer time scales. While P has only one stable isotope, P in most organic and inorganic P forms is strongly bonded to oxygen (O), which has three stable isotopes, providing a system to track phosphorus cycling and transformations using the stable isotopes of O in phosphate ($\delta^{18}O_p$).

Since the pioneering study of Longinelli and Nuti (1973) and several subsequent publications (Fricke et al. 1998; Longinelli et al. 1976; Longinelli 1984; Luz et al. 1984; Luz and Kolodny 1985; Shemesh et al. 1983, 1988), oxygen isotope ratios of bioapatite in teeth and bones have been widely used as paleoenvironmental proxies. The oxygen isotope paleothermometer is based on an empirical equation that is assumed to represent equilibrium fractionations between phosphate and water as a function of temperature as follows (Longinelli and Nuti 1973):

$$T\ (^{\circ}C) = 111.4 - 4.3(\delta^{18}O_P - \delta^{18}O_w) \quad (21.1)$$

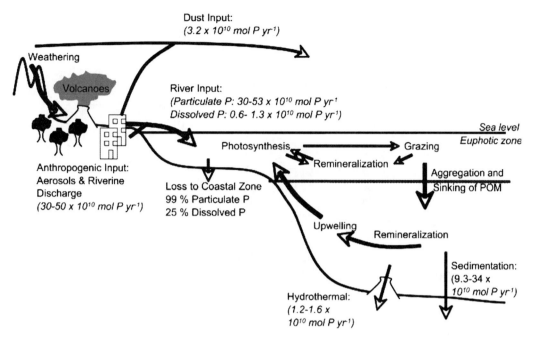

Fig. 21.1 The marine phosphorus cycle. Fluxes are given in italics. Flux data are from Benitez-Nelson (2000) and Follmi (1995). Continental weathering is the primary source of phosphorus to the oceanic phosphorus cycle. Most of this phosphorus is delivered via rivers with a smaller portion delivered via dust deposition. In recent times, anthropogenic sources of phosphorus have become a large fraction of the phosphorus delivered to the marine environment, effectively doubling the pre-anthropogenic flux. The primary sink for phosphorus in the marine environment is loss to the sediments. Much of the particulate flux from rivers is lost to sediments on the continental shelves, and a smaller portion is lost to deep-sea sediments. Hydrothermal systems constitute an additional small sink for P. Figure modified from Paytan and McLaughlin (2007)

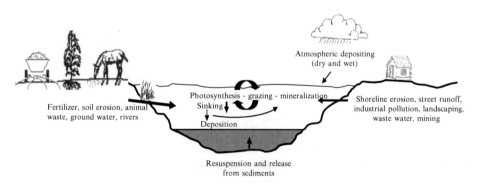

Fig. 21.2 Phosphate sources to lakes include fertilizers, animal waste, soil erosion, industrial and mining waste water input as well as atmospheric deposition. Phosphate enters lakes through rivers, groundwater, direct disposal and runoff. Plants and algae utilize the phosphate as a nutrient. Phosphate is transferred through the food web and some of this particulate matter is remineralized in the water column. Some phosphate is deposited in the sediment. Under anoxic conditions phosphate from the sediments may be recycled back into the water

where $\delta^{18}O_P$ and $\delta^{18}O_W$ are the oxygen isotopic composition of phosphate and water, respectively, in equilibrium with environmental temperature T (°C). Importantly, at most earth-surface temperatures (<80°C) and pressures, the P-O bond is resistant to inorganic hydrolysis and does not exchange O without biological mediation (Blake et al. 1997; Lecuyer et al. 1996) preserving the signature of the temperature and water isotope ratio of the solution from which the minerals precipitated.

However, due to the large sample size required for isotope analysis and the low concentrations of

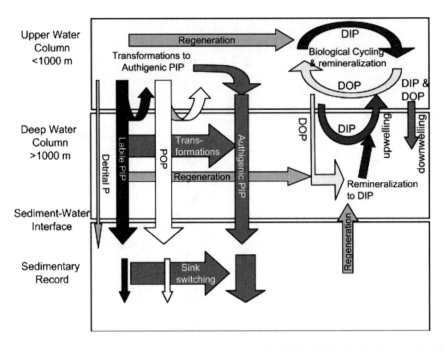

Fig. 21.3 Transformations between P pools in the water column and sediments. Abbreviations are as follows: PIP, particulate inorganic phosphorus; POP, particulate organic phosphorus; DIP, dissolved inorgranic phosphorus; DOP, dissolved organic phosphorus. Particulate phosphorus forms can undergo transformations throughout the water column and within sediments. Particulate phosphorus forms may also undergo regeneration into dissolved forms. Particulate phosphorus is lost from surface waters via sinking. Biological cycling and remineralization are the primary mechanisms of tranformations of the dissolved phases and are dominant in surface waters, though microbial remineralization continues at depth. Dissolved phosphorus forms are lost from surface waters via downwelling and biological uptake (into POP) and are returned to surface waters via upwelling and mixing. Regeneration form sediment can add more dissolved phosphate to deep water. Figure modified from Paytan and McLaughlin (2007)

dissolved phosphate in most water bodies, the oxygen isotopic composition of phosphate, $\delta^{18}O_p$, has only recently been applied systematically for tracking dissolved phosphate in water bodies. Pioneering work by Longinelli et al. (1976) found no variation in the $\delta^{18}O_p$ of dissolved phosphate in seawater with either depth or latitude in the Atlantic and Pacific Oceans, although there was a significant difference between the two ocean basins. The $\delta^{18}O_p$ values were thought to reflect kinetic–biological isotopic fractionation. Longinelli et al. (1976) extracted and concentrated P from large volumes of water without pre-filtration using iron-coated fibers that absorb inorganic and organic P indiscriminately. Analysis of mixed organic and inorganic P samples may have confounded interpretation of the results (Blake et al. 2005) and because of the analytical limitations few attempts to follow up on this work have been made for over a decade (Paytan 1989). These complications have been overcome with current technologies, several detailed protocols for isolating, purifying and precipitating small quantities of phosphate from complex matrix solutions such as fresh and ocean waters were published and this system has now been applied to various water bodies including oceans (Colman et al. 2005; McLaughlin et al. 2006b, 2011), estuaries (McLaughlin et al. 2006a, d) and lakes (Elsbury et al. 2009; Markel et al. 1994). In addition, extensive and innovative laboratory studies have been conducted to carefully determine the fractionation associated with various biogenic and abiotic transformations of P (Blake et al. 1997, 1998, 2001, 2005; Liang 2005; Liang and Blake 2006a, b, 2007, 2009).

21.1.2 Isotopic Signatures of Potential Phosphate Sources to Aquatic Systems

Identifying point and non-point nutrient sources is important for understanding ecosystem health, and has implications for designing best management practices, industry regulation and allocation of water discharge permits. P sources can be separated into point sources, such as sewage and industrial discharge sites, and non-point sources like urban and agricultural run-off (Young et al. 2009). Phosphate oxygen isotope tracer studies in natural environments are limited. However, recent field studies have demonstrated the utility of $\delta^{18}O_p$ as a tracer of various phosphate sources to lakes, rivers, estuaries and the coastal ocean (Coleman 2002; McLaughlin et al. 2006b, d). A wide range of $\delta^{18}O_p$ values from 6 to 27‰ has been documented in these various studies (Fig. 21.4a). A significant portion of these samples are not in isotopic equilibrium with the surrounding water, indicating that complete intracellular biological cycling of the orthophosphate had not taken place, and a source signature may have been partially retained. In addition, the $\delta^{18}O_p$ of some potential end-member sources (wastewater treatment plant effluent, fertilizers, soaps,

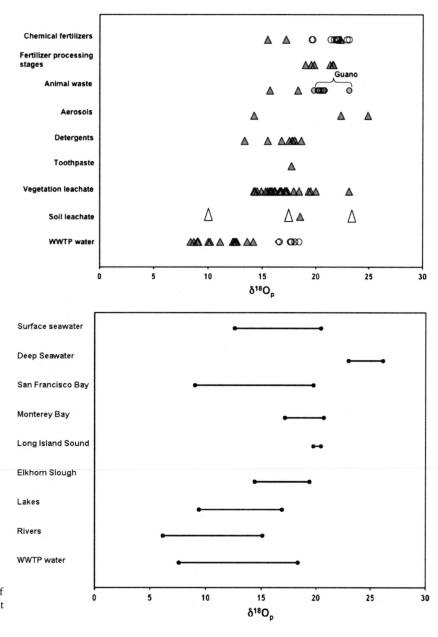

Fig. 21.4 (a) $\delta^{18}O_p$ of some potential end-member sources. *Full triangles* from Young et al. (2009); *Open circles* from Gruau et al. (2005); *Full circles* from Avliffe et al. (1992); *Diamonds* from Coleman (2002); *Empty triangles* from Zohar et al. (2010a, b). Figure modified from Young et al. (2009). (b) Range of $\delta^{18}O_p$ values observed in different water systems

soil extracts, etc.) has been published (Young et al. 2009) (Fig. 21.4b).

A considerable range of $\delta^{18}O_p$ values has been measured in various P sources and the differences observed among sources are much larger than the analytical precision (± 0.3‰) associated with this technique. Although there is considerable overlap in $\delta^{18}O_p$ measured in the various groups of samples, these results indicate that in specific geographic regions, different P source types may span a narrower range and have distinct signatures, and in these cases, the $\delta^{18}O_p$ could be useful for identifying the contribution of the different sources. For example, while the entire range of reported $\delta^{18}O_p$ values for worldwide wastewater treatment plant effluent overlaps with the values measured for multiple types of detergents, organic fertilizers, and chemical fertilizers, all measured $\delta^{18}O_p$ values for the Palo Alto Regional Water Quality Control Plant are significantly lower than any of the measured fertilizers and detergents (Young et al. 2009). Thus, if phosphate is not heavily cycled within an ecosystem such that the source signature is reset, $\delta^{18}O_p$ can be used to identify isotopically distinct phosphate sources and/or the extent of phosphate cycling in aquatic systems (i.e. the deviation from the isotopic composition of the source towards the expected equilibrium value).

21.1.3 Isotope Fractionations Involved in P Cycling

Isotope fractionations associated with several of the important reactions and transformations operating in the P cycle have been determined in controlled laboratory experiments (Table 21.1). This information provides the basis for interpretation of isotope data ($\delta^{18}O_p$) obtained from phosphate in the natural environment. In the absence of biological activity at ambient temperatures, pH, and pressure, isotope exchange between phosphate oxygen and water (or other solutions) is slow and can be considered negligible for the time scales of concern of most environmental applications (Blake et al. 1997; Longinelli and Nuti 1973, Longinelli et al. 1976; O'Neil et al. 2003). Studies of precipitation

Table 21.1 Isotope fractionation effects associated with various biogeochemical processes

Process	Fractionation (Δ or ε)	Reference
Precipitation/dissolution of P minerals (apatite)	+0.7‰ to +1‰ Heavy isotope in mineral phase	Blake et al. (1997)
Adsorption/desorption of P to/from mineral surfaces	~+1‰ Heavy isotope in mineral phase	Liang and Blake (2007)
Precipitation with sesquioxides and hydroxides	~+1‰ Heavy isotope in mineral phase	Jaisi et al. (2009)
Abiotic hydrolysis of polyphosphate (O:P = 3.33), pyrophosphate (O:P = 3.5), phosphonates (O:P = 3.0), monoesters (O:P = 3.0) and diesters (O:P = 2.0)	No fractionation or temperature effect, however incorporation of oxygen from water during formation of PO_4 (O:P = 4) occurs	McLaughlin et al. (2006a)
Transport by water or air	No fractionation or temperature effect	Longinelli (1965)
Assimilation by phytoplankton	Light isotopes preferentially utilized, enrichment of the residual solution ($\varepsilon = -3$‰)	Blake et al. (2005)
Intracellular processing such as inorganic pyrophosphatase (PPase) catalysis	Equilibrium isotopic exchange T and $\delta^{18}O_w$ impact (21.2)	Blake et al. (2005)
Alkaline phosphatase (APase) hydrolization of phosphomonoesterase (extracellular)	Kinetic isotope effects $\varepsilon = -30$‰ effecting only the newly incorporated oxygen	Liang and Blake (2006a, b)
5′-nucleotidase hydrolization (extracellular)	Kinetic isotope effects $\varepsilon = -10$‰ effecting only the newly incorporated oxygen	Liang and Blake (2006a, b)
First step of DNAse hydrolization	Kinetic isotope effects $\varepsilon = -20$‰ effecting only the newly incorporated oxygen	Liang and Blake (2009)
First step of RNAse hydrolization	Kinetic isotope effects $\varepsilon = +20$‰ effecting only the newly incorporated oxygen	Liang and Blake (2009)
Transport from roots to leaves (by transporters)	Enrichment in the process foliage heavier than roots	

and dissolution of various P bearing minerals and studies of P adsorption and desorption onto/from mineral surfaces indicate that the fractionation associated with these processes (given equilibration time of more than a few hours) is small – in the range of 1‰ (Jaisi et al. 2009; Liang 2005; Liang and Blake 2006b). Typically the heavier isotopes in these reactions are associated with the mineral phase while the solution retains phosphate with lighter isotopes. Precipitation or dissolution of apatite minerals (inorganically) will be accompanied by a small oxygen isotope fractionation in the range of +0.7 ‰ to +1‰ (Blake et al. 1997). Similarly, adsorption or precipitation with sesquioxides and hydroxides imprints a small positive isotope effect (Jaisi et al. 2009). In contrast, enzyme mediated biological activity could break the P-O bond in processes that involve large isotopic fractionation. Intracellular as well as extracellular enzymes are expressed by various organisms for the uptake and utilization of P and may play a role in determining the oxygen isotopic composition of phosphate in aquatic systems. Different enzymatic processes induce different isotopic fractionations (Table 21.1). The most dominant enzymatic process controlling $\delta^{18}O_p$ in the environment is the intracellular activity of pyrophosphatase (PPase) (Blake et al. 2005), which involves equilibrium isotopic exchange. Blake et al. (2005) found that this enzymatic activity results in isotopic equilibrium of oxygen in phosphate similar to that described by Longinelli and Nuti (1973). The equation for phosphate extracted from microbial cultures was described by Blake et al. (1997):

$$T(°C) = 155.8 - -6.4(\delta^{18}O_p - \delta^{18}O_w) \quad (21.2)$$

These equilibrium relations have been observed in tissues of a variety of organisms, including fish, mammals (Kolodny et al. 1983), bacteria and algae (Blake et al. 1997, 2005; Paytan et al. 2002). Results of an algae culture experiment indicate that intracellular oxygen isotope exchange between phosphorus compounds and water within cells is very rapid (Paytan et al. 2002). These processes are expected to occur in all organisms and phosphate released from cells to the environment will carry this equilibrium signature and impact dissolved phosphate $\delta^{18}O_p$ values leading to equilibrium values. Extracellular remineralization and hydrolization of organic P (P_o) compounds by phosphohydrolase enzymes such as alkaline phosphatase (APase) and 5′-nucleotidase, involves incorporation of one or more oxygen atoms from the ambient water with an isotope fractionation of −30 and −10‰, respectively (Liang and Blake 2006b). A summary of published fractionation values to date is given in Table 21.1. The resulting phosphate from such processes will reflect the fractionation and would typically shift $\delta^{18}O_p$ towards values that are lower than equilibrium. Work by several groups is currently ongoing to determine the isotope fractionation associated with additional enzymes, and will enable better interpretation of field data. Uptake and utilization (assimilation) of phosphate by aquatic plants, algae, and microorganisms is also associated with isotope fractionation. The phosphate with lighter isotopes is preferentially utilized, a process that could enrich the residual solution with phosphate that has heavy isotopes (Blake et al. 2005).

The isotopic composition of dissolved phosphate and particularly the degree of isotope equilibrium or deviation from equilibrium of phosphate in various aquatic systems has been used for deciphering the extent of biological utilization and turnover of phosphate in aquatic systems (Colman et al. 2005; Elsbury et al. 2009; McLaughlin et al. 2006b, d, 2011). This application is based on the assumption that extensive recycling and turnover will lead to isotopic equilibrium while deviation from equilibrium may reflect source signatures or other processes that do not result in isotopic equilibrium such as expression of extracellular enzymes or phosphate uptake (Fig. 21.5). The following sections will describe the methodology (sample preparation and analysis), give examples of application of this system in various settings and address the needs for future progress in this field.

21.2 Materials and Methods

For analysis of $\delta^{18}O_P$ by isotope ratio mass spectrometry (IRMS), it is necessary to convert the phosphate into a pure solid phase without isotopic alteration. The purification steps are of great importance, since the presence of oxygen sources other than phosphate compromises the results (Weidemann-Bidlack et al. 2008). The final compound analyzed should be non-hygroscopic, stable under laboratory conditions, and should decompose to form carbon monoxide (CO) at temperatures attainable in a lab furnace. Silver phosphate (Ag_3PO_4) has been proven a convenient phase for this purpose

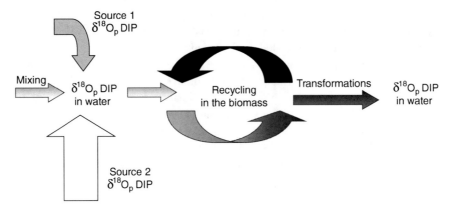

Fig. 21.5 A graphical representation of P mixing and cycling in the water illustrating the utility of $\delta^{18}O_p$ for identifying sources if biological transformations do not erase the source signatures or the degree of intracellular biological cycling and turnover by determining the difference between the source signature and expected equilibrium values (see McLaughlin et al. 2006a, b, c, d)

(Firsching 1961; O'Neil et al. 1994) and has gradually substituted the earlier hazardous fluorination technique (Kolodny et al. 1983; Tudge 1960). Ag_3PO_4 is reduced with carbon in an oxygen free atmosphere at high temperature (>1,300°C) using a thermal combustion elemental analyzer (TCEA) to yield carbon monoxide for analysis by IRMS (Coleman 2002, McLaughlin et al. 2004, O'Neil et al. 1994). The TCEA and mass spectrometer are linked via a continuous flow interface, and the CO gas is measured instantaneously after formation (Kornexl et al. 1999). Prior to mass spectrometric analysis phosphate has to be concentrated (phosphate concentrations are low in many environments), isolated, purified, and precipitated as Ag_3PO_4.

Several detailed protocols for isolating, purifying and precipitating small quantities of phosphate from complex matrix solutions such as fresh and ocean waters have been published (Coleman 2002; Goldhammer et al. 2011; Jaisi and Blake 2010, McLaughlin et al. 2004; Tamburini et al. 2010, Weidemann-Bidlack et al. 2008; Zohar et al. 2010a). Most of these procedures involve a concentration step to collect sufficient amounts of phosphate and remove some of the dissolved organic phosphate and interfering ions from the sample. This is done through a series of precipitations and/or resin treatments followed by a final precipitation as Ag_3PO_4 (Table 21.2). It is important to ensure that the concentration and preparation process does not introduce any isotopic fractionation and all of the above methods report that authentic signatures are preserved. Problems with the final precipitation of silver phosphate have been experienced when working with water samples containing very high concentrations of dissolved organic matter. Several promising approaches for addressing this problem have been explored, including UV radiation of the sample (Liang and Blake 2006b), passing the sample through phosphate-free activated carbon (Gruau et al. 2005), using resins such as DAX-8 to remove organics (Tamburini et al. 2010), precipitation of humic acids (Zohar et al. 2010a) and treatment with H_2O_2 (Goldhammer et al. 2011; Zohar et al. 2010a). Published procedures report that these methods to remove organic matter retain the original isotopic signature of phosphate.

As mentioned above it is very important that the only source of oxygen analyzed (as Ag_3PO_4) originates from the "authentic" phosphate in the sample. There are however two separate processes that may compromise this requirement. If not all of the organic matter is removed or if other minerals that contain oxygen (such as $AgNO_3$) precipitate along with the Ag_3PO_4 (e.g. the Ag_3PO_4 is not pure), then the oxygen contributing to the CO gas will not reflect that of phosphate. Data has to be monitored to ensure that this does not occur. This is done by monitoring the oxygen yield (peak area compared to the pure silver phosphate standards) expected based on the weight of the Ag_3PO_4 sample. The oxygen content per unit weight of Ag_3PO_4 is 15.3% and samples which deviate from this value particularly towards higher oxygen yield should be suspected of contamination. Plotting the oxygen yield (or peak area) of analyzed pure Ag_3PO_4 standards along with the samples should yield a linear relation with weight (Fig. 21.6). It is also advised to include a step to remove tightly sorbed

Table 21.2 Published procedures for the concentration and purification of phosphate from water samples and the precipitation of Ag_3PO_4 for analysis of $\delta^{18}O_p$

McLaughlin et al. (2004, 2006a, b, c, d); (Elsbury et al. 2009); (Young et al. 2009)	Colman (2002); Colman et al. (2005); Goldhammer et al. (2011)	Tamburini et al. (2010); Tudge (1960); Kolodny et al. (1983); Paytan et al. (2002); Liang and Blake (2007)
Magnesium-induced coprecipitation (MagIC, Karl and Tien 1992)	Magnesium-induced coprecipitation (MagIC, Karl and Tien 1992)	Magnesium-induced coprecipitation (MagIC, Karl and Tien 1992)
Dissolution in acetic and nitric acids and buffering at pH 5.5 with 1M potassium acetate	Dissolution in 0.1 M HNO_3	Dissolution in 1 M HCl
Precipitation as cerium phosphate	Anion removal (AG1X8) in $NaHCO_3$ form	Precipitation as ammonium phosphomolybdate
Rinses to remove chloride	HCO_3 removal in acid	Dissolution in citric-acid NH_4OH
Dissolution in 0.2 M nitric acid	Cation removal (AG50X8)	Precipitation of magnesium ammonium phosphate
Cation removal (AG50X8)	Volume reduction by evaporation at 60°C	Rinse and dissolve in 0.5 M HNO_3
Ag_3PO_4 fast precipitation	Ag_3PO_4 slow micro precipitation in $P:Ag:NO_3:NH_4OH$ molar ratios of 1:10:30:75	Cation removal (AG50X8)
		Ag_3PO_4 slow precipitation in $P:Ag:NO_3:NH_4OH$ molar ratios of 1:100:300:750

Note that in water rich in dissolved organic matter (DOM) a step to remove DOM either from the water before the MagIC step or right after dissolution of the magnesium hydroxide is needed. This could be done be repeat MagIC co-precipitation (Goldhammer et al. 2011), DAX-8 Amberlite resin (Tamburini et al. 2010), activated char (Gruau et al. 2005), or precipitation (Zohar et al. 2010a, b)

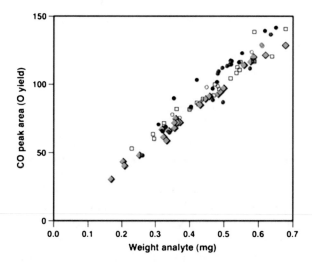

Fig. 21.6 CO peak area for silver phosphate standards (*green diamonds*) and various samples (*other symbols*) relative to sample weight introduced into the mass spectrometer. The expected oxygen yield from pure silver phosphate is 15.3%. If samples fall off the line defined by the standards the sample is likely contaminated by an external source of oxygen and might not represent the oxygen isotope ratios in phosphate. Figure modified from Tamburini et al. (2010)

water molecules from silver phosphate. This can be done by heating the silver phosphate samples to ~450°C to get strongly adsorbed water off.

Another potential process by which data could be compromised is contribution of phosphate which is hydrolyzed from condensed forms or organic forms of phosphate for which the O:P ratio is less than 4 during sample processing (this is an analytical artifact) (McLaughlin et al. 2006c). Using ^{18}O-labled and unlabeled reagents on replicates of the same sample these artifacts could be monitored and corrected. If hydrolysis takes place, oxygen from the acid solution is incorporated into the phosphate group, and because the phosphate in the labeled acid solution will have a higher isotope value than phosphate in the unlabeled solution it could be tracked (McLaughlin et al. 2006c). In this case, the use of a simple equation allows the correction and determination of the isotope value of the extracted phosphate (McLaughlin et al. 2006c).

While all of the concentration, purification, separation and precipitation methods published (Table 21.2) were tested for this potential artifact and report that any impact, if exists, is below analytical error, it is important to note that because of the vast array of organic P compounds in nature and the huge variability in their concentration and relative abundance in different environmental samples each new set/type of samples should be tested to ensure that such artifacts do not compromise the data.

For mass spectrometric analysis about 200–600 μg of Ag_3PO_4 should be weighed into silver capsules. Some laboratories also add a small amount of finely powdered glassy carbon or nickel-carbide, to improve the reaction between the silver phosphate and carbon during pyrolysis. The samples are introduced into the TCEA via a zero blank autosampler. The TCEA furnace is kept at a constant and consistent temperature (1,375 and 1,450°C have been used). The furnace itself consists of a ceramic tube filled with glassy carbon chips encased in a glassy carbon tube. The produced reaction gases are carried by constantly flushing with a high purity helium stream through a GC column held at fixed temperature (e.g. 80°C) to purify the sample from trace contaminants. The gas is admitted to the mass-spectrometer via a Conflow interface. Some systems also include a copper tube which removes oxygen from the helium carrier gas. The ion currents of masses m/z 28, 29 and 30 are registered on the Faraday cups and converted to $\delta^{18}O$ values relative to a carbon monoxide standard gas for which $\delta^{18}O$ has been calculated relative to SMOW. Each sample is run for 300 s with a CO reference peak preceding the sample peak.

Calibration and corrections for instrumental drifts are accomplished by repeated measurements of internal standards. The standard deviation of the analysis based on repeated measures of the standards is typically less than ±0.4‰. In order to capture instrumental drift with time, delta value linearity, and sample size variability, working standards with known $\delta^{18}O$ values are weighed out in a range of sizes and analyzed along with the samples during each run (for example at ten sample increments). Raw $\delta^{18}O$ values are then corrected to the range of standards for drift and off set of the delta values and sample-size linearity.

The oxygen isotopic composition of phosphate is reported in standard delta notation ($\delta^{18}O$), which is calculated using the following equation:

$$\delta^{18}O = \left[\frac{R_{sample}}{R_{VSMOW}} - 1\right] \times 1,000 \quad (21.3)$$

where R_{sample} is the ratio of $^{18}O/^{16}O$ in a sample and R_{VSMOW} is the ratio of $^{18}O/^{16}O$ in the isotopic standard for O, Vienna Standard Mean Ocean Water (VSMOW).

Currently there are no certified international Ag_3PO_4 standards and various laboratories use different "home-made" internal standards for which the $\delta^{18}O_p$ has been determined via fluorination (McLaughlin et al. 2004; Vennemann et al. 2002).

21.3 Applications

The use of $\delta^{18}O_p$ of dissolved inorganic phosphate (DIP) to study phosphate sources and cycling is relatively new and it is not yet widely used. In the past decade it has been applied in a variety of aquatic systems including estuaries, coastal water, lakes, rivers, and the open ocean. A brief summary of representative examples is given below. These examples demonstrate the great utility of this system and it is likely that now that the methodology has been worked out extensive application of this tool will take place.

21.3.1 Use of $\delta^{18}O_p$ as a Tracer for Phosphate Sources is Estuaries

In a study of North San Francisco Bay, McLaughlin et al. (2006d) used $\delta^{18}O_p$ to assess mixing of dissolved inorganic phosphate (DIP) sources along an estuarine flow path. Due to different sources of phosphate, temperatures, and $\delta^{18}O_w$ the $\delta^{18}O_p$ signatures of oceanic and riverine phosphate sources are distinct. Based on salinity and $\delta^{18}O_w$, waters in the North San Francisco Bay can be described as a two end-member mixing system between Pacific Ocean waters and the freshwaters of the San Joaquin and Sacramento Rivers (Ingram et al. 1996; McLaughlin et al. 2006d). This mixing can be used to calculate an expected mixing line for $\delta^{18}O_p$. Such a trend will be observed if phosphate in the Bay is not being cycled extensively through the biomass or affected by processes that may alter the source $\delta^{18}O_p$ signatures. Deviations from the mixing-line are observed and attributed to contribution of phosphate with unique $\delta^{18}O_p$ signatures at various locations along the estuary (from point and non-point sources) such as the discharge points of tributaries or wastewater treatment plants inputs.

The general lack of isotopic equilibrium in DIP throughout the Bay indicates that phosphate cycling is not rapid compared to phosphate input (low utilization rate, short residence time), and that source $\delta^{18}O_p$ contributed to the observed signature at most, if not all, stations. The deviations from the $\delta^{18}O_p$ mixing model have been interpreted to represent inputs of phosphate from local point sources within the North Bay (e.g. at the Napa River confluence) (Fig. 21.7).

At another estuary, Elkhorn Slough (McLaughlin et al. 2006a), the phosphate $\delta^{18}O_p$ within the main channel also indicates variability in phosphate sources throughout the channel, which are related to the surrounding land use. Trends in $\delta^{18}O_p$ show high values near the mouth reflecting phosphate of an oceanic origin, a minimum value near Hummingbird Island in the central slough reflecting phosphate input from groundwater, and high values near the head of the slough reflecting fertilizer input. A clear change in the relative contribution of these sources is observed and linked to water mixing during changing tidal conations at the mouth of the Slough.

In these studies, McLaughlin et al. (2006a, d) demonstrated that it is possible to use $\delta^{18}O_p$ to identify point and non-point source phosphate inputs to aquatic systems and suggest that this may be applied in other impacted systems to identify specific anthropogenic sources, such as fertilizer and sewage phosphate, or to trace natural sources of phosphate. This information is crucial for mitigation of pollution impacts and successful restoration of estuaries and other aquatic systems.

21.3.2 Phosphate Sources and Cycling in Lakes

Phosphorous loading in freshwater lakes has been identified as one of the leading causes for eutrophication and thus linked to hypoxia, harmful algal blooms and other adverse impacts (Schlesinger 1991; Sharp 1991). Despite the clear and wide spread impacts of phosphate loading, relatively few studies have used $\delta^{18}O_p$ to track sources and learn about P cycling in lakes. A study by Markel et al. (1994) focuses on sediments and suspended matter in Lake Kinneret, Israel. The isotope data show that about 70% of the particulate phosphate come to the lake from a basaltic source ($\delta^{18}O_p = 6‰$) with the balance being of sedimentary/anthropogenic origin ($\delta^{18}O_p = 18–25‰$).

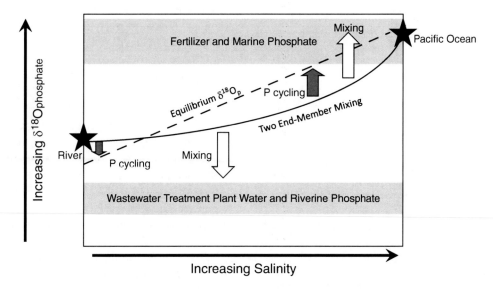

Fig. 21.7 Diagram indicating two end-member mixing (*black line*) and the expected equilibrium line (*dashed line*). Values below both the two end-member and the equilibrium line (*white* down facing *arrows*) indicate mixing with either riverine or wastewater treatment plant effluent. Deviations which move off the two end-member mixing line in the direction of equilibrium line be indicative of phosphate cycling, though they may also represent mixing with fertilizer phosphate. Deviations which fall off the two-end member mixing line in the direction of equilibrium, but in excess of equilibrium are indicative of mixing with fertilizer phosphate or treatment plant effluent depending on location along the salinity gradient

This study also alludes to some internal cycling (precipitation and dissolution) of phosphate in the lake. A study by Elsbury et al. (2009) records the distribution of $\delta^{18}O_p$ in water samples from the western and central basins of Lake Erie along with several potential sources (rivers, waste water treatment plants, atmospheric deposition). $\delta^{18}O_p$ of lake water is largely out of equilibrium with ambient conditions, indicating that source signatures may be discerned. $\delta^{18}O_p$ values in the lake range from +10 to +17‰, whereas the equilibrium value is expected to be around +14‰ and riverine weighted average $\delta^{18}O_p$ value is +11‰ (Fig. 21.8). Therefore, they conclude that some of the lake $\delta^{18}O_p$ values could not be explained by any known source or process. This indicates that there must be one or more as yet uncharacterized source(s) of phosphate with a high $\delta^{18}O_p$ value. In this study the authors speculate that a likely source may be the release of phosphate from sediments under reducing conditions that are created during anoxic events in the hypolimnion of the central basin of Lake Erie.

21.3.3 Phosphate Sources and Cycling in Riverine Systems

The range of potential $\delta^{18}O_p$ values for DIP in riverine systems is much greater than the range expected for open-ocean and coastal waters due to the wider range of temperatures, $\delta^{18}O$ water values, and phosphate sources found in riverine systems. Furthermore, land use patterns are thought to have a significant impact on nutrient stoichiometry and concentrations in riverine environments (Harris 2001; Lehrter 2006; Neill et al. 2001), thus, differences in land use could provide unique $\delta^{18}O_p$ signatures with which to trace the relative influence of specific sources to receiving waters. Common phosphate sources for rivers include wastewater treatment effluent, agricultural and urban runoff, manure, leaking septic systems, and natural rock and soil weathering. In addition, river discharge can be viewed as a source of phosphate in relation to other systems; for example, tributaries entering larger rivers, lakes, estuaries, or coastal waters. Although the $\delta^{18}O_p$ of river water will usually be controlled by a complex combination of source inputs, if the $\delta^{18}O_p$ of the sources are known, this can be used to trace the phosphate as it moves down the river's flow path. There are not many data sets for $\delta^{18}O_p$ of rivers but the few that are available report values that do not represent equilibrium and thus most likely reflect changes in source contribution along the river flow path (Fig. 21.9).

Water samples collected from the San Joaquin River (SJR), a hypereutrophic river in the major agricultural region of the California Central Valley, span a range $\delta^{18}O_p$ values greater than the analytical error, and only one sample fell along the expected equilibrium line. The samples do not show a consistent offset from equilibrium, indicating that the $\delta^{18}O_p$ at least partially reflects inputs of phosphate sources with different $\delta^{18}O_p$ signatures, rather than full biological cycling and complete oxygen exchange with water (Young et al. 2009).

Rivers flowing into Lake Erie range in isotope values from +10.5 to +15.2‰. These values range from 4.0‰ lower than the expected isotopic equilibrium to 1.0‰ higher, with two samples falling within range of the expected equilibrium (~ 14‰) and in general are lower than lake values (Elsbury et al. 2009). Tributaries to Lake Tahoe, CA, are also not at equilibrium with values from 8.2 to 12.2‰ (equilibrium ~11‰). While more research is needed, the

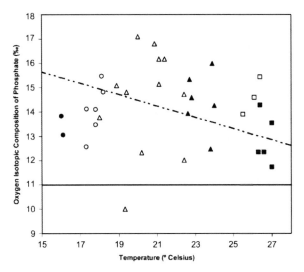

Fig. 21.8 Oxygen isotopic composition of phosphate ($\delta^{18}Op$) of Lake Erie surface water. Symbols refer to different sampling trips. The line at +11‰ represents the weighted average riverine $\delta^{18}Op$. The *dashed line* represents the expected $\delta^{18}Op$ value calculated based on the temperature for each sample and at the average lake surface $\delta^{18}Ow$ of −6.78‰ (standard deviation 0.3‰). Samples plotting between the river and equilibrium lines could be explained by P cycling a process that would tend to erase source signature and bring the $\delta^{18}Op$ values towards equilibrium. Lines above the equilibrium line suggest a source with $\delta^{18}Op$ higher than 17‰. Figure modified from Elsbury et al. (2009)

Fig. 21.9 Offset from isotopic equilibrium of various river samples. *Gray bar* represents the range of values that would be at equilibrium considering analytical error and calculation errors associated with determining the equilibrium value. Figure modified from Young et al. (2009)

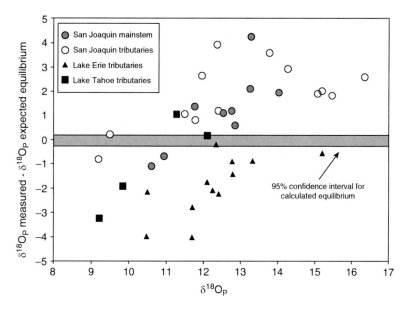

results of river studies in California and the Lake Erie area (Michigan and Ohio) demonstrate two important factors for using $\delta^{18}O_p$ as a source tracer in river systems. The $\delta^{18}O_p$ value of the majority of water samples are not in isotopic equilibrium, indicating that source signatures are not being rapidly overprinted by equilibrium signatures within the river, and in several instances, certain tributaries had $\delta^{18}O_p$ values that are distinct from those of other tributaries, indicating that the contribution of phosphate from specific tributaries to the receiving water body could be identified using this isotope tracing approach.

21.3.4 Phosphorous Cycling in a Coastal Setting

Phosphate in many coastal systems is not the limiting nutrient for productivity, yet is heavily utilized, thus it is expected that the source signature will be at least partially overprinted and that the $\delta^{18}O_p$ will shift towards equilibrium values. If this is indeed the case the degree of deviation from the source signature could be used as a measure of phosphate turnover rate relative to new phosphate input. This principle has been used in California coastal waters (Monterey Bay) (Fig. 21.10). In this system, $\delta^{18}O_p$ tracks seasonal changes in phosphate cycling through the biomass (e.g. phosphate utilization rates) with the greatest phosphate oxygen isotope exchange occurring during the upwelling season (McLaughlin et al. 2006b). Spatially the greatest percent of phosphate oxygen exchange, and thus the greatest phosphate utilization relative to input, occurs at the locus of upwelling. Episodes of higher phosphate turnover occurs simultaneously throughout the upper 200 m of the water column and on a broad spatial scale. $\delta^{18}O_p$ data also suggest that deep water (~500 m) may be a source of phosphate to the euphotic zone in Monterey Bay.

The degree of P cycling differs among different coastal systems. Colman (2002) concluded that the large deviations in $\delta^{18}O_p$ between riverine and coastal waters in the Long Island Sound reflects extensive equilibration with local coastal water and indicates that in this geographic area rapid microbial cycling overprints source $\delta^{18}O_p$ values on a timescale of weeks.

21.3.5 Phosphorous Cycling in Open Ocean Settings

Dissolved inorganic phosphorus (DIP) concentrations in the open ocean have a typical nutrient profile with low concentrations in surface water due to extensive uptake by primary producers and increasing concentration with depth resulting from regeneration of DIP from sinking particulate matter. The deep Pacific has higher DIP concentration than the deep Atlantic due to

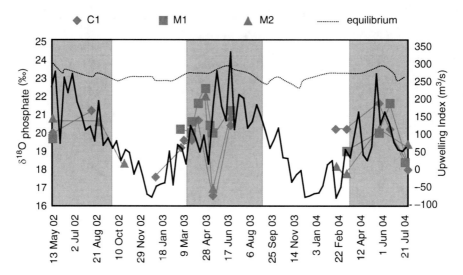

Fig. 21.10 Observed phosphate $\delta^{18}O_p$ variability from May 2002–August 2004. *Solid line* is the 10-d running mean of the NOAA upwelling index as a function of time; $\delta^{18}O_p$ is for samples collected at 10-m depth at three monitoring stations (C1, M1, M2); *dashed line* is the expected equilibrium phosphate $\delta^{18}O_p$; The $\delta^{18}O$ of the phosphate source from deep water upwelling is ~17‰. Data from McLaughlin et al. (2006a, b, c, d)

aging of the water along the circulation pathway (Broecker and Peng 1982). It would thus be expected that the $\delta^{18}O_p$ in open ocean waters be primarily a function of biological turnover with potentially some impact of circulation. Colman et al. (2005) measured the $\delta^{18}O_p$ depth distributions in the Atlantic and Pacific Oceans. At both basins $\delta^{18}O_p$ values were close to, but slightly offset from, the expected equilibrium values (calculated from equation (21.1) and the seawater temperature and $\delta^{18}O_w$). Because seawater values at intermediate depths approaches the equilibrium isotopic composition, intracellular cycling at these depths is suggested as the main process affecting the isotopic signatures. The offset at depth is attributed to differences between the deep water temperature and high latitude surface water temperatures, where DIP is equilibrated and transported along the circulation path (Colman et al. 2005) (Fig. 21.11).

In oligotrophic systems, such as the surface waters of the Sargasso Sea, DIP concentrations are extremely low. Consequently, P is thought to limit or co-limit primary productivity in this region. McLaughlin et al. (2011) investigated the biogeochemical cycling of P in the Sargasso Sea, utilizing multiple techniques including $\delta^{18}O_p$, alkaline phosphatase enzyme-labeled fluorescence (ELF), and ^{33}P uptake derived phosphate turnover rates. Results from these studies indicate that dissolved organic phosphorus (DOP) is utilized by phytoplankton and bacteria to supplement cellular requirements for this vital nutrient. They show that remineralization of the DOP pool is most extensive above the thermocline, as indicated by a large fraction of eukaryotes producing alkaline phosphatase, rapid phosphorus turnover times, and a large deviation from equilibrium of $\delta^{18}O_p$ towards lighter values. These data suggest that DOP remineralization by extracellular enzymes is prevalent and that DOP can account for up to 60% of P utilized and support a corresponding amount of primary production. Below the thermocline, alkaline phosphatase expression is reduced, turnover times increase, and $\delta^{18}O_p$ values approach equilibrium, all of which are indicative of intracellular phosphate cycling and slower turnover of the DOP pool. This study highlights the importance of bioavailable organic P to primary productivity in oligotrophic systems and has implications for the global carbon cycle.

21.4 Future Directions

For a more rigorous interpretation of $\delta^{18}O_p$ data from environmental samples several gaps in our understanding of how phosphate oxygen is fractionated in aquatic systems must be addressed. Particularly, characterization of the isotopic fractionation of phosphate oxygen associated with additional processes including

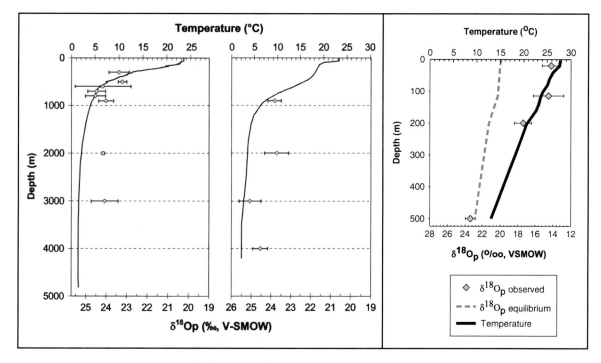

Fig. 21.11 *Left panel* represented the depth profile of $\delta^{18}O_p$ in the Pacific (**a**) and Atlantic (**b**). *Solid lines* represent the temperature depended equilibrium values and *open circles* are measured values. *Error bars* represent 95% confidence intervals based on replicate mass spectrometric analyses of single samples. Note the approach to equilibrium values at intermediate depth and off sets in the deep ocean (modified from Colman et al. 2005). On the right is a depth profile in the upper 500 m of the oligotrophic Sargasso Sea (data from McLaughlin et al. 2011). *Circles* are measured values and the *solid line* represents the expected equilibrium $\delta^{18}O_p$ calculate based on the oxygen isotope value of seawater and the temperature at the respective depth using the equation for equilibrium relation

those associated with different enzymes as well as inorganic processes (desorption from particles and sediment regeneration). Specifically, there is a dearth of data on the fractionation associated with freshwater periphyton (soft algae and diatoms) and freshwater heterotrophic bacteria. Research has suggested that bacteria are superior competitors for phosphate in aquatic systems compared to phytoplankton (Currie and Kalff 1984); however, differences in isotopic fractionation associated with bacterial cycling of phosphorus compared to algal cycling have not been fully defined. More research is needed to understand how various organisms fractionate phosphate oxygen under a variety of temperature and phosphorus concentration regimes.

Isotopic fractionation associated with sorption onto particulate matter and in co-precipitation of phosphate with various minerals must also be further explored. Phosphate interactions with sediments and co-precipitates in lakes and streams have been found to be an important factor in controlling the dissolved phosphate pool (Fox 1989; House 2003). Such effects are assumed to be negligible in most systems but could potentially play a role in hardwater systems where co-precipitation of phosphate can result in the removal of up to 30% of the dissolved P pool (House 2003). Finally, fractionation associated with remineralization and sedimentary fluxes also needs to be defined.

Procedures for the analysis of oxygen isotopes in organic phosphate compounds have not been fully tested. The only published procedure is of McLaughlin et al. (2006c) in which an $\delta^{18}O$ enriched isotope spike is used to correct for incorporation of reagent oxygen during hydrolysis of organic phosphate. While data presented in this paper is promising, fractionation effects associated with hydrolysis have not been fully evaluated and work on natural samples is limited. The signatures of dissolved organic phosphate compounds and plant material may be important yet these values are mostly unknown.

The database characterizing source signatures is also relatively limited. Specifically, groundwater, atmospheric deposition and agriculture and urban runoff which are known sources to many aquatic systems have not been measured and only limited information regarding isotope signatures of phosphate regenerated from sediments is available. Similarly the observation that waste water treatment effluents from different locations and treatment plants have different $\delta^{18}O_p$ values warrants further work to determine how specific treatment protocols impact the isotope ratio. Thus a more extensive data base at a wide range of locations and settings is needed.

In addition, it is vital that a certified international silver phosphate standard be prepared, characterized and distributed to a wider scientific community. This would permit comparison of results among laboratories and will enable laboratories to establish better QA/QC protocols.

Finally, it could be interesting to extend the utility of this isotope system to other areas of environmental research including environmental forensics, climate research, and to study phosphorous cycling in vegetation and soils. Indeed a few attempts to move in that direction have taken place (Jaisi and Blake 2010; Tamburini et al. 2010; Zohar et al. 2010b). The development is rapid and it is expected that this system will see a great expansion in application to a broad range of problems in the near future.

Acknowledgments The authors would like to thank Megan Young, Steve Silva, and Carol Kendall at the USGS Menlo Park, CA, and Federica Tamburini from ETH, Zurich for sharing their experience and knowledge of sample preparation and analyses of oxygen isotopes in phosphate.

References

Ayliffe LK, Veeh HH, Chivas AR (1992) Oxygen isotopes of phosphate and the origin of island apatite deposits. Earth Planet Sci Lett 108(1–3): 119–129
Beman JM, Arrigo KR, Matson PA (2005) Agricultural runoff fuels large phytoplankton blooms in vulnerable areas of the ocean. Nature 434:211–214
Benitez-Nelson CR (2000) The Biogeochemical Cycling of Phosphorus in Marine Systems. Earth Sci. Rev., 51, 109–135
Benitez-Nelson CR, Buessler KO (1998) Measurement of cosmogenic ^{32}P and ^{33}P activities in rainwater and seawater. Analytical Chem 70:64–72
Benitez-Nelson CR, Buesseler KO (1999) Variability of inorganic and organic phosphorus turnover rates in the coastal ocean. Nature 398:502–505
Benitez-Nelson CR, Karl DM (2002) Phosphorus cycling in the North Pacific Subtropical Gyre using cosmogenic 32P and 33P. Limnol Oceanogr 47:762–770
Blake RE, O'Neil JR, Garcia GA (1997) Oxygen isotope systematics of biologically mediated reactions of phosphate: I. Microbial degradation of organophosphorus compounds. Geochim Cosmochim Acta 61:4411–4422
Blake RE, O'Neil JR, Garcia GA (1998) Effects of microbial activity on the $\delta^{18}O$ of dissolved inorganic phosphate and textural features of synthetic apatites. Am Mineralog 83:1516–1531
Blake RE, Alt JC, Martin AM (2001) Oxygen isotope rations of PO4: an inorganic indicator of enzymatic activity and P metabolism and a new biomarker in the search for life. Proc Natl Acad Sci USA 98:2148–2153
Blake RE, O'Ncil JR, Surkov AV (2005) Biogeochemical cycling of phosphorus: insights from oxygen isotope effects of phosphoenzymes. Am J Sci 305:596–620
Bothwell ML (1985) Phosphorus limitation of lotic periphyton growth rates: an intersite comparison using continuous-flow troughs (Thompson River System, British Columbia). Limnol Oceanogr 30:527–542
Broecker WS, Peng TH (1982) Tracers in the Sea. Columbia University, Lamont-Doherty Geological Observatory
Carpenter SR, Caraco NF, Correl DL et al (1998) Nonpoint pollution of surface waters with phosphorus and nitrogen. Ecol Appl 8:559–568
Coleman A (2002) The oxygen isotope composition of dissolved inorganic phosphate and the marine phosphorus cycle. Geology and Geophysics, Yale
Colman A (2002) The oxygen isotope composition of dissolved inorganic phosphate and the marine phosphorus cycle. Geology and Geophysics, Yale
Colman AS, Blake RE, Karl DM et al (2005) Marine phosphate oxygen isotopes and organic matter remineralization in the oceans. Proc Natl Acad Sci USA 102:13023–13028
Currie DJ, Kalff J (1984) A comparison of the abilities of freshwater algae and bacteria to acquire and retain phosphorus. Limnol Oceanogr 29:298–310
Elsbury KE, Paytan A, Ostrom NE, Kendall C, Young MB, McLaughlin K, Rollog ME, and Watson S (2009) Using Oxygen Isotopes of Phosphate To Trace Phosphorus Sources and Cycling in Lake Erie. Environmental Science and Technology 43:3108–3114
Emsley J (2000) The Shocking History of Phosphorus. Macmillan, London
Firsching FH (1961) Precipitation of silver phosphate from homogeneous solution. Analytical Chem 33:873–874
Föllmi KB (1995) 160 m.y. record of marine sedimentary phosphorus burial: Coupling of climate and continental weathering under greenhouse and icehouse conditions: Geology 23:859–862
Fox LE (1989) Model for inorganic control of phosphate concentrations in river waters. Geochim Cosmochim Acta 53:417–428
Fricke HC, Clyde WC, O'Neil JR et al (1998) Evidence for rapid climate change in North America during the latest Paleocene thermal maximum: oxygen isotope compositions of biogenic

phosphate from the Bighorn Basin (Wyoming). Earth Planet Sci Lett 160:193–208

Goldhammer T, Max T, Brunner B et al (2011) Marine sediment pore-water profiles of phosphate oxygen isotopes using a refined micro-extraction technique. Limno. Oceanogr Methods (submitted)

Gruau G, Legeas M, Riou C et al (2005) The oxygen isotopic composition of dissolved anthropogenic phosphates: a new tool for eutrophication research? Water Res 39:232–238

Harris GP (2001) Biogeochemistry of nitrogen and phosphorus in Australian catchments, rivers and estuaries: effects of land use and flow regulation and comparisons with global patterns. Mar Freshwater Res 52:139–149

Hecky RE, Kilham P (1988) Nutrient limitation of phytoplankton in freshwater and marine environments: a review of recent evidence on the effects of enrichment. Limnol Oceanogr 33:796–822

House WA (2003) Geochemical cycling of phosphorus in rivers. Appl Geochem 18:739–748

Howarth RW (1988) Nutrient limitation of primary production in marine ecosystems. Annu Rev Ecol Syst 19:89–110

Ingram BL, Conrad ME, Ingle JC (1996) Stable isotope and salinity systematics in estuarine waters and carbonates: San Francisco Bay. Geochim Cosmochim Acta 60:455–467

Jaisi DP, Blake RE, Kukkadapu RK (2009) Fractionation of oxygen isotopes in phosphate during its interactions with iron oxides. Geochim Cosmochim Acta 74:1309–1319

Jaisi DP, Blake RE (2010) Tracing sources and cycling of phosphate in Peru Margin sediments using oxygen isotopes in authigenic and detrital phosphates. Geochim Cosmochim Acta 74:3199–3212

Karl DM, Tien G (1992) MAGIC: A sensitive and precise method for measuring dissolved phosphorus in aquatic environments. Limnol and Oceanogr 37:105–116

Karl DM, Tien G (1997) Temporal variability in dissolved phosphorus concentrations in the subtropical North Pacific Ocean. Mar Chem 56:77–96

Karl DM, Bjorkman KM, Dore JE et al (2001) Ecological nitrogen-to-phosphorus stoichiometry at station ALOHA. Deep-Sea Res II 48:1529–1566

Kolodny Y, Luz B, Navon O (1983) Oxygen isotope variations in phosphate of biogenic apatites, I. Fish bone apatite – rechecking the rules of the game. Earth Planet Sci Lett 64:398–404

Kornexl BE, Gehre M, Hofling R et al (1999) On-line $\delta^{18}O$ measurement of organic and inorganic substances. Rapid Comm Mass Spect 13:1685–1693

Krom MD, Kress N, Brenner S et al (1991) Phosphorus limitation of primary productivity in the eastern Mediterranean Sea. Limnol Oceanogr 36:424–432

Lal D, Chung Y, Platt T et al (1988) Twin cosmogenic radiotracer studies of phosphorus recycling and chemical fluxes in the upper ocean. Limnol Oceanogr 33:1559–1567

Lal D, Lee T (1988) Cosmogenic ^{32}P and ^{33}P uses as tracers to study phosphorus recycling in the upper ocean. Nature 333:752–754

Lecuyer C, Grandjean P, Emig CC (1996) Determination of oxygen isotope fractionation between water and phosphate from living lingulids: potential application to palaeoenvironmental studies. Paleogeogr Paleoclim Paleoecol 126:101–108

Lee T, Barg E, Lal D (1991) Studies of vertical mixing in the Southern California Bight with cosmogenic radionuclides ^{32}P and ^{7}Be. Limnol Oceanogr 36:1044–1053

Lehrter JC (2006) Effects of land use and. land cover, stream discharge, and interannual climate on the magnitude and timing of nitrogen, phosphorus, and organic carbon concentrations in three coastal plain watersheds. Water Environ Res 78:2356–2368

Liang Y (2005) Oxygen isotope studies of biogeochemical cycling of phosphorus. Ph.D. Thesis. Department of Geology and Geophysics, Yale University, New Haven, CT, USA

Liang Y, Blake RE (2006a) Oxygen isotope composition of phosphate in organic compounds: isotope effects of extraction methods. Organic Geochem 37:1263–1277

Liang Y, Blake RE (2006b) Oxygen isotope signature of Pi regeneration from organic compounds by phosphomonoesterases and photooxidation. Geochim Cosmochim Acta 70:3957–3969

Liang Y, Blake RE (2007) Oxygen isotope fractionation between apatite and aqueous-phase phosphate: 20–45°C. Chem Geol 238:121–133

Liang Y, Blake RE (2009) Compound- and enzyme-specific phosphodiester hydrolysis mechanisms revealed by $\delta^{18}O$ of dissolved inorganic phosphate: implications for the marine P cycling. Geochim Cosmochim Acta 73:3782–3794

Longinelli A (1965) Oxygen isotopic composition of orthophosphate from shells of living marine organisms. Nature 207:716–719

Longinelli A (1984) Oxygen isotopes in mammal bone phosphate: a new tool for paleohydrological and paleoclimatological research. Geochim Cosmochim Acta 48:385–390

Longinelli A, Bartelloni M, Cortecci G (1976) The isotopic cycle of oceanic phosphate: I. Earth Planet Sci Lett 32:389–392

Longinelli A, Nuti S (1973) Revised phosphate-water isotopic temperature scale. Earth Planet Sci Lett 19:373–376

Luz B, Kolodny Y, Kovach J (1984) Oxygen isotope variations in phosphate of biogenic apatites III. Conodonts. Earth Planet Sci Lett 69:255–262

Luz B, Kolodny Y (1985) Oxygen isotope variation in phosphate of biogenic apatites IV. Mammal teeth and bones. Earth Planet Sci Lett 75:29–36

Markel D, Kolodny Y, Luz B et al (1994) Phosphorus cycling and phosphorus sources in Lake Kinneret: tracing by oxygen isotopes in phosphate. Isr J Earth Sci 43:165–178

McLaughlin K, Kendall C, Silva S et al (2004) A precise method for the analysis of d18O of dissolved inorganic phosphate in seawater. Limnol Oceanogr Methods 2:202–212

McLaughlin K, Cade-Menun BJ, Paytan A (2006a) The oxygen isotopic composition of phosphate in Elkhorn Slough. A tracer for phosphate sources. Estuarine, Coastal and Shelf Science, California. doi:10.1016/j.ecss.2006.06.030

McLaughlin K, Chavez FP, Pennington JT et al (2006b) A time series investigation of the oxygen isotopic composition of dissolved inorganic phosphate in Monterey Bay. Limnol Oceanogr 51:2370–2379

McLaughlin K, Kendall C, Silva SR et al (2006c) Oxygen isotopes of phosphatic compounds – application for marine particulate matter, sediments and soils. Mar Chem 98:148–155

McLaughlin K, Paytan A, Kendall C et al (2006d) Phosphate oxygen isotopes as a tracer for sources and cycling of phosphate in North San Francisco Bay. Journal of Geophysical Research-Biogeosciences 111:G03003. doi:10.1029/2005JG000079

McLaughlin K, Sohm J, Cutter G et al (2011) Insights into the sources and cycling of phosphate in the Sargasso Sea: a multi-tracer approach. Global Biogeochemical Cycles (submitted)

Neill C, Deegan LA, Thomas SM et al (2001) Deforestation for pasture alters nitrogen and phosphorus in small Amazonian Streams. Ecol Appl 11:1817–1828

Nicholson D, Dyhrman S, Chavez F et al (2006) Alkaline phosphatase activity in the phytoplankton communities of Monterey Bay and San Francisco Bay. Limnol Oceanogr 51:874–883

O'Neil JR, Roe LJ, Reinhard E et al (1994) A rapid and precise method of oxygen isotope analysis of biogenic phosphate. Isr J Earth Sci 43:203–212

O'Neil JR, Vennemann TW, McKenzie WF (2003) Effects of speciation on equilibrium fractionations and rates of oxygen isotope exchange between $(PO_4)_{aq}$ and H_2O. Geochim Cosmochim Acta 67:3135–3144

Paytan A (1989) Oxygen isotope variations of phosphate in aquatic systems, Master's Thesis. Hebrew University, Jerusalem

Paytan A, Kolodny Y, Neori A et al (2002) Rapid biologically mediated oxygen isotope exchange between water and phosphate. Global Biogeochem Cycles 16:1013

Paytan A, McLaughlin K (2007) The Oceanic Phosphorus Cycle. Chemical Reviews 107:563–576

Rabalais NN, Turner RE, Wiseman WJ (2002) Gulf of Mexico hypoxia, a.k.a. "The dead zone". Annu Rev Ecol Syst 33:235–263

Schlesinger WH (1991) Biogeochemistry. Academic Press, San Diego, An Analysis of Global Change

Sharp JH (1991) Review of carbon, nitrogen, and phosphorus biogeochemistry. Rev Geophys Suppl 29:648–657

Shemesh A, Kolodny Y, Luz B (1983) Oxygen isotope variations in phosphate of biogenic apatites. II. Phosphorite rocks. Earth Planet Sci Lett 64:405–416

Shemesh A, Kolodny Y, Luz B (1988) Isotope geochemistry of oxygen and carbon in phosphate and carbonate phosphorite francolite. Geochim Cosmochim Acta 52:2565–2572

Smith REH, Kalff J (1983) Competition for phosphorus among co-occuring freshwater phytoplankton. Limnol Oceanogr 28:448–464

Smith SV (1984) Phosphorus versus nitrogen limitation in the marine environment. Limnol Oceanogr 29:1149–1160

Sundareshwar PV, Morris JT, Koepfler EK et al (2003) Phosphorus limitation of coastal ecosystem processes. Science 299:563–565

Tamburini F, Bernasconi SM, Angert A et al (2010) A method for the analysis of the $\delta^{18}O$ of inorganic phosphate in soils extracted with HCl. Eur J Soil Sci 61:1025–1032

Tilman D, Fargione J, Wolff B et al (2001) Forcasting agriculturally driven global environmental change. Science 292:281–284

Tudge AP (1960) A method of analysis of oxygen isotopes in orthophosphate – its use in the measurement of paleotemperatures. Geochim Cosmochim Acta 18:81–93

Vaccari DA (2009) Phosphorus famine: the treat to our food supply. Sci Am 300:54–59

Vennemann TW, Fricke HC, Blake RE et al (2002) Oxygen isotope analysis of phosphates: a comparison of techniques for analysis of Ag_3PO_4. Chem Geol 185:321–336

Weidemann-Bidlack FB, Colman AS, Fogel ML (2008) Stable isotope analyses of phosphate oxygen from biological apatite: a new technique for microsampling, microprecipitation of Ag_3PO_4, and removal of organic contamination. Rapid Commun Mass Spectrom 22:1807–1816

Wu JF, Sunda W, Boyle EA et al (2000) Phosphate depletion in the western North Atlantic Ocean. Science 289:759–762

Young MB, McLaughlin K, Kendall C et al (2009) Charaterizing the oxygen isotopic composition of phosphate sources to aquatic systems. Environ Sci Tech 43:5190–5196

Zohar I, Paytan A, Shaviv A (2010a) A method for the analysis of oxygen isotopic composition of various soil phosphate fractions. Environ Sci Tech 44(19):7583–7588

Zohar I, Shaviv A, Young M et al (2010b) Phosphorus dynamics in soil following irrigation, a study using oxygen isotopic composition of phosphate. Geoderma 159:109–121

Chapter 22
Isotopic Tracing of Perchlorate in the Environment

Neil C. Sturchio, John Karl Böhlke, Baohua Gu, Paul B. Hatzinger, and W. Andrew Jackson

Abstract Isotopic measurements can be used for tracing the sources and behavior of environmental contaminants. Perchlorate (ClO_4^-) has been detected widely in groundwater, soils, fertilizers, plants, milk, and human urine since 1997, when improved analytical methods for analyzing ClO_4^- concentration became available for routine use. Perchlorate ingestion poses a risk to human health because of its interference with thyroidal hormone production. Consequently, methods for isotopic analysis of ClO_4^- have been developed and applied to assist evaluation of the origin and migration of this common contaminant. Isotopic data are now available for stable isotopes of oxygen and chlorine, as well as ^{36}Cl isotopic abundances, in ClO_4^- samples from a variety of natural and synthetic sources. These isotopic data provide a basis for distinguishing sources of ClO_4^- found in the environment, and for understanding the origin of natural ClO_4^-. In addition, the isotope effects of microbial ClO_4^- reduction have been measured in laboratory and field experiments, providing a tool for assessing ClO_4^- attenuation in the environment. Isotopic data have been used successfully in some areas for identifying major sources of ClO_4^- contamination in drinking water supplies. Questions about the origin and global biogeochemical cycle of natural ClO_4^- remain to be addressed; such work would benefit from the development of methods for preparation and isotopic analysis of ClO_4^- in samples with low concentrations and complex matrices.

22.1 Introduction

Perchlorate (ClO_4^-) is a stable oxyanion consisting of four O^{2-} ions bonded in tetrahedral coordination with a central Cl^{7+} ion. It is ubiquitous in the environment at trace concentrations, and has natural and anthropogenic sources. Natural ClO_4^- is present in precipitation (generally <0.1 μg L^{-1}), in soil (generally <10 μg kg^{-1}), and in groundwater at concentrations ranging from background levels of about 0.01 to >100 μg L^{-1} in some arid regions where perchlorate has been concentrated by evaporation (Dasgupta et al. 2005; Jackson et al. 2005a; Rao et al. 2007; Rajagopalan et al. 2006, 2009; Parker et al. 2008).

A major production mechanism for natural perchlorate apparently involves reactions of atmospheric Cl species with ozone (O_3) (Simonaitis and Heicklen 1975; Jaegle et al. 1996; Bao and Gu 2004; Dasgupta et al. 2005; Kang et al. 2008; Catling et al. 2010; Rao et al. 2010). Anthropogenic ClO_4^- salts are synthesized by electrolysis of NaCl brines (Schumacher 1960) in quantities on the order of 10^7 kg $year^{-1}$ (DasGupta et al. 2006) for military, aerospace, and other industrial applications, e.g., solid rocket fuel, explosives, fireworks, road flares, electroplating solutions.

Perchlorate salts (NH_4ClO_4, $KClO_4$, $NaClO_4$) are soluble in water and some organic solvents. The

N.C. Sturchio (✉)
University of Illinois at Chicago, Chicago, IL 60607, USA
e-mail: sturchio@uic.edu

J.K. Böhlke
U.S. Geological Survey, Reston, VA 20192, USA

B. Gu
Oak Ridge National Laboratory, Oak Ridge, TN 37831, USA

P.B. Hatzinger
Shaw Environmental, Inc., Lawrenceville, NJ 08648, USA

W.A. Jackson
Texas Tech University, Lubbock, TX 79409, USA

ClO_4^- ion in aqueous solution is non-complexing and unreactive at low temperature, resistant to O exchange with H_2O (Hoering et al. 1958), and adsorbs weakly to solids present in soils and aquifers. It is not readily removed by conventional water treatment processes, but special ion exchange resins have been developed to remove ClO_4^- from drinking water (Gu et al. 2007, 2011). Perchlorate can be reduced by microbes under anoxic conditions after available nitrate (NO_3^-) is depleted (Coates and Achenbach 2004), and this biodegradation process has been utilized for both in situ and ex situ treatment of water (Hatzinger 2005). Its chemical properties give ClO_4^- high mobility and general persistence in surface waters, vadose zone environments, and shallow oxic aquifers.

The widespread occurrence of ClO_4^- in drinking water and food supplies (Gullick et al. 2001; Kirk et al. 2003; Jackson et al. 2005b; Sanchez et al. 2005, 2006) has attracted attention because of the potential deleterious health effects of ClO_4^- ingestion caused by interference with human thyroidal hormone production (NRC 2005). Pathways of natural and synthetic perchlorate through the environment, and potential natural attenuation processes, are thus of great interest with respect to human health and the protection of environmental quality.

Stable isotope ratio analyses of oxygen (O) and chlorine (Cl), along with ^{36}Cl abundance measurements, have shown that different ClO_4^- sources have distinct isotopic compositions that provide multiple isotopic tracers of the sources and behavior of ClO_4^- in the environment (Bao and Gu 2004; Böhlke et al. 2005; Sturchio et al. 2006, 2009; Jackson et al. 2010). In this chapter we review the development and current status of sampling and isotopic analytical methods for ClO_4^-, along with a survey of published data. Remaining questions about the origin and behavior of natural ClO_4^- are outlined.

22.1.1 Isotope Abundances and Notation

There are three stable isotopes of O (^{16}O, ^{17}O, and ^{18}O) having natural abundances of approximately 99.76, 0.04, and 0.2%, respectively (Coplen et al. 2002; Hoefs 2009). Variations in O isotope ratios are reported as $\delta^{18}O$ and $\delta^{17}O$, defined as the difference between the $^{18}O/^{16}O$ or $^{17}O/^{16}O$ atom ratio, respectively, of a sample and that of Vienna Standard Mean Ocean Water (VSMOW):

$$\delta^{18}O = \left[\left(^{18}O/^{16}O\right)_{sample} / \left(^{18}O/^{16}O\right)_{VSMOW} - 1 \right] \quad (22.1)$$

$$\delta^{17}O = \left[\left(^{17}O/^{16}O\right)_{sample} / \left(^{17}O/^{16}O\right)_{VSMOW} - 1 \right] \quad (22.2)$$

These δ values are normally reported in parts per thousand (‰) following multiplication of both sides of (22.1) and (22.2) by 1,000. In systems where isotopic fractionation is strictly mass-dependent, $\delta^{17}O \cong 0.52 \cdot \delta^{18}O$ and is not normally reported. However, natural ClO_4^- commonly has ^{17}O in excess of this relationship (Bao and Gu 2004; Böhlke et al. 2005) and the ^{17}O anomaly commonly is reported as a deviation from the abundance expected for mass-dependent fractionation, according to the approximation (Thiemens 2006):

$$\Delta^{17}O = \delta^{17}O - 0.52 \cdot \delta^{18}O \quad (22.3)$$

Alternatively (Miller 2002), and in this paper:

$$\Delta^{17}O = \left[\left(1 + \delta^{17}O\right)/\left(1 + \delta^{18}O\right)^{0.525} \right] - 1 \quad (22.4)$$

The $\Delta^{17}O$ value is normally reported in parts per thousand (‰) following multiplication of both sides of (22.3) and (22.4) by 1,000.

There are two stable isotopes of Cl (^{35}Cl and ^{37}Cl) having natural abundances of approximately 75.76 and 24.24%, respectively (Coplen et al. 2002; Hoefs 2009). Chlorine stable isotope ratios are reported as $\delta^{37}Cl$, defined as the difference between the $^{37}Cl/^{35}Cl$ atom ratio of a sample and that of Standard Mean Ocean Chloride (SMOC) (Long et al. 1993; Godon et al. 2004):

$$\delta^{37}Cl = \left[\left(^{37}Cl/^{35}Cl\right)_{sample} / \left(^{37}Cl/^{35}Cl\right)_{SMOC} - 1 \right] \quad (22.5)$$

Values of $\delta^{37}Cl$ are normally reported in parts per thousand (‰) following multiplication of both sides of (22.5) by 1,000.

Chlorine-36 is a long-lived radioactive isotope (half-life = 301,000 years) produced largely by cosmic-ray interactions with atmospheric Ar as well as by thermal neutron capture on ^{35}Cl in the terrestrial subsurface environment. Chorine-36 abundances are measured by accelerator mass spectrometry (AMS) and are expressed as the atom ratio of $^{36}Cl/Cl_{total}$. This ratio ranges from about 10^{-15} to 10^{-11} in terrestrial materials (Phillips 2000).

22.2 Methods

All high-precision stable isotope data ($\delta^{37}Cl$, $\delta^{18}O$, $\Delta^{17}O$) reported for ClO_4^- to date have been measured by isotope-ratio mass spectrometry (IRMS). Methods published by Ader et al. (2001) and Sturchio et al. (2003) are suitable only for determination of Cl isotope ratios in ClO_4^-, and these have been replaced in routine analysis by the methods developed and applied to isotopic analysis of both O and Cl by Böhlke et al. (2005); Sturchio et al. (2006); Hatzinger et al. (2009), and Jackson et al. (2010), which are described below in Sect. 22.2.3. Such measurements begin with a pure alkali-ClO_4 salt, preferably $KClO_4$, $RbClO_4$, or $CsClO_4$. A typical isotopic analysis of ClO_4^- by IRMS requires at least 0.2 mg of ClO_4^-. Although synthetic ClO_4^- reagents are readily available in large amounts, the concentration of ClO_4^- in most environmental materials is low (µg/kg) and requires elaborate methods of preconcentration, extraction, and purification prior to isotopic analysis (Gu et al. 2011).

22.2.1 Sampling

The most successful and widely used method for isotopic sampling of ClO_4^- from environmental materials has involved the use of a highly ClO_4-selective bifunctional ion-exchange resin for preconcentration (Gu et al. 2001, 2007, 2011). This resin was developed initially for large-scale water treatment and is commercially available as Purolite® A530E. Its use for preconcentration requires that the ClO_4^- be contained in an aqueous solution. Thus A530E is suitable for preconcentration of ClO_4^- directly from natural waters or from aqueous solutions obtained by leaching ClO_4-bearing solid materials such as soil, caliche, plants, or industrial products.

In practice, the A530E resin (20–30 mesh size) is normally packed into 100-mL size columns fabricated from 38 mm (1.5 in.) diameter clear polyvinyl chloride (PVC) tubing. Water is passed through the column at an optimal rate of 1–2 L per minute to preconcentrate ClO_4^- on the resin, with the objective of obtaining at least 3 mg for isotopic analysis. A prefilter (5–10 µm) may be used to remove particles from the water before passing through the column. Common anions such as Cl^-, NO_3^-, and SO_4^{2-}, along with humic material, if present in sufficient quantities, tend to interfere with uptake of ClO_4^- on the A530E resin. This interference increases with salinity and results in lower efficiency of ClO_4^- adsorption and early breakthrough, thus it is advisable to pass a larger volume of water through the column than that required by the assumption of 100% efficiency. Further details about the performance of A530E resin in ClO_4^- removal are given by Gu et al. (2001, 2007, 2011).

22.2.2 Extraction and Purification

When ClO_4^- has been concentrated on the A530E resin column, there are normally substantial quantities of other common anions present. The first step in recovering ClO_4^- is to flush the column with four to five bed volumes of 4 M HCl. This displaces a large portion of the adsorbed NO_3^-, SO_4^{2-}, carbonates, and organic anions, but adds substantial Cl^- to the resin. The ClO_4^- is then eluted using a solution of 1 M $FeCl_3$ in 4 M HCl, and the ClO_4^- is displaced from the A530E resin by the $FeCl_4^-$ anion (Gu et al. 2001). After ClO_4^- has been eluted from the column, there are multiple options for Fe^{3+} and Cl^- removal and further purification (Gu et al. 2011). Depending on the bulk sample composition, various issues may arise in ClO_4^- purification, and certain samples are more difficult to purify than others. Generally, lower-concentration samples ($ClO_4^- < \sim 1$ µg/L) are more challenging. Ultimately, the purified ClO_4^- is normally precipitated as $CsClO_4$, the purity of which is verified by ion chromatography and/or Raman spectroscopy, for isotopic analysis.

22.2.3 Isotopic Analysis

22.2.3.1 Oxygen Isotopes

Oxygen isotope ratios in ClO_4^- may be analyzed by two IRMS methods. First, $\delta^{18}O$ values may be determined by reaction with glassy carbon at 1,325°C (or higher) to produce CO, which is transferred in a He carrier through a molecular-sieve gas chromatograph to an isotope-ratio mass spectrometer and analyzed in continuous-flow (CF) mode by monitoring peaks at m/z 28 and 30 (this method is referred to as CO-CFIRMS, for CO continuous-flow isotope-ratio mass spectrometry). Yields of O (as CO) typically are $100 \pm 2\%$ for pure ClO_4^- reagents and samples. The 1σ analytical precision of $\delta^{18}O$ values ranges from ± 0.1 to 0.3‰, based on replicate analyses of samples and isotopic reference materials.

For the second IRMS method, both $\delta^{18}O$ and $\delta^{17}O$ values are measured on O_2 produced by in vacuo decomposition of the $CsClO_4$ at 600–650°C, according to the reaction:

$$CsClO_4 \rightarrow CsCl + 2O_2 \quad (22.6)$$

Decomposition can be done in sealed quartz or Pyrex glass tubes (Böhlke et al. 2005; Sturchio et al. 2007) or in a vacuum system equipped with an O_2 trap (Bao and Gu 2004). The O_2 gas is admitted to an isotope-ratio mass spectrometer and analyzed in dual-inlet (DI) mode by measurements at m/z 32, 33, and 34 (this method is referred to as O2-DIIRMS, for O_2 dual-inlet isotope-ratio mass spectrometry). Yields of O (as O_2) by the sealed-tube method are typically within $\pm 5\%$ of those expected from processed sample amounts for pure ClO_4^- reagents and samples and measured aliquots of tank O_2. Partial exchange of O between O_2 and glass may occur during this decomposition reaction, as a function of sample size, temperature, and time.

Oxygen isotope analyses of ClO_4^- are calibrated using a pair of $KClO_4$ isotopic reference materials (USGS37, USGS38) with isotopic compositions that differ by more than the samples being analyzed. To monitor drift, analyses include an internal laboratory reference gas (either CO or O_2, see below), against which all samples and isotopic reference materials are analyzed in the mass spectrometer during a single batch of analyses in which all samples and reference materials are the same size (in terms of O) and are treated identically. The isotopic reference materials consist of reagent-grade $KClO_4$ salts (USGS37, USGS38) that were prepared specifically for calibration of ClO_4^- isotopic analyses. The $\delta^{18}O$ scale is based on CO-CFIRMS analyses of the ClO_4^- isotopic reference materials against international water, nitrate, and sulfate isotopic reference materials as described by Böhlke et al. (2003) and all data are referenced to the conventional VSMOW-SLAP (Standard Light Antarctic Precipitation) scale. The $\Delta^{17}O$ scale for ClO_4^- is based provisionally on the assumption that the normal reagent $KClO_4$ reference material (USGS37) has $^{18}O{:}^{17}O{:}^{16}O$ ratios that are related to those of VSMOW by normal mass dependent relations ($\Delta^{17}O = 0.0$‰) with exponent $\lambda = 0.525$ (see (22.4) above) (Miller 2002; Böhlke et al. 2005). Typical 1σ reproducibility of the $\Delta^{17}O$ measurements is approximately ± 0.1‰ after normalization, based on replicate analyses of samples and isotopic reference materials.

22.2.3.2 Chlorine Isotopes

Stable isotope ratios of Cl in purified ClO_4^- samples are analyzed using IRMS techniques, whereas $^{36}Cl/Cl$ ratios are determined using AMS (Sturchio et al. 2007, 2009). For $\delta^{37}Cl$ determinations, ClO_4^- salts are first decomposed at 600–650°C in evacuated glass tubes to produce alkali chloride salts, which are then analyzed according to well established methods (Sturchio et al. 2007). The alkali chloride salts produced by ClO_4^- decomposition are dissolved in warm 18.2 MΩ deionized water, and Cl^- is precipitated as AgCl by addition of $AgNO_3$. The resulting AgCl is recovered by centrifugation, washed in dilute HNO_3, dried, and reacted in a sealed glass tube with excess CH_3I at 300°C for 2 h to produce CH_3Cl. The resulting CH_3Cl is purified using gas chromatography, cryo-concentrated, and then admitted to an isotope-ratio mass spectrometer and analyzed in either continuous-flow or dual-inlet mode (depending on sample size) by measurements at m/z 50 and 52.

Chlorine isotope ratio analyses are normalized using the same ClO_4^- isotopic reference materials described for O analysis in Sect. 22.2.3.1 (USGS37, USGS38). To monitor drift, analyses include an internal laboratory reference gas (CH$_3$Cl), against which all samples and reference materials are analyzed in the mass spectrometer during a single batch of analyses. The δ^{37}Cl scale is based on isotopic analyses of the USGS ClO$_4^-$ isotopic reference materials against SMOC. After normalization, the 1σ analytical precision of δ^{37}Cl values is approximately ±0.2‰, based on replicate analyses of samples and isotopic reference materials.

Analysis of ^{36}Cl in ClO$_4^-$ is performed by AMS (Sturchio et al. 2009). Prior to AMS, the CsClO$_4$ is decomposed in a sealed tube as described above, then the chloride is recovered as CsCl and precipitated as AgCl by addition of excess AgNO$_3$ solution. The AgCl is dissolved in dilute NH$_4$OH solution and the Cl$^-$ is purified by anion chromatography on analytical grade 1-X8 resin (using an unpublished protocol provided by the PRIME Lab of Purdue University) to ensure removal of trace amounts of S that might cause isobaric interference (by ^{36}S) at mass 36. An alternative method for purification of Cl$^-$ involves cation-exchange chromatography (Jiang et al. 2004). Purified Cl$^-$ is then precipitated as AgCl for AMS measurement. Analysis of seawater Cl$^-$ provides a reference datum of ^{36}Cl/Cl = 0.5 × 10^{-15} (Argento et al. 2010).

22.3 Results and Discussion

22.3.1 Stable Isotopic Composition of Synthetic Perchlorate

Stable isotope data for samples of synthetic ClO$_4^-$ reagents and other commercial ClO$_4^-$-bearing products have been published by Ader et al. (2001), Bao and Gu (2004), Böhlke et al. (2005), and Sturchio et al. (2006). Data for samples of synthetic ClO$_4^-$ in which both δ^{18}O and δ^{37}Cl isotope ratios were measured are shown in Fig. 22.1. Published δ^{37}Cl values for synthetic ClO$_4^-$ range from −3.1 to +2.3‰ and have a mean value near +0.6‰. This mean value is within the range of common industrial NaCl sources, such as halite from Phanerozoic bedded evaporites having δ^{37}Cl values in the range of 0.0 ± 0.9‰ (Eastoe et al. 2007), and indicates that relatively little (<1‰) isotopic fractionation of Cl accompanies ClO$_4^-$ syn-

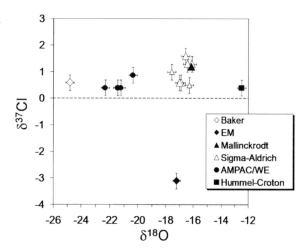

Fig. 22.1 δ^{37}Cl (‰) versus δ^{18}O (‰) for representative samples of synthetic ClO$_4^-$ grouped by source (see legend). Analytical errors shown at ±0.3‰. *Dashed line* is δ^{37}Cl reference value of 0.0‰ for Standard Mean Ocean Chloride. Data from Sturchio et al. (2006)

thesis. Published δ^{18}O values of synthetic ClO$_4^-$ range from −24.8 to −12.5‰, presumably reflecting a range in composition of the local water sources used in production and O isotopic fractionation that occurs during ClO$_4^-$ synthesis (Sturchio et al. 2006). In contrast, Δ^{17}O values of synthetic ClO$_4^-$ are consistently 0.0 ± 0.1‰, indicating little or no mass-independent isotopic fractionation in the ClO$_4^-$ synthesis process. All stable isotopic data generated to date for primary synthetic ClO$_4^-$ (as of September, 2010) are within the ranges of δ^{18}O and δ^{37}Cl shown in Fig. 22.1. Preliminary measurements on ClO$_4^-$ produced by disproportionation reactions in commercial NaOCl (bleach) solutions, however, have indicated some anomalously low values of δ^{18}O and anomalously high values of δ^{37}Cl (e.g., δ^{37}Cl = +14.0‰; (Sturchio et al. 2009)), yielding a potential means by which to distinguish this source of ClO$_4^-$ from that produced by electrochemical synthesis.

22.3.1.1 Isotopic Fractionation of Perchlorate by Microbial Reduction

Kinetic isotopic fractionations are caused by mass dependent differences in reaction rates of different isotopic species during fast, incomplete, or unidirectional reactions such as diffusion, evaporation, and microbial respiration, where reaction products are generally enriched in the lighter isotopes (Criss 1999). The isotopic fractionation factor, α, is defined as

$$\alpha = R_A/R_B \quad (22.7)$$

where R is an isotope ratio, and A and B are two substances (in the present case, product and reactant, respectively). For O and Cl isotope ratios, R represents $^{18}O/^{16}O$ and $^{37}Cl/^{35}Cl$, respectively. Values of α can be obtained from the experimental results by assuming the exponential function

$$R/R_0 = f^{\alpha-1} \quad (22.8)$$

where R and R_0 are the O or Cl isotope ratios ($^{18}O/^{16}O$ or $^{37}Cl/^{35}Cl$) of the residual ClO_4^- and the initial, unreacted ClO_4^-, respectively, and f is the fraction of ClO_4^- remaining. In terms of the δ values of the ClO_4^-, as defined in (22.1), (22.2), and (22.5),

$$(\delta + 1)/(\delta_0 + 1) = f^{\alpha-1} \quad (22.9)$$

where δ is the isotopic composition of the ClO_4^- at any value f, and δ_0 is the isotopic composition at $f = 1$. The value of α can be obtained by fitting data with the natural log of (22.8):

$$\alpha - 1 = \ln(R/R_o)/\ln f \quad (22.10)$$

This describes the mass-dependent Rayleigh-type isotopic fractionation that accompanies a variety of natural processes (Broecker and Oversby 1971). Isotopic fractionation factors are commonly expressed in terms of ε, where

$$\varepsilon = \alpha - 1 \quad (22.11)$$

with ε normally expressed in parts per thousand (‰).

Isotopic fractionation of Cl and O accompanying microbial reduction of ClO_4^- have been investigated in several laboratory experiments and in a field study involving an in situ aquifer push-pull test. Two independent liquid culture experiments published in 2003 used the bacterial species *Azospira suillum* with acetate as the electron donor and ClO_4^- as the sole electron acceptor to investigate the Cl kinetic isotope effect (Coleman et al. 2003; Sturchio et al. 2003). Coleman et al. (2003) maintained the temperature of their cultures at 37°C, whereas Sturchio et al. (2003) maintained the temperature of their cultures at 22°C, during incubation. Coleman et al. (2003) reported $\varepsilon^{37}Cl$ values of -15.8 ± 0.4 and -14.8 ± 1.3‰ obtained from two separate cultures, both exhibiting nearly complete ClO_4^- reduction in about 90 min. Sturchio et al. (2003) reported comparable $\varepsilon^{37}Cl$ values of -16.6 and -12.9‰ obtained from two separate cultures with complete ClO_4^- reduction times of 18 days and 5.5 h, respectively, noting that less fractionation was observed in the experiment in which ClO_4^- was consumed more rapidly. A subsequent presentation and re-evaluation of the Coleman et al. (2003) data concluded a single and more precise $\varepsilon^{37}Cl$ value of -14.98 ± 0.15‰, following a statistical refinement of the complete experimental data set (Ader et al. 2008).

Further work by Sturchio et al. (2007) produced data on both Cl and O isotope effects of microbial ClO_4^- reduction, this time using two different bacterial genera (*A. suillum* JPLRND and *Decholospirillum* sp. FBR2) in liquid cultures at temperatures of 22 and 10°C, and including an experiment using ^{18}O-enriched water to test for O exchange between ClO_4^- and H_2O. This set of experiments resulted in composite $\varepsilon^{37}Cl$ and $\varepsilon^{18}O$ values (regressed from the combined data of the five separate experiments, with all points weighted equally) of -13.2 ± 0.5 and -33.1 ± 1.2‰, respectively. There was no evidence (within analytical error) for dependence on either bacterial strain or temperature, and no evidence for O exchange between ClO_4^- and H_2O, during ClO_4^- reduction. A key result of this work was the observation that, despite a range in apparent ε values of the individual experiments, the ratio $\varepsilon^{18}O/\varepsilon^{37}Cl$ showed a constant value of 2.50 ± 0.04 for the combined experimental results over a range of f from 1.00 to 0.01 (Fig. 22.2).

The applicability of laboratory isotopic fractionation factors in the field was explored by Hatzinger et al. (2009), who performed a push-pull test in 2006 at a site where a soybean oil emulsion had been injected three years earlier in a demonstration of its potential to stimulate ClO_4^- bioremediation (Borden 2007). In the single injection push-pull test, 405 L of contaminated upgradient groundwater (containing 5.5 mg L^{-1} ClO_4^- and 7.7 mg L^{-1} NO_3^-) was amended with NaBr (as a tracer for groundwater mixing) and injected at a rate of 30 L min^{-1} into the zone containing residual emulsified oil (the monitoring well was

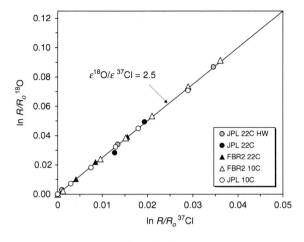

Fig. 22.2 Relation of $\varepsilon^{18}O$ and $\varepsilon^{37}Cl$ in combined results of five liquid culture experiments of microbial ClO_4^- reduction (Sturchio et al. 2007). A constant slope of 2.50 ± 0.04 was observed, independent of bacterial strain and temperature. The experiment labeled "HW" was done in ^{18}O-enriched water ($\delta^{18}O = +198‰$) and indicates no measurable O exchange between ClO_4^- and H_2O during microbial reduction under these conditions

screened at depths of 1.8–4.9 m, where total well depth was 5 m below ground surface). Nine water samples of 1–30-L volume (increasing with time) were removed periodically over a period of 30 h for chemical and isotopic analysis. Apparent in situ isotopic fractionation factors for both O and Cl in ClO_4^- were only about 0.3–0.4 times the values reported for pure culture studies (Sturchio et al. 2007; Ader et al. 2008); similarly, apparent fractionation factors for N and O in NO_3^- were only 0.2–0.6 times those reported in laboratory culture studies. The relatively small apparent isotopic fractionation factors observed in the push-pull test may be attributed to physical and chemical heterogeneity in the aquifer, as similar effects have been observed and modeled for NO_3^- reduction in groundwater and surface water (Mariotti et al. 1988; Lehman et al. 2003; Green et al. 2010). Despite the relatively small apparent in situ values of $\varepsilon^{18}O$ and $\varepsilon^{37}Cl$ measured for ClO_4^- in the push-pull test, the ε ratio $\varepsilon^{18}O/\varepsilon^{37}Cl$ was 2.63, in excellent agreement with the laboratory-determined value of 2.50 ± 0.04 (Sturchio et al. 2007). This indicates that the fundamental process of ClO_4^- reduction may have been the same in laboratory and field settings, but the heterogeneity of the field setting was such that it caused substantial underestimation of the extent of ClO_4^- reduction when using an isotopic approach based on laboratory fractionation factors.

22.3.2 Stable Isotopic Composition of Natural Perchlorate

22.3.2.1 Atacama Desert, Chile

The most well-known occurrence of naturally occurring ClO_4^- is in the extensive NO_3^--rich "caliche"-type salt deposits of the hyperarid Atacama Desert of northern Chile, which contain up to 0.6 wt.% ClO_4^- (Ericksen 1981). The O isotopic composition of Atacama ClO_4^- was first measured by Bao and Gu (2004) who reported $\delta^{18}O$ values ranging from -24.8 to $-4.5‰$ and $\Delta^{17}O$ values ranging from $+4.2$ to $+9.6‰$. The elevated ^{17}O abundance was interpreted by Bao and Gu (2004) as a reflection of the atmospheric origin of ClO_4^- through photochemical reactions of atmospheric Cl species with O_3. Additional isotopic data for Atacama ClO_4^- were reported by Böhlke et al. (2005), Sturchio et al. (2006), and Jackson et al. (2010), extending the O isotopic results of Bao and Gu (2004) and, in addition, documenting extreme ^{37}Cl-depletion with $\delta^{37}Cl$ values ranging from -14.5 to $-11.8‰$. An atmospheric origin for ClO_4^- is consistent with N and O isotopic evidence for the atmospheric origin of NO_3^- in the Atacama salt deposits and soils (Böhlke et al. 1997; Michalski et al. 2004; Ewing et al. 2007), where both compounds may have accumulated for millions of years (Ericksen 1981; Rech et al. 2010; Ewing et al. 2006). There is growing interest in the origin and isotopic composition of Atacama ClO_4^- for several reasons: (1) billions of kg of Atacama NO_3^- deposits have been imported to the U.S. for use in agricultural fertilizer (DasGupta et al. 2006), and the distinctive isotopic composition of Atacama ClO_4^- has been found in aquifers across the U.S., indicating past use of Atacama nitrate fertilizer is a source of ClO_4^- in U.S. groundwater (Böhlke et al. 2005, 2009; Sturchio et al. 2011); and (2) the recent discovery of ClO_4^- in Martian and Antarctic soils has led to renewed efforts to understand the mechanism of atmospheric ClO_4^- formation and the planetary geochemical cycles for ClO_4^- on both Earth and Mars (Hecht et al. 2009; Ming et al. 2010; Catling et al. 2010; Kounaves et al. 2010).

22.3.2.2 Southwestern United States

The occurrence of indigenous natural ClO_4^- (unrelated to imported Atacama ClO_4^-) throughout the United States, especially in the arid southwestern regions, has become increasingly apparent through surveys of ClO_4^- concentrations in precipitation, soil, and groundwater (Plummer et al. 2006; Jackson et al. 2005a; Rajagopalan et al. 2006, 2009; Rao et al. 2007; Parker et al. 2008). The stable isotopic composition of indigenous natural ClO_4^- analyzed to date (hereafter referred to as SW ClO_4^-) is distinct from both Atacama ClO_4^- and synthetic ClO_4^- (Jackson et al. 2010) (Fig. 22.3).

The three types of ClO_4^- shown in Fig. 22.3 can be distinguished completely by their $\delta^{18}O$ values. Two types of ClO_4^- have overlapping $\delta^{37}Cl$ values: synthetic ClO_4^- has a range from about -3 to $+2‰$, whereas SW ClO_4^- has a range from about -3 to $+6‰$. In contrast, Atacama ClO_4^- has a uniquely low range in $\delta^{37}Cl$ from about -15 to $-9‰$. The reason for such a wide range in $\delta^{37}Cl$ values for natural ClO_4^- is not yet understood, but may be related to regional or hemispheric differences in the source(s) of atmospheric Cl reactants or the mechanism(s) involved in ClO_4^- production.

The $\Delta^{17}O$ values of synthetic ClO_4^- are uniformly $0.0 \pm 0.2‰$. Natural ClO_4^- shows a significant ^{17}O excess, with Atacama ClO_4^- showing a relatively narrow range in $\Delta^{17}O$ values from $+8.1$ to $+10.5‰$, whereas SW ClO_4^- shows a much wider range from about $+0.2‰$ in groundwaters from the Southern High Plains to as high as $+18.4‰$ for an unsaturated zone salt sample from the Zabriskie area (Ericksen et al. 1988) near Death Valley, CA. This mass-independent isotopic fractionation effect (Thiemens 2006) is diagnostic of all natural ClO_4^- analyzed to date. The $\Delta^{17}O$ values of the SW ClO_4^- summarized in Fig. 22.3 display a bimodal distribution, with one group (labeled SHP, for Southern High Plains) having a range from

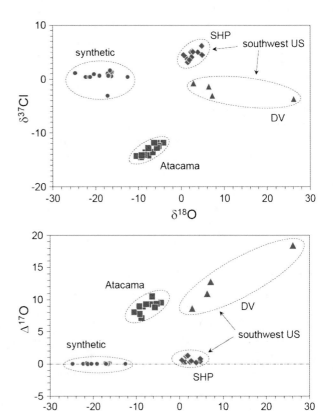

Fig. 22.3 $\delta^{37}Cl$ (per mil) versus $\delta^{18}O$ (per mil) values (upper diagram) and $\Delta^{17}O$ (per mil) versus $\delta^{18}O$ (per mil) values (lower diagram) for samples of synthetic ClO_4^-, Atacama ClO_4^-, and SW ClO_4^-. Southwest ClO_4^- is subdivided into Southern High Plains (SHP) and Death Valley (DV) as described in text. Data from Böhlke et al. (2005); Sturchio et al. (2006); Jackson et al. (2010)

about +0.2 to +2.6‰, and a second group (labeled DV, for Death Valley) having a range from +8.6 to +18.4‰. This distribution may be related in part with local climate and the mode of ClO_4^- occurrence, as the lower range of values represents ClO_4^- extracted from a number of groundwater samples and an unsaturated zone soil leachate from the Southern High Plains and Middle Rio Grande Basin (Texas and New Mexico), whereas the higher values are from unsaturated zone salts from the vicinity of Death Valley, California. The ranges of O isotopic compositions could imply either (1) there is a primary, high-$\Delta^{17}O$ atmospheric ClO_4^- signature that is retained for thousands of years in the most arid unsaturated zone profiles (and for millions of years in the hyperarid Atacama Desert) but diminished by slow isotopic exchange with O in slightly wetter environments with time, or (2) there is another production mechanism (e.g., involving UV but not O_3) that results in lower $\Delta^{17}O$ values for ClO_4^- (Jackson et al. 2010).

22.3.3 Chlorine-36 Abundance in Perchlorate

The radioactive isotope ^{36}Cl (half-life = 301,000 years) provides an additional tracer for ClO_4^-. Chlorine-36 is produced in the atmosphere largely by cosmic-ray interactions with Ar, and also in the subsurface by thermal neutron capture on ^{35}Cl (Lehmann et al. 1993). The range in measured $^{36}Cl/Cl$ ratios of Cl^- in pre-anthropogenic groundwater across the continental U.S. is from ~10×10^{-15} near the coasts to as high as $1,670 \times 10^{-15}$ in the central Rocky Mountains (Davis et al. 2003). The lower ratios near the coasts reflect dilution by marine aerosols in which $^{36}Cl/Cl = 0.5 \times 10^{-15}$ (Argento et al. 2010). Testing of thermonuclear bombs in the Pacific Ocean during 1952–1958 injected a large amount of ^{36}Cl, produced by neutron irradiation of seawater Cl^-, into the stratosphere, resulting in worldwide ^{36}Cl fallout for several years thereafter (Phillips 2000). The presence of this bomb-pulse Cl^- may be identified from its anomalously high $^{36}Cl/Cl$ ratio and its association with relatively high tritium activity. The highest measured $^{36}Cl/Cl$ ratio reported for bomb-affected groundwater Cl^- is $12,800 \times 10^{-15}$ (Davis et al. 2003). The $^{36}Cl/Cl$ ratio of Cl^- in Long Island (NY) rainwater sampled in 1957 (during the peak of the nuclear bomb testing era in the Pacific) was as high as $127,000 \times 10^{-15}$ (Schaeffer et al. 1960). Arctic deposition in 1957 also had anomalously high $^{36}Cl/Cl$ ($28,600 \times 10^{-15}$) as observed in the Dye-3 ice core (Synal et al. 1990).

A reconnaissance survey of ^{36}Cl in ClO_4^-, including a variety of synthetic and natural ClO_4^- samples, showed a range of $^{36}Cl/Cl$ ratios exceeding four orders of magnitude (Sturchio et al. 2009) (Fig. 22.4). Synthetic ClO_4^- samples had relatively low $^{36}Cl/Cl$ ratios from $\leq 2.5 \times 10^{-15}$ to 40×10^{-15}, reflecting those expected from ancient, marine-derived NaCl sources

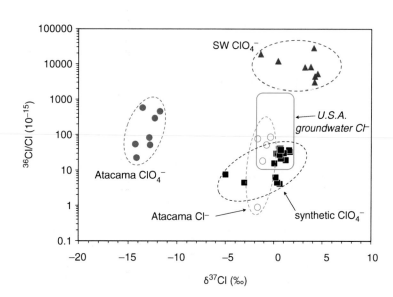

Fig. 22.4 $^{36}Cl/Cl$ (atom ratio) versus $\delta^{37}Cl$ (‰) in representative samples of synthetic ClO_4^-, Atacama ClO_4^- and associated Cl^-, and SW ClO_4^- (adapted from Sturchio et al. 2009)

used in the electrochemical synthesis, such as bedded evaporites and salt domes. Southwest ClO_4^- samples had extremely high $^{36}Cl/Cl$ ratios in the range $3{,}130 \times 10^{-15}$ to $28{,}800 \times 10^{-15}$. Although the presence of bomb ^{36}Cl could not be ruled out for some of these samples, $^{36}Cl/Cl$ ratios of $3{,}130 \times 10^{-15}$ to $12{,}300 \times 10^{-15}$ were measured in ClO_4^- extracted from groundwaters having no detectable tritium, indicating groundwater recharge prior to bomb testing and implying that the high $^{36}Cl/Cl$ ratios in these ClO_4^- samples are unrelated to the bomb tests. The fact that these ratios are much higher than any others measured for Cl^- from pre-anthropogenic groundwater implies a largely stratospheric source for the ClO_4^-, as any near-surface ClO_4^- production mechanism should yield $^{36}Cl/Cl$ ratios in the range of near-surface Cl^- (i.e. $<2{,}000 \times 10^{-15}$).

Atacama ClO_4^- had an intermediate range of $^{36}Cl/Cl$ ratios from 22×10^{-15} to 590×10^{-15}, and Cl^- from two samples of the Atacama caliche and one sample of an Atacama commercial $NaNO_3$ product had relatively low $^{36}Cl/Cl$ ratios identical within error to those of the co-occurring ClO_4^-. The Atacama data could be consistent with initially high $^{36}Cl/Cl$ ratios (similar to those of SW ClO_4^- samples) that have decreased by radioactive decay. In those samples where ClO_4^- and Cl^- have identical $^{36}Cl/Cl$ ratios, ClO_4^- and Cl^- have apparently approached secular equilibrium with the local radiation environment (i.e., a condition in which the rate of radioactive decay of ^{36}Cl is equal to its rate of production within the sample environment). There has been adequate time for secular equilibrium to be attained, as many of the Atacama caliche deposits began forming at least 9.4 million

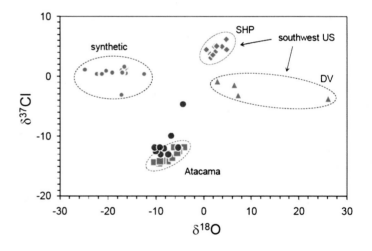

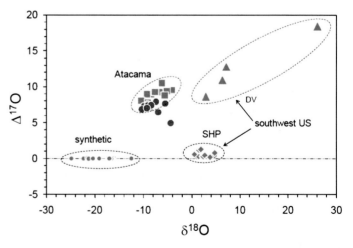

Fig. 22.5 $\delta^{37}Cl$ (per mil) versus $\delta^{18}O$ (per mil) values (*upper diagram*) and $\Delta^{17}O$ (per mil) versus $\delta^{18}O$ (per mil) values (*lower diagram*) showing stable isotope data for ClO_4^- in groundwater samples from the Chino Basin, California (*red symbols*) in comparison to the principal known ClO_4^- source types in the region as identified in Fig. 22.3. A dominantly Atacama source is inferred for the Chino Basin samples (Sturchio et al. 2008)

22 Isotopic Tracing of Perchlorate in the Environment

years ago (Rech et al. 2006), whereas it takes only about two million years for ^{36}Cl to reach a condition of >99% secular equilibrium.

22.3.4 Tracing Perchlorate in Contaminated Aquifers

The isotopic data accumulated for ClO_4^- during the past decade have created opportunities for tracing ClO_4^- in contaminated aquifers with relatively robust results. The three principal sources of ClO_4^- found in contaminated groundwaters of the US (i.e., synthetic, Atacama, and SW ClO_4^-) are isotopically distinct with respect to stable isotopes of O and Cl (Bao and Gu 2004; Böhlke et al. 2005; Sturchio et al. 2006) (Jackson et al. 2010) as well as ^{36}Cl (Sturchio et al. 2009). Furthermore, the low rate of O exchange between ClO_4^- and H_2O (Hoering et al. 1958) could maintain the integrity of ClO_4^- source O isotopic ratios over at least several decades under normal groundwater conditions (Böhlke et al. 2009). Biodegradation of ClO_4^- produces systematic stable isotope enrichments that do not significantly affect $\Delta^{17}O$ values or $^{36}Cl/Cl$ ratios, thus potentially allowing detection of the biodegradation without masking the initial source(s) of ClO_4^- (Sturchio et al. 2007; Hatzinger et al. 2009).

Several studies involving the application of stable isotope measurements for tracing ClO_4^- sources in contaminated aquifers have been published with supporting information from other chemical and isotopic tracers. Increasing concentrations of ClO_4^- detected in municipal water supplies of Chino, Ontario, and Pomona, California during 2004 prompted an isotopic investigation to determine the source of this increase. The data showed a dominantly agricultural source

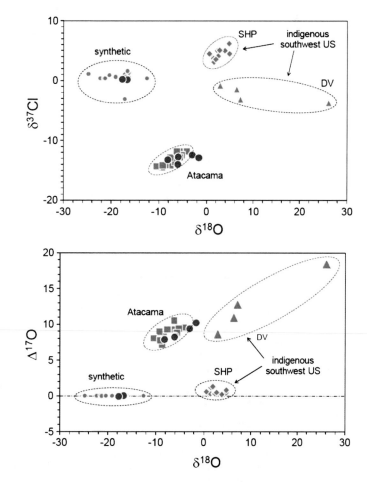

Fig. 22.6 $\delta^{37}Cl$ (per mil) versus $\delta^{18}O$ (per mil) values (*upper diagram*) and $\Delta^{17}O$ (per mil) versus $\delta^{18}O$ (per mil) values (*lower diagram*) showing stable isotope data for ClO_4^- in groundwater samples (labeled *red symbols*) from Long Island, New York (Böhlke et al. 2009), in comparison to the principal known ClO_4^- source types in the region as identified in Fig. 22.3

with isotopic compositions resembling those of Atacama ClO_4^-, with apparent minor contributions of synthetic and SW ClO_4^- (Fig. 22.5). The northern portion of the Chino Basin was formerly used for citrus cultivation and there is anecdotal evidence for extensive use of Chilean NO_3^- fertilizer (Sturchio et al. 2008).

Böhlke et al. (2009) demonstrated that ClO_4^- originating from known synthetic and agricultural sources in Long Island, New York could be traced from stable isotopic compositions (Fig. 22.6). A history of groundwater contamination by Atacama ClO_4^- in an agricultural area in eastern Long Island was inferred from depth profiles of environmental tracer ages along with concentration and isotopic data for ClO_4^- and NO_3^- (Böhlke et al. 2009). The possible sources of ClO_4^- in the groundwater of Pasadena, California also were investigated using multiple chemical and isotopic tracers, and evidence was found for at least three distinct sources of ClO_4^-, of which at least two were synthetic (Slaten et al. 2010).

The most intensive local isotopic investigation of ClO_4^- sources in groundwater to date was performed in the southeastern San Bernardino Basin, California. Discrete, intersecting ClO_4^- plumes were found to have distinct sources (Sturchio et al. 2011). One plume containing synthetic ClO_4^- emanated from a former rocket-testing site that had been used subsequently as an artificial recharge basin (labeled "site" at the right side of Fig. 22.7), and another plume containing Atacama-type ClO_4^- emanated from a predominantly agricultural area. Where these plumes intersect, some groundwater wells contain mixtures of synthetic and Atacama ClO_4^- (Figs. 22.7 and 22.8).

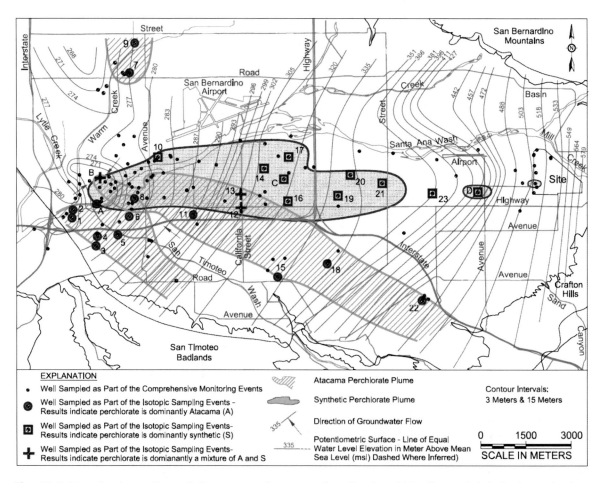

Fig. 22.7 Map showing outlines of discrete groundwater ClO_4^- plumes in Bunker Hill Basin southwest of San Bernardino, California in relation to groundwater potentiometric surface. Map legend identifies symbols indicating predominant ClO_4^- source type. Adapted from Sturchio et al. (2011)

Fig. 22.8 $\delta^{37}Cl$ (per mil) versus $\delta^{18}O$ (per mil) (*upper diagram*) and $\Delta^{17}O$ (per mil) versus $\delta^{18}O$ (per mil) (*lower diagram*) for samples of groundwater ClO_4^- plumes in the southeastern San Bernardino Basin, California (*red symbols*) in comparison to the three principal known ClO_4^- source types in the region as identified in Fig. 22.3. One plume has a synthetic source, and the other has a predominantly agricultural (i.e. Atacama-type) source, with some apparent mixing between sources (data from Sturchio et al. 2011)

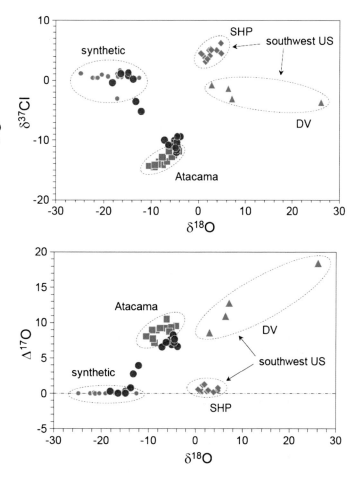

22.4 Summary and Research Opportunities

Isotopic data can be used for tracing the origin and behavior of ClO_4^- in the environment. Four independently varying parameters have been measured on individual ClO_4^- samples for this purpose: $\delta^{37}Cl$, $\delta^{18}O$, $\Delta^{17}O$, and $^{36}Cl/Cl$. At least three distinct types of ClO_4^- have been identified isotopically (i.e., synthetic, Atacama, and SW ClO_4^-), and these distinctions have proven to be useful in forensic applications. Additional data for indigenous, natural ClO_4^- are needed, however, to obtain a global picture of its isotopic variations. Improved methods for sample preparation and isotopic analysis with much better sensitivity would be helpful for measuring ClO_4^- isotopic variations in some sample types such as aerosols and precipitation as well as foodstuffs and body fluids, which have been precluded by the impracticality of obtaining the currently required milligram amounts of ClO_4^-. Further experimental and theoretical investigations of atmospheric ClO_4^- production mechanisms may lead to improved explanations of observed isotopic variations in natural samples.

Acknowledgements Much of the work reviewed in this chapter was supported by contracts from the Strategic Environmental Research and Development Program and the Environmental Security Technology Certification Program of the U.S. Department of Defense, and by the National Research Program of the U.S. Geological Survey. Use of product or trade names in this paper is for identification purposes only and does not constitute endorsement by the U.S. government. We thank Hans Eggenkamp, Stephanie Ewing, Doug Kent, and an anonymous referee for constructive reviews of the manuscript.

References

Ader M, Coleman ML, Doyle SP, Stroud M, Wakelin D (2001) Methods for the stable isotopic analysis of chlorine in chlorate and perchlorate compounds. Anal Chem 73: 4946–4950

Ader M, Chaudhuri S, Coates JD, Coleman M (2008) Microbial perchlorate reduction: a precise laboratory determination of the chlorine isotope fractionation and its possible biochemical basis. Earth Planet Sci Lett 269:604–612

Argento DC, Stone JO, Fifield LK, Tims SG (2010) Chlorine-36 in seawater. Nucl Instr Meth Phys Res B 268:1226–1228

Bao H, Gu B (2004) Natural perchlorate has its unique oxygen isotope signature. Environ Sci Technol 38:5073–5077

Böhlke JK, Ericksen GE, Revesz K (1997) Stable isotope evidence for an atmospheric origin of desert nitrate deposits in northern Chile and southern California. Chem Geol 136:135–152

Böhlke JK, Mroczkowski SJ, Coplen TB (2003) Oxygen isotopes in nitrate: new reference materials for $^{18}O:^{17}O:^{16}O$ measurements and observations on nitrate-water equilibration. Rapid Comm Mass Spec 17:1835–1846

Böhlke JK, Sturchio NC, Gu B, Horita J, Brown GM, Jackson WA, Batista JR, Hatzinger PB (2005) Perchlorate isotope forensics. Anal Chem 77:7838–7842

Böhlke JK, Hatzinger PB, Sturchio NC, Gu B, Abbene I, Mroczkowski SJ (2009) Atacama perchlorate as an agricultural contaminant in groundwater: isotopic and chronologic evidence from Long Island, New York. Environ Sci Technol 43:5619–5625

Borden RC (2007) Anaerobic bioremediation of perchlorate and 1,1,1-trichloroethane in an emulsified oil barrier. J Contam Hydrol 94:13–33

Broecker WS, Oversby VM (1971) Chemical equilibrium in the earth. McGraw-Hill, New York

Catling DC, Clair MW, Zahnle KJ, Quinn RC, Clarc BC, Hecht MH, Kounaves SP (2010) Atmospheric origins of perchlorate on Mars and in the Atacama. J Geophys Res 115:E00E11. doi:10.1029/2009JE003425

Coates JD, Achenbach LA (2004) Microbial perchlorate reduction: rocket-fuelled metabolism. Nat Rev Microbiol 2:569–580

Coleman ML, Ader M, Chaudhuri S, Coates JD (2003) Microbial isotopic fractionation of perchlorate chlorine. Appl Environ Microbiol 69:4997–5000

Coplen TB, Hopple JA, Böhlke JK, Peiser HS, Rieder SE, Krouse HR, Rosman KJR, DingT, Vocke RD, Jr, Revesz KM, Lamberty A, Taylor P, De Bievre P (2002) Compilation of minimum and maximum isotope ratios of selected elements in naturally occurring terrestrial materials and reagents. U.S. Geology Survey Water Resources Investigation Report 01-4222

Criss RE (1999) Principles of stable isotope distribution. Oxford University Press, New York

Dasgupta PK, Martinelango PK, Jackson WA, Anderson TA, Tian K, Tock RW, Rajagopalan S (2005) The origin of naturally occurring perchlorate: the role of atmospheric processes. Environ Sci Technol 39:1569–1575

DasGupta PK, Dyke JV, Kirk AB, Jackson WA (2006) Perchlorate in the United States. Analysis of relative source contributions to the food chain. Environ Sci Technol 40:6608–6614

Davis SN, Moysey S, Cecil LD, Zreda M (2003) Chlorine-36 in groundwater of the United States: empirical data. Hydrogeol J 11:217–227

Eastoe CJ, Peryt TM, Petrychenko OY, Geisler-Cussy D (2007) Stable chlorine isotopes in Phanerozoic evaporites. Appl Geochem 22:575–588

Ericksen GE (1981) Geology and origin of the Chilean nitrate deposits. Professional Paper 1188. U.S. Geological Survey, Washington, DC

Ericksen GE, Hosterman JW, Amand S (1988) Chemistry, mineralogy, and origin of the clay-hill nitrate deposits Amaragosa River Valley, Death Valley region, California, U.S.A. Chem Geol 67:85–102

Ewing SA, Sutter B, Owen J, Nishiizumi K, Sharp W, Cliff SS, Perry K, Dietrich W, McKay CP, Amundsen R (2006) A threshold in soil formation at Earth's arid-hyperarid transition. Geochim Cosmochim Acta 70:5293–5322

Ewing SA, Michalski G, Thiemens M, Quinn RC, Macalady JL, Kohl S, Wankel SD, Kendall C, McKay CP, Amundson R (2007) The rainfall limit of the N cycle on earth. Glob Biogeochem Cycles 21:GB3009. doi:10.1029/2006GB002838

Godon A, Jendrzejewski N, Eggenkamp HGM, Banks DA, Ader M, Coleman ML, Pineau F (2004) An international cross calibration over a large range of chlorine isotope compositions. Chem Geol 207:1–12

Green CT, Böhlke JK, Bekins B, Phillips S (2010) Mixing effects on apparent reaction rates and isotope fractionation factors during denitrification in a heterogeneous aquifer. Water Resour Res 46:W08525. doi:10.1029/2009WR008903: 1-19

Gu B, Brown GM, Maya L, Lance MJ, Moyer BA (2001) Regeneration of perchlorate (ClO_4^-)-loaded anion exchange resins by novel tetrachloroferrate ($FeCl_4^-$) displacement technique. Environ Sci Technol 35:3363–3368

Gu B, Brown GM, Chiang CC (2007) Treatment of perchlorate-contaminated groundwater using highly selective, regenerable ion-exchange technologies. Environ Sci Technol 41: 6277–6282

Gu B, Böhlke JK, Sturchio NC, Hatzinger PB, Jackson WA, Beloso AD Jr, Heraty LJ, Bian Y, Brown GM (2011) Removal, recovery and fingerprinting of perchlorate by ion exchange processes. In: SenGupta AK (ed) Ion exchange and solvent extraction: a series of advances, vol 20. CRC, Boca Raton, FL

Gullick RW, Barhorst TS, LeChevallier MW (2001) Occurrence of perchlorate in drinking water sources. J Am Water Works Assoc 93:66–77

Hatzinger PB (2005) Perchlorate biodegradation for water treatment. Environ Sci Technol 39:239A–247A

Hatzinger P, Böhlke JK, Sturchio NC, Gu B, Heraty LJ, Borden RC (2009) Fractionation of stable isotopes in perchlorate and nitrate during in situ biodegradation in a sandy aquifer. Environ Chem 6:44–52

Hecht MH, Kounaves SP, Quinn RC, West SJ, Young SMM, Ming DW, Catling DC, Clark BC, Boynton WV, Hoffman J, DeFlores LP, Gospodinova K, Kapit J, Smith PH (2009) Detection of perchlorate and the soluble chemistry of Martian soil at the Phoenix Lander site. Science 325:64–67

Hoefs J (2009) Stable isotope geochemistry, 6th edn. Springer, Berlin

Hoering TC, Ishimori FT, McDonald HO (1958) The oxygen exchange between oxy-anions and water, 11. Chlorite, chlorate and perchlorate ions. J Am Chem Soc 80:3876

Jackson WA, Anandam SK, Anderson T, Lehman T, Rainwater K, Rajagopalan S, Ridley M, Tock R (2005a) Perchlorate occurrence in the Texas southern high plains aquifer system. Ground Water Monit Rem 25:137–149

Jackson WA, Preethi J, Laxman P, Tian K, Smith PN, Yu L, Anderson TA (2005b) Perchlorate accumulation in forage and edible vegetation. J Agric Food Chem 53:369–373

Jackson WA, Böhlke JK, Gu B, Hatzinger PB, Sturchio NC (2010) Isotopic composition and origin of indigenous perchlorate and co-occurring nitrate in the southwestern United States. Environ Sci Technol 44:4869–4876

Jaegle L, Yung YL, Toon GC, Sen B, Blavier J (1996) Balloon observations of organic and inorganic chlorine in the stratosphere: the role of $HClO_4$ production on sulfate aerosols. Geophys Res Lett 23:1749–1752

Jiang S, Lin Y, Jiang H (2004) Improvement of the sample preparation method for AMS measurement of ^{36}Cl in natural environment. Nucl Instr Meth Phys Res B 223–224: 318–322

Kang N, Jackson WA, Dasgupta PK, Anderson TA (2008) Perchlorate production by ozone oxidation of chloride in aqueous and dry systems. Sci Total Environ 405:301–309

Kirk AB, Smith EE, Tian K, Anderson TA, Dasgupta PK (2003) Perchlorate in milk. Environ Sci Technol 37:4979–4981

Kounaves SP, Stroble ST, Anderson RM, Moore Q, Catling DC, Douglas S, McKay CP, Ming DW, Smith PH, Tamppari LK, Zent AP (2010) Discovery of natural perchlorate in the Antarctic Dry Valleys and its global implications. Environ Sci Technol 44:2360–2364

Lehman MF, Reichert P, Bernasconi SM, Barbieri A, McKenzie JA (2003) Modeling nitrogen and oxygen isotope fractionation during denitrification in a lacustrine redox-transition zone. Geochim Cosmochim Acta 67:2529–2542

Lehmann BE, Davis SN, Fabryka-Martin JT (1993) Atmospheric and subsurface sources of stable and radioactive nuclides used for groundwater dating. Water Resour Res 29:2027–2040

Long A, Eastoe CJ, Kaufmann RS, Martin JG, Wirt L, Finley JB (1993) High-precision measurement of chlorine stable isotope ratios. Geochim Cosmochim Acta 57:2907–2912

Mariotti A, Simon B, Landreau A (1988) ^{15}N isotope biogeochemistry and natural denitrification process in groundwater: application to the chalk aquifer of northern France. Geochim Cosmochim Acta 52:1869–1878

Michalski G, Böhlke JK, Thiemens M (2004) Long term atmospheric deposition as the source of nitrate and other salts in the Atacama Desert, Chile: new evidence from mass-independent oxygen isotopic compositions. Geochim Cosmochim Acta 68:4023–4038

Miller MF (2002) Isotopic fractionation and the quantification of ^{17}O anomalies in the oxygen three-isotope system: an appraisal and geochemical significance. Geochim Cosmochim Acta 66:1881–1889

Ming DW, Smith PH, Tamppari LK, Zent AP (2010) Discovery of natural perchlorate in the Antarctic Dry Valleys and its global implications. Environ Sci Technol 44:2360–2364

NRC (2005) Committee to Assess the Health Implications of Perchlorate Ingestion, National Research Council. Health implications of perchlorate ingestion. National Academy, Washington, DC. http://www.nap.edu/catalog/11202.html

Parker DR, Seyfferth AL, Reese BK (2008) Perchlorate in groundwater: a synoptic survey of "pristine" sites in the conterminous Unites States. Environ Sci Technol 42:1465–1471

Phillips FM (2000) Chlorine-36. In: Cook P, Herczeg A (eds) Environmental tracers in subsurface hydrology. Kluwer, Boston, pp 299–348

Plummer LN, Böhlke JK, Doughten MW (2006) Perchlorate in Pleistocene and Holocene groundwater in north-central New Mexico. Environ Sci Technol 40:1757–1763

Rajagopalan S, Anderson TA, Fahlquist L, Rainwater KA, Ridley M, Jackson WA (2006) Widespread presence of naturally occurring perchlorate in high plains of Texas and New Mexico. Environ Sci Technol 40:3156–3162

Rajagopalan S, Anderson TA, Cox S, Harvey G, Cheng Q, Jackson WA (2009) Perchlorate in wet deposition across North America. Environ Sci Technol 43:616–622

Rao B, Anderson TA, Orris GJ, Rainwater KA, Rajagopalan S, Sandvig RM, Scanlon BR, Stonestrom DA, Walvoord MA, Jackson WA (2007) Widespread natural perchlorate in unsaturated zones of the Southwest United States. Environ Sci Technol 41:4522–4528

Rao B, Anderson TA, Redder A, Jackson WA (2010) Perchlorate formation by ozone oxidation of aqueous chlorine/oxychlorine species: role of Cl_xO_y radicals. Environ Sci Technol 44:2961–2967

Rech JA, Currie BS, Michalski G, Cowan AM (2006) Neogene climate change and uplift in the Atacama Desert, Chile. Geology 34:761–764

Rech JA, Currie BS, Shullenberger ED, Dunagan SP, Jordan TE, Blanco N, Tomlinson AJ, Rowe HD, Houston J (2010) Evidence for the development of the Andean rain shadow from a Neogene isotopic record in the Atacama Desert, Chile. Earth Planet Sci Lett 292:371–382

Sanchez CA, Crump KS, Krieger RL, Khandaker NR, Gibbs JP (2005) Perchlorate and nitrate in leafy vegetables of North America. Environ Sci Technol 39:9391–9397

Sanchez CA, Krieger RL, Khandaker NR, Valentin-Blasini L, Blount BC (2006) Potential perchlorate exposure from citrus irrigated with contaminated water. Anal Chim Acta 567:33–38

Schaeffer OA, Thompson SO, Lark NL (1960) Chlorine-36 radioactivity in rain. J Geophys Res 65:4013–4016

Schumacher J (1960) Perchlorates: their properties, manufacture, and uses, American Chemical Society Monograph Series146. Reinhold, New York

Simonaitis R, Heicklen J (1975) Perchloric acid: possible sink for stratospheric chlorine. Planet Space Sci 23: 1567–1569

Slaten S, Fields KA, Santos S, Barton A, Rectanus HV, Bhargava M (2010) Integrated environmental forensics approach for evaluating the extent of dissolved perchlorate originating from multiple sources. Environ Forensics 11:72–93

Sturchio NC, Hatzinger PB, Arkins MD, Suh C, Heraty LJ (2003) Chlorine isotope fractionation during microbial reduction of perchlorate. Environ Sci Technol 37:3859–3863

Sturchio NC, Böhlke JK, Gu B, Horita J, Brown GM, Beloso AD Jr, Hatzinger PB, Jackson WA, Batista JR (2006) Stable isotopic compositions of chlorine and oxygen in synthetic and natural perchlorates. In: Gu B, Coates JD (eds) Perchlorate environmental occurrences, interactions, and treatment. Springer, New York, pp 93–109

Sturchio NC, Böhlke JK, Beloso AD Jr, Streger SH, Heraty LJ, Hatzinger PB (2007) Oxygen and chlorine isotopic fractionation during perchlorate biodegradation: laboratory results and implications for forensics and natural attenuation studies. Environ Sci Technol 41:2796–2802

Sturchio NC, Beloso Jr AD, Heraty LJ, LeClaire J, Rolfe T, Manning KR (2008) Isotopic evidence for agricultural perchlorate in groundwater of the western Chino Basin, California. In: Sixth international conference on remediation of chlorinated and recalcitrant compounds, Monterey, CA, 18–22 May 2008

Sturchio NC, Caffee MR, Beloso AD Jr, Heraty LJ, Böhlke JK, Gu B, Jackson WA, Hatzinger PB, Heikoop JR, Dale M (2009) Chlorine-36 as a tracer of perchlorate origin. Environ Sci Technol 43:6934–6938

Sturchio NC, Hoaglund JR III, Marroquin RJ, Beloso Jr AD, Heraty LJ, Bortz SE, Patterson TL (2011) Isotopic mapping of groundwater perchlorate plumes. Ground Water. DOI:10.1111/j.1745-6584.2011.00802.x

Synal HA, Beer J, Bonani G, Suter M, Wölfi W (1990) Atmospheric transport of bomb-produced ^{36}Cl. Nucl Meth Instr Phys Res B52:483–488

Thiemens MH (2006) History and applications of mass-independent isotope effects. Annu Rev Earth Planet Sci 34:217–262

Chapter 23
The Isotopomers of Nitrous Oxide: Analytical Considerations and Application to Resolution of Microbial Production Pathways

Nathaniel E. Ostrom and Peggy H. Ostrom

Abstract Nitrous oxide (N_2O) is an important greenhouse gas and the most important substance involved in ozone depletion. The N_2O molecule consists of two N atoms that differ in the nature of their covalent bonds and consequently tend to acquire distinct isotopic compositions. Thus, N_2O contains both bulk ($\delta^{15}N$ and $\delta^{18}O$) and site specific isotopic information. Site preference (SP) is defined as the difference in $\delta^{15}N$ between the central (α) and outer (β) N atoms in N_2O. SP has emerged as a potential conservative tracer of microbial N_2O production because (1) it is independent of the isotopologue composition of the substrates of nitrification and denitrification and (2) does not exhibit fractionation during production. In pure microbial culture distinct SP values for N_2O production from bacterial denitrification, including nitrifier denitrification (-10 to 0 ‰), relative to hydroxylamine oxidation and fungal denitrification (33–37 ‰) provides a basis to resolve production pathways in the natural environment. Future directions for isotopomer research include (1) improvements in calibration and mass overlap corrections, (2) evaluation of SP signals of microbial production pathways not yet studied including heterotrophic nitrification, codenitrification, and dissimilatory reduction of nitrate to ammonium, and (3) evaluation of the potential for biological production to produce a ^{17}O anomaly in atmospheric N_2O.

23.1 Introduction

Nitrous oxide (N_2O) is a long lived greenhouse gas that is also involved in ozone depletion. Although the atmospheric abundance of N_2O is approximately one thousandth that of CO_2, relative to CO_2 its radiative efficiency (Wm^{-2} ppb^{-1}) is 216 times greater. Thus, the contribution of N_2O to atmospheric global warming is substantial (6 %) (Forster et al. 2007). Currently N_2O is the most important ozone depleting substance emitted by human activity but it is not regulated by the Montreal protocol (Ravishankara et al. 2009). The fact that N_2O production is strongly linked to fertilizer application has important implications for management of emissions of this gas (Forster et al. 2007). With expectations of an increase in human population from a current estimate of 7–9 billion by 2050 and the associated escalation in fertilizer production, N_2O emissions are anticipated to rise (Mosier and Kroeze 2000; Tilman et al. 2001; Galloway et al. 2003). While there is compelling need to reduce emissions, we are unable to achieve a lower atmospheric N_2O concentrations simply by controlling anthropogenic sources as the majority of N_2O is not directly derived from human activity. Rather microbial activity produces N_2O and is stimulated upon disturbance of soils, fertilizer application and nitrogen loading to aquatic ecosystems (Seitzinger et al. 2006; Groffman et al. 2006; Forster et al. 2007). Consequently, mitigation of N_2O emissions will not only be dependent upon reducing fertilizer use but also dependent upon measures that regulate microbial activities associated with N_2O production.

Identifying the microbial origin of N_2O in soils and aquatic environments is a long-standing research objective that has remained surprisingly elusive

---

N. E. Ostrom (✉) and P. H. Ostrom
Department of Zoology, Michigan State University, 204 Natural Sciences Building, East Lansing, MI 48824, USA
e-mail: ostromn@msu.edu

(Mosier et al. 1998a, b; Robertson et al. 2004; Naqvi et al. 2006). A focal point of this objective has been the determination of the importance of nitrification and denitrification in the production of this gas. Nitrification involves oxidation of ammonium to nitrite and/or nitrate in the presence of oxygen. During this process, N_2O is formed as a byproduct of hydroxylamine oxidation. When oxygen is limiting, denitrification invokes the successive reduction of nitrate, nitrite, nitric oxide, and N_2O with dinitrogen gas (N_2) as the most common final product (Firestone and Davidson 1989). Owing to the sensitivity of these reactions to oxygen, the activity of nitrifiers and denitrifiers varies in response to different environmental controls such as changes in pH and the concentration of nitrate and organic carbon (Robertson and Groffman 2006). Thus, microbial controls of N_2O production are expected to depend on management strategies that impact the soil and water environments (Matson et al. 1989; Seitzinger et al. 2006). However, managing soil N_2O emissions is not as simple as regulating nitrification and denitrification. Additional transformation pathways are influential. Nitrifying bacteria produce N_2O by two pathways, hydroxylamine oxidation and reduction of nitrite (Wrage et al. 2001; Fig. 23.1), there is increasing awareness that fungal denitrification is an important mechanism of N_2O production in soils (Laughlin and Stevens 2002; Crenshaw et al. 2008) and other nitrogen cycling pathways such as heterotrophic nitrification, dissimilatory reduction of nitrate to ammonia (DRNA) and anaerobic ammonium oxidation (anammox) may contribute to N_2O production (Scott et al. 2008; Wrage et al. 2001; Kampschreur et al. 2008).

Traditional approaches for apportioning N_2O production to nitrification and denitrification include acetylene inhibition and/or isotope labeling approaches that inherently alter microbial activity (Madsen 1998; Groffman et al. 2006). Acetylene inhibition or ^{15}N tracer approaches have been widely used in both soil and aquatic environments (Groffman et al. 2006). The use of acetylene to define specific production pathways can be problematic owing to the potential for (1) incomplete diffusion of acetylene into the soil and incomplete inhibition of N_2O reduction, (2) alteration of microbial activity as acetylene is a metabolizeable source of carbon, (3) inhibition of nitrification and its associated supply of nitrate, and (4) production of N_2O from novel pathways such as fungal denitrification and codenitrification that may not respond to inhibitors (Madsen 1998; Groffman et al. 2006; Wrage et al. 2004, 2005; Laughlin and Stevens 2002; Crenshaw et al. 2008). As a result, there is a strong desire to avoid the use of acetylene and other inhibitors in N cycling studies (Wrage et al. 2004, 2005; Groffman et al. 2006). Application of ^{15}N enriched nitrate or ammonium is a common approach to evaluate N_2O production from nitrification and denitrification, however, this approach may be susceptible to incomplete diffusion of the tracer and stimulation of activity by the addition of the ^{15}N enriched substrates (Groffman et al. 2006). Consequently the challenges to resolving the microbial origins of N_2O are considerable and isotopologue data offer important advantages in the determination of production pathways relative to inhibitor and tracer studies. These include a lack of concern regarding incomplete diffusion of tracers or inhibitors, lack of alteration of microbial activity due to the addition of tracers or inhibitors, and an ability to apply results across broad temporal and spatial scales that is rarely feasible using incubation approaches.

In 1999 two research groups argued that the analysis of the intramolecular distribution of ^{15}N in N_2O could offer additional information regarding the origins of this greenhouse gas relative to the bulk isotope measurements $\delta^{15}N$ and $\delta^{18}O$ (Toyoda and Yoshida

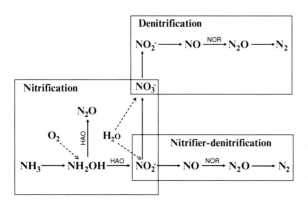

Fig. 23.1 Diagram of microbial N cycling showing pathways leading to N_2O production. The enzyme hydroxylamine oxidoreductase (HAO) is involved in the oxidation of ammonia to hydroxylamine and N_2O with the incorporation of an O atom from O_2. Nitrous oxide reductase (NOR) is involved in the reduction of nitric oxide to N_2O in both nitrifier-denitrification and denitrification and involves addition of O from water. Adapted from Wrage et al. (2001)

1999; Brenninkenmeijer and Röckman 2000; Yoshida and Toyoda 2000). Consequently, distinct terminology is needed to distinguish changes in the average isotopic composition of the N_2O molecule from isotopic variation that is positional dependent. We use the term "isotopomer" to indicate molecules of the same mass in which trace isotopes are arranged differently. "Isotopologue" is a more general term that indicates molecules that differ in their isotopic composition. The N atoms in N_2O have distinct covalent bonds with a central N atom bonded to the terminal N and O atoms. The central and outer N atoms are termed α and β, respectively (Toyoda and Yoshida 1999) or 2 and 1, respectively (Brenninkenmeijer and Röckman 2000). Alternatively, Yung and Miller (1997) designate the isotopomers $^{15}N^{14}N^{16}O$ and $^{14}N^{15}N^{16}O$ as 546 and 456. In this chapter, we utilize the α and β terminology as it is common in the biological literature. As the nature of covalent bonds between the α and β positions in N_2O differ there is ample opportunity for equilibrium, photochemical and biological isotope effects. As a result ^{15}N becomes unevenly distributed between the two N atoms within the N_2O molecule (Miller and Yung 2000; Sutka et al. 2006). Consequently, the N_2O molecule contains positional dependent ($\delta^{15}N^\alpha$ and $\delta^{15}N^\beta$) in addition to bulk isotopic information $\delta^{15}N$ and $\delta^{18}O$. Positional dependent isotope values of N_2O are commonly expressed as site preference (SP) which is the difference in $\delta^{15}N$ between the α and β positions.

23.2 Analytical Methods

23.2.1 Gas Purification and Mass Spectrometry

The primary instrument used for determining the bulk and positional dependent isotope values of N_2O is the magnetic sector isotope ratio mass spectrometer equipped with 3–5 detectors (collectors). These instruments are commercially available and commonly have one narrow center and two wide axial collectors that enable analysis of the isotopic composition of CO_2, N_2 and O_2. Because analysis of the bulk $\delta^{15}N$ and $\delta^{18}O$ of N_2O requires monitoring of the same masses as CO_2 (44, 45 and 46 amu), it is readily accomplished without modification to commercial instruments. A primary challenge for isotopic analysis of N_2O is its low concentration in the atmosphere, 319 ppv (as of 2007; Forster et al. 2007). In comparison, the atmospheric concentration of CO_2 is 1,000 times greater. Because of its low concentration, analysis of the isotope composition of N_2O often required extraction of N_2O from large volumes of air or water. For example, Yoshida and Matsuo (1983) concentrated N_2O from 60 L of air onto molecular sieve 5A. This approach also required the removal of water and CO_2 via a 6 m trap of drierite and a 3 m trap consisting of ascarite, respectively. N_2O was extracted from the molecular sieve by heating and cryogenic purification under vacuum. Quantitative removal of CO_2 is particularly critical owing to the mass overlap with N_2O. Similarly, early analyses of ocean water were accomplished by extracting N_2O from 100 L of seawater using a large stainless steel tank provided from Asahi Breweries, Ltd. (Yoshida et al. 1984).

The use of a gas chromatograph interface to the mass spectrometer has greatly decreased the minimum sample size required for isotope analysis of N_2O. An early approach used vacuum to pass 500 mL of air through ascarite and LiOH traps for removal of CO_2 and water, respectively, and then through a liquid N_2 cooled trap where N_2O was concentrated. Non-condensable gases passed through these traps and were removed by vacuum. The purified N_2O was thawed and separated from remaining traces of CO_2 using a GS-Q gas chromatographic column (J&W Scientific, Inc.) prior to introduction to a mass spectrometer (Bergsma et al. 2001). This approach enabled analysis of samples containing as little as 5 nmol of N_2O. Commercial gas chromatographic systems are currently available from both Elementar Inc. (TraceGas) and Thermo Scientific Inc. (Precon) that transfer the sample to a cryogenic trap via He flow instead of vacuum, utilize a cryofocusing step to further increase sensitivity by approximately a factor of 5, and a gas chromatographic column such as Porapak Q to isolate N_2O from any remaining gases (Brand 1995; Sutka et al. 2003; Röckmann et al. 2003). Röckmann et al. (2003) observed several peaks that eluted well after N_2O and minimized these peaks by optimizing temperature and the column flow rate. Despite this, long eluting peaks carried over from one analysis into the next.

This phenomenon was eliminated by use of a switching valve and by splitting the GC column into pre- and analytical columns. Alternatively, long eluting peaks have been eliminated by preconditioning the column at high temperature (200°C) for several hours after each day of analysis (personal observation).

Estimates of bulk and positional isotope values require monitoring the molecular ion of N_2O^+ with collectors position at m/z 44, 45 and 46 as well as the fragment ion, NO^+, with collectors positioned at m/z 30 and 31. This enables subsequent determination of ^NR, the ratio of the trace to abundant isotope in which N is the mass of the trace isotope (e.g. the m/z ratios 45/44, 46/44 and 31/30 are denoted ^{45}R, ^{46}R and ^{31}R, respectively). As will be discussed below these ratios are required for determination of bulk and positional isotope values. Although the use of fragment ions in stable isotope analysis was pioneered over 60 years ago, it was not applied to natural samples until the late 1990s (Friedman and Bigeleisen 1950; Brenninkenmeijer and Röckman 2000; Toyoda and Yoshida 1999; Yoshida and Toyoda 2000). Analysis of the five m/z values required for obtaining N_2O isotope values can be conducted on a traditional isotope ratio mass spectrometer with three collectors. The sample, however, must be analyzed twice; initially for ^{45}R and ^{46}R and secondly for ^{31}R. It is now common for instruments to be constructed with five collectors for the simultaneous monitoring of m/z 30, 31, 44, 45 and 46 (Brenninkenmeijer and Röckman 2000; Toyoda and Yoshida 1999; Sutka et al. 2003).

23.2.2 Determining Bulk and Positional Isotope Values

Bulk isotope values are expressed as:

$$\delta^N E = \left(\left[^N R_{sample}/^N R_{standard}\right] - 1\right) \times 1{,}000 \quad (23.1)$$

$R_{standard}$ is V-SMOW and air-N_2 for $\delta^{18}O$ and $\delta^{15}N$, respectively. Bulk isotope values derive from the analysis of m/z 44, 45, and 46 associated with the molecular ion, N_2O^+. $\delta^{15}N^\alpha$ and $\delta^{15}N^\beta$ are derived from a series of (23.2)–(23.7) and also require analysis of the fragment ion NO^+ at m/z 30 and 31. Owing to the complexity of the topic, a complete explanation of the equations required for estimating bulk and positional isotope values is beyond the scope of this chapter. Thus, we provide a foundation with brief explanations of the equations below and refer the reader to Toyoda and Yoshida (1999), Kaiser et al. (2004a) and Westley et al. (2007) where a more expansive discussion can be found.

Determination of the $\delta^{15}N$, $\delta^{18}O$, $\delta^{15}N^\alpha$ and $\delta^{15}N^\beta$ of N_2O requires an estimate of ^{15}R, $^{15}R^\alpha$, ^{17}R and ^{18}R that are ultimately derived from analysis of the molecular and fragmentation ions, N_2O^+ and NO^+. The two stable isotopes of N and three stable isotopes of O in N_2O can combine to produce 12 possible isotopomers (Miller and Yung 2000). Of these, eight isotopomers have m/z values of 44, 45, or 46 and can, therefore, be detected within a triple collector mass spectrometer:

$^{14}N^{14}N^{16}O$ m/z 44
$^{15}N^{14}N^{16}O$ m/z 45
$^{14}N^{15}N^{16}O$ m/z 45
$^{14}N^{14}N^{17}O$ m/z 45
$^{14}N^{14}N^{18}O$ m/z 46
$^{14}N^{15}N^{17}O$ m/z 46
$^{15}N^{14}N^{17}O$ m/z 46
$^{15}N^{15}N^{16}O$ m/z 46

The isotopomers of the fragment ion NO^+ are with m/z values of 30 and 31 are:

$^{14}N^{16}O$ m/z 30
$^{15}N^{16}O$ m/z 31
$^{14}N^{17}O$ m/z 31

In the isotopologue analysis of N_2O m/z 30, 31, 44, 45 and 46 are monitored and therefore the isotopomers shown above will contribute to these collectors. Additional collectors can be added however, isotopomers beyond m/z 32 and 46 for NO^+ and N_2O^+, respectively, are exceedingly rare and mass 32 suffers from interference from trace levels of O_2 in the mass spectrometer. The goal of monitoring these m/z is to determine ^{15}R, $^{15}R^\alpha$, $^{15}R^\beta$ and ^{18}R. As shown above, however, ^{15}R cannot be obtained directly from the ratio of m/z 45 to m/z 44 (^{45}R) because m/z 45 also receives a contribution from the $^{14}N^{14}N^{17}O$ isotopomer. Thus an expression for ^{45}R is:

$$^{45}R = {}^{15}R + {}^{17}R \quad (23.2)$$

where

$$^{15}R = (^{15}R^\alpha + {}^{15}R^\beta)/2 \quad (23.3)$$

Traditional mass spectrometric analysis with three collectors at m/z 44, 45 and 46 can only determine ^{15}R, which (23.3) indicates is the average of $^{15}R^\alpha + {}^{15}R^\beta$. Thus only bulk $\delta^{15}N$ can be determined from m/z 44, 45 and 46. Positional isotope abundances require additional monitoring of m/z 30 and 31 as it is the ratio of these masses that contains N from the α position only.

The predominant isotopomer that contributes to m/z 46 is $^{14}N^{14}N^{18}O$. However, m/z 46 also receives a contribution from $^{14}N^{15}N^{17}O$, $^{15}N^{14}N^{17}O$, and $^{15}N^{15}N^{16}O$. Although the probability of N_2O^+ containing two trace isotopes is small, these isotopomers are sufficiently abundant that they cannot be ignored. Consequently, ^{46}R is described as:

$$^{46}R = {}^{18}R + (^{15}R^\alpha + {}^{15}R^\beta) * {}^{17}R + {}^{15}R^\alpha + {}^{15}R^\beta \quad (23.4)$$

$\delta^{15}N^\alpha$ is obtained from the 31 to 30 ion current ratio with a correction for interference from ^{17}O. $\delta^{15}N$ is the average of $\delta^{15}N^\alpha$ and $\delta^{15}N^\beta$ and $\delta^{15}N^\beta$ is determined by difference. $\delta^{18}O$ is obtained from the 46:44 ion current ratio. ^{31}R reflects contributions of ^{15}N from the α position and ^{17}O:

$$^{31}R = {}^{15}R^\alpha + {}^{17}R \quad (23.5)$$

Equations (23.2) through (23.4) are robust given a random or statistical isotope distribution of the N and O isotopes in N_2O. In other words, the presence of one trace isotope does not increase or decrease the probability of the occurrence of a second trace isotope. Theoretical and experimental measurements support this contention (Röckmann et al. 2003; Kaiser et al. 2003); however as we discuss later, this assumption does not extend to the fragmentation and rearrangement processes.

Equations (23.2) and (23.4) can be solved numerically with the assumption of a mass-dependent relationship between $\delta^{17}O$ and $\delta^{18}O$ as described by Santrock et al. (1985):

$$\delta^{17}O = 0.516 * \delta^{18}O \quad (23.6)$$

Or as presented in Kaiser et al. (2003):

$$^{17}R = 0.00937035 \left(^{18}R\right)^{0.516} \quad (23.7)$$

The mass-dependent relationship between the oxygen isotopes is the fundamental basis for the widely used correction for the overlap of the $^{12}C^{16}O^{17}O$ isotopomer with $^{13}C^{16}O^{16}O$ in determination of $\delta^{13}C$ that was initially developed by Harmon Craig and has been followed with more recent improvements (Craig 1957; Santrock et al. 1985; Kaiser 2008; Kaiser and Röckmann 2008; Brand et al. 2010). Indeed a simple correction for the interference of ^{17}O with the $\delta^{15}N$ of N_2O was proposed by Brand (1995) which simply involved applying the CO_2 correction to N_2O multiplied by 0.75. This is convenient as most commercial mass spectrometers are provided with software that includes mass dependent corrections for the $\delta^{13}C$ of CO_2 but not for N_2O.

23.2.3 Mass-Dependence Between $\delta^{17}O$ and $\delta^{18}O$

While the mass dependent relationship between $\delta^{17}O$ and $\delta^{18}O$ is robust for the majority of materials on Earth, photochemical reactions in the atmosphere may alter the slope of (23.6) markedly (Thiemens 2006). Most notable is stratospheric ozone that exhibits a $\delta^{17}O$ to $\delta^{18}O$ relationship of approximately 0.62 (Lammerzahl et al. 2002). Further, the precise value of the slope for materials on Earth may vary from 0.5000 to 0.5305 (Matsuhisa et al. 1978; Assonov and Brenninkmeijer 2001). The $\delta^{17}O$ of N_2O is infrequently determined directly and, consequently, corrections for mass overlap depend strongly on the assumption of the mass-dependent relationship between $\delta^{17}O$ and $\delta^{18}O$. Tropospheric N_2O, however, is slightly enriched in ^{17}O relative to the mass-dependent line by approximately 0.9 ‰ (Cliff et al. 1999; Cliff and Thiemens 1997; Kaiser et al. 2003, 2004b). As we discuss later, the precise cause of the ^{17}O anomaly in tropospheric N_2O is not known and may reflect an influence from photochemical reactions and/or, possibly, biological production (Kaiser et al. 2004b). At current values of precision commonly obtained in mass spectrometric analysis of N_2O (~0.1–1 ‰) any error introduced to the mass interference correction from the ^{17}O isotopomer at m/z 45 is minor; 0.1 ‰ or less (Toyoda and Yoshida 1999).

Accurate determination of both $\delta^{17}O$ and $\delta^{18}O$ values of N_2O is not possible by analysis of the N_2O^+ and NO^+ ions owing to mass overlap and requires decomposition of N_2O to O_2. One approach for decomposition includes high temperature decomposition of N_2O to both O_2 and N_2 in the presence of a gold catalyst followed by cryogenic purification of each product (Cliff and Thiemens 1994) or under He flow (Kaiser et al. 2007). Alternatively N_2O can be decomposed to O_2 at high temperature in the presence of BrF_5 (Brenninkenmeijer and Röckman 2000). While both of these approaches can be used to accurately constrain the presence or absence of a ^{17}O anomaly, any site specific $\delta^{15}N$ data will be lost.

23.2.4 Isotope Rearrangement

A unique challenge to the isotopic analysis of N_2O is the rearrangement or scrambling of N atoms in the NO^+ fragment between the α and β positions during ionization. Scrambling was first observed by Friedman and Bigeleisen (1950) and has since been verified in a number of studies based on the analysis of N_2O enriched in ^{15}N in either the α or β position. These studies have shown that between 7 and 11.7 % of α N atoms may be found in the β position (Friedman and Bigeleisen 1950; Brenninkenmeijer and Röckman 2000; Toyoda and Yoshida 1999; Sutka et al. 2004; Westley et al. 2007). The precise causes of scrambling are not certain as the NO^+ ion may be formed by several pathways from an excited N_2O^+ molecule. These include collision induced decomposition and decay of metastable ions in transit from the source to collectors (Mark et al. 1981; Westley et al. 2007). As these processes are affected by source conditions (such as acceleration voltage) and are instrument dependent, it is strongly recommended that the degree of scrambling or rearrangement factor be determined on a regular basis and whenever source conditions change markedly (such as filament replacement) (Westley et al. 2007). Nonetheless, the rearrangement factor, γ, is stable under routine operating conditions and provides a basis to correct the observed value for $^{15}R^\alpha$ for rearrangement (Toyoda and Yoshida 1999):

$$^{15}R^\alpha{}_{obs} = (1 - \gamma)^{15}R^\alpha + \gamma^{15}R^\beta \qquad (23.8)$$

In addition to (23.4), (23.5) can be solved to correct $^{15}R^\alpha$ for the contribution from ^{17}O. However, early correction routines involved two critical assumptions: (1) that different isotopomers of N_2O produce the NO^+ fragment ion at equal rates and (2) that the fraction of NO^+ produced from ionization of $^{15}N^{14}N^{16}O$ is identical to that produced upon ionization of $^{14}N^{14}N^{16}O$ (Toyoda and Yoshida 1999; Kaiser et al. 2004a). Both of these assumptions have recently been shown to be incorrect (Westley et al. 2007). Differences in bond strength and differences in zero point energies between molecules with and without a trace isotope are the fundamental controls on isotope fractionation. Consequently, it is not surprising that $^{15}N^{14}N^{16}O$ and $^{14}N^{14}N^{16}O$ would cleave to produce the NO^+ fragment at different rates. The mass correction and calibration routine developed by Kaiser et al. (2004a) involved mixtures of isotopically enriched N_2O with a natural abundance standard and was based on the assumption that $^{15}N^{15}N^{16}O$ yields $^{15}N^{16}O$ at that same rate that $^{14}N^{16}O$ is produced from $^{14}N^{14}N^{16}O$. This has been shown to be invalid and calibration based on this assumption yields different results for the same gas analyzed by different mass spectrometers (Westley et al. 2007). The assumption of equal fragmentation rates for different isotopomers in calibration is the likely reason for the nearly 30 ‰ difference in SP reported for tropospheric N_2O by two research groups (Toyoda and Yoshida 1999; Yoshida and Toyoda 2000; Kaiser et al. 2004a). In contrast to mixtures of ^{15}N enriched and natural abundance standards, calibrations based on the thermal decomposition of ammonium nitrate to produced N_2O are robust (Westley et al. 2007). In thermal decomposition, nitrate and ammonium contribute uniquely to the α and β positions in N_2O, respectively with small and symmetrical fractionation effects (Friedman and Bigeleisen 1950; Toyoda and Yoshida 1999; Westley et al. 2007).

23.2.5 Spectroscopic Analysis of N_2O Isotopomers

An important advantage of spectroscopic approaches in N_2O isotopomer analysis relative to mass spectrometry is that the absorption spectra of each isotopomer are unique and consist of distinct spectral lines such that mass overlap is less of a concern. Thus spectroscopic data provides a truly independent approach to mass spectrometry. Spectroscopic measurements of N_2O isotopologues in environmental samples are relatively rare (Wahlen and Yoshinari 1985; Yoshinari and Wahlen 1985; Esler et al. 2000; Pérez et al. 2001; Griffith et al. 2000; Turatti et al. 2000). This likely reflects the considerable challenges of spectroscopic approaches for obtaining isotopomer data on N_2O at near atmospheric abundances of approximately 319 ppbv (as of 2007; Forster et al. 2007).

Wahlen and Yoshinari (1985) and Yoshinari and Wahlen (1985) determined the $\delta^{18}O$ of N_2O in wastewater treatment plants using a high resolution infrared tunable diode laser with a precision better than 1 ‰ for concentrations from ambient to 50 ppmv. Recently, Picarro Instruments (www.picarro.com) reported precision better than 1 ‰ for both $\delta^{15}N$ and $\delta^{18}O$ using a cavity ring down spectroscopic instrument with an effective cell path length of 20 km. Determination of site specific isotope values in N_2O have, thus far, been restricted to Fourier Transform Infrared (FT-IR) spectroscopy which may reflect the challenge and time consuming nature of the measurement (Esler et al. 2000; Pérez et al. 2001; Griffith et al. 2000; Turatti et al. 2000). With continued improvements in design broader application of spectroscopic instruments in isotopomer studies will undoubtedly increase. Spectroscopic measurements have proven quite useful in the calibration of internal laboratory standards and in providing accurate data on the SP of tropospheric N_2O (Toyoda and Yoshida 1999; Röckmann et al. 2003; Griffith et al. 2009).

23.2.6 Recommendations for Calibration and Mass Interference Corrections

At the present time, the most accurate approach for calibration and mass overlap corrections of N_2O isotopomers is presented in Westley et al. (2007). Calibration using mixtures of isotopically enriched and natural abundance N_2O standards has not proven successful; however, this approach is essential for determination of the scrambling or rearrangement factor. As the rearrangement factor is dependent upon source conditions it needs to be determined for each instrument and after significant changes to the mass spectrometer ion source (e.g. cleaning and filament replacement). The isotopomer research community suffers from the lack of an international isotopically characterized N_2O standard. Internal laboratory standards may best be calibrated by thermal decomposition of ammonium nitrate for which the $\delta^{15}N$ of the ammonium and nitrate are determined by independent approaches (Toyoda and Yoshida 1999; Westley et al. 2007). There is a compelling need for the research community to develop a standard that could be made available from an international agency such as the IAEA.

Recent efforts support a SP of tropospheric N_2O of 18.7 ± 2.2 ‰ (Toyoda and Yoshida 1999; Westley et al. 2007) which is consistent with that obtained by FTIR of 19.8 ± 2.1 ‰ (Griffith et al. 2009). Consequently, all laboratories can analyze the SP of tropospheric N_2O to validate calibrations. Mass interference corrections will undoubtedly continue to be revised. For example, in their corrections, Westley et al. (2007) assumed that $^{14}N^{14}N^{17}O$ has the same rate of fragmentation within the ion source as $^{14}N^{14}N^{16}O$. Yet, they demonstrated that this assumption is not valid for other isotopomers. The impact of this assumption is likely negligible at present levels of precision but could become important in high precision applications or if the sample of N_2O analyzed was far from mass dependence. As the mass interference corrections are not trivial and likely to evolve there is value in using common methods for these corrections (e.g. a downloadable spreadsheet). Similar spreadsheets are currently available for download from the IAEA for spectroscopic analysis of the isotopic composition of waters (http://www-naweb.iaea.org/napc/ih/index.html). Because correction routines are complex and changing there may be a need to publish (at least electronically) raw ion current ratios (^{31}R, ^{45}R, and ^{46}R) as well as the isotopomer values of the internal laboratory standard used to enable future revisions of mass interference corrections to be applied to existing data.

A regrettable outcome of the use of mixtures of ^{15}N enriched and natural abundance N_2O for calibration has been the publication of SP values that cannot be considered accurate. These include not only the SP values for tropospheric N_2O but for other natural samples. A simple offset correction cannot be used because the approach based on mixtures of ^{15}N enriched and natural abundance standards will yield different results for the same gas analyzed on different mass spectrometers (Westley et al. 2007).

Consequently, the reader needs to verify the calibration protocol before referencing specific SP values; especially for studies prior to the publication of Westley et al. (2007). It may be possible for the authors of earlier papers to publish corrected values.

23.3 Application of Isotopomers to Resolve Biological Production Pathways

23.3.1 Overview

The isotopologue composition of any material records two kinds of information: its origin and/or process of formation. The former is dependent upon a unique isotopologue signature for the material that is retained during transport or transformation in the environment. This is considered conservative behavior. Non-conservative behavior occurs with isotopic fractionation; the tendency for a molecule with the light isotope to react at a faster rate than the same molecule retaining the heavy isotope. In general, the behaviors of molecules that contain the isotopes of nitrogen are rarely conservative. This is a consequence of (1) the wide range of valence states characteristic of this element and the associated wide range of covalent bonds that can be formed and (2) the fact that N is intimately involved in a large number of different biological reactions and is often a limiting nutrient to primary production and metabolism. In rare circumstances, isotopologue values may retain both source and process information (e.g. Ostrom et al. 2002), but most often source information is lost once fractionation occurs. The behavior of the isotopologies of N_2O is no exception. The observation, however, that during its production in pure culture SP (1) is independent of the substrate's isotopic composition and (2) is constant during the course of production even when $\delta^{15}N$ and $\delta^{18}O$ changes markedly, provides strong support for the use of SP as a conservative tracer (Sutka et al. 2003, 2006, 2008; Toyoda et al. 2005). In the remainder of this review, we will discuss the application of isotopologues to resolve microbial N_2O production pathways and how both conservative and non-conservative behavior strengthen or limit interpretations of its origins and processes of formation.

23.3.2 Non-Conservative Behavior of the $\delta^{15}N$ of N_2O: An Example from Fungal Denitrification

A challenge to the application of bulk isotopes values to resolve origins of N_2O is the wide range of fractionation factors that are associated with production from different pathways. In this paper, we describe fractionation factors as constants that are specific to a single reaction or reaction step. We follow the convention of Mariotti et al. (1981) by defining the magnitude of isotopic fractionation during a single unidirectional reaction as the fractionation factor α,

$$\alpha = k_2/k_1 \qquad (23.9)$$

where k_1 and k_2 are the reaction rates for the light and heavy isotopically substituted compounds, respectively (although some authors use the inverse of this ratio). We further define an isotopic enrichment factor, ε:

$$\varepsilon = (\alpha - 1) * 1{,}000 \qquad (23.10)$$

The value ε is convenient because it is approximately equal to the difference in the isotopic composition of the substrate (δ_s) and product (δ_p) or Δ at any point during the reaction:

$$\varepsilon \approx \Delta = \delta_s - \delta_p \qquad (23.11)$$

Values for ε during production of N_2O from nitrification and denitrification vary between -45 and -111 ‰ for nitrification and -13 and -45 ‰ for denitrification (Yoshida 1988; Barford et al. 1999; Pérez et al. 2006). As we explain below, variable isotopic enrichment factors are a consequence of processes in which isotope fractionation may occur at multiple steps. Denitrification, for example, consists of a series of steps that involves diffusion of nitrate into the cell followed by the sequential reduction of nitrate to nitrite to nitric oxide to N_2O and finally N_2 (Davidson 1993). Each step in the denitrification reaction sequence is undoubtedly associated with a unique and constant fractionation factor. Although fractionation occurs at multiple points, it is common to describe fractionation as the difference in the isotope value of the substrate and product (e.g. difference in $\delta^{15}N$ between nitrate and N_2). This "observed" fractionation is the net effect of several isotope effects (several cases of fractionation) whose relative importance may vary depending upon environmental conditions. For this reason, we prefer to describe fractionation factors in biological reactions as net isotope effects (NIE) and replace ε with η. Following this practice, ε is a constant specific to fractionation for a single reaction whereas η is the NIE. We explore this concept below using production of N_2O by fungal denitrification as an example.

Fractionation factors can be determined from a Rayleigh distillation equation that describes fractionation as a function of changes in the abundance and isotopic composition of the substrate and product for a uni-directional reaction within a closed system. The Rayleigh distillation equation has been written in numerous forms, e.g. (23.13), and is amenable to data of Sutka et al. (2008) on N_2O production by fungal denitrification because it allows η to be expressed as a function of the isotopic composition of the accumulating product. This form of the Rayleigh equation is:

$$\delta^{15}N_p = \delta^{15}N_{so} - \eta_{p/s}[(f \ln f)/(1-f)] \quad (23.12)$$

where f is defined as the fraction of substrate (NO_2^-) remaining (determined as the ratio of the calculated NO_2^- concentration at any point in time divided by the initial concentration), $\eta_{p/s}$ is the NIE for any substrate (s) converted to a product (p), and (so) refers to the substrate prior to initiation of the reaction (Mariotti et al. 1982). When (23.12) is plotted in coordinates of $-\dfrac{f \ln f}{1-f}$ vs. $(\delta^{15}N_p - \delta^{15}N_{so})$ the slope of a linear regression equation equates to $\eta_{p/s}$ (Fig. 23.2). Values of η for the change in $\delta^{15}N$ during fungal denitrification are not constant and range from −6.6 to −74.7 ‰ across separate cultures and species (Fig. 23.2).

Evidence for departure from Rayleigh behavior is pronounced for replicate A of *Cylindrocarpon tonkinese*, e.g. in Fig. 23.2 as a curvilinear relationship is observed ($r^2 = 0.97$). The curvilinear nature of the data indicates that η is not constant but variable over the course of fungal denitrification. Therefore, η must be the net result of fractionation during at least two steps.

A frequently cited equation describing η during photosynthesis is (Oleary 1981; Farquhar et al. 1982):

$$\eta = \varepsilon_a + (\varepsilon_b - \varepsilon_a)K_1/(K_2 + K_3) \quad (23.13)$$

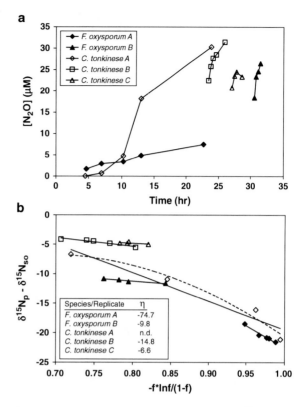

Fig. 23.2 Changes in the concentration (**a**) and $\delta^{15}N$ (**b**) of N_2O produced during fungal denitrification for two replicate cultures of *Fusarium oxysporum* (A, B) and three cultures of *Cylindrocarpon tonkinese* (A–C) (data from Sutka et al. 2008). The *symbols* shown in (**a**) also apply to (**b**). The slopes of the regression lines through each culture are equivalent to η as determined by (23.11) and shown for each species within the *inset* in (**b**). A curvilinear relationship is evident in the data for *C. tonkinese* A indicating that the net isotope effect is changing over the course of N_2O production

Here the constants ε_a and ε_b are the isotopic enrichment factors for diffusion of CO_2 in air (−4.4 ‰) and reduction by the enzyme Rubisco (−29.4 ‰), respectively (Guy et al. 1993). Rate constants for diffusion into the cell, out of the cell and for the enzyme activity are given by K_1, K_2, and K_3, respectively. By (23.13), variation in the relative importance of these rate constants determines whether the fractionation factor for diffusion or for enzymatic reduction is the predominant control on η. When $K_1 \ll (K_2 + K_3)$, then $K_1/(K_2 + K_3)$ approaches zero and, η approaches ε_a. In this case η is small because it is largely a function of the small fractionation factor for diffusion. In contrast, when diffusion does not limit the supply of N_2O to the enzyme, $K_1/(K_2 + K_3)$

approaches 1 and η approaches ε_b. Enzymatic fractionation factors tend to be quite large relative to those associated with diffusion. Thus (23.13) demonstrates why η associated with many biological reactions tends to be variable while ε is constant and why this is the case even though both η and ε are determined using the same equations. Further, a consideration of the relative magnitude of the rate constants indicates that η should be small when diffusion limits the supply of substrate to the enzyme and large when diffusion is not limiting to the enzymatic process.

Equation (23.13) can be applied to a suite of biological reactions and has been proposed as an explanation for the wide range of NIE values observed during denitrification, N_2O reduction, and consumption of O_2 during respiration (Ostrom et al. 2002, 2005, 2007). Variation in η for fungal denitrification by the same microbial species most likely reflects depletion of the nitrite substrate, over time. A temporal decline in substrate concentration is a common occurrence in "batch" cultures and slows the rate of substrate diffusion into the cell. This then causes $K_1/(K_2 + K_3)$ to shift toward zero and reduces the magnitude of η. Observations of constant fractionation factors during multi-step reactions can be explained as the result of two types of experimental conditions: (1) when depletion of the substrate is minor such that the relationship between $-\frac{f \ln f}{1-f}$ and $(\delta^{15}N_p - \delta^{15}N_{so})$ is approximated by a line or (2) "steady-state" conditions that are offered, for example, by chemostats (Barford et al. 1999). Even within steady-state systems η varies with the relative importance of the reaction rate constants for diffusion and for the enzyme. Ideally, fractionation factors are best constrained by determining the isotope effect associated with each step within a reaction sequence. This can often be accomplished by work with pure enzymes (Guy et al. 1993).

The wide variation in η challenges the use of bulk isotope approaches to resolve the origins of N_2O. The range of published values of η for nitrification, bacterial denitrification, and fungal denitrification, −45 to −111 ‰, −13 to −45 ‰, and −6.6 to −74.7 ‰, respectively, overlap (Yoshida 1988; Barford et al. 1999; Pérez et al. 2006; Sutka et al. 2008). Therefore, $\delta^{15}N$ values cannot be accurately attributed to a specific microbial production pathway. Bulk N isotope values may be insightful if certain processes are known to be inoperative, such as anoxic conditions that prohibit nitrification, or with the use of inhibitors to isolate specific pathways (but see Sect. 23.1 for challenges associated with inhibitor approaches). While $\delta^{15}N$ values may not provide precise resolution of microbial production pathways they still give insight into sources of N_2O to the atmosphere (e.g. terrestrial vs. oceanic) as well as changes in such sources over time (Kim and Craig 1993; Dore et al. 1998). For example, changes in the $\delta^{15}N$ and $\delta^{18}O$ of ice core N_2O from pre-industrial times to present are consistent with increased emissions from agricultural activities (Stein and Yung 2003; Sowers et al. 2003; Bernard et al. 2006).

23.3.3 The Use of Bulk $\delta^{18}O$ Values in Assessing the Origins of N_2O

$\delta^{18}O$ has been proposed as a tracer of microbial N_2O production in a variety of studies (Ostrom et al. 2000; Mandernack et al. 2000, 2009; Frame and Casciotti 2010). The use of $\delta^{18}O$ requires that different N_2O production pathways have unique fractionation factors or incorporate elemental O from isotopically distinct sources (e.g. water or O_2) (Ostrom et al. 2000).

During nitrification, the first O atom added to ammonium to form hydroxylamine is from O_2 where as those added to form nitrite and nitrate are from water (Dua et al. 1979; Hollocher et al. 1981; Andersson and Hooper 1983; Kumar et al. 1983) (Fig. 23.1). These sources of O in the ocean and soil environments typically differ by greater than 20 ‰ (Kroopnick and Craig 1976). Ostrom et al. (2000) observed that the difference in $\delta^{18}O$ between O_2 and N_2O changed markedly with depth in the water column in the North Pacific Ocean and concluded that these shifts reflected changes in the proportion of N_2O derived from hydroxylamine oxidation or reduction of nitrite. Thus changes in the $\delta^{18}O$ of N_2O and O_2 were used to infer whether N_2O was produced by via hydroxylamine oxidation or by reduction of nitrite (Fig. 23.1). The $\delta^{18}O$ of O_2 is not commonly measured concurrent with that of N_2O particularly in soil environments. Consumption of O_2 during respiration invokes a large isotope effect that can then cause increases in the $\delta^{18}O$ of N_2O upon incorporation of O_2 during nitrification (Mandernack et al. 2009). Consequently, changes in the $\delta^{18}O$ of N_2O may not always reflect changes in sources but incorporation of a respiration isotope effect from O_2.

There is ample evidence that elemental O in N_2O exchanges with that in water during the microbial generation of this gas (Kool et al. 2007). Casciotti et al. (2002) found exchange to be low during N_2O production by denitrification in *Pseudomonas aereofaciens* (<10 %) but as high as 78 % in *Pseudomonas chlororaphis*. In incubation of soils with acetylene to inhibit N_2O reduction, Wrage et al. (2005) observed no evidence of exchange, however, in the absence of acetylene exchange was nearly 100 %. Exchange of 1 to 25 % between O in N_2O and water has been shown for a variety of microorganisms during the first step of nitrification, ammonia oxidation (Casciotti et al. 2010). Exchange with water is not expected during N_2O production via hydroxylamine oxidase (HAO) but enzyme studies have shown that exchange occurs during N_2O production by nitrite reductase (NiR) and nitric oxide reductase (NOR) (Ye et al. 1991; Kool et al. 2007). The potential for exchange to occur during microbial N_2O production has led numerous studies to conclude that $\delta^{18}O$ is a poor tracer of microbial origins (Kool et al. 2007, 2009a, b; Ostrom et al. 2007, 2010). Nonetheless, Frame and Casciotti (2010) recently reported a strong relationship between $\delta^{18}O$ and SP in microbial cultures where O_2 levels were varied to favor production from hydroxylamine oxidation or nitrifier-denitrification and concluded that $\delta^{18}O$ is a potential tracer of these pathways. This observation implies that water exchange must be limited during nitrifier-denitrification and hydroxylamine oxidation. The challenge to studies addressing the origins of N_2O on the basis of $\delta^{18}O$ is to know when exchange of O between N_2O and water occurs and to what extent. In this regard, incubations with ^{18}O-enriched water are a viable approach to resolve this issue.

23.3.4 The Use of SP to Resolve Microbial N_2O Production Pathways

A key advantage of SP in tracing origins of N_2O is that, in contrast to bulk isotope values, SP is independent of the isotopic composition of the substrates of nitrification or denitrification (O_2, water, nitrate, nitrite or ammonium) (Toyoda et al. 2005; Sutka et al. 2006). Further, in pure microbial cultures the SP of N_2O produced by nitrifying and denitrifying bacteria did not exhibit fractionation ($\eta = 0$ ‰) even though bulk, α and β $\delta^{15}N$ values were characterized by non-zero η values (Fig. 23.3). These characteristics, lack of dependence on the isotopic composition of the substrates and

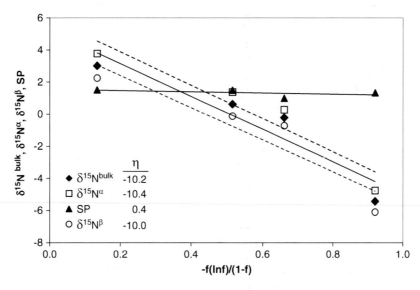

Fig. 23.3 Isotopomer values for N_2O production via bacterial denitrification by *Pseudomonas chlororaphis* as a function of the change in the extent of the reaction, f. Linear regression lines for $\delta^{15}N^{bulk}$ and SP are shown with *solid lines* and for $\delta^{15}N^\alpha$ and $d^{15}N^\beta$ with *dashed lines*. Based on (23.3), the slope of the regression equation for each isotopomer data set is equivalent to η. Values for η are indicated to the *left* of each isotopologue value in the legend. These data demonstrate that while fractionation is evident during N_2O production by bacterial denitrification for $\delta^{15}N_{bulk}$, $\delta^{15}N^\alpha$ and $\delta^{15}N^\beta$ fractionation in SP was essentially zero. Data are from Sutka et al. (2006)

lack of fractionation during production, provide strong support for the use of SP as a conservative tracer of microbial production. Critical to this approach, however, is that different sources of N_2O have unique SP values. Recently Sutka et al. (2006) found unique SP values for N_2O derived from hydroxylamine oxidation (nitrification) and nitrite or nitrate reduction (denitrification and nitrifier denitrification) in bacterial cultures and suggested that SP could be used to apportion microbial sources of this gas. Production of N_2O via nitrite or nitrate reduction, whether by nitrifying or denitrifying bacteria, resulted in an average SP values of 0 ‰ compared to 33 ‰ for N_2O produced from hydroxylamine oxidation (Sutka et al. 2006). Toyoda et al. (2005) similarly reported a SP value for N_2O produced from one species of denitrifying bacteria of -5.1 ‰. A second species produced N_2O with a SP of 23.3 ‰; however, this culture occurred with a low growth rate that may have been influenced by inorganic production. Sutka et al. (2008) further demonstrated that fungal denitrification yields N_2O with a SP of 37 ‰, which is not distinct from that associated with nitrification but clearly distinct from that produced during bacterial denitrification. Collectively these pure culture studies provide compelling evidence that a SP of 0 ‰ or less is indicative of N_2O produced by denitrifying bacteria whereas values of 33–37 ‰ indicates production from hydroxylamine oxidation or fungal denitrification.

Soil incubations directed toward defining the SP values associated with nitrification and denitrification have found values distinct from those obtained in pure culture. Although widely cited, the study by Pérez et al. (2006) was based on the calibration approach of Kaiser et al. (2004a) and reported SP values need to corrected. Well et al. (2008) recently reported a range of 18.3–36.3 ‰ for N_2O produced by nitrification in soils. The higher values are in agreement with the SP values obtained via hydroxylamine oxidation in pure culture (Sutka et al. 2006). The lower values are consistent with production from a mixture of hydroxylamine oxidation and nitrifier-denitrification which has been shown to have a SP in pure culture near 0 ‰ (Sutka et al. 2006). Thus the variable SP signals attributed to nitrification are quite readily explained by a mixture of N_2O derived from hydroxylamine oxidation and nitrifier denitrification. Two incubation studies using temperate soils yielded variable SP values for N_2O production from denitrification of 3.1–8.9 ‰ (Well and Flessa 2009) and 2.3–16 ‰ (Well et al. 2006). The lower values are similar to those obtained in pure culture (-5 to 0 ‰; Toyoda et al. 2005; Sutka et al. 2006). The higher values could reflect production from fungal denitrification (Sutka et al. 2008) or, when acetylene was not used, an increase in SP due to reduction of N_2O (Ostrom et al. 2007; Jinuntuya-Nortman et al. 2008). These studies indicate that a significant challenge to soil incubation studies is to isolate specific production pathways in a complex soil community and the results obtained thus far are not inconsistent with SP endmember values obtained in pure culture.

The ultimate control on the SP of N_2O may not be the bacterial species or pathways but the enzymes involved in its production, in particular, NIR and NOR (Stein and Yung 2003; Schmidt et al. 2004). HAO catalyzes the four electron transfer of hydroxylamine to nitrite although N_2O is given off in a branched reaction (Kostera et al. 2008). There are three bacterial and one fungal type of NOR enzymes reconized: CNOR, qNOR, and qCuNor and P450NOR, respectively. cNOR is associated with bacterial denitrification while qNOR and qCuNor are associated with detoxification of NO and may not be environmentally relevant (Hendriks et al. 2000). Traditionally production of N_2O via P450NOR has not been considered important but most fungi lack N_2O reductase and, therefore, all nitrate reduction by fungi yields N_2O whereas most reduction in bacteria yields N_2 (Shoun et al. 1992). Further, recent studies applying antibiotics to soils have implicated fungi as primary sources of N_2O (Crenshaw et al. 2008). The three bacterial NOR mainly differ in electron donors but the active site structures of NOR are considered highly homologous; therefore, they are expected to have identical active sites and behave very similarly (Hendriks et al. 2000; Tavares et al. 2006). The NOR enzymes used in nitrifier-denitrification are nearly identical to those found in denitrifying bacteria and likely were obtained by lateral gene transfer (Garbeva et al. 2007; Kool et al. 2007).

Distinctions and similarities in the SP of N_2O during various production pathways can be understood by consideration of the enzyme pathways in addition to the specific microbial species involved. Sutka et al. (2006) cultured two species of *Pseudomonas* sp. Each species possessed one of the two known types of dissimilatory NIR. A similar SP was observed in N_2O produced by both species (0 ‰) even though the type of NIR enzyme involved differed. This result provides support for the suggestion that during bacterial denitrification NOR is more important in controlling SP than NIR (Stein and Yung 2003). Highly contrasting SP values have been demonstrated for denitrifying fungi (37 ‰) and bacteria (0 ‰) consistent with a control of NOR on SP (Sutka et al. 2006, 2008). As the NIR and NOR enzymes in denitrifiers and nitrifiers are structurally and genetically related (Casciotti and Ward 2001, 2005; Garbeva et al. 2007) it is not surprising that the SP for N_2O from bacterial denitrification (−2.5 to 1.8 ‰) and nitrifier-denitrification (−4.0 to 1.9 ‰) were found to be similar (Sutka et al. 2006). Bacteria using qNOR have not been cultured to evaluate SP. As this pathway is primarily used in detoxification of environmental NO it likely constitutes a small portion of N_2O production in soils (Hendriks et al. 2000). Thus the SP associated with bacterial denitrification (including nitrifier-denitrification) via cNOR (−5 to 0 ‰) is distinct from other pathways (nitrification via hydroxylamine oxidation and fungal denitrification (33–37 ‰)) and has the potential, therefore, to act as a tracer of its production.

Frame and Casciotti (2010) recently proposed revision of the SP endmember values to −10 and 37 ‰, respectively, for nitrite reduction and hydroxylamine oxidation. Collectively pure culture studies provide compelling evidence that a SP of −10 to 0 ‰ is indicative of N_2O produced by denitrifying bacteria whereas values of 33–37 ‰ indicates production from hydroxylamine oxidation and/or fungal denitrification. These revised ranges of SP values can be used to apportion production of N_2O to (1) bacterial denitrification and (2) hydroxylamine oxidation/fungal denitrification. For example, SP values for soil-derived N_2O were measured throughout a growing season in a temperate agricultural field varied from −12.5 to 17.6 ‰ (Fig. 23.4). The majority of values are consistent with an origin from bacterial denitrification. When each value is weighted by the N_2O flux to the atmosphere at that time, the flux weighted SP for this field is −1.3 ‰. Assuming endmember values of 0 and 33 ‰ or −10 to 37 ‰ for bacterial denitrification and hydroxylamine oxidation/fungal denitrification, respectively, the proportion of N_2O derived from bacterial denitrification is between 73 and 100 %. This result indicates that management activities at this particular location should focus on actions that minimize

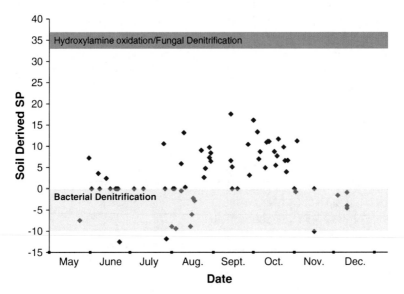

Fig. 23.4 SP values for soil-derived N_2O from an agricultural field planted in wheat in 2007 at the Kellogg Biological Station, Hickory Corners, Michigan. The field received application of urea-ammonium nitrate (28%) fertilizer in the spring at a rate of 246 kg N ha^{-1}. Shown in *yellow* and *blue* are the SP values associated with N_2O production from bacterial denitrification and hydroxylamine oxidation/fungal denitrification. Static flux chambers were used to collect all samples which initially (at time zero) includes N_2O from the atmosphere. Consequently, soil-derived values are calculated by mass balance assuming concentration and isotopomer values for atmospheric N_2O (see Opdyke et al. 2009). N. Ostrom, P. Ostrom, N. Millar, G.P. Robertson, unpublished data

denitrification. Such activities might include, for example, the avoidance of anaerobic conditions resulting from excessive irrigation. Currently, there is not a definitive means on the basis of SP to distinguish the two pathways of N_2O production by nitrifying bacteria; however, some studies have indicated that $\delta^{18}O$ values may hold promise (Mandernack et al. 2000, 2009; Wrage et al. 2005; Frame and Casciotti 2010) and with improvements in precision, $\delta^{17}O$ may be insightful as well (see Sect. 23.3.7).

23.3.5 Isotopomer Effects Associated with Reduction of N_2O

While the SP values of N_2O have the potential to reveal microbial production pathways, consumption (or reduction) of N_2O during denitrification can impart an isotope effect that could bias or obscure source isotope values (Tilsner et al. 2003; Wrage et al. 2004; Yamagishi et al. 2005). Rates of N_2O reduction independent of gross denitrification are rarely determined in soils or in the ocean; hence it is difficult to know the importance of this process a priori (e.g. Bandibas et al. 1994; Cavigelli and Robertson 2001). Denitrification results in cleavage of the N and O covalent bond in N_2O, which based on kinetic isotope theory, is expected to result in an increase in the ^{15}N content of the α position and, therefore, an increase in SP (Popp et al. 2002; Yamagishi et al. 2005). The increase in SP in response to reduction would result in a shift away from values associated with bacterial denitrification (−10 to 0 ‰) towards those associated with hydroxylamine oxidation/fungal denitrification (33–37 ‰) (Sutka et al. 2006; Frame and Casciotti 2010) and result in an underestimate of the importance of bacterial denitrification as a source of N_2O. Consequently, the importance of the N_2O reduction isotope effect on SP values should, ideally, be established prior to interpretation of the microbial origins of N_2O.

A number of studies have addressed isotopologue effects associated with N_2O reduction. Ratios of $\delta^{18}O$ to $\delta^{15}N$ and $\eta^{18}O$ to $\eta^{15}N$ are strongly linear during N_2O reduction in pure microbial cultures and soils and to cluster within a narrow range of 2.4–2.6 ‰ (Menyailo and Hungate 2006; Ostrom et al. 2007; Vieten et al. 2007; Jinuntuya-Nortman et al. 2008). Similarly, within pure microbial cultures and soil incubations the relationship between $\eta^{18}O$ and the NIE for SP has been reported to be 1.7 and 1.9 ‰, respectively (Ostrom et al. 2007; Jinuntuya-Nortman et al. 2008). Consequently, these mass dependent relationships between the isotopologues and η values are diagnostic of N_2O reduction. Studies of the η for SP during N_2O reduction have been considerably fewer than those for bulk isotopes. Values for ηSP have been reported as −6.4 ‰ from one ocean study, −6.8 to −5.0 ‰ in microbial culture, and −4.5 to −2.9 ‰ in a study of temperate soils (Yamagishi et al. 2005; Ostrom et al. 2007; Jinuntuya-Nortman et al. 2008). Jinuntuya-Nortman et al. (2008) found a strong correlation between ηSP and the degree of water filled pore space in soils which was consistent with an increase in the importance of diffusion in controlling η (by (23.13)) as soils become more saturated. Overall the magnitude of the η for SP is small relative to that for bulk $\delta^{15}N$ and $\delta^{18}O$ values but substantial enough that the impact of N_2O reduction on SP should not be ignored. The mass dependent relationships between η for SP and bulk $\delta^{15}N$ and $\delta^{18}O$ for N_2O reduction are strongly diagnostic of this process and unique relative to isotopologue shifts that occur in response to mixing of soil and oceanic N_2O with atmospheric N_2O. Consequently, Ostrom et al. (2007) suggested that the small η for N_2O reduction and observation of a slope of less than 1 for a plot of $\delta^{18}O$ vs. $\delta^{15}N$ indicated that shifts in SP during N_2O reduction were relatively minor whereas those greater than 1 implied that a N_2O reduction had altered SP appreciably. Under the latter condition, a correction to SP is needed before using SP to determine the relative proportion of N_2O derived from bacterial denitrification and hydroxylamine oxidation/fungal denitrification as described in preliminary models (Yamagishi et al. 2005; Ostrom et al. 2007; Opdyke et al. 2009).

23.3.6 Correction of SP Values for N_2O Reduction

All fractionating processes alter the isotopic composition of the substrate and when this occurs the source isotope values are altered and potentially loss. While soils are well established as a source of N_2O to the atmosphere it is also recognized that N_2O is reduced in soils and that reduction alters $\delta^{15}N$, $\delta^{18}O$ and SP (Chapuis-Lardy et al. 2007; Ostrom et al. 2007; Jinuntuya-Nortman et al. 2008). Once isotopic alteration has occurs the proportion of N_2O derived from bacterial denitrification and hydroxylamine oxidation/fungal denitrification determined from SP will be incorrect unless the isotope effect during reduction can be quantified. Correcting for fractionation during reduction and determining the SP value of N_2O prior to reduction is challenged by the fact that the NIE for reduction is variable. Nonetheless, the mass dependent relationships between $\delta^{15}N$ and $\delta^{18}O$ and $\delta^{15}N$ and SP during N_2O reduction are unique and diagnostic of this process (Menyailo and Hungate 2006; Ostrom et al. 2007; Vieten et al. 2007; Jinuntuya-Nortman et al. 2008). While not quantitative, a simultaneous production and reduction model that describes both isotopologue shifts during production and fractionation during reduction has been developed to estimate the maximum likely shift in SP due to reduction within soil trace gas flux chambers (Ostrom et al. 2007; Opdyke et al. 2009).

Within soil flux chambers the isotopologue composition of N_2O will change with time owing to (1) mixing of atmospheric and soil derived N_2O and (2) N_2O reduction. The isotopologue shifts due to mixing are described by:

$$\delta_{mix} = (\delta_{atm}C_{atm} + \delta_{SD}C_{SD})/C_{mix} \quad (23.14)$$

where, δ and C refer to the isotopic composition ($\delta^{15}N$, $\delta^{18}O$ or SP) and concentration of N_2O, respectively. The subscripts atm and SD refer to atmospheric and soil derived N_2O, respectively, and mix indicates the resulting mixture of these two sources. Equation (23.14) is shown graphically on a plot of $\delta^{18}O$ vs. $\delta^{15}N$ as the "Atm-Soil N_2O Mixing" line in Fig. 23.5. Once a soil flux chamber is closed and soils release N_2O

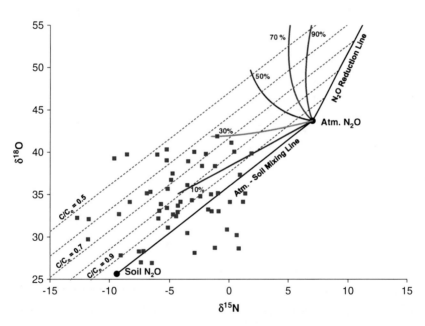

Fig. 23.5 Oxygen and nitrogen isotope data from soil flux chambers in an agricultural field planted in wheat in 2007 at the Kellogg Biological Station, Hickory Corners, Michigan (*squares*) in comparison to modeled results of simultaneous production and reduction. The isotopic composition of flux weighted "Soil N_2O" for the entire season is shown as closed circle. In the absence of N_2O reduction, atmospheric N_2O (also shown as a *closed circle*) mixes with soil derived N_2O and values would lie on a line (Atm. − soil mixing) connecting these two endmembers. The N_2O reduction line indicates where isotope values would lie if reduction was the only process affecting N_2O isotopologues (Ostrom et al. 2007; Jinuntuya-Nortman et al. 2008). The *curvilinear lines* represent isotopologue values resulting from simultaneous production (mixing of air and soil N_2O) and reduction at difference percentages of the rate of reduction relative to the rate of production (10, 30, 50, 70 and 90%). The distance along any of the simultaneous production and reduction lines away from atmospheric N_2O values indicates a greater degree of N_2O reduction as described by the observed concentration (C) relative to that which would have been present in the absence of reduction (C_o). The *dashed lines* connect all points in which the extent of reduction, C/C_o, is the same. Data points below the mixing line likely represent variation in the isotopologue values of the soil N_2O endmember or non-conservative behavior such as exchange of O with water during production

to the headspace, the isotopologue values will move from those of the atmosphere towards those of soil derived N_2O shown in Fig. 23.5 as the "atm-soil N_2O mixing" line.

The change in isotopologue values resulting from reduction of N_2O can be determined from a Rayleigh fractionation model as the difference between the isotope value of the residual substrate (δ_s) and that prior to any reduction (δ_o) (Mariotti et al. 1981):

$$\Delta = \delta_s - \delta_o = -\eta \ln(C/C_o) \quad (23.15)$$

Here, η is the fractionation factor for N_2O reduction and C/C_o is the ratio of the concentration of the residual N_2O relative to that which was present prior to reduction. Λ is defined as the difference in isotopic composition of N_2O at any point (δ_s) in time and prior to reduction (δ_o). During reduction of N_2O, the ratio of $\eta^{18}O$ and $\eta^{15}N$ has been shown to be constant at a value of 2.6 (Ostrom et al. 2007; Jinuntuya-Nortman et al. 2008). Therefore, if reduction occurred in the absence of production the concentration of N_2O would decline from atmospheric levels and $\delta^{18}O$ and $\delta^{15}N$ values would increase along a slope of 2.6 (shown as the "N_2O Reduction" line in Fig. 23.5).

The situation in which both production and reduction occur simultaneously can be described by combining (23.14) and (23.15). To effect this we equate the isotopologue and concentration values of the mixture of atmospheric and soil derived N_2O (δ_{mix} and C_{mix}, respectively) in (23.14) to those of the initial substrates in (23.15) (δ_o and C_o, respectively). Thus (23.14) becomes:

$$\delta_o = (\delta_{atm}C_{atm} + \delta_{SD}C_{SD})/C_o \quad (23.16)$$

δ_o from (23.16) is substituted into (23.15) and solved for δ_s

$$\delta_s = (\delta_{atm}C_{atm} + \delta_{SD}C_{SD})/C_o + \eta \ln(C/C_o) \quad (23.17)$$

Equation (23.17) can be written in terms of any isotopologue:

$$\delta^{18}O_s = (\delta^{18}O_{atm}C_{atm} + \delta^{18}O_{SD}C_{SD})/C_o$$
$$+ \eta^{18}O \ln(C/C_o) \quad (23.18)$$

and

$$\delta^{15}N_s = (\delta^{15}N_{atm}C_{atm} + \delta^{15}N_{SD}C_{SD})/C_o$$
$$+ \eta^{15}N \ln(C/C_o) \quad (23.19)$$

Equations (23.18) and (23.19) describe the changes in isotopologue values of N_2O when production occurs simultaneously with consumption as a function of the isotopic composition of soil derived and atmospheric N_2O, the fractionation factors for reduction, and the concentration prior to (C) and after reduction (C_o). A desirable outcome of these equations is the term C/C_o as this would provide a basis to determine the shift in SP (Δ) due to reduction by (23.15). If C/C_o can be determined, Δ for SP can be determined and subtracted from measured SP values to provide a corrected estimate of the fraction of N_2O derived from bacterial denitrification vs. that derived collectively from hydroxylamine oxidation and fungal denitrification. Because C_o appears in both the linear and logarithmic portions of (23.5) and (23.6) an explicit solution is not possible, however, an implicit relationship can be generated graphically using theoretical scenarios of simultaneous production and reduction as shown in Fig. 23.5. An arbitrary production rate is chose and the rate of reduction is set to a percentage of the production rate (10, 30, 50, 70, and 90 % in Fig. 23.5). The following terms are used in (23.18) and (23.19):

$\delta^{15}N_{atm} = 7.0$ ‰ (Yoshida and Toyoda 2000)
$\delta^{18}O_{atm} = 43.7$ ‰ (Yoshida and Toyoda 2000)
$C_{atm} = 315$ ppbv (Yoshida and Toyoda 2000)
$\delta^{15}N_{SD}$ and $\delta^{18}O_{SD}$ (taken as the flux weighted average for the entire data set)
C_{SD} (difference between the concentration at t_1 and t_0 (C_{atm}))
$\delta^{15}N_s$ and $\delta^{18}O_s$ (measured at t_1)
$\eta^{15}N$ (between -9.2 and -1.8 ‰) (Ostrom et al. 2007)
$\eta^{18}O$ (calculated from $\eta^{18}O$ to $\eta^{15}N$ constant relationship of 2.6) (Jinuntuya-Nortman et al. 2008)

The minimum value for $\eta^{15}N$ observed at this study site (-9.2 ‰) was chosen to create a "worse-case-scenario" in which the greatest Δ SP value can be expected. Production and reduction are allowed to occur for multiple time points over several hours and the results are plotted as simultaneous production and reduction (SPR) curved lines shown in Fig. 23.5. All of the SPR curves originate at the isotope values for atmospheric N_2O (no production or reduction and, therefore, a C/C_o value of 1). As the rate of production outpaces that of reduction the C/C_o term decreases with time. Thus, as the isotope values along any SPR curve move away from the atmospheric endmember C/C_o decreases. Common C/C_o values between the SPR curves are connected to create the parallel "C/C_o" lines shown in Fig. 23.5. Once this theoretical graph is generated field data (from Fig. 23.4) is overlaid. The location of each data point corresponds to a value of C/C_o and a production to reduction ratio. Thus, the data presented has maximum values of the production to reduction ratio of slightly greater than 30 % and C/C_o slightly greater than 0.5. This latter value can be entered into (23.15) to determine the maximum possible Δ value expected. Based on the maximum value for C/C_o and the largest fractionation factor in SP observed during N_2O reduction in soils at this location (−4.5 ‰) (Jinuntuya-Nortman et al. 2008), a maximum shift in SP of 4.0 ‰ can be expected. Given SP values of 0 ‰ for production from denitrification and 37 ‰ from other sources, a shift in SP of 4 ‰ corresponds to an underestimate of production from bacterial denitrification of 10.8 %. Consequently, the maximum expected shifts in SP due to N_2O reduction in this study are small but should not be ignored.

There are limitations to the SPR model that prevent us from determination of Δ values quantitatively. Constant and invariant isotopologue values for the soil derived endmember are assumed (determined as the flux weighted isotope values for the entire data set); however, the soil derived endmember values likely varies in time and space. Further, constant values for $\eta^{15}N$ and $\eta^{18}O$ during reduction are assumed but these values can be expected to vary as well. Thus, the results of the SPR model are considered estimates of the shift in SP during reduction under a "worse-case" scenario. Nonetheless, the results presented in Fig. 23.5 suggest (1) that as the ηSP values are small, reduction has to be quite extensive (>50 % the rate of production) to alter SP value substantially and (2) if uncorrected, the shift in SP due to reduction would bias the results in favor of production from hydroxylamine oxidation/fungal denitrification. Thus the observation of a predominance of SP values near the −10 to 0 ‰ endmember for production from bacterial denitrification (Fig. 23.4) is compelling evidence that this process dominates N_2O production as all other processes for which SP is known (hydroxylamine oxidation, fungal denitrification and N_2O reduction) will result in higher SP values.

23.3.7 Resolution of the Causes of the ^{17}O Anomaly in N_2O

It is well known that all biological and most geochemical processes on Earth produce a consistent mass dependent relationship between $\delta^{17}O$ and $\delta^{18}O$ such that the slope of the line relating these terms is constant and equal to 0.516 (Thiemens 2006). Tropospheric N_2O is slightly enriched in ^{17}O relative to the "mass-dependent" line by approximately 0.9 ‰ (Cliff and Thiemens 1997; Cliff et al. 1999; Kaiser et al. 2003, 2004b). Photochemical reactions in the stratosphere are known to produce "mass-independent" relationships and most notable is stratospheric ozone that is defined by $\delta^{17}O$ to $\delta^{18}O$ relationship of approximately 0.62 (Lammerzahl et al. 2002).

The unusual ^{17}O enrichment in tropospheric N_2O has long been assumed to result from photochemical reactions and exchange of N_2O between the stratosphere and troposphere. Prasad and Zipf (2008) recently proposed that the anomaly is a consequence of photochemical production of N_2O. If correct, then the IPCC is overestimating the biological sources of N_2O by approximately 7 % and the global budget of N_2O would be unbalanced. A recent assessment of laboratory experiments and models, however, suggested that only part of the "anomaly" can be explained by photochemical reactions. Biological production has been hypothesized as the cause of the remaining ^{17}O anomaly (Kaiser et al. 2004b; Kaiser and Röckmann 2005). If the anomaly is biological, this would be the first documentation of a biological mass-independent effect. The issue of the ^{17}O anomaly remains unresolved because the $\delta^{17}O$ and $\delta^{18}O$ composition of N_2O produced photochemically has not been determined and by the fact that "no measurements of the oxygen triple isotopic composition [^{16}O, ^{17}O and ^{18}O] of biogenic N_2O exist" (Kaiser and Röckmann 2005). The key limitation has been the size of the anomaly, 0.9 ‰ relative to current analytical precision (0.1–0.5 ‰).

Biological production itself is unlikely to result in mass independence, however, the source of elemental O in N_2O can be either O_2 or water and there is a small but measurable difference in the slopes of the mass-dependent relationships for O_2 and water (Meijer and Li 1998; Angert et al. 2003). Until recently, the relationship between $\delta^{17}O$ and $\delta^{18}O$ was described by a slope with a value of 0.52 that has been termed the "terrestrial fractionation line" (Young et al. 2002). With this in mind, the ^{17}O anomaly, $\Delta^{17}O$, can be defined as:

$$\Delta^{17}O = \delta^{17}O - 0.52 * \delta^{18}O \qquad (23.20)$$

which describes the magnitude of the enrichment or depletion in ^{17}O relative to what the sample would have if it were in "mass dependence" (Cliff and Thiemens 1997; Kaiser et al. 2004b). A $\Delta^{17}O$ value of 0 ‰ would normally indicate that the sample followed the terrestrial fractionation line and was mass-dependent. As mentioned, the primary causes of non-zero or "mass-independent" $\Delta^{17}O$ values are photochemical reactions in the stratosphere and subsequent incorporation into tropospheric materials (Thiemens 2006). As values for the slope are dependent on the isotopic composition of the reference gas, a more accurate way to represent slope values is:

$$\Delta^{17}O = (\delta^{17}O + 1) - \lambda*\ln(\delta^{18}O + 1) \qquad (23.21)$$

where λ is the slope of the trend line in a $\ln(\delta^{17}O +1)$ vs. $\lambda*\ln(\delta^{18}O + 1)$ plot and approximately equivalent to the slope from (23.20) (note that values are reported in ‰, however the factor of 1,000 has been omitted for convenience) (Luz and Barkan 2005).

With recent improvements in precision, λ values have been defined to the third decimal and it is now apparent that there are slight variations in the these values between different materials (Luz and Barkan 2005). For example, λ values of 0.518, 0.528 and 0.525 have been reported for respiration, natural waters and rocks, respectively (Meijer and Li 1998; Miller 2002; Angert et al. 2003). For atmospheric N_2O, a slope of 0.516 has been reported (Cliff and Thiemens 1997; Kaiser et al. 2004b). Because different mass-dependent properties may be described by different λ values, it is possible to obtain a non-zero $\Delta^{17}O$ value for some compounds without invoking mass-independent fractionation. N_2O may be such a compound. Recently it was proposed that if during the biological production of N_2O via denitrification O exchange with water occurred, then N_2O could be produced following the mass-dependent slope of 0.528 for water rather than the value of 0.516 for atmospheric N_2O (Kaiser et al. 2004b). Exchange with water during the production of N_2O by two different species of dentrifying bacteria has been shown and the extent of exchange was species dependent (Casciotti et al. 2002). With this foundation, Kaiser and Röckmann (2005) proposed that at least 0.4 ‰ of the 0.9 ‰ ^{17}O anomaly present in the atmosphere could be due to biological production. Exchange with water during biological production (Mechanism 1) is only one of three ways that a biological ^{17}O anomaly may become expressed in N_2O.

A second means by which a ^{17}O anomaly may become expressed in N_2O depends upon the pathway of O incorporation during nitrification (Mechanism 2). During nitrification, the first O atom added is derived from O_2 whereas the second is derived from water (Dua et al. 1979; Andersson and Hooper 1983; Kumar et al. 1983) (Fig. 23.1). The $\delta^{18}O$ of water is generally 20 ‰ greater than that of O_2 (Kroopnick and Craig 1976); therefore, N_2O produced via oxidation of hydroxylamine (in which O is derived from O_2) should be enriched in N_2O relative to that derived from reduction of nitrite (50:50 O_2 and water) and nitrate (1/3 O_2, 2/3 water). With this foundation shifts in the $\delta^{18}O$ of N_2O with depth in the ocean have been attributed to shifts in the relative importance of hydroxylamine oxidation or nitrite or nitrate reduction (Ostrom et al. 2000). Since O_2 and water are also characterized by different mass-dependent slopes N_2O derived from oxidation of hydroxylamine or reduction of nitrite or nitrate should acquire distinct ^{17}O anomalies. In fact, as the proportion of elemental O derived from O_2 and water in nitrite and nitrate produced by nitrification differ, the ^{17}O anomaly and measured λ values may enable distinction of N_2O produced by reduction of these two compounds. This would be a significant advance as the SP produced upon reduction of nitrite via nitrifier-denitrification is no different than that produced upon reduction of nitrate by denitrification (Sutka et al. 2006). Consequently, the ^{17}O anomaly may provide a novel tracer of microbial production and may help resolve the importance of nitrifier-denitrification in N_2O reduction relative to other production pathways (Wrage et al. 2001, 2005). Exchange with water may potentially erase the distinction in $\delta^{17}O$ between N_2O derived from reduction of nitrite or nitrate. Nonetheless exchange with water will result in a distinct ^{17}O anomaly in N_2O relative to N_2O produced from oxidation of hydroxylamine which introduces O from O_2.

A third means for introduction of a ^{17}O anomaly is a function of the origin of nitrate denitrified to produce

N$_2$O (Mechanism 3). Atmospheric nitrate has recently been recognized to have a strong ^{17}O anomaly that ranges between 21 and 35 ‰ and reflects incorporation of a signal from photochemical reactions (Michalski et al. 2003, 2004; Morin et al. 2007, 2009; Kunasek et al. 2008; Michalski et al. 2010; Savarino and Morin 2011). Michalski et al. (2003) proposed that the biological production of N$_2$O from the denitrification of atmospheric nitrate could be an important source of the ^{17}O anomaly in atmospheric N$_2$O. This pathway is not likely to be important in many terrestrial environments in which atmospheric deposition is a small fraction of nitrate inputs relative to that applied as fertilizer or released following soil mineralization/nitrification (e.g. Ostrom et al. 1998) but could be important in organic poor soils (e.g. deserts) and oligotrophic waters. Incorporation of a ^{17}O anomaly via mechanism 3 would also potentially be moderated to some extent if exchange of water during N$_2$O production from denitrification is significant. Resolution of which mechanism is predominant is dependent on an improvement in precision beyond 0.1 ‰ and an analytical approach that measures the δ^{17}O and δ^{18}O of N$_2$O without mass interference.

23.4 Synthesis and Future Directions

The pioneering research of Toyoda and Yoshida (1999) and Brenninkenmeijer and Röckman (2000) provided strong foundations for the application of SP to evaluate the biogeochemical origins and sinks of N$_2$O. Because the N$_2$O molecule contains two elements with a total of 5 stable isotopes a large number of isotopomers of N$_2$O exist which greatly complicates corrections for mass interference in the mass spectrometer. Refinement of mass interference corrections has been challenging and will undoubtedly continue (e.g. Westley et al. 2007; Frame and Casciotti 2010). The considerable complexity and lack of a consistent correction routine is a serious impediment to the widespread determination of the SP of N$_2$O even though all modern mass spectrometers are provided with three collectors appropriately configured for its analysis (^{31}R, ^{45}R and ^{46}R can be determined with three collectors by analyzing the same sample twice). Because of this it is recommend that all published manuscripts include (1) the isotopologue values of the internal laboratory standard used relative to international standards (VSMOW and Air-N$_2$), (2) the correction routine used, and (3) ^{31}R, ^{45}R and ^{46}R values obtained (perhaps as an electronic annex). In this manner, if revisions to the SP of tropospheric N$_2$O occur, an international standard is developed, and if improvements to correction routines become available, published data can be corrected and revised to newer scales. Mass interference correction models should be made available to the international community, perhaps posted to a web site, to enable widespread access, consistent reporting of isotopologue data, and improvement to mass correction routines.

There is a current lack of an international standard for calibration and need for the research community to collaborate on its development. At the present time, calibration may be best accomplished by exchange of standards with the laboratory of N. Yoshida at the Tokyo Institute of Technology, analysis of the SP of tropospheric air to obtain a value of 18.7 ‰, and/or by spectroscopic approaches. FT-IR is the only spectroscopic approach to date that has provided data on the SP of tropospheric N$_2$O and is advantageous given that the analysis avoids the mass interference issues that challenge mass spectrometry (Griffith et al. 2009). Applications of FTIR to obtain field data have been limited but with improvements in methodology and technology spectroscopic approaches (including tunable diode laser and cavity ring down spectroscopy) will undoubtedly increase.

Additional studies are needed using pure microbial cultures, co-cultures, and natural soil and water microbial communities to better understand SP signals and the importance of and extent of water exchange in controlling the δ^{18}O values of N$_2$O as well as SP values associated with production by anammox, DNRA and codenitrification. The use of inhibitors may be insightful but results should be interpreted with caution. N$_2$O production from fungal denitrification, for example, has recently been cited as important based on the use of antibiotics (Laughlin and Stevens 2002), however the inhibitor used does not inhibit archaea and archaea have been implicated in both nitrification and denitrification (Cabello et al. 2004; Leininger et al. 2006). Thus N$_2$O production in the presence of inhibitors could potentially have been the result of both fungal and archaeal activity. More research needs to be done to evaluate SP signals from novel N_2O production pathways such as archaeal nitrification and denitrification, codenitrification, annamox and DNRA. Research evaluating isotopologue signals associated with purified enzymes is an important avenue of research for obtaining precise ε values in the absence of other isotope effects such as diffusion.

Accurate evaluation of the ^{17}O anomaly in N_2O is a decisive research direction that will help resolve the global N_2O budget and potentially production pathways, such as nitrifier-denitrification, that cannot be resolved on the basis of SP. Accurate determination of the $\delta^{17}O$ of N_2O cannot be determined based on the isotopic analysis of the molecular and fragmentation ions with the mass spectrometer as this approach requires assumption of mass dependence. Evaluation of the $\delta^{17}O$ of N_2O will require decomposition of N_2O to O_2 (e.g. Thiemens 2006; Kaiser et al. 2007) but it also requires an improvement in analytical precision to resolve causes of the small (0.9 ‰) mass independent signal in N_2O. Decomposition of N_2O to O_2 followed by high precision analysis of O_2 (e.g. Barkan and Luz 2003) is potentially a viable approach for obtaining per meg (1/meg = 0.001 ‰) levels of precision for $\delta^{15}N$, $\delta^{17}O$ and $\delta^{18}O$ in N_2O.

Efforts to refine analytical approaches and data interpretation associated with the isotopomers of N_2O have an important payoff in the arena of global climate change. There is a large uncertainty in estimates of N_2O to GWP even though (1) increases in N_2O emissions primarily derive from a single source (agriculture: 70–81 %), (2) estimates for N_2O emissions from fertilized cropland and grasslands have been recently published (3.3 and 1.8 Tg N_2O-N year^{-1}, respectively) and (3) there has been a significant increase in the number of N_2O measurements from agricultural since 2002 (Barton et al. 2008). The magnitude of our uncertainty is illustrated by the large range in the 85 % confidence interval for global N_2O emissions from agricultural soils; −51 to +107 % (Barton et al. 2008). The ability to refine these uncertainties, however, will not reduce global warming. Instead, we need to reduce N_2O emissions. This is where data on N_2O isotopomers and ^{17}O anomalies will be significant. We know that N_2O emissions primarily derived from soil microbial processes occurring in agricultural soils and that management strategies (e.g. tillage) influence environmental conditions that control microbial N_2O production pathways. Thus, agriculture is a setting where management practices can be adjusted to minimize N_2O emissions. Effective management strategies will require knowledge of the microbial pathways that control N_2O flux effective. This is, perhaps, the most compelling call for additional and more accurate isotopomer and ^{17}O anomaly data.

Acknowledgements This manuscript has benefited greatly from conversations over the past decade with a number of researchers and students including Tim Bergsma, Sourendra Bhattacharya, Hasand Gandhi, Eric Hegg, Matthew Opdyke, Malee Jinuntuya-Nortman, Jay Lennon, Adam Pitt, G. Phil Robertson, Robin Sutka, and Joe von Fisher. We greatly appreciate the assistance of Sakae Toyoda, Naohiro Yoshida, David Griffith and Stephen Parkes in the calibration of our internal laboratory standard. We thank two reviewers of this manuscript for their insightful comments. We thank Phil Robertson, Neville Millar, and James Humpula for assistance in the collection and/or analysis of data presented in Figs. 23.4 and 23.5.

References

Andersson KK, Hooper AB (1983) O_2 and H_2O are each the source of one O in NO_2^- produced from NH3 by nitrosomonas – N-15-NMR evidence. FEBS Lett 164:236–240

Angert A, Barkan E, Barnett B, Brugnoli E, Davidson EA, Fessenden J, Maneepong S, Panapitukkul N, Randerson JT, Savage K, Yakir D, Luz B (2003) Contribution of soil respiration in tropical, temperate, and boreal forests to the ^{18}O enrichment of atmospheric O_2. Global Biogeochem Cy 17:1089. doi:10.1029/2003GB002056

Assonov SS, Brenninkmeijer CAM (2001) A new method to determine the ^{17}O isotopic abundance in CO_2 using oxygen isotope exchange with a solid oxide. Rapid Commun Mass Spectrom 15:2426–2437

Bandibas J, Vermoesen A, De Groot CJ, Cleemput OV (1994) The effect of different moisture regimes and soil characteristics on nitrous oxide emission and consumption by different soils. Soil Sci 158:106

Barford CC, Montoya JP, Altabet MA, Mitchell R (1999) Steady-state nitrogen isotope effects of N_2 and N_2O production in Paracoccus denitrificans. Appl Environ Microbiol 65: 989

Barkan E, Luz B (2003) High-precision measurements of $^{17}O/^{16}O$ and $^{18}O/^{16}O$ of O_2 and O_2/Ar ratio in air. Rapid Commun Mass Spectrom 17:2809–2814

Barton L, Kiese R, Gatter D, Butterbach-Bahl K, Buck R, Hinz C, Murphy DV (2008) Nitrous oxide emissions from a cropped soil in a semi-arid climate. Global Change Biol 14:177–192

Bergsma TT, Ostrom NE, Emmons M, Robertson GP (2001) Measuring simultaneous fluxes from soil of N_2O and N_2 in the field using the ^{15}N-Gas "nonequilibrium" technique. Environ Sci Technol 35:4307–4312

Bernard S, Röckmann TR, Kaiser J, Barnola JM, Fischer H, Blunier T, Chappellaz J (2006) Constraints on N_2O budget changes since pre-industrial time from new firn air and ice core isotope measurements. Atmos Chem Phys 6:493–503

Brand WA (1995) PreCon: a fully automated interface for the Pre-Gc concentration of trace gases on air for isotopic analysis. Isot Environ Health Stud 31:277–284

Brand WA, Assonov SS, Coplen TB (2010) Correction for the ^{17}O interference in $\delta(^{13}C)$ measurements when analyzing CO_2 with stable isotope mass spectrometry (IUPAC technical report). Pure Appl Chem 82:1719–1733

Brenninkenmeijer CAM, Röckman T (2000) Mass spectrometry of the intramolecular nitrogen isotope distribution of environmental nitrous oxide using fragment-ion analysis. Rapid Commun Mass Spectrom 13:2028–2033

Cabello P, Roldan MD, Moreno-Vivian C (2004) Nitrate reduction and the nitrogen cycle in archaea. Microbiology 150:3527

Casciotti KL, Ward BB (2001) Dissimilatory nitrite reductase genes from autotrophic ammonia-oxidizing bacteria. Appl Environ Microbiol 67:2213–2221

Casciotti KL, Ward BB (2005) Phylogenetic analysis of nitric oxide reductase gene homologues from aerobic ammonia-oxidizing bacteria. FEMS Microbiol Ecol 52:197–205

Casciotti KL, Sigman DM, Hastings MG, Bohlke JK, Hilkert A (2002) Measurement of the oxygen isotopic composition of nitrate in seawater and freshwater using the denitrifier method. Anal Chem 74:4905–4912

Casciotti KL, McIlvin M, Buchwald C (2010) Oxygen isotopic exchange and fractionation during bacterial ammonia oxidation. Limnol Oceanogr 55:753–762

Cavigelli MA, Robertson GP (2001) Role of denitrifier diversity in rates of nitrous oxide consumption in a terrestrial ecosystem. Soil Biol Biochem 33:297–310

Chapuis-Lardy L, Wrage N, Metay A, Chotte JL, Bernoux M (2007) Soils, a sink for N_2O? A review. Global Change Biol 13:1–17

Cliff SS, Thiemens MH (1994) High-precision isotopic determination of the $^{18}O/^{16}O$ ratios in nitrous-oxide. Anal Chem 66:2791–2793

Cliff SS, Thiemens MH (1997) The $^{18}O/^{16}O$ and $^{17}O/^{16}O$ ratios in atmospheric nitrous oxide: a mass-independent anomaly. Science 278:1774–1776

Cliff SS, Brenninkmeijer CAM, Thiemens MH (1999) First measurement of the O-18/O-16 and O-17/O-16 ratios in stratospheric nitrous oxide: a mass-independent anomaly. J Geophys Res Atmos 104:16171–16175

Craig H (1957) Isotopic standards for carbon and oxygen and correction factors for mass-spectrometric analysis of carbon dioxide. Geochim Cosmochim Acta 12:133–149

Crenshaw CL, Lauber C, Sinsabaugh RL, Stavely LK (2008) Fungal control of nitrous oxide production in semiarid grassland. Biogeochemistry 87:17–27

Davidson EA (1993) Soil water content and the ratio of nitrous oxide to nitric oxide emitted from soil. In: Oremland R (ed) Biogeochemistry of global change. Chapman Hall, New York, pp 369–386

Dore JE, Popp BN, Karl DM, Sansone FJ (1998) A large source of atmospheric nitrous oxide from subtropical North Pacific surface waters. Nature 396:63–66

Dua RD, Bhandari B, Nicholas DJD (1979) Stable isotope studies on the oxidation of ammonia to hydroxylamine by Nitrosomonas europaea. FEBS Lett 106:401–404

Esler MB, Griffith DWT, Turatti F, Wilson SR, Rahn T, Zhang H (2000) N_2O concentration and flux measurements and complete isotopic analysis by FTIR spectroscopy. Chemosphere Global Change Sci 2:445–454

Farquhar GD, O'Leary MH, Berry JA (1982) On the relationship between carbon isotope discrimination and the intercellular carbon dioxide concentration in leaves. Aust J Plant Physiol 9:121–137

Firestone MK, Davidson EA (1989) Microbiological basis of NO and N_2O production and consumption in soil. In: Andreae MO, Schimel DS (eds) Exchange of trace gases between terrestrial ecosystems and the atmosphere. Wiley, New York

Forster P et al (2007) Changes in atmospheric constituents and in radiative forcing. In: Solomon S (ed) Climate change 2007: the physical science basis. Contribution of working group I to the fourth assessment report of the intergovernmental panel on climate change. Cambridge University Press, New York

Frame C, Casciotti K (2010) Biogeochemical controls and isotopic signatures of nitrous oxide production by a marine ammonia-oxidizing bacterium. Biogeosci Discuss 7:3019–3059

Friedman L, Bigeleisen J (1950) Oxygen and nitrogen isotope effects in the decomposition of ammonium nitrate. J Chem Phys 18:1325–1331

Galloway JN, Aber JD, Erisman JW, Seitzinger SP, Howarth RW, Cowling EB, Cosby BJ (2003) The nitrogen cascade. Bioscience 53:341–356

Garbeva P, Baggs EM, Prosser JI (2007) Phylogeny of nitrite reductase (nirK) and nitric oxide reductase (norB) genes from Nitrosospira species isolated from soil. FEMS Microbiol Lett 266:83–89

Griffith DWT, Toon GC, Sen B, Blavier JF, Toth RA (2000) Vertical profiles of nitrous oxide isotopomer fractionation measured in the stratosphere. Geophys Res Lett 27:2485–2488

Griffith DWT, Parkes SD, Haverd V, Paton-Walsh C, Wilson SR (2009) Absolute calibration of the intramolecular site preference of N-15 fractionation in tropospheric N_2O by FT-IR spectroscopy. Anal Chem 81:2227–2234

Groffman PM, Altabet MA, Bohlke JK, Butterbach-Bahl K, David MB, Firestone MA, Giblin AE, Kana TA, Nielsen LP, Voytek MA (2006) Methods for measuring denitrification: diverse approaches to a difficult problem. Ecol Appl 16:2091–2122

Guy RD, Fogel ML, Berry JA (1993) Photosynthetic fractionation of the stable isotopes of oxygen and carbon. Plant Physiol 101:37–47

Hendriks J, Oubrie A, Castresana J, Urbani A, Gemeinhardt S, Saraste M (2000) Nitric oxide reductases in bacteria. Biochim Biophys Acta 1459:266–273

Hollocher TC, Tate ME, Nicholas DJ (1981) Oxidation of ammonia by Nitrosomonas europaea. Definite 18O-tracer evidence that hydroxylamine formation involves a monooxygenase. J Biol Chem 256:10834

Jinuntuya-Nortman M, Sutka RL, Ostrom PH, Gandhi H, Ostrom NE (2008) Isotopologue fractionation during microbial reduction of N_2O within soil mesocosms as a function of water-filled pore space. Soil Biol Biochem 40:2273–2280

Kaiser J (2008) Reformulated ^{17}O correction of mass spectrometric stable isotope measurements in carbon dioxide and a critical appraisal of historic "absolute" carbon and oxygen isotope ratios. Geochim Cosmochim Acta 72:1312–1334

Kaiser J, Röckmann T (2005) Absence of isotope exchange in the reaction of $N_2O + O(^1D)$ and the global $\Delta^{17}O$ budget of nitrous oxide. Geophys Res Lett 32. doi:10.1029/2005GL023199

Kaiser J, Röckmann T (2008) Correction of mass spectrometric isotope ratio measurements for isobaric isotopologues of O_2, CO, CO_2, N_2O and SO_2. Rapid Commun Mass Spectrom 22:3997–4008

Kaiser J, Röckmann T, Brenninkmeijer CAM (2003) Complete and accurate mass spectrometric isotope analysis of tropospheric nitrous oxide. J Geophys Res Atmos 108:17

Kaiser J, Park S, Boering KA, Brenninkmeijer CAM, Hilkert A, Röckmann T (2004a) Mass spectrometric method for the absolute calibration of the intramolecular nitrogen isotope distribution in nitrous oxide. Anal Bioanal Chem 378:256–269

Kaiser J, Röckmann T, Brenninkmeijer CAM (2004b) Contribution of mass-dependent fractionation to the oxygen isotope anomaly of atmospheric nitrous oxide. J Geophys Res Atmos 109. doi:10.1029/2003JD003613

Kaiser J, Hastings MG, Houlton BZ, Röckmann T, Sigman DM (2007) Triple oxygen isotope analysis of nitrate using the denitrifier method and thermal decomposition of N_2O. Anal Chem 79:599–607

Kampschreur MJ, van der Star WRL, Wielders HA, Mulder JW, Jetten SM, van Loosdrecht MCM (2008) Dynamics of nitric oxide and nitrous oxide emission during full-scale reject water treatment. Water Res 42:812–826

Kim KR, Craig H (1993) ^{15}N and ^{18}O characteristics of nitrous-oxide – a global perspective. Science 262:1855–1857

Kool DM, Wrage N, Oenema O, Dolfing J, Van Groenigen JW (2007) Oxygen exchange between (de)nitrification intermediates and H_2O and its implications for source determination of NO_3^- and N_2O: a review. Rapid Commun Mass Spectrom 21:3569–3578

Kool DM, Muller C, Wrage N, Oenema O, Van Groenigen JW (2009a) Oxygen exchange between nitrogen oxides and H_2O can occur during nitrifier pathways. Soil Biol Biochem 41:1632–1641

Kool DM, Wrage N, Oenema O, Harris D, Van Groenigen JW (2009b) The ^{18}O signature of biogenic nitrous oxide is determined by O exchange with water. Rapid Commun Mass Spectrom 23:104–108

Kostera J, Youngblut MD, Slosarczyk JM, Pacheco AA (2008) Kinetic and product distribution analysis of NO reductase activity in Nitrosomonas europaea hydroxylamine oxidoreductase. J Biol Inorg Chem 13:1073–1083

Kroopnick P, Craig H (1976) Oxygen isotope fractionation in dissolved oxygen in deep-sea. Earth Planet Sci Lett 32:375–388

Kumar S, Nicholas DJD, Williams EH (1983) Definitive ^{15}N-NMR evidence that water serves as a source of O during nitrite oxidation by Nitrobacter agilis. FEBS Lett 152:71–74

Kunasek S, Alexander B, Steig E, Hastings MG, Gleason D, Jarvis J (2008) Measurements and modeling of $\Delta^{17}O$ of nitrate in snowpits from Summit, Greenland. J Geophys Res 113. doi:10.1029/2008JD010103

Lammerzahl P, Röckmann T, Brenninkmeijer CAM, Krankowsky D, Mauersberger K (2002) Oxygen isotope composition of stratospheric carbon dioxide. Geophys Res Lett 29. doi:10.1029/2001GL014343

Laughlin RJ, Stevens R (2002) Evidence for fungal dominance of denitrification and codenitrification in a grassland soil. Soil Sci Soc Am J 66:1540–1548

Leininger S, Urich T, Schloter M, Schwark L, Qi J, Nicol GW, Prosser JI, Schuster SC, Schleper C (2006) Archaea predominate among ammonia-oxidizing prokaryotes in soils. Nature 442:806–809

Luz B, Barkan E (2005) The isotopic ratios $^{17}O/^{16}O$ and $^{18}O/^{16}O$ in molecular oxygen and their significance in biogeochemistry. Geochim Cosmochim Acta 69:1099–1110

Madsen EL (1998) Epistemology of environmental microbiology. Environ Sci Technol 32:429–439

Mandernack KW, Rahn T, Kinney C, Wahlen M (2000) The biogeochemical controls of the $\delta^{15}N$ and $\delta^{18}O$ of N_2O produced in landfill cover soils. J Geophys Res Atmos 105:17709–17720

Mandernack KW, Mills CT, Johnson CA, Rahn T, Kinney C (2009) The $\delta^{15}N$ and $\delta^{18}O$ values of N_2O produced during the co-oxidation of ammonia by methanotrophic bacteria. Chem Geol 267:96–107

Mariotti A, Germon JC, Hubert P, Kaiser P, Letolle R, Tardieux A, Tardieux P (1981) Experimental-determination of nitrogen kinetic isotope fractionation – some principles – illustration for the denitrification and nitrification processes. Plant Soil 62:413–430

Mariotti A, Germon JC, Leclerc A (1982) Nitrogen isotope fractionation associated with the NO_2^- to N_2O step of denitrification in soils. Can J Soil Sci 62:227–241

Mark E, Mark TD, Kim YB, Stephan K (1981) Absolute electron impact ionization cross section from threshold up to 180 eV for N2O+e -> N2O+ + 2e and the metastable and collision induced dissociation of N2O+. J Chem Phys 75:4446–4453

Matson P, Vitousek P, Schimel D (1989) Regional extrapolation of trace gas flux based on soils and ecosystems. In: Andreae M, Schimel D (eds) Exchange of trace gases between terrestrial ecosystems and the atmosphere. Wiley, New York, pp 97–108

Matsuhisa Y, Goldsmith JR, Clayton RN (1978) Mechanisms of hydrothermal crystallization of quartz at 250 C and 15 kbar. Geochim Cosmochim Acta 42:173–182

Meijer H, Li W (1998) The use of electrolysis for accurate $\delta^{17}O$ and $\delta^{18}O$ isotope measurements in water. Iso Env Health Stud 34:349–369

Menyailo OV, Hungate BA (2006) Stable isotope discrimination during soil denitrification: production and consumption of nitrous oxide. Global Biogeochem Cy 20. doi:10.1029/2005GB002527

Michalski G, Scott Z, Kabiling M, Thiemens MH (2003) First measurements and modeling of $\delta^{17}O$ in atmospheric nitrate. Geophys Res Lett 30. doi:10.1029/2003GL017015

Michalski G, Bohlke JK, Thiemens M (2004) Long term atmospheric deposition as the source of nitrate and other salts in the Atacama Desert, Chile: new evidence from mass-independent oxygen isotopic compositions. Geochim Cosmochim Acta 68:4023–4038

Michalski G, Bhattacharya SK, Mase DF (2010) Oxygen isotopes in atmospheric nitrate. In: Baskaran M (ed) Handbook of environmental isotope geochemistry. Springer, Berlin

Miller MF (2002) Isotopic fractionation and the quantification of ^{17}O anomalies in the oxygen three-isotope system: an appraisal and geochemical significance. Geochim Cosmochim Acta 66:1881–1889

Miller CE, Yung YL (2000) Photo-induced isotopic fractionation of stratospheric N_2O. Chemosphere Global Change Sci 2:255–266

Morin S, Savarino J, Bekki S, Gong S, Bottenheim JW (2007) Signature of Arctic surface ozone depletion events in the isotope anomaly ($\Delta^{17}O$) of atmospheric nitrate. Atmos Chem Phys 7:1451–1469

Morin S, Savarino J, Frey MM, Domine F, Jacobi HW, Kaleschke L, Martins JMF (2009) Comprehensive isotopic composition of atmospheric nitrate in the Atlantic Ocean boundary layer from 65 degrees S to 79 degrees N. J Geophys Res Atmos 114:19

Mosier A, Kroeze C (2000) Potential impact on the global atmospheric N_2O budget of the increased nitrogen input required to meet future global food demands. Chemosphere Global Change Sci 2:465–473

Mosier AR, Duxbury JM, Freney JR, Heinemeyer O, Minami K (1998a) Assessing and mitigating N_2O emissions from agricultural soils. Clim Change 40:7–38

Mosier A, Kroeze C, Nevison C, Oenema O, Seitzinger S, van Cleemput O (1998b) Closing the global N_2O budget: nitrous oxide emissions through the agricultural nitrogen cycle – OECD/IPCC/IEA phase II development of IPCC guidelines for national greenhouse gas inventory methodology. Nutr Cycl Agroecosyst 52:225–248

Naqvi SWA, Naik H, Pratihary A, D'Souza W, Narvekar PV, Jayakumar DA, Devol AH, Yoshinari T, Saino T (2006) Coastal versus open-ocean denitrification in the Arabian Sea. Biogeosciences 3:621–633

Oleary MH (1981) Carbon isotope fractionation in plants. Phytochemistry 20:553–567

Opdyke MR, Ostrom NE, Ostrom PH (2009) Evidence for the predominance of denitrification as a source of N_2O in temperate agricultural soils based on isotopologue measurements. Global Biogeochem Cy 23. doi:10.1029/2009GB003523

Ostrom NE, Knoke KE, Hedin LO, Robertson GP, Smucker AJM (1998) Temporal trends in nitrogen isotope values of nitrate leaching from an agricultural soil. Chem Geol 146:219–227

Ostrom NE, Russ ME, Popp BN, Rust TM, Karl D (2000) Mechanisms of nitrous oxide production in the subtropical North Pacific based on determinations of the isotopic abundances of nitrous oxide and di-oxygen. Chemosphere Global Change Sci 2:281–290

Ostrom NE, Hedin LO, von Fischer JC, Robertson GP (2002) Nitrogen transformations and NO_3^- removal at a soil-stream interface: a stable isotope approach. Ecol Appl 12:1027–1043

Ostrom NE, Russ ME, Field A, Piwinski L, Twiss MR, Carrick HJ (2005) Ratios of community respiration to photosynthesis and rates of primary production in Lake Erie via oxygen isotope techniques. J Great Lakes Res 31:138–153

Ostrom NE, Pitt A, Sutka R, Ostrom PH, Grandy AS, Huizinga KM, Robertson GP (2007) Isotopologue effects during N_2O reduction in soils and in pure cultures of denitrifiers. J Geophys Res Biogeosci 112. doi:10.1029/2006JG000287

Ostrom N, Sutka R, Ostrom P, Grandy S, Huizinga K, Gandhi H, von Fischer J, Robertson G (2010) Isotopologue data reveal bacterial denitrification as the primary source of N_2O during a high flux event following cultivation of a native temperate grassland. Soil Biol Biochem 42:499–506

Pérez T, Trumbore SE, Tyler SC, Matson PA, Ortiz-Monasterio I, Rahn T, Griffith DWT (2001) Identifying the agricultural imprint on the global N_2O budget using stable isotopes. J Geophys Res 106:9869–9878

Pérez T, Garcia-Montiel D, Trumbore S, Tyler S, Camargo P, Moreira M, Piccolo M, Cerri C (2006) Nitrous oxide nitrification and denitrification ^{15}N enrichment factors from Amazon forest soils. Ecol Appl 16:2153–2167

Popp BN, Westley MB, Toyoda S, Miwa T, Dore JE, Yoshida N, Rust TM, Sansone FJ, Russ ME, Ostrom NE, Ostrom P (2002) Nitrogen and oxygen isotopomeric constraints on the origins and sea-to-air flux of N_2O in the oligotrophic subtropical North Pacific gyre. Global Biogeochem Cy 16. doi:10.1029/2001GB001806

Prasad SS, Zipf EC (2008) Atmospheric production of nitrous oxide from excited ozone and its potentially important implications for global change studies. J Geophys Res Atmos 113. doi:10.1029/2007JD009447

Ravishankara AR, Daniel JS, Portmann RW (2009) Nitrous oxide (N_2O): the dominant ozone-depleting substance emitted in the 21st century. Science 326:123–125

Robertson G, Groffman P (2006) Nitrogen transformations. In: Paul E, Clark F (eds) Soil microbiology, ecology and biochemistry. Springer, New York, pp 341–364

Robertson GP, Broome JC, Chornesky EA, Frankenberger JR, Johnson P, Lipson M, Miranowski JA, Owens ED, Pimentel D, Thrupp LA (2004) Rethinking the vision for environmental research in US agriculture. Bioscience 54:61–65

Röckmann T, Kaiser J, Brenninkmeijer CA, Brand WA (2003) Gas chromatography/isotope-ratio mass spectrometry method for high-precision position-dependent ^{15}N and ^{18}O measurements of atmospheric nitrous oxide. Rapid Commun Mass Spectrom 17:1897–1908

Santrock J, Studley SA, Hayes JM (1985) Isotopic analyses based on the mass spectra of carbon dioxide. Anal Chem 57:1444–1448

Savarino J, Morin S (2011) The N, O, S isotopes of oxy-anions in ice cores and polar environments. In: Baskaran M (ed) Handbook of environmental isotope geochemistry. Springer, Berlin

Schmidt HL, Werner RA, Yoshida N, Well R (2004) Is the isotopic composition of nitrous oxide an indicator for its origin from nitrification or denitrification? A theoretical approach from referred data and microbiological and enzyme kinetic aspects. Rapid Commun Mass Spectrom 18:2036–2040

Scott JT, McCarthy MJ, Gardner WS, Doyle RD (2008) Denitrification, dissimilatory nitrate reduction to ammonium, and nitrogen fixation along a nitrate concentration gradient in a created freshwater wetland. Biogeochemistry 87:99–111

Seitzinger S, Harrison JA, Bohlke JK, Bouwman AF, Lowrance R, Peterson B, Tobias C, Van Drecht G (2006)

Denitrification across landscapes and waterscapes: a synthesis. Ecol Appl 16:2064–2090

Shoun H, Kim DH, Uchiyama H, Sugiyama J (1992) Denitrification by fungi. FEMS Microbiol Lett 94:277–281

Sowers T, Alley RB, Jubenville J (2003) Ice core records of atmospheric N_2O covering the last 106,000 years. Science 301:945–948

Stein LY, Yung YL (2003) Production, isotopic composition, and atmospheric fate of biologically produced nitrous oxide. Annu Rev Earth Planet Sci 31:329–356

Sutka RL, Ostrom NE, Ostrom PH, Gandhi H, Breznak JA (2003) Nitrogen isotopomer site preference of N_2O produced by Nitrosomonas europaea and Methylococcus capsulatus Bath. Rapid Commun Mass Spectrom 17:738–745

Sutka RL, Ostrom NE, Ostrom PH, Breznak JA, Gandhi H (2004) Erratum: nitrogen isotopomer site preference of N_2O produced by *Nitrosomonas europaea* and *Methylococcus capsulatus* Bath. Rapid Commun Mass Spectrom 18:1–2

Sutka RL, Ostrom NE, Ostrom PH, Breznak JA, Gandhi H, Pitt AJ, Li F (2006) Distinguishing nitrous oxide production from nitrification and denitrification on the basis of isotopomer abundances. Appl Environ Microbiol 72:638–644

Sutka R, Adams G, Ostrom NE, Ostrom P (2008) Isotopologue fractionation during N_2O production by fungal denitrification. Rapid Commun Mass Spectrom 22:3989–3996

Tavares P, Pereira AS, Moura JJG, Moura I (2006) Metalloenzymes of the denitrification pathway. J Inorg Biochem 100:2087–2100

Thiemens MH (2006) History and applications of mass-independent isotope effects. Annu Rev Earth Planet Sci 34:217–262

Tilman D, Fargione J, Wolff B, D'Antonio C, Dobson A, Howarth R, Schindler D, Schlesinger WH, Simberloff D, Swackhamer D (2001) Forecasting agriculturally driven global environmental change. Science 292:281–284

Tilsner J, Wrage N, Lauf J, Gebauer G (2003) Emission of gaseous nitrogen oxides from an extensively managed grassland in NE Bavaria, Germany II. Stable isotope natural abundance of N_2O. Biogeochemistry 63:249–267

Toyoda S, Yoshida N (1999) Determination of nitrogen isotopomers of nitrous oxide on a modified isotope ratio mass spectrometer. Anal Chem 71:4711–4718

Toyoda S, Mutobe H, Yamagishi H, Yoshida N, Tanji Y (2005) Fractionation of N_2O isotopomers during production by denitrifier. Soil Biol Biochem 37:1535–1545

Turatti F, Griffith DWT, Wilson SR, Esler MB, Rahn T, Zhang H, Blake GA (2000) Positionally dependent ^{15}N fractionation factors in the UV photolysis of N_2O determined by high resolution FTIR spectroscopy. Geophys Res Lett 27:2489–2492

Vieten B, Blunier T, Neftel A, Alewell C, Conen F (2007) Fractionation factors for stable isotopes of N and O during N_2O reduction in soil depend on reaction rate constant. Rapid Commun Mass Spectrom 21:846–850

Wahlen M, Yoshinari T (1985) Oxygen isotope ratios in N_2O from different environments. Nature 313:780–782

Well R, Flessa H (2009) Isotopologue signatures of N_2O produced by denitrification in soils. J Geophys Res Biogeosci 114. doi:10.1029/2008JG000804

Well R, Kurganova I, de Gerenyu VL, Flessa H (2006) Isotopomer signatures of soil-emitted N_2O under different moisture conditions – a microcosm study with arable loess soil. Soil Biol Biochem 38:2923–2933

Well R, Flessa H, Xing L, Ju XT, Romheld V (2008) Isotopologue ratios of N_2O emitted from microcosms with NH_4^+ fertilized arable soils under conditions favoring nitrification. Soil Biol Biochem 40:2416–2426

Westley MB, Popp BN, Rust TM (2007) The calibration of the intramolecular nitrogen isotope distribution in nitrous oxide measured by isotope ratio mass spectrometry. Rapid Commun Mass Spectrom 21:391–405

Wrage N, Velthof GL, van Beusichem ML, Oenema O (2001) Role of nitrifier denitrification in the production of nitrous oxide. Soil Biol Biochem 33:1723–1732

Wrage N, Velthof GL, Oenema O, Laanbroek HJ (2004) Acetylene and oxygen as inhibitors of nitrous oxide production in Nitrosomonas europaea and Nitrosospira briensis: a cautionary tale. FEMS Microbiol Ecol 47:13–18

Wrage N, van Groenigen JW, Oenema O, Baggs EM (2005) A novel dual-isotope labelling method for distinguishing between soil sources of N_2O. Rapid Commun Mass Spectrom 19:3298–3306

Yamagishi H, Yoshida N, Toyoda S, Popp BN, Westley MB, Watanabe S (2005) Contributions of denitrification and mixing on the distribution of nitrous oxide in the North Pacific. Geophys Res Lett 32. doi:10.1029/2004GL021458

Ye RW, Torosuarez I, Tiedje JM, Averill BA (1991) H_2O-^{18}O isotope exchange studies on the mechanism of reduction of nitric oxide and nitrite to nitrous oxide by denitrifying bacteria – evidence for an electrophilic nitrosyl during reduction of nitric oxide. J Biol Chem 266: 12848–12851

Yoshida N (1988) ^{15}N-depleted N_2O as a product of nitrification. Nature 335:528–529

Yoshida N, Matsuo S (1983) Nitrogen isotope ratio of atmospheric N_2O as a key to the global cycle of N_2O. Geochem J 17: 231–239

Yoshida N, Toyoda S (2000) Constraining the atmospheric N_2O budget from intramolecular site preference in N_2O isotopomers. Nature 405:330–334

Yoshida N, Hattori A, Saino T, Matsuo S, Wada E (1984) $^{15}N/^{14}N$ ratio of dissolved N_2O in the eastern tropical Pacific Ocean. Nature 307:442–444

Yoshinari T, Wahlen M (1985) Oxygen isotope ratios in N_2O from nitrification at a wastewater treatment facility. Nature 317:349–350

Young ED, Galy A, Nagahara H (2002) Kinetic and equilibrium mass-dependent isotope fractionation laws in nature and their geochemical and cosmochemical significance. Geochim Cosmochim Acta 66:1095–1104

Yung YL, Miller CE (1997) Isotopic fractionation of stratospheric nitrous oxide. Science 278:1778–1780

Chapter 24
Using Cosmogenic Radionuclides for the Determination of Effective Surface Exposure Age and Time-Averaged Erosion Rates

D. Lal

Abstract We discuss the cosmogenic method of determining exposure ages and erosion rates of exposed surfaces. This method is recognized to date as the only reliable method of quantifying important processes in geochronology, provided the exposure history of the surfaces can be successfully characterized using two or more cosmogenic nuclides produced in situ in the surfaces of interest. The success of the cosmogenic technique is based on two important characteristics of the secondary cosmic ray beam interacting with the exposed surfaces: (1) cosmic ray labeling is sensitively dependent on the geometry of exposure of the solid, and (2) the mean distance characterizing the absorption of cosmic ray beam in typical rocks is only about 50 cm. These facts make it possible to determine the past exposure history of the exposed surfaces; this task can be put on much firmer ground if information on the exposure history can be constrained based on geological information.

We summarize the early suggestions of Lal (Earth Planet Sci Lett 104:424–439, 1991) for using cosmogenic nuclides in geomorphology, as well as the complex exposure geometries considered subsequently in a paper by Lal and Chen (Earth Planet Sci Lett 236(3–4):797–813, 2005). These methods combined with the ongoing steady improvements in the AMS sensitivity should considerably widen the scope of applications of cosmic ray produced nuclides in geomorphology.

24.1 Introduction

Comic ray interactions in the earth's atmosphere produced several radionuclides which are used as tracers to study the motion of the atmospheric gases within the stratosphere and the troposphere, and exchange of gases between the stratosphere and the troposphere (Lal and Baskaran 2011). These tracers also provide valuable data on the movement of the stratospheric aerosols to the troposphere and their removal from the troposphere by wet precipitations and gravitational settling.

Cosmic rays continue to produce radionuclides in situ in the hydrosphere, cryosphere and the lithosphere. In the case of the latter, one can measure cosmogenic effects down to underground depths of ~300 m. They continue to be useful as tracers for studying the movement of fluid in the hydrosphere but they also serve in many different ways, e.g., to date sediments in the hydrosphere and ice layers in the cryosphere. In the lithosphere and in the cryosphere, in-situ produced radionuclides serve to provide valuable information on geomorphologic processes and solar activity, respectively. *In this chapter we would focus on how cosmic ray interactions can be used to delineate geomorphic processes, in particular surface exposure ages and surficial erosion rates* (Lal 1991, 2007; Lal and Chen 2005).

The basis of application of cosmogenic radionuclides in geomorphology is that the continuous rate at which "new" nuclides are "created" in matter, depends on the geometry and duration of irradiation. One of the most remarkable and pervasive natural phenomena which occurs on the Earth, everywhere in the Solar System, and likely in the remotest parts of the Universe, is the occurrence of nuclear interactions produced by energetic particles of the cosmic radiation,

---

D. Lal (✉)
Geosciences Research Division, Scripps Institution of Oceanography, 9500 Gilman Drive, La Jolla, CA 92093-0244, USA
e-mail: dlal@ucsd.edu

which fill up the galaxies as a result of occasional particle acceleration by type II supernovae. The mean kinetic energies of the primary cosmic ray particles far exceed the binding energy of nucleons in nuclei of ordinary matter, by more than 3 orders of magnitude. The lifetime of cosmic rays in our galaxy is more than 4 orders of magnitude greater than the frequency of occurrence of supernovae in which the cosmic rays are accelerated. These facts translate to a continuous nuclear bombardment of matter in the universe by cosmic ray particles, at an essentially constant rate. Two earlier papers (Lal 1991; Lal and Chen 2005) discuss this essential feature of cosmic ray labeling of erosion surfaces and consider viable models for studying geomorphic processes (Stone 2000; Gosse and Phillips 2001). Lal (1991) discussed the essentials of cosmic ray production rates of several nuclides in *terrestrial solids*, and considered first order models of exposure of surfaces eroding at constant rates, which was followed by more complex exposure models (Lal and Chen 2005).

Before proceeding further, it must be stated that in passage through the atmosphere, the cosmic ray energy and flux are both appreciably reduced at sea level; the rate of production of radioisotopes is also correspondingly appreciably reduced (Fig. 24.1 which gives the rate of nuclear disintegrations in the atmosphere at different geomagnetic latitudes). The field of cosmic ray geomorphology had therefore to wait improvements in the sensitivity of measurement of cosmogenic radioisotopes. This advance occurred with the development of the accelerator mass spectrometry (AMS), which allows direct identification and measurement of the number of atoms of a nuclide in a sample, for several nuclides of interest (Bennett et al. 1977; Nelson et al. 1977; also see reviews by Lal 1988; Tuniz et al. 1998; Gosse and Phillips 2001). Special cases here are the long-lived nuclei, such as ^{14}C (half-life = 5730 years), ^{36}Cl (half-life = 0.3 My), ^{26}Al (half-life = 0.7 My), ^{10}Be (half-life = 1.5 My) and ^{129}I (half-life = 160 My), for which the decay counting method adopted so far required working with large samples, requiring 10^9–10^{13} atoms for a measurement. The improvement in sensitivity with the AMS for these nuclides allows convenient study of their atoms in the range of (10^5–10^6) atoms/sample, whereby one can now design studies of cosmogenic effects in the troposphere, and at underground depths, thereby addressing new questions in planetary science,

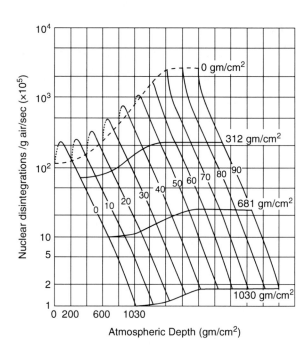

Fig. 24.1 The total rate of nuclear disintegrations in the atmosphere, with energy release of ≥40 MeV, is plotted as a function of altitude and geomagnetic latitude in intervals of 10°. Curves for different latitudes have been displaced along the abscissa by 200 g cm^{-2} (based on Lal and Peters 1967)

which could not be addressed earlier. There are two other obvious, highly significant advantages: (1) a reduction in the size of samples (resulting in both economy and simpler chemical processing), and (2) the rapidity with which cosmogenic effects can now be measured. Combination of these improvements allow time-series and/or synoptic measurements of cosmogenic effects in the geospheres.

24.2 Expected Cosmogenic In Situ Radionuclide Concentrations in Solids, for Eroding Surfaces

The time for which a surface has been exposed to cosmic radiation depends on its exposure history and the history of cosmic ray flux during the exposure of the surface. The cosmic ray flux at the top of the Earth's atmosphere is appreciably influenced by two parameters: (1) Geomagnetic field of the Earth, and (2) Solar activity.

1. Cosmic ray flux and nuclide production during different geomagnetic fields

If the past geomagnetic field is known, one can calculate the expected cosmogenic nuclide production rate at different geomagnetic latitudes (Lal et al. 1985; Nishiizumi et al. 1989), basing on the present day field (cf. nuclide production rates for different nuclides as given by Lal and Peters 1967). The vertical cut-off rigidity in the cosmic ray spectrum at the top of the atmosphere, R_c (=pc/Ze), is given by the following relation:

$$R_c = 14.9(M/M_0) \cos^4(\lambda)$$

where p is the particle momentum, c is the speed of light, Z is the particle's atomic number, e is the electronic charge; M and M_0 are the dipole fields during an epoch in the past and during the present time, respectively, for which cosmogenic production rates in the atmosphere have been calculated (Lal and Peters 1967), and λ is the geomagnetic latitude. The corresponding geomagnetic latitude for the past, λ_M, when the dipole field was M, is given by the relation (Lal et al. 1985; Nishiizumi et al. 1989):

$$\cos(\lambda_M) = \left(\frac{M_0}{M}\right)^{1/4} \cos(\lambda_0) \quad (24.1)$$

where λ_0 is the geomagnetic latitude for the present field, M_0.

2. Cosmic ray flux and nuclide production during periods of different solar activity

The primary galactic cosmic ray (GCR) flux incident at the top of the Earth's atmosphere is anti-correlated with solar activity (with some phase lag). See Gleeson and Axford (1968) for a convection-diffusion theoretical framework for the transport of GCR particles through the heliosphere, and see Castagnoli and Lal (1980) and Papini et al. (1996) for the essential features of the experimental data on solar modulation of primary cosmic ray flux. In terms of the solar modulation formulism (Gleeson and Axford 1968), the effective change in the energy of a charged particle moving from a point P on the boundary along a dynamical trajectory to an interior point Q is $Ze\phi$, irrespective of its path to Q. If the measured kinetic energy of a charged particle at Q is E_k/nucleon, then it follows (Randall and Van Allen 1986) that its kinetic energy/nucleon at P must have been ($E_k + Ze\phi$). As shown by Randall and Van Allen (1986), this leads to the following simple equation relating differential fluxes at points P and Q:

$$\frac{j_Q(T)}{T(T+2mc^2)} = \frac{j_P\left(T + \frac{Ze\phi}{A}\right)}{\left(T + \frac{Ze\phi}{A}\right)\left(T + \frac{Ze\phi}{A} + 2mc^2\right)} \quad (24.2)$$

In (24.2), A is the mass number; T is defined as the kinetic energy per nucleon, E_k; the total kinetic energy = A E_k. If the point P is taken at the heliosphere boundary, and the point Q at 1 AU then the charged particle is modulated through a potential difference of ϕ MeV/nucleon.

For more than four decades, the solar modulation of cosmic ray flux has been studied using satellites/spacecraft in orbit around the Earth, and more recently on deep space probes. The range of observed variations in the modulated near-Earth differential flux of GCR protons until the late 1970s are shown in Fig. 24.2 for protons as described by different curves designated by the modulation parameter, ϕ (Randall and Van Allen 1986; Castagnoli and Lal 1980). The hypothetical curve, $\phi = 0$, corresponds to the predicted shape of the interstellar GCR proton flux, i.e. the GCR flux outside the heliosphere. Information about near Earth changes in flux have come largely from the charged particle monitoring experiment (CPME) aboard the IMP 8 satellite since 1973. For a recent summary of these data reference is made to Simunac and Armstrong (2004).

Our knowledge of modulation of GCR through the outer heliosphere comes from deep space probes (Lockwood and Webber 1995) which transmitted in situ cosmic ray data up to distances of more than 70 AU (Pioneers 10 and 11, Voyagers 1 and 2). For details see Lal (2007). Nuclide production rates for different geomagnetic field intensities and modulation parameters, ϕ, have been calculated earlier (cf. Lal 1992).

3. Present day nuclide production rates at different latitudes and at altitudes ≤ 10 km

If the production rate of a nuclide in a rock/soil is known at some latitude and altitude (≤ 10 km), then

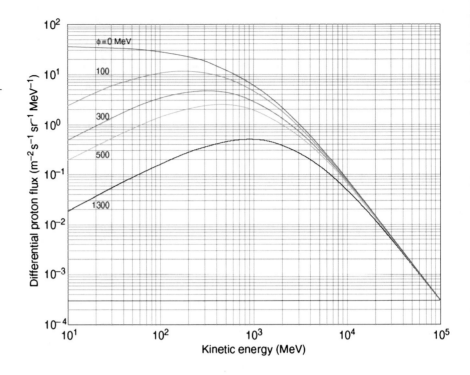

Fig. 24.2 Estimated proton spectra for different solar modulation intensities as defined by the parameter, ϕ. The curve labeled $\phi = 0$ corresponds to the unmodulated proton spectra outside the heliosphere

Table 24.1 Nuclear disintegration rates (Lal and Peters 1967) in the atmosphere (g^{-1} year^{-1})

Geomagnetic latitude (degrees)	Polynomial coefficients[a]			
	a1	a2	a3	a4
0	3.307×10^2	2.559×10^2	9.843×10^1	2.050×10^1
10	3.379×10^2	2.521×10^2	1.110×10^2	2.073×10^1
20	3.821×10^2	2.721×10^2	1.325×10^2	2.483×10^1
30	4.693×10^2	3.946×10^2	9.776×10^1	4.720×10^1
40	5.256×10^2	5.054×10^2	1.420×10^2	5.887×10^1
50	5.711×10^2	5.881×10^2	1.709×10^2	7.612×10^1
60–90	5.634×10^2	6.218×10^2	1.773×10^2	7.891×10^1

[a]The polynomials (24.3) are valid for 0–10 km altitudes

the production rate at other latitudes and altitudes can be derived using the scaling factors given by Lal and Peters (1967). The scaling factors are primarily based on thermal neutron measurements and take into account the thresholds for nuclide production from the targets of interest (Lal 1958). Third degree polynomials are presented by Lal (1991) for nuclear disintegrations (g^{-1} air year^{-1}) in the atmosphere in Table 24.1, based on Lal and Peters (1967), and for ^{10}Be and ^{26}Al in quartz (g^{-1} year^{-1}), based on measurements in quartz from glacially polished rocks (Nishiizumi et al. 1989). The nuclear disintegration rates or the production rates of ^{10}Be, and ^{26}Al in quartz are given by the corresponding polynomial coefficients, P_1, P_2, P_3 and P_4 for q (L, y):

$$q(L, y) = P_1(L) + P_2(L)\,y + P_3(L)\,y^2 + P_4(L)\,y^3 \quad (24.3)$$

where q(L,y) is the nuclear disintegration rate in the atmosphere or ^{10}Be or ^{26}Al production rate in quartz (g^{-1} SiO$_2$ year^{-1}) at latitude L, and altitude y (km). The polynomial coefficients are given separately for latitudes: 0, 10, 20, 30, 40, 50, 60–90°, in Tables 24.1 and 24.2.

For completeness we give here the third degree polynomial for converting altitude (y km) to pressure, P (g cm^{-2}), based on the ICAO standard atmosphere (International Civil Aviation Organization 1964):

$$P = p1 + p2 \cdot y + p3 \cdot y^2 + p4 \cdot y^3 \quad (24.4)$$

Table 24.2 Production rates of ^{10}Be and ^{26}Al in quartz (g^{-1} SiO$_2$ year^{-1})

Geomagnetic latitude (degrees)	Polynomial coefficients[a]							
	a		b		c		d	
	^{10}Be	^{26}Al	^{10}Be	^{26}Al	^{10}Be	^{26}Al	^{10}Be	^{26}Al
0	3.511	21.47	2.547	15.45	0.9512	5.751	0.18608	1.1154
10	3.596	22.00	2.522	15.32	1.0668	6.444	0.188229	1.1287
20	4.0607	24.84	2.734	16.61	1.2673	7.652	0.2253	1.3504
30	4.994	30.55	3.904	23.67	0.9739	5.911	0.42671	2.5563
40	5.594	34.21	4.946	29.92	1.3817	8.361	0.53176	3.1853
50	6.064	37.08	5.715	34.57	1.6473	9.955	0.68684	4.1138
60–90	5.994	36.67	6.018	36.38	1.7045	10.3	0.71184	4.2634

[a]The polynomials (24.3) are valid for 0–10 km altitudes

where p1 = 1032.92; p2 = −121.95; p3 = 5.657; p4 = −0.1095.

For scaling factors for ^{14}C, reference is made to Miller et al. (2006) and Lifton et al. (2001). Earlier scaling factors for ^{3}He and ^{20,21,22}Ne are given in Lal (1991) but they have since been considerably improved as a result of efforts made by scientists working jointly in the CRONUS-Earth Project (Cosmic ray produced systematics on Earth Project; Stuart and Dunai 2009).

4. Expected nuclide concentrations in exposed rocks for first order exposure models

As expected, the principal obstacle in determining surface exposure ages or erosion rates is the lack of knowledge regarding the rock's surface exposure history. For example, the rock may have been covered with sediment or have had undergone exfoliation during its prior exposure history. Further, the erosion rate may have varied with climate in the past. Clearly, any evidence from the geologic history of the rock is of paramount importance.

Since the production of cosmogenic nuclides within the rock is a sensitive function of the geometry of the rock, use of two or more cosmogenic nuclides can help constrain the exposure history of the rock. In an earlier paper, Lal (1991) considered the effects of changes in exposure geometry, and also considered first order exposure models. In a subsequent paper, Lal and Chen (2005) considered more complicated exposure geometries and considered possibilities of constraining exposure histories using cosmogenic data alone.

In the following, we will discuss first order exposure models and develop simple equations to bring out salient features of the cosmogenic method, for instance, the constraints on exposure history that are possible using two or more cosmogenic nuclides.

For simplicity, we will consider a horizontal rock surface; complex rock geometries have been considered by Lal and Chen (2005). As the surface of a rock horizon is exposed, nuclear interactions of cosmic rays produce nuclides at the exposed surface and at depths. The number of atoms of radio-nuclides at time t, N(x,t) within the rock at any depth, x(t) are given by the following differential equation:

$$\frac{dN(x,t)}{dt} = -N(x,t) + P(x,t) \quad (24.5)$$

where λ is the disintegration constant of the nuclide and P(x,t) is the nuclide production rate at depth x (cm), and time t (s). If one considers nuclide production by nucleons, then the nuclide production rate, P(x,t) is fairly well described by the equation:

$$P(x,t) = P(0,t) \, e^{-\mu \, \rho \, x(t)} \quad (24.6)$$

where μ (cm^{-2} g^{-1}) is the absorption coefficient (inverse of μ is defined as the absorption mean free path (g cm^{-2}) in the rock). The absorption mean free path is altitude and latitude dependant (Table 24.3).

The value of x(t) in (24.2) is given by the integral:

$$x(t) = \int_0^t \varepsilon(t) \, dt \quad (24.7)$$

(a) Special case of fixed erosion rate, fixed cosmic ray production rate and uniform erosion history of the surface

If one assumes a constant erosion rate, ε, and constant cosmic ray nuclide production rate, P_0 at the rock

Table 24.3 Absorption mean free paths in the atmosphere for nuclear disintegrations of energy release >40 MeV

Altitude range (km)	Mean free path (g cm^{-2})		
	0°	30°	60°
0–5	160	150	140
5–10	175	160	150

surface, then the solution of (24.1) nuclide concentration at time t, at depth x can also be written in the integral form:

$$N(x, T) = \int_0^T P_0 \, e^{-\mu\rho\,(x+\varepsilon t)} \cdot e^{-\lambda(T-t)} \, dt \quad (24.8)$$

integrating backwards in time, setting the present time at 0, and $t = T$, when the rock was first exposed to cosmic radiation, as T, the total duration of the exposure.

Solving (24.5) or (24.8) gives the nuclide concentration, $N(x, T)$ in the rock at depth, x:

$$N(x, T) = \frac{P_0}{\lambda + \mu\varepsilon} \, e^{-\mu x} \left(1 - e^{-T(\lambda+\mu\varepsilon)}\right) \quad (24.9)$$

If however, the rock sample had some initial concentration of the radionuclide at the start of the irradiation, $N(x,0)$, then an additional term, $N(x,0) \exp(-\lambda T)$ should be added to (24.9).

Equation (24.9) states that if a rock surface undergoes *steady state erosion*, the in situ radionuclides attain secular equilibrium concentration corresponding to an effective disintegration constant, $\lambda + \mu\varepsilon$. The effective irradiation time, T_{eff} for the top surface of the rock is then given by:

$$T_{eff} = N\left(0, T \gg \frac{1}{\lambda + \mu\varepsilon}\right)/P_0 = \frac{1}{\lambda + \mu\varepsilon} \quad (24.10)$$

The validity of a steady state model can be tested by studying two or more radionuclides of different half-lives in the rock surface. The model steady state erosion at the surface is then given by, based on (24.5):

$$\varepsilon = \frac{1}{\mu}\left[\frac{P(0)}{N(0,T)} - \lambda\right] = \frac{1}{\mu}\left[\frac{1}{T_{eff}} - \lambda\right] \quad (24.11)$$

If the rock had been undergoing steady state erosion for periods of the order of $\geq$(4–5) times T_{eff}, it follows from (24.11) that the quantities below are invariant for different radionuclides:

$$\left(\frac{P(0)}{N(0,T)}\right)_1 - \lambda_1 = \left(\frac{P(0)}{N(0,T)}\right)_2 - \lambda_2$$
$$= \left(\frac{P(0)}{N(0,T)}\right)_3 - \lambda_3 \quad (24.12)$$

Note that the validity of (24.8) does not necessarily imply that surface has been undergoing erosion on a micro-scale. However it does imply that if the rock has undergone macro-erosion, then the chip-off distances have always been much $\ll$ than the distance $1/\mu$.

The suitability of a radionuclide for determining the erosion rate of a rock horizon is determined by the condition that λ and $\mu\varepsilon$ should be of the same order; in other words T_{eff} and the mean half-life of the radionuclide, λ, should be of the same order. If $\lambda \gg \mu\varepsilon$, the radionuclide would be built to its secular equilibrium value ($T_{eff} = 1/\lambda$) and no information can be obtained from its study about the erosion rate. Else if $\mu\varepsilon \gg 1$, the nuclide concentrations would build up to the same concentration independent of the half-life of the radionuclide. In this case, the nuclide decay during the build-up period is unimportant and the nuclide behaves similar to a stable nuclide. Also, in this case ($\mu\varepsilon \gg 1$), the apparent surface exposure age, $T_{eff} = 1/\mu\varepsilon$, i.e. it is the time during which erosion removes a rock depth equivalent to one absorption mean free path for cosmic rays, $1/\mu$, ~50 cm in common rocks.

Under steady state, the nuclide build up is shown in Fig. 24.3 for five radionuclides, ^{39}Ar, ^{14}C, ^{36}Cl, ^{26}Al and ^{10}Be, and a stable nuclide. These curves are dependent on the values of ε and λ, and provide useful insight if the exposure history of the rock deviates from the assumed steady state erosion (24.10).

If two radionuclides are used, e.g., ^{10}Be and ^{26}Al, their steady state concentrations are well described for all erosion rates by the shadowed curve in Fig. 24.4, labelled "steady state erosion island", The temporal evolution of the curves is given by the following:

$$Ratio\,(^{26}Al/^{10}Be)_t$$
$$= \frac{P(0)_{26Al}}{P(0)_{10Be}} \cdot \frac{\lambda_{10Be} + \mu\varepsilon}{\lambda_{26Al} + \mu\varepsilon} \frac{\left(1 - e^{-t(\lambda_{26Al}+\mu\varepsilon)}\right)}{\left(1 - e^{-t(\lambda_{10Be}+\mu\varepsilon)}\right)} \quad (24.13)$$

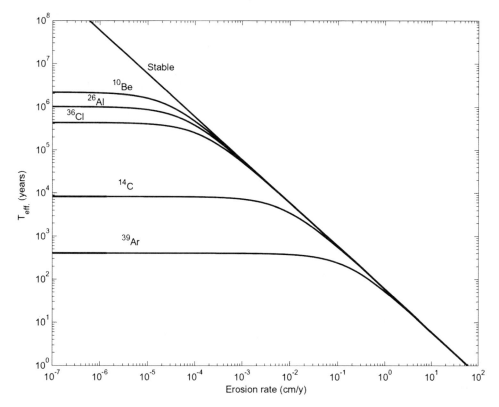

Fig. 24.3 Effective steady state surface exposure ages, T_{eff} (24.6), as a function of erosion rates for five radionuclides, ^{39}Ar, ^{14}C, ^{36}Cl, ^{26}Al and ^{10}Be, and a stable nuclide

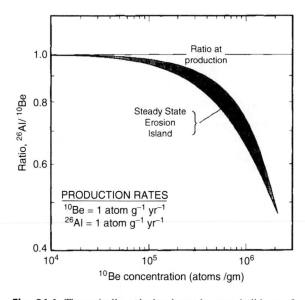

Fig. 24.4 Theoretically calculated steady state build up of ^{26}Al/^{10}Be ratio and ^{10}Be concentration for hypothetical production rates of 1 atom g^{-1} rock for both ^{10}Be and ^{26}Al. The *steady state erosion island* includes the temporal build up of radionuclide concentrations as given by (24.9) for all erosion rates. In steady state all regions above the *shaded curve* are forbidden

with the ratio lying between two limits:

$$\left[\frac{P(0)_{26Al}}{P(0)_{10Be}}\right] \quad and \quad \left[\frac{P(0)_{26Al}}{P(0)_{10Be}} \cdot \frac{\lambda_{10Be}}{\lambda_{26Al}}\right] \quad (24.14)$$

corresponding to large and small erosion rates, respectively.

(b) Complex cosmic ray irradiation histories

If the exposure history of the rock is simple as so far considered: a flat rock surface, eroding at a constant rate, curves like that shown in Fig. 24.4 for ^{26}Al and ^{10}Be can be used to ascertain its veracity. One can list a number of complex rock irradiation histories: for instance (1) rock's of different geometrical shapes, (2) the top surface spalls off sometime in the past durting irradiation, and (3) the top surface is buried under a layer of sediment. The first case has been considered in some detail by Lal and Chen (2005), who have cosidered a sphroidal target, a rectangular geometry with a flat top, a sloping surface and a beach terrace with a flat top. The second case of a rock

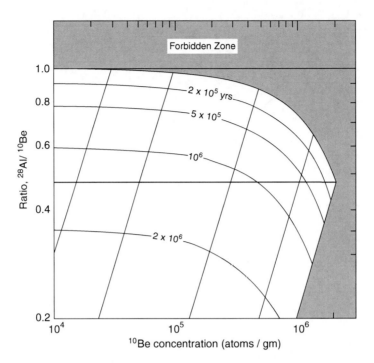

Fig. 24.5 The upper bound of the curve corresponds to temporal evolution for the case of ε = 0, as also in Fig. 24.4. The *straight lines* denote the path traced by the rock surface after the top surface was shielded by a thick sediment or rock fragments. The production rates are fixed at 1 atom g^{-1} ^{10}Be and ^{26}Al in the rock

surface spalling of has been considered in detail by Lal (1991), who have considered two cases as examples: the effective exposure age of the rock is (1) small compared to the mean life of the radionuclide, and (2) comparable to the mean life. The third case was in fact discovered in a real case of a soil profile based on studies of ^{14}C and ^{10}Be (Lal et al. 1996). Reference is also made to Miller et al. (2006), who studied ice sheet erosion and complex exposure histories in Bafin Island, Arctic Canada, based on ^{14}C, ^{26}Al and ^{10}Be.

We will not consider here the cases of complex rock geometries since these have been dealt with in detail by Lal and Chen (2005), and also the case (2) of rock spall during cosmic ray exposure (Lal 1991). The third case is an interesting one since in some cases the rock surface may be shielded by a thick sediment or rock layer. In this case, if the shielding of rock surface is substantial, the shorter lived radioisotope would decay much faster than the longer lived radionuclide (Lal 1991), providing an information on the time for which the rock surface was shielded. As a case in point we consider the special case of a rock surface which was exposed under zero erosion rate evolving finally to a ratio, ^{26}Al/^{10}Be:

$$\text{Ratio}\left(^{26}Al/^{10}Be\right)_{\text{Steady State and } \varepsilon=0} = \frac{P(0)_{26Al}}{P(0)_{10Be}} \cdot \frac{\lambda_{10Be}}{\lambda_{26Al}} \quad (24.15)$$

corresponding to the top curve in Fig. 24.5. If during the evolution of this curve, the rock surface is overlain by a thick deposit of sediment or rock fragments, the ratio ^{26}Al/^{10}Be and the ^{10}Be concentration will evolve along the example straight lines due to faster decay of ^{26}Al compared to ^{10}Be. These points would *generally* lie outside the "steady state erosion island" in Fig. 24.4, and can be used to find out if the rock surface was part of the time or continuously shielded on the top surface by a thick sediment layer or rock fragments.

24.3 Future Directions

In this chapter, we have discussed that cosmic ray labeling of erosion surfaces allows one to determine both cosmic ray exposure ages and surface erosion rates. Several radionuclides are produced in situ in rocks/soils from major target elements, e.g., ^{39}Ar, ^{14}C, ^{36}Cl, ^{26}Al and ^{10}Be. This task is accomplished fairly well using two or more radionuclides produced in situ in the rock by cosmic rays.

The single most salient feature of cosmic ray labeling of erosion surfaces is the high sensitivity of the rate of in situ nuclear interactions on the geometry of the irradiation. Note also that the mean absorption distance for cosmic radiation in typical rocks is ~50 cm only. The text in this chapter along with ideas discussed by Lal (1991) and Lal and Chen (2005) cover most of the ideas, which allow one to put severe constraints on the "unknown" exposure history of the rock horizon. As an example: Use of the shorter half-life radionuclides, ^{39}Ar, ^{14}C and ^{36}Cl, in conjunction with ^{10}Be allow one to put severe constraints on the "unknown" rock exposure histories. This task can be put on much firmer ground if information on the exposure history can be constrained based on geological information.

24.4 Conclusion

The cosmogenic nuclear method described in this chapter, bases itself on the fact that several radionuclides are produced in situ in rocks/soils from major target elements. As stated earlier, the single most salient feature of cosmic ray labeling of erosion surfaces is the high sensitivity of the rate of in situ nuclear interactions on the geometry of the irradiation, coupled with the fact that the mean absorption distance for cosmic radiation in typical rocks is ~50 cm.

In the absence of information on the exposure histories of the rock surfaces studied, several ideas to constrain exposure histories were discussed by Lal (1991) and Lal and Chen (2005); these allow one to put severe constraints on the "unknown" exposure history of the rock horizon. As an example, use of the *shorter* half-life radionuclides, ^{39}Ar, ^{14}C and ^{36}Cl, in conjunction with ^{10}Be allow one to put severe constraints on the "unknown" rock exposure histories.

Estimation of exposure ages and erosion requires an accurate knowledge of nuclide production rates. Realizing the importance of determining accurate time scales in earth sciences, several scientists have joined a project "Cosmic ray produced nuclide systematics on Earth project (CRONUS-Earth Project)"

for improving rates of production of cosmogenic nuclides in targets exposed to cosmic radiation under different conditions. These improvements combined with the steady improvements in the AMS sensitivity (cf. Galindo-Uribarri et al. 2007) should considerably widen the scope of applications of cosmic ray produced nuclides in geomorphology.

References

Bennett CL, Beukens RP, Clover MR, Gove HE et al (1977) Radiocarbon dating using electrostatic accelerators: negative ions provide the key. Science 198:508–510

Castagnoli G, Lal D (1980) Solar modulation effects in terrestrial production of carbon 14. Radiocarbon 22:133–158

Galindo-Uribarri A et al (2007) Pushing the limits of accelerator mass spectrometry. Nucl Instr Meth Phys Res B 299:123–130

Gleeson LJ, Axford WI (1968) Solar modulation of galactic cosmic rays. Astrophys J 423:426–431

Gosse JC, Phillips FM (2001) Terrestrial in situ cosmogenic nuclides: theory and application. Quat Sci Rev 20:1475–1560

International Civil Aviation Organization (1964) Manual of the ICAO standard atmosphere, 2nd edn. International Civil Aviatory Organization, Montreal

Lal D (1958) Investigations of nuclear interactions produced by cosmic rays. Ph.D. Thesis, Bombay University

Lal D (1988) In-situ produced cosmogenic isotopes in terrestrial rocks. Ann Rev Earth Planet Sci Lett 16:355–388

Lal D (1991) Cosmic ray tagging of erosion surfaces: in situ production rates and erosion models. Earth Planet Sci Lett 104:424–439

Lal D (1992) Expected secular variations in the global terrestrial production rate of radiocarbon. In: Bard E, Broecker WS (eds) The last deglaciation: absolute and radiocarbon chronologies. NATO ASI series, vol 12. Springer, Berlin, pp 113–126

Lal D (2007) Cosmic ray interactions in minerals. In: Elias SA (ed) Encyclopedia of quaternary science. Elsevier, Oxford, pp 419–436

Lal D, Baskaran M (2011) Applications of cosmogenic-isotopes as atmospheric tracers. In: Baskaran M (ed) Handbook of environmental isotope geochemistry. Springer, Heidelberg

Lal D, Chen J (2005) Cosmic ray labeling of erosion surfaces II: special cases of exposure histories of boulders, soils and beach terraces. Earth Planet Sci Lett 236(3–4):797–813

Lal D, Peters B (1967) Cosmic ray produced radioactivity on the earth. In: Flugge S (ed) Handbook der Physik, vol 46/2. Springer, Berlin, pp 551–612

Lal D, Arnold JR, Nishiizumi K (1985) Geophysical records of a tree: new application for studying geomagnetic field and solar activity changes during the past 10^4 years. Meteoritics 20(2):403–414

Lal D, Pavich M, Gu ZY, Jull AJT et al (1996) Recent erosional history of a soil profile based on cosmogenic in-situ radionuclides ^{14}C and ^{10}Be. In: Basu A, Hart S (eds) Earth processes reading the isotopic code. Geophysics monograph series, vol 95, pp 371–376

Lifton NA, Jull AJT, Quade J (2001) A new extraction technique and production rate estimate for in situ cosmogenic 14C in quartz. Geochim Cosmochim Acta 65:1953–1969

Lockwood JA, Webber WR (1995) An estimate of the location of modulation boundary for E >70 MeV galactic cosmic rays using voyager and pioneer spacecraft data. Astrophys J 442:852–860

Miller GH, Briner JP, Lifton NA, Finkel RC (2006) Limited ice-sheet erosion and complex exposure histories derived from in situ cosmogenic ^{10}Be, ^{26}Al and ^{14}C on Baffin Island, Arctic Canada. Quat Geochronol 1:74–85

Nelson DE, Koertling RG, Stott WR (1977) Carbon-14: direct detection at natural concentrations. Science 198:507–508

Nishiizumi K, Winterer EL, Kohl CP, Lal D et al (1989) Cosmic ray production rates of ^{10}Be and ^{26}Al in quartz from glacially polished rocks. J Geophys Res 94:17907–17916

Papini P, Grimani C, Stephens SA (1996) An estimate of the secondary proton spectrum at small atmospheric depths. Nuovo Cimento 19C(3):367–388

Randall BA, Van Allen JA (1986) Heliocentric radius of the cosmic ray modulation boundary. Geophys Res Lett 13:628–631

Simunac KDC, Armstrong TP (2004) Solar cycle variations in solar and interplanetary ions observed with interplanetary monitoring platform. J Geophys Res 109:A10101

Stone JO (2000) Air pressure and cosmogenic isotope production. J Geophys Res 105(10):23753–23759

Stuart GM, Dunai TJ (2009) Advances in cosmogenic isotope research from CRONUS-EU. Quat Geochronol 4(6):435–436

Tuniz C, Bird JR, Fink D, Herzog GF (1998) Accelerator mass spectrometry. CRC Press, Washington, p 371

Chapter 25
Measuring Soil Erosion Rates Using Natural (^{7}Be, ^{210}Pb) and Anthropogenic (^{137}Cs, ^{239,240}Pu) Radionuclides

Gerald Matisoff and Peter J. Whiting

Abstract This chapter examines the application of natural (^{7}Be and ^{210}Pb) and anthropogenic fallout radionuclides (^{134}Cs, ^{137}Cs, ^{239,240}Pu) to determine soil erosion rates. Particular attention is given to ^{137}Cs because it has been most widely used in geomorphic studies of wind and water erosion. The chapter is organized to cover the formation and sources of these radionuclides; how they are distributed in precipitation and around the globe: their fate and transport in undisturbed and tilled soils; and their time scales of utility. Also discussed are methods for soil collection, sample preparation for ^{137}Cs analysis by gamma spectroscopy, and the selection of standards and instrument calibration. Details are presented on methods for calculating soil erosion, including empirical methods that are related to the Universal Soil Loss Equation (USLE), box models that compare ^{137}Cs activities in a study site to a reference site, and time dependent methods that account for the temporal inputs of ^{137}Cs and precipitation induced erosion. Several examples of recent applications, including the combination of radionuclides with other techniques or measurements, are presented. The chapter concludes with suggestions for future work: the value of new methods and instrumentation to allow for greater spatial resolution of rates and/or greater accuracy; the need to incorporate migration of radionuclides in the time-dependent models; the opportunities to concurrently use the global and Chernobyl signals to better understand temporal variation soil erosion processes and rates; and the importance of the use of these tracers to characterize C storage and cycling.

G. Matisoff (✉) and P.J. Whiting
Department of Geological Sciences, Case Western Reserve University, Cleveland, OH 44106-7216, USA
e-mail: gerald.matisoff@case.edu; peter.whiting@case.edu

25.1 Introduction

25.1.1 Soil Erosion; Nature of the Problem

Soil is among our most fundamental resources and soil processes help regulate atmospheric composition and climate. Soil anchors and sustains the vegetation that provides sustenance for animals and humans and provides fibers and material used in everything from cotton for clothing to lumber for homes to biomass for energy. The soil itself can be mined for key materials, minerals and metals, and energy. The foundations of most human structures – homes, buildings, and roads – are built on soil. Soil and soil processes filter water, reduce toxicity of airborne pollutants delivered to the land surface, and store carbon and nutrients. The value of soil in terms of ecosystem function and service has been estimated in the hundreds of billions of dollars per year (Pimental et al. 1995).

A comprehensive understanding of material fluxes on the earth surface and its effects on geochemical cycles (hydrologic, C, and N), atmospheric composition and climate, and ocean chemistry depends upon an understanding of soil and soil movement on the landscape including erosion, transport, and deposition. Soils sequester C and N from the atmosphere and retain certain metals during the weathering of rocks, but soil erosion either moves those materials to places of long-term storage or exposes soils to greater reactivity. Soils hold 2,300 Gt of carbon, about four times as much carbon as is in the atmosphere (Lal 2003). It has been suggested that if carbon on the landscape lost by erosion is replaced by new vegetative growth, then intermediate storage in fluvial systems of the eroded carbon represents a net removal of carbon from the atmosphere and may be the "missing" anthropogenic carbon (Harden et al. 1992; Stallard 1998). Others note that oxidation of a portion of the carbon in transport may produce 0.8–1.2 GtC per year. Thus anthropogenically enhanced soil erosion may reinforce global warming.

Soil is moved by a variety of processes including water (splash, sheetwash, rills), wind, ice (freeze-thaw, glaciers, periglacial), gravity (dry ravel, creep, toppling, debris flows, earthflows), tillage, and bioturbation. Erosion is often accelerated by disturbance (clearing, fire, plowing, overgrazing, compaction, or desiccation) that disrupts soil structure and removes vegetative covering. Oldeman (1994) estimated that 1,094 Mha (1 ha = 10^4 m^2) are affected by water erosion and 549 Mha by wind erosion. These numbers represent 12 and 6% of agricultural land areas, respectively. Total erosion of these areas is approximately 75 billion tons/year (Pimental et al. 1995).

The net loss of soil has both on-site and off-site consequences as summarized by Pimental et al. (1995). In croplands, the diminished fertility due to topsoil erosion requires fertilization or results in diminished yields, creates pressure to deforest new areas as fertility of existing cropland decreases, and results in the loss in water holding capacity of soils. Fertilization, in turn, often has its own consequences. Most fertilizers rely on fossil fuels to create, ship, and apply the material and the applied fertilizer has the potential for creating downstream water quality concerns. The additional water use required because of diminished soil retention taxes another critical resource. In forestlands, soil loss can change species composition, diminish water-holding capacity, and speed desertification. In suburban and urban areas, soil loss can reduce the ability of soils to sustain vegetative cover and trees helpful in addressing air, water, heat, and sound pollution.

Fine sediments derived from erosion of soil are disproportionately responsible for degradation of surface waters (Nelson and Logan 1983; Dong et al. 1984). Eroded soil impairs water quality (Sekely et al. 2002; Sharpley et al. 1994; Pote et al. 1996) to the point that drinking water supplies, aquatic environments, and opportunities for recreation are threatened. Eroded soil often harms aquatic environments by inhibiting light penetration (Yamada and Nakamura 2002), by siltation of rivers (Reiser 1998) and reservoirs (Williams and Wolman 1984), by eutrophication of waterways, lakes, and seas (Rabalais et al. 1999), and by contamination (Tarras-Wahlberg and Lane 2003). In 2000, the US Environmental Protection Agency reported that siltation debilitated 12% of the stream reaches assessed by states and tribes and was responsible for 33% of impairments to beneficial use (USEPA 2000). In areas where wind is an important process of erosion, the transported fine material can be a health problem, foul equipment, and cause abrasion requiring the repainting of structures (Lyles 1985).

History shows that civilization can collapse as the soil resource is depleted (Montgomery 2007; Diamond 2005; Hyams 1952). Plato ascribed the poor soils of his native Attica to erosion after land clearing and his view of the causative factors of poor soil was shared by Aristotle (Montgomery 2007). Loss of production associated with soil loss and degradation ultimately affected the stability of the Greek civilization as it did the Romans later. Lowdermilk (1953) describes a trail of societies from Judea to Syria to China where poor stewardship of the land and resulting erosion led indirectly to conquest or societal discord. More recent examples of societal dislocations (famine and migration) associated with soil erosion and land degradation include the Dustbowl of the 1930s, the Sahel of the 1970s, and Haiti.

25.1.2 Tools for Measuring Soil Erosion

Critical to the understanding and quantification of soil erosion are tools for its measurement. Erosion pins, sediment accumulated in reservoirs, measured sediment concentration in streamflow, photographic techniques, and soil tracers each have their usefulness and limitations. Sediment budgets (Dietrich and Dunne 1978) are often a basis for quantifying the various processes and paths that move soil on the landscape and result in local loss (erosion) and local gain (deposition) of soil.

A particularly useful tool for measuring soil erosion is a conservative tracer of the soil particles, especially when the tracer is relatively easy to measure. Important considerations in the use of a tracer are that the concentration of the tracer is relatively uniform; adsorption of the tracer to soil is strong and quick; variation in adsorption to various sizes or mineralogic/organic constituents is minor or can be accounted for; and methods exist to measure the tracer.

The best known of the tracers for estimating soil erosion are natural and anthropogenic radionuclides. The anthropogenic radionuclides found on the landscape were produced largely by atmospheric nuclear bomb testing and the fallout was distributed globally. The list of fission products is extensive, although many of these radionuclides are too short-lived to be useful tracers of soil erosion. Of the longer-lived fission products, the best known is ^{137}Cs, but the list of other useful tracers includes ^{134}Cs, ^{238,239,240}Pu, and ^{241}Am as minute solid particles or sorbed to soil particles; and ^{3}H and ^{90}Sr as soluble tracers. The naturally-occurring radionuclides are produced by various nuclear reactions, or uranium or thorium decay chains (Porcelli and Baskaran 2011) and include ^{7}Be, ^{210}Pb and a few others. ^{137}Cs, ^{7}Be, and ^{210}Pb are each suitable as particle tracers because they have a global distribution, adsorb efficiently to soil particles and thus move with soil, and are relatively easily measured.

^{137}Cs is the most widely used radionuclide tracer for soil erosion (Ritchie and McHenry 1990). For years, Ritchie and Ritchie (2008) maintained a bibliography of publications that utilized ^{137}Cs in the study of soils and sedimentation. Figure 25.1, redrafted from Ritchie and Ritchie (2008) and updated here to include papers published after December 15, 2008, illustrates how widespread the use of ^{137}Cs as a tracer has been. There are a total of about 4,500 Cs references with the vast majority of the papers following the Chernobyl accident in 1986. In comparison, there are about 2,700 references to the use of ^{210}Pb in studies of soils and sedimentation. The use of ^{210}Pb in such studies now exceeds the use of ^{137}Cs. The use of ^{7}Be as a tracer in the context of soils and sediment is relatively new and has resulted in about 90 papers to date. It should be noted that substantially less than half of the total number of papers have used the respective tracer to quantify soil erosion.

25.1.3 Summary of Contents of Book Chapter and the Approach Used

This chapter focuses primarily upon the use of anthropogenic and naturally-occurring radionuclides to study soil erosion processes and to quantify rates of soil erosion. Many other processes affect soil and sediment transport and deposition and other radionuclides used in those studies are detailed in other chapters of this collection. Here we devote much of our attention to the anthropogenic radionuclide ^{137}Cs but we also look at other radionuclide tracers in part to show which radionuclides may be the most suitable for given applications. Specifically, we describe the source of the radionuclides; their characteristics, deposition, sorption, and transport; the methods of measurement; the assumptions associated with their use as

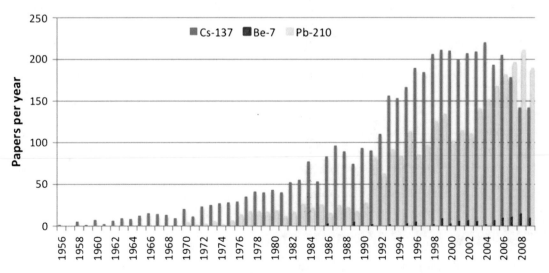

Fig. 25.1 The annual number of papers utilizing ^{137}Cs, ^{210}Pb, and ^{7}Be in chronologic, geomorphic and sedimentologic studies. The total number of ^{137}Cs papers is now about 4,500. Modified and updated from Ritchie and Ritchie (2008)

tracers; the models used to estimate erosion with the tracers; the relative merits of different models and tracers for investigating soil erosion; recent applications of radionuclides in erosion studies; and foreseeable changes in the use of tracers and the tracers used.

25.2 Background

25.2.1 Radionuclides in the Environment

25.2.1.1 Sources of ^{137}Cs, ^{7}Be, ^{210}Pb, and ^{239,240}Pu

Radionuclides that have been used in studies of soil erosion on decadal time scales or less include ^{134}Cs, ^{7}Be, ^{210}Pb, and ^{239,240}Pu but the most widely known of these is ^{137}Cs ($t_{1/2}$ = 30.1 years). ^{134}Cs ($t_{1/2}$ = 2.06 years), ^{137}Cs, ^{239}Pu ($t_{1/2}$ = 4,110 years) and ^{240}Pu ($t_{1/2}$ = 6,537 years) are present on the global landscape largely as a result of anthropogenic fallout from thermonuclear weapons testing primarily during the period 1954–1968 (Fig. 25.2). In parts of Europe, ^{137}Cs fallout from the nuclear reactor accident at Chernobyl in 1986 is superimposed on the global signal of fallout and the Pu isotopic signatures from stratospheric fallout and from Chernobyl are different from each other (Ketterer et al. 2011; Hong et al. 2011). Other accidental releases of Cs isotopes have created local zones of contamination. ^{210}Pb ($t_{1/2}$ = 22.3 years) is delivered to the landscape as a result of the decay of gaseous ^{222}Rn in the ^{238}U decay chain. In that chain, ^{226}Ra in soils and rock decays to ^{222}Rn ($t_{1/2}$ = 3.8 days) which is released to the atmosphere and undergoes a series of short-lived decays to ^{210}Pb which, as a particulate, is delivered to the landscape by wet and dry fallout. This ^{210}Pb is termed excess ^{210}Pb (^{210}Pb$_{xs}$) (Fig. 25.2). However, some ^{222}Rn that occurs in the soil is trapped in mineral matter or unable to escape to the atmosphere during its 3.8 day half-life, and its decay builds the supported pool of ^{210}Pb in the soil (supported ^{210}Pb). ^{7}Be ($t_{1/2}$ = 53.3 days) is produced by cosmic ray spallation of nitrogen (90%) and oxygen (10%) in the troposphere and stratosphere (Kaste et al. 2002; Kaste and Baskaran 2011) and then it is removed from the atmosphere and is delivered to the landscape often during large thunderstorms (Dibb 1989).

^{137}Cs atmospheric fallout is at present very small and the negligible amount, if any, is due to eolian resuspension of near-surface soil particles with adsorbed ^{137}Cs. ^{210}Pb concentration is highest in air originating over continental regions (Turekian et al. 1983; Paatero and Hatakka 2000) and the activity of ^{210}Pb in the atmosphere decreases with altitude (Kownacka 2002) consistent with the sources of the radionuclide in soil and rock at the earth surface. ^{7}Be is produced through spallation interactions of atmospheric O and N nuclei and the nucleonic component of the atmospheric cascade induced by galactic cosmic rays (Dorman 2004) (Fig. 25.2). Sugihara et al. (2000) suggest that 70% is produced in the stratosphere. ^{7}Be is also added to the troposphere from the stratosphere by mixing along the polar front and the subtropical jet and by mixing in large synoptic storms (mid-latitude cyclones and hurricanes). ^{7}Be concentration increases with altitude given the substantial reservoir in the stratosphere where concentrations may be an order of magnitude higher than in the upper troposphere

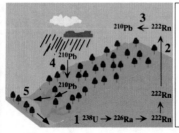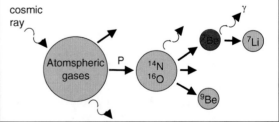

Fig. 25.2 Schematic illustration of fallout radionuclide sources used in erosion studies of decadal or shorter time scales. *Left figure* is a photograph of an atmospheric bomb test, the main source of global radioactive Cs and Pu fallout. Photo from: http://www.nv.doe.gov/news&pubs/photos&films/atm.htm. The *middle figure* illustrates excess ^{210}Pb fallout. The figure on the *right* illustrates the production of ^{7}Be by cosmic ray spallation of oxygen and nitrogen in the stratosphere

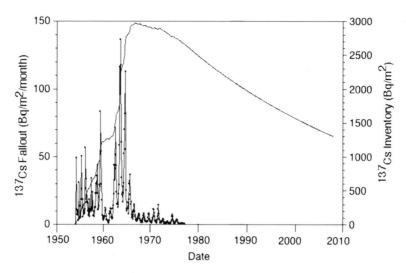

Fig. 25.3 ^{137}Cs fallout (Bq m^{-2} month^{-1}) calculated as 1.45 times the average ^{90}Sr monthly fallout at the US DOE monitoring sites (Birmingham, AL; Los Angeles, CA; San Francisco, CA; Denver, CO; Coral Gables/Miami, FL; Argonne, IL; International Falls, MN; Columbia, MO; Helena, MT; New York, NY; Williston, ND; Wooster, OH; Tulsa, OK; Medford, OR; Columbia, SC; Vermillion, SD; and Green Bay, WI). Data from the Health and Safety Laboratory (1977). Also shown is the calculated ^{137}Cs inventory (Bq m^{-2}) assuming that all the fallout to the soil surface remains in the soil and is subject only to radioactive decay. Chernobyl fallout is not included in this calculation

(Lal and Peters 1967; Dutkiewicz and Husain 1985; Lal and Baskaran 2011). There appears to be some seasonality of the concentrations of ^{210}Pb and ^{7}Be in the atmosphere with the highest values in late spring and summer (Turekian et al. 1983; Baskaran et al. 1993; Caillet et al. 2001). Valles et al. (2009) show a 2–3-fold variation in monthly concentrations of ^{7}Be and ^{210}Pb. Greater emissions of ^{222}Rn from snow-free areas and dry soils are thought to increase atmospheric concentrations of ^{210}Pb during summer (Olsen et al. 1985; Caillet et al. 2001). The seasonal peak in ^{7}Be concentration in late spring and summer (e.g. El-Hussein et al. 2001) is thought to be caused by enhanced mixing of stratospheric air into the troposphere. Longer term fluctuation in ^{7}Be fallout has been tied to variations in the flux of cosmic galactic primary radiation caused by the 11-year sunspot cycle (Azahra et al. 2003; Kikuchi et al. 2009).

^{137}Cs (+), ^{7}Be (2+), and ^{210}Pb (2+, 4+) are strongly chemically active, so they rapidly become associated with aerosols and particles (0.7–2 μm; Ioannidou and Papastefanou 2006) and are delivered from the atmosphere to the earth surface predominantly in precipitation (Olsen et al. 1985). Snow is more efficient than rain at removing radionuclides from air (McNeary and Baskaran 2003; Ioannidou and Papastefanou 2006) and wet precipitation is more efficient than dry precipitation.

Residence times in the troposphere are estimated to be 22–48 days (Durana et al. 1996) for ^{7}Be and just a few days for the aerosols carrying the ^{210}Pb (Tokieda et al. 1996). ^{137}Cs also has a short residence time partly because the source of the radionuclide is now resuspension of soil particles that are coarser (Papastenfanou et al. 1995; Ioannidou and Papastefanou 2006). During the weapons testing era, ^{137}Cs residence time in the atmosphere was longer (~1–10 years) than it is today because of its injection into the stratosphere (Joshi 1987).

Although the first nuclear detonation was in 1945, ^{137}Cs was first detected in fallout in 1951 and over most of the globe it was below detection prior to about 1954, peaked in about 1963, and was effectively below detection by 1983 (Cambray et al. 1985) (Fig. 25.3). The data in Fig. 25.3, for the most part, are not based on actual ^{137}Cs fallout data because ^{137}Cs was not routinely monitored during the 1950–1983 time frame. Instead, ^{90}Sr fallout was usually monitored and ^{137}Cs fallout is inferred by assuming a ^{137}Cs/^{90}Sr ratio in the bomb fallout (Health and Safety Laboratory 1977; Robbins 1985). This ratio has been estimated to range from 1.4 to 1.65 and is characteristic of production rates of the isotopes and is independent of location. The $^{239+240}$Pu depositional fluxes are less well known because certain data (i.e., the ratios of plutonium to ^{137}Cs, ^{90}Sr) are still classified by the U.S. Government. More recently, for example during the monitoring of Chernobyl fallout, deposition of ^{131}I was monitored and the deposition of all other radionuclides was calculated from the ^{131}I deposition density values using the relationships calculated by Hicks (1982). The overall geographic patterns of ^{90}Sr and $^{239+240}$Pu differ slightly from those for ^{137}Cs and ^{131}I primarily due to the differences in the nuclear fuel used in different tests, the size of the particles associated with the radionuclides, and the directions of travel of the clouds of radioactive particles from each test. The overall deposition of ^{90}Sr was very similar to that of ^{137}Cs. For most of the regions in the United States, the activity of ^{137}Cs was 10–20 times the activity of $^{239+240}$Pu deposited (Beck 1999; CDC 2006).

Because of the temporal variability with a peak ^{137}Cs fallout activity in 1963, that peak is often used as a time horizon in sediments to indicate the time of deposition of that layer of sediment. Linear sedimentation rates (cm year^{-1}) are then simply calculated as the depth of the layer (cm) divided by the time in years since 1963. The mass flux deposition rates (g cm^{-2} year^{-1}) are calculated as the cumulative dry mass of sediment above the 1963 time horizon divided by the cross sectional area of the sediment core divided by the time in years since 1963 until the date of coring (Walling and He 1997a, b; Goodbred and Kuehl 1998).

Inspection of Fig. 25.3 shows that the ^{137}Cs depositional flux was highly variable, reflecting the times of atmospheric testing. There are peaks in fallout observed each year between 1954 and 1959, and much larger peaks in 1962, 1963 and 1964. The maximum global fallout of radioactive nuclides occurred in Spring 1964, but because most field data cannot resolve the 1962 to 1964 fallout peaks, the position of the maximum in ^{137}Cs is often assigned a date of 1963 (1964 in the southern hemisphere). In the late 1960s to mid-1970s, fallout decreased considerably, but the seasonal cycle reflecting precipitation can be seen through the 1970s.

As a result of the nuclear accident at Chernobyl in 1986, additional ^{137}Cs fallout occurred in some areas of Europe. It is estimated that the Chernobyl accident released from 10 to 16% as much ^{137}Cs to the environment as was emitted from all nuclear weapons tests (Flavin 1987). Little Chernobyl fallout occurred over North America (Roy et al. 1988). Note the much higher levels of ^{137}Cs deposition from Chernobyl than from stratospheric fallout (compare Figs. 25.3–25.5). For example, the *lowest* contours on the Europe map following Chernobyl are ~2 kBq m^{-2} (light yellow in Fig. 25.4) De Cort et al. (1998) whereas in Fig. 25.3 the *largest* fallout values in the US following bomb testing are ~3,000 Bq m^{-2} (3 kBq m^{-2}). Monthly deposition values are 20–30 times less than the lowest fluxes deposited from Chernobyl. The highest Chernobyl fluxes are ~1,480 kBq m^{-2}, a value 10,000× larger than the peak monthly fallout from atmospheric weapons testing (~140 Bq m^{-2} month^{-1}) and a value at least 100× larger than the inventory in US soils (8–13 kBq m^{-2}, Fig. 25.5). Delivery of ^{137}Cs from the atmosphere is again currently near zero (Quine 1995). The much higher deposition of Chernobyl-derived ^{137}Cs over northern Europe is significant, because in these locations it has swamped the prior global fallout signature and has had the effect of "resetting" the ^{137}Cs soil erosion clock to 1986 because downcore soil inventories and soil activities near the soil surface are now dominated by Chernobyl-derived ^{137}Cs. This is not the case over most of the rest of the world, as the ^{137}Cs fallout from Chernobyl was minimal (Department of National Health and Welfare 1986). Consequently, the current distribution of ^{137}Cs in soils in northern Europe is dominated by Chernobyl fallout and in the US by stratospheric fallout. This resulted because Chernobyl was characterized by the relatively intermittent release of a full range of radionuclides at relatively low temperatures with very heavy local fallout from tropospheric transport. Weapons testing fallout was at a high temperature with more uniform stratospheric transport, longer residence times, and with much less pronounced local fallout (Joshi 1987). Pu atom ratios can be used to distinguish between Chernobyl and stratospheric fallout. The ^{240}Pu/^{239}Pu atom ratio was about 0.38 in Chernobyl fallout and 0.18 in stratospheric fallout (Ketterer et al. 2011). Further, the distribution of Pu from Chernobyl was much more localized than was ^{137}Cs. Pu isotopes were specifically associated with non-volatile fuel particles (Mietelski and Was 1995) whereas ^{137}Cs volatilized in the reactor accident and was more widely dispersed over much of northern Europe and Russia. ^{7}Be and ^{210}Pb deposition on the other hand continues as natural processes continue to produce these radionuclides.

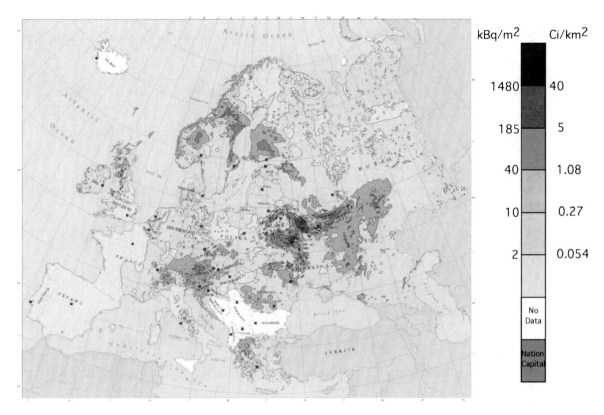

Fig. 25.4 ^{137}Cs fallout deposition over Europe following the Chernobyl accident. Modified from EC/IGCE, Roshydromet (Russia)/Minchernobyl (Ukraine)/Belhydromet (Belarus) (1998)

25.2.1.2 ^{137}Cs, ^{7}Be and ^{210}Pb in Precipitation

Deposition of radionuclides occurs both by dry and wet fall. The vast majority of ^{7}Be falls in association with rainfall. Workers have reported dry fall as making up 3–10% of total fallout (Wallbrink and Murray 1994; McNeary and Baskaran 2003; Salisbury and Cartwright 2005; Ioannidou and Papastefanou 2006; Sepulveda et al. 2008). Dry deposition of ^{210}Pb appears to be more variable than ^{7}Be because of the importance of resuspended soil and dust in contributing to the flux (Todd et al. 1989). The monthly atmospheric depositional flux (wet plus dry) of both ^{7}Be and ^{210}Pb varies by about a factor of 5 over the course of a year (Matisoff et al. 2005) but exhibits a maximum in the spring (Turekian et al. 1983; Olsen et al. 1985; Dibb 1989; Todd et al. 1989; Robbins and Eadie 1991; Koch et al. 1996; Baskaran et al. 1997) and represents the variability in the quantity of precipitation (Turekian et al. 1983; Koch et al. 1996; Baskaran et al. 1997) and, to a lesser extent, the seasonality in stratospheric ^{7}Be production and troposphere-stratosphere exchange (Turekian et al. 1983; Todd et al. 1989; Koch et al. 1996). ^{7}Be and ^{210}Pb deposition during precipitation events is well correlated to precipitation amount (Caillet et al. 2001; Ciffroy et al. 2003; Su et al. 2003) reflecting the fact that wet fall is a dominant delivery mechanism. Caillet et al. (2001) found that ^{7}Be deposition was slightly better correlated with precipitation than ^{210}Pb deposition ($r^2 = 0.66$ for ^{7}Be vs. 0.55 for ^{210}Pb). The better correlation with ^{7}Be is due to the fact that more of the ^{210}Pb comes from dry deposition and ^{7}Be is often derived from the stratosphere during large thunderstorms. Quine (1995) reported that ^{137}Cs delivery from global fallout was also highest in spring and early summer. Callender and Robbins (1993) observed an annual cycle of ^{137}Cs deposition in the high sedimentation environment of Oahe Reservoir (South Dakota, USA).

Given the correlation between fallout and precipitation amounts during events, it is not surprising that the annual fallout of ^{137}Cs, ^{7}Be and ^{210}Pb is correlated to

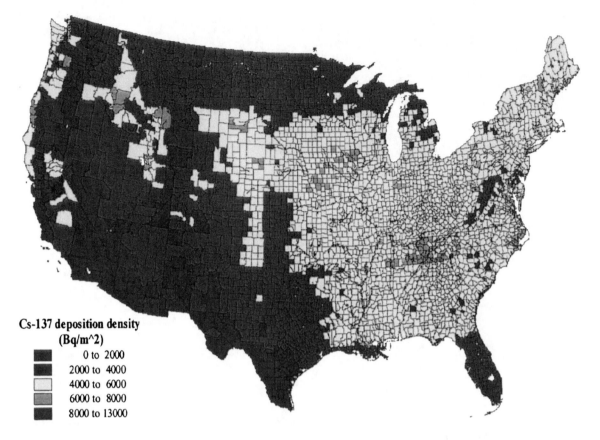

Fig. 25.5 ^{137}Cs depositional flux in the continental United States from global fallout (CDC 2006)

annual precipitation. In most places, the precipitation-normalized annual depositional fluxes have remained constant (Baskaran 1995). Ritchie and McHenry (1978) noted that the variation in ^{137}Cs inventory in the northcentral US was best explained by mean annual precipitation. Rowan (1995) found that variation in mean annual precipitation (900–2,000 mm) explained 75% of the variation in ^{137}Cs inventory in the Exe basin in the United Kingdom. Lance et al. (1986) found a linear relationship between the inventory of ^{137}Cs in soils and average annual precipitation in the southern USA. Baskaran et al. (1993) observed that ^{210}Pb fluxes increased with precipitation and Gallagher et al. (2001) found that ^{210}Pb inventory was significantly higher on the wetter west coast of Ireland than on the east coast. Our survey of annual delivery of ^{7}Be reported by workers (Turekian et al. 1983; Dibb 1989; Papastefanou and Ionannidou 1991; Baskaran et al. 1993; Caillet et al. 2001; Ayub et al. 2009) as compared to annual precipitation shows a strong correlation: $r^2 = 0.72$ (Fig. 25.6).

The fallout of ^{137}Cs over the globe as indicated by soil inventories is higher in the mid-latitudes than equatorial areas, and higher in the Northern Hemisphere than in the Southern Hemisphere (Stokes and Walling 2003; Walling et al. 2003). Collins et al. (2001) found inventories that were almost an order of magnitude higher in mid-latitudes of the Northern Hemisphere. Figure 25.7 illustrates the general pattern of inventories by latitude across the globe and Fig. 25.8 illustrates the reconstructed global fallout of ^{137}Cs as of 1970 indicating both the localized fallout from the test sites and the global fallout. McHenry et al. (1973), summarizing several studies, noted that fallout in the USA increased to the north and to the east (Fig. 25.5). Sarmiento and Gwinn (1986) developed a semi-empirical model for the deposition of ^{90}Sr (and hence ^{137}Cs) that is based on latitude, time since 1954, and the monthly precipitation. The distribution

Fig. 25.6 Linear relationship between ^{7}Be depositional flux and annual precipitation

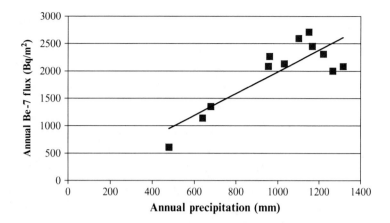

Fig. 25.7 Variation in bomb-derived ^{90}Sr (a surrogate for ^{137}Cs) by latitude band (from Stokes and Walling 2003)

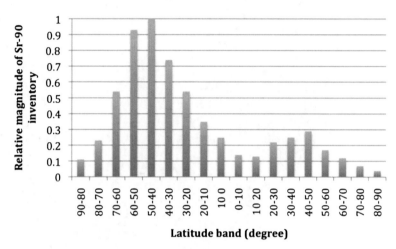

of bomb-derived ^{137}Cs fallout over the globe is relatively smooth in contrast to the Chernobyl fallout (compare Figs. 25.4 and 25.5) because the global fallout signal is due to a series of nuclear explosions that reached into the stratosphere allowing ^{137}Cs to be well mixed in the atmosphere. In contrast, the plume of radioactive debris from Chernobyl did not extend higher than a few km into the atmosphere and thus prevailing winds, and then rainout, determined the pattern of fallout. ^{239,240}Pu was also vaporized during bomb testing and its global distribution is similar to that of ^{137}Cs. However, this did not occur during Chernobyl, so the Pu was delivered as fine particulates and therefore its distribution was not widespread (Mietelski et al. 1996; Ketterer et al. 2004; Brudecki et al. 2009). Global fallout flux of ^{210}Pb is summarized in Baskaran (2011). The inventories of ^{210}Pb are lower in the Southern Hemisphere than the Northern Hemisphere because of the smaller percentage of land area.

The radon flux from the oceans is negligible compared to the continents (Turekian et al. 1977) thus with less continental area in the Southern Hemisphere there will be a lower atmospheric concentration and less deposition of ^{210}Pb. ^{7}Be inventories should be similar in both hemispheres because of its stratospheric source.

25.2.1.3 ^{137}Cs, ^{7}Be and ^{210}Pb in Soils

Once the radionuclides ^{137}Cs, ^{7}Be and ^{210}Pb reach the earth surface, they rapidly sorb to the mineral grains and organic materials of soil particles (Tamura and Jacobs 1960). Cs fixation to soil material depends strongly on soil composition and mineralogy. There is a complicated pattern of preferred sorption of radionuclides to finer particle sizes and to organic materials (He and Walling 1996; Wallbrink and Murray 1996; Motha et al. 2002) and clays (Hawley et al. 1986;

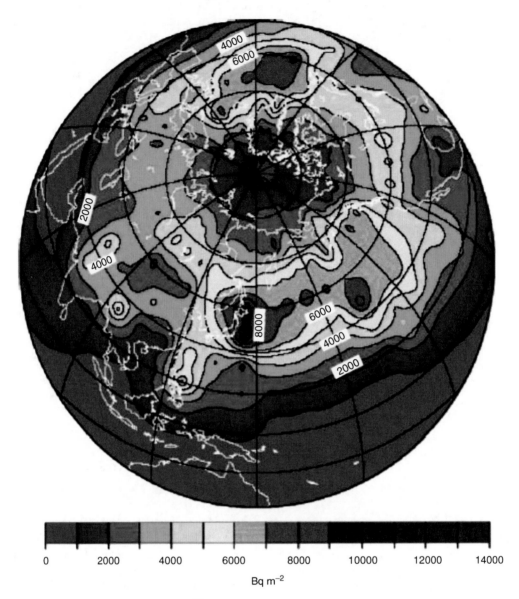

Fig. 25.8 Global distribution of bomb-derived ^{137}Cs fallout illustrating high fallout near testing sites and a strong latitudinal distribution (from Aoyama et al. 2006)

Wang and Cornett 1993; Balistrieri and Murray 1984). ^{137}Cs is a 1+ cation and its sorption exceeds that of all other alkali ions (Schultz et al. 1959). Lomenick and Tamura (1965) suggested that the adsorption is almost non-exchangable. Riise et al. (1990) estimated that less than 10% of Cs was leachable. ^{7}Be reaches the earth surface as a Be^{2+} cation with high charge density which makes it prone to adsorption. ^{210}Pb likewise reaches the surface as a 2+ or 4+ cation and is rapidly adsorbed. Partition coefficients, K_d, for these radionuclides are ~10^5 (Olsen et al. 1986; You et al. 1989; Hawley et al. 1986; Steinmann et al. 1999) indicating that these radionuclides are sufficiently particle-bound that they are suitable for tracing erosion, transport, and deposition of soil.

There are, however environments in which ^{137}Cs might fail as a conservative particle tracer. Some previous studies of radiocaesium in soils have failed because K_d values have been derived under conditions very different from those in situ. The partitioning of Cs in organic-rich soils is reversible and the partition coefficients can be low. Where the soils are peaty or podzolic, the mobility of cesium is considerably greater than in other soils (Sanchez et al. 1999). In acidic soils in coniferous forests in N. Europe Dorr and Munnich (1989) found that Cs migrated faster than ^{210}Pb and suggested that chemical exchange was occurring between the organic and hydrous phases. It is generally believed that radiocaesium retention in soils and sediments is due to the presence of a small number of highly selective sites. Cremers et al. (1988) found that strong Cs adsorption by the solid phase was regulated by the availability of frayed edge sites on illite and is inhibited by the presence of competing poorly hydrated alkali cations (K^+ or NH_4^+). Where geochemical migration of Cs is significant, workers should turn to alternatives such as Pu isotopes.

In soil, cesium has a low mobility; the majority of cesium ions are retained in the upper 20 cm of the soil surface and usually they do not migrate below a depth of 40 cm (Korobova et al. 1998; Matisoff et al. 2002a). For example, vertical migration patterns of ^{137}Cs in four agricultural soils from southern Chile indicated that approximately 90% of the applied cesium was retained in the top 40 cm of soil and as much as 100% was found in the upper 10 cm (Schuller et al. 1997).

^{137}Cs uptake by plants also affects its downcore migration. Clay and zeolite minerals strongly bind cesium cations in interlayer positions of the clay particles and therefore reduce the bioavailability of ^{137}Cs and its uptake by plants (Paasikallio 1999). Plant uptake of ^{134}Cs in a peat soil was decreased by a factor of 8 when zeolites were added (Shenber and Johanson 1992).

The kinetics of sorption are less understood, but Baskaran and Santschi (1993) reported that approximately 80% of ^{7}Be became associated with particulates within an hour of a rainfall event. In a study where ^{134}Cs was applied in solution to the surface of soils, it was adsorbed in the upper 2.5 cm of the soil (Owens et al. 1996). These results largely mimic the findings of Rogowski and Tamura (1970) who applied ^{137}Cs to the soil surface and also found penetration of only a few cm. This modest penetration during infiltration suggests a timeframe for adsorption of up to 10 min if typical rates of infiltration for these silty sand soils are used.

Some proportion of the radionuclide fallout is retained on vegetation. Doering et al. (2006) observed that 18% of ^{7}Be was retained on vegetation whereas only about 1% of ^{210}Pb$_{xs}$ was retained on vegetation. These differences almost surely reflect the much shorter halflife of ^{7}Be than any differences in affinity for organic material. Very little ^{137}Cs is typically on vegetation today because of the negligible fallout. Plants may also take up radionuclides from soil. Coughtrey and Thorne (1983) supposed that plant uptake could result in a 0.2% loss in ^{137}Cs per year. Numerous studies have shown that radioactive contaminants move through the food chain and can exceed health standards (Davis 1986; Revelle and Revelle 1988). For example, mushrooms have been reported to uptake as much as 50% of the ^{137}Cs inventory in the soil and moose, caribou, sheep and milk consumption have all had radioactive residues that restrict their consumption (Korky and Kowalski 1989).

25.2.1.4 Radionuclide Profiles in Undisturbed Soils

The simplest profile of the set of radionuclides most commonly used in soil erosion studies may be that of ^{7}Be (Fig. 25.9; see also Kaste and Baskaran 2011). From a peak concentration at the surface, it decreases exponentially down core to non-detectable values below 2–3 cm (Bonniwell et al. 1999; Wilson et al. 2003; Doering et al. 2006). In the case of ^{7}Be, some fraction is associated with the vegetation growing on the surface (Doering et al. 2006 reported 18%) or associated with organic litter. The half-life of ^{7}Be is just 53.3 days thus there is a relatively brief period to mix/move the radionuclide downward before the radionuclide decays (Walling et al. 2009). Only very rarely has ^{7}Be been measured deeper than a couple of cm in the soil. Kaste et al. (1999a, b) noted a second down-profile peak in activity that they hypothesized might be caused by subsurface pipeflow quickly taking the ^{7}Be cation to depth before sorption.

The next simplest profile may be that of ^{210}Pb (Figs. 25.9 and 25.10). As with ^{7}Be, ^{210}Pb$_{xs}$ is delivered by wet and dry fallout to the surface and mixed downward in the soil profile. In the case of ^{210}Pb however, there is both the atmospherically derived ^{210}Pb$_{xs}$ and the steady background ^{210}Pb activity (called supported ^{210}Pb) maintained by continuous in

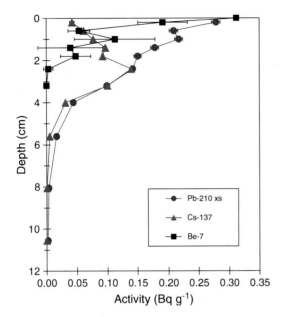

Fig. 25.9 The distribution of radionuclides ^{7}Be, ^{137}Cs, and ^{210}Pb$_{xs}$ in this upland soil from near Corwin Springs, Montana, USA is characteristic of undisturbed soils. ^{7}Be activity is highest at the surface and little is found below 3 cm. ^{137}Cs displays a subsurface peak in activity, in this case between 2 and 3 cm of depth, and no radionuclide activity below 10 cm. ^{210}Pb$_{xs}$ displays a surface peak in activity. ^{210}Pb$_{xs}$ activity is negligible below 10 cm (from Whiting et al. 2005)

situ decay of ^{226}Ra in soil and rock (Goldberg and Koide 1962). The surface maximum in activity decreases exponentially downcore to the value of the supported ^{210}Pb. Most ^{210}Pb$_{xs}$ is found in the upper 10 cm of soil (Bonniwell et al. 1999; Doering et al. 2006). The greater penetration of ^{210}Pb$_{xs}$ than ^{7}Be is due to its greater half-life of 22.3 years. With more time to operate, downward migration extends further.

The ^{137}Cs profile is typically more complicated (Figs. 25.9 and 25.10). Over much of the globe, the undisturbed ^{137}Cs profile features a subsurface maximum and then an exponential decrease below that. Very little ^{137}Cs is found below 20 cm (Owens and Walling 1996) in most locales. Unlike the other two radionuclides, the delivery of ^{137}Cs was not steady (Fig. 25.3). It was first detected in fallout in 1951, peaked in about 1963, and was negligible by the early 1980s. In the absence of constant replenishment to the surface, the peak in concentration has migrated below the surface by several cm. In areas receiving significant Chernobyl fallout after the nuclear plant accident in 1986, the distribution of ^{137}Cs from stratospheric fallout is swamped by the Chernobyl fallout so that global fallout can no longer be identified. This interpretation is supported by Pu isotope data which show "hot"

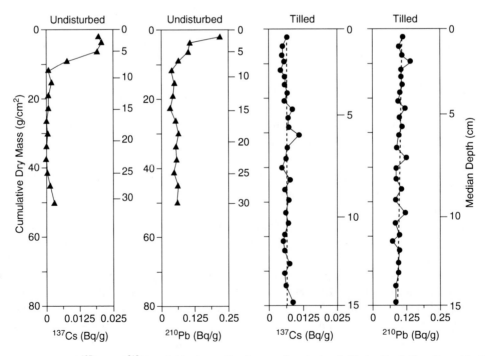

Fig. 25.10 Comparison of ^{137}Cs and ^{210}Pb activities in profiles from undisturbed and tilled soils. Soils collected in Ohio, USA. After Matisoff et al. (2002a, b)

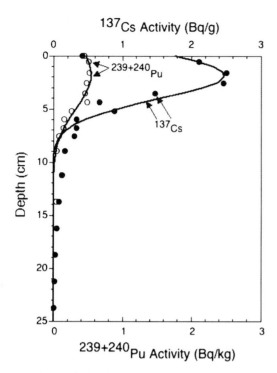

Fig. 25.11 ^{137}Cs and $^{239+240}$Pu data from a core collected in 2008 in Skogsvallen, Sweden and a non-local bioturbation model fit to the data. Note that the Pu is derived from global fallout (as determined from Pu isotopes, not shown here), but the ^{137}Cs is derived primarily from Chernobyl fallout (as determined by its relatively high activity)

Chernobyl-derived Pu particles at depth while stratospheric Pu is found both above and below the Chernobyl Pu and the ^{137}Cs profile mimics the Pu profile (Matisoff et al. 2011). Additional support can be seen in Fig. 25.11, where the Pu profile is from global fallout (as determined from Pu isotopes, not shown here), but the ^{137}Cs is derived primarily (as determined by its relatively high activity) from Chernobyl fallout. (Note the much higher activity of ^{137}Cs in Fig. 25.11 compared to that in Figs. 25.9 or 25.10 indicating a Chernobyl source to the soil collected in Sweden in Fig. 25.11 and a global fallout source to the soils collected in the USA shown in Figs. 25.9 and 25.10.)

25.2.1.5 Downcore Migration of Radionuclides in Soils

The radionuclides delivered to the soil surface are moved downward by soil processes including bioturbation, leaching, diffusion, and translocation. Movement can be enhanced by disturbance such as plowing. The depth of penetration of each radionuclide into the soil is determined by the rate of this downward movement and the half-life of each radionuclide. DeJong et al. (1986) argue for low leachability of ^{137}Cs but some workers have found substantial mobility, particularly in peaty soils. Kaste et al. (1999a, b) report that as pH increases, ^{7}Be can be desorbed.

Since the Chernobyl accident there have been a number of studies on the migration of ^{137}Cs and other radionuclides into the soil. Several authors have reported that ^{137}Cs penetration decreases exponentially downcore (Miller et al. 1990; Walling and Woodward 1992; Wallbrink and Murray 1996; Doering et al. 2006). Several studies have described the downcore transport of ^{137}Cs in terms of a diffusion-advection equation for a sorbed radionuclide (Konshin 1992; Bossew and Kirchner 2004; Shimmack et al. 1997; Rosen et al. 1999). These studies have concluded that the model fails to describe the rapid downward migration of the radionuclides shortly after fallout so that a non-linearity or irreversible sorption process leads to a decrease in the migration rate over time. This may help explain why the ^{239,240}Pu and ^{137}Cs profiles match in Fig. 25.11 even though the ^{239,240}Pu was delivered from stratospheric fallout whereas the ^{137}Cs was delivered from Chernobyl fallout.

Recently it has been suggested that ^{137}Cs transport is facilitated by colloidal transport (Flury et al. 2004; Chen et al. 2005), but colloidal transport in a natural, unsaturated media is small (Honeyman and Ranville 2002; Lenhart and Saiers 2002). Février and Martin-Garin (2005) conducted laboratory experiments that demonstrated retention of anionic radionuclides by microbial activity. Bundt et al. (2000) noted enrichment of radionuclides of Pb, Pu, and Cs at depth in preferential flow paths.

The transport of radionuclides in soils also can be facilitated by burrowing organisms such as earthworms, ants, termites and pocket gophers. Some of these bioturbating organisms are present in soils in numbers that can approach tens of millions per square meter and they actively mix the soil to depths ranging from a few centimeters to a few meters (Johnson et al. 1990; Müller-Lemans and van Dorp 1996; Gabet 2000; Gabet et al. 2003). Darwin (1881) first reported the role of earthworms in facilitating the downward migration of objects by ingesting soil and/or plant matter at depth and depositing it onto the surface as casts. This mixing style incorporates both random (diffusive) and directed (advective) processes and has been termed "conveyor-belt" feeding (Rhoads and Stanley 1964; Rhoads 1974; Powell 1977) or "non-local" mixing (Boudreau 1986a, b; Robbins 1986). Earthworms have been suggested to be the most important factor controlling the vertical transport of radionuclides in central European soils, with the potential to turn over the top layer of soil within 5–20 years (Müller-Lemans and van Dorp 1996). It is possible to model mixing by bioturbation in soils as a diffusive process and burial as an advection velocity. Kaste et al. (2007) used this approach and reported bio-diffusion mixing coefficients of 1–2 cm^2 $year^{-1}$ in a grassland in California (USA) and from a forested landscape in Australia, and inferred that the more actively bioturbated soils exhibit higher erosion rates. This approach treats the downward migration as a burial velocity and while it may simulate well the downcore profiles, it does not account for the directed recycling of tracer from depth (non-local mixing) as demonstrated by Robbins et al. (1979) using aquatic oligochaetes. Alternatively it is possible to model mixing by bioturbation as a diffusive process and the downward migration as a burial caused by non-local feeding in which the organisms place soil from depth onto the soil surface (Jarvis et al. 2010). Figure 25.11 is an application of the non-local bioturbation model to Chernobyl fallout at a site in Sweden and yields a bioturbation coefficient of 0.1 cm^2 $year^{-1}$ and a feeding rate of 0.002–0.003 $year^{-1}$ over the top 20 cm. In other words, this bioturbation model can account for the burial of the radionuclides by recycling 0.2–0.3% of the top 20 cm of the soil column each year.

25.2.1.6 Effects of Tillage on Soil Profiles

The radionuclide profiles in tilled soils exhibit considerable differences from the profiles in undisturbed soils (Matisoff et al. 2002b). Undisturbed soils show either a surface maximum (^7Be, ^{210}Pb) or a near surface peak (^{137}Cs) followed at depth by a decrease to a constant activity (^{210}Pb) or to near zero (^7Be, ^{137}Cs) (Fig. 25.9). Plowing of agricultural soils will mix soils to depths of 10–30 cm creating a relatively uniform concentration over the plow depth (Wise 1980 in Martz and deJong 1985; Owens et al. 1996). The distribution of radionuclides in the tilled soils is largely homogeneous because plowing mixes surface soils, which are enriched in radionuclides, with deeper soils, which are depleted in both ^{210}Pb and ^{137}Cs (Fig. 25.10).

Figure 25.12 illustrates a generalized model of the distribution of ^7Be, ^{137}Cs and ^{210}Pb in a soil profile. All three isotopes are deposited on the surface through wet and dry fallout. Each radionuclide is distributed differently in the soil because of differences in half-lives, delivery rates, delivery histories, and land use. An undisturbed soil will exhibit higher radionuclide activities near the soil surface, which reflects their surficial input and slow downward transport. The shorter half-life of ^7Be compared to ^{210}Pb results in less downward migration of ^7Be. Hence ^7Be is found only at the soil surface but some ^{210}Pb will migrate down core. In addition, some ^{210}Pb is produced in the soil by in situ decay of ^{222}Rn resulting in some ^{210}Pb activity at all depths in the soil. Because ^{137}Cs had its peak delivery in the early 1960s and almost no delivery before 1951 or since 1975 (or had its peak delivery in 1986 in areas affected by Chernobyl fallout) its activities have a distinct peak at some depth (~10 cm) in the soil. Plowing homogenizes ^{210}Pb and ^{137}Cs within the plowed layer, but because of its short half-life and constant input, ^7Be activities are highest at the surface and are homogeneous only immediately after plowing. Because ^{137}Cs fallout occurred at a distinct instance of time, its distribution will remain homogeneous within the soil profile, even after the cessation of plowing. On the other hand, ^{210}Pb, like ^7Be, is continuously deposited on the land surface. Its distribution will remain homogeneous if the soil is plowed annually, but it will accumulate at the surface and slowly rebuild a profile with decreasing activity with depth if tillage ceases.

Because ^7Be, ^{137}Cs, and ^{210}Pb have different distributions in the soil profile, erosion of the soil to different depths will yield an assemblage of radionuclides in the eroded material that is characteristic of only one depth. Shallow erosion produces proportionally larger amounts of ^7Be and ^{210}Pb because these radionuclides are concentrated near the surface. Deeper incision yields progressively less additional ^7Be and no additional ^7Be below about 1 cm. The

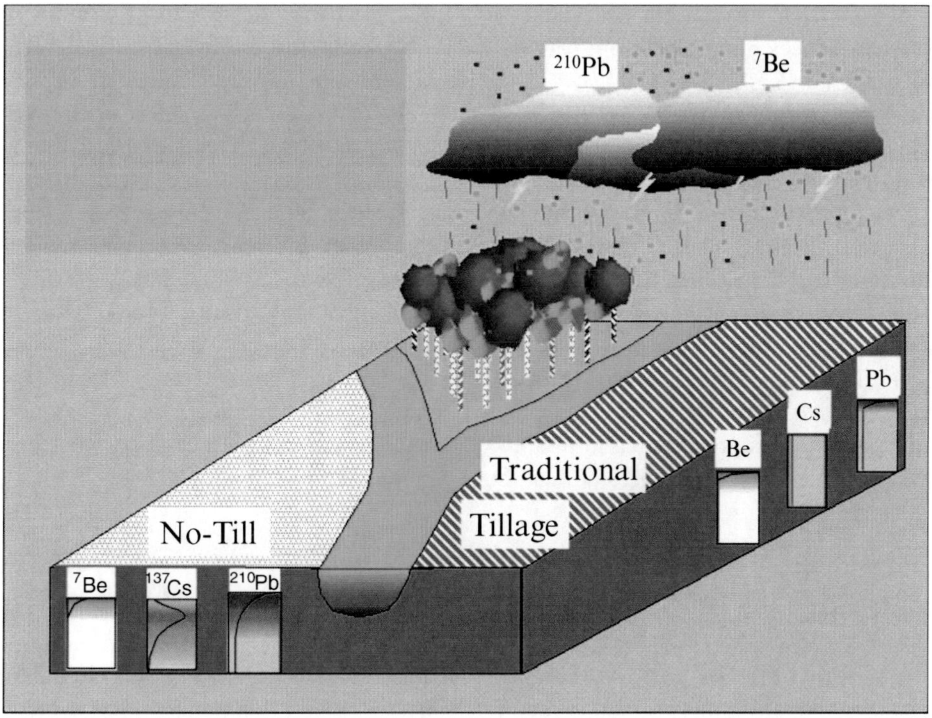

Fig. 25.12 Schematic illustration of soil profiles of ^{7}Be, ^{137}Cs, and ^{210}Pb in undisturbed soils (no-till) and in soils that have been mixed by plowing (traditional tillage)

proportional contribution of ^{137}Cs increases to the depth of its maximum. Erosion below that depth yields incrementally more ^{210}Pb and ^{137}Cs but at progressively decreasing rates. Below some depth in the soil (~15–30 cm), deeper erosion yields little if any additional ^{137}Cs. ^{210}Pb yield continues to grow in correspondence to the constant supported ^{210}Pb activity.

Sediment eroded from a soil will have a radionuclide signature corresponding to the tillage practice and the depth of erosion. Thus radionuclide signatures in suspended sediments can provide a means of tracing particles eroded from the landscape and can identify soil sources and be used to quantify the erosion (Whiting et al. 2001). The distinct distributions of radionuclides permit in principle the use of multiple mass balances to quantitatively estimate the areal extent of rill and sheet erosion and the characteristic depth of erosion associated with each mechanism (Wallbrink et al. 1999; Whiting et al. 1999, 2001). In essence, it is possible to infer the "recipe" for erosion – for example, 1 part sheet erosion to 10 parts rill erosion – on the basis of the total yield of radionuclides and sediment.

25.2.2 Concept of Inventory

The inventory, or standing stock, of a radionuclide represents the total amount of a radionuclide per unit area and is determined by atmospheric delivery, radioactive decay, gain, and loss. Gain typically reflects deposition and loss typically reflects erosion or leaching. In the case of ^{137}Cs, local inventories at stable locations (neither eroding or depositing) are steadily dropping because contemporary fallout to replenish the stock is essentially zero (Fig. 25.3). In 2010, ^{137}Cs inventories are less than half of what they were in 1974, except in areas affected by Chernobyl fallout. ^{210}Pb inventories at stable sites are more or less steady because annual contributions (fallout) to the stock and variation in the delivery from year-to-year are both small compared to the total inventory. For instance, fallout of ^{210}Pb in a typical year represents no more than 3% of the stock of ^{210}Pb$_{xs}$. ^{7}Be inventories at stable sites tend to vary over the year due to the variation in delivery with the seasons and the short half-life of 53 days. Individual storms can deliver

>50% of the ^{7}Be inventory particularly when the precipitation event arrives after a period of drought lasting weeks to months.

To use these tracers to quantify erosion or deposition, the amount of the radionuclides in a vertical column of soil (the measured inventory) is compared to the expected amount in the soil profile based upon fallout (the expected inventory). The fundamental idea is that the inventory (stock) of radionuclides will be changed only by additions of radionuclides as wet- and dry-fallout or by sediment deposition or by losses by decay and soil erosion. If the measured inventory is less than the expected inventory, then erosion has occurred and the soil loss is proportional to the ratio of the inventories. If the inventory is greater than the expected inventory, deposition has occurred.

25.2.2.1 The Spatial Variability of Inventory

Erosion and deposition are the primary causes of variation in radionuclide inventory. But other factors also influence the inventory and these include: (1) random spatial variability, (2) systematic spatial variability, (3) sampling variability, and (4) precision in measurement (Owens and Walling 1996). Random spatial variability can be tied to local soil properties, microtopography (i.e. Lance et al. 1986), and localized vegetation distributions. Systematic variation is caused by gradients in precipitation, slope, aspect, windfields, and soil types. Sampling variability is tied to the area of collection with large areas of collection featuring lower variability (Loughran et al. 1988). Owens and Walling (1996) and Foster et al. (1994) estimate variation associated with sampling area is approximately 5%. Finally, measurement precision is an important source of variability and is typically about 10%. Owens and Walling (1996) concluded that local variation in ^{137}Cs inventory was relatively large thus single measurements were unlikely to be appropriate measures of activity. Their summary of their own work and the work of others showed a range in coefficient of variation in Cs inventories of 5–47% with an average value of 29%. For ^{7}Be, Wilson et al. (2003) found a coefficient of variation of 33% in inventories. It might be expected that with the shorter halflife that local variation might add substantially to variability, but that does not appear to be the case.

The change in the inventory of a radionuclide tracer is the basis for estimating soil erosion. In using a radionuclide to quantify erosion, one assumes that the erosion that removes soil also removes the radionuclide tracer. Ideally there is equivalence between the mass loss relative to the potentially erodible material and the radionuclide loss relative to the potentially lost material (inventory). However, radionuclides can still be used to quantify erosion in the absence of equivalence if the tracer loss:soil loss relationship can be quantified. For instance, preferential erosion of certain sized particles enriched in radionuclides might appear to invalidate the assumption of equivalence, but the ability to describe this enrichment will permit determination of erosion.

25.2.3 Time Scales of Utility

Fundamental to the use of tracers to measure rates of soil erosion is the time span over which the loss is calculated. In other words, to determine the rate of erosion using radionuclides, the denominator – time in this case – must be established. Using ^{137}Cs as an example, delivery to the earth surface began around about 1954, peaked in about 1963, and was effectively below detection by 1983 (except in parts of Europe). If we ignore for the moment various other issues, we might reasonably calculate the amount of erosion by measuring the amount of Cs in the soil lost relative to the amount we would expect based upon fallout (accounting for decay). To determine a rate, we do not know anything about the timing of erosion in this interval since deposition so we will assume that the erosion occurred over the period since 1954 when deposition began. In such a scenario, the erosion rate is determined for a period of 56 years (assuming 2010 as the measurement date).

What sets the shortest and longest scales for the use of a radionuclide to estimate erosion? If erosion rate is being determined retrospectively, these are the considerations for the timescale of utility. In the case of a retrospective analysis, the scientist is typically measuring one time the inventory of the radionuclide at a site where erosion occurred and comparing this inventory to the inventory at a reference site without erosion or to the inventory expected given the estimated fallout. The changes in inventory are converted to soil loss and the rate is determined by the time since deposition.

If repeated measurements of inventory are made and made sufficiently far apart in time, the erosion of soil over shorter time periods can be examined. What is "sufficiently far apart in time?" To estimate erosion a difference in inventory must be measurable. It is typical that analytical errors in activity are at least 10%. In the case of ^{137}Cs, a 10% uncertainty in measured activity when compared to the decay rate, corresponds to a resolution in time no finer than 5 years (~10% of the mean life of the ^{137}Cs). Longer time scales, at least a decade (Walling and Quine 1990), are typically needed to produce measurable changes in ^{137}Cs inventories. In the case of Pb, this finest resolution is about 3 years (~10% of the mean life of ^{210}Pb). For ^{7}Be, the variable flux means that under certain circumstances, the finest resolution is a matter of days or at the scale of individual precipitation events. One factor influencing the timeframe over which changes in inventory can be recognized is the distribution of the tracer in the soil. If the bulk of the tracer is at the surface, significant amounts of the tracer can be removed with a given amount of erosion. Less of the tracer will be removed by the same amount of erosion if the same inventory is distributed more deeply in the soil. In summary, the amount of erosion; the half-life; the history of deposition, erosion, and mixing of the surface fallout into the soil profile; and the precision in measurement are the keys to the timescale of utility for the various radionuclides. Section 25.4, which summarizes the major classes of models for estimating erosion rates using ^{137}Cs and their assumptions, will treat this subject further.

25.3 Collection and Measurement of Samples

25.3.1 Soil Collection

In order to determine soil radioactivity inventories it is necessary to collect intact soil cores and measure vertical profiles of the radioactivity. Since the radioactivity of ^{137}Cs has often migrated downcore to depths of 20 cm or more, push cores, for example the types used by golf courses to examine turf quality, are too short. Accordingly, hammer-driven cores and soil pits are used to obtain deeper cores. If high resolution profiling of the cores is desired, for example to obtain a ^{7}Be profile, then the diameter of the tube needs to be large enough to collect sufficient sample to measure the radionuclide activities in a reasonable counting time. Typical push or hammer cores obtain a 1-in. diameter tube, and by using extension rods the tubes can be collected to a depth of 60-cm or more. However, this type of coring method causes significant (often ~20–30%) compaction of the soil, so that the true depths are poorly known even if they are compaction corrected. Compaction can be minimized by using a larger diameter tube, but it is more difficult to obtain deeper cores with a larger diameter tube. High-resolution soil profiles are easier to obtain from soil pits (Wilson et al. 2003) but the digging and sampling of soil pits is tedious and slower. Alternatively, thin layers of soil may be scraped from the surface of a large area to obtain high resolution vertical profiles (Walling et al. 1999a, b) – a method that is also quite tedious to construct. Regardless of the sampling technique used, it is necessary to know the surface area sampled since radionuclide inventories are expressed as activity per unit area.

After collection, the soil samples need to be dried, ground with a mortar and pestle to a fine texture, placed in standardized geometries, and analyzed by gamma spectroscopy for their radionuclide activities.

25.3.2 ^{137}Cs Measurement

The measurement of ^{137}Cs is accomplished by gamma spectroscopy. ^{137}Cs undergoes β^- (negatron) decay and emits gamma energies at 31.82 keV (relative intensity = 1.96%), 32.19 keV (relative intensity = 3.61%), 36.40 keV (relative intensity = 1.31%), and 661.66 keV (relative intensity = 85.21%). The two lowest energy photopeaks usually cannot be baseline resolved, so they are not used. The photopeak at 36.40 keV has been used, but it is not an ideal choice for measurement because the weak gamma in that energy range undergoes significant sample self absorption for which a large correction is necessary (Cutshall et al. 1983), its relative intensity is fairly small, and, depending on what other isotopes are being analyzed in the same spectrum, that portion of the spectrum may not have good photopeak separation. The photopeak at 661.66 keV is usually used because the relative intensity is much larger, the spectrum background and the efficiency are relatively constant in that part of the spectrum, and there are no other peaks that overlap or interfere. However, the detector efficiency for solid-state detectors is low at that energy requiring large sample sizes or long counting times to acquire a suitable peak for quantitation.

There are several different types of detectors capable of measuring ^{137}Cs. While Si detectors are commonly used for X-ray analyses, their efficiencies are too low at the higher energies needed for ^{137}Cs detection. The most common detectors that can measure the 661.66 kcV photopeak of ^{137}Cs include NaI and Germanium (HPGe) semiconductors, although it is possible to use other detectors, such as Cd-Te, Cd-Zn-Te, and Hg-I. These other detectors have an advantage of reasonably good photopeak resolution and peak to Compton ratio while not require liquid nitrogen cooling. Their disadvantage is that for most applications they are too small and have a very low efficiency at 661.66 keV and therefore require counting times that are too long for practical applications. However, larger crystals are being developed and they may eventually replace NaI and HPGe detectors. NaI detectors have the advantages of being relatively inexpensive, have the highest efficiency of all the detectors discussed here, and requiring no liquid nitrogen for cooling. However, the photopeak energy resolution for NaI detectors is very poor and in most samples the ^{137}Cs photopeak is not visually detectable. Quantitation of a non-visual photopeak requires complex peak separation software, and since the peaks are not visually obvious, the use of NaI detectors to measure ^{137}Cs in soils and sediments seems a "bit like magic" or at the very least results in a lack of confidence in the results. High purity germanium detectors (HPGe) provide the easiest, and most accurate quantitation of the 661.66 keV photopeak of ^{137}Cs. These detectors provide excellent energy resolution and peak to Compton ratios over most of the applicable gamma-ray spectrum, so that ^{210}Pb, ^{7}Be, ^{134}Cs, ^{137}Cs and ^{40}K and other gamma-emitting isotopes may all be determined at the same time from a single spectrum. However, they are more expensive than the other types of semiconductors discussed here and they require liquid nitrogen cooling. Other analytical techniques, such as alpha and beta spectroscopy for the determination of ^{210}Pb and ICP-MS for the determination ^{239,240}Pu are discussed in more detail in other chapters (Ketterer et al. 2011).

25.3.3 Calibration and Standards

Calibration of a detector for ^{137}Cs analyses consists of both an energy calibration and an efficiency calibration. The purpose of the energy calibration is to assign the correct photopeak to its appropriate energy. This is accomplished by counting a series of known radionuclides (energies) to determine their channel positions on the energy spectrum. The efficiency calibration is required to relate measured counts to the absolute values of the sample activities and it depends on kind and size of the detector, instrument settings, sample geometry (i.e., the size and shape of the container), sample volume, and shielding (background activities). Consequently, it is necessary to calibrate the energy and efficiency of each machine for every amplifier setting and sample geometry that will be used. Typical efficiencies for the low energy ^{210}Pb photopeak (46.52 keV) are ~0.5% and the efficiencies are about a factor of 3–5 less at the higher energies of ^{7}Be (477.59 keV) and ^{137}Cs (661.66 keV). Because of these low efficiencies and the low activities in most samples (some Chernobyl samples are an exception) counting times of about 1 day are needed to obtain enough counts to reduce the counting error to <10%. Typical self-absorption correction factors (Cutshall et al. 1983) at the low energy of ^{210}Pb can range from 10 to 80% depending on the size and shape of the sample while the self-absorption correction factor at the higher energy of ^{137}Cs is negligible.

Determining the efficiency requires the use of a standard for which the activity is known. Unfortunately there are no commercially available standards of soils or sediments that are suitable for the efficiency determination of the entire suite of radionuclides commonly measured (^{210}Pb, ^{7}Be, ^{137}Cs, ^{40}K) although there are a couple of samples that can be used as Quality Assurance samples or as a laboratory standard for efficiency calibration for ^{137}Cs, such as NIST 4350B Columbia River Sediment and International Atomic Energy Agency Reference Material IAEA-375. A list of primary standards available for the calibration of instruments and chemical procedures is given in Baskaran et al. (2009). In particular, RGU-1 (IAEA – 400 µg g^{-1} with ^{238}U concentration with all its daughter products in secular equilibrium) is a suitable standard for the calibration of ^{210}Pb. Alternatively, it is possible to prepare a standard by spiking a "clean" soil or sediment with the appropriate radionuclides of known activities (for example, Amersham plc QCY44 or NG4 Mixed Radionuclide Solutions) and then placing those prepared soils in the appropriate container for the efficiency calibration of that geometry.

25.4 Methods for Calculating Soil Erosion

Soil erosion can be calculated by the change in the inventory of radionuclides. While the same basic approaches can be used with any of the radionuclides, we detail the use of ^{137}Cs in erosion studies because it was for ^{137}Cs that the methods were originally developed.

Estimation of the erosion rate using ^{137}Cs inventories is typically addressed as an inverse problem with known ^{137}Cs inventories, cultivation and precipitation histories. The simplest method to estimate erosion rates is by a comparison of the ^{137}Cs inventory in soil cores with a reference value from a nearby non-eroded site. The non-eroded site represents the expected baseline fallout to the local geographic area, so that an eroded site would be expected to have less inventory compared to the non-eroded site. Various models have been applied to ^{137}Cs inventories where the erosion rate is assumed to be correlated to the percent of the ^{137}Cs inventory lost from a site (Ritchie et al. 1974; Spomer et al. 1985; Brown et al. 1981; Kachanoski and deJong 1984; DeJong et al. 1986; Lowrance et al. 1988; Soileau et al. 1990; Montgomery et al. 1997; Walling et al. 1999a, b; Fornes et al. 2005). These methods have been recently reviewed (Walling et al. 2002; Poreba 2006) in more detail than we present below and an analysis of these methods led Zapata et al. (2002) to conclude that there are uncertainties associated with the model selection used to estimate erosion rates, that most of the models do not address short-term changes in erosion rates

such as those related to changes in land use and management practices, and that the methodology needs further standardization of the protocols for its general application worldwide.

25.4.1 Empirical, Non-Linear Model

Studies conducted in the 1960s and early 1970s in which ^{137}Cs inventory loss was compared with soil loss estimated using the Universal Soil Loss Equation (Wischmeier and Smith 1978) showed a strong, non-linear empirical relationship at several sites following the equation

$$E = 0.0087 P^{1.18} \quad (25.1)$$

where E is erosion rate (g cm^{-2} year^{-1}) and P is the percent ^{137}Cs inventory lost from the study site (Ritchie et al. 1974). Percent radionuclide inventory lost is defined as

$$P = (I_{ref} - I_{site})/I_{ref} * 100 \quad (25.2)$$

where I_{ref} is the ^{137}Cs inventory from an undisturbed reference site (Bq m^{-2}) and I_{site} is the ^{137}Cs inventory for the study site. This led to the development of a number of models to calculate erosion rate on the basis of percent inventory lost. In essence, the model assumes a uniform distribution of ^{137}Cs throughout the soil profile so that erosion removes a proportional fraction of the inventory. This approach can result in significant error in the estimation of erosion rate in areas where land use, conservation practices, precipitation or other environmental conditions are not constant throughout the sampling period. For example, if atmospheric ^{137}Cs fallout were confined to the upper layer of soil prior to being eroded or distributed within the cultivation layer, a singular erosion event occurring just after the atmospheric fallout peak in 1964 (following the peak in stratospheric fallout) or in 1986 (following Chernobyl fallout in Europe) would remove a disproportionate amount of ^{137}Cs inventory and result in an overestimate of the soil erosion rate throughout the sampling period. Conversely, a major erosion event today will remove only a small fraction of the inventory because of the downward migration of ^{137}Cs in the intervening time period and may result in an underestimate of the erosion rate. The use of a reference site does not solve this problem.

25.4.2 Linear Box Model Methods

There are two major classes of models that linearly correlate the percent ^{137}Cs inventory lost with erosion rate. One approach assumes that the total ^{137}Cs inventory is eligible for erosion (Linear 1-Box model; Brown et al. 1981; Spomer et al. 1985) and those in which only the ^{137}Cs inventory within a surface or tillage layer is eligible for erosion (Linear-2Box model; DeJong et al. 1986; Lowrance et al. 1988; Soileau et al. 1990; Montgomery et al. 1997). The key difference between the Linear-1Box and Linear-2Box models lies in the vertical distribution of ^{137}Cs. The Linear-1Box model assumes (sometimes incorrectly) that the entire ^{137}Cs inventory is confined to the tillage layer, whereas the Linear-2Box model allows for some portion of the inventory to reside beneath the tillage layer where it is unavailable for erosion. Spomer et al. (1985) used the Linear-1Box model to describe soil erosion during the period 1954–1974, so that

$$E = DP_L K/T \quad (25.3)$$

where D is the average tillage mixing depth (cm), K is the dry bulk density of the soil (g cm^{-3}), and T is the time since the onset of ^{137}Cs deposition (y). P_L is defined as

$$P_L = (I_{ref} - I_{site})/I_{ref} \quad (25.4)$$

The Linear-2Box model (i.e., tillage depth < ^{137}Cs penetration depth) described soil erosion with the relationship

$$E = DP_{till} K/T \quad (25.5)$$

P_{till} is analogous to P_L, but the inventory is confined to the tillage layer or to the same depth as the tillage layer at reference sites. Comparisons of results obtained with both the Linear 1-Box and Linear 2-box models indicate little differences in derived erosion rates at Spomer et al.'s (1985) field site (Fornes et al. 2005).

25.4.3 Time-Dependent Model Methods

Some approaches (detailed in Ritchie et al. 1974) are limited in their use because the relationship between erosion rate and ^{137}Cs inventory is incorrectly assumed to be independent of time or cultivation, even failing to account for the radioactive decay of ^{137}Cs (Kachanoski and deJong 1984). These various approaches produced a wide range of erosion rates at a single site (Fornes et al. 2000) confirming that model assumptions significantly influence ^{137}Cs-derived soil erosion rates. Thus, these models are limited because they are single time-step models that do not consider the time-dependent nature of ^{137}Cs fallout (Kachanoski and deJong 1984; Spomer et al. 1985; Walling et al. 1999a, b) or the downcore migration of ^{137}Cs (Figs. 25.9–25.11).

It is possible to obtain improved estimates of soil erosion rates based upon changes in ^{137}Cs inventories by addressing issues associated with model assumptions and changes in soil conservation practices by modeling the time-dependent ^{137}Cs fallout, precipitation and soil cultivation (Kachanoski and deJong 1984; Walling and He 1997a, b; Zhang et al. 1999; Fornes et al. 2005). Using this approach, the equation describing the rate of change of total ^{137}Cs inventory is

$$\partial I/\partial t = F(t) - EC(t) - \lambda I \quad (25.6)$$

where I is ^{137}Cs inventory (Bq cm^{-2}), E is erosion rate (g cm^{-2} year^{-1}), C(t) is the concentration of ^{137}Cs (Bq g^{-1}), F(t) is the time-dependent atmospheric fallout of ^{137}Cs (Bq cm^{-2}), and λ is the radioactive decay constant (year^{-1}). Temporal variation of ^{137}Cs fallout, F(t), can be derived from the measured fallout data (Fig. 25.3) with the magnitude adjusted to the local ^{137}Cs reference inventory (Health and Safety Laboratory 1977; Cambray et al. 1989). This model can be kept compatible with the Linear-1Box and Linear-2Box models, by assuming steady-state erosion and cultivation. Soil erosion rates can also be calculated under more detailed constraints such as precipitation-dependent erosion and ephemeral homogenization of the cultivated layer (Walling and He 1997a, b; Zhang et al. 1999; Fornes et al. 2005) for more accurate and realistic erosion rates.

Figure 25.13 illustrates the differences in calculated erosion rates that model assumptions governing the

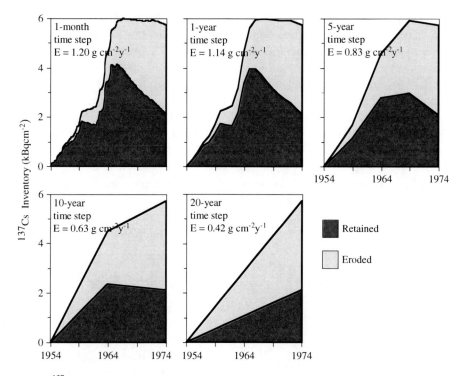

Fig. 25.13 Calculated ^{137}Cs inventories in a soil subject to atmospheric deposition of ^{137}Cs, periodic (annual) cultivation, and erosion over the time period 1954–1974. Note that inclusion of monthly data results in an erosion rate about 3 times larger than that calculated from the soil inventory only at the end of the 20-year time period. After Fornes et al. (2005)

choice of time step exert on the calculated erosion rates. In these simulations of Fornes et al. (2005), the erosion rate decreased by a factor of three as the model time step was increased from 1 month to 20 years. With a single 20-year time step, erosion rates are virtually the same as those estimated by Spomer et al. (1985) because Spomer et al. (1985) used a Linear-1Box model to describe soil erosion during a single time interval of 1954–1974. The modest differences between the 20-year time step and those reported by Spomer et al. (1985) can be explained by noting that the cultivation depth used in the Fornes et al. (2005) simulations was 10 cm, but Spomer et al. (1985) used a 15 cm cultivation depth. Deposition, cultivation, and erosion in the Spomer et al. (1985) model occur in a single step beginning c. 1954 and ending when the core was collected in 1974. The Fornes et al. (2005) model incorporates a monthly time step (Fig. 25.13) so the 20-year interval is approximated by 240 monthly time steps of deposition and erosion. These results indicate that ^{137}Cs-derived erosion rates are highly sensitive to the length of the time step used in the model because of the combination of the timing of the fallout, cultivation, and erosion. These results suggest that models that fail to account for temporal variations in atmospheric deposition and/or cultivation practices can grossly miscalculate erosion rates.

25.4.4 Tillage Erosion

While the radionuclide inventory in a core and/or the downcore profile of a radionuclide in that core may indicate erosion, the models described above do not necessarily inform the cause of that erosion. The pattern of radionuclide inventories in a field may be strongly influenced by tillage translocation rather than water erosion. For example, poor agreement has been reported between spatially-distributed ^{137}Cs-derived soil redistribution rates with those derived from water erosion (Quine 1999). Reduced inventories on hillslope convexities and deposition in hollows have been attributed to soil loss caused by tillage redistribution rather than by water erosion

(DeJong et al. 1983; Lobb et al. 1995). Consequently, several workers have developed models which attempt to distinguish the two drivers of radionuclide redistribution and thereby obtain an estimate of the erosion caused solely by water erosion (Govers et al. 1994, 1996; Lobb and Kachanoski 1999). Van Oost et al. (2006) provide a review of the literature on tillage erosion and conclude that based on a global data set tillage erosion rates are comparable to or higher than water erosion rates. They note that because of the widespread use of tillage practices, the high translocation rates resulting from tillage, and the effects of tillage on soil properties, that tillage erosion should be considered in soil landscape studies.

The most widespread used tillage model treats tillage erosion as a diffusion-type process (Govers et al. 1994). The model of Govers et al. (1994) relates the rate of soil translocation to the soil bulk density, the average soil translocation distance in the direction of tillage, and the depth of tillage. They treat translocation distances resulting from a single tillage pass to be linearly and inversely related to slope and that multiple passes in opposing directions results in a net downslope transport. Additional assumptions are that the tillage depth and soil bulk density do not vary in space, tillage soil translocation can be expressed as a linear function of the slope gradient, and tillage is conducted in opposing directions. Using the continuity equation they determine that the tillage erosion may be written as

$$E = -\frac{\partial Q_s}{\partial x} = -D\rho_b b \frac{\partial h}{\partial x} = k_{til} \frac{\delta^2 h}{\delta x^2} \quad (25.7)$$

where E is the tillage erosion rate (kg m^{-2} a^{-1}); Q_s is the rate of soil translocation in the direction of tillage (kg m^{-1} a^{-1}); D is the tillage depth (m); ρ_b is the soil bulk density (kg m^{-3}); x is the distance (positive downslope) (m); d is the average soil translocation distance in the direction of tillage (m a^{-1}) (=a + bS where a and b are regression constants (m a^{-1}) and S is the slope tangent (positive upslope; negative downslope) (dimensionless)); h is the height at a given point of the hillslope (m); and k_{til} (=$-D\rho_b b$) is the tillage transport coefficient (kg m^{-2} a^{-1}).

This model has been applied to a number of studies to compare different plowing directions (Van Muysen et al. 2002; St Gerontidis et al. 2001; De Alba 2001; Quine and Zhang 2004; Heckrath et al. 2006), and erodability of the landscape (Lobb et al. 1999) as affected by implement characteristics (tool shape, width, length) and operational parameters (tillage depth, speed, tillage direction). Reported implement erosivities as characterized by the tillage transport coefficient are fairly consistent and range from 400 to 800 kg m^{-2} year^{-1} for mechanized plowing and from 70 to 260 kg m^{-2} year^{-1} for non-mechanized agriculture (Van Oost et al. 2006). Van Oost et al. (2006) also report that decreasing tillage depth and plowing along contour lines substantially reduce tillage erosion rates and can be considered as effective soil conservation strategies.

25.4.5 Wind Erosion

Compared to water there has been relatively little work using ^{137}Cs to study aeolian processes of erosion and deposition. Recently, however, the application of ^{137}Cs to study wind erosion has received some attention (Sutherland and deJong 1990; Sutherland et al. 1991; Chappell 1996, 1998; Yan and Zhang 1998; Yan et al. 2001; Yan and Shi 2004; Hu et al. 2005). The determination of the wind erosion rate is important in assessing the extent and intensity of desertification and the effectiveness of counter-measures.

Calculation of wind erosion rates has been based on the same proportional inventory models that are sometimes used to estimate erosion by water (Sutherland and deJong 1990; Walling and Quine 1993), although it is recognized that these models sometimes suffer from the same issues affecting their use to quantify water erosion – failure to account for surface enrichment of ^{137}Cs and ^{137}Cs dilution by tillage. Mass balance models (Kachanoski and deJong 1984; Zhang et al. 1990) are more suitable for assessing soil erosion in soils where ^{137}Cs is distributed uniformly, such as croplands. In settings where the ^{137}Cs profile is not homogeneous, such as grasslands, the profile distribution model is more appropriate. The calculation of wind erosion loss in these models (Yan et al. 2001; Hu et al. 2005) is estimated by

$$E = CPR * Bd * DI * 10^4/T \quad (25.8)$$

where E is the net wind erosion rate of the sample site (Mg ha^{-1} a^{-1}); Bd is the bulk density of the soil (Mg m^{-3}); DI is the sampling depth increment (assumed = plow depth in farmlands, Sutherland and deJong 1990; Walling and Quine 1993) and in undisturbed soils it is approximately the depth over which ^{137}Cs is found (~0.1–0.3 m); T is the time period between the year of initial ^{137}Cs fallout (assumed to be the year of maximum fallout, 1963) and the sampling year of the study; CPR is the percentage residual at a sampling point in the field relative to the native control area (%):

$$CPR = (CPI - k * CRI) * 100/(k * CRI) \quad (25.9)$$

where CPI is the ^{137}Cs inventory at the sampling site (Bq m^{-2}); k is a coefficient of ^{137}Cs redistribution caused by snow-blown and vegetation removal (sometimes set =1; =0.95 in Yan et al. 2001); and CRI is the ^{137}Cs inventory at the reference site (Bq m^{-2}).

Studies using this model to quantify wind erosion yield rates that range from 300 to 8,400 Mg m^{-2} a^{-1} depending on the vegetative cover (Yan et al. 2001; Yan and Shi 2004; Hu et al. 2005) where grasslands exhibit the least erosional loss and croplands and dunelands the highest. In one study (Li et al. 2005), an attempt was made to estimate the relative magnitudes of both wind and water erosion in the same study area, and the authors report that wind erosion can account for at least 18% of the total soil loss.

25.5 Recent Applications

In this section, we describe new applications of ^{137}Cs and other radionuclides in studies of surface erosion and deposition.

25.5.1 Single Event Erosion Measurement

One limitation of the longer-lived radionuclides ^{137}Cs and ^{210}Pb is their inability to quantify erosion over short time periods; for instance, the erosion occurring during a single storm. ^{137}Cs and ^{210}Pb are unsuitable for such studies because the change in inventory due to erosion during a single event is small compared to the total inventory. Walling and Quine (1990) estimated that at least a decade of erosion is necessary to produce a large enough change in ^{137}Cs inventory (t$_{1/2}$ = 30.1 years) that the difference is measurable. Following this reasoning, at least 5–8 years of erosion would be required to recognize erosion using changes in ^{210}Pb$_{xs}$ inventory (t$_{1/2}$ = 22.3 years). ^{7}Be has a sufficiently short halflife (t$_{1/2}$ = 53 days) that Blake et al. (1999), Walling et al. (1999a, b) and Wilson et al. (2003) used the radionuclide to estimate erosion occurring during single rainfall events. The workers determined the pre-storm and post-storm inventory of ^{7}Be as well as the delivery with precipitation. The erosion during the single event studied by Wilson et al. (2003) corresponded to 23% of the average annual erosion rate determined by ^{137}Cs and the mass-balance approach yielded an erosion amount that matched the erosion determined from sediment flux off the field.

Vitko (2007) used ^{7}Be to recognize redistribution of plowed soil from ridges into adjacent furrows finding that fields plowed along contour had greater retention of ^{7}Be than fields plowed up-down the slope presumably due to less soil erosion and/or greater deposition.

25.5.2 Sediment Sources

The radionuclides ^{7}Be, ^{137}Cs, and ^{210}Pb have been used individually and in tandem to quantify the relative contributions of different processes of erosion to sediment. Nagle et al. (2007) used the activity of ^{137}Cs in suspended sediments to determine the provenance of sediment. Collins and Walling (2007) utilized ^{137}Cs and several other constituents in a multivariate sediment mixing model to identify principle sources of fine sediment. He and Owens (1995) used ^{137}Cs, ^{210}Pb$_{xs}$, and ^{226}Ra in tandem to provide radionuclide fingerprints of cultivated land, uncultivated land, and stream banks and their contributions to the sediment pool. Walling and Woodward (1992) used ^{7}Be, ^{137}Cs, ^{210}Pb$_{xs}$ to distinguish sediment from surficial and channel sources and the trend in ^{7}Be during the studied event allowed them to recognize changing sources. Brigham et al. (2001) used ^{210}Pb$_{xs}$, ^{7}Be, and ^{137}Cs as separate tracers to determine the relative importance of surficial and streambank erosion. There was substantial variation in the estimated percentage of sediment from surface soil erosion – ^{137}Cs = 34%, ^{210}Pb$_{xs}$ = 71%, ^{7}Be = 91%. Gellis and Landwehr (2006) used ^{137}Cs and unsupported ^{210}Pb, stable isotopes (del ^{13}C and del ^{15}N), total carbon, nitrogen, and phosphorus to successfully partition sources of fine-grained suspended sediment in the Pocomoke River watershed from cropland, forest, channel and ditch banks, and ditch beds. Devereux et al. (2010) used concentrations of 63 elements and two radionuclides to fingerprint fine sediment sources by both physiographic province (Piedmont and Coastal Plain) and source locale (streambanks, upland and street residue).

Here we consider several specific examples of the use of tracers in source identification. Whiting et al. (2005) established the contribution of streambanks to the suspended sediment flux with a two-component mixing model that used the distinctive radionuclide signatures of the streambanks and of the landscape soils to define the end members of the mixing model (Fig. 25.14). Sediment delivered to the channel by erosion of the soil surface has relatively high activities of ^{7}Be and ^{210}Pb because of the atmospheric delivery of the radionuclides to the soil surface (Fig. 25.12). Sediment delivered to the channel by bank erosion (largely toppling) has much lower radionuclide activities because the low-activity material from deeper in the bank dilutes the high activity material from atop the bank. The height of the collapsed bank is so much larger than the depth of penetration of ^{7}Be into the soil that the activity of ^{7}Be in the eroded bank material is effectively zero. The depth to which the bank material is removed corresponds to that part of the soil profile where only the low-activity supported ^{210}Pb is found. Consequently, the ^{210}Pb activity of the bank material is much lower than the surface activity and is similar to the constant value that reflects in situ decay of soil minerals.

In another study, Whiting et al. (2001) used the distinct distribution of ^{7}Be, ^{137}Cs, and ^{210}Pb with depth in soils on an agricultural plot (Fig. 25.12) and the measured radionuclide flux in runoff in multiple mass balances to quantitatively estimate the areal extent of rill and sheet erosion and the characteristic depth of erosion associated with each mechanism. They examined 15.5 million possible combinations of the depth and areal extent of rill and sheet erosion and found that the best solution to the mass balances corresponded to rills eroding 0.38% of the basin to a depth of 35 mm and sheetwash eroding 37% of the basin to a depth of 0.012 mm. Rill erosion produced 29 times more sediment than sheet erosion.

The identification of sources and depositional locations identified by the radionuclide ratios as discussed herein can be the basis for quantifying a sediment budget. Blake et al. (2002) used ^{7}Be and ^{137}Cs to quantify various elements of a sediment budget – soil erosion and remobilization from hillslopes, erosion and deposition in river channels, and floodplain deposition. ^{137}Cs has been a common tool for characterizing soil redistribution on landscapes (Walling and Quine 1992; Owens et al. 1997; Walling et al. 2003; Kaste et al. 2006).

25.5.3 Pu as a Tracer

^{239,240}Pu may be a promising substitute for ^{137}Cs measurements. Stratospheric ^{239,240}Pu is also bomb-derived, behaves similarly once delivered to the earth surface, and gives equivalent information. In fact, Figs. 25.11 and 25.15 show how similar the profiles for the two radionuclides can be and Fig. 25.16 illustrates the strong correlation between ^{137}Cs and ^{239,240}Pu inventories. There are several reasons why ^{239,240}Pu may become used more often than ^{137}Cs. With the development of a method for using inductively coupled plasma mass spectrometry (ICPMS)

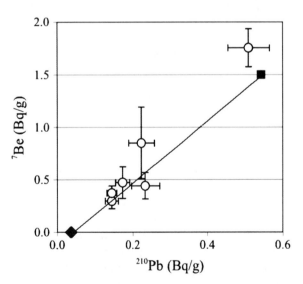

Fig. 25.14 The suspended sediment samples (o) along Soda Butte Creek in Yellowstone National Park, Wyoming, USA, are intermediate between two endmember sources for fine sediment: bank material (*diamond*) and surficial soils (*square*)

Fig. 25.15 Comparison of ^{137}Cs and $^{239+240}$Pu profiles in undisturbed prairie soils from Konza Prairie, KS. Plutonium data courtesy of Michael Ketterer

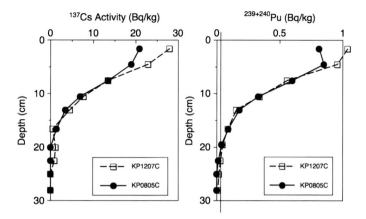

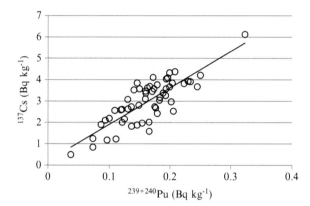

Fig. 25.16 ^{137}Cs inventory vs. ^{239,240}Pu inventory for soil auger samples from 54 sites in the Murder Creek, Georgia USA watershed (from Stubblefield et al. ms)

and high sample through-put, the use of ^{239,240}Pu may be more cost effective (Ketterer and Szechenyi 2008). Moreover, ^{137}Cs half-life ($t_{1/2} = 30.1$ years) is such that peak inventories were observed in the mid- to late-1960s, and by 2010 inventories are only ~40% of original values and dropping. The halflife of both Pu isotopes is several thousand years. Finally, the ^{240}Pu/^{239}Pu values in fallout in the 1950s and after 1965 differ which may allow partial differentiation of the time frame of erosion.

25.5.4 Use with Other Techniques and in Other Settings

Recently we have seen the use of the fallout radionuclides in conjunction with other erosion measurement techniques. Fondran (2007) combined the use of multiple mass-balances of radionuclides and a laser line scanner with sub-millimeter vertical resolution. With the laser line scanner, Lorimor was able to measure changes in topography (erosion and deposition) and compare these results to the estimates of the depth and areal extent of rills and sheetwash as determined from the multiple mass balances. In the same study, rare earth element tagged soils were deployed to provide information on the proportion of sediment delivered from the top of the slope compared to the middle and bottom of the slope (Stubblefield et al. 2006). O'Farrell et al. (2007) used pond sediment volumes, ^{137}Cs and ^{210}Pb activities, and ^{10}Be and ^{26}Al activities to examine different approaches and timeframes for estimating landscape denudation. Walling et al. (1999a, b) used ^{7}Be and ^{137}Cs respectively to evaluate event and decadal erosion rates. The conjunctive use of fallout radionuclides (Mabit et al. 2008) especially with other radionuclides, promises to provide key information on variation in denudation rates over time driven by changes in climate and landuse.

Yet other applications include using radionuclide based sedimentation estimates to test predictions of floodplain sedimentation (Siggars et al. 1999) and to compare aerial photography-derived and radionuclide-derived meander migration rates (Black et al. 2010) or accretion rates (Provansal et al. 2010). Various workers have used ^{7}Be and/or ^{210}Pb to investigate resuspension in rivers and estuaries (Fitzgerald et al. 2001; Wilson et al. 2003, 2005, 2007; Jweda et al. 2008), erosion after forest harvesting (Schuller et al. 2006), and wildfire (Blake et al. 2009).

While the use of radionuclides in surficial studies has focused on tracing fine materials, Salant et al. (2006) used ^{7}Be to recognize and track medium and coarse sand released from an impoundment. Fisher et al. (2010) likewise used ^{7}Be to constrain the timescales of sediment storage associated with large woody debris.

25.6 Future Directions

Although there are about 5,000 papers in the literature devoted to the radionuclides discussed in this chapter there is still room for improvement in the methodologies and applications. From a technological perspective, improvements in instrumentation and detection and lowering of costs should enable greater numbers of samples to be collected and analyzed. For example, the development of larger, low cost gamma detectors that do not require liquid nitrogen cooling will enable a more routine use of the equipment, field deployment and more detailed spatial resolution. This could lead to watershed scale or even landscape scale studies. Another technological improvement could be the development of ICP-MS methods for the measurement of ^{137}Cs and ^{210}Pb. Measurement of gamma decays is inefficient because it measures only those atoms that undergo decay during the counting period. But the vast majority of the atoms of interest do not undergo decay during the counting period, especially for the longer lived isotopes. Instead, mass measurements have the advantage of measuring all the atoms that are present in the sample. While ICP-MS methods have been developed for Pu isotopes (Ketterer et al. 2002) it appears that natural levels of ^{137}Cs and ^{210}Pb cannot be detected by current instrumentation. However, with improvements in ICP-MS technology this may change and permit cost-effective analyses of much larger numbers of samples. As with the development of larger low cost gamma detectors, the development of these techniques could enable much higher spatial resolution studies. For example studies comparing ^{239,240}Pu-derived erosion rates as a surrogate for ^{137}Cs-derived erosion rates can be done now, since there are already ICP-MS methods for ^{239,240}Pu and larger numbers of samples can be run for ^{239,240}Pu than for ^{137}Cs.

Earlier in this chapter we demonstrated that the derived erosion rate is model dependent, and in particular, also depends on the time interval used to evaluate erosion. But these models do not account for the steady, downward migration of the radionuclides. Further progress will require the incorporation of the downward migration into the time dependent erosion models. Furthermore, these models will need to be more comprehensive in their treatment of the downward migration process. Particle mixing by bioturbation as well as solute transport with adsorption will both need to be included in models to enable a better description of the vertical profiles as a function of time. Furthermore, there is evidence that the downward migration process is not steady, and may be affected by non-linear adsorption and/or flow through macropores or channels. A more rigorous evaluation and treatment of these processes needs investigation and incorporation into the models.

It may be possible to use the much better ^{137}Cs soil data sets collected following Chernobyl than the older soil profiles collected a couple of decades after stratospheric fallout to constrain erosion models, quantify soil properties, and improve estimates of soil transport and sources. Even more intriguing is the possibility of combining from the same soil column ^{137}Cs profiles that are dominated by Chernobyl fallout with ^{239,240}Pu profiles derived from global fallout because this will enable comparison of two different tracers over two different time periods within the same soil profile. Furthermore, it may be possible to find locations where the ^{137}Cs signal from stratospheric fallout is comparable in magnitude to the ^{137}Cs fallout from Chernobyl. In such a situation it may be possible to identify and separate the two overlapping signals which will help constrain soil processes and model coefficients.

Finally, soil fertility and C sequestration in soils remain global issues. Thus, accurately determining soil erosion rates, which clearly impacts soil fertility and affects estimates of C sequestration and C removal, remains an important component in research on these topics. Similar studies could also be done to relate past and current erosion rates to land usage to provide both estimates of the historical impacts of land usage on erosion and also future impacts of land usage on C storage and cycling.

Acknowledgements This chapter is dedicated to the memory of Jerry Ritchie.

We would like to thank our former students for their contributions to our research, some of which is highlighted in this chapter: undergraduates Rita Cabral, Joel Saylor, Derek Smith,

Natalie Vajda, and Lauren Vitko; graduate students Chris Bonniwell, Travis Bukach, Carol Fondran, and Chris Wilson; post-doctoral research associates Bill Fornes and Andrew Stubblefield; and colleagues Michael Ketterer, J. Wojciech Mietelski, Klas Rosén, Fred Soster, Louis Thibodeaux, and their students. James Kaste and an anonymous reviewer provided comments that improved the manuscript.

References

Aoyama M, Hirose K, Igarashi Y (2006) Re-construction and updating our understanding on the global weapons tests ^{137}Cs fallout. J Environ Monitor 8:431–438

Ayub JJ, Di Gregorio DE, Velasco H, Huck H, Rizzotto LF (2009) Short-term seasonal variability in Be-7 wet deposition in a semiarid ecosystem of central Argentina. J Environ Radioact 100:977–981

Azahra M, Camacho-Garcia A, Gonzalez-Gomez C, Lopez-Penalver JJ, El Bardouni T (2003) Seasonal Be-7 concentrations in near-surface air of Granada (Spain) in the period 1993–2001. Appl Radiat Isot 59:159–164

Balistrieri LS, Murray JW (1984) Marine scavenging: trace metal adsorption by interfacial sediment from MANOP site H1. Geochim Cosmochim Acta 48:921–929

Baskaran M (1995) A search for the seasonal variability on the depositional fluxes of ^{7}Be and ^{210}Pb. J Geophys Res 100:2833–2840

Baskaran M (2011) Po-210 and Pb-210 as atmospheric tracers and global atmospheric Pb-210 fallout: a review. J Environ Radioact 102:500–513

Baskaran M, Santschi PH (1993) The role of particles and colloids in the transport of radionuclides in coastal environments of Texas. Mar Chem 43:95–114

Baskaran M, Coleman CH, Santschi PH (1993) Atmospheric depositional fluxes of Be-7 and Pb-210 at Galveston and College Station, Texas. J Geophys Res 98:20555–20571

Baskaran M, Ravichandran M, Bianchi TS (1997) Cycling of ^{7}Be and ^{210}Pb in a high DOC, shallow, turbid estuary of southeast Texas. Estuar Coast Shelf Sci 45:165–176

Baskaran M, Hong G-H, Santschi PH (2009) Radionuclide analysis is seawater. In: Wurl O (ed) Practical guidelines for the analysis of seawater. CRC Press, Boca Raton, pp 259–304

Beck HL (1999) External radiation exposure to the population of the continental U.S. from Nevada weapons tests and estimates of deposition density of radionuclides that could significantly contribute to internal radiation exposure via ingestion. Report to the National Cancer Institute, P.O. #263-MQ-909853. New York

Black E, Renshaw CE, Magilligan FJ, Kaste JM, Dade WB, Landis JD (2010) Determining lateral migration rates using fallout radionuclides. Geomorphology 123:364–369

Blake WH, Walling DE, He Q (1999) Fallout beryllium-7 as a tracer in soil erosion investigations. Appl Radiat Isot 51:599–605

Blake WH, Walling DE, He Q (2002) Using cosmogenic Beryllium-7 as a tracer in sediment budget investigations. Geogr Ann 84:89–102

Blake WH, Wallbink PJ, Wilkinson SN, Humphreys GS, Doerr SH, Shakesby RA, Tomkins KM (2009) Deriving hillslope sediment budgets in wildfire affected forests using fallout radionuclide tracers. Geomorphology 104:105–116

Bonniwell EC, Matisoff G, Whiting PJ (1999) Fine sediment residence times in rivers determined using fallout radionuclides (^{7}Be, ^{137}Cs, ^{210}Pb). Geomorphology 27:75–92

Bossew P, Kirchner G (2004) Modeling the vertical distribution of radionuclides in soil. Part 1. The convection-dispersion equation revisited. J Environ Radioact 73:127–150

Boudreau BP (1986a) Mathematics of tracer mixing in sediments: I. Spatially-dependent, diffusive mixing. Am J Sci 286:161–198

Boudreau BP (1986b) Mathematics of tracer mixing in sediments: II. Nonlocal mixing and biological conveyor-belt phenomena. Am J Sci 286:199–238

Brigham ME, McCullough CJ, Wilkinson P (2001) Analysis of suspended-sediment concentrations and radioisotope levels in the Wild Rice River basin, northwestern Minnesota, 1973–98. US Geological Survey Water Resources Investigation Report 2001-4192

Brown RB, Kling GF, Cutshall NH (1981) Agricultural erosion indicated by ^{137}Cs redistribution: II. Estimates of erosion rates. Soil Sci Soc Am J 45:1191–1197

Brudecki K, Suwaj J, Mietelski JW (2009) Plutonium and ^{137}Cs in forest litter: the approximate map of plutonium from Chernobyl deposition in Northeastern and Eastern Poland. Nukleonika 54(3):199–209

Bundt M, Albrecht A, Froidevaux BP, Fluhler H (2000) Impact of preferential flow on radionuclide distribution in soil. Environ Sci Technol 34:3895–3899

Caillet S, Arpagaus P, Monna F, Dominik J (2001) Factors controlling Be-7 and Pb-210 atmospheric deposition as revealed by sampling individual rain events in the region of Geneva, Switzerland. J Environ Radioact 53:241–256

Callender E, Robbins JA (1993) Transport and accumulation of radionuclides and stable elements in a Missouri River reservoir. Water Resour Res 29:1787–1804

Cambray RS, Playford K, Lewis GNJ (1985) Radioactive fallout in air and rain: results to the end of 1984. In: Report *AERE-R11915*, Harwell

Cambray RS, Playford K, Carpenter RC (1989) Radioactive fallout in air and rain: results to the end of 1988. Report no. AERE-R 10155, U.K. Atomic Energy Authority

CDC (2006) Report on the feasibility of a study of the health consequences to the American population from nuclear weapons tests conducted by the United States and Other Nations. http://www.cdc.gov/nceh/radiation/fallout/. Accessed date: March 17, 2010

Chappell A (1996) Modelling the spatial variation of processes in redistribution of soil: digital models and ^{137}Cs in the south-west Niger. Geomorphology 17:249–261

Chappell A (1998) Using remote sensing and geostatistics to map ^{137}Cs-derived net soil flux in the south-west Niger. J Arid Environ 39:441–455

Chen G, Flury M, Harsh JB, Lichtner PC (2005) Colloid-facilitated transport of cesium in variably saturated Hanford sediments. Environ Sci Technol 39:3435–3442

Ciffroy P, Reyss J, Siclet F (2003) Determination of the residence time of suspended particles in the turbidity maximum of the Loire estuary by ^{7}Be analysis. Estuar Coast Shelf Sci 57:553–568

Collins AL, Walling DE (2007) The storage and provenance of fine sediment on the channel bed of two contrasting lowland permeable catchments, UK. River Res Appl 23: 429–450

Collins AL, Walling DE, Sichingabula HM, Leeks GJL (2001) Using 137Cs measurements to quantify soil erosion and redistribution rates for areas under different land use in the Upper Kaleya River basin, southern Zambia. Geoderma 104:229–323

Coughtrey PJ, Thorne MC (1983) Radionuclide distribution and transport in terrestrial and aquatic ecosystems: a critical review. A.A. Balkema, Rotterdam

Cremers AE, De Preter P, Maes A (1988) Quantitative analysis of radiocaesium retention in soils. Nature 335:247–249

Cutshall NH, Larsen IL, Olsen CR (1983) Direct analysis of ^{210}Pb in sediment samples; self adsorption correction. Nucl Instr Meth 206:309–312

Darwin C (1881) The formation of vegetable mould through the action of worms. London: J. Murray (Facsimilies republished in 1982 and 1985, University of Chicago Press, Chicago)

Davis L (1986) Concern over Chernobyl-tainted birds. Sci News 130:54

de Alba S (2001) Modeling the effects of complex topography and patterns of tillage on soil translocation by tillage with mouldboard plough. J Soil Water Conserv 56:335–345

De Cort M, Dubois G, Fridman SD, Germenchuk MG, Izrael YA, Janssens A, Jones AR, Kelly GN, Kvasnikova EV, Matveenko IL, Nazarov IM, Pokumeiko YuM, Sitak VA, Stukin ED, Tabachny LY, Tsaturov YS, Avdyushin SI (1998) Atlas of caesium deposition on Europe after the Chernobyl accident, EUR 16733 EC. Office for Official Publications of the European Communities, Luxembourg

DeJong E, Begg CBM, Kachanoski RG (1983) Estimates of soil-erosion and deposition for some Saskatchewan soils. Can J Soil Sci 63:607–617

DeJong E, Wang C, Rees HW (1986) Soil redistribution on three cultivated New Brunswick hillslopes calculated from ^{137}Cs measurement, solum data and the USLE. Can J Soil Sci 66:721–730

Department of National Health and Welfare (1986) Environmental radioactivity in Canada (Radiological monitoring annual report). Department of National Health and Welfare, Ottawa

Devereux OH, Prestegaard KL, Needleman BA, Gellis AC (2010) Suspended-sediment sources in an urban watershed, Northeast Branch Anacostia River, Maryland. Hydrol Process (www.interscience.wiley.com). doi:10.1002/hyp.7604

Diamond J (2005) Collapse: how societies choose to fail or succeed. Viking, New York

Dibb JE (1989) Atmospheric deposition of beryllium-7 in the Chesapeake Bay region. J Geophys Res 94:2261–2265

Dietrich WE, Dunne T (1978) Sediment budget for a small catchment in mountainous terrain. Z Geomorphol Suppl Bd 29:191–206

Doering C, Akber R, Heijnis H (2006) Vertical distributions of Pb-210 excess, Be-7 and Cs-137 in selected grass covered soils in Southeast Queensland, Australia. J Environ Radioact 87:135–147

Dong A, Chesters G, Simsiman GV (1984) Metal composition of soil, sediments, and urban dust and dirt samples form the Menomonee River watershed, Wisconsin, USA. Water Air Soil Pollut 22:257–275

Dorman L (2004) Cosmic rays in the Earth's atmosphere and underground. Kluwer Academic, Dordrecht

Dorr H, Munnich KO (1989) Downward movement of soil organic matter and its influence on trace-element transport (^{210}Pb, ^{137}Cs) in the soil. Radiocarbon 31:655–663

Durana L, Chudy M, Masarik J (1996) Investigations of Be-7 in the Brastislava atmosphere. J Radioanal Nucl Chem 207:345–356

Dutkiewicz VA, Husain L (1985) Stratospheric and tropospheric components of ^{7}Be in surface air. J Geophy Res 90:5783–5788

EC/IGCE (1998) Roshydromet (Russia)/Minchernobyl (Ukraine)/Belhydromet (Belarus)

El-Hussein A, Mohamemed A, Abd El-Hady M, Ahmed AA, Ali AE, Barakat A (2001) Diurnal and seasonal variation of short-lived radon progeny concentration and atmospheric temporal variations of Pb 210 and Be-7 in Egypt. Atmos Environ 35:4305–4313

Environmental Protection Agency (2000) National water quality inventory, 2000 report. URL:http://www.epa.gov/305b/2000report/. Accessed date: March 19, 2010

Février L, Martin-Garin A (2005) Biogeochemical behaviour of anionic radionuclides in soil: evidence for biotic interactions. Radioprotection 40:S79–S86

Fisher GB, Magilligan FJ, Kaste JM, Nislow KH (2010) Constraining the timescales of sediment sequestration with large woody debris using cosmogenic ^{7}Be. J Geophys Res 115: doi:10.129/2009JF001352

Fitzgerald SA, Klump JV, Swarzenski PW, Mackenzie RA, Richards KD (2001) Beryllium-7 as a tracer of short-term sediment deposition and resuspension in the Fox River, Wisconsin. Environ Sci Technol 35:300–305

Flavin C (1987) Reassessing nuclear power: the fallout from Chernobyl. Worldwatch paper 75, Washington

Flury MS, Czigany S, Chen G, Harsh JB (2004) Cesium migration in saturated silica sand and Hanford sediments as impacted by ionic strength. J Contam Hydrol 71:111–126

Fondran CL (2007) Characterizing erosion by laser scanning. M.A. Thesis, Case Western Reserve University, 60 pp

Fornes WL, Matisoff G, Wilson CG, Whiting PJ (2000) Cs-137-derived soil erosion rates under changing tillage practices. EOS Transactions, Amer. Geophys. Union Fall Meeting, San Francisco, CA

Fornes WL, Matisoff G, Wilson CG, Whiting PJ (2005) Erosion rates using Cs-137; the effects of model assumptions and management practices. Earth Surf Process Landforms 30:1181–1189

Foster IDL, Dalgleish H, Deargin JA, Jones ED (1994) Quantifying soil erosion and sediment transport in drainage basins; some observations on the use of ^{137}Cs. In: Variability in stream erosion and sediment transport. IAHS Press, Wallingford

Gabet EJ (2000) Gopher bioturbation: field evidence for non-linear hillslope diffusion. Earth Surf Process Landforms 25:1419–1428

Gabet EJ, Reichman OJ, Seabloom EW (2003) The effects of bioturbation on soil processes and sediment transport. Ann Rev Earth Planet Sci 31:249–273

Gallagher D, McGee EJ, Mitchell PI (2001) A recent history of C-14, Cs-137 and Pb-210 accumulation in Irish peat bogs: an east versus west coast comparison. In: RadioCarbon conference, Jerusalem, June 2000

Gellis AC, Landwehr JM (2006) Identifying sources of fine-grained suspended-sediment for the Pocomoke River, an Eastern Shore tributary to the Chesapeake Bay. In: Proceedings of the 8th federal interagency sedimentation conference, Reno, Nevada, 2–6 April 2006, Reno, NV, Paper 5C-1 in CD_ROM file ISBN 0-9779007-1-1, 9

Goldberg ED, Koide M (1962) Geochronological studies of deep sea sediments by the ionium/thorium method. Goeochim Cosmochim Acta 26:417–450

Goodbred SL, Kuehl SA (1998) Floodplain processes in the Bengal Basin and the storage of Ganges-Brahmaputra river sediment: an accretion study using ^{137}Cs and ^{210}Pb geochronology. Sediment Geol 121:239–258

Govers G, Vandaele K, Desmet P, Poesen J, Bunte K (1994) The role of tillage in soil redistribution on hillslopes. Eur J Soil Sci 45:469–478

Govers G, Quine TA, Desmet PJ, Walling DE (1996) The relative contribution of soil tillage and overland flow erosion to soil redistribution on agricultural land. Earth Surf Process Landforms 21:929–946

Harden JW, Sundquist ET, Stallard RF, Mark RK (1992) Dynamics of soil carbon during deglaciation of the Laurentide ice sheet. Science 258:1921–1924

Hawley N, Robbins JA, Eadie BJ (1986) The partitioning of ^{7}beryllium in fresh water. Geochim Cosmochim Acta 50:1127–1131

He Q, Owens P (1995) Determination of suspended sediment provenance using caesium-137, unsupported lead-210 and radium-226; a numerical mixing model approach. In: Gurnell A, Webb B, Foster I (eds) Sediment and water quality in river catchments. London, Wiley, pp 207–227

He Q, Walling DE (1996) Interpreting particle size effects in the adsorption of ^{137}Cs and unsupported ^{210}Pb by mineral soils and sediments. J Environ Radioact 30:117–137

Health and Safety Laboratory (1977) Final tabulation of monthly ^{90}Sr fallout data: 1954–1976. USERDA Report HASL-329

Heckrath G, Halekoh U, Djurhuus J, Govers G (2006) The effect of tillage direction on soil redistribution by mouldboard ploughing on complex slopes. Soil Tillage Res 88:225–241

Hicks HG (1982) Calculation of the concentration of any radionuclide deposited on the ground by off-site fallout from a nuclear detonation. Health Phys 42:585–600

Honeyman BD, Ranville JF (2002) Colloid properties and their effects on radionuclide transport through soils and groundwater. In: Zhang P-C, Brady PV (eds.) Geochemistry of soil radionuclides. SSSA special publication 59; Soil Science Society of America, Madison, pp 131–163

Hong G-H, Hamilton TF, Baskaran M, Kenna TC (2011) Applications of anthropogenic radionuclides as tracers to investigate marine environmental processes. In: Baskaran M (ed) Handbook of environmental isotope geochemistry. Springer, Heidelberg

Hu Y, Liu J, Zhuang D, Cao H, Yan H, Yang F (2005) Distribution characteristics of ^{137}Cs in wind-eroded soil profile and its use in estimating wind erosion modulus. Chin Sci Bull 50:1155–1159

Hyams E (1952) Soil and civilization. Thames and Hudson, London

Ioannidou A, Papastefanou C (2006) Precipitation scavenging of Be-7 and Cs-137 radionuclides in air. J Environ Radioact 85:121–136

Jarvis NJ, Taylor A, Larsbo M (2010) Modeling the effects of bioturbation on the re-distribution of ^{137}Cs in an undisturbed grassland soil. Eur J Soil Sci 61:24–34

Johnson DL, Keller EA, Rockwell TK (1990) Dynamic pedogenesis – new views on some key soil concepts, and a model for interpreting quaternary soils. Quat Res 33:306–319

Joshi SR (1987) Early Canadian results on the long-range transport of chernobyl radioactivity. Sci Total Environ 63:125–137

Jweda J, Baskaran M, van Hees E, Schweitzer L (2008) Short-lived radionuclides (Be-7 and Pb-210) as tracers of particle dynamics in a river system in southeast Michigan. Limnol Oceanogr 53:1934–1944

Kachanoski RG, deJong E (1984) Predicting the temporal relationship between soil cesium-137 and erosion rate. J Environ Qual 13:301–304

Kaste JM, Baskaran M (2011) Meteoric ^{7}Be and ^{10}Be as process tracers in the environment. In: Baskaran M (ed) Handbook of environmental isotope geochemistry. Springer, Heidelberg

Kaste JM, Fernandez IJ, Hess CT, Norton SA, Pillerin BA (1999a) Sinks and mobilization of cosmogenic beryllium-7 and bedrock-derived beryllium-9 in forested watersheds in Maine, USA. Geol Soc Am Abst Prog 31:305

Kaste JM, Fernandez IJ, Hess CT, Norton SA (1999b) Delivery of cosmogenic beryllium-7 to forested watersheds in Maine, USA. Geol Soc Am Abst Progr 31:305

Kaste JM, Norton SA, Hess CT (2002) Environmental chemistry of beryllium-7. Rev Mineral Geochem 50:271–289

Kaste JM, Heimsath AM, Hohmann M (2006) Quantifying sediment transport across an undisturbed prairie landscape using cesium-137 and high resolution topography. Geomorphology 76:430–440

Kaste JM, Heimseth AM, Bostick BC (2007) Short-term soil mixing quantified with fallout radionuclides. Geology 35:243–246

Ketterer ME, Szechenyi SC (2008) Determination of plutonium and other transuranic elements by inductively coupled plasma mass spectrometry: a historical perspective and new frontier in the environmental sciences. Spectrochim Acta 63:719–737

Ketterer ME, Watson BR, Matisoff G, Wilson CG (2002) Rapid dating of recent aquatic sediments using Pu activities and ^{240}Pu/^{239}Pu as determined by quadrupole inductively coupled plasma mass spectrometry. Environ Sci Technnol 36:1307–1311

Ketterer ME, Hafer KM, Mietelski JW (2004) Resolving Chernobyl vs. global fallout contributions in soils from Poland using plutonium atom ratios measured by inductively coupled plasma mass spectrometry. J Environ Radioact 73:183–201

Ketterer ME, Zheng J, Yamada M (2011) Source tracking of transuranics using their isotopes. In: Baskaran M (ed) Handbook of environmental isotope geochemistry. Springer, Heidelberg

Kikuchi S, Sakurai H, Gunji S, Tokanai F (2009) Temporal variation of Be-7 concentrations in atmosphere for 8 y from 2000 at Yamagata, Japan: solar influence on the Be-7 time series. J Environ Radioact 100:515–521

Koch DM, Jacob DJ, Graustein WC (1996) Vertical transport of tropospheric aerosols as indicated by ^{7}Be and ^{210}Pb in a chemical tracer model. J Geophys Res 101:18651–18666

Konshin OV (1992) Mathematical model of ^{137}Cs migration in soil: analysis of observations following the Chernobyl accident. Health Phys 63:301–306

Korky JK, Kowalski L (1989) Radioactive cesium in edible mushrooms. J Agric Food Chem 37:568–569

Korobova E, Ermakov A, Linnak V (1998) ^{137}Cs and ^{90}Sr mobility in soils and transfer in soil-plant systems in the Novozybkov district affected by the Chernobyl accident. Appl Geochem 13:803–814

Kownacka L (2002) Vertical distributions of beryllium-7 and lead-210 in the tropospheric and lower stratospheric air. Nukleonika 47:79–82

Lal D, Baskaran M (2011) Applications of cosmogenic-isotopes as atmospheric tracers. In: Baskaran M (ed) Handbook of environmental isotope geochemistry. Springer, Heidelberg

Lal D, Peters B (1967) Cosmic ray produced radioactivity on the Earth, p. 551–612. In Encyclopedia of physics. V. 46/2. Springer

Lal R (2003) Soil erosion and the global carbon budget. Environ Internat 29:437–450

Lance J, McIntyre SC, Naney JW, Rousseva SS (1986) Measuring sediment movement at low erosion rates using cesium 137. Soil Sci Soc Am J 50:1303–1309

Lenhart JJ, Saiers JE (2002) Transport of silica colloids through unsaturated porous media: experimental results and model comparisons. Environ Sci Technol 36:769–777

Li M, Li ZB, Liu PL, Yao WY (2005) Using Cesium-137 technique to study the characteristics of different aspect of soil erosion in the wind-water erosion crisscross region on Loess Plateau of China. Appl Radiat Isot 62:109–113

Lobb DA, Kachanoski RG (1999) Modeling tillage translocation using step, linear-plateau and exponential functions. Soil Till Res 51:317–330

Lobb DA, Kachanoski RG, Miller MH (1995) Tillage translocation and tillage erosion on shoulder slope landscape positions measured using Cs-137 as a tracer. Can J Soil Sci 75:211–218

Lobb DA, Kachanoski RG, Miller MH (1999) Tillage translocation and tillage erosion in the complex upland landscapes of southwestern Ontario, Canada. Soil Tillage Res 51:189–209

Lomenick TF, Tamura T (1965) Naturally occurring fixation of cesium-137 on sediments of lacustrian origin. Soil Sci Soc Am Proc 29:383–386

Loughran J, Elliott GL, Campbell BL, Shelly DJ (1988) Estimation of soil erosion from caesium-137 measurements in a small, cultivated catchment in Australia, Applied Radiation and Isotopes A39, pp. 1153–1157

Lowdermilk WC (1953) Conquest of the land through 7000 years. Agricultural Info Bull No. 99

Lowrance R, McIntyre S, Lance C (1988) Erosion and deposition in a field/forest system estimated using cesium-137 activity. J Soil Water Conserv 43:195–199

Lyles L (1985) Predicting and controlling wind erosion. Agric Hist Soc 59:205–214

Mabit L, Benmansour M, Walling DE (2008) Comparative advantages and limitations of the fallout radionuclides Cs-137, Pb-210(ex) and Be-7 for assessing soil erosion and sedimentation. J Environ Radioact 99:1799–1807

Martz LW, de Jong E (1985) The relationship between land surface morphology and soil erosion-deposition rates in a small Saskatchewan basin. Proceedings of the Canadian Society for Civil Engineering, Annual Conference, Hydrotechnical Division, July 1985, Saskatoon, Sask., 1–19

Matisoff G, Bonniwell EC, Whiting PJ (2002a) Soil erosion and sediment sources in an Ohio watershed using Beryllium-7, Cesium-137, and Lead-210. J Environ Qual 31:54–61

Matisoff G, Bonniwell EC, Whiting PJ (2002b) Radionuclides as indicators of sediment transport in agricultural watersheds that drain to Lake Erie. J Environ Qual 31:62–72

Matisoff G, Wilson GC, Whiting PJ (2005) ^{7}Be/^{210}Pb ratio as an indicator of suspended sediment age or fraction new sediment in suspension. Earth Surf Process Landforms 30:1191–1201

Matisoff G, Ketterer ME, Rosén K, Mietelski JW, Vitko LF, Persson H, Lokas E (2011) Downward Migration of Chernobyl-derived Radionuclides in Soils in Poland and Sweden. Applied Geochemistry 26:105–115

McHenry JR, Ritchie JC, Gill AC (1973) Accumulation of fallout cesium 137 in soils and sediments in selected watersheds. Water Resour Res 9:676–686

McNeary D, Baskaran M (2003) Depositional characteristics of Be-7 and Pb-210 in southeastern Michigan. J Geophys Res Atmos 108:4210

Mietelski JW, Was B (1995) Plutonium from Chernobyl in Poland. Appl Radiat Isot 46:1203–1211

Mietelski JW, Jasińska M, Kozak K, Ochab E (1996) The method of measurements used in the investigation of radioactive contamination of forests in Poland. Appl Radiat Isot 47:1089–1095

Miller KM, Kuiper JL, Helfer IK (1990) ^{137}Cs fallout depth distributions in forest versus field sites: implications for external gamma dose rates. J Environ Radioact 12:23–47

Montgomery DR (2007) Dirt: the erosion of civilizations. University of California Press, California

Montgomery JA, Busacca AJ, Frazier BE, McCool DK (1997) Evaluating soil movement using cesium-137 and the revised universal soil loss equation. Soil Sci Soc Am J 61:571–579

Motha JA, Walbrink PJ, Hairsine PB, Grayson RB (2002) Tracer properties of eroded sediment and source material. Hydrol Process 16:1983–2000

Müller-Lemans H, van Dorp F (1996) Bioturbation as a mechanism for radionuclide transport in soils: relevance of earthworms. J Environ Radioact 31:7–20

Nagle GN, Fahey T, Ritchie JC, Woodbury PB (2007) Variations in sediment sources in the Finger Lakes and Western Catskills regions of New York. Hydrol Process 21:828–838

Nelson DW, Logan TJ (1983) Chemical processes and transport of phosphorus. In: Schaller FW, Bailey GW (eds) Agricultural management and water quality. Iowa State University Press, Iowa

O'Farrell CR, Heimsath AM, Kaste JM (2007) Quantifying hillslope erosion rates and processes for a coastal California landscape over varying timescales. Earth Surf Process Landforms 32:544–560

Oldeman LR (1994) The global extent of soil degradation. In: Greland DJ, Szacolcs I (eds) Soil resilience and sustainable land use. CAB International, Wallingford, pp 99–118

Olsen CR, Larsen IL, Lowry PD, Cutshall NH, Todd JF, Wong GTF, Casey WH (1985) Atmospheric fluxes and marsh-soil inventories of ^{7}Be and ^{210}Pb. J Geophys Res 90:10487–10495

Olsen CR, Larsen IL, Lowry PD, Cutshall NH (1986) Atmospheric fluxes and marsh-soil inventories of ^{7}Be and ^{210}Pb. J Geophys Res 90:10487–10495

Owens PN, Walling DE (1996) Spatial variability of caesium-137 inventories at reference sites: an example from two contrasting sites in England and Zimbabwe. Appl Radiat Isot 47:699–707

Owens PN, Walling DE, He Q (1996) The behavior of bomb-derived caesium-137 fallout in catchment soils. J Environ Radioact 32:169–191

Owens PN, Walling DE, He Q, Shanahan J, Foster I (1997) The use of caesium-137 measurements to establish a sediment budget for the Start catchment, Devon, UK. Hydrol Sci 42:405–423

Paasikallio A (1999) Effect of biotite, zeolite, heavy clay, bentonite and apatite on the uptake of radiocesium by grass from peat soil. Plant Soil 206:213–222

Paatero J, Hatakka J (2000) Source areas of airborne Be-7 and Pb-210 measured in Northern Finland. Health Phys 79:691–696

Papastefanou C, Ioannidou A (1991) Depositional fluxes and other physical characteristics of atmospheric Beryllium-7 in the temperate zones (40N) with a dry (precipitation free) climate. Atmos Environ 25:2335–2343

Papastefanou C, Ioannidou A, Stoulos S, Manolopoulou M (1995) Atmospheric deposition of cosmogenic ^{7}Be and ^{137}Cs from fallout of the Chernobyl accident. Sci Total Environ 170:151–156

Pimental D, Harvey C, Resosuarmo P, Sinclair K, Kurz D, McNair M, Crist S, Shpritz L, Fitton L, Saffourni R, Blair R (1995) Environmental and economic costs of soil erosion and conservation benefits. Science 267:1117–1123

Porcelli D, Baskaran M (2011) An overview of isotope geochemistry in environmental studies. In: Baskaran M (ed) Handbook of environmental isotope geochemistry. Springer, Heidelberg

Poreba GJ (2006) Caesium-137 as a soil erosion tracer: a review. Geochronometria 25:37–46

Pote DH, Daniel TC, Sharpley AN, Moore PA, Edwards DR, Nichols DJ (1996) Relating extractable soil phosphorus to phosphorus losses in runoff. Soil Sci Soc Am J 60:855–859

Powell EN (1977) Particle size selection and sediment reworking in a funnel feeder, *Leptosynapta benvis* (Holothuroidea, Synaptidae). Int Rev Gesamten Hydrobiol 62:385–408

Provansal M, Villiet J, Eyrolle F, Raccasi G, Gurriaran R, Antonelli C (2010) High-resolution evaluation of recent bank accretion rate of the managed Rhone: a case study by multi-proxy approach. Geomorphology 117:287–297

Quine TA (1995) Estimation of erosion rates from Caesium-137 data; the calibration question. In: Foster IDL et al (eds) Sediment and water quality in river catchments. Chichester, Wiley, pp 307–329

Quine TA (1999) Use of caesium-137 data for validation of spatially distributed erosion models: the implications of tillage erosion. Catena 37:415–430

Quine TA, Zhang Y (2004) Re-defining tillage erosion: quantifying intensity – direction relationships for complex terrain. (1) Derivation of an adirectional soil transport coefficient. Soil Use and Management 20(2):114–123

Rabalais NN, Turner RE, Dubravko J, Dortsch Q, Wisman WJ (1999) Characterization of hypoxia: topic 1 report for the integrated assessment on hypoxia in the Gulf of Mexico. NOAA Coastal Ocean Program Decision Analysis Series No. 17, NOAA Coastal Ocean Office, Silver Spring

Reiser DW (1998) Sediment in gravel bed rivers. In: Klingeman PC, Beschta RL, Komar PD, Bradley JB (eds) Ecological and biological considerations. Water Resource Publications, Highland Ranch

Revelle P, Revelle C (1988) The environment. Jones and Bartlett, Boston

Rhoads DC (1974) Organism-sediment relations on the muddy sea floor. Oceanogr Mar Biol Annu Rev 12:263–300

Rhoads DC, Stanley DJ (1964) Biogenic graded bedding. J Sediment Petrol 35:956–963

Riise G, Bjornstad HE, Lien HN, Oughton DH, Saibu B (1990) A study on radionuclide association with soil components using a sequential extraction procedure. J Radioanal Nucl Chem 142:531–538

Ritchie JC, McHenry JR (1978) Fallout Cs-137 in cultivated and noncultivated North Central United States watersheds. J Environ Qual 7:40–44

Ritchie JC, McHenry JR (1990) Application of radioactive fallout Cesium-137 for measuring soil erosion and sediment accumulation rates and patterns – a review. J Environ Qual 19:215–233

Ritchie JC, Ritchie CA (2008) Bibliography of publications of ^{137}cesium studies related to erosion and sediment deposition. http://web.cena.usp.br/apostilas/Osny/BiblioCs137December2008.pdf. Accessed 21 Jan 2010

Ritchie JC, Spraberry JA, McHenry JR (1974) Estimating soil erosion for the redistribution of fallout ^{137}Cs. Soil Sci Soc Am J 38:137–139

Robbins JA (1985) Great Lakes regional fallout source functions. NOAA Technical Memorandum ERL GLERL-56, Ann Arbor, MI

Robbins JA (1986) A model for particle-selective transport of tracers in sediments with conveyor-belt deposit feeders. J Geophys Res 91(C7):8542–8558

Robbins JA, Eadie BJ (1991) Seasonal cycling of trace elements ^{137}Cs, ^{7}Be, and $^{239+240}$Pu in Lake Michigan. J Geophy Res 96:17081–17104

Robbins JA, McCall PL, Fisher JB, Krezoski JR (1979) Effect of deposit feeders on migration of ^{137}Cs in lake sediment. Earth Planet Lett 42:277–287

Rogowski AS, Tamura T (1970) Environmental mobility of cesium-137. Radiat Bot 10:35–45

Rosen K, Oborn L, Lonsjo H (1999) Migration of radiocaesium in Swedish soil profiles after the Chernobyl accident, 1987–1995. J Environ Radioact 46:45–66

Rowan JS (1995) The erosional transport of radiocaesium in catchment systems: a case study of the Exe basin, Devon. In: Foster IDL et al (eds) Sediment and water quality in river catchments. Wiley, Chichester

Roy JC, Cote JE, Mahfoud A, Villeneuve S, Turcotte J (1988) On the transport of Chernobyl radioactivity to eastern Canada. J Environ Radioact 6:121–130

Salant NL, Renshaw CE, Magilligan FJ, Kaste JM, Nislow KH, Heimsath AM (2006) The use of short-lived radionuclides to quantify transitional bed material transport in a regulated river. Earth Surf Process Landforms 32:509–524

Salisbury RT, Cartwright J (2005) Cosmogenic Be-7 deposition in North Wales: Be-7 concentrations in sheep faeces in relation to altitude and precipitation. J Environ Radioact 78:353–361

Sanchez AL, Wright SM, Smolders E, Naylor C, Stevens PA, Kennedy VH, Dodd BA, Singleton DL, Barnett CL (1999) High plant uptake of radiocesium from organic soils due to Cs mobility and low soil K content. Environ Sci Technol 33:2752–2757

Sarmiento JL, Gwinn E (1986) Strontium-90 fallout prediction. J Geophys Res 91:7631–7646

Schuller P, Elliers A, Kirchner G (1997) Vertical migration of fallout ^{137}Cs in agricultural soils from Southern Chile. Sci Total Environ 193:197–205

Schuller P, Iroume A, Walling DE, Mancilla IIB, Castillo A, Trumper RE (2006) Use of beryllium-7 to document soil redistribution following forest harvest operations. J Environ Qual 35:1756–1763

Schultz RK, Overstreet R, Barshad I (1959) On the soil chemistry of cesium-137. Soil Sci 89:19–27

Sekely AC, Bauer DW, Mulla DJ (2002) Streambank slumping and its contribution to the phosphorus and suspended sediment loads of the Blue Earth River, Minnesota. J Soil Water Conserv 57:243–250

Sepulveda A, Schuller P, Walling DE, Castillo A (2008) Use of Be-7 to document soil erosion associated with a short period of extreme rainfall. J Environ Radioact 99:35–49

Sharpley AN, Daniel TC, Edwards DR (1994) Phosphorus movement in the landscape. J Product Agric 6:492–500

Shenber MA, Johanson KJ (1992) Influence of zeolite on the availability of radiocaesium in soil to plants. Sci Total Environ 113:287–295

Shimmack W, Flessa H, Bunzl K (1997) Vertical migration of Chernobyl-derived radiocesium in Bavarian Grassland soils. Naturvissenschaften 84:204–207

Siggars GB, Bates PD, Anderson MG, Walling DE, He Q (1999) A preliminary investigation of the integration of modeled floodplain hydraulics with estimates of overbank floodplain sedimentation derived form Pb-210 and Cs-137 measurements. Earth Surf Process Landforms 24: 211–231

Soileau JM, Hajek BF, Touchton JT (1990) Soil erosion and deposition evidence in a small watershed using fallout cesium-137. Soil Sci Soc Am J 54:1712–1719

Spomer RG, McHenry JR, Piest RF (1985) Sediment movement and deposition using cesium-137 tracer. Trans Am Soc Agric Eng 28:767–772

St Gerontidis DV, Kosmas G, Detsis G, Marathianou M, Zafirios T, Tsara M (2001) The effect of mouldboard plow on tillage erosion along a hillslope. J Soil Water Conserv 56:147–152

Stallard RF (1998) Terrestrial sedimentation and the carbon cycle: coupling weathering and erosion to carbon burial. Global Biogeochem Cy 12:231–257

Steinmann P, Billen T, Loizeau J-L, Dominik J (1999) Beryllium-7 as a tracer to study mechanisms and rates of metal scavenging from lake surface waters. Geochim Cosmochim Acta 63:1621–1633

Stokes S, Walling DE (2003) Radiogenic and isotopic methods for the direct dating of fluvial sediments. In: Kondolf GM, Piegay H (eds) Tools in fluvial geomorphology. Wiley, Chichester

Stubblefield AP, Fondran C, Ketterer ME, Matisoff G, Whiting PJ (2006) Radionuclide and rare earth element tracers of erosional processes on the plot scale. In: Joint Eighth Federal Interagency Sedimentation Conference and Third Federal Interagency Modeling Conference, Reno, Nevada

Stubblefield AP, Whiting PJ, Matisoff G, Fondran C, Wilson C, Calhoun FC, ms. Delivery of nutrients by rill and sheet erosion in agricultural settings. In preparation

Su CC, Huh CA, Lin FJ (2003) Factors controlling atmospheric fluxes of Be-7 and Pb-210 in northern Taiwan. Geophys Res Lett 30:2018

Sugihara S, Momoshima N, Maeda Y, Osaki S (2000) Variation of atmospheric ^{7}Be and ^{210}Pb depositions at Fukuoka, Japan. In: 10th international congress of the international radiation protection association. Hiroshima, Japan 10–19 May 2000

Sutherland RA, deJong E (1990) Estimation of sediment redistribution within agricultural fields using caesium-137. Crystal Springs, Saskatchewan, Canada. Appl Geograph 10:205–213

Sutherland RA, Kowalchuk T, deJong E (1991) Caesium-137 estimates of sediment redistribution by wind. Soil Sci 151:387–396

Tamura T, Jacobs DG (1960) Structural implications in cesium sorption. Health Phys 2:391–398

Tarras-Wahlberg NH, Lane SN (2003) Suspended sediment yield and metal contamination in a river catchment affected by El Nino events and gold mining activities; the Puyango River Basin, Southern Ecuador. Hydrol Process 17:3101–3123

Todd JF, Wong GRF, Olsen CR, Larsen IL (1989) Atmospheric depositional characteristics of Beryllium 7 and Lead 210 along the southeastern Virginia Coast. J Geophys Res 94:11106–11116

Tokieda T, Yamanaka K, Harada K, Tsunogai S (1996) Seasonal variation of residence time and upper atmospheric contribution of aerosols studied with Pb-210, Po-210 and Be-7. Tellus 48:690–702

Turekian KK, Nozaki Y, Benninger LK (1977) Geochemistry of atmospheric radon and radon products. Ann Rev Earth Planet Sci 5:227–255

Turekian KK, Benninger LK, Dion EP (1983) ^{7}Be and ^{210}Pb total deposition fluxes at New Haven, Connecticut and at Bermuda. J Geophys Res 88:5411–5415

Valles I, Camacho A, Ortega X (2009) Natural and anthropogenic radionuclides in airborne particulate samples collected in Barcelona (Spain). J Environ Radioact 100:102–107

Van Muysen W, Govers G, Van Oost K (2002) Identification of important factors in the process of tillage erosion: the case of mouldboard tillage. Soil Till Res 65:77–93

Van Oost K, Govers G, de Albe S, Quine TA (2006) Tillage erosion: a review of controlling factors and implications for soil quality. Progr Phys Geog 4:443–466

Vitko L (2007) Evaluating soil erosion on agricultural plots using radionuclides. BA Thesis, Case Western Reserve University

Wallbrink PJ, Murray AS (1994) Fallout of ^{7}Be in south Eastern Australia. J Environ Radioact 25:213–228

Wallbrink PJ, Murray AS (1996) Distribution and variability of Be-7 in soils under different surface cover conditions and its potential for describing soil redistribution processes. Water Resour Res 32:467–476

Wallbrink PJ, Murray AS, Olley JM (1999) Relating suspended sediment to its original soil depth using fallout radionuclides. Soil Science Society of America Journal 63:369–378

Walling DE, He Q (1997a) Investigating spatial patterns of overbank sedimentation on river floodplains. Water Air Soil Pollut 99:9–20

Walling DE, He Q (1997b) Models for converting ^{137}Cs measurements to estimates for soil redistribution on cultivated and uncultivated soils (including software for model implementation). In: International Atomic Energy Agency (IAEA) coordinated research programmes on soil erosion (D1.50.05) and sedimentation (F3.10.01). University of Exeter, Exeter

Walling DE, Quine TA (1990) Calibration of cesium-137 measurements to provide quantitative erosion rate data. Land Degrad Rehab 2:161–175

Walling DE, Quine TA (1992) The use of caesium-137 measurements in soil erosion surveys. In: Erosion and sediment transport monitoring programmes in River Basins, IAHS publications 210, Wallingford

Walling DE, Quine TA (1993) Use of caesium-137 as a tracer of erosion and sedimentation: Handbook for the application of the caesium-137 technique. U.K Overseas Development Administration Research Scheme R4579, Department of Geography, University of Exeter, Exeter, United Kingdom, 196 p

Walling DE, Woodward JC (1992) Use of radiometric fingerprints to derive information on suspended sediment sources. In: Erosion and sediment transport monitoring programmes in River Basins, IAHS publications 210, Wallingford

Walling DE, He Q, Blake W (1999a) Use of Be-7 and Cs-137 measurements to document short- and medium term rates of water-induced soil erosion on agricultural land. Water Resour Res 35:3865–3874

Walling DE, Owens PN, Leeks GJL (1999b) Fingerprinting suspended sediment sources in the catchment of the River Ouse, Yorkshire, UK. Hydrol Process 13:955–975

Walling DE, He Q, Appleby PG (2002) Conversion models for use in soil-erosion, soil-redistribution and sedimentation investigations. In: Zapata F (ed) Handbook for the assessment of soil erosion and sedimentation using environmental radionuclides. Kluwer Academic Publishers, Dordrecht, pp 111–164

Walling DE, Collins AL, Sichingabula HM (2003) Using unsupported lead-210 measurements to investigate erosion and sediment delivery in a small Zambian catchment. Geomorphology 52:193–213

Walling DE, Schuller P, Zhang Y, Iroume A (2009) Extending the timescale for using beryllium 7 measurements to document soil redistribution by erosion. Water Resour Res 45: W02418

Wang K, Cornett RJ (1993) Distribution coefficients of ^{210}Pb and ^{210}Po in laboratory and natural aquatic systems. J Paleolimnol 9:179–188

Whiting PJ, Bonniwell EC, Matisoff G (1999) A Mass Balance Method for Determining the Depth and Area of Rill and Sheetwash Erosion of Soils using Fallout Radionuclides (^{7}Be, ^{137}Cs, ^{210}Pb): EOS Transactions, American Geophysical Union Fall Meeting. v. 80. p. F445

Whiting PJ, Bonniwell EC, Matisoff G (2001) Depth and areal extent of sheet wash and rill erosion from radionuclides in soils and suspended sediment. Geology 29:1131–1134

Whiting PJ, Matisoff G, Fornes W, Soster F (2005) Suspended sediment sources and transport distances in the Yellowstone basin. Geol Soc Am Bull 117:515–529

Williams GP, Wolman MG (1984) Downstream effects of dams on alluvial rivers. US Geological Survey Professional Paper 1286, p 83

Wilson CG, Matisoff G, Whiting PJ (2003) Short-term erosion rates from a Be-7 inventory balance. Earth Surf Process Landforms 28:967–977

Wilson CG, Matisoff G, Whiting PJ (2005) Transport of fine sediment through the Old Woman Creek, OH, wetland using radionuclide tracers. J Great Lakes Res 31:56–67

Wilson CG, Matisoff G, Whiting PJ (2007) The use of ^{7}Be and ^{210}Pb$_{xs}$ to differentiate suspended sediment sources in South Slough, OR. Estuar Coasts 30:348–358

Wischmeier WH, Smith DD (1978) Predicting rainfall erosion losses – a guide to conservation planning. In: Agricultural handbook No. 537. US Department of Agricultural Science and Education Administration, Washington

Wise SW (1980) Caesium-137 and lead-210: a review of the technique and application to geomorphology, pp. 109–127. In: Cullingford RA, Davidson RA, Lewin J (eds) Timescales in geomorphology, John Wiley and Sons, New York

Yamada H, Nakamura F (2002) Effect of fine sediment deposition and channel works on periphyton biomass in the Makomanai River, Northern Japan. River Res Appl 18:481–493

Yan P, Shi P (2004) Using the ^{137}Cs technique to estimate wind erosion in Gonghe Basin, Qinghai Province, China. China Soil Sci 169:295–305

Yan P, Zhang XB (1998) Prospects of caesium-137 used in the study of aeolian processes. J Desert Res 18:182–187

Yan P, Dong Z, Dong G, Zhang X, Zhang Y (2001) Preliminary results of using ^{137}Cs to study wind erosion in the Qinghai-Tibet plateau. J Arid Environ 47:443–452

You CF, Lee T, Li YH (1989) The partition of Be between soil and water. Chem Geol 77:105–118

Zapata F, Garcia-Agudo E, Ritchie JC, Appleby PG (2002) Introduction. In: Zapata F (ed) Handbook for the assessment of soil erosion and sedimentation using environmental radionuclides. Kluwer Academic Publishers, Dordrecht, pp 1–14

Zhang XB, Higgitt DL, Walling DE (1990) A preliminary assessment of the potential for using cesium-137 to estimate rates of soil-erosion in the Loess Plateau of China. Hydrol Sci 35:243–252

Zhang XB, Walling DE, He Q (1999) Simplified mass balance models for assessing soil erosion rates and cultivated land using cesium-137 measurements. Hydrol Sci 44:33–45

Chapter 26
Sr and Nd Isotopes as Tracers of Chemical and Physical Erosion

Gyana Ranjan Tripathy, Sunil Kumar Singh, and S. Krishnaswami

Abstract The applications of radiogenic isotopes to investigate chemical and physical erosion processes, particularly in river basins of the Himalaya, have led to interesting inferences on the relationship between tectonics, weathering and climate. The chemical weathering studies rely more on Sr isotopes because of their widely different ratios in various end members, their uniform distribution in the oceans and the availability of continuous and robust record of marine $^{87}Sr/^{86}Sr$ through much of the geological past. The record for the Cenozoic shows steady increase in $^{87}Sr/^{86}Sr$; one of the hypotheses suggested to explain this is enhanced continental silicate weathering due to the uplift of the Himalaya. This hypothesis linking tectonics-weathering-climate, based on $^{87}Sr/^{86}Sr$ as an index of silicate weathering, however, is being challenged by the recent observations that there are a variety of carbonates in the river basins of the Himalaya with $^{87}Sr/^{86}Sr$ similar to that of silicates which have the potential to contribute significantly to the high $^{87}Sr/^{86}Sr$ of rivers such as the Ganga-Brahmaputra. Further, the non-stochiometric release of Sr isotopes during chemical weathering of minerals and rocks, the imbalance of Sr isotope budget in the oceans and temporal variations in riverine fluxes due to impact of glaciations all have compounded the problem.

Studies on the provenance of sediments and physical erosion pattern employ both Sr and Nd isotopes under the assumption that their source signatures are preserved in sediments. Though there are concerns on how well this assumption is satisfied especially by the Sr isotope system, both Sr and Nd systems are being used to learn about physical erosion in the Himalaya, its variability and causative factors. The results show that at present the major source of sediments to the Ganga plain and the Bay of Bengal is the Higher Himalayan Crystallines and that physical erosion among the various sub-basins is very heterogeneous with maximum rates in regions of intense precipitation and high relief. There are three such "hot-spots", one each in the basins of the Ganga, Brahmaputra and the Indus, which unload huge amount of sediments promoting rapid uplift of regions surrounding them and enhance chemical weathering by exposing fresh rock surfaces. The pattern of physical erosion and its temporal variations shows that it is influenced by climate change both on ky and My time scales though during the latter periods the erosion regime has been by and large stable. This article reviews investigations on the present and past chemical and physical erosion in river basins of the Himalaya using Sr and Nd isotope systematics in water and sediments.

26.1 Introduction

The Earth's continental surface is subject to continuous physical and chemical weathering and erosion by wind, water and glaciers. Physical weathering is the process of breakdown of rocks into finer fragments without affecting their composition, whereas chemical weathering is a complex process of conversion of rocks into its soluble and insoluble secondary products through water-rock interactions. The process of removal of soluble and particulate weathering

G.R. Tripathy, S.K. Singh, and S. Krishnaswami (✉)
Geosciences Division, Physical Research Laboratory, Ahmedabad 380009, India
e-mail: swami@prl.res.in

products from the site of their formation to their final repository is erosion, though weathering and erosion are often used interchangeably. Erosion therefore plays an important role in exposing "fresh" rock and mineral surfaces in a drainage basin for weathering, a key factor regulating their chemical weathering rates (Stallard and Edmond 1983; Kump et al. 2000). A major driver of chemical weathering is carbonic acid formed by the solution of CO_2 in natural waters. Therefore, studies of chemical weathering are intimately linked to global carbon cycle. In recent years there have been several studies to determine the chemical erosion rates of major global river basins and the various factors regulating it, especially the role of tectonics and climate (Negrel et al. 1993; Derry and France-Lanord 1997; Edmond and Huh 1997; Gaillardet et al. 1997, 1999; Galy and France-Lanord 1999; Huh and Edmond 1999; Krishnaswami et al. 1999; Kump et al. 2000; Millot et al. 2002; Blum and Erel 2003; Dessert et al. 2003; Bickle et al. 2005; Das et al. 2005; Singh et al. 2005; West et al. 2005; Wu et al. 2005; Moon et al. 2007; Peucker-Ehrenbrink et al. 2010; Tripathy and Singh 2010). One of the key topics of interest in this area is silicate weathering of river basins, a major sink for atmospheric CO_2 on million year time scales. This makes the determination of silicate erosion rates (SER) of river basins and its temporal variations an important component of studies on atmospheric CO_2 balance through time and its impact on long term global change, the tectonics-weathering-climate link (Walker et al. 1981; Berner et al. 1983; Raymo et al. 1988; Raymo and Ruddiman 1992; Ruddiman 1997; Kump et al. 2000). These studies though rely primarily on the chemical composition of dissolved phase of rivers, radiogenic isotopes, particularly $^{87}Sr/^{86}Sr$ and $^{143}Nd/^{144}Nd$, have provided important additional insights into mineral and rock weathering processes that have implications to isotope geochemistry of rivers and oceans and contemporary and paleo-silicate erosion of continents. The unique radiogenic isotope composition of minerals and rocks and the near conservative behavior of the isotope ratios in solution (unlike major elements which are prone to removal from solution by precipitation, ion-exchange and biological cycling) make these isotope techniques very useful to investigate kinetics of rock-water interactions in natural systems where the reactions often proceed at very slow rates (Blum and Erel 2003; Gaillardet 2008).

In contrast to studies on chemical erosion which rely mainly on the chemical and isotopic composition of dissolved phases, investigations on physical erosion (e.g. provenance studies) depend on chemical and isotopic signatures of sediments. Construction of the physical erosion pattern of sedimentary basins and its controlling factors requires tracking the source of sediments. Sr and Nd isotopes have been extensively used as provenance indicators of river and ocean sediments (Dia et al. 1992; Allegre et al. 1996; Pierson-Wickmann et al. 2001; Singh and France-Lanord 2002; Clift 2006; Colin et al. 2006; Singh et al. 2008; Viers et al. 2008) ever since the pioneering study of Dasch (1969) on tracing the source of deep-sea surface sediments of the Atlantic using Sr isotopes. These isotope systems have become increasingly popular in such studies because their signatures among the various sources (end members) are often markedly different and easily distinguishable. Further, the application of these isotopes as provenance tracers is promoted by the property of the sediments to preserve in them the isotope composition of sources. There have been a number of Sr and Nd isotopic studies in river and ocean sediments from different basins to tag their provenances (e.g. Derry and France-Lanord 1997; Colin et al. 1999; Tutken et al. 2002; Clift et al. 2008; Singh et al. 2008) and their temporal variations to infer about factors controlling physical erosion in the past, particularly its link to regional changes in climate and/or tectonics (France-Lanord et al. 1993; Walter et al. 2000; Li et al. 2003; Clift et al. 2008; Rahaman et al. 2009; Galy et al. 2010). During the past decade, in addition to Sr-Nd, the applications of other isotope systems such as U series disequilibrium (Chabaux et al. 2008, Vigier and Bourdon 2011), Lu-Hf (Ma et al. 2010) and some of the non-traditional isotopes, e.g. Si, Mg and Ca (Tipper et al. 2008; Reynolds 2011) are also being explored to investigate chemical and physical erosion.

This chapter reviews selected studies on the applications of Sr and Nd isotopes as tracers to investigate chemical and physical erosion with emphasis on river basins of the Himalaya. The topics addressed include chemical weathering of minerals and rocks, silicate erosion in river basins and sediment provenance studies and the relation among erosion, tectonics and climate.

26.2 Chemical Weathering in River Basins

26.2.1 Overview of Sr, Nd in Rocks and Their Isotope Systematics

Strontium is an alkaline earth element with chemical properties similar to that of Ca. It occurs as a minor constituent in rock-forming minerals often replacing Ca or K. Sr has four naturally occurring isotopes, of these ^{87}Sr is radiogenic, produced from the radioactive decay of ^{87}Rb ($t_{1/2} = 4.88 \times 10^{10}$ years). Rb is an alkali element with properties similar to K, therefore it is more abundant in K-rich minerals. As a result, minerals with high K/Ca (~Rb/Sr) ratio develop more radiogenic Sr with time. Thus in a rock suite with minerals of the same age, those with high Rb and low Sr (e.g. biotite, muscovite) will have higher ^{87}Sr/^{86}Sr (refers to ^{87}Sr/^{86}Sr ratio) compared to minerals with low Rb and high Sr (e.g. plagioclase, apatite). During chemical weathering, Sr is generally more mobile than Rb; therefore, the Rb/Sr ratio of weathered residues is higher than that in parent rocks. The results of Dasch (1969) and subsequent studies on Rb/Sr abundances in weathering profiles developed on basalts and granites confirm this behavior with the weathered residues having higher Rb/Sr than the parent rocks. Preferential adsorption of Rb on clays can also contribute to the higher Rb/Sr in the weathered residues.

The rare earth elements (REEs) occur as trace elements in major rock forming minerals and in higher concentrations in the accessory minerals, e.g. zircon, apatite, monazite and allanite. The concentrations of REEs vary widely in accessory minerals; they are often a minor constituent in zircon and apatite, but are a major component of REE bearing minerals such as monazite and allanite. The REEs are classified broadly into two groups, the light rare earth elements (LREE; La-Sm) and the heavy rare earth elements (HREE; Eu-Lu). Major rock forming minerals are relatively enriched in one of these two groups, for example feldspars and apatite are more abundant in LREE whereas pyroxenes and garnet prefer HREE. The concentrations of Sm and Nd in rock-forming minerals increase in the sequence in which they crystallize from magma, however due to the close similarity in the chemical properties between Sm and Nd their fractionation during geological processes is limited. As a result, the Sm/Nd abundance ratio in terrestrial rocks and minerals is within a narrow range (0.1–0.5), in contrast to Rb/Sr which varies widely, from 0.005 to 3 (Faure 1986).

Neodymium is a light rare-earth element having seven naturally occurring isotopes. Among these, ^{142}Nd and ^{143}Nd are radiogenic produced by the α-decay of ^{146}Sm ($t_{1/2} = 1.03 \times 10^8$ years) and ^{147}Sm ($t_{1/2} = 1.06 \times 10^{11}$ years), respectively. ^{146}Sm, because of its short half-life is now an extinct radionuclide; its signature however has been detected in meteorites through measurements of ^{142}Nd (Amelin and Rotenberg 2004; Anderasen and Sharma 2006). The ^{143}Nd/^{144}Nd ratio of minerals and rocks increases with time due to production of ^{143}Nd from ^{147}Sm, their ^{143}Nd/^{144}Nd, therefore depends on their Sm/Nd ratios and ages. The isotopic composition of Nd (^{143}Nd/^{144}Nd) is expressed in epsilon units (ε, DePaolo and Wasserburg 1976), which represents the relative deviation of ^{143}Nd/^{144}Nd in the sample from ^{143}Nd/^{144}Nd of the chondritic uniform reservoir (CHUR) in units of 10^4. ε is expressed as,

$$\varepsilon_{Nd}(0) = \left(\frac{\left(^{143}Nd/^{144}Nd\right)_{sample}(0)}{\left(^{143}Nd/^{144}Nd\right)_{CHUR}(0)} - 1 \right) \times 10^4$$

where (^{143}Nd/^{144}Nd)$_{CHUR}$(0) and (^{143}Nd/^{144}Nd)$_{sample}$(0) are the present day ^{143}Nd/^{144}Nd ratio of CHUR (= 0.512638; Jacobsen and Wasserburg 1980) and the sample, respectively. The normalization of Nd isotopic ratios of rocks with chondritic value provides information on the relative Sm/Nd ratio of their sources. For example, a negative epsilon value indicates that the rock is derived from sources lower in Sm/Nd than CHUR (Faure 1986). Further, as the Nd isotopic variations in natural systems are quite small, the ε notation provides a convenient approach to express and appreciate these small variations.

26.2.2 Behavior of Sr and Nd Isotopes During Chemical Weathering of rocks and Mineral Weathering Kinetics

26.2.2.1 $^{87}Sr/^{86}Sr$ Release During Mineral and Rock Weathering: Laboratory and Field Studies

The major source of dissolved Sr to rivers is chemical weathering of silicates and carbonates in their drainage basins, with minor contribution from other lithologies such as evaporites. The silicates of bed rocks are an assemblage of minerals, which have different chemical and mineralogical composition and weathering properties. Each of these minerals has their own distinct $^{87}Sr/^{86}Sr$ corresponding to their age and Rb and Sr concentrations. During chemical weathering of silicate rocks, Sr isotopes from various minerals are released depending on their weathering rates and accessibility to weathering. As a result, Sr from easily weatherable minerals is expected to be released to solution initially from freshly exposed rock surfaces. The exposure of fresh rock surfaces is facilitated by physical erosion, tectonics and glaciations.

Studies on the release of Sr and $^{87}Sr/^{86}Sr$ during weathering of continental rocks and their subsequent behavior are being carried out (1) to determine their release pattern from minerals in the parent rock and its implications to mineral weathering rates and $^{87}Sr/^{86}Sr$ of rivers, (2) to assess the relative roles of silicate and carbonate sources to their budget in river/ground water systems and to explore the possibility of using $^{87}Sr/^{86}Sr$ in rivers as a proxy for silicate weathering and (3) to interpret Sr isotope records in marine archives. The $^{87}Sr/^{86}Sr$ of the oceans, as recorded in carbonate shells of marine sediments, is known to be steadily increasing since the Cenozoic (Burke et al. 1982; Veizer 1989; Richter et al. 1992). Delineating the sources responsible for this increase in terms of silicate/carbonate weathering can prove useful in deciphering the weathering history of continents and its role in atmospheric CO_2 drawdown.

Erel et al. (2004) conducted laboratory studies of granitoid weathering to determine trends in Sr isotope release with time and its implications to mineral weathering. These authors based on combined studies of major elements and Sr isotopes in minerals of a granitoid rock from Elat, Israel and in the solution from its weathering observed that during initial stages the granitoid dissolution is dominated by contributions from easily weatherable trace phases, calcites/apatites, followed by biotites in the intermediate stages and plagioclase during the later stages of weathering (Fig. 26.1). By comparing these laboratory experimental data with Sr isotope distribution in soil chronosequences from the Wind River and Sierra Nevada mountains, Erel et al. (2004) were able to deduce the temporal pattern of Sr release during granitoid weathering under natural conditions and from it the mineral dissolution sequence for the granite. The results showed that the easily weatherable trace phases in the granitoid, calcites and apatites are the major source of elements to solution from rock weathering during the initial few hundred years, biotites determining the supply from rock surfaces exposed to weathering a few hundreds to ~10,000 years and plagioclase accounting for much of the cations from surfaces that have been weathered for more than ~100,000 years.

The distribution of Sr isotopes in different components of soil chronosequences provided independent evidences for the non-stochiometric release of Sr during rock weathering. Measurements of Sr isotopes

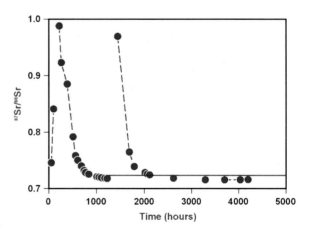

Fig. 26.1 The release pattern of $^{87}Sr/^{86}Sr$ during dissolution of a granitoid rock with time. The low $^{87}Sr/^{86}Sr$ during the early stage of dissolution is a result of preferential weathering of calcites; this is followed by a sharp peak in Sr isotopic ratio due to weathering of biotites. The straight line parallel to the x-axis is the $^{87}Sr/^{86}Sr$ of the whole rock. The weathered residue of the granitoid rock after ~1,200 h of experiment was dried and subjected to further dissolution; this resulted in the second $^{87}Sr/^{86}Sr$ peak. Figure replotted from Erel et al. (2004)

(Blum and Erel 1995, 1997) in the exchangeable fractions and bulk soils from a granitic soil chronosequence developed on the Wind River Mountains (0.4–300 ky) showed very significant and systematic decrease in $^{87}Sr/^{86}Sr$ of the exchangeable pool with soil age (Fig. 26.2). The primary source of exchangeable Sr in the soil profile is that released during chemical weathering of rocks. This Sr, retained in the exchangeable sites of soils, can be released to solution by ion exchange with suitable reagents. Therefore the decrease in $^{87}Sr/^{86}Sr$ (from 0.7947 to 0.7114; Fig. 26.2) of the exchangeable pool with age suggests that $^{87}Sr/^{86}Sr$ released during the early stages of weathering was far more radiogenic than that released during the later stages. This was interpreted in terms of preferential weathering of biotites from freshly exposed rock surfaces (<20 ky), underscoring its importance as a source of highly radiogenic Sr to waters draining young rocks rich in biotite. These granites/gneisses were devoid of even trace amounts of calcite (Blum and Erel 1995) precluding it as a source for Sr in the early stages of weathering (c.f. Erel et al. 2004). The data also suggested that biotites weather much faster relative to plagioclase in young soils (~8 times) and that the impact of its (biotite) weathering becomes less as the soil gets older, primarily due to depletion in the abundances of unweathered biotites (Blum and Erel 1995). More importantly, the non-stochiometric release of $^{87}Sr/^{86}Sr$ due to differences in mineral weathering rates led to the suggestion that exposure of new weathering surfaces (e.g. by glaciation and/or mountain uplift) can make the $^{87}Sr/^{86}Sr$ of rivers significantly more radiogenic for periods of ~20 ky following their exposure and thus elevate the riverine $^{87}Sr/^{86}Sr$ input to the oceans during such periods. This result also provides a mechanism to link global glaciations with Sr isotope evolution of the oceans (Armstrong 1971; Blum and Erel 1995, 1997; Zachos et al. 1999). More studies on Sr isotopes in granitic soil chronosequence from other locations however showed that the composition of the exchangeable Sr depended on the history of the profile. For example, in a profile from central Sierra Nevada (Bullen et al. 1997), developed on granitic alluvium the exchangeable $^{87}Sr/^{86}Sr$ was dominated by supply from plagioclase and K-feldspars and not biotite as was observed by Blum and Erel (1997). This difference was attributed to the absence of fresh biotites in the soil profile of Sierra Nevada, further attesting to the importance of biotite weathering from freshly exposed rock surfaces in contributing to highly radiogenic Sr to solutions.

Analogous to field studies of rock weathering, mineral weathering experiments also brought out the role of non-stochiometric release of Sr isotope ratios during their early stages of dissolution. Brantley et al. (1998) in their dissolution experiments of feldspars observed that in the initial stages the release of Sr was not stoichiometric and its isotope composition was different from that of the host mineral. This difference was explained in terms of contribution of Sr either from fluorite and zeolites, trace secondary phases in the feldspars and/or to preferential leaching of Sr from more easily weatherable sites of the primary mineral. The isotopic composition of the solution attained steady-state and became close to that of the host mineral with progressive dissolution. These results have led to the suggestion that non-stochiometric release of Sr and $^{87}Sr/^{86}Sr$ can also occur in natural systems, but only for short time periods (<10^3 years) following the exposure of fresh mineral/ rock surfaces for weathering. Taylor et al. (2000) also have reported similar non-stochiometric release of Sr and $^{87}Sr/^{86}Sr$ during early stages of dissolution of biotites and phlogopites (Fig. 26.3). These results were interpreted in terms of preferential dissolution of trace

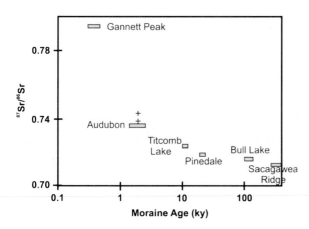

Fig. 26.2 Variations in the Sr isotopic ratio in the exchangeable fractions of a granitoid soil chronosequence developed on glacial moraines in the Wind River Range. The negative correlation between exchangeable $^{87}Sr/^{86}Sr$ and moraine age indicates release of more radiogenic $^{87}Sr/^{86}Sr$ from biotites during the early stage of weathering. The $^{87}Sr/^{86}Sr$ values for two stream water samples (marked as +) draining the Audubon till are also shown in the figure. Figure redrawn from Blum and Erel (1995)

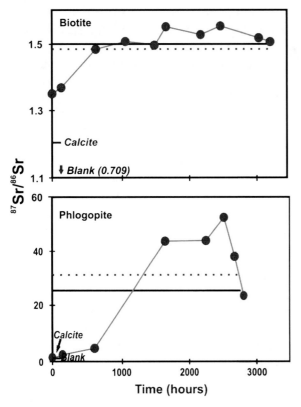

Fig. 26.3 The release pattern of $^{87}Sr/^{86}Sr$ in laboratory leaching experiments of biotite and phlogopite. The low Sr isotopic ratio during the initial stages of the experiment is a result of Sr release from trace calcite contained in the minerals. The *solid* (*black*) and *dotted* (*red*) *lines* in the figure represent the pre-experiment and post-experiment values of the minerals respectively. Figure redrawn from Taylor et al. (2000)

calcite in the primary minerals and materials from their interlayer sites, the later being more radiogenic as they concentrate Rb. Based on the results of these experiments the authors calculated the dissolution rate of ^{87}Sr to be marginally higher than that of ^{86}Sr in biotites and that steady state with respect to release of Sr isotopes would be reached in a few thousand years.

The importance of trace calcic phases (e.g. calcite, apatite, and bytownite) in dominating the supply of Ca, Sr and Sr isotopes during granite weathering has also been demonstrated through field studies (Blum et al. 1998; Jacobson et al. 2002; Oliva et al. 2004). Results of Sr isotope measurements in rivers draining small catchments of the Himalaya suggest that vein calcites in granites of the basin can be an important source of Ca and Sr with high $^{87}Sr/^{86}Sr$ (Blum et al. 1998). The researches of Oliva et al. (2004) on chemical weathering of high elevation Estibere granitic watershed showed that trace calcic phases, epidote, bytownite, prehnite and apatite, contributed significantly to the Ca, Sr and Sr isotope fluxes. However, unlike in the crystallines of the Himalaya, the majority of trace calcic phases in the Estibere watershed is silicates and therefore contributes to silicate weathering and atmospheric CO_2 drawdown. Similarly, the studies of Aubert et al. (2001) in the Strengbach watershed (France) showed that the Sr and Nd isotope composition of streams and spring waters of the region can be interpreted in terms of two end member mixing, apatite and plagioclase. The authors also inferred that biotites and K-feldspars only have a "weak influence" on the Sr isotope budget of these waters (c.f. Blum and Erel 1995; Bullen et al. 1997). All these field and laboratory studies bring out the important role of trace calcic inclusions such as calcites, apatites and silicates (epidote, bytownite, prehnite) often contained in silicate rocks in supplying Ca and Sr and $^{87}Sr/^{86}Sr$ to streams during rock weathering. These results have vital implications on the use of Sr isotopes to track silicate weathering (see Sect. 26.2.4).

The impact of mineral weathering kinetics is also evident in the seasonal variations of $^{87}Sr/^{86}Sr$ of rivers. Tipper et al. (2006a) observed that during the monsoon season the $^{87}Sr/^{86}Sr$ of the headwater tributaries of the Marsyandi, a sub-tributary of the Ganga was less radiogenic compared to that during non-monsoon periods (Fig. 26.4) similar to that noted earlier by Krishnaswami et al. (1999) and Bickle et al. (2003) for the headwaters of the Ganga. The Sr isotopic composition of rivers in the Himalaya is determined by the mixing proportion of contributions from two major sources, marginally radiogenic sedimentary carbonates ($^{87}Sr/^{86}Sr$~0.715; Singh et al. 1998) and radiogenic silicates (with associated trace phases; $^{87}Sr/^{86}Sr$ ~0.75–0.80, Krishnaswami et al. 1999). The difference in weathering kinetics of silicates and carbonates and its climatic dependence is an important cause contributing to the seasonal variations in $^{87}Sr/^{86}Sr$ of rivers. The weathering rates of carbonates are significantly higher than silicates, this enhances their relative contribution to rivers during high runoff. Considering that Sr in sedimentary carbonates is generally far less radiogenic relative to Sr in silicates, higher proportion of Sr from such carbonates to rivers during monsoon would make the riverine Sr lower in $^{87}Sr/^{86}Sr$. In contrast during dry periods when more time becomes

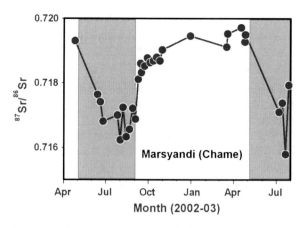

Fig. 26.4 Seasonal variations in $^{87}Sr/^{86}Sr$ of the Marsyandi River, Nepal, a sub-tributary of the Ganga river. The data show significantly lower $^{87}Sr/^{86}Sr$ ratio during monsoon (*shaded zone*) due to relatively more contribution of unradiogenic Sr from carbonates. Figure replotted from Tipper et al. (2006a)

available for water–rock interactions the contribution from silicates is relatively more than that during monsoon. Similar seasonal trends in riverine silicate cations and $^{87}Sr/^{86}Sr$ have been documented in a few other rivers (Moon et al. 2007; Rai and Singh 2007; Tripathy et al. 2010).

Thus the laboratory experiments and field data provide evidences to suggest that during weathering, particularly in early stages, Sr isotope composition of the solution can be significantly different from that of the parent material primarily due to differences in mineral weathering kinetics. Such non-stochiometric release of Sr isotopes can result in increasing $^{87}Sr/^{86}Sr$ of rivers, if minerals with highly radiogenic Sr isotope composition (e.g. biotite) are freshly exposed for weathering. These studies also establishes a link between dissolved fluxes of rivers and glaciation/mountain uplift as the exposure of fresh rock surfaces for weathering is promoted by tectonics, glaciation and intense physical erosion. Two other consequences of non-stochiometric release of Sr isotopes pertain to their applications in weathering and provenances studies. The finding that trace calcites and apatites can be important sources of Ca, Sr and radiogenic $^{87}Sr/^{86}Sr$ to rivers can constrain the use of $^{87}Sr/^{86}Sr$ as a proxy of silicate weathering if such contributions are major on a basin wide scale. Similarly the non-stochiometric release of $^{87}Sr/^{86}Sr$ to solution, if significant on a basin scale, it can result in the formation of sediments with $^{87}Sr/^{86}Sr$ different from that of parent material, a result that can challenge the use of $^{87}Sr/^{86}Sr$ as a provenance tracer.

26.2.2.2 REE (Sm/Nd) and $^{143}Nd/^{144}Nd$ Release During Mineral and Rock Weathering: Laboratory and Field Studies

Compared to Rb-Sr and $^{87}Sr/^{86}Sr$, there are only a few investigations on the behavior of Sm-Nd and $^{143}Nd/^{144}Nd$ during progressive weathering of rocks. Generally, most of the REEs in rocks, ~70–90%, is present in accessory minerals (e.g. allanite, monazite, sphene, zircon and apatite) and the reminder distributed among the primary rock forming minerals (Barun et al. 1993; Harlavan et al. 2009). Therefore studies on the release of REEs during rock weathering serve as a probe to investigate their fractionation and to infer the weathering characteristics of the accessory minerals. Early investigations (Nesbitt 1979) based on the abundances and distribution of LREEs and HREEs in soil profiles developed on granodiorite demonstrated that during chemical weathering they are leached from the upper layers of soil depleting their concentrations relative to parent rocks. During their downward transport they get sequestered in the weathering products in the deeper layers enriching their abundances. The mobilization and fractionation behavior of the two groups of REEs during chemical weathering however, was found to be different and that it was determined by (1) the chemistry of soil water, pH and dissolved organic matter concentration. This is consistent with the subsequent studies that these two properties of water play an important role in regulating the concentration and mobilization of dissolved Nd in rivers (Sect. 26.2.3.2), (2) the abundance and weathering pattern of different accessory minerals in rocks hosting the REEs and (3) the formation of secondary phases in the soil that sequester the REEs.

Subsequent studies based on different components of soil profiles and laboratory leaching experiments on granitoid weathering have by and large attested to the above findings (Nesbitt and Markovics 1997; Aubert et al. 2001; Harlavan and Erel 2002; Ma et al. 2007; Harlavan et al. 2009); solubilization of REEs from accessory minerals in the upper layers of the soil and their uptake in secondary phases in the deeper layers causing their redistribution within the soil column.

The solubilization and transport of REEs was linked to the availability of organic matter.

The accessory minerals involved in the REE release were identified and their weathering sequence determined through inter-element/isotope association. For example, in the soil profile from Vosges Mountains, France (Aubert et al. 2001), the correlation of REEs with P and Th suggested apatite and monazite to be the main phases determining the mobilization and budget of REEs. Similarly, Harlavan and Erel (2002) in their laboratory experiments on the release of Pb isotopes, REEs and major elements during granitoid weathering observed that dissolution of allanite dominated REEs supply to solution initially, followed by weathering of allanite, apatite and sphene in the later stages and contributions from dissolution of feldspar becoming prominent during the final stages. These results also provided weathering sequences for the accessory minerals as allanite > apatite > sphene. These findings were independently confirmed by Harlavan et al. (2009) based on their studies on REE abundances and Pb isotopes in the labile pool of soil chronosequences. These latter studies, as they were based on soil chronosequences, also provided time constraints on the processes governing REEs mobilization and distribution in soils. Investigations on the mobilization and redistribution of REEs during intense weathering of basalts (Ma et al. 2007) also reveal similar pattern with extensive removal from the upper layers of soil and uptake in the deeper layers. The mobilization results in fractionation of LREEs from HREEs, the former group being relatively less mobile. As in earlier studies, the availability of organic matter was found to be a key factor in the mobilization of REEs and the formation of secondary phases such as phosphates and Fe-Mn oxy/hydroxides for their sequestration in deeper layers.

Information on the behavior of Sm/Nd during weathering is derived mainly from studies of their distribution in soil profiles (Fig. 26.5). These results however do not show any consistent pattern on their relative mobilities. For example, Nesbitt and Markovics (1997) did not observe significant fractionation between Sm and Nd as a function of weathering intensity in the Toorango soil profile, Australia (Fig. 26.5a) relative to the parent granitoid rock though both elements were mobilized by chemical weathering. These results led to

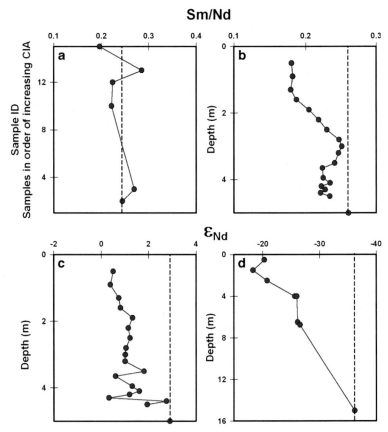

Fig. 26.5 The behavior of Sm/Nd and ε_{Nd} during chemical weathering as recorded in soil profiles. The results show different patterns. (**a**) (Toorongo granodiorite, Australia; Nesbitt and Markovics 1997) indicates Sm/Nd remain nearly constant during weathering; whereas (**b**, **c**) (Neogene basalts from South China; Ma et al. 2010) and (**d**) (granodiorite from southern Cameroon; Viers and Wasserburg 2004) depict effects of preferential mobilization of Sm. The *bottom* and *top* sides of the figures show the least and most weathered rocks; the *dashed straight line* in the figure represents the bed rock value

the conclusion that in this profile chemical weathering has not affected the Sm-Nd chronology. In contrast, a number of studies (e.g. MacFarlane et al. 1994; Ohlander et al. 2000; Ma et al. 2010) report significant fractionation in Sm/Nd during chemical weathering. Ohlander et al. (2000) observed that during weathering of till, Nd is preferentially released over Sm from minerals with lower Sm/Nd (allanite, monazite) than the bulk soil leading to enhancement in the ε_{Nd} value, by up to four units (~ -22 to -18), in the weathered till. MacFarlane et al. (1994) also observed preferential release of LREEs and fractionation of Nd isotopes and $^{147}Sm/^{144}Nd$ in two weathering profiles preserved between flows of the Mt. Roe basalts in Western Australia. Investigations of Sm/Nd and ε_{Nd} in soil profiles formed by intensive weathering of basalts (Ma et al. 2010) also showed effects of Sm/Nd fractionation with corresponding changes in ε_{Nd} values (Fig. 26.5b, c). Viers and Wasserburg (2004) investigated the behavior of Sm/Nd and Nd isotope composition during weathering of granite by analyzing a lateritic soil profile in a tropical watershed from Cameroon along with surface waters collected from the region. The results showed major changes in ε_{Nd}, from ~ -36 in the parent rock to -18 near the top of the soil profile (Fig. 26.5d), without any major shifts in Sm/Nd. These results were attributed to preferential dissolution of minerals such as feldspars and apatite with least radiogenic Nd.

The above studies thus suggest that an important source of REEs to solution during chemical weathering of rocks is dissolution of accessory minerals. The differences in mineral weathering rates can cause the Nd isotope composition of the solution and residual solids to be different from that of the parent rock. However, the changes observed in Nd isotopic ratios due to chemical weathering in the basins are often less compared to the variations observed among global sediments (Ma et al. 2010).

26.2.3 Dissolved Sr, $^{87}Sr/^{86}Sr$, Nd and ε_{Nd} in Rivers

26.2.3.1 Sr and $^{87}Sr/^{86}Sr$

The concentration of dissolved Sr and its isotopic composition have been measured in a large number of global rivers to learn about its behavior during weathering and transport, its flux to the oceans and its marine budget (Goldstein and Jacobsen 1987; Palmer and Edmond 1989; Trivedi et al. 1995; Gaillardet et al. 1999; Vance et al. 2009; Peucker-Ehrenbrink et al. 2010). The dissolved Sr concentration of major global rivers varies widely, by more than an order of magnitude, from ~0.21 to ~7.46 µM (Table 26.1). Among the major rivers (Table 26.1) Sr concentration is the highest for the Yellow and the lowest for the Orinoco, with the Yangtze delivering the maximum amount of Sr to the ocean, ~7% of the global riverine Sr flux compared to its contribution of ~2.5% to water discharge. The global discharge-weighted average concentration of Sr based on data in Table 26.1 (which make up ~44% of global water discharge and ~33% of drainage area) is ~0.94 µM; this corresponds to an annual flux of 36.5×10^9 mol/y consistent with some of the earlier estimates (Palmer and Edmond 1989; Vance et al. 2009; Table 26.1). This is expected considering that much of the data in Table 26.1 are from earlier compilations. More recently, Peucker-Ehrenbrink et al. (2010) have calculated following a different approach (by averaging Sr concentration on the basis of large scale drainage regions and extrapolating them) the contemporary supply of dissolved Sr to the ocean to be 47×10^9 mol/y, ~30% higher than the present and some of the earlier estimates (Table 26.1).

The major uncertainty in the flux estimates arises from the use of the available Sr data (Table 26.1) which are often based on a single or a few measurements in a river during a year, as the annual mean Sr concentration. Such an approach can be subject to significant errors because of short-term/seasonal variations in Sr abundance of rivers, in general it seems to show a decreasing trend with increasing water discharge. For example, Sr in the Brahmaputra measured at biweekly intervals over a period of ~10 months shows that it varies by a factor of ~2 (Rai and Singh 2007), similar to the range reported by Galy et al. (1999) and for some of the other major global rivers such as the Orinoco, Yukon, Mississippi and the Ganga (Palmer and Edmond 1989; Xu and Marcantonio 2007; Rai et al. 2010; Table 26.2). It is apparent from these data that the determination of Sr flux based on a single measurement in a river can differ from the annual average value by as much as ±50%. Other potential sources of uncertainty in the flux estimates are (1) inter-annual variation in water discharge. The

Table 26.1 Dissolved Sr concentration and $^{87}Sr/^{86}Sr$ ratios of global rivers

River	Water discharge (km³/y)	Drainage area (10⁶ km²)	Sr (nM)	Sr flux (10⁹ mol/y)	$^{87}Sr/^{86}Sr$	Reference
Amazon	6,590	6.112	310	2.04	0.71165	Peucker-Ehrenbrink et al. (2010)
Zaire (Congo)	1,200	3.698	313	0.38	0.71550	Palmer and Edmond (1989)
Orinoco	1,135	1.1	210	0.24	0.71830	Palmer and Edmond (1989)
Mississippi	580	2.98	2,130	1.24	0.70957	Xu and Marcantonio (2007)
Parana	568	2.783	520	0.30	0.71390	Gaillardet et al. (1999)
Lena	525	2.49	1,100	0.58	0.71048	Peucker-Ehrenbrink et al. (2010)
Tocantins	372	0.757	380	0.14	0.72067	Gaillardet et al. (1999)
Amur	344	1.855	500	0.17	0.70923	Moon et al. (2009)
St. Lawrence	337	1.02	1,200	0.40	0.70962	Yang et al. (1996)
Mackenzie	308	1.787	2,740	0.84	0.71138	Millot et al. (2003)
Columbia	236	0.669	982	0.23	0.71210	Goldstein and Jacobsen (1987)
Danube	207	0.817	2,760	0.57	0.70890	Gaillardet et al. (1999)
Yukon	200	0.849	1,590	0.32	0.7137	Gaillardet et al. (1999)
Niger	154	1.2	250	0.04	0.71400	Palmer and Edmond (1989)
Fraser	112	0.22	913	0.10	0.71200	Palmer and Edmond (1989)
Rhine	69	0.224	6,227	0.43	0.70920	Palmer and Edmond (1989)
HT rivers						
Yangtze (Chang Jiang)	928	1.808	2,830	2.63	0.71032	Noh et al. (2009)
Brahmaputra	510	0.58	730	0.37	0.71920	Krishnaswami et al. (1992)
Ganga	493	1.05	560	0.28	0.72910	Galy et al. (1999)
Irrawaddy[a]	486	0.41	-	-	0.71010	Tipper et al. (2006b)
Mekong	467	0.795	3,080	1.44	0.71035	Noh et al. (2009)
Pearl	363	0.437	767	0.28	0.71190	Palmer and Edmond (1989)
Indus	238	0.47	3,689	0.88	0.71110	Pande et al. (1994)
Salween	211	0.325	1,330	0.28	0.71405	Noh et al. (2009)
Hong (Red)	123	0.12	1,158	0.14	0.71284	Moon et al. (2007)
Huang He (Yellow)	41	0.752	7,458	0.31	0.71110	Palmer and Edmond (1989)
Peninsular Indian rivers						
Godavari	105	0.313	1,375	0.14	0.7152	Trivedi et al. (1995)
Mahanadi	66	0.132	704	0.05	0.71930	Trivedi et al. (1995)
Narmada	39	0.102	2,156	0.08	0.71140	Trivedi et al. (1995)
Krishna	30	0.259	3,742	0.11	0.71420	Trivedi et al. (1995)
Kaveri (Cauvery)	21	0.088	2,690	0.06	0.71498	Pattanaik et al. (2007)
Discharge-weighted Sr concentration and $^{87}Sr/^{86}Sr$						
HT Rivers	3,860	6.75	1,956	7.55	0.71179	
Non-HT rivers	1,3199	29.5	641	–	0.71163	
Global river	3,8857	110	939	36.5	0.71171	
Earlier estimates of global river Sr and $^{87}Sr/^{86}Sr$						
Goldstein and Jacobsen (1987)			705	–	0.7101	
Palmer and Edmond (1989)			890	33.3	0.7119	
Vance et al. (2009)			-	33.7	0.7114	
Peucker-Ehrenbrink et al. (2010)			1,220	47	0.7111	

Average discharge-weighted Sr and $^{87}Sr/^{86}Sr$ values for global river are estimated based on the data in the table. The discharge-weighted average Sr for rivers listed in the table is taken to be the same for the mean global river, this value is multiplied by global water discharge to estimate global riverine Sr flux.

[a]For flux calculations, the Sr concentration of Irrawaddy is assumed to be the same as the mean of HT rivers.

Hydrological parameters are from Pande et al. (1994); Gaillardet et al. (1999); Peucker-Ehrenbrink (2009). For the HT and Non-HT rivers, average Sr and $^{87}Sr/^{86}Sr$ are estimated based on the data compiled in this study from the citations in the table.

Table 26.2 Temporal variations in Sr and $^{87}Sr/^{86}Sr$ in rivers

River	Location	Sr, nM	$^{87}Sr/^{86}Sr$	Reference
Ganga	Patna	1,017–2,270	0.7224–0.7249	Krishnaswami et al. (1992)
	Rajshahi	560–2,710	0.7235–0.7291	Galy et al. (1999)
	Rajmahal	1,122–2,157	–	Rai et al. (2010)
Brahmaputra	Goalpara	670–1,082	0.7187–0.7197	Krishnaswami et al. (1992)
	Chilmari	320–900	0.7218–0.7413	Galy et al. (1999)
	Guwahati	604–1,392	0.7159–0.7180	Rai and Singh (2007)
Mekong	Da Hai	3,080–5,770	0.7098–0.7104	Noh et al. (2009)
Orinoco	–	132–312	0.7175–0.7199	Palmer and Edmond (1989)
Yukon	–	1,116–2,390	0.7136–0.7139	Palmer and Edmond (1989)
Red	Phu Tho	1158–1413	0.7121–0.7128	Moon et al. (2007)

Ganga, for example shows a spread of ±20% in water discharge during decadal time scales (1950–1960; 1965–1973; http://www.grdc.sr.unh.edu)). This variation however is much higher, a factor of ~2 on a year to year basis and (2) validity of calculating global flux by extrapolating Sr data from available measurements (Table 26.1) which represents only 40–50% of global river discharge and about a third of drainage area. Such a calculation requires that the measured data is globally representative in terms of different factors that include discharge (climate), areal coverage and lithological distribution of the river basins. In this context, the available data (Table 26.1) may not have adequate representation of rivers from volcanic islands and island arcs.

The concentration of dissolved Sr in rivers is determined largely by the lithology of the drainage basin and the intensity of chemical weathering. In general, silicate and carbonates are the major lithologies of river basins with minor occurrence of evaporites and sulfides. The weathering kinetics of carbonates is much faster than that of silicates (Drever 1997); this can make carbonates an important source of Sr to rivers even if their areal exposure in river basins is relatively less and their Sr abundances comparable to that in silicates. The role of carbonates and silicates in determining the budget of Sr in rivers is borne out from the linear co-variation of dissolved Sr/Na with Ca/Na in several large and medium size global rivers (Gaillardet et al. 1999), attributable to mixing between a high Ca/Na, Sr/Na end member (carbonates) and a low Ca/Na, Sr/Na end member (silicates). Model calculations to constrain the silicate and carbonate derived Sr in rivers have met with challenges due to large variations in elemental ratios of end members and precipitation of calcite from water (e.g. Galy et al. 1999; Krishnaswami et al. 1999; Bickle et al. 2005). Some of these challenges, particularly those pertaining to end member values have been addressed through the use of inverse model (Negrel et al. 1993; Gaillardet et al. 1999; Moon et al. 2007; Tripathy and Singh 2010). The application of such a model (Tripathy and Singh 2010) to the headwaters of the Ganga system rivers in the Himalaya show that ~70% of dissolved Sr in them is derived from carbonates and ~20% from silicates, despite relatively low aerial exposure of carbonates in the catchment (Amiotte Suchet et al. 2003; Tripathy and Singh 2010). The uncertainties in the estimates of Sr supply from silicates and carbonates and insufficient knowledge on the release ratios of Sr to (Na, K, Mg and Ca) from the major lithologies of the basin restricts the use of dissolved Sr concentration in rivers as an index to derive silicate and carbonate erosion rates of their basins.

The $^{87}Sr/^{86}Sr$ of the major rivers ranges from 0.7089 to 0.7291 (Table 26.1) with the highest value for the Ganga and the lowest for the Danube. The $^{87}Sr/^{86}Sr$ of rivers is determined by the mixing proportion of Sr contributed predominantly by silicates and carbonates in the catchment. The Sr isotopic composition of sedimentary carbonates is fairly well established and is generally unradiogenic, in the range of 0.705–0.709 (Veizer 1989; Allegre et al. 2010). Metamorphic alteration of carbonates however, can make their $^{87}Sr/^{86}Sr$ far more radiogenic, as has been observed in some of the carbonates from the Himalaya (Singh et al. 1998; Bickle et al. 2001). The limited range in $^{87}Sr/^{86}Sr$ of carbonates is in contrast to the Sr isotope composition of silicates which show a very wide range, from unradiogenic volcanic rocks to highly radiogenic granites and gneisses such as those from the Himalaya (Singh et al. 1998, 2008; Dalai et al. 2003). The $^{87}Sr/^{86}Sr$ of silicates in the drainage basin is often reflected in the isotopic composition of dissolved Sr in rivers draining them. The recent compilation of $^{87}Sr/^{86}Sr$ of dissolved Sr from large scale drainage regions (Peucker-Ehrenbrink et al. 2010) shows an overall increasing trend with bed rock ages of the region.

The Sr flux weighted $^{87}Sr/^{86}Sr$ for the global river is 0.7117 (based on data in Table 26.1), close to the value of 0.7119 (Palmer and Edmond 1989) and is marginally more radiogenic than the values of 0.7114 and 0.7111 reported by Vance et al. (2009) and Peucker-Ehrenbrink et al. (2010). Analogous to Sr concentration, $^{87}Sr/^{86}Sr$ of rivers also show significant seasonal variations (Table 26.2). This can result from variations in the relative contribution of Sr from silicates and carbonates and/or from different tributaries. Whatever may be the cause for such variations, they highlight the need for long term monitoring of rivers along with their discharge to obtain representative global riverine Sr concentration and $^{87}Sr/^{86}Sr$.

The Sr elemental and isotopic data for the major rivers listed in Table 26.1 show distinctly high $^{87}Sr/^{86}Sr$ with moderate Sr concentration in the Ganga-Brahmaputra rivers compared to the others. Identifying the source for the high $^{87}Sr/^{86}Sr$ in these rivers has been a topic of investigation and debate over the last few decades (Sect. "Can $^{87}Sr/^{86}Sr$ be a Proxy for Silicate Erosion?"). The discharge weighted Sr concentration for the nine Himalayan-Tibetan rivers listed in Table 26.1 is 1.96 μM with a flux weighted $^{87}Sr/^{86}Sr$ of 0.7118. These rivers account for ~21% of global riverine Sr flux, disproportionately higher compared to their contribution to water discharge (~10%).

26.2.3.2 Nd and ε_{Nd}

Studies on the concentration of dissolved Nd in rivers (0.20–0.45 μm filtered) and its isotopic geochemistry are limited compared to those of Sr. Initial studies of Nd concentration in river waters showed that it varies widely, from 0.02 to 21.8 nM, with a discharge-weighted average value of 0.28 nM for the global river (Goldstein and Jacobsen 1987). The concentration of Nd in these rivers was found to be inversely correlated with their pH and (Na + Ca) abundances; relationships that underscore the importance of water chemistry in determining the Nd concentration of rivers. Further, the measurements of Sm in these rivers indicated that its geochemical behavior is similar to that of Nd, though there was a minor fractionation between them. The Sm/Nd in solution was marginally higher (~10%) compared to that in suspended phases, suggestive of preferential release of Sm to solution. There was also a hint that the fractionation increased with pH, likely due to greater stability of Sm-carbonate complexes. The importance of pH in determining the abundances and fractionation of REEs in the dissolved load of rivers was also brought out in the studies of Gaillardet et al. (1997) in rivers of the Amazon basin and Tricca et al. (1999) for mature rivers of the Rhine valley. Some of the subsequent studies have provided an alternative explanation for the pH–REE relationship. Ingri et al. (2000) in their investigations of Kalix River observed that La concentrations are strongly correlated with DOC (dissolved organic carbon) and that high La and DOC abundances are associated with lower pH. These results led to the suggestion that DOC is the primary factor governing REE abundances in filtered river waters and that the REE-pH relation is a result of pH–DOC correlation. Subsequent studies have both supported (e.g. Johannesson et al. 2004; Shiller 2010) and challenged (e.g. Steinmann and Stille 2008) the REE–DOC relationship.

The role of colloids in contributing to REE abundances in filtered river water and its impact on REE behavior in rivers and estuaries became a topic of investigation following advances in analytical techniques for the study of colloids. Measurement of REEs in filtered river waters and their colloidal fractions (Ingri et al. 2000; Andersson et al. 2001) suggest that REEs are largely associated with colloids, with less than 5% in dissolved phase (<3 kD fraction). Both organic rich and inorganic (Fe oxyhydroxides) colloids have been suggested as potential "carriers" of REEs. Barroux et al. (2006) and Steinmann and Stille (2008) have invoked the important role of REE-colloid interactions in determining the concentrations and fractionation of REEs in rivers. The observation that the Nd concentration in the Amazon increases linearly with water discharge led Barroux et al. (2006) to suggest that enhanced mobilization of particles and colloids during rain events can be a cause for the discharge–REE relationship. More recently, Steinmann and Stille (2008) have explained the steady increase in LREE depletion along the course of small rivers in Massif Central (France) in terms of precipitation of colloidal Fe-oxyhydroxides that preferentially scavenge LREEs.

The behavior of dissolved Nd in estuaries determines the significance of rivers in contributing to the budget of Nd and its isotopes in the ocean. Investigations on the abundance and distribution of dissolved REEs in estuaries show their widespread removal in the low salinity regions due to coagulation and settling of colloids, the major carrier phase of dissolved Nd (Sholkovitz 1995; Sholkovitz and Szymczak 2000; Frank 2002; Jeandel et al. 2007; Porcelli et al. 2009). The budget of dissolved Nd and its isotopic composition in the oceans, therefore has to be supported by input from other sources such as desorption from river particulates and atmospheric dust and boundary exchange at the continental margins, among these the boundary exchange has been suggested as the dominant mode of supply (Tachikawa et al. 2003; Jeandel et al. 2007).

The ε_{Nd} values for rivers also show a wide range (−44 to +7) with a global Nd flux weighted riverine average of −8.4 (Goldstein and Jacobsen 1987). The ε_{Nd} of river water is often found to be similar to that of the bedrocks in their basin, and therefore correlates with the age of the rocks (Goldstein and Jacobsen 1987; Peucker-Ehrenbrink et al. 2010). There are however, minor differences in the ε_{Nd} values of dissolved and suspended loads of rivers, most likely a result of preferential weathering of phases with different Sm/Nd and ε_{Nd}. The ε_{Nd} of rivers shows an inverse relation with $^{87}Sr/^{86}Sr$ resulting from Nd and Sr isotope composition of rocks being weathered and mixing of various end members with different ε_{Nd} and $^{87}Sr/^{86}Sr$.

26.2.4 Silicate Erosion in River Basins

26.2.4.1 Contemporary Silicate Erosion in the Ganga-Brahmaputra System

Silicate weathering on land is a fundamental process that determines the supply of materials derived from crustal silicates to the oceans. This process is driven primarily by CO_2, making silicate weathering an important regulator of CO_2 budget over million year time scales and a key component of global carbon cycle. Therefore, changes in silicate weathering rates on long time scales can affect the atmospheric CO_2 budget and hence global climate. Recognizing the importance of silicate weathering on the exogenic cycles of elements, global carbon cycle and climate, there have been a number of investigations to determine silicate erosion rates (SER) and the factors regulating it. These investigations have been based both on small streams which by and large drain mono-lithologic terrains and large river systems which integrate contributions from their tributaries draining different lithologies. The studies on small streams are an approach to derive mineral weathering rates in natural settings, whereas larger river basins provide estimates of regional silicate erosion. The studies on large river basins, particularly those draining young orogenic belts such as the Himalaya have also been motivated by the hypothesis (Raymo et al. 1988; Raymo and Ruddiman 1992; Ruddiman 1997) that enhanced silicate weathering in this mountain belt can be the driver of global cooling during the Cenozoic, as the conducive monsoon climate, high relief and intense physical erosion of the region all can significantly enhance silicate weathering and associated atmospheric CO_2 drawdown. The steady increase in the Sr isotope composition of the oceans during the Cenozoic (Fig. 26.6) has been suggested as a major support for the tectonics-weathering-climate hypothesis (Raymo and Ruddiman 1992; Richter et al. 1992. Based on the present day Sr concentration and its isotopic composition of rivers draining the

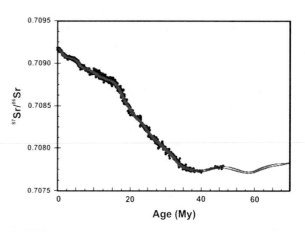

Fig. 26.6 Evolution of seawater $^{87}Sr/^{86}Sr$ since the last 70 My. Intense chemical weathering in the Himalaya is suggested as a potential cause for the steady increase in $^{87}Sr/^{86}Sr$ since 40 My (Raymo and Ruddiman 1992). Figure modified from Ravizza and Zachos (2003)

Himalayan-Tibet region, it is argued that the observed steady increase in the marine $^{87}Sr/^{86}Sr$ can result from silicate weathering in this region. This hypothesis invokes the use of Sr isotope composition of seawater as a proxy of silicate weathering on the continents. This suggestion has been a topic of debate as diverse and at times controversial sources have been proposed for Sr and its isotopes in rivers of the Himalaya (Edmond 1992; Krishnaswami et al. 1992; Palmer and Edmond 1992; Blum et al. 1998; Jacobson and Blum 2000; Bickle et al. 2001). If, however, the use of Sr isotopes as a proxy of silicate weathering is validated, it would also serve as a tool to investigate its past variations which are difficult to determine using other approaches.

The determination of contemporary SER relies mainly on the chemical composition of rivers. The chemistry of rivers is dominated by contributions from various lithologies of river basins and the kinetics of rock–water interaction. From the measured concentrations, the contributions from silicates, carbonates and evaporites are deduced using either the forward or the inverse model (Negrel et al. 1993; Singh et al. 1998; Gaillardet et al. 1999; Galy and France-Lanord 1999; Krishnaswami et al. 1999; Bickle et al. 2005; Singh et al. 2005; Wu et al. 2005; Hren et al. 2007; Tripathy and Singh 2010). In the forward model, Na* (Na corrected for Cl; Na* = Na_{riv} − Cl_{riv}, where "riv" refers to measured concentration in rivers) is used as an index of silicate weathering. This coupled with knowledge of release ratios of (K + Mg + Ca) to Na* from silicates in the basin to rivers is used to derive SER. The release ratios would depend on the silicate lithology of the basin and the nature of weathering (mineral specific/stochiometric) and therefore can have a wide range. In contrast, in the inverse model likely elemental ratios for various end members contributing to major cation abundances to rivers is assigned a priori and following the method of iteration and material balance considerations the best end member ratios that can generate the measured river water composition is derived. The removal of Ca from rivers by calcite precipitation can be a source of uncertainty in this approach. Both the forward and inverse methods have been widely used to derive contemporary SER and associated CO_2 consumption rates for various major rivers including the Ganga-Brahmaputra. The contemporary SER (calculated from the sum of cations derived from silicates and SiO_2) for the headwaters of the Ganga, Bhagirathi and Alaknanda range between 10 and 15 tons $km^{-2} y^{-1}$ (Krishnaswami and Singh 2005), a factor of ~2–3 higher than the global average value of ~5.5 tons $km^{-2} y^{-1}$ (Gaillardet et al. 1999). The higher SER in the headwater basins of the Ganga suggests more intense silicate erosion in these Himalayan basins (Galy and France-Lanord 1999; Krishnaswami et al. 1999). Independent estimates of SER have been obtained by France-Lanord and Derry (1997) and McCauley and DePaolo (1997) based on the deficiency in Na, K, Ca, and Mg in the Bay of Bengal sediments relative to that in crystallines of the Himalaya; these estimates generally are time-averaged values, unlike that based on river water data which are based on snap shot sampling.

Can $^{87}Sr/^{86}Sr$ be a Proxy for Silicate Erosion?

Generally rivers draining old silicate rocks have lower Sr with higher $^{87}Sr/^{86}Sr$ compared to rivers weathering marine limestones/evaporites. The large difference in the $^{87}Sr/^{86}Sr$ among the different lithologies in the drainage basin promotes the use of Sr isotopes as a natural tracer to determine Sr contribution to rivers from various end members.

Figures 26.7 (a, b) are examples of mixing plots for Sr isotopes in two river systems, the Orinoco and the Yamuna headwaters. The plots show an overall linear trend albeit some scatter, suggestive of two component mixing (Palmer and Edmond 1992; Dalai et al. 2003). From the mixing array, the Sr contribution to the rivers from the two end members can be determined if their Sr isotope composition are known and from it the SER of the basin if (Na + K + Ca + Mg)/Sr release ratios are known. Estimates of these release ratios are available based on different approaches and models (e.g. Krishnaswami et al. 1999), however as mentioned earlier they are not well constrained. More importantly, such calculations generally assume that the high $^{87}Sr/^{86}Sr$ end member is silicates; ascertaining the validity of this assumption is critical to the application of $^{87}Sr/^{86}Sr$ as proxy for silicate weathering. Krishnaswami et al. (1992) and Dalai et al. (2003) based on the widespread occurrence of crystallines with highly radiogenic Sr isotope composition (~0.75–1.0) in the drainage basins of the headwaters of the Yamuna and the Ganga and the strong correlation between "silicate indices" and $^{87}Sr/^{86}Sr$ of these

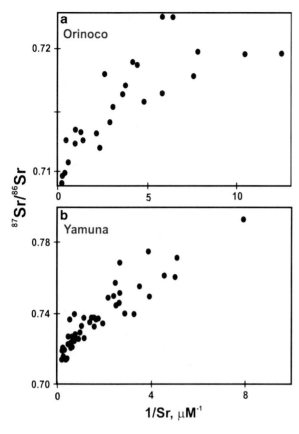

Fig. 26.7 Sr mixing plots for two major rivers, the Orinoco and the Yamuna headwaters. The near linear trend is suggestive of a two component mixing, the high $^{87}Sr/^{86}Sr$ and low Sr silicates with the low $^{87}Sr/^{86}Sr$ and high Sr carbonates. Figure redrawn from Palmer and Edmond (1989) and Dalai et al. (2003)

rivers (Fig. 26.8) suggested that the high $^{87}Sr/^{86}Sr$ end member is indeed silicates of the drainage basin. However, the occurrence of metamorphic carbonates and vein calcites in these river basins with highly radiogenic $^{87}Sr/^{86}Sr$ similar to that of the crystallines and considering that carbonates are more easily weatherable compared to silicates has challenged the assignment of high $^{87}Sr/^{86}Sr$ end member to silicates. These studies instead suggest that the metamorphosed carbonates and vein calcites to be the dominant sources of high $^{87}Sr/^{86}Sr$ to the rivers in the Himalaya (Palmer and Edmond 1992; Blum et al. 1998; Quade et al. 1997; Harris et al. 1998; Jacobson and Blum 2000; Oliver et al. 2003). Attempts to test the different components of this hypothesis have resulted in contrasting conclusions. Singh et al. (1998) measured Sr isotopic composition of Pre-Cambrian carbonates from the Lesser Himalaya to assess their role in contributing to the high $^{87}Sr/^{86}Sr$ of the Ganga-Yamuna source waters. Their results, though brought out the important role of metamorphism in considerably elevating the $^{87}Sr/^{86}Sr$ of these sedimentary carbonates locally, suggested that their weathering is unlikely to be a major source for the high $^{87}Sr/^{86}Sr$ to the headwaters of the Ganga on a basin wide scale. The results of Bickle et al. (2005) on the Sr isotopic composition of the Ganga headwater basins also show that it is dominated by contribution from weathering of silicates (~50% of total Sr) with the balance from trace calcites (~10%) and sedimentary carbonates (~35%). In contrast, the

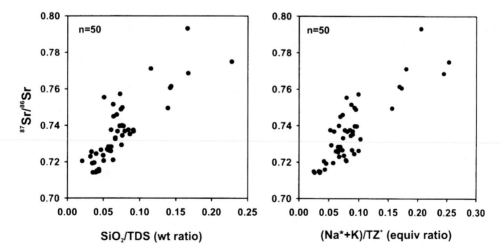

Fig. 26.8 Scatter diagrams of $^{87}Sr/^{86}Sr$ with other "silicate indices" (SiO_2/TDS and $(Na^* + K)/TZ^+$) in the Yamuna headwaters. The good correlation between them has been interpreted to suggest that the high $^{87}Sr/^{86}Sr$ end member in these rivers is of silicate origin and that $^{87}Sr/^{86}Sr$ can serve as a proxy of silicate weathering. Figure from Dalai et al. (2003)

work of Oliver et al. (2003) brought out the major role of metasedimentary carbonates as the source of elevated $^{87}Sr/^{86}Sr$ to the sub-tributaries of the Ganga (Bhote Kosi-Sun Kosi, Nepal). Their work based on downstream variations in the chemistry and $^{87}Sr/^{86}Sr$ of rivers and mass balance considerations led to infer that the Paleo-Proterozoic impure carbonates with highly radiogenic Sr isotope composition can be significant contributor to the high $^{87}Sr/^{86}Sr$ of the Ganga river, accounting for a major fraction of their supply to the marine $^{87}Sr/^{86}Sr$. Thus, the available results on Sr isotope composition of metamorphosed carbonates from the Himalaya and their interpretation have led to diverging views on their role as a major source of high $^{87}Sr/^{86}Sr$ to the Ganga river on a basin wide scale.

In addition to metamorphosed carbonates, the significance of disseminated carbonates and vein calcites as sources of highly radiogenic Sr to the Ganga river also have been investigated. For example, Sr isotope studies (Jacobson et al. 2002) on glacial moraine chronosequences from the Himalaya containing trace carbonates demonstrated that their weathering can be a significant source of Ca, Sr and radiogenic Sr to rivers for thousands of years subsequent to their exposure. In addition, as mentioned earlier, there are also results which suggest that vein calcites in granites of the Himalaya can be an important source of Ca and Sr with high $^{87}Sr/^{86}Sr$ (Blum et al. 1998).

It is thus evident from the available results and interpretation that both silicates and carbonates can contribute to the high $^{87}Sr/^{86}Sr$ of the Ganga-Brahmaputra rivers. However, their relative significance to the Sr isotope budget remains to be established quantitatively. This information is essential to decide on the applicability of Sr isotopes as a proxy of silicate erosion, one of the basis of the tectonics-weathering-climate hypothesis.

26.2.4.2 Seawater $^{87}Sr/^{86}Sr$ Record During the Cenozoic: Impact of Weathering in the Himalaya and Deccan Traps

Chemical weathering of continental crust and the supply of its products to the ocean is a key process regulating its chemical and isotopic evolution through time. The chemical and isotopic composition of the authigenic and biogenic components of marine sediments hold clues to temporal variations in continental inputs to the sea. Among these the Sr isotope records hold promise to retrieve continental weathering history.

The evolution of seawater Sr isotopic composition can be expressed using a balance equation involving the various Sr supply and removal terms. Early attempt in this direction was made by Brass (1976) to explain the variations in the oceanic $^{87}Sr/^{86}Sr$ over the past ~400 My based on changes in $^{87}Sr/^{86}Sr$ input from weathering of silicates and carbonates on continents. These mass balance calculations showed that the variations in seawater $^{87}Sr/^{86}Sr$ since the last 200 My can be attributed to changes in the $^{87}Sr/^{86}Sr$ supplied to the ocean from weathering of silicates on land; the ratio being dependent on the proportion of radiogenic Sr from old acidic rocks and unradiogenic Sr from young basic rocks. The mixing proportion of Sr from these two silicate lithologies can vary depending on their exposure to weathering which in turn is governed by processes such as tectonics/glaciation.

The Sr isotope evolution of the ocean was investigated in great detail during 1970–1990 which resulted in a robust and dense record of $^{87}Sr/^{86}Sr$ in marine carbonates since the Cenozoic (Fig. 26.6; Veizer 1989). Richter et al. (1992) based on these data developed a model to test the hypothesis (Raymo et al. 1988; Raymo and Ruddiman 1992) that silicate weathering in the Himalaya-Tibet (HT) is the primary factor contributing to the steady increase in the Sr isotope composition of the oceans during this period. In this model, the marine budget of Sr and its isotopes are determined by their supply from rivers, hydrothermal sources and diagenetic alteration/dissolution of carbonates from sediments and their removal to ocean floor *via* calcareous skeletons. Among the three input sources, Sr flux from diagenetic alteration/dissolution of carbonates is much less than the fluxes from the other two sources (Richter et al. 1992; Banner 2004) and therefore as an approximation this term is often neglected from the budget calculations. The mass balance equation for the rate of change of marine Sr isotope ratio is:

$$N\frac{dR_{SW}}{dt} = J_r(R_r - R_{SW}) + J_h(R_h - R_{SW}) \quad (26.1)$$

where N is total Sr (in moles) in the ocean. J_r and J_h are the flux (moles/y) of Sr from the riverine and hydrothermal sources. R_{sw}, R_r and R_h are the $^{87}Sr/^{86}Sr$ of seawater, river and hydrothermal sources respectively. Equation (26.1) suggests that the $^{87}Sr/^{86}Sr$ of seawater is governed by the balance between supply of radiogenic Sr from rivers and unradiogenic mantle like Sr derived from hydrothermal sources. In the above equation the temporal changes in J_r, R_r and J_h are not known, though J_h is often assumed to be proportional to ocean floor production (Richter et al. 1992). The variations in either J_r or R_r or both can bring about changes in R_{sw} (Edmond 1992; Krishnaswami et al. 1992; Richter et al. 1992; Derry and France-Lanord 1997; McCauley and DePaolo 1997; Kump et al. 2000; Gaillardet 2008). Increase in J_r (riverine flux) would imply more intense continental weathering (Raymo et al. 1988) whereas increasing R_r would suggest change in the source material being weathered (Edmond 1992; Derry and France-Lanord 1997; Kump et al. 2000). Therefore to determine variations in the intensity of continental weathering, information on changes in R_r is required. These data are generally unavailable and therefore the equation is solved using reasonable assumptions. Richter et al. (1992) by solving equation (26.1) with the then available data on contemporary fluxes and ratios (Table 26.3) showed that the increase in the $^{87}Sr/^{86}Sr$ of seawater since the past ~40 My is due to enhanced supply of Sr and $^{87}Sr/^{86}Sr$ by rivers and that increase in both J_r and R_r are required to balance the evolution of R_{sw}. In their calculations they also evaluated the impact of temporal variations in J_h on the evolution of seawater $^{87}Sr/^{86}Sr$ and concluded that though changes in J_h could explain about half the change in R_{sw}, the evolutionary trend was not consistent with the observations.

To further evaluate the role of Himalayan-Tibetan (HT) rivers on $^{87}Sr/^{86}Sr$ of oceans, the river water flux and Sr isotopic ratio terms in the above equation is split into two parts, one representing global rivers excluding those from HT and the other representing the extra flux (J_{r1}) with $^{87}Sr/^{86}Sr$, R_{r1} required to explain the observed variation in seawater Sr evolution (Richter et al. 1992). The modified equation is:

$$N\frac{dR_{SW}}{dt} = J_{ro}(R_{ro} - R_{SW}) + J_{r1}(R_{r1} - R_{SW}) + J_h(R_h - R_{SW}) \quad (26.2)$$

where, J_{ro} and R_{ro} are the Sr flux and $^{87}Sr/^{86}Sr$ from the global rivers excluding those from the HT. The calculations showed that the Sr flux (J_{r1}) required to reproduce the sea water $^{87}Sr/^{86}Sr$ curve using present day value of R_{r1} is roughly consistent with that measured indicating that the HT rivers can be a dominant source to account for the marine Sr isotope evolution during the past ~40 My. These calculations and inferences derived from them rely on the compilation of riverine Sr flux and $^{87}Sr/^{86}Sr$ (c.f. Table 26.1), which as mentioned earlier can be subject to uncertainties arising from intra and inter-annual variations in Sr concentration and $^{87}Sr/^{86}Sr$ of rivers, water discharge and inadequate sampling of rivers from various major lithologies. These results and therefore the conclusions based on them maybe subject to revision as more robust data on Sr concentration of rivers and its isotopic composition become available. For example, subsequent to the work of Richter et al. (1992) there have been more measurements of Sr and $^{87}Sr/^{86}Sr$ in the HT rivers, particularly in the Ganga and the Brahmaputra during various seasons including peak discharge. These results show that the dissolved Sr concentration of the Ganga is much lower during peak discharge (0.56 μM; Galy et al. 1999) than the value used by Richter et al. (1992) in their model (1.58 μM), the corresponding $^{87}Sr/^{86}Sr$ values also being significantly different (0.7257–0.7291). The Brahmaputra at its mouth also exhibits a similar difference in its Sr concentration with a value of 0.32 μM (Galy et al. 1999) during peak flow compared to 0.93 μM used in the model (Richter et al. 1992). Considering that much of the annual water discharge of the Ganga (~75%) is during its peak flow, the Sr concentration measured during this period is likely to be more representative of its annual average

Table 26.3 Numerical (present day) values of various parameters used in seawater Sr budget model (Richter et al. 1992)

Parameter		
J_r	Riverine flux	3.3×10^{10} mol/y
J_h	Hydrothermal alteration flux	0.82×10^{10} mol/y
J_d	Digenetic flux	~0
J_{ro}	Riverine flux excluding HT rivers	2.2×10^{10} mol/y
J_{r1}	HT riverine flux	7.7×10^9 mol/y
N	Total Sr in ocean	1.25×10^{17} mol
R_r	Riverine $^{87}Sr/^{86}Sr$	0.711
R_h	Hydrothermal $^{87}Sr/^{86}Sr$	0.7030
R_d	Digenetic $^{87}Sr/^{86}Sr$	0.7084
R_{ro}	$^{87}Sr/^{86}Sr$ global rivers excluding HT rivers	0.710
R_{r1}	$^{87}Sr/^{86}Sr$ of HT rivers	0.7127

concentration. This value for the Sr concentration of the Ganga yields a Sr flux weighted ^{87}Sr/^{86}Sr ratio of 0.7118 for the HT rivers, very close to ^{87}Sr/^{86}Sr of global rivers 0.7117 and lower than 0.7127, used in the model calculations (Richter et al. 1992). The use of the revised ^{87}Sr/^{86}Sr of the HT rivers in the model calculation suggest that the Sr flux (J_{r1}) required from the HT rivers would be significantly higher than the estimates of Richter et al. (1992) and their contemporary measured flux (Table 26.1). These observations underscore the importance of regular monitoring of rivers to obtain representative values for the annual average Sr concentration and ^{87}Sr/^{86}Sr.

The inferences of Richter et al. (1992) can be used to deduce the weathering history of continents if data on temporal variations of either J_r or R_r are available. Derry and France-Lanord (1996) obtained ^{87}Sr/^{86}Sr data for the Ganga during the past ~20 My based on Sr isotope record of the Ganga floodplain as preserved in the pedogenic clays of the Bay of Bengal sediments. The results showed significant temporal changes in the Sr isotopic ratio of the Ganga with highly radiogenic ratios during 1–7 My (~0.74) compared to ~0.72 during 0–1 and 7–20 My periods. The higher ^{87}Sr/^{86}Sr during 1–7 My requires a decrease in the riverine Sr flux during this period to balance the Sr isotope data of the oceans. This led to the suggestion that increase in ^{87}Sr/^{86}Sr of oceans need not be driven by a concomitant increase in riverine Sr flux and therefore continental weathering. Quade et al. (1997) also observed that the Sr isotopic composition of Himalayan rivers (Ganga, Indus) had varied with time based on ^{87}Sr/^{86}Sr of shells and paleosol carbonates from the Siwaliks, further attesting to the idea that the Sr isotope evolution of the oceans can be a result of variations in source composition.

Another recent development on Sr budget in the oceans is the finding that its hydrothermal flux is much less than that estimated earlier (~1.0×10^{10} mol/y; Palmer and Edmond 1989) and that the current estimate can support only about a third or less of Sr required to balance its budget in the oceans (Davis et al. 2003). This imbalance in the marine Sr budget therefore, requires that there has to be additional major source (s) of unradiogenic Sr to the oceans. Investigations on chemical weathering rates, Sr concentration and ^{87}Sr/^{86}Sr of rivers, groundwater and hydrothermal sources in volcanic islands and island arcs (Rad et al. 2007; Allegre et al. 2010) suggest that supply from these regions can be the "missing source" of unradiogenic Sr to the oceans. Budget model for present day oceanic ^{87}Sr/^{86}Sr shows that hydrothermal sources contribute only ~27% of mantle-like unradiogenic Sr to the sea compared to ~73% from island arcs and volcanic islands (Allegre et al. 2010). Temporal variation in the Sr concentration from these sources to the sea can also impact the Sr isotope evolution of the ocean; however there is no data to evaluate this.

Analogous to the studies on the role of HT rivers on the ^{87}Sr/^{86}Sr of the oceans, the impact of weathering Deccan trap basalts was also assessed using data from rivers flowing through them and suitable models (Dessert et al. 2001; Das et al. 2006). The results of Dessert et al. (2001) showed an increase in seawater ^{87}Sr/^{86}Sr from 0.70782 to 0.70789 over a period of a few My after the Deccan eruption followed by a steady decrease. The initial increase was explained in terms of enhanced continental weathering by CO_2 (and hence temperature) released during the emplacement of the Deccan basalts. The subsequent decrease in the seawater ^{87}Sr/^{86}Sr is attributed to a combined effect of contribution of unradiogenic Sr from the weathering of Deccan basalts and decrease in atmospheric CO_2 which decreases continental weathering. Das et al. (2005, 2006) carried out major ion and Sr isotope studies on several rivers exclusively draining the Deccan traps to investigate the role of their weathering on the evolution of marine ^{87}Sr/^{86}Sr at around K-T boundary and early Tertiary. Their results and model calculations showed that the supply of unradiogenic Sr from the Deccan basalt weathering caused the decline in seawater ^{87}Sr/^{86}Sr during the early Tertiary. As basalts are a source of unradiogenic Sr to the oceans, changes in their Sr contribution relative to that from old sialic rocks can modify the riverine ^{87}Sr/^{86}Sr which in turn can affect the Sr isotope composition of the oceans (Brass 1976; Taylor and Lasaga 1999). Such a relative change in weathering of silicate lithology can also bring about changes in oceanic ^{87}Sr/^{86}Sr without invoking changes in the intensity of chemical weathering on continents (Taylor and Lasaga 1999).

Thus, many of the above studies bring out the potential of Sr isotopes in oceans and rivers to provide information on contemporary and past global chemical weathering patterns and the role of weathering in the Himalaya and Deccan traps in contributing to the Sr isotope evolution of the oceans since the Cenozoic. The potential of Sr isotopes as a proxy of silicate weathering in the Himalaya however awaits quantitative information on the role of radiogenic carbonates as a source riverine ^{87}Sr/^{86}Sr and temporal variations in riverine ^{87}Sr/^{86}Sr.

26.3 Physical Erosion in River Basins

Physical erosion is a key process that determines the geomorphology of the earth surface and promotes chemical weathering intensity of river basins. The records of physical erosion contained in sediments hold clues to the spatial and temporal erosion pattern of the source region and its dependence on climatic and tectonic factors (e.g., Dia et al. 1992; Asahara et al. 1999; Burbank et al. 2003; Singh et al. 2008; Rahaman et al. 2009; Galy et al. 2010). Over the years there have been a number of studies both in river and marine basins to decipher these patterns by tracking the sediment provenances through their geochemical, mineralogical and isotopic composition. The successful application of these proxies require their source signatures to be well preserved in the sediments, a requirement that has met with challenges because of chemical weathering of source rocks and size sorting of sediments. Many of the initial studies on sediment provenances were based on clay mineralogy. The clay minerals provide information more on their weathering provenances than their geological provenances (Dasch 1969) as they are the end products of chemical weathering of minerals present in source rocks.

Historically, the use of Sr isotopes as a proxy for sediment provenance was recognized and demonstrated during the 1960's (Dasch 1969). The detailed study of spatial distribution of ^{87}Sr/^{86}Sr in surface sediments of the Atlantic (Dasch 1969; Fig. 26.9) showed large variations which were interpreted in terms of differences in their provenances. During the last few decades, radiogenic isotope ratios particularly ^{87}Sr/^{86}Sr and ε_{Nd} of sediments have often found applications as tracers to track the provenance of sediments (Goldstein and O'Nions 1981; France-Lanord et al. 1993; Winter et al. 1997; Pierson-Wickmann et al. 2001; Banner 2004; Yang et al. 2007; Singh et al. 2008; Viers et al. 2008; Ahmad et al. 2005, 2009; Galy et al. 2010). These isotope systems serve as reliable proxies for sediment provenance studies because of their distinctly different isotopic composition in different sources and the near preservation of the source isotope composition during weathering and transport. Processes that can affect the preservation of source isotope composition in sediments are chemical weathering and sediment transport. These processes are likely to influence the preservation of ^{87}Sr/^{86}Sr more than that of ^{143}Nd/^{144}Nd because (1) generally the isotopic composition of Nd is relatively more uniform among the minerals in the parent rock compared to ^{87}Sr/^{86}Sr as the fractionation between Sm and Nd is significantly less compared to Rb and Sr during mineral formation. Therefore, preferential dissolution of specific minerals during chemical weathering though can influence the preservation of source isotope composition of both Sr and Nd, the impact is likely to be more on ^{87}Sr/^{86}Sr and (2) during transportation of sediments size sorting is common; if this results in mineralogical differences in the various size fractions, then the isotope composition of the sediment may differ from that of the source rock. Here again, the influence is likely to be more on ^{87}Sr/^{86}Sr because of its wider range among various minerals present in a parent rock compared to ε_{Nd}. Studies of soil profiles (c.f. Sect. 26.2.2) and sediments (Tutken et al. 2002) have provided evidences for alteration of isotopic composition of source rock during weathering and transport (Fig. 26.10). These data indicate that between ^{87}Sr/^{86}Sr and ε_{Nd}, the later is less susceptible for alteration during weathering and transportation making it a more robust tracer for fingerprinting the provenance of sediments (Walter et al. 2000). Such differences in the behavior of Sr and Nd isotopes during weathering and transport have led to combine studies of Sr-Nd isotope systematics in sediments to derive more detailed information on sediment provenances and related sedimentary processes (e.g. Colin et al. 1999; Walter et al. 2000; Tutken et al. 2002; Singh et al. 2008).

Sr and Nd isotope compositions of riverine and marine sediments though are being used often to track their provenances, there are lingering questions about how well the source isotope signatures are preserved in them. Therefore to make the interpretation more robust preference should be for the use of multi-tracer approaches involving isotopic and chemical/mineralogical composition of sediments. Such approaches can also help in determining the factors contributing to the changes in the isotopic,

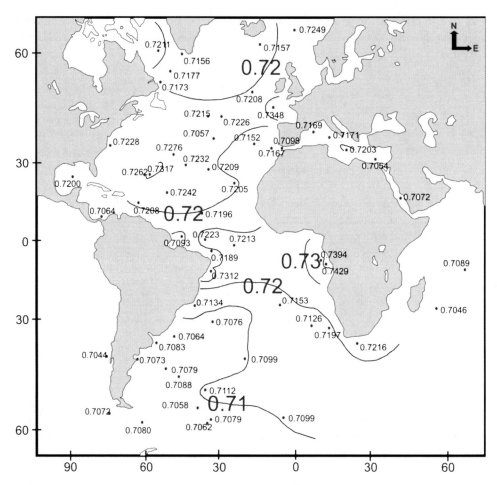

Fig. 26.9 $^{87}Sr/^{86}Sr$ of Atlantic surface sediments. The ratio shows significant spatial variations which has been used to infer their provenances. Figure redrawn from Dasch (1969)

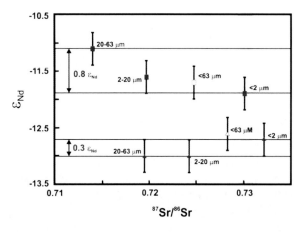

Fig. 26.10 Sr and Nd isotopic ratios in different size fractions of two sections from an Arctic sediment core. The $^{87}Sr/^{86}Sr$ shows significant variation with size whereas ε_{Nd} is relatively constant. Figure modified from Tutken et al. (2002).

chemical and mineralogical composition of sediments (France-Lanord et al. 1993; Colin et al. 1999; Singh and France-Lanord 2002; Viers et al. 2008).

26.3.1 Contemporary Physical Erosion Pattern of River Basins in the Himalaya

The rate of physical erosion of river basins of the Himalaya, particularly the Ganga and the Brahmaputra is a topic of investigation in recent years because of their importance to the sediment budget of the ocean, the global carbon and geochemical cycles of various elements and to understand the physical erosion pattern of their sub-basins and their controlling factors.

Singh and France-Lanord (2002) and Singh et al. (2008) investigated the Sr-Nd isotopic composition of bank and suspended sediments of the Brahmaputra and the Ganga plain along with those of their tributaries to determine the sources of sediments to the mainstream and their relative contributions. The Brahmaputra mainstream receives sediments from three sources, the Trans-Himalayan plutonic belt, the Lesser and the Higher Himalaya. The Sr-Nd isotopic composition of the Brahmaputra sediments is nearly uniform all along its course in the plain downstream Pasighat, in spite of contributions from the Himalayan tributaries with a wide range of Sr and Nd isotope composition (Fig. 26.11). This uniformity in ε_{Nd} and $^{87}Sr/^{86}Sr$ suggests that the sediment contributions from the Siang, draining the Eastern Syntaxis of the Himalaya and southern Tibet (Tsangpo) overwhelm that from other sources. Material balance calculations based on isotope data show that about half of the sediments of the Brahmaputra at its outflow are from the Siang resulting from intense localised erosion. Thus the sediment flux of the Brahmaputra is dominated by the Eastern Syntaxis "hot spot" where the physical erosion is controlled by runoff and relief (Singh and France-Lanord 2002; Singh 2006, 2007).

Similar studies in the Ganga mainstream sediments and its tributaries show that their $^{87}Sr/^{86}Sr$ and ε_{Nd} values are close to or by and large within the range reported for the Higher Himalayan Crystallines (HH, Fig. 26.12). The Sr and Nd isotope mass balance of these sediments on the basis of a two component (Higher and Lesser Himalayan Crystallines) mixing suggests that the HH account for more than two thirds of the contemporary Ganga Plain sediments (Fig. 26.12), consistent with the results from sediments of the Bay of Bengal (France-Lanord et al. 1993; Derry and France-Lanord 1997; Galy et al. 2010). More interestingly, the Sr isotopic composition of the Ganga sediments showed significant spatial variation along the plain with a drastic decrease from 0.768 to 0.755 immediately following the confluence of the Ganga with the Gandak (Fig. 26.13). A similar feature was also recorded in the spatial distribution of ε_{Nd}. Budget calculations based on the isotopic composition of these sediments show that the Gandak sub-basin which occupies only ~5% of the total drainage

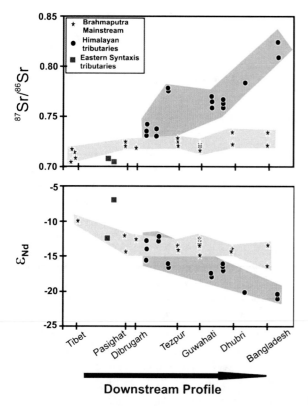

Fig. 26.11 Spatial variations in $^{87}Sr/^{86}Sr$ and ε_{Nd} of sediments of the Brahmaputra river system. The Sr-Nd isotopic signatures of the Brahmaputra mainstream sediments downstream Pasighat remains nearly uniform. Figure redrawn from Singh and France-Lanord (2002)

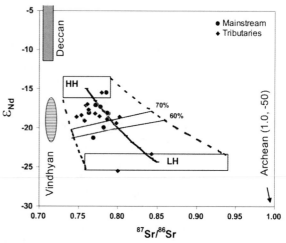

Fig. 26.12 Sr-Nd isotope mixing plot of river sediments from the Ganga plain. The isotopic composition of these sediments is interpreted in terms of a two component mixing, HH and LH. The isotopic compositions of various other end members which can contribute to the Ganga sediments are also shown. Figure modified from Singh et al. (2008)

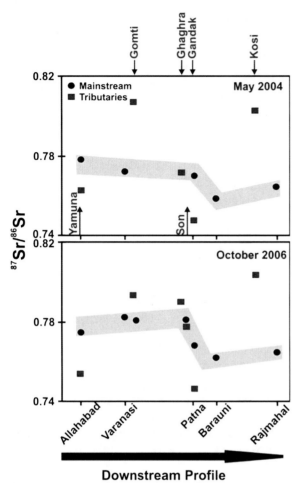

Fig. 26.13 Spatial variations (along the Ganga mainstream) in $^{87}Sr/^{86}Sr$ and ε_{Nd} of sediments from the Ganga and its tributaries. A major dip in the Sr isotopic ratio of the Ganga sediments is seen after the confluence of the Gandak, suggesting dominant supply of sediments to the Ganga mainstream from this tributary. Figure modified from Singh et al. (2008)

area of the Ganga, accounts for about half of the Ganga sediments at its outflow. These results bring out the intense physical erosion of the Gandak basin due to the high relief and focused precipitation over its headwater basins in the Himalaya (Singh et al. 2008). This observation highlights the coupling between high rainfall and relief with high physical erosion rate of river basins, linking climate (rainfall) and tectonics (relief) with contemporary erosion pattern of the Himalaya.

The Sr and Nd isotope studies of sediments from various sub-basins of the three major global rivers draining the Himalaya, the Brahmaputra, the Ganga and the Indus show that physical erosion rates are quite high in sub-basins characterized by intense precipitation over regions of high relief, the Gandak, the Siang, and the Nanga Parbat (Leland et al. 1998; Clift et al. 2002; Singh and France-Lanord 2002; Singh 2006; Singh et al. 2008) and that the sediment fluxes of these major rivers are dominated by contributions from these "hot spots" of physical erosion. Such high and focused erosion in these hot spots is unloading huge amount of sediments from the Himalaya causing regions around them to uplift more rapidly compared to other regions (Molnar and England 1990; Montgomery 1994; Zeitler et al. 2001). Importantly, these results underscore the coupling among tectonics, climate and physical erosion. Further as physical and chemical erosion are linked (Bluth and Kump 1994; Edmond and Huh 1997; Gaillardet et al. 1999; Millot et al. 2002; Singh et al. 2005) these hotspots could also facilitate intense chemical weathering by continuously exposing "fresh" rock surfaces.

26.3.2 Paleo-Erosion in Selected River Basins of the Himalaya and Its Controlling Factors

26.3.2.1 Shorter Time (ky) Scales

One of the key topics of interest in studies of physical erosion is its temporal variations and factors contributing to it. In this regard there have been a number of investigations in recent years to decipher the physical erosion history of the Himalaya and its link to tectonics and climate using river and marine sediments of Himalayan origin.

Rahaman et al. (2009) derived temporal variations in the physical erosion pattern of the Ganga basin from depth profiles of Sr and Nd isotopes in the silicate fraction of a dated sediment core from the Ganga Plain. Both $^{87}Sr/^{86}Sr$ and ε_{Nd} exhibited significant variations with depth in the core, the range in their values, however, were bound by those for the Higher and the Lesser Himalayan Crystallines, the two major sediment sources to the Ganga plain. This led to the suggestion that the observed depth variations were due to differences in the mixing proportion of sediments from the Higher and Lesser Himalaya. Furthermore, the results also showed two major excursions in the

Sr-Nd isotope composition at approximately 20 and 70 ky (Fig. 26.14), which were attributed to a decrease in the proportion of sediments from the Higher Himalaya caused by decrease in monsoon precipitation and an increase in glacial cover both of which result from lower solar insolation. Thus, the temporal variations in the relative sediment flux from the Higher and Lesser Himalaya were attributed to climate variability.

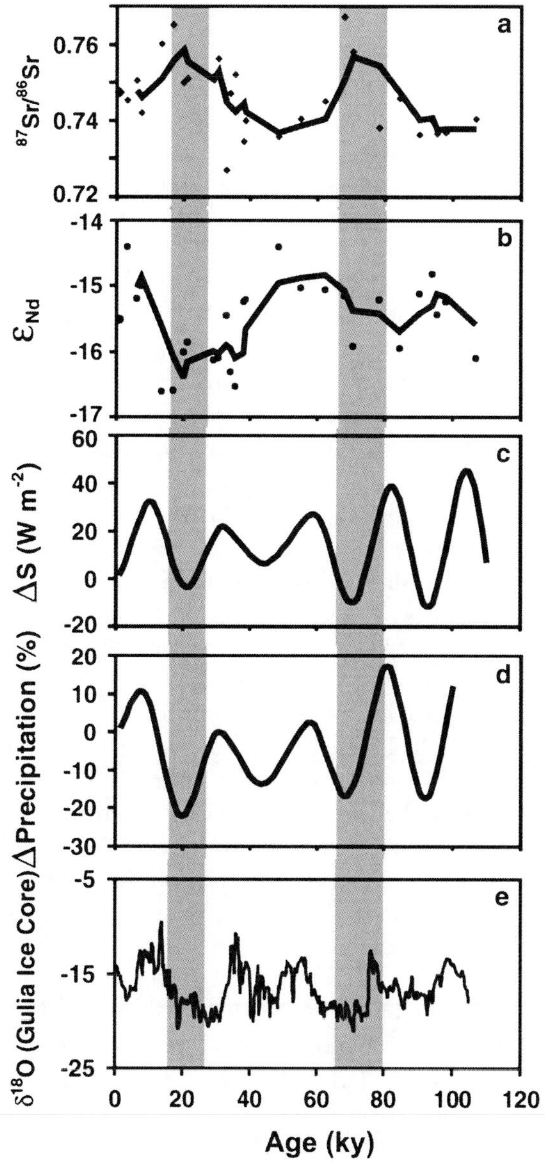

Colin et al. (1999, 2006) used Sr and Nd isotope composition along with clay mineralogy to constrain sediment sources to the Bay of Bengal and the Andaman Sea, to reconstruct the erosional history of the Indo–Burman ranges and to assess the role of climate changes (monsoon rainfall variations) on erosion. The geographical (spatial) distribution of Nd and Sr isotope composition in the Bay of Bengal and the Andaman Sea suggest that these basins receive sediments from several sources; the Ganga-Brahmaputra rivers dominating the supply to the western Bay of Bengal, the Himalayan and Indo-Burman ranges to the eastern Bay and the Irrawaddy River to the Andaman Basin. The sediments of the Andaman Sea and the Bay of Bengal deposited during the last glacial maximum (LGM) are characterized by more radiogenic $^{87}Sr/^{86}Sr$ compared to interglacial sediments with roughly the same ε_{Nd} (Fig. 26.15). The lack of concomitant changes in the Sr and Nd isotope composition led to interpret the $^{87}Sr/^{86}Sr$ results in terms of variations in chemical weathering intensity rather than to provenance changes (see Sect. 26.2.2). The more radiogenic $^{87}Sr/^{86}Sr$ during glacial periods at the core site was attributed to efficient transport of unaltered Rb-rich

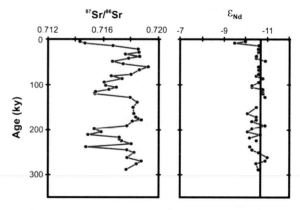

Fig. 26.14 Temporal variations in Sr and Nd isotopic ratios (*panel* **a** and **b**) of sediments from the Ganga plain during the past ~100 ky. These variations suggest relative changes in provenances (HH and LH) of the sediments and coincide with regional climatic changes (*panel* **c**: Solar insolation, *panel* **d**: Monsoon precipitation and *panel* **e**: Oxygen isotopic record of a Tibetan ice core), bringing out the role of climate on physical erosion of the Himalaya. Figure modified from Rahaman et al. (2009)

Fig. 26.15 Glacial-interglacial variations in Sr and Nd isotopic ratios of Andaman sea sediments. $^{87}Sr/^{86}Sr$ data show significant temporal variations, whereas ε_{Nd} is nearly uniform with depth. The straight line through ε_{Nd} data is the present-day ε_{Nd} value of the Irrawaddy river sediment. The lack of concomitant changes in $^{87}Sr/^{86}Sr$ and ε_{Nd} suggests that the variations in $^{87}Sr/^{86}Sr$ are a result of fractionation during weathering and transport rather than changes in provenance. Data Source: Colin et al. (2006)

minerals exposed in the headwater basin of the Irrawaddy river by glacier grinding.

The impact of climate variations on physical erosion in the Irrawaddy Basin was also brought out in the Sr-Nd isotope composition of sediments from a core raised from the Ninety East Ridge of the Bay of Bengal (Ahmad et al. 2005). The results showed concomitant temporal changes in $^{87}Sr/^{86}Sr$ and ε_{Nd} and that these changes overlapped with the records of Heinrich events. This led to suggest that these fluctuations in the isotope records are probably a result of relatively higher sediment contribution from the Irrawaddy river caused by intensification of NE monsoon in response to the cold events of the North Atlantic via an atmospheric teleconnection.

Analogous to the studies in the Bay of Bengal which receives sediments predominantly from the central and eastern Himalaya, there have also been investigations in the Indus delta and the Arabian Sea to decipher the erosional history of the western Himalaya. In one such study, Clift et al. (2008) observed significant temporal variations in the Nd isotope composition of sediments from the Indus delta with lesser radiogenic Nd in sediments younger than 14 ky. This suggested relatively increased sediment flux to these regions from sources with less radiogenic Nd, the Lesser and/or Greater (Higher) Himalaya compared to other sources (e.g. primitive arc rocks of the Indus suture and Trans-Himalaya) with LH dominating the contribution. This change in provenance of Indus sediments is attributed to change in the monsoon strength during these periods highlighting the importance of climate in regulating the erosion over western Himalaya on millennial time scales.

26.3.2.2 Longer Time (My) Scales

The Himalaya is one of the young and active mountain ranges and ranks first in the supply of sediments to the ocean (Milliman and Meade 1983; Milliman and Syvitski 1992). The temporal variation in the sediment supply from the Himalaya on million year time scales, particularly from its different sub-basins (e.g. Higher and Lesser Himalaya) and the factors regulating it have implications to global carbon cycle and therefore to climate. This has prompted studies on the pattern of physical erosion in the Himalaya and its variations over My time scales based mainly on Sr and Nd isotopic composition of sedimentary archives preserved in the Bay of Bengal and the Arabian Sea. In one such study, France-Lanord et al. (1993) investigated variations in the Sr and Nd isotopic ratios, together with stable isotopes and clay mineral abundances in a sediment core from the Bay of Bengal spanning 17 My time period. The results showed that the dominant source of these sediments is the Higher Himalayan Crystallines. The narrow range in Sr-Nd isotopic values of these sediments over their 17 My depositional history further suggested that their source remained nearly invariant despite significant changes in sedimentation rate, tectonic history and climate. In a more recent study, Galy et al. (2010) observed from the Sr and Nd isotopic composition of sediments deposited in the deep Bengal Fan that they were mainly derived from the High Himalaya Crystallines with sub-ordinate contributions from the Lesser Himalaya and Trans-Himalayan batholiths. Further, it was observed that the isotopic composition of sediments deposited during the recent 20 ky was very similar to that deposited during the past 12 My confirming the generally stable erosion regime of the Himalaya during this long time period. This stable erosion regime argues against any major impact on Himalayan erosion by glacier advance and retreat. The overall covariation between Sr and Nd isotope data further suggested that the isotope systematics were by and large, controlled by sediment provenances and not by changes in chemical weathering intensity. The near synchronicity between minor changes in erosion pattern during late Miocene and at the Plio-Pleistocene transition and climate parameters, precipitation and glacier development, brought out yet another example of climate-erosion link.

Clift and Blusztajn (2005) reconstructed the long-term erosional history of the Indus river system based on Nd isotope composition of the Arabian Sea sediments deposited during 30 My to present. The ε_{Nd} values showed a range from $\sim(-11)$ to $\sim(-9.5)$ during the period 30–5 My ago. These values suggest a provenance dominated by Karakoram (ε_{Nd}: -10 to -11), in contrast to a Himalayan source which is characterized by less radiogenic ε_{Nd} (~-25 to ~-15). The ε_{Nd} values during 5 My to Late Pleistocene (~300 ky ago) showed a trend towards more negative values, from $\sim(-10$ to $\sim-13)$ that require enhanced relative contribution from the Himalayan source at the expense of supply from the Karakoram or the Indus suture zone.

The negative shift in ε_{Nd} has been explained in terms of capture of the ancestral Himalayan tributaries of the Ganga by the Indus. Currently much of the Himalayan sediments of the Indus river is supplied by its four large tributaries, the Sutlej, Ravi, Chennab and Jellum. It is suggested that ~5 My ago these rivers were part of the Ganga river system and not of the Indus drainage.

The above examples demonstrate the successful applications of Sr-Nd isotopes to reconstruct the erosion patterns of different sediment basins at various spatial and temporal scales. These studies also have brought out the links between the patterns of physical erosion and tectonic/climate changes.

26.4 Future Directions

This review addressed the applications of Sr and Nd isotope systems for investigating contemporary and past chemical (silicate) and physical erosion in river basins, particularly those from the Himalaya. Available results show that both these isotope systems, especially $^{143}Nd/^{144}Nd$, have been successfully used to track sediment sources and characterize physical erosion patterns, which have brought out the coupling between tectonics, erosion and climate. The use of Sr isotopes as a proxy for silicate weathering in the Himalaya has been controversial because of the presence of carbonates with highly radiogenic Sr similar to that of silicates, which makes it difficult to clearly identify the sources of radiogenic Sr to the rivers; resolving this issue would be an important challenge for the future. An approach to pursue this problem could be through a multi-tracer systematic study of rivers (major ions, Sr isotopes and other potential proxies of silicate/carbonate weathering, e.g. Ge, $^{234}U/^{238}U$, Ca and Mg isotopes) draining small and medium sized watersheds of the Himalaya with well characterized basin properties, lithology, mineralogy and temporal variations in water discharge. There have been a few investigations in this direction. For example, Chabaux et al. (2001) through measurements of U, Sr isotopes and major ions in selected Himalayan rivers concluded that "U activity ratios, in association with Sr isotope ratios, can be used to trace the sources of dissolved fluxes carried by these rivers". However, analogous to Sr isotopes, the role of trace phases in rocks in contributing to dissolved U in rivers can frustrate the interpretation of results. Another group of isotope tracers that may help in quantification of sources of alkaline earth elements to rivers is Ca and Mg isotopes (Tipper et al. 2008). Studies of these isotopes in rivers of the Himalayan-Tibetan Plateau indicate that in small rivers these isotopes hold promise to provide information on the sources of Ca and Mg whereas in large rivers the lithological control on the isotopic composition seem obscured by isotope fractionation during weathering and calcite precipitation. Better characterization of the impact of these processes on the isotope composition of Ca and Mg may provide a way forward for their application to quantify their sources to rivers.

Additional complexities on the use of Sr isotopes as silicate weathering proxy arise from their supply to rivers via hot springs. Studies on the role of hot springs in regulating river water chemistry and on the Sr isotopic budget of the Marsyandi river, a sub-tributary of the Ganga (Evans et al. 2001, 2004) have shown that they can be an important source of dissolved major ions and Sr with highly radiogenic $^{87}Sr/^{86}Sr$ to this river. An interesting aspect of this study has been the application of Ge/Si as a tracer to evaluate the hot spring contribution to the water flux and alkalinity of river. More detailed studies are required to quantitatively evaluate the role of hot springs in determining the budget of major ions and Sr isotopes on basin-wide and global scales.

The marine budget of strontium shows an imbalance between its supply and removal (Davis et al. 2003; Vance et al. 2009). Uncertainties in the estimates of hydrothermal and riverine fluxes and recent enhancement in riverine flux due to weathering of finely ground rock left behind by glaciers (Vance et al. 2009) can all contribute to this imbalance. Therefore, obtaining better estimates of mean Sr flux and $^{87}Sr/^{86}Sr$ to the ocean through temporally and spatially (lithology based) representative sampling of rivers from basins with different climate and physical features should provide better understanding of the Sr isotope budget of the oceans and their potential to yield information on continental weathering history. In this context, the roles of island arcs and ocean islands in contributing unradiogenic Sr to the oceans and the release of Sr during enhanced chemical weathering of fresh and finely ground rocks produced by glaciers are being increasingly recognized. Similarly, ascertaining the relative variation in riverine fluxes between glacial and interglacial periods through investigations of weathering indices in dated sediments from selected basins can also contribute to better understanding of the marine geochemistry and budget of Sr.

Another source of dissolved Sr to the oceans which is not included in its marine budget calculations is sub-marine groundwater discharge (SGD), though its importance is getting increasingly recognized. For example, available results for the Bay of Bengal (Basu et al. 2001) indicate that SGD can be a very significant source of Sr to the bay, with moderately radiogenic Sr ($^{87}Sr/^{86}Sr$ ~0.715–0.720). More data for other coastal regions are needed to quantify the global significance of this input and its role in Sr isotope evolution of the oceans.

Recently, questions have been raised about the uplift-erosion-climate hypothesis based on depositional trends of dissolved $^{10}Be/^{9}Be$ in ocean sediments and Fe-Mn crusts during the past 12 My (Willenbring and Blanckenburg 2010). These trends do not show any significant temporal variation leading to infer that global weathering fluxes have not been interrupted by mountain building pulses during the late-Cenozoic era. A topic for future study could be reconciliation of these findings with the marine Sr isotope record which shows a steady increase during this period.

26.5 Summary

Continental erosion, both physical and chemical, regulates global geochemical cycles, shapes Earth's surface morphology and influences its climate pattern. The global significance of erosion processes has led to several studies on them and their variations in space and time to learn about the evolutionary history of our planet. A topic of considerable interest in recent years in this field has been to understand the coupling between tectonics-erosion-climate. Radiogenic isotopes, particularly $^{87}Sr/^{86}Sr$ has proven to be a powerful tool in these investigations.

Chemical weathering of silicate minerals on continents is an important process that serves as a major sink for atmospheric CO_2 on My time scales. An important development on this topic has been the suggestion that enhanced silicate weathering in young orogenic belts such as the Himalaya is the driver of global cooling during the Cenozoic. This led to detailed investigations on silicate weathering, both present and past, particularly its relation with tectonics and climate. One of the key requirements in these studies is the availability of suitable proxies that can track changes in silicate weathering in the past. Sr isotope ratios recorded in carbonates of deep sea sediments hold clues to these variations, however deciphering them is challenging because of multiple sources supplying Sr to the sea and the assumptions required to interpret the Sr isotope record. For example, the Cenozoic increase in marine $^{87}Sr/^{86}Sr$ can result either from enhanced riverine Sr flux of constant $^{87}Sr/^{86}Sr$ or changes in its $^{87}Sr/^{86}Sr$ or both. Earlier studies argued in favor of increasing the riverine Sr flux and therefore intensity of silicate weathering to reproduce the marine Sr isotope record, whereas some of the later work suggested changing the riverine $^{87}Sr/^{86}Sr$ and therefore the source composition rather than the intensity of silicate weathering to explain the increase in seawater $^{87}Sr/^{86}Sr$. Sr isotope studies of clays and river carbonates formed during the past several My seem to support the later view, thereby challenging the proposition of enhanced silicate weathering during the Cenozoic. Another issue pertaining to the Sr isotope problem is the source of highly radiogenic Sr in rivers of the Himalaya; there have been both laboratory and field studies which demonstrate that trace calcites contained in granites and metamorphosed sedimentary carbonates dispersed in the drainage basins can also supply high radiogenic Sr to rivers. If the major source of radiogenic Sr to the Himalayan rivers is indeed trace calcites and metamorphosed carbonates, it would have serious repercussions on the use of Sr isotopes as a silicate proxy. Thus the multiple variables associated with the supply of Sr and $^{87}Sr/^{86}Sr$ to rivers and oceans have raised new challenges on the use of Sr isotopes to determine silicate weathering rates.

Physical erosion is another key process that influences climate by regulating the transport and burial of organic carbon and determining chemical erosion rates. This has led to a number of studies on the physical erosion pattern of river basins, their variability in space and time and the factors controlling them. These studies rely mainly on Sr and Nd isotopes to determine the sediment provenances, under the assumption that the isotope signatures of sources are well preserved in the sediments. This assumption though is generally valid, there are evidences based both on laboratory and field studies that suggest non-stochiometric release of Sr and Nd isotopes during different stages of rock weathering; the release being governed by the weathering sequence of minerals in the rock. For example, it has been demonstrated that during early stages of granite weathering Sr isotopes are released preferentially from minerals such as calcites, apatites, bytownite and biotite contained in them.

The non-stochiometric release of Sr isotopes from rocks during their weathering though has raised concerns about its application as a provenance tracer, it has been used extensively often with Nd isotopes to investigate spatial and temporal variations in physical erosion pattern of river basins and factors controlling them. Nd being less mobile than Sr during weathering and transport its source composition is expected to be preserved better in sediments, therefore concomitant changes in both Sr and Nd isotopic composition of sediments is generally interpreted in terms of source variations. In recent years there have been many studies, using Sr and Nd isotope proxies, on the physical erosion pattern of river basins from the Himalaya because of their dominance in sediment flux to the ocean and major role in global carbon cycle. These studies identified HH to be the major source of sediments depositing at present and over the past several My in the Ganga Plain and the Bay of Bengal and that the erosional regime has been generally stable during this period. The primary factors controlling contemporary physical erosion in the Himalaya are relief and climate and that most of the erosion occurs in "hot spots" characterized by intense precipitation over regions of high relief. These hot spots unload huge amounts of sediments promoting more rapid uplift of regions around them thereby coupling tectonics and erosion. The past erosion pattern of the Himalaya deduced from the Sr and Nd isotope records in sediments from the Ganga Plain, the Bay of Bengal and the Arabian Sea all show that it was linked to climate changes both on ky and My time scales, however there was only limited impact of glacier advance and retreat on erosion. There are also evidences to demonstrate that capture of Ganga tributaries by the Indus contributed to changes in its erosion pattern during the past ~5 My. It is evident from the above examples that Sr and Nd isotope studies of contemporary and ancient sediments hold valuable information on their sources and controlling agents of physical erosion in respective basins. These studies also have underscored the importance of climate and tectonics in regulating continental erosion.

Acknowledgements SK thanks the Indian National Science Academy, New Delhi for Senior Scientistship and the Director, PRL for logistical support. Reviews and comments from Prof. M. Baskaran and two anonymous reviewers have helped improve the article.

References

Ahmad S, Babu G, Padmakumari V et al (2005) Sr, Nd isotopic evidence of terrigenous flux variations in the Bay of Bengal: implications of monsoons during the last 34,000 years. Geophys Res Lett 32:1–4

Ahmad S, Padmakumari V, Babu G (2009) Strontium and neodymium isotopic compositions in sediments from Godavari, Krishna and Pennar rivers. Curr Sci 97:1766–1769

Allegre C, Dupre B, Negrel P et al (1996) Sr–Nd–Pb isotope systematics in Amazon and Congo River systems: constraints about erosion processes. Chem Geol 131:93–112

Allegre C, Louvat P, Gaillardet J et al (2010) The fundamental role of island arc weathering in the oceanic Sr isotope budget. Earth Planet Sci Lett 292:51–56

Amelin Y, Rotenberg E (2004) Sm-Nd systematics of chondrites. Earth Planet Sci Lett 223:267–282

Amiotte Suchet P, Probst J, Ludwig W (2003) Worldwide distribution of continental rock lithology: implications for the atmospheric/soil CO_2 uptake by continental weathering and alkalinity river transport to the oceans. Glob Biogeochem Cycles 17:1038. doi:10.1029/2002GB001891

Anderasen R, Sharma M (2006) Solar Nebula heterogenity in p-process Samarium and Neodymium isotopes. Science 314:806–809

Andersson P, Dahlqvist R, Ingri J et al (2001) The isotopic composition of Nd in a boreal river: a reflection of selective weathering and colloidal transport. Geochim Cosmochim Acta 65:521–527

Armstrong RL (1971) Glacial erosion and the variable isotopic composition of strontium in seawater. Nature 230:132–134

Asahara Y, Tanaka T, Kamioka H et al (1999) Provenance of the north Pacific sediments and process of source material transport as derived from Rb–Sr isotopic systematics. Chem Geol 158:271–291

Aubert D, Stille P, Probst A (2001) REE fractionation during granite weathering and removal by waters and suspended loads: Sr and Nd isotopic evidence. Geochim Cosmochim Acta 65:387–406

Banner J (2004) Radiogenic isotopes: systematics and applications to earth surface processes and chemical startigraphy. Earth Sci Rev 65:141–194

Barroux G, Sonke J, Boaventura G et al (2006) Seasonal dissolved rare earth element dynamics of the Amazon River main stem, its tributaries, and the Curuaí floodplain. Geochem Geophys Geosyst 7:Q12005. doi:10.1029/2006GC001244

Barun JJ, Pagel M, Herbillon A et al (1993) Mobilization and redistribution of REEs and thorium in a syenitic lateritic profile: a mass balance study. Geochim Cosmochim Acta 57:4419–4434

Basu A, Jacobsen S, Poreda R et al (2001) Large groundwater strontium flux to the oceans from the Bengal basin and the marine strontium isotope record. Science 293:1470–1473

Berner R, Lasaga A, Garrels R (1983) The carbonate–silicate geochemical cycle and its effect on atmospheric carbon dioxide over the past 100 million years. Am J Sci 283:641–683

Bickle M, Harris N, Bunbury J et al (2001) Controls on the $^{87}Sr/^{86}Sr$ ratio of carbonates in the Garhwal Himalaya, Headwaters of the Ganges. J Geol 109:737–753

Bickle M, Bunbury J, Chapman H et al (2003) Fluxes of Sr into the headwaters of the Ganges. Geochim Cosmochim Acta 67:2567–2584

Bickle M, Chapman H, Bunbury J et al (2005) Relative contributions of silicate and carbonate rocks to riverine Sr fluxes in the headwaters of the Ganges. Geochim Cosmochim Acta 69:2221–2240

Blum JD, Erel Y (1995) A silicate weathering mechanism linking increases in marine Sr-87/Sr-86 with global glaciation. Nature 373:415–418

Blum JD, Erel Y (1997) Rb-Sr isotope systematics of a granitic soil chronosequence: the importance of biotite weathering. Geochim Cosmochim Acta 61:3193–3204

Blum JD, Erel Y (2003) Radiogenic isotopes in weathering and hydrology. In: Holland HD, Turekian KK (eds) Surface and ground water, weathering and soils, vol. 5, Treatise on geochemistry. Elsevier-Pergamon, Oxford, pp 365–392

Blum JD, Gazis CA, Jacobson A et al (1998) Carbonate versus silicate weathering rates in the Raikot watershed within the High Himalayan crystalline series. Geology 26:411–414

Bluth G, Kump L (1994) Lithologic and climatologic controls of river chemistry. Geochim Cosmochim Acta 58:2341–2359

Brantley S, Chesley J, Stillings L (1998) Isotopic ratios and release rates of strontium measured from weathering feldspars. Geochim Cosmochim Acta 62:1493–1500

Brass GW (1976) The variation of the marine $^{87}Sr/^{86}Sr$ ratio during Phanerozoic time; interpretation using a flux model. Geochim Cosmochim Acta 40:721–730

Bullen T, White A, Blum A et al (1997) Chemical weathering of a soil chronosequence on granitoid alluvium: II mineralogic and isotopic constraints on the behavior of strontium. Geochim Cosmochim Acta 61:291–306

Burbank D, Blythe A, Putkonen J et al (2003) Decoupling of erosion and precipitation in the Himalayas. Nature 426:652–655

Burke W, Denison R, Hetherington E et al (1982) Variation of seawater $^{87}Sr/^{86}Sr$ throughout Phanerozoic time. Geology 10:516–519

Chabaux F, Riotte J, Clauer N et al (2001) Isotopic tracing of the dissolved U fluxes of Himalayan rivers: implications for present and past U budgets of the Ganges-Brahmaputra system. Geochim Cosmochim Acta 65:3201–3217

Chabaux F, Bourdon B, Riotte J (2008) U-series geochemistry in weathering profiles, river waters and lakes. In: Krishnaswami S, Cochran JK (eds) U/Th series radionuclides in aquatic systems, vol 13, Radioactivity in the environment. Elsevier, New York, NY, pp 49–104

Clift P (2006) Controls on the erosion of Cenozoic Asia and the flux of clastic sediment to the ocean. Earth Planet Sci Lett 241:571–590

Clift P, Blusztajn J (2005) Reorganization of the western Himalayan river system after five million years ago. Nature 438:1001–1003

Clift P, Lee J, Hildebrand P et al (2002) Nd and Pb isotope variability in the Indus river system: implications for crustal heterogeneity in the western Himalya. Earth Planet Sci Lett 200:91–106

Clift P, Giosan L, Blusztajn J et al (2008) Holocene erosion of the Lesser Himalaya triggered by intensified summer monsoon. Geology 36:79–82

Colin C, Turpin L, Bertaux J et al (1999) Erosional history of the Himalayan and Burman ranges during the last two glacial-interglacial cycles. Earth Planet Sci Lett 171:647–660

Colin C, Turpin L, Blamart D et al (2006) Evolution of weathering patterns in the Indo-Burman ranges over 280 kyr: effects of sediment provenance on $^{87}Sr/^{86}Sr$ ratios tracer. Geochem Geophys Geosyst 7:Q03007. doi:10.1029/2005GC000962

Dalai T, Krishnaswami S, Kumar A (2003) Sr and $^{87}Sr/^{86}Sr$ in the Yamuna River System in the Himalaya: sources, fluxes, and controls on Sr isotope composition. Geochim Cosmochim Acta 67:2931–2948

Das A, Krishnaswami S, Sarin M et al (2005) Chemical weathering in the Krishna Basin and Western Ghats of the Deccan Traps, India: rates of basalt weathering and their controls. Geochim Cosmochim Acta 69:2067–2084

Das A, Krishnaswami S, Kumar A (2006) Sr and $^{87}Sr/^{86}Sr$ in rivers draining the Deccan Traps (India): implications to weathering, Sr fluxes and marine $^{87}Sr/^{86}Sr$ record around K/T. Geochem Geophys Geosyst 7:Q06014. doi:10.1029/2005GC001081

Dasch E (1969) Strontium isotopes in weathering profiles, deep-sea sediments, and sedimentary rocks. Geochim Cosmochim Acta 33:1521–1552

Davis A, Bickle M, Teagle D (2003) Imbalance in the oceanic strontium budget. Earth Planet Sci Lett 211:173–187

DePaolo D, Wasserburg G (1976) Nd isotopic variations and petrogenetic models. Geophys Res Lett 3:249–252

Derry L, France-Lanord C (1996) Neogene Himalayan weathering history and river $^{87}Sr/^{86}Sr$: impact on the marine Sr record. Earth Planet Sci Lett 142:59–74

Derry L, France-Lanord C (1997) Himalayan weathering and erosion fluxes: climate and tectonic controls. In: Ruddiman WF (ed) Tectonic uplift and climate change. Plenum, New York, pp 290–312

Dessert C, Dupre B, Francois L et al (2001) Erosion of Deccan Traps determined by river geochemistry: impact on the global climate and the $^{87}Sr/^{86}Sr$ ratio of sea water. Earth Planet Sci Lett 188:459–474

Dessert C, Dupre B, Gaillardet J et al (2003) Basalt weathering laws and the impact of basalt weathering on the global carbon cycle. Chem Geol 20:1–17

Dia A, Dupré B, Allègre C (1992) Nd isotopes in Indian Ocean sediments used as a tracer of supply to the ocean and circulation paths. Mar Geol 103:349–359

Drever J (1997) The geochemistry of natural waters, 3rd edn. Prentice Hall, NJ, p 436

Edmond JM (1992) Himalayan tectonics, weathering processes, and the strontium isotope record in marine limestone. Science 258:1594–1597

Edmond J, Huh Y (1997) Chemical weathering yields from basement and orogenic terrains in hot and cold climates. In: Ruddiman WF (ed) Tectonic Uplift and Climate Change. Plenum Press, New York, pp 330–351

Erel Y, Blum JD, Roueff E et al (2004) Lead and strontium isotopes as monitors of experimental granitoid mineral dissolution. Geochim Cosmochim Acta 68:4649–4663

Evans M, Derry L, Anderson S et al (2001) Hydrothermal source of radiogenic Sr to Himalayan rivers. Geology 29:803–806

Evans M, Derry L, France-Lanord C (2004) Geothermal fluxes of alkalinity in the Narayani river system of central Nepal. Geochem Geophys Geosyst 5:Q08011. doi:10.1029/2004GC000719

Faure G (1986) Principles of Isotope Geology, 2nd edn. Wiley, Hoboken, N.J

France-Lanord C, Derry L (1997) Organic carbon burial forcing of the carbon cycle from Himalayan erosion. Nature 390:65–67

France-Lanord C, Derry L, Michard A (1993) Evolution of the Himalaya since Miocene time: isotopic and sedimentologic evidence from the Bengal Fan. In: Treloar PJ, Searle M (eds) Himalayan tectonics, vol 74, Geological Society of London Special Publication. Geological Society of London, London, pp 603–621

Frank M (2002) Radiogenic isotopes: tracers of past ocean circulation and erosional input. Rev Geophys 40(1):1001. doi:10.1029/2000RG000094

Gaillardet J (2008) Isotope geochemistry as a tool for deciphering kinetics of water-rock interaction. In: Brantley S, Kubicki J, White A (eds) Kinetics of water-rock interaction, Chap. 12. Spinger, New York, pp 611–674

Gaillardet J, Dupre B, Allegre C (1997) Chemical and physical denudation in the Amazon river basin. Chem Geol 142:141–173

Gaillardet J, Dupre B, Louvat P et al (1999) Global silicate weathering and CO_2 consumption rates deduced from the chemistry of large rivers. Chem Geol 159:3–30

Galy A, France-Lanord C (1999) Weathering processes in the Ganges-Brahmaputra basin and the riverine alkalinity budget. Chem Geol 159:31–60

Galy A, France-Lanord C, Derry L (1999) The strontium isotopic budget of Himalayan Rivers in Nepal and Bangladesh. Geochim Cosmochim Acta 63:1905–1925

Galy V, France-Lanord C, Peucker-Ehrenbrink B et al (2010) Sr–Nd–Os evidence for a stable erosion regime in the Himalaya during the past 12 Myr. Earth Planet Sci Lett 290:474–480

Goldstein SL, Jacobsen SB (1987) The Nd and Sr isotopic systematics of river-water dissolved material: implications for the sources of Nd and Sr in the seawater. Chem Geol 66:245–272

Goldstein SL, O'Nions RK (1981) Nd and Sr isotopic relationships in pelagic clays and ferromanganese deposits. Nature 292:324–327

Harlavan Y, Erel Y (2002) The release of Pb and REE from granitoids by the dissolution of accessory phases. Geochim Cosmochim Acta 66:837–848

Harlavan Y, Erel Y, Blum JD (2009) The coupled release of REE and Pb to the soil labile pool with time by weathering of accessory phases, Wind River Mountains, WY. Geochim Cosmochim Acta 73:320–336

Harris N, Bickle M, Chapman H et al (1998) The significance of Himalayan rivers for silicate weathering rates: evidence from the Bhote Kosi tributary. Chem Geol 144:205–220

Hren M, Chamberlain C, Hilley G et al (2007) Major ion chemistry of the Yarlung Tsangpo-Brahmaputra river: chemical weathering, erosion, and CO_2 consumption in the southern Tibetan Plateau and eastern syntaxis of the Himalaya. Geochim Cosmochim Acta 71:2907–2935

Huh Y, Edmond J (1999) The fluvial geochemistry of the rivers of Eastern Siberia: III Tributaries of the Lena and Anbar draining the basement terrain of the Siberian Craton and the Trans-Baikal Highlands. Geochim Cosmochim Acta 63:967–987

Ingri J, Widerlund A, Land M et al (2000) Temporal variations in the fractionation of the rare earth elements in a boreal river; the role of colloidal particles. Chem Geol 166:23–45

Jacobsen S, Wasserburg G (1980) Sm-Nd isotopic evolution of chondrites. Earth Planet Sci Lett 50:139–155

Jacobson AD, Blum JD (2000) The Ca/Sr and $^{87}Sr/^{86}Sr$ geochemistry of disseminated calcite in Himalayan silicate rocks from Nanga Parbat: influence on river water chemistry. Geology 28:463–466

Jacobson AD, Blum JD, Chamberlain CP et al (2002) The Ca/Sr and Sr isotope systematics of a Himalayan glacial chronosequence: carbonate versus silicate weathering rates as a function of landscape surface age. Geochim Cosmochim Acta 66:13–27

Jeandel C, Arsouze T, Lacan F et al (2007) Isotopic Nd compositions and concentrations of the lithogenic inputs into the ocean: a compilation, with an emphasis on the margins. Chem Geol 239:156–164

Johannesson K, Tang J, Daniels J et al (2004) Rare earth element concentrations and speciation in organic-rich blackwaters of the Great Dismal Swamp, Virginia. USA Chem Geol 209:271–294

Krishnaswami S, Singh SK (2005) Chemical weathering in the river basins of the Himalaya. India Curr Sci 89:841–849

Krishnaswami S, Trivedi JR, Sarin MM et al (1992) Strontium isotopes and rubidium in the Ganga-Brahmaputra river system: weathering in the Himalaya, fluxes to the Bay of Bengal and contributions to the evolution of oceanic $^{87}Sr/^{86}Sr$. Earth Planet Sci Lett 109:243–253

Krishnaswami S, Singh SK, Dalai TK (1999) Silicate weathering in the Himalaya: role in contributing to major ions and radiogenic Sr to the Bay of Bengal. In: Somayajulu BLK (ed) Ocean science, trends and future directions. Indian National Science Academy and Akademia International, New Delhi, pp 23–51

Kump LR, Brantley SL, Arthur MA (2000) Chemical weathering, atmospheric CO_2 and climate. Annu Rev Earth Planet Sci 28:611–667

Leland J, Reid MR, Burbank DW et al (1998) Incision and differential bedrock uplift along the Indus River near Nanga Parbat, Pakistan Himalaya, from ^{10}Be and ^{26}Al exposure age dating of bedrock straths. Earth Planet Sci Lett 154:93–107

Li XH, Wei GJ, Shao L et al (2003) Geochemical and Nd isotopic variations in sediments of the South China Sea: a response to Cenozoic tectonism in SE Asia. Earth Planet Sci Lett 211:207–220

Ma J, Wei G, Xu Y et al (2007) Mobilization and re-distribution of major and trace elements during extreme weathering of basalt in Hainan Island, South China. Geochim Cosmochim Acta 71:3223–3237

Ma J, Wei G, Xu Y et al (2010) Variations of Sr–Nd–Hf isotopic systematics in basalt during intensive weathering. Chem Geol 269:376–385

MacFarlane A, Danielson A, Holland H et al (1994) REE chemistry and Sm-Nd systematics of late Archean weathering profiles in the Fortescue Group, Western Australia. Geochim Cosmochim Acta 58:1777–1794

McCauley S, DePaolo D (1997) The marine $^{87}Sr/^{86}Sr$ and $\delta^{18}O$ records, Himalayan alkalinity fluxes and Cenozoic climate records. In: Ruddiman W (ed) Tectonics uplift and climate change. Plenum, New York, pp 428–467

Milliman J, Meade R (1983) World-wide delivery of river sediment to the oceans. J Geol 91:1–21

Milliman JD, Syvitski PM (1992) Geomorphic/Tectonic control of sediment discharge to the ocean: the importance of small mountainous rivers. J Geol 100:525–544

Millot R, Gaillardet J, Dupre B et al (2002) The global control of silicate weathering rates and the coupling with physical erosion: new insights from rivers of the Canadian Shield. Earth Planet Sci Lett 196:83–98

Millot R, Gaillardet J, Dupre B et al (2003) Northern latitude chemical weathering rates: clues from the Mackenzie River basin, Canada. Geochim Cosmochim Acta 67:1305–1329

Molnar P, England P (1990) Late Cenozoic uplift of mountain ranges and global climate change: chicken or egg? Nature 346:29–34

Montgomery DR (1994) Valley incision and the uplift of mountain peaks. J Geophys Res 99:913–921

Moon S, Huh Y, Qin J et al (2007) Chemical weathering in the Hong (Red) River basin: rates of silicate weathering and their controlling factors. Geochim Cosmochim Acta 71:1411–1430

Moon S, Huh Y, Zitsev A (2009) Hydrochemistry of the Amur River: weathering in a northern temperate basin. Aquat Geochem 15:497–527

Negrel P, Allegre CJ, Dupre B et al (1993) Erosion sources determined by inversion of major and trace element ratios and strontium isotopic ratios in river water: the Congo Basin case. Earth Planet Sci Lett 120:59–76

Nesbitt H (1979) Mobility and fractionation of rare earth elements during weathering of a granodiorite. Nature 279:206–210

Nesbitt HW, Markovics G (1997) Weathering of granodioritic crust, long-term storage of elements in weathering profiles, and petrogenesis of silicate minerals. Geochim Cosmochim Acta 61:1653–1670

Noh H, Huh Y, Qin J et al (2009) Chemical weathering in the three rivers region of Eastern Tibet. Geochim Cosmochim Acta 73:1857–1877

Ohlander B, Ingri J, Land M et al (2000) Change of Sm-Nd isotope composition during weathering of till. Geochim Cosmochim Acta 64:813–820

Oliva P, Dupre B, Martin F et al (2004) The role of trace minerals in chemical weathering in a high-elevation granitic watershed (Estibere, France): chemical and mineralogical evidence. Geochim Cosmochim Acta 68:2223–2243

Oliver L, Harris N, Bickle M et al (2003) Silicate weathering rates decoupled from the $^{87}Sr/^{86}Sr$ ratio of the dissolved load during Himalayan erosion. Chem Geol 201:119–139

Palmer M, Edmond J (1989) The strontium isotope budget of the modern ocean. Earth Planet Sci Lett 92:11–26

Palmer MR, Edmond JM (1992) Controls over the strontium isotope composition of river water. Geochim Cosmochim Acta 56:2099–2111

Pande K, Sarin MM, Trivedi JR et al (1994) The Indus river system (India-Pakistan): major-ion chemistry, uranium and strontium isotopes. Chem Geol 116:245–259

Pattanaik J, Balakrishnan S, Bhutani R et al (2007) Chemical and strontium isotopic composition of Kaveri, Palar and Ponnaiyar rivers: significance to weathering of granulites and granitic gneisses of southern Peninsular India. Curr Sci 93:523–531

Peucker-Ehrenbrink B (2009) Land2Sea database of river drainage basin sizes, annual water discharges, and suspended sediment fluxes. Geochem Geophys Geosyst 10:Q06014. doi:10.1029/2008GC002356

Peucker-Ehrenbrink B, Miller MW, Arsouze T et al (2010) Continental bedrock and riverine fluxes of strontium and neodymium isotopes to the oceans. Geochem Geophys Geosyst 11:Q03016. doi:10.1029/2009GC002869

Pierson-Wickmann AC, Reisberg L, France-Lanord C et al (2001) Os-Sr-Nd results from sediments in the Bay of Bengal: implications for sediment transport and the marine Os record. Paleoceanography 16:435–444

Porcelli D, Anderson P, Baskaran M et al (2009) The distribution of neodymium isotopes in Arctic Ocean basins. Geochim Cosmochim Acta 73:2645–2659

Quade J, Roe L, DeCelles P et al (1997) The late neogene $^{87}Sr/^{86}Sr$ record of lowland Himalayan rivers. Science 276:1828–1831

Rad S, Allegre C, Lovat P (2007) Hidden erosion on volcanic islands. Earth Planet Sci Lett 262:109–124

Rahaman W, Singh SK, Sinha R et al (2009) Climate control on erosion distribution over the Himalaya during the past 100 ka. Geology 37:559–562

Rai SK, Singh SK (2007) Temporal variation in Sr and $^{87}Sr/^{86}Sr$ of the Brahmaputra: implications for annual fluxes and tracking flash floods through chemical and isotope composition. Geochem Geophys Geosyst 8:Q08008

Rai SK, Singh SK, Krishnaswami S (2010) Chemical weathering in the plain and peninsular sub-basins of the Ganga: impact on major ion chemistry and elemental fluxes. Geochim Cosmochim Acta 74:2340–2355

Ravizza G, Zachos J (2003) Records of Cenozoic chemistry. In: Holland HD, Turekian KK (eds) The oceans and marine chemistry, vol. 6, Treatise on geochemistry. Elsevier-Pergamon, Oxford, pp 551–582

Raymo ME, Ruddiman WF (1992) Tectonic forcing of late Cenozoic climate. Nature 359:117–122

Raymo ME, Ruddiman WF, Froelich PN (1988) Influence of late Cenozoic mountain building on ocean geochemical cycles. Geology 16:649–653

Reynolds B (2011) Silicon isotopes as tracers of terrestrial processes. In: Baskaran M (ed) Handbook of environmental isotope geochemistry. Chapter 6. Springer, Heidelberg

Richter FM, Rowley DB, DePaolo DJ (1992) Sr isotope evolution of seawater: the role of tectonics. Earth Planet Sci Lett 109:11–23

Ruddiman W (1997) Tectonic Uplift and Climate Change. Plenum, New York, p 535

Shiller A (2010) Dissolved rare earth elements in a seasonally snow-covered, alpine/subalpine watershed, Loch Vale, Colorado. Geochim Cosmochim Acta 74:2040–2052

Sholkovitz E (1995) The aquatic chemistry of rare earth elements in rivers and estuaries. Aquat Geochem 1:1–34

Sholkovitz E, Szymczak R (2000) The estuarine chemistry of rare earth elements: comparison of the Amazon, Fly, Sepik and the Gulf of Papua systems. Earth Planet Sci Lett 179:299–309

Singh SK (2006) Spatial variability in erosion in the Brahmaputra basin: causes and impacts. Curr Sci 90:1271–1276

Singh SK (2007) Erosion and weathering in the Brahmaputra river system. In: Gupta A (ed) Large rivers. Wiley, Chichester, pp 373–393

Singh SK, France-Lanord C (2002) Tracing the distribution of erosion in the Brahmaputra watershed from isotopic compositions of stream sediments. Earth Planet Sci Lett 202:645–662

Singh SK, Trivedi JR, Pande K et al (1998) Chemical and strontium, oxygen, and carbon isotopic compositions of carbonates from the Lesser Himalaya: implications to the strontium isotope composition of the source waters of the Ganga, Ghaghara, and the Indus Rivers. Geochim Cosmochim Acta 62:743–755

Singh SK, Sarin MM, France-Lanord C (2005) Chemical erosion in the eastern Himalaya: major ion composition of the Brahmaputra and $\delta^{13}C$ of dissolved inorganic carbon. Geochim Cosmochim Acta 69:3573–3588

Singh SK, Rai SK, Krishnaswami S (2008) Sr and Nd isotopes in river sediments from the Ganga basin: sediment provenance and spatial variability in physical erosion. J Geophys Res 113:F03006. doi:10.1029/2007JF000909

Stallard RF, Edmond JM (1983) Geochemistry of the Amazon 2. J Geophys Res 88:9671–9688

Steinmann M, Stille P (2008) Controls on transport and fractionation of the rare earth elements in stream water of a mixed basaltic-granitic catchment basin (Massif Central, France). Chem Geol 254:1–18

Tachikawa K, Athias V, Jeandel C (2003) Neodymium budget in the modern ocean and paleo-oceanographic implications. J Geophys Res 108:3254. doi:10.1029/1999JC000285

Taylor AS, Lasaga AC (1999) The role of basalt weathering in the Sr isotope budget of the oceans. Chem Geol 161:199–214

Taylor AS, Blum JD, Lasaga AC et al (2000) Kinetics of dissolution and Sr release during biotite and phlogopite weathering. Geochim Cosmochim Acta 64:1191–1208

Tipper E, Bickle M, Galy A et al (2006a) The short term climatic sensitivity of carbonate and silicate weathering fluxes: insight from seasonal variations in river chemistry. Geochim Cosmochim Acta 70:2737–2754

Tipper E, Galy A, Gaillardet J et al (2006b) The magnesium isotope budget of the modern ocean: constraints from riverine magnesium isotope ratios. Earth Planet Sci Lett 250:241–253

Tipper E, Galy A, Bickle M (2008) Calcium and magnesium isotope systematics in rivers draining the Himalaya-Tibetan-Plateau region: lithological or fractionation control? Geochim Cosmochim Acta 72:1057–1075

Tricca A, Stille P, Steinmann M et al (1999) Rare earth elements and Sr and Nd isotopic compositions of dissolved and suspended loads from small river systems in the Vosges mountains (France), the river Rhine and groundwater. Chem Geol 160:139–158

Tripathy GR, Singh SK (2010) Chemical erosion rates of river basins of the Ganga system in the Himalaya: reanalysis based on inversion of dissolved major ions, Sr, and $^{87}Sr/^{86}Sr$. Geochem Geophys Geosyst 11:Q03013. doi:10.1029/2009GC002862

Tripathy GR, Goswami V, Singh SK et al (2010) Temporal variations in Sr and $^{87}Sr/^{86}Sr$ of the Ganga headwaters: estimate of dissolved Sr flux to the mainstream. Hydrol Process 24:1159–1171. doi:10.1002/hyp. 7572

Trivedi J, Pande K, Krishnaswami S et al (1995) Sr isotopes in rivers of India and Pakistan: a reconnaissance study. Curr Sci 69:171–178

Tutken T, Eisenhauer A, Wiegand B et al (2002) Glacial-interglacial cycles in Sr and Nd isotopic composition of Arctic marine sediments triggered by the Svalbard/Barents Sea ice sheet. Mar Geol 182:351–372

Vance D, Teagle D, Foster G (2009) Variable Quaternary chemical weathering fluxes and imbalances in marine geochemical budgets. Nature 458:493–496

Veizer J (1989) Strontium isotopes in sea water through time. Annu Rev Earth Planet Sci 17:141–168

Viers J, Wasserburg GJ (2004) Behavior of Sm and Nd in a lateritic soil profile. Geochim Cosmochim Acta 68:2043–2054

Viers J, Roddaz M, Filizola N et al (2008) Seasonal and provenance controls on Nd–Sr isotopic compositions of Amazon river suspended sediments and implications for Nd and Sr fluxes exported to the Atlantic Ocean. Earth Planet Sci Lett 274:511–523

Vigier N, Bourdon B (2011) Constraining rates of chemical and physical erosion using U-series radionuclides. In: Baskaran M (ed) Handbook of environmental isotope geochemistry, Chapter 27. Springer, Heidelberg

Walker J, Hays P, Kasting J (1981) A negative feedback mechanism for the long-term stabilization of Earth's surface temperature. J Geophy Res 86:9776–9782

Walter HJ, Hegner E, Diekmann B et al (2000) Provenance and transport of terrigenous sediment in the south Atlantic Ocean and their relations to glacial and interglacial cycles: Nd and Sr isotopic evidence. Geochim Cosmochim Acta 64:3813–3827

West AJ, Galy A, Bickle M (2005) Tectonic and climatic controls on silicate weathering. Earth Planet Sci Lett 235:211–228

Willenbring JK, Blanckenburg FV (2010) Long-term stability of global erosion rates and weathering during late-Cenozoic cooling. Nature 465:211–214

Winter B, Johnson C, Clark D (1997) Strontium, neodymium and lead isotope variations of authigenic and silicate sediment components from the Late Cenozoic Arctic Ocean:

implications for sediment provenance and source of trace metals in sea water. Geochim Cosmochim Acta 61: 4181–4200

Wu L, Huh Y, Qin J et al (2005) Chemical weathering in the Upper Huang He (Yellow River) draining the eastern Qinghai-Tibet Plateau. Geochim Cosmochim Acta 69:5279–5294

Xu Y, Marcantonio F (2007) Strontium isotope variations in the lower Mississippi River and its estuarine mixing zone. Mar Chem 105:118–128

Yang C, Telmer K, Veizer J (1996) Chemical dynamics of the "St. Lawrence" riverine system: δD_{H2O}, $\delta^{18}O_{H2O}$, $\delta^{13}C_{DIC}$, $\delta^{34}S_{sulfate}$, and dissolved $^{87}Sr/^{86}Sr$. Geochim Cosmochim Acta 60:851–866

Yang S, Jiang S, Ling H et al (2007) Sr-Nd isotopic compositions of the Changjiang sediments: implications for tracing sediment sources. Sci China Ser D-Earth Sci 50:1556–1565

Zachos J, Opdyke B, Quinn T et al (1999) Early Cenozoic glaciation, Antarctic weathering, and seawater $^{87}Sr/^{86}Sr$: is there a link? Chem Geol 161:165–180

Zeitler P, Koons P, Bishop M et al (2001) Erosion, Himalayan geodynamics, and the geomorphology of metamorphism. GSA Today 11:4–9

Chapter 27
Constraining Rates of Chemical and Physical Erosion Using U-Series Radionuclides

Nathalie Vigier and Bernard Bourdon

Abstract This chapter relates recent developments concerning the use of several U-series nuclides, in particular ^{234}U-^{238}U and ^{230}Th-^{238}U disequilibria, for constraining physical and chemical erosion rates and sediment age. Indeed, the ability to measure these disequilibria with an extremely high precision, even in samples with low concentrations such as natural waters, has opened new avenues for investigating erosional processes. This chapter is articulated in three main parts: a brief introduction and presentation of modern technical methods is followed by a description of how ^{234}U-^{238}U and ^{230}Th-^{238}U disequilibria measured in dissolved (water) and solid (sediment) river phases can be used to provide quantitative constraints on physical and chemical erosion rates at the basin scale. In parallel, recoil effects occurring during sediment formation can now be modelled and used to estimate the residence time of a sediment within a basin. Finally, the last part of this chapter presents the latest findings concerning the study of weathering profiles and the modelling of U and Th migration within an aquifer system.

N. Vigier (✉)
CRPG-CNRS, Nancy Université, 15 rue Notre Dame des Pauvres, 54500 Vandoeuvre les Nancy, France
e-mail: nvigier@crpg.cnrs-nancy.fr

B. Bourdon
Institute of Geochemistry and Petrology, Clausiusstrasse 25, ETH Zurich, Zurich 8092, Switzerland

27.1 Introduction

27.1.1 U-Series as a Tool for Studying Environmental Processes

Across a broad range of fields within the Earth sciences, U-series nuclides have emerged as a set of powerful tools to constrain the rates and timescales of geological processes. Following the discovery of radioactivity by Becquerel (1896), it was soon realized that the radioactivity of natural uranium was due to other elements contained in minute quantities. The most famous decay product was radium, discovered in 1898 by the Curies. In the following years many other elements were identified and the concept of a decay chain was elaborated by Rutherford and Soddy (1903). In the Earth sciences, it took many decades before the applications of these nuclides to Earth processes would flourish. As fully summarized in Ivanovich and Harmon (1992), there have been some pioneering applications of U-series nuclides to understand weathering processes based on α-counting measurements (Plater et al. 1992, 1994; Rosholt et al. 1966; Latham and Schwarz 1987) but the expansion of the field was somewhat limited by the precision of measurements, as well as the relatively large quantities of samples needed for analysis. With the advent of mass spectrometric techniques in the 1990 (Edwards et al. 1987; Goldstein and Stirling 2003), the variations of the longer lived U- series nuclides could then be fully investigated. Recent reviews of U-series applications to studying environmental processes can be found in Chabaux et al. (2003a); Dosseto et al. (2008a); Chabaux et al. (2008).

27.1.2 Properties of U-Series Nuclides in the Weathering Environment

In what follows we summarize the geochemical and nuclear properties of U-series nuclides that are relevant to the weathering environment. First, these nuclides are radioactive with half lives that range between 1,600 yr and 244 kyr, therefore providing a range of timescales relevant for the study of chemical weathering. Second, the geochemical properties of U-Th-Ra and Pa are not identical. During the dissolution of minerals and water-rock interaction, U and Ra are considered as mobile elements (U > Ra) while Th and Pa are considered as immobile elements. The geochemical properties of these elements have been reviewed in detail in Ivanovich and Harmon (1992); Bourdon et al. (2003); or Chabaux et al. (2008). Essentially the fractionation of U-series nuclides in the weathering environment is controlled by their speciation in aqueous solution, their nuclear properties and their interaction with mineral surfaces and with organic material present as solid load or as dissolved load. In oxidized environments, U^{6+} reacts with the water molecule and is in the form of UO_2^{2+}, which is a dissolved cation that is often complexed by carbonates or possibly other ligands if they are present in solution. The adsorption of U onto mineral surfaces has been studied rather extensively and shows that uranium can interact with carbonates, iron oxides or clay minerals (e.g., Duff et al. 2002; Bruno et al. 1995; Giammar and Hering 2001; Singer et al. 2009; Ames et al. 1983).

In contrast, Th is far more insoluble and reacts readily with mineral surfaces and with organic substances to a greater extent (Chen and Wang 2007). Thorium will easily adsorb to many mineral surfaces and its own solubility is extremely low (Dähn et al. 2002; Degueldre and Kline 2007; Geibert and Usbeck 2004; Rojo et al. 2009). Thus, it is expected that under weathering conditions, U concentrations in waters will be greatly enhanced relative to that of Th.

In comparison, there have been fewer studies on the relative mobility of Ra. As an alkali earth, Ra is more mobile than Th although its intrinsic solubility in aqueous solutions is low. In general, the complexation of Ra in solution is more limited than that of Th, but one should not underestimate the ability of Ra to interact with negatively charged surfaces of clay minerals in mildly acidic to neutral conditions (Lazar et al. 2008; Hidaka et al. 2007; Willett and Bond 1995).

In summary, upon congruent dissolution of silicate minerals, all U-series nuclides should be released into solution. In general, their concentrations in solution is not high enough to allow direct precipitation of $Th(OH)_4$, UO_2 or $RaSO_4$ but they will definitely interact to a variable degree with mineral surfaces. A complete database of sorption behavior in the presence of relevant ligands is currently not available. However, it can be predicted that the level of interaction of U-Th-Ra-Pa with minerals under oxic conditions follows the order: Th ≈ Pa > U > Ra. For the purpose of this review, it suffices to mention that in general natural river waters will show a larger abundance of U and Ra relative to Th and Pa. There could be an exception to this case if waters are organic rich (dissolved or colloidal organic matter), which enhances the solubility of Th in waters (Dosseto et al. 2006c). As the affinity of Th for organic acids is extremely high, the presence of dissolved organic matter can result in a lower adsorption of Th to solids and hence a higher effective concentration in solutions (Reiller et al. 2002).

The second property of the U-series nuclides that must be considered is the effect of recoil. As detailed in Bourdon et al. (2003), radioactive decay inside minerals leads to the emission of α-particles with a high energy (4–8 MeV) (Fleischer 1980; Fleischer and Raabe 1978; Huang and Walker 1967; Kigoshi 1971). A non-negligible part of this energy is given to the daughter nuclide upon decay (e.g. ^{234}U, ^{230}Th or ^{226}Ra). This energy termed "recoil energy" is in the form of kinetic energy, which means that the daughter nuclide will be displaced from its original site by a distance that depends on the magnitude of this energy and the properties of the mineral. The theoretical recoil distance is around 30 nm for silicate and carbonate minerals with the consequence that, depending on the grain size of minerals, and how far from the grain boundary the parent nuclides are located, a variable fraction of the daughter nuclides is ejected from minerals into surrounding solutions. The recoil effect can significantly contribute to the fractionation between ^{234}U-^{238}U observed during erosion processes, or perhaps, in some cases, be the dominant cause of this fractionation. As the geochemical contrast between U and Th or Ra and Th is large, it is generally considered that the main cause for fractionation between these nuclides is chemical rather than being due to recoil effects (e.g. Bourdon et al. 2003; DePaolo et al. 2006).

27.2 Material and Methods

The measurement of U-series nuclides, and in particular of U and Th isotopes, in weathering products (sediments and soil samples), as well as in river and ground waters, requires specific procedures. The main reasons are: (1) solid phases sampled in soils may contain U or Th rich accessory phases particularly resistant to dissolution procedures (2) natural waters are often characterized by very low levels of U and Th, and the precise measurement of their isotope composition requires the use of highly sensitive mass spectrometers, as well as a specific chemical separation allowing negligible external contamination.

Soil and regolith samples are usually collected using drilling techniques and hand augers (Ma et al. 2010; Hubert et al. 2006; Bourdon et al. 2009). Soil and river sediments are generally sieved before being crushed, in order to remove pebbles and gravels (greater than sand-sized particles). In order to avoid contamination, the crushing is generally done using agate material. Powders are then dissolved in mixtures of concentrated acids (generally HF, HNO_3 and HCl, with $HClO_4$ in the case of organic matter) (e.g. Ma et al. 2010; Granet et al. 2007). In some cases, samples are ashed at 550°C overnight before dissolution (Dosseto et al. 2008a, b). In order to ensure complete dissolution and to avoid fluoride precipitation that may easily retain elements such as thorium, boric acid may be added (e.g. Vigier et al. 2001).

Since most natural water samples have very low levels in U-series nuclides, it is often necessary to perform a pre-concentration procedure. This can be done through the evaporation of a significant volume of water, but evaporation of large volumes of water may result in substantial external contamination. As a consequence, another pre-concentration technique has been developed (Chen et al. 1986; Vigier et al. 2001), which consists of co-precipitating radionuclides with Fe hydroxides from 15–20 L of filtered water.

Filtration is usually performed in the field, the same day of the sampling, using 0.2 or 0.45 μm filters (e.g. Hubert et al. 2006, Vigier et al. 2001). Sampling techniques specific for aquifer environments are given in Hubert et al. (2006) and in Maher et al. (2006a).

Before isotope composition measurements, U and Th need to be fully separated from the rest of the sample matrix. This is classically performed by ion chromatography, using anion exchange resin columns, in a clean laboratory (Manhès 1981; Luo et al. 1997; Turner et al. 1997; Aciego et al. 2009).

Over the past few years, U and Th isotope compositions (^{234}U, ^{238}U, ^{230}Th and ^{232}Th) have typically been measured using new generation mass spectrometers, either by thermal ionisation mass spectrometer (TIMS) (e.g. TRITON) or by multicollector-inductively coupled plasma mass spectrometer (MC-ICP-MS) (e.g. Nu Instruments, Neptune or Isoprobe) (Ma et al. 2010; Granet et al. 2007, 2010; Pelt et al. 2008; Dosseto et al. 2008a, b; Hubert et al. 2006; Bourdon et al. 2009; Maher et al. 2006a, b; Vigier et al. 2006; Andersen et al. 2004, 2008; Ball et al. 2008; Fietzke et al. 2005; Hellstrom 2003). For U isotopes, both types of mass spectrometers require a quantity of 10–100 ng of uranium, depending on the sensitivity of the instrument. Both ^{234}U/^{238}U and ^{230}Th/^{232}Th ratios are very low in nature (usually of the order of 10^{-5} to 10^{-6} respectively), which requires first that they cannot be measured with the same type of collector, and second that the abundance sensitivity of the mass spectrometer is highly optimized (it typically needs to be below 0.1 ppm). The measured isotope ratio needs to be corrected for instrumental mass bias (more significant with MC-ICP-MS), and for the relative gain between the ion counting system (IC) used for the low abundance isotope, and Faraday cup, used for the high abundance isotope. Bourdon et al. (2009) report specific procedures used to perform both types of corrections. The most recent internal precision and published reproducibility (at the 2σ level) for ^{234}U/^{238}U ranges between 1 and 3‰. The accuracy is checked using uranium reference materials, either synthetic solutions (e.g. NBS 960, now CRM-145 or NBL 112a) or rocks (HU1, BeN, AthO, UB-N, BHVO-1).

For Th isotope measurements, the amount of Th used is highly variable and mostly depends on the Th concentration of the sample. Solids are generally enriched in Th, and isotope analyses require sample size that generally contain 100–500 ng Th. In that case, similar corrections as for uranium isotopes are made (see e.g. Turner et al. 2001; Sims et al. 2008), and both published internal and external errors are around 5‰ (±2‰). For samples with much lower concentrations, such as natural waters, highly sensitive MC-ICP-MS are preferred. In that specific case, errors obtained are directly correlated to the amount of available Th for isotope analyses (see Vigier et al. 2006 for more details). The accuracy of Th isotope measurements is checked using Th synthetic solutions (e.g. Th-105, Th-S1, IRMM 35 and 36), and rock standards (generally TML and BeN).

27.3 Chemical and Physical Erosion Dynamics at Large Scale

The application of U-series nuclides to study the dynamics of chemical and physical erosion in large river basins is relatively recent although some earlier work by Plater et al. (1992) had hinted at some potential applications. More recently, the studies of Porcelli et al. (1997, 2001) highlighted the behaviour of U-series nuclides in organic-rich northern-high-latitude waters. Following the study of Vigier et al. (2001) many studies have exemplified how U-series can be used to constrain the timescales of physical and chemical erosion in large rivers. Recent reviews of the approach are given in Chabaux et al. (2003b, 2008); Dosseto et al. (2008a) In this section we review critically the most recent literature and further examine the great potential of U-series for constraining chemical and physical erosion rate and timescales.

27.3.1 Rates of Chemical and Physical Erosion

As explained in the Introduction, the different chemical and nuclear properties of U-series nuclides leads to fractionation in the U-series decay chain. First, the greater mobility of U and Ra relative to Th leads to enhancement of U and Ra in the dissolved fraction while Th is enriched in solids. Overall, river waters show ^{230}Th/^{238}U and ^{230}Th/^{226}Ra activity ratios in the dissolved loads smaller than 1, while in the suspended loads these activity ratios are greater than 1 (Fig. 27.1). Part of the fractionation can be related to adsorption of U-series nuclides either in the soil environment or

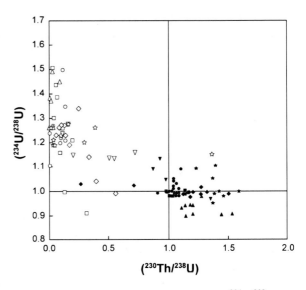

Fig. 27.1 Compilation of literature data of ^{234}U-^{238}U and ^{230}Th-^{238}U disequilibria measured in river filtered waters (*open symbols*) and corresponding river sediments (*black symbols*). *Triangles*: Mackenzie Basin rivers (Vigier et al. 2001). *Squares*: Bolivian Andes rivers (Dosseto et al. 2006b). *Stars*: Deccan Traps rivers (Vigier et al. 2005). *Inverse triangles*: Murray-Darling River (Dosseto et al. 2006c). *Circles*: Icelandic rivers (Vigier et al. 2006). *Diamonds*: Amazon rivers (Dosseto et al. 2006a)

during transport in rivers. In principle, adsorption of Th onto solids would be greater than that of U and Ra, leading to even greater ^{230}Th/^{238}U and ^{230}Th/^{226}Ra activity ratios. If the fractionation in ^{230}Th/^{238}U and ^{226}Ra/^{230}Th activity ratios is related to the degree of mineral alteration, then, in principle, the measurements of these ratios can be used to estimate the timescales and the rates of chemical weathering.

Another important type of fractionation that has been widely observed is the fractionation between ^{234}U and ^{238}U. There can be two distinct processes for creating ^{234}U-^{238}U fractionation. First, the α-decay of ^{238}U to ^{234}Th and then to ^{234}U leads to direct ejection of the daughter nuclide from its current site by a distance that depends on the properties of the surrounding medium. In minerals, this distance should be approximately 30 nm. For grain that are smaller than 10 μm, the recoil can lead to measurable depletion of the solids over timescales of the order of magnitude of the half life of ^{234}U (Fig. 27.2). Another possibility is the preferential leaching of minerals leading to the enhancement of ^{234}U in the solution, as was first documented by Kigoshi (1971). This effect has been further documented in a number of laboratory

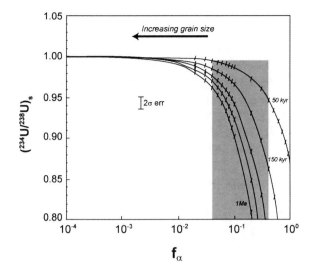

Fig. 27.2 Evolution of $(^{234}U/^{238}U)$ activity ratio in a sediment as a function of the fraction of recoiled ^{234}Th (f_α, see text for more details). Each curve correspond to a specific comminution age. If f_α can be estimated independently for a size fraction of sediment, a corresponding precise measurement of $(^{234}U/^{238}U)$ can lead to an estimation of the comminution age. The *grey shadow* corresponds to published range for f_α (DePaolo et al. 2006; Maher et al. 2006a, b; Hubert et al. 2006; Dosseto et al. 2010; Bourdon et al. 2009)

experiments (Hussain and Lal 1986; Andersen et al. 2009; Bourdon et al. 2009) for silicate and carbonate lithologies. In both cases, significant ^{234}U-^{238}U fractionation can be observed. More specifically, it is likely that the degree of ^{234}U-^{238}U fractionation actually depends on the nature of mineral phases. In what follows, we describe a number of applications of U-series nuclides for understanding the dynamics of weathering, rates and timescales of chemical weathering.

27.3.1.1 Dynamics of Erosion

An intrinsic property of landscape evolution is that it is in general a slow process, relative to human timescales, such that the direct measurement of erosion rates or soil formation rates is difficult. For dynamic systems that evolve slowly it is tempting to assume steady-state erosion. In this particular context, "steady-state" implies that the "weathering zone" in a given basin has reached a constant mass (M_w) or in other words that the output fluxes (mass being exported per unit of time as solid and dissolved loads in rivers, ϕ_{out}) is balanced by the input fluxes (new bedrock material being weathered per unit of time, ϕ_{in}). The equation describing steady-state of erosion can be written as:

$$\frac{dM_w}{dt} = \phi_{in} - \phi_{out} = 0 \quad (27.1)$$

One can then write a similar equation for each of the nuclides of the U-series decay chain (e.g. C_i):

$$\frac{dM_w C_i}{dt} = \phi_{in} C_i^{in} - \phi_{out} C_i^{out} = 0 \quad (27.2)$$

The output flux can be estimated based on both the dissolved (d) and the suspended solids or particles (sol) exported by rivers:

$$\frac{dM_w C_i}{dt} = \phi_{in} C_i^{in} - \phi_{sol} C_i^{sol} - \phi_d C_i^d = 0 \quad (27.3)$$

After rearranging the equations, it can be shown that (Vigier et al. 2001):

$$\left(\frac{^{234}U}{^{238}U}\right)_d X_d + \left(\frac{^{234}U}{^{238}U}\right)_{sol} (1 - X_d) = 1 \quad (27.4)$$

where X_d represents the mass fraction of dissolved uranium relative to the total U export flux. Thus, using measured U activity ratios in the dissolved and suspended load of a basin river, one can evaluate if erosion operates in steady-state or not at the basin scale. This can be done by comparing the value of X_d deduced from U isotopes with that estimated based on direct measurements of U content in the dissolved and solid load (U_w and U_s), and of the total suspended solids (TSS), thereby providing an important constraint on the dynamics of erosion. Indeed, X_d is an inverse function of TSS ($X_d = U_w/(U_w+TSS. U_s)$), and therefore of the physical erosion rate (Φ = TSS × discharge/area). In the end, this approach can be used to assess whether soils are being destroyed or be constructed at the scale of a watershed or a large basin. When the physical erosion rate (or sediment yield) based on U-series is slower than the observed erosion rate, it would indicate that the soils are being destroyed, while if it is faster, then the soils should be building up (Fig. 27.3). Thus, the U-series can provide important constraints on the dynamics of soils in river basins.

In principle, the attainment of steady-state erosion should be observed when the forcing parameters that

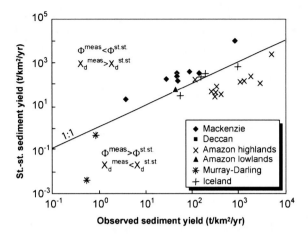

Fig. 27.3 Comparison of erosion rates (sediment yield or Φ = TSS × discharge/area) based on direct measurements of suspended load in rivers, and that inferred from U-series measurements (assuming a steady-state of erosion at the scale of the basin). See text for details. If the measured sediment yield is lower than the steady-state one, this can result either from a recent decrease of physical erosion rates, or alternatively from a recent increase of the proportion of dissolved uranium (X_d, see (27.4)), which is expected to be related to an increase of chemical erosion rate. The Mackenzie Basin (Canada) and the Amazon highlands are the regions the furthest from steady-state. Corresponding references are (Dosseto et al. 2006a, b, c; Vigier et al. 2001, 2005, 2006). Graph modified from Dosseto et al. (2008a)

affect the rates of erosion (due to climate, tectonic activity and/or human activity) have been varying slowly enough to allow the rate of soil production to adjust to new conditions. The parameters that control the rate of soil production are not precisely known and it is even more difficult to constrain the rate at which soil productions can adjust to new conditions. However it has been shown that the rate of soil production is lower when soils are thicker (Heimsath et al. 1997). The study of several large river basins has revealed that the erosion process in such basins is not always at steady-state (Vigier et al. 2001, 2005; Dosseto et al. (2006a, 2008a) (Fig. 27.3). In most cases, the absence of steady-state can be explained by recent changes in natural or human forcing parameters. For example, in the case of the Mackenzie river basins, this was attributed to recent deglaciation after the LGM. This concept has been further utilized by Vance et al. (2009) to explain variations in Sr isotopes in the marine record. Studies of rivers draining the Deccan trap (Vigier et al. 2005) have shown that agriculture was likely responsible for the destruction of soils in this region of India, which is characterized by intense deforestation. In the Andes, Dosseto et al. (2008a, b) have shown that accelerated erosion rates due to increased rainfall during the Holocene led to destruction of soils. However, in contrast, in Iceland, where chemical and physical erosion rates display a large range, erosion processes were found to operate at steady-state at the scale of the island, although this region was strongly affected by the last glaciation, and maximum soil age was estimated to be 10 kyr (Vigier et al. 2006).

There have been criticisms of this approach (e.g. Granet et al. 2007) and it is quite clear that these studies have to be refined to accommodate them. Indeed, although equation (27.4) may seem rather simple, it hides some complexities that need to be assessed. In order to have a robust mass balance, one needs to make sure that there is no additional flux of dissolved uranium that is coming from another weathering zone, in particular from deep groundwaters. It is possible that in some cases, there is a significant contribution of uranium from deep aquifers, as suggested by Durand et al. (2005). This aspect was also studied by Bourdon et al. (2009) who have analyzed surface and groundwater samples in a small chalk river basin. This study showed that up to 40% of the uranium budget could be coming from aquifers in this carbonate environment. This might be an extreme case because the dissolution of carbonate is relatively fast. In the Amazon river basin, even by considering a contribution of up to 20% uranium from aquifers, the conclusion that erosion is not at steady-state is not affected. Given that the degree of disequilibrium of erosion processes is high for the Amazon basins, the conclusions are overall robust. Second, the potential for U or Th redistribution during transport also needs to be examined carefully. These questions have been discussed at some length in Dosseto et al. (2006a, b, 2008a) and it was shown that in most cases, the consideration of these effects did not greatly affect the conclusions, although they should not be neglected. For example, the adsorption of Th onto particles could decrease the $^{230}Th/^{238}U$ ratio measured in the dissolved load while the $^{230}Th/^{238}U$ ratio in the suspended load would increase. In fact it has been shown in several studies that the Th in the dissolved load of rivers is mostly associated with colloids (Viers et al. 1997; Dosseto et al. 2006c) and that its abundance in river waters is generally much higher than what the solubility would predict. Thus, given the association of Th with colloids it is not clear that adsorption of Th onto suspended particles (size >0.2 μm) is taking place during transport.

Another important pitfall of this approach (that is in fact not specific to the U-series) is the representativeness of the analysis with respect to the possibility that the composition of the suspended load is not homogeneous or that the bedload also needs to be taken into account. The study of Granet et al. (2010) shows that there is sometimes a significant variability in the ^{230}Th/^{238}U ratios of the suspended load of rivers (2–50% in the most extreme case). Similarly, Dosseto et al. (2006c) have shown that the ^{230}Th/^{238}U and ^{234}U/^{238}U ratios of various size fractions in rivers is highly variable and that these size fractions can be represented by a mixture of a true dissolved load and suspended loads. However, the technique that consists of filtering samples at 0.2 μm rather than determining the true endmember of the dissolved load yielded similar results in terms of mass balance.

In the end, the study of erosion dynamics based on U-series will need some refinement and a careful assessment of the pitfalls and potential complexities need to be made to assess the robustness of conclusions. In particular, further detailed sampling will certainly be needed in the future.

27.3.1.2 Rates and Timescale of Chemical Weathering

In principle, U-series nuclides should be one of the best suited tools to determine the rates and timescales of chemical weathering in river basins. In comparison, the ^{10}Be method has been used to determine physical erosion rates by analyzing ^{10}Be production in quartz transported in rivers (Hewawasam et al. 2003; von Blanckenburg 2006). Physical and chemical erosion rates using Nd and Sr isotopes are discussed in Tripathy et al. (2011).

The basic principles for deriving the rates and timescales of chemical weathering are based on some simple assumptions about the origin of U-series fractionation. Upon mineral alteration, the U-series nuclides are fractionated due either to differences in chemical behaviour or due to recoil effects. If one can estimate the degree of fractionation based on some simple laws, it is then possible to derive a timescale by measuring the fractionation between ^{234}U-^{238}U or ^{230}Th-^{238}U in the weathered solids (as in river suspended particles) and in solutions (river waters). Furthermore, if we can make the additional assumption regarding the initial state of the system prior to the inception of chemical weathering, one can derive the rates of chemical and physical weathering. For the U-series nuclides, one can write the following equations. The equation of decay for a given number of atoms, N, of a given nuclide can be written as follows:

For the parent nuclide (N_1):

$$\frac{dN_1}{dt} = -\lambda_1 N_1 \qquad (27.5)$$

For all the intermediate nuclides i of interest:

$$\frac{dN_i}{dt} = -\lambda_i N_i + \lambda_{i-1} N_{i-1} \qquad (27.6)$$

where the decay constant, λ is related to the half-life by $\lambda = \ln 2 / t_{1/2}$ and subscript i-1 refers to the next nuclide higher up in the decay chain (i.e. parent nuclide). For timescales greater than 5–6 times the half life of the longest lived daughter, the decay chain can be shown to be in secular equilibrium, i.e. all activities are equal:

$$\lambda_{i-1} N_{i-1} = \lambda_i N_i \qquad (27.7)$$

We assume that the U-series decay chain is initially in secular equilibrium in minerals, i.e. the U decay chain has not been disturbed over the past million year and all activities are equal. The main consequence of this assumption is that the net effect of chemical weathering can be assessed by considering the measured activity ratios in weathering products. However, it is arguable whether this assumption is completely valid. This can be assessed by considering measurements of U-series nuclides in boreholes that extend below the typical weathering zone in order to examine to what extent U-series nuclides are disturbed in the underground, below the regolith. Several studies suggest that where there is no large crustal fluid flow, the U-series decay chain is generally in secular equilibrium at least for the ^{234}U-^{238}U or ^{230}Th-^{238}U systems. This has been checked in the case of carbonates (Bourdon et al. 2009) or in the case of granites (Noseck et al. 2008), or even in the case of clays (Pekala et al. 2009). It turns out that for a number of lithologies, although there are small deviations from secular equilibrium, overall the assumption of initial secular equilibrium for bulk rocks can be considered as valid.

The second condition for deriving the rates and timescale of weathering concerns the assumption that one can make about the laws describing chemical weathering. This has been the subject of some discussion for a number of reasons. Following the early study of Plater et al. (1992), a number of authors have assumed that the release of U-series nuclides follow first order kinetics (Plater et al. 1992; Vigier et al. 2001; Dosseto et al. 2006a, b, c; Chabaux et al. 2003a, b). There are a number of obvious reasons for which such a law is likely to be incorrect, simply because U-series nuclides are prone to redistribution rather than simple release. As U, Th and Ra often interact with mineral surfaces or with organic matter, it is not unreasonable to consider that these elements can be redistributed in weathering profiles (see e.g. Dequincey et al. 2002). Such a redistribution is particularly likely in climate where iron oxides or hydroxides form due to iron oxidation and transport. In such a context, the adsorption of U or Th is likely and the first order law for chemical weathering is probably inadequate and an additional term for describing the evolution of U-series nuclides fractionation is justified (e.g. Granet et al. 2010). However, in other cases, as argued by Chabaux et al. (2008), for systems that are largely undersaturated, first-order kinetics are justified. If the system reaches saturation in a secondary phase, then it is expected that the rates of chemical weathering will be different (Ganor et al. 2007; Maher et al. 2009).

By analyzing the dissolved and suspended load in rivers it is possible to infer both the rate of U release and the time over which this release has taken place. If the timescale of transport is relatively short, then the timescale over which U release has taken place should be equivalent to the residence time of U in soils (Dosseto et al. 2008a, b). As described in Vigier et al. (2001), the process of chemical weathering for the U-series decay chain in the solid (p) can be described by the following equation:

$$\frac{d^{238}U_p}{dt} = -k_8{}^{238}U_p \qquad (27.8)$$

Where $^{238}U_p$ is the number of atoms of ^{238}U in river or soil particles. For ^{234}U, the equation can be written as:

$$\frac{d^{234}U_p}{dt} = \lambda_8{}^{238}U_p - (k_4 + \lambda_4){}^{234}U_p \qquad (27.9)$$

Where $^{234}U_p$ represents the number of ^{234}U atoms, λ_4 the decay constant of ^{234}U and k_4 the rate of release of ^{234}U into water of nuclide i. For the dissolved load, it is possible to integrate over the soil thickness, or in other words over the timescale since the inception of weathering in the soils:

$$^{238}U_d = \int_0^\tau k_8{}^{238}U_p(t)dt \qquad (27.10)$$

In the case of ^{238}U-^{234}U-^{230}Th system, a system of four non-linear equations with four unknowns (time, k_8, k_4 and k_0) can be solved. Some more complex forms of this equations have been used in the literature to derive timescales (Dequincey et al. 2002; Vigier et al. 2005; Granet et al. 2007) although the introduction of more complex models is not always warranted. For example, based on the earlier models of Ghaleb et al. (1990), Dequincey et al. (2002) have proposed that the release of U can be described by the following equation:

$$\frac{d^{234}U_p}{dt} = \lambda_8{}^{238}U_p - (k_4 + \lambda_4){}^{234}U_p + F_4 - F_8 \quad (27.11)$$

Where F_i represent a zeroth order additional flux due to input of U per unit of time. This form of model adds an additional level of complexity, thereby providing additional degrees of freedom without having an obvious external constraint to solve the equations. Nevertheless, in some cases, a reasonable solution, albeit non unique can be derived (Granet et al. 2007).

If dissolution is congruent, then the constant k_i should be directly related to the dissolution rate (Maher et al. 2004). Here the dissolution rate is reported in yr^{-1}, while the usual unit for dissolution rates are in mol m^{-2} y^{-1} (r_{diss}). The relationship between the two dissolution rates can be described by the following equations:

$$r_{diss} = \frac{k_i}{M.S_{BET}} \qquad (27.12)$$

Where S_{BET} is the BET surface area in m^2/g, and M the molar mass of the minerals. A compilation of dissolution rates obtained for large river basins based on U-series showed that there is a negative correlation between the rate of dissolution and the timescale of weathering (Fig. 27.4). A similar correlation has also been reported by Maher et al. (2004) and White and Brantley (2003), although the trend based on U-series was offset from that based on dissolution experiments for silicate minerals. The general explanation for this trend is still unclear although several explanations have been proposed. Over long timescales, it could be that the nature of weathering changes due to modification of mineral surface area or composition. For example, the soils with older ages could be characterized by a greater proportion of secondary clay minerals that should dissolve slower. Alternatively, longer soil residence times are expected to correspond to greater soil thickness, which could ultimately lead to slower weathering rates. In principle, the presence of thicker soils is in general an indicator of areas with stable tectonics that should be characterized by a transport-limited erosion regime. Thermodynamic effects (i.e. fluids equilibrating with the solids) and the potential change in hydraulic properties (e.g. reduced permeability) may also play key roles (e.g. Maher 2010).

In such a regime it is expected that the soil production is actually inversely correlated with soil thickness. Thus, the negative correlation shown in Fig. 27.4 could simply reflect the slower weathering rates expected from thicker soils.

The U-series nuclide data for large river basin can also be used to infer the timescale of chemical weathering, i.e. the average soil residence time at the scale of the whole basin (if the following transport time is short). Although there could be some pitfalls in this approach (see above), a compilation of the data reported so far in the literature lends some confidence in their reliability. First, the soil residence times are positively correlated with independant estimations of soil thickness (not shown here), which would make sense because one would expect that the time spent by a given solid particle in a soil should scale directly with soil thickness. Furthermore, a consideration of several large river basins shows that the sediment residence time is often directly related to the past climates or geological history: mountain areas are generally characterized by short timescale, while stable cratons with constant climatic conditions show long residence time (Dosseto et al. 2006a). The recently glaciated basin of the Mackenzie river (Vigier et al. 2001) is perhaps an exception as it shows relatively short soil residence time. In this particular case, it is expected that the Laurentian ice sheet that covered this area removed existing soils and prevented soil formation until the onset of deglaciation.

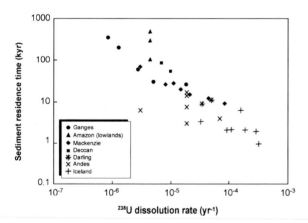

Fig. 27.4 ^{238}U dissolution or leaching rate as a function of the timescale of weathering deduced from U-series disequilibria measured in river waters and/or sediments. The negative trend is close to the one obtained based on the comparison of field and laboratory experiments (White and Brantley 2003). Corresponding references are: Ganges (Granet et al. 2007, 2010); Amazon (Dosseto et al. 2006a, b, c); Mackenzie (Vigier et al. 2001); Deccan (Vigier et al. 2005); Andes (Dosseto et al. 2006a, b, c); Iceland (Vigier et al. 2006). Modified from Dosseto et al. (2008a, b)

27.3.2 Sediment Residence Time and Transport Time

The timescales over which sediments can be transported from their source region to the oceans and the possible implications on the evolution of landscapes is still a key question in geomorphology, despite recent theoretical or analytical developments (e.g. von Blanckenburg 2006). Based on U-series radionuclide measurements, new approaches to derive timescales of sediment production and transport have been proposed and the principles are based again on the fractionation of the U-series decay chain. There have been two main approaches used recently to derive the timescale of sediment production and transport.

27.3.2.1 Comminution Ages of Sediment

The grain size of sediment is in general variable with a generally decreasing size away from the source region. Starting from a massive bedrock, physical and chemical erosion processes will lead to the destruction of the cohesive structure of rocks, thereby leading to the observed grain size distribution (Turcotte 1986). As the grain size approaches micrometers, the fraction of recoiled ^{234}U increases so as to yield a measurable depletion in the grain (more than a few permil). If one is able to estimate reliably the fraction of ^{234}Th ejected due to recoil (f_α), then one can estimate the time since the grains reached their current size, provided uranium in the mineral is uniformly distributed. DePaolo et al. (2006) have coined this time the "comminution age" of the sediment. The methodology has been applied to a number of sediments and shown to yield reasonable estimates for the age of comminution (DePaolo et al. 2006; Dosseto et al. 2010; Lee et al. 2010). For example, Dosseto et al. (2010) have shown that the comminution age of sediments in a semi-arid Australian catchment has been variable over the past 120 kyr and could be explained by variations in the source of sediments, ultimately due to climatic change.

One key aspect of this methodology is to obtain a reliable estimate of the parameter f_α (the fraction of recoiled nuclide), in order to calculate an age. The various approaches have been summarized for example in Maher et al. (2006a, b). One can use the ratio of surface to volume with the equation:

$$f_\alpha = \frac{3\lambda_r L}{2d_P} \quad (27.13)$$

L is the recoil distance of ^{234}Th, d_p is the diameter of the particle, and λ_r is a correction for surface roughness. An alternative is to use the surface area based on BET surface area measurements:

$$f_\alpha = \frac{1}{4} L.S.\lambda_r n \rho_S \quad (27.14)$$

(S being for the specific surface area, and ρ_s is the density, and similar symbols as for 27.13).

This method overestimates the value of f_α because the lengthscale of recoil is much greater than the size of individual N_2 molecules used in the BET experiments. As shown by Bourdon et al. (2009), this effect can be corrected for by using the fractal dimension of the surface and the theoretical model of Semkow (1991). In this case:

$$f_\alpha = \frac{1}{4} \left[\frac{2^{D-1}}{4-D} \right] \left[\frac{a}{R} \right]^{D-2} R.S_{BET}\rho_s \quad (27.15)$$

where D is the fractal dimension of the surface, a the diameter of the adsorbate molecule (N_2), R the recoil length, ρ_S the density of the solid, and S_{BET} the measured BET surface area.

This latter approach has been further validated by the calculated f_α values used for the dating of ice-core samples (Aciego et al. 2009).

Another aspect of this methodology is that the sediment often needs to be leached prior to U-series analysis and this leaching could lead to some spurious ^{234}U-^{238}U fractionation in the grain. It is known that leaching can produce excess of ^{234}U in the solution (see above), therefore the leaching procedure needs to be more thoroughly evaluated. It is currently not known whether the leaching with HCl produces spurious fractionation or only removes carbonates or nuclides adsorbed at the mineral surfaces. Last, the determination of comminution ages neglects the possible effect of preferential release of ^{234}U into solution during alteration/dissolution, as highlighted experimentally. If there is preferential release of ^{234}U, the comminution ages probably need to be adjusted (DePaolo et al. 2006; Lee et al. 2010).

Overall, this new method for deriving comminution ages looks very promising and should be further developed and calibrated while paying specific attention to its analytical methodology.

27.3.2.2 Sediment Transport Time

As suggested by Dosseto et al. (2006b), it is possible to investigate the transport time of sediments in rivers based on U-series nuclide fractionation. By measuring ^{238}U-^{230}Th in the suspended or bed load in rivers along a given river, it is possible to determine the timescale of transport since the ^{238}U-^{230}Th fractionation is time-dependent. However, some assumptions are required in order to apply this methodology. First, one needs to ascertain the origin of particles; ideally they should come from the same source region, as particles from multiple sources with various chemical compositions would make this method untractable. Second, one needs to show that the observed ^{238}U-^{230}Th variation is indeed due to transport or weathering during transport. If this is the case, then a sediment transport time can be calculated. For example Dosseto et al. (2006b) investigated the suspended load of the Amazon river and were able to show that there was a negligible contribution of particles from the plain river while most of the suspended load was derived from the Andes. They also showed that the variation in ^{238}U-^{230}Th was directly correlated with the depletion in Na relative to Sm as particles were being transported to the river mouth, and that the transport time was increasing up to approximately 20 kyr along the river. This was interpreted as a result of mixing of suspended and remobilized particles. Granet et al. (2007, 2010) also investigated the time scale of sediment transport in Himalayan rivers both on the suspended and bedload fractions and found that the transport time of the bedload was on the order of several 100 kyrs while that of the suspended load was less than 20–25 kyr, which was consistent with the results found by Dosseto et al. (2006b). However, there might be some additional inherent complexities and further tests in other basins would be useful. A complete characterization of the sediments could probably help to ensure that all the conditions for obtaining reliable timescale are fulfilled.

27.4 Small Scale Studies

27.4.1 Soil Age and Production Rate

At a smaller scale, at the scale of a soil or a weathering profile, it is also possible to use U-series disequilibria for estimating the timescale and rate of chemical weathering. Over the last couple of years, this tool has even been investigated for estimating soil production rate.

The pioneering work of Rosholt et al. (1966, 1983) have early established the following observations, based on a systematic study of ^{234}U-^{238}U and ^{230}Th-^{238}U in soil samples developed on loess, shales and trachytes, and also by comparing fresh and weathered granites. First, in petrologically fresh granites, U-series nuclides of concern are in secular equilibrium, within analytical uncertainties. Second, the preferential leaching of ^{234}U occurs before soil development, in the first step of rock alteration. In the following stages of weathering, there is a significant leaching of U relative to Th at the beginning of soil formation, associated with little ^{234}U-^{238}U fractionation. In contrast, significant ^{234}U-^{238}U fractionation can be observed in soil horizons where the most mobile U has already been leached away. Third, there can also be uptake of U in the upper soil horizons, that seems to be linked to the occurrence of organic matter and sometimes iron oxides, and that has been interpreted as adsorption of dissolved U which is often enriched in ^{234}U (Fig. 27.1).

From these first set of studies, two main results already emerged: the time dependence of ^{234}U-^{238}U and ^{230}Th-^{238}U disequilibria systematics found in solid weathered products, and the relationship between ^{234}U-^{238}U fractionation and the degree of chemical erosion, responsible for U release from soils. These fundamental observations are the basis for the modelling of U-series in soils that were developed more recently, and which were aimed at constraining (1) the age of the soil and (2) soil or saprolite production rates.

Most U-Th data for soils developed on silicate lithologies lie very near the equiline in the classical ^{238}U-^{230}Th isochron diagram (e.g. Fig. 27.5a). This implies that, in order to constrain timescales as precisely as possible from U-series disequilibria, isotope ratios need to be measured with high precision techniques, which has been made possible only with the advent of mass spectrometric techniques (see Sect. 27.2).

Various sets of soils developed on granites and granodiorites yield positive correlations passing through the secular equilibrium point in a ^{234}U-^{238}U versus ^{230}Th-^{238}U diagram (Fig. 27.5b). This pattern is similar for a series of weathering profiles developed in different climatic and geographical contexts: for example the Kaya laterite, located in Northern Burkina Faso (Dequincey et al. 2002), and soils from the Bega River Basin (SE Autralia, Dosseto et al. 2008b). In contrast, various regoliths developed on US shales display a negative trend in the same diagram (Ma et al. 2010).

The model presently used for determining the soil age, or the age of the most recent U mobilisation in a weathering profile is based on equations detailed in Dequincey et al. (2002), and are partly summarized by equation (27.11). In this model, in addition to U release from the soils, U inputs are also taken

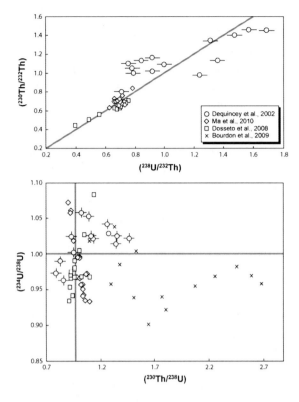

Fig. 27.5 Compilation of recent literature data of U-series disequilibria measured in soil profiles (**a**) reported in a isochron-type diagram (**b**) reported in the diagram of $(^{234}U/^{238}U)$ versus $(^{230}Th/^{238}U)$ diagram, often used in soil studies (see Chabaux et al. 2003b for a review). Data from Dequincey et al. (2002), Ma et al. (2010) and Dosseto et al. (2008b) correspond to soils developed on silicate lithologies. The Bourdon et al. (2009) data is a profile developed on carbonates

and k_j for the rate of release into the water of nuclide f (in yr^{-1}) (for simplifying we use k_4, k_8, k_0, k_2 and f_4, f_8, f_0 and f_2 for k and f coefficients of ^{234}U, ^{238}U, ^{230}Th and ^{232}Th respectively). In this model, the f_f coefficient represents the rate at which nuclide addition occurs, through chemical illuviation in saprolite, or in soils through dust deposition and from soils located upslope. In all cases, there are more unknowns than input parameters (usually measured activity ratios and U and Th concentrations). Consequently, some or several simplifications need to be made for solving the entire set of equations: (1) no leaching of ^{230}Th (Dequincey et al. 2002); (2) identical k_4/k_8 and f_4/f_8 for several saprolites located along a single slope (Dosseto et al. 2008b); (3) $k_0/k_2 = 1$ and $f_0/f_2 = 1$, i.e. there is no differential fractionation of thorium isotopes during leaching and input (Dosseto et al. 2008b); and (4) constant input and output coefficients (f_8, f_4, f_0, k_8, k_4, k_0) for all samples of a single soil profile (Ma et al. 2010).

Soil ages that could be calculated with this method cover a large range, between 7 ka (± 1) and more than 6 Ma (± 0.9) (for the Australian saprolite). The two studies concerning a series of profiles located along a (~300 m) hillslope display roughly the same systematics: small residence time close to the ridge and an increase of the soil and saprolite age as a function of the distance from the ridge top (Fig. 27.6). For a saprolite developed from granodiorite, Dosseto et al. (2008b) interpreted these results as an increase in the duration of weathering with distance from the ridge.

into account. These inputs may come either from atmospheric deposition (Hirose 2011), from soils located upslope, or simply from vertical transfer within a profile. Ages for the African Kaya laterites (Dequincey et al. 2002), regoliths developed on shales (Ma et al. 2010) and saprolite and soils developed on Australian granodiorite (Dosseto et al. 2008a, b) could be determined, using the following equation, describing the time variation of a nuclide j in a soil sample (N_j):

$$\frac{dN_j}{dt} = \lambda_i N_i - \lambda_j N_j + f_j N_{j,0} - k_j N_j \qquad (27.16)$$

where λ_i is the decay constant of the parent nuclide, N_i the abundance of the parent nuclide in the sample, λ_j the decay constant of nuclide j, f_j an input coefficient for nuclide j (in yr^{-1}), $N_{j,0}$ the initial nuclide abundance

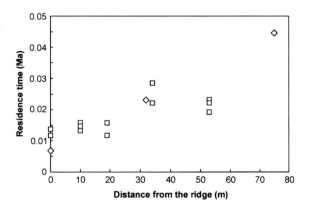

Fig. 27.6 Soil residence time as a function of the distance from the ridge for soils of two different locations (in Australia, Dosseto et al. 2008a, b, *squares*), and in central Pennsylvania (USA, Ma et al. 2010, *diamonds*). Both series of profiles show the same trend

In addition, U dissolution rates (k_8) were also found to decrease with increasing distance from the ridge. Two laterites of similar thickness located in the Amazonian craton, and in Northern Burkina Faso respectively yielded ages greater than 300 ka.

With this methodology, the formation rate of saprolite and soils (ε_s) could also be estimated. Indeed, using the age calculated from U-series measurements (τ_s) and the height of a sample, either above the bedrock-saprolite boundary or above the soil-saprolite boundary respectively (h_s), the production rate is:

$$\varepsilon_s = h_s / \tau_s \qquad (27.17)$$

By considering the parameter h_s, (i.e. the difference between the sample depth and the bottom limit of the profile), the production rate is found to be independent of physical erosion rate that removes material from the top of the profile. Consequently, the production rate estimated with this methodology corresponds to the rate at which the profile deepens within the bedrock, and there is no need of assuming a balance between physical and chemical erosion (such as for a soil in steady-state which would have remained at a constant thickness during its estimated age). If instead of h_s, the whole profile thickness is considered, then calculated production rates are valid only if the soil rapidly reached a steady-state since the onset of its formation, i.e. only if the integrated soil production has counterbalanced the integrated denudation rates, over the lifetime of the profile.

The few recent studies using U-series for determining profile production rate have shown that soils, saprolite and regolith all display a range of production rate that is relatively large and relatively similar (from a few mm/kyr to ~60 mm/kyr). This is surprising since the various profiles also have significantly different thicknesses and weathering timescales. However, soils, regoliths – as well as saprolites and laterites – are characterized by a systematic inverse trend as a function of the profile age: more "recent" weathering profiles exhibit the largest production rates, in soils and also within saprolite (Fig. 27.7). This is consistent with larger scale studies, where it is found that dissolution rate or uranium leaching coefficient is decreasing as a function of time, or of the sediment age (Fig. 27.4). Chemical erosion is thus thought to be more intensive at the beginning of a profile formation, when the proportion of fresh minerals is still high. These deductions are in agreement with initial observations of Rosholt et al. (1966), although more

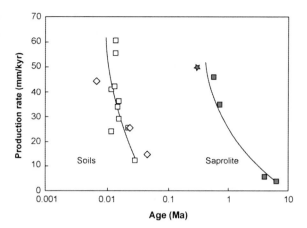

Fig. 27.7 Compilation of production rate and age of soil and saprolite profiles, both determined using U-series disequilibria. *White symbols* correspond to soil samples (Dosseto et al. 2008b; Ma et al. 2010), and saprolite data are displayed in *grey* (Mathieu et al. 1995; Dosseto et al. 2008b)

studies of weathering profile are now needed, that should benefit from recent advances in the field.

27.4.2 Weathering in Aquifers

Over the past few years, an effort has also been made to use U-series nuclides for determining weathering rates and fluid flow within aquifers. Groundwaters are characterized by U-series activity ratios that are generally distinct from river water signatures. For example, (^{234}U/^{238}U) ratios can be extremely high in some aquifers (see review in Ivanovich and Harmon 1992; Osmond and Coward 2000), and this can help their contribution to river waters to be quantified (e.g. Durand et al. 2005; Riotte and Chabaux 1999; Chabaux et al. 2001; Riotte et al. 2003). This was the case in a systematic study of river and ground waters of the Rhine graben (Durand et al. 2005). The coupling of ^{234}U-^{238}U disequilibria and ^{87}Sr/^{86}Sr show that all the stream samples can be explained by a mixing of three types of waters (deep, shallow and alluvial).

In parallel with these studies focusing on the origin of elements carried in river waters, transport modelling has been developed and applied in more complete and sophisticated ways to single aquifer systems (see for a detailed review Porcelli and Swarzenski 2003; Porcelli 2008). Since the early 1980, simple mathematical treatments of aquifer models have been developed for constraining rates of weathering, adsorption/desorption, and migration of some specific radionuclides (e.g. Krishnaswami et al. 1982; Davidson and Dickson 1986; Ku et al. 1992; Luo et al. 2000). These modelling approaches could recently be further constrained due to improved precision on isotope ratios, and the ability to measure substantial number of low level samples – using MC-ICP-MS. Indeed, until a decade ago, most of the data of U-series nuclides in aquifers were obtained using counting techniques, and were associated with significant uncertainties. High precision measurements were developed over the past ten years, and only a few aquifer systems have been studied thus far. These studies have shown that it was possible to provide, for a few case studies, some quantitative constraints on weathering rates within the aquifer or the vadose zone, and/or other parameters such as fluid flow and the age of the system. For example, based on the distribution of U-Th series radionuclides in an unconfined aquifer, Tricca et al. (2001) proposed that the source of ^{222}Rn is mainly derived from the precipitation of ^{232}Th and ^{230}Th on surface coating since the formation of the aquifer. This surface coating is found to account for ~10% of the total Th content of the rock.

In summary, different approaches have been developed, based on the same general equation, of a one dimensional advective transport along a groundwater flow path:

$$\left(\frac{\partial C^i_w}{\partial t}\right) + v\left(\frac{\partial C^i_w}{\partial x}\right) = R_\alpha + R_{diss} + R_{des} \\ - R_{ads} + \lambda_p C^p_w - \lambda_i C^i_w \quad (27.18)$$

In this equation, C^i_w represents the concentration of a nuclide i in water, and C^p_w the concentration of its direct parent within its U-decay series. R_α represents the α-recoil loss rate. It is therefore a function of C^p_s, i.e. the concentration of the parent nuclide in the solid, and f_α (see Sect. 27.3.2.1). R_{diss} is the dissolution term, and is function of the dissolution coefficient for the nuclide (i) and C^i_s, the concentration of (i) in the solid. In Tricca et al. (2000, 2001) and Reynolds et al. (2003), for simplification purposes, the coefficient of dissolution is assumed to be the same for various nuclides such as ^{234}U, ^{238}U and ^{232}Th. The desorption term (R_{des}) is a function of a desorption coefficient and C^i_s, and, similarly, the adsorption term (R_{ads}) is a function of an adsorption coefficient (and C^i_w). The two other terms in the right hand side of equation (27.18) represent input from decay of the radioactive parent present in water, and output due the radioactive decay of nuclide (i) of concern.

The differences between the various published models reside first in the number of aquifer phases taken into account. In most cases, except in Hubert et al. (2006), the solid phase is not analysed. However, a fine layer located at the grain surface is now included in the modelling. This surface layer is considered to "collect" the nuclides ejected by the solid phase due to α-recoil, and the dissolved nuclides that get adsorbed. Tricca et al. (2001) estimated the residence time of Th in the surface coating to be ~3,000 yr while in the water it is ~1 h. The second key point is the modelling of the recoil fraction, as described in detail in Sect. 27.3.2.1.

Even when considering the three aquifer phases (solid, surface layer and water), the number of unknowns always exceeds the number of measured parameters (i.e. activity ratios and nuclide concentrations). As a consequence, the resolution of such models necessarily requires some basic assumptions, and/or, as recently developed by Maher et al. (2006a, b), the coupling with other isotope systems (i.e. Sr isotopes). The first set of studies assumed a steady-state: it was supposed that no evolution of the nuclide concentration as a function of time occurs at the time of the water sampling (Tricca et al. 2000; Reynolds et al. 2003). The number of unknowns was also reduced by considering that all nuclides were released into water at the same rate (w). In other words, no ^{234}U-^{238}U fractionation occurred due to leaching/dissolution. Finally, it was assumed that all the isotope and chemical signatures of the waters are the result of in-situ weathering, with no significant input from the vadose zone for example. Under these conditions, (^{234}U/^{238}U) activity ratio in the water is directly related to the ratio of uranium dissolution coefficient (w_8) and the α-recoil degree (f_α = fraction of rock ^{234}Th ejected into the water). If f_α can be estimated independently (from geometrical consideration for example, see Sect. 27.3.2.1), then w_8 can be determined:

$$\left(^{234}U/^{238}U\right)_w = 1 + f_\alpha\, \lambda_4/w_8 \quad (27.19)$$

From this equation, it appears that particularly high (^{234}U/^{238}U)$_w$ ratios, such as those found in the aquifer of Ojo Alamo Aquifer (New Mexico), located in an

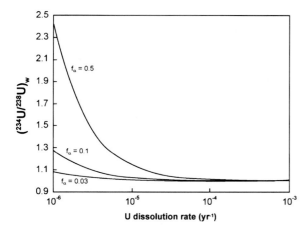

Fig. 27.8 Theoretical dependence of groundwater ($^{234}U/^{238}U$) as a function of the U dissolution rate assuming a steady-state and no ^{234}U-^{238}U fractionation during dissolution (see equation 19). High ($^{234}U/^{238}U$) in old aquifer waters can thus be explained by low dissolution rates and high recoil effects (high f_α values)

The study of the aquifer solid phase can provide important constraints on the modelling of the U-series nuclides. First, considering that the solid phase has also reached a steady-state, Maher et al. (2006a, b) have shown that its ($^{234}U/^{238}U$) is roughly equal to (1-f_α), assuming no ^{234}U-^{238}U fractionation linked to dissolution. Consequently, for old sediments (typically >1 Ma), a measurement of their ($^{234}U/^{238}U$) leads to a direct estimation of f_α. However, this technique is applicable only if f_α is significantly different from 1 (within uncertainties), i.e. for grains lower than a certain size (estimated by the authors to be ~100μm). If steady-state has not been reached, ($^{234}U/^{238}U$) of the solid is simply a function of time and of the α-recoil term (f_α):

$$\left(\frac{^{234}U}{^{238}U}\right)_s = (1 - f_\alpha)(1 - e^{-\lambda_4 t}) + e^{-\lambda_4 t} \qquad (27.20)$$

arid environment (Reynolds et al. 2003), can result from low weathering rates (resulting in low w_8), and strong α-recoil effects (Fig. 27.8).

However, over short distances, ($^{234}U/^{238}U$)$_w$ is expected to evolve as a function of the distance from the recharge zone. In that case, if one still assumes a steady-state, ($^{234}U/^{238}U$)$_w$ is found to be a function of w_i/v (the dissolution rate over the water velocity ratio). Thus, further constraints on the dissolution rate can be obtained either by a direct estimation of the water infiltration flux, or by coupling with other isotope systems, such as Sr isotopes (see Maher et al. 2006a, b; Singleton et al. 2006). Here, the high ($^{234}U/^{238}U$)$_w$ values found in aquifers were explained by low infiltration rates. When the water velocity increases, the ($^{234}U/^{238}U$) of the pore water is expected to decrease.

In the case of a non steady-state approach, such as developed in Hubert et al. (2006), ($^{234}U/^{238}U$)$_w$ is found to vary as a function of the water velocity and f_α, but also as a function of time, the initial activity ratio of the solid (at t = 0, x = 0), the distance from catchment divide, and the uranium distribution coefficient (resulting from adsorption and desorption processes). It can be seen that activity ratios measured in aquifer waters are not significantly affected by the activity of nuclides in the surface layer. Indeed, this layer is in steady-state regardless of adsorption and desorption processes (see Hubert et al. 2006 for more details), having therefore no influence on the water composition.

assuming that no uranium isotope fractionation occurs during dissolution, and that the initial solid (at t = 0) was in secular equilibrium (Hubert et al. 2006). Again, the role of the surface layer is negligible, since the fraction of this surface is very small compared to the typical grain size. Equation (27.20) shows that, if an independent estimation of f_α can be obtained, the age of the aquifer can be determined (Fig. 27.9).

Overall, recent studies of aquifers in carbonate (Bourdon et al. 2009) and silicate lithologies show that, with a few assumptions, it is possible to estimate

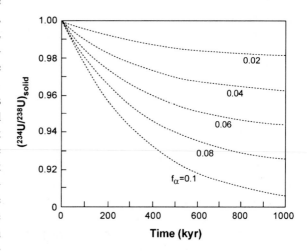

Fig. 27.9 Theoretical evolution of ($^{234}U/^{238}U$) in an aquifer rock matrix versus the age of the rock since the beginning of water percolation. Model curves are labeled with f_α values. Modified from Hubert et al. (2006)

the chemical erosion rate, based on U-series nuclides measured in water and solid samples, even in the case of a non steady-state system. It appears that, as well as for all studies concerning U-series nuclides in surperficial environment, there is now a strong need for independent and precise estimates of the α-recoil term (f_α), and for a better understanding of nuclide fractionation during chemical erosion and leaching of minerals.

27.5 Conclusion and Future Directions

Over the past few years, several nuclides of the U-series have emerged as powerful tracers of chemical erosion at various scales, including those pertinent to continents, basins, soils and aquifers. This is essentially due to new mass spectrometric developments permitting high precision measurements of low concentration samples (typically water samples). Furthermore, these analytical methods have triggered the development of new and more sophisticated transport and erosion models. It must be pointed out, however, that recent studies all highlight that in the future it will be necessary to have a better determination of recoil effects at the mineral scale, and to have a better knowledge of nuclide fractionation and redistribution during chemical alteration and transport. To this end, experimental studies should help to quantify key parameters such as the leaching coefficients. In natural systems, coupling of U-series disequilibria with other tools such as ^{10}Be, or Sr and Nd isotopes for example, has been little investigated thus far, nevertheless this is a promising approach in the sense that both methods provide independent and complementary quantitative information.

Despite the uncertainties highlighted, some systematics are already apparent, such as the major role of recent climatic variations and land use on physical erosion rates presently estimated for some basins. This has been observed in North American cratons, but also in mountaineous areas such as the Andes. Another concept that has emerged is the link between the age of a profile and its production rate. Chemical erosion rates are high at the beginning of soil formation. A similar relationship is found for sediments carried by rivers: old sediments are located in areas where chemical erosion rates (and therefore U dissolution rates) are lower. Since high precision data are still quite limited, these trends must of course be refined. Nevertheless, if they are confirmed they could greatly improve our understanding/modelling of the recent evolution of continental chemical erosion and its effect on the transfer of weathered material to the sea, and in consequence on ocean chemistry.

Acknowledgements We thank the two reviewers, Kate Maher and Anthony Dosseto, for their helpful comments on this chapter. We are also grateful for the constructive comments of the Editor, Mark Baskaran.

References

Aciego SM, Bourdon B, Lupker M, Rickli J (2009) A new procedure for separating and measuring radiogenic isotopes (U, Th, Pa, Ra, Sr, Nd, Hf) in ice cores. Chem Geol 266:194–204

Ames LL, McGarrah JE, Walker BA, Salter PF (1983) Uranium and radium sorption on amorphous ferric oxyhydroxide. Chem Geol 40:135–148

Andersen MB, Stirling CH, Potter EK, Halliday AN (2004) Toward epsilon levels of measurement precision on U-234/U-238 by using MC-ICPMS. Int J Mass Spec 237:107–118

Andersen MB, Stirling CH, Potter EK, Halliday AN, Blake SG, McCulloch MT, Ayling BF, O'Leary M (2008) High-precision U-series measurements of more than 500,000 year old fossil corals. Earth Planet Sci Lett 265:229–245

Andersen MB, Erel Y, Bourdon B (2009) Experimental evidence for ^{234}U-^{238}U fractionation during granite weathering with implications for ^{234}U/^{238}U in natural waters. Geochim Cosmochim Acta 73:4124–4141

Ball L, Sims KWW, Schwieters J (2008) Measurements of ^{234}U/^{238}U and ^{230}Th/^{232}Th in volcanic rocks using the Neptune MC-ICP-MS. JAAS 23:173–180

Becquerel AH (1896) On the invisible rays emitted by phosphorescent bodies. CR Séances Acad Sci 122:501–503

Bourdon B, Turner S, Henderson GM, Lundstrom CC (2003) Introduction to U-series geochemistry. In: Bourdon B, Henderson GM, Lundstrom CC, Turner SP (eds) Uranium-series geochemistry, vol 52. Geochemical Society – Mineralogical Society of America, Washington, DC, pp 1–21

Bourdon B, Bureau S, Andersen MB, Pili E, Hubert A (2009) Weathering rates from top to bottom in a carbonate environment. Chem Geol 258:275–287

Bruno J, de Pablo J, Duro L, Figuerola E (1995) Experimental study and modeling of the U(VI)-Fe(OH)$_3$ surface precipitation/coprecipitatopn equilibria. Geochim Cosmochim Acta 59:4113–4123

Chabaux F, Riotte J, Clauer N, France-Lanord C (2001) Isotopic tracing of the dissolved U fluxes of Himalayan rivers: implications for present and past U budgets of the Ganges-Brahmaputra system. Geochim Cosmochim Acta 65(19):3201–3217

Chabaux F, Dequincey O, Leveque JJ, Leprun JC, Clauer N, Riotte J, Paquet H (2003a) Tracing and dating recent chemical transfers in weathering profiles by trace-element geochemistry and U-238-U-234-Th-230 disequilibria: the example of the Kaya lateritic toposequence (Burkina-Faso). CR Geosci 335:1219–1231

Chabaux F, Riotte J, Dequincey O (2003b) U-Th-Ra fractionation during weathering and river transport. In: Bourdon B, Henderson GM, Lundstrom CC, Turner SP (eds) Uranium-series geochemistry, vol 52. Geochemical Society – Mineralogical Society of America, Washington, DC, pp 533–576

Chabaux F, Bourdon B, Riotte J (2008) U-series geochemistry in weathering profiles, river waters and lakes. In: Krishnaswami S, Cochran JK (eds) U/Th series radionuclides in aquatic systems, vol 13, Radioactivity in the environment. Elsevier, New York, NY, pp 49–104

Chen JH, Edwards RL, Wasserburg GJ (1986) ^{238}U, ^{234}U and ^{232}Th in seawater. Earth Planet Sci Lett 80:241–251

Chen C, Wang X (2007) Sorption of Th (IV) to silica as a function of pH, humic/fulvic acid, ionic strength, electrolyte type. Appl Radiat Isot 65:155–163

Dähn R, Scheidegger AM, Manceau A, Curti E, Baeyens B, Bradbury MH, Chateigner D (2002) Th uptake on montmorillonite: a powder and polarized extended X-ray absorption fine structure (EXAFS) study. J Colloid Interface Sci 249:8–21

Davidson MR, Dickson BL (1986) A porous flow model for steady-state transport of radium in ground waters. Water Resour Res 22:34–44

DePaolo DJ, Maher K, Christensen JN, McManus J (2006) Sediment transport time measured with U-series isotopes: results from ODP North Atlantic drift site 984. Earth Planet Sci Lett 248:394–410

Degueldre C, Kline A (2007) Study of thorium association and surface precipitation on colloids. Earth Planet Sci Lett 264:104–113

Dequincey O, Chabaux F, Clauer N, Sigmarsson O, Liewig N, Leprun JC (2002) Chemical mobilizations in laterites: evidence from trace elements and ^{238}U-^{234}U-^{230}Th disequilibria. Geochim Cosmochim Acta 66:1197–1210

Dosseto A, Hesse PP, Maher K, Fryirs K, Turner S (2010) Climatic and vegetation control on sediment dynamics during the last glacial cycle. Geology 38:395–398

Dosseto A, Bourdon B, Turner S (2008a) Uranium-series isotopes in river materials: Insights into the timescales of erosion and sediment transport. Earth Planet Sci Lett 265:1–17

Dosseto A, Turner SP, Chappell J (2008b) The evolution of weathering profiles through time: new insights from uranium-series isotopes. Earth Planet Sci Lett 274:359–371

Dosseto A, Bourdon B, Gaillardet J, Allègre CJ, Filizola N (2006a) Timescale and conditions of chemical weathering under tropical climate: study of the Amazon basin with U-series. Geochim Cosmochim Acta 70:71–89

Dosseto A, Bourdon B, Gaillardet J, Maurice-Bourgoin L, Allègre CJ (2006b) Weathering and transport of sediments in the Bolivian Andes: time constraints from uranium-series isotopes. Earth Planet Sci Lett 248(3–4):759–771

Dosseto A, Turner S, Douglas GB (2006c) Uranium-series isotopes in colloids and suspended sediments: timescale for sediment production and transport in the Murray-Darling River system. Earth Planet Sci Lett 246:418–431

Durand S, Chabaux F, Rihs S, Duringer P, Elsass P (2005) U isotope ratios as tracers of groundwater inputs into surface waters: Example of the Upper Rhine hydrosystem. Chem Geol 220:1–19

Duff MC, Coughlin JU, Hunter DB (2002) Uranium co-precipitation with iron oxide minerals. Geochim Cosmochim Acta 66:3533–3547

Edwards RL, Chen JH, Wasserburg GJ (1987) ^{238}U-^{234}U-^{230}Th-^{232}Th systematics and the precise measurement of time over the past 500,000 years. Earth Planet Sci Lett 81:175–192

Fietzke J, Liebtrau V, Eisenhauer A, Dullo C (2005) Determination of uranium isotope ratios by multi-static MIC-ICP-MS: method and implementation for precise U- and Th-series isotope measurements. JAAS 20:395–401

Fleischer RL (1980) Isotopic disequilibrium of uranium – alpha-recoil damage and preferential solution effects. Science 207 (4434):979–981

Fleischer RL, Raabe OG (1978) Recoiling alpha-emitting nuclei – mechanisms for uranium-series disequilibrium. Geochim Cosmochim Acta 42(7):973–978

Giammar DE, Hering JG (2001) Timescale for sorption-desorption and surface precipitation of uranyl on goethite. Environ Sci Technol 35:3332–3337

Goldstein SJ, Stirling CH (2003) Techniques for measuring uranium series nuclides: 1992–2002. In: Bourdon B, Henderson GM, Lundstrom CC, Turner SP (eds) Uranium-series geochemistry, vol 52. Geochemical Society – Mineralogical Society of America, Washington, DC, pp 23–57

Ganor J, Lu P, Zheng Z, Zhu C (2007) Bridging the gap between laboratory measurements and field estimations of weathering using simple calculations. Environ Geol 53:599–610

Geibert W, Usbeck R (2004) Adsorption of thorium and protactinium onto different particle types: experimental findings. Geochim Cosmochim Acta 68:1489–1501

Ghaleb B, Hillaire-Marcel C, Causse C, Gariepy C, Vallières S (1990) Fractionation and recycling of U and Th isotopes in a semiarid endoreic depression of central Syria. Geochim Cosmochim Acta 54:1025–1035

Granet M, Chabaux F, Stille P, France-Lanord C, Pelt E (2007) Time-scales of sedimentary transfer and weathering processes from U-series nuclides: clues from the Himalayan rivers. Earth Planet Sci Lett 261:389–406

Granet M, Chabaux F, Stille P, Dosseto A, France-Lanord C, Blaes E (2010) U-series disequilibria in suspended river sediments and implication for sediment transfer time in alluvial plains: the case of the Himalayan rivers. Geochim Cosmochim Acta 74:2851–2865

Heimsath AM, Dietrich WE, Nishiizumi K, et al. (1997) The soil production function and landscape equilibrium. Nature 388:358–361

Hellstrom H (2003) Rapid and accurate U/Th dating using parallel ion-counting multi-collector ICP-MS. JAAS 18:1346–1351

Hewawasam T, von Blanckenburg F, Schaller M, Kubik PW (2003) Increase of human over natural erosion rates in tropical highlands constrained by cosmogenic nuclides. Geology 31:597–600

Hidaka H, Horie K, Gauthier-Lafaye F (2007) Transport and selective uptake of radium into natural clay minerals. Earth Planet Sci Lett 264:167–176

Huang WH, Walker RM (1967) Fossil alpha-particle recoil tracks: a new method of age determination. Science 155:1103–1106

Hubert A, Bourdon B, Pili E, Meynadier L (2006) Transport of radionuclides in an unconfined chalk aquifer inferred from U-series disequilibria. Geochim Cosmochim Acta 70:5437–5454

Hussain N, Lal D (1986) Preferential solution of ^{234}U from recoil tracks and ^{234}U/^{238}U radioactive disequilibrium in natural waters. Proc Ind Acad Sci (Earth Planet Sci) 95:245–263

Ivanovich M, Harmon RS (1992) Uranium-series disequilibrium: applications to earth, marine, and environmental sciences. Oxford University Press, Oxford, p 910

Kigoshi K (1971) Alpha recoil thorium-234: dissolution into water and uranium-234/uranium-238 disequilibrium in nature. Science 173:47–49

Krishnaswami S, Graustein WC, Turekian KK, Dowd JF (1982) Radium, thorium and radioactive lead isotopes in groundwaters: application to in situ determination of adsorption desorption rate constants and retardation factors. Water Res 18:1663–1675

Ku T-L, Luo S, Leslie BW, Hammond DE (1992) Decay-series disequilibria applied to the study of rock-water interaction and geothermal system. In: Ivanovich M, Harmon RS (eds) Uranium-series disequilibrium: applications to earth, marine, and environmental sciences. Clarendon, Oxford, pp 631–668

Latham AG, Schwarz HP (1987) On the possibility of determining rates of removal of uranium from crystalline igneous rocks using U-series disequilibria-I: a U-leach model and its applicability to whole-rock data. App Geochem 2:55–65

Lazar B, Weinstein Y, Paytan A, Magal E, Bruce B, Kolodny Y (2008) Ra and Th adsorption coefficients in lakes – Lake Kinneret (Sea of Galilee) "natural experiment. Geochim Cosmochim Acta 72:3446–3459

Lee VE, DePaolo DJ, Christensen JN (2010) Uranium-series comminution ages of continental sediments: case study of a Pleistocene alluvial fan. Earth Planet Sci Lett 296:244–254

Luo SD, Ku T-L, Roback R, Murrell M, McLing TL (2000) In-situ radionuclide transport and preferential groundwater flows at INEEL (Idaho): decay-series disequilibrium studies. Geochim Cosmochim Acta 64:867–881

Luo X, Rehkamper M, Lee DC, Halliday AN (1997) High precision ^{230}Th/^{232}Th and ^{234}U/^{238}U measurements using energy-filtered ICP magnetic sector multiple collector mass spetrometry. Int J Mass Spectrom Ion Process 171:105–117

Ma L, Chabaux F, Pelt E, Blaes E, Jin L, Brantley S (2010) Regolith production rates calculated with uranium-series isotopes at Susquehanna/Shale Hills Critical Zone Observatory. Earth Planet Sci Lett 297:211–225

Maher K, DePaolo DJ, Lin C-F (2004) Rates of silicate dissolution in deep-sea sediment: in situ measurement using ^{234}U/^{238}U of pore fluids. Geochim Cosmochim Acta 68:4629–4648

Maher K, DePaolo DJ, Christensen JN (2006a) U-Sr isotopic speedometer: fluid flow and chemical weathering rates in aquifers. Geochim Cosmochim Acta 70:4417–4435

Maher K, Steefel CI, DePaolo DJ, Viani BE (2006b) The mineral dissolution rate conundrum: insights from reactive transport modeling of U isotopes and pore fluid chemistry in marine sediments. Geochim Cosmochim Acta 70:337–363

Maher K, Steefel CI, White AF et al. (2009) The role of reaction affinity and secondary minerals in regulating chemical weathering rates at the Santa Cruz soil chronosequence, California. Geochim Cosmochim Acta 73:1804–2831

Maher K (2010) The dependence of chemical weathering rates on fluid residence time. Earth Planet Sci Lett 294:101–110

Manhès G (1981) Développement de l'ensemble chronomtrique U-Th-Pb: contribution à la chronologie initiale du système solaire Thèse d'état, Université Paris 6, Paris

Mathieu D, Bernat M, Nahon D (1995) Short-lived U and Th isotope distribution in a tropical laterite derived from granite (Pitinga river basin, Amazonia, Brazil): application to assessment of weathering rate. Earth Planet Sci Lett 136:703–714

Noseck U, Brasser T, Suksi J, Havlová V, Hercik M, Denecke MA, Förster H-J (2008) Identification of uranium enrichment scenarios by multi-method characterisation of immobile uranium phases Phys. Chem Earth 33:969–977

Osmond JK, Coward JB (2000) U-series nuclides as tracers in groundwater hydrology. In: Cook P, Herczeg A (eds) Environmentaltracers in subsurface hydrology. Kluwer, Boston, pp 290–333

Pekala M, Kramers JD, Waber HN, Gimmi T, Alt-Epping P (2009) Transport of ^{234}U in the Opalinus Clay on centimetre to decimetre scales. App Geochem 24:138–152

Pelt E, Chabaux F, Innocent C, Navarre-Sitcher A, Sak PB, Brantley SL (2008) Uranium-Thorium chronometry of weathering rinds: rock alteration rate and paleo-isotopic record of weathering fluids. Earth Planet Sci Lett 276:98–105

Plater AJ, Ivanovich M, Dugdale RE (1992) Uranium series disequilibrium in river sediments and waters: the significance of anomalous activity ratios. Appl Geochem 7:101–110

Plater AJ, Dugdale RE, Ivanovich M (1994) Sediment yield determination using uranium-series radionuclides: the case of The Wash and Fenland drainage basin, eastern England. Geomorphology 11(1):41–56

Porcelli D, Andersson PS, Baskaran M, Wasserburg GJ (2001) Transport of U- and Th-series nuclides in a Baltic Shield watershed and the Baltic Sea. Geochim Cosmochim Acta 65:2439–2459

Porcelli D, Andersson PS, Wasserburg GJ, Ingri J, Baskaran M (1997) The importance of colloids and mires for the transport of uranium isotopes through the Kalix River watershed and Baltic Sea. Geochim Cosmochim Acta 61:4095–4113

Porcelli D, Swarzenski PW (2003) The behavior of U- and Th-series nuclides in groundwater. In: Bourdon B, Henderson GM, Lundstrom CC, Turner SP (eds) Uranium-series geochemistry, vol 52. Mineralogical Society of America – Geochemical Society, Washington, DC, pp 317–361

Porcelli D (2008) Investigating groundwater processes using U- and Th-series nuclides. In: Krishnaswami S, Cochran JK (eds) U-Th series nuclides in aquatic systems, vol 13. Elsevier, New York, NY, pp 105–154

Reiller P, Moulin V, Casanova F, Dautel C (2002) Retention behaviour of humic substances onto mineral surfaces and consequences upon thorium (IV) mobility: case of iron oxides. Appl Geochem 17:1551–1562

Reynolds BC, Wasserburg GJ, Baskaran M (2003) The transport of U- and Th-series nuclides in sandy confined aquifers. Geochim Cosmochim Acta 67:1955–1972

Riotte J, Chabaux F (1999) ^{234}U/^{238}U) Activity ratios in freshwaters as tracers of hydrological processes: the Strengbach watershed (Vosges, France. Geochim Cosmochim Acta 63:1263–1275

Riotte J, Chabaux F, Benedetti M, Dia A, Gerard M, Boulegue J, Etame J (2003) Uranium colloidal transport and origin of the U-234-U-238 fractionation in surface waters: new insights from Mount Cameroon. Chem Geol 202:365–381

Rojo I, Seco F, Rovira M, Giménez J, Cervantes G, Martí V, de Pablo J (2009) Thorium sorption onto magnetite and ferrihydrite in acidic conditions. J Nucl Mater 385:474–478

Rosholt JN, Doe BR, Tastumoto M (1966) Evolution of the isotopic composition of uranium and thorium in soil profiles. Geol Soc Am Bull 77:987–1004

Rosholt JN (1983) Isotopic composition of uranium and thorium in cristalline rocks. J Geophys Res 88:7315–7330

Rutherford E, Soddy F (1903) The radioactivity of uranium. Philosop Mag 5:441–445

Semkow TM (1991) Fractal model of radon emanation from solids. Phys Rev Lett 66:3012–3015

Sims KW, Gill JB, Dosseto A, Hoffmann DL, Lundstrom CC, Williams RW, Ball L, Tollstrup D, Turner S, Prytulak J, Glessner JJ, Standish J, Elliott T (2008) An Inter-Laboratory Assessment of the Thorium Isotopic Composition of Synthetic and Rock Reference Materials. Geostand Geoanal Res 32:65–91

Singer DM, Maher K, Brown GE Jr (2009) Uranyl-chlorite sorption/desorption: evaluation of different U(VI) sequestration processes. Geochim Cosmochim Acta 73:5989–6007

Singleton MJ, Maher K, DePaolo DJ, Conrad ME, Dresel PE (2006) Dissolution rates and vadose zone drainage from strontium isotope measurements of groundwater in the Pasco Basin, WA unconfined aquifer. J Hydrol 321:39–58

Tricca A, Porcelli D, Wasserburg GJ (2000) Factors controlling the groundwater transport of U, Th, Ra and Rn. Proc Ind Acad Sci (Earth Planet Sci) 109:95–108

Tricca A, Wasserburg GJ, Porcelli D, Baskaran M (2001) The transport of U- and Th-series nuclides in a sandy unconfined aquifer. Geochim Cosmochim Acta 65:1187–1210

Tripathy GR, Singh SK, Krishnaswami S (2011) Sr and Nd Isotopes as Tracers of Chemical and Physical Erosion, Handbook of environmental Isotope Geochmistry, Springer, Ed: M. Baskaran, ch. 26, this book.

Turcotte DL (1986) Fractals and fragmentation. J Geophys Res 91:1921–1926

Turner S, Van Calsteren P, Vigier N, Thomas L (2001) Determination of thorium and uranium isotope ratios in low-concentration geological materials using a fixed multi-collector-ICP-MS. JAAS 16:612–615

Turner S, Hawkesworth C, Rogers N, Bartlett J, Worthington T, Hergt J, Pearce J, Smith I (1997) ^{238}U-^{230}Th disequilibria, magma petrogenesis, and flux rates beneath the depleted Tonga-Kermadec, island arc. Geochim Cosmochim Acta 61:4855–4884

Vance D, Teagle DAH, Foster GL (2009) Variable Quaternary chemical weathering fluxes and imbalances in marine geochemical budgets. Nature 458:493–496

Viers J, Dupre B, Polve M, Schott J, Dandurand JL, Braun JJ (1997) Chemical weathering in the drainage basin of a tropical watershed (Nsimi-Zoetele site, Cameroon): comparison between organic-poor and organic-rich waters. Chem Geol 140:181–206

Vigier N, Bourdon B, Turner S, Allègre CJ (2001) Erosion timescales derived from U-decay series measurements in rivers. Earth Planet Sci Lett 193:549–563

Vigier N, Bourdon B, Turner S, Van Calsteren P, Subramanian V, Dupré B, Allègre CJ (2005) Parameters influencing the duration and rates of weathering deduced from U-series measured in rivers: the Deccan Trap region (India). Chem Geol 219:69–91

Vigier N, Burton KW, Gislason SR, Rogers NW, Duchene S, Thomas L, Hodge E, Schaefer B (2006) The relationship between riverine U-series disequilibria and erosion rates in a basaltic terrain. Earth Planet Sci Lett 249(3–4):258–273

Von Blanckenburg F (2006) The control mechanisms of erosion and weathering at basin scale from cosmogenic nuclides in river sediment. Earth Planet Sci Lett 242:224–239

White AF, Brantley SL (2003) The effect of time on the weathering of silicate minerals: why do weathering rates differ in the laboratory and field? Chem Geol 202:479–506

Willett IR, Bond WJ (1995) Sorption of manganese, uranium, and radium by highly weathered. Soils J Environ Qual 24:834–845

Hirose K (2011) Uranium, thorium and anthropogenic radionuclides as atmospheric tracers. In: Baskaran M (ed) Handbook of environmental isotope geochemistry. Springer, Heidelberg

Tripathy GR, Singh SK, Krishnaswami S (2011) Sr and Nd Isotopes as tracers of chemical and physical erosion. In: Baskaran M (ed) Handbook of environmental isotope geochemistry. Springer, Heidelberg